Principles of Medical Physiology

Sabyasachi Sircar, M.D.

Department of Physiology
University College of Medical Sciences
Delhi

795 Illustrations

127 Tables

19 Text boxes

Thieme
Stuttgart • New York

Library of Congress Cataloging-in-Publication Data is available from the publisher.

Important note: Medicine is an ever-changing science undergoing continual development. Research and clinical experience are continually expanding our knowledge, in particular our knowledge of proper treatment and drug therapy. Insofar as this book mentions any dosage or application, readers may rest assured that the authors, editors, and publishers have made every effort to ensure that such references are in accordance with **the state of knowledge at the time of production of the book**.

Nevertheless, this does not involve, imply, or express any guarantee or responsibility on the part of the publishers in respect to any dosage instructions and forms of applications stated in the book. **Every user is requested to examine carefully** the manufacturers' leaflets accompanying each drug and to check, if necessary in consultation with a physician or specialist, whether the dosage schedules mentioned therein or the contraindications stated by the manufacturers differ from the statements made in the present book. Such examination is particularly important with drugs that are either rarely used or have been newly released on the market. Every dosage schedule or every form of application used is entirely at the user's own risk and responsibility. The authors and publishers request every user to report to the publishers any discrepancies or inaccuracies noticed. If errors in this work are found after publication, errata will be posted at www.thieme.com on the product description page.

© 2008 Georg Thieme Verlag,
Rüdigerstrasse 14, 70469 Stuttgart, Germany
http://www.thieme.de
Thieme New York, 333 Seventh Avenue,
New York, NY 10001, USA
http://www.thieme.com

Some of the product names, patents, and registered designs referred to in this book are in fact registered trademarks or proprietary names even though specific reference to this fact is not always made in the text. Therefore, the appearance of a name without designation as proprietary is not to be construed as a representation by the publisher that it is in the public domain.

Typesetting by Vitasta Publishing Pvt Ltd, Delhi, India
Printed in India

ISBN 978-3-13-144061-7 (TPS, Rest of World)
ISBN 978-1-58890-572-7 (TPN, The Americas)

1 2 3 4 5 6

Dedicated to the memory of my parents

who were the best teachers I have ever known

Contents

Section 1   General Physiology   1

1. Principles of Physics in Physiology   3
2. Principles of Physical Chemistry in Physiology   19
3. Principles of Control System in Physiology   25
4. The Cell   29
5. Cell Membrane   36
6. Cell Cycle   44
7. Applied Genetics   49

Section 2   Nerve and Muscle Physiology   55

8. Functional Anatomy of the Neuromuscular system   57
9. Degeneration and Regeneration of Nerve and Muscle   66
10. Resting Membrane Potential   73
11. Membrane Excitation and Action Potential   80
12. Electrophysiology of Ion Channels   85
13. Conduction of Nerve Impulses   90
14. Neuromuscular Transmission   96
15. Mechanism of Muscle Contraction   100
16. Characteristics of Muscle Contraction   104
17. Muscle Elasticity   111
18. Electromyography and Electroneurography   114
19. Muscle Mechanics   120
20. Smooth Muscles   125
21. Cardiac Muscle   130

Section 3   Blood and Immune System   135

22. Body Fluids and Blood   137
23. Red Blood Cells   141
24. Hemoglobin   146
25. Hematinic Factors   154
26. Blood Grouping and Transfusion   161
27. Blood Platelets and Hemostasis   166
28. Hemostatic Balance   173
29. Granulocytes   179
30. Agranulocytes and Lymphoid Organs   184
31. Immunity, Tolerance and Hypersensitivity   189

32. Immune Mechanisms   195
33. Hemopoiesis   202

Section 4   Cardiovascular System   207

34. Cardiac Excitation and the Electrocardiogram   209
35. Abnormalities of Cardiac Excitation   219
36. Cardiac Cycle   226
37. Cardiac Output   232
38. Circulatory Pathway and Hemodynamics   243
39. Capillary Exchange and Lymphatic Circulation   257
40. Chemical Control of Cardiovascular System   262
41. Neural Control of Cardiovascular System   267
42. Blood Pressure Regulation   273
43. Circulatory Shock   280
44. Coronary Circulation   285
45. Cerebral Circulation   289
46. Pulmonary and Pleural Circulation   296
47. Cutaneous, Muscle, and Splanchnic Circulation   300

Section 5   Respiratory System   307

48. Functional Anatomy of the Respiratory System   309
49. Mechanics of Pulmonary Ventilation   314
50. Measurement of Pulmonary Ventilation   328
51. Alveolar Ventilation, Perfusion, and Gaseous Exchange   336
52. Transport of Gases   343
53. Neural Control of Respiratory Rhythm   348
54. Chemical Control of Pulmonary Ventilation   352
55. High and Low Pressure Breathing   359
56. Pulmonary Function Tests and Respiratory Disorders   363

Section 6   Renal System   371

57. Functional Anatomy of the Kidney   373
58. Glomerular Filtration and Tubular Reabsorption   379

59.	Renal Handling of Sodium	383
60.	Renal Regulation of Urine Volume and Osmolarity	388
61.	Body Fluid and Electrolyte Balance	392
62.	Renal Regulation of Acid-Base Balance	398
63.	Renal Regulation of Potassium Balance	404
64.	Renal Handling of Miscellaneous Substances	407
65.	Hormones Acting on the Kidney	410
66.	Quantification of Renal Functions	413
67.	Urine Analysis and Renal Function Tests	417
68.	Renal Syndromes	421
69.	Urinary Bladder and Cystometry	425

Section 7   Gastrointestinal System   431

70.	Events in the Mouth and Esophagus	433
71.	Events in the Stomach	437
72.	Events in the Duodenum	443
73.	Events in the Small Intestine	449
74.	Events in the Colon	453
75.	Gastrointestinal Hormones	456
76.	Gastrointestinal Disorders	460

Section 8   Metabolism and Nutrition   467

77.	Dietary Nutrients and Fibers	469
78.	Nutritional Assessment and Dietary Planning	476
79.	Metabolic Pathways	480
80.	Metabolic States and the Liver	488

Section 9   Endocrine System   497

81.	Mechanism of Hormonal Action	499
82.	Hypothalamic and Pituitary Hormones	504
83.	Thyroid Hormones	510
84.	Calcitropic Hormones	517
85.	Adrenocortical Hormones	525
86.	Adrenomedullary Hormones	534
87.	Pancreatic Hormones	537

Section 10   Reproductive System   545

88.	Testicular and Ovarian Hormones	547
89.	Puberty and Gametogenesis	552
90.	Menstrual Cycle	558
91.	Sperm Transport and Fertilization	562
92.	Sexual Differentiation of the Fetus	571
93.	Pregnancy	576
94.	Parturition and Lactation	581

Section 11   Central Nervous System   585

95.	Anatomy of the Central Nervous System	587
96.	Spinal Cord and Brainstem	597
97.	Cerebellum	605
98.	Diencephalon	612
99.	Basal Ganglia	616
100.	Cerebral Cortex	621
101.	Autonomic Nervous System	630
102.	Synaptic Mechanisms and Neurotransmitters	635
103.	Sensory Mechanisms	645
104.	Regulation of Muscle Length and Tone	660
105.	Motor Planning, Programming, and Execution	667
106.	Thought, Emotion, and Conation	679
107.	Electroencephalogram and Epilepsies	685
108.	Sleep	692
109.	Regulation of Body Temperature	697
110.	Regulation of Food Intake	703
111.	Memory and Learning	707
112.	Language and Speech	715

Section 12   Special Senses   719

113.	Functional Anatomy of the Eye	721
114.	Visual Optics	726
115.	The Retina and Phototransduction	735
116.	Visual Pathways and Image Processing	741
117.	Oculomotor Mechanisms	750
118.	Auditory Mechanisms	756
119.	Vestibular Mechanisms	766
120.	Olfactory and Gustatory Mechanisms	770
	Photo Acknowledgments	775
	Index	777

Preface

The real voyage of discovery consists not in seeking new landscapes but in having new eyes.

Marcel Proust

About a decade ago, I set out to write a book of human physiology that undergraduate medical students could read at bedtime for relaxation, and the result is now in your hands! The four principal considerations that have shaped this book are utility, lucidity, brevity, and contextual relevance.

Every sentence in the text has been carefully weighed for its utility before it was granted admission to this book. The utility could be its clinical relevance, conceptual elegance, or its importance in examinations. For deciding on the utility, I have used the simple test of time: Surely, any physiological fact that I fail to recall myself despite teaching physiology for more than two decades need not be taught to the student either. Admittedly though, I have erred and included in this book a lot more than what I can recall effortlessly.

Lucidity has been attempted through numerous full-color illustrations, flow charts and 3D-schematic diagrams (all drawn by me), tables, analogies, cartoons, anecdotes, and caveats, and by restricting the text to the basics.

Brevity has been achieved, not by compromising on the elucidation of important facts and concepts but through a rigorous economy of words, elimination of introductory passages to chapters, avoiding repetitions, and substituting wherever appropriate a well-labeled picture in place of "a thousand words."

Ensuring contextual relevance has been the toughest of all and is also by far the most distinctive feature of this book. I have spent months planning as to what goes in where and mulling over the rubrics. Sundry facts that could not be woven into a flowing narrative have been omitted unless deemed especially important. Some topics do not fit nicely anywhere but are too small to be presented as a separate chapter. Others seem to be appropriate at too many places. There are still others that can be presented as a separate chapter

only at the cost of extensive repetition. Hence, topics like exercise physiology or fetal physiology are not presented as separate chapters but are discussed at several places throughout the book, wherever relevant.

Conspicuous by their paucity or absence in this book are the references to the work of scientists, and the normal range of laboratory values. I do not discount the importance of these but leave them for students to find out from reference books. I have witnessed with dismay heated arguments among teachers on what should be considered the "normal" range of physiological parameters, often sidetracking vital conceptual questions, and would not like to join the fray! Rather, I would like to impress on the readers that there can be no global range of normal biological values and that in all such considerations, the context is important, as was strikingly brought out by the comment of the copy-editor of this book, "Are these data real?" referring to the 60 kg weight of a "reference man" (Table 78.3) that is well below western norms. Hence all "normal" values mentioned in this book should be taken as more illustrative than authoritative as they are drawn from myriad sources and altered slightly to make them easy to remember.

I hope this book ushers in happier times for students interested in understanding medical physiology and doing well in competitive examinations. I invite the comments of teachers and students for incorporation into future editions. They may be sent to me through the publisher or directly to me at the following address:

Sabyasachi Sircar
Department of Physiology
University College of Medical Sciences
Delhi–110 095

August, 2007

Contributors

In writing some of the chapters in this textbook, I have sought and obtained assistance from my colleagues in other specialties. These contributions are listed below.

Chapter 5 Cell Membrane. Sections: Membrane Composition and Membrane Structure Chapter 7 Applied Genetics	Gautam Sarkar, Professor, Department of Biochemistry, A.V.Medical College, Pondicherry.
Chapter 19 Muscle Mechanics and the Skeletal Lever System	Sanjay Wadhwa, Professor, Department of Physical Medicine and Rehabilitation, Postgraduate Institute of Medical Education and Research, Chandigarh.
Chapter 44 Coronary Circulation. Section: Coronary Insufficiency	M. Abid Geelani, Professor, Department of Cardiothoracic Surgery, GB Pant Hospital, Maulana Azad Medical College, New Delhi.
Chapter 56 Respiratory Disorders	Mrinal Sircar, Senior Consultant Pulmonologist and Intensivist, Fortis Escorts Hospital and Research Center, Faridabad.
Chapter 78 Nutritional Assessment and Dietary Planning	Sanjay Chaturvedi, Professor, Department of Preventive and Social Medicine, University College of Medical Sciences, Delhi.
Chapter 81 Mechanism of Hormonal Action	Rafat Ahmed, Reader, Department of Biochemistry, University College of Medical Sciences, Delhi.
Chapter 94 Parturition and Lactation. Section: Parturition	Amita Suneja, Professor, Department of Obstetrics and Gynecology, University College of Medical Sciences, Delhi.
Chapter 113 Functional Anatomy of the Eye	Zia Chaudhuri, Assistant Professor of Ophthalmology, Maulana Azad Medical College, New Delhi.

Acknowledgments

Acknowledgements are due to the Thieme team at Stuttgart, notably Mr. Malik Norbert Lechelt, Vice President, International Business Development, Thieme, and Stefanie Langner, Production Manager, Thieme, not least for making available to me a large number of illustrations of stunning quality from their various atlases. In particular, I would like to thank Mr. Lechelt for his steadfast support extended to me throughout the long months of production of this book, especially when the chips were down, and Mr. Santosh Kumar Verma, CEO, Vitasta Publishing Pvt Ltd, who coordinated this project before Mr. Lechelt stepped in. Acknowledgements are also due to the Vitasta team in India for giving their best in typesetting what they felt was a difficult manuscript. I particularly thank Ms. Renu Kaul Verma, Managing Director, Vitasta Publishing Pvt Ltd, for her indulgence in allowing me to breathe down the neck of the typesetters to have the page layouted exactly as I wanted, and to Mr. K. K. Raman, Production Manager, Vitasta Publishing Pvt Ltd, for superbly coordinating the typesetting work to ensure its timely completion. I would also give due credits to the artists, Mr. Akhilesh Kumar and Mr. Abhimanyu Sinha, for their excellent work with sketches and cartoons. Thanks are also due to several of my peers in and outside Delhi who, on hearing that I was working on a book of Medical Physiology, posted me their questions and doubts which I could work upon. My search for their answers often yielded salutary results, as in the case of "What is Einthoven's equation?". Special thanks are due to Mr. Arun Kumar Sharma, retired Technical Assistant, who took excellent care of my laboratory while I concentrated on this book, and also went through the manuscript thoroughly, making numerous intelligent remarks that led to its considerable improvement. I am also indebted to my former student, Dr. Vineet Sahay, now in University of New South Wales, Sydney, for promptly sending me several scientific papers that I was finding difficult to procure. Last but not least, I must thank my wife Madhumita who cheerfully bore the brunt of my negligence of household chores through the long years of my devotion to this book, though I wish she would remind me of this less often!

Section 1

General Physiology

1. Principles of Physics in Physiology

2. Principles of Physical Chemistry in Physiology

3. Principles of Control System in Physiology

4. The Cell

5. Cell Membrane

6. Cell Cycle

7. Applied Genetics

Principles of Physics in Physiology

Several areas of physiology cannot be appreciated without an elementary knowledge of physics. If only to emphasize the point, selected principles of physics that are important in physiology are outlined below. Also discussed are mathematical principles that are inseparable from physics.

Mathematics

Vectors

Vectors are quantities that have a magnitude as well as a direction. A vector is represented with an arrow that indicates its direction. The length of the arrow gives a measure of the magnitude of the vector. A vector can be resolved into components. Conversely, two or more vectors can be combined into a resultant vector. Physiological applications of vectors are found in biomechanics, electrocardiography, motor control, and vestibular mechanisms, among others. The basic concepts of vector resolution are explained with the following examples.

Suppose a person travels 150 km from point O to point P (Fig. 1.1A). In doing so, he travels 138 km to the south, 60 km to the east and 140 km to the southeast. The distance he travels in a given direction can be calculated simply by dropping a perpendicular from point P onto the axis representing that direction. This basic method can be applied to any vector quantity. For example, if the person travels from point O toward point P with a velocity of 150 km h^{-1}, then he moves south with a velocity of 138 km h^{-1}, toward east with a velocity of 60 km h^{-1} and toward southeast with a velocity of 140 km h^{-1}. Conversely, if it is known at what velocity he is moving toward south and east, the actual velocity and direction of his movement can be calculated graphically. A vector does not have any component perpendicular to it. This is self-evident; a person traveling due south does not move toward east or west. The calculation of the mean electrical axis of the heart from the recorded voltages in the ECG leads (see Fig. 34.**11**) is based on the same principle. It explains why an ECG wave that has the highest amplitude in lead I would have the lowest amplitude in lead aVF which is perpendicular to lead I. Vector principles also explain why the effect of gravity on circulation gets nullified when the body is supine. In the supine position, the direction of gravity is perpendicular to the direction of blood flow in most of the large vessels.

A special case of the vector principle is illustrated in Figure 1.1B. Suppose a person is standing at point O situated at the center of an equilateral triangle formed by points A, B, and C. When the person moves toward point P with a velocity of (say) 70 km h^{-1}, he is moving toward point A at 51 km h^{-1}, toward point B at 16 km h^{-1} and toward point C at –67 km h^{-1}. (The minus sign denotes that he is actually moving away from point C.) On adding the three velocities, we get zero. This principle forms the basis of the zero potential obtained by interconnecting the vertices of Einthoven's triangle of electrocardiography.

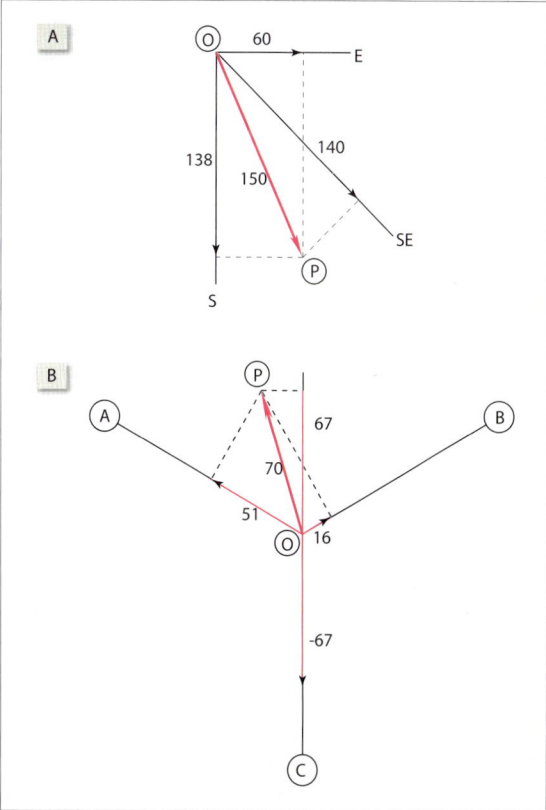

Fig. 1.1   Vector principles. [A] Resolution of a vector into its components. [B] Vector principles underlying Einthoven's hypothesis of electrocardiography.

Calculus

There are two broad branches of calculus: differential and integral. Both find applications in several areas of physiology.

Differential calculus makes it possible to calculate the slope of a curve at any point. The slope of a line is expressed mathematically as $\delta y/\delta x$, where δy is the small vertical distance through which a point on the line moves up as it shifts a small distance δx to the right. In the case of a straight line, this ratio remains the same anywhere along the line but in the case of a curved line, this ratio is different for every small segment of the line (Fig. 1.2A). Clearly, a precise description of the slopes of various parts of a curve requires that we break up the curve into infinitesimally small segments. The slope of each segment, which is small enough to be called a point, is denoted by the expression dy/dx.

Integral calculus makes it possible to calculate the area under the curve in a Cartesian plane. The area under the curve can be calculated by breaking up the area into rectangular blocks. However, if the curve is irregular, the blocks are not perfect rectangles and the result obtained is inaccurate. Greater accuracy is possible only if the width of blocks is infinitesimally small (dx) so that each block becomes rectangular. The area of the curve (A) is then given by:

$$A = \int y \, dx$$

Or stated simply, A is the sum of areas of an infinite number of rectangles with an infinitesimally small width (dx) so that the area of each rectangle is given by y × dx. It should be readily apparent from Figure 1.2B that $\int y \, dx$ is not the same as $\int x \, dy$. The graphic calculation of the work done by the lung and heart, the pressure-time index of the heart and mean blood pressure, and the amount of dye flowing in the indicator dilution method of blood flow estimation are based on the principles of integral calculus.

Logarithms

The logarithm of a number is the exponent to which 10 must be raised so as to equal the number. Expressed mathematically, if $10^a = N$, then log N = a. For example, log 1000 = 3 and log 0.001 = –3. It may be noted here that 1000 and 0.001 are reciprocal numbers. In other words, if log N = a, then log 1/N = –a. Logarithms appear in the Nernst equation. In accordance with the logarithmic property just explained, the equation can be written in either way as follows:

$$Em = \frac{61}{z} \log \frac{[C_o]}{[C_i]}$$

or

$$Em = \frac{-61}{z} \log \frac{[C_i]}{[C_o]}$$

An important application of logarithms is the logarithmic scale which is a convenient scale for describing certain types of data. The decibel scale of sound (see below) and the pH scale of acidity are examples of logarithmic scale.

Consider the sound pressure from various sources (Table 1.1). The inadequacy of a linear scale in representing these data graphically is obvious from Figure 1.3. On the other hand, if the data are converted into their logarithms and then plotted, the advantage is instantly apparent. The logarithmic scale of sound intensity is called the *decibel scale*.

Mechanics

Laws of motion

Newton's first law of motion states that a body continues to be stationary or to move in a straight line with uniform velocity until it is acted upon by an external force. The law helps in defining force itself (see below). The law is also known as the *law of inertia*. Because of inertia, a stationary body cannot start moving on its own and a moving body cannot stop on its own. The

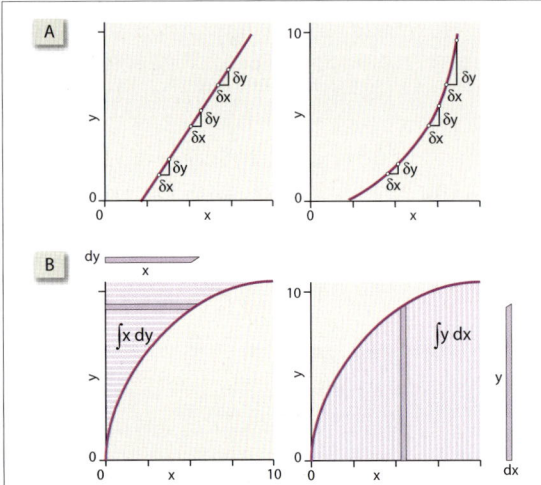

Fig. 1.2 [A] (*Left*) The slope ($\delta y/\delta x$) is the same in all segments of a straight line. (*Right*) For the same δx, the δy varies in different segments of the curve. Hence the slope ($\delta y/\delta x$) is different in all segments. [B] Areas defined by the terms $\int X dy$ (*left*) and $\int Y dx$ (*right*).

Table 1.**1**   Sound pressures at auditory threshold and in different situations

	Sound pressure in dynes cm^{-2}	Sound pressure as a multiple of threshold	Log of the multiple
Threshold	0.002	1	0
Conversation	0.02	10	1
Factory noise	2	1000	3
Discomfort	200	100 000	5

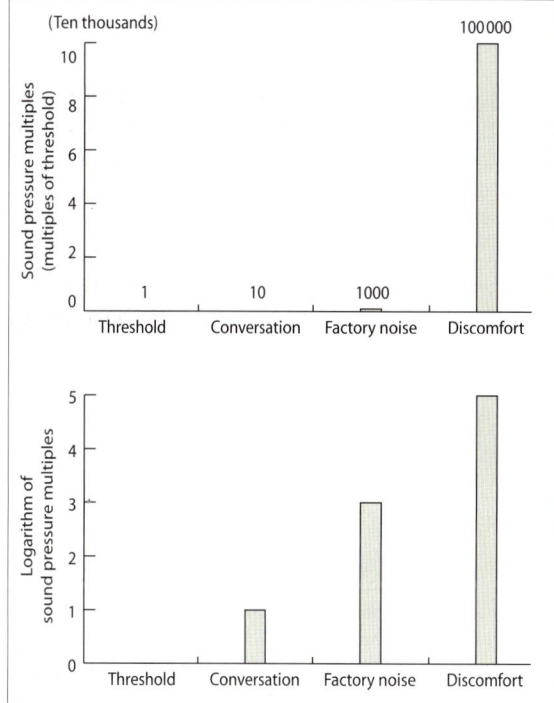

Fig.1.**3**   Sound pressures expressed as multiples of hearing threshold are depicted graphically on a linear scale (*above*) and a logarithmic scale (*below*).

concept of inertia helps in understanding the effect of coup and counter-coup injuries and the working of the semicircular canals and otolith organs of the vestibular apparatus.

Newton's second law of motion tells us that the acceleration or retardation of a body is directly proportional to the force applied on it and inversely proportional to the mass of the body. The product of the mass and acceleration of a body gives the force acting on it.

Newton's third law of motion states that every action has an equal and opposite reaction. The third law has been put to use in ballistocardiography. It also helps in understanding muscle action (see Fig. 49.**3**).

Force

Force is anything that changes or tends to change the uniform motion of a body in straight line. The SI unit of force is the Newton (N). A force of 1 N, when applied to a 1 kg mass placed on a horizontal frictionless surface, produces an acceleration of 1 m s^{-2}.

A shearing force is a force that is directed tangentially to the surface of a body (Fig. 1.**4**). Under the effect of a shearing force, the component laminae of a body move through different distances. An understanding of the nature of shearing force is important in the context of otolith organs. Shearing force is also exerted by blood flow on the capillary endothelium.

Gravitational force, as applicable on earth, is a special kind of force that has two remarkable features: (1) It is always directed toward the center of the earth and thereby, helps us in defining the vertical. (2) It is directly proportional to the mass of the body. The acceleration produced by gravitational force is called *acceleration due to gravity*. A heavier body is pulled with a greater gravitational force (in accordance with Newton's law of gravitation). However, for a given force, a heavier body has less acceleration (in accordance with Newton's second law of motion). Therefore the acceleration due to gravity does not vary with the mass of a body and remains absolutely constant at 9.8 m s^{-2}. This acceleration due to gravity, denoted by g, gives us the feeling of weight.

Under certain conditions, we feel heavier or lighter than usual due to what may be called the *apparent g,* which is denoted as Gz. Any change in Gz has important effects on cardiac output and blood pressure. Gz differs from *g* only when the body is subjected to other vertical accelerative forces in addition to the gravitational force.

The body is often subjected to accelerative forces in other directions too. Accordingly, the force is called Gx

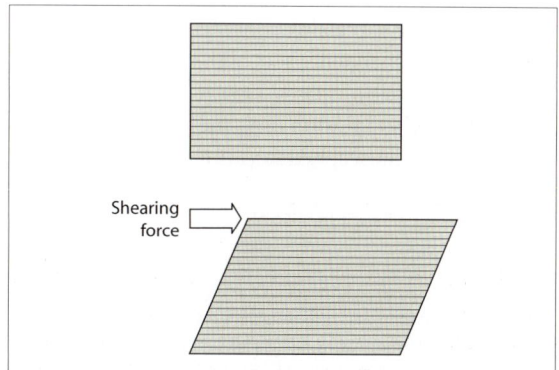

Fig. 1.**4**   Shearing force.

or Gy depending on whether it is directed anteroposteriorly or laterally, in relation to the subject. Together, the Gx, Gy, and Gz forces are called *G forces.*[1] The organs for detecting G forces are the otolith organs.

High and low Gz are experienced when there is acceleration or deceleration along the long axis of the body. Very high or low Gz occurs during the take off and landing of space rockets. More commonly, such changes in Gz are experienced during looping of airplanes and during parachute jumping when the parachute is suddenly opened out after a period of free fall (the opening shock load). A slight feel of high and low Gz is obtained in an elevator (lift) when it starts or stops.

Zero Gz is perceived by astronauts in orbiting satellites where the gravitational forces are precisely counterbalanced by the centrifugal forces generated by the orbiting satellite. On earth, zero Gz is experienced during free-fall under the effect of gravity. Stated mathematically

$$Gz = g - a$$

where:

Gz = apparent acceleration due to gravity
a = actual acceleration in the direction of gravity
g = 9.8 ms^{-2}

When a = g, Gz becomes 0. In other words, a body falling under the effect of gravity is weightless (see Fig. 38.**15**).

Pressure

Pressure is the force exerted per unit area. The pressure of a column of fluid (hydrostatic pressure) depends only on its vertical height; neither the width of the column, nor its inclination makes any difference to the pressure exerted (Fig. 1.**5**A). This concept is of importance in understanding the measurement of jugular venous pressure and the direct manometry of blood pressure. The dependence of fluid pressure on the height of the fluid column also explains why the venous pressure is higher in the dependent parts of the body and why the atmospheric pressure decreases at high altitudes. Its knowledge helps in estimating the pressure at different depths of the sea, which is important in deep-sea diving.

The SI unit of pressure is Newton m^{-2}, which is also

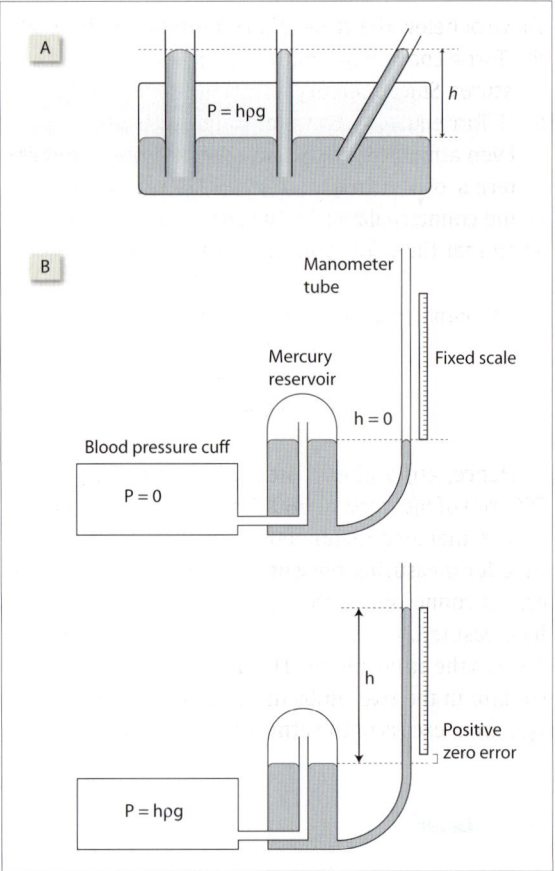

Fig. 1.**5** [A] Hydrostatic pressure depends only on the vertical height of a fluid column. It remains unaffected by the diameter and inclination of the tube. [B] A mercury manometer connected to a rubber cuff is used in sphygmomanometry. (*Above*) The cuff pressure is zero and the mercury levels in the reservoir and manometer tube are equal. (*Below*) As the cuff pressure increases, the mercury level in the tube rises while the mercury level in the reservoir falls. The difference h in the mercury levels gives the cuff pressure. Due to the fall in mercury level in the reservoir, the fixed scale gives a false-high reading due to positive zero error. The error can be reduced somewhat by scaling down the calibration. It can be verified that in a sphygmomanometer, the 1 cm calibration is actually a little less than 1 cm.

called 1 *Pascal* (Pa). The atmospheric pressure at sea level is taken as 101.29 kilopascals (kPa). This pressure is also known as 1 *atmospheric pressure*. The value of 1 atmospheric pressure, when rounded off to 100 kPa, is called 1 *bar*. However, fluid pressure is often expressed simply in terms of the vertical height of a fluid column because the pressure of a fluid column is given by hρg where h is the height of the fluid column and ρ is the density of the fluid. The fluid of reference is usually mercury but can be water, saline or even blood, whichever is convenient. The pressure exerted by 1 mm of mercury column is called 1 *Torr*. Nearly all fluid pressures inside the body are a few millimeters of mercury above or below the atmospheric pressure, which makes the Torr a convenient unit for expressing physiological pressures. Since mercury is 13.6 times denser than water, 1 Torr equals 13.6 mm of water pressure.

Even atmospheric pressure is commonly expressed in terms of the height of the mercury column that would counterbalance it. The height can be calculated, given that the density of mercury is 13.6×10^3 kg m^{-3}.

$$\text{Atmospheric pressure} = h \times \rho \times g$$
$$101.29 \times 10^3 = h \times (13.6 \times 10^3) \times 9.8$$
$$\therefore \quad h = 0.76 \text{ m}$$
$$= 760 \text{ mm}$$

Hence, atmospheric pressure is also expressed as 760 mm of mercury, or 760 Torr.

The mercury manometer (Fig. 1.**5**B) is commonly used for measuring pressure. One limb of the manometer is connected to the system whose pressure is to be measured. The other limb of the manometer is left open to the atmosphere. The difference in the mercury column in the two limbs indicates the pressure of the system in excess of the atmospheric pressure.

Levers

A lever is a rigid bar that is acted on by forces that tend to rotate the bar about its pivot or fulcrum. Levers are of three types depending on the relationship between the fulcrum, weight, and force applied. If the fulcrum is central, it is a *class I lever*. If weight is central, it is *class II*. If force is central, it is called *class III*.

The *mechanical advantage* of a lever is the ratio of the load force to the effort force. The ratio is proportional to the ratio of effort arm to load arm (Fig. 1.**6**). The greater the mechanical advantage, the less will be the force required to move a load.

$$\text{Mechanical advantage} = \frac{\text{Load}}{\text{Effort}}$$

The *velocity ratio* is the ratio of the distance moved by the effort to that moved by the load. The greater the velocity ratio, the greater is the distance through which an effort has to move for lifting a load.

$$\text{Velocity ratio} = \frac{\text{Distance moved by effort}}{\text{Distance moved by load}}$$

In a frictionless lever, mechanical advantage equals velocity ratio. Therefore, in all three classes, what is gained in excursion of the load is lost in the effective force acting on the load and *vice versa*. Class-II levers always produce force-gains, class-III levers always produce excursion gains and Class-I levers can produce

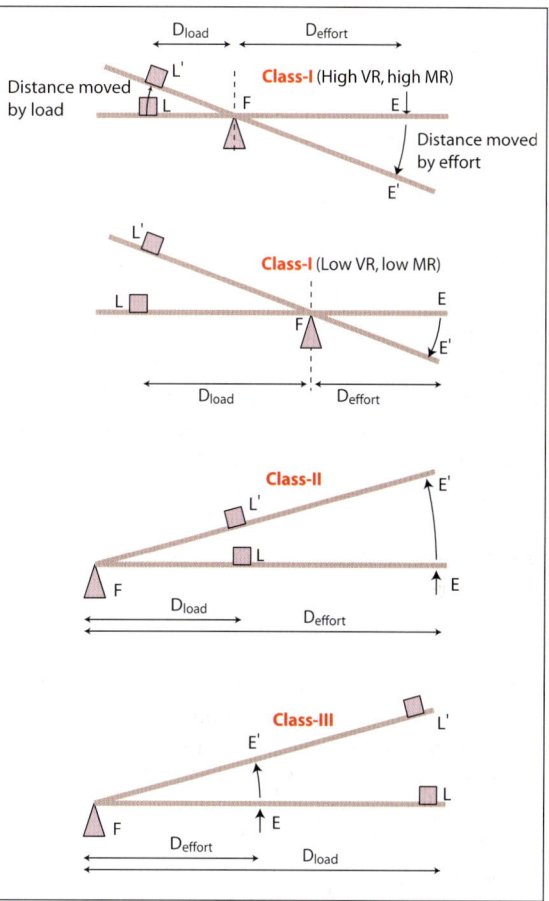

Fig. 1.**6**   The physics of a lever (explanation in text).

either of the two. The principles of levers are essential to the understanding of the biomechanics of the musculoskeletal system (Chapter 19).

Work and energy

Work is done when the point of application of a force (F) moves through a certain distance (D). Stated mathematically,

$$\text{Work done (W)} = F \times D$$

The SI unit of work is the *Joule*. 1 Joule of work is done when a force of 1 N moves a body through 1 meter.

Energy is the capacity for doing work and the unit of energy is the same as that of work, i.e., Joule. The energy associated with motion is called *kinetic energy* (KE) and is given by the formula:

$$KE = \tfrac{1}{2}mv^2$$

where *m* is the mass of the body and *v* is its velocity. The amount of kinetic energy that a body can gain by falling under the effect of gravity gives a measure of its potential energy (PE) and is given by the formula:

$$PE = mgh$$

where *m* is the mass of the body, *g* is acceleration due to gravity and *h* is the height through which the body can fall.

When work is done by a body by expending its own energy, the work done is said to be positive. For example, a body that falls through a certain height expends its potential energy and therefore the work done *by the body* is positive. Conversely, a body that is lifted against gravity gains in potential energy and therefore the work done *on the body* is negative.

When a fluid is compressed by application of pressure P, the fluid gains in energy, which is at least partly in the form of heat. The work done *on* the fluid is therefore negative and is given by:

$$Work\ done\ (W) = -\int P\ dV$$

where dV is the small decrement in volume through which it has been compressed. Conversely, for expanding against an incumbent pressure, the fluid has to expend energy and therefore it loses energy. The work done *by* the fluid is therefore positive and is given by:

$$Work\ done\ (W) = +\int P\ dV$$

These concepts help us understand why the work done by the lungs or the heart is given by the area enclosed in its pressure-volume loop.

In the case of the lungs, the work done on the lungs is negative during inspiration because the lungs themselves do not expend any energy; rather, they are made to inflate by the inspiratory muscles. During inspiration, the lungs gain in elastic recoil energy. This recoil energy stored in the lungs is expended during expiration as the lungs deflate. Therefore the work done by the lungs during expiration is positive. The work done on and by the lungs during inspiration and expiration are shown graphically in Figure 1.7A. When the work done during inspiration and expiration are added, we are left with a small amount of negative work done that is represented by the area enclosed within the *pressure volume loop* (called the *hysteresis loop*). The negative work signifies that at the end of one breathing cycle, work has been done *on* the lungs *by* the respiratory muscles which have expended their energy. The lungs have not spent energy of their own; rather, they have gained some energy. This energy is the heat energy that has been generated by the frictional (viscous) forces inside the lungs.

In the same way, the area inside the ventricular pressure-volume loop of the cardiac cycle gives the positive work done *by* the ventricle in overcoming viscous resistance of blood flow (Fig. 1.7B).

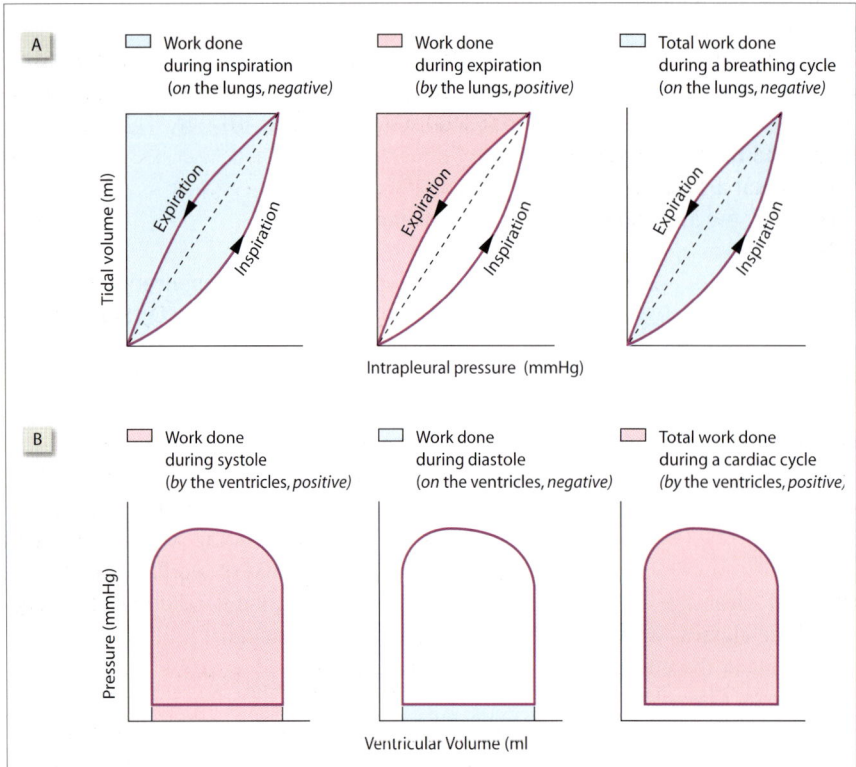

Fig. 1.7 (*Above*) Work done on the lung during breathing. (*Below*) Work done by the ventricle during a cardiac cycle.

Buoyancy

When a body is immersed in a fluid, the fluid displaced pushes the body up so that the body loses weight. The loss in weight is called *buoyancy* and it is equal to the weight of the fluid displaced. Providing buoyancy to the brain is an important function of the cerebrospinal fluid.

Surface tension

Surface tension is the property of a liquid surface which causes it to behave like a stretched membrane. Because of surface tension, a drop of fluid tries to minimize its surface area. It therefore assumes a spherical shape because the surface to volume ratio is minimum for a sphere. The fact that a bubble does not collapse completely is due to the fact that the air inside it provides a distending pressure that exactly balances the collapsing pressure of the fluid shell.

The **law of Laplace** states that tension (T) in the wall of a cylinder is equal to the product of the transmural pressure (P) and the radius (R). The formula has several variations (Table 1.2) depending on the geometry of the surface (spherical or cylindrical) and its composition (liquid drop, liquid bubble or air bubble) (Fig. 1.8). Applications of Laplace's law in physiology assume P = 2T/R as an approximation, which is the formula for a spherical air bubble in air.

The law of Laplace helps explain several interesting phenomena. (1) It explains how the thin-walled capillaries are able to withstand an internal pressure as high as 25 mmHg, which is the normal capillary hydrostatic pressure. This is possible because capillaries have a very small radius R. Even though the wall tension is low, the small value of R in the denominator makes a very high P possible. (2) The law of Laplace puts the dilated heart at a disadvantage. When the radius R increases, the wall tension T must go up proportionately if the ventricular pressure P is to be maintained. (3) In accordance with the Laplace law, the lower the functional residual capacity of the lungs, the more difficult it is to inflate it. (4) Because of Laplace's law, the greater the gastric filling, the lower is the pressure that causes gastric emptying. This is obviously beneficial to the process of digestion. (5) Laplace's law explains why reduction in detrusor muscle tension prevents rise of intravesical pressure even as the bladder fills up to greater volume.

Viscosity

Viscosity is fluid friction. When fluid moves along a tube, it does so in the form of multiple layers or laminae that slip on one another, moving at different velocities due to friction between the layers. The lamina at the middle of the blood vessel moves fastest while the one closest to the vessel wall does not move at all. Thus, there is a velocity gradient of laminae, which is called the *shear rate*. A fluid has 1 *poise* of viscosity if there is a frictional force of 1 dyne cm^{-2} between its layers when flowing at a shear rate of 1 cm s^{-1} cm^{-1}. A Newtonian fluid is one in which the viscosity is independent of the shear rate. Blood is not a Newtonian fluid.

Just as friction affects the velocity of a body, viscosity affects the velocity of fluid flow. In a long narrow tube of uniform radius, the relation of the flow rate (Q) with the pressure gradient ($P_A - P_B$) i.e., the pressure difference at the two ends of the tube, fluid viscosity (η), tube radius (R), and tube length (L) is given by the *Poiseuille-Hagen formula*.

$$Q = (P_A - P_B) \times \frac{\pi}{8} \times \frac{1}{\eta} \times \frac{R^4}{L}$$

This law is important in the understanding of hemodynamics. Since resistance varies inversely with the fourth power of the radius, even small changes in the vessel diameter cause large variations in blood flow through it, enabling effective regulation of blood flow

Table 1.2   Formulas defining Laplace's law

	Spherical	Cylindrical
Liquid drop in air	2T/R	T/R
Air bubble in liquid	4T/R	2T/R
Air bubble in air	2T/R	T/R

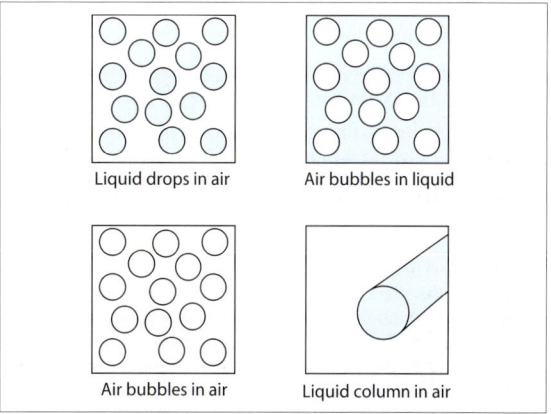

Liquid drops in air   Air bubbles in liquid

Air bubbles in air   Liquid column in air

Fig. 1.8   The Laplace law is applicable to any of the above (see Table 1.2).

through vascular beds. The relevance of the Poiseuille-Hagen formula is best brought out by the differences in the blood flow in the renal cortex and the renal medulla, where blood viscosity, capillary length and capillary diameter all have a role.

At high velocities, the flow of fluid becomes turbulent and does not remain streamlined. The probability of turbulence is related to the diameter of the vessel and the viscosity of the blood. This probability is expressed by the *Reynolds number.*

$$Re = \frac{\rho DV}{\eta}$$

where, Re is the Reynolds number, ρ is the density of the fluid; D is the diameter of the tube (in cm); V is the velocity of the flow (in cm s^{-1}) and η is the viscosity of the fluid (in poise). The higher the value of Re, the greater the probability of turbulence. When Re < 2000, flow is usually not turbulent whereas if Re > 3000, turbulence is almost always present. Turbulence of blood is responsible for cardiac murmurs and the Korotkov sounds.

The Poiseuille-Hagen formula can be rewritten as:

$$(P_A - P_B) = Q \times \frac{8}{\pi} \times \eta \times \frac{L}{R^4}$$

Thus when the flow rate, fluid viscosity, and tube radius remain constant, the pressure drop ($P_A - P_B$) along a tube is directly proportional to the length of the tube and inversely proportional to the tube radius, as explained in Figure 1.**9**.

■ Bernoulli's principle

When a constant amount (Q) of fluid flows through a tube, the total fluid energy, i.e., the sum of its kinetic energy, pressure energy, and potential energy, remains constant (Fig. 1.**9**). This is known as *Bernoulli's principle* and it helps in understanding several important hemodynamic principles. It explains, for example, why the fluid pressure is low in a blood vessel at places where its radius is less[2] (Fig. 1.**9**). It also explains the suction effect of venous blood flow on the thoracic duct terminating in the vein (see Fig. 39.**7**).

---

[2] The fact that both the Poiseuille-Hagen formula and Bernoulli's principle give the effect of tube diameter on fluid pressure need not be cause for confusion. The two formulas deal with different aspects of fluid energy. The Poiseuille-Hagen formula deals with the conversion of fluid pressure into heat and the resultant loss of fluid pressure as the fluid flows through a long tube of uniform diameter. It does not deal with the other forms of fluid energy like kinetic energy or potential energy. On the other hand, Bernoulli's principle deals with the interconversion of the three different forms of fluid energy and not with its conversion into heat energy.

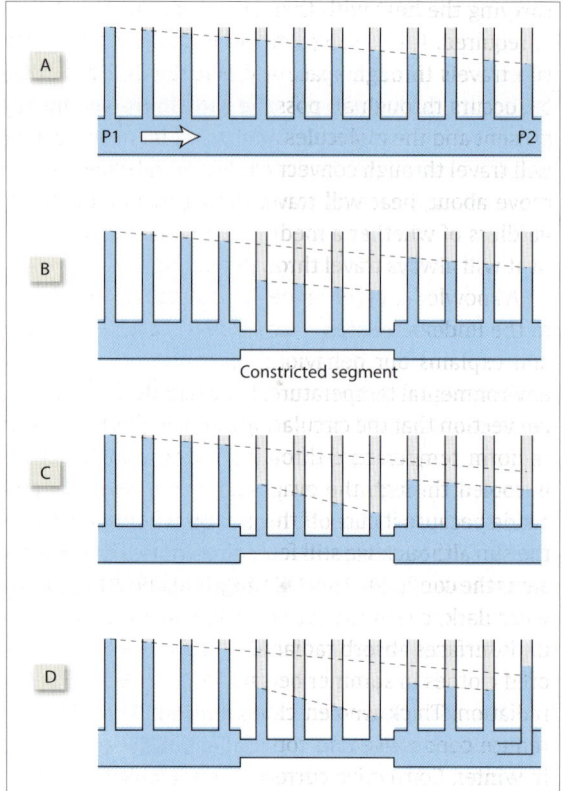

Fig. 1.**9**   A horizontal tube is fitted with uniformly spaced manometers for measuring the pressure drop. [A] The pressure falls uniformly with distance in accordance with the Poiseuille-Hagen formula. [B] The pressure falls in the constricted segment in accordance with Bernoulli's principle. Beyond the constricted segment, the pressure rises again. The depiction is not entirely correct: the correct pressures are depicted immediately below. [C] The pressure in the constricted segment is low in accordance with Bernoulli's principle and the pressure drop along the segment is steeper in accordance with the Poiseuille-Hagen formula. [D] The fluid rises higher in the last manometer because it captures a part of the fluid's kinetic energy and converts it into pressure energy in accordance with Bernoulli's principle.

▬ Heat

Heat always flows from a higher temperature to a lower temperature. When the temperatures of two bodies are equal, no heat transfer occurs between them. In *conduction*, transfer of heat occurs through a medium whose molecules do not move about freely. Thus, when we hold an iron rod in fire, the iron atoms vibrate and transfer the heat from the fire to the hand through the rod. However, the iron atoms do not move from their fixed places. In *convection*, the molecules of the medium transfer heat by actually moving about, carrying the heat with them. Evaporative cooling is an example: the water molecules vaporize and move away, carrying the heat with them. In *radiation*, no medium is required. It is through radiation that heat from the sun travels through space and reaches us. Heat transfer occurs through all possible modes. If a medium is present and the molecules are free to move about, heat will travel through convection. If the molecules cannot move about, heat will travel through conduction. Regardless of whether a medium is present or not, some heat will always travel through radiation.

A knowledge of the modes of heat transfer is relevant to the understanding of thermoregulation in the body and explains our behavioral responses to changes in environmental temperature. For example, it is through convection that the circulating blood maintains a fairly uniform temperature throughout the body. The body is cooled through the evaporation of sweat. We seek shade because it cuts off the direct radiation heat from the sun although we still feel the convective heat of hot air or the conductive heat of the ground we lie upon. We wear dark, coarse-textured clothes in winter because dark surfaces absorb radiated heat. We wear light-colored clothes in summer because they reflect back heat radiation. Thick woolen clothes that trap air in them reduce conductive and convective losses of body heat in winter. Convective currents are employed in the caloric stimulation test for testing vestibular functions.

Light

Light bends as it passes from air into a denser medium or when it emerges out of the medium into air. This bending is known as *refraction* and the extent to which it bends while passing in and out of the medium is given by the refractive index (μ) of the medium. The refractive index of water is 1.33. The refractive indices of the various compartments of the eye are shown in Figure 114.**1**.

Refraction of light underlies the formation of an image by the lens. A knowledge of how lenses work helps in understanding the refraction in the eye. The basic formula for lenses is:

$$\frac{1}{f} = \frac{1}{v} - \frac{1}{u}$$

where *u* is the distance of the object from the lens, *v* is the distance of the image from the lens, and *f* is the focal length of the lens. The distances *u*, *v*, and *f* may be positive or negative, depending on whether the measurements are made along the direction of the light or against it (Fig. **1.10**). The focus is the point where parallel rays passing through the lens would converge (in the case of a convex lens) or appear to diverge (in the case of a concave lens). The focal length of a convex lens is positive while the focal length of a concave lens is negative.

The **power of a lens** is expressed in *diopter* (D) and is the reciprocal of its focal length in meters. Thus the power of a convex lens with focal length of +25 cm is +4 diopter whereas the power of a concave lens of –50 cm is –2 diopter. The power of the intraocular lens can be increased by contracting some of the intraocular muscles. This phenomenon is called *accommodation*. When two coaxial lenses are placed together, their dioptric powers get added algebraically. This forms the basis of correction of refractory errors by using external lenses.

The physics of image formation is best understood by integrating it with the physiology of sensory perception, because whatever we see around us is essentially the *sensory projection* of images formed on our retina. Consider a point object (O) and its point image I formed by a convex lens L (Fig. **1.11**). A point image is formed when all the rays diverging from a point object are made to converge at a single point using a convex lens. In Figure **1.11**A, the image I_r is formed in front of the eye. After converging at I_r, the rays diverge again and therefore the point image (the point of convergence of rays) behaves like a point object (the point of divergence of rays). Hence the image is called a *real image*. The diverging rays from I_r converge on the retina after passing through the convex lens of the eye. Past sensory experience of the observer's brain tells it that the rays forming the retinal image are originating from point I_r and therefore, the brain 'sees' an image at point I_r. This 'calculated guess' made by the brain is known as

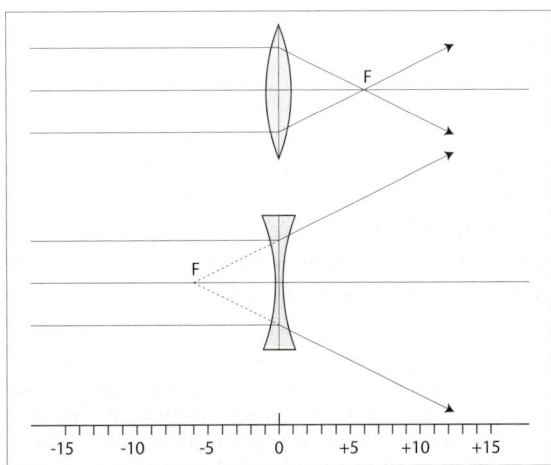

Fig. 1.10 Ray diagrams showing the focal point of a convex lens (*above*) and a concave lens (*below*). In the examples illustrated here, the focal length of the convex lens is +6 cm and the focal length of the concave lens is –6 cm.

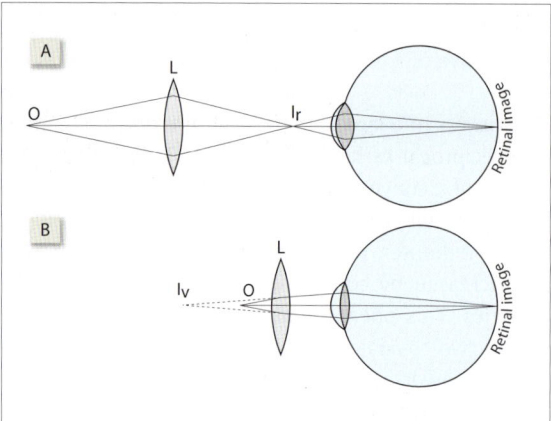

Fig. 1.**11** The optics of image formation of point-objects.

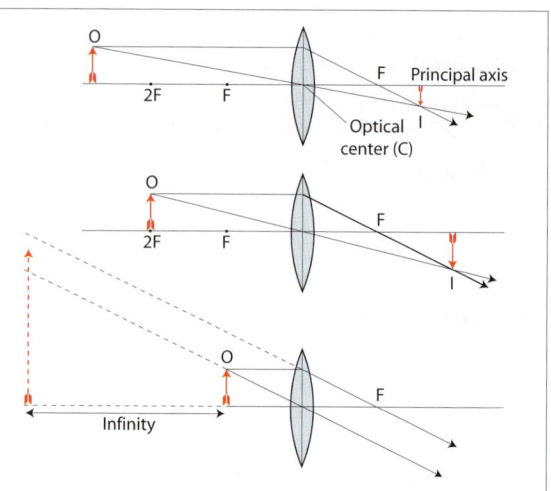

Fig. 1.**12** The optics of image formation of objects placed greater than two focal lengths away (*above*), at twice the focal length (*middle*) and at focal length (*below*).

sensory projection. In making the projection, the brain takes into consideration, among other factors, the extent of accommodation required by the intraocular muscles.

In Figure 1.**11**B, the convex lens L bends the divergent rays from the object but not enough to make them converge in front of the eye. It is the intraocular lens that finally converges the rays on to a point on the retina. As in the previous case, the brain projects the retinal image to the point where the divergent rays incident on the eye seem to be originating from. Hence, the brain 'sees' an image at point I_v. Unlike I_r, no rays are actually diverging from I_v; they only appear to do so due to sensory projection. Hence, I_v is called a *virtual image* (virtual = amounting to). Stated simply, an image is real if it is formed by the convergence of rays. The image is virtual if it is visible at a site where there is no convergence of rays. Going by this definition, the image formed on the retina of the eye is a real image.

The images formed when the object is located at different distances from the lens are shown in Figure 1.**12**. These ray diagrams can be easily constructed by remembering that (1) rays parallel to the principal axis converge at the focus after passing through the lens, and (2) the rays passing through the optical center (C) of the lens do not deviate.

How big does an object appear to the observer? The answer is that the nearer an object is to the eye, the bigger is the retinal image and the bigger it appears to the observer. However, how close we can bring an object to our eyes is limited by the power of our accommodation. Short-sighted people (myopes) see bigger images because they can bring the object nearer to the eye and still see it clearly. Another question arises here. Given two objects of different sizes placed at different distances from the eye, which one will appear bigger? The answer lies in the *visual angle* subtended by the two objects (Fig. 1.**13**) which forms the basis of

the testing of visual acuity. The visual angle subtended by an object is given by the formula:

$$\text{Visual angle} = \tan^{-1} \frac{\text{Height of the object}}{\text{Distance of the object from the eye}}$$

Thus, if an 8.75 cm high object[3] is kept at a distance of 60 m (i.e., 6000 cm), the visual angle subtended by it will be:

$= \tan^{-1} [8.65 / 6000]$

$= \tan^{-1} [0.00145]$

$= 0.0835°$

$= 0.0835 \times 60$ minutes

$= 5.01$ minutes

The retina is normally situated at the focus of the intraocular lens (see Fig. 114.**4**) and therefore when there is no accommodation, only objects kept at infinity (e.g., the stars in a night sky) will be clearly visible. Anything nearer appears hazy. This can be verified by putting atropine drops on the eye, which abolishes the power of accommodation. For the same reason, the best images are seen by our eyes when the object is kept at the focus of a magnifying glass and its virtual image is formed at infinity.[4] Due to sensory projection, parallel rays falling on the eye result in a virtual image that is infinitely large and is formed at infinity. How big does an

---

[3] In the Snellen chart, the topmost letter is 8.75 cm high, which is readable by a normal person from a distance of 60 m (see Fig. 114.**9**).

[4] This point is stressed here because several textbooks of high school physics mention that when an object is kept at the focus of a convex lens, the rays become parallel and therefore no image is formed. That is incorrect.

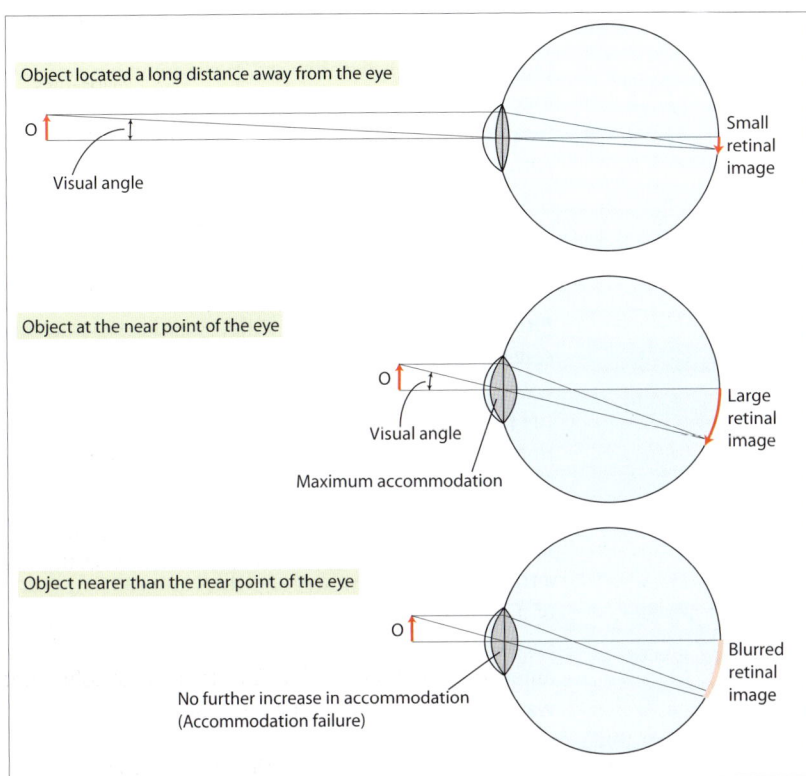

Fig. 1.**13** The visual angle increases as the object is brought nearer to the eye. Note that the greater the visual angle, the larger is the retinal image. When the object comes nearer than the near point, the retinal image becomes blurred.

Fig. 1.**14** Red eye (Bruckner's) reflex in a photograph shot in the dark using flashlight. The principle of direct ophthalmoscopy is similar.

infinitely large image located at infinity appear to the eye? As already explained, the answer depends on the visual angle subtended at the eye by the virtual image.

Because the eyes do not have to be strained (accommodated) for seeing a virtual image at infinity, all optical instruments (microscopes, telescopes) are so designed that the image formed by them is virtual and is situated at infinity. A commonly observed virtual image at infinity is the *red eye* well known to photographers: The pupil appears red in a photograph that has been shot in dim light using flashlight (Fig. 1.**14**). Because the pupils are dilated in dim light, the red choroid layer beneath the retina gets illuminated by

the flashlight and gets photographed. The illuminated choroid, which is situated at the focus of the intraocular lens, sends out parallel rays from the pupil. These parallel rays get captured by the camera even from a distance of several meters. Image formation when the object is at focus is also of importance in understanding direct ophthalmoscopy and retinoscopy.

Figure 1.**12** also shows that the nearer an object is to the focal point F, the greater the distance between the image and the lens. In the eye, the distance between the lens and the retina is fixed at 17 mm and the maximum power of the lens that can be attained after full accommodation is 69 diopters which is equivalent to a focal length of 14.5 mm. Substituting these values in the lens formula, we get $u = 10$ mm, which is as close the object can get to our eyes without blurring the retinal image and is called the *near point*.

Waves and Sound

A wave propagates through a medium when the particles of the medium oscillate rhythmically about their mean position. The wave is called *transverse* or *longitudinal* depending on whether the plane of oscillation of the particles is perpendicular or parallel to the direction of wave motion. A transverse wave has alternate crests and troughs while a longitudinal wave has alternate zones of condensation and rarefaction (Fig. 1.**15**).

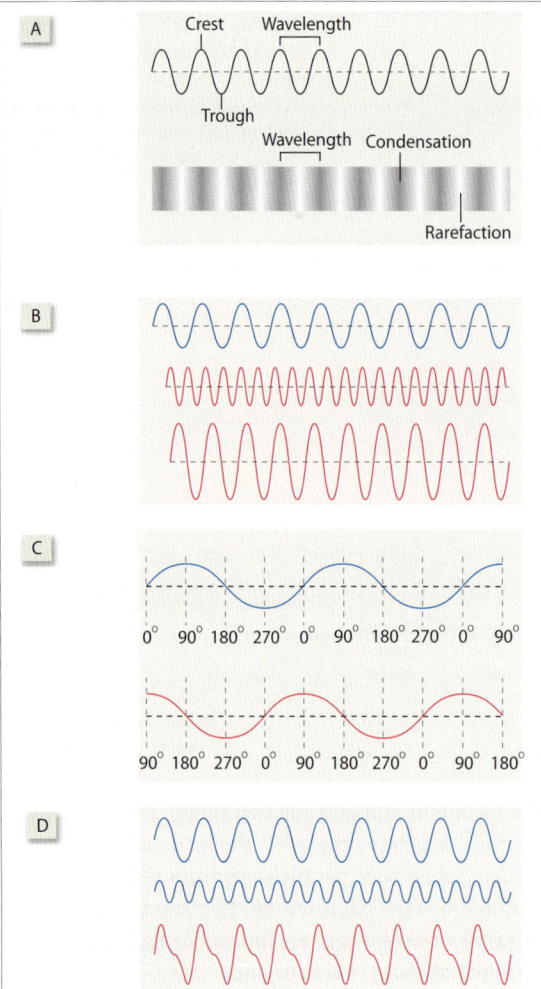

Fig. 1.15 Characteristics of wave motion. [A] Transverse and longitudinal waves. [B] Waves with different frequency and amplitude. [C] Waves with a phase difference. [D] Wave summation. Note that the wave in red color is formed by the summation of the two waves shown in blue, one of which has a lower frequency (fundamental wave) and the other has double the frequency (harmonic wave). The shape of the summated wave is different from the component waves, and gives the wave a characteristic sound quality or timbre.

The ocean waves or the ripples in a pool of water or the fluttering of a flag in the wind are examples of transverse waves. Transverse waves are also set up on the basilar membrane of the cochlea (see Fig. 118.**6**). Sound waves are longitudinal waves.

The *frequency* of a wave is the number of waves produced in 1 second and is expressed in *Hertz* (Hz). It is inversely related to its wavelength. Two waves with the same frequency can have a phase difference (Fig. 1.**15**C). For example, if at any point of time, one wave is at its crest (i.e., at 90°) and the other is at its trough (i.e., at 270°), the two waves are 180° out of phase. The term *phase difference* or *phase lag* has been used in this

book in the context of fluctuations in blood gases in Cheyne-Stokes breathing and in the ionic basis of cardiac pacemaker potentials. Two waves with different frequencies and amplitudes can summate to produce a complex wave. Such complex waves form the basis of the quality or timbre of sound waves (Fig. 1.**15**D). The characteristics of a sound wave, namely, its pitch, loudness, and timbre are discussed below.

Pitch   The pitch (or tone) of the sound is what is perceived by the human ear as higher or lower musical notes: It is directly related to its frequency. The greater the frequency, the higher is the tone. Pitch discrimination by the human ear is best in the range of 1000–3000 Hz.

Intensity and loudness   A sound wave is essentially a traveling pressure wave and therefore the amplitude of a sound wave is expressed as the maximum pressure variation in it. The intensity of a sound wave is the amount of energy (in Joule) transported by the sound wave in unit time (1 s) per unit area (1 m²), across a surface perpendicular to the direction of propagation. The intensity of a sound wave is directly proportional to the square of its pressure amplitude. The pressure amplitude of the faintest sound that can be heard is about 3×10^{-5} Pa and the corresponding intensity is 10^{-12} J s⁻¹ m⁻². A more convenient unit for expressing sound intensity is the decibel (dB).

$$\text{Number of dB} = 10 \log \frac{\text{Intensity of the sound}}{\text{Intensity of barely audible sound}}$$

The term loudness refers to the listener's subjective perception of the magnitude of a sound sensation. Loudness increases with intensity, and the relation between the two is empirical, which is given by the Weber–Fechner law or any of its modifications (see Fig. 103.17).

Since the decibel scale is a logarithmic scale, every 10 decibels indicate a tenfold increase in sound intensity. The faintest audible sound has an intensity of zero decibel. A sound is 10 times louder than threshold at 10 dB, 100 times louder at 20 dB, a million times louder at 60 dB and 10 billion times louder at 100 dB. The usual loudness of speech is 65 dB. A 100 dB sound damages the peripheral auditory apparatus while 120 dB causes pain. 150 dB causes permanent damage to the hearing apparatus.

Timbre or quality   Most musical sounds are made of complex waves formed by the summation of a high-amplitude wave (fundamental frequency) with several other waves of smaller amplitudes whose frequencies are multiples of the fundamental frequency (harmonics or overtones). The fundamental frequency determines the pitch of the sound while the overtones give the sound its characteristic timbre or quality. Variations in timbre enable us to distinguish tones of the same pitch produced by different musical instruments.

A mathematical theorem called the *Fourier Transform* makes it possible to express any wave with a given frequency and amplitude as the sum of a fundamental frequency and its numerous harmonics. Fourier transform has tremendous application in engineering. A well-known example is the musical synthesizer in which the tone of any musical instrument can be synthesized by mixing several multiples of a fundamental frequency in appropriate proportions of their amplitudes. Fourier transform has been applied to biological signals too, e.g., for data reduction in electroencephalogram. Finally, it is interesting to note that the basic function of the basilar membrane of the cochlea is to perform a Fourier transform on the sound waves: The sound wave breaks up into its component frequencies on the basilar membrane and each component registers its maximum amplitude at a different site of the basilar membrane. This information is conveyed separately to the brain which is then able to 'synthesize' the quality of the sound.

Electricity

Electrical charge   Atoms are made of protons, neutrons, and electrons. A proton has a unit positive charge, an electron has a unit negative charge, and neutrons do not have any charge on them. A body containing equal numbers of protons and electrons is electrostatically neutral. When the number of electrons in a body exceeds the number of protons in it, the body develops a negative charge. Conversely, it develops a positive charge when the number of electrons is less than the number of protons in it. The SI unit of charge is the *Coulomb* which equals the charge of 6.242×10^{18} electrons.

Current   The flow of charged particles is called an electric current. The SI unit of current is the *Ampere*. A flow of 1 Coulomb of charge every second is 1 Ampere current.

Potential   The potential of a body is the work done when a charge of +1 Coulomb approaches the body from infinity under the effect of its attractive or repulsive force. If the work done is 1 Joule, the potential of the body is said to be 1 *Volt*, which is the SI unit of potential. Two charged bodies have 1 Volt potential difference if 1 Joule of work is done in moving a charge of +1 Coulomb from one body to the other.

Resistance   The resistance of a conductor determines how much current it allows to pass through it when a potential difference is applied across it: for a given potential difference across a conductor, the current flowing through the conductor is inversely proportional to its resistance. This is known as the *Ohm's law*. If the current is 1 Ampere when the potential difference is 1 Volt, the resistance is said to be 1 Ohm, which is the SI unit of resistance. Conductance is the reciprocal of resistance and its SI unit is Mho.

$$\text{Voltage} = \text{Current} \times \text{Resistance}$$

An important derivation of Ohm's law is that when resistances (R1, R2, R3...) are added up in series (Fig. 1.**16**A), the total resistance (R) is given by:

$$R = R1 + R2 + R3 \ldots\ldots$$

When R1, R2, and R3 are connected in parallel (Fig. 1.**16**B), the total resistance is given by the formula:

$$1/R = 1/R1 + 1/R2 + 1/R3 \ldots\ldots$$

The resistance of a conductor varies directly with its length and inversely with its cross-sectional area. This principle is important in the understanding of nerve conduction and explains why nerve conduction velocity increases with axon diameter.

Capacitance   The potential of a body is directly proportional to the amount of charge it contains. The greater the amount of charge a body contains, the greater is its potential. However the extent to which

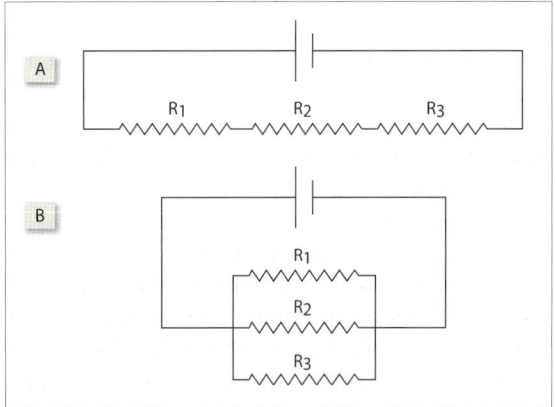

Fig. 1.**16**   Resistances in series [A] and parallel [B].

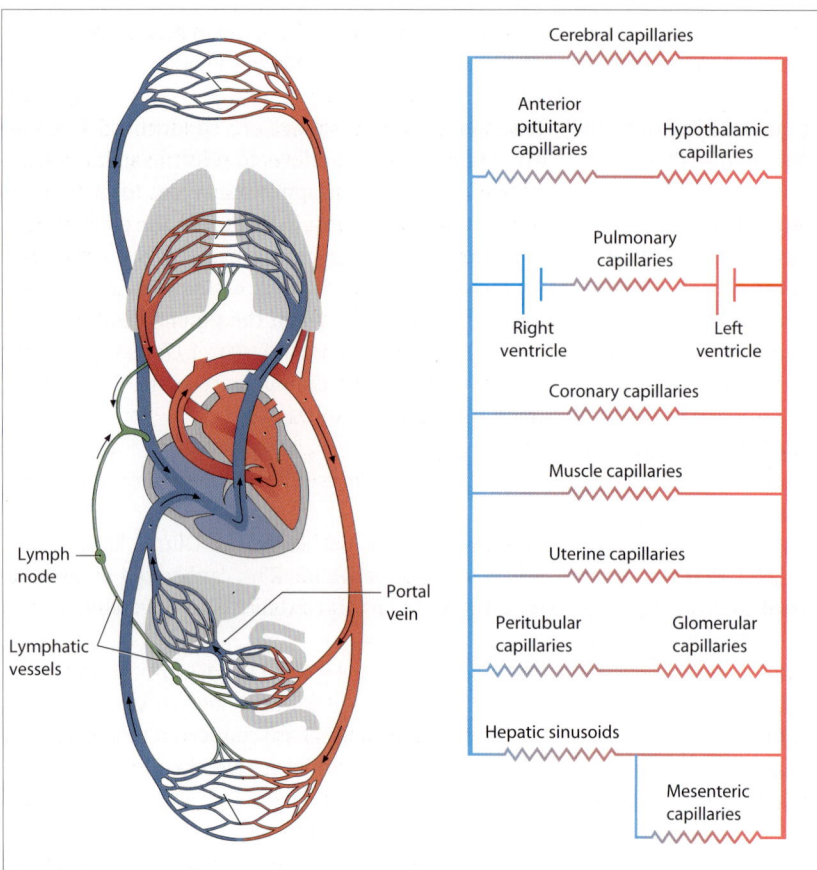

Fig. 1.17 Circulatory beds in series and parallel with the heart. Note that the capillary beds that are in series constitute a portal system.

Cerebral capillaries

Anterior pituitary capillaries

Hypothalamic capillaries

Pulmonary capillaries

Right ventricle

Left ventricle

Coronary capillaries

Muscle capillaries

Uterine capillaries

Peritubular capillaries

Glomerular capillaries

Hepatic sinusoids

Mesenteric capillaries

Lymph node

Portal vein

Lymphatic vessels

the potential rises when a given amount of charge is imparted to it will differ for different bodies. A body that can hold a large amount of charge without much rise in its potential is said to have a high capacitance. Stated mathematically,

$$\text{Capacitance} = \frac{\text{Charge}}{\text{Potential}}$$

Larger bodies have higher capacitance and therefore, the earth has almost an infinite capacitance. Thus the potential of earth (ground potential) remains unchanged no matter how much charge enters it. This stable ground potential is assumed to be zero and serves as a convenient reference for all electrical measurements. For the same reason, any charged body, when grounded, loses all its charge to the earth and its potential becomes zero.

A device that can store a large amount of charge is called a *capacitor*. A capacitor can hold a lot of charge and yet its potential remains low. On the other hand, a body with a low capacitance develops a high potential when it is charged. At high potentials, charge tends to 'leak out' and hence a body with low capacitance cannot hold much charge. A capacitor can be built by sandwiching a thin sheet of dielectric (nonconducting

material) between two parallel metal plates. Such a capacitor is called a *parallel plate capacitor*. When a potential difference is applied to the plates, a large amount of charge gets stored in the plates. The nerve membrane behaves like a dielectric, and when both its surfaces are layered with ions of opposite charges, it acts like a parallel plate capacitor (see Fig. 10.1). Membrane capacitance is an important determinant of membrane excitability and conduction velocity of nerve membranes.

The concepts of electricity are relevant to fluid dynamics too and help in understanding several important hemodynamic phenomena. Conversely, simple analogies of fluid mechanics help in understanding certain electrical phenomena like capacitance. These analogies are explained below.

Fluid pressure is analogous to electric potential and fluid usually flows from high-pressure to low-pressure just as electric current flows from high to low potential. The amount of fluid flowing in unit time is analogous to electric current. Fluid resistance is quite similar to electric resistance. The relationship between flow rate, pressure drop, and resistance in the blood vessels is similar to the relationship between the current, potential difference, and resistance:

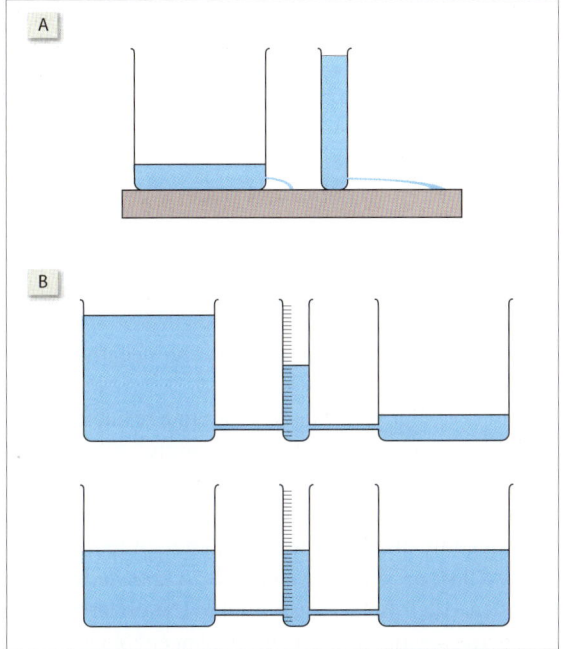

Fig. 1.18 Fluid analogy of electricity. [A] Fluid pressure and capacity are analogous to electric potential and capacitance. Shown here are two containers containing the same volume of fluid and with both having a small hole near their base. The container on the left has a broader base and therefore the hydrostatic pressure is low and leakage is much less. The container on the right has a narrow base. Hence the hydrostatic pressure is high and leakage is more. [B] A graduated tube interconnects two fluid-filled containers and is equidistant from both. The fluid pressure in the tube gives the mean of the fluid pressures in the two containers. However, the pressures in the two containers soon equalize unless the connecting limbs have high resistance to fluid flow.

Pressure drop = Flow Rate × Resistance

A similar formula is employed for defining blood flow resistance in vascular beds in terms of flow rate and perfusion pressure. Even the formulas for electrical resistances in series and parallel are valid for fluid dynamics and have important implications in hemodynamics. For example, most capillary beds are disposed in parallel to the heart (Fig. 1.17). The portal beds are however in series. The physics of parallel circuits explains the phenomena of circulatory 'steal' like coronary steal or subclavian steal: the electrical analogy is that when multiple resistors are connected in parallel, more current flows through resistors with lower resistance and *vice versa*. The electrical analogy also helps in understanding why the blood pressure falls when the resistance in any of the capillary beds decreases.

The electrical capacitance of a body can be compared with the fluid capacity of a container. A container with a large base is similar to a capacitor because it can store a large amount of fluid (analogous to charge) without its hydrostatic pressure (analogous to potential) increasing too much. On the other hand, the pressure in a narrow tube rises quickly when filled with fluid. For the same reason, leakage of charge is higher when the capacity is low (Fig. 1.18A).

Another similarity between electricity and fluid dynamics pertains to the measurement of potential. Consider two containers with fluid filled to different levels. For measuring their average potential, the two containers are connected to a graduated measuring tube as shown in Figure 1.18B. The height of the fluid column

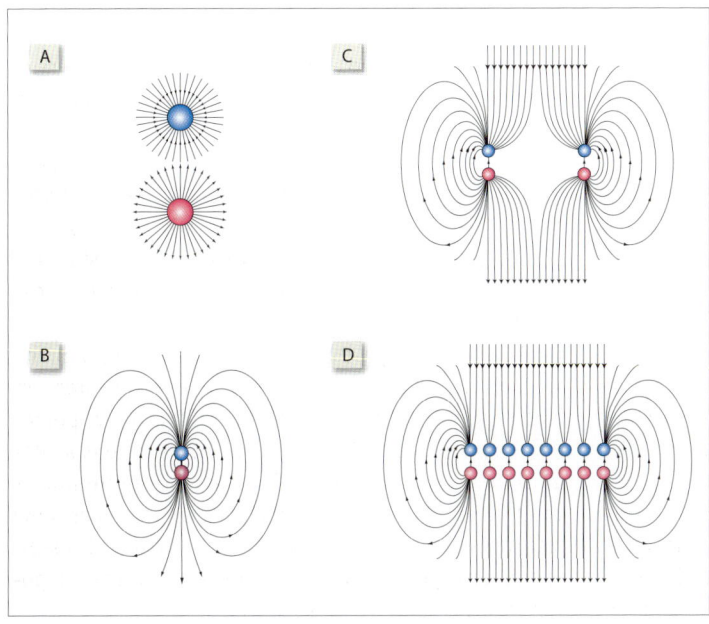

Fig. 1.19 [A] The electric field of a positive charge (red) and a negative charge (blue). Each line is a vector representing the force and direction of the repulsive or attractive force exerted on a charge of +1 Coulomb. [B] The lines of force around a dipole. [C] The lines of force when two dipoles are placed side-by-side. [D] When several dipoles are lined up, the current flowing along the long axis of the dipoles summate and get recorded by a distant electrode.

in the graduated tube gives the average fluid pressure in the two containers. The problem with this method is that since the two containers are interconnected, their pressures equalize very soon. The problem can be tackled by increasing the resistance of the connecting limbs. This analogy helps in understanding why it is important in electrocardiography to interconnect the three limb electrodes only through high (5000 Ohm) resistors for obtaining zero potential (see Fig. 34.**4**).

■   **Electric field of dipoles**

Electric field refers to the area of influence of an electric charge (or charges). The electric field of a single charged particle is depicted in Figure 1.**19**A. Each vector in the diagram represents the direction and force with which a positive charge of 1 Coulomb moves (or tends to move) under the effect of the electric field. If an electric field is applied to a medium containing mobile charged particles, the charged particles start moving along the lines of the electric field, resulting in the flow of an electric current. This happens in body fluid where an electric field is set up by the heart.

A **dipole** is a pair of opposite charges lying close together. The electric field set up by a dipole is more complex. Each dipole results in the flow of current between the opposite charges along multiple pathways as shown in Figure 1.**19**B. Although the conventional current flows from the positive to the negative charge, to a recording electrode placed at a distance, the current appears to flow along a vector directed from the negative to the positive charge.

Multiple dipoles located along a line result in a large resultant current. Each dipole generates a large field around it as shown in Figures 1.**19**C and 1.**19**D. Only currents that travel straight along the axis of the dipole get added up and become strong enough to be picked up by a distant recording electrode. A knowledge of dipoles and their fields is essential to the understanding of electrocardiography and electroencephalography.

2 Principles of Physical Chemistry in Physiology

Gases

Gas laws

Universal gas law The volume of a gas is directly proportional to the absolute temperature (T) of the gas (Charles' law) and inversely proportional to the pressure (P) applied upon it (Boyle's law). The volume is also directly proportional to the number of moles of gas (n). Stated mathematically:

$$V(volume) = n \times \frac{T}{P}$$

or

$$PV = nRT$$

where R is the proportionality constant and is called the *universal gas constant*.

The variability in gas volume with temperature and pressure makes it imperative to express all gas volumes at standard temperature and pressure (STP). The standard temperature is 0°C (273 Kelvin) and the *standard pressure* is 1 atmosphere (760 mmHg).

In physiology, gas volumes are often expressed at body temperature (37°C) and 1 atmosphere pressure, or BTP for short.

Dalton's law of partial pressure When two or more gases are mixed, the total pressure exerted by the mixture is equal to the sum of the pressures exerted by the component gases when occupying the same volume.

Suppose 500 ml of O_2 and 500 ml of N_2 at 0°C and 760 mmHg are mixed together to make a 1 L gaseous mixture (Fig. 2.1). What will be the pressure exerted by the mixture? Since the final volume of the gaseous mixture is 1 L, both the gases will come to occupy 1 L each (and not 500 ml each). In accordance with Boyle's law, the pressure of each gas will drop to 380 mmHg. Finally, in accordance with Dalton's law of partial pressure, the pressure of the mixture will be 380 + 380 = 760 mmHg.

In the above example, the pressure of 380 mmHg will be called the *partial pressure* of each of the constituent gases. It is calculated by the formula:

$$\text{Partial pressure of gas (380 mmHg)} = \frac{\text{Total pressure}}{\text{(760 mm Hg)}} \times \frac{\text{Volume of the gas at STP (500 ml)}}{\text{Volume of the mixture at STP (1 L)}}$$

$$= \text{Total pressure} \times \frac{\text{Concentration of a gas in a gaseous mixture}}{}$$

Taking another example:

The concentration of O_2 in atmospheric air = 21%. Therefore, its partial pressure in atmospheric air will be 760 × 21% = 160 mmHg

The partial pressure of a gas dissolved in a liquid is equal to the partial pressure of the gas in a gaseous mixture that is in equilibrium with the liquid.

Aqueous tension When water vapor saturates a gas or a gas mixture, it exerts an additional pressure of 47 mmHg at 37°C. This pressure is called the *saturated vapor pressure* or the *aqueous tension*. When air is saturated with water vapor, the concentration of all the constituent gases decreases. However, the percentage composition of a gas mixture is always expressed assuming that there is no moisture in it. For example, when we say that O_2 constitutes 21% of atmospheric

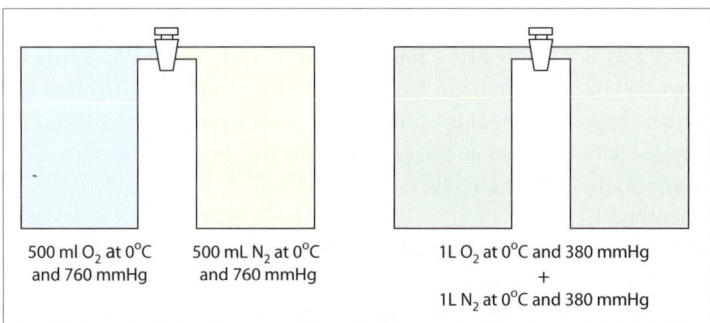

500 ml O_2 at 0°C and 760 mmHg   500 mL N_2 at 0°C and 760 mmHg   1L O_2 at 0°C and 380 mmHg
+
1L N_2 at 0°C and 380 mmHg

Fig. 2.1 An example of Dalton's law of partial pressures.

air, we actually mean 21% of 'dry' atmospheric air. If the air is saturated with water vapor, the saturated vapor pressure has to be deducted from the atmospheric pressure before calculating the partial pressure of constituent gases individually. For example, the partial pressure of O_2 (Po_2) in humidified atmospheric air (at 37°C) is $(760 - 47) \times 21\% = 150$ mmHg.

In physiology, it is a common practice to record the volume of gases saturated with water vapor at body temperature and at atmospheric pressure. Volumes thus recorded are denoted as BTPS ('S' for saturated).

Solutions

Diffusion of solutes

Simple diffusion   If there are two solutions of different solute concentrations separated by a membrane, the solutes diffuse from higher to lower concentration. This is because the molecules in solution, due to their constant random motion, spread out to fill all of the available volume. The greater the concentration of a substance in solution, the greater is its tendency to spread out. Hence, although solutes in both the solutions are diffusing farther apart, there is a net flux of solutes from   higher to lower concentration. The magnitude of the diffusing tendency from one region to another is directly proportional to the cross-sectional area across which the diffusion takes place and the concentration gradient, which is the difference in concentration of the diffusing substance divided by the thickness of the boundary *(Fick's law of diffusion)*. Thus,

$$J = -DA \frac{\Delta c}{\Delta x}$$

where J is the net rate of diffusion, D is the diffusion coefficient, A is the area of the partition that separates the two solutions, $\Delta c/\Delta x$ is the concentration gradient, i.e., the rate of change of concentration with distance. The minus sign is a sign-convention that indicates the direction of diffusion. When solutes diffuse from higher to lower concentration, $\Delta c/\Delta x$ is negative and therefore multiplying by $-DA$ gives a positive value.

Nernst potential   When two ionic solutions A and B of different concentrations (C_A and C_B) are separated by a permeable membrane, the ions tend to diffuse along their concentration gradient. Since ions are charged particles, their diffusion can be stopped by an appropriate electrical potential (E) applied across the membrane. The magnitude and polarity of the potential that must be applied to side A of the membrane for stopping the diffusion of ions (E_A) is given by the *Nernst equation*:

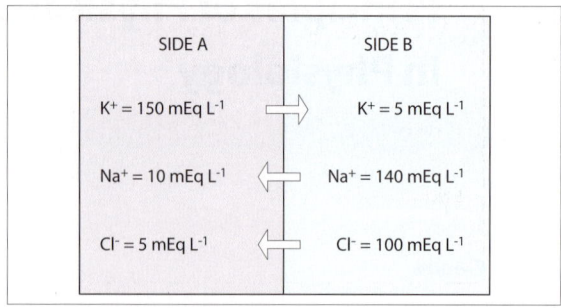

Fig. 2.**2**   Ionic distribution in two compartments for illustrating the principles of Nernst potential.

$$E_A = \frac{61}{z} \log \frac{[C_B]}{[C_A]}$$

where z is the valence of the ion. This equation can be also written as:

$$E_A = \frac{-61}{z} \log \frac{[C_A]}{[C_B]}$$

In the example illustrated in Figure 2.**2**, the application of Nernst equation to the concentration of K+ on side A (150 mEq L^{-1}) and side B (5 mEq L^{-1}) gives a value of -90 mV. This potential is called the *Nernst potential* for K+ (E_K).

$$E_K = \frac{-61}{+1} \log \frac{150}{5}$$

It has *two* implications. (1) It means that -90 mV applied to side A will prevent any outward diffusion of the K+. (2) It also means that if the K+ concentrations are equal on both sides of the membrane, a potential difference of -90 mV applied to the membrane will produce the same rate of diffusion as an ionic concentration ratio of 30 (i.e., $150 \div 5$).

Similarly, the diffusion of Na+ will be prevented by a potential of $+70$ mV (E_{Na}) applied to side A and the diffusion of Cl- (E_{Cl}) will be prevented by applying a potential of -80 mV to side A.

$$E_{Na} = \frac{-61}{+1} \log \frac{10}{140}$$

$$E_{Cl} = \frac{-61}{-1} \log \frac{5}{100}$$

Thus every ion distributed asymmetrically across the membrane has its own Nernst potential (E) that will prevent its diffusion. In other words, the Nernst potential of an ion gives the electrical equivalent of its diffusion energy.

Gibbs–Donnan equilibrium   If two compartments are separated by a semipermeable membrane and one of the compartments contains diffusible ions, the ions

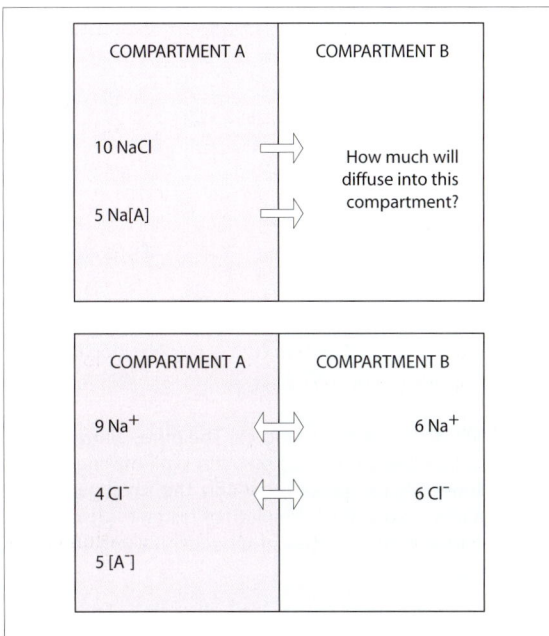

Fig. 2.**3**   Ionic distribution in two compartments for illustrating the principles of Gibbs–Donnan equilibrium. [A⁻] denotes an impermeable anion.

Table 2.**1**   Ionic distribution at Gibbs–Donnan equilibrium.

	Inside	**Outside**
Total +ve charge	9(Na^+)	6(Na^+)
Total –ve charge	9(4 Cl^- +5 A^-)	6(Cl^-)
Product of diffusible ions [Na^+] × [Cl^-]	36(9 Na^+ × 4 Cl^-)	36(6 Na^+ × 6 Cl^-)
Total number of ions	18(9 + 4 + 5)	12(6 + 6)

up, the microscopic appearance of which is known to pathologists as the cloudy swelling.

■ Diffusion of solvent

Osmosis is the movement of solvent across a semipermeable membrane that separates two solutions of different solute concentrations and restricts the movement of solutes across it. For osmosis to occur, it is not necessary for the membrane to be totally impermeable to the solute. A solute with limited permeability also produces osmosis, though to a lesser extent. The solute permeability is given by the *reflection coefficient* (s), which varies between 0 (freely permeable) and 1 (totally impermeable). The reflection coefficient is the probability of a solute molecule reflecting back from the membrane instead of passing through it. The reflection coefficient of a solute is not an absolute constant but varies with the type of membrane. It will be zero for all solutes if the membrane has very large pores in it. Osmosis is proportional to the reflection coefficient as well as to the concentration gradient. Osmosis does not occur when the reflection coefficient is zero, i.e., when the membrane is freely permeable to the solute.

Osmotic pressure   The osmosis from a dilute to a concentrated solution can be prevented by increasing the hydrostatic pressure of the concentrated solution. The hydrostatic pressure necessary to prevent the osmosis is called the osmotic pressure of the solution (Fig. 2.**4**).

Osmotic pressure, like freezing-point depression and boiling-point elevation, is a *colligative property*, i.e., it depends on the number rather than the type of particles in a solution. It is given by the *Van't Hoff relationship*:

$$\pi = RT \times \phi \, nC$$

where π is the osmotic pressure, R is the universal gas constant, T is the absolute temperature, ϕ is the osmotic coefficient, n is the number of ions yielded by the dissociation of a solute molecule, C is the molar concentration (moles L^{-1}) of the solute.

The Van't Hoff equation shows that at a constant

---

will soon distribute themselves equally across the two compartments. If, however, one of the compartments (compartment A) contains impermeable ions (Fig. 2.**3**), the typical equilibrium (with equal numbers of ions on both sides of the membrane) as observed in simple diffusion will not occur. Instead, a different type of equilibrium called the *Gibbs–Donnan equilibrium* will occur in which two conditions must be satisfied: (1) Both the compartments must be electroneutral, and (2) the product of *diffusible* ions (anions and cations) must be equal in both compartments. The solution is shown in Table 2.**1**.

At equilibrium, compartment A will have a larger concentration of solutes (18) than compartment B (12). As a result, water will move from B to A due to osmosis and thereby, decrease the solute concentration in compartment A. However, since the impermeable anion continues to be present in compartment A, the Gibbs–Donnan equilibrium is re-established. As a result, compartment A would continue to have a higher solute concentration, no matter how much osmosis occurs!

A living cell contains impermeable proteins anions. This imposes the Gibbs–Donnan equilibrium, so that there are more ions inside the cell than outside it. As a result, water continuously moves into the cell by osmosis, which will inevitably rupture the cell. To survive, the cell must continuously pump out excess ions. It does so with the help of the ATP-driven Na^+-K^+ pump. When a cell dies, the pump stops and the cell swells

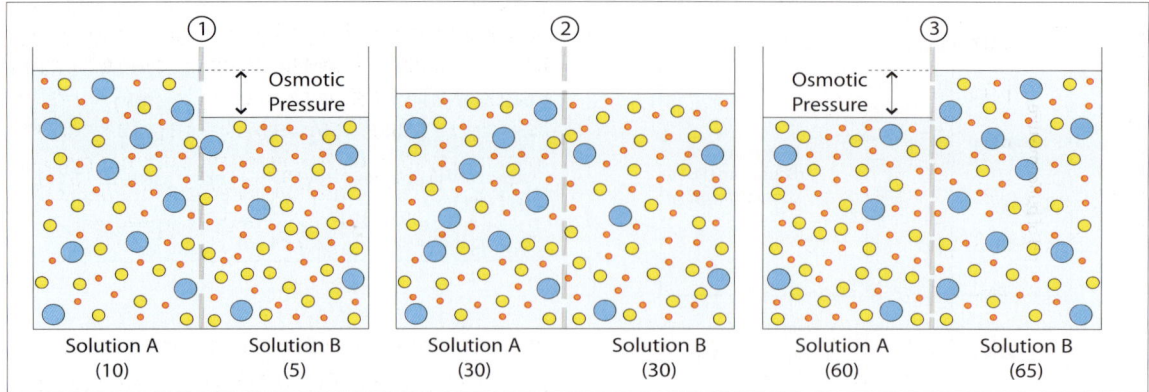

Fig. 2.4 A membrane separates two solutions, A and B, containing ions of three different sizes. Solution A contains 60 particles: 10 large, 20 medium-sized, and 30 small. Solution B contains 65 particles: 5 large, 25 medium-sized, and 35 small. The tonicity of the solutions depends on the membrane pore size. The figures below each solution indicate the number of osmotically active particles in it. (1) The pores allow only the medium and small particles to pass through, making solution A hypertonic. (2) The pores allow only the small particles to pass through, making the two solutions isotonic. (3) The pores are impermeable to particles of all sizes, making solution B hypertonic.

temperature, osmotic pressure is proportional to the number of particles in solution per unit volume of solution (given by n × C in the formula). The *osmotic coefficient* (ϕ) in the formula is not the same as the reflection coefficient. The osmotic coefficient does not depend on membrane characteristics: it is constant for a solute molecule and is assumed to be 1.0 in all approximate calculations.

Osmoles, osmolarity and osmolality   One mole[1] of osmotically active particles is called one *osmole* (Osm). Thus, a molar solution of glucose contains 1 osmole, a molar solution of NaCl contains 2 osmoles (1 mole of Na^+ and 1 mole of Cl^-) while a molar solution of $CaCl_2$ contains 3 osmoles (1 mole of Ca^{2+} and 2 moles of Cl^-). The osmolar concentration of a solution in Osm L^{-1} is called *osmolarity*. When expressed in Osm kg^{-1} of solution, it is called *osmolality*. Osmolarity is affected by temperature, which changes the volume of solution. Also, dissolution of solutes is associated with a slight rise in the volume of the solution. This increase is different for different solutes. Hence, when 1 mole of glucose is dissolved in 1 L of water, the osmolarity will be slightly less than 1 Osm L^{-1}. Osmolality however is unaffected by temperature changes or the increase volume of the solution that accompanies dissolution.

The measurement of osmolarity (nC) is based on the principle that the freezing point of a solution is depressed in proportion to the number of osmoles present in it. It is given by the formula:

$$nC = \Delta T_f \div 1.86$$

where ΔT_f is the reduction in the freezing temperature. The freezing point of normal human plasma averages –0.54°C, which corresponds to an osmolal concentration in plasma of 290 mOsm L^{-1}. This is equivalent to an osmotic pressure against pure water of 7.3 atmospheres.

Two solutions having identical osmolarity are called *isoosmolar*. They exert the same osmotic pressure and are therefore also called *isoosmotic*. If one of the two has greater osmolarity, it is said to be *hyperosmolar* or *hyperosmotic* in comparison to the other, which is called *hypoosmolar* or *hypoosmotic*. Except immediately after a sudden change in composition, all fluid compartments of the body are in osmotic equilibrium.

Tonicity   When a membrane with unknown characteristics separates two solutions with different osmolarity, it does not necessarily mean that there will be osmosis from the hypoosmolar to the hyperosmolar solution. This is made clear by the example illustrated (Fig. 2.4) in which the solutions have solutes of three different sizes: large, medium, and small.

The three solutes are present in different concentrations. Solution A contains 10, 20, and 30 particles, per unit volume, of the large, medium, and small solutes respectively. Solution B contains 5, 25, and 35 particles per unit volume. It is seen that the magnitude

---

[1] A 'mole' is the name of a quantity. Just as a dozen equals 12 and a score equals 20, a mole equals 6.023×10^{23}. The weight of one mole of atoms or molecules of any element or compound is exactly equal to the atomic or molecular weight of the element or compound expressed in grams. For example, the molecular weight of NaCl is 58.5. Therefore, 6.023×10^{23} molecules (i.e., 1 mole) of NaCl weighs exactly 58.5 grams.

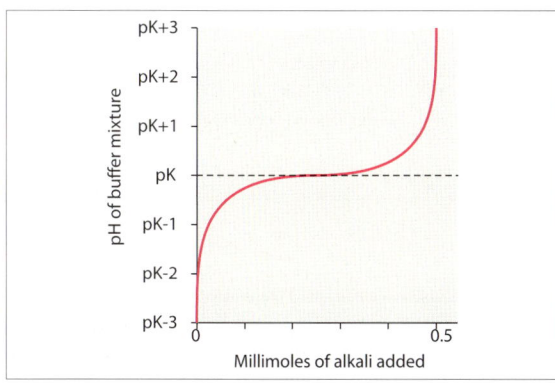

Fig. 2.**5**   The titration curve of a buffer mixture with alkali. The buffering power is maximum at pK.

and direction of osmosis cannot be predicted without knowing the pore-size of the membrane. (1) If the pores exclude only the largest particles but allow the medium and small particles to pass through, solution A will be hypertonic with respect to solution B although the latter is hyperosmolar. This is because the number of osmotically active particles in solution A is 10 while solution B has only 5. (2) If the pores exclude the large and medium sized particles but allow the small particles to pass through, the two solutions will become isotonic and no osmosis will occur. This is because the total number of osmotically active particles will be 30 in both solutions: 10 + 20 in solution A and 5 + 25 in solution B. (3) If the pores do not permit any solute to pass through, solution B will be hypertonic with respect to solution A, i.e., there will be osmosis from compartment A to B. This is because the total number of osmotically active particles will be 60 in solution A and 65 in solution B.

In real-life situations, the complexity of biological membranes makes it impossible to accurately predict whether a solute particle will pass through it. Moreover, body fluids contain innumerable types of solutes, which is another reason why their tonicity cannot be predicted. The only way, therefore, of estimating the tonicity is to determine it experimentally. In clinical parlance, the word tonicity always refers to the tonicity of a solution with respect to an erythrocyte. In other words, it is the erythrocytic cell membrane across which the tonicity is tested. If the erythrocyte shrinks in a solution by losing water through osmosis, the solution is *hypertonic*. If the erythrocyte swells up in a solution by gaining water through osmosis, the solution is *hypotonic*. If the erythrocyte neither shrinks nor swells in the solution, the solution is called *isotonic*.

The fluid used in clinics for intravenous transfusion is both isotonic and isoosmolar to the plasma. A 0.9% saline solution is most often used for the purpose. The

0.9% NaCl solution is roughly isoosmolar (308 mOsm L^{-1}) to the body fluids (290 mOsm L^{-1}) as can be calculated easily.[2] A 5% glucose solution is also isotonic initially when infused intravenously, but as the glucose is metabolized, the solution gradually becomes hypotonic. An isoosmolar solution of urea will not be isotonic since urea rapidly diffuses into the erythrocytes.

■ Buffers

The pH of a solution is maintained at near constant levels with the help of buffers (Fig. 2.**5**). Most body buffers comprise weak acids [HA] with their conjugate bases [NaA]. By the law of mass action:

$$k = \frac{[H^+][A^-]}{[HA]}$$

where k is the dissociation constant.

$$[H^+] = k \times \frac{[HA]}{[A^-]}$$

$$-\log [H^+] = -\log k - \log \frac{[HA]}{[A^-]}$$

$$pH = pK + \log \frac{[A^-]}{[HA]}$$

The above equation is known as the *Henderson–Hasselbalch* equation.

The *buffering power* of the system is the number of moles of acid or base that must be added to 1 mole of the buffer to change its pH by 1 unit. The maximum buffering power of any buffer is 0.575. The buffering power of a buffer is maximal when the pH of the solution is identical with that of the pK value, i.e., when pH = pK, [A⁻] = [HA]. It also depends on the concentration of the buffer.

■ Chemical equilibrium

All chemical reactions are potentially reversible but they tend to proceed unidirectionally if one of the reactants or products is removed from the field of reaction.

---

[2] Molecular weight of NaCl = 58.5.

∴ 58.5 g of NaCl contain 1 mole of NaCl molecules.

∴ A solution containing 58.5 g NaCl will contain 2 osmoles (1 osmole of Na^+ and 1 osmole of Cl^-)

∴ A solution containing 9 g of NaCl contains $(2 \times 9)/58.5$ moles = 0.308 osmoles (or 308 millismoles)

∴ The osmolarity of a 9 g L^{-1} (0.9 g%) solution = 308 mOsm L^{-1}

Consider the reaction shown below.

$$CO_2 + H_2O \xrightleftharpoons[\text{anhydrase}]{\text{Carbonic}} H_2CO_3 \leftrightarrow H^+ + HCO_3^-$$

The reaction will proceed in forward direction if at least one of the products is quickly removed, e.g., if the H^+ is buffered. On the other hand, if the CO_2 escapes from the reaction mixture, the reaction will proceed in the backward direction.

■ Electrophoresis

Electrophoresis is the migration of charged solutes or particles in a liquid medium under the influence of an electrical field (Fig. 2.**6**). Positively charged ions (cations) migrate towards the cathode and negatively charged ions (anions) migrate towards the anode. An *ampholyte* (earlier called *zwitterion*) takes on a positive charge in an acidic solution and negative charge in a basic solution. Depending on its pH, therefore, an ampholyte will migrate towards the anode or cathode. At a particular pH called the *isoelectric pH*, an ampholyte will be neutral and therefore will not move in an electric field. Because proteins contain ionizable amino (NH_2) and carboxyl (COOH) groups, they behave as ampholytes in solutions. Nucleic acids too are ampholytes.

The rate of electrophoretic migration depends on (1) the net electrical charge of the molecule, (2) the

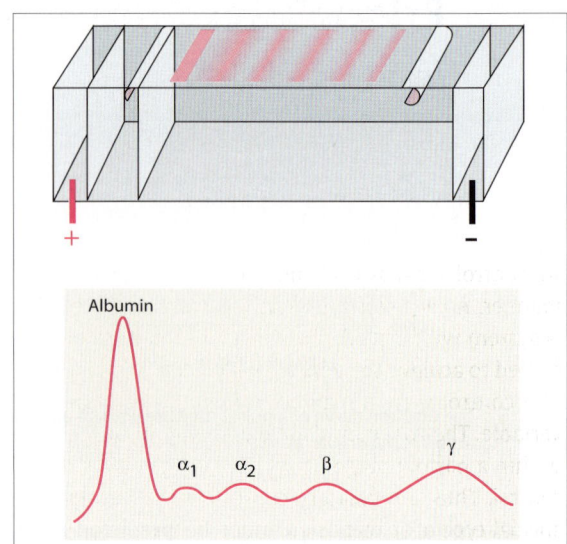

Fig. 2.**6** (*Above*) Electrophoresis apparatus. (*Below*) Major components of plasma proteins separated out through electrophoresis.

size and shape of the molecule, (3) electrical field strength, (4) properties of the supporting medium, and (5) temperature of operation. Electrophoretic mobility is defined as the rate of migration (cm s^{-1}) per unit field strength (volts cm^{-1}). Since electrophoretic mobility depends, among other things, on the size and shape of the molecule, electrophoresis is used for the separation of different proteins in a solution.

3 Principles of Control Systems in Physiology

To control means to change something in a desired manner. A *control system* can therefore be defined as 'a system with a goal'. The things that have to be controlled to achieve the goal are called *variables*. The goal of a control system may be to increase or decrease a variable. The goal may also be to maintain the variable within a narrow range, in which case it is a *regulatory control*. Thus, a regulatory control can be defined as a special type of control meant for maintaining the constancy of a variable. On the other hand, a *servo control* sets the variable at any predetermined level that is not necessarily within a narrow range.

The above concepts can be understood easily by considering the example of a fan-regulator. The regulator controls (increases or decreases) the speed of the fan (a variable), which in turn controls the amount of evaporation (another variable) of our sweat. The final result is the regulation of our body temperature (a third variable) that is maintained within physiological limits. The regulation is achieved through the servo control of the fan. Similarly, there is a control system in our body that controls the cardiac output (increases or decreases it) in order to regulate (maintain constant) tissue oxygenation. Other physiological examples are the regulation of blood pressure, regulation of blood glucose, regulation of body temperature, regulation or control of posture, control of heart rate, and control of breathing.

A special type of regulation is *homeostasis* which refers to the regulation of the internal environment or *milieu intérieur* of the body. The importance of homeostasis becomes apparent when we consider the environmental adversities faced by a unicellular organism like the ameba with the cozy environment of any human cell. The human cell enjoys a steady supply of oxygen at 100 mmHg partial pressure and glucose at a concentration of 100 mg dl^{-1}. Its metabolites are promptly washed away from its immediate vicinity. It basks in the cozy comfort of 37°C and is secure in the protective cover of an elaborate defense system. All this is made possible by the cooperation of all the cells that constitute a multicellular organism in which each cell plays its part and together, ensure the maintenance of a steady internal environment. For example, when the body temperature falls it is sensed by a group of cells in the brain that signals to the muscles to step up

their activity so that the temperature can be restored. Likewise, each parameter that is held at a constant level by the body is the result of a unique homeostatic mechanism.

Regulatory Mechanisms

Several broad principles of regulatory system can be appreciated by considering the example of an air-conditioned hall (Fig. 3.1). The temperature inside the air-conditioned hall is regulated so that it remains constant (say, at 25°C). Whenever the temperature of the hall rises, the message is signaled to the control room. In a few minutes, an extra blast of cold air cools the hall. Conversely, when the hall temperature falls below 25°C, the cooling power of the air conditioner is reduced. All the essential elements of a regulatory system are present in the above example. There is a sensor or receptor that senses the variable (temperature inside the hall). There is an input that signals the information to the control center and there is an output (electrical power supply to the air conditioning plant) that acts on the motor or the effector (the pump delivering the cold blast) that restores the variable to its original level.

Closed loop (guided) control   The control system described above is an example of a closed loop control. The loop begins with the sensing of any change in the variable and ends with the correction of the same variable. In other words, a closed loop control system begins and ends with the controlled variable. Such a

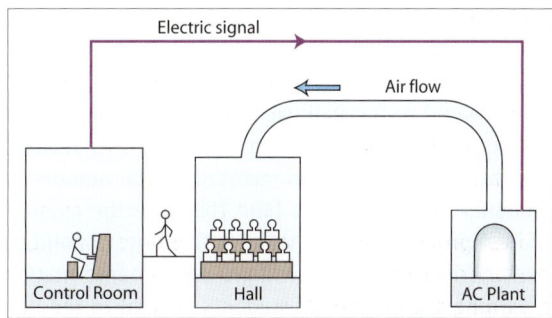

Fig. 3.1   A manually controlled temperature-regulatory system in a large hall can explain several principles of a control system.

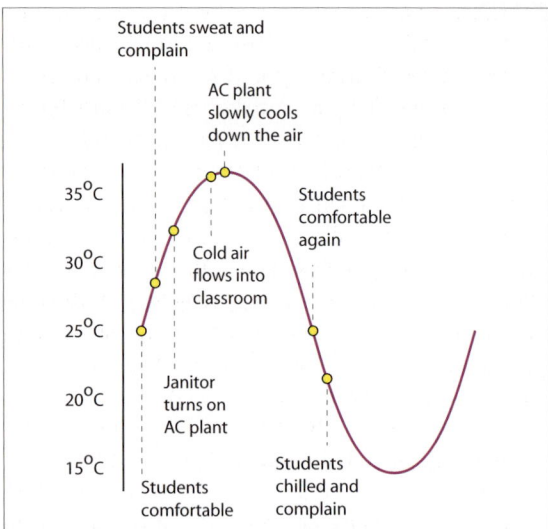

Fig. 3.**2** Genesis of temperature oscillations in an air-conditioned hall. The temperature regulatory system is shown in Fig. 3.**1**.

system can be a crude one, manned by an untrained janitor who would only know when to increase or decrease the cooling power but not by how much. Hence, it is likely that each time the temperature drifts from the set point, it gets overcorrected resulting in temperature oscillations (Fig. 3.**2**). Despite its simplicity and its obvious propensity for oscillations, this system guarantees through repeated corrections that the hall temperature will hover around the desired temperature. It is called *guided control* and requires virtually no knowledge of the factors that affect the variable.

Open loop (parametric, anticipatory) control Not all control systems have closed loops. Consider the same example of an air-conditioned hall. A brilliant engineer decides that he will not let the hall temperature drift at all so that there is no need whatsoever of correcting the temperature. So he derives a complex formula for calculating the amount of heat lost from the hall each minute and the amount of body heat produced by the students present in the hall. From this formula, he calculates the exact amount of cold air that is to be delivered to the hall each minute so that the hall temperature does not change. The engineer realizes that his formula needs to incorporate the number of students present the hall (and therefore the amount of heat produced in the hall). He therefore appoints a person (the sensor) who will provide him information regarding the number of students present in the hall during a lecture. This type of control system is an open loop control system. Here, the controlled variable (hall temperature) is *not* the same as the feedback (number

of students) provided by the sensor. The control system tries to cool the hall *in anticipation* even before it starts warming up. Although the system is extremely prompt, it does not have any temperature sensors to judge if the hall temperature has been properly corrected. The open-loop control system is also called a *parametric control system* since it is heavily dependent on parameters and calculations based on them. Parametric control might appear to be only a theoretical proposition that is impracticable. That is not true. The vestibuloocular reflex (p 751) is an example of parametric control in which the vestibular apparatus (sensor) detects jerky head movements and reflexly corrects ocular gaze (controlled variable).

Normally, a parametric control system does not depend on information related to the controlled variable. However, if the same information is provided to it, parametric control becomes more efficient and is called *parametric feedforward control*. However, it still will never match the accuracy of a guided control system. This can be explained by extending the example of the air-conditioned hall as follows. Suppose at the end of a lecture, the students present in the hall complain to the engineer that the temperature of the hall was warmer than it should have been. The engineer realizes that he needs to revise his temperature-correcting formula. He therefore goes back to the hall and does a detailed survey of the hall to find out the reasons for the failure of his formula. He realizes that the projector used by the professor was generating a lot of heat and therefore incorporates this factor too in his new revised formula which turns out to be better and the students are satisfied. It is important to realize that the feedback regarding the controlled variable was not a continuous feedback and therefore, is essentially different from guided control. In parametric feedforward control, the control system learns and becomes wiser with time and the control formula becomes more and more accurate.

What happens if the engineer is unable to detect the cause of failure to maintain the hall-temperature at the set-point? After all his efforts to detect the cause have failed, he will probably increase the cooling power a little above that suggested by his formula. He does so without understanding why his formula has failed, which could be an undetected leak in the airconditioning duct. Such a system is called *parametric feedback control*.

Combined control Finally, there will always be certain factors, known or unknown, that are so unpredictable and random that no adjustments in the formula are possible for them. Such disturbances are called *noise* which is an unavoidable component of any control

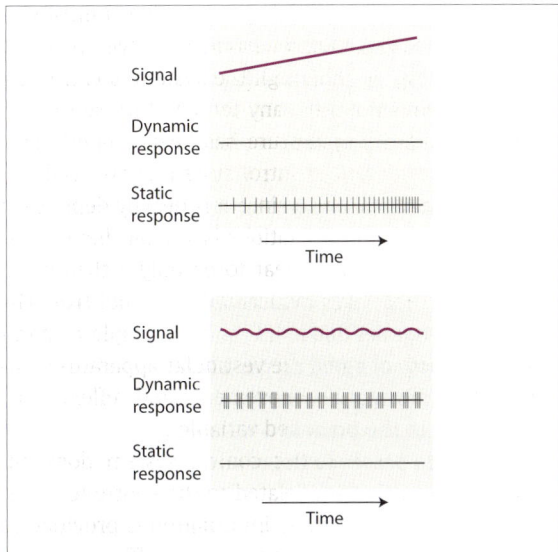

Fig. 3.3 Static versus dynamic sensitivity of receptors. (*Above*) A signal that rises slowly to a high amplitude remains undetected by receptors with dynamic sensitivity but stimulates receptors with static sensitivity. (*Below*) Low amplitude fluctuations go undetected by receptors with static sensitivity but cannot escape detection by receptors with dynamic sensitivity. Note that the density of the vertical lines (action potentials) indicates the intensity of receptor stimulation.

system. Parametric control systems have no way of dealing with noise. Only guided control systems can deal with noise. Hence, parametric and guided control systems can be integrated into a remarkably efficient control system in which parametric control ensures quick and near-perfect correction of the variable to its set point while guided control slowly but surely corrects any residual error that is attributable to noise.

Static versus dynamic sensors   In the air-conditioning example, it is important to know how much the temperature has drifted from the set point if feedback correction is to be instituted. Another crucial piece of information required for the correction is the rate at which the hall temperature is changing. If the hall temperature is increasing at an alarming rate, the air-conditioning plant has to be turned on full-blast. On the other hand, if the hall temperature is rising slowly, the blast needs only to be gently turned on or, the temperature will overshoot. To acquire this information, it is important to install in the hall two types of sensors: *static sensors* that will signal the extent of temperature drift, and *dynamic sensors* that will signal the rate of temperature change.

A static sensor is ideal for detecting very small, slow changes in the variable, whereas a dynamic sensor is suited for detecting rapid fluctuations in the variable.

If receptors with dynamic sensitivity are absent, regulation will fail whenever there are rapid fluctuations in the variable. Conversely, in the absence of receptors with static sensitivity, a large change in the variable can go undetected if the change occurs extremely slowly (Fig. 3.3). It is therefore advantageous to adjust the static and dynamic sensitivity of the receptors depending on how the variable tends to change. If the variable is changing extremely rapidly, the dynamic sensitivity needs to be increased. If the variable is changing extremely slowly, the static sensitivity needs to be increased. Where the rate of change in the variable is assuredly slow, dynamic sensitivity of receptors is not required. This is true for the chemical composition of body fluids and therefore chemoreceptors do not show dynamic sensitivity. However mechanoreceptors mostly have two subtypes with static and dynamic sensitivity (see muscle spindle, p 661).

The total time taken by the afferent sensory information to reach the control center and the time taken by the efferent corrective measures to take effect constitute the *feedback correction time*. An excessive delay in the feedback correction makes the regulatory system susceptible to oscillations (*see below*). The *gain* of feedback control may be defined as the ratio of the intensity of the corrective response to the intensity of the sensory input to the control center. In simpler words, when the control center overreacts to the sensory information, the gain is high; when it underreacts, the gain is low. If the gain is low, the regulatory system will fail when there are rapid fluctuations in the variable. Conversely, a high gain will overcorrect a slowly changing variable, making the regulatory system prone to oscillations.

$$\text{Gain} = \frac{\text{Intensity of corrective response}}{\text{Intensity of sensory information}}$$

Oscillations in a regulatory system

A regulatory mechanism is sometimes prone to overcorrection, as is exemplified by the air conditioning-plant analogy. When the hall warms up, the lecture-hall attendant goes to the control center to cool it down. But by the time he returns to the hall, he might find all the students shivering with cold, especially, if the control has a high gain. The attendant again sets off for the air-conditioning plant to turn down its cooling power. The oscillations resulting from such a regulatory system are depicted in Figure 3.2. The oscillations can be prevented if the change in the variable can be detected early and the corrective measure is instituted quickly. In other words, the receptor sensitivity has

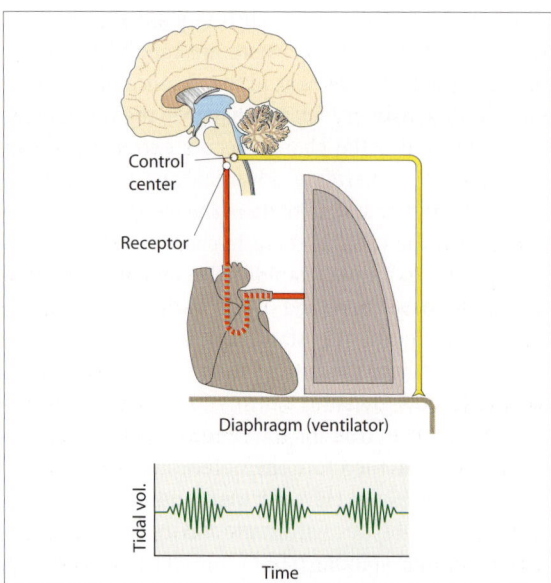

Fig. 3.**4**   Cheyne–Stokes breathing. (*Above*) The circulatory and neural pathways that underlie the oscillations in the respiratory regulatory system. These oscillations are analogous to the oscillations that would occur in an air-conditioning regulatory system shown in Figure 3.**1**. (*Below*) The spirogram in Cheyne–Stokes breathing.

to be high and the feedback correction time has to be short. A third factor that will dampen oscillation is the size of the hall. A large hall will have a fairly stable temperature: it will not warm up or cool down quickly.

A physiological example of oscillations is seen in Cheyne–Stokes breathing. The depth of quiet breathing is normally uniform. However, in Cheyne–Stokes breathing, the depth of quiet breathing shows rhythmic oscillations as shown in Figure 3.**4**. Its causes are briefly mentioned here for illustrating the factors that make a regulatory system oscillate in a physiological situation.

In the chemical control of breathing, the partial pressure of CO_2 in alveolar air is regulated by controlling the rate and depth of breathing. When breathing slows down, Pco_2 rises in alveolar air. Simultaneously, the blood Pco_2 rises due to its equilibration with alveolar air. The flow of blood acts as a slow messenger, conveying the information on alveolar Pco_2 to the receptors located in the brain in a little over 7 seconds. The message then passes to the control center in the brain itself. The control center initiates corrective changes in pulmonary ventilation in order to bring down the alveolar Pco_2. The regulatory system is shown schematically in Figure 3.**4**.

This regulatory system starts oscillating in three types of situations. (1) Any prolongation of the time taken by the blood to flow from the lungs to the brain increases the feedback correction time. This allows the alveolar Pco_2 to rise to high levels before the information reaches the brain receptors, and thereby sets off oscillations. (2) A reduction in the sensitivity of the receptors in the brain to blood Pco_2 allows the alveolar Pco_2 to rise to high levels before correction is instituted, and the system starts oscillating. (3) A reduction in the *functional residual capacity* of the lung results in rapid changes in Pco_2. This predisposes to oscillations in ventilation.

The Cell

Cell Organelles

An *organelle* is a membrane-limited subcellular entity that can be isolated by centrifugation at high speeds. It includes the nucleus, mitochondria, endoplasmic reticulum, Golgi apparatus, peroxisome, lysosome, and the plasma membrane. The nucleolus, ribosomes, and cytoskeleton proteins are not organelles because they are not membrane-bound. Although not organelles, the nucleolus and ribosome are considered here with the nucleus for functional reasons.

The various subcellular fractions can be separated by differential centrifugation. *Low-speed* centrifugation precipitates the nuclei. *Intermediate-speed* centrifugation separates out the mitochondria, lysosomes, and peroxisomes. *High-speed* centrifugation precipitates the free ribosomes and endoplasmic reticulum, both smooth and rough. These precipitates are collectively called *microsomes*, a term of common clinical usage. For example, the commonly used term *hepatic microsomal enzymes* refers to the enzymes contained in smooth endoplasmic reticulum in liver cells. These enzymes are important for several important physiological reactions and for detoxification of drugs and poisons.

Apparatus for protein synthesis

The **nucleus** consists mainly of the chromosomes. Except in germ cells, the chromosomes occur in pairs, one originally from each parent. Each chromosome consists of a long strand of DNA (see Fig. 7.1). It is wrapped at intervals around a core of histone proteins to form a nucleosome (Fig. 4.1). Thus, a chromosome is like a string of beads. The whole complex of DNA and proteins is called chromatin. When the cell is not dividing, only clumps of chromatin can be discerned in the nucleus. During cell division, the coiling around histones is loosened, probably by acetylation of the histones, and the chromosomes become distinct.

The nucleus is enclosed in a double membrane bearing pores (nuclear pores) through which mRNA can pass out and some of the proteins synthesized in the cytosol can enter back into the nucleus. The nuclear pores are guarded with two rings that open and close to regulate the passage of large molecules. The transport through the nuclear pore requires proteins called importins and exportins.

Nucleolus   The nucleus of most cells, especially the growing cells, contains one or more nucleoli. The nucleolus is the site of ribosome synthesis. It contains the genes for ribosome synthesis along with a considerable amount of RNA and proteins representing ribosomes in various stages of production. Cells that are active in protein synthesis have prominent nucleoli.

Ribosomes are the sites of protein synthesis (Fig. 4.1). They are small granules of RNAs. Ribosomes may be present free in the cytosol or on the surface of the rough endoplasmic reticulum. The *free ribosomes* synthesize cytoskeletal proteins and other cytoplasmic proteins such as hemoglobin. The proteins synthesized by them

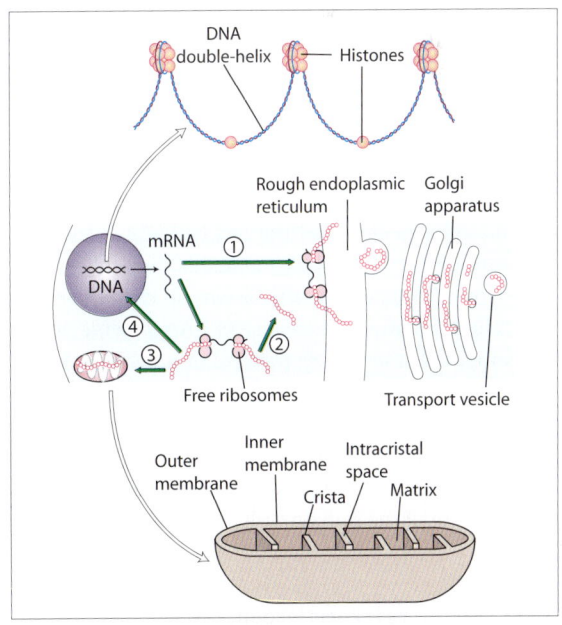

Fig. 4.1   Peptides synthesized on the rough endoplasmic reticulum mostly end up in the Golgi apparatus (1) while those synthesized by the free ribosomes end up either as cytoplasmic proteins (2), or find their way into the mitochondria (3) or the nucleus (4). The DNA is magnified above, showing it as a 'string of beads'. DNA forms the strings while nucleosomes form the beads. The mitochondrion is magnified below, showing the outer and inner membranes, the matrix, and the intracristal spaces.

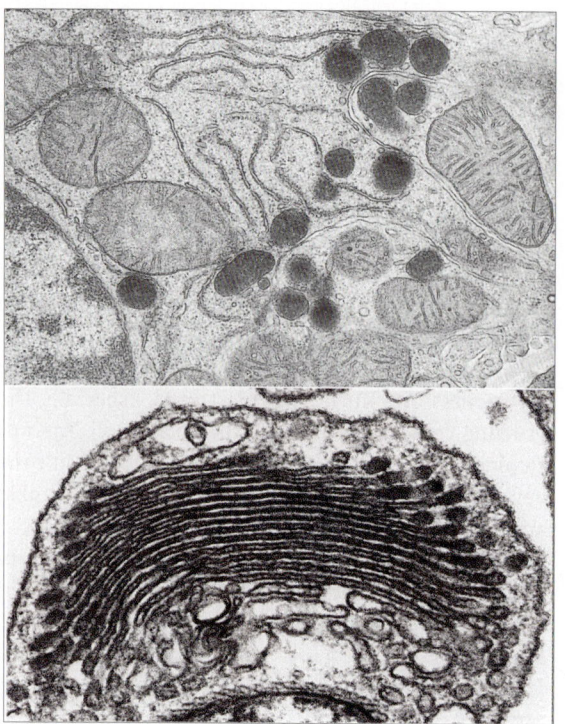

Fig. 4.**2** (*Above*) Mitochondria, rough endoplasmic reticulum and peroxisomes (× 10 000). (*Below*) Golgi body (× 33 000).

also find their way into the nucleus, mitochondria, and peroxisomes (Fig. 4.**2**). The *bound ribosomes* are located on the rough ER. They synthesize all membrane proteins and most of the proteins destined for secretion. The proteins synthesized are extruded into the rough ER. Bound and free ribosomes are structurally identical and interchangeable, and the cell can adjust the relative number of each as its metabolism changes. Both types of ribosomes usually occur in clusters called *polyribosomes* attached to one mRNA molecule, an arrangement that increases the rate of polypeptide synthesis. The ribosomes of eukaryotes and prokaryotes are entirely dissimilar. Hence in bacterial infections, drugs like tetracycline and streptomycin are able to selectively inhibit the prokaryotic ribosomes of the bacteria but not of the human cells.

Secretory (exocytic) apparatus

Rough endoplasmic reticulum The endoplasmic reticulum (ER), whether rough or smooth, is made of a series of membranous tubules in the cytoplasm of the cell (Figs. 4.**1** and 4.**2**). The rough ER has ribosomes attached to its cytoplasmic side. It is concerned with protein synthesis and the initial folding of polypeptide chains. After the polypeptide chain is formed, it is modified (posttranslational modification) to the final protein by chemical modifications of the amino acid residues, cleavage of the polypeptide chain, and folding and packaging of the protein into its ultimate, often complex configuration.

Most proteins synthesized in the ribosomes have at their amino terminal a special sequence of 15–30 amino acids called the signal peptide. As the protein enters the rough ER, the signal peptide remains at the leading end and hence, it is also called the leader sequence. The signal peptide, and with it, the rest of the peptide, is guided into the ER by the signal recognition particle (SRP) present in the cytosol. The process is analogous to a needle guiding the thread. The SRP binds to receptors present on the surface of the ER and thereby anchors the signal peptide to the ER. Thereafter, the signal peptide dissociates from the SRP and binds to the protein (translocon) that forms the pores of the endoplasmic reticulum. The signal peptide is then cleaved from the rest of the peptide by a signal peptidase. In the case of hormones, removal of the signal peptide results in the conversion of preprohormone to prohormone. Before secretion, additional amino acid residues are removed in the Golgi apparatus, resulting in the formation of the hormone.

The final step in posttranslational modification is protein folding. The way the protein gets folded is determined by the sequence of the amino acids in the polypeptide chain. Special protein molecules called molecular chaperones are sometimes required for ensuring a proper protein folding. A fully formed polypeptide chain may fold up and therefore fail to enter the ER. Such failure does not occur because the polypeptide chain enters the ER well before it is completely synthesized. Thus, while the signal protein is entering the ER, the other end of the polypeptide is still lengthening due to continuing translation. The process is called cotranslational insertion. The polypeptide chain anchors the ribosome on which it is synthesized to the ER. The ER is called a rough ER when a large number of ribosomes are anchored to it.

The **Golgi apparatus** is a stack of six or more membrane-enclosed sacs (cisterns). The cell contains one or more Golgi apparatuses, usually near the nucleus. Each Golgi apparatus is a polarized structure with a *cis* (convex) side and a *trans* (concave) side. Membranous vesicles containing newly synthesized proteins bud off from the rough ER and fuse with the *cis* side of Golgi apparatus. The proteins are then passed from cistern to cistern, finally reaching the *trans* side, from which vesicles bud off into the cytoplasm and are finally exocytosed (Figs. 4.**1** and 4.**2**).

Secretory vesicles budding off from the trans-Golgi are released to the cell exterior through exocytosis. The passage of the vesicle from the Golgi body to the exterior can be along nonconstitutive or constitutive pathways. In the *constitutive* or *unregulated pathway*, the contents of the secretory vesicles are exocytosed immediately. The exocytosis results not only in secretion but in the incorporation of bits of new membrane into the existing cell membrane. Vesicles are also pinched off from the cell membrane by endocytosis. These vesicles, called endosomes, eventually become lysosomes (see below). In the *nonconstitutive or regulated pathway*, the secretory vesicles are released in a regulated fashion only on appropriate external stimulation, as in glandular cells and neuronal endings.

Apparatus for metabolic reactions

Mitochondria are sausage-shaped structures made of an outer membrane and an inner membrane that is folded to form shelves (*cristae*). The space between the two membranes is called the intracristal (or intermembrane) space, and the space inside the inner membrane is called the matrix space (Figs. 4.**1** and 4.**2**). The *outer membrane* of each mitochondrion is studded with the enzymes concerned with biologic oxidations, providing raw materials for the reactions occurring in the matrix space. The *matrix space* contains enzymes of the citric acid cycle and enzymes required for β-oxidation of fatty acids. The *inner membrane* contains succinate dehydrogenase and ATP synthetase. The *intracristal*

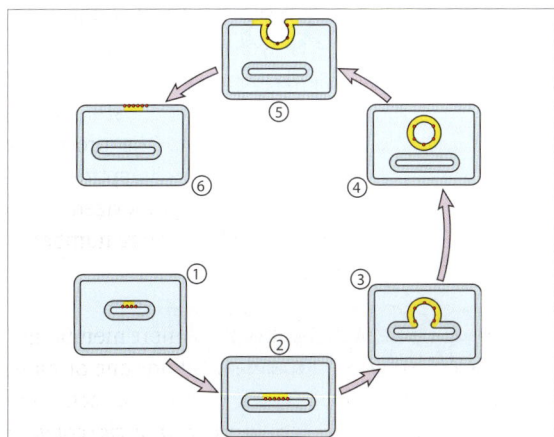

Fig. 4.**3**   Phospholipid transfer from vesicle to membrane. The smooth endoplasmic reticulum contains enzymes for phospholipid synthesis on its cytoplasmic surface (1). The phospholipids synthesized are inserted into the membrane of the endoplasmic reticulum (2) and subsequently bud off (3) to form lipid vesicles (4) that can get inserted into other membranes, e.g., into the cell membrane (5) and (6).

space contains adenyl kinase and creatine kinase. The enzymes in the matrix, inner membrane, and intermembrane space are *marker enzymes* for mitochondria, i.e., they are found nowhere else.

Mitochondria have their own genome, which suggests that mitochondria were once autonomous organisms that came to develop a symbiotic relation with eukaryotic cells. Compared to the nuclear genome, the mitochondrial genome contains much less DNA. The mitochondrial DNA codes certain key enzymes of oxidative phosphorylation.

All the mitochondria in the zygote come from the ovum and therefore the inheritance of mitochondrial DNA is entirely maternal. Mitochondrial DNA mutates far more frequently than nuclear DNA. Mutations in mitochondrial DNA cause a large number of relatively rare diseases.

Peroxisomes are cellular storehouses of metabolic enzymes. The peroxisome matrix contains nearly 50 enzymes that catalyze a variety of metabolic reactions. Peroxisomes in the liver detoxify alcohol and other harmful compounds. The marker enzymes for peroxisomes are catalase and urate oxidase.

Peroxisome membrane also has an elaborate mechanism for importing several enzymes into its matrix. An autosomal mutation that produces a defect in peroxisome membrane transport is the cause of Zellweger syndrome, which is fatal in infancy. Conversely, clofibrate, the drug used for lowering blood lipids, causes an increase in the number of peroxisomes. It is only one of the several *peroxisome proliferators* that are known to increase the number of peroxisomes.

The **smooth endoplasmic reticulum** is the site of steroid synthesis. Steroid-secreting cells are rich in smooth ER. The enzymes responsible for the synthesis of membrane phospholipids reside in the cytoplasmic surface of the cisterns of the ER. As phospholipids are synthesized at that site, they self-assemble into bimolecular layers, thereby expanding the membrane and promoting the detachment of *lipid vesicles* from it. These vesicles then travel to other sites, donating their lipids to other membranes (Fig. 4.**3**).

Smooth ER is also the site of detoxification of drugs and poisons in other cells, especially liver cells. Hepatic detoxification has important implications in pharmacology and medicine. Liver cell ER is also the site where glucose-6-phosphate is stripped of its phosphate radical before releasing glucose into the blood. Glucose-6-phosphatase is the marker enzyme for smooth ER. In striated (skeletal and cardiac) muscle, the smooth ER is called the sarcoplasmic reticulum. It stores Ca^{2+} in high concentration and plays an important role in muscle contraction.

■ Digestive apparatus

Lysosomes are membranous bags containing hydro-lytic enzymes. The cell employs lysosomal enzymes for digesting large molecules of proteins, polysaccharides, fats, and nucleic acids. Lysosomes bud off from the *trans* face of the Golgi apparatus. Lysosomal enzymes are synthesized in the rough ER and then processed by the Golgi apparatus. The marker enzyme for lysosome is acid phosphatase.

The lysosomal membrane has a hydrogen-ion pump that pumps in H⁺ from the cytosol and lowers the ly-sosomal pH to 5.0, which is the optimum pH for the lysosomal enzymes. The characteristic granules of the granulocytic leukocytes are actually lysosomes. Phago-cytic vacuoles fuse with the lysosome, whose enzymes digest the phagocytosed matter. The products of enzy-matic digestion are absorbed into the cytosol while the rest is exocytosed. The lysosomes also engulf worn-out components of the cell in which they are located, form-ing autophagic vacuoles, and return the digested prod-ucts for reuse by the cell. When a cell dies, lysosomal enzymes cause autolysis of the remnants. Apoptosis (p 47) by lysosomal enzymes is often important in the process of development. For example, in the embryon-ic stage, the hands are webbed till lysosomes digest the tissues between the fingers.

Small amounts of enzymes that leak out of lyso-somes into the cytosol normally get inactivated due to the absence of acidic pH. However, large leakages cause destruction of the cell. In gout, phagocytes ingest uric acid crystals and such ingestion triggers the extracel-lular release of lysosomal enzymes that contribute to the inflammatory response in the joints.

When a lysosomal enzyme is congenitally absent, it results in a rare but well-known group of disorders called the lysosomal storage disorders. In this, the lyso-somes become engorged with indigestible substrates. For example, in Tay–Sachs disease, a lipid-digesting enzyme is missing. It leads to brain dysfunction due to engorgement of brain cells with lipids.

■ Cytoskeletal apparatus

The cytoskeleton comprises three types of filaments: microtubules, microfilaments, and intermediate fila-ments. All the three are made of protein subunits. They not only maintain the structure of the cell but also per-mit it to change shape and move.

Microtubules and microfilaments are long polymers that move about by depolymerization at one end (the minus end) and polymerization at the other (the plus end). Both serve as 'railroads' on which certain proteins called *molecular motors* can 'walk' (Fig. 4.**4**B). Both lend some amount of structural solidity to the cell. However, there are differences too.

Microtubules are tubular structures, about 15-20 nm in diameter. A microtubule is made of α-tubulin and β-tubulin subunits that form stacks of rings con-taining 13 subunits (Fig. 4.**4**A). The molecular motors for microtubule are *kinesin* and *dynein*. Microtubules

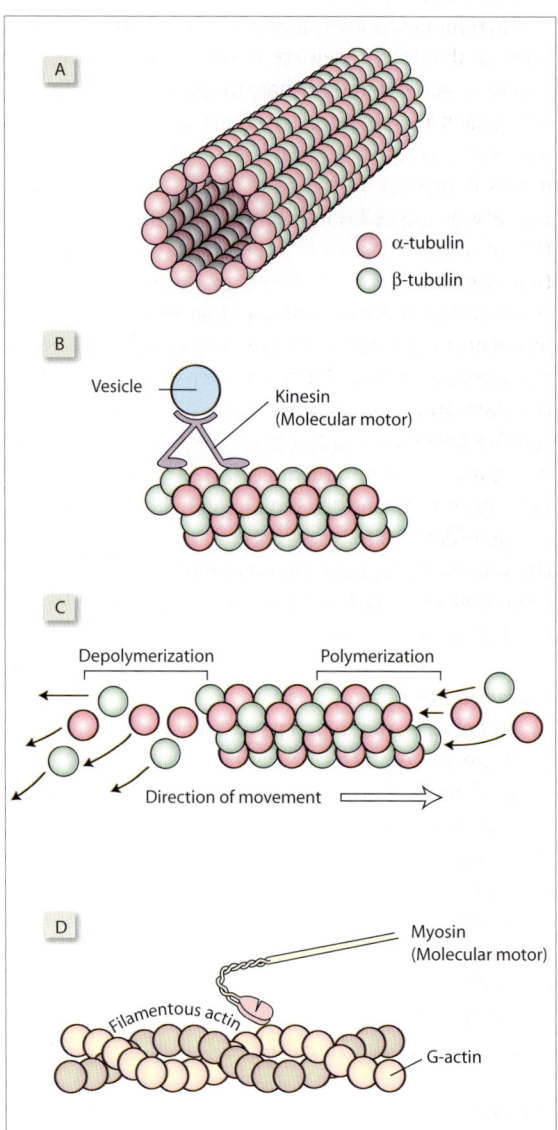

Fig. 4.**4** Microtubules and microfilaments. [A] A microtubule is made of α-tubulin and β-tubulin. Thirteen of these subunits form rings that stack up to form a hollow tubule. [B] A kinesin molecule carrying a vesicle is moving on a microtubule. [C] The microtubules themselves can move about in the cytoplasm by depolymerizing at one end and polymerizing at the other. [D] Two strands of F-actin are intertwined and are moved by the myosin molecule.

are generally concerned with movements of cellular components, transporting secretory granules, vesicles, and mitochondria from one part of the cell to another. One end of its molecular motor binds to the transport vesicle while the other end 'walks' on the microtubule. They also form mitotic spindles, which move the chromosomes in mitosis. The anticancer drug paclitaxel binds to microtubules and makes them so stable that organelles cannot move and mitotic spindles cannot form, leading to cell death. Microtubule assembly is prevented by colchicine and vinblastine.

Microfilaments are filamentous structures, about 3-5 nm in diameter. They are made of two F-actin (filamentous actin) strands that are coiled helically (Fig. 4.4D). Each F-actin strand is a polymer of G-actin (globular actin) subunits. The molecular motor of microfilaments is myosin. Microfilaments are concerned with motility of the cell itself. The most well-known function of microfilaments is their role in muscle contraction: the actin filaments in muscle are microfilaments. Microfilaments are anchored to the plasma membrane by anchoring proteins. Microfilaments and the anchoring proteins together form a dense matrix beneath the cell membrane, which is particularly abundant at the *zonulae adherentes* (intercellular junctions). Microfilaments are abundant in lamellipodia, the processes that cells put out when they crawl along surfaces. Bundles of microfilaments are also present in the core of microvilli, where they lend structural support. Also, when a cell divides, it is pinched into two by a constricting band of microfilaments.

Intermediate filaments have a diameter (8–10 nm) that is between that of microtubule and microfilament. They are more abundant than either microtubule or microfilament. They are made of cytokeratin. Unlike microtubule or microfilaments, these proteins are very stable and remain mostly polymerized. They serve as the 'bones' of the cell, giving structural strength to it. For example, the nucleus sits in a cage made of intermediate filaments.In axons, the intermediate filaments are called *neurofilaments*: they maintain the axonal diameter. Intermediate filaments are also present in desmosomes (intercellular junctions).

The **centrosome** is located near the nucleus. It is made of two centrioles which are short cylinders arranged at right angles to each other (Fig. 4.5). Microtubules in groups of three run longitudinally in the walls of each centriole. There are nine of these triplets spaced at regular intervals around the circumference.

The centrioles are surrounded by a small amount of γ-tubulin. Microtubules grow out of this γ-tubulin in the pericentriolar material and hence, the centrosome

is also called the *microtubule-organizing center*. When a cell divides, the centrosomes duplicate themselves and the pairs move apart to form the poles of the mitotic spindles, which are made of microtubules.

Cilia and flagella are motile processes of cell. Cilia usually occur in large numbers on cell surface while flagella are usually limited to one or a few per cell. Both have the same diameter but flagella are about ten times longer. They also differ in their pattern of beating. Dynein is the molecular motor responsible for the beating of flagella and cilia. The propulsion of a sperm is an example of flagellar action. The mucokinesis in the respiratory tract is brought about by ciliary action.

Both flagella and cilia resemble the centriole in having an array of nine tubular structures in their wall, but they have an additional pair of microtubules in the center (Fig. 4.5). Also, there are two rather than three microtubules in each of the nine circumferential structures. The microtubule assembly of a cilium or flagellum is anchored in the cell by a *basal body*, which is structurally identical to the centriole. It is from the basal body that the cilium or the flagellum starts to grow.

A congenital form of ciliary inactivity is *Kartagener's syndrome*, in which the axonemal dynein, the molecular motor that produces ciliary beating, is absent. Patients with this condition suffer from bronchiectasis, an irreversible, chronic dilatation of the bronchioles. They often have situs inversus, presumably because the cilia necessary for rotating the viscera are nonfunctional during embryonic development. Male patients are infertile because they lack motile sperm. However female patients are able to conceive normally, proving that ciliary motility is not essential for transport of the ovum to the site of fertilization.

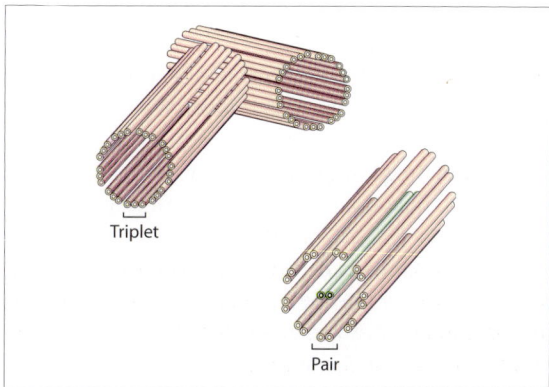

Fig. 4.5   (*Above*) A pair of centrioles lying at right angles to each other. Each centriole is made of nine circumferential triplets of microtubules. (*Below*) A cilium or a flagellum is made of nine circumferential pairs and one central pair of microtubules.

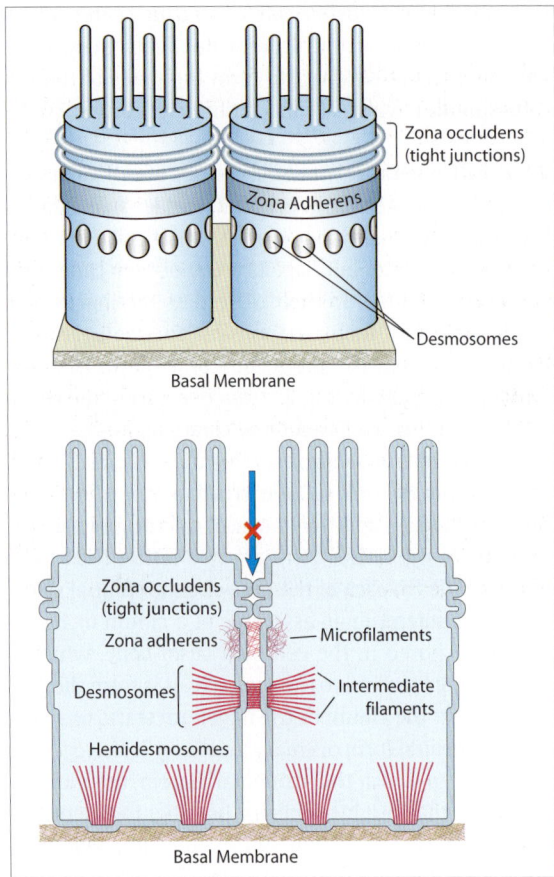

Fig. 4.**6**   Intercellular junctions. *(Above)* The zona occludens and zona adherens form a continuous ring around the cell. The desmosomes form an interrupted ring around the cell. *(Below)* Longitudinal section of two cells showing intercellular junctions. The zona occludens forms the tightest junctions and usually does not permit movement of electrolytes across them.

Intercellular Junctions

Intercellular junctions (Fig. 4.**6**) fasten cells to each other and to the basement membrane. They are of four types: zona occludens, zona adherens, desmosomes and gap junction. Gap junctions not only fasten cells together but more importantly, they permit passage of water, electrolytes and ionic currents through them.

The **zona occludens or tight junction** is a band-like ridge just below the apex of a cell. The membranes of the adjacent ridges are apposed tightly, leaving little space in between. Some are so tight that they do not allow water and ions to pass through them and are therefore called *'tight'* tight junctions. Others that are more permeable are called *'leaky'* tight junctions. These have important functions in the renal tubules.

Tight junctions also help to maintain polarity of the cells. Membrane proteins in the apical region of the cell cannot float across the band of zona occludens.

The **zona adherens** is another band immediately below the zona occludens. However, the band is formed not by a ridge, but by microfilaments running under the surface of the membrane. The membranes of adjacent cells are apposed less tightly here, leaving a gap of about 20 nm. Like zona occludens, they strongly bind adjacent cells and also lend polarity to the cell by preventing the drift of integral membrane proteins. However, they are quite permeable to water and ions.

The **desmosome or macula adherens** is not a band but a row of junctional patches below zona adherens (*macula = spot*). Like zona adherens, these spot-junctions leave 20 nm clefts between adjacent membranes. Intermediate filaments of the two adjacent cells converge on their respective side of the junction. The desmosome therefore not only attaches adjacent cells but also lends structural stability to the entire epithelium by anchoring together the cytoskeleton of the adjacent cells. A *hemidesmosome* (one-half of a desmosome) attaches a cell to the basement membrane.

The **gap junction** consists of a pair of *hemichannels* or *connexons* inserted into the membrane of adjacent cells (Fig. 4.**7**). Each connexon is made of six identical protein subunits called *connexins* which enclose a

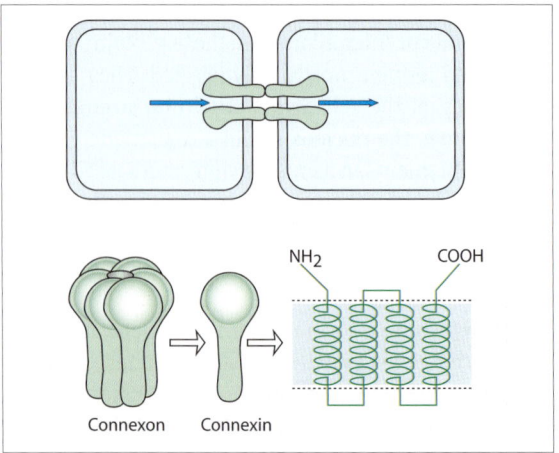

Fig. 4.**7**   Gap junctions connecting adjacent cells are designed to permit free communication through them. A gap junction is made of two connexons, each of which is made of 6 connexins. Each connexin is made of a long peptide chain that traverses the membrane four times.

central channel. When the corresponding connexons of the adjacent cell link up end-to-end, they form a continuous channel that permits substances to pass through from cell to cell. Each connexin is made of four transmembrane peptide segments. Mutant connexons produce the Charcot–Marie–Tooth disease, a form of peripheral neuropathy.

At gap junctions, the intercellular space narrows down to 3 nm, thereby helping in binding the cells together. However, their real physiological significance lies in allowing ions to flow through them, i.e., they conduct ionic current. This enables electrical excitation to spread from cell to cell, as in smooth and cardiac muscles. The pore size of a gap junction decreases when intracellular Ca^{2+} is high or pH is low, both of which are commonly associated with cell damage. Closure of gap junctions in response to these stimuli isolates damaged cells so that the Ca^{2+} and H^+ do not spread from the damaged to the normal cells.

Cell adhesion molecules

Cell adhesion molecules (CAMs) are surface glycoproteins present on the cell membrane surface. They are important for binding to other cells as well as to the extracellular matrix where they bind to laminins (see Fig. 8.**8**B). Apart from holding the tissues together, CAMs have important roles in inflammation and wound healing, embryonic development, axonal growth, and tumor metastasis. The four important groups of CAMs are selectins, integrins, cadherins, and the IgG superfamily of immunoglobulins. *Selectins* are required for capturing the free-flowing neutrophils and making them roll along the endothelial wall (see Fig. 29.**2**). *Integrins* and *IgG superfamily immunoglobulins* are present on the endothelium. They bring about a firm adhesion of the neutrophil with the endothelium. Neutrophilic migration through the endothelial wall (diapedesis) requires dissociation of intercellular *cadherin* contacts.

5   Cell Membrane

Membrane Composition

Membranes are complex structures composed of lipids, proteins, and carbohydrates. The cell membrane contains proteins and lipids in a 50:50 ratio. This ratio refers to the ratio of their masses and not numbers. An average membrane protein is several times larger than the average lipid molecule but lipid molecules are about 50 times more numerous than protein molecules. The ratio is not absolute and varies from membrane to membrane. The exact ratio between the two varies with the function of the cell. For example, the myelin sheath of nerves has about 75% lipids and 25% proteins, whereas membranes involved in energy transduction, such as the inner mitochondrial membrane, have 75% proteins and 25% lipids.

The major membrane lipids are phospholipids, glycosphingolipids, and cholesterol. Membrane phospholipids (Fig. 5.1) are broadly of two types: the phosphoglycerides, which are relatively more abundant, and the sphingomyelins, which are prominent in myelin sheath. Glycosphingolipids present in the membrane include cerebrosides and gangliosides. Both are derivatives of sphingosine. Cholesterol exists mainly in the plasma membranes of cells. It is also found in small quantities in mitochondria, Golgi complexes, and nuclear membranes.

The membrane proteins comprise enzymes, transport proteins, structural proteins, antigens (e.g., for histocompatibility) and receptors for various molecules. The external side of membrane proteins mostly has oligosaccharide chains (carbohydrates) attached to them. The plasma membrane contains over a hundred different proteins.

Membrane Structure

Lipid bilayer

Membrane lipids are amphipathic, i.e., they contain both hydrophobic and hydrophilic regions. The hydrophilic (polar) region is their globular head while the hydrophilic (nonpolar) regions are their fatty acid tails. The membrane lipids are organized into a closed bilayer (Fig. 5.2A) in which the hydrophobic regions of the phospholipids are protected from the aqueous environment while the hydrophilic regions are immersed in water. Proteins are found inserted into this lipid bilayer and are classified into integral proteins and peripheral proteins.

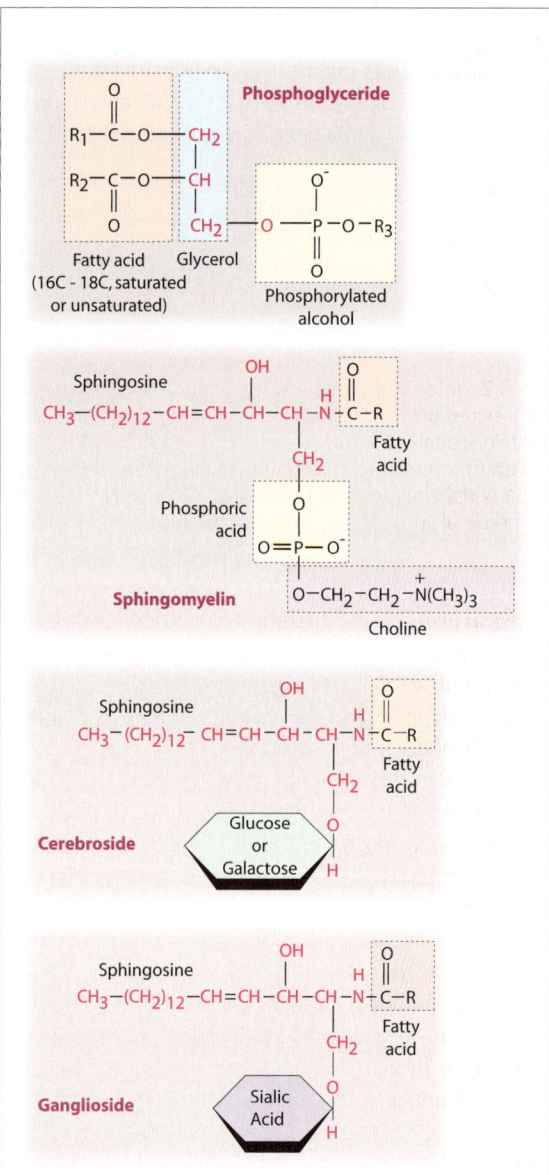

Fig. 5.1   Chemical structure of membrane phospholipids and glycosphingolipids.

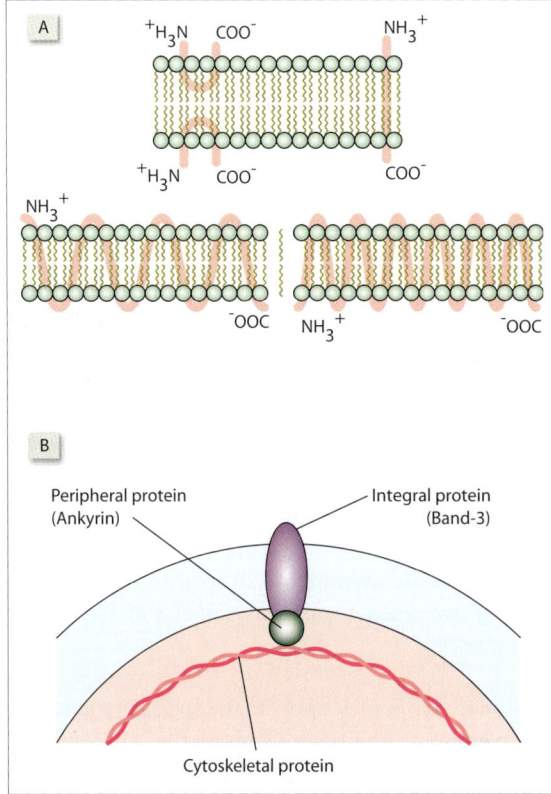

Fig. 5.**2**   Integral proteins. [A] Integral proteins with their polar residues projecting out. Also shown are a molecule of G-protein spanning the membrane seven times, and a glucose transporter molecule, spanning the membrane twelve times. [B] A peripheral protein anchoring the integral protein to the cytoskeletal protein of an erythrocyte.

Integral proteins are anchored to membranes through a direct interaction with the lipid bilayer. Some of them run the entire thickness of the membrane, often several times through. Others are located more on the outside or inside of the membrane (Fig 5.**2**A).

Integral proteins are amphipathic, consisting of two hydrophilic ends separated by an intervening hydrophobic region that traverses the hydrophobic core of the bilayer. The hydrophilic external region of the integral protein is found outside the membrane: either on its external or internal surface. Integral proteins serve as: (1) *channels*, which permit the passage of selected ions through the membrane; (2) *carriers* (or transporters), which ferry substances across the membrane by binding to them; (3) *pumps*, which are carriers that split ATP and use the energy derived for membrane transport of substrates; (4) *receptors* (located on the outside), which bind to specific molecules and send a chemical signal to the cell interior, initiating intracellular reactions, and (5) *enzymes* catalyzing reactions at the membrane surfaces, both outer and inner.

Peripheral proteins do not interact directly with the phospholipids in the bilayer. They are weakly bound to the hydrophilic regions of integral proteins. They are located on both surfaces of the membrane. Many hormone receptor molecules are integral proteins and therefore the polypeptide hormones that bind to these receptor molecules are considered as peripheral proteins. Peripheral proteins serve as cell adhesion molecules (CAMs) that anchor cells to neighboring cells and to the basal lamina. They also contribute to the cytoskeleton when present on the cytoplasmic side of the membrane. For example ankyrin, a peripheral protein located on the inside of the membrane, anchors spectrin (a cytoskeletal protein in the erythrocyte) to band-3 (an integral protein of erythrocyte membrane). Ankyrin plays an important role in the maintenance of the biconcave shape of the erythrocyte (Fig. 5.**2**B).

■   Fluid mosaic

The fluid mosaic model of membrane structure has been likened to icebergs (membrane proteins) floating in a sea of predominantly phospholipid molecules (Fig. 5.**3**A). Phospholipids too float about in the plane of the membrane. This diffusion, termed translational diffusion, can be as rapid as several micrometers per second. Several membrane transport processes and enzyme activities depend on the optimum fluidity of the cellular membrane. As membrane fluidity increases, there is a rise in membrane permeability to water and small hydrophilic solutes. The fluidity of a cell membrane depends on the lipid composition of the membrane, the density of integral proteins, and the temperature.

Role of fatty acids   A lipid bilayer made up of only one type of phospholipid changes from a liquid state to a rigid crystalline state (gel state) at a characteristic freezing point. This change in state is known as phase transition and the temperature at which it occurs is called the phase transition temperature (T_m). The T_m is higher (fluidity is low) when the constituent fatty acid chains are long and mostly saturated (without double bonds). Long chains have greater interactions among themselves, making the membrane stiffer. Saturated fatty acids have straight tails, whereas unsaturated fatty acids have kinked tails. As more kinks are inserted in the tails, the membrane becomes less tightly packed and therefore its fluidity increases (Fig. 5.**3**B).

Role of cholesterol   The presence of cholesterol in the membrane makes it possible for the cell membrane to maintain its fluidity across a wide range of temperatures. The number of cholesterol molecules in

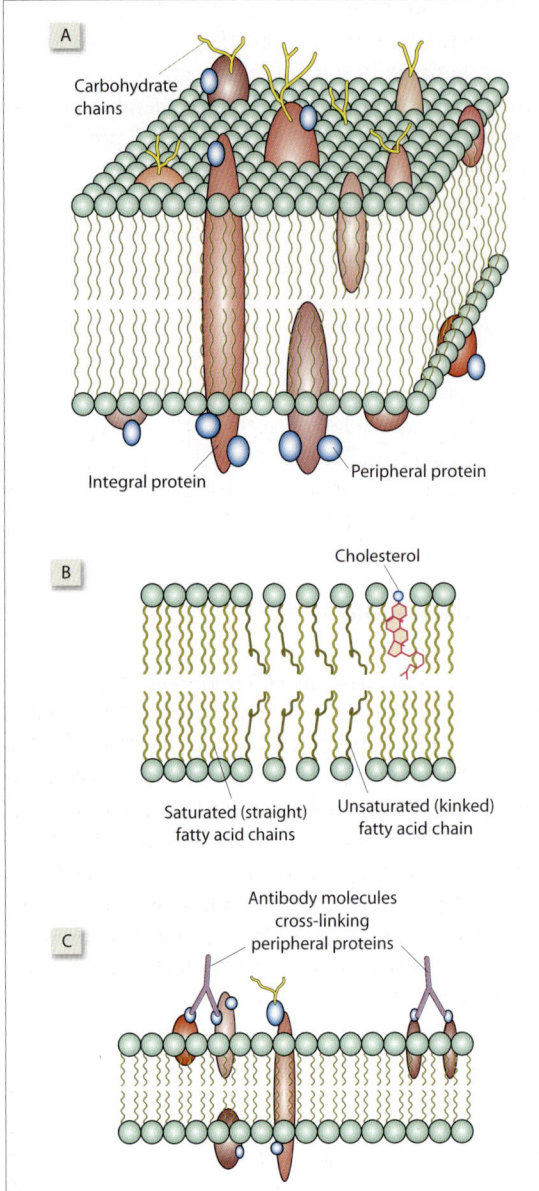

Fig. 5.**3** Membrane structure. [A] The cell membrane show-ing integral and peripheral proteins. [B] Intercalation of kinked unsaturated fatty acid chains and cholesterol molecules spaces out the phospholipid molecules and affects membrane fluidity. [C] Antibody molecules cross-linking peripheral proteins. This reduces the mobility of integral proteins and reduces mem-brane fluidity.

the membrane can be as high as the number of phos-pholipids. At high cholesterol:phospholipid ratios, the transition temperature is abolished altogether, i.e., the membrane always remains fluid. Cholesterol interca-lates among the phospholipids of the membrane with its hydroxyl group at the aqueous interface and the remainder of the molecule, among the fatty acid tails of phospholipids (Fig. 5.**3**B). At temperatures above

the T_m, cholesterol partially immobilizes those por-tions of the fatty acid chains that lie adjacent to it and thus makes the membrane stiffer. At temperatures be-low the T_m it minimizes the mutual interaction of the hydrocarbon tails of fatty acids and thereby increases membrane fluidity.

Membrane areas having a high density of integral proteins have low membrane fluidity due to protein–protein interaction. Some of the protein–protein inter-actions taking place within the plane of the membrane may be mediated by interconnecting peripheral pro-teins, such as cross-linking antibodies (Fig. 5.**3**C). These peripheral proteins may then restrict the mobility of integral proteins within the membrane.

Membrane asymmetry

Membranes are asymmetric structures. This asymme-try is of two types: regional asymmetry and inside-out asymmetry.

Regional asymmetry refers to the specialization of the cell membrane at different sites on the cell. For exam-ple, in renal tubules and intestinal mucosal cells, only the membrane facing the lumen is thrown into folds, forming microvilli. Similarly, only the membranes that are contiguous with adjacent cells show specializa-tions for intercellular junctions.

Inside–outside (transverse) asymmetry refers to the structural differences through the thickness of the cell membrane. For example, the phospholipids are not symmetrically disposed across the membrane thick-ness. The choline-containing phospholipids (lecithin and sphingomyelin) are located mainly in the outer molecular layer, while the aminophospholipids (phos-phatidylserine and cephalin) are preferentially located in the inner layer. Cholesterol is generally present in larger amounts on the outside than on the inside. Gly-colipids lie exclusively on the outside of the membrane. Proteins too are differentially located in the outer, inner or middle parts of the membrane. The carbohydrates are attached only to the membrane proteins on the outer surface. In addition, specific enzymes are located exclusively on the outside or inside of membranes, as in the mitochondrial and plasma membranes.

Membrane disorders

Mutation of membrane proteins affects their func-tion as receptors, transporters, ion channels, enzymes, and structural components. For example, in *hereditary spherocytosis*, there is mutation in the genes encoding spectrin, resulting in the tendency of the cell to become spherical rather than biconcave. Membrane proteins are also affected by antibodies against them. Autoantibodies to the acetylcholine receptor in skeletal muscle cause *myasthenia gravis*. Ischemia can quickly affect the integrity of various ion channels in membranes. Excess of cholesterol in the membrane can also affect membrane function, e.g., in *familial hypercholesterolemia*. The fragility of red cells is critically dependent on the protein:cholesterol ratio.

Membrane Transport

Simple diffusion

Since simple diffusion involves no expenditure of energy, it can occur only from a region of high solute concentration to low solute concentration. Simple diffusion is bidirectional, i.e., it occurs in both directions across the membrane. However, the diffusion along the concentration gradient is much higher than the diffusion against the gradient. Hence, the net diffusion is always along the concentration gradient. The rate of simple diffusion is directly proportional to the concentration gradient across the membrane and the permeability of the membrane to the solute. The membrane permeability to a substance depends on its size, lipid solubility, and electrical charge.

Gases such as O_2, CO_2, and N_2 are small molecules with little interaction with solvents. They readily diffuse through the hydrophobic regions of the membrane. Hydrophobic molecules like steroid hormones and weak organic acids and bases dissolve in the lipid bilayer and cross it with ease.

The permeability of the lipid bilayer to water is much more than that predicted by its size and lipid solubility. This is because water molecules can pass between adjacent phospholipid molecules without actually dissolving in the region occupied by the fatty acid chains.

Diffusion of charged particles, i.e., ions through the lipid bilayer is very slow. However, in biological membranes, the membrane permeability to ions like Na^+, K^+, Cl^-, and Ca^{2+} is high due to the presence of membrane channels that allow these ions to pass through in their hydrated form. Except through channels, the lipid bilayer of the membrane prevents diffusion of ions.

Large uncharged hydrophilic molecules such as glucose have extremely slow rates of simple diffusion. Most biological membranes employ carrier-mediated transport to speed up the diffusion of glucose.

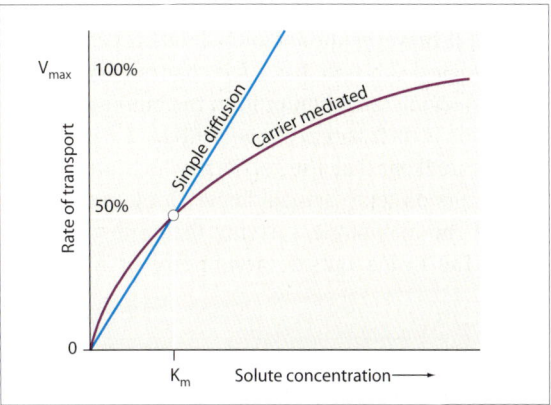

Fig. 5.**4**   Chemical kinetics of simple diffusion and carrier-mediated membrane transport. Note that while the amount of carrier mediated transport plateaus off at high solute concentration, there is no such limit to simple diffusion.

Carrier-mediated transport

Membrane transport can also occur through carrier-mediated transport, which employs certain integral membrane proteins as carriers or transporters for specific substrates. It may occur without any energy expenditure (facilitated diffusion) or may involve energy expenditure (active transport). When a carrier transports only a single substance, it is called *uniport*. When it transports more than one substance in the same direction, it is called *cotransport* or *symport*. When it transports two substances in opposite directions, it is called *countertransport* or *antiport*.

The rate of carrier-mediated transport depends on the concentration gradient across the membrane, the amount of carrier available (which is the key control step) and the rapidity of bonding and dissociation of the carrier with its substrate. The rate cannot exceed a certain maximum, called the V_{max}. At V_{max} all substrate-binding sites on the carrier are saturated. The substrate concentration at which the transport is 50% of the maximum is called the binding constant (K_m) of the carrier (Fig. 5.**4**).

Carrier-mediated transport can be blocked by inhibitors that bear structural similarity to the physiological substrate of the carrier. These inhibitors compete with the physiological substrate for a place on the carrier. Moreover, once they bind to the carrier these inhibitors do not dissociate easily, thereby blocking the transport mechanism. For example, the carrier-mediated transport of glucose is blocked by phloridzin.

The inhibitors are often classified as competitive and noncompetitive. If an inhibitor binds irreversibly to the carrier leaving no chance for the substrate to compete for a place on the carrier, the inhibition is called *noncompetitive*. The carrier in effect gets inactivated. If however the inhibitor binds reversibly, the physiological substrate has a fair chance of competing and dislodging the inhibitor from the binding site. The inhibition is then said to be *competitive*.

It is unlikely that the carriers, which are integral membrane proteins, actually move through the thickness of the membrane, carrying their substrate with them. The inside–outside asymmetry of membrane

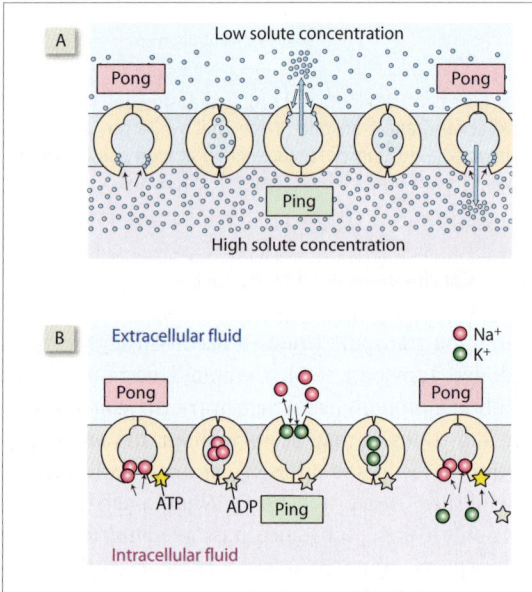

Fig. 5.**5**   The ping-pong model of carrier-mediated transport. [A] Facilitated diffusion. Note that transport of solute occurs in both directions. However, transport is greater from higher to lower solute concentration. [B] Active transport. Receptors for Na⁺ are present only at the inner side while receptors for K⁺ are present only at the outer side of the membrane carrier. This ensures that Na⁺ can only move out while K⁺ can only move in.

proteins is too stable to permit such movements. Rather, a *ping-pong mechanism* is proposed. In this model, the carrier protein exists in two principal conformations: ping and pong. In the pong state, it is exposed to high concentrations of solute and the molecules of the solute bind to specific sites on the carrier protein. Transport occurs when a conformational change to the ping state exposes the carrier to a lower concentration of solute. The transition between the ping and pong states is powered by the bond energy released when the substrate binds to the solute. This is true for all carrier-mediated transport. In active transport, the binding of ATP to the carrier provides additional energy (Fig. 5.**5**).

While integral membrane proteins employ the ping-pong mechanism for transporting ions, certain microbes synthesize small organic molecules called *ionophores* that transport ions by actually traveling across the membrane. These ionophores contain hydrophilic centers that bind specific ions and a hydrophobic exterior that allows them to dissolve in the membrane and diffuse across it.

Passive carrier-mediated transport can transport substrates only from a high concentration to low concentration. It is also called *facilitated diffusion*. Like simple diffusion, it is bidirectional, i.e., it occurs in both directions. However, when the concentration on one side is higher than the other, the sheer difference in the kinetics of solute–carrier interaction ensures that there is a net solute movement from high to low concentration.

Glucose and other large uncharged hydrophilic molecules have extremely slow rates of simple diffusion across the lipid bilayer. They cross the membrane much faster through facilitated diffusion. Examples of facilitated diffusion are the cotransport of Na⁺ with

Fig. 5.**6**   The boys with roller skates (carriers) are able to roll faster (facilitated diffusion) down the slope (concentration gradient) compared to those who are sliding down (simple diffusion). However, the number of boys who can roll down is limited by the availability of the roller skates.

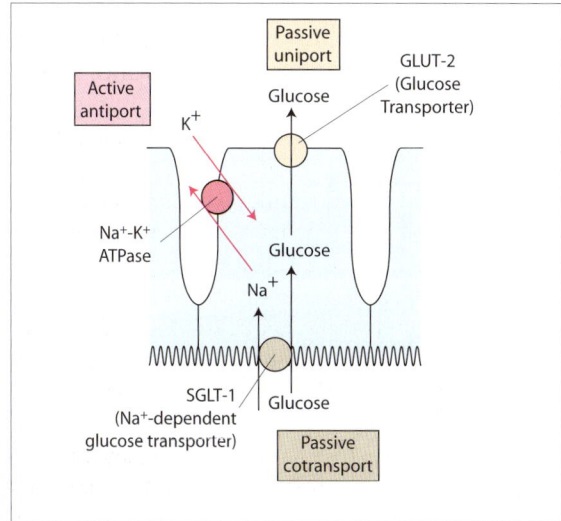

Fig. 5.**7**   Secondary active transport is a form of facilitated diffusion in which the concentration gradient is built up through active transport.

monosaccharides and amino acids in renal tubular cells and intestinal mucosal cells. An example of facilitated counter-transport is the Cl^-–HCO_3^- antiport found in renal tubular cells and gastric parietal cells.

Facilitated diffusion is faster than simple diffusion but the amount transported by facilitated diffusion is limited by the availability of the carrier. (Fig. 5.**6**). Hormones can regulate facilitated diffusion by changing the number of carriers available. For example, insulin increases glucose transport into cells by moving glucose transporters from an intracellular reservoir to the membrane surface.

Primary-active carrier-mediated transport involves energy expenditure. The energy comes mostly from ATP that is hydrolyzed by the carrier protein itself, which also acts as an ATPase. Unlike passive transport, active transport can transport substrates against a concentration gradient. The best-known example of a carrier ATPase is the Na^+-K^+ ATPase. It has binding sites for both ATP and Na^+ on the cytoplasmic side of the membrane but the K^+ binding site is located on the extracellular side of the membrane (Fig. 5.**5**). This asymmetry of location of the binding sites explains why, unlike facilitated diffusion, primary-active transport can occur only in one direction. Ouabain or digitalis inhibits this ATPase by binding to the extracellular site on the transporter.

Secondary-active carrier-mediated transport represents a combination of primary-active transport and facilitated diffusion. It is exemplified by glucose transport across renal tubular cells and intestinal mucosal

cells (Fig. 5.**7**). The basolateral border of the cell lowers the intracellular Na^+ concentration through primary active transport of Na^+ ions to the exterior. The low intracellular Na^+ concentration provides the necessary concentration gradient for Na^+ to diffuse in passively at the luminal border through facilitated cotransport with glucose. Thus, the Na^+-K^+ ATPase indirectly powers the movement of glucose into the cell, and glucose can move in against a concentration gradient so long as Na^+ diffuses in along a concentration gradient.

▪ Endocytosis

Endocytosis is the process by which cells take up macromolecules and large particles. The process requires ATPase, Ca^{2+}, and microfilaments. Endocytosis occurs by invagination of the plasma membrane so as to enclose a small droplet of extracellular fluid and its contents. The invagination gets pinched off at its neck to form an endocytotic vesicle. The vesicle then transports its contents to other organelles by fusing with their membranes.

Alternatively, it can fuse back with the plasma membrane. Depending on what is endocytosed, endocytosis is called phagocytosis or pinocytosis. Endocytosis of cells, bacteria, viruses or debris is called phagocytosis. Endocytic vesicles containing these fuse with primary lysosomes to form secondary lysosomes where the ingested particles are digested. Endocytosis of water, nutrient molecules and parts of the cell membrane is called pinocytosis.

Fluid-phase pinocytosis is a nonselective process in which the cell takes up fluid and all its solutes indiscriminately. Vigorous fluid-phase pinocytosis is associated with internalization of considerable amounts of the plasma membrane. To avoid reduction in the surface area of the membrane, the membrane is replaced simultaneously by exocytosis of vesicles. In this way, the plasma membrane is constantly recycled.

Absorptive pinocytosis is also called receptor-mediated pinocytosis. It is responsible for the uptake of selected macromolecules for which the cell membrane bears specific receptors. Such uptake minimizes the indiscriminate uptake of fluid or other soluble macromolecules. The vesicles formed during absorptive pinocytosis are derived from invaginations (pits) that are coated on the cytoplasmic side with a filamentous material called clathrin, a peripheral membrane protein. Such pits are called *coated-pits*.

An example of absorptive pinocytosis is provided by the endocytosis of low-density lipoprotein (LDL) molecules. The molecules bind with their receptor on the plasma membrane and the receptor-LDL complexes are internalized by means of coated pits. The endocytotic vesicles fuse with lysosomes in the cell. The receptor is released and recycled back to the cell surface membrane but the LDL is metabolized.

Endocytosed hormone receptors can trigger intracellular events after being endocytosed. Following pinocytosis, these hormone receptors form *receptosomes*, vesicles that avoid lysosomes and deliver their contents to other intracellular sites such as the Golgi body.

Receptor-mediated endocytosis can at times be self-defeating. Viruses causing hepatitis, poliomyelitis, and AIDS gain access into the cell through this mechanism. Iron toxicity also begins with excessive uptake through endocytosis.

Exocytosis

Exocytosis is the process for release of macromolecules to the exterior. Exocytosis is associated with an increase in the area of the plasma membrane. Working in tandem with fluid-filled endocytosis, this process is involved in remodeling of the membrane. Exocytosed molecules are of three types. Some attach to the cell surface and become peripheral proteins, e.g., antigens. Some become part of the extracellular matrix, e.g., collagen and glycosaminoglycans. Some enter extracellular fluid and serve as hormones and neurotransmitters. These are exocytosed only when the cell is stimulated.

Membrane Channels

Ion channels are integral membrane proteins that enclose a central pore. They traverse the entire thickness of the cell membrane, projecting a little at both the outer and inner membrane surfaces.

The structural characteristics of some ionic channels are shown in Figure 5.**8** and summarized in Table 5.**1**. A voltage-sensitive Na^+ ion channel, for example, is made of 4 subunits designated as α, β, γ, and δ. Each subunit in turn is made of 6 coiled transmembrane segments designated S1–S6. In general, channels have variable numbers of subunits and transmembrane domains having different designations.

Ion channel specificity

Ion channels are highly specific for certain ions. Ions, regardless of their size, tend to pass through their own channels. While it is understandable that a larger ion

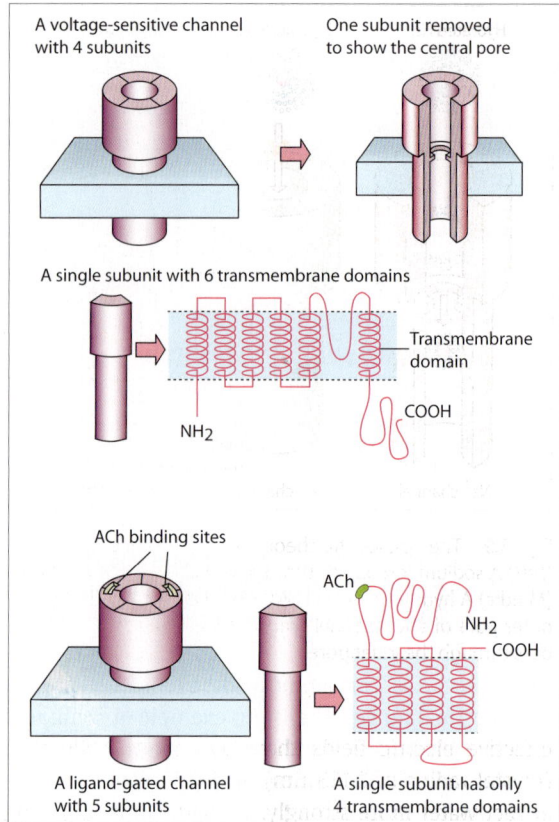

Fig. 5.**8** (*Above*) Structure of a voltage-gated channel. It is made of 4 subunits with each subunit made of 6 transmembrane domains. The subunits enclose a central pore. (*Below*) Structure of a ligand-gated cation channel. It is made of 5 subunits, each with 4 transmembrane domains. The subunits enclose a central pore. The binding sites for acetylcholine are present on the exterior at the junctions between two adjacent subunits.

Table 5.**1** Structural characteristics of ionic channels.

	Number of subunits	Number of transmembrane domains per subunit
Voltage-gated Na^+ and K^+ channels	4 subunits. All are α-type and are designated I, II, III, and IV	6 (S1, S2, S3, S4, S5, and S6)
Ligand-gated cation channel	5 subunits. 2 are α-type, 1 each are β, γ, and δ.	4 (M1, M2, M3, and M4)

cannot pass through a smaller channel, it requires some explanation as to why smaller ions do not pass through larger channels. This is explained by the *closest-fit hypothesis*. The hydration of ions is an important consideration in this hypothesis.

The smaller an ion, the more highly localized is its charge and stronger is its electric field. Smaller ions such as Na^+ (crystal radius of 0.095 nm) have stronger

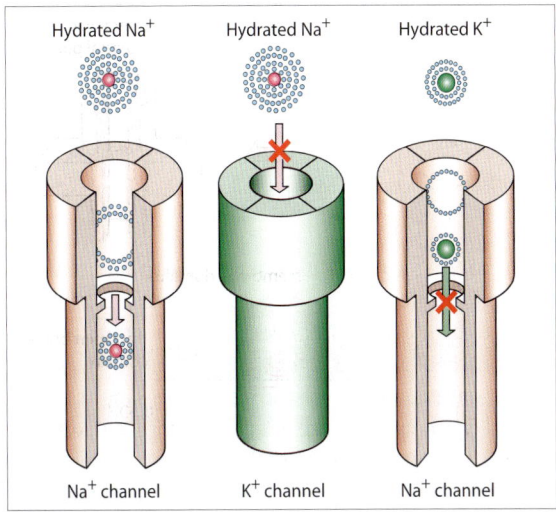

Fig. 5.**9** The 'closest-fit theory' of ion channel specificity. (*Left*) A sodium ion passes through the outer and inner pores. (*Middle*) A hydrated sodium ion is unable to pass through the outer pore of a K⁺ channel. (*Right*) A hydrated K⁺ is unable to pass through the inner pore.

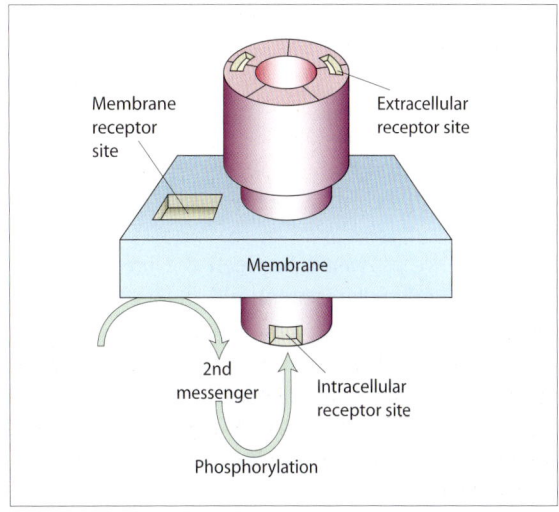

Fig. 5.**10**   Ligand-binding sites may be located on the outer membrane surface, or they may be present on the outer or inner parts of a membrane channel.

effective electric fields than larger ions such as K^+ (crystal radius of 0.133 nm). As a result, smaller ions attract water more strongly. Thus, the strong electrostatic attraction for water causes Na^+ to have a larger water shell.

For passing completely through a channel, an ion has to negotiate two barriers: an *outer pore*, and another pore called the *selectivity filter* located midway inside the channel (Fig. 5.**9**). The ion enters the outer pore with its complete water of hydration. Once it arrives at the inner selectivity filter, the ion sheds most of its water shell and forms a weak electrostatic bond with the polar (carboxyl) residues of the amino acids that line the channel wall. This electrostatic bond must release adequate energy for stripping the ion of its water shell. The energy released is maximum when the unhydrated ion fits closely in the channel and is less if it floats loosely within the channel.

The hydrated Na^+ ion is larger than the hydrated K^+ ion and as would be expected, a Na^+ channel has a larger diameter than a K^+ channel. A hydrated Na^+ ion is a little too large to pass through the outer pore of a K^+ channel. The hydrated K^+ ion is slightly less in diameter than the Na^+ channel and yet it is unable to pass through the Na^+ channel because the selectivity filter of the Na^+ channel is oversized for a K^+ ion bereft of its water shell. Hence, the energy released due to the electrostatic attraction between the K^+ ion and the wall of the Na^+ selectivity filter is inadequate for stripping the K^+ ion of its water shell. With its shell intact, the K^+ is unable to negotiate the selectivity filter of Na^+ channel.

▪ Types of channels

What makes channels unique as compared to other mechanisms of membrane transport is that they can be gated precisely. Depending on the factors that produce their opening and closing (gating), ion channels are classified into four types.

Voltage-gated channels are gated by changes in membrane potential. Examples are voltage-gated Na^+, K^+, and Ca^{2+} channels.

Ligand-gated channels are regulated by ligands, i.e. chemicals that bind to them. The ligand affects channel permeability in different ways (Fig. 5.**10**) (1) It can bind to the receptor channel protein at an extracellular site, e.g., in acetylcholine receptors. (2) It can bind to the channel at an intracellular site (e.g., pronase, local anesthetics). (3) It can bind to membrane receptors and activate a second messenger cascade leading to phosphorylation of channel protein.

Mechanically gated channels have pores that respond to mechanical stimuli like stretch.

Resting channels are not gated at all. Resting channels make a substantial contribution to the membrane potential.

Apart from these physiological channels, membrane pores may be created in pathological situations. Diphtheria toxin and activated serum complement components like the C5b-C6-C7-C8-C9 fragment can produce large pores in cellular membranes and thereby provide macromolecules with direct access to the cell interior.

6   Cell Cycle

The sequence of events between the birth of a cell (through division of its parent cell) and its division into daughter cells is called the cell cycle. A cell roughly doubles its cytoplasm and organelles, and duplicates its DNA. Thereafter, the cell nucleus divides through a process called mitosis and its cytoplasm is apportioned (cytokinesis) into two daughter cells that have identical genomes. The mitosis and cytokinesis together constitute the M phase (mitotic phase) of the cell cycle. The remaining part of the cell cycle is called interphase and constitutes 90% of the cell cycle.

The interphase is further subdivided into three phases (Fig. 6.1). The phase that immediately follows the M phase is the G_1 phase (G_1 for the first gap). In this phase, there is rapid growth and metabolic activity in the cell, and its centrioles replicate. The next phase is the S phase in which DNA replication occurs. The S phase is followed by the G_2 phase (second gap) in which there is further growth and final preparation for the impending mitosis. Somewhere within the G_1 phase is a stage called the restriction point where the cell is at a

crossroads: from there it either goes into a nondividing state called the G_0 phase or goes through the entire cell cycle all over again. This 'checkpoint function' is under the control of the *p53* gene.

An understanding of the cell-cycle kinetics is essential for the proper use of anticancer (antineoplastic, cytotoxic) drugs. Many of the potent anticancer agents act by damaging DNA and their toxicity is greater during the S phase. Others block the formation of mitotic spindles in the M phase. These agents have activity only against cells that are in the process of division. Accordingly, neoplasms with a high percentage of cells undergoing division are highly susceptible to chemotherapeutic measures. For the same reason, normal tissues that proliferate rapidly (bone marrow, hair follicles, and intestinal epithelium) are susceptible to these drugs. On the other hand, slowly growing tumors are often irresponsive to cytotoxic agents. Cells of slow growing tumors may remain in the G_0 state for prolonged periods, only to reenter the cell cycle at a later time. Damaged cells that reach the restriction point undergo apoptosis (see below) if the *p53* gene is intact and it exerts its normal checkpoint function. If the *p53* gene is mutated and the checkpoint function fails, damaged cells will not be diverted to the apoptotic pathway. These cells will proceed through S phase and some will emerge as a drug-resistant population.

Mitosis

Mitosis (Greek *mitos* = thread, fiber) is the type of cell division that occurs in the somatic cells of the body during growth. In mitosis, each of the two daughter cells receives exactly the same number and the same type of chromosomes contained in the parent cell. Mitosis occurs in five stages: interphase, prophase, metaphase, anaphase, and telophase.

In *interphase*, the nucleus is well defined and bound by the nuclear membrane. It contains one or more nucleoli. Just outside the nucleus are two pairs of centrioles, formed earlier by the replication from a single pair. The two centrioles at first lie side-by-side just outside the nucleus. Around each centriole pair, microtubules are formed in a radial array called an aster (star). During early interphase, the cell is diploid, i.e., its nucleus

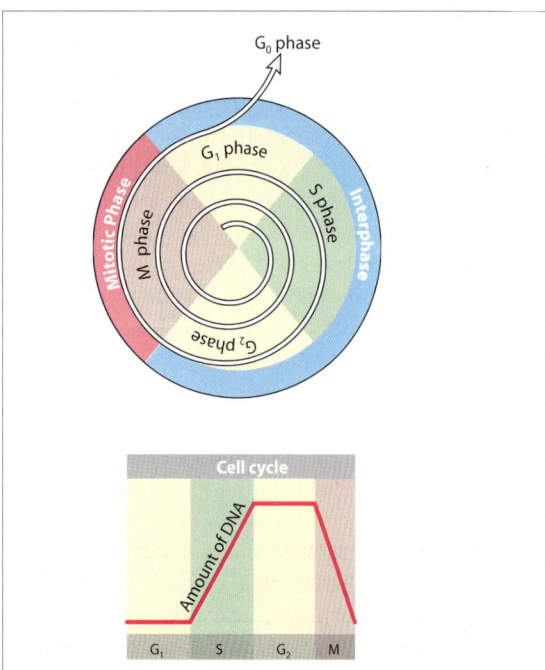

Fig. 6.1   The cell cycle. (*Above*) Phases of the cell cycle. (*Below*) Amount of DNA in different phases.

contains 22 pairs of autosomal chromosomes and a pair of sex chromosomes (Fig. 6.**2**). The chromosomes duplicate by the middle of the interphase, making the cell tetraploid. Till this stage the chromosomes cannot be distinguished individually as they are still in the form of loosely packed chromatin fibers.

During *prophase*, changes occur in both the nucleus and the cytoplasm. In the nucleus, the nucleoli disappear. The chromatin fibers become more tightly coiled and folded into discrete chromosomes. Each duplicated chromosome now appears as two identical sister chromatids joined at the centromere. Since each of the two sister chromatids contains a double-stranded DNA, the structure is called a tetrad and contains a tetraploid genome.

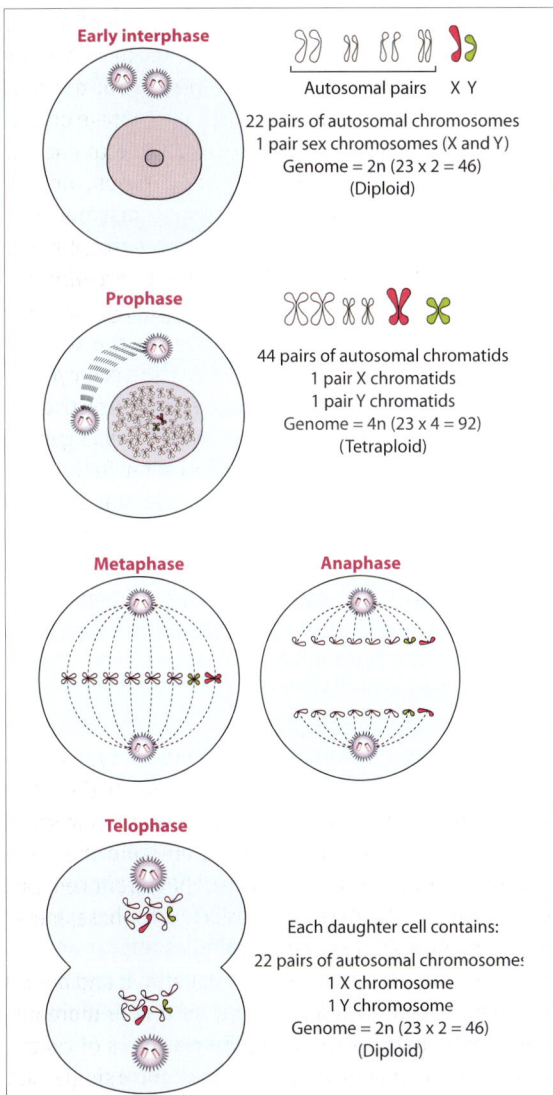

Early interphase

Autosomal pairs   X   Y

22 pairs of autosomal chromosomes
1 pair sex chromosomes (X and Y)
Genome = 2n (23 x 2 = 46)
(Diploid)

Prophase

44 pairs of autosomal chromatids
1 pair X chromatids
1 pair Y chromatids
Genome = 4n (23 x 4 = 92)
(Tetraploid)

Metaphase   **Anaphase**

Telophase

Each daughter cell contains:

22 pairs of autosomal chromosomes
1 X chromosome
1 Y chromosome
Genome = 2n (23 x 2 = 46)
(Diploid)

Fig. 6.**2**   Stages of mitosis. Note that the cell is diploid in interphase, becomes tetraploid in prophase due to DNA replication, and finally, becomes diploid again in telophase.

In the cytoplasm, the mitotic spindles form. They are made of microtubules extending between the two pairs of centrioles. During prophase, the two centrioles move away from each other, pushed by the elongating bundles of microtubules between them. Late in prophase, the nuclear membrane fragments. The microtubules of the spindle invade the nucleus and interact with the chromosomes, which have become even more condensed. Bundles of microtubules called polar fibers extend from each pole toward the equator of the cell.

During *metaphase*, the centriole pair is located at the opposite poles of the cell. The chromosomes line up in the equatorial region of the cell.

Anaphase begins when the paired centromeres of each chromosome move apart, liberating the sister chromatids from each other. Each chromatid is now representative of a fully-fledged chromosome. By the end of anaphase, the two poles of the cell have the complete (diploid) set of chromosomes.

At *telophase*, the daughter nuclei begin to form at the two poles of the cell where the chromosomes have gathered. The nuclear membrane forms, the nucleoli reappear and the chromosomes uncoil. Cytokinesis occurs shortly thereafter and the cytoplasm is apportioned into the two daughter cells.

■ Mitotic chromosomal nondisjunction

Chromosomal nondisjunction (nonseparation of chromosomes) results in abnormal mitosis. It usually occurs immediately after zygote formation and results in mosaicism. In mosaicism, the number of chromosomes is not the same in all the cells of the body. For example, approximately 50% of the body cell population may have a certain number of chromosomes while the remaining 50% have a different number of chromosomes. If half the cells in the body are 45,X (i.e., 44+X) and the remaining cells are 47,XYY (i.e., 44+XYY), the mosaicism is denoted symbolically as XY/XYY. Its mechanism is depicted diagrammatically in Figure 6.**3**.

■ Meiosis

Meiosis (Greek *meioum* = diminish) occurs during gamete formation (spermatogenesis and oogenesis). Like mitosis, it is preceded by the duplication of the chromosomes during *interphase*, but this duplication is followed by two successive cell divisions called meiosis-I and meiosis-II. Meiosis-II is almost identical to mitosis. The result is the formation of four haploid daughter cells (Fig. 6.**4**).

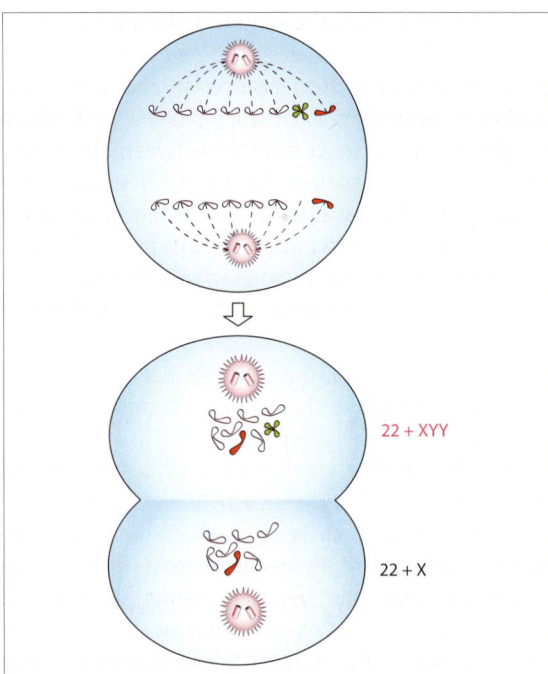

Fig. 6.3 Mitotic nondisjunction. The Y chromosomes fail to separate. Therefore one of the daughter cells gets two Y chromosomes while the other gets none.

Prophase-I takes 90% of the time required for meiosis-I. During this phase, the homologous chromosomes derived from each parent come together. Since each chromosome is made of two sister chromatids, the genome is tetraploid. The two homologous chromosomes are apposed in such a way that each gene is juxtaposed with its homologous gene on the opposite chromatid. Corresponding segments of adjacent nonsister chromatids may exchange by breaking and reattaching to the other chromatid. This event is called crossing over and it ensures that genetic traits of the homologous chromosomes get randomly mixed up. The nonsister chromatids remain attached at the sites of *crossing over*, which are called chiasmata (singular: chiasma). Other changes during prophase-I are the fragmentation of the nuclear membrane, dispersion of the nucleoli, and migration of the centrioles to the opposite poles.

During *metaphase-I*, the tetrads line up in the equatorial region of the cell in such a way that the homologous chromosomes are on opposite sides of the equatorial line.

Anaphase-I begins when the homologous chromosomes in each tetrad move apart. The sister chromatids in each chromosome, however, remain attached by the centromere. By the end of anaphase, each pole of the cell has a haploid set of chromosomes. Since each chromosome still has two chromatids, the genome is diploid.

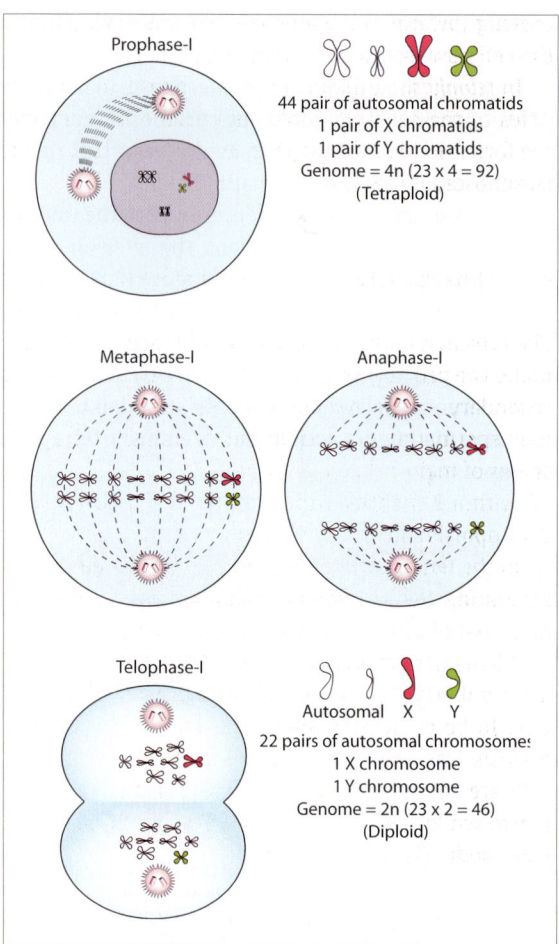

Fig. 6.4 Stages of meiosis. During prophase-I, the cell becomes tetraploid due to DNA replication. In telophase-I, the two daughter cells become diploid, not due to separation of chromatids as in mitosis, but due to the apportioning of homologous chromosomes into two cells. Hence, one of the daughter cells gets the X chromosome while the other gets the Y chromosome. Each chromosome still contains two chromatids. Meiosis-II results in the formation of four haploid daughter cells due to separation of chromatids to the daughter cells.

In *telophase-I*, cytokinesis occurs and the cytoplasm gets apportioned into the two daughter cells. Nuclear membrane formation and reappearance of nucleoli may or may not occur depending on the duration of the interval (interkinesis) between telophase-I and prophase-II. If the interkinesis is brief, neither occurs.

In *prophase-II*, the mitotic spindles appear and the chromosomes move towards the equator. If the nuclear membrane and nucleoli appear during telophase-I, they disintegrate again during this phase.

During *metaphase-II*, the haploid number of chromosomes line up along the equatorial line with their centromeres aligned.

During *anaphase-II*, the paired centromeres of each chromosome move apart with the sister chromatids moving towards the opposite poles. The two poles of the cell now have a haploid genome each.

In *telophase-II*, the nuclei begin to form at opposite poles of the cell and cytokinesis occurs. The result is the formation of four daughter cells, each with haploid chromosomes (Fig. 6.**4**).

■ Interkinesis

The typical meiosis described above is seen only in the male. The primary spermatocyte of the male forms two secondary spermatocytes after meiosis-I and finally four spermatids after meiosis-II (see Fig. 89.**3**). Both stages of meiosis begin after puberty and are completed within 2 months. However, the prophase-I takes a disproportionately long time of about 22 days.

In the female, there are long intervals (interkinesis) separating the different phases of meiosis (Fig. 6.**5**). Meiosis-I of ovarian germ cells begins immediately after birth but gets arrested at prophase, only to resume after puberty. Meiosis-II gets arrested in metaphase, only to be completed after fertilization. Moreover, in meiosis-I as well as in meiosis-II, the two daughter cells are not uniform in size: one gets most of the cytoplasm while the other gets little of it and is called the *polar body*. The polar bodies degenerate.

■ Meiotic chromosomal nondisjunction

Meiotic chromosomal nondisjunction results in abnormal gametogenesis. In the classical form of Klinefelter syndrome (see Fig. 92.6), chromosomal nondisjunction during meiosis results in formation of an abnormal ovum. When fertilized, the abnormal ovum results in a zygote with the chromosomal configuration of XXY. The mechanism through which the abnormal ovum is formed is depicted diagrammatically in Figure 6.**6**.

■ Apoptosis

Apoptosis (dropping off, like autumn leaves) is genetically programmed cell death that does not significantly affect neighboring cells and tissues. It is distinct from necrosis, which is an unregulated destruction of larger tissue areas. If necrosis is cell murder, apoptosis is cell suicide. Apoptosis makes an interesting contrast with mitosis. Mitosis is the manifestation of the cell's will to divide and live on while apoptosis represents its decision to call it quits!

Apoptosis is responsible for (1) death of intestinal mucosal cells as they reach the tip of the villus to be sloughed off; (2) death of epidermal cells as they reach the surface; (3) death of large number of neurons in the central nervous system that do not make appropriate synaptic contact with their target organs; (4) death of clones of lymphocytes that are likely to react with

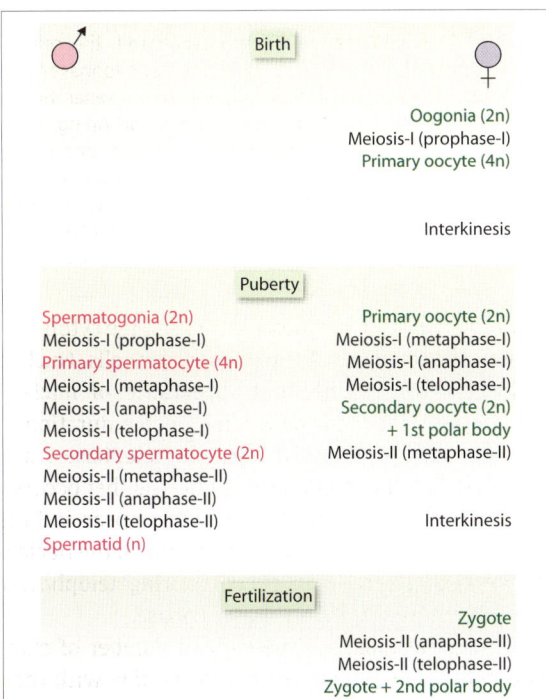

Fig. 6.**5**   Male versus female meiosis. Meiosis in females shows two long intervals called the interkineses.

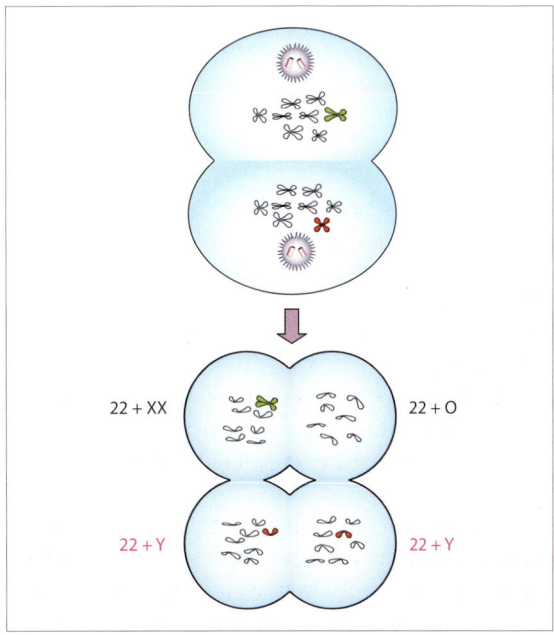

Fig. 6.**6**   Meiotic nondisjunction resulting in four haploid germ cells of which two have abnormal genomes 22 + XX and 22 + 0. See also Fig. 92.5.

'self'; (5) removal of the webs between the fingers in fetal life; (6) regression of one or the other of Wolffian duct and Mullerian duct systems in the course of sexual differentiation in the fetus; (7) regression of incompletely developed Graafian follicles in ovary after ovulation; (8) cyclic breakdown of the endometrium, leading to menstruation; (9) crystalline lens protein that is derived from the remnants of apoptotic cells.

Apoptosis is completed in four phases: induction, initiation, execution and disposal.

The *induction phase* is the stage of gene activation. Several endogenous and exogenous stimuli activate the apoptotic genes through signal transduction. Two well-known activating factors are Fas-antigen and tumor necrosis factor (TNF). Another antigen, the p53-antigen, which is the product of the tumor suppressor gene, exerts an anticancer activity by activation of apoptosis. Mitochondrial factors also play an important role in apoptosis. Several proapoptotic factors, like ultraviolet light, oxidants, cytokines, neurotoxins, etc. damage the mitochondrial membranes causing release of apoptosis-inducing factor (AIF).

In the *initiation phase*, the activated gene initiates a proteolytic cascade involving a family of at least ten proteases named *caspases* (an abbreviation for the word cysteine aspartase). The enzymes have cysteine in their active site and cleave their target proteins at aspartic acids.

The *execution phase* completes the death program. The dying cell shrinks and loses its contact with neighboring cells. The cell DNA gets fragmented and the cytoplasm and chromatin condense. The nucleus and cytoplasm eventually break up into small cell remnants called apoptotic bodies.

In the *disposal phase*, the apoptotic bodies are either phagocytosed or are lost from the epithelial surfaces.

7 Applied Genetics

Nucleic Acids

The characteristics of a cell are determined by its structural proteins and the proteinaceous enzymes that control its function. The structure and functions of proteins are determined by their amino acid sequence, which in turn is determined by the chemical composition of the nucleic acids in the cell's nucleus. A change in the composition of the nucleic acid alters the protein synthesized, sometimes with deleterious consequences.

The basic subunits of nucleic acids are the nucleotides (Fig. 7.1). A nucleotide is made of a nitrogenous base, a pentose sugar molecule and up to three phosphate groups. The pentose sugar may be deoxyribose or ribose. Accordingly, the nucleic acid is either deoxyribonucleic acid (DNA) or ribonucleic acid (RNA). DNA is found mostly in the chromosomes but it is present in the mitochondria too. RNA, though synthesized in the nucleus, is present mostly in the cytosol. In both DNA and RNA, the nucleotides are linked to each other by phosphodiester bonds, forming long chains of nucleotides. The codes for the synthesis of different types of amino acids reside in the sequence of the different nucleotides in the DNA. There are four types of nucleotides in DNA depending on the type of their nitrogenous base, which are adenine, guanine, cytosine and thymine. Similarly, the nucleotides in RNA contain adenine, guanine, cytosine or uracil. The nucleotide chain has two free ends, a 5' (read five prime) end and a 3' end. The sequence of nucleotides in a segment of DNA or RNA is conventionally stated from the 5' end to the 3' end.

DNA has a double helical structure, with its two strands interconnected by hydrogen bonds. RNA usually does not assume a double helical structure but often folds upon itself to form various shapes (See tRNA, Fig. 7.**3**). There are three major types of RNA: ribosomal RNA (rRNA), transfer RNA (tRNA), and messenger RNA (mRNA). The rRNA, along with several proteins, forms the ribosomes upon which protein synthesis takes place. The tRNA transports specific amino acids from the cellular amino acid pool to the ribosomes which are the sites of protein synthesis. The mRNA acts as a template of the genetic code.

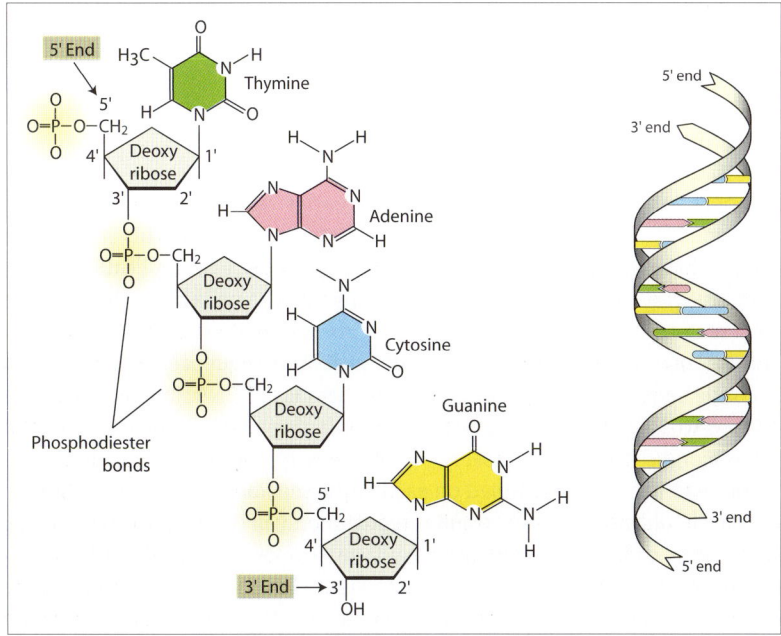

Fig. 7.**1** (*Left*) Nucleotides of DNA. (*Right*) The double helix structure of DNA.

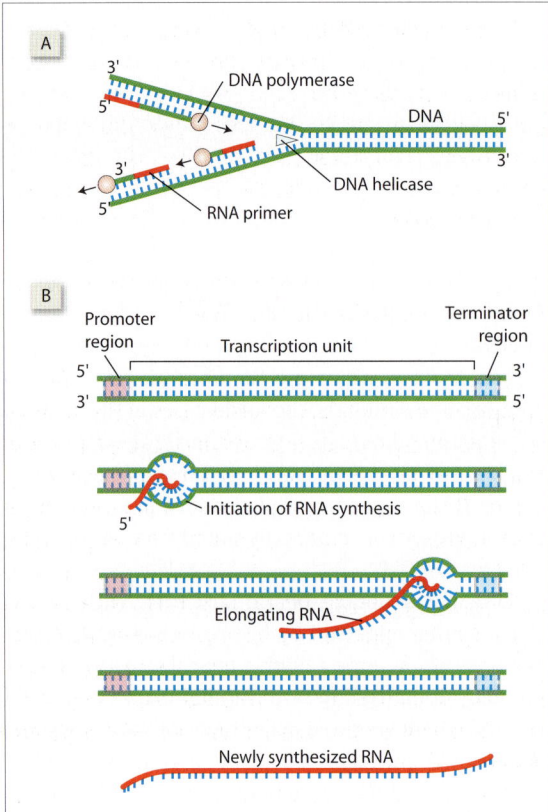

Fig. 7.**2** [A] DNA replication showing the replication fork. [B] Transcription (RNA synthesis).

Table 7.**1**   Codons for amino acids and STOP codons.

Glycine	GGU, GGC, GGA, GGG
Alanine	GCU, GCC, GCA, GCG
Valine	GUU, GUC, GUA, GUG
Leucine	UUA, UUG, CUU, CUC, CUA, CUG
Isoleucine	AUU, AUC, AUA
Serine	UCU, UCC, UCA, UCG, AGU, AGC
Threonine	ACU, ACC, ACA, ACG
Cysteine	UGU, UGC
Methionine	AUG
Aspartic acid	GAU, GAC
Aspartine	AAU, AAC
Glutamic acid	GAA, GAG
Glutamine	CAA, CAG
Arginine	CGU, CGC, CGA, CGG, AGA, AGG
Lysine	AAA, AAG
Histidine	CAU, CAC
Phenylalanine	UUU, UUC
Tyrosine	UAU, UAC
Tryptophan	UGG
Proline	CCU, CCC, CCA, CCG
STOP codons	UAA, UAG, UGA

Genetic Code

The genetic code is made of three-lettered words (triplets), each letter representing an RNA nucleotide. Each triplet codon always codes for the synthesis of a specific amino acid. It is possible to arrange 4 nucleotides into 64 ($4^3 = 64$) different triplets. Of these 61 code for amino acids and the remaining three (UAG, UGA, UAA) code for stop codons. Since the number of codons (61) exceed the number of amino acids to be coded for (20), several amino acids have more than one codon coding for them (Table 7.**1**). The reverse is not true, i.e., one codon never represents more than one amino acid.

Not all stretches of DNA in the chromosome code for proteins. For example, if the nucleotide sequence in a stretch of RNA is 5'-AUG-CCA-UUG-GAU-UAA-3', then the polypeptide coded by it is methionine-proline-leucine-aspartic acid. UAA does not code for any amino acid: rather, it signals that the chain should stop growing further. UAA is therefore called a 'stop' codon. Long stretches of DNA in eukaryotic cells do not code for protein. These noncoding regions are known as introns. The intervening stretches that code for proteins are known as exons.

Mutation refers to a permanent change in the genetic code. Such changes can have lethal effects on an organism due to alterations in the protein encoded by the mutated segment of the DNA.

DNA replication

During cell division, each strand of the double stranded DNA creates an identical copy of itself. This is known as DNA replication. In eukaryotes, DNA replication begins at multiple sites along the DNA helix but in prokaryotes, the replication begins at a single site. The site(s) must contain a unique nucleotide sequence. At the site(s) of replication, the two strands of DNA unwind, forming a replication fork. Unwinding requires the enzyme DNA helicase. A complementary nucleotide sequence is assembled on each of the separated DNA strands. Since a DNA strand is read out only in the 3' to 5' direction, the complementary DNA strands grow in opposite directions as shown in Figure 7.**2**A. Replication requires the enzyme DNA polymerase. The initiation of replication requires an RNA primer, a short RNA segment attached to the 3' end of the DNA strands.

Transcription

Transcription is the process through which RNA is synthesized using one of the DNA strands as the template. The part of the template bearing the nucleotide sequence TATAAT or TTGACA (read in the 5' to 3' direction) signals the initiation of transcription. These sequences are called the promotor regions of the DNA (Fig. 7.2B).

The enzyme RNA polymerase binds to the DNA at the promotor region. Transcription begins a few nucleotides downstream from the promoter site. This produces a growing chain of RNA bearing complementary nucleotides. Transcription stops when the RNA polymerase reaches an area on the DNA known as the termination region.

Translation

Translation is the process of mRNA-guided protein synthesis. The mRNA is formed in the nucleus. It leaves the nucleus to enter the cytosol and reaches the ribosome. Protein synthesis takes place on the ribosomal surface. Translation requires the availability of amino acids that are to be incorporated into the peptide, the mRNA bearing the codons, the tRNA bearing the anticodons and having a binding site for one amino acid, and ribosomes on which the protein assembly can take place. Also required are GTP and ATP as energy sources and several protein factors that are necessary for the initiation, elongation, and termination of the peptide chain.

Initiation of translation begins with the binding of the mRNA with the ribosome so that the initiation codon of the mRNA (the first codon that is to be translated) is located at the P-site (P for peptide) of the ribosome and its second codon is located at the A-site (A for amino acid) of the ribosome (Fig. 7.3). Next, a tRNA carrying methionine binds to the initiation codon AUG in the mRNA. Protein molecules called initiation factors bind to the complex, lending it stability.

Elongation of the peptide chain occurs when a second tRNA carrying the corresponding amino acid (say, tryptophan) binds to the next codon (UGG) on the mRNA at the A-site of the ribosome. This binding requires GTP and an elongation factor. Next, the methionine molecule binds to the tryptophan molecule to form a dipeptide. The reaction is catalyzed by the enzyme peptidyltransferase and an rRNA molecule. The tRNA left behind by methionine dissociates from the P-site. The tRNA now moves to the vacant P-site, carrying its dipeptide and the second codon (UGC) with it. A fresh tRNA carrying a third amino acid now binds to the A-site and the cycle repeats all over again, resulting in a growing peptide chain.

Termination of translation occurs when a stop codon (say, UAA) on the mRNA comes to occupy the A-site. There are no tRNAs for stop codons and therefore the peptide chain cannot grow further. Instead, the peptide

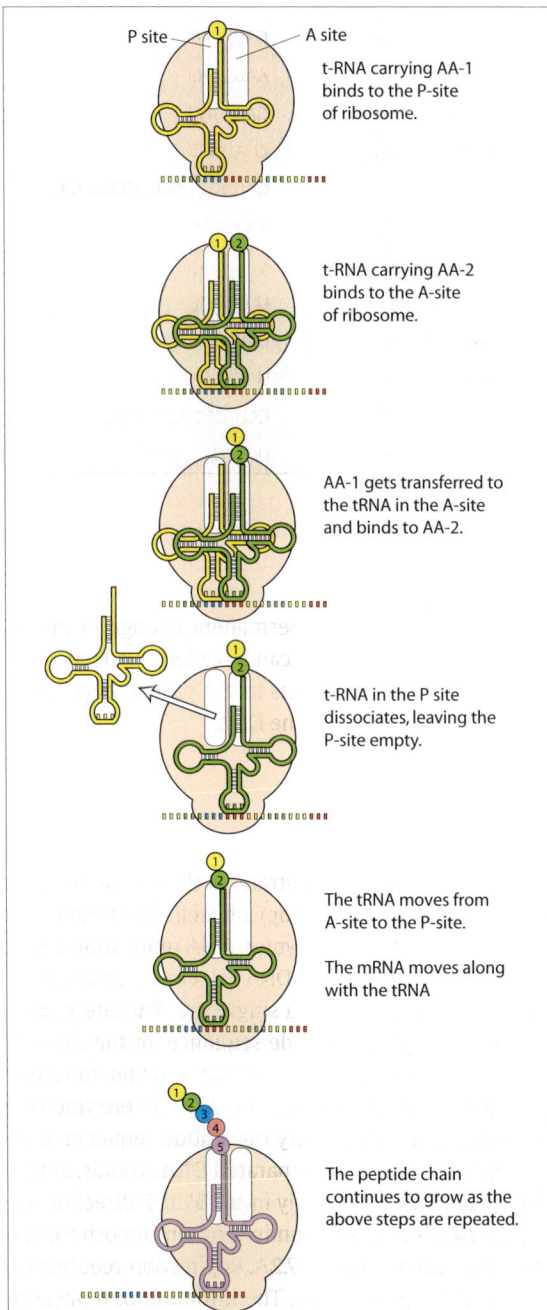

P site — A site

t-RNA carrying AA-1 binds to the P-site of ribosome.

t-RNA carrying AA-2 binds to the A-site of ribosome.

AA-1 gets transferred to the tRNA in the A-site and binds to AA-2.

t-RNA in the P site dissociates, leaving the P-site empty.

The tRNA moves from A-site to the P-site.

The mRNA moves along with the tRNA

The peptide chain continues to grow as the above steps are repeated.

Fig 7.**3**   Steps in translation. AA = Amino acid

chain is released from the tRNA carrying it through hydrolysis. The hydrolysis requires peptidyltransferase, releasing factors, and GTP.

Molecular Techniques

Southern blotting

The Southern blotting technique helps to detect the presence or absence of stretches of DNA with a specific nucleotide sequence in an entire genome or in a given DNA sample. The principle of the technique is outlined below.

The DNA isolated from eukaryotic (including human) cells is cut into multiple fragments called restriction fragments with the help of enzymes called restriction endonucleases. There are many different types of restriction endonucleases, each having a specific restriction site, i.e., the site where it cuts the DNA. For example, the restriction enzyme EcoRI recognizes the sequence GAATTC and cuts the DNA wherever this nucleotide sequence occurs on the DNA.

The DNA, after isolation and restriction digestion, is subjected to electrophoresis. The rate of migration of DNA fragments is inversely proportional to the size of the fragment. DNA fragments are not visible to the naked eye. To make them visible, the agarose gel containing the DNA is soaked in ethidium bromide solution. The DNA binds to ethidium bromide and becomes visible when exposed to ultraviolet light. The electrophoretically separated DNA fragments are blotted (transferred by blotting or soaking) onto an artificial membrane made of nitrocellulose or nylon, either mechanically or electrically (electroblotting), producing an exact replica of the original bands. The replica of DNA on the blotting membrane is called a *blot*. The transfer makes it easier for the experimenter to work with the DNA, as the artificial membrane is stronger and thus easier to handle than the fragile gel membrane. One or more of the several DNA bands that are blotted onto the artificial membrane may contain the nucleotide sequence of interest. The DNA fragment of interest is detected using a probe, a small piece of single-stranded radiolabelled DNA having a nucleotide sequence that is complementary to the DNA fragment of interest. The binding of the radioactive probe to the DNA fragment of interest is called *hybridization*. The band containing the DNA of interest is now radioactive: it can be radiographed by apposing the artificial membrane containing the DNA bands to an X-ray plate.

Similar techniques (electrophoresis following hybridization with specific probes), when applied to RNA fragments is called *northern blotting* and when applied to proteins, is called *western blotting*. For proteins, the probes used are specific antibodies directed against the proteins.

DNA fingerprinting

Most of the noncoding regions of DNA, i.e., the introns, consist of multiple repeats called *tandem repeats* of certain similar or identical DNA sequences. These repetitive sequences of DNA are either contiguous or are dispersed randomly along the length of the DNA. An example of dispersed repetitive DNA is the *Alu* element. It is a 300-base-pair DNA sequence that occurs randomly in the genome as many as 300000 to 500000 times. The location and number of these repetitive DNA is unique and is thus a molecular autograph of an individual. If the genomic DNA is digested with a restriction endonuclease, the length of the restriction fragments varies from individual to individual. This polymorphism (broad genotypic variations within a species) is known as *restriction fragment length polymorphism* and is made use of in DNA fingerprinting. The restriction fragments are subjected to Southern hybridization using the probe for one of the tandem repeats of DNA nucleotides. Due to restriction fragment length polymorphism, the blot shows bands whose distribution is unique for an individual. This technique can be utilized to confirm or rule out the molecular identity of an individual. The DNA required for the purpose can be extracted from any biologic material like blood or semen stain, or a hair.

DNA cloning

Multiple copies of a DNA fragment can be produced by incorporating it into a vector and inserting the vector into an actively multiplying cell. A vector is a molecule of DNA which replicates itself, when inserted into a cell. Plasmids and viruses are examples of cloning vectors.

Polymerase chain reaction

The polymerase chain reaction (PCR) is an alternative to DNA cloning. It allows rapid synthesis of a large number of copies of any specific DNA segment, provided that short DNA sequences on both sides of the DNA segment, called flanking sequences, are known. PCR consists of three steps: denaturation (separating the DNA strands) by heating, annealing (cooling the DNA and allowing the primers to get attached to the flanking DNA sequences), and elongation of the primer

Chapter 7   Applied Genetics   53

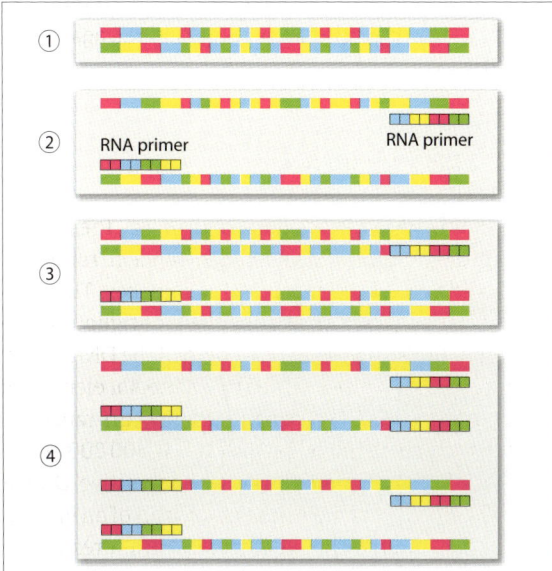

Fig. 7.4   Steps in a polymerase chain reaction. (1) A double-stranded DNA (2) DNA strands separated and attached to RNA primers. (3) DNA replicates by elongation of the complementary strand. (4) The strands of the replicated DNA separate and get attached to primers, repeating the above cycle all over again.

(by allowing the assembly of base pairs, starting from the primers). The variations in temperature required for the process are usually achieved with the help of an instrument known as thermocycler.

These three steps are repeated several times, leading to the synthesis of enormous number of copies of the given DNA. If the DNA is subjected to 30 cycles, a billion copies of the original DNA are generated. PCR can be used to identify the presence of bacteria and viruses, even if only one organism is present. PCR is also used in forensic medicine to detect the identity of a suspect by using any biologic material. It is used in research laboratories to detect and amplify vanishingly small quantities of DNA.

Section 2

Nerve and Muscle

8. Functional Anatomy of the Neuromuscular system

9. Degeneration and Regeneration of Nerve and Muscle

10. Resting Membrane Potential

11. Membrane Excitation and Action Potential

12. Electrophysiology of Ion Channels

13. Conduction of Nerve Impulses

14. Neuromuscular Transmission

15. Mechanism of Muscle Contraction

16. Characteristics of Muscle Contraction

17. Muscle Elasticity

18. Electromyography and Electroneurography

19. Muscle Mechanics

20. Smooth Muscles

21. Cardiac Muscle

8   Functional Anatomy of the Neuromuscular System

Neuron

The neuron or nerve cell (Fig. 8.**1**) is unique in that it can conduct electric impulses in which are codified various types of motor and sensory information. A neuron has a cell body called the soma, and two kinds of processes, namely, the axon and the dendrites.

The neuron contains most of the organelles present in a typical cell. However, the centriole, which is necessary for mitosis, is absent in the mature neuron which does not undergo mitosis. The cisternae of the rough endoplasmic reticulum are disposed in parallel

arrays called *Nissl bodies*. Due to their ribosomic RNA content, Nissl bodies stain intensely with chromophilic dyes and are very conspicuous. Polyribosomes are also found free in the cytoplasm. Nissl bodies are present in the soma and dendrites but are usually absent from the axon hillock and the axon. Smooth endoplasmic reticulum is present in all parts of the neuron.

The *cytoskeleton* of the neuron is made of microtubules, microfilaments, and intermediate filaments. In neurons, intermediate filaments are called *neurofilaments*. Under the light microscope, a network of *neurofibrils* (~2 μm in diameter) is seen in the perikaryon, extending into the dendrites and the axon. Neurofibrils are actually bundles of neurofilaments and microtubules. In certain degenerative diseases like Alzheimer's disease, the neurofilament protein gets altered, resulting in the formation of characteristic lesions called *neurofibrillary tangles*.

The **axon**[1] arises as a single long process from the *axon hillock*, which is a conical extension of the cell body. The axon is thinner and much longer than dendrites of the same cell. In some neurons, the axon may be myelinated. In certain very small local-circuit neurons, for example, the amacrine cells of the retina, there may be no axon at all. The part of the axon between the axon hillock and the beginning of the myelin sheath is called the *initial segment*. In response to the various synaptic inputs on the dendrites, the axon hillock and initial segment give rise to an action potential and hence, are together called the *spike trigger zone*. The axon conducts the action potential over long distances.

Dendrites are the multiple cell processes which branch extensively immediately after taking root from the soma. The dendritic tree so formed provides most of the surface for receiving signals from other neurons via their synapses with axon terminals. Dendrites often look 'thorny' due to numerous minute projections called *spines* present on their surface. The spines are

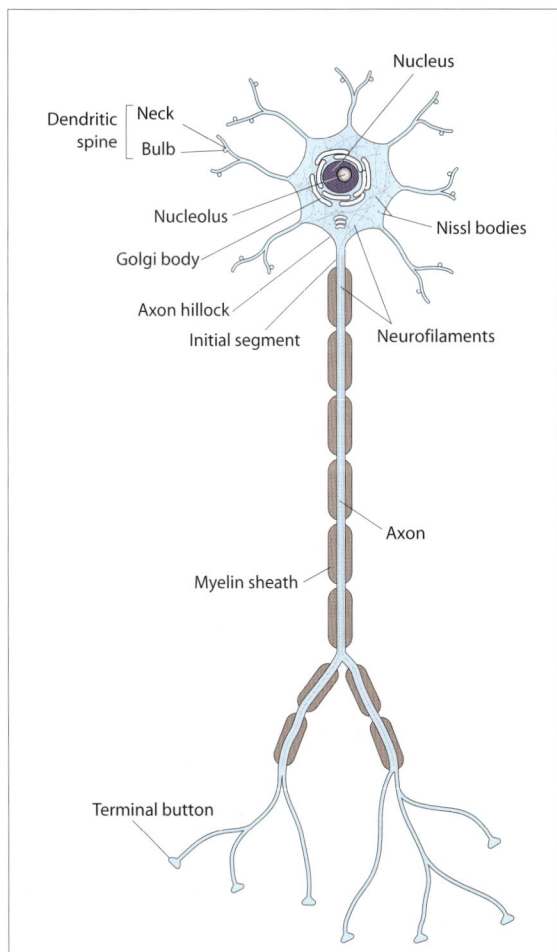

Fig. 8.**1**   Structure of a neuron. Note the bulb and the narrow neck of the dendritic spines.

---

[1] The axon is sometimes defined as the process that carries signals away from the soma. Similarly, dendrites are sometimes defined as the processes that carry impulses towards the soma. These definitions are fallacious. In a pseudounipolar cell or a bipolar cell, the axon carries information both towards and away from the cell.

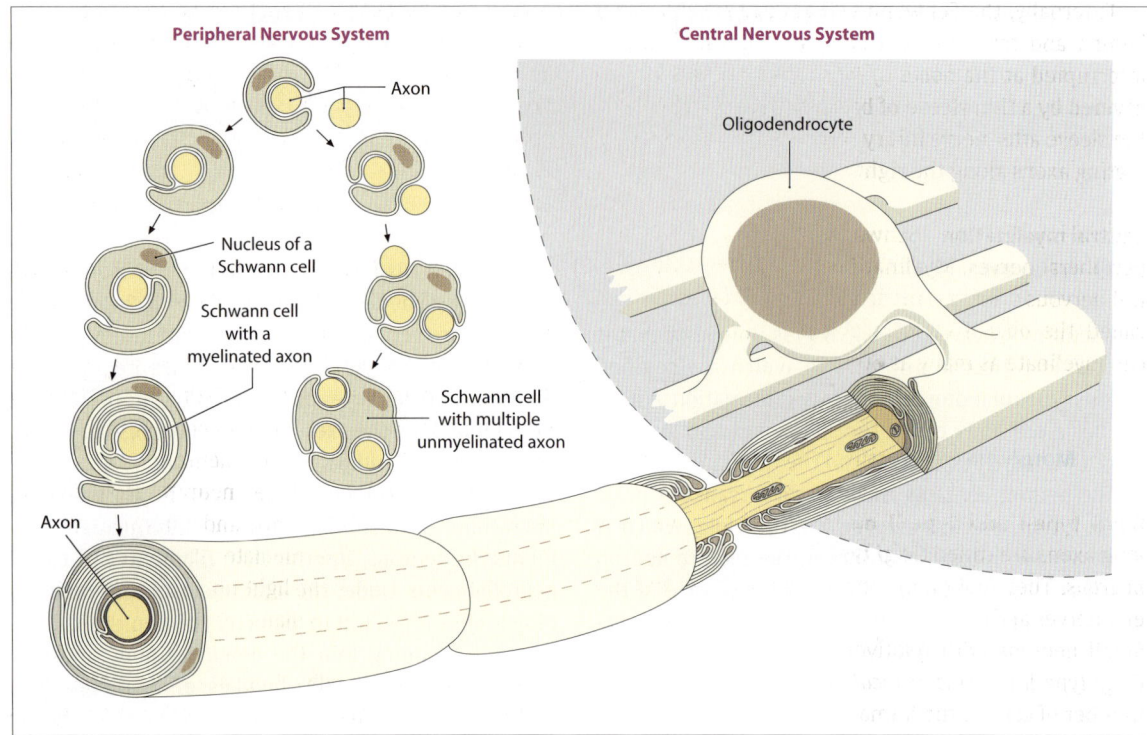

Fig. 8.**2**   (*Left*) Myelination of peripheral neurons by Schwann cells. (*Right*) Myelination of central neurons by oligodendrocytes.

the sites of synaptic contact, connected to the main shaft by a thin neck and ending in a more bulbous head. Each spine forms at least one synapse. During synaptic transmission, Ca^{2+} concentration rises in the spine. The thin neck of the spine prevents the rise in Ca^{2+} concentration in the spine from spreading to the main shaft, increasing the holding capacity of the dendrites (p 643).

Spines decrease in number after neuronal deafferentation or nutritional deprivation and exhibit structural changes in older people and in individuals with certain chromosomal abnormalities (trisomies 13 and 21). The synaptic inputs produce local potentials that get algebraically summated on the surface of dendrites. Dendrites usually do not produce action potentials. They mostly conduct graded potentials. Those that are very long like axons may conduct action potentials.

■ Myelination of neurons

Myelination of axons increases their speed of conduction but greatly increases their diameter. In motor nerves, speed is important. Hence, motor neurons to muscles and sensory fibers from muscles (proprioceptive fibers) are heavily myelinated. Inside the brain, neurons travel short distances and hence speed is

less important. Rather, the space available inside the cranium is limited. Hence, it is advantageous to have thinly myelinated or unmyelinated neurons that occupy less space, allowing more neurons to be accommodated inside the cranium.

Myelination begins at about 4 months *in utero* and is complete by about 5 years. Motor nerve roots are largely myelinated at birth, but the optic nerve and sensory roots lag behind by 3 to 4 months. The corticospinal tracts require a year for full myelination, and commissural axons of the cerebral hemispheres, 7 years or more. It is only after the completion of myelination that a neuron becomes fully functional. In the infant, functions like vision, walking, and performance of directed movements are possible only after the completion of myelination.

Peripheral myelination   The myelin sheath is produced by glial cells called *Schwann cells* that encircle the axon forming around it a thin sleeve called the *sheath of Schwann*. The myelin sheath is composed of several layers of cell membrane wrapped spirally around the axon (Fig. 8.**2**). The myelin gets compacted by the elimination of the intervening cytoplasm. The so-called 'unmyelinated' axons are also covered by a single layer of membrane and the cytoplasm of the enveloping Schwann cell. However, the myelin is not prominent.

Externally, the Schwann cell is covered by a basal lamina and the endoneurium. The myelin sheath is interrupted at the *nodes of Ranvier* where the axon is covered by a thin sleeve of basal lamina. Persistence of the sleeve after nerve injury serves to guide the regenerating axons along the right path.

Central myelination   Schwann cells are found only in peripheral nerves. Myelination of neurons in the central nervous system is made by a type of neuroglial cell called the *oligodendrocyte*. A single oligodendrocyte can myelinate as many as 60 axons.

Morphological classification of neurons

Golgi type-I and type-II neurons   Large cells with long axons are called *Golgi type-I neurons* or *projection neurons*. They include the neurons that form peripheral nerves and long tracts of the brain and spinal cord. Small neurons with relatively short axons are called *Golgi type-II neurons* or *local circuit neurons*. The total number of cells in the human nervous system is about 1 trillion, most of which are local circuit neurons that have proliferated markedly in the course of evolution. The evolutionary increase in the number of sensory neurons has been much less. The number of motor neurons has remained relatively small at about 2 million.

Pseudopolar, bipolar, and multipolar neurons   Except in early embryonic stages, *true unipolar neurons* are not found in vertebrates. However, the primary sensory neurons (neurons conveying impulses from the sensory receptors to the spinal cord) are *pseudounipolar*. During early development, these neurons are bipolar but later, the roots of the two processes shift around the perikaryon and fuse into a single stem. The stem and both its branches are myelinated axons. There may be short dendritic processes at the end of the distal branch, often in association with some kind of receptor structure. In *bipolar neurons*, an axon projects from each end of a fusiform cell body. Cells of this type are found in the vestibular and cochlear ganglia and in the nasal olfactory epithelium. Most vertebrate neurons, especially in the central nervous system, are *multipolar*. The dendrites branch profusely to form the dendritic tree. There is usually a single axon (Fig. 8.**3**).

Organization of neurons

The nervous system is divided into the central nervous system (CNS) constituted by the brain and the spinal cord, and the *peripheral nervous system* (PNS)

constituted by neurons lying outside the central nervous system. An aggregation of neuronal cell bodies located inside the CNS is called a *nucleus*. An aggregation of neuronal cell bodies located outside the CNS is called a ganglion. A compact bundle of axons located inside the CNS is called a tract. A compact bundle of axons located outside the CNS is called a *nerve*.

Axoplasmic Transport

Axoplasmic transport is vital to nerve cell functions. There are two forms of axonal transport: fast and slow. Fast axoplasmic transport has a velocity of 20 to 400 mm per day and may be anterograde or retrograde in direction. Slow axoplasmic transport occurs at 0.2–0.4 mm per day and is always anterograde.

Fast axoplasmic transport

Fast anterograde transport occurs along microtubules. It occurs from the soma to axon terminal and is driven by the molecular motor *kinesin*. It transports membrane-bounded organelles like short segments of the endoplasmic reticulum, mitochondria, small vesicles, actin, myosin, and the clathrin used in recycling of synaptic vesicle membrane.

Fast retrograde transport occurs from the nerve terminals to the cell body, returning materials for degradation or reuse. These materials include proteins and small molecules picked up by the axon terminal.

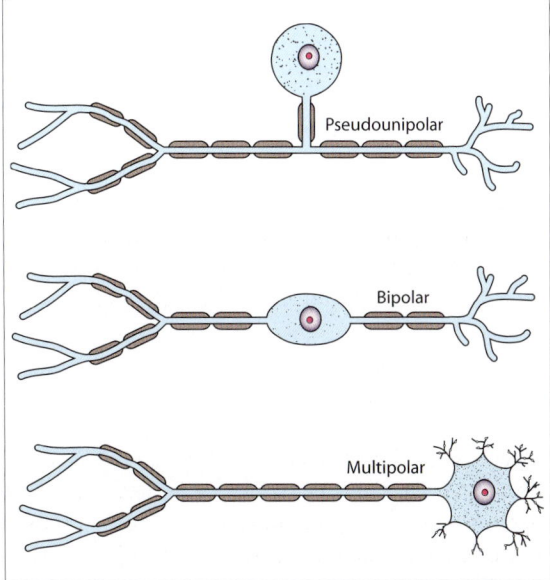

Fig. 8.**3**   Pseudounipolar, bipolar, and multipolar neurons

It employs the molecular motor *dynein*, and has one-third the velocity of anterograde transport. Retrograde transport is responsible for carrying tetanus toxin and neurotropic viruses such as herpes simplex and rabies directly to cell bodies in the CNS. It has been employed by neuroanatomists for charting out neural pathways.

■ Slow axoplasmic transport

Slow transport carries protein subunits of neurofilaments, tubulins of the microtubules, and soluble enzymes. It also moves some cytosol with it. The mechanism of slow axonal transport is not clear: Probably the cytoskeleton moves as a whole due to the continual polymerization at the leading end and depolymerization at the trailing end, and the axoplasmic matrix moves with it.

■ Neuroglia

Neuroglia, meaning 'nerve glue', is the connective tissue of the brain. In the central nervous system, neuroglial cells are ten times more numerous than neurons. These cells provide structural and trophic support to neurons and depending on their type, serve other important functions. Although they are interconnected through gap junctions, glial cells do not conduct action potentials. The major neuroglial cells are astrocytes, oligodendroglia, and microglia (Fig. 8.**4**). The Schwann cells of peripheral nerves and the ependymal cells of cerebral ventricles are also considered to be neuroglial cells.

Astrocytes are star-shaped glial cells. They form the 'skeleton' of the central nervous system. In the developing nervous tissue, astrocytes provide the framework along which young neuronal cells migrate to their final positions. They form an insulation around synapses so as to prevent chaotic spread of nerve impulses. These astrocytes around synapses have an important role in reestablishing synaptic contact following disruption of synapses. Astrocytes proliferate and form scars after neuronal damage. They clear the neural tissues of K^+ and excess neurotransmitters that accumulate extracellularly following neural activity. Through their processes and footplates on the walls of blood vessels, astrocytes contribute to the formation of the blood-brain barrier. Astrocytes also store glycogen. When stimulated by neurotransmitters like norepinephrine, they convert the glycogen into glucose and release it for the consumption of the neurons.

Oligodendrocytes resemble astrocytes but as their name indicates, have fewer dendrites. They myelinate neurons of the central nervous system.

Microglial cells resemble oligodendrocytes but as the name suggests, are much smaller in size. They are phagocytic cells that clear up myelin and cellular debris in injured areas.

■ Muscle fibers

Muscles are contractile tissues whose cells are traditionally called 'fibers'. *Histologically*, some muscle fibers show cross striations while others do not. *Functionally*, some muscles conduct ephaptically, with the action potential spreading from cell to cell. Such muscle fibers contract spontaneously. In others, ephaptic conduction is absent. These muscle cells contract only when stimulated through nerve fibers. Accordingly, there are four types of muscle fibers based on their histological and functional characteristics (Table 8.**1**).

Skeletal muscle fibers are striated cylindrical and multinucleate (Fig. 8.**5**). They are 10 to 30 cm in length—long enough to justify the term 'fiber'. They are formed

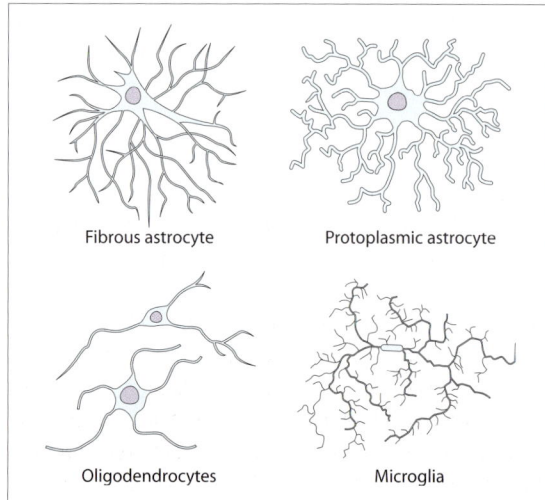

Fig. 8.**4**   Neuroglial cells.

Table 8.**1**   Classification of muscle fibers.

	Striated	**Non-striated**
Ephaptic conduction present	Cardiac muscle	Single-unit smooth muscle
Ephaptic conduction absent	Skeletal muscle	Multi-unit smooth muscle

Fibrous astrocyte

Protoplasmic astrocyte

Oligodendrocytes

Microglia

by fusion of several smaller cells into a multinucleate syncytium. Skeletal muscles are so named because they are attached to bones by tendons and move these bones and the loads borne by them. They are the only muscles under voluntary control.

Smooth muscle cells are non-striated fusiform (spindle-shaped) and uninucleate (Fig. 8.**5**). They are 0.02 to 0.5 mm in length. *Single-unit* (or *visceral*) *smooth muscles* are present in walls of the gastrointestinal and genitourinary tract. All the fibers in a single-unit smooth muscle contract together, near-simultaneously or sequentially, due to the presence of cell-to-cell conduction of electrical excitation. *Multi-unit* (or *motor-unit*) *smooth muscles* are found in the intrinsic muscles of the eye, piloerector muscles, vas deferens and in the walls of large elastic arteries. Fibers of multi-unit smooth muscles contract in response to nerve stimulation in much the same way as skeletal muscle fibers. However, they are innervated by autonomic fibers that are not under voluntary control.

Cardiac muscle cells are striated, about 0.1-mm-long cylindrical fibers, some of which are branched (Fig. 8.5). The cells are joined end-to-end forming *intercalated disks* that provide low-resistance bridges for passage of electrical excitation from cell to cell. Cardiac muscle is found only in the heart, and hence its name.

■   Contractile apparatus of a striated muscle

Like all other cells, the striated muscle fiber also has a cell membrane (called the sarcolemma), smooth endoplasmic reticulum (called sarcoplasmic reticulum), cytoplasm (called sarcoplasm) and cytoskeletal proteins which are of three types: contractile, regulatory, and anchoring proteins. The *contractile proteins* are myosin and actin. These two proteins interact to generate the contractile force in a muscle. The *regulatory proteins* are tropomyosin and troponin. Also called the 'relaxation proteins', these regulate the interaction between actin and myosin. The *anchoring proteins* are α-actinin, titin, nebulin, and dystrophin. These proteins anchor the cytoskeletal proteins to each other as well as to the sarcolemma and the extracellular matrix.

The cytoskeletal proteins are disposed in a remarkably orderly way, because of which the muscle fiber shows transverse *striations* when viewed under the light microscope. With special techniques, the striations are seen to correspond with dark and light bands (see Fig. 15.2). A line called the Z-line extends through the middle of the light band. The part of the muscle fiber that extends between two consecutive Z-lines

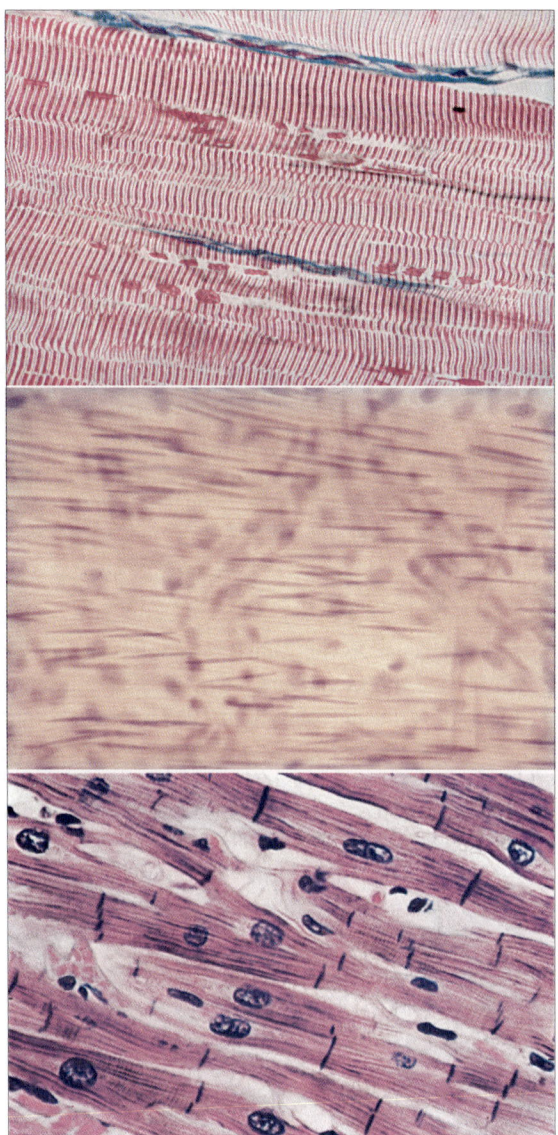

Fig. 8.**5**   Muscle cells. (*Above*) Skeletal muscle fibers. (*Middle*) Smooth muscle fibers. (*Below*) Cardiac muscle cells.

is called a *sarcomere*, which is the contractile unit of muscle. During muscle contraction, the Z-lines come closer together and the sarcomeres shorten.

Sarcoplasmic reticulum   In a muscle cell, the sarcoplasmic reticulum is disposed in a highly geometrical way: some run parallel to the length of the fiber (the *longitudinal* or *L-tubules*) while others are disposed radially, coursing from the surface towards the center of the fiber (the *transverse* or *T-tubules*). The transverse tubules are transverse invaginations of the sarcolemma into the cell. They typically occur at the A-I junctions

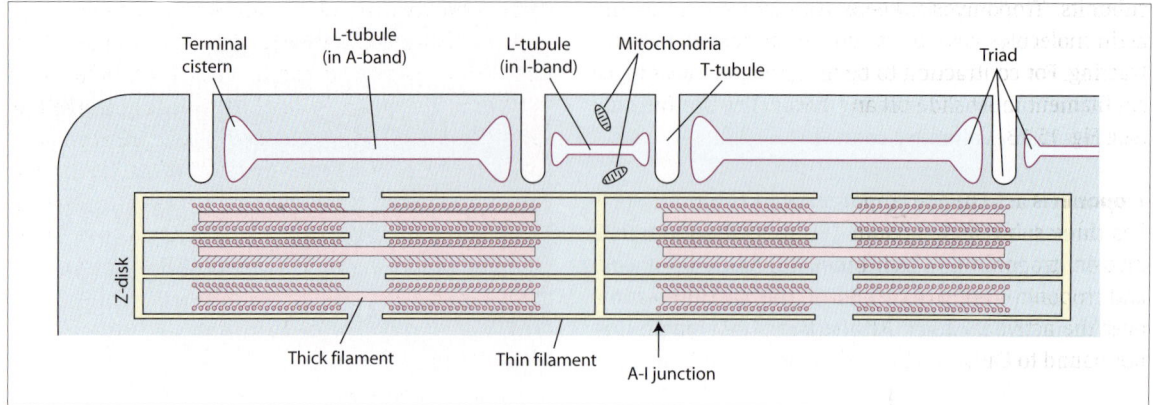

Fig. 8.**6**   The transverse and longitudinal tubules. Note that the L-tubules in the I-band are shorter and less regular than the L tubules in the A-band. This is because the I-band region is crowded with mitochondria. Moreover, the width of the I-band changes during muscle contraction subjecting the L-tubule to compression. The T-tubule and two lateral cisterns on its sides constitute a triad.

in mammals and at the Z-disk in amphibians. Being extensions of the sarcolemma, the transverse tubules conduct the *action potentials* from the sarcolemma into the interior of the cell along its membranes. The longitudinal tubules are disposed longitudinally along the entire length of the A and I bands. They end in dilated sacs called *lateral cisterns* (Fig. 8.**6**).

Myosin filaments are thick and disposed longitudinally at the center of a sarcomere (Fig. 8.**6**). A myosin filament is made of two intertwined heavy H-chains. The heavy chain has a globular head and a long tail. The tails are helically intertwined while the heads remain separate. Each globular head contains two additional light L-chains. In the sarcomere, the myosin tails are grouped into a bundle from which the globular heads called myosin heads project out (Fig. 8.**7**A).

Digestion with trypsin generates two fragments of the myosin molecule: the heavy meromyosin (HMM) and the light meromyosin (LMM). The HMM contains the globular head as well a part of its fibrous tail. It can be split further by papain into two parts: the globular HMM S1, which has all the ATPase activity and actin-binding ability, and the fibrous HMM S2, which has none of it. The LMM comes entirely from the fibrous tail of the myosin molecule. It does not have any ATPase activity or actin-binding ability (Fig. 8.**7**B).

Actin filament is a double helical filament (F-actin) (Fig. 8.**8**). It is made of globular subunits called G-actin. In a sarcomere, actin filaments are attached at one end to the Z-disk. There are specific sites on the actin filament which bind to the myosin head during muscle contraction. These sites are called *active sites*.

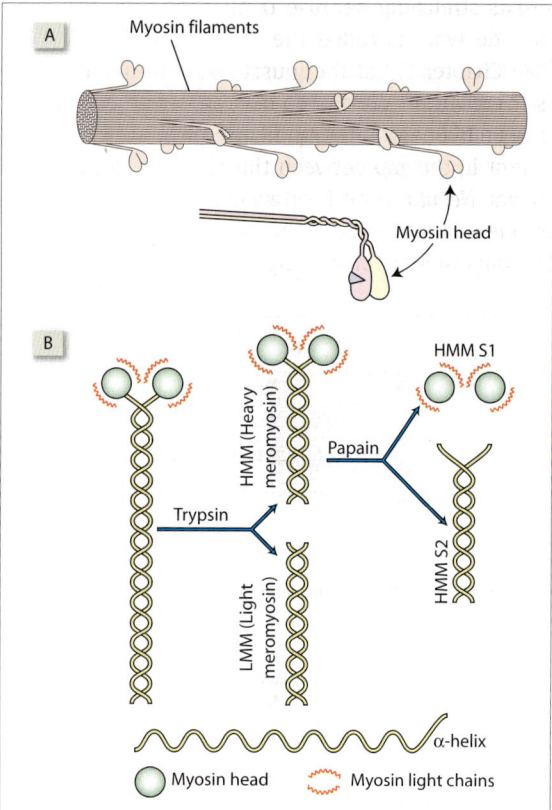

Fig. 8.**7**   Myosin filaments. [A] Organization of myosin filaments. [B] Structure of myosin filaments.

Tropomyosin is a filamentous molecule that consists of two chains, α and β. The tropomyosin filament lies in the groove between the two filaments of actin, and each tropomyosin filament spans seven G-actin subunits. Tropomyosin keeps the active sites of the actin molecules covered when the muscle is not contracting. For contraction to be initiated, the tropomyosin filament must slide off and uncover the active sites (see Fig. 15.1B).

Troponin is attached to tropomyosin (see Fig. 15.1B). It has three subunits: troponin-T which binds to tropomyosin, troponin-C which binds to four ions of Ca^{2+}, and troponin-I which holds the tropomyosin filament over the active sites on actin so long as troponin-C is not bound to Ca^{2+}.

■ Anchoring proteins of a striated muscle

The α-*actinin* is located in the Z-band. It cross-links the actin filaments in the Z-band. *Titin* (earlier called connectin or gap filament) is a large elastic filament that interconnects the Z-disks (Fig. 8.8A). The sarcomere resists stretching because of the titin filaments that provide what is called the series elastic component (see Chapter 17) of the muscle. When the sarcomere is stretched till there is no overlap between actin and myosin filaments, the titin filaments become more apparent in the gap between the actin and myosin filaments. *Nebulin* is an inextensible filament connected at one end to the α- actinin in the Z-disk and to the tropomyosin-troponin complex at regular intervals.

Dystrophin-associated glycoproteins include some proteins that are extracellular (dystroglycan and laminin), some located on the inner side of sarcolemma (dystrophin, syntrophin, and utrophin) and some that span the membrane thickness (sarcoglycan, sarcospan). Dystrophin, the best known of these, anchors actin to the membrane through syntrophin, and also to the basal lamina matrix through dystroglycan and laminin (Fig. 8.8B). Genetic defects in dystrophin molecule produce Duchenne's muscular dystrophy.

■ Smooth muscles

Smooth muscles have functionally important anatomic differences with skeletal muscles (Fig. 8.9). (1) The sarcolemma show short invaginations into the cytoplasm called caveoli, which are analogous to the T-tubules in skeletal fibers. (2) There are several fusiform densities present in the cytoplasm (cytoplasm dense bodies) and along the inner surface of the sarcolemma (subsarcolemmal dense plaque). The cytoplasmic dense bodies contain the protein α-actinin and are the equivalent of the Z-disks in skeletal muscle fibers. Subplasmalemmal dense plaques contain vinculin and talin. Extending between cytoplasmic dense bodies are three types of filaments: actin, myosin, and intermediate filaments. Intermediate filaments are made of desmin (in vascular smooth muscle, it is made of vimentin) that provides a cytoskeletal framework. (3) The ratio of actin

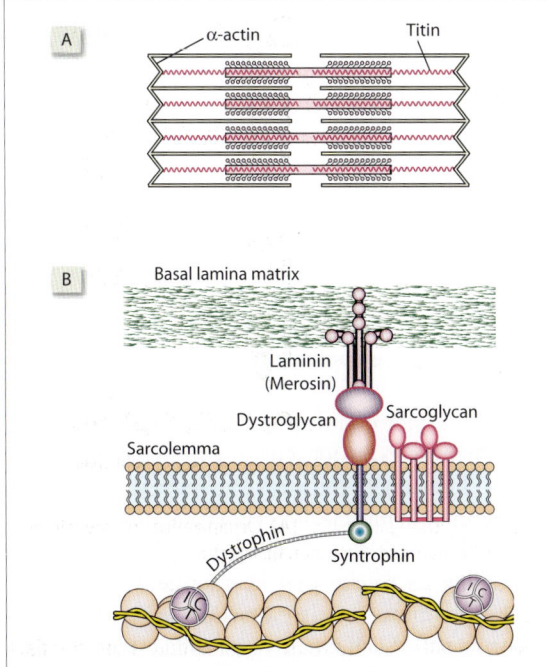

Fig. 8.8   Anchoring proteins in muscle. [A] Location of α-actinin and tinin filaments in a sarcomere. [B] Dystrophin-associated glycoproteins.

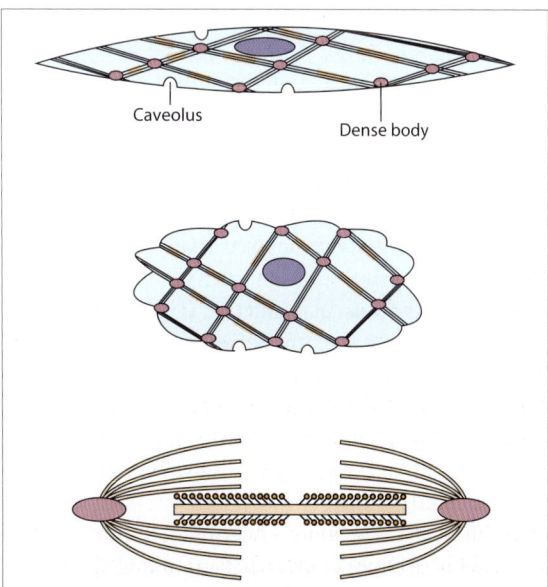

Fig. 8.9   Smooth muscle fiber. (*Above*) Structure of a smooth muscle fiber. (*Middle*) A smooth muscle fiber in the contracted state. (*Below*) Actin fibers radiate from the dense bodies, with several actin filaments surrounding a single myosin filament.

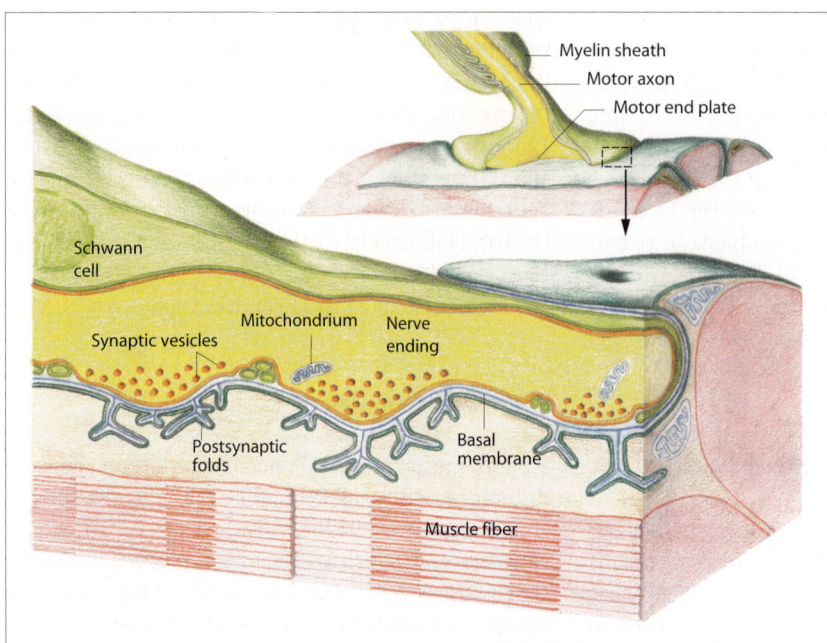

Fig. 8.**10**   Neuromuscular junction.

Labels in figure:
Myelin sheath
Motor axon
Motor end plate
Schwann cell
Synaptic vesicles
Mitochondrium
Nerve ending
Postsynaptic folds
Basal membrane
Muscle fiber

and myosin filaments is ~12:1 in smooth muscles compared to ~2:1 in skeletal muscles. The myosin filaments are different in skeletal and smooth muscles. In smooth muscles, the projecting heads of the myosin molecules form cross-bridges with the actin filaments along their entire length, whereas the thick filaments of skeletal muscle have a bare central segment that is devoid of cross-bridges. (4) The regulatory proteins tropomyosin and troponin that are present in skeletal muscle are absent in smooth muscle. Instead, smooth muscles contain calmodulin, which is analogous to troponin.

Neuromuscular Junction

A junction between two neurons that permits the passage of an electrical impulse across it is called a synapse. The neuromuscular junction (Fig. 8.10) is essentially a synapse that exists between an axonal ending and a muscle fiber. Neuromuscular junction in skeletal muscle As the axon supplying a skeletal muscle fiber approaches its termination, it loses its myelin sheath. The axis cylinder then branches into a number of bulb-shaped endings called *terminal buttons*, each of which innervates a muscle fiber. The terminal buttons contain many small, *synaptic vesicles* that contain acetylcholine, which is the neurotransmitter at neuromuscular junctions. The sites of the prejunctional membrane that are specialized for vesicular release of neurotransmitter are called the *active zones*.

The terminal buttons come in close proximity with a thickened trough on the muscle membrane called the *motor endplate* that bears receptors for acetylcholine (ACh). The motor endplate is thrown into folds called the *junctional folds*. The acetylcholine receptors are clustered at the crest of each junctional fold, which are positioned opposite the active zones of the prejunctional membrane. The 100-nm-wide space between the terminal button and the motor endplate is called the *myoneural cleft*. The cleft is not uniformly wide. It is less than 1 nm in width at the tips of the junctional folds. Within the cleft is a basement membrane or the *basal lamina* made of collagen and other matrix proteins. The enzyme *acetylcholinesterase* (AChE) is anchored to the collagen fibrils of the basement membrane.

Synapse en passant in smooth muscle

Unlike the motor endplate of a skeletal muscle fiber, smooth muscles do not show any specialization at their site of contact with the axon. Moreover, the axon innervating smooth muscle does not end at the site of innervation. Rather, it shows multiple swellings or varicosities along its course that liberate neurotransmitters. Each axon forms multiple junctions with muscle cells along its path. Such synapses are called *synapse en passant* (Fig. 8.**11**). This is unlike in skeletal muscle where the neuron must branch and terminate to innervate multiple fibers. Finally, the neurotransmitters released

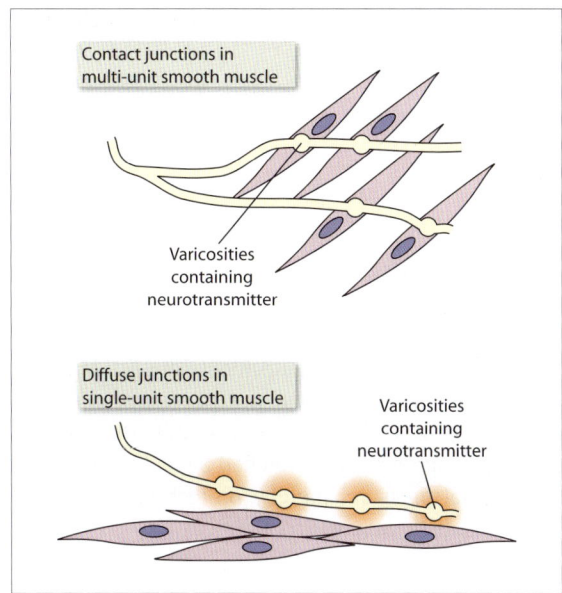

Fig. 8.**11**   Synapse en passant. (*Above*) Contact junctions. (*Below*) Diffuse junctions.

from the varicosities (the *junctional neurotransmitters*) can be either acetylcholine or norepinephrine. This is unlike the axonal ending in skeletal muscle that releases only acetylcholine.

Contact junctions   In multi-unit smooth muscles, the varicosities come in close contact with individual cells forming contact junctions. Here, the muscle membrane is separated from the varicosity by a gap of roughly the same width as in a neuromuscular junction. Hence, the junctional delay (i.e., time taken for the impulse to travel from the neuron to the muscle membrane) is comparable to that in a neuromuscular junction of a skeletal muscle.

Diffuse junction   In single-unit smooth muscle, the varicosities do not come in close contact with any cell. Neurotransmitters from the varicosities diffuse away to reach all the muscle cells. This arrangement has been called a diffuse junction. The junction delay is considerably more in diffuse junctions.

9 Degeneration and Regeneration of Nerve and Muscle

Nerve

Peripheral nerves are enclosed in three connective tissue sheaths: the epineurium, perineurium, and endoneurium (Fig. 9.**1**).

The **endoneurium** is bounded internally by the basal lamina around the Schwann cells and externally by the relatively impermeable inner basal lamina of the perineurium. The interstitial fluid of the endoneurium does not exchange freely with the capillary blood flowing through it because the capillaries have tightly adherent endothelial cells. This barrier represents an extension of the blood–brain barrier. It is important for maintaining the appropriate environment for the axons and for protecting them from harmful agents. When a nerve gets infected, the sleeve of endoneurium provides a conduit through which the invading bacteria or neurotropic viruses can reach the central nervous system.

The **perineurium** is the middle connective tissue sheath, which is bounded by its internal and external basal lamina. The cells of the perineurium are tightly adherent and act as a barrier to passage of particulate tracers, dye molecules or toxins into the endoneurium.

The **epineurium** is the outermost connective sheath that sends extensions into the nerve, compartmentalizing the nerve into fascicles. It limits the extent to which the nerve can be stretched by body movements or external pressure, thereby protecting the delicate axons inside the nerve.

Nerve growth

This **growth cone** is the expanded tip of an elongating axon. It has three regions: a central core which is rich in mitochondria, the *filopodia* which are slender projections of the central core, and the *lamellopodia* which are the 'webs' between the filopodia (Fig. 9.**2**).

The motility of the filopodia is made possible by the reorganization of the cytoskeleton inside it. The actin microfilaments polymerize near the tip of the filopodia and depolymerize at the other end. The filopodial membrane also has receptors for various molecules present

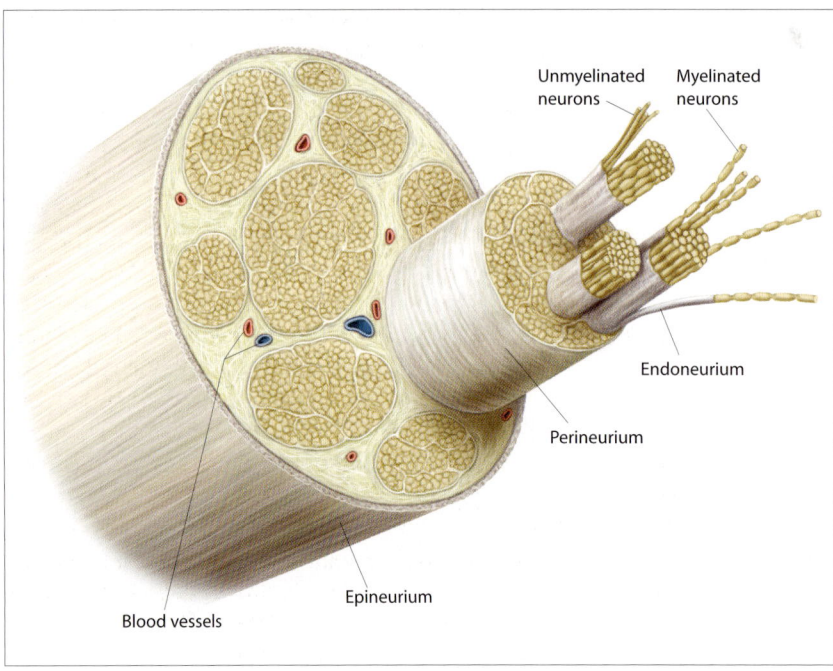

Fig. 9.**1**   Structure of a peripheral nerve.

Unmyelinated neurons

Myelinated neurons

Endoneurium

Perineurium

Epineurium

Blood vessels

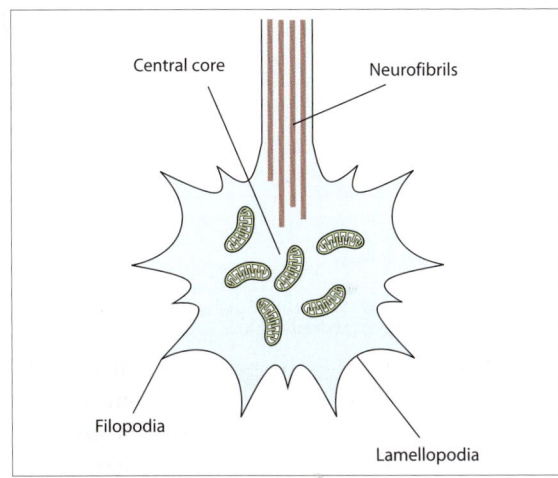

Fig. 9.**2**   The growth cone of a neuron.

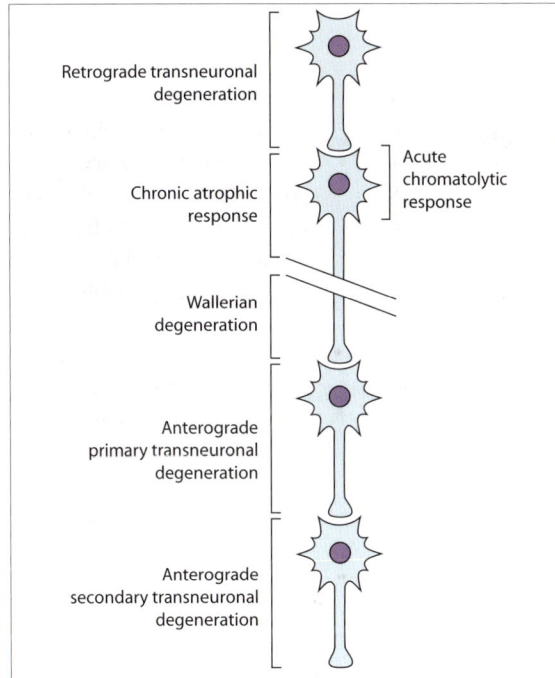

Fig. 9.**3**   Types of neuronal degeneration.

in its environment. When certain key molecules bind to the filopodial receptors, they activate a second messenger system that alters filopodial motility.

Trophic factors affecting axonal growth   *Neurotrophins* are secreted by the target cells. They promote nerve growth. The nerve growth factor (NGF) is the best known neurotrophin and also the first one to be discovered. Neurotrophins are essential for the survival of neurons. In their absence, the neuron perishes through apoptosis. Neurotrophins induce the formation of

certain proteins that inhibit apoptosis of neurons. Hence neurons heading towards neurotrophin-secreting target cells survive while those straying away perish. Laminins are growth-promoting molecules present in the extracellular matrix that attract the growth cone towards them. Integrins are the complementary molecules present on the growth cone that bind to laminins and other growth-promoting molecules in the extracellular matrix. Cell-adhesion molecules (CAMs), which are known for their ability to bring about cell-to-cell adhesions, also stimulate axonal growth. They are present in the extracellular matrix. CAMs are of two types: those which are Ca^{2+}-dependent, e.g., cadherin, and those which are not, e.g., neural cell-adhesion molecule (NCAM).

Tropic factors affecting axonal growth   The word tropic means 'turning toward or from a stimulus'. It is different from the word trophic which means 'nourishing or providing nutrition'. Neurons growing towards the source of a trophic factor are likely to survive while those growing away from it generally perish. Hence, trophic factors also function as tropic factors although the reverse is not true. Netrins are chemoattractant molecules that attract the growth cone towards them. Ephrins and semaphorins provide *repulsive axon guidance signals* (RAGS), i.e., they direct the advancing growth cone away from them.

Nerve degeneration

When an axon is crushed or severed, it results in degeneration of the neuron both distal (anterograde) and proximal (retrograde) to the site of injury (Fig. 9.**3**).

Anterograde degenerative changes can affect both the injured neuron as well as those neurons that suffer no direct damage but are functionally associated with the damaged cell. Such transneuronal degeneration is well known in lesions of the retina, which result not only in the degeneration of the ganglion cells (primary transneuronal degeneration) but also of the lateral geniculate cells (secondary transneuronal degeneration).

Anterograde degeneration in the injured neuron is commonly called *Wallerian degeneration* (Fig. 9.**4**). Within 24 hours, neurofilaments break up and the axon fragments into short lengths. Within 10 days, the myelin sheath breaks down into lipid droplets around the axon. Within 30 days, the myelin is denatured chemically. Within 3 months, macrophages from the endoneurium invade the degenerating myelin sheath and axis cylinder. The macrophages phagocytose the debris, leaving behind only the endoneural tube. In the central nervous system, nerve fibers undergo a slower Wallerian degeneration than those in the peripheral nervous system.

Retrograde degenerative changes affect the dendritic tree, the cell body, and the part of the axon proximal to the lesion. Retrograde changes can affect the injured neuron or may be transneuronal. An example of retrograde transneuronal degeneration is seen following surgical ablation of the cerebral cortex. It results in degeneration not only of the neurons projecting from the anterior thalamic nuclei to the cortex but also of the neurons that project to the anterior thalamic nuclei from other thalamic nuclei (retrograde transneuronal degeneration). Most of the time, the retrograde degeneration extends only up to the first node of Ranvier proximal to the injury.

The degeneration in the injured neuron may be of the acute or the chronic type. *Acute retrograde (chromatolytic) response* starts after 24 hours and is characterized by the disintegration of Nissl substance into fine dust (Fig. 9.**5**). Nissl substance represents inactive, aggregated RNA. Following chromatolysis, the RNA disperses in the cytoplasm, enabling the neuron to mobilize the protein synthesis required for regeneration. *Chronic retrograde atrophic response* follows the acute response only occasionally. It involves the atrophy of the nerve cell body with all its processes.

Retrograde chromatolysis begins 24–48 hours after axonal injury and reaches its peak in about 2 weeks when it loses its staining reaction. It begins near the axon hillock around the nucleus and spreads to other parts of the cell body (central chromatolysis). Chromatolysis can also occur in certain infectious or degenerative diseases of the nervous system such as poliomyelitis and progressive muscular atrophy. Such chromatolysis usually fades inward away from the cell periphery toward the nucleus (peripheral chromatolysis) rather than outward from the nucleus, as is the case in retrograde chromatolysis. The degree of chromatolysis depends on the neuron type (it is more in motor neurons) and the proximity of the site of injury to the nerve cell (the more the axoplasm is detached from the cell body, the greater is the retrograde chromatolysis). The other acute retrograde changes that follow an axonal injury are: (1) Swelling of the cell. (2) Displacement of the nucleus from its normal central position to one at the periphery, away from the axon hillock. It may even be completely extruded, in which case the cell atrophies and completely disappears. (3) Fragmentation and reduction of the Golgi apparatus. (4) Disappearance of neurofibrils. (5) Withdrawal of the synapsing axon from the cell or its separation from the cell by the interposition of a glial cell (synaptic stripping).

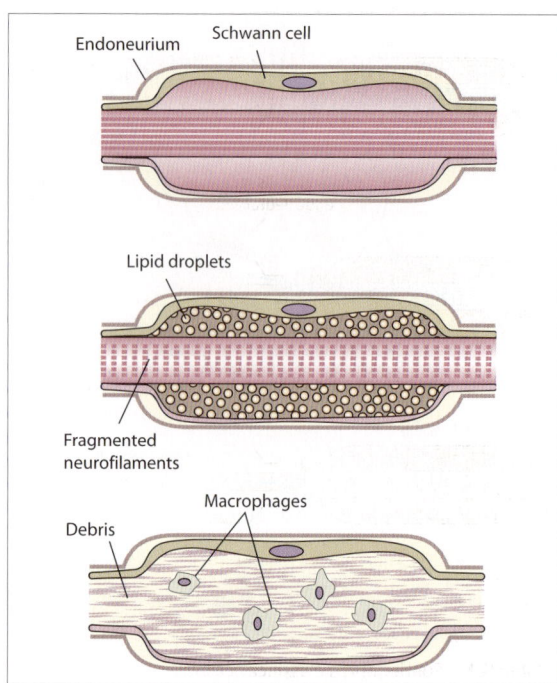

Fig. 9.**4**   Wallerian degenerative changes in a nerve fiber.

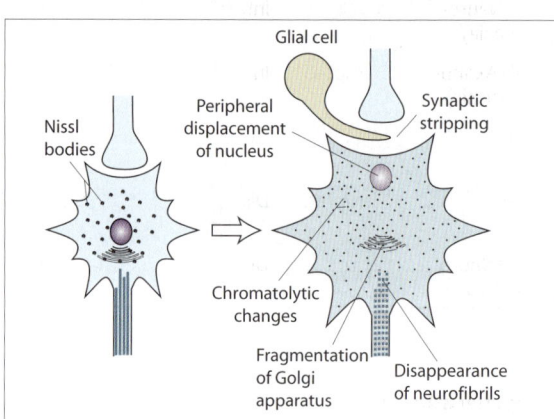

Fig. 9.**5**   Chromatolytic changes in the soma of a nerve.

Neuroanatomists can cut a nerve or tract and locate its origin by looking for neurons exhibiting chromatolysis. Until the advent of techniques exploiting orthograde and retrograde axonal transport, a time-consuming search for chromatolysis and tracing Wallerian degeneration were the only methods available for working out the neural pathway in the central nervous system.

Nerve regeneration

Repair begins about 20 days after nerve section and is complete in 3 months. Regenerative changes usually occur simultaneously in the cell and the axon.

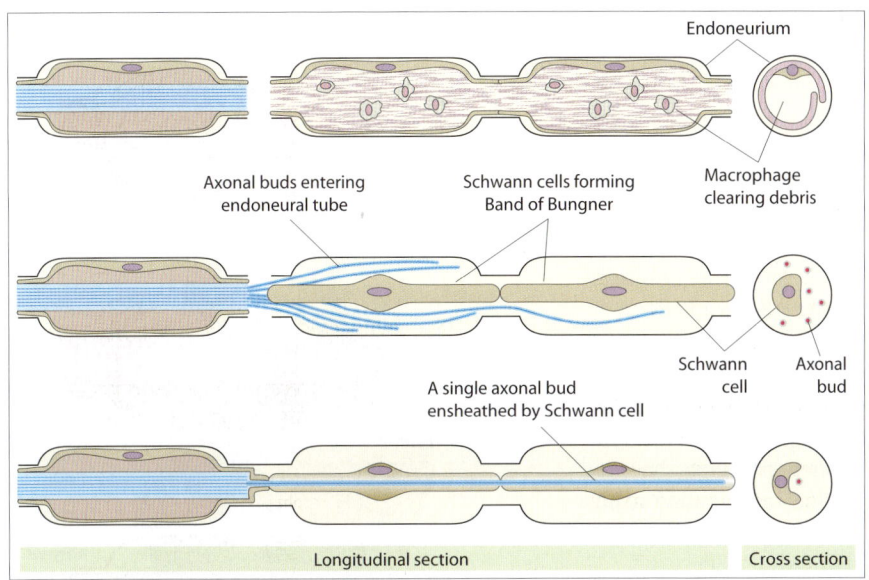

Fig. 9.**6**   Regenerative changes in the axon of a neuron.

Table 9.**1**   Sunderland's classification of nerve injury.

Type	Axon	Basal lamina	Endoneurium	Perineurium	Epineurium	Recovery	Tinel sign
I (Neuro-praxia)	Intact	Intact	Intact	Intact	Intact	Full recovery in hours	Tinel sign absent
II (Axono-temesis)	Disrupted	Intact	Intact	Intact	Intact	Full recovery in 3 months	Advancing Tinel sign
III	Disrupted	Disrupted	Disrupted	Intact	Intact	Incomplete recovery at 3 months	Advancing Tinel sign
IV	Disrupted	Disrupted	Disrupted	Disrupted	Intact	No recovery	Tinel sign stationary
V (Neuro-temesis)	Disrupted	Disrupted	Disrupted	Disrupted	Disrupted	No recovery	Tinel sign stationary

Regenerative changes in the axon The Schwann cells at the site of injury multiply and grow at the rate of up to 1 mm per day, forming a solid cord of elongated cells (band of Bungner) within the endoneural tube (Fig. 9.**6**). The plasmalemma of the Schwann cells and adjacent basal lamina separate, creating an annular compartment between the Schwann cells and the endoneurium. Up to 100 axonal sprouts, each containing a neurofibril, grow out in all directions from the proximal axon. A variable number (up to 25) grow into the distal annular compartment. The daily rate of growth is up to 4 mm in the peripheral stump. Eventually all but one axonal sprout degenerate. The surviving fibril enlarges to fill the distal tube. The Schwann cells in the Band of Bungner form myelin sheath around the reinnervating axonal sprout. The sheath begins to develop in about 15 days and is complete within a year. Increase in fiber diameter takes place very slowly. Regenerated fibers attain a fiber diameter of up to 80 percent of normal.

Regenerative changes in the cell body The Nissl substance and Golgi apparatus gradually reappear. Sometimes, they reappear in greater than the original amounts, reflecting the extra metabolic effort required to replenish large amounts of the axoplasm. The cell regains its normal size and the nucleus returns to its central position.

Classification of nerve injury

Seddon's classification describes three types of nerve injury, viz., neuropraxia, axonotemesis, and neurotemesis, while Sunderland's classification describes five types of injury (Table 9.**1**). *Neuropraxia* occurs due to minor nerve stretch or pressure causing ischemic injury to the nerve. It results in conduction block without causing any structural damage. The common experience of a limb 'going to sleep' is due to neuropraxia.

Axonotemesis occurs due to excessive stress injury to the nerve. It results in Wallerian degeneration. The basal lamina of Schwann cells and other sheaths are all intact. Before the advent of antitubercular drugs, axonotemesis was carried out therapeutically in tuberculosis, crushing the phrenic nerve so as to paralyze one half of the diaphragm. Decreased ventilation of the lung, it was hoped, would kill the aerobic tubercle bacilli. *Neurotemesis* occurs as a result of penetrating injury to the nerve. All the sheaths are disrupted.

Tinel sign   The damaged nerve is tested by lightly percussing on its course with a patella hammer from the distal to the proximal end. The Tinel sign refers to the tingling sensation that is experienced when the site of regeneration is reached. As the regenerating axon grows, the site of the tingling sensation also shifts. This is called the *advancing Tinel sign*.

Skeletal Muscle

Each skeletal muscle fiber is surrounded by a thin layer of connective tissue called the *endomysium* or the basal lamina. Parallel fibers are grouped into fascicles that are surrounded by the *perimysium*. The whole muscle is enclosed in a tough connective tissue sheath called *epimysium* from which thin septa extend to join the perimysium. Blood vessels ramify in the epimysium and penetrate via the septa of the perimysium to form a rich capillary network in the endomysium (Fig. 9.7).

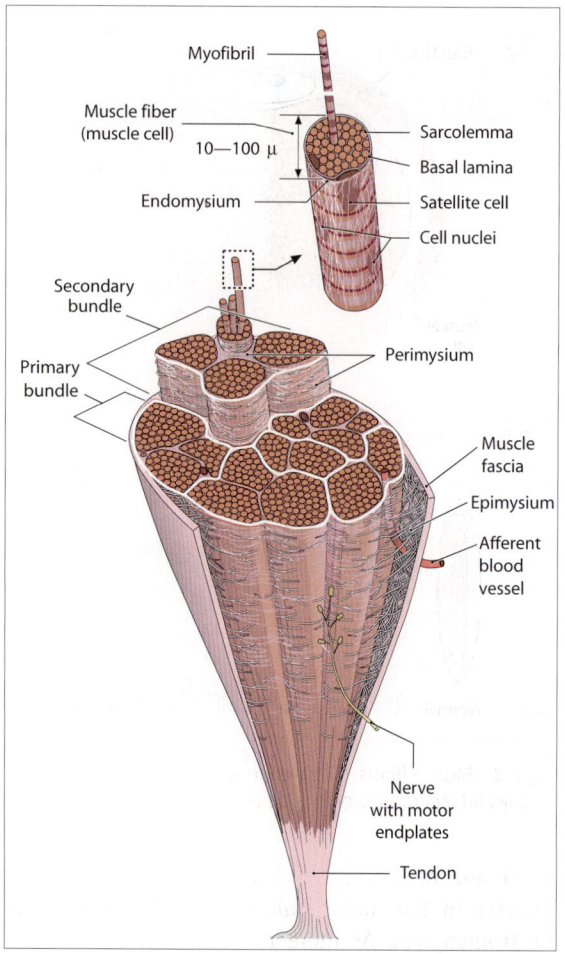

Fig. 9.**7**   Structure of a whole muscle, shown in cross section.

Muscle growth

Skeletal muscle fibers develop from uninucleate precursors called *myoblasts*. These proliferate and fuse to form long multinucleate cells called *myotubes*, which subsequently develop in their cytoplasm myofibrils and transform into myocytes or muscle cells. Small, flattened mononucleate cells called *myosatellite cells* are located between the sarcolemma and the basal lamina (Fig. 9.8A). These cells are myoblasts that fail to fuse into myotubes. They get activated later during muscle hypertrophy or regeneration and provide new myocytes.

Growth of skeletal muscle involves either an increase in the number of muscle fibers (hyperplasia), an increase in the diameter of the muscle fibers (hypertrophy) or an increase in the length of the muscle fibers (lengthening). Fiber length increases if the freshly synthesized myofibrils are added in series, while fiber diameter increases if the myofibrils are

added in parallel (Fig. 9.8B). Muscle lengthening allows increased muscle shortening during contraction. Muscle hyperplasia and hypertrophy increase muscle force during contraction.

Effect of exercise   *Dynamic exercise* (see Fig. 42.2) leads to an increase in the rate of oxygen utilization and is associated with an increase in the number of mitochondria and muscle capillary density. There is little change in muscle fiber diameter or the ratio of slow and fast muscle fibers (see Table 116.1). *Static exercise* causes muscle hypertrophy and thereby increases the force-generating capacity. The hypertrophy occurs even when the load is applied on a denervated muscle.

Effect of aging   With age, sporadic death of motor neurons becomes frequent. The denervated muscle fibers become innervated by the surviving motor neurons. As a result, the motor units (p 107) become

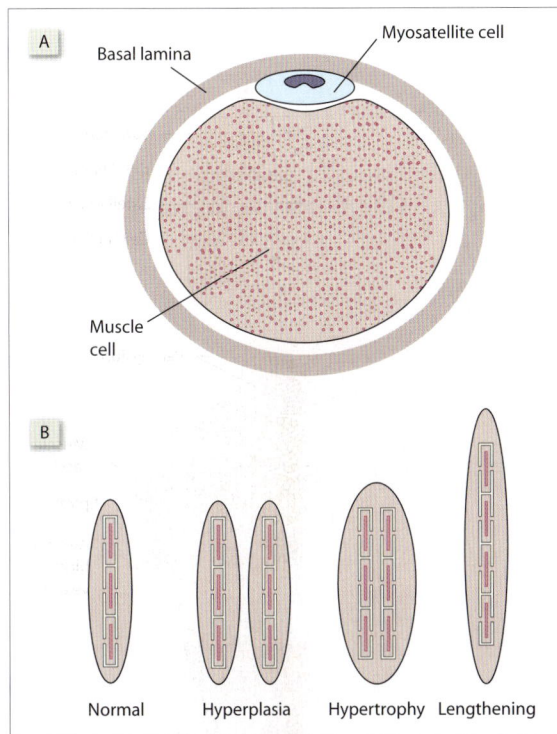

Fig. 9.**8**   Muscle fibers. [A] Satellite cells in a muscle. [B] Hypertrophy and hyperplasia of muscle cells.

larger and motor control becomes less precise. The increase in the motor unit size is apparent in the electromyogram. As more motor neurons die, more muscle fibers remain denervated causing loss of muscle power. Eventually, some muscles become completely denervated.

Muscle degeneration

Muscle injury can result in degeneration of muscle. Sarcolemmal damage results in entry of Ca^{2+} into the cell with consequent activation of Ca^{2+}-dependent enzymes like proteases and endogenous phospholipases. If the vascular supply to the muscle is intact, the degenerating muscle fiber is invaded by phagocytes from blood. The phagocytes cross the intact basal lamina sheaths and ingest muscle fiber debris. Within a few days, the bulk of this debris is removed by the phagocytes, which then depart from the site.

Denervation supersensitivity of muscle   Each motor endplate contains about 50 million acetylcholine (ACh) receptors, which are highly concentrated near the tips of the postjunctional folds. Few ACh receptors (AChRs) are located elsewhere on the muscle membrane.

Following a section of the motor nerve the AChR increase more than 10-fold in number and become dispersed over the entire surface of the sarcolemma. The muscle cell then becomes sensitive to ACh applied anywhere on its surface, a phenomenon known as *denervation supersensitivity*. Denervation also lowers the membrane potential to about -60mV, making it prone to *fibrillations*. After the motor nerve regenerates and functional innervation of the muscle is reestablished, the accumulation of AChRs decreases throughout the sarcolemma and the sensitivity of the nonjunctional portion of the sarcolemma diminishes to normal levels. The normal resting potential is restored and the fibrillation disappears.

Muscle regeneration

Even before the degeneration of muscle is complete, satellite cells within the basal lamina sheath begin to multiply and fuse to form multinucleate myotubes within the basal lamina tube of the degenerated muscle fiber. The regenerated myotube later forms its own basal lamina. Hence, a regenerated muscle fiber usually shows duplication of its basal lamina. If the nerve supply is intact, the myotubes get innervated within 2 days. It takes up to 3 weeks for the reinnervated muscle to grow to normal size. Cardiac muscle does not possess satellite cells and is therefore unable to regenerate.

Synaptogenesis

The neuromuscular junction is formed when the growing axon contacts a muscle fiber, the growth cone differentiates into the prejunctional apparatus, and the muscle membrane in contact with the nerve terminal differentiates into the postjunctional apparatus. The formation of the junction (synaptogenesis) between the axonal ending and the muscle membrane is guided by several chemical signals emanating from both and influencing each other's development.

Role of muscle in prejunctional differentiation   *Laminins* are major components of all basal lamina and are potent promoters of axonal outgrowth from the neuron. Several other factors located on the muscle (e.g., cadherin and NCAM) are involved in prejunctional differentiation.

Role of nerve in postjunctional differentiation   *Agrin* is a proteoglycan released from nerve terminals and causes clustering of AChRs. It does not bind directly to AChR. Rather, it binds to a membrane-associated enzyme called MuSK (Muscle-Specific tyrosine Kinase), triggering a second-messenger cascade that brings about clustering of rapsyn, a protein associated with AChR.

Neuregulin is secreted by the motor axon. It binds to specific tyrosine kinases (called erb-kinases) present on the postjunctional membrane and triggers a cascade reaction that ultimately increases the synthesis of AChR by activating transcription of AChR genes. Also, the high sarcoplasmic Ca^{2+} associated with muscle contraction activates a cascade of protein-kinase reactions culminating in the depression of AChR gene transcription in the in nonsynaptic areas.

10 Resting Membrane Potential

The conduction of a nerve impulse is an electrical phenomenon. The inner surface of the nerve membrane is charged negative with respect to its outer surface, giving the nerve membrane a definite polarity. When this polarity is briefly reversed, it serves as a signal that is conducted over long distances along the nerve membrane. This signal is called the action potential. However, before discussing action potential (Chapter 11), or its conduction over the nerve membrane (Chapter 13), it is necessary to appreciate the origin of the membrane polarity. This is discussed below in detail.

Diffusion Potential

The inner surface of a cell membrane is usually charged negative with respect to its outer surface. The potential of the inner surface of the membrane with respect to its outer surface is called the *membrane potential*. The potential is called a membrane potential because it exists essentially on the membrane. The cations and anions form two layers on the outer and inner membrane surfaces respectively. The membrane potential is produced by the passive diffusion of ions. Hence it is also called a *diffusion potential*.

The membrane potential changes markedly when the membrane is stimulated. The potential recorded intracellularly when a cell is *not* stimulated is called the *resting membrane potential* (RMP) and the membrane is said to be *polarized*. The unmyelinated axonal membrane has an RMP of –70 mV. The membrane potential is more (–90 mV) in striated muscles and less (–50 mV) in certain smooth muscles.[1] The general principles of membrane potential however are the same in all excitable tissues. Most of the discussion on membrane potential in the following passages assume a membrane potential of –80 mV. When the inside of the membrane becomes more negative, it is said to be *hyperpolarized*. When the inside of the membrane becomes less negative, it is said to be *depolarized*. When following depolarization, the membrane returns to the polarized state, it is said to be *repolarized*.

Origin of diffusion potential

The activity of the Na^+-K^+ pump builds up a large concentration of Na^+ in the extracellular fluid (ECF) and a large concentration of K^+ in the intracellular fluid (ICF). It also indirectly contributes to the build-up of a large Cl^- concentration in the ECF. The membrane potential develops when these ions leak back along their respective concentration gradients, i.e., when Na^+ leaks back from ECF to ICF or K^+ leaks out from ICF to ECF. The streams of ions that leak back from high to low concentration are called *leak currents* or *channel currents* to differentiate them from the *pump current*[2] produced by the Na^+-K^+ pump. In its resting state, the membrane is quite leaky to K^+ but not so leaky to Na^+. Hence, the amount of K^+ that leaks out in a given time is much more than the amount of Na^+ that leaks in through the membrane in the same time. As a result, an excess of K^+ accumulates in the ECF, the amount of which is too small to change the ECF or ICF concentration of K^+ but large enough to make the exterior of the membrane positive with respect to its interior by several millivolts.

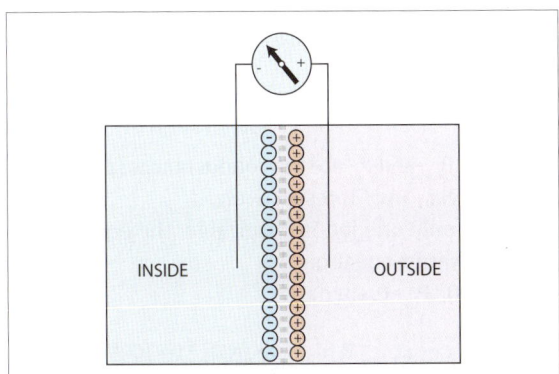

Fig. 10.**1** A potential of about –80 mV is measurable across the cell membrane. The potential exists essentially at the membrane, with anions and cations layered on the opposite surfaces of the cell membrane.

---

[1] The terms 'low' and 'high' membrane potential cause considerable confusion, and understandably so. These terms refer to the absolute values of membrane potential. When the potential changes from –70 to –50 mV, it is decreasing. However, when it changes from +10 to + 30 mV, it is increasing.

[2] The pump current is mostly ignored in discussions on membrane potential. Its role is discussed at the end of this chapter.

Diffusion potential of a single ion   Before trying to understand the diffusion potential of an ion, it is important to recall that whenever an ion is distributed asymmetrically across a membrane, there exists a unique potential E_{ion} that will prevents its diffusion. This potential is called the Nernst potential of the ion. It can be calculated using the Nernst equation. If the ions are allowed to diffuse, they produce a diffusion potential. The following passages explain why the diffusion potential produced by an ion is exactly equal to its Nernst potential.

Take the case illustrated in Figure 10.2. Assume that the membrane is permeable only to K^+ and impermeable to Na^+ and Cl^-. Assume also that the membrane potential E_m is 0 mV to start with. The following sequence of events will occur: (1) K^+ diffuses out of the cell along its concentration gradient. The resultant decrease in intracellular K^+ concentration is so small (< 10 picoEq dl^{-1}) that it hardly produces any change in the K^+ concentration gradient or its rate of outward diffusion. (2) Since the membrane is impermeable to both Na^+ and Cl^-, the diffusion of K^+ produces an intracellular negativity which is not nullified by the diffusion of Na^+ in the opposite direction, or by the diffusion of Cl^- in the same direction. (3) The intracellular negativity reduces the rate of subsequent outward diffusion of K^+. (4) As the outward diffusion of K^+ continues, the membrane becomes increasingly negative inside, and the outward diffusion of K^+ keeps decreasing. (5) When the membrane potential reaches about –90 mV (which is the Nernst potential for K^+), the outward K^+ diffusion stops. Hence, the diffusion potential produced by a single ion will always equal its own Nernst potential (Fig. 10.3).

The intracellular negativity pulls back the extracellular cations to the external surface of the membrane. Similarly, the external positivity attracts the intracellular anions to the internal surface of the membrane. Hence, opposite charges line up on the opposite surfaces of the membrane.[3] Hence, the membrane potential resides essentially at the membrane and can even be abolished temporarily by simply scraping the membrane surface.

Diffusion potential of multiple ions   When multiple ions are diffusing across the membrane, each ion tries to drive the membrane potential towards its own Nernst potential. The resultant diffusion potential (E_m)

---

[3] Together, the membrane and the opposite charges on its opposite surfaces form a parallel-plate capacitor in which the membrane is the dielectric. The capacitance of a parallel plate capacitor is known to be inversely proportional to the thickness of the dielectric. Because the membrane is very thin, the capacitance of the membrane is quite high. Membrane capacitance is an important consideration in the conduction of nerve impulses (Chapter 13).

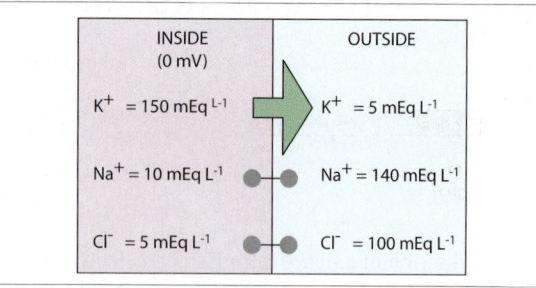

Fig. 10.**2**   Distribution of ions across a semipermeable membrane that is permeable only to potassium ions. Potassium ions will diffuse out resulting in a diffusion potential that will be equal to the Nernst potential of K^+.

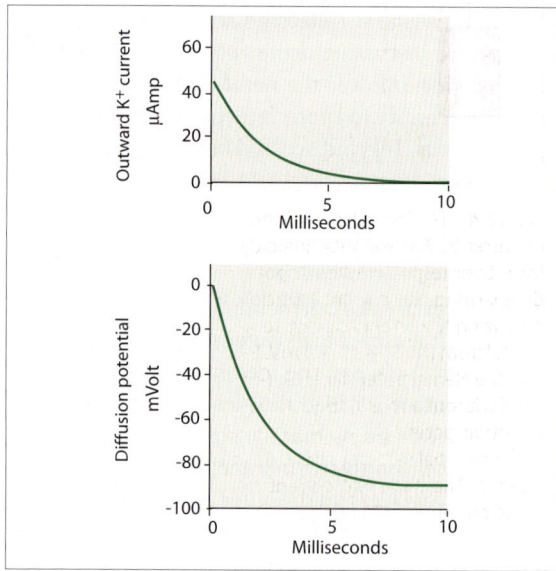

Fig. 10.**3**   Consequences of the ionic distribution shown in Figure 10.**2**. The outward diffusion of potassium ions declines with the build-up of a diffusion potential. The diffusion stops when the diffusion potential equals the Nernst potential of K^+.

is called the *equilibrium potential of the membrane*. It is given by the Goldman equation:

$$E_m = 61 \log \frac{P_K[K_0] + P_{Na}[Na_0] + P_{Cl}[Cl_i]}{P_K[K_i] + P_{Na}[Na_i] + P_{Cl}[Cl_0]}$$

where P_K, P_{Na}, and P_{Cl} are the conductances (permeability) of K^+, Na^+, and Cl^- respectively.

When only one ion is permeable, the equation reduces to Nernst equation.

If $P_{Na} = 0$, $P_{Cl} = 0$, then:

$$E_m = 61 \log \frac{P_K[K_0] + 0 \times [Na_0] + 0 \times [Cl_i]}{P_K[K_i] + 0 \times [Na_i] + 0 \times [Cl_0]}$$

$$E_m = 61 \log \frac{P_K[K_0]}{P_K[K_i]}$$

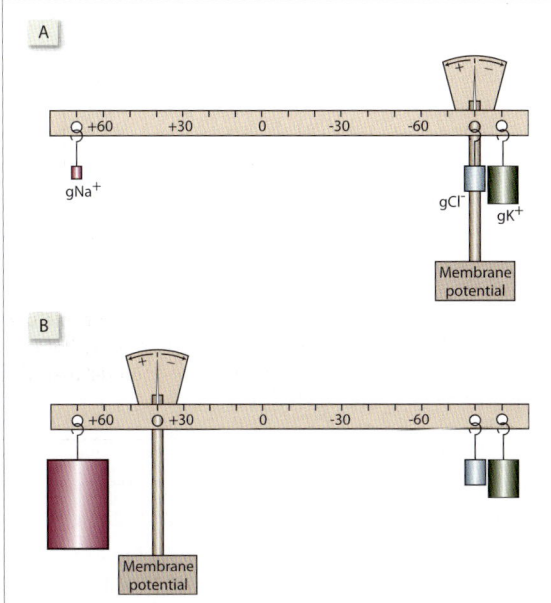

Fig. 10.**4** [A] The balance-of-moments analogy of the balance of currents. The weights (ionic conductances) are suspended from their respective Nernst potential. A balance between the depolarizing Na^+ current (anticlockwise moment) and hyperpolarizing K^+ current (clockwise moment) is established when the fulcrum (RMP) is at –80 mV. Since the membrane potential is at the Nernst potential of Cl^- (–80 mV), the chloride ion finds itself without a role in this balance of currents. [B] At the peak of action potential, the Na^+ conductance increases massively and a new balance of currents is established, with the RMP at +40mV. The huge Na^+ current is now balanced by the combined currents of K^+ and Cl^-.

$$E_m = 61 \log \frac{[K_0]}{[K_i]}$$

In other words, when the permeability of one ion far exceeds that of the others, the equilibrium potential of the membrane will approach the Nernst potential of the highly permeable ion. The implications are as follows: (1) There is a balance of currents (Fig. 10.**4**), with each diffusing ion trying to drive the equilibrium potential towards its own Nernst potential. For example, Na^+ tries to drive the RMP towards +70 mV, K^+ tries to drive the RMP towards –90 mV, and Cl^- tries to drive the RMP towards –80 mV. (2) The ion that is more permeable and more distant from the equilibrium potential is more effective in driving the equilibrium potential towards its own Nernst potential. (3) The equilibrium potential or the RMP settles at –80 mV. At this RMP, both Na^+ and K^+ are moderately effective and balance each other. The effectiveness of the Na^+ ion comes from the distance of its Nernst potential from the RMP. On the other hand, the effectiveness of the K^+ ion comes from its moderate permeability. The Nernst potential of Cl^- is equal to the RMP; hence it is unable to influence the RMP.

Membrane currents

In the case illustrated in Figure 10.**5**, the outward diffusion of the positively charged K^+ ions will tend to hyperpolarize the membrane. In other words, K^+ carries a hyperpolarizing current.[4] Since K^+ is flowing out, it constitutes an outward hyperpolarizing current. Similarly, Na^+ carries an inward depolarizing current and Cl^- carries an inward hyperpolarizing current.

The number of ions flowing out per second per square centimeter of the membrane (current density) can be calculated from Ohm's Law. Since the reciprocal of resistance is conductance, Ohm's Law can be rewritten as:

Current (I) = Potential (E) × Conductance (g)

Outward current per unit membrane area ($\mu Amp\ cm^{-2}$)

$$= \left[\begin{matrix} \text{Membrane} \\ \text{potential} \\ \text{(mV)} \end{matrix} - \begin{matrix} \text{Nernst potential} \\ \text{of an ion} \\ \text{(mV)} \end{matrix} \right] \times \begin{matrix} \text{Ionic conductance} \\ \text{per unit membrane area} \\ \text{(mmho cm}^{-2}) \end{matrix}$$

Thus,

$$I_{Na} = (E_m - E_{Na}) \times g_{Na}$$
$$I_K = (E_m - E_K) \times g_K$$
$$I_{Cl} = (E_m - E_{Cl}) \times g_{Cl}$$

where I_{Na}, I_K, and I_{Cl} are the membrane currents, E_{Na}, E_K, and E_{Cl} are the Nernst potentials, and g_{Na}, g_K, and g_{Cl} are the membrane conductances for Na^+, K^+, and Cl^- ions respectively. These formulas explain why the membrane current of an ion is directly proportional to the difference between the membrane potential and the Nernst potential of the ion. This principle has already been enunciated in the balance-of-moments analogy (Fig. 10.**4**). The point is further illustrated through the following example.

Figure 10.**5** shows the intracellular and extracellular concentrations of Na^+, K^+, and Cl^- and their membrane conductances. Based on these data, the membrane currents of these ions at two different membrane potentials, 0 mV and –80 mV, have been calculated[5] and tabulated (Table 10.**1**).

When membrane potential is at 0 mV, K^+ will carry a strong outward hyperpolarizing current (54 µAmp),

---

4 The diffusion rate of ions can be expressed as the number of ions flowing out per second. A more convenient way is to express the ionic diffusion as a current, which is defined as the rate of flow of electric charge. One ampere current is the flow of 1 coulomb of charge per second. Since each monovalent ion carries 1.6×10^{-19} coulombs of electric charge, 1 ampere of K^+ current is the flow of 6.25×10^{18} K^+ ions per second.

5 In these calculations, it has been assumed that the membrane conductances of the ions do not change with changes in membrane potential. The assumption is not correct, as explained in Chapters 11 and 12.

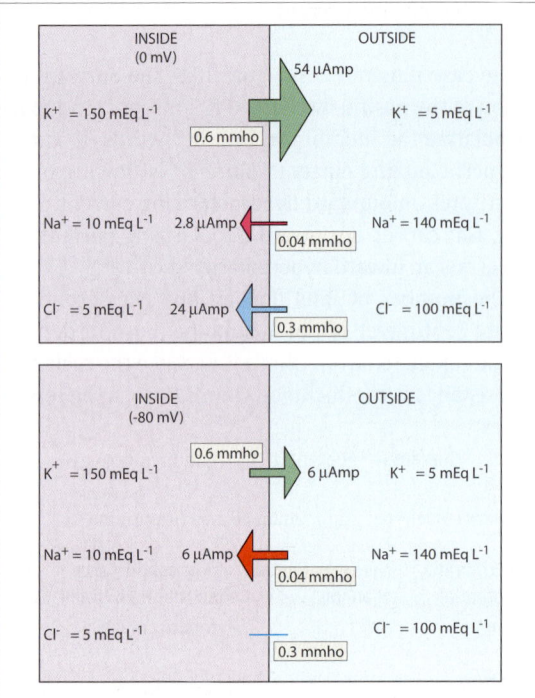

Fig. 10.**5**   (*Above*) The membrane potential is 0 mV. The total hyperpolarizing current (78 μAmp) carried by K⁺ and Cl⁻ far exceeds the depolarizing Na⁺ current (2.8 μAmp). (*Below*) The membrane potential is –80 mV. The outward K⁺ current exactly balances the inward Na⁺ current.

Table 10.1   Calculation of ionic currents (I_{ion}) at two different membrane potentials, 0 mV and –80 mV, using the formula: $I_{ion} = (E_m - E_{ion}) \times g_{ion}$

Membrane potential (E_m mV)	Nernst potential of (E_{ion} mV)	Membrane conductance of (g_{ion} mmho)	Membrane current of ion (I_{ion} μAmp)
0 mV	$E_K = -90$	$g_K = 0.6$	$I_K = 54$
0 mV	$E_{Na} = +70$	$g_{Na} = 0.04$	$I_{Na} = 2.8$
0 mV	$E_{Cl} = -80$	$g_{Cl} = 0.3$	$I_{Cl} = 24$
–80 mV	$E_K = -90$	$g_K = 0.6$	$I_K = 6$
–80 mV	$E_{Na} = +70$	$g_{Na} = 0.04$	$I_{Na} = 6$
–80 mV	$E_{Cl} = -80$	$g_{Cl} = 0.3$	$I_{Cl} = 0$

Na⁺ will carry a weak inward depolarizing current (2.8 μAmp), and Cl⁻ will carry a strong inward hyperpolarizing current (24 μAmp). Thus the total hyperpolarizing current (78 μAmp), carried by K⁺ and Cl⁻ far exceeds the depolarizing Na⁺ current (2.8 μAmp). The intracellular potential therefore becomes negative.

With increasing negativity, the depolarizing Na⁺ current increases and the hyperpolarizing currents of K⁺ and Cl⁻ decrease. When membrane potential reaches –80 mV, K⁺ will carry a moderate outward hyperpolarizing current, Na⁺ will carry a moderate inward

depolarizing current and Cl⁻ current stops, since –80 mV equals E_{Cl}. At this stage, the hyperpolarizing and depolarizing currents are equal and the membrane potential does not change any further. The potential at which an equilibrium is established between the depolarizing and hyperpolarizing membrane currents denotes the resting membrane potential of the membrane. Expressed mathematically,

$$I_{Na} + I_K + I_{Cl} = 0$$

$$\therefore \ \{(E_m - E_{Na}) \times g_{Na}\} + \{(E_m - E_K) \times g_K\} + \{(E_m - E_{Cl}) \times g_{Cl}\} = 0$$

By expanding and rearranging, we get

$$E_m = \frac{(E_K \times g_K) + (E_{Na} \times g_{Na}) + (E_{Cl} \times g_{Cl})}{g_K + g_{Na} + g_{Cl}}$$

This is the Hodgkin–Huxley equation for calculation of membrane potential. More accurate values are given by the Goldman equation, the derivation of which is beyond the scope of this book. Students should however satisfy themselves that for the same set of ionic concentrations and conductances, the Hodgkin–Huxley equation and the Goldman equation give slightly different values.

Chloride current   It should be clear from the foregoing discussions that Na⁺ always carries an inward depolarizing current and K⁺ always carries an outward hyperpolarizing current. The Cl⁻ current is somewhat different. Unlike Na⁺ or K⁺ ions, Cl⁻ ions do not have a Cl⁻ pump. Hence, the Cl⁻ gradient is created entirely by the RMP, which pushes out Cl⁻ ions from the cell. Hence, the Nernst potential of Cl⁻ equals the RMP. Consequently, there is no significant Cl⁻ current across the resting membrane. Several types of nerve cells, however, do pump out Cl⁻ through secondary active transport using K⁺-Cl⁻ cotransport. In these cells, the E_{Cl} is slightly more negative than the RMP, and Cl⁻ carries a weak, inward hyperpolarizing current.

Another situation in which the E_{Cl} and RMP are different is when the g_{Na} increases massively, as during an action potential. In this situation, the Cl⁻ sets up a moderately strong inward hyperpolarizing current as shown in Figure 10.**4**B.

■   **Effect of changes in ECF ionic concentrations**

Extracellular ionic concentrations are much more prone to perturbations as compared to intracellular concentrations. Any variation in ECF ionic concentration alters the Nernst potential of the ion. However, the change in Nernst potential is less for ions that have high concentration in the ECF (Na⁺, Cl⁻) and more for K⁺,

Table 10.**2** Effect of increasing the ECF concentration of ions (m Eq L^{-1}) on the Nernst potential (mV).

Nernst potential for ions when ECF concentration are normal.

Ion	ICF conc.	ECF conc.	Nernst potential
Na$^+$	10	140	69.9
K$^+$	150	5	–90.1
Cl$^-$	5	100	–79.4

Nernst potential for ions when ECF concentrations are increased by 5 mEq L^{-1}.

Ion	ICF conc.	ECF conc.	Nernst potential
Na$^+$	10	145	70.8
K$^+$	150	10	–71.7
Cl$^-$	5	105	–80.7

which has a low ECF concentration. Table 10.**2** shows that for a given amount of change in ECF concentration (e.g., 5 mEq L^{-1}), the percentage change in [C$_o$] / [C$_i$] is much greater for K$^+$ (100%) than for Na$^+$ (3.6%) or for Cl$^-$ (5%). The changes in the Nernst potentials are correspondingly larger. Hence, perturbations in extracellular K$^+$ concentration have large effects on membrane potential. As shown in the above example, a 5 mEq L^{-1} increase in the ECF concentration of Cl$^-$ will hyperpolarize the membrane slightly as its Nernst potential changes from –79.4 mV to –80.7 mV. However, a similar increase in ECF K$^+$ will depolarize the membrane markedly as its Nernst potential changes from –90.1 mV to –71.7 mV. Hence, addition of KCl to the bathing solution of an excitable tissue depolarizes the cells (Fig. 10.**6**A).

Pump Current

The role of the Na$^+$-K$^+$ pump in the development of membrane potentials is mostly misunderstood by students. This is unfortunate, since membrane potential is one of the best understood areas of physiology. Before considering the role of the Na$^+$-K$^+$ pump in membrane potential, two facts that are discussed below must be understood.

A strong pump that is working slowly The first thing that must be realized is that the Na$^+$-K$^+$ pump works very feebly in the cells. It serves to pump out the Na$^+$ that diffuses back into the cell and pumps in the K$^+$ that diffuses out of the cell. In the resting membrane, these diffusions are negligible. However, during an action potential, large amounts of Na$^+$ diffuse in and equally

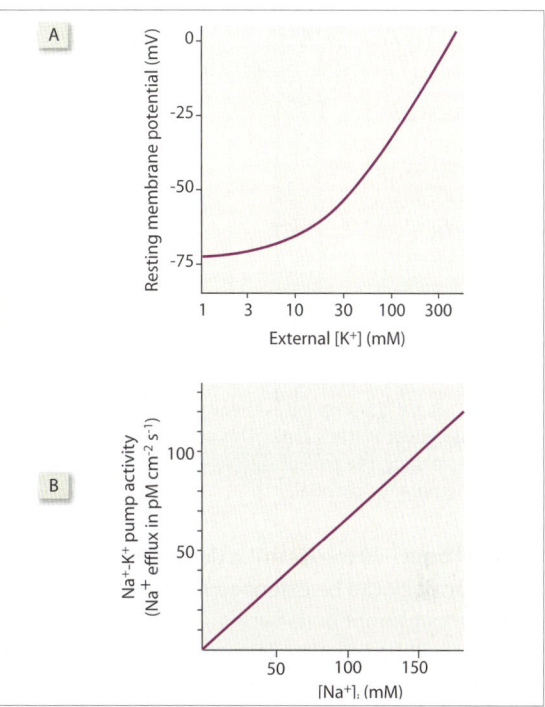

Fig. 10.**6** [A] Effect of increasing the extracellular K$^+$ concentration of the membrane potential. [B] Dependence of Na$^+$-K$^+$ pump activity on the intracellular Na$^+$ concentration.

large amounts of K$^+$ diffuse out. These are pumped back to the extracellular and intracellular compartments respectively for the maintenance of cell homeostasis.

Let us imagine the original situation when the ECF and ICF concentrations were equal and the Na$^+$-K$^+$ pump just got into action. The pump activity must have been very high initially because the intracellular Na$^+$ concentration was high. However, as more and more Na$^+$ was pumped to the exterior, the pump current got reduced. This is because the activity of Na$^+$-K$^+$ pump is directly proportional to the intracellular Na$^+$ concentration (Fig. 10.**6**B). At the usual intracellular Na$^+$ concentration of 10 mEq L^{-1}, the pump current is feeble and adds only about –3 mV to the diffusion potential.

The analogy of the overhead tank for domestic water supply provides a useful insight (Fig. 10.**7**). A water pump (analogous to the Na$^+$-K$^+$ pump) fills up the overhead water tank and creates a hydrostatic pressure head (analogous to the ionic concentration gradient). When the water tank is full, the pump cannot run fast. Nor is it required to, since it needs only to replenish the water that is flowing out continuously to the houses. Switching off the pump therefore will not immediately affect the pressure of domestic water supply (analogous to RMP) which is maintained by the pressure head in the water tank. This is analogous to the situation in neurons where inhibition of the pump reduces

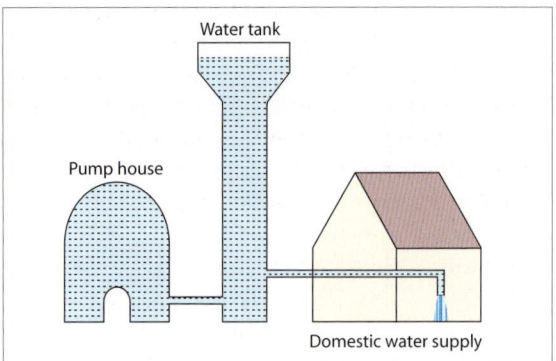

Fig. 10.**7**   A water pump raises the water head in an overhead tank for domestic water supply. This provides a useful analogy for understanding the role of the Na+-K+ pump in the generation of membrane potential.

the RMP from –70 to –67 mV, a decrease of only 3 mV.

The analogy can be extended further. If the domestic water consumption is very high, the water head in the tank will be low. The low water head will enable the pump to run faster. A new equilibrium is reached in which the pump works very hard to replenish the large amount of water that is consumed by domestic users. However, the equilibrium is attained at a lower pressure head in the overhead tank. If the pump is switched off now, the pressure head in the tank will fall considerably. This is analogous to the situation in smooth muscles where the RMP is low (–50 mV) and inhibition of the pump brings down the potential by another 20 mV.

An electrogenic pump rendered electroneutral The second important thing that must be realized is that although the Na+-K+-pump is potentially electrogenic (because of pumping out three Na+ ions for every two K+ ions that are pumped in), at no stage is the pump able to build up a significant membrane potential. This is because, as soon as the pump creates a negative potential inside the cell, chloride ions rush out of the cell and restore electroneutrality. Thus in effect, the Na+-K+ pump pumps out three Na+ ions and one Cl– ion for every two K+ ions it pumps in. The real role of the Na+-K+ pump is that it builds up a high K+ concentration within the cell and a high Na+ concentration outside the cell, setting the stage for the development of a diffusion potential.

Restoration of cell homeostasis

During the resting state, a cell with a polarized membrane continuously loses K+ to the exterior and gains Na+ through their respective channels. The membrane

potential does not change because the outward K+ current balances the inward Na+ current. However, due to these continuous channel currents, the intracellular and extracellular concentrations of these ions tend to equalize and the membrane potential disappears over a period of time. However, that does not happen due to the Na+-K+ pump which pumps out three Na+ ions for every two K+ ions transported inside.

Balancing pump current with channel current Suppose, in a resting nerve cell, the channel currents produce Na+ influx and K+ efflux in a ratio of 1:1. Yet, the Na+-K+ pump which produces Na+ efflux and K+ influx in a ratio of 3:2. is able to restore cell homeostasis. Before understanding how the same is possible, it is important to recall that the activity of the Na+-K+ pump (and the resultant pump current) increases when the intracellular concentration of Na+ rises.

Let us take the example shown in Figure 10.**8**. When the pump is inactive, the leak currents of Na+ and K+

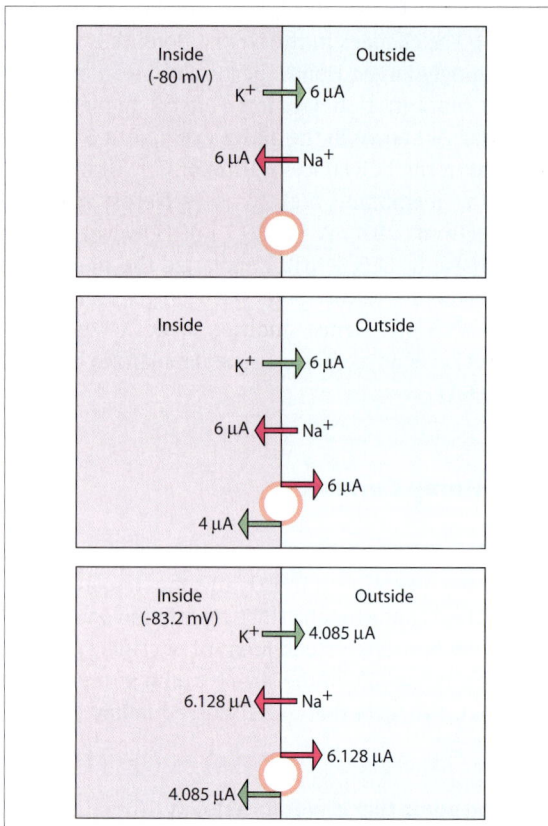

Fig. 10.**8**   Effect of pump current on cell homeostasis and membrane potential. (*Above*) Membrane potential is –80 mV. Pump is not operational. The leak currents balance each other. (*Middle*) Pump is operational, resulting an unbalanced outward K+ current of 2 µA. (*Below*) Membrane potential is –83.2 mV. Leak currents and pump currents get readjusted accordingly.

balance each other (both are 6 μA) so that the membrane potential does not change. However, cell homeostasis is threatened as the intracellular concentration of K$^+$ and extracellular concentration of Na$^+$ decrease continuously. To prevent these losses, the Na$^+$-K$^+$ pump becomes operational, generating an outward Na current of 6 μA and an inward K$^+$ current of 4 μA. (Recall that the Na$^+$ and K$^+$ pump currents are in a ratio of 3:2.) The pump current of Na$^+$ exactly balances the leak current of Na$^+$. However, the K$^+$ leak current is not fully balanced by the K$^+$ pump current and there remains an unbalanced K$^+$ leak current of 2 μA. This outward K$^+$ current hyperpolarizes the membrane slightly, taking the membrane potential to –83.2 mV. The hyperpolarization reduces the K$^+$ leak current to 4.085 μA and increases the Na$^+$ leak current to 6.128 μA. The increase in Na$^+$ leak current stimulates the pump activity so that the pump currents for K$^+$ and Na$^+$ become 4.085 μA and 6.128 μA respectively. The pump currents now completely balance the leak currents and cell homeostasis is no longer threatened.

In the new equilibrium, the following may be noted: (1) The Na$^+$ current (6.128 μA) and K$^+$ current (4.085 μA) are in a ratio of 3:2. This is true for both leak currents as well as pump currents. (2) The activity of the Na$^+$-K$^+$ pump has made a difference of only 3.2 mV.

11   Membrane Excitation and Action Potential

Action Potential

The action potential (nerve impulse) is the signal that is conducted along the axon over long distances without any reduction in amplitude. Any quick and brief change in the RMP (depolarization or hyperpolarization, small or large in magnitude) constitutes a signal. However, except the action potential, all other signals fade out after traveling short distances along a membrane. The signal of action potential consists of a brief reversal of membrane polarity followed by a quick restoration of the normal polarity. The whole process lasts a few milliseconds.

All-or-none law   An action potential is usually full-sized with a fixed amplitude of about 110 mV (from –70 mV to +40 mV). Subthreshold stimuli cannot trigger action potentials. Once triggered, an action potential runs its entire course producing a fully-fledged spike. This is known as the *all-or-none* law.

The law does not rule out the possibility of a small or medium-sized action potential that may be produced, depending on the prevailing conditions of membrane potential and excitability, ionic concentrations, and temperature. It only emphasizes that the magnitude of the spike does not increase in a *graded* manner with the stimulus strength: it is either maximum (all) or it is not triggered at all (*none*) (Fig. 11.**1**). Instead, a change in stimulus strength alters the frequency of action potentials generated.[1]

Phases of action potential   The action potential, which is recorded using an intracellular electrode (Fig. 11.**2**), has five phases, viz. prepotential, depolarization, repolarization, after-depolarization, and hyperpolarization (Fig. 11.**2**). (1) The *prepotential*, also called the 'foot' of the action potential, is a local membrane potential that drifts slowly towards –55 mV, which is called the *firing level*. Strictly speaking, it not a part of the action potential. (2) During *depolarization*, the potential shoots up

to +40 mV in less than a millisecond. (3) During *repolarization*, the potential drops to near resting levels, i.e., to about –40 mV. This phase also lasts less than a millisecond. The depolarization and repolarization phases of the action potential together constitute the *spike potential*. Although the total duration of the action potential is about 40 ms, the spike lasts less than 2 ms. It is the spike potential rather than the entire action potential that can truly be termed as the nerve impulse. (4) During *after-depolarization*, the rate of repolarization slows down and gradually reaches the resting potential of –70 mV. This slow phase of repolarization lasts about 2 ms. (5) During *after-hyperpolarization*, the intracellular negativity overshoots the normal resting value of –70 mV to reach a more negative value of about –75 mV. It lasts for nearly 40 milliseconds before slowly returning to the normal resting potential of –70 mV.

Membrane excitability during action potential   An action potential is triggered only if the stimulus raises the membrane potential above the firing level of -55mV.

Fig. 11.**1**   The (f)all or none law! A boulder being pushed up a slope exemplifies a graded response: the greater the push, the higher the boulder moves. When the boulder is pushed over the cliff, it symbolizes the 'all-or-none' law: The push has to exceed a certain minimum for the boulder to fall into the sea or else it rolls back on to land. The strength of the push does not determine the height through which the boulder falls before splashing into the sea. The height of the fall may vary though, as the sea rises and falls with the tide.

---

[1] In telecommunication parlance, transmission of signals by altering the frequency of the carrier waves is called frequency modulation or FM transmission. FM transmission is better than AM transmission, which is more susceptible to external disturbances. The transmission of action potential is a form of FM transmission.

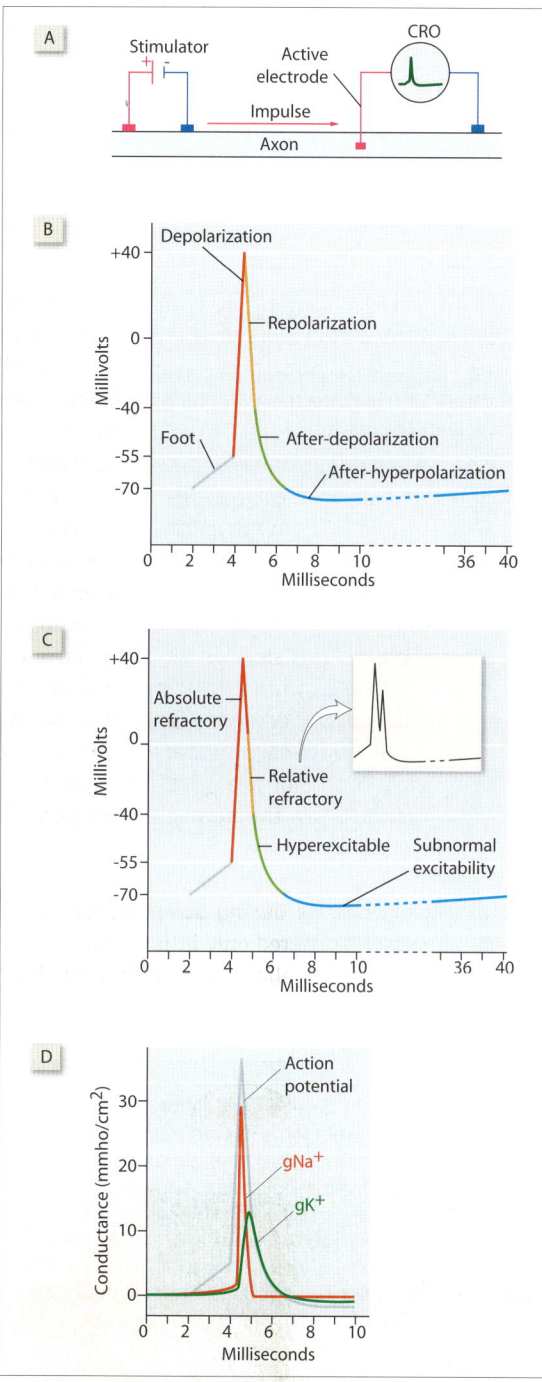

Fig. 11.**2** [A] Set up for the recording of an action potential. [B] Phases of an action potential. [C] Refractory periods of an action potential. Inset shows a second, smaller action potential triggered during the relative refractory period. [D] Ionic conductances during action potential.

The minimum strength of stimulus that can trigger an action potential is called a *threshold stimulus*.

The membrane is *refractory* (i.e., not responsive) to a second stimulus during most of the spike (Fig. 11.**2**C). From the spike onset till repolarization is about

one-third complete, the membrane is *absolutely refractory*, i.e., no stimulus howsoever strong can elicit a response. Thereafter, till the onset of after-depolarization, the membrane is *relatively refractory*, i.e., a sufficiently high stimulus can elicit a response. During after-depolarization, the membrane is hyperexcitable. Having just come out of the refractory period, the membrane is excitable and being closer to the firing level makes it *hyperexcitable*. During hyperpolarization, the membrane excitability is low but slowly returns to normal. The *effective refractory period* (ERP) includes the absolute refractory period and the early part of the relative refractory period. At the end of ERP, the membrane is able to conduct physiologically produced action potentials. If the membrane is in the relative refractory period, the actions potentials are smaller and are conducted slowly.

Origin and spread of action potential   During the depolarization phase of action potential, ionic currents flow between the depolarized membrane and the adjacent polarized areas, which then get depolarized too. When the depolarization of the adjacent area exceeds 15mV, a fresh action potential is triggered in the adjacent area. In this way, the action potential travels along the membrane as a wave of depolarization.

If every action potential is triggered by another action potential in its immediate vicinity, how is the first action potential stimulated? It stands to reason that ultimately, an action potential must be triggered by other types of electrical signals. These electrical signals are of different types, e.g., the excitatory post-synaptic potential (EPSP), end-plate potential (EPP), receptor potential, etc. These potentials are formed by the transduction of light, sound, heat, chemical or mechanical stimuli into electrical signals. Action potentials can also be triggered experimentally by a pair of stimulating electrodes: the action potential gets triggered at the cathode.

Ionic conductances during action potential   All phases of the action potential except the prepotential are the direct consequence of the permeability (conductance) changes of Na$^+$ and K$^+$, which are due to channel activation and inactivation. An increase in Na$^+$ conductance depolarizes the membrane and an increase in K$^+$ permeability repolarizes (or hyperpolarizes) the membrane. The actual membrane potential at any instant is determined by the relative magnitudes of Na$^+$ and K$^+$ permeability (Fig. 11.**2**D).

The prepotential is induced by some external source of negative potential. The source may be an external electrode (a cathode) or a depolarized area on the membrane in the immediate vicinity. Depolarization

Fig. 11.**3**   The hare and tortoise story epitomizes the ionic changes during action potential. The hare (Na⁺ permeability) races ahead but soon goes off to sleep, allowing the tortoise (K⁺ permeability) to overtake it !

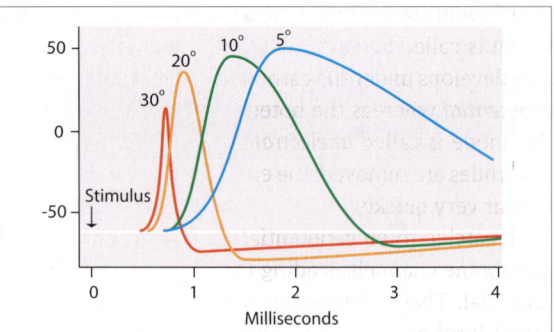

Fig. 11.**4**   Effect of temperature on action potential. As the temperature falls, both the duration and the amplitude of the action potential increase.

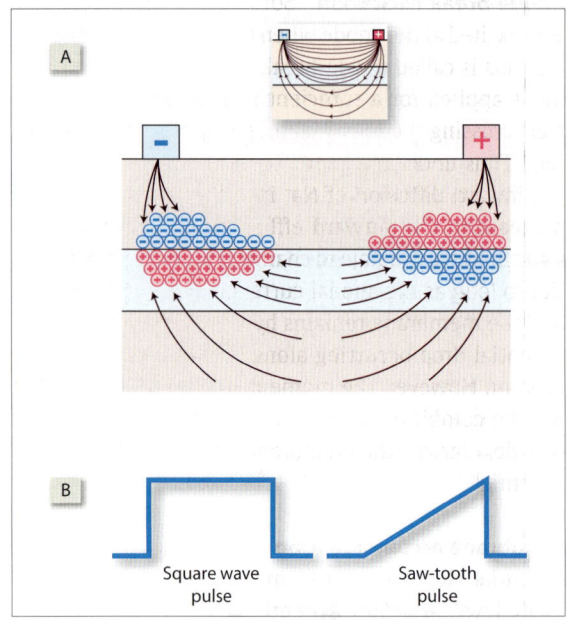

Fig. 11.**5**   [A] Current flowing from anode to cathode, intersecting the axonal membrane in its path. Beneath the cathode, anions accumulate outside and cations accumulate inside the membrane. Beneath the anode, there is accumulation of cations outside and anions inside the membrane. [B] A square-pulse stimulus rises sharply to its peak level and effectively triggers an action potential whereas a saw-tooth pulse, which rises to its peak slowly, often fails to trigger an action potential.

occurs due to a steep rise in Na⁺ conductance. Repolarization occurs due to two reasons, viz., the rapid rise of K⁺ conductance shortly after the surge in Na⁺ conductance, and the spontaneous inactivation of the Na⁺ channels which open during the depolarization phase (Fig 11.**3**). During after-depolarization, the K⁺ conductance is slightly more than Na⁺ conductance. Hence, the potential drifts back slowly towards the resting value. Hyperpolarization occurs because K⁺ conductance is still high although the Na⁺ conductance has returned to normal.

Effect of temperature on action potential   Temperature affects the rate at which membrane channels open and close. With a fall in temperature, the depolarization phase becomes less steep due to slower opening of Na⁺ channels. Also, the overshoot goes higher due to a slower onset of Na⁺ channel inactivation and a delayed rise in K⁺ conductance (Fig. 11.**4**).

■   **Membrane stimulation**

In clinical situations, both the positive and negative electrodes are placed extracellularly, usually on the surface of the skin. When current is passed through the electrodes, the part of the nerve below the cathode gets excited, and the part of the nerve below the anode gets hyperpolarized. In the electrodes, the current is carried by electrons while in the tissues, the current is carried by ions, both positive and negative. The direction of the current is conventionally the direction of flow of positive ions.

Cathodal stimulation   When the stimulating electrodes are turned on, the anions and cations start flowing along the electric field set up between the electrodes. Initially, the ions are unable to flow across the membrane at sites where the field intersects the membrane. Hence at these sites, the ions pile up on both surfaces of the membrane. These ionic currents are called *capacitive currents*. Immediately below the cathode, the extracellular anions flowing away from the cathode line up on the outer membrane surface while the axoplasmic cations flowing towards the cathode adhere to the inner membrane surface (Fig. 11.**5**A).

These events result in a change in membrane potential, which is called an *electrotonic potential*. The potential that develops under the cathode is called *catelectrotonic potential* whereas the potential that develops under the anode is called *anelectrotonic potential*. When the electrodes are removed, the electrotonic potentials disappear very quickly.

A catelectrotonic potential results in opening of membrane channels, leading to changes in membrane potential. These changes in membrane potential are called *local potentials* and they last a little longer after the electrodes are removed. If the local potential is strong enough, it triggers the action potential.

Anode-break excitation   Sometimes, the membrane gets excited at the anode when the stimulus is switched off. This is called anode-break excitation. If the stimulus is applied for a sufficient length of time, the ions start crossing the membrane, setting up a *resistive current*. Thus under the anode there is a large increase in the inward diffusion of Na+ into the axon and a large decrease in the outward efflux of K+. However, these ionic fluxes are unable to change the membrane potential so long as the anodal current is passing through it and the membrane remains hyperpolarized due to the potential drop occurring along the path of the anodal current. However, the moment the anode is switched off, the combination of high Na+ influx and low K+ efflux depolarizes the membrane and triggers an action potential.

Membrane accommodation   When a catelectrotonic potential depolarizes the membrane rapidly to the firing level, an action potential is triggered. However, when the catelectrotonic potential depolarizes the membrane slowly to the firing level, the action potential often fails to be triggered. This phenomenon is called membrane accommodation and is due to Na+ channel inactivation. When the membrane depolarizes, both Na+ and K+ channels open up. If the depolarization occurs very slowly, more and more Na+ channels open up, only to get quickly inactivated. The K+ channels remain open and tend to restore the membrane potential. By the time the potential touches the firing level, most of the Na+ channels have got inactivated while the K+ channels remain open as long as the membrane remains depolarized. Therefore, the explosive depolarization fails to occur. If however the membrane depolarizes rapidly to the firing level, a large number of Na+ channels open all at once. Before they get inactivated, they are able to trigger off the action potential by overcoming the effect of K+ channels. For the same reason, the square pulse is more effective than the saw-tooth pulse in stimulating the membrane (Fig. 11.**5**B).

Strength-duration curve   The ease with which a membrane can be stimulated depends on two factors, viz., the strength of the stimulus and the duration for which the stimulus is applied. As the strength of stimulus increases, the time required to excite the membrane decreases and *vice versa*. The strength-duration curve is obtained by plotting various combinations of the stimulus strength and the time it takes to stimulate a membrane (Fig. 11.**6**). Any index of membrane excitability must take into account both the factors. Two such indices of membrane excitability are rheobase and chronaxie. *Rheobase* is the theoretical minimum current of electrical stimulus which when applied for an adequate length of time will excite the membrane.

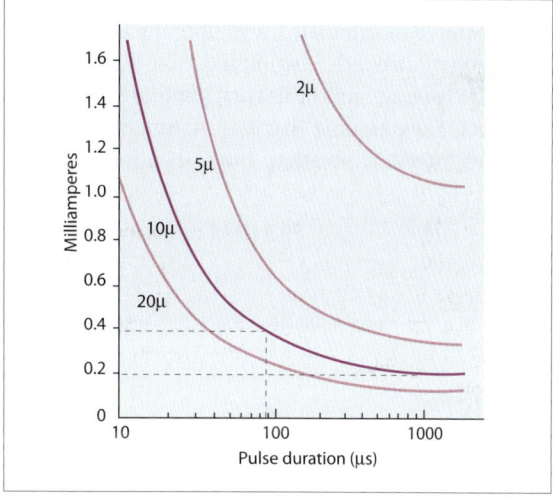

Fig. 11.**6**   Strength-duration curves of nerve fibers of different diameters. The 10µ fiber has a rheobase of 0.2 mA, utilization time of nearly 1000 µs and a chronaxie of 100 µs (0.1 ms).

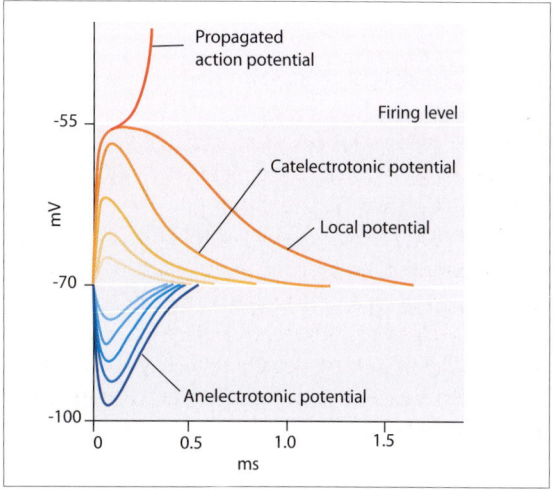

Fig. 11.**7**   Electrotonic and local potentials.

Table 11.**1**   Differences between graded potential and action potential.

Graded potential	Action potential
Amplitude proportionate to stimulus strength and can get summated	Amplitude constant for all suprathreshold stimuli and cannot be summated
Can be a depolarization or hyperpolarization	Always a depolarization
Conduction is associated with reduction in magnitude	Conducted without reduction in magnitude
Can be generated spontaneously in response to physical or chemical stimuli	Generated only in response to membrane depolarization

Utilization time is the time taken by a stimulus of rheobase strength to excite the membrane. *Chronaxie* is the time required to stimulate the membrane using double the rheobase strength of stimulus.

Both rheobase and utilization time are theoretical concepts. They assume that the membrane is stimulated for indefinite length of time. Such long duration of weak stimulus results in membrane accommodation and therefore can never be used for practical determination of either the rheobase or the utilization time. Hence, chronaxie is a practical index of membrane excitability. Its value ranges from 0.02 ms in the largest peripheral neurons (Aα) to a maximum of 1.5 ms in the smallest C-fibers. An increase in chronaxie indicates a reduction in membrane excitability. Serial measurements of chronaxie provide useful information regarding the progress of nerve healing following injury and surgical repair.

Graded Potentials

Unlike action potentials, local potentials do not obey the all-or-none law. They are proportional to the stimulus strength and hence are also called *graded potentials*. Other examples of graded potentials include the receptor potential and the end-plate potential. Some of the differences between graded potentials and action potentials are listed in Table 11.**1**.

Channel Dynamics

Effect of membrane potential on channel permeability

Both Na⁺ channels and K⁺ channels are voltage-sensitive. They open (activate) and close (deactivate) with changes in the membrane potential. Individual channels obey the all-or-none law, i.e., at any instant, they remain either fully open or fully closed. When the membrane depolarizes, the probability of a channel remaining open increases. Stated simply, as the membrane depolarizes, more and more channels change from the fully closed to the fully open state.

When the membrane depolarizes, both Na⁺ and K⁺ channels open up, resulting in measurable increases in their conductances (i.e., membrane permeability). The increase in conductance is roughly proportional to the change in membrane potential.

Sodium channel inactivation   Though both Na⁺ and K⁺ conductance increase with membrane depolarization, the increase in Na⁺ conductance does not last long. In other words, if the membrane is kept depolarized for a certain period of time, the Na⁺ channels that initially get activated start closing down spontaneously after a few milliseconds (Fig. 12.1A). This process is called *Na⁺ channel inactivation*. They recover from inactivation only after the membrane returns to the normal RMP. Na⁺ channel activation has two important consequences, viz., the *refractory period* and *membrane accommodation*.

Na⁺ channels are guarded by two gates: the activation and inactivation gates.[1] During depolarization, the activation gate opens quickly followed slowly by the closure of the inactivation gate. This results in a brief

interval during which the channel remains open (Fig. 12.1B). During repolarization, the activation gate closes first followed by the opening of the inactivation gate so that the channel remains obliterated throughout repolarization.

Effect of channel permeability on membrane potential

Stabilizing effect of potassium channels   When the membrane potential becomes less negative, the K⁺ channels open up. Opening of the K⁺ channels increases the outward hyperpolarizing current and thereby makes the membrane more negative. Conversely, if the membrane potential becomes more negative, a large

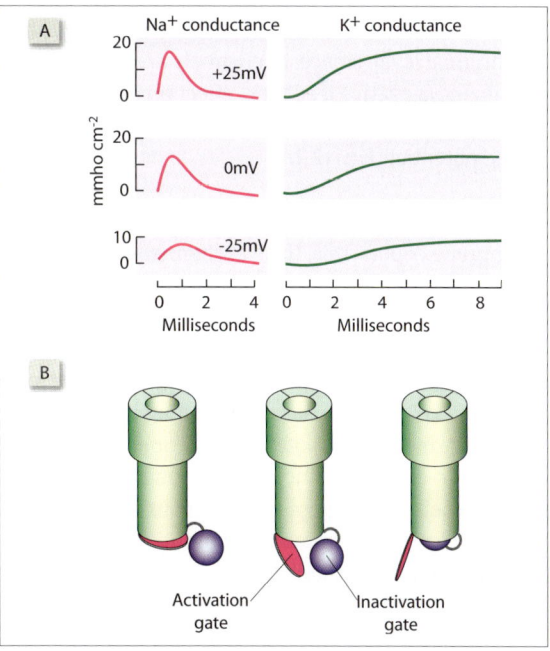

Fig. **12.1**   [A] Effect of membrane depolarization on membrane permeability to potassium and sodium ions over a period of 4 to 8 milliseconds. The effects of three different grades of depolarization are shown here, as the membrane is depolarized to +25 mV (*above*), 0 mV (*middle*), and –25 mV (*bottom*). Note that the changes in membrane permeability of Na⁺ are transient while those of K⁺ are sustained as long as the membrane remains depolarized. [B] Activation (m) and inactivation (h) gates of a sodium channel.

---

[1] The activation and inactivation gates are also called the m gates and h gates respectively. The origins of the names 'm' and 'h' are to be found in the empirical equations developed in the 1950s by Hodgkin and Huxley to describe the time course of voltage changes during an action potential. They inferred quite correctly that the changes in Na⁺ conductance occurred due to two processes: a rapid process that increases channel permeability (represented by the constant m) and a slower process that reduces channel permeability (represented by the constant h). The physical evidence of the presence of activation and inactivation gates such as the identification of the protein chains came several years later as a testimony to the power of mathematical modeling.

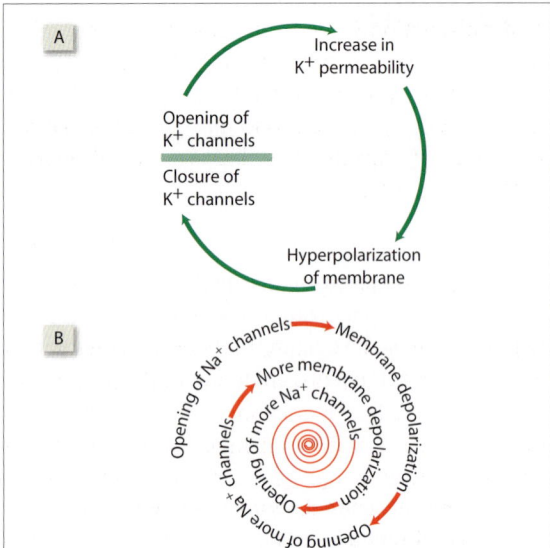

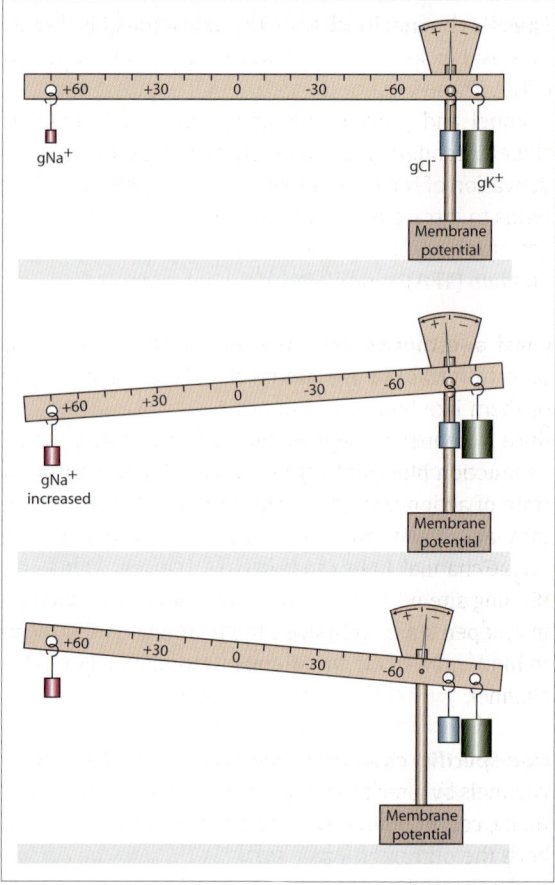

Fig. 12.**2** Effect of channel permeability on membrane potential. [A] Stabilizing effect of K⁺ ions on membrane potential. [B] The Hodgkin cycle, explaining the destabilizing effect of Na+ ions on membrane potential.

Fig. 12.**3** The ionic basis of the firing level. (*Above*) Membrane at resting potential of –80 mV. All ionic conductances are resting conductances. (*Middle*) Sodium conductance (g_{Na}) increases slightly, resulting in a depolarizing tendency. (*Below*) Due to the increase in g_{Na}, the membrane depolarizes to –70 mV. Immediately, g_{Cl} and g_K start exerting a strong hyperpolarizing effect. This happens even when there is no change in g_{Cl} and g_K (see also Fig. 10.**4**).

number of K⁺ channels close down and the membrane potential drifts to a less negative value. Thus, whenever there is a change in membrane potential, the activated K⁺ channels tend to restore the membrane potential to the original level. Hence, K⁺ channels are said to have a stabilizing (negative feedback) effect on the membrane potential, i.e., they tend to prevent any change in membrane potential (Fig. 12.**2**A).

Destabilizing effect of sodium channels When the membrane depolarizes, the Na⁺ channels start opening up. The opening of the Na⁺ channels increases the inward depolarizing current and thereby depolarizes the membrane further, leading to the opening of greater numbers of Na⁺ channels. Thus, there is a positive feedback spiral (the *Hodgkin cycle*) resulting in a very rapid change in membrane potential and opening of nearly all the Na⁺ channels. Since the opening of even a few Na⁺ channels can trigger off this vicious cycle resulting in large changes in membrane potential, Na⁺ channels are said to have a destabilizing effect on membrane potential (Fig. 12.**2**B).

Firing level It is clear from Figure 12.**1**A that when the membrane depolarizes, the rise in Na⁺ conductance is much quicker than the rise in K⁺ conductance. Hence, even a small depolarization should destabilize the membrane potential and set off the Hodgkin cycle. What actually happens is that the membrane quickly repolarizes following a small depolarization. Why is it so?

The question is best answered by considering the 'balance of moments' analogy (Fig. 12.**3**) which explains why the slightest depolarization of the membrane is immediately opposed by g_K and g_{Cl} even when these conductances remain unchanged. If additionally the K⁺ channels open up, the g_K increases and the opposition to membrane depolarization is even greater. Thus, a fairly large increase in g_{Na} is required before the Hodgkin cycle can be set off. The membrane potential at which this occurs is called the firing level.

■   **Agents altering channel function**

Several naturally occurring substances affect the permeability of Na⁺ or K⁺ channels. These compounds have been used extensively in research for studying channel properties using the voltage clamp technique.

Specific channel blockers   The activation of Na⁺ channels is blocked by *tetrodotoxin* (TTX) and *saxitoxin* (STX). These bind to the extracellular sites on the Na⁺ channel and prevent the activation of Na⁺ channels, thereby rendering the membrane inexcitable. The inactivation of Na⁺ channels is blocked by *pronase*, which binds to intracellular sites on the Na⁺ channel. The inactivation of the K⁺ channel is blocked by *tetraethylammonium* (TEA). It prolongs the action potential.

Local anesthetics are lipid-soluble substances that bind to intracellular sites on the Na⁺ channels. Some of them like lidocaine and procainamide tend to produce a frequency-dependent block, i.e., they produce conduction block only after the membrane conducts a train of action potentials. This is probably because the local anesthetic molecule binds to an intracellular site on the channel. In its charged form, it gains access to its binding site within the pore only when the channel is in an open state. Moreover, the local anesthetic seems to bind more tightly to the inactivated form of the Na⁺ channel.

Non-specific channel modulators   Ca²⁺ ions block channels by altering the charge on the membrane. Normally, cell membranes carry a net negative charge on both the outer and inner surfaces as a result of negatively charged groups in the membrane phospholipids. Positive cations, especially multivalent cations like Ca²⁺, are attracted to the negative charges on the membrane and repel other cations (Na⁺ and K⁺). Lowering the concentrations of extracellular Ca²⁺ can cause spontaneous impulse generation, which can be reversed by the addition of Ca²⁺ or other bivalent ions, probably because of their stabilizing effects on the channel. Spontaneous impulse generation in nerves due to low serum Ca²⁺ results in widespread muscle spasms. The condition is called tetany and is commonly seen in hypoparathyroidism.

Channel Characterization

Voltage clamp technique

Measuring the electrical conductance of a membrane involves measuring membrane currents at different voltages. A major hurdle in these efforts is the explosive change in membrane potential that occurs once the voltage touches the threshold. This led to the search for a technique that would 'clamp' the membrane potential at any desired value despite the surge in current.

In the voltage clamp technique, the membrane potential can be stepped rapidly to a predetermined level of depolarization (called the command potential) by passing current across the cell membrane. Every time the membrane potential tends to change from the command potential, the clamp circuit opposes the change in potential by injecting a precisely calculated feedback current (Fig. 12.**4**).

Patch clamp technique

Before the advent of the patch clamp technique, the voltage clamp technique was used to determine conductance of a whole membrane with a mixed population of different types of channels. The conductance characteristics of a single type of channel were deduced by blocking the other known channels present in the membrane. For example, Figure 12.**5** shows the records of membrane conductance before and after application of TTX. The conductance measured after applying TTX gives the K⁺ conductance provided the membrane has only Na⁺ and K⁺ channels.

With patch clamp technique, it is possible to record the conductance of a single channel. A small fine polished glass micropipette with a tip diameter of ~1 μm is pressed against the membrane. The pipette is filled with a solution having a composition like the extracellular fluid. A metal electrode in contact with the electrolyte in the micropipette connects the pipette to a special electric circuit that measures the current flowing through channels in the membrane under the pipette tip. A small amount of suction to the patch

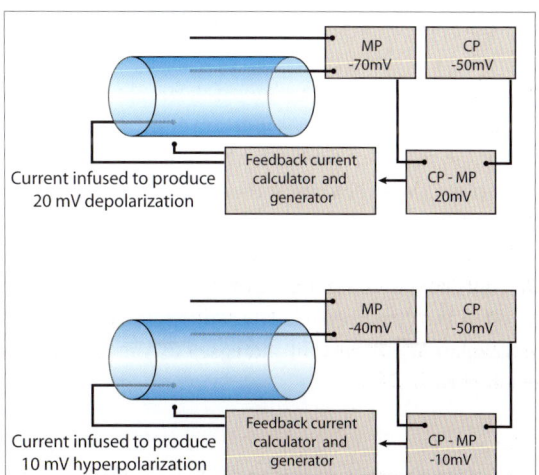

Fig. 12.**4**   (*Above*) Membrane potential (MP = −70 mV) is made equal to the command potential (CP = −50 mV) by injecting a calculated amount of current. (*Below*) If membrane potential tries to change from −50 to −40 mV, it is immediately corrected by injecting a current calculated to produce 10 mV hyperpolarization.

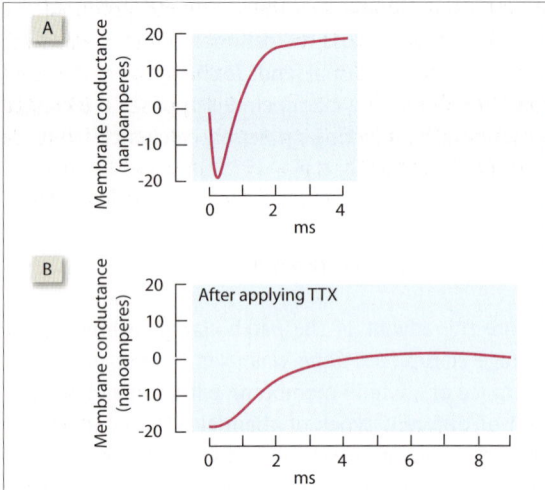

Fig. 12.**5** [A] Conductance of a membrane that contains only Na and K+ channels. [B] Conductance of the same membrane after blocking the Na channels with TTX.

pipette greatly increases the tightness of the seal between the pipette and the membrane. The result is a seal with extremely high resistance (the gigaohm seal) between the inside and outside of the pipette.

Patch clamping offers three types of experimental possibilities (Fig. 12.**6**A): (1) The experimenter can apply various stimuli from within the pipette and measure the behavior of the trapped channels. (2) The experimenter can detach the membrane from the cell, thereby exposing the cytoplasmic mouth of the channels. (3) The membrane patch can be ruptured without breaking the gigaohm seal so that the experimenter can alter the constituents of the living cell's cytoplasm.

Channel currents recorded by patch clamp One striking observation made possible through patch-clamp studies of single channels is that channels obey the all-or-none law: they are either fully open (conducting a maximum current of about 3 picoAmperes) or are fully closed (Fig. 12.**6**B,C). Also, channels are usually not continuously open but switch frequently between the open and closed states. The probability of a channel being open at any instant increases as the membrane depolarizes.

Channel Structure

The molecular structures of several channels have been characterized and the sites responsible for specific channel characteristics have been identified. The structure of an ion channel has been described on p 42. Some of the details of a voltage-sensitive Na+ channel

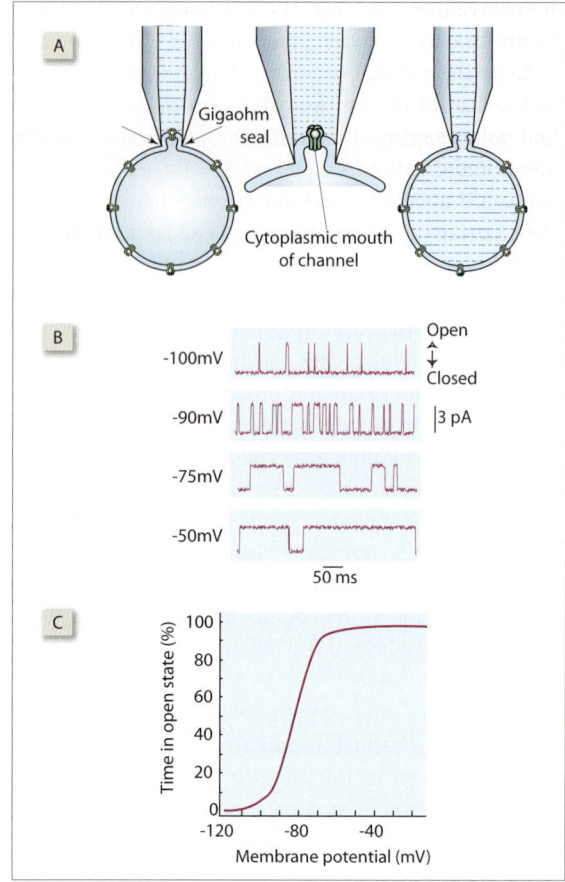

Fig. 12.**6** [A] Three methods of the patch clamp technique. [B] Channel current of a single channel recorded through patch clamp. The channel current obeys the all-or-none law: it is either 3 pA (maximum) or 0 pA (minimum). As the membrane is depolarized in steps (–100 mV, –90 mV, –75 mV, and –50 mV), the channel remains open for longer durations. [C] Graph showing the relation between membrane potential and the time spent by the channel in the open state.

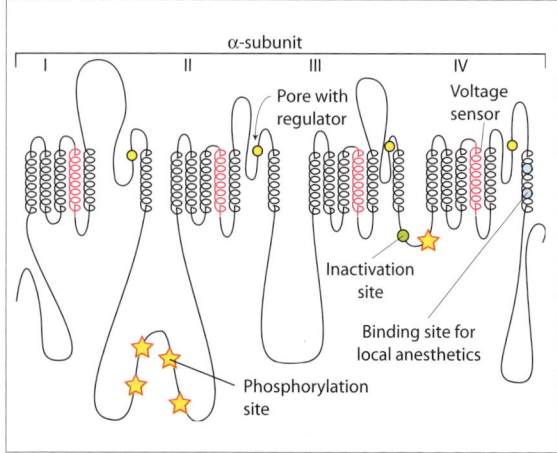

Fig. 12.**7** The ultrastructure of the α–subunit of the sodium channel showing its functionally important sites.

are shown in Figure 12.**7**. The S-4 transmembrane segments in each homologous domain of the α-subunit contain the voltage sensors. The positively charged amino acid in the third position within S-4 is the actual voltage sensor. S-5 and S-6 transmembrane segments and the short membrane-associated P loop (segments SS1 and SS2) enclose between them the channel pore. Certain amino acids in the P loop are critical for determining conductance and ion selectivity of the Na^+ channel and its ability to bind TTX and saxitoxin. The short intracellular loop connecting homologous domains III and IV serves as an inactivation gate. The loops between I and II, and between III and IV are susceptible to phosphorylation. Phosphorylation at these two sites impairs channel activation and inactivation respectively.

13   Conduction of Nerve Impulses

Before considering the mechanism of nerve conduction, it will be useful to consider the analogy of a model for transfer of heat from one end of a rod to the other. Consider two rods, one wooden and the other made of steel. One end of each rod is placed over a flame with a temperature of 600°C. Which will transfer the heat more effectively from one end to the other? The answer seems too obvious: the steel rod of course. However, there is another way of looking at it. The far end of the steel rod can never get heated to 600°C because of heat losses. On the other hand, the wooden rod burns itself out and transfers the flame to its far end, which then reaches a temperature of 600°C. This novel method of heat transfer with a high fidelity may be called the traveling flame mechanism (Fig. 13.1A). However, the wooden rod can conduct the flame only once, after which a fresh wooden rod has to be provided for another flame to pass over it. Moreover, it is very slow. However, where the fidelity of the 'temperature-signal' is a priority, this mechanism cannot be ignored.

A hybrid model incorporating both steel and wood incorporates the advantages of both: steel gives speed and wood gives fidelity (Fig. 13.1B). The role of steel is to transfer enough heat to its far end so that the wood catches fire. The burning wood restores the temperature to 600°C. For achieving high speeds of heat transfer, the length of the steel rods should be long. The wooden chips should be few and far between because it is here that the heat transfer slows down.

Making the steel rods too long is however fraught with the risk of heat transfer failure. If the temperature at the far end of the steel rod falls below the ignition temperature of wood, the heat transfer stops there. This imposes a limit to the length of steel rods that can be interposed between the wooden chips. This limit can be maximized by using insulated steel rods so that there is little heat loss in the intervening segments. This analogy will not be extended any farther although there can be numerous corollaries to it. For example, how does the flame temperature affect the velocity of the traveling flame?

Mechanism of Nerve Conduction

Electrotonic conduction

Local currents When an action potential is in its depolarization phase, local currents flow from the depolarized membrane to the adjacent polarized areas and depolarize them too. In the axoplasm, the current flows away from the depolarized area. The current then flows out through the axonal membrane and returns through the extracellular fluid. This passive flow of local currents to adjacent membrane areas is called *electrotonic conduction*. The local current circuit can be resolved into three components: (1) the internal longitudinal current or axoplasmic current which flows through the axoplasm, (2) the radial or membrane current which flows out through the membrane, and (3) the external longitudinal current which flows through the extracellular fluid. The effect of local currents becomes feebler with distance (Fig. 13.2A). In an unmyelinated neuron, a fully-fledged action potential of 110 mV can produce a depolarization of 15 mV at a distance of 1.0 mm but a depolarization of only 2.0 mV at a distance of 2.0 mm.

Factors affecting electrotonic conduction Electrotonic conduction is determined by four physical factors: (1) The axoplasmic resistance (R_i) offered by the axoplasm to the flow of axoplasmic current. The thinner the axon, the higher is the axoplasmic resistance

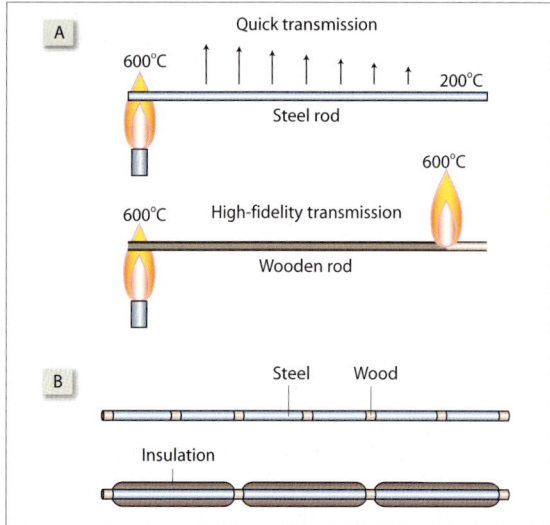

Fig. 13.1   The traveling flame analogy of nerve conduction. [A] A steel rod transfers the heat faster but the wood transfers the flame itself. [B] A hybrid wood-and-steel rod for speeding up flame transfer.

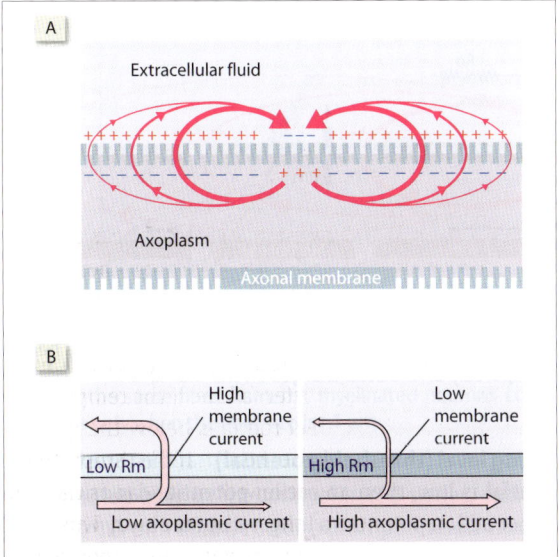

Fig. 13.**2**   [A] In unmyelinated membrane, local currents cannot travel far because of high axoplasmic resistance (Ri), low membrane resistance (Rm), and high membrane capacitance (Cm). High Cm is shown as a dense distribution of membrane charge. [B] Membrane resistance and currents. As membrane resistance rises, membrane current decreases and axoplasmic current rises.

and the weaker is the electrotonic conduction. (2) The resistance (Ro) offered by the extracellular fluid to the outer longitudinal current. The greater this resistance, the weaker is the electrotonic conduction. Ro usually remains constant. (3) The resistance (Rm) offered by the membrane to the transmembrane current. When Rm is high, the membrane current falls resulting in higher axoplasmic current and stronger electrotonic conduction. Rm is high when the membrane is thick, as in the myelinated segment of the neuron (Fig. 13.**2**B). (4) The electrical capacitance (Cm) of the membrane. The greater the membrane capacitance, the poorer is the electrotonic conduction.[1] Cm is low when the membrane is thick, as in the myelinated segment of the neuron.

Propagation of action potential

Local currents from one action potential spread to the adjacent membrane area and depolarize it to the threshold level, triggering another action potential and the process continues. Thus, there are two essential components of action potential propagation: (1) The flow of local currents, which depolarize the adjacent membrane, is called the *conductive component* of impulse propagation. This electrotonic conduction of membrane depolarization is associated with progressive decrease in the magnitude of depolarization.

(2) The triggering of a fresh action potential in the adjacent membrane is called the *regenerative component* of impulse propagation. The new action potential depolarizes the membrane maximally and thereby restores the depolarization to its original magnitude of 110 mV.

Without the conductive component, an action potential cannot be conducted at all. Without the regenerative component, an action potential can be conducted only up to a limited distance beyond which it will fade out. Before the local currents fade out completely, a fresh action potential must be generated for its continued propagation. The conductive component is very fast and depends on the electrical characteristics (the cable properties) of the membrane. The regenerative component is much slower. Frequent action potentials along the neuron tend to slow down the conduction velocity.

Nerve Conduction Velocity

Nerve conduction velocity can be affected by factors influencing electrotonic conduction (conductive component), the action potential (regenerative component), or both.

Axon diameter   An axon with a larger diameter allows easier passage to axoplasmic current.[2] When the electrotonic conduction is stronger, an action potential can quickly trigger another action potential a long distance away. In an unmyelinated fiber, the speed of propagation is directly proportional to the square root

---

[1]   An electrical impulse is conducted by inducing a potential change in its adjacent area. The greater the capacitance of a body, the more it resists any change in its potential. Hence, if the capacitance of the adjacent area is higher, it takes longer to induce a potential change in that area and therefore, electrotonic conduction slows down.

[2]   The resistance of a conductor is inversely related to its cross-sectional area.

A nerve impulse does not get conducted from one cell to another unless there are low electrical bridges connecting the two cells. Hence, special mechanisms are required for the impulse to travel from one neuron to another (synaptic transmission), or from a neuron to the muscle cell (neuromuscular transmission). The mechanisms of synaptic transmission and neuromuscular transmission are quite similar. The broad steps in neuromuscular transmission are summarized in Figure 14.1A. The structure of the neuromuscular junction is discussed on p 64.

The steps of neuromuscular transmission can be summed up as follows: (1) An action potential is conducted down the motor axon to the prejunctional axon terminal. (2) Depolarization of the terminal buttons opens up voltage-gated Ca^{2+} channels in its membrane. Ca^{2+} moves into the terminal along an electrochemical gradient. (3) Elevated Ca^{2+} concentration in the terminal button causes exocytosis of synaptic vesicles into the myoneural cleft. (4) The ACh released from synaptic vesicles diffuses across the myoneural cleft and binds to specific acetylcholine receptors on the motor endplate. (5) The binding of ACh to the ACh receptor increases the conductance of the postjunctional membrane to Na^+ and K^+, resulting in a transient depolarization of the postjunctional membrane. This depolarization is called the endplate potential (EPP). (6) The EPP is transient because acetylcholine is quickly hydrolyzed by the enzyme acetylcholinesterase (AChE) into choline and acetate. AChE is present in high concentration in the junctional cleft. (7) The postjunctional membrane cannot generate action potentials. However, the EPP depolarizes the adjacent muscle membrane by electrotonic conduction. When the depolarization exceeds the threshold, an action potential is triggered in the adjacent muscle membrane.

■ Acetylcholine as a neurotransmitter

Synthesis of acetylcholine Acetylcholine (ACh) is synthesized in the cytosol of the motor nerve terminal in the presence of the enzyme choline O-acetyltransferase.

$$Acetyl\text{-}CoA + Choline \xrightarrow[\text{acetyltransferase}]{\text{Choline O-}} Acetylcholine + CoA$$

Nervous tissue cannot synthesize choline, which is obtained from diet. Significant amounts of choline are taken up from the junctional cleft into the nerve terminal through Na^+-coupled secondary active transport. After synthesis, ACh is incorporated into the synaptic vesicles. Some vesicles originate in the soma and are transported distally by axoplasmic flow. Some are formed in the axon while some are formed in the terminal button itself.

Release of acetylcholine Acetylcholine is released into the myoneural cleft by exocytosis of synaptic vesicles. The exocytosis is triggered by the rise of Ca^{2+} in the terminal button. The release of acetylcholine occurs in quanta or 'packets'. Each quantum corresponds to the content of one synaptic vesicle, i.e., 10 000 molecules of ACh. A few quanta of acetylcholine are also released spontaneously.

Inactivation of acetylcholine ACh is removed from its site of release partly by diffusion but largely by

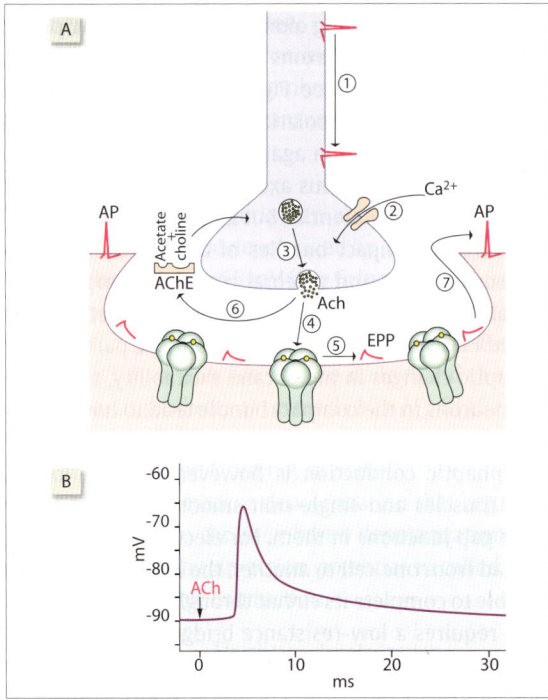

Fig. 14.1 [A] Major steps in neuromuscular transmission. [B] An endplate potential (EPP).

hydrolysis into acetate and choline by the enzyme acetylcholinesterase (AChE) present in the basal lamina. Choline is taken up from the synaptic cleft and reused for the synthesis of acetylcholine.

Acetylcholine receptors   ACh has to diffuse 1 μm to reach the end plate, which it does in less than a millisecond. ACh combines with ACh receptors located on the ACh-gated cation channels (see Table 5.**1**) present on the post-junctional membrane. The binding is short lasting because as the ACh concentration in the cleft falls, ACh quickly dissociates from its receptor, in accordance with the laws of chemical kinetics. Formation of an ACh-receptor complex opens up the channels to both Na⁺ and K⁺ and as a result, the postjunctional membrane gets depolarized. According to the Goldman equation the membrane potential (Vm) is given by:

$$Vm = 61 \log \frac{P_K[K_0] + P_{Na}[Na_0] + P_{Cl}[Cl_i]}{P_K[K_i] + P_{Na}[Na_i] + P_{Cl}[Cl_0]}$$

Assuming $P_K = P_{Na}$, and ignoring P_{Cl} (since both P_K and P_{Na} increases markedly in comparison to P_{Cl}), the above formula reduces to:

$$Vm = 61 \log \frac{[K_0] + [Na_0]}{[K_i] + [Na_i]}$$

or

$$Vm = 61 \log \frac{5 + 140}{150 + 10}$$

$$= -2.6 \text{ mV}$$

Thus the simultaneous increase in Na⁺ and K⁺ permeability drives the membrane potential towards 0 mV. However, the presence of ACh in the cleft is normally extremely short-lived. Hence, the ligand-gated cation channels remain open for a very short time, permitting only a very small local depolarization (0.5 mV) of the membrane. Larger depolarization is possible if ACh remains attached to the channels for longer durations. This knowledge is used in the treatment of myasthenia gravis (see below) in which the channels are kept open for longer periods.

Post-Junctional potentials

Miniature endplate potential (MEPP)   Each quantum of ACh released into the synaptic cleft produces a small depolarization called a miniature endplate potential (MEPP), which is 0.5 mV in amplitude and not enough to trigger an action potential in the adjacent muscle plasma membrane. MEPPs occur spontaneously even when the neuron is not stimulated, with an average frequency of about 1 s⁻¹. Such sporadic MEPPs do not have any physiologic effect. A large number of MEPPs

must be produced synchronously to bring about a substantial change in the potential of the motor endplate.

Endplate potential (EPP)   Each action potential reaching the axonal ending results in the release of about 60 quanta of ACh, with each quantum containing 10000 ACh molecules. The 600000 ACh molecules bind to 300000 ACh-R (2 on each) and produce 300000 MEPPs, each having a strength of 0.5 mV. These MEPPs summate spatially, taking the membrane potential from –90 mV to –65 mV (Fig. 14.**1**B). This 25 mV change in membrane potential is called the endplate potential (EPP). The EPP gets conducted electrotonically from the endplate to the adjacent muscle fiber membrane, which then gets depolarized to its firing level and generates an action potential. No action potential is generated in the endplate itself. Only 6 quanta of ACh are necessary for the transmission of the action potential across the neuromuscular junction but as many as 60 are released. This provides a 10-fold safety factor for neuromuscular transmission.

Drugs Affecting Neuromuscular Transmission

Drugs affecting acetylcholine release

Amidopyridines   increase the duration of the action potential in the nerve terminal, probably by blocking the voltage-gated K⁺ channels on the nerve terminals. Prolongation of action potential results in greater Ca²⁺ influx and release of more quanta of ACh. These drugs have found some limited use in LEMS (see below).

Botulinum toxin produces muscular paralysis by inhibiting the exocytosis of synaptic vesicles. Although best known for causing food poisoning, in recent years, botulinum toxin (Botox) has found therapeutic use as a long-acting muscle relaxant whose effect lasts 3–4 months. Botox has been used in the treatment of squint, facial wrinkles, achalasia (to relax the lower esophageal sphincter), anal fissure (to relax the anal sphincter) and in skeletal deformities that occur due to muscle spasm.

Alpha-Latrotoxin is the toxin of the black widow spider *Lactrodectus mactans*. It binds to receptors on the presynaptic membrane. The binding results in massive Ca²⁺ influx into the presynaptic terminal with consequent secretion of large amounts of ACh, leading to muscle spasm. The spider is commonly found in North America, where it causes the greatest number of deaths from arthropod bites.

Hemicholiniums are drugs that block the choline transport system and inhibit choline uptake. Prolonged treatment with hemicholiniums depletes the neuronal store of transmitter and ultimately decreases the ACh content of the quanta.

Drugs inhibiting acetylcholinesterase

Anticholinesterases (anti-ChE) are drugs that bind to AChE, thereby preventing it from hydrolyzing ACh (Fig. 14.**2**A). The AChE is freed only after it hydrolyzes the anti-ChE bound to it. By inactivating AChE, anti-ChE drugs permit large amounts of ACh to accumulate in the myoneural cleft. As a result, the EPP is larger (more depolarization) and dramatically prolonged. Initially the depolarization results in persistent muscle contraction (spasm) but after a few seconds, makes the motor endplate inexcitable due to channel desensitization. Channel desensitization is not the same as channel inactivation. Desensitization occurs in a ligand-gated channel when the ligand remains attached to the channel receptor for several milliseconds. The molecular mechanism of channel desensitization is not well understood. Anti-ChE drugs are of two types: reversible and irreversible.

Reversible inhibitors Reversible anti-ChE drugs like neostigmine are readily hydrolyzed by AChE. Therefore in a few hours, the AChE becomes available once again for inactivating ACh. Reversible inhibitors are used in the treatment of myasthenia gravis and in curare poisoning.

Irreversible inhibitors Organophosphate compounds like Malathion and Baygon (insecticides), diisopropyl phosphofluoridate (DFP, Dyflos) and nerve gas used in chemical warfare are poorly hydrolyzed by AChE, and therefore remain attached to the AChE molecule for several weeks. AChE is therefore unable to bind and hydrolyze ACh. The persistence of ACh in the myoneural cleft causes desensitization of the channels in the motor endplate. Hence, irreversible anti-ChE produces deadly paralysis.

Drugs blocking acetylcholine receptors

Drugs that bind to acetylcholine receptors are of two types: those that do not have any intrinsic activity (nondepolarizing blocker) and those that mimic the action of acetylcholine (depolarizing blocker). Drugs that have affinity for acetylcholine receptors but do not have any intrinsic activity produce nondepolarizing neuromuscular

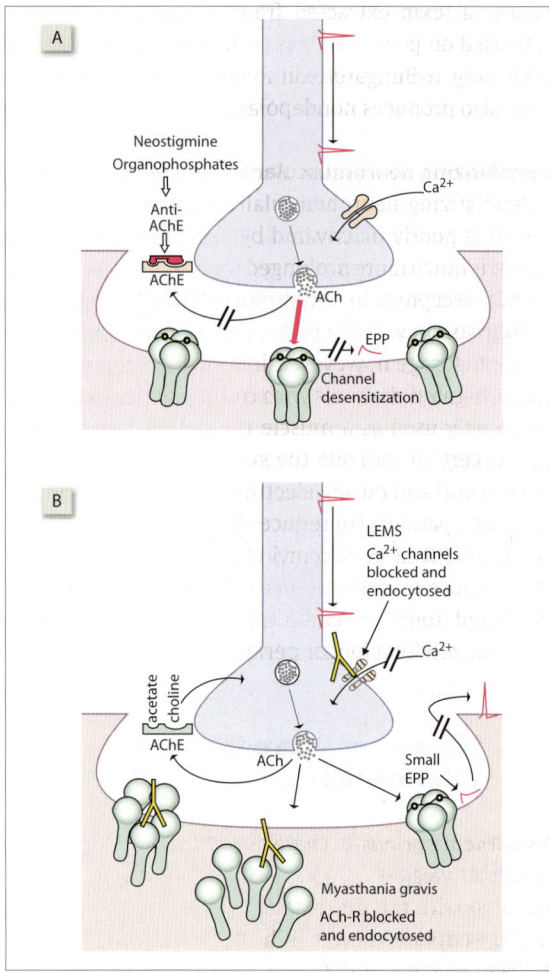

Fig. 14.**2**   [A] Effect of anti-AChE drugs. [B] Pathophysiology of Myasthenia gravis and LEMS.

block. They do not depolarize the endplate but keep the ACh receptors blocked so that ACh is unable to bind to them. Drugs which mimic the action of acetylcholine produce depolarizing neuromuscular block.

Nondepolarizing neuromuscular blocker An example of a nondepolarizing neuromuscular blocker is d-tubocurarine. It has no effect of its own but prevents ACh from producing neuromuscular transmission, causing muscle paralysis that can be fatal due to the paralysis of respiratory muscles. An ACh receptor will bind more to the agonist that is present in higher concentration. Hence, if the concentration of ACh in the cleft is increased by administering anti-AChE, the ACh molecules are able to dislodge d-tubocurarine from the ACh receptors, making them functional again. Hence, the neuromuscular block produced by tubocurare is said to be reversible. D-tubocurarine is obtained from curare, a toxin extracted from certain plants, which was used on poison arrows by Red Indians to paralyze their prey. α-Bungarotoxin found in the venom of the krait also produces nondepolarizing block.

Depolarizing neuromuscular blocker An example of a depolarizing neuromuscular blocker is succinylcholine. It is poorly inactivated by AChE and therefore its action is much more prolonged than that of ACh. It binds to ACh receptors in the motor end plate and causes muscle spasm when administered in moderate dosage. In high dosage however, it induces desensitization of the ACh-gated channels with consequent paralysis. It is commonly used as a muscle relaxant during abdominal surgery (to provide the surgeon easier access into the viscera) and during electroconvulsion treatment of psychotic patients (to reduce violent movements). The block produced by succinyl choline is called irreversible because even if the succinylcholine molecules are dislodged from the ACh receptors, the channels remain desensitized for a longer period.

Diseases of Neuromuscular Transmission

Myasthenia gravis is characterized by severe skeletal muscular weakness and rapid onset of fatigue. It is associated with the destruction of ACh receptors on the motor endplate. Fewer ACh receptors result in fewer MEPPs and smaller EPP.

Myasthenia gravis is caused by autoantibodies to ACh receptors on the motor endplate. The antibodies block the ACh-receptors so that ACh cannot bind to them. The antibodies also cross-link the ACh-receptors, triggering their removal by endocytosis (Fig. 14.**2**B). Alternatively, it has been suggested that the disease is associated with benign tumor of the thymus (thymoma) and a hypersecretion of thymopoietin which binds to the ACh receptors to inactivate them.

Neostigmine, an anti-ChE drug, produces striking symptomatic improvement, each dose acting for several hours. When AChE is administered, ACh persists in the myoneural cleft for a longer period, resulting in larger MEPPs that add up to produce an EPP large enough to trigger an action potential.[1] Treatment also includes suppression of autoimmunity by immunosuppressants, thymectomy, and plasmapheresis (p 140).

In **Lambert-Eaton myasthenic syndrome (LEMS)**, autoantibodies damage the voltage-gated Ca^{2+} channels present in the nerve endings at the neuromuscular junction (Fig. 14.**2**). This decreases the Ca^{2+} influx that causes acetylcholine release during neuromuscular transmission and results in muscle weakness. Symptoms of LEMS get alleviated with continued muscular activity due to the progressive accumulation of Ca^{2+} in the terminal buttons, with consequent improvement of ACh release from the nerve ending. This is in contrast to the symptoms of myasthenia gravis which get aggravated with continued muscular activity due to the progressive reduction of ACh release from the nerve endings.

[1] If the ACh receptors are damaged and fewer in number in patients of myasthenia gravis, why should a rise in ACh concentration in the myoneural cleft improve neuromuscular transmission? It is so because the presence of ACh in the cleft is normally extremely short-lived. Hence, the ligand-gated cation channels remain open for a very short time, permitting a very small MEPP. Larger MEPPs are possible if ACh remains attached to the channels for longer duration. If the channels are reduced in number, a larger MEPP at every channel can compensate to produce a normal endplate potential.

15   Mechanism of Muscle Contraction

Contractile Mechanism

The shortening of a muscle fiber occurs due to the sliding of actin filaments on myosin filaments. The repetitive events that bring about this shortening are called the cross-bridge cycling, and can be briefly summarized as follows (Fig. 15.1A).

Cross-bridge cycling

ATP binds to myosin ATPase present on the myosin head and splits into ADP and P_i. The energy released activates the myosin head, which is now ready to bind to actin. The activated myosin head binds with the active sites of actin filaments forming actomyosin. Simultaneously, the myosin head flexes at its hinge. As a result, the actin filament slides on the myosin filament, bringing the Z-disks closer together and thereby shortening the sarcomere. As the myosin head flexes, the ADP and P_i present on it are cast off, making way for a fresh molecule of ATP. When ATP binds to myosin ATPase, the myosin head detaches from actin, and the cross-bridge cycle repeats all over again.

Power stroke   When the muscle is in a relaxed state, the myosin head is in the high-energy 90° conformation. The high-energy state is brought about by the energy released when an ATP molecule binds to myosin ATPase (present on the S-1 head) and is hydrolyzed into ADP and P_i. The activated myosin head binds to one of the several active sites on the actin filament to form the actin-myosin-ADP-P_i complex. Formation of the complex triggers two near-simultaneous events: (1) the flexion of the myosin head which now changes to the low-energy 45° conformation, and (2) the release of the P_i and ADP from the complex.

The *flexion of the myosin head* from the high-energy 90° conformation to the low-energy 45° conformation generates mechanical force and is called the 'power stroke'. It has either or both of the following effects (Fig. 15.1C). If the load on the muscle is small, then the actin filament slides over the myosin filament, producing muscle shortening (isotonic contraction). If the load on the muscle is large, then the actin and ADP + P_i myosin filaments are unable to slide over each other (isometric

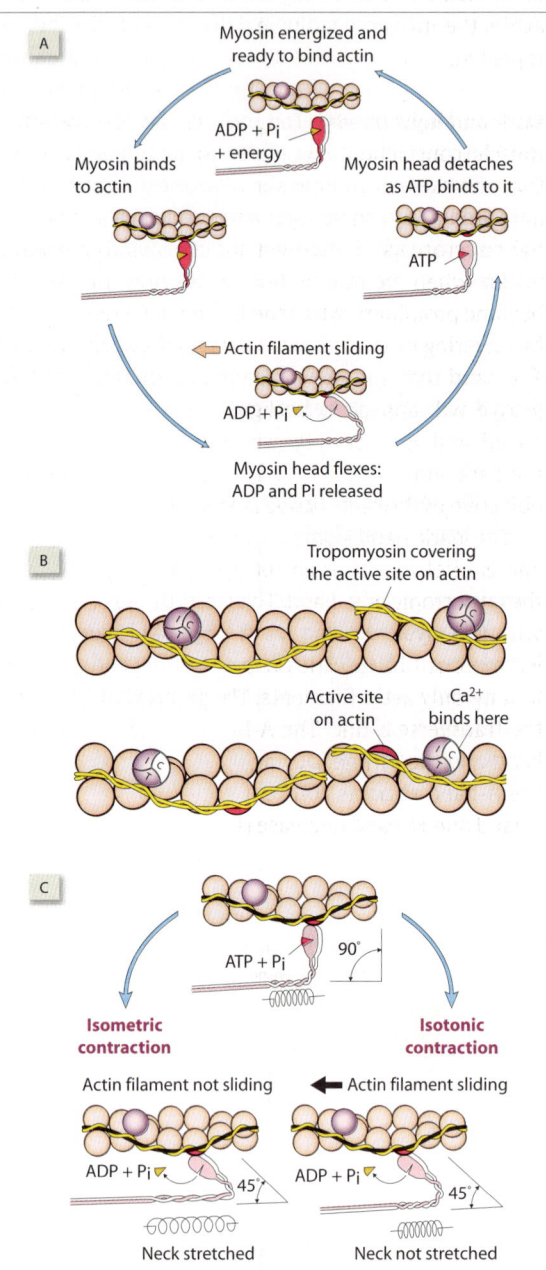

Fig. 15.**1**   [A] The cross-bridge cycle in a skeletal muscle. [B] Binding of calcium ions to troponin-C shifts the tropomyosin molecule, uncovering the active sites on actin that bind to myosin heads. [C] The power stroke in isometric and isotonic contraction. In isometric contraction, the actin filament cannot slide, resulting in the stretching of the neck of the actin filament.

contraction). Flexion of the myosin head will instead stretch the elastic neck of the myosin molecule (see Series elastic component, p 111).

The *release of ADP and P$_i$* allows a fresh ATP molecule to bind to the myosin head. The myosin-ATP complex has a low affinity for actin and therefore it results in the dissociation of the myosin head from the actin filament. The freshly bound ATP molecule is split again, the myosin head is reactivated, and the cycle is repeated.

Dark and light bands   Long before the mechanism of muscle contraction was understood, it was observed that the skeletal muscle showed cross striations that appeared as alternate light and dark bands under the light microscope. These light and dark bands are barely visible when the muscle fiber is perfectly focused but become prominent when the fiber is defocused slightly by lowering or raising the objective of the microscope. The band that appears to be dark on lowering the objective will appear to be light when the objective is raised, and vice versa. By convention, all references to the dark and light bands are those observed when the objective of the microscope is lowered.

The dark band contains highly refractile material and is birefringent or anisotropic and hence, is called the anisotropic or A-band. The A-bands are coextensive with the myosin filaments. The light bands are called isotropic or I-bands, and are present in the region containing only actin filaments. The I-band is bisected by the transverse Z-line. The A-band is bisected by the H band and the M-line. During contraction, the width of the A-band remains constant while the widths of the I-band and H-band decrease (Fig. 15.**2**).

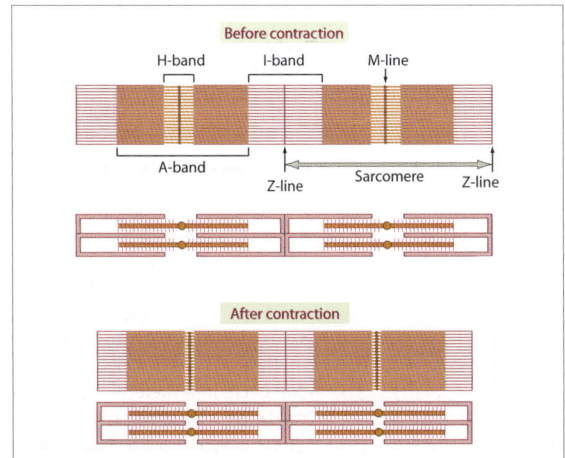

Fig. 15.**2**   Dark and light bands in a striated muscle. Contraction reduces the widths of the I-band and H-band but the width of the A-band remains unchanged.

Fig. 15.**3**   The handshake between the fat myosin and the thin actin is prevented by the tropomyosin gate with a troponin lock. The calcium ion is the key.

The above observations served as a lodestar for a century of research on the mechanism of muscle contraction. Any theory of muscle contraction had to account for the above observations. The sliding filament theory satisfactorily explains them all. The A-band is birefringent because it is the zone in which the actin and myosin fibers overlap.

■ Initiation and termination of cross-bridge cycling

Tropomyosin   The cross-bridge cycling is switched off or on by the tropomyosin molecule which slides on the actin molecule to cover or uncover the active sites on it. Tropomyosin covers the active sites on actin when the sarcoplasmic Ca^{2+} concentration is low and uncovers them when it rises (Fig. 15.**1**B).

Troponin   The regulatory action of Ca^{2+} on tropomyosin is mediated by troponin (Tp), a Ca^{2+}-binding protein. When the sarcoplasmic Ca^{2+} concentration rises, four ions of Ca^{2+} bind to TpC. The TpC-4Ca^{2+} complex induces changes in TpI and TpT, which in turn brings about a shift of tropomyosin away from the active sites on actin. When the sarcoplasmic Ca^{2+} falls, Ca^{2+} dissociates from TpC and tropomyosin slides back on the actin filament to cover the active sites.

Sarcoplasmic calcium concentration   The Ca^{2+} concentration in the sarcoplasm is quite low. The concentration in the L-tubule is about a thousand times higher because Ca^{2+} is continuously pumped from the sarcoplasm into the L-tubule by Ca^{2+}-ATPase. These calcium pumps are located on the membranes along the entire length of the L-tubule. Most of the Ca^{2+} that enters the

L-tubule moves to the terminal cisterns where it is bound in large amounts to a calcium-binding protein called *calsequestrin*.

When the sarcolemma depolarizes, the depolarization spreads to the T-tubules too. When the T-tubule depolarizes, the Ca^{2+} present in the lateral cisterns diffuses out into the sarcoplasm through special Ca^{2+} channels called the *ryanodine receptor* (RYR) channels. The released Ca^{2+} is taken up almost immediately and pumped back into the L-tubules by Ca^{2+}-ATPase. Hence, the Ca^{2+} concentration rises in the sarcoplasm for a very brief period and has been aptly called the Ca^{2+}-*pulse*. During the Ca^{2+} pulse, the sarcoplasmic Ca^{2+} concentration rises a thousand fold, from about 10^{-7} to 10^{-4} moles L^{-1}.

The sarcoplasmic Ca^{2+} concentration is determined by a dynamic balance between (1) the rate at which Ca^{2+} enters it from the lateral cisterns and (2) the rate at which Ca^{2+} is pumped back into the L-tubules. Whenever the frequency of action potentials increases, the sarcoplasmic Ca^{2+} increases. Conversely, a decrease in the frequency of action potentials results in a fall in sarcoplasmic Ca^{2+}.

Role of ATP   ATP has three roles in muscle contraction and relaxation. (1) It provides the energy for the power stroke of the myosin head. (2) It brings about a dissociation of the myosin head from the actin filament. (3) It brings about muscle relaxation by pumping out Ca^{2+} from the sarcoplasm into the L-tubules.

Muscle relaxation occurs when the cross-bridge cycle is interrupted. It is a common misconception that ATP causes muscle relaxation by disengaging the actin and myosin filaments. Disengagement of actin and myosin by ATP does not stop the cross-bridge cycle; rather, it keeps the cycle going. Disengagement of actin and myosin by ATP is followed immediately by a fresh interaction between actin and myosin, and results in another power stroke. Hence for the cycle to stop, the dissociation of myosin heads from actin must be accompanied by a lowering of sarcoplasmic Ca^{2+} and the consequent covering of the active sites on actin. When a muscle becomes fatigued, its ATP content decreases. Depletion of ATP slows down the pumping of Ca^{2+} into sarcoplasmic tubules and therefore increases the sarcoplasmic Ca^{2+} concentration. The active sites on actin remain uncovered (due to high sarcoplasmic Ca^{2+}) and bound to the myosin heads (because ATP is also required for the dissociation of myosin heads from actin). The muscle therefore fails to relax completely and remains in a partially contracted state called *contraction remainder*. For identical reasons (i.e., depletion of ATPs), the muscles become rigid after death, a state known as *rigor mortis*.

Excitation–Contraction Coupling

The term *excitation–contraction coupling* refers to the events between the generation of sarcolemmal action potential and the outpouring of Ca^{2+} from L-tubule cisterns into the sarcoplasm. For years, researchers have tried to understand how the depolarization of the T tubules results in an outpouring of Ca^{2+} from the lateral cisterns although there is no continuity between them. Presently, the favored theory is that of mechanically interlocked receptors of the T-tubule and lateral cistern, as explained below.

The RMP of skeletal muscle is about –90 mV. The action potential runs along the T-tubule to enter deep into the muscle fiber. When the action potential reaches up to the tip of the T-tubule, it activates certain voltage-gated receptors called *dihydropyridine receptors* (DHP) located on the T-tubule membrane. Activated DHP receptors trigger the opening of the RYR Ca^{2+} release channels located on the lateral cisterns. This is possible because the lateral cisterns are located very close to the tips of the transverse tubules, and the protein chains of DHP and RYR are mechanically interlocked. When the DHP is activated by the depolarization of the T-tubule, it undergoes a conformational change, which results in the RYR being pulled open mechanically (Fig. 15.**4**).

Muscle Energetics

Energy source for muscle contraction

The ATP store of a muscle cell is exhausted in the first 3 seconds of exercise. Thereafter, the ATP stores are replenished continuously by the dephosphorylation of creatine phosphate reserves of the muscle fiber.

$$\text{Creatine phosphate} + \text{ADP} \rightarrow \text{creatine} + \text{ATP}$$

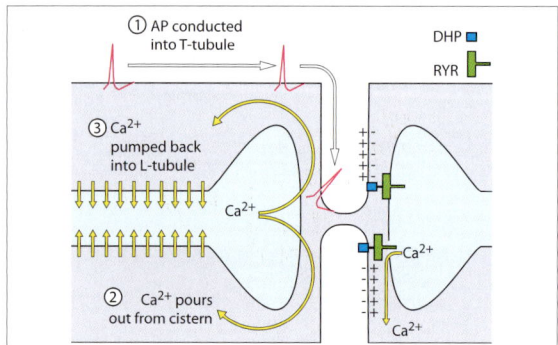

Fig. 15.**4** Excitation–contraction coupling. The T-tubule above is shown as polarized while below, it is shown as depolarized. Depolarization of the T-tubule results in outpouring of Ca^{2+} from the lateral cistern.

In another 5 seconds, the creatine phosphate reserves too get depleted and further supply of energy for ATP replenishment comes from glycolysis. However, the lactic acid accumulation associated with glycolysis makes it difficult for muscle contraction to continue beyond 1 minute. Continuous exercise for several hours is possible only when the ATP is replenished continuously through the Krebs cycle. The efficiency of the muscles under aerobic conditions is about 20%.

Despite being an inefficient and short-term process of ATP generation, the value of glycolysis lies in that unlike the Krebs cycle, its peak rate is not limited by the oxygen supply to the muscle. It is therefore ideal for short but severe exercise, e.g., a 100-meter sprint. The Krebs cycle on the other hand provides a steady supply of ATP at a rate limited by the oxygen uptake of the tissues, and is indispensable for a marathon.

■ Muscle heat

Early researchers carefully recorded the heat produced in muscle in order to understand the mechanism of muscle contraction. These details corroborate the current concepts of the contractile and elastic properties of the muscle and also indicate the timings of the physical and chemical processes that underlie muscle contraction.

Resting heat is the heat produced in the unstimulated muscle. It is the energy released due to the basic cellular metabolism and activity of the Na^+-K^+ pump.

Activation heat is produced at the onset of muscle contraction, before any generation of tension and/or shortening. Activation heat is associated with the release of Ca^{2+} from sarcoplasmic reticulum and the activity of myosin ATPase activity during cross-bridge cycling. Repeated activation of the muscle fiber, as occurs during muscle tetanus, results in the summation of activation heat. The total activation heat released during the course of a sustained muscle contraction is called the *maintenance heat*.

Shortening heat occurs during the actual process of shortening. It is absent in isometric contraction. Shortening heat is the energy released in overcoming the viscous forces inside the muscle while shortening. The sum of activation (or maintenance) heat and shortening heat is called the *initial heat*.

Recovery heat is the heat liberated due to the activity of Ca^{2+}-ATPase as it pumps Ca^{2+} back into the sarcoplasmic reticulum and also regeneration of ATP and other energy substrates. Recovery heat begins almost immediately after the onset of contraction and continues for several minutes after the cessation of contraction, indicating that increased Ca^{2+}-ATPase activity and ATP regeneration begin soon after contraction starts.

Relaxation heat is the amount of heat produced in addition to the recovery heat. It is released when an isotonically contracted muscle is stretched back by the load to its original length and is attributable to the viscous resistance to isotonic muscle relaxation.

The structural unit of a skeletal muscle is a muscle fiber but its functional unit is the motor unit (see below). While most of the properties of the whole muscle are also present in each muscle fiber, some of its properties can be related only to the behavior of its motor unit. Hence, the gross characteristics of muscle contraction are discussed under two headings, viz., the single fiber characteristics and the motor unit characteristics.

Single Muscle Fiber

Although discussed as the characteristics of a single muscle fiber, most of the experiments discussed in the following passages can also be performed with the whole muscle. How can we be sure that the experimentally observed properties of a whole muscle are also valid for the muscle fiber? For example, we know that when a train of stimuli are applied, the muscle generates more tension than it does in response to a single stimulus. There could be two reasons for this increase in muscle tension. One possibility is that multiple stimuli activate more muscle fibers. The other possibility is that multiple stimuli increase the contractility of each muscle fiber. In practice, we rule out the first possibility by employing a maximal stimulus for stimulating the muscle. When a maximal stimulus is applied to a muscle, all the muscle fibers contract. Hence, the increase in muscle tension cannot be attributed to stimulation of a grater number of fibers. Rather, it must be the response of individual muscle fibers. Most experiments on whole muscle employ maximal stimulus for stimulating the muscle.

The all-or-none law   A single muscle fiber obeys the all-or-none law: it either contracts maximally or not at all, depending on whether the stimulus is of threshold intensity or subthreshold. Although muscle fibers obey the all-or-none law, the whole muscle does not. As the stimulus strength increases, more and more fibers in the muscle are stimulated, and the muscle shows a graded response to the stimuli.

The actin-myosin filaments themselves do not obey the all-or-none law. This can be shown in skinned fibers in which the sarcolemma of the muscle fiber is removed. The filaments show graded contraction proportional to the Ca^{2+} concentration of the bathing fluid. Why then does the whole fiber obey the all-or-none law? It is because the sarcolemma depolarizes in an all-or-none fashion. If it were possible to produce graded depolarization of the sarcolemma, the muscle fiber too would show graded contractions.

Single muscle twitch

The single muscle twitch (commonly called the simple muscle twitch) may be recorded in the isotonic or isometric condition. In isotonic recording, the muscle is made to contract against minimal load and the shortening is recorded using a system of levers. In isometric recording, the contracting muscle is not allowed to shorten and the tension (force) developed is recorded using a tension transducer.

Recording a perfectly isotonic contraction requires that the load on the muscle is zero. Recording of a perfectly isometric contraction requires that the load is completely immovable. Both are technically impossible. For recording isotonic muscle shortening, the muscle has to move a writing lever. The recording lever itself has a finite weight, which acts as a load on the muscle. Hence, it is not possible to record a perfectly isotonic muscle contraction.

On the other hand, for recording isometric increase in muscle tension, the muscle is made to produce a very small displacement of a very heavy load. The displacement, which is measured after magnification, gives an indirect measure of the degree of muscle tension. If no displacement is produced in the recording device, then the muscle tension cannot be assessed. Hence, no recording can be perfectly isometric.

As a practical approximation, isotonic recording is made using a very light isotonic lever that is moved easily by the muscle. Isometric recording is made using a tension transducer, which serves two purposes. (1) It acts like a heavy load, allowing only minimal muscle shortening in response to large contractile forces. (2) It produces a measurable current that is proportional to the muscle shortening. The current produced is therefore proportional to the rise in muscle tension.

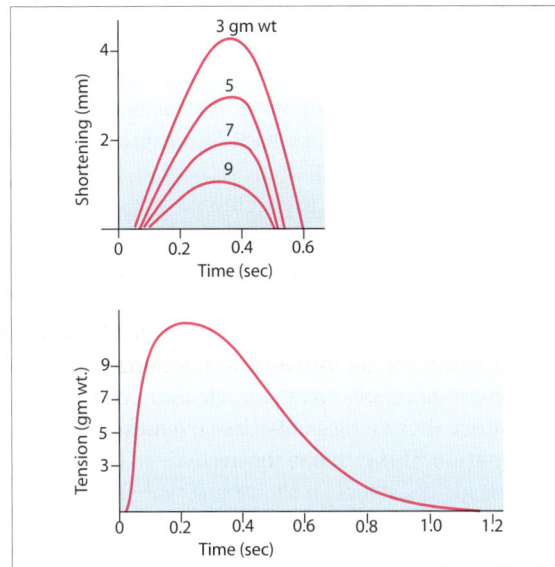

Fig. 16.**1**   (*Above*) Single muscle twitches recorded isotonically from frog sartorius muscle. (*Below*) Single muscle twitches recorded isometrically from frog sartorius muscle.

Phases of a muscle twitch   A typical isotonic twitch recorded from a frog's gastrocnemius-sciatic preparation takes about 0.1s and shows three phases (Fig. 16.**1**). (1) The *latent phase* is the time taken by the impulse to travel along the nerve to the neuromuscular junction, neuromuscular transmission, excitation-contraction coupling, the initial isometric phase of contraction and the inertia of the recording lever. The latent period is longer when the distance between the point of nerve stimulation and the neuromuscular junction is more, or when a heavier load prolongs the initial isotonic contraction phase. (2) *Isotonic contraction* is the phase in which the muscle shortens by up to 20% of its resting length. The greater the load, the shorter is the isotonic contraction phase, and the less is the shortening. (3) *Isotonic relaxation* is the phase in which the muscle gets stretched back to its original length by the dead weight suspended from the recording lever. The relaxation period is prolonged when the dead weight is lighter. If the load on the muscle is zero, the muscle will not regain its original length: a muscle that has contracted isotonically remains shortened till it is restored to its original length by stretching.

Isotonic versus isometric twitch   There are salient differences between the time courses of isotonic and isometric twitches. Isotonic contraction has a longer latent period. This is because no contraction is entirely isotonic: all isotonic contractions begin with a brief phase of isometric contraction, which accounts for a part of the latent period of isotonic contractions. For the same reason, the total duration of isotonic shortening is less than the duration of rise in isometric tension. The peak isotonic shortening occurs after peak isometric tension.

Slow and fast twitch muscles   Depending on the duration of a single twitch, muscle fibers are categorized into slow twitch and fast twitch muscle fibers (Fig. 16.**2**A). Slow twitch fibers rely on oxidative metabolism and are red due to their myoglobin content. Fast twitch fibers are pale in color because they lack myoglobin. Most muscles contain a varying admixture of both types of fibers and are called pale muscles, examples of which are the gastrocnemius muscle, extraocular muscles, and the dorsal interossei. They contract quickly but are easily fatigable. Muscles that contain only red fibers are called red muscles, examples of which are the soleus and the lumbricals. Their contraction is slow and sustained, and are not easily fatigued.

Summation of twitches and muscle tetanus   When two stimuli are so timed that the second twitch starts before the completion of the first, the second twitch records a larger shortening (or tension) than the first. This is called *summation of twitches*. As the interval between the two stimuli is decreased, individual peaks of the two twitches become less discernable till only a single large twitch is observed. When multiple stimuli are delivered in quick succession to produce summation of twitches, the muscle gets *tetanized*, i.e., it remains contracted and does not relax. If the stimuli are spaced sufficiently close, the individual peaks fuse to produce a complete tetanus and the contraction reaches a near-perfect plateau. If the peaks of the individual twitches are discernable, it indicates the presence of brief relaxation between peaks, and the tetanus is said to be incomplete (Fig. 16.**2**B).

Tetanic tension is about four times the twitch tension. There are two theories for this higher tension generated during muscle tetanus: One theory assumes that during a single twitch, the amount of Ca^{2+} released into the sarcoplasm is not enough to produce tetanic tension. When the muscle is stimulated in rapid succession, Ca^{2+} comes out into the sarcoplasm with each stimulus and there is a progressive accumulation of Ca^{2+} in the sarcoplasm. Tetanic tension is reached when sarcoplasmic Ca^{2+} levels reach their maximum. Studies with the Ca^{2+}-sensitive photoprotein aequorin show that sarcoplasmic Ca^{2+} does increase on stimulation in rapid succession. On the other hand, there are experiments to suggest that even during a single twitch, enough Ca^{2+} is released into the sarcoplasm to cause complete shortening of its sarcomeres. However, the Ca^{2+} starts moving back into sarcoplasmic reticulum

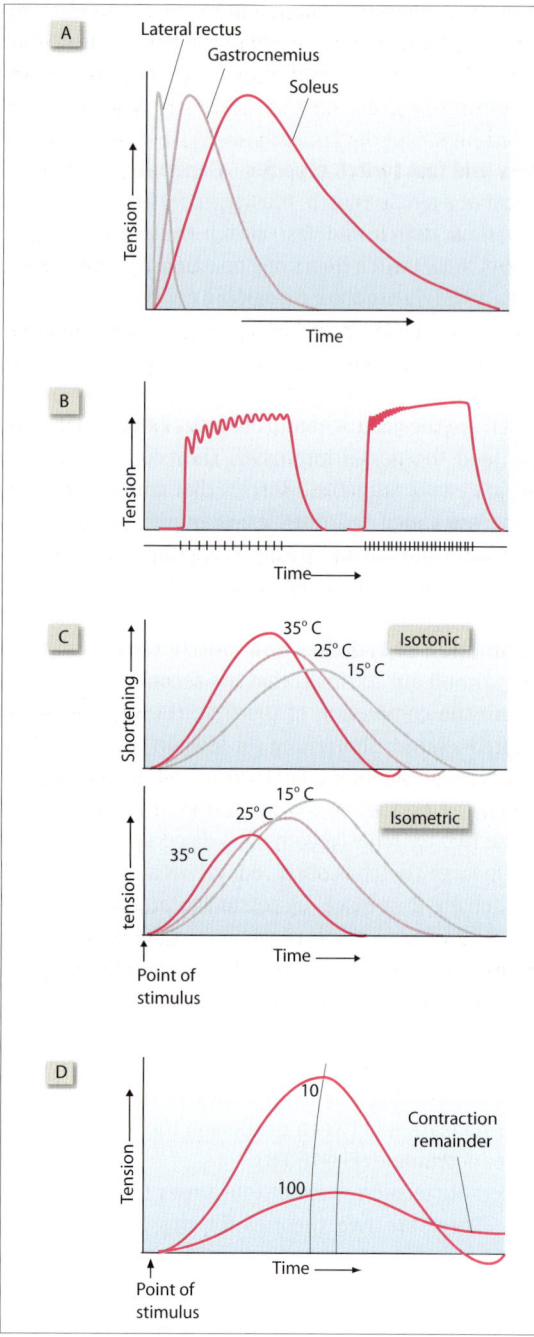

Fig. 16.**2** [A] Isometric twitches of inferior rectus, gastrocnemius, and soleus. [B] Subtetanic and tetanic contractions. [C] Effect of temperature on isotonic contractions and isometric contractions. [D] Muscle fatigue. The figure shows the tenth twitch and the 100th twitch obtained when the muscle is repeatedly stimulated in rapid succession.

Effect of temperature on muscle contraction   A rise in temperature reduces the isometric twitch tension of muscles (Fig. 16.**2**C). A rise in temperature (within physiological limits) promotes the activity of Ca^{2+}-ATPase which pumps Ca^{2+} back faster into the sarcoplasmic reticulum and hastens relaxation. Contraction is quicker too, due to faster diffusion of Ca^{2+} from reticulum to sarcoplasm. Since the duration of the Ca^{2+} pulse decreases with rise in temperature, less time is available for the rise of twitch tension. Hence, the strength of an isometric muscle twitch falls at higher temperature. For the same reason, temperature does not have much effect on tetanic tension.

Isotonic shortening of muscles increases with rise in temperature. This is due to the decrease in the internal viscoelastic resistance to shortening which more than compensates for the reduction in muscle tension.

Muscle fatigue is associated with a fall in muscle tension and an increase in relaxation time (Fig. 16.**2**D). The fatigue occurs at least partly due to ATP-depletion in the muscle itself and is most prominently revealed by its incomplete relaxation. In a large motor unit, fatigue can also occur at the neuromuscular junction.

However, the depletion of acetylcholine occurs only at prolonged, high frequency (>50 Hz) stimulation. In real life, 'psychological fatigue' occurs in the central synapses of the brain earlier than in the muscle itself. This fatigue can however be overcome by adequate encouragement and motivation.

Length-tension relationship   The length of a muscle, when it is detached from its bony attachments, is called its *equilibrium length*. If the muscle is stimulated after stretching it passively, the contractile force developed by it will vary depending on the amount of passive stretch. In other words, the contractile force developed by a muscle depends on its initial length. The force of muscle contraction recorded at various initial lengths of the muscle is shown in Figure 16.**3**A. The force recorded is the total force, i.e., it is the sum of the active contractile force and the passive recoil of the muscle in response to stretching. If the muscle is stretched but not stimulated, the force recorded will give the passive recoil force. The active contractile force generated by the muscle can be obtained by subtracting the passive recoil force from the total tension. It is seen that as the muscle is stretched from its equilibrium length, the active tension first increases to a maximum and then decreases. The length at which the muscle generates the maximum contractile force is called its *resting length*. The resting length of a sarcomere is 2.0 to 2.25 μ. The rise and fall of active tension occurs due to the varying degree of cross-bridge overlap (Fig. 16.**3**B).

well before the muscle tension is able to rise to tetanic levels. During tetanus, Ca^{2+} is continuously present in the sarcoplasm and therefore the muscle gets adequate time to reach tetanic tension.

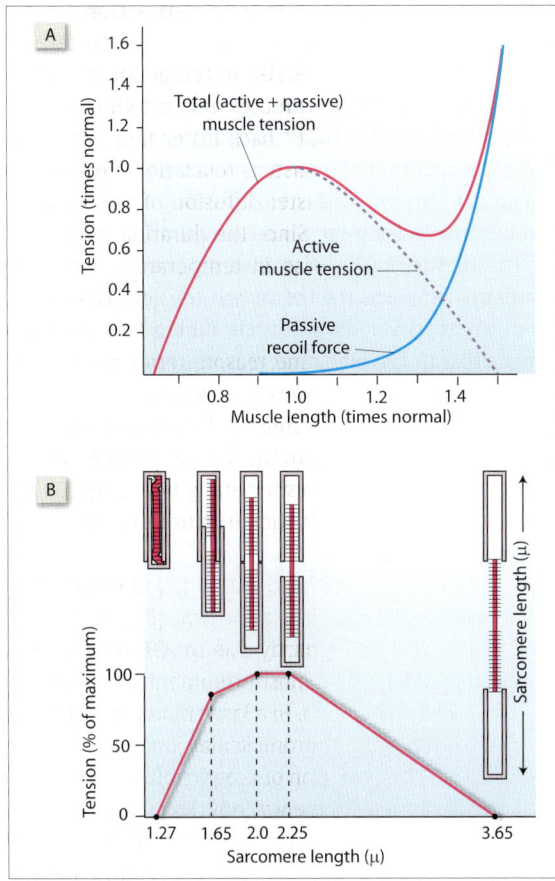

Fig. 16.3 Length-tension relationship in a skeletal muscle fiber. [A] Active, passive, and total tension in a muscle which is stimulated at different preloaded conditions. [B] Cross-bridge overlap at different sarcomere length.

Muscle contractility Since the strength of muscle contraction varies with muscle length, it is important that studies on muscle contractility are made on isometric contraction. An increase in muscle contractility is associated with a rise in peak tension, an increase in the rate of rise and fall in tension, and a shortening of the duration of contraction. Thus the rise in peak tension that occurs with a fall in temperature (Fig. 16.2C) does not signify a rise in contractility because the contraction and relaxation periods are prolonged. Muscle contractility is an important consideration in cardiac muscle physiology.

Preload versus afterload *Preload* is the load placed on a muscle before the muscle contracts. It serves to stretch the muscle sarcomeres, thus producing a passive tension in the muscle. This passive tension increases muscle contraction in two ways: (1) It adds an elastic recoil force to the muscle during its contraction. (2) It stretches the muscle to its resting length, producing the optimum length-tension relationship

(described below) for active force generation. In real life situations, it is a common practice to prestretch a muscle using antagonist muscles. Through experience, we learn how much force our muscles generate at different lengths and unconsciously, we adjust the muscle length before initiating a movement to develop the power we want.

Afterload is the load that the muscle encounters after it starts shortening. The contraction of an afterloaded muscle is comparatively less. In real life situations, after-loaded contractions tend to occur in novel situations when there are unexpected perturbations in load.

A muscle can also be partly preloaded and partly afterloaded (Fig. 16.4). The concept of preload and afterload is also applicable to the heart which is made of striated muscle. A heart can withstand a greater preload than afterload without going into cardiac failure.

▄▄ Motor Unit

A solitary motor neuron with all its peripheral branches and innervated extrafusal muscle fibers is called a *motor unit*. The term motor unit is sometimes extended to multi-unit smooth muscles too. The motor unit is the functional unit of muscle contraction *in vivo*.

The motor unit obeys the *all-or-none law*, i.e., when the nerve fiber of the motor unit is stimulated, all the fibers of the motor unit will either contract maximally or not contract at all, depending upon whether the stimulus is of threshold or subthreshold intensity.

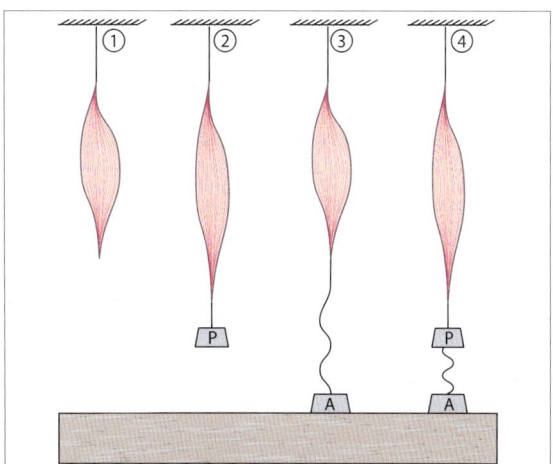

Fig. 16.4 Different types of muscle loading. (1) Unloaded muscle (ignoring the weight of the muscle itself). (2) Preloaded muscle. (3) Afterloaded muscle. (4) preloaded and afterloaded muscle.

Innervation ratio The number of muscle fibers innervated by a single neuron is called the innervation ratio of the motor unit. The ratio is low in muscles concerned with precision movement and high in those requiring power rather than precision. For example, the extraocular muscles have the lowest innervation ratio of less than 6 per axon, while the gastrocnemius muscle has an innervation ratio of up to 2000 per axon.

Motor unit territory In the transverse section of a muscle, the area occupied by a single motor unit is called the motor unit territory. For example, the territory of a motor unit in biceps ranges from 2–15 mm in diameter. There are extensive overlaps of the motor unit territories of different motor units, and muscle fibers belonging to different motor units intermingle singly (Fig. 16.5A). It is uncommon to find two or more motor fibers of the same motor unit lying adjacent to each other. The extensive intermingling of motor fibers from different motor units results in smoothing of the muscle contraction.

Endplate zone The motor endplates of all the muscle fibers in a motor unit are aligned in a narrow band called the *endplate zone* or *innervation zone*. It is located midway between the ends of the muscle fibers (Fig. 16.5B). However, the endplate zones of different motor units are not always aligned.

Motor recruitment

When a muscle begins to contract, only a few units contract first. If the power generated is inadequate, more units are 'recruited'. This process of employing progressively greater number of motor units for muscle contraction is called motor recruitment.

Henneman (size) principle When a muscle contracts against a load, the smallest motor units (which have only a few muscle fibers in them) are recruited first. If the force generated is insufficient, then the larger motor units are recruited. This order of recruitment from the smaller to the larger motor units is called the

Henneman principle or the *size principle*. In fact, the largest motor units in a muscle are idle most of the time since such large muscle forces are rarely required (Fig. 16.6). On stimulation, the larger motor units are found to be more excitable than the smaller ones. Hence, the orderly recruitment from smaller to the larger units occurs due to the way the motor system is organized and not because of their intrinsic excitability.

The Henneman principle brings out an interesting similarity between the motor and the sensory system. Since progressively larger motor units are recruited, it means that as the load increases, the number of motor

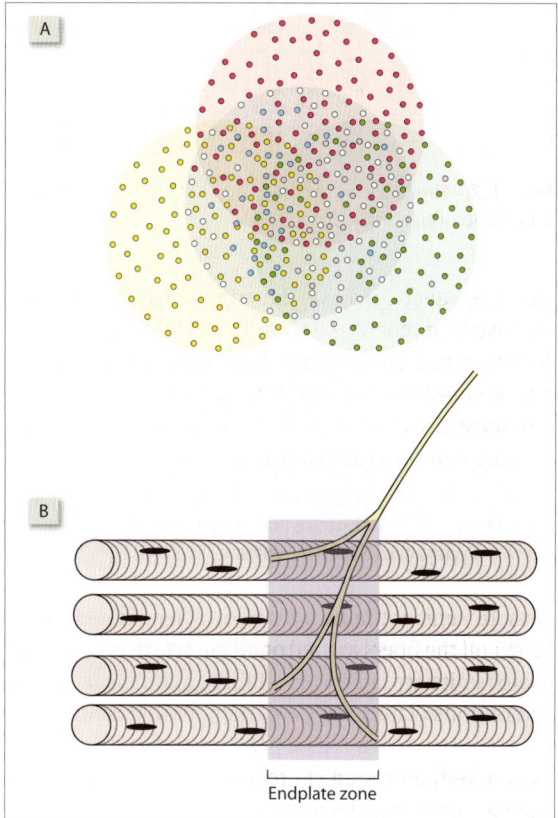

Fig. 16.**5** [A] Motor unit territories of four motor units (shown in different colors) seen in cross section. Fibers intermingle singly, as shown in the central overlapping area. [B] Endplate zone of a muscle.

Fig. 16.**6** The size principle!

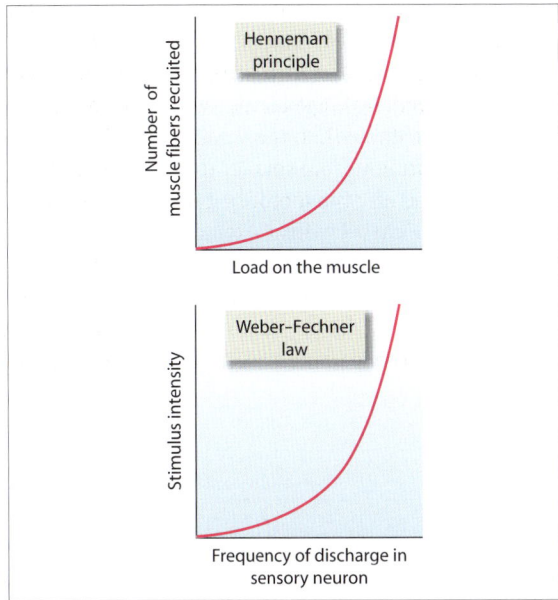

Fig. 16.**7** Similarity between Henneman (motor) principle and Weber–Fechner (sensory) law.

fibers recruited increases exponentially. This exponential rise in motor strength in response to a linear rise in motor neuron discharge is reminiscent of the Weber–Fechner law, which states that the stimulus intensity must rise exponentially to produce a linear rise in sensory neuronal discharge (Fig. 16.**7**).

■ Type-I and Type-II motor units

A motor unit contains only one type of muscle fiber: either all the fibers are red or all are white. Accordingly, motor units are called type-I (red) and type-II (white) motor units (Table 16.**1**).

Type-I (red) motor units are adapted for slow and sustained (tonic) contractions without fatigue. They are smaller and are the first to be recruited whenever a muscle contracts. These units are therefore almost continuously (i.e., tonically) active during routine activity. This is also the reason why all muscles must have at least a few type-I motor units in them. Since type-I units contract continuously for prolonged periods without fatigue, they have to be dependent on aerobic metabolism, and are endowed with a high capillary density for ensuring adequate oxygen supply. However, the capillaries get squeezed between contracting muscle fibers and blood flows through the capillaries only during intermittent muscle relaxation. Hence, red fibers contain myoglobin for storing oxygen and ensuring adequate oxygen supply during tonic contractions. Myoglobin

gets rapidly oxygenated during the brief phases of relaxation that occur in between contractions. Type-I units are present mostly in postural muscles which do not require speedy contractions. In keeping with the requirements, axons of these units have smaller diameter and slow conduction velocity.

Type-II (white) motor units are adapted for brief (phasic) bursts of powerful contractions. They are inactive most of the time and are recruited much later when the Type-I motor units fail to move the load. They contract only in short bursts (i.e., phasically) when a brief but powerful contraction is needed. In keeping with their function, axons of Type II units have larger diameter with greater conduction velocity.

Table 16.**1** Comparison between Type-I and Type-II motor units.

Motor unit	Type I	Type IIb
Names	Slow oxidative	Fast glycolytic
	Red fibers	White fibers
	Tonic fibers	Phasic fibers
	S (slow)	F (fast)
Metabolism	Aerobic	Anaerobic
Glycolytic capacity	Low	High
Oxidative capacity	High	Low
Fatigability	Little or none	Rapid
Fiber length	Small	Large
Fiber diameter	Small	Large
Glycogen content	Low	High
Mitochondria	High	Low
Sarcoplasmic reticulum	Normal	Extensive
Ca^{2+} pumping into SR	Moderate	High
Capillary density	High	Low
Blood supply	High	Normal
Myoglobin content	High	Low
Myosin ATPase activity	Low	High
Phosphorylase	Low	High
Succinic dehydrogenase	High	Low
NADH dehydrogenase	High	Low
Number of units in a muscle	Many	Few
Number of terminals per axon	Few	Many
Axon diameter	Small	Large
Conduction velocity	Slow	Fast
Order of recruitment	Early	Later
Twitch duration	Long	Brief
Tetanic tension	Small	Large

For contracting and relaxing quickly, white muscle fibers have more extensive sarcoplasmic reticulum with a higher capacity for pumping Ca^{2+}. They also have a faster isoenzyme of myosin ATPase. Since these muscles are fast muscles, they have little time for taking up O_2 and glucose from blood. Hence white fibers depend on anaerobic metabolism and store adequate glycogen in them. However, the lactic acidosis resulting from anaerobic metabolism makes the white fibers fatigue-prone.

17 Muscle Elasticity

Physical Models of Muscle

The contractile properties of the muscle explain how the muscle shortens actively in response to a stimulus. However, it does not explain: (1) how the muscle regains its original length after it is stretched passively, and (2) how the muscle is able to contract even when its external length does not change. These characteristics of muscle are attributable to its elasticity. To explain how the elastic and contractile components are arranged in the muscle, the two-compartment model and later, the three-compartment model of muscle were proposed. Later models have incorporated several more compartments in order to explain all the observed characteristics of gross muscle contraction. However, for most purposes, the three-compartment model suffices.

The two-compartment model

The two-compartment model assumes that the muscle has a contractile component (CC) and an elastic component (EC). The CC represents the actin and myosin filaments. It is considered to be plastic, i.e., when stretched, it produces no elastic recoil. The CC is therefore unable to return to its original length after it has been stretched. The EC represents the elastic element which resists stretch and restores a contracted muscle to its resting length.

In the two-compartment model (Fig. 17.1A), there is only one elastic component which may be disposed either in series or in parallel with the contractile component. In either case, it is unable to explain all the observed phenomena related to muscle elasticity. If an elastic element is inserted in series with the contractile component, the CC would continue to elongate when the muscle is stretched and would ultimately snap, while the *series elastic component* (SEC) would remain unstretched. However, the SEC does explain how the muscle is able to contract without a change in length and why the muscle reverts to its resting tension at the end of isometric contraction. This is because during isometric contraction, the SEC gets stretched and during relaxation, it recoils and stretches out the CC.

To resist muscle stretching, an elastic component must be present in parallel with the CC. However, a *parallel elastic component* (PEC) cannot explain how a muscle regains its original length after isometric contraction. A parallel elastic element will simply fold up during active shortening of the CC and therefore, will be unable to provide the necessary elastic recoil for restoration of the original muscle length. Nonetheless, it does explain why the muscle resists passive stretching.

The three-compartment model

The inadequacy of the two-compartment model led to the proposition of the three-compartment model of muscle elasticity (Fig. 17.1B,C) in which the presence of two elastic components of muscle elasticity was posited, one in series with the CC of the muscle (SEC) and the other in parallel (PEC). SEC explains how the muscle is able to contract even when its external length does not change. It also explains how it regains its original length after contracting isometrically. PEC explains why the muscle regains its original length after it is passively stretched. The SEC resides in the elastic neck of myosin filaments and, in the case of a whole muscle, in the tendon. The PEC resides in the sarcolemma and the gap filaments.

Effect of Load

Phases of muscle contraction

A muscle that contracts tetanically against a load shows three phases which differ in the extent of shortening and the amount of tension generated (Fig. 17.2A).

Initial isometric contraction phase All muscle contractions begin with this phase in which the shortening of the CC merely stretches the SEC. The load does not move and there is no change in the external length of the muscle (hence the name isometric). Stretching of the SEC is associated with a rise in muscle tension. As the shortening of the CC continues, the SEC gets stretched more and more. Stretching of the SEC results

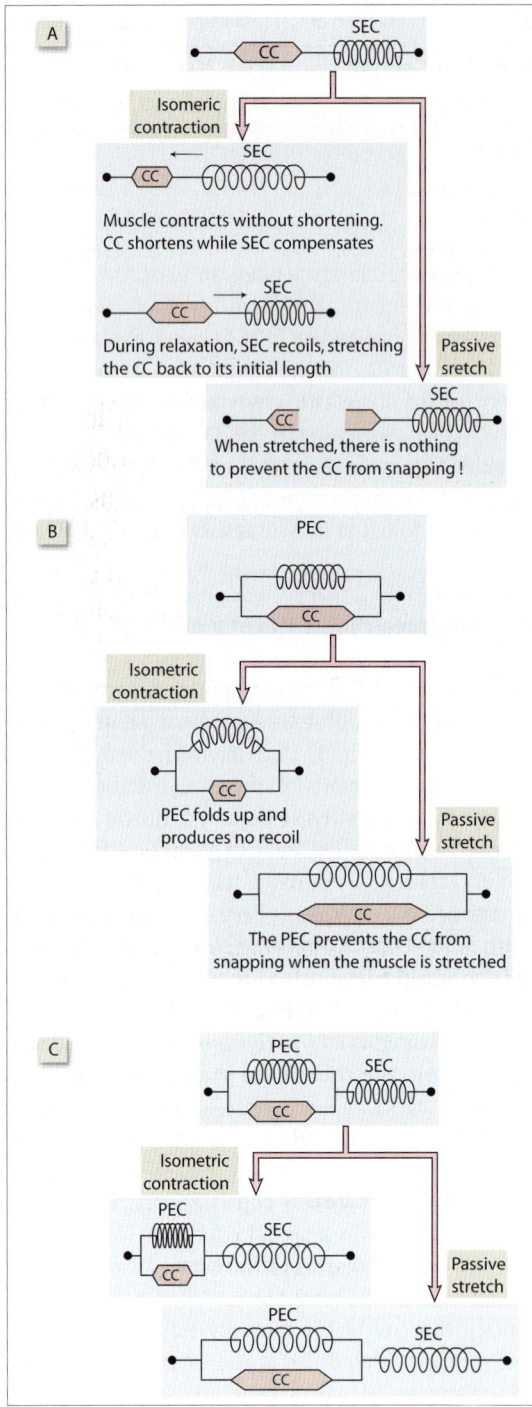

Fig. 17.**1** [A] Two-compartment model with the elastic component in series (SEC). The model explains the mechanism of isometric contraction but fails to explain the effect of passive stretch. [B] Two-compartment model with the elastic component in parallel (PEC). The model explains the mechanism of recoil of muscles following passive stretch but fails to explain the mechanism of isometric contraction. [C] The three-compartmental model of the muscle showing the CC, PEC, and SEC. When the muscle contracts without external shortening, the shortening of CC is compensated by the stretching of SEC. When the muscle passively stretched, all the components are stretched. The recoil of the PEC restores the CC to its original length.

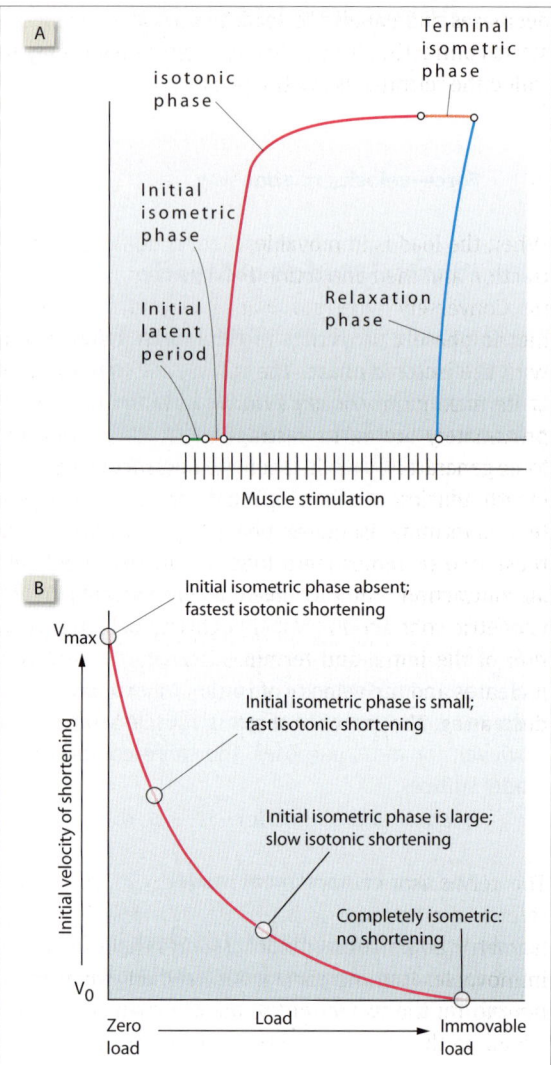

Fig. 17.**2** [A] The three phases of a tetanic contraction. [B] Force–velocity relationship in a skeletal muscle.

in rise of muscle tension till the tension equals the load. This point marks the end of the initial isometric phase.

Intermediate phase of isotonic contraction This phase commences when the muscle tension exceeds the load and the load starts moving. There is no further stretching of the SEC and no further increase in muscle tension (tone) during this phase.

Terminal isometric contraction phase As the muscle shortens, the force generated by its CC changes in accordance with the length–tension relationship (see Fig. 16.**3**). Once the muscle becomes shorter than the resting length, any further shortening is associated with a decrease in tension. When the tension generated decreases and equals the load, the muscle once again starts contracting isometrically, entering what may be called the terminal isometric phase.

Force–velocity relationship

When the load is immovable, there is no isotonic contraction and the contraction becomes entirely isometric. Conversely, when the load is zero, the initial isometric phase disappears, and the contraction begins with the isotonic phase. The muscle initially shortens at its maximum velocity (Vmax) and shortens to approximately 60% of its resting length. Thereafter, the force generated by its CC is zero (in accordance with the length–tension relationship) and the contraction enters its terminal isometric phase (Fig. 17.**2**B). Between these two extremes (zero load and immovable load), all contractions have variable durations of isotonic and isometric contraction. With increasing load, the duration of the initial and terminal isometric contraction increases and the velocity of isotonic shortening keeps decreasing. The extent of shortening also decreases.

Isotonic and isometric contractions

The terms isotonic and isotonic contraction usually refer to the extreme ends of the force–velocity graph. Isometric contractions occur while trying to move the immovable. Isotonic contraction is rarely seen in its pure form: the weight of the muscle itself constitutes a load on the muscle. The differences are summarized in Table 17.**1**.

Energy expenditure in isometric contraction In isometric contraction, there is no external shortening of the muscle and therefore, the external work done (force × distance moved) is zero. Yet, sustained isometric contraction is associated with fatigue, indicating that the contraction requires continuous

Table 17.**1**   Comparison of isotonic and isometric contraction.

Isometric contraction	Isotonic contraction
Shortening of the CC is compensated by stretching of the SEC.	Shortening of CC results in shortening of muscle. SEC not stretched.
Tension rises due to stretching of SEC.	Since SEC not stretched, tension remains unchanged.
No shortening occurs and therefore no external work is done.	Shortening occurs and external work is done.
Occurs at the beginning and end of all contractions.	Occurs in the middle of a contraction.
Isometric phase increases when load increases.	Isotonic phase decreases when load increases.
Heat released is less and therefore more energy-efficient.	Heat released is more (due to the release of heat of shortening) and therefore, is less energy-efficient.
An isometric twitch has a shorter latent period, shorter contraction period, and longer relaxation period.	An isotonic twitch has a longer latent period, longer contraction period, and shorter relaxation period.
A rise in temperature decreases isometric twitch tension.	A rise in temperature increases isotonic twitch shortening.

energy expenditure. The energy is utilized for doing 'internal' work, i.e., for stretching the SEC. It may be argued that the SEC needs to be stretched but once, and thereafter no further energy expenditure should be required. That is not correct. The SEC continuously tends to recoil back each time the myosin head dissociates from actin filament, and it requires continuous energy-intensive cross-bridge cycling to keep the SEC stretched. The situation is analogous to a car kept stationary on an incline with its engine running, by adjusting the clutch and the accelerator pedals. The car continuously burns fuel and yet does no external work. The car can also be kept stationary by switching off the engine and applying brakes. The situation then becomes analogous to the latch phenomenon of smooth muscles discussed on p 126.

The action potential shown in Figure 11.2 is recorded intracellularly (see also Fig. 11.1). In clinical practice, action potentials are recorded extracellularly. The extracellularly recorded action potentials can be monophasic, biphasic or triphasic (Fig. 18.1).

An action potential recorded by a pair of recording electrodes placed on an axon or a muscle fiber is usually *biphasic potential*. If a part of the axon is crushed so that it remains permanently depolarized, and the reference electrode is kept on the crushed area, the active electrode kept some distance away will record a *monophasic action potential*. If the recording electrodes are not in direct contact with the tissue, they only record the electrical current set up in the surrounding conductive medium as the action potential travels along the tissue. The electrode records a positive potential when the current flows towards it, a negative potential when the current flows away from it, and zero potential when the current flows perpendicular to it. The result is a *triphasic action potential*.

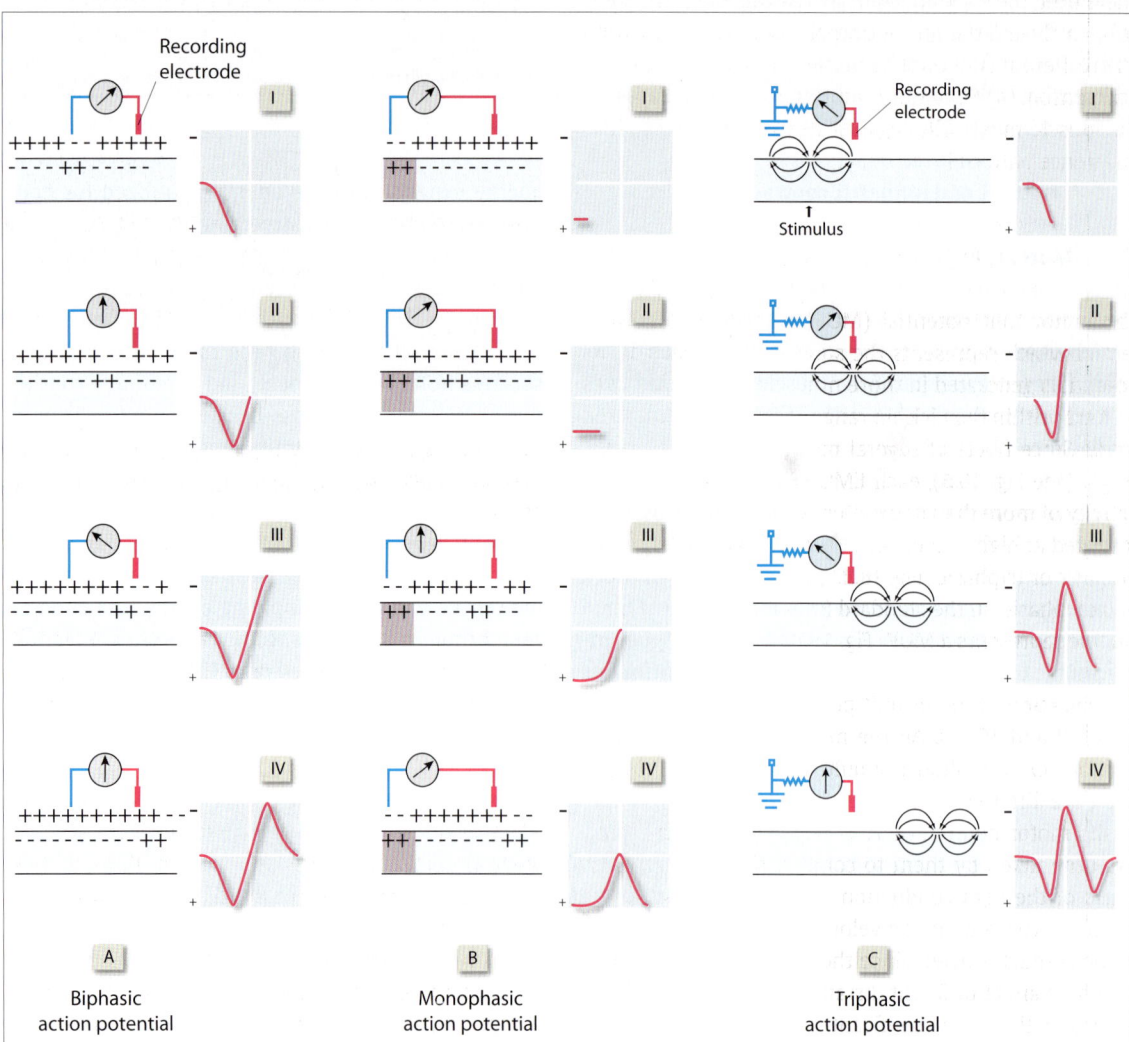

Fig. 18.**1** EMG recording. [A] Monophasic action potential, [B] Biphasic action potential, and [C] Triphasic action potential

Electromyography

Electromyography is the recording of the electrical activity of the muscle using surface electrodes placed on the muscle or more commonly, needle electrodes inserted into the muscle. The needle electrode consists of a pointed steel cannula through which runs a fine silver, steel, or platinum wire that is insulated except at its tip. The potential difference between the outer cannula and inner wire is recorded and the patient is grounded by a separate surface electrode. The ground lead is attached to the same limb as the muscle to be examined.

The standard protocol for EMG recording is as follows. (1) The electrode is inserted into the muscle while it is relaxed and the *insertion activity* is noted. (2) The muscle is then explored systematically with the electrode for the presence of any spontaneous activity. Normally, there is no spontaneous activity. (3) Thereafter, the subject is asked to contract the muscles voluntarily and the motor unit potentials are recorded from different sites on the muscle at different grades of contraction. (4) Finally, the subject is asked to contract the muscle maximally (against resistance) and the interference pattern is recorded.

Motor unit potential

The motor unit potential (MUP) is a compound potential which represents the sum of individual action potentials generated in those muscle fibers of the unit that are within the pick-up range of the recording electrode. Since fibers of several motor units intermingle singly (see Fig. 16.**5**), each EMG electrode records the activity of more than one motor unit. When the EMG is recorded at high speed, each MUP is usually seen to be biphasic or triphasic (Fig. 18.**2**A). Sometimes, they may be polyphasic. In the standard EMG record, each vertical line represents a MUP (Fig. 18.**3**).

The **duration** of motor unit potentials is normally between 2 and 15 ms. All the muscle fibers of a motor unit are excited almost simultaneously. Even though there are differences in the length of different branches of the motor neuron, there is negligible difference in the time taken by them to conduct the impulses because of the high conduction velocity of Aα neurons. However, the conduction velocity of the muscle membrane is much slower. Since the endplates of the muscle fibers are at different distances from the recording electrode, there are significant differences in the time taken by the action potentials to travel from these endplates to the recording electrode (Fig. 18.**2**B). Hence,

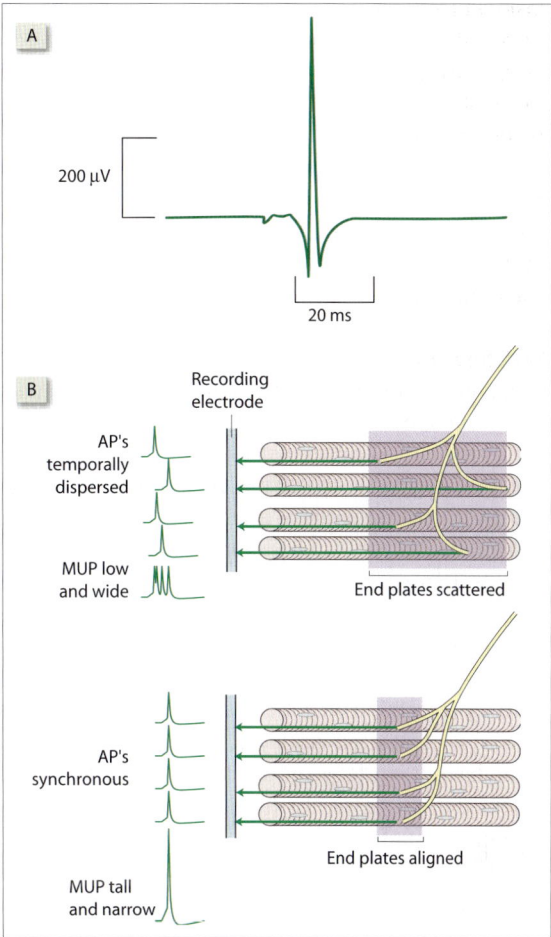

Fig. 18.**2**   [A] A solitary motor unit potential. The one shown above is triphasic. [B] The anatomic scatter of the motor endplate affects both the duration and amplitude of the motor unit potential.

the duration of the motor unit potentials is related to the anatomic scatter of endplates of those muscle fibers in the units that are within the pick-up zone of the recording electrode.

The **amplitude** of motor unit potentials is usually between 0.3 and 3 mV. It is high when there are actively contracting muscle fibers very close to the recording electrode. Conversely, the amplitude is low if the actively contracting fibers are located some distance away from the electrode. The amplitude of the MUP is also affected by the number and size of active fibers lying close to the electrode and the anatomic scatter of the endplates: the MUP is taller when the endplate scatter is less (Fig. 18.**2**B).

▪ Normal electromyogram

Insertion activity   Electrical activity usually cannot be recorded outside the endplate region of healthy muscle at rest, except immediately after insertion or movement of the needle recording electrode. The activity related to the electrode movement, called insertion activity, is due to mechanical stimulation or injury of the muscle fibers and usually stops within 3 seconds of the movement (Fig. 18.**3**A). Insertion activity is prolonged in denervated muscle and is absent when muscle tissue is not viable.

Endplate noise   After the insertion activity ceases, spontaneous activity may be found only in the endplate region, and nowhere else. Such activity is called endplate noise. Endplate noise consists of monophasic negative potentials that have an irregular, high frequency discharge pattern, a duration of between 0.5 and 2.0 ms, and an amplitude that is usually less than 0.1 mV. These potentials correspond to the miniature endplate potentials (MEPP) of the neuromuscular junctions (Fig. 18.**3**B).

Activity associated with muscle contraction   When a muscle is contracted weakly, a few small motor units begin firing irregularly and at a low rate. As the force of contraction increases, the firing rate of the small motor units increase until it reaches a certain frequency. Thereafter, larger units are recruited. The motor unit potentials of the larger units can be recognized by their greater amplitude. Eventually, so many units are recruited that the baseline is interrupted continuously by MUPs and therefore the baseline cannot be seen clearly. The resulting appearance of the EMG is called the *interference pattern* (Fig. 18.**3**C). The features to be noted in analyzing motor recruitment patterns are: (1) the frequency at which the motor unit potentials first appear (onset-frequency), (2) the frequency at which a particular unit must fire before another is recruited (recruitment-frequency), and (3) the density of the interference pattern.

▪ Abnormal EMG

▪ Spontaneous EMG during muscle relaxation

Fasciculation potentials   Fasciculation is the involuntary contraction of single motor units. It occurs in patients with chronic partial denervation, especially when it is due to spinal cord lesion. Fasciculation produces jerky, visible twitchings of a group of muscle fibers. The motor unit potentials that occur

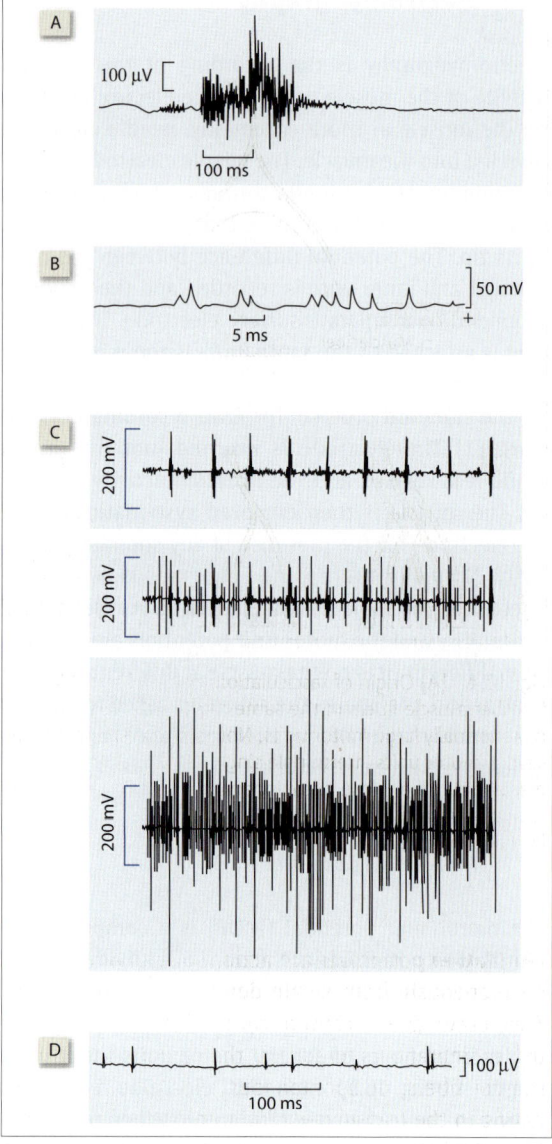

Fig. 18.**3**   [A] Insertion activity. [B] Endplate noise. [C] Recruitment patterns in slight (above), moderate (middle), and maximal (below) contraction. [D] Fibrillation potentials.

during fasciculation are called fasciculation potentials. Fasciculations can occur in normal persons too, and fasciculation potentials resemble normal MUPs. The MUPs recorded from pathological fasciculations are more polyphasic and have durations greater than 15 ms. Fasciculation potentials occur due to instability of membrane potential. They originate spontaneously at multiple sites along the diseased motor neuron and its immature sprouts. Potentials originating lower down the neuron spread antidromically by axonal reflex to other nerve terminals of the motor units (Fig. 18.**4**A).

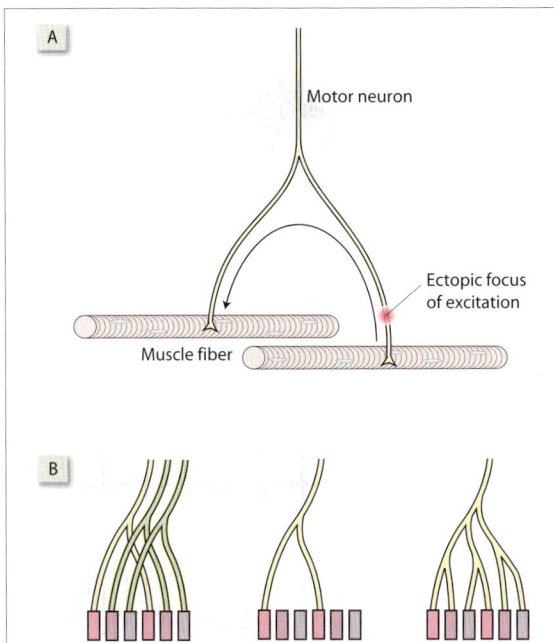

Fig. 18.**4** [A] Origin of fasciculation potential and its spread to other muscle fibers in the same motor unit. [B] Formation of abnormally large motor units. Normally, muscle fibers from several motor units intermingle singly. When denervated, several muscle fibers lose their innervation (orphaned). During reinnervation, the orphaned fibers get innervated by axonal branches of the same neuron.

Fibrillation potentials are action potentials that arise spontaneously from single denervated muscle fibers. They occur due to rhythmical oscillations of the resting membrane potential in the denervated skeletal muscle fibers, or spontaneous depolarizations originating in the transverse tubular system of the muscle fiber. Fibrillation potentials usually have an amplitude between 20 and 300 µV (much smaller than a MUP), a duration of less than 5 ms, and a firing rate of 2 to 20 s^{-1} (Fig. 18.**3**D).

Abnormal EMG during muscle contraction

Sparse interference pattern In patients with neuropathic weakness, the number of functional motor units is reduced and therefore fewer motor units are available for recruitment. Hence, there is a decrease in the density of the interference pattern during maximal voluntary contraction. In severe cases, the interference pattern is not produced at all and individual MUPs can be recognized. Since the number of muscle fibers in each motor unit is normal, the amplitude of the MUP is normal.

Low MUP amplitude In *myopathic disorders*, the number of functional motor units remains unchanged until an advanced stage of the disorder but the number of fibers per motor unit is reduced. Hence, the number of motor units available for recruitment is not reduced but the amplitude of each MUP is considerably reduced. The motor unit firing rates remain normal and the interference pattern remains full. The interference density is often higher than normal for a given degree of voluntary activity because more motor units are recruited to compensate for the reduced tension that individual units are able to generate.

High MUP amplitude is often recordable in denervated muscles that have subsequently been reinnervated, as occurs with aging (p 70). These muscles have large motor units (Fig. 18.**4**B) whose muscle fibers do not intermingle. (Normally, there is intermingling of muscle fibers from different motor units as shown in Figure 16.**5**A.) An electrode placed in such a muscle picks up the potentials from a large number of muscle fibers belonging to the same motor unit and therefore records a huge MUP.

Electroneurography

Measurement of nerve conduction velocity

Compound action potential A peripheral nerve is made of several neurons with different conduction velocities. When a nerve trunk is strongly stimulated, all the neurons in it get excited. A monophasic record obtained from the nerve surface some distance away shows multiple peaks, each peak corresponding to the action potentials of a group of neurons having similar conduction velocities. The record so obtained is called the compound action potential.

Suppose the time interval between the stimulus and the appearance of the first peak (i.e., Aα) is t_1. If the distance between the stimulating and recording electrodes is d_1, the conduction velocity (V_1) of the nerve fiber represented by that peak (i.e., Aα) is given by the formula:

$$V_1 = d_1 / t_1$$

Similarly, the conduction velocity (V_2) of Aβ will be d_2/t_2, and so on. Conduction velocities in infants are roughly half of the adult values, reaching the adult range by 5 to 8 years. They decline by 1 m s^{-1} per decade between 20 and 55 years and thereafter, at the rate of 3 m s^{-1} per decade. Nerve fibers have been classified on the basis of their conduction velocities (Table 18.**1**).

Table 18.**1**   Classification of nerve fibers based on their diameter and conduction velocity.

	Function	Fiber diameter (μ)	Conduction velocity* (ms⁻¹)	Spike duration (ms)
Aα	Somatic motor Proprioception	12–20	100	0.5
Aβ	Touch, pressure	5–12	60	0.5
Aγ	Fusimotor	3–6	40	0.5
Aδ	Pain, temperature Crude touch	2–5	20	0.5
B	Preganglionic autonomic (thinly myelinated)	<2	10	1.0
C	Postganglionic autonomic, pain, temperature (all are unmyelinated)	<1	2	2.0

* range: ± 20%

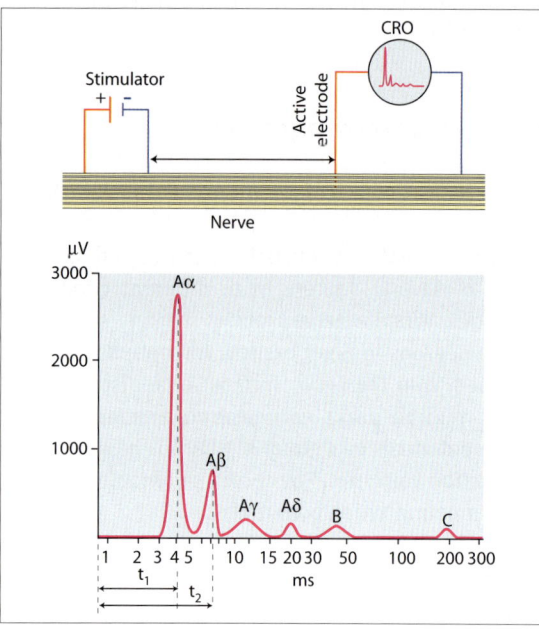

Fig. 18.**5**   (*Above*) Set up for recording a compound action potential. (*Below*) A compound action potential.

■   M response, F response, and H reflex

The simplest method of estimating nerve conduction velocity is to measure the time taken by an impulse to travel a short distance along the axon, between the point of stimulation and the point of recording. The disadvantage associated with such recording is that only a short segment of the nerve is studied. The values

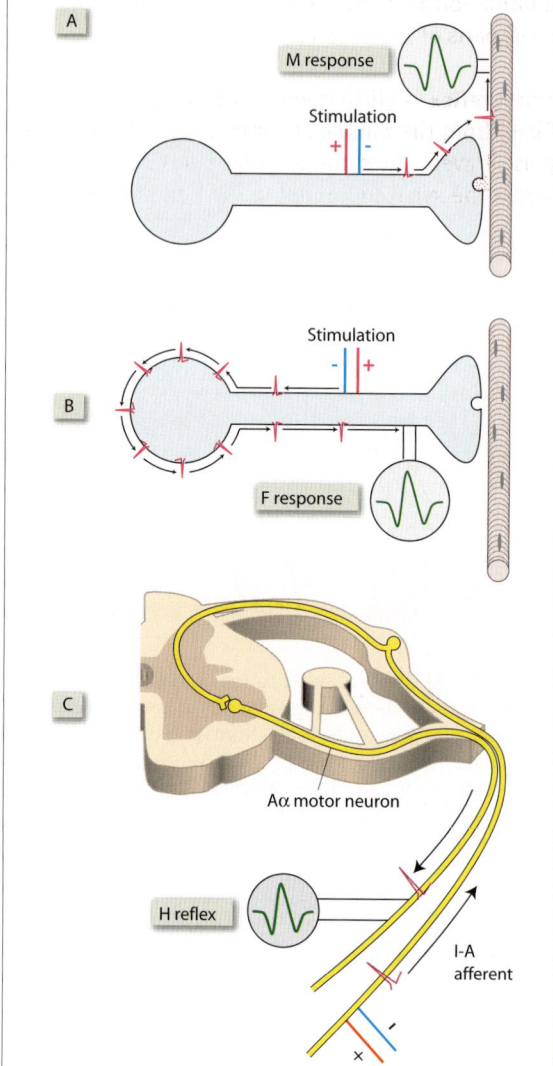

Fig. 18.**6**   The M response, F response, and H reflex.

obtained are not very precise and any abnormality in other parts of the axon goes undetected. The recording of the M and F responses and the H reflex are more informative (Fig. 18.**6**).

The **M response** or M wave is recordable from a muscle by stimulating the motor nerve. This ensures that the most peripheral part of a motor nerve is studied.

F response   When a motor nerve is stimulated, the nerve impulse travels not only orthodromically toward the muscle but also antidromically towards its cell body located in the ventral horn of the spinal cord. After depolarizing the cell body, roughly 1 out of 10 impulses return orthodromically along the same axon as the F-wave or the F-response. Thus, the conduction velocity recorded from the F-response will give information about the most central (i.e., near the spinal cord) parts of motor axons.

The **H reflex** is elicited when the proprioceptive (I-A) fibers from the muscle are stimulated. The nerve impulse travels along the I-A fibers to the spinal cord, across the synapse located in the ventral horn of the spinal cord and finally orthodromically along the motor (Aα) neuron. Thus, the conduction velocity recorded from the H response gives information about the most central parts of both motor as well as sensory axons. The height of the H wave indicates the motor neuron excitability, which depends on supraspinal facilitation and inhibition.

19 Muscle Mechanics

Muscle Tension versus Excursion

The external work done by a muscle is observable either as the change in its length (*muscle excursion*) and/or in its stiffness (*muscle tension*). The excursion of a muscle depends on the length of the muscle, i.e., the number of sarcomeres in series. The tension developed by the muscle is proportional to the physiological (not anatomical) cross-sectional area of the muscle, i.e., the number of sarcomeres disposed in parallel. The *anatomical cross-sectional area* is the cross-sectional area of the muscle at its thickest parts. The *physiological cross-sectional area* is the sum of the cross-sectional areas of all the muscle fibers at their thickest part. Based on the relative amounts of tension and excursion, muscle contraction has been classified into three types, viz., isometric, eccentric, and concentric. Eccentric contractions generate the maximum amounts of tension followed by isometric and isotonic contractions.

In **isometric contraction** there is a rise in muscle tension but there is no muscle excursion. The excursion is zero if the load is immovable, or if an antagonistic muscle contracts with equal force. Also called static or holding contractions, the functional role of isometric contractions is mostly to stabilize the joints.

In **eccentric contraction** there is simultaneous increase in muscle length and tension. This occurs when a contracting muscle is subjected to an external force greater than the active tension generated. It is also called *lengthening contraction*. Eccentric contractions decelerate body segments and provide shock absorption, e.g., when walking or landing from a jump. Elbow flexors contract eccentrically when the glass of water is lowered to the table. Another example of eccentric contraction is the quadriceps muscle when the body is being lowered to sit on a chair. The tension of an eccentric contraction increases as the speed of active lengthening increases.

In **concentric contraction** the muscle shortens while the muscle tension may increase (auxotonic), decrease (meiotonic) or remain unchanged (isotonic). *Auxotonic contraction* is associated with a continuous rise in muscle tension and is seen while pulling a spring. *Meiotonic contraction* is associated with a fall in tension. It is seen when the muscle acts on a device like the clasp knife, which offers resistance that is initially high but drops suddenly. *Isotonic contraction* is not associated with any change in muscle tension and is seen when the load remains unchanged throughout contraction. It is seen when a load is lifted by flexing the elbow. The speed of shortening in a concentric contraction is inversely proportional to the tension produced.

Internal architecture of muscle

The body requires a wide range of excursion and tension, and muscles tend to specialize in one or the other. The relative amounts of muscle tension and excursion produced by a muscle are determined by its internal architecture (Fig. 19.**1**, Table 19.**1**).

In **strap and fusiform muscles** the fibers run through the entire length of the muscle. Each fiber therefore

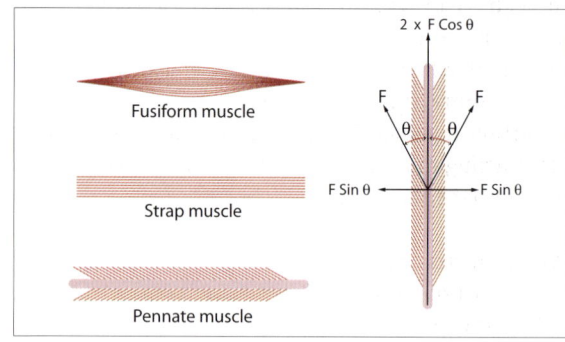

Fig. 19.**1** Internal structure of muscle. (*Left*) Three types of muscles based on the orientation of muscle fibers. (*Right*) Vector analysis showing how fiber orientation affects the force exerted by a pennate muscle.

Table 19.**1** Muscle excursion and tension of different types of muscles.

	Excursion	Tension
Strap-like	High	Low
Fusiform	Moderate	Moderate
Pennate	Low	High

must be at least as long as the muscle itself. Hence for a given volume of muscle, only a limited number of fibers can be accommodated in parallel. Fusiform muscles have larger physiological cross-sectional area than strap muscles, which are generally thin and long. Accordingly, strap muscles (e.g., sartorius, gracilis, and semitendinosus) can quickly contract through a large range of excursion but develop relatively weaker tension as compared to fusiform muscles.

In **pennate muscles** the length of the fibers is much smaller than the length of the muscle. Hence, a very large number of fibers can be accommodated in a relatively small volume of muscle. This reduces the excursion but greatly augments the tension generated. Most of the muscles in the body are of pennate structure (e.g., the vasti of the quadriceps, soleus, and gastrocnemius). In pennate muscles, both the excursion and the tension contributed by individual fibers to the whole muscle get reduced by a factor of Cos θ due to the oblique disposition of fibers. θ is the angle of insertion of fibers.

Muscle Action

Laws of muscle action   The outcome of muscle contraction in terms of the movements in the skeletal lever system is known as muscle action. Muscles rarely contract alone; rather, several muscles contribute to produce the desired force and resulting motion. Although the resultant effect is often quite complex, they abide by certain elementary principles known as the *laws of muscle action*. The *law of approximation* states that a contracting muscle tends to bring its origin and insertion together. The *law of detorsion* states that a contracting muscle tends to bring its origin and insertion into the same plane.

Components of muscle action   The muscle action at the joint can be resolved into three components (Fig. 19.**2**). The *swing component* tends to change the angulation of the joint. The *shunt component* tends to compress the articular surfaces. The *spin component* tends to rotate the bone about its long axis. The swing component is discussed further with the skeletal lever system (see below).

Integrated muscle action

The functional role of a muscle in any movement can be classified into three different types: agonist, antagonist, and synergist. The synergist in its turn may act as an assistant mover, neutralizer or a stabilizer.

An **agonist** is a contracting muscle or a muscle group that is the principal muscle producing a joint motion. The agonist always contracts actively to produce a concentric, isometric or eccentric contraction. An agonist is also called a *prime mover* indicating that the muscle provides the most important force creating a particular torque.

An **antagonist** is a muscle (or muscle group) that possesses the opposite anatomic action of the agonist. The antagonist is usually a noncontracting muscle that neither assists nor resists motion but that passively elongates or shortens to permit the motion to occur. For example, during elbow flexion, the biceps brachii muscle is an agonist, while the triceps brachii is an antagonist. When the agonist and antagonist contract simultaneously, the result is isometric contraction of both with consequent stiffening of the muscle and immobilization of the joint. The antagonist then functions as a stabilizer (see below). Body builders show off their biceps by simultaneously contracting the triceps (see Fig. 105.**7**).

A **synergist** contracts at the same time as the agonist. The action of the synergist may be nearly identical to that of the agonist, partially antagonistic to the agonist, or neither. Accordingly, a synergist may serve three different roles as an assistant mover, neutralizer or a stabilizer. An *assistant mover* is a servile synergist muscle that can aid the prime mover to act as an emergency muscle either when great force is required or when paralysis has occurred. For example, during elbow flexion, the brachioradialis acts as the assistant mover with the brachialis serving as the prime mover. The *neutralizer* is a partially antagonistic synergist muscle that opposes an unwanted action of the prime mover. For example, the pronator teres prevents the supination action of the biceps brachii during elbow flexion. Another example is the wrist extensors preventing wrist flexion when long flexors of the fingers contract to close the fist. A *stabilizer* is a synergist muscle that steadies or supports

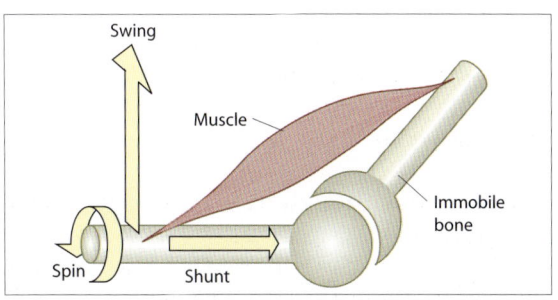

Fig. 19.**2**   Components of muscle action at the joint.

a proximal joint so that another muscle may act effectively at a distal joint. This contraction is generally isometric. For example, the wrist extensors often act as stabilizers so that the hand can be used effectively. To reach out with the hand, the scapula must be stabilized against the thorax. This is done by the rhomboids.

The relationships of muscles as agonists, antagonists, and synergists are situational and not absolute. When a person in a sitting position flexes his elbow to lift a load in hand, the flexors contract concentrically and are called agonists. The extensors are the antagonists and they are relatively relaxed and elongate to permit the elbow flexion motion. In order to lower the load, the elbow is extended. In this situation, the flexors perform an eccentric contraction and are still called agonists, whereas the extensors remain relatively inactive and are still called antagonists. But when the person is placed in the supine position with the shoulder in 90 degrees flexion and is asked to perform the same motion of elbow flexion and extension, the agonist–antagonist relation is reversed. Now the elbow extensors are the agonists for elbow extension (concentric contraction) and for elbow flexion (eccentric contraction), while the flexors are the antagonists for both these motions.

Actions of two-joint muscles

Many muscles in the human body cross two or more joints. Examples are the biceps brachii, the long head of the triceps brachii, the hamstrings, the rectus femoris, and a number of muscles crossing the wrist and finger joints. These muscles affect motion simultaneously at both or all the joints over which they pass.

Active insufficiency During joint movements, the passive tensions in two-joint muscles change much more than that in one-joint muscles. During certain joint movements, two-joint muscles become excessively slack, thereby failing to produce active tension (*active insufficiency*). An example of active insufficiency is seen in the finger flexors, which cannot produce a tight fist when the wrist is in flexion. A tight fist is possible only when the wrist is in a neutral position.

Passive insufficiency During certain other movements, two joint muscles get stretched excessively, thereby restricting the range of joint movements (*passive insufficiency*). An example of passive insufficiency is seen in the gastrocnemius, which restricts ankle dorsiflexion when the knee is flexed. A larger range of ankle dorsiflexion is possible when the knee is in flexion due to the change in the tightness of the gastrocnemius.

Both active and passive insufficiency can occur at the same time in a pair of antagonists. A good example is the combined movement of hip extension and knee flexion (Fig. 19.**3**). This movement produces active insufficiency of the hamstrings and passive insufficiency of the rectus femoris. Conversely, simultaneous hip flexion and knee extension causes passive insufficiency of the hamstrings and active insufficiency of the rectus femoris.

Skeletal Lever System

A joint represents the fulcrum (F) of a lever system (see Fig. 1.**6**) in which the swing component of muscle action provides the effort (E) for moving the external

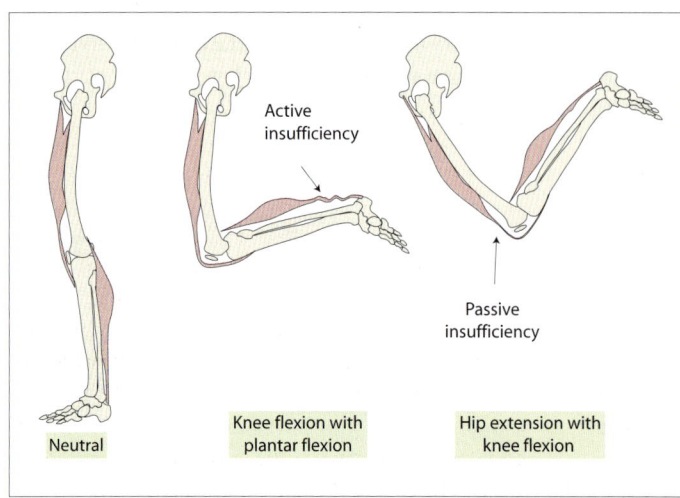

Neutral

Active insufficiency

Passive insufficiency

Knee flexion with plantar flexion

Hip extension with knee flexion

Fig. 19.**3** Action at two-joint muscle. Active insufficiency of gastrocnemius occurs due to excessive slackness when there is knee flexion with plantar flexion. Passive insufficiency of rectus femoris occurs due to excessive stretching when there is knee flexion with hip extension.

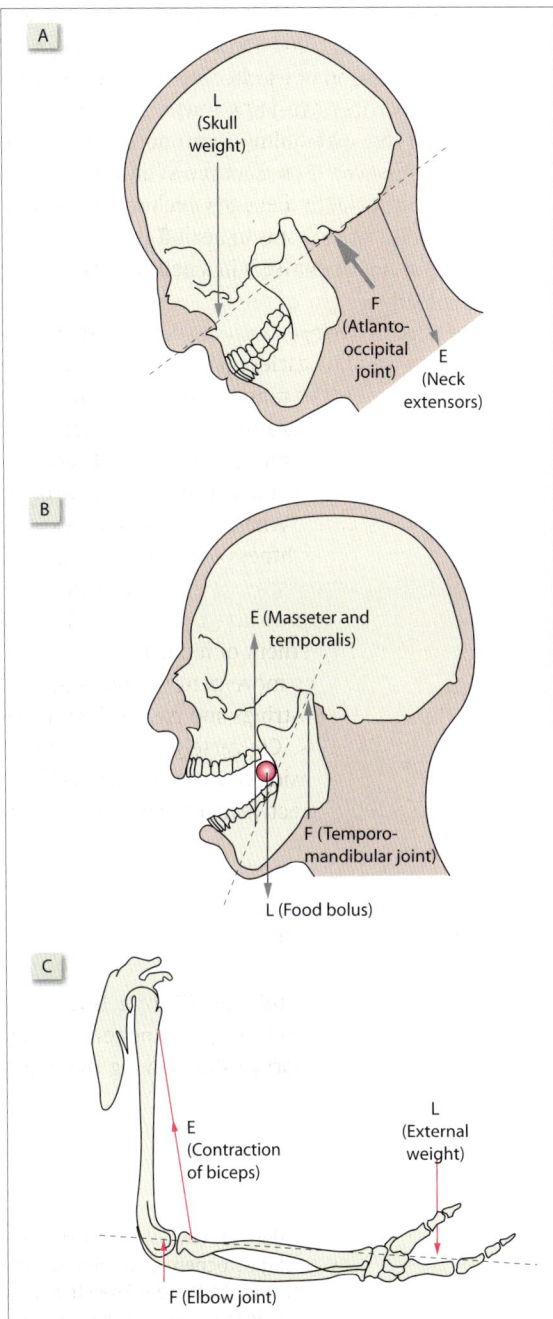

Fig. 19.**4**   Examples of levers in the musculoskeletal system. [A] Class-I lever. [B] Class-II lever. [C] Class-III lever.

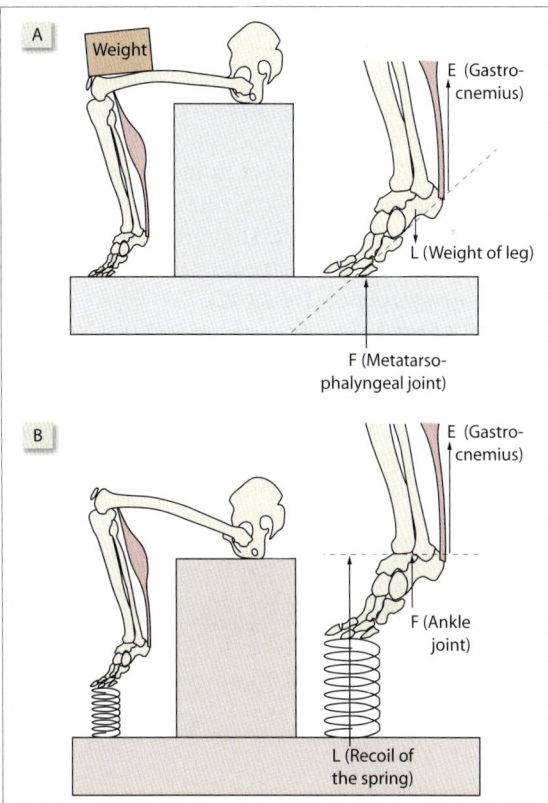

Fig. 19.**5**   [A] Plantar-flexion against resistance can be viewed as a Class-I lever. [B] Plantar-flexing to raise the heel can be viewed as a Class-II lever.

load (L). Most body levers are of class-III. They magnify the swing component. The skeletal lever system delivers muscle power to the load. The lever does not magnify power, which remains unchanged. It merely alters the ratio of force transmitted to the load and the excursion of the load. In all three classes of lever, what is gained in excursion is lost in force, and what is gained in force is lost in excursion. Class-II levers always produce force-gains (high mechanical advantage), while class- III levers always produce excursion gains (low velocity ratio).

Class-I lever   Examples of a class-I lever include the atlantooccipital joint where the weight of the head is balanced by neck extensor muscle force. A head that tips forwards in sleep is extended by the neck extensors when the person wakes up (Fig. 19.**4**A). Conversely, in full extension, the center of gravity of the head lies behind the axis so that the flexor muscles now act to pull the head forward. The same leverage system works to balance the trunk at the hip joints in standing.

Class-II lever   There are fewer examples of class-II levers in our body. The anterior fibers of the masseter muscle usually pass in front of the back molar teeth. Therefore as the mandible is elevated, the food crushed between these teeth lies closer to the axis of the temporo-mandibular joint than the anterior part of the muscle. Class-II levers provide a force-advantage such that large weights can be supported or moved by a smaller force (Fig. 19.**4**B).

Class-III lever The vast majority of lever systems within the body are class-III levers, with the joint closer to the muscle attachment than to the load. Contraction of the brachialis, an elbow flexor, causes motion at the fulcrum of the elbow joint, which results in elbow flexion, i.e., raising the object in the hand as well as the weight of the forearm. Class-III levers give the advantage of higher speed and larger excursion. They are ideal for most muscles that contract through a small range with great force (Fig. 19.**4**C).

Lever systems, though, can be considered in several ways, depending on the points of reference chosen and the axis about which movement occurs. For example, during plantar-flexion of the foot against resistance by contracting the triceps surae, the whole foot acts as a class-I lever if the ankle joint is considered as the axis (Fig. 19.**5**A). However, if the foot is on the ground with the subject sitting with a weight on the bent knees, then contraction of the same muscles can be seen to cause rotation about the metatarsophalangeal joints, raising the load (weight of the leg) acting at the ankle joint. This is a class-II lever system (Fig. 19.**5**B).

20 Smooth Muscles

There are several differences between smooth and skeletal muscles. The structural differences have been discussed on p 60 and are briefly listed in Table 20.1. The functional characteristics of smooth muscles are discussed in this chapter.

Compared to skeletal muscles, smooth muscles have delayed onset of contraction and relaxation: the contraction begins about 200 ms after the peak of a spike and ends 500 ms after excitation is over. Smooth muscles have lower energy-requirements for contraction and can maintain a high tension without actively contracting (latching). Smooth muscles have a higher percentage of shortening. They contract up to 70% of their initial length. Finally, smooth muscles are able to readjust their resting length, i.e., the length at which they generate maximum active tension (plasticity).

■ Mechanism of smooth muscle contraction

The cross-bridge cycling in smooth and skeletal muscles are identical and each cross-bridge cycling of a myosin head generates the same amount of shortening (1 nm) and tension (5×10^{-12} Newton). However, its regulation is different. Smooth muscle does not contain tropomyosin or troponin. One of the light chains of the myosin filament located in the neck region, called the *regulatory chain of myosin*, serves the function of tropomyosin. Similarly, a Ca^{2+}-binding protein called *calmodulin* serves the role of troponin.

Table 20.1 Structural differences between skeletal and smooth muscle fibers.

Skeletal muscle fiber	Smooth muscle fiber
Is large, cylindrical, and multinucleate	Is small, spindle shaped, and uninucleate
Has well-developed T-tubules	Has caveoli which are rudimentary T-tubules
Has Z-disks	Has dense bodies that are analogous to Z-disks
Actin and myosin filaments are roughly equal in number	Actin filaments far outnumber the myosin filaments
Actin filaments are parallel	Actin filaments radiate from the dense bodies

Phosphorylation of myosin cross-bridges When sarcoplasmic Ca^{2+} rises, the Ca^{2+} binds to calmodulin. The Ca^{2+}-calmodulin complex activates the enzyme *myosin light chain kinase* (MLCK), which in turn phosphorylates the myosin regulatory chain. Phosphorylation of the regulatory chain (often called *cross-bridge phosphorylation*) permits actin–myosin interaction and the cross-bridge cycling starts. The cycling stops when another enzyme, called *myosin phosphatase*, dephosphorylates the regulatory chain.

Variability of the average cycling rate Cross-bridge cycling occurs only when the myosin cross-bridge is phosphorylated. Dephosphorylation prevents both attachment as well as the detachment of cross-bridges. In a smooth muscle cell, some cross-bridges may be cycling while others are halted at any one instant, depending on whether or not the cross-bridge is phosphorylated. Hence, the average cycling rate of all the cross-bridges in a sarcomere at any instant can be anything from zero (when none of the cross-bridges is phosphorylated) to maximum (when all the cross-bridges are phosphorylated).

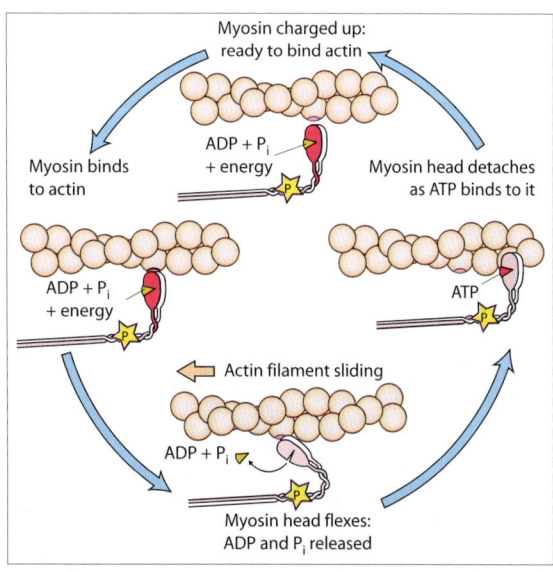

Fig. 20.1 Cross-bridge cycling in smooth muscle.

Latch mechanism   When the cycling rate is high, the work done is high too. The shortening speed and the contractile force rise. When the average cycling rate is low, work done is also low. The shortening speed is low, and so is the active tension generated. However, the resistance to passive tension in the muscle increases! This is because as the cycling rate decreases, more and more myosin heads remain *latched* to the actin filament.

When the rate of cross-bridge cycling nears zero, the energy consumption is minimal and the muscle goes into a state that is analogous to the contraction remainder of a fatigued skeletal muscle: Such a muscle cannot generate active tension but can effectively resist passive stretching. This suits the smooth muscle well, since in most instances, it has to resist stretch rather than actively move a load. Smooth muscles are mostly found in the walls of hollow viscera that must resist excessive stretching, e.g., the arterial walls. Rarely is smooth muscle required to move a heavy load.

> **Load Handling**
>
> Moving a load and resisting a load are two different things. An analogy of how a heavy load is handled differently by a strong and a weak man (Fig. 20.**2**) illustrates the point. The strong man has the muscles to lift a heavy load. The frail man is not doing any muscular work. He is only preventing the load from falling by latching on to a fixed bar. The maximum load that the frail man can resist in this way is determined by the structural strength of his ligaments and muscles that hold his bones together. Similarly, the amount of load that can be resisted by a smooth muscle depends upon the strength of its actin–myosin bonds.

Plasticity   The smooth muscle defies the usual length–tension relationship that is valid for striated muscles. Unlike a striated muscle fiber which can contract to roughly 60% of its initial length, a smooth muscle fiber can contract to almost 30% of its initial length. Similarly, when stretched passively, the passive tension it develops gradually

Fig. 20.**2**   Resisting a load by latching on to a pole. Smooth muscles employ a similar mechanism for resisting load.

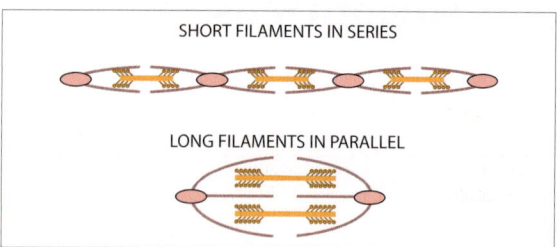

Fig. 20.**3**   Forty-eight cross-bridges disposed in series (*above*) and parallel (*below*).

reduces to the prestretch level. In other words, whether actively contracted or passively stretched, the smooth muscle fiber tends to behave as if it is always at its resting length! This phenomenon is called plasticity and is possible due to a process in which the thick filaments dissolve and reorganize themselves so that at longer fiber lengths, the thick filaments are shorter but more numerous, and are disposed in series. Conversely, at shorter lengths, the thick filaments are longer and disposed in parallel.

■ Excitation and inhibition of smooth muscle

Based on how smooth muscles are stimulated or inhibited, smooth muscle cells are classified into single-unit and multi-unit types. *Multi-unit smooth muscles* are stimulated only through nerves, quite like skeletal muscle fibers. *Single-unit smooth muscles* show ephaptic conduction, i.e., the excitation spreads from cell to cell, resulting in the contraction of all the fibers in a unit. Single-unit smooth muscles are excited in several ways that include: (1) spontaneous excitation through pacemaker, (2) ephaptic excitation from adjacent cells, (3) stimulation through autonomic nerves, i.e., by neurotransmitters, (4) stimulation by hormones, (5) stimulation by stretch, and (6) stimulation by cold temperature.

Some hormones and neurotransmitters even inhibit smooth muscle contraction. For inhibition to occur, the muscle must already be contracting in response to some other excitatory stimulus. For example, a smooth muscle excited through a pacemaker is inhibited by a nerve secreting inhibitory neurotransmitter. Smooth muscles that remain contracted most of the time and relax only in response to inhibitory stimuli are called *tonic smooth muscles*. Examples of tonic smooth muscles are the smooth muscles of gastrointestinal and urogenital sphincters. Smooth muscles located in the walls of blood vessels and airways also remain partially contracted. On the other hand, muscles forming the walls of gastrointestinal and urogenital tracts remain

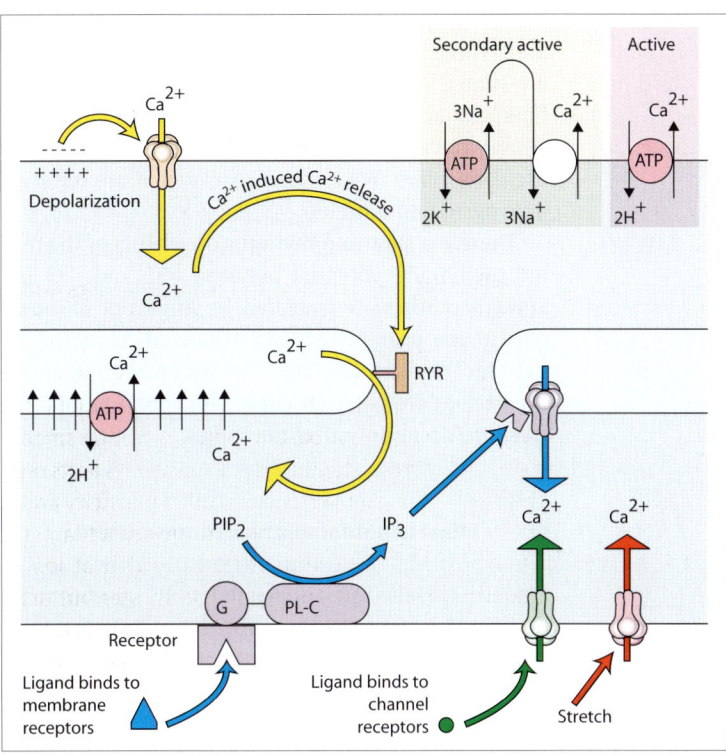

Fig. 20.**4**   Excitation–contraction coupling in smooth muscles.

mostly relaxed and contract only in response to excitatory stimuli. These are called *phasic smooth muscles*.

Mechanisms of excitation-contraction coupling

With so many types of stimuli exciting a smooth muscle, the excitation-contraction coupling in smooth muscle is not unexpectedly a diverse phenomenon. There are at least three different ways in which the smooth muscle excitation can be coupled to its contraction and two sources from which Ca²⁺ can be mobilized: extracellular and intracellular (Fig. 20.**4**).

In **electromechanical coupling**, the smooth muscle is excited through sarcolemmal depolarization. When the membrane depolarizes, voltage-gated Ca²⁺ channels present on the sarcolemma open up and Ca²⁺ moves into the sarcoplasm from the extracellular fluid. This Ca²⁺ stimulates the release of more Ca²⁺ from the sarcoplasmic reticulum. This is called *Ca²⁺-induced Ca²⁺ release* (CICR). A fully-fledged action potential is not always required for electromechanical coupling. In a multi-unit smooth muscle fiber, even subthreshold electrotonic depolarizations can raise the sarcoplasmic Ca²⁺ sufficiently to bring about excitation-contraction coupling.

A fall in temperature also excites the smooth muscles. It does so by inhibiting the Na⁺-K⁺ pump, thereby depolarizing the sarcolemma. This is possible because, unlike in striated muscles, the Na⁺-K⁺ pump makes a substantial direct contribution to membrane potential in smooth muscle.

In **pharmacomechanical coupling**, the muscle is excited by chemical agents in the absence of any membrane depolarization. There are two mechanisms of pharmacomechanical coupling: (1) Neurotransmitters and hormones bind to membrane receptors to activate group-II hormonal mechanisms (see Chapter 81) to release Ca²⁺ from the sarcoplasmic reticulum. (2) Neurotransmitters and hormones bind directly to ligand-gated Ca²⁺ channels on the sarcolemma and open them up, letting in extracellular Ca²⁺.

In **mechanomechanical coupling**, smooth muscles are excited by stretch, which opens up stretch-sensitive Ca²⁺ channels on the sarcolemma, letting in extracellular Ca²⁺.

Resting membrane potential

Unlike in the skeletal muscle, the RMP of the smooth muscle cell is less, approximately –50 mV. In single-unit

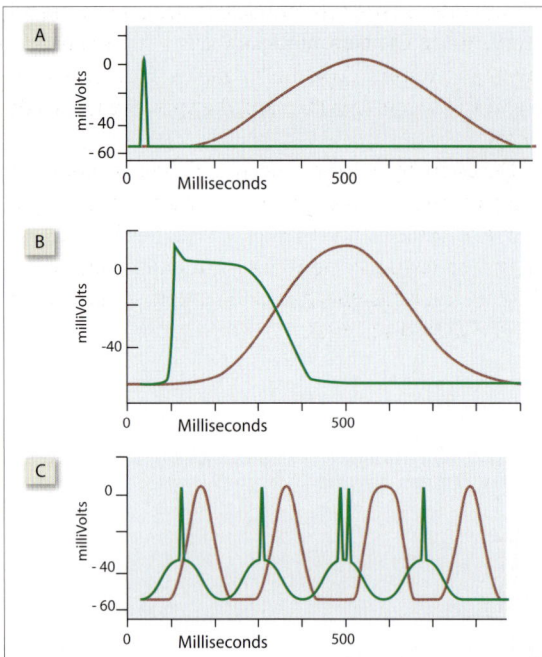

Fig. 20.**5** Action potentials (green) and the resulting contractions (yellow) in smooth muscles. [A] A single spike potential producing a single muscle twitch. [B] Action potential with a plateau. It is associated with a prolonged contraction. [C] Pacemaker potential with superimposed spikes.

smooth muscles, the RMP is often unstable, oscillating between –55 and –35 mV. These oscillations are called *pacemaker potentials*. They occur due to rhythmic changes in either Ca^{2+} channel permeability and/or the activity of the Na^+-K^+ pump. In skeletal muscle, the Na^+-K^+ pump contributes very little to the RMP. In smooth muscle, the Na^+-K^+ pump makes a significant contribution (~20 mV) to the RMP (see water-tank analogy, Fig. 10.**7**).

Action potentials

In multi-unit smooth muscle, even a subthreshold depolarization without an action potential is associated with some amount of contraction. Action potentials are however essential for single-unit smooth muscle to contract, because only an action potential can excite several smooth muscle fibers through ephaptic conduction. A subthreshold depolarization will fade out after traveling a short distance. In single-unit smooth muscle, the excitation needs to spread from cell to cell. Not unexpectedly, subthreshold depolarizations have no role in the contraction of single-unit smooth muscle.

The action potentials in single-unit smooth muscles are of three types (Fig. 20.**5**), (1) spike potentials, which are similar to the action potential in a skeletal muscle fiber, (2) action potentials with plateaus, which are similar to cardiac action potentials, and (3) spikes on oscillatory pacemaker potentials, which are observed only in smooth muscles.

There is a slow and rhythmic oscillation of the resting membrane potential between –55 and –35 mV. Spike potentials are triggered, in singles or in bursts, only at the peak of the oscillations when the potential touches the firing level. The pacemaker potentials themselves are not associated with contractions. Contractions occur only after the spikes.

Effect of autonomic neurotransmitters

Smooth muscles are innervated only by autonomic nerves. The potentials generated in smooth muscle when they are stimulated through autonomic nerves are called *junctional potentials*. They may be excitatory junctional potentials (EJP) or inhibitory junctional potentials (IJP) depending on whether the neurotransmitter secreted at the junction depolarizes or hyperpolarizes the smooth muscle membrane.

Acetylcholine binds to ligand-gated ACh receptors present on smooth muscles and depolarizes the membrane. The ACh receptors present on smooth muscles are of the muscarinic type. In multi-unit smooth muscles, the depolarization triggers action potentials. In single-unit smooth muscles with spontaneous action potentials, the depolarization results in a higher frequency of action potentials. The action potentials themselves become slightly smaller and wider. The contractions become stronger and more frequent.

The effect of **catecholamines** on smooth muscle depends on the type of receptor stimulated. Adrenergic receptors are of two types, α and β. In general, β adrenoceptors are inhibitory, causing relaxation while α-adrenoceptors may be either excitatory or inhibitory. Inhibitory α and β adrenoceptors are widely distributed in gastrointestinal smooth muscles. *Excitatory α-adrenoceptors* are present mainly in vascular and urogenital smooth muscles and gastrointestinal sphincters. They produce membrane depolarization by increasing Na^+ permeability. *Inhibitory α-adrenoceptors* induce relaxation mainly through membrane hyperpolarization caused by an increase in K^+ permeability. The inhibition is quicker and short-lasting. *Inhibitory β-adrenoceptors* induce relaxation mainly by suppressing spontaneous pacemaker activity, with or without hyperpolarization of the membrane. The spike characteristics are not affected much.

Effect of ovarian hormones

The effects of ovarian hormones on uterine smooth muscle are complicated by the type of excitation-contraction coupling, hormonal interactions and differences in the circular and longitudinal uterine smooth muscles. In general, estrogens stimulate and progesterone relaxes uterine smooth muscles. Stimulation generally occurs when the muscle depolarizes and its spike-frequency increases. Conversely, inhibition occurs when the muscle hyperpolarizes and the spike frequency decreases.

Occasionally, both estrogen and progesterone violate this general rule, possibly because they employ both electromechanical and pharmacomechanical coupling. For example, they may stimulate contractions through pharmacomechanical coupling while inhibiting the spike discharges.

21 Cardiac Muscle

Cardiac muscle fibers are of two types: nonautomatic and automatic. The nonautomatic fibers of the atria and ventricles are meant primarily for generating contractile force. The automatic fibers are spontaneously excitable and constitute the conducting system of the heart. They are relatively less contractile as they have fewer myofibrils and mitochondria and have poorly developed sarcoplasmic reticulum.

Cardiac muscle has functional resemblances with both smooth muscle and skeletal muscle. Cardiac muscle resembles visceral smooth muscle in that it shows automaticity and ephaptic conduction, and its contractility is affected by hormones. Cardiac muscle resembles skeletal muscle in that it contains regular sarcomeres delimited by Z-disks, shows similar length–tension relationship, and its contraction is regulated by troponin–tropomyosin complex.

Cardiac Action Potentials

Cardiac action potentials resemble those of the smooth muscle. The action potentials of automatic cardiac fibers comprise single spikes on the peaks of an oscillating membrane potential. Action potentials of nonautomatic cardiac fibers have a characteristic plateau. This prolonged plateau increases the duration of the cardiac action potential.

Action potential in nonautomatic cardiac fibers

The action potential in a nonautomatic cardiac fiber shows a characteristic plateau. In all, it has five phases (Fig. 21.1). The absolute refractory period of cardiac action potential includes phases 0, 1, and 2, while in phase 3, the cardiac muscle is in the relative refractory period. The cardiac action potential is thus refractory for most of its duration.

Phase 0 is the phase of depolarization that occurs due to the opening of Na^+ channels, resulting in a *fast inward depolarizing current*.

Phase 1 is the partial repolarization that occurs due to Na^+ channel inactivation and a transient increase in K^+ permeability (the *transient outward K^+ current*).

Phase 2 is the plateau during which the membrane repolarizes either very slowly, or not at all. It occurs due to (1) the opening of sarcolemmal *L-type Ca^{2+} channels* (L for long-lasting) that lets in a slow, inward depolarizing current and (2) closure of a distinct set of K^+ channels called the *inward-rectifying K^+ channels*.

The K^+ channels normally open up when the membrane depolarizes. In cardiac muscles, some K^+ channels (the inward-rectifying K^+ channels) do exactly the opposite, that is, they remain open when the membrane is polarized and close down with depolarization. Hence, at the resting membrane potential, they make a substantial contribution to the total K^+ current. Conversely, during the sustained depolarization of the plateau phase, they close down. If they remained open during depolarization, large amounts of K^+ would flow out of the cell throughout the plateau phase, a loss that would be difficult to replenish.

Phase 3 is the phase of complete repolarization. It occurs due to two outward K^+ currents through two different types of K^+ channels: (1) The *delayed outward rectifyied K^+ current* flows through voltage gated channels that are activated slowly. (2) The *Ca^{2+}- activated K^+ current* flows through K^+ channels that are activated by the elevated sarcoplasmic Ca^{2+} levels.

Phase 4 is the resting phase with a potential of about –90 mV. It is maintained by a *resting K^+ current*, the largest contributor to which is the inward rectifying K^+ current mentioned above.

Action potential in automatic cardiac fibers

Phases of action potential The action potential of an automatic cardiac fiber shows salient differences from that of a nonautomatic fiber (Fig. 21.2). (1) The RMP is less negative (–70 mV) than in nonautomatic cardiac fibers (–90 mV). (2) Phase 4 of the action potential shows

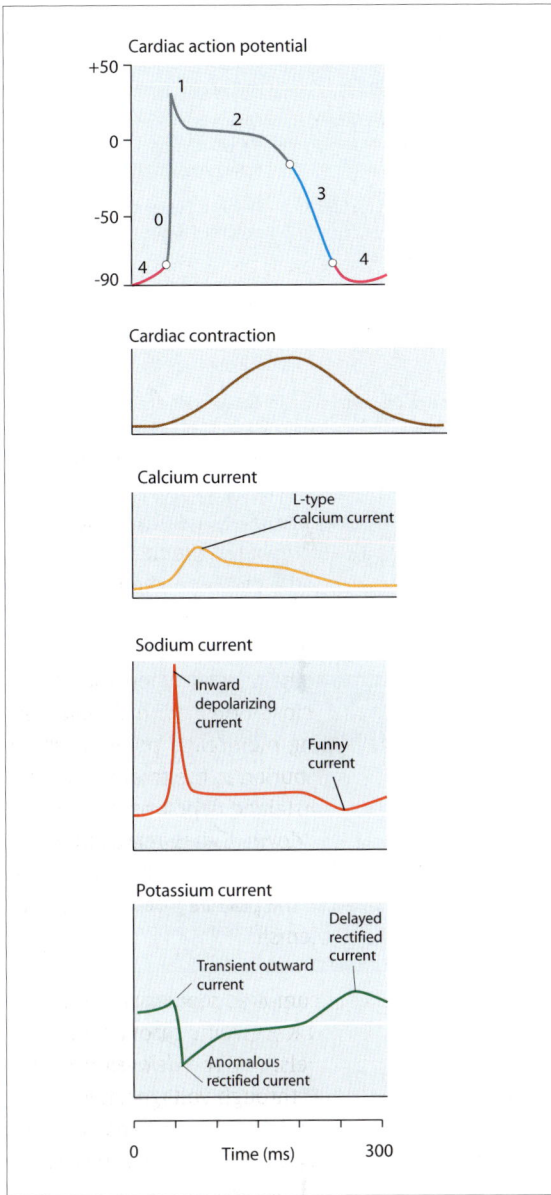

Fig. 21.**1** Cardiac action potential and the resulting cardiac contraction. Also shown are Ca^{2+}, Na^+, and K^+ currents associated with the action potential.

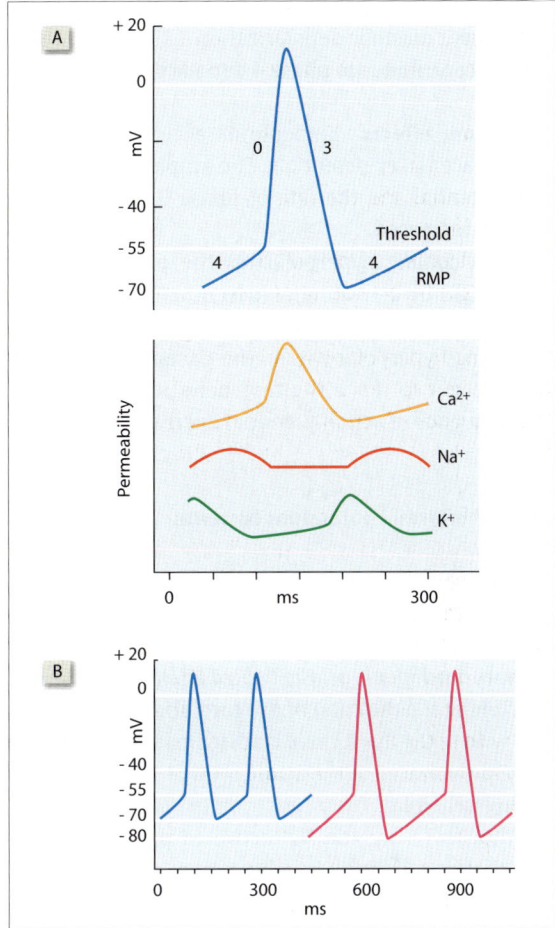

Fig. 21.**2** [A] Cardiac pacemaker potential and the associated ionic permeability changes. [B] Effect of acetylcholine on cardiac pacemaker.

a slow drift towards the threshold potential till phase 0 is triggered. This characteristic phase 4 is also called the *pacemaker potential* or the *diastolic depolarization*. (3) Phase 0 is less steep and its peak is less sharp (more rounded) than phase 0 in non-automatic fibers. This is because the depolarization in phase 0 occurs due to an increase in Ca^{2+} permeability. This is unlike phase 0 of nonautomatic cells which is produced by an increase in Na^+ permeability. (4) In the absence of the plateau (phase 2), phases 1 and 3 blend into a single phase of repolarization called phase 3.

Ionic basis of pacemaker potential Permeability changes in all the three ions, i.e., Na^+, K^+, and Ca^{2+} seem to be responsible for the slow diastolic depolarization (Table 21.**1**). (1) The *'funny' Na^+ current* is carried by a population of funny Na^+ channels that open up when the membrane hyperpolarizes beyond −50 mV (instead of closing down like the regular Na^+ channels). The resulting Na^+ influx depolarizes the membrane towards threshold. (2) The *outward (delayed) rectifier K^+ current* is carried by channels that open up with depolarization and close down as the membrane repolarizes (just like ordinary K^+ channels). However, the opening and closing of channels is somewhat slow or delayed and therefore, there is a slight phase lag between the membrane potential and the K^+ permeability changes. As a result, when the membrane is fully depolarized, the K^+ permeability is still increasing. Conversely, when the membrane is fully repolarized, the K^+ permeability is still decreasing. This delayed decrease in outward K^+

current following full repolarization contributes to the subsequent diastolic depolarization. (3) An *inward Ca²⁺ current* contributes to phase 4 depolarization.

Autonomic effects Epinephrine accelerates phase 4 of the pacemaker potentials. The amplitude of the action potential and the rate of phase 0 depolarization are also increased.

Acetylcholine hyperpolarizes the pacemaker cells (Fig. 21.**2**B) by activating an *acetylcholine-activated K⁺ channel*. The activation is mediated by a G-protein. Following hyperpolarization, the pacemaker potential takes longer to touch the threshold potential. Hence, the frequency of action potential decreases.

■ Abnormal potassium currents

Certain types of K⁺ channels assume importance in cardiac disorders (Table 21.**1**). Arachidonic acid and other fatty acids that are liberated from ischemic cardiac cells activate *arachidonic acid-activated K⁺ channels*, thereby shortening the duration of the cardiac action potential and with it, the duration of cardiac systole. The effect is more pronounced at the acidic pH that is often present during ischemia.

Also, there are the *Na⁺-activated potassium channels* that get activated in a heart overloaded with Na⁺ ions, e.g., following digitalis administration. Hence, the Q-T interval of ECG is shortened in a patient receiving digitalis.

Table 21.**1** Normal and abnormal cardiac K⁺ currents of clinical importance.

Normal K⁺ currents	Role
Transient outward K⁺ current	Opens briefly immediately after depolarization. Contributes to early repolarization.
Anomalous (inward) rectifer K⁺ current	Responsible for the resting potential. Closes with depolarization. Thus, opposes depolarization and prolongs the plateau.
Delayed (outward) rectified K⁺ current	Opens at the end of plateau. Initiates repolarization.
Ca²⁺-activated potassium current	Activated by the rise in sarcoplasmic Ca²⁺ concentration that occurs at the onset of contraction. Accelerates repolarization and thereby shortens the action potential.
ACh-activated K⁺ current	Activated in response to vagal stimulation. Hyperpolarizes the resting cell, thereby slowing the SA node and shortening the atrial action potential.
Abnormal K⁺ currents	**Role**
Na⁺-activated K⁺ current	Activated by high sarcoplasmic Na⁺ concentrations. Accelerates repolarization in a Na⁺-overloaded heart.
ATP-sensitive K⁺ current	Normally inhibited by ATP. Opens in energy-starved heart.
Arachidonic acid-activated K⁺ current	Activated by arachidonic acid and other fatty acids, especially at acidic pH.

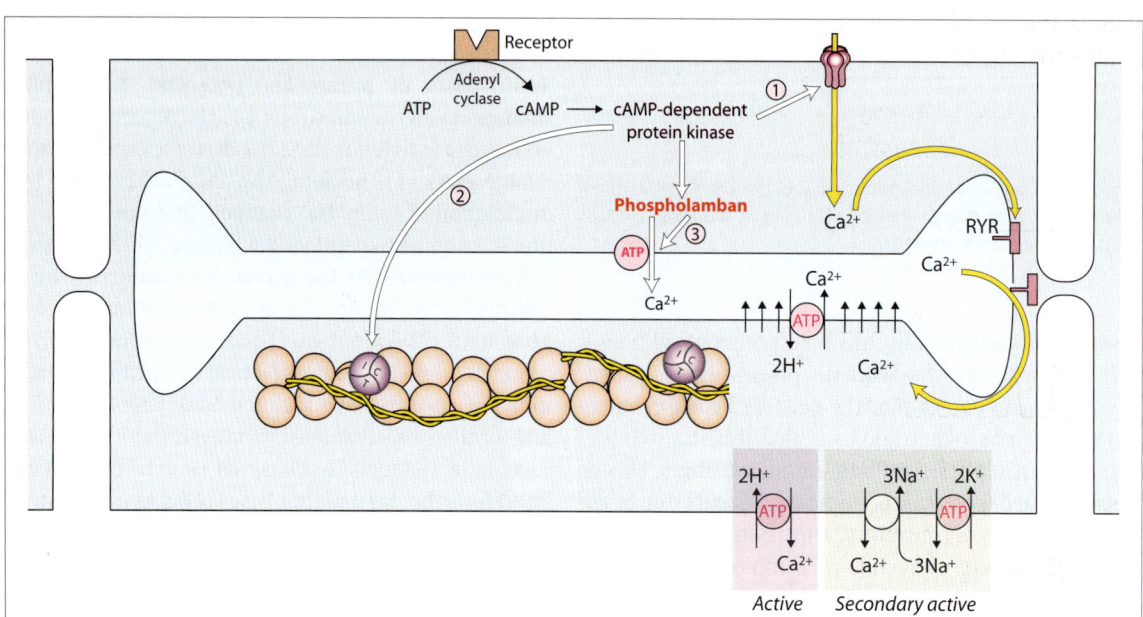

Fig. 21.**3** Excitation–contraction coupling in cardiac muscle and the modulation of cardiac contractility.

Cardiac Contraction

Excitation–contraction coupling

Like in smooth muscle, the sarcoplasmic Ca^{2+} concentration in the cardiac muscle is kept low by two types of Ca^{2+} transport. One is a primary active transport in which a Ca^{2+}-$2H^+$ antiport occurs using Ca^{2+}-ATPase for splitting ATP. The other is a secondary active transport in which a passive Ca^{2+}-$3Na^+$ antiport is coupled to an active Na^+-K^+ antiport (Fig. 21.**3**).

The sarcoplasmic Ca^{2+} pulse is produced predominantly by mobilization of intracellular Ca^{2+} when the ryanodine channels on lateral cisterns open up. This mobilization is triggered by a smaller amount of extracellular Ca^{2+} that enters through the sarcolemma when the membrane is excited, and binds to the ryanodine receptors. This process is called *calcium-induced calcium release* (CICR). Although cardiac fibers show T-tubules as in skeletal muscle fibers, these T-tubules do little else other than to let in extracellular Ca^{2+} like the rest of the sarcolemma. The contraction starts just after depolarization and lasts until about 50 ms after repolarization is completed (Fig. 21.**1**).

Cross-bridge cycling and sliding of filaments

The cross-bridge cycling in cardiac muscle is no different from that of skeletal or smooth muscle. However, its regulation shares the features of both. Like the skeletal muscle, its onset and termination are controlled by the troponin-tropomyosin complex. Like the smooth muscle, the contractility of cardiac muscle is sensitive to phosphorylation.

Modulation of cardiac muscle contractility

Phosphorylation   At least three components of the cardiac excitation-contraction coupling machinery are susceptible to phosphorylation (Fig. 21.**3**). (1) Phosphorylation of sarcoplasmic Ca^{2+} channels results in their remaining open for longer durations. Sarcoplasmic Ca^{2+} concentration therefore rises and results in quicker and stronger contraction. (2) Phosphorylation of troponin-I inhibits binding of Ca^{2+} to troponin-C, thereby facilitating relaxation. (3) *Phospholamban* is a regulatory protein present only in cardiac muscles. It controls the activity of the sarcoplasmic Ca^{2+} pump. When phosphorylated, phospholamban stimulates the Ca^{2+} pump in the L-tubules and lowers sarcoplasmic Ca^{2+} concentration, thereby accelerating relaxation. The overall effect of activation of the intracellular phosphorylation system is to increase the strength of contraction (positive inotropic effect) and speed of contraction (positive chronotropic effect) as well as to quicken relaxation. Phosphorylation of intracellular regulatory proteins is induced by catecholamines, which bind to sarcolemmal β_1 receptors and activate the group-IIa hormonal mechanism.

Membrane pump and exchanger   The Na^+-K^+ pump on the sarcolemma lowers intracellular Na^+. This promotes the entry of extracellular Na^+ into the cell in exchange for Ca^{2+} efflux. In other words, the sarcolemma pumps out Ca^{2+} through secondary active transport (Fig. 21.**3**). Inhibition of this active transport, e.g., by digitalis or other cardiac glycosides (a group of structurally similar compounds that increase cardiac contractility) raises the intracellular Ca^{2+} concentration and thereby increases myocardial contractility.

Fig. 21.**4**   Summation of cardiac contraction. Positive staircase effect occurs when a quiescent heart starts beating, or when there is a sudden increase in heart rate. Negative staircase effect occurs following a sudden decrease in heart rate.

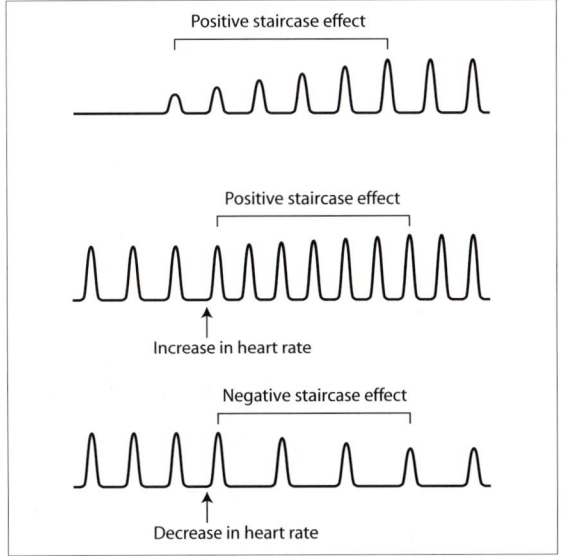

■ Characteristics of cardiac muscle contraction

Due to ephaptic conduction between cardiac myocytes, the contraction of cardiac muscle obeys the *all-or-none law*, i.e., it either contracts completely or does not contract at all, depending on whether the strength of the stimulus is threshold or subthreshold.

Cardiac muscle also shows summation of stimuli. When a quiescent cardiac muscle starts contracting spontaneously (as occurs at the onset of idioventricular rhythm), its first few contractions show a progressive increase in height. The phenomenon is known as the positive staircase effect and occurs due to the cumulative increase in sarcoplasmic calcium concentration. A similar *positive staircase effect* is seen over a few beats immediately after an increase in heart rate. Conversely, a *negative staircase effect* (progressive reduction in strength of contraction) is observed for a few seconds immediately after a decrease in heart rate.

Although cardiac contractions show the staircase effect, they never fuse together because the cardiac muscle cannot be stimulated in quick succession due to its long refractory period. Thus unlike skeletal muscle, cardiac muscle cannot be tetanized.

Section 3

Blood and Immune System

22. **Body Fluids and Blood**

23. **Red Blood Cells**

24. **Hemoglobin**

25. **Hematinic Factors**

26. **Blood Grouping and Transfusion**

27. **Blood Platelets and Hemostasis**

28. **Hemostatic Balance**

29. **Granulocytes**

30. **Agranulocytes and Lymphoid Organs**

31. **Immunity, Tolerance and Hypersensitivity**

32. **Immune Mechanisms**

33. **Hemopoiesis**

Body Fluids

The percentage of body weight that is made of water is about 62% in males, 52% in females, and 72% in infants. Females have the least proportion of body water because they have more adipose tissue than males. In both sexes, water constitutes approximately 72% of the lean (adipose-free) body mass. In a man weighing 60 kg, the volume of total body water (TBW) is approximately 36 L. This water is divided into various compartments as shown in Figure 22.**1**.

Extracellular fluid (ECF) is present outside the cells. Of the 12 L of ECF, 3 L is present inside the blood vessels as intravascular fluid, i.e., plasma while the remainder is present around the cells as interstitial fluid and is separated from the intravascular fluid only by the walls of the blood vessels. Some of the extracellular fluid is present in specialized compartments, and is called transcellular fluid. Examples include cerebrospinal fluid, synovial fluid, pleural fluid, pericardial fluid, peritoneal fluid, intraocular fluid, gastrointestinal secretions, and urine.

The volume of the various body compartments can be measured by injecting into them an indicator substance and estimating its volume of distribution from the degree of its dilution. The indicator substance used must fulfill certain criteria. For measurement of body fluid compartments, the most relevant consideration is that the dye should remain confined to the compartment whose volume is to be measured and should get uniformly diluted by the fluid in the compartment

(Table 22.**1**). Other considerations are that the dye should be non-toxic, should not change (pharmacologically or otherwise) the fluid volume, should not be metabolized, altered or excreted in significant amounts in a short time and should be easy to estimate in the laboratory.

Two spaces that are calculated indirectly are the intracellular fluid volume (TBW volume – ECF volume) and the extravascular fluid volume (ECF volume – plasma volume).

Example

100 mg of inulin was injected intravenously into a 60 kg man. After 30 minutes, the inulin concentration was found to be 0.75 mg dl⁻¹. Also, 25 mg was excreted in urine during that period. Calculate the extracellular volume.

Solution

The ECF will be given by the volume of distribution of inulin, which is equal to:

$$\frac{100 - 25}{0.75} \times 100 = 10000 \text{ ml } (10 \text{ L})$$

Body fluid osmolarity   The normal osmolarity of (extracellular) body fluids is about 290 mOsm L⁻¹. Some ions, like K^+, Mg^{2+}, and PO_4^{3-} are predominantly intracellular. Others, like Na^+, Cl^-, and Ca^+, are predominantly extracellular and are important contributors to ECF osmolarity.

It should be clear from Table 22.**2** that the major contributors to the body osmolarity are Na^+ and Cl^-. The total body osmolarity can be calculated by adding up the millimolar concentrations of Na^+ (140), K^+ (5), Cl^- (100), HCO_3^- (25), glucose (5), and urea (5)

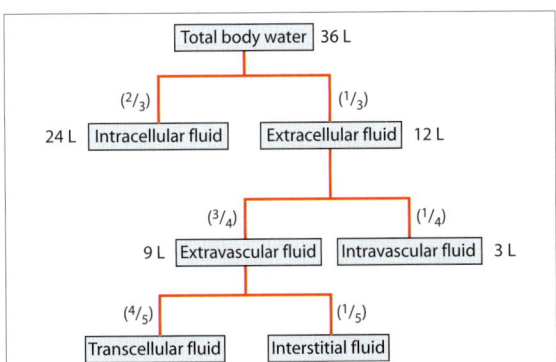

Fig. 22.**1**   Body fluid compartments. The volumes indicated are in reference to a 60 Kg man.

Table 22.**1**   Dyes for measuring body fluid compartments.

Total body water (TBW) volume	Deuterium oxide (heavy water), tritium oxide, aminopyrine
Extracellular fluid (ECF) volume	Sodium thiosulfate, sucrose, inulin, mannitol
Intravascular space	Evans Blue (T-1824), ¹³¹I₂ (these bind to plasma albumin, and therefore cannot escape from the blood vessels)

Table 22.**2**   Important contributors to body osmolarity.

Constituent	Concentration
Na$^+$	135–155 mmole L^{-1}
Cl$^-$	90–110 mmole L^{-1}
HCO$_3^-$	25 mmole L^{-1}
K$^+$	3.5–5.0 mmole L^{-1}
Glucose	60–100 mg dl^{-1} (3.4–5.6 mmole L^{-1})
Blood Urea Nitrogen	10–20 mg dl^{-1} (3.6–7.1 mmole L^{-1})

Table 22.**3**   Normal values for the cellular elements in human blood.

Red blood cells	Males	5.5 ($\pm$1.0) million ml^{-1}
	Females	4.8 ($\pm$1.0) million ml^{-1}
White blood cells		
Total leukocytic count		4000–11 000 ml^{-1}
Differential leukocytic count	Neutrophils	3000–6000 (50–70%)
	Eosinophils	150–300 (1–4%)
	Basophils	0–100 (0–1%)
	Lymphocytes	1500–4000 (20–40%)
	Monocytes	300–600 (2–8%)
Platelets		200 000–500 000 ml^{-1}

Table 22.**4**   Constituents of plasma.

Constituent	Examples
Electrolytes	Na$^+$, K$^+$, Cl$^-$, HCO$_3^-$
Proteins	Albumin, globulin, amino acids
Carbohydrates	Glucose
Lipids	Cholesterol, fatty acids
Minerals	Ca^{2+}, PO$_4^{3-}$
Enzymes	Alkaline phosphatase, amylase
Metabolites	Bilirubin, urea, uric acid, creatinine

white blood cells (WBC) or leukocytes and the platelets or thrombocytes (Table 22.**3**).

The plasma contains innumerable substances, of which proteins (the plasma proteins) are the major constituents. The major constituents of plasma are summarized in Table 22.**4**. Plasma contains, among other substances, coagulation proteins. Hence like blood, plasma too clots on standing. Plasma without the coagulation proteins is called serum. It is obtained by allowing whole blood to clot. As the clot hardens, it shrinks and extrudes the serum. Serum has essentially the same composition as plasma except that its clotting factors have been consumed and it has a higher serotonin content due to the breakdown of platelets during clotting.

The **specific gravity** of blood is between 1.055 and 1.060 and is determined mainly by the hemoglobin and plasma protein concentration. It therefore serves as a quick screening test for the overall quality of blood and is often performed for screening candidates for blood donation. A drop of blood obtained from the donor is gently dropped into a column of CuSO$_4$ solution with a specific gravity of 1.050. The drop should sink at least halfway down the column. If the drop floats, its specific gravity is low and the blood donation should not be accepted.

and adding another 10 for other less abundant ions. Alternatively, it can be quickly calculated by the formula (2 × Na$^+$ concentration) + glucose (concentration) + urea (concentration). This formula is convenient for quickly calculating body fluid osmolarity when it rises alarmingly in hyperglycemia or uremia. Proteins do not contribute significantly to body fluid osmolarity because their molar concentration is low.

Blood

The body contains about 5 L of blood. The salient functions of blood are transport of gases (O$_2$ and CO$_2$), nutrients, metabolites, and hormones. It provides defense against infection and has a role in the maintenance of body temperature, acid-base balance, and fluid-electrolyte balance. It has an inbuilt mechanism (hemostasis) for preventing its own loss from the body.

Composition of blood   Blood is made of formed (cellular) elements (45%) and plasma (55%). The two can be separated by centrifuging whole blood at ~1000 g (usually equivalent to ~3000 rpm in a small centrifuge). The bulk of the cellular elements of blood are made of red cells or erythrocytes. Others include the

Erythrocytic sedimentation   When anticoagulated blood is allowed to stand in a long vertical tube its erythrocytes settle down leaving clear plasma at the top of the tube. The rate at which the erythrocytes settle down in a tube of standard dimensions (2.5 mm inner diameter, 200 mm height) is called the erythrocytic sedimentation rate (ESR) and is one of the oldest hematological tests that is still in use.

The most important factor that determines ESR is the extent of rouleaux formation by erythrocytes (see Fig. 23.**1**). Rouleaux formation is determined mainly by the nature of plasma. Increased rouleaux formation occurs when plasma contains increased amounts of fibrinogen (as in pregnancy) and serum globulin (as in inflammatory diseases). Red cell characteristics also affect rouleaux formation and therefore affect the ESR too. Red cells with higher MCHC (mean corpuscular hemoglobin concentration) tend to fall more slowly in plasma than those with normal or low MCHC. Excessive nonuniformity of shape (poikilocytosis) or size (anisocytosis) of the red cells reduces ESR.

Two well-known pathological causes of elevated ESR are tuberculosis and rheumatoid arthritis. However, ESR is elevated in so many diseases that it has little diagnostic utility. It is elevated in almost all inflammatory disorders and collagen diseases. Its main utility is in prognosis, i.e., prediction of the probable course of a disease in an individual and the chances of recovery. Thus, during a 6-month course of tuberculosis treatment, serial measurements of ESR will indicate if the patient is improving. It has similar use in certain malignancies, especially Hodgkin's disease.

Identification test for blood   Blood has obvious forensic importance. Hence it is important to distinguish bloodstains from other red colored stains. Two microchemical tests for hemoglobin are sensitive and confirmatory for blood. In the Teichman test, a few crystals of chemically pure NaCl crystals are placed on the stain extract over a glass slide. A cover slip is put over it. Two to three drops of glacial acetic acid are then added to the side of the cover slip. The slide is then gently heated by passing over a flame and then cooled. On examination under a high power microscope, yellow or dark brown rhombic crystals of hemin or hematin are seen in clusters. On addition of H_2O_2 over the crystals, gas bubbles are seen coming out which confirms that the stain is blood.

In the Takayama test, two drops of Takayama reagent are added to the stain extract, which is placed over a glass slide under a cover slip and warmed a little to hasten the reaction. On observation under microscope, pink, feathery crystals of hemochromogen or reduced alkaline hematin arranged in clusters or sheaves are seen in 1 to 6 minutes. The test is declared negative if no crystals are formed within 30 minutes. It is positive even in very old stain. However, it is more time consuming and less sensitive than the Teichman test.

Plasma Proteins

Properties of plasma proteins   Plasma proteins are large molecules with molecular weights ranging mostly from 50 000 to 300 000 Daltons. With the exception of albumin, nearly all plasma proteins are glycoproteins containing oligosaccharides. The oligosaccharide

Table 22.**5**   Functions of plasma proteins.

Function	Plasma protein
Nutrients	Lipoproteins
Enzymes	Amylase, alkaline phosphatase
Hormones	Anterior pituitary hormones, angiotensin
Antibodies	Gamma globulin
Clotting and fibrinolytic factors	Fibrinogen, prothrombin, fibrinolysin
Carriers	Albumin, cerruloplasmin, transferrin
Scavengers	Gel-solin, Gc protein

chains are responsible for certain properties of plasma proteins like solubility, viscosity, charge, denaturation, etc. Like most other proteins, their charged residues tend to be located on the surface. Many plasma proteins exhibit polymorphism.[1] Plasma proteins showing polymorphism are haptoglobin, transferrin, cerruloplasmin, and immunoglobulins.

Certain properties of plasma protein molecules are attributable to their large size and high molecular weight. For example, unlike electrolytes or other smaller molecules, they can be separated from the plasma by ultracentrifugation. They are unable to pass across the capillary membrane and consequently exert an oncotic pressure of about 25 mmHg. Owing to their size and particularly their shape, they greatly contribute to blood viscosity. The plasma protein fibrinogen is a significant contributor to blood viscosity.

Another set of properties of plasma proteins is attributable to the presence of polar residues like NH_3^+ and COO^- on their molecular surface. Thus, plasma protein molecules are soluble in water and show electrophoretic mobility. They are amphoteric in nature because the polar residues comprise both positive (NH_3^+) and negative (COO^-) groups. The plasma proteins act as efficient buffers by virtue of their amphoteric nature. Their polar residues bind easily with, and therefore serve as good carriers for, metallic ions and steroids.

Functions of plasma proteins   The two general functions of plasma proteins are (1) holding back the fluid portion of blood inside the blood vessels by virtue of the osmotic pressure they exert, and (2) buffering body fluids. Besides these general functions, different plasma proteins have specific functions: they can act variously as nutrients, enzymes, hormones, antibodies,

---

[1] Polymorphism is a Mendelian trait that exists in the population in at least two phenotypes, neither of which is rare.

clotting and fibrinolytic factors, and as carrier molecules (Table 22.**5**).

The carrier function of plasma proteins helps in prolonging the half-life of hormones in blood. Binding the hormones to plasma proteins prevents the rapid filtration of hormones through the glomeruli. The protein-bound hormone also acts as a reservoir of the hormone. When the free hormone level falls, the bound hormone readily dissociates from the carrier protein and restores the free hormone level.

The scavenger function of plasma proteins comes into play after cell necrosis which is associated with the release of large amounts of actin (present in the cytoskeleton of cells) into the circulation. These are scavenged by gel-solin, which depolymerizes F-actin, and Gc protein (also called the vitamin D-binding protein), which binds to G-actin.

Types of plasma proteins   Only two types of plasma proteins, albumin, and globulin, were originally separated using the salting out technique. Electrophoretic techniques later showed the globulin fraction to be highly heterogeneous. The term globulin now signifies a large group of plasma proteins (designated as α_1, α_2, β, and γ), each with a specified range of electrophoretic mobility (Fig. 2.**6**). The albumin fraction is more homogeneous. The term albumin has therefore been retained to denote the largest single protein in the albumin fraction and the remaining proteins in that fraction have been given other names, e.g., prealbumin.

The plasma concentration of total proteins in adults is about 7 g dl^{-1} of which albumin accounts for more than half, i.e., 4 g dl^{-1}. Albumin molecules far outnumber globulin molecules and therefore albumin is the principle contributor to plasma oncotic pressure. When plasma albumin falls, the plasma oncotic pressure decreases and results in edema.

Synthesis of plasma proteins   The γ globulins (immunoglobulins) are produced by plasma cells. Most other plasma proteins are synthesized in the liver. However contributions to plasma proteins also come from macrophages (complement), intestinal cells (apoproteins), and endothelial cells (coagulation factors).

The relation of plasma proteins to diet has been studied in dogs rendered hypoproteinemic by repeated plasmapheresis.[2] It was noted that the 10 essential amino acids have to be provided in the diet for the satisfactory synthesis of plasma proteins. Usually, dietary proteins of animal origin favor albumin synthesis while those of plant origin favor globulin synthesis.

Hypoproteinemia due to increased protein losses occurs in nephrotic syndrome and in protein-losing enteropathy. Hypoproteinemia due to reduced protein synthesis occurs in liver diseases, malabsorption, malnutrition, prolonged starvation, and chronic inflammation (due to inhibition of hepatic synthesis of albumin by inflammatory mediators). The mid-pregnancy fall in total plasma protein concentration is largely due to hemodilution and occurs despite the increase in hepatic synthesis of globulin.

Hyperproteinemia is seen in acute inflammation and in multiple myeloma. Globulins increase sharply during any acute inflammation. These are called acute phase proteins. They include C-reactive proteins (CRP), so called because it reacts with C-polysaccharide of pneumococci. Other acute phase proteins are α-antitrypsin, haptoglobin, von Willebrand factor, and fibrinogen. C-reactive proteins increase in chronic inflammation and malignancy too. These acute phase proteins are important for non-specific immunity of the body. In multiple myeloma, plasma cells secrete large amounts of immunoglobulins resulting in hypergammaglobulinemia.

[2] Plasmapheresis is the procedure for collecting plasma from donors without depleting their blood cells. Whole blood is withdrawn from the donor, the plasma is separated from the cells by centrifugation, and the cells are returned to the donor's circulatory system without fluid replacement. The term plasmapheresis is often used interchangeably with the term therapeutic plasma exchange, the procedure used for removing excess antibodies from the blood in certain immunological disorders. In plasma exchange, the blood cells are reconstituted in saline, plasma substitutes or donor plasma before being returned to the patient.

23 Red Blood Cells

The red blood cells or *erythrocytes* are biconcave, circular disk-shaped cells without a nucleus, mitochondria or ribosomes. The average life span of a red cell in circulation is 120 days. The red cell count is highest on the first day of life. The changes in red cell count with age parallel the changes in hemoglobin concentration (see Fig. 24.**1**). The red cell counts in adult males and females are given in Table 22.**4**.

An abnormally high red cell count is called *polycythemia*, and is seen in hypoxic conditions. The malignant form of polycythemia is called *polycythemia vera*. Whether benign or malignant, polycythemia is associated with high blood viscosity and increased resistance to blood flow, resulting in stagnant hypoxia and peripheral cyanosis (p 355).

The sole importance of the red cell lies in the presence of hemoglobin in its cytoplasm. The advantages of carrying hemoglobin within a cell as against in the free form in the plasma are threefold. (1) It prevents the rapid destruction and elimination of the hemoglobin. (2) It prevents the marked increase in the plasma viscosity that would occur if hemoglobin existed as a plasma protein. (3) It prevents blood from exerting a high osmotic pressure across the capillary wall.

The biconcave shape of the red cell gives it a large surface-to-volume ratio, which represents the most efficient shape for rapid gaseous exchange. The red cell is flexible and it is readily distorted during its passage in the circulation; thus as it passes through capillaries, it assumes a parachute-like configuration (Fig. 23.**1**A). Erythrocytes are mostly round but a few of them are slightly oval. The term *poikilocytosis* refers to an excessive variation in red cell shape. The mean diameter of red cells as measured directly in a film is 7.2 μm. The disk thickness is 2 μm. Excessive variation in red cell size is called anisocytosis. It may be due to an increase in the number of small or large cells or both.

Rouleaux formation Within a blood vessel, in the absence of significant flow, the red cells tend to form rouleaux, i.e., they tend to align themselves side by side, forming stacks (Fig. 23.**1**B). It is a reversible phenomenon. The protein coating of red cells plays a major role in rouleaux formation. Generally all large molecules that are not spherical are capable of inducing rouleaux formation, especially fibrinogen. Albumin and globulins also induce rouleaux formation, but only at concentrations much higher than normal.

Chromicity Red cell staining is deeper at the periphery and fades at the center (the central pallor). Cells that stain normally are assumed to have a normal concentration of hemoglobin and are called *normochromic*. The term hypochromia is used to describe a decrease in the intensity of staining. Hypochromia is nearly always associated with a decrease in the mean cell hemoglobin concentration (MCHC), i.e., a decrease in the concentration of hemoglobin in the red cells (see below). However, cells that are thinner than normal (as in thalassemia) may appear slightly hypochromic even though the MCHC is normal. The term *hyperchromia* is used to describe an increase in the intensity of staining of the red cell, in which the cell stains more deeply and the central pallor is not apparent. Such an appearance is usually due to an increase in the thickness of the cell, and not to an increase in MCHC.

Spherocytes show both an increase in thickness and an elevated MCHC. They certainly are hyperchromic. In megaloblastic anemia, the red cell diameter is increased but MCHC remains normal. Whether or not the cells should be called hyperchromic depends on what is meant by the term hyperchromia. Considering the

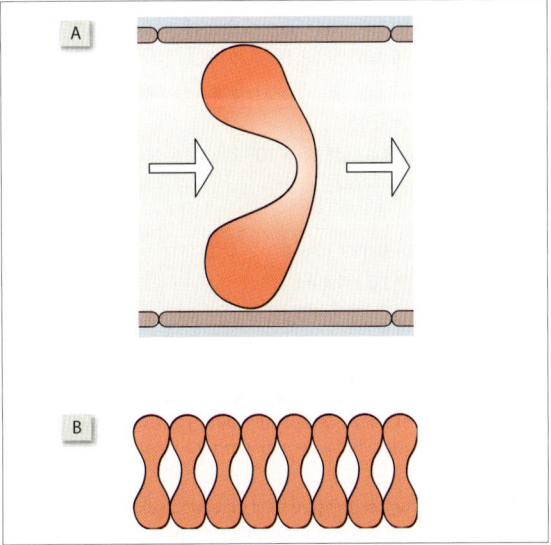

Fig. 23.**1** [A] A red cell assumes a parachute-like shape for squeezing through a capillary. [B] Rouleaux formation.

ambiguity of the term hyperchromia, it is best abandoned in favor of the more objective terms like MCH and MCHC.

Volume of packed red cells (VPRC) is total volume of packed red cells[1] in 100 ml of blood. It is determined by centrifuging the blood at a high speed in a graduated centrifuge tube.[2] The centrifugation results in the separation of three layers: a supernatant layer of plasma, a sediment layer of red cells, and a thin buffy coat separating the two. The *buffy coat* is made of leukocytes and platelets. The volume of red cell sediment gives the VPRC. The VPRC is a useful clinical test that is used for detecting anemia and polycythemia. However, a normal red cell count or VPRC does not rule out an absolute deficiency or excess of red cells in the body: it might be a case of hemodilution (as in pregnancy) or hemoconcentration (as in dehydration).

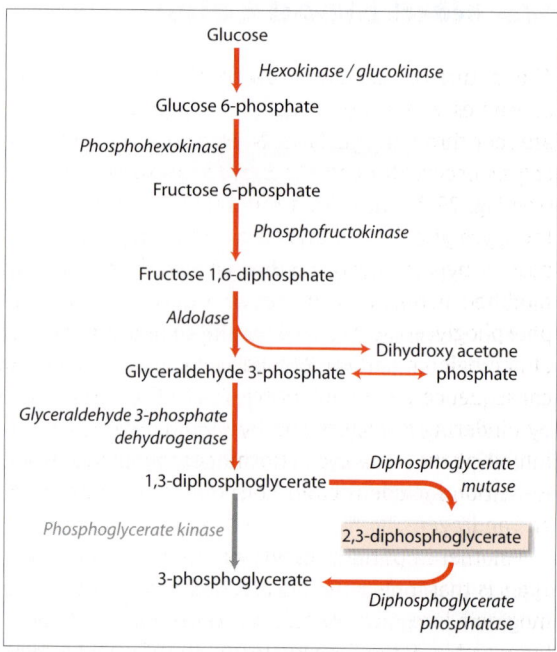

Fig. 23.**2** The Embden Meyerhof pathway in erythrocytes. Compare it with Figure 79.**3**.

■ Red cell indices

The red cell indices refer to the mean corpuscular volume (MCV), the mean corpuscular hemoglobin (MCH) and the mean corpuscular hemoglobin concentration (MCHC). These indices are calculated from VPRC, the hemoglobin concentration, and the red cell count. The three indices are related by the formula:

$$MCHC = \frac{MCH}{MCV}$$

The **mean cell volume** (MCV) is the average volume of the red cells. It is calculated by dividing the packed cell volume (VPRC) by the red cell count. The result is expressed in femtoliters (fl) or cubic micrometers (μ^3).

Example

$$VPRC = 0.45$$
$$\text{Red cell count} = 5 \times 10^{12} \text{ L}^{-1}$$
$$MCV = \frac{0.45}{5 \times 10^{12} \text{ L}^{-1}}$$
$$= 0.09 \times 10^{-12} \text{ L}$$
$$= 90 \times 10^{-15} \text{ L}$$
$$\text{Normal MCV} = 85 \pm 8 \text{ fl}$$

---

[1] The term VPRC emphasizes that only the volume of red cells, and not other cells like leukocytes and platelets, is to be measured. It is therefore preferable to the term packed cell volume (PCV) that was used earlier.

[2] When hematological measurements are made electronically, VPRC is computed from the product of red cell count and the mean red cell volume. When thus computed, the term VPRC becomes a misnomer since no 'packing' of red cell is involved. The term hematocrit is therefore currently preferred over VPRC.

The **mean cell hemoglobin** (MCH) is the average mass of hemoglobin (in picograms) contained in each red cell. It is increased in spherocytosis and megaloblastic anemia. The MCH is calculated by dividing the amount of hemoglobin in 1 L blood by the number of red cells present in 1 L of blood (the red cell count).

Example

$$\text{Hemoglobin concentration} = 15 \text{ g dl}^{-1} (150 \text{ g L}^{-1})$$
$$\text{Red cell count} = 5 \times 10^{12} \text{ L}^{-1}$$

$$MCH = \frac{150 \text{ g L}^{-1}}{5 \times 10^{12} \text{ L}^{-1}}$$

$$= 30 \times 10^{-12} \text{ g}$$
$$\text{Normal MCH} = 30 \pm 2 \text{ pg}$$

The **mean cell hemoglobin concentration** (MCHC) is the average concentration of hemoglobin in an erythrocyte. MCHC is increased in spherocytosis and sickle cell anemia. MCHC is calculated by dividing the amount of hemoglobin in 1 L of blood by the volume of packed cells in 1 L blood (VPRC).

Example

$$\text{Hemoglobin concentration} = 15 \text{ g dl}^{-1}$$
$$VPRC = 0.45$$
$$MCHC = \frac{15 \text{ g dl}^{-1}}{0.45}$$
$$= 33.33 \text{ g dl}^{-1}$$
$$\text{Normal MCHC} = 33 \pm 2 \text{ g dl}^{-1}$$

Red cell metabolism

The mature red cell has a low respiratory activity and consumes very little oxygen. Its energy requirements are met through glycolysis. Ninety percent of the glycolysis occurs through the *Embden Meyerhof pathway* (see Fig. 79.2) but with a difference: In erythrocytes, the glycolytic step catalyzed by phosphoglycerate kinase is bypassed (Fig. 23.**2**). The significance of this modified pathway is the production of 2,3 DPG (diphosphoglycerate) that influences the oxygen affinity of hemoglobin and thereby has important physiologic consequences. Acidemia decreases DPG in the red cell by hindering glycolysis and hypoxia increases DPG by inhibiting the Krebs cycle. Hormones that increase DPG formation include thyroid hormone, growth hormone and androgens.

Another important feature of glycolysis in erythrocytes is that it generates NADH + H$^+$. Normally methemoglobin is formed continually by the autooxidation of hemoglobin. Simultaneous reduction of methemoglobin by NADH + H$^+$ keeps the proportion of methemoglobin below 1% of the hemoglobin content.

Ten percent of glycolysis occurs through the *pentose phosphate pathway*, also called the hexose monophosphate (HMP) shunt. The significance of this pathway is that it generates NADPH + H$^+$ which is required for the reduction of glutathione. Reduced glutathione protects the sulphydryl groups of hemoglobin against oxidation. NADPH + H$^+$ is also needed in some way for the maintenance of normal red cell fragility.

Red Cell Turnover

Erythropoiesis

In the adult, the erythrocytes are produced in the bone marrow. The erythrocytic precursor cells are the proerythroblast, erythroblasts (early, intermediate, and late), and reticulocytes.

Erythropoietin   The production of erythrocytes is regulated by a hormone called *erythropoietin*. Erythropoietin is a glycoprotein secreted mainly by the kidney (85%) and liver (15%). In the kidney, erythropoietin-secreting cells are found in the interstitium of the inner cortex and outer medulla, lying just outside the tubular basement membrane.

Erythropoietin increases erythropoiesis by acting on the bone marrow as well as the fetal yolk sac, liver, and spleen where it promotes: (1) differentiation of erythropoietic stem cells to proerythroblasts, (2) proliferation of committed stem cells, (3) hemoglobin synthesis, by increasing globulin synthesis and potentiating δ-amino levulinic acid synthetase, and (4) release of erythrocytes from bone marrow.

The factors that promote erythropoietin secretion are: (1) hypoxia, (2) alkalosis, (3) hormones, like the growth hormone, prolactin, thyroid hormones, catecholamines, corticosteroids, and androgens, and (4) hemolysates (products, mainly nucleotides released following red cell destruction) like cAMP, NAD$^+$, and NADP$^+$. Stimulation of erythropoietin by hypoxia is a physiological feedback mechanism through which the number of erythrocytes is increased whenever there is increased destruction of red cells resulting in the release of hemolysates and causing O$_2$ deficiency in the tissues. The increase in red cell count tends to improve tissue oxygenation.

Nutritional requirements for erythropoiesis   Vitamin B$_{12}$ and folate are essential for DNA synthesis. Deficiency of either of them results in megaloblastic anemia. Iron is essential for hemoglobin synthesis. Deficiency of iron, occurring mostly due to inadequate intake, is one of the commonest causes of anemia. An adequate supply of high-quality protein in the diet is essential for supplying amino acids for the synthesis of the globin of hemoglobin. Hemoglobin gets a high priority for available protein and hence, the protein deficiency must be very marked before hemoglobin synthesis is impaired.

Red cell destruction

About 1% of the red cells, the older and the abnormal ones, are removed daily from circulation by macrophages lining the walls of sinusoids where the blood flow is slow, as in the liver, spleen, and bone marrow. Inside the macrophage, hemoglobin is degraded into bilirubin after separating the iron from it. The iron and bilirubin are released into circulation.

Red cell fragility   Due to certain pathological changes in its membrane or in its contents, the red cell sometimes becomes mechanically more fragile, i.e., more vulnerable to deforming stresses than normal red cells. Such cells are removed from circulation by macrophages in larger numbers than normal cells. A convenient method for testing red cell fragility is to test for its *osmotic fragility*. Erythrocytes are suspended in a series of saline solutions with strengths ranging from 0.9 to 0.3%. Red cells with normal osmotic fragility start showing hemolysis (red cell rupture with loss of hemoglobin) in 0.5% saline and are completely hemolyzed in 0.35% saline (Fig. 23.**3**). However, a normal or low osmotic fragility does not rule out the possibility

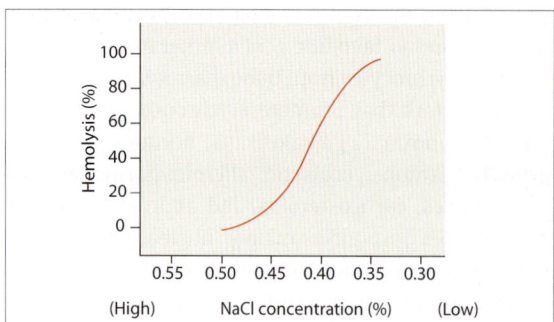

Fig. 23.**3**   Osmotic fragility curve. The curve shifts to the left when osmotic fragility of red cells increases.

of a high mechanical fragility. In sickle cell anemia the sickle-shaped erythrocytes have a high mechanical fragility. However, the osmotic fragility of the sickle cells is normal or even low.

In **hereditary spherocytosis**, the spherocytes (spherical erythrocytes) hemolyze more readily, i.e., they start hemolyzing at higher saline concentration. This is because spherocytes, as the name suggests, are rather more spherical than diskoid which leaves them with little margin for swelling in hypotonic solutions. Spherocytes are removed by the spleen in large numbers, resulting in hemolytic anemia (see below). Spherocytosis is caused by abnormalities of the protein network that maintains the shape and flexibility of the red cell membrane (see Fig. 5.**2**B). Ankyrin, a peripheral protein located on the inside of the membrane, anchors spectrin (a cytoskeletal protein in the red cell) to band-3, (an integral protein of erythrocyte membrane). Defects in band-3, spectrin, and ankyrin have all been observed in the spherocytes.

G6PD deficiency   The deficiency of the enzyme glucose 6-phosphate dehydrogenase (G6PD), a key enzyme in the pentose phosphate pathway, reduces the production of NADPH + H$^+$. In the absence of adequate NADPH + H$^+$, the susceptibility of red cells to hemolysis increases. Severe G6PD deficiency also inhibits the ability of granulocytes to kill bacteria and thereby predisposes to severe infections.

Anemias

Anemia is defined as a reduction in the concentration of hemoglobin in the peripheral blood below the normal for the age and sex of the patient. Thus, an adult male is said to be anemic when his hemoglobin falls below 13.0 g dl^{-1} and an adult female, when her hemoglobin falls below 11.5 g dl^{-1}. Somewhat arbitrarily, anemia is called moderate when it falls below 9 g dl^{-1} and severe when it falls below 6 g dl^{-1}. The fall of the hemoglobin concentration below normal values is usually, but not always, accompanied by a fall in the red cell count. Thus occasionally, especially in the hypochromic microcytic anemia of iron deficiency, the red cell count is normal although the hemoglobin is significantly reduced due to the low hemoglobin content of individual cells.

There are two main classifications of anemia, the etiological classification, based on the cause of the anemia and the laboratory classification, based on the characteristics of the red cell.

The **laboratory classification** is based on MCV and MCHC. In normocytic anemias, the MCV is within the normal range. Most *normocytic anemias* are also normochromic with a normal MCHC, but mild hypochromia may occur occasionally. If the erythrocytes are normocytic and normochromic, why should there be anemia at all? The anemia in these cases results from a low red cell count which reduces the hemoglobin concentration of blood. In *microcytic anemias*, the MCV is reduced. Microcytic anemias are mostly hypochromic with the MCHC reduced. In *macrocytic anemias*, the MCV is increased. Most macrocytic anemias are normochromic but a mild hypochromia (reduced MCHC) may occur in some cases.

The **etiological classification** of anemias is based on the cause of anemia (Fig. 23.**4**). Broadly, they can be due to deficiency of hemopoietic factors (deficiency anemia), excessive breakdown of erythrocytes (hemolytic anemia) or excessive blood loss (hemorrhagic anemia). *Deficiency anemias* occur due to impaired red cell formation, which can occur when there is deficiency of iron (iron-deficiency anemia), or due to vitamin B$_{12}$ or folate deficiency (megaloblastic anemia). *Hemolytic anemias* are relatively uncommon. Red cells are destroyed in large numbers either due to some defect in the red cell itself (intracorpuscular defect), or due to some defect in its immediate environment (extracorpuscular defect). *Hemorrhagic anemia* is a common form of anemia and can be due to acute or chronic blood loss. Anemia following acute blood loss (acute post-hemorrhagic anemia) results in hemodilution. This is because the plasma is restored much faster than the cellular elements. The result is a normocytic, normochromic anemia. Anemia due to chronic blood loss (chronic posthemorrhagic anemia) tends to show features of iron deficiency, i.e., a microcytic, hypochromic anemia. *Hemodilutional anemia* occurs when there is hypervolemia due to fluid retention. The physiological anemia of pregnancy occurs due to the hemodilution caused by progesterone.

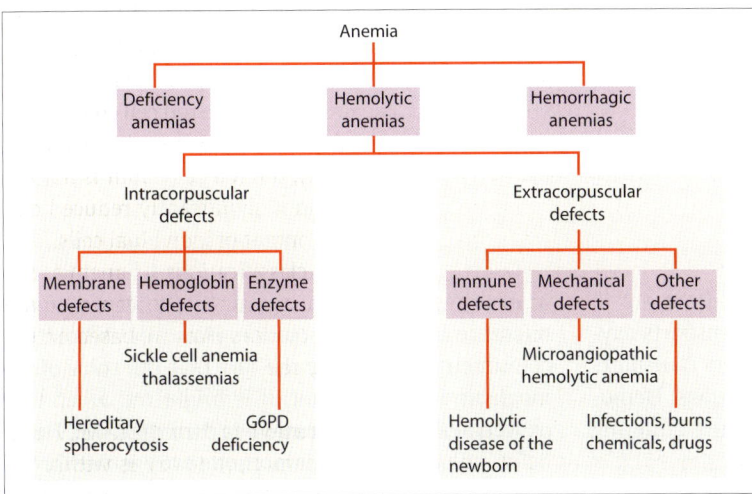

Fig. 23.**4**   Etiological classification of anemias.

Signs and symptoms Anemia results in reduced oxygen-carrying capacity of blood, resulting in tissue hypoxia. The hypoxia causes symptoms like easy fatigability (due to muscle hypoxia) and faintness (due to cerebral hypoxia), especially on exertion. The hypoxia brings about several compensatory responses. Although these responses alleviate tissue hypoxia, they also produce symptoms like *breathlessness* due to compensatory stimulation of the respiratory center, and *palpitation* due to compensatory increase in cardiac output. The raised cardiac output creates turbulence of blood during its passage through the cardiac valves, resulting in clinical signs like *cardiac murmurs*. The reduction in red cell count reduces blood viscosity and thereby contributes to the turbulence and the resultant murmurs (see Reynolds number, p 10).

Blood picture In *megaloblastic anemia*, there is marked macrocytosis, anisocytosis, and poikilocytosis. There is neutropenia with aged, hypersegmented neutrophils, and thrombocytopenia. This is because the production of other blood cells is also affected due to impaired DNA synthesis.

In *iron-deficiency anemia*, the erythrocytes are typically microcytic and appear hypochromic in blood film. Both MCH and MCHC are reduced. The red cell count is also reduced.

In *hemolytic anemias*, the reticulocyte count is increased due to the appearance of larger reticulocytes which are less mature. These are released prematurely from bone marrow into circulation under the effect of erythropoietin in response to increased demand. Hence these reticulocytes are called *shift reticulocytes*. A few nucleated red cells (erythroblasts released prematurely into circulation) may also be seen.

Bone marrow picture The state of bone marrow reflects either the cause of the anemia (as in deficiency anemias) or the effect of anemia (as in hemolytic and hemorrhagic anemias).

In *megaloblastic anemia*, the proerythroblasts and erythroblasts show megaloblastic changes and are respectively called promegaloblasts and megaloblasts. Compared to their erythroblastic counterparts, megaloblasts have the following features: (1) The cell is larger with a larger nucleus and more cytoplasm. (2) The chromatin is more reticular (loose mesh) than the erythroblast which has a more clumped chromatin. Clumping of chromatin is a sign of cell maturity. (3) Hemoglobinization of the cytoplasm proceeds normally. Thus, nuclear maturation (indicated by the state of chromatin) can be said to be lagging behind cytoplasmic maturation (indicated by hemoglobinization). (4) Due to maturation arrest, the primitive precursor cells (like promegaloblast) are more numerous than usual.

In *iron deficiency anemia*, the bone marrow shows proliferation of the precursor cells (erythroid hyperplasia) with a larger proportion of the mature forms. Some of the precursor cells show scanty, polychromatic cytoplasm (a sign of cytoplasmic immaturity) with a pyknotic nucleus (a sign of nuclear maturity). It indicates that cytoplasmic maturation lags behind nuclear maturation.

In *hemolytic anemias*, the red cell precursors show excessive proliferation (erythroid hyperplasia). Normally, white cell precursors are 3 to 5 times more numerous than red cell precursors, i.e., the myeloid to erythroid ratio is 3–5:1. In erythroid hyperplasia, their numbers become roughly equal (1:1). The marrow-space in the bone widens, resulting in detectable bone changes in radiographs.

24 Hemoglobin

Hemoglobin is an oxygen-binding protein present in the cytoplasm of the red blood cells. It transports oxygen from the lungs to the tissues and also transports carbon dioxide from the tissues to the lungs. Hemoglobin is also the largest contributor to the buffering capacity of blood.

Hemoglobin in the newborn The hemoglobin concentration of blood from the umbilical cord averages 16.5 g dl^{-1}. The cord blood is representative of the infant's blood before birth. Shortly after birth the hemoglobin value of normal infants increases rapidly, the hemoglobin on the first day of life being 18.5 g dl^{-1}. This is due to two reasons. (1) There is transfusion of red cells from the placenta to the infant. Hemoglobin values are significantly higher in infants in whom the cord is not tied for a few minutes after delivery, allowing extra placental blood to pass into the infant. (2) There is rapid reduction of plasma volume in the neonate, resulting in hemoconcentration. The reduction in plasma volume allows the infant to accommodate the extra maternal blood cells obtained through transfusion. After the first two days, there is a marked fall in hemoglobin concentration for two weeks, stabilizing by the third month at 12.0 g dl^{-1}. It starts rising again at the end of the first year towards the adult level. An adolescent spurt occurs in boys but not in girls. The adult values of hemoglobin are 15.5 ± 2.5 g dl^{-1} (for men) and 14.0 ± 2.5 g dl^{-1} (for women).

Estimation of hemoglobin Being a colored substance, the hemoglobin concentration of a blood sample can be easily estimated by comparing its color to that of a reference hemoglobin solution. However, difficulty in colorimetry is posed by the fact that the color of hemoglobin is not constant; it is bright red when fully oxygenated and bluish red when deoxygenated. Hence before colorimetry, it is important to oxygenate the hemoglobin completely (the oxyhemoglobin method) or to denature it using acid (acid-hematin method), alkali (alkali-hematin method) or cyanide (cyanmethemoglobin method).

Structure of hemoglobin Hemoglobin (molecular weight 64,450 Daltons) is a globular molecule comprising four subunits, each subunit consisting of a polypeptide chain and *heme*, an iron-containing porphyrin. The four polypeptide chains present in a hemoglobin molecule together constitute the protein called *globin*. One gram of hemoglobin contains 3.4 mg of iron and can carry up to 1.34 ml of oxygen.

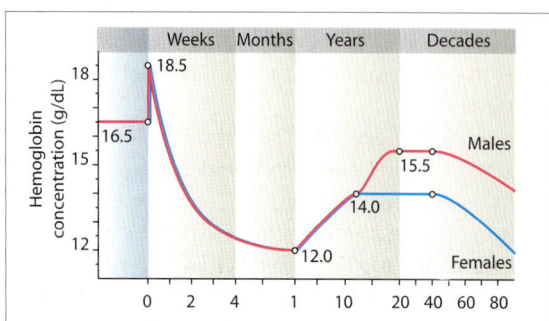

Fig. 24.**1** Hemoglobin concentration at different ages.

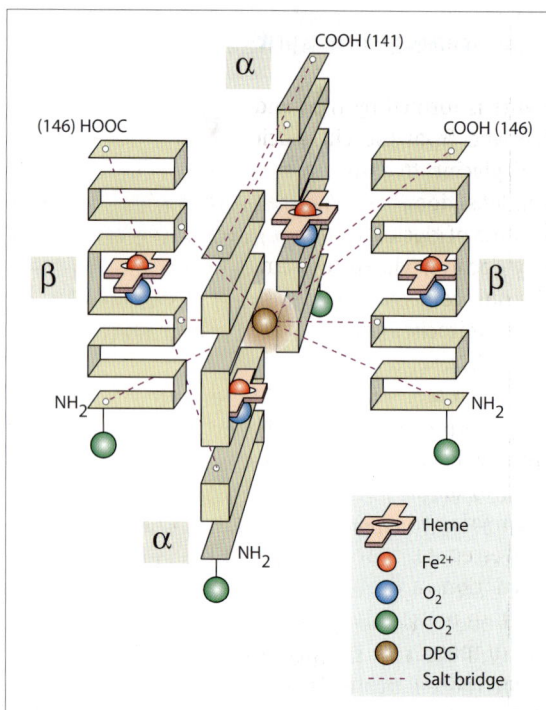

Fig. 24.**2** Structure of hemoglobin shown schematically.

Different forms of hemoglobin have different polypeptide chains in them. Hemoglobin A, which comprises 97% of the hemoglobin in adults, is made of two α chains, each containing 141 amino acid residues and two β chains, each containing 146 amino acid residues. It is symbolically written as HbA ($\alpha_2\beta_2$). There are salt links (noncovalent, electrostatic bonds) between the terminal carboxyl residues (COO⁻) and the intermediate amino residues (NH₃⁺) of the four polypeptide chains. The iron atom of heme is attached between the histidine residues of the α chain at positions 58 and 87 (also called His-E7 and His-F8 respectively). In the β chain, the iron atom is attached between histidine residues at positions 63 and 92. In the deoxygenated state, the iron atom is located a little outside the plane of the ring in the direction of His-F8 (Fig. 24.**2**).

Fetal hemoglobin or HbF ($\alpha_2\beta_2$) is the main form of hemoglobin in the fetus. It has greater affinity for oxygen. It is more resistant to alkali denaturation and therefore the routine alkali-hematin method fails to estimate HbF unless the solution is heated in a water bath for 4 minutes. HbF disappears from the red cells of normal infants by the age of 6 months although very small amounts (less than 2%) can be detected in the red cells of most children and adults.

HbA₂ ($\alpha_2\beta_2$) is present in small amounts (< 3% in normal adults). Hb-Bart's (γ_4) is present in the fetus in small amounts. Hb-Gower 1($\zeta_2\epsilon_2$), Hb-Gower 2($\alpha_2\epsilon_2$), and Hb-Portland ($\xi_2\epsilon_2$) are embryonic forms of hemoglobin.

Hemoglobin Synthesis

Heme is formed by the condensation of succinyl-CoA (derived from the citric acid cycle in mitochondria) and glycine to form α-amino-β-ketoadipic acid. The condensation requires pyridoxal phosphate for the activation of glycine. α-amino-β-ketoadipic acid gets rapidly decarboxylated to δ-amino levulinate (ALA). It is the rate-controlling step in heme synthesis and occurs in the mitochondria in the presence of ALA synthetase (Fig. 24.**3**).

Two molecules of ALA condense in the cytosol to form porphobilinogen in the presence of ALA dehydratase. Four molecules of porphobilinogen condense to form uroporphyrinogen-III in the presence of uroporphyrinogen synthetase. Uroporphyrinogen-III is converted to coproporphyrinogen-III by the decarboxylation of the acetate groups (A), which changes them to methyl (M) groups. The reaction is catalyzed by uroporphyrinogen decarboxylase. Coproporphyrinogen-III then enters the mitochondria, where it is converted to protoporphyrinogen-III by the mitochondrial enzyme coproporphyrinogen oxidase which

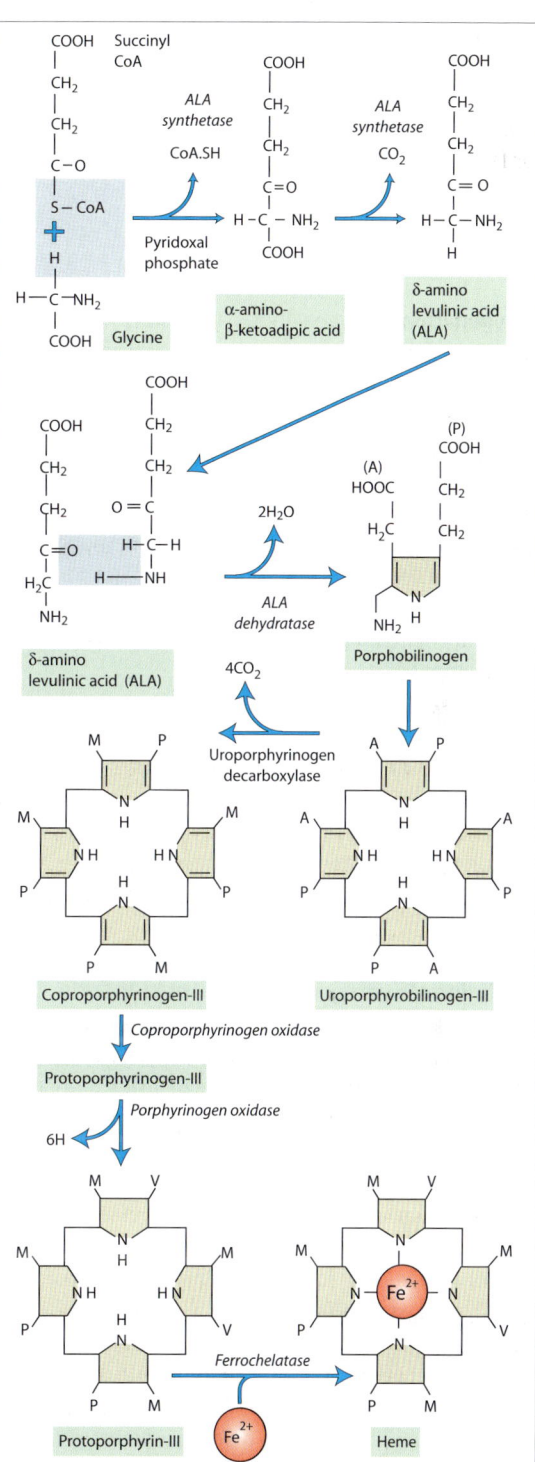

Fig. 24.**3** Steps in hemoglobin synthesis. A=Acetate, P=propionate, M=Methyl, V=Vinyl

catalyzes the decarboxylation and oxidation of two propionic acid chains into vinyl groups. Another mitochondrial enzyme porphyrinogen oxidase converts protoporphyrinogen-III to protoporphyrin-III. Protoporphyrin is a component of a number of cellular enzymes (e.g., cytochromes, catalase, and P-450 microsomal enzymes). The final step in heme synthesis involves the incorporation of ferrous iron into protoporphyrin in a reaction catalyzed by heme synthetase or ferrochelatase, another mitochondrial enzyme.

Reactions of Hemoglobin

Oxygenation Oxygen binds reversibly with hemoglobin to form oxyhemoglobin. The oxygen molecule occupies the sixth coordinate position of the iron atom, which remains in the ferrous state. The insertion of the O_2 molecule has the following effects. The iron atom moves into the plane of the heme molecule. As the iron atom moves, it pulls the His-F8 along it, distorting the polypeptide chains. The distortion of the polypeptide chains results in the rupture of the salt links between the chains and the release of H^+ ions. The quaternary structure of hemoglobin changes from the taut (T) state to the relaxed (R) state.

A total of four O_2 molecules bind to a molecule of hemoglobin, one each to a heme molecule. The insertion of the first molecule of O_2 ruptures all the salt links. Subsequent molecules of oxygen therefore find it easier to bind to hemoglobin. This phenomenon is called *cooperative binding kinetics*. It results in the sigmoid O_2-dissociation curve of hemoglobin (see Fig. 52.01).

Buffering of hydrogen ions H^+ ions bind to the NH_2 of the intermediate histidine residues to form NH_3^+. These NH_3^+ groups interact with the terminal COO^- groups to form salt links. Formation of salt-links reduces the binding of O_2 to hemoglobin. Conversely, binding of O_2 to hemoglobin breaks the salt links and releases the H^+ ions. Thus, oxygenation of hemoglobin reduces the buffering capacity of hemoglobin. Conversely, the buffering of H^+ by hemoglobin decreases its oxygenation. The buffering of H^+ ions is also important for transport of CO_2 in plasma as bicarbonates.

Formation of carbamino hemoglobin CO_2 binds avidly to the terminal αNH_2 groups of valine in the β polypeptide chain and less avidly, to the ones in the α polypeptide chain.

$$R-\underset{\underset{H}{|}}{N}-H \quad + \quad CO_2 \quad = \quad R-\underset{\underset{H}{|}}{N}-COO^- \quad + \quad H^+$$

The reaction produces a carbamate and releases H^+, which is buffered by the intermediate histidine residues of hemoglobin. These reactions favor the formation of salt links that make oxygenation more difficult. Thus, whenever CO_2 is transported in blood,

whether as carbamino hemoglobin or as bicarbonates, it is associated with a reduction in the oxygenation capacity of hemoglobin. This phenomenon is called the *Bohr effect*.

Reaction with 2,3 diphosphoglycerate (DPG) Only a single molecule of DPG binds to a molecule of hemoglobin. The DPG molecule occupies the central cavity enclosed by the four polypeptide chains of hemoglobin. It is held in place by salt bridges with three NH_3^+ groups on each β polypeptide chain. One is the NH_3^+ group of the terminal valine residue while the two others belong to the intermediate lysine and histidine residues. Formation of the salt links makes it difficult for O_2 to bind to hemoglobin and shifts the O_2 dissociation curve to the right (see Fig. 52.**2**).

Formation of carboxyhemoglobin Carbon monoxide reacts with hemoglobin to form carboxyhemoglobin. The affinity of hemoglobin for CO is 200 times higher than its affinity for O_2. It binds exactly where O_2 binds to heme.

Formation of methemoglobin The ferrous iron of hemoglobin is susceptible to oxidation by superoxides and other oxidizing agents, forming methemoglobin (hemoglobin with its iron in ferric form) which cannot carry oxygen. The red cell possesses an effective *methemoglobin reductase system* for reducing Fe^{3+} back to Fe^{2+}. The system comprises $NADH + H^+$, cytochrome b5, and cytochrome b5 reductase.

$$Hb-Fe^{3+} + Cyt\ b5\ red \rightarrow Hb-Fe^{2+} + Cyt\ b5\ ox$$

$$Cyt\ b5\ ox + NADH + H^+ \xrightarrow{Cyt\ b5\ reductase} Cyt\ b5\ red + NAD^+$$

Methemoglobinemia can occur due to genetic deficiency of methemoglobin reductase. It is also caused by the ingestion of certain drugs like sulphonamides. When more than 10% of the normal hemoglobin changes to methemoglobin, it causes a dusky discoloration of the skin resembling cyanosis.

Glycosylation of hemoglobin Hemoglobin is nonenzymatically glycosylated by the glucose that enters the erythrocytes. The amount of glycosylated hemoglobin (HbA_{1c}) is proportionate to the blood glucose concentration and is normally about 5%. The HbA_{1c} level reflects the average blood glucose concentration over the preceding 6 to 8 weeks and serves as an index of long-term control of diabetes mellitus. Thus a patient who is mostly careless about controlling his hyperglycemia and takes an insulin injection only before visiting the physician would have a normal blood glucose but his HbA_{1c} would remain elevated.

Hemoglobin Disorders

Hemoglobinopathies

Hemoglobinopathies refer to abnormalities in the amino acid sequence of the polypeptide chains of hemoglobin. Examples of hemoglobinopathies are HbS, HbC, HbE, and HbD Punjab. They occur due to a defect in the gene that directs the synthesis of these polypeptide chains. Several abnormal hemoglobins have no harmful effects but HbS results in sickle cell anemia, a potentially fatal disorder.

Sickle cell anemia In HbS, the α chains are normal but the β chains are abnormal, with a valine residue at position 6 instead of the usual glutamic acid. This substitution results in the formation of certain reactive sites called *sticky patches* on the β chain, which bind to *sticky receptors* on the α chains of HbS. This binding causes polymerization of HbS (Fig. 24.**4**) forming long, fibrous precipitates. The precipitates distort the erythrocyte, making it sickle shaped. Since polymerization results in fewer molecules of hemoglobin, the osmotic pressure exerted by hemoglobin decreases and the sickle cell becomes dehydrated. For the same reason, the sickle cell can imbibe more water without bursting and its osmotic fragility is lower than normal cells. The mechanical fragility of sickle cells is however higher because polymerization of hemoglobin deforms the membrane and makes it rigid. The membrane also gets altered chemically, favoring the deposition of IgG and complement on its surface. Sickling of red cells causes two types of problems.

(1) Sickle cells are hemolyzed in large numbers, resulting in hemolytic anemia. The hemolysis is both intravascular and extravascular. *Intravascular hemolysis* occurs due to cell lysis caused by complement activation. It also occurs due to the fragmentation of the rigid cells due to mechanical stress as it passes through narrow capillaries. *Extravascular hemolysis* occurs due to excessive phagocytosis of the IgG-coated sickle cells by macrophages in splenic sinusoids.

(2) Sickle cells also obstruct the capillary microcirculation. The reasons include increased rigidity of the cell, increased tendency of sickle cells to adhere to the endothelium and a general increase in blood viscosity. The occlusion causes stagnant hypoxia and severe ischemic pain (vasoocclusive crisis). The hypoxia causes sickling of more cells and sets off a vicious cycle.

Sickling occurs only in deoxygenated hemoglobin. The conformational change that occurs with oxygenation hides the sticky receptors in the deeper regions of the molecules so that they are not present on the surface and no sickling can occur. Sickling is also reduced

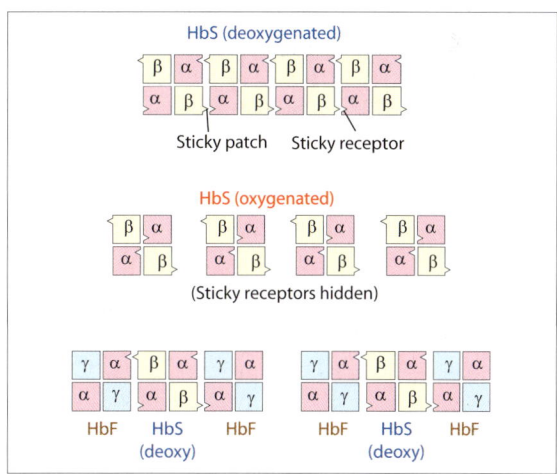

Fig. 24.**4**   Sticky receptors and sticky patches in sickle cell anemia. (*Above*) Polymerization of deoxygenated HbS. (*Middle*) Lack of polymerization in oxygenated HbS. (*Below*) Lack of polymerization of deoxygenated HbS in the presence of HbF. Only trimers are formed.

in cells that have more HbF because HbF has sticky receptors but no sticky patches. Hence in effect, HbF blocks the sticky patches of HbS. As a result, they prevent the formation of long polymers of HbS (Fig. 24.**4**).

Thalassemias

Thalassemias refer to the reduced synthesis or absence of a pair of polypeptide chains of hemoglobin. The affected polypeptide has a normal sequence of amino acids. It occurs due to defects in the regulatory portion of the globin genes. Decreased or absent α and β polypeptides are called α and β thalassemia respectively. Decreased polypeptide synthesis has two effects: (1) a general reduction in the amount of hemoglobin synthesized, resulting in anemia, and (2) an imbalance between the amounts of different α chains. In α thalassemia for example, in the absence of α chains, four β chains aggregate to form β$_4$, called *HbH*. These aggregates precipitate in the cytoplasm and cause membrane damage. Red cells with these aggregates are removed in large numbers by splenic macrophages, causing hemolytic anemia.

Hemoglobin Breakdown

Formation of bilirubin   Aged erythrocytes are phagocytosed by macrophages, mostly those lining the sinusoids in the liver, bone marrow and spleen. Inside the macrophage, the heme molecule is cleaved off

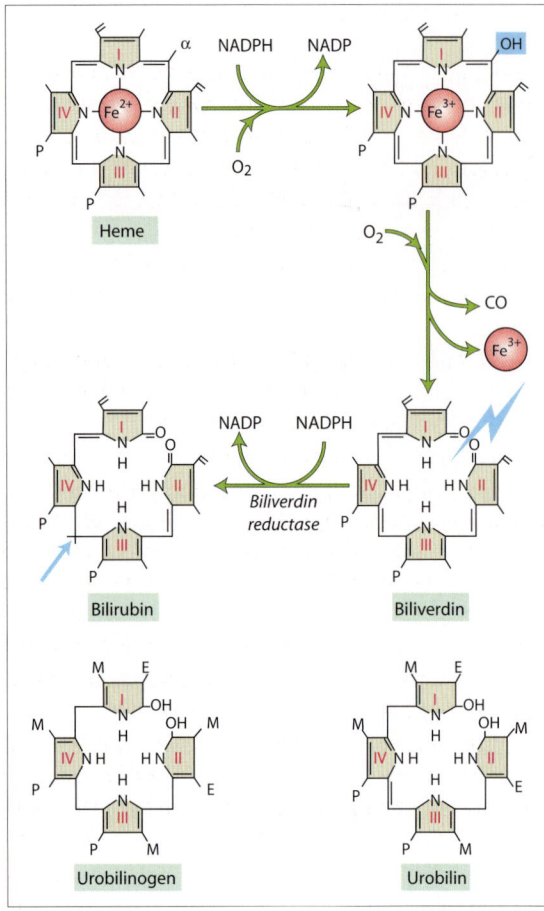

Fig. 24.**5** Steps in hemoglobin degradation.

reticulum of the liver cells). The reaction occurs in two stages as shown below. The conjugated bilirubin is excreted into the hepatic canaliculi through active transport. This active transport is the rate-limiting step in the entire process of bilirubin excretion.

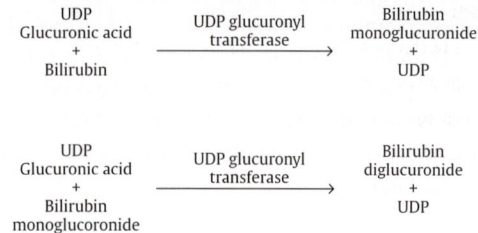

Formation and excretion of urobilinogen The conjugated bilirubin excreted in bile is acted upon by intestinal bacteria in the terminal ileum and the colon. The bacterial enzyme β-glucuronidase splits off the glucuronide and converts bilirubin into *urobilinogen* (also called stercobilinogen), a colorless compound. About 20% of the urobilinogen formed is reabsorbed from the intestine and reexcreted by the liver into the intestine through bile. This constitutes the *enterohepatic cycle* of urobilinogen. A relatively small amount of urobilinogen that is absorbed from the intestine is excreted in urine. Most of the urobilinogens (colorless) formed in the intestine are further oxidized to *urobilins* (colored compounds) and excreted in feces. The darkening of feces on standing in air is due to conversion of the fecal urobilinogens to urobilins.

globin. Globin is degraded into amino acids and reused. The Fe^{2+} in the porphyrin is converted into Fe^{3+}. The ring is broken to release the iron and carbon monoxide, resulting in the formation of *biliverdin*, a green pigment. These reactions are catalyzed by heme oxygenase, an enzyme complex present in the microsomes of macrophages. The fate of the released iron is discussed on p 155. The biliverdin is converted into *bilirubin*, a yellow pigment, by biliverdin reductase which reduces the methenyl bridge between pyrrole III and pyrrole IV to a methylene group (Fig. 24.**5**).

Excretion of bilirubin Bilirubin is poorly soluble in water and is transported in plasma bound to albumin. In the hepatic sinusoids, bilirubin dissociates from albumin to enter the liver cells through facilitated diffusion. Bilirubin is lipid soluble and tends to persist inside the cell. The hepatocytes conjugate bilirubin with *uridine diphosphate glucuronic acid*, making it water soluble so that it can be excreted in bile. The reaction is catalyzed by the enzyme *glucuronyl transferase* present in hepatic microsomes (i.e., the smooth endoplasmic

Hyperbilirubinemia

The normal serum bilirubin level ranges from 0.3 mg–1.0 mg dl^{-1}. Most of this bilirubin is in the unconjugated form (0.2–0.7 mg dl^{-1}). The serum level of conjugated bilirubin is 0.1–0.3 mg dl^{-1}.

The term *jaundice* or *icterus* refers to the yellowish discoloration of the skin and mucous membranes caused by hyperbilirubinemia. Clinical jaundice is apparent when serum bilirubin exceeds 2 mg dl^{-1}. The excess bilirubin in blood may be mostly conjugated or mostly unconjugated. If more than 50% of the bilirubin is conjugated, it is called *predominantly conjugated* hyperbilirubinemia. If less than 15% is conjugated, it is called *predominantly unconjugated* hyperbilirubinemia (Table 24.**1**).

In hypothyroid states, serum carotene is elevated and the skin becomes yellow. This skin condition differs from that observed in jaundice in that the sclera of the eye is not yellow. The yellowing of sclera is prominent in jaundice because the elastase in sclera has a high affinity for bilirubin.

Table 24.**1**   Conjugated versus unconjugated bilirubinemia.

Conjugated bilirubinemia	Unconjugated bilirubinemia
Jaundice is choluric	Jaundice is acholuric
Jaundice is mostly due to posthepatic causes. Hepatic causes include infective hepatitis.	Jaundice is mostly due to prehepatic causes. Hepatic causes include neonatal jaundice.
Kernicterus can never occur	Kernicterus can occur
Stool may be clay colored	Stool is never clay colored
Van den Berg reaction is direct	Van den Berg reaction is indirect.

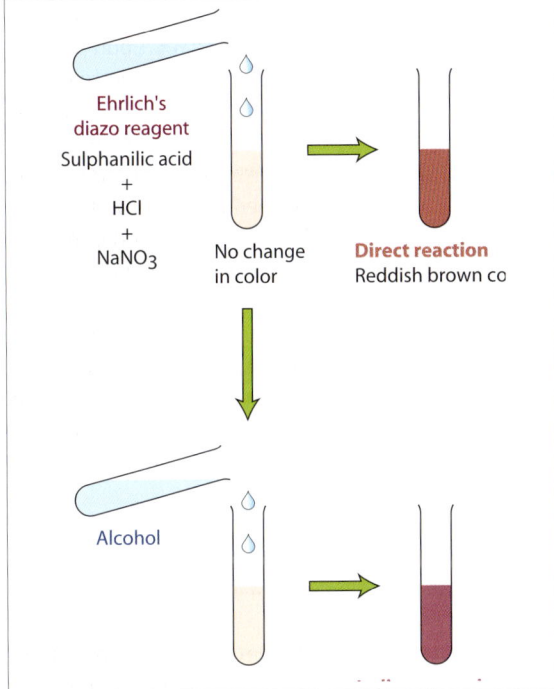

Fig. 24.**6**   The Van den Berg reaction.

The **Van den Berg test** is used for estimating conjugated as well as unconjugated bilirubin levels in plasma. When Ehrlich's diazo reagent is added to conjugated bilirubin, a reddish brown coloration is obtained. It is called the *direct Van den Berg reaction*. Unconjugated bilirubin, which is water-insoluble, has to be dissolved in methanol to complete the Van den Berg reaction. Hence, if the direct reaction is absent, methanol is added to the solution to detect the presence of unconjugated bilirubin. A reddish violet coloration obtained on adding methanol confirms the presence of unconjugated bilirubin and is called the *indirect Van den Berg reaction*.

◼ Unconjugated bilirubinemia

Unconjugated bilirubinemia occurs when bilirubin is produced in excess of what can be conjugated in the liver. It is also called *retention hyperbilirubinemia*. Unconjugated bilirubin is insoluble in water: it is transported in plasma in bound form with albumin. Since albumin is not filtered into urine, unconjugated bilirubin does not appear in the urine (Fig. 24.**7**A). Hence in unconjugated bilirubinemia, the jaundice is acholuric (no bile in urine).

Albumin has two binding sites for bilirubin: a high-affinity site and a low-affinity site. The high-affinity sites on albumin molecules can bind unconjugated bilirubin up to a concentration of 20 mg dl^{-1}. Unconjugated bilirubin in excess of this amount is bound to the low-affinity site on albumin and dissociates readily. Being lipid soluble, unconjugated bilirubin can cross the blood–brain barrier. It is unable to penetrate the adult brain in large amounts but in infants, hyperbilirubinemia in excess of 20 mg dl^{-1} results in a neurological disorder called *kernicterus* due to deposition of bilirubin in the lipid-rich basal ganglia.

Unconjugated bilirubinemia can occur due to *prehepatic causes* like excessive hemolysis or due to *hepatic causes* like decreased hepatic conjugation of bilirubin. Gilbert Syndrome (mild reduction in UDP glucuronyl transferase), Crigler–Najjar Syndrome Type II (moderate reduction in UDP glucuronyl transferase), and Crigler–Najjar Syndrome Type I (UDP glucuronyl transferase absent) are hepatic causes of unconjugated bilirubinemia that have genetic etiology. A more common cause is neonatal jaundice.

In **neonatal jaundice**, also called *physiologic jaundice of the newborn*, up to 5 mg dl^{-1} of hyperbilirubinemia may be present normally. It appears within 2 to 5 days of birth and lasts for about a week. In the fetus, bilirubin is removed from circulation by the placenta. At birth, the newborn has to suddenly excrete its own bilirubin but its hepatic conjugation of bilirubin is still inadequate due to reduced UDP-glucuronyl transferase activity. As a result, hyperbilirubinemia occurs. The hyperbilirubinemia subsides when the newborn's liver matures. The jaundice can be prevented by administration of *phenobarbital* to the pregnant mother or to the newborn. Phenobarbital belongs to a group of drugs called hepatic microsomal enzyme inducers. Microsomal inducers increase the activity of glucuronyl transferase. Neonatal jaundice can be reduced by *phototherapy*. Exposure of the skin to white or blue lights causes photoisomerization of bilirubin to water-soluble lumirubin which is rapidly excreted in bile without requiring any conjugation.

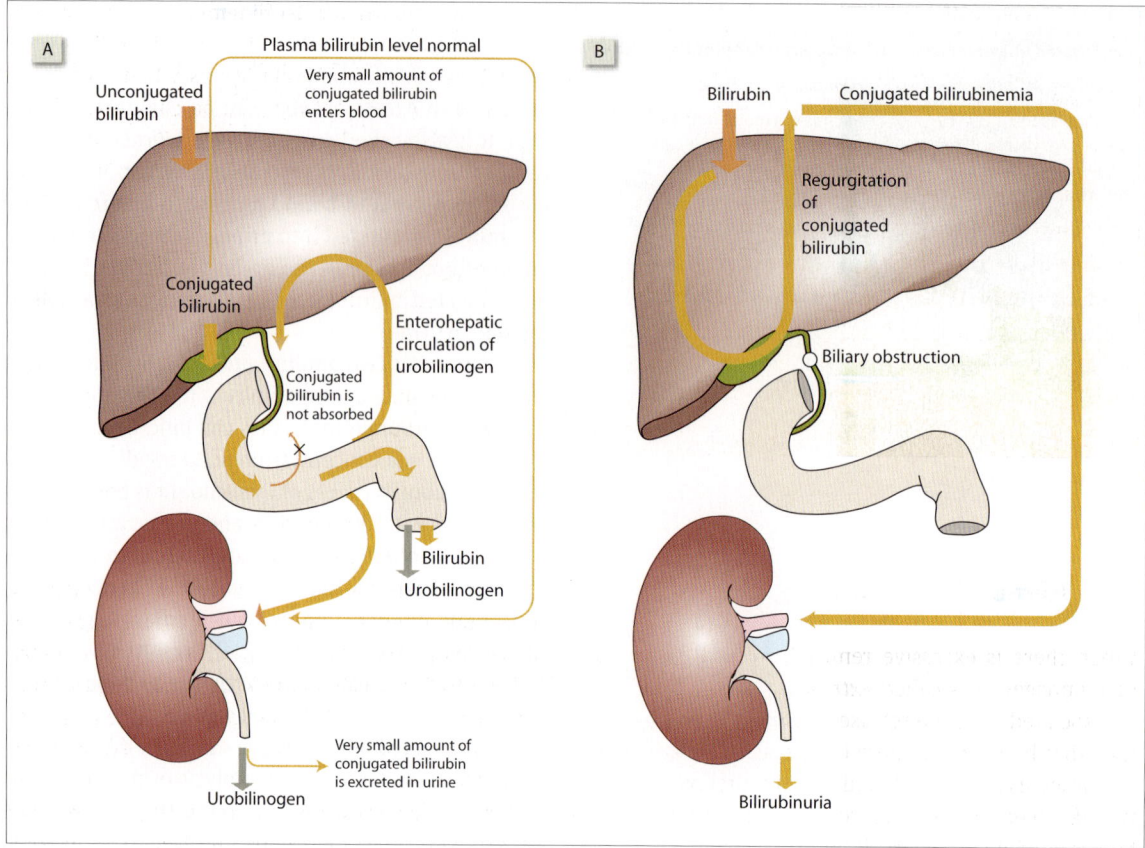

Fig. 24.**7**   Hyperbilirubinemia. [A] Unconjugated bilirubinemia. [B] Conjugated bilirubinemia.

Conjugated bilirubinemia

Conjugated bilirubinemia occurs when there is impaired hepatic excretion of bilirubin and the conjugated bilirubin regurgitates back into circulation. It is also called *regurgitation hyperbilirubinemia*. Conjugated bilirubin is water soluble and is present in plasma in dissolved form.[1] It gets easily filtered into urine (Fig. 24.**7**B). Hence, in conjugated bilirubinemia, the jaundice is *choluric* (bile pigment present in urine). Conjugated bilirubinemia can occur due to posthepatic or hepatic impairment of bilirubin excretion.

Posthepatic causes of hyperbilirubinemia (obstructive jaundice) include obstruction of bile ducts or intrahepatic biliary canaliculi. In these conditions, the conjugated bilirubin formed in the liver cells cannot flow out of the biliary canaliculi easily and therefore regurgitates into the hepatic sinusoids, resulting in conjugated hyperbilirubinemia. Since the bile does not

enter the intestine, the stool is pale (clay-colored). No urobilinogen is formed either and therefore none is excreted in the urine.

Hepatic causes include hepatocellular diseases and certain genetic defects in bilirubin excretion like Dubin–Johnson Syndrome and Rotor Syndrome. In hepatocellular diseases, all the three functions of the liver cells are impaired: uptake of bilirubin, conjugation of bilirubin, and excretion of bilirubin. However, it is the excretion that is most affected, leading to conjugated rather than unconjugated bilirubinemia. Hepatocellular jaundice is therefore choluric and is associated with pale stools. The urine urobilinogen levels first increase and then decrease with increasing severity of hepatocellular jaundice. The *initial rise* in urine urobilinogen is due to a reduction in hepatic reuptake of the urobilinogen absorbed from intestine into blood. Hence, the urobilinogen is eliminated in larger amounts in urine. The *subsequent fall* in urine urobilinogen is due to the reduced excretion of bilirubin into the intestine with consequent reduction in the intestinal urobilinogen synthesis. For the same reason, a rising urinary urobilinogen level in a patient with jaundice under treatment is a good sign.

---

[1]   Some amount of conjugated bilirubin binds to albumin following prolonged hyperbilirubinemia. During recovery, this albumin-bound bilirubin is responsible for the persistence of yellowish coloration of skin even after the free conjugated bilirubin in plasma returns to normal.

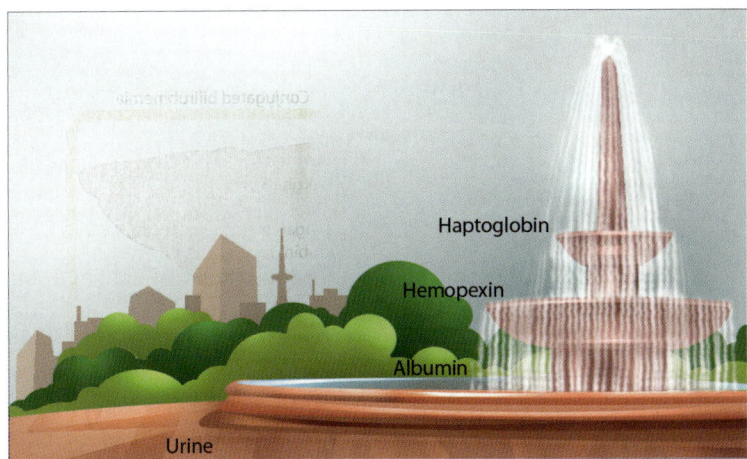

Fig. 24.**8**   The symbolic cascade of hemoglobinemia, showing that the hemoglobin released in blood is successively captured by haptoglobulin, hemopexin, and albumin before overflowing into urine.

Hemoglobinemia

When there is excessive removal of erythrocytes by macrophages, it is called *extravascular hemolysis* and is associated with the release of bilirubin in blood. On the other hand, when there is excessive breakdown of erythrocytes inside the blood vessels, it is called *intravascular hemolysis*, and is associated with the release of free hemoglobin into circulation.

When hemoglobin is released into circulation in large amounts, plasma proteins bind to it so that iron is not lost from the body into the urine. The complexes formed are taken up by hepatocytes and the iron is recycled. Free hemoglobin released into circulation first combines with *haptoglobin*. When the binding capacity of haptoglobin is exceeded, the hemoglobin gets converted into methemoglobin. The ferriheme (heme containing Fe^{3+}) dissociates from globin and binds to *hemopexin*. When the binding capacity of hemopexin too is exceeded, the ferriheme is bound to plasma *albumin* to form methemalbumin. Low plasma haptoglobin

level is an early indicator of excessive hemoglobin breakdown. Low plasma hemopexin or the appearance of methemalbumin in plasma indicates severe increase in hemoglobin breakdown (Fig. 24.**8**). When the binding capacities of all the three plasma proteins are exceeded, hemoglobin starts circulating in blood in free form (hemoglobinemia).

Freely circulating hemoglobin is taken up by macrophages and broken down, resulting in hyperbilirubinemia. Hemoglobin is also filtered into the renal tubules. Small amounts of hemoglobin are removed from the tubules through endocytosis by tubular cells. The iron is stored in renal tubular cells as *hemosiderin*. Hemosiderinuria occurs when these tubular cells are desquamated into tubules. When a large amount of hemoglobin is filtered into the tubules, it is passed into the urine (hemoglobinuria). If the urine is alkaline, the hemoglobin is oxygenated to oxyhemoglobin (pink) in the urinary tract. If the urine is acidic, the hemoglobin is oxidized to methemoglobin (dark brown). The urine color varies accordingly.

25   Hematinic Factors

Iron

Iron balance is unique in that it is achieved by control of absorption rather than by control of excretion. Iron is not actively excreted: It is lost from the body only when iron-laden cells are lost, especially epithelial cells from the gastrointestinal tract.

Mechanism of iron absorption

Iron absorption occurs almost entirely in the duodenum (Table 25.1). Only 5% of the ingested iron is normally absorbed. A nonvegetarian diet contains iron mostly in the heme form while a vegetarian diet contains iron

Table 25.1   Nutritional aspects of iron.

Dietary form	Ferric (Fe^{3+})
Daily requirements	10–20 mg in diet (only 10% of the dietary intake is absorbed).
Sources	Meat, liver, egg, leafy vegetables, whole wheat, and jaggery.
Absorption	Fe^{3+} is reduced to Fe^{2+} by ferric reductase present on enterocytes. Reducing substances like vitamin C enhance absorption. Phosphates and phytates (present in cereals) decrease absorption by forming insoluble complexes with iron.
Site of absorption	Duodenum and upper jejunum.
Body resources	5 g (in the liver, spleen, bone marrow, lymph nodes, and the RES cells).
Storage forms	2/3 as ferritin and 1/3 as hemosiderin.
Transport form	As Fe^{2+} bound to transferrin; as ferritin (ferric hydrophosphate) bound to apoferritin.
Role	Synthesis of hemoglobin, myoglobin, and cytochromes (intracellular enzymes).
Causes of deficiency	Decrease in dietary intake or absorption; Increased demand (pregnancy) or losses (hemorrhage).
Effects of deficiency	Iron deficiency anemia.
Effects of excess	Hemosiderosis and hemochromatosis.
Diagnosis of deficiency	Serum iron level (normal = 120 μg dl^{-1}) and Total iron binding capacity (normal = 250-450 μg dl^{-1}).
Therapeutic form	Ferrous sulfate.

in the nonheme form. The absorption of nonheme and heme iron have different mechanisms.

Dietary nonheme iron is absorbed mainly in the form of insoluble salts like $Fe(OH)_3$ and $FePO_4$, and insoluble complexes with dietary substances like phytates (present in grains and bran), polyphenols (present in legumes, tea, coffee, and wine), egg white, and bovine milk protein. In its insoluble form, iron is unable to diffuse in adequate amounts to the intestinal brush border.

Gastric acid is necessary for iron absorption because it dissolves the insoluble ferric salts and complexes. The dissolved Fe^{3+} forms complexes with reducing substances in the diet like ascorbic acid and cysteine (present in meats) and is reduced to Fe^{2+}. The intestinal brush border can absorb only Fe^{2+}. Hence, iron absorption is impaired in subjects with gastrectomy or achlorhydria.

The enzyme *ferric reductase*, present on the intestinal brush border, reduces any remaining Fe^{3+} to Fe^{2+}. Ferrous iron is taken up into the intestinal mucosal cells (enterocytes) through facilitated diffusion across the luminal border. Once inside, the Fe^{2+} is oxidized to Fe^{3+} by the enzyme *ferroxidase* and bound to apoferritin to form *ferritin*. At the basolateral border of the enterocyte, the ferritin dissociates to release Fe^{3+}, which diffuses out into blood through facilitated diffusion.

Dietary heme iron is freed from its apoprotein (globin) by exposure to gastric acid and proteases. The free heme, containing iron in the Fe^{2+} form, is transported into the enterocyte intact by a specific *heme transporter*. Inside the enterocyte the enzyme *heme oxygenase* releases Fe^{2+} from heme and adds it to the free Fe^{2+} pool in the enterocyte. The subsequent fate of heme and nonheme Fe^{2+} are the same. Absorption of heme iron is relatively unaffected by the composition of the diet.

Regulation of iron absorption

The intestinal mucosal cell absorbs iron in proportion to the body's requirement, especially, the rate of erythropoiesis. As mentioned above, the iron absorbed into the mucosal cell binds to apoferritin to form ferritin.

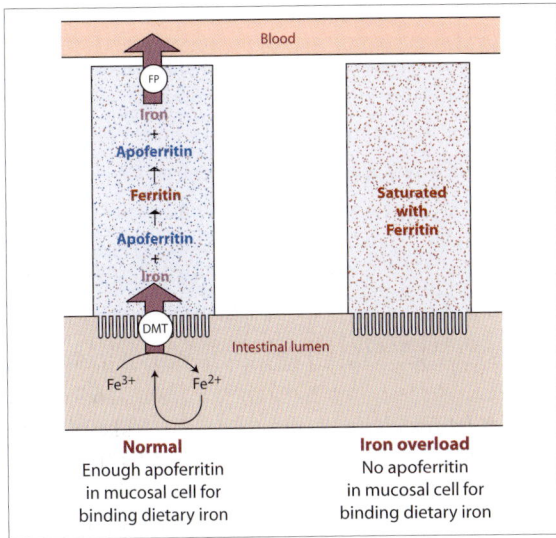

Fig. 25.1   Regulation of iron absorption DMT = dimethyltrypt-amine; FP = ferroportin.

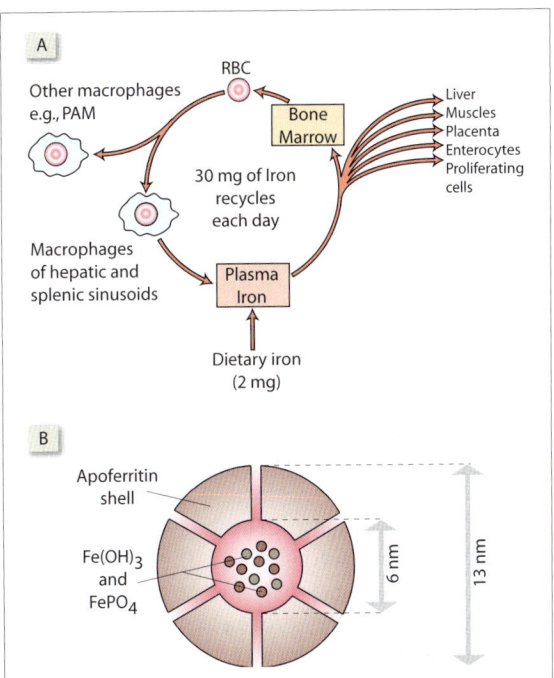

Fig. 25.2   [A] The iron cycle. [B] Structure of ferritin.

The ferritin then releases iron into the bloodstream and reverts back to apoferritin. The *mucosal block theory* (Fig. 25.**1**) says that when the body iron stores are high, mucosal ferritin does not release iron into the blood-stream (in accordance with the dynamics of chemical equilibrium). Consequently, the mucosal cells remain loaded with ferritin. With all the apoferritin converted into ferritin, the mucosal cell is not able to take up any more luminal iron. Mucosal cells have a lifespan of 3–4 days at the end of which they get sloughed off into the intestine. Mucosal ferritin iron is also lost with them, ridding the body of its unwanted iron.

Iron cycle

The term iron cycle (Fig. 25.**2**A) refers to the uptake of iron from plasma into erythrocytes and its eventual return to the plasma after the death of the erythro-cytes. Each day about 30 mg of iron goes through this cycle. About 2 mg iron enters the cycle from diet and the same amount leaves the cycle into the tissues. Iron travels from plasma to the bone marrow where it is incorporated into hemoglobin. At end of their life span erythrocytes are engulfed by macrophages. After phagocytosis, the red cell membrane is lysed, heme is released from hemoglobin and iron is liberated by the enzyme heme oxygenase.

Macrophages lining the sinuses of liver, spleen, and bone marrow release the iron into the circula-tion, where it gets bound to transferrin, completing the cycle. Other macrophages like the pulmonary alveolar macrophages (PAM) lack the ability to return the iron to the circulation; they store the iron. These iron-laden macrophages may remain in the tissues, or may enter the circulation only to leave the body subsequently through the intestine. Less than 2 mg iron is taken up each day by many tissues for the synthesis of nonhe-moglobin, iron-containing enzymes, and proteins like cytochromes, catalases, and myoglobin. These tissues include: (1) liver, which takes up the largest amount of iron next only to the red cells; (2) placenta, which transfers the iron to the fetus; (3) intestinal mucosal cells, which store iron as ferritin, thereby regulating iron absorption; (4) skeletal muscles, which produce myoglobin; and (5) proliferating cells that require ex-cess cytochromes and catalases.

Iron transport in plasma

The plasma protein that transports iron is called *trans-ferrin*. It is a glycoprotein and it binds up to two mol-ecules of Fe^{3+}. Transferrin is synthesized chiefly in the liver. Its rate of synthesis is inversely related to the amount of iron stored in the body. Hence, in iron de-ficiency, plasma transferrin increases. Clinically, the amount of transferrin is expressed in terms of the amount of iron it will bind, i.e., the *total iron-binding capacity* (TIBC). Transferrin-mediated transport en-sures delivery of iron only to those cells that need iron, i.e., the red cell precursors. Unbound iron becomes distributed to all cells equally, irrespective of their iron requirement. The importance of transferrin is apparent in *congenital atransferrinemia* where erythrocytes show signs of iron deficiency but tissues are loaded with iron.

Regulation of cellular uptake of iron

Transferrin delivers its iron to the red cells and other tissues by binding to specific *transferrin receptors*. The rate of iron uptake by a cell is related to the number of transferrin receptors on its surface. During erythropoiesis, the number of transferrin receptors reaches its peak in intermediate normoblasts when hemoglobin first makes its appearance. Cells shed their receptors as their iron requirement decreases. The shed receptors can be found in plasma in a concentration that correlates with the rate of erythropoiesis.

When plasma iron concentration is high, the cells inhibit the synthesis of transferrin receptors and increase the synthesis of apoferritin, thereby limiting further acquisition of iron and facilitating storage of any excess. In contrast, when plasma iron concentration is low, the cells increase the synthesis of transferrin receptors and reduce the synthesis of apoferritin, maximizing iron entry and utilization while minimizing the diversion of iron into stores.

Iron storage proteins

The two iron storage compounds are ferritin and hemosiderin. Iron exchanges mostly with the ferritin. The exchange with the hemosiderin is much slower.

Ferritin (Fig. 25.2B) is made of an iron-free apoferritin (protein) shell with a hollow central cavity containing a complex of $Fe(OH)_3$ and $FePO_4$. The cavity communicates with the surface by six channels, through which iron can enter and leave. A fully saturated ferritin molecule can hold up to about 4500 iron atoms.

Hemosiderin represents a more stable form of storage iron than ferritin. It is formed in the macrophages and hepatocytes when a large amount of iron accumulates in them. Hemosiderin is a dense aggregate of ferritin crystals. It is formed inside the lysosome and is therefore usually membrane bound. The lysosomal enzymes digest the proteins of the ferritin. As a result, the iron increases in concentration and polymerizes. Lipids and other substances are added to the aggregate, making the composition of hemosiderin quite variable. The

Table 25.**2**   Comparison of ferritin and hemosiderin.

Ferritin	Hemosiderin
Composition fairly consistent	Composition highly variable
Not enclosed by membrane	May be enclosed by membrane
Iron content lesser (~20%)	Iron content greater (~30%)
Protein content higher	Protein content lower
Iron in the form of $Fe(OH)_3$ and $FePO_4$	Iron in the form of Fe_2O_3
Water-soluble	Water-insoluble
Stains with Prussian Blue	Does not stain with Prussian Blue

iron in hemosiderin is mostly in the form of Fe_2O_3 (Table 25.**2**)

Iron storage disorders

Siderotic granules are granules of nonheme iron in the red cell. The granules represent the iron present in excess of requirements for hemoglobin synthesis. *Siderocytes* are red cells containing siderotic granules, while precursor cells (normoblasts and reticulocytes) containing siderotic granules are called *sideroblasts*. Sideroblasts are normally present in the bone marrow. Siderocytes appear in the blood in significant numbers after splenectomy. This is because normally, the spleen temporarily sequesters reticulocytes released from the bone marrow till they complete heme synthesis, using up all their iron (siderotic) stores. Following splenectomy, reticulocyte maturation occurs in the peripheral blood and thus, siderocytes appear in blood.

Hemosiderosis is the deposition of hemosiderin in the tissues. The term *siderosis* is broader and includes the deposition of other forms of iron too, like ferritin. *Hemochromatosis* is a hereditary condition in which the iron overload occurs due to increased iron absorption from the gut. However, the term is also used synonymously with hemosiderosis. To avoid confusion, the hereditary form of hemosiderosis is called primary hemochromatosis while the other forms are called secondary hemochromatosis.

Hemosiderosis is of four types: local hemosiderosis, hemolytic or transfusional hemosiderosis, nutritional hemosiderosis and primary hemochromatosis. In local hemosiderosis, the iron deposition is localized to a small part of the body. In others, the hemosiderin deposits are found throughout the body.

Local hemosiderosis develops when there is hemorrhage into the tissues. Macrophages phagocytose the red cells and store their iron as ferritin and hemosiderin. If the hemorrhage is small, the macrophages migrate away from the site but if the hemorrhage is large, the hemosiderin-laden macrophages remain at the site.

Hemolytic hemosiderosis is seen in hemolytic anemias. It is also called transfusional hemosiderosis because it occurs following repeated blood transfusion. Initially, the hemosiderin accumulates in the MPS cells (p 184) but in prolonged and severe cases, the parenchymal cells, like those of the liver and myocardium, are also overloaded with hemosiderin.

Nutritional hemosiderosis occurs due to increased intake of iron. It is sometimes caused by excessive consumption of red wine which has a high iron content and because alcohol increases the absorption of iron.

Primary hemochromatosis is a genetic disease that results in increased absorption of iron from the gut. Its exact mechanism is not known.

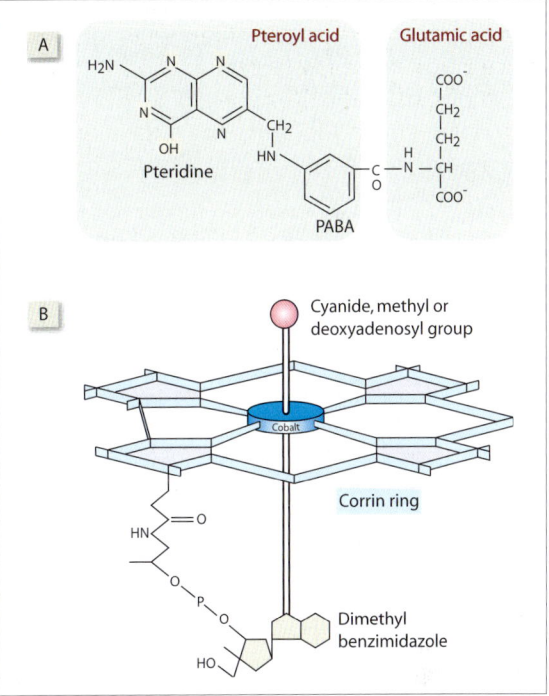

Fig. 25.**3** [A] Structure of folic acid. [B] Structure of vitamin B$_{12}$.

Folic Acid

Folic acid, or *pteroylglutamic acid*, consists of the base pteridine attached to one molecule each of *p-aminobenzoic acid* (PABA) and *glutamic acid* (Fig. 25.3A). Animals cannot synthesize folates and therefore, require folate in their diet.[1] In plants, folic acid is present as a *polyglutamate conjugate* consisting of a polypeptide chain of seven glutamate residues. In liver (consumed as food), the major folate is a *pentaglutamyl conjugate*. These *polyglutamyl folates* in the diet are broken down by intestinal enzymes to *monoglutamyl folates* for absorption. Most of this is reduced to *tetrahydrofolate* (THF) in the intestinal cell by the enzyme *folate reductase*. THF is the active form of folic acid (Table 25.**3**).

Action of folic acid

Folic acid is important for DNA synthesis. Tetrahydrofolate acts as a carrier for activated one-carbon units like *methyl, methylene, methenyl, formyl* and *formimino*. By transferring a methyl group, it helps in DNA synthesis (Fig. 25.**4**) which requires methylation of uridine to thymidine. Serine provides the methyl group required. *Tetrahydrofolate* (THF) transfers the methyl group from serine to uridine in two steps: (1) Serine transfers the methylene group to THF to form glycine and *methylene-THF.* (2) *Methylene-THF* transfers the methyl group to uridine, and in the process, gets converted to *dihydrofolate* (DHF).

---

[1] While animals are able to assimilate dietary folate, bacteria must synthesize their own folate because folic acid cannot diffuse into bacterial cells. Sulphonamides, which are structurally similar to PABA, inhibit the bacterial synthesis of folic acid and thereby inhibit their multiplication without affecting the host cells.

The regeneration of THF from DHF occurs in the presence of *folate reductase*. If the THF is not regenerated, it results in folate deficiency, which can be overcome only by excess dietary intake. *Methotrexate*, an anti-cancer drug, is an inhibitor of folate reductase. By preventing the regeneration of THF, it impairs DNA synthesis. Hence, it especially impairs the function of vigorously multiplying cells that have to synthesize a lot of DNA. Some of the methylene-THF formed gets converted (trapped) into *methyl-THF*. It is called the *methyl trap* or the *folate trap*. Recovery of THF from methyl-THF requires vitamin B$_{12}$.

Folate deficiency

The common causes of folate deficiency are: (1) *Inadequate intake*, mostly due to unbalanced diet (common in alcoholics, teenagers with food fads and addicted to junk foods), (2) *Increased requirements*, as in pregnancy, infancy, malignancy, and increased hematopoiesis (chronic hemolytic anemias), (3) *Malabsorption due to sprue*, and (4) *Folate inhibitors* like methotrexate.

Because folic acid is important for DNA synthesis and cell division, especially affected are the cells that multiply rapidly, like the hemopoietic cells. Erythropoiesis, leukopoiesis, and thrombopoiesis—all are affected. Defective erythropoiesis results in anemia. The

Table 25.**2**   Nutritional aspects of vitamin B$_{12}$ and folic acid.

	Folic acid	Cobalamin
Dietary form	Polyglutamates	Deoxyadenosyl-cobalamin
Daily requirements	200 µg	2 µg
Sources	Liver, leafy vegetables, yeast, and pulses	Animal products (meat, liver, fish, eggs, and milk)
Absorption	Carboxypeptidases reduce poly- to mono-glutamates. Jejunal mucosal cells convert monoglutamates to absorbable methyl-THF	Intrinsic factor secreted by the parietal cells binds to B$_{12}$ as well as to the ileal receptors thereby facilitating diffusion
Site of absorption	Ileum	Ileum
Body resources	500 mg (in liver, erythrocytes, and leukocytes).	5 mg (in liver).
Storage forms	5-methyl THF	Deoxyadenosyl-cobalamin
Transport form	5-methyl THF	As methylcobalamin bound to transcobalamin
Role	DNA synthesis in blood cells	DNA synthesis in blood cells
Causes of deficiency	Decreased dietary intake or absorption; increased demand (pregnancy)	Low intake / absorption; high demand pregnancy) or losses (fish tapeworm infestation).
Effects of deficiency	Megaloblastic anemia	Megaloblastic anemia
Effects of excess	Water soluble. Excess is excreted	Water soluble. Excess is excreted
Diagnosis of deficiency	Serum folic acid level (normal = 200 ng dl^{-1}); FIGLU Test (increased urinary excretion of formiminoglutamic acid)	Serum B$_{12}$ level (Normal = 50 µg dl^{-1}); Methylmalonic acid test (increased urinary excretion in deficiency)
Therapeutic form	Pteroylglutamic acid	Hydroxycobalamin

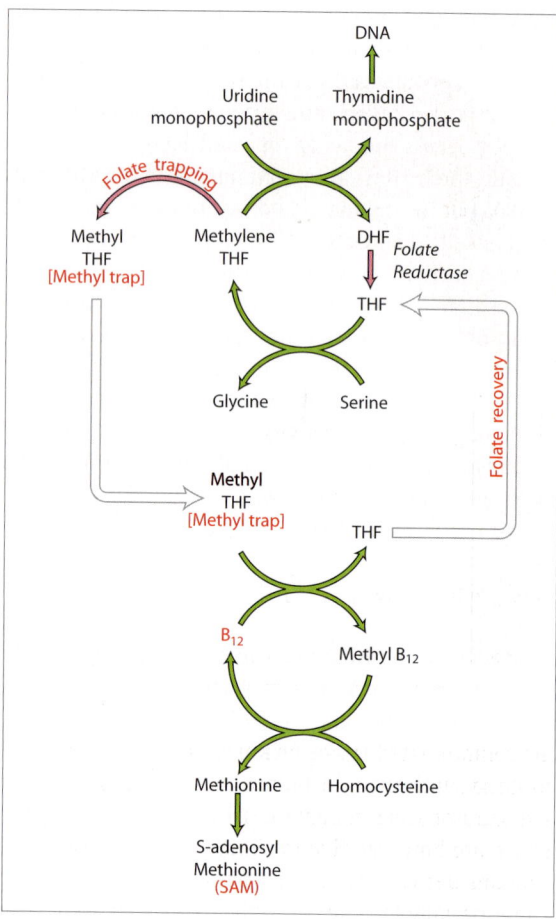

Fig. 25.**4**   The role of folate in DNA synthesis, the folate trap, and the role of vitamin B$_{12}$ in folate recovery.

red cell precursors like the normoblasts grow in size but fail to divide in time, resulting in large cells called *megaloblasts*. The anemia that is associated with the presence of megaloblasts in the bone marrow is called megaloblastic anemia. The erythrocytes that appear in circulation are also large (macrocytes).

Test for folate deficiency   *Formiminoglutamate* (FIGLU), an intermediate product of histidine metabolism, transfers its formimino group to THF to form *formimino-THF*. This reaction is employed for detection of folate

deficiency (the *FIGLU test*). In folate deficiency, there is a marked rise in the urinary excretion of FIGLU after a large oral dose of histidine. The test is however not specific for folates; it is positive in vitamin B$_{12}$ deficiency too. This is because in the absence of vitamin B$_{12}$, a lot of THF remains trapped in the methyl trap causing a deficiency of THF in the body.

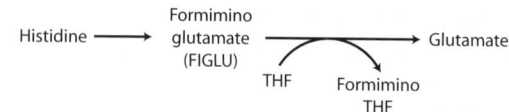

Vitamin B$_{12}$

Vitamin B$_{12}$ (*cobalamin*) is a corrin ring (similar to a porphyrin ring) with a cobalt ion at its center (Fig. 25.**3**B). The vitamin is synthesized entirely by bacteria and certain fungi. It is absent from plants unless they

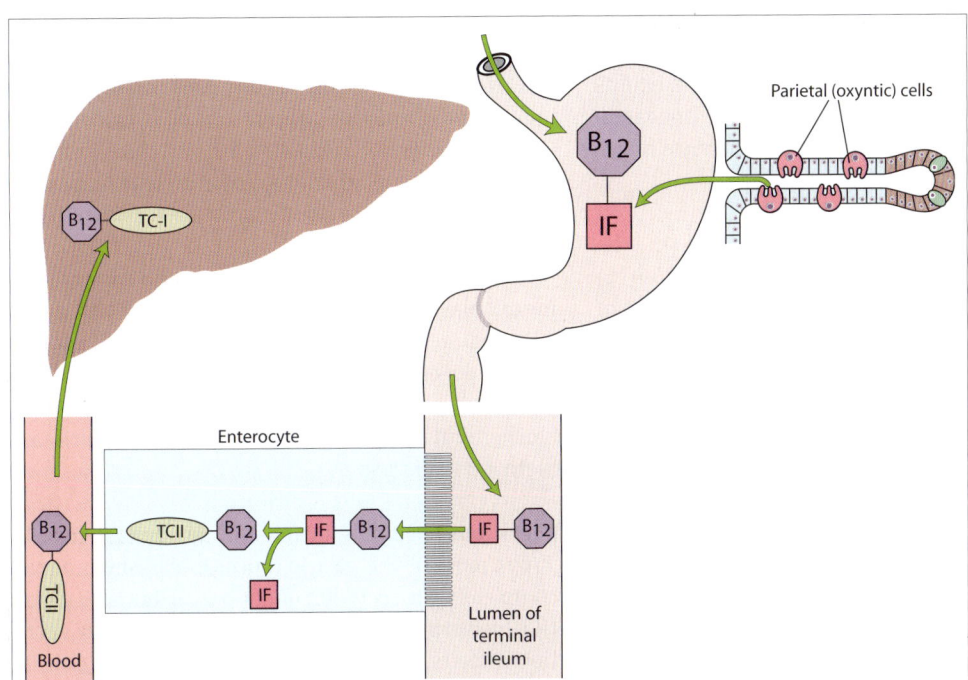

are contaminated by bacteria or fungi. Although colonic bacteria synthesize cobalamin, the colonic cobalamins are unabsorbable in human. Animals, especially ruminants, are however able to absorb colonic cobalamins. Humans derive their cobalamins from the meat, milk, and especially the liver of these animals. The commercial preparation is *cyanocobalamin*. The stability of this compound and therefore its shelf-life is much higher than other forms of cobalamin (Table 25.**3**).

Absorption and Storage

The intestinal absorption of vitamin B$_{12}$ is possible only if the vitamin B$_{12}$ is bound to the *intrinsic factor*, a glycoprotein secreted by parietal cells of the gastric mucosa. The intrinsic factor–cobalamin complex then binds to specific receptors in the ileum and is absorbed by endocytosis. Inside the mucosal cells, cobalamin is transferred from intrinsic factor to transcobalamin II, another cyanocobalamin-binding protein. Transcobalamin II transports cobalamin in blood (Fig. 25.**5**). Cobalamin is stored in liver, bound to transcobalamin I.

The **Schilling test** is used for estimating vitamin B$_{12}$ absorption. In this test, a small dose of radiolabeled vitamin B$_{12}$ is administered orally and the amount of B$_{12}$ absorbed is assessed by measuring the radioactivity in urine. If the absorption of vitamin B$_{12}$ appears to be reduced, the test is repeated after supplementing intrinsic factor along with oral radioactive vitamin B$_{12}$.

Actions of vitamin B$_{12}$

Inside the cells, cobalamin is converted into its active forms: methylcobalamin and deoxyadenosylcobalamin. *Methylcobalamin* is required for THF recovery from methyl trap and S-adenyl methionine (SAM) production. *Deoxyadenosylcobalamin* is a coenzyme for an important gluconeogenetic reaction.

THF recovery and SAM production The recovery of THF from methyl trap and SAM production are interlinked (Fig. 25.**4**). Cobalamin removes the methyl group from methyl-THF and in the process gets converted into methylcobalamin. Methylcobalamin then transfers the methyl group to homocysteine, which gets converted into methionine. Cobalamin gets regenerated in the process. One of the products of methionine is SAM, which is important for maintenance of myelin in neurons.

The above reactions explain an observation of great clinical importance. Vitamin B$_{12}$ deficiency produces megaloblastic anemia and also produces neurological signs and symptoms. Megaloblastic anemia occurs because the folate that gets trapped in the methyl trap is not recovered, resulting in folate deficiency. Hence, an increase in dietary folate improves the megaloblastic anemia caused by vitamin B$_{12}$ deficiency. The neurological deficiencies however are not improved by folate administration. It is therefore important not to administer folate supplements indiscriminately. If there is a deficiency of vitamin B$_{12}$, the folate supplements prevent the occurrence of megaloblastic anemia. In the absence of megaloblastic anemia, the vitamin B_{12} deficiency tends to go undiagnosed although the neurological changes associated with it continue unabated. The above reaction also explains the occurrence of *homocysteinuria* in vitamin B_{12} deficiency.

Gluconeogenesis Deoxyadenosylcobalamin is the coenzyme for conversion of methylmalonyl-CoA to succinyl-CoA. This is a key reaction in the pathway of conversion of propionate (an amino acid) to succinyl–CoA (a member of the citric acid cycle) and is therefore of significance in the process of gluconeogenesis. The urinary excretion of methylmalonyl-CoA is increased (methylmalonic aciduria) in vitamin B_{12} deficiency and has been used as a test for vitamin B_{12} deficiency.

■   **Vitamin B_{12} deficiency**

Dietary deficiency of vitamin B_{12} is uncommon even in vegetarians who obtain adequate vitamin B_{12} in milk. Most cases of vitamin B_{12} deficiency are due to a reduction in intestinal absorption of vitamin B_{12}, the causes of which can be: (1) *Inadequate intrinsic factor* (IF), as in pernicious anemia and gastrectomy; (2) *Disorders* (*sprue, enteritis*) or *resection of terminal ileum*; (3) *Competition for cobalamin* by fish tapeworm and bacteria (in the blind loop syndrome) which consume most of the cobalamin in the intestine, leaving little for absorption.

The main effects of vitamin B_{12} deficiency are: (1) megaloblastic anemia, (2) glossitis (sore tongue), (3) *neurological manifestations* due to demyelination and axonal degeneration of peripheral nerves and posterolateral columns of the spinal cord (p 599), and (4) biochemical consequences like homocysteinuria and methylmalonic aciduria.

Pernicious anemia is an autoimmune disease that destroys most of the gastric mucosa, abolishing almost completely the secretion of not only IF but also gastric HCl (*achlorhydria*) and pepsin. Abolition of IF secretion impairs vitamin B_{12} absorption producing all the signs of vitamin B_{12} deficiency, notably megaloblastic anemia.

Blood Grouping and Transfusion

The outcome of indiscriminate transfusion of blood from one individual (the donor) to another (the recipient) can be fatal. This happens when certain immunoglobulins (called agglutinins) present in the recipient's plasma react with certain antigenic proteins (called agglutinogens) on the donor's erythrocytes. The reaction results in the lysis of the donor's erythrocytes with lethal consequences. Both agglutinogens and agglutinins are of several different types and each agglutinogen has a specific agglutinin with which it reacts. By ensuring that the recipient's plasma does not have the agglutinins that are specific for the agglutinogens on the donor erythrocytes, fatal transfusion can be avoided. This led to the system of blood grouping.

Blood Group Systems

Blood is grouped on the basis of the type of agglutinogen present on its erythrocytes. Thus, blood group A has A agglutinogen on its erythrocytes while blood group M has agglutinogen-M on its erythrocytes. All blood groups obey, in whole or in part, *Landsteiner's law* which states that: (1) when the blood contains a particular agglutinogen, its corresponding agglutinin is always absent in that blood, and (2) when a particular agglutinogen is absent in the blood, its corresponding agglutinin is always present in the blood. The first clause of the law is always true but the second clause is valid only for the ABO blood groups.

While innumerable agglutinogens have been deciphered in the blood, the important ones are those which are widely prevalent in the population and those which cause the worst transfusion reactions. These are called the *major blood group systems*, e.g., the ABO and the Rhesus (CDE) systems. Some blood groups are found only in a small proportion of the population and occasionally produce mild transfusion reactions. These are called the *minor blood group systems*, e.g., MN, P, etc. In addition to the major and minor blood groups, there are *familial blood groups* such as the Kell, Duffy, Diego, Lewis, Lutheran, Kidd, and many others that are named after individuals, mostly women, whose blood groups were detected during childbirth. These blood agglutinogens are prevalent only in a few families.

The **ABO system** comprises two agglutinogens A and B whose corresponding agglutinins are α and β. Accordingly, there are 4 blood groups in the ABO system: *Group A*, which has A agglutinogen, *group B* which has B agglutinogen, *group AB* having both, and *group O* having neither. Group A has α agglutinin, group B has β agglutinin, group AB has neither, and *group O* has both (Table 26.**1**). Both α and β agglutinins are immunoglobulin-M (IgM), which is very effective in causing agglutination (clumping) of the red cells.

In India, about 22% of the population have A group, 33% have B group and 40% have O group blood. Only 5% have AB group blood. 85% of Caucasians are D+. Among Asians, over 99% are D+.

The **Rhesus blood group** agglutinogens were first discovered in the erythrocytes of rhesus monkeys, and hence the name. Rhesus blood group comprises a system of 3 agglutinogens: C, D, and E. However, for all practical purposes, the term Rhesus agglutinogen refers to the D agglutinogen which produces the worst transfusion reactions. Accordingly, the Rhesus system comprises only two blood groups: the Rhesus positive (Rh positive or D+) and the Rhesus negative (Rh negative or D–) blood groups depending on the presence or absence of D agglutinogen (Table 26.**1**).

Unlike in the ABO system, there are no natural antibodies to rhesus agglutinogens. Anti-D antibodies develop only when a D– person is transfused with D+ blood. Once produced, these antibodies persist in blood for years and can produce serious reactions during a second transfusion.

Anti-D agglutinins are predominantly immunoglobulin G (IgG) and partly immunoglobulin M (IgM).

Table 26.**1** ABO and Rhesus blood groups

Group		Agglutinogen	Agglutinin
ABO system	A	A	α
	B	B	β
	AB	A and B	–
	O	–	α and β
Rhesus system	Rh +	D	–
	Rh –	–	–

Unlike IgM which is very effective in agglutinating agglutinogen-bearing red cells, IgG does not agglutinate the red cells although they do react with the agglutinin. Such immunoglobulins which do not cause agglutination are called *incomplete antibodies*. Although they do not agglutinate red cells, IgG-coated red cells still get lysed due to the activation of complement on their surface (see Complement system, p 200).

Blood grouping

For *ABO blood grouping*, the test sample of blood or erythrocyte suspension is reacted with sera containing α and β (called antiserum-A and antiserum-B). The sample is grouped according to the serum that agglutinates its red cells (Fig. 26.**1**).

Rhesus blood grouping can be done in the same way as ABO grouping if the anti-D agglutinin used is of the IgM type. If the anti-D agglutinin used is IgG, the D+ red cells will get coated with anti-D agglutinin but there will be no agglutination of the cells. The coated red cells will agglutinate only on subsequent addition of Coombs' (anti-immunoglobulin) serum (Fig. 26.**2**). Agglutination will also occur if the IgG anti-D is potentiated by adding albumin to it.

Genotypes and inheritance

The *ABO phenotypes* are controlled by a pair of codominant alleles A and B. An individual who has inherited A-agglutinogen from one parent and B agglutinogen from the other parent will have the AB blood group. Similarly, an individual whose phenotypic blood group is B may have either the genotype BB (homozygous) or BO (heterozygous), as shown in Table 26.**2**.

The *Rh phenotypes* are controlled by three sets (C, D, and E) of two alternative alleles (dominant and recessive). Each phenotype has a variable number of possible genotypes. For example, cde has only one possible genotype, i.e., ccddee. CDE on the other hand can have eight possible genotypes, viz., CCDDEE, CCDDEe, CCDdEE, CCDdEe, CcDDEE, CcDDEe, CcDdEE, and CcDdEe.

Table 26.**2**   Genotypes of ABO blood groups

Blood group	Possible genotype(s)
A	AA, AO
B	BB, BO
AB	AB
O	OO

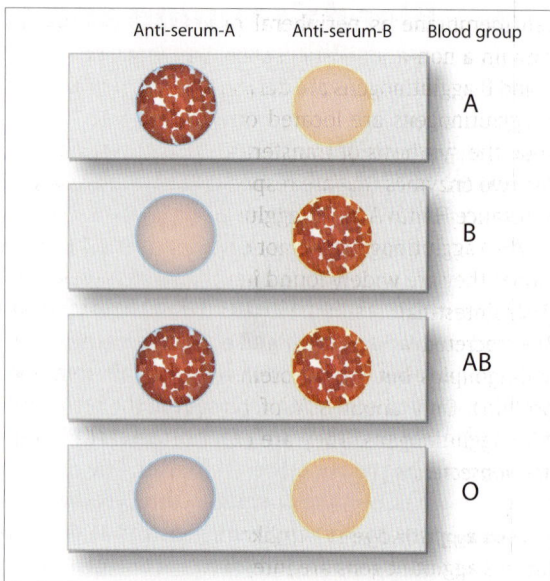

Fig. 26.**1**   The agglutination test for blood grouping.

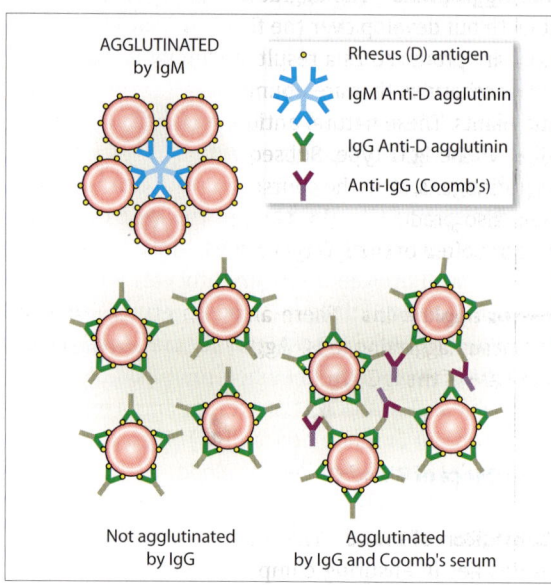

Fig. 26.**2**   Rhesus agglutinogens and agglutinins. (*Above*) Rhesus (D) positive red cells agglutinated by IgM anti-D agglutinins. (*Below, left*) D-positive red cells coated but not agglutinated by IgG anti-D agglutinins. (*Below, right*) IgG anti-D coated red cells agglutinated by anti-IgG which bind to IgG molecules by their stem (Fc).

Agglutinogens and agglutinins

The **ABO agglutinogens** represent only a few of the approximately one million agglutinogens present on an erythrocyte. The ABO agglutinogens are glycosphingolipids (oligosaccharide plus sphingolipid). The antigenicity of the agglutinogens resides in the oligosaccharide moiety. The ABO agglutinogens are present on the red cell membrane as peripheral proteins. O group cells contain a non-antigenic H substance from which both A and B agglutinogens are derived. The genes for A and B agglutinogens are located on chromosome 9. They code the synthesis of transferase-A and transferase-B, the two enzymes that are responsible for conversion of substance H into A and B agglutinogens.

ABO agglutinogens are not confined to erythrocytes alone; they are widely found in the secretory glands of gastrointestinal, respiratory, and genitourinary tracts. The secreted agglutinogens are however not glyco-sphingolipids but glycoproteins (oligosaccharide plus protein). Only about 80% of the population secretes ABO agglutinogens. They are called *secretors*. The rest are *nonsecretors*.

Rhesus agglutinogens   Unlike the ABO agglutinogens, Rhesus agglutinogens are integral membrane proteins. They are not found anywhere other than on red cells.

ABO agglutinins   The agglutinins α and β are absent at birth but develop over the first 3 to 6 months of life. They are produced as a result of exposure to ABO-like polysaccharides that are abundant in microbes, seeds, and plants. These natural antibodies are immunoglobulins of the IgM type. Subsequent exposures to ABO agglutinogens, as in the course of mismatched transfusion, also produce agglutinins. Such immune agglutinins are often of the IgG type.

Rhesus agglutinins   There are no natural antibodies to Rhesus agglutinogens. Agglutinins formed against them are of the IgG type.

Importance of Blood Groups

Transfusion of blood   The main importance of blood groups lies in ensuring compatible blood transfusion. This is discussed in detail below.

Medicolegal importance   When the blood types of the parents are known, it can be stated with certainty which of the blood groups cannot be present in the offspring. This knowledge is useful in exclusion of pretenders in cases of disputed paternity. It is however not possible to state conclusively that a certain person is indeed the parent of a child. The predictive value is increased if several blood group systems are considered. With the use of DNA fingerprinting, the exclusion rate for paternity rises close to 100%.

Association with diseases   The incidence of certain diseases is related to the blood group. For example

duodenal ulcers are twice as common in group O non-secretors than in group A or B secretors, and tumors of the salivary glands, stomach, and pancreas are more common in group A than in group O individuals.

Blood Transfusion

Autologous blood transfusion

Autotransfusion is completely free from the risks of transfusion reactions and transfusion-transmitted diseases. In this, the patient's own blood is withdrawn in advance of elective surgery and then transfused back if needed during the surgery. Up to 1.5 L of blood is withdrawn over a 3-week period while the person receives iron supplements.

Blood grouping and cross-matching

Autotransfusion is only occasionally possible. Mostly, the patient needs a blood donor. Preferably, the blood groups of the donor and the recipient should be the same. In emergency situations, there may be no time for finding out the blood group of the patient and even when known, the blood of the same group may not be available. In such situations, blood group O– may be transfused indiscriminately to all patients in dire need of transfusion. This is because O– group blood has no agglutinogens and the chances of fatal reactions occurring following a mismatched transfusion (O– group blood donated to persons with A+, B+ or AB+ blood groups) are the lowest. Persons with O– blood group are therefore called *universal donors*. In the same way, persons with AB blood group are *universal recipients*. In emergency situations, they can be transfused with any of the ABO blood groups. This is because AB group blood has neither α nor β agglutinins.

The idea of a universal donor is not always safe. Normally, the α and β agglutinins present in the transfused O group blood are greatly diluted by the recipient's plasma and therefore, are unable to lyse the recipient's erythrocytes.[1] However, the O group donor may have very high titers of α and β agglutinins, and these may cause hemolysis of the recipient's erythrocytes. Such O group donors are called *dangerous universal donors*.

Transfusion of blood of the same group does not

---

[1]   It may be argued that the donor's agglutinogens also get diluted. That, however, is not true. The effective concentration of agglutinogens is determined by their concentration on the surface of the erythrocyte, which remains unchanged even after the dilution of erythrocytes in the recipient's plasma.

guarantee a reaction-free transfusion. The donor and recipient's blood may be ABO and Rhesus compatible but the donor might have P agglutinogens and the recipient might have anti-P agglutinins. Since there are innumerable minor and familial blood groups that are never ascertained, the donor and recipient's blood have to be directly tested (cross-matched) against each other. Cross matching may be major or minor. *Major cross-matching* involves testing the donor's erythrocytes against the recipient's serum. *Minor cross-matching* involves testing the recipient's erythrocytes against the donor's serum.

Blood grouping, however, does help in narrowing down the search for compatible blood. For example, if the recipient is B+, the blood bank technician needs to test only a few samples of B+ blood for the perfect compatibility. Without blood grouping, a much larger number of random blood samples would have to be cross-matched.

Complications of transfusion

Fatal *hemolytic reactions* can occur in mismatched transfusion. Rapid hemolysis results in liberation of free hemoglobin into the plasma, often resulting in severe jaundice and renal tubular damage. When reactions occur, transfusion must be stopped immediately and the patient intravenously injected with rapid-acting corticosteroids. Febrile reactions occur due to destruction of leukocytes and platelet by antibodies against them.

Circulatory overload can develop if transfusion is too rapid. The rate of transfusion therefore should not exceed 1ml per kilogram body weight per hour.

Hemosiderosis is caused by repeated blood transfusions, as in thalassemic patients. There is iron deposition in, and consequent damage to, several organs like the liver, heart, and endocrine organs.

Electrolyte disturbances, especially hyperkalemia and hypocalcaemia, are not uncommon. *Hyperkalemia* occurs because in stored blood the erythrocytes leak out intracellular K^+ into the plasma. *Hypocalcemia* occurs because stored blood contains citrates as anticoagulant. When transfused into a recipient the citrates are metabolized. However, if the rate of transfusion exceeds the rate of citrate metabolism, the citrates chelate Ca^{2+} in recipient's blood causing hypocalcemia.

Anemia hypoxia (p 355) can be a problem in patients receiving a large transfusion of stored blood. Red cells in stored blood have very low amounts of 2,3 diphosphoglycerate (DPG) in them. Hence, stored blood has high affinity for O_2 and consequently, tends to give off less O_2 to the tissues.

Transmission of diseases like hepatitis B or C and AIDS constitutes a serious risk.

Hemolytic Disease of the Newborn

Erythroblastosis fetalis

If a Rhesus negative mother bears a Rh positive baby in two consecutive pregnancies, the second baby is prone to develop hemolytic disease. The hemolytic disease can also occur in the first baby if the Rh negative mother is sensitized to Rhesus agglutinogens by a prior Rh positive blood transfusion.

During the first pregnancy, the mother is sensitized to Rh agglutinogens of the fetus due to leakage of fetal erythrocytes into the maternal circulation. The leak occurs late in the third trimester or during parturition when the maternal and fetal bloods come in contact for the first time. However, agglutinins are not formed quickly or in significant titers on first exposure and the first D+ fetus usually escapes unharmed. A second D+ fetus however evokes rapid formation of anti-D agglutinins in the third trimester.

The anti-D formed is almost entirely of the IgG type which can cross the placental barrier. The agglutinins therefore cross over to the fetal circulation, hemolyzing the fetal red cells. The hemolysis triggers compensatory hyperactivity of the fetal erythropoietic organs, resulting in the appearance of immature erythroblasts in the fetal circulation, which gives this disorder its name.

Depending on its severity, erythroblastosis fetalis takes one of the following forms. In severe cases, the fetus looks like a bag of waters due to edema and often dies *in utero*. The condition is called *hydrops fetalis*. Less severe cases result in jaundice of the newborn and the condition is called *icterus neonatorum gravis*. The jaundice is much less severe till birth when the mother conjugates and excretes most of the fetal bilirubin load but exacerbates immediately after birth. The mildest cases result only in anemia and are called the *congenital anemia of the newborn*.

Kernicterus is a complication of icterus neonatorum gravis. It is a neurological syndrome in which unconjugated bilirubin is deposited in the basal ganglia. The condition is uncommon in adults but infants are vulnerable to it because their blood–brain barrier is more permeable. Moreover, since the liver is not mature enough in infancy, there is a greater rise in the plasma concentration of unconjugated bilirubin, which is lipid soluble and crosses the blood–brain barrier more easily.

The diagnosis of erythroblastosis fetalis is confirmed only if anti-D agglutinins are detected on fetal red cells or in maternal blood. Anti-D coated fetal red cells agglutinate on adding Coombs' serum. The test is called the *direct Coombs' test*. For detecting anti-D agglutinin in maternal blood, the agglutinin must first be adsorbed on 'carrier' red cells. Subsequent addition of Coombs' serum will cause the red cells to agglutinate. The test is called the *indirect Coombs' test*.

Prevention of the condition is possible by not allowing the mother to get sensitized to Rh agglutinogens during the first pregnancy. This is done by administering a single dose of anti-Rh serum during the postpartum period in the first 72 hours. The anti-Rh antibodies promptly destroy any Rh agglutinogens that might gain access to maternal circulation and prevent the mother from developing active immunity against the Rh agglutinogen. Treatment consists of exchange transfusion in which the hemolyzed blood of the newborn is withdrawn from a suitable peripheral artery and fresh Rhesus negative blood is transfused simultaneously into a convenient peripheral vein.

Fetal ABO hemolytic disease

Fetal ABO hemolytic disease, which has a similar mechanism to that described above, is surprisingly mild. There are at least four reasons for the mildness. (1) Fetal erythrocyte membrane possesses fewer A and B agglutinogenic sites. (2) Anti-A and anti-B agglutinins do not bind complement on the fetal erythrocyte. (They do so in adults, causing severe hemolysis in ABO incompatibility.) (3) Anti-A and anti-B are mostly IgM, which do not cross the placenta. (4) The small amounts of IgG anti-A and anti-B that do cross the placenta bind to several types of cells other than the red cell. Consequently, their effect on erythrocytes is diluted.

27 Blood Platelets and Hemostasis

Blood Platelets

Blood platelets or *thrombocytes* are thin biconvex anucleate disks 2–4 μm in diameter. They are produced in the bone marrow by fragmentation of very large nucleated cells called *megakaryocytes*. They are released into the blood where they have a lifespan of about 10 days.

Blood platelets have at least four functions: (1) When the endothelium is disrupted, the breach is closed by a mass of platelets called the platelet thrombus or the platelet plug. (2) Phospholipids present on the surface of platelets are essential in blood coagulation. Some of the key reactions take place only on the surface of platelets. (3) Platelet granules release several prohemostatic and antihemostatic substances as well as factors that are important in tissue healing. (4) Platelets have a weak phagocytic activity.

Platelet structure

In stained blood smears, platelets exhibit two concentric zones: a peripheral zone called the hyalomere and a central zone called the granulomere. The *hyalomere* has a circumferential bundle of microtubules that maintains the shape of the platelet. It also has actin and myosin proteins. When the intracellular calcium concentration rises, the myosin light chain kinase is activated and the myosin interacts with actin filaments, resulting in platelet contraction. Platelet contraction

was earlier thought to be mediated by a special contractile protein called thrombosthenin, which is now known to be nothing but actin. The *granulomere* contains a canalicular system, secretory granules, one or two mitochondria, and scattered particles of glycogen. However, it lacks a nucleus and is therefore unable to synthesize proteins.

The **platelet canalicular system** is of two types. The *surface-connected canalicular system* is analogous to the T tubules of skeletal muscle cell. It is open on the platelet surface at several sites and is the major pathway for the discharge of secretory products upon activation of the platelets. The *dense tubular system* is analogous to the L tubule of skeletal muscle cell. It maintains a high concentration of Ca^{2+} inside it. This Ca^{2+} is released into the platelet cytosol when the platelet is activated.

The **platelet granules** are of three types: α-granules, dense granules, and lysosomal granules. The contents of α-granules and dense granules are secreted on platelet activation while the lysosomal granules are secreted when the platelet begins to disintegrate.

α-platelet granules secrete numerous substances, most of which are already present in plasma and therefore do not have any physiological role. An example is the von Willebrand factor (vWF) which is important for platelet adhesion. Adequate amounts of vWF come from endothelial secretions anyway and the presence or absence of vWF in platelets does not affect platelet adhesion. The important α-granule proteins are those that are absent from plasma until secreted by activated platelets, and are as follows. (1) *Platelet factor 4* inhibits the antithrombin III–heparin system. (2) *Thrombospondin* binds to the glycoprotein receptors on the platelet surface and brings about platelet aggregation. (3) *Platelet-derived growth factor* (PDGF) and *transforming growth factor-β* (TGF-β) are chemoattractants for leukocytes, smooth muscle cells, and fibroblasts. They also stimulate mitosis of these cells. Thus, they help in inflammation and wound healing. (The term platelet-derived growth factor is inappropriate because these factors are secreted not only by platelets but also by monocytes, macrophages, and endothelin cells.) (4) *Fibronectin* helps in adhesion of platelets to the site of injury.

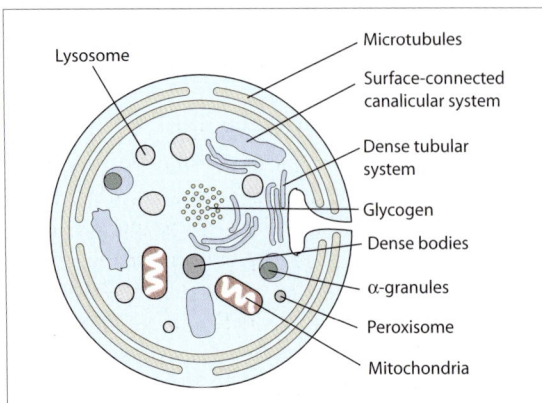

Fig. 27.**1**   Structure of a thrombocyte.

Dense platelet granules secrete *ADP* which amplifies platelet activation. Patients with hereditary disorders that prevent the storage of adequate quantities of ADP in the dense granules have moderate bleeding disorders. Dense granules also secrete *serotonin* which constricts arterioles, thereby helping in hemostasis.

Platelet integrins   The platelet bears on its surface integrins which are cell adhesion molecules. Chemically, these molecules are glycoproteins (GP) and have various subtypes. GP Ib is present on all platelets. GP IIb–IIIa is formed only after platelet activation. GP I is important for platelet adhesion while GP IIb–IIIa is important for platelet aggregation.

Platelet count

The normal platelet count is 150,000–400,000 mm^{-3} of blood. **Thrombocytopenia** is the fall in platelet count below 150,000 mm^{-3}. It results in *purpura*, a bleeding disorder. When the purpura occurs without any obvious cause, it is called primary thrombocytopenic purpura. Secondary thrombocytopenic purpura (where the cause is identifiable) occurs following drug administration (e.g., aspirin), in malignancies (e.g., leukemias, aplastic anemia, and bone marrow infiltration), and in hypersplenism in which an abnormally hyperactive spleen destroys platelets and other blood cells in larger than usual numbers. Life-threatening thrombocytopenia occurs in hemorrhagic dengue fever, a viral disease.

Thrombocytosis is a rise of the platelet count above 400,000 mm^{-3}. Individuals with thrombocytosis are predisposed to thrombotic events. The common causes of thrombocytosis are acute hemorrhage, surgery, and trauma, particularly bone fractures. Splenectomy causes thrombocytosis by reducing the number of platelets that are removed from circulation.

Hemostasis

The term hemostasis means the stoppage of bleeding. It occurs in three major steps: constriction of the damaged blood vessels, formation of a hemostatic plug, and clot retraction and dissolution (Fig. 27.2).

Constriction of the damaged vessel slows down the bleeding. Unless the blood flow slows down, any clot formed is washed away. The immediate constriction occurs mainly due to the direct response of vascular smooth muscle to injury. A little later, further

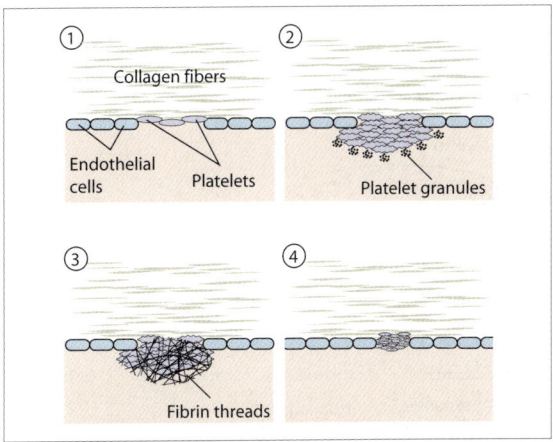

Fig. 27.**2**   Formation of a platelet plug. (1) A few platelets adhere to the collagen fibers below the disrupted endothelium. (2) Several more platelets aggregate at the site, forming a platelet plug. (3) Fibrin threads are deposited in the plug, making it firm. (4) Platelet plug contracts, pulling together the edges of the disrupted endothelium.

constriction is induced by the serotonin released from platelets that adhere to the site of injury.

Formation of a platelet plug arrests the bleeding by plugging the breach in the endothelium. The platelet plug is formed by a large number of tightly packed platelets.

Consolidation of the platelet plug occurs through the formation of fibrin threads that run criss-cross through it. This involves the enzymatic conversion of the plasma protein fibrinogen into fibrin through an elaborate process called coagulation or clotting. The fibrin threads bind to the integrins on platelets and anchor the platelets tightly together into a thrombus or clot.

Clot Retraction (shrinking of the clot) pulls the edges of the wound together, making wound healing easier. Clot retraction occurs due to the contraction of platelets. Integrins are also required for clot retraction since they anchor platelets to fibrin threads.

Fibrinolysis (dissolution of the clot) is necessary for the restoration of normal blood flow to the healed tissue. Fibrinolysis is produced by a substance called *plasmin* which splits fibrin enzymatically. Plasmin is formed from plasminogen by *tissue plasminogen activator* (tPA) secreted by endothelial cells. Newly formed clots, which are rich in platelets, are less susceptible to fibrinolysis than older clots. This is because platelets release *plasminogen activator inhibitor*. This ensures that clots are lysed only after sufficient time has elapsed during which tissue healing can occur.

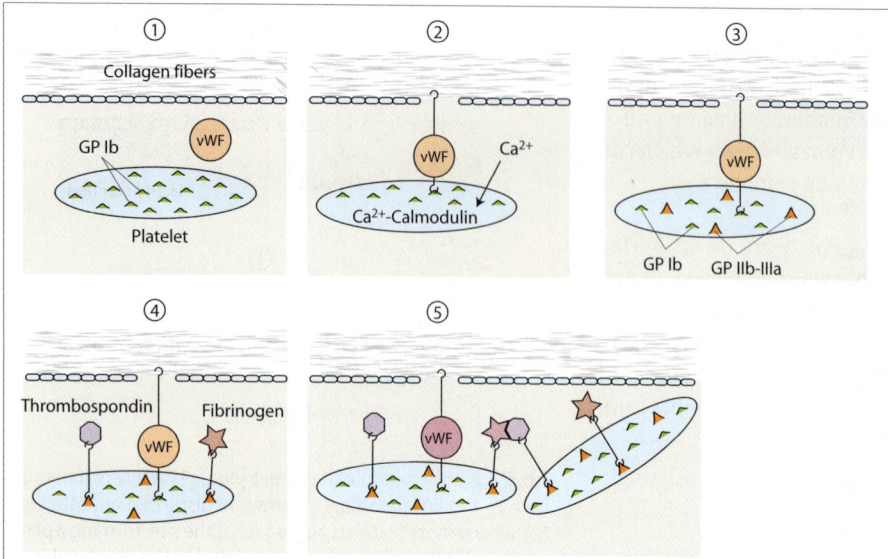

Fig. 27.**3** Schematic illustration of platelet adhesion and aggregation.

Platelet plug formation

The formation of platelet plug begins when platelets come in contact with the subendothelial collagen fibers (collagen fibers lying under the endothelium of blood vessels) that get exposed as a result of injury. The platelet plug is formed in four stages: platelet adhesion, platelet activation, platelet aggregation, and platelet contraction (Fig. 27.**3**).

Platelet adhesion requires the presence of the *von Willebrand factor* (vWF), a glycoprotein secreted by the vascular endothelial cells and present in plasma. When the endothelium is disrupted, the circulating vWF binds to the exposed subendothelial collagen fibers. On binding with collagen, vWF gets altered so that it is now able to bind with GP Ib present on the platelet surface. The vWF thereby anchors the platelet to subendothelial collagen.

Platelet activation refers to the rise in intracellular Ca^{2+} concentration that occurs in the platelet after it gets anchored to collagen through vWF. The contact with collagen activates group II-C hormonal mechanisms (see Chapter 81) in the platelet. The IP_3 (the second messenger of group II-C hormones) formed triggers the release of Ca^{2+} from the dense tubular system of platelets into the platelet cytosol. Activated platelets contract (due to activation of actin-myosin interaction), develop pseudopodia (due to reorganization of microtubules), discharge their granules (due to Ca^{2+}-mediated exocytosis), and develop another type of integrin on its surface called GP IIb-IIIa (due to reorganization of the platelet membrane).

The rise in Ca^{2+} during platelet activation also leads to the formation of thromboxane A_2 inside the platelet (Fig. 27.**4**). Thromboxane A_2 leads to further rise in cytosolic Ca^{2+}. Thus, there is a positive feedback cycle of platelet activation. The ADP secreted by the dense granules contributes to platelet activation by inducing thromboxane A_2.

Platelet aggregation   The GP IIb-IIIa on the platelet surface binds to *thrombospondin*, an adhesive protein secreted by the platelet itself. GP IIb-IIIa also binds to fibrinogen present in plasma. Aggregation of platelets occurs due to the cross-linking of thrombospondin with fibrinogen.

Blood coagulation

Blood coagulation is brought about through a cascade system in which one activated factor activates another factor. The cascade ends with the formation of fibrin. The coagulation factors are listed in Table 27.**1**.

The coagulation process is initiated by two different mechanisms, the extrinsic and the intrinsic. The two pathways soon converge on to a common pathway that begins with the activation of factor IX and ends with the formation of fibrin threads (Fig. 27.**5**).

The extrinsic pathway is initiated by the *tissue factor*, a lipoprotein released by the injured tissues. The intrinsic pathway is initiated by contact of blood platelets with negatively charged surfaces like glass. It is called 'intrinsic' because all the necessary factors required in this pathway are present in the plasma itself. It is through the intrinsic pathway that blood clots

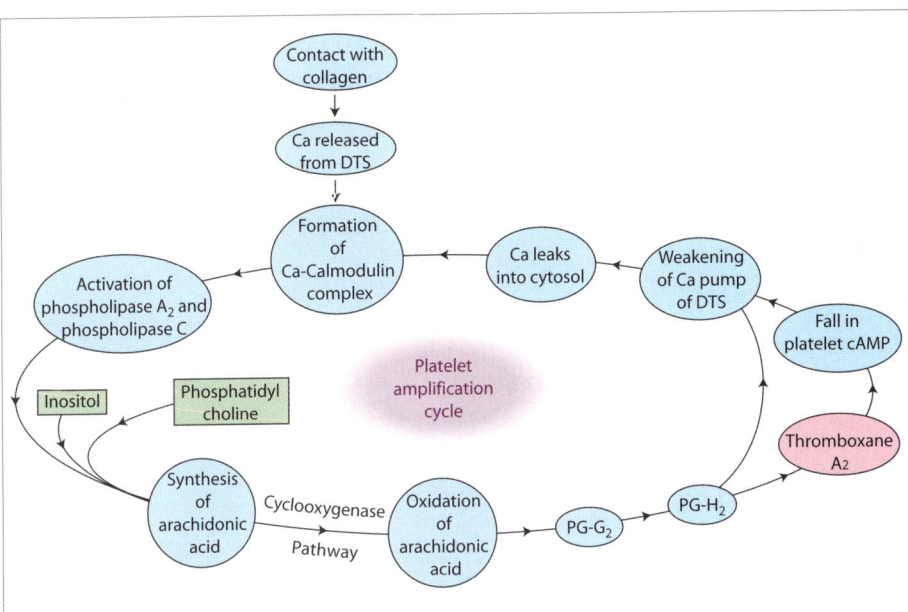

Fig. 27.**4**   The amplification cycle for platelet activation. See also Fig. 28.**2**.

Table 27.**1**   Coagulation factors

I	Fibrinogen
II	Prothrombin
III	Thromboplastin
IV	Calcium ions
V	Labile factor
VI	Has been dropped. Earlier called accelerin
VII	Stable factor. Earlier called preaccelerin
VIII	Antihemophilic globulin
IX	Christmas factor
X	Stuart-Prower factor
XI	Plasma thromboplastin antecedent (PTA)
XII	Hageman factor
XIII	Fibrin stabilizing factor (Laki Lorand factor)
HMWK	High molecular weight protein
Pre-K	Prekallikrein
Ka	Kallikrein
PPL	Platelet phospholipids

with a negatively charged surface. XIIa in its turn activates XI to XIa. The formation of XIIa is greatly enhanced through feedback activation by prekallikrein (Fig. 27.**7**): XIIa activates prekallikrein to kallikrein which in turn activates more XII to XIIa.

HMWK is responsible for attracting prekallikrein and factor XI (both are bound to HMWK in plasma) to the site of reaction with factor XII. This is possible because HMWK, like factor XII, is attracted toward negatively charged surfaces which provide the site of reaction.

Initiation of the extrinsic pathway occurs when tissue factor (TF), a glycoprotein, is released from fibroblasts and smooth muscle cells of the vessel wall. TF is the cofactor of both factor VII and VIIa. The binding of TF with factor VII or VIIa has to occur in the presence of membrane phospholipids. Hence, the reaction occurs mainly on the fibroblasts of the vessel wall, with the fibroblast cell membrane providing the phospholipids necessary for the reaction.

When TF binds to factor VII, it promotes its activation to VIIa. When TF binds to factor VIIa, the TF-VIIa complex catalyzes the activation of factor IX (Fig. 27.**5**) and also, somewhat weakly, of factor X (Fig. 27.**6**).

Activation of factors VII, VIII, and V (Fig. 27.**7**) does not occur in the regular course of the cascade. Factor VII is activated by several factors, the most important ones being TF and factor IXa. It is also activated by VIIa (autoactivation). Factor V and factor VIII are activated only through feedback activation. In the coagulation cascade, activation of factor VIII and factor V occur before

when allowed to stand in a glass test tube. Injured blood vessels too can trigger the intrinsic pathway because injury exposes the collagen fibers in the vessel wall. Collagen provides the negative surface that triggers the intrinsic pathway.

Initiation of the intrinsic pathway requires four contact activation factors: factor XII, factor XI, prekallikrein, and high molecular weight kininogens (HMWK). Factor XII is activated to XIIa when it comes in contact

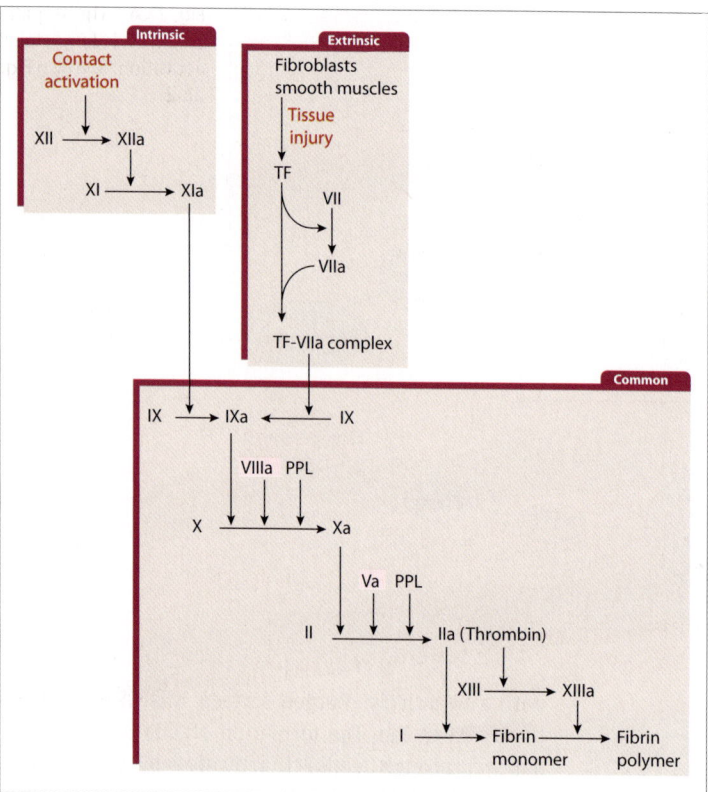

Fig. 27.**5** The coagulation cascade. All steps require calcium ions (not shown). TF = tissue factor; PPL = platelet phospholipids (present on platelet surface).

The Beginning of the Common Pathway

The common pathway is sometimes described as beginning with the activation of factor X. This is partly true but the activation of factor X through the extrinsic pathway is a rather weak one. If tissue factor and VIIa could adequately activate factor X, the deficiency of factors IX and VIII would not have caused such serious bleeding disorders as they are known to (Fig. 27.**6**).

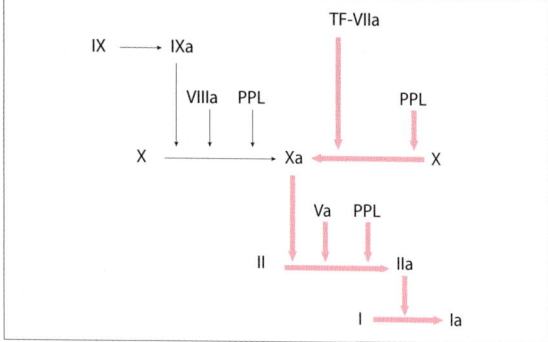

Fig. 27.**6** A coagulation pathway showing a strong, direct activation of factor X by factor VII and tissue factor. Such a pathway would not have any role for factors VIII and IX. The direct activation of factor X by tissue factor and factor VII does exist but is very weak.

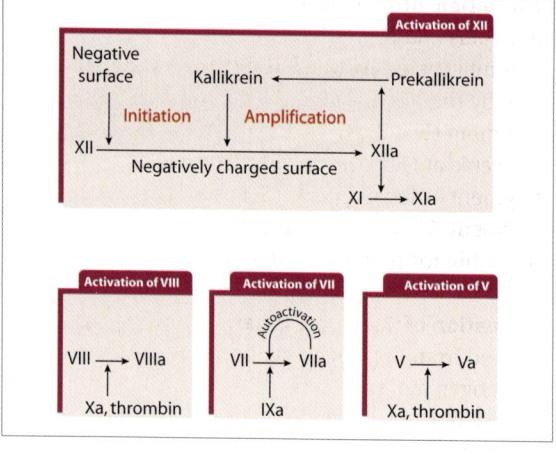

Fig. 27.**7** Activation of factors XII, VIII, VII, and V.

Xa and thrombin may be available for the activation of factors V and VIII. However, as coagulation proceeds, more and more factor Xa and thrombin are generated and are available for activation of factors V and VIII.

Activation of factor X requires platelet phospholipids and therefore it occurs on the platelet surface. A complex of factors X, IXa, VIIIa, platelet phospholipids, and calcium ions (X-IXa-VIIIa-PPL-Ca^{2+} complex), called *tenase*, forms on the platelet surface (Fig. 27.**8**) and results in the activation of Factor X to Xa.

activation of factor X, and much before the formation of thrombin. However, the activation of factors V and VIII requires Xa and thrombin. Initially, only traces of

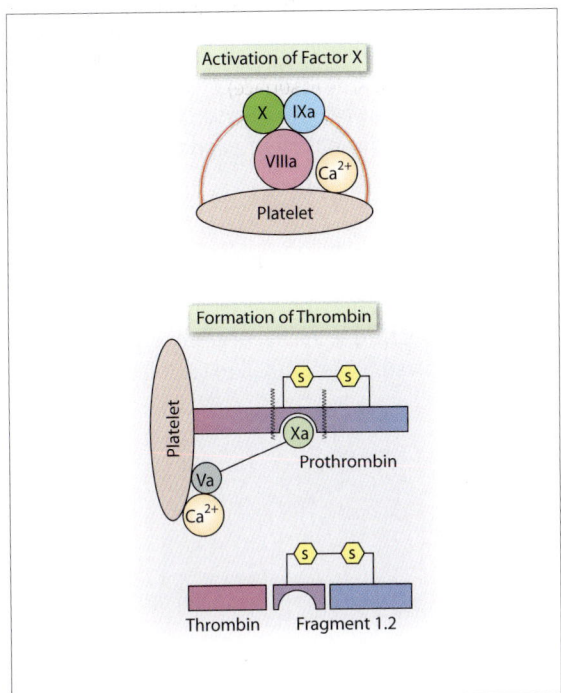

Fig. 27.**8**   (*Above*) The IXa-VIIIa-PPL-Ca²⁺ complex (tenase) that activates factor X into Xa. (*Below*) The Xa-Va-PPL-Ca²⁺ complex that splits prothrombin into thrombin.

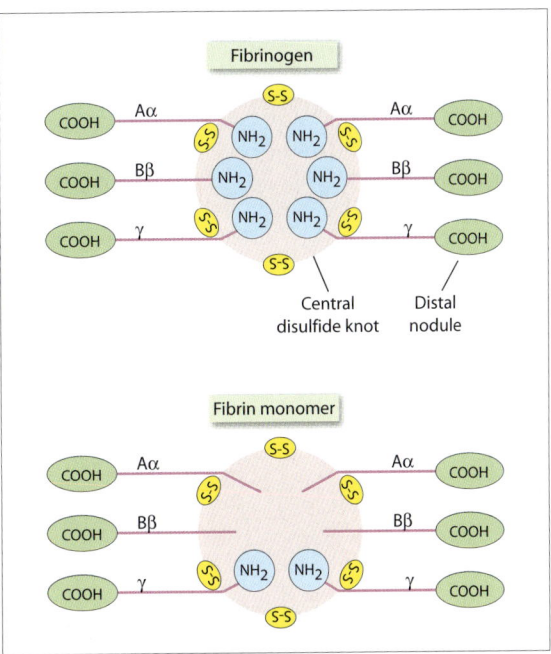

Fig. 27.**9**   Structure of fibrinogen (*above*) and fibrin monomer (*below*).

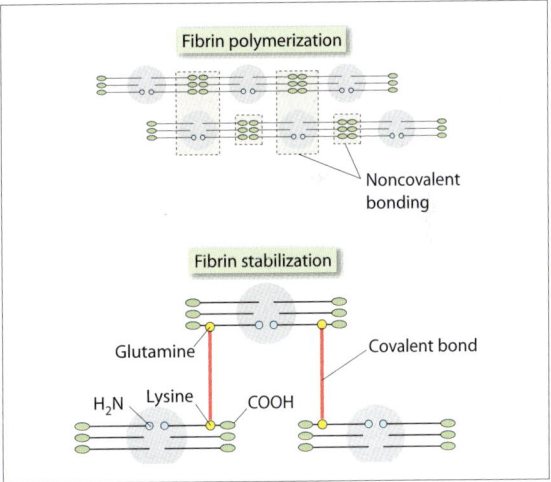

Fig. 27.**10** (*Above*) Polymerization of fibrin molecules through non-covalent bonding. (*Below*) Fibrin stabilization by factor XIII. Shown here are the peptide bonds formed between the glutamine and lysine residues on the γ-chains of the fibrin monomers.

Formation of thrombin too requires platelet phospholipids and therefore occurs on the platelet surface. Thrombin is cleaved off from the prothrombin molecule by the Xa-Va-PPL-Ca²⁺ complex that binds to prothrombin (Fig. 27.**8**). Factor Xa splits the prothrombin molecule at two sites, forming thrombin and a residual fragment called *fragment 1.2*. In hypercoagulable states, fragment 1.2 serves as a useful marker for excessive thrombin formation.

Formation of fibrin monomers   Fibrin monomers are formed from fibrinogen by the action of thrombin. The fibrinogen molecule is made of three pairs of nonidentical chains, viz., the Aα, Bβ, and γ chains (Fig. 27.**9**). The three chains are disposed symmetrically with their NH₂ groups linked together by disulfide bonds into a central nodule (or the disulfide knot). Thrombin cleaves off short segments from the NH₂ terminals of Aα and Bβ chains to form fibrin monomer.

Polymerization of fibrin monomers   The fibrin monomers initially polymerize through weak noncovalent bonds to form fibrin strands. The bonds are formed both end-to-end (between the distal nodules) as well as side-to-side (between the central nodule and the distal nodule). In the central nodule, the bonds are formed at the active sites formed by the detachment of the NH₂ groups from Aα and Bβ chains (Fig. 27.**10**).

Stabilization of fibrin polymers   The fibrin polymers are formed through noncovalent bonding of fibrin monomers and are therefore mechanically weak. It is made stronger and more resistant to fibrinolysis by factor XIIIa which polymerizes fibrin by catalyzing the formation of peptide bonds between fibrin monomers (Fig. 27.**10**).

What is Factor III?

There is some ambiguity about the term *thromboplastin* which is understandable in the context of the history of coagulation research. The term thromboplastin or factor III was originally assigned to the putative activator of factor II (prothrombin). Since the tissue factor was earlier thought to be directly responsible for the conversion of prothrombin to thrombin, the tissue factor was called tissue thromboplastin. In light of our present knowledge, however, it is the Xa-Va-PPL-Ca^{2+} complex that should be designated as thromboplastin. In the thromboplastin generation test for blood coagulation, the word thromboplastin actually refers to the Xa-Va-PPL-Ca^{2+} complex. Because of these ambiguities, the terms factor-III and thromboplastin are best avoided.

Factor XIII is cleaved by thrombin into its active form XIIIa. Following activation, factor XIII undergoes a conformational change so that its cysteine molecules get exposed at the surface. It is these cysteine molecules that catalyze the formation of peptide bonds between fibrin monomers.

Role of factor IV　Calcium is required for all the steps of the coagulation cascade. Clinical hypocalcemia does not impair coagulation for which Ca^{2+} has to fall to very low levels. Death due to hypocalcemic tetany occurs well before plasma Ca^{2+} can fall to such levels.

Anti-Hemostatic Mechanisms

The coagulating tendency of blood is balanced in vivo by endogenous antihemostatic mechanisms (Fig. 28.1) that tend to prevent clotting inside the blood vessels and to dissolve any clots that do form. In addition, there are exogenous antihemostatic mechanisms that have been exploited for therapeutic purposes.

Factors inhibiting platelet aggregation

Normally, a delicate balance exists between thromboxane A_2 which promotes platelet aggregation and prostacyclin (PGI_2) which inhibits platelet aggregation. Thromboxane A_2 is produced by platelets while PGI_2 is produced by endothelial cells. Both are produced by the cyclooxygenase pathway of arachidonic acid oxidation (Fig. 28.2). Prostacyclin stimulates membrane adenyl cyclase and raises cAMP levels in platelets. cAMP stimulates Ca-ATPase and thereby increases the pumping of Ca^{2+} into the dense tubular system of the platelet with consequent lowering of cytosolic Ca^{2+}. Prostacyclin thereby prevents platelet activation. On the other hand, thromboxane A_2 reduces the cAMP levels in platelets, resulting in the opposite effect.

Aspirin irreversibly (for the remaining life of the platelet) inhibits the cyclooxygenase pathway of arachidonic acid oxidation. By interrupting the amplification circuit of platelet activation (see Fig. 27.4) aspirin minimizes platelet activation and aggregation. This makes aspirin a valuable drug for the prevention of thrombosis. Aspirin also inhibits arachidonic acid oxidation in endothelial cells, thereby reducing prostacyclin secretion. However, the effect is comparatively weaker. Ingestion of certain fish oils prolongs the bleeding time: Certain fatty acids found in fish oils decrease arachidonic acid release from platelet membrane phospholipids and thus limit the synthesis of thromboxane A_2.

Factors inhibiting coagulation

Protein C pathway Three protein factors, viz., protein C, thrombomodulin, and protein S constitute an important negative feedback pathway that keeps blood coagulation under control (Fig. 28.3). *Protein C is activated by factors Xa and thrombin. The activated*

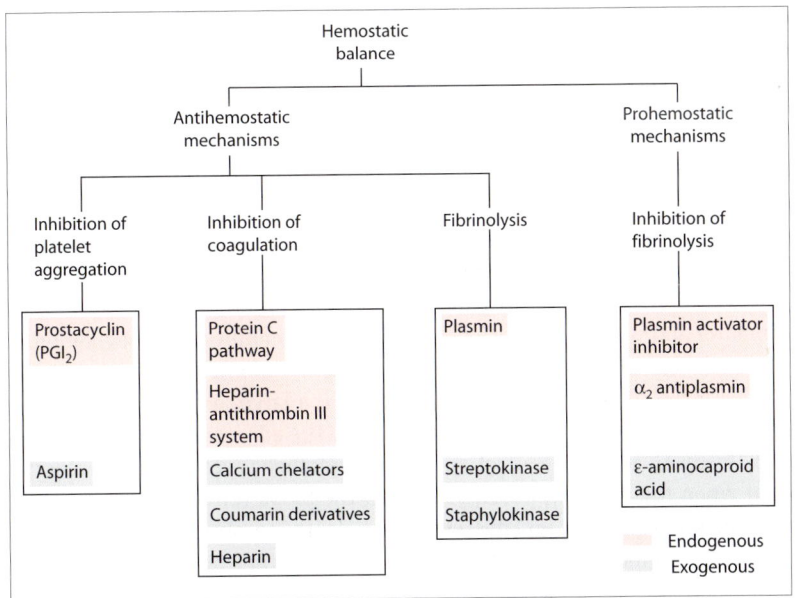

Fig. 28.**1** Antihemostatic and prohemostatic mechanisms.

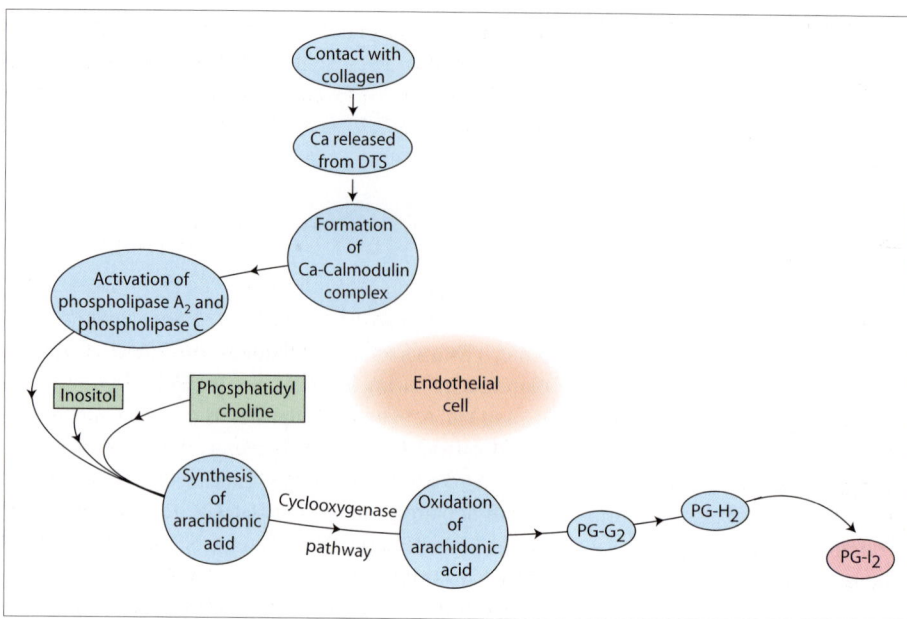

Fig. 28.**2** Prostacyclin synthesis in endothelial cells. Compare with Figure 27.**4**.

protein C inactivates factors VIIIa and Va, the two key factors responsible for the formation of thrombin and Xa. The inactivation requires the presence of two cofactors, *protein S* and *thrombomodulin*. Protein S and protein C are plasma proteins. Thrombomodulin is a protein present on vascular endothelium.

Antithrombin-heparin system *Antithrombin III* is present in plasma as well as in vascular endothelium. It inactivates a number of coagulation factors including thrombin. *Heparin sulfate* is present only in the vascular endothelium. It inhibits both the production of thrombin (from prothrombin) as well as the action of thrombin (on fibrinogen). Heparin potentiates antithrombin III.

Exogenous anticoagulants *In vivo*, a plasma Ca^{2+} level low enough to impair blood clotting is incompatible with life. However, clotting can be prevented *in vitro* if Ca^{2+} is removed from blood by the addition of substances such as oxalates, which form insoluble salts with Ca^{2+} or by the addition of chelating agents which bind Ca^{2+}.

Coumarin derivatives such as *dicoumarol* and *warfarin* are effective anticoagulants. These compounds inhibit the reduction of vitamin K to its active form vitamin KH_2 resulting in the impaired synthesis of factors II, VII, IX, and X (VKDP, see below). However, coumarin derivatives are not administered when the anticoagulant effect needs to be reversed quickly. Such a situation arises, for example, during dialysis of blood, when the blood drawn from the patient's vein must be treated with an anticoagulant before passing it through the dialysis machine and the anticoagulant must be quickly neutralized before reinfusing the blood into the patient's vein. Heparin is the anticoagulant of choice in such situation, because its effect is quickly neutralized by *protamine*, a highly basic protein that forms an irreversible complex with heparin.

■ Factors causing fibrinolysis

Plasmin (fibrinolysin) is the principal fibrinolytic factor. It is formed from a circulating glycoprotein called plasminogen. Plasmin lyses both fibrin and fibrinogen. Exogenous plasmin activators include the bacterial enzymes *streptokinase* and *staphylokinase*. These are used in the treatment of early myocardial infarction for lysing clots in coronary arteries.

The main endogenous activator of plasmin is the *tissue-type plasminogen activator* (t-PA), which is present in the endothelial cells. In its absence, there is extensive spontaneous fibrin deposition. There are also defects in growth and fertility, since plasmin not only lyses clots but also plays a role in cell motility and in ovulation.

Adrenaline stimulates endothelial cells to secrete t-PA. In violent deaths, a large amount of adrenaline is released into blood. The adrenaline causes rapid release of t-PA from endothelial cells, causing massive fibrinolysis. Hence, the blood remains fluid and incoagulable even after death, an observation of forensic importance. The catalytic efficiency of t-PA increases several hundred-fold when it binds to fibrin. This is an

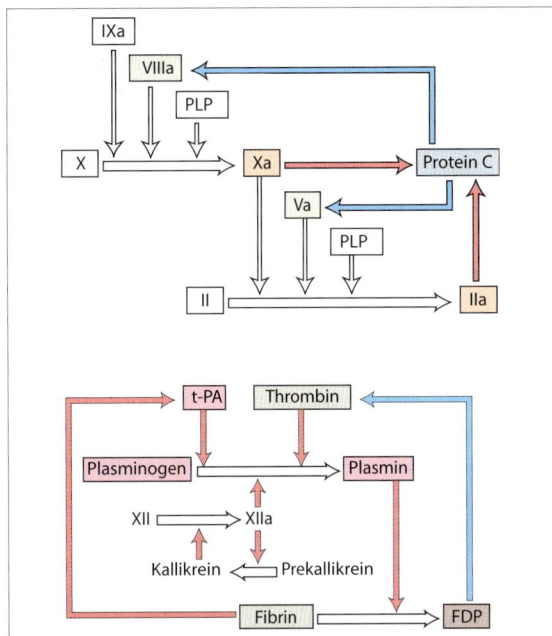

Fig. 28.**3**   (*Above*) The protein C pathway for inhibition of coagulation. (*Below*) Plasminogen activation. FDP = Fibrin degradation products. Red: stimulation; Blue: inhibition.

example of autoinactivation wherein fibrin indirectly brings about its own degradation. Human t-PA is now produced by recombinant DNA techniques and is available for clinical use. It lyses clots in the coronary arteries if given to patients soon after the onset of myocardial infarction.

Another activator of plasmin is thrombin. However, the products of fibrinogen breakdown (called *fibrinogen degradation products*) inhibit the activity of thrombin. Thus there is a negative feedback that controls plasmin generation. Plasmin is also activated by kallikrein and factor XIIa (Fig. 28.**3**). It is interesting to note that fibrinolysis is initiated by the same factors that initiate coagulation (kallikrein and factor XIIa) or are involved in the final stages of coagulation (thrombin, fibrinogen).

Pro-Hemostatic Mechanisms

Fibrinolysis Inhibitors

Exogenous inhibitors of fibrinolysis include drugs like *ε-aminocaproic acid*. Dentists sometimes apply it locally during tooth extraction.

Endogenous inhibitors of fibrinolysis include *plasminogen activator inhibitors* (PAI) which inhibit the activation of plasmin, and α_2-antiplasmin which inhibits plasmin directly. During clotting, the fibrin threads

bind to both plasminogen activators like t-PA as well as to plasminogen activator inhibitors like α_2-antiplasmin. Hence, both fibrinolytic and antifibrinolytic substances are present inside the thrombus.

Hemorrhagic Disorders

Hemorrhagic disorders are of three types: vascular disorders due to increased capillary fragility, platelet disorders that result in inadequate formation of platelet plugs, and coagulation disorders (Fig. 28.**4**) that result from weakness of the platelet plug.

The characteristics of bleeding due to vascular and platelet disorders are similar. They tend to cause spontaneous bleeding, resulting in purplish patches on the skin or mucous membranes. The condition is called *purpura*. The term purpura comes from the word purple which denotes the color of the bleeding patches. Coagulation disorders do not cause purpura. Spontaneous bleeding, when it does occur in severe coagulation disorders, occurs in joints and muscle. Bleeding mostly occurs following injury and does not stop easily. Although the bleeding characteristics give some clue regarding their cause (vascular/platelet or coagulation disorders), the final diagnosis requires confirmatory tests.

Tests for vascular / platelet disorders

The **tourniquet test** is positive in most cases of thrombocytopenia. It is performed by placing the sphygmomanometer cuff around the upper arm and raising the pressure to halfway between the arterial and venous blood pressures. The idea is to permit arterial flow but occlude the venous return so as to severely increase the capillary hydrostatic pressure distal to the cuff. The pressure is kept elevated for 5 minutes. After the cuff is deflated and the congestion disappears, the number of petechiae in the cubital fossa is counted. If more than 20 petechiae are present in an area of 3 cm diameter, the test is positive and suggests thrombocytopenia.

The **bleeding time** is estimated by pricking the finger or the earlobe to a measured depth and mopping the blood that flows out at regular intervals. The normal bleeding time is 2 to 6 minutes. Bleeding time increases in thrombocytopenia.

Clot retraction   Fresh blood is taken in a test tube and incubated in a water bath at 37°C. Clot retraction is indicated by the pulling away of the clot from the sides of the tube. Normally, clot retraction should begin in 1 hour and be complete in 24 hours. Also, the volume of serum left behind after complete clot retraction is about half the volume of blood. The clot retraction time is prolonged in thrombasthenia.

■ Tests for coagulation disorders

Although platelet phospholipid is required for clotting, the coagulation time is not affected by moderate thrombocytopenia. This is the reason why a platelet count is not considered to be a test for coagulation disorder. However, the coagulation time may be affected in severe thrombocytopenia.

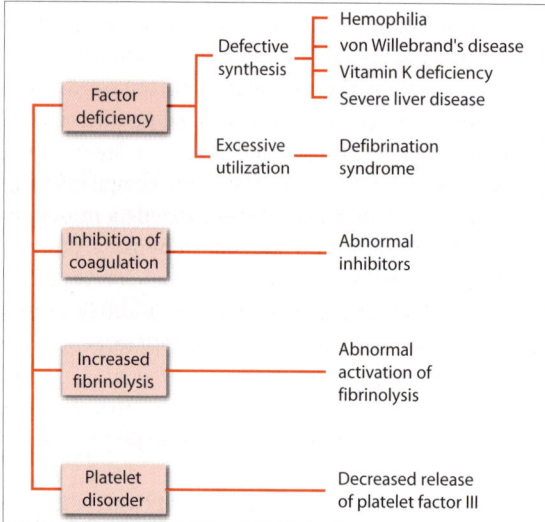

Fig. 28.**4**   Types of coagulation disorders.

Table 28.**1**   Clinical differences between vascular or platelet disorders and coagulation disorders.

Vascular/platelet disorders	Coagulation disorders
Bleeding usually confined to the skin.	Bleeding usually in deeper tissues.
Bleeding usually takes the form of confluent petechiae and small ecchymosis.	Bleeding usually takes the form of large ecchymoses.
Spontaneous bleeding is common.	Spontaneous bleeding is uncommon.
Wound bleeding:	Wound bleeding:
a. is excessive;	a. is less profuse;
b. is immediate;	b. is delayed for several hours;
c. stops quickly on application of local pressure;	c. does not stop quickly on application of local pressure;
d. lasts < 48 hours;	d. continues > 48 hours;
e. rarely recurs.	e. tends to recur.

Whole blood clotting time tests the intrinsic and common pathways. It is the time taken for blood to clot spontaneously in a glass tube. Normal clotting time is 5 to 11 minutes. The test is not sensitive: Clotting time is prolonged only in severe coagulation disorders. However, it remains a useful and simple test for controlling the dose of heparin during anticoagulant therapy.

Partial thromboplastin time (PTT) assesses the intrinsic and common pathways of coagulation. In this test, the intrinsic coagulation pathway is activated by incubating the plasma with kaolin (a contact factor) in the presence of Ca^{2+} and platelet lipid substitute. Normally, the plasma clots within 45 seconds. If the clotting time is prolonged by more than 10 seconds, the result is considered abnormal.

Prothrombin time (PT) assesses the extrinsic and common pathways of coagulation. In this test, the extrinsic coagulation pathway is activated by incubating the plasma with tissue factor (usually an extract of brain tissue) and Ca^{2+}. Since the time taken varies with the type of tissue factor used, the test is always performed simultaneously on the patient's plasma and normal plasma. If the time taken for the patient's plasma to clot is 20% more than that of a normal subject, it is considered abnormal.

Thrombin time estimates the fibrinogen concentration in plasma and therefore, tests the common pathway. It is the time taken by plasma to clot after addition of thrombin to it. With a standardized thrombin solution, plasma clots in 15 seconds. Prolongation to 18 seconds or more is regarded as abnormal.

Thromboplastin generation test (TGT)   It is a test for the intrinsic pathway. It is done in two stages. In stage I, thromboplastin is produced by reacting together factors XII, XI, IX, VIII, V, platelet phospholipids and Ca^{2+}. Stage II estimates the amount of thromboplastin formed in stage I by adding it to normal plasma (containing prothrombin and fibrinogen) and noting the time taken for fibrin formation.

■ Platelet disorders

Bleeding due to a reduction in platelet count is called *thrombocytopenic purpura*. Bleeding disorders can also be due to defects in platelet adhesion or aggregation. Platelet adhesion is impaired in von Willebrand's disease in which there is deficiency of vWF. von Willebrand's disease is also associated with defective coagulation. Platelet aggregation is impaired in *thrombasthenia*, a rare hereditary disorder in which the GP IIb-IIIa receptors are not formed following platelet activation.

Vascular disorders

Bleeding disorders from vascular causes are often called *athrombocytopenic purpura*. Strictly speaking, the term 'athrombocytopenic' should also include functional disorders of platelets like thrombasthenia. However, since these are rare conditions, the term usually refers to vascular disorders. Two well known vascular causes of bleeding are *scurvy* (see Table 77.**2**) which causes failure of collagen formation associated with impaired hydroxyproline synthesis, and *Cushing's syndrome*, which causes loss of perivascular supporting tissues. Other causes include severe infections, drugs, and senile purpura (due to atrophy of collagen in old age).

Coagulation disorders

Hemophilia occurs due to a genetic deficiency of factor VIII (hemophilia A or classical hemophilia) or of factor IX (hemophilia B or Christmas disease). Hemophilia A is three times more common, occurring in 1 in 10,000 male births. Hemophilia is inherited as a sex-linked (on X chromosome) disorder mostly transmitted by females who themselves have no symptoms. To have frank hemophilia, a female has to be homozygous for the hemophilic gene. Female carriers of hemophilia, who are heterozygotes, usually produce sufficient factor VIII for normal hemostasis.

Hemophilia is characterized by bleeding into soft tissues, muscles, and weight-bearing joints. Hematuria and epistaxis are common. Hemophilic bleeding can result in large collections of partially clotted blood putting pressure on adjacent tissues. Despite frequent bleeding, severe iron-deficiency anemia is uncommon because most of the bleeding is internal and the iron released is recycled.

A prolonged PTT is considered diagnostic for hemophilia. Hemophilia is treated with factor VIII concentrate. ε-aminocaproic acid (EACA) has been used in hemophilia, especially during dental extractions. It is a potent antifibrinolytic agent that inhibits plasminogen activators present in oral secretions and stabilizes clot formation in oral tissue.

In **von Willebrand's disease**, patients bleed excessively due to a congenital deficiency of vWF. Its incidence is about 1 in 1000 and if one considers subclinical cases too, it is as high as 1 in 100. In addition to promoting platelet adherence, vWF serves as a carrier protein for factor VIII in plasma and increases its plasma half-life. In the absence of vWF, factor VIII cannot be maintained at adequate levels in the plasma. An analogue of vasopressin, *desamino-arginine-vasopressin*, is often given to these patients just before surgery to induce endothelial cells to release their stores of vWF.

Vitamin K deficiency   Vitamin K is obtained in part from food, especially green leafy vegetables (as vitamin K_1 or *phylloquinone*), and partly from the bacterial flora in the intestine which synthesizes the vitamin (as *menaquinone*). When one source is deficient, the other compensates. Dietary vitamin K_1 is fat-soluble and requires bile salts for its absorption. Bacterial vitamin menaquinone is water-soluble and is absorbed even in the absence of bile.

The physiologically active form of vitamin K is vitamin KH_2 (reduced hydroquinone). In the liver, vitamin K is reduced to vitamin KH_2. Liver cells store enough vitamin K to last a month and therefore inadequate dietary intake is not a common cause of vitamin K disorder. There are three major causes of vitamin K deficiency: (1) *Antibiotics* eliminate intestinal bacteria and reduce the synthesis of menaquinone. (2) *Intestinal malabsorption* of vitamin K can occur in obstructive jaundice. Due to absence of bile, absorption of fats and fat-soluble vitamins are impaired. (3) *Hepatocellular diseases* cause vitamin K deficiency by impairing its conversion to the active form.

Vitamin K deficiency results in low plasma levels of both procoagulants as well as some anticoagulants. These proteins are called *vitamin K-dependent proteins* (VKDP). The procoagulants are factors II, VII, IX, and X. The anticoagulants are protein C and protein S. The plasma level of factor VII is the first to decrease.

The vitamin K dependent proteins undergo some post-translational processing before they are secreted. This involves carboxylation of about ten of their terminal glutamic acid residues. Vitamin KH_2 serves as a cofactor in the carboxylation reaction. In the process, vitamin KH_2 gets oxidized to vitamin K oxide. Vitamin KH_2 is subsequently regenerated from vitamin K oxide by the enzyme epoxide reductase. This is called the *vitamin K cycle* (Fig. 28.**5**). The commonly used anticoagulants warfarin and dicoumarol are inhibitors of the enzyme epoxide reductase and prevent the regeneration of the physiologically useful vitamin KH_2.

Thrombotic disorders

Formation of clots inside blood vessels is called *thrombosis* to distinguish it from the normal extravascular

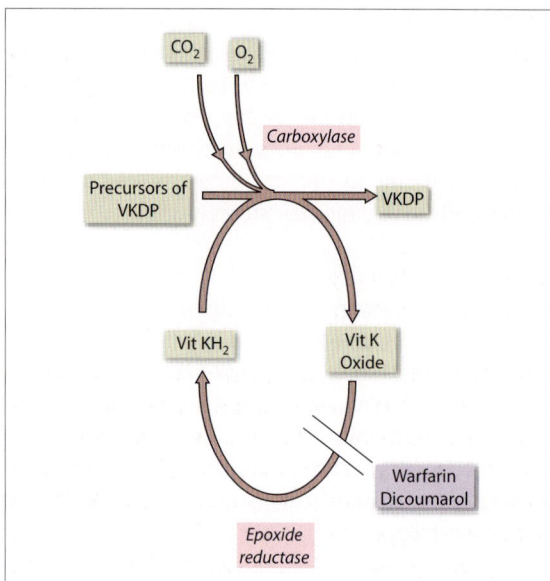

Fig. 28.**5**   Vitamin K cycle.

clotting of blood. Thrombi can occlude the arterial supply to the organs in which they form. Bits of thrombus (emboli) sometimes break off and travel in the bloodstream to distant sites, blocking the blood supply to other organs.

Three factors predispose to thrombosis: (1) *Hemodynamic factors*: Thrombosis is prone to occur where blood flow is sluggish, e.g., in the veins of the legs after operations and delivery, because the slow flow permits activated clotting factors to accumulate instead of being washed away. (2) *Vascular factors*: Thrombosis tends to occur where the arterial intima is damaged by atherosclerotic plaques. (3) *Blood factors*: A congenital absence of protein C leads to massive intravascular coagulation and usually death in infancy. A genetic abnormality in factor V, making it resistant to inactivation by

protein C is a commoner cause of thrombosis. Hypercoagulability of blood can occur due to rise in plasma levels of vWF. It is an acute phase protein and its plasma level rises in inflammatory states. Plasma vWF level also rises during the third trimester of pregnancy.

■ Disseminated intravascular coagulation

Disseminated intravascular coagulation (DIC) has two phases, the thrombotic phase, and the fibrinolytic phase.

In the **thrombotic phase**, there is formation of numerous small thrombi and emboli throughout the microvasculature, causing blockage of circulation and ischemic organ damage. The clots also use up most of the coagulation factors and platelets, resulting in bleeding tendencies. Hence, the condition is also called *defibrination syndrome* or *consumption coagulopathy*.

In the **fibrinolytic phase**, there is fibrinolysis of the clots. The fibrin degradation products that are formed have antihemostatic effect and aggravate bleeding.

DIC usually occurs in *metastatic malignancies* (malignancies that spread to various parts of the body) and massive trauma because malignant or necrotic tissues release tissue factor into the circulation, triggering coagulation.

DIC also occurs in gram-negative septicemia because the endotoxin from *gram-negative bacteria* activates factor XII and stimulates the secretion of tissue factor. DIC is commoner during *pregnancy* because pregnancy is associated with a rise in plasma fibrinogen level and vWF. DIC also occurs following *snake bite*: The venom of Malaysian pit viper has a direct effect on fibrinogen, converting it to fibrin.

29 Granulocytes

Leukocytes are classified into granulocytes (neutrophils, eosinophils, and basophils) and agranulocytes (monocytes and lymphocytes). The cytoplasm of granulocytes contains secretory granules that readily take up Romanowsky-type stains.[1] Monocytes and lymphocytes do not have granules in their cytoplasm and are therefore called agranulocytes. The granules that distinguish the different types of granulocytes are called *specific* or *secondary granules*. They make their appearance only in the myelocyte stage (p 204). Another set of granules called the *primary* or *azurophilic granules* are common to all granulocytes. They appear in the promyelocyte stage but get somewhat obscured by the appearance of the specific granules in the myelocyte stage. All cytoplasmic granules, primary or secondary, contain biologically active substances that are involved in inflammatory and allergic reactions.

Neutrophils approach, ingest, and kill bacteria. Eosinophils attack parasites that are too large to be engulfed by phagocytosis. They enter tissues and are especially abundant in the mucosa of the respiratory, lower urinary, and gastrointestinal tracts. They increase in number in allergic diseases. Basophils, which resemble but are not identical to mast cells, release histamine and other inflammatory mediators. They are important in immediate-type hypersensitivity reactions. The relative and absolute counts of leukocytes are given in Table 22.**3**.

Neutrophils

The neutrophil (Fig. 29.**1**) is 10–14 μm in diameter. It is circular in shape but the shape keeps changing due to ameboid movements. Its nucleus is multilobed with up to six lobes and hence the neutrophil is also called a *polymorphonuclear leukocyte*. The number of lobes is related to the age of the neutrophils, with the younger ones having a single-lobed horseshoe shaped nucleus and the older ones having a multilobed nucleus. The frequency distribution of neutrophils with different numbers of lobes is called the *Arneth count* and is done to assess the average age of neutrophils in circulation. Nearly 45% of the neutrophils are trilobed and in the right stage of maturity for optimum functioning, 20% are overaged with 4 or 5 lobes, and 35% are underaged with fewer than 3 lobes. A preponderance of immature cells is called a 'shift to the left', while a preponderance of the older cells is called a 'shift to the right'.

Of the total number of neutrophils present in blood, only 50% are circulating in the blood at any instant, while another 50% remain marginated (sidelined) on vessel walls or sequestered (isolated) in closed capillaries. Margination of neutrophils is the first step in their migration to the tissues, which is their ultimate destination. After migration into tissues, they never return to the bloodstream and survive in tissues only for a few days. The average half-life of neutrophils in the circulation is only 6 hours. Up to 100 ml of packed neutrophils are eliminated daily, mainly in the intestine and also in respiratory secretions. Neutrophils that die in tissues are taken up by macrophages.

For every neutrophil in circulation, there are about 100 mature neutrophils held in the bone marrow as a reserve. These stored neutrophils enter the bloodstream when they are stimulated by cortisol or by a granulocyte-inducing factor derived from dead leukocytes. Thus, an increase in the blood neutrophilic count can occur in three ways: (1) Mobilization of marginated or sequestered neutrophils from blood vessels. Adrenaline, exercise, and corticosteroids produce transient neutrophilia in this way. (2) Release of stored neutrophils from bone marrow. This too produces a transient neutrophilia. (3) Stimulation of neutrophil production in the bone marrow.

[1] The most commonly used stains in the hematology laboratory belong to the Romanowsky group of stains. A Romanowsky stain is defined as any stain that contains methylene blue or its oxidation product, and a halogenated fluorescein dye, usually eosin. Romanowsky and Malachowski first described the stain that used oxidized methylene blue solution with eosin. It was refined by Leishmann who used methyl alcohol in the stain. Other alcohol-based modifications of the Romanowsky stain include the Wright and Giemsa stains. Romanowsky stains are also called polychrome stains because they can impart typical colors to different cell components. Methylene blue is a basic stain that stains the nucleus and some cytoplasmic structures a blue or purple color. These stained structures are thus basophilic (e.g., DNA or RNA). Eosin is an acidic stain that stains some cytoplasmic structures an orange-red color. The orange-red staining structures are acidophilic (e.g., proteins with amino groups). When both the basic and the acidic components of the mixture stain a cytoplasmic structure, a pink or lilac color develops. This is referred to as a neutrophilic reaction.

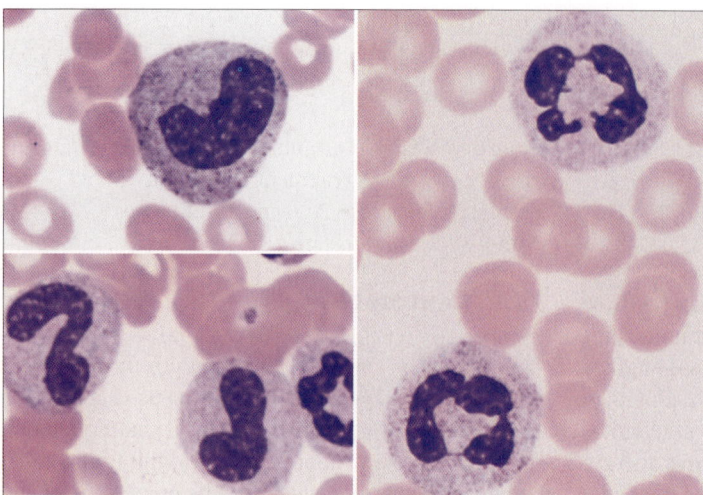

Fig. 29.**1** Neutrophils. (*Above left*) Transitional form between a metamyelocyte and a band cell. (*Below left*) Two band cells. (*Right*) Segmented neutrophilic granulocytes.

Neutrophilic granules Like all granulocytes, neutrophils have two types of granules: the azurophilic granules that are also present in other granulocytes, and the specific granules that are peculiar to neutrophils alone and give them their name.

Azurophilic granules are released into phagocytic vacuoles containing ingested microbes. Well known among the azurophilic granule contents are myeloperoxidase and defensins. *Myeloperoxidase* is a protein that catalyzes the production of hypochlorite (OCl⁻) from chloride and hydrogen peroxide produced by the oxygen burst. It imparts a greenish color to the pus. *Defensins* are cationic proteins that kill a variety of bacteria, fungi, and viruses.

Although the contents of the azurophilic granules are mostly secreted into phagocytic vacuoles, some enzymes may escape into surrounding tissues. This happens when the ingested organism is large and the phagocytic vacuole cannot be completely sealed off. The enzymes, particularly elastase, cause considerable damage to the surrounding tissues. Such damage underlies several disorders including rheumatoid arthritis and pulmonary emphysema.

Specific granules are largely released into the extracellular space. They are released early, when the neutrophil moves toward the site of inflammation. The major constituent of specific granules is an iron-binding protein called *apolactoferrin*. By binding iron, lactoferrin exerts a bactericidal effect by depriving bacteria of their necessary iron. It also contains *collagenase* which facilitates neutrophilic mobility by hydrolyzing the extracellular matrix. The membrane of specific granules bears integrin molecules. When the granules fuse with the cell membrane, these integrin molecules appear on the cell membrane where they perform an important role in chemotaxis.

Neutrophil count An increase is called *neutrophilia* and a decrease is called *neutropenia*. Since neutrophils comprise the bulk of leukocytes, neutrophilia is associated with leukocytosis (an increase in the total leukocytic count) and neutropenia is associated with leukopenia (a decrease in the total leukocytic count). Conversely, most cases of leukocytosis are due to neutrophilia and most cases of leukopenia are due to neutropenia.

At birth, there is leukocytosis (25,000 mm⁻³) that is mainly neutrophilic. It returns to normal after 1 week. In adults, the neutrophilic count can show a transient increase due to mobilization of neutrophils from the marginal pool. The phenomenon is called the *shift-leukocytosis*, and is mostly due to the stimulation of glucocorticoid secretion. Marked shift-leukocytosis with neutrophilia occurs after strenuous exercise. During pregnancy, the leukocytosis increases till term and peaks at parturition, with the count sometimes doubling. Leukocytosis also occurs in anxiety or stress of any kind. Finally, there is a diurnal factor, with the leukocytic count increasing slightly in the afternoon (the afternoon-tide).

Neutrophilia occurs in: (1) Infections and septicemia especially with pyogenic cocci (like *Staphylococci*) and bacilli (like *Escherichia coli*) and also with non-pyogenic organisms (like diphtheria and cholera bacilli); (2) Hemorrhage and trauma, as in surgery, fractures, crush injuries, and burns; (3) Malignancies like myeloid leukemia; (4) Cardiac disorders like myocardial infarction; (5) Metabolic disturbances like renal failure and diabetic coma; and (6) Medications like adrenaline.

Neutropenia commonly occurs in: (1) Viral infections like influenza, measles, and infective hepatitis; (2) Bacterial infections like typhoid fever; (3) Aplastic anemia; and (4) Hypersplenism.

Neutrophilic functions

The major role of neutrophils is to protect the host against infectious agents. To fulfill this role, neutrophils are endowed with the capability to sense infection, migrate to the site of the infecting organism, and then destroy the infectious agent.

Chemotaxis Neutrophils are attracted toward the site of infection by chemoattractant molecules, which are detected by specific receptors on the neutrophilic membrane. These chemoattractants are either molecules released by the degradation of bacteria and damaged tissue cells or molecules formed through the interaction of bacteria with the host defense system. Notable among the latter are complement C5 (see Table 32.**1**), leukotriene B4 (which is the most potent chemoattractant molecule known yet), secretions of platelet α-granules (platelet factor 4, platelet-derived growth factor), and secretions of mast cell granules (see below).

Margination When a neutrophil flowing inside a capillary approaches an area of inflammation, it becomes marginated, i.e., it is attracted to the capillary endothelium and starts rolling along its surface (Fig. 29.**2**). Margination usually occurs in a postcapillary venule. The margination is caused by the binding of selectins (cell adhesion molecules) present on the endothelial cells with the carbohydrate molecules present on the surface of neutrophils. Endothelial selectins are more numerous in areas where there is inflammation. However, they are present normally too, which is the reason why nearly half the circulating neutrophils normally remain marginated.

Activation While rolling along the endothelial surface, if the neutrophil comes in contact with chemoattractant molecules, it becomes activated. The activation is mediated by G-protein. Activation of neutrophils is associated with degranulation of the specific granules to the neutrophil cell surface and a marked increase in neutrophilic adhesion to endothelial cells, resulting in the cessation of rolling. This occurs due to the appearance of integrins (cell adhesion molecules) on the surface of the neutrophil which bind with its receptors on endothelial cells.

Diapedesis Neutrophils next insinuate themselves through the walls of the capillaries between endothelial cells by a process called diapedesis. The neutrophils then migrate to the tissue along the chemoattractant gradient. During diapedesis, the neutrophil elongates and develops a broad 'head' called a lamellipodium

which contains the bulk of the cytoplasm, and a thin forked 'tail' called the pseudoflagellum. There is degranulation of specific granules from the lamellipodium.

Phagocytosis Neutrophils have on their surface receptors for the Fc fragment of immunoglobulin molecules. When these Fc receptors bind to the IgG and complement proteins present on bacterial surface (see opsonization, p 198), it triggers increased neutrophilic motility, exocytosis, and respiratory burst, thereby facilitating phagocytosis. These responses are mediated by G-protein.

During phagocytosis, the granulocyte membrane extends pseudopodia around the particle. The pseudopodia fuse to form a phagocytic vacuole and enclose the particle in it. The primary (lysosomal) and secondary (specific) granules fuse with the phagocytic vacuole and discharge into it enzymes like lysozyme, myeloperoxidase, cathepsin G, elastase, lactoferrin, and cationic proteins called defensins. Following the respiratory burst, superoxide anion and its metabolites H_2O_2 and hypohalites diffuse into the phagocytic vacuole and initiate oxygen-dependent microbial killing.

Respiratory burst Within seconds of stimulation, neutrophils sharply increase their oxygen uptake, a phenomenon known as the respiratory burst. The increased oxygen uptake is utilized for oxidation of glucose via the hexose monophosphate shunt, resulting in the production of NADPH and reduced glutathione (GSH). NADPH reduces molecular oxygen to superoxide ion (O_2^-) and then to H_2O_2 while GSH helps in detoxifying the extra H_2O_2 produced.

The reduction of O_2^- to H_2O_2 inside the neutrophil is catalyzed by the enzyme superoxide dismutase (SOD).

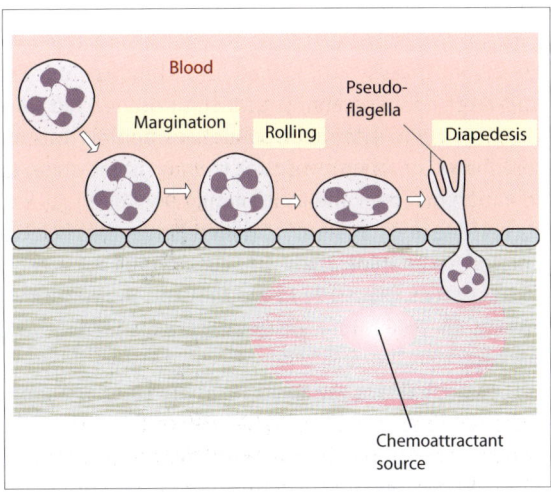

Fig. 29.**2** Neutrophilic margination, rolling, and diapedesis.

The O_2^- that escapes from neutrophils into extracellular space is reduced to H_2O_2 by cerruloplasmin, an acute phase protein. Neither O_2^- nor H_2O_2 have significant bactericidal activity though the H_2O_2 formed in the tissues damages cells and the connective tissue matrix. However, in the presence of the azurophilic granule enzyme myeloperoxidase, H_2O_2 oxidizes halide ions (Cl^-, Br^-, I^-) to form hypohalites (HOCl, HOBr, HOI) which are powerful microbicidal agents.

$$2O_2 + NADPH = 2O_2^- + NADP^+ + H^+$$
(weak bactericidal)

$$2O_2^- + 2H^+ \xrightarrow[\text{Cerruloplasmin (extracellular)}]{\text{Superoxide dismutase (intracellular)}} H_2O_2 + O_2$$

$$H_2O_2 + 2Cl^- \xrightarrow{\text{Myeloperoxidase}} 2HClO$$
(strong bactericidal)

◾ Disorders of neutrophilic functions

Abnormal polymerization of actin filaments in neutrophils or a congenital deficiency of leukocytic integrins results in *neutrophil hypomotility*. In a group of rare hereditary disorders known collectively as *chronic granulomatous disease*, the inability of neutrophils and monocytes to generate O_2^- leads to recurrent bacterial infections. Severe *congenital glucose 6-phosphate dehydrogenase (G6PD) deficiency* causes susceptibility to multiple infections due to failure to generate the NADPH necessary for O_2^- production. In *congenital myeloperoxidase deficiency*, microbial killing power is reduced because hypohalite ions are not formed.

◼ Eosinophils

Eosinophils (Fig. 29.**3**) contain a bilobed nucleus and large coarse granules that stain orange-red with Wright's stain. Like neutrophils, eosinophils too are distributed in an intravascular compartment made up of a marginal pool and a circulating pool of equal size. Its half-life in circulation is about 8 hours. For each circulating eosinophil, about 100 eosinophils are present in the tissues, primarily in the skin and in the submucosa of the respiratory, gastrointestinal, and genitourinary tracts.

Eosinophils are attracted to tissue sites of allergic reactions by chemotactic factors secreted by mast cells. At the site of allergic reaction, eosinophils degrade mast cell products and thereby decrease the clinical manifestations of allergic responses.

Eosinophilic granules   The predominant constituent of eosinophilic granules is a material called *major basic protein* (MBP) which plays a key role in the eosinophil's ability to damage the helminthic larvae. Also present in the large granules is a potent bactericidal called *eosinophilic cationic protein*, a neurotoxic protein, and an *eosinophil peroxidase* with properties different from the myeloperoxidase of neutrophils. The smaller eosinophilic granules contain *aryl sulfatase B*, an enzyme that can inactivate the sulfur-containing leukotrienes liberated by mast cells in immediate hypersensitivity reactions. Eosinophils secrete *lysophospholipase* which forms crystals called Charcot-Leyden crystals in the pulmonary secretions of patients with asthma.

Eosinophilic count   Glucocorticoids cause margination and sequestration of circulating eosinophils, lowering the eosinophilic count. The eosinophilic count shows a diurnal variation that varies inversely with the diurnal fluctuations in glucocorticoid levels (see Fig. 85.**6**). It is lowest at 8 am and highest at midnight. Emotional stress decreases it. The eosinophil count decreases progressively during the intermenstrual period.

Eosinophilia occurs in: (1) Allergic disorders like asthma, drug allergy, and food sensitivity; (2) Parasitic infestations like hookworm, tapeworm, hydatid, and filaria. In general, eosinophilia is more pronounced with parasites causing tissue infection than with those causing intestinal infestations; (3) Skin diseases like eczema, dermatitis, and scabies; and (4) Pulmonary eosinophilia, a group of diseases associated with pulmonary infiltration. Their cause is unknown but could be related to filarial infestation.

Eosinopenia occurs in: (1) Endocrine disorders like Cushing's disease, and (2) Stress, as in acute infection, traumatic shock, surgical operation, severe exercise, burns, acute emotional stress, or exposure to cold.

▰ Basophils and Mast Cells

◾ Basophils

The mature basophil (Fig. 29.**3**) has a lighter staining nucleus than the neutrophil. It seldom contains more than two lobes. The cytoplasm is pink and contains a varying number of large, deeply-staining basophilic granules. The granules are usually not very numerous, and although they do not pack the cytoplasm as do eosinophil granules, they overlie the nucleus and tend to obscure its detail.

Basophils remain in the blood only for a few hours after which they move into tissues. Their fate in the

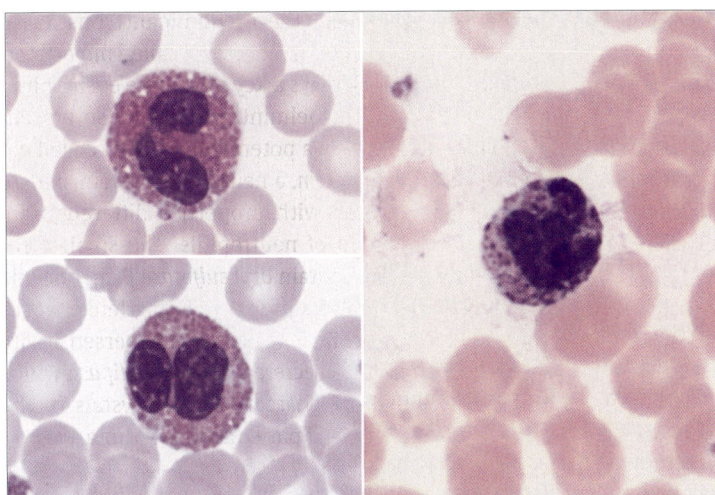

Fig. 29.**3** Eosinophils and basophils. (*Left, above and below*) Eosinophils with corpuscular, orange-stained granules; (*Right*) Basophils with corpuscular granules that stain deep blue to violet.

Table 29.**1** Comparison of basophil and mast cell.

Both basophil and mast cell	
• are derived from marrow stem cells	
• are stimulated to grow by the same interleukins	
• have surface receptors for IgE and degranulate when an antigen interlinks the IgE on their surfaces.	

Basophil	Mast cell
• is present in the blood stream	• is present in tissues
• has a multilobed nucleus	• has a round nucleus
• granules contain chondroitin sulfate	• granules contain heparin

tissues is uncertain. Contrary to earlier belief, basophils do not transform into mast cells in tissues. The basophil surface membrane contains receptors for the Fc fragment of IgE molecules. Basophil degranulates when an antigen reacts with IgE bound to basophils and thereby interconnects the IgE molecules on the basophil surface (see Fig. 31.**2**, Immediate hypersensitivity). Basophilic granules contain histamine which mediates allergic reactions. Basophils also contain the proteoglycan chondroitin sulfate whose function in the basophil is not known.

Basophilia occurs in chronic myeloid leukemia and hypersensitivity states. *Basopenia* occurs in Cushing's disease and prolonged corticosteroid therapy. It also occurs when there is neutrophilic leukocytosis, as in infections.

■ Mast cells

Mast cells are cells similar to basophils. They are found wandering in connective tissues, especially, beneath epithelial surfaces. Mast cells are most abundant at those places where the body comes in direct contact with the environment, e.g., in the lung, skin, lymphoid tissues, and the submucosal layers of the digestive tract. The similarities and differences between basophils and mast cells are summarized in Table 29.**1**.

Mast cells contain inflammatory mediators like histamine, leukotrienes, prostaglandins, and chemotactic factors. Excessive mast cell degranulation produces clinical manifestations of allergy and even anaphylaxis. Leukotrienes released by the mast cell trigger the bronchospasm and the mucosal edema of bronchial asthma. The slow reacting substance of anaphylaxis or SRS-A is a mixture of leukotrienes C4, D4, and E4 (see prostaglandins, p 264). Mast cells also participate in immune response and tissue repair. They release tumor necrosis factor (TNF) in response to bacterial products by an antibody-independent mechanism, thus contributing to the nonspecific natural immunity. Mast cells also contain the Charcot–Leyden crystal protein and small amounts of major basic protein.

Monocytes

Monocytes (Fig. 30.1) are spherical cells. When suspended in isotonic fluid, their diameter is 9–12 μm but when flattened in dried blood smears, they measure up to 14–18 μm in diameter. They have a relatively abundant dull blue cytoplasm with a few azurophilic granules. The nucleus is round or kidney-shaped and is eccentric in position. Its chromatin stains less intensely than that of lymphocytes and there are one or two nucleoli.

Monocytes originate in the bone marrow and circulate in blood for three days before migrating through the walls of postcapillary venules into the connective tissue of various organs where they differentiate into tissue macrophages. Tissue macrophages do not re-enter circulation but persist in the tissues for another three months. Circulating monocytes perform no important function and are only a mobile reserve of cells capable of migrating into tissues and developing into macrophages.

Monocytosis (an increase in monocyte count, see Table 22.3) occurs in: (1) bacterial infections like tuberculosis and typhoid; (2) protozoal infections like malaria and kala-azar; (3) rickettsial infections like the oriental sore; (4) neoplastic diseases like infectious mononucleosis, Hodgkin's disease, and monocytic leukemia; and (5) chronic inflammatory diseases like rheumatoid arthritis, collagen diseases, ulcerative colitis, and regional enteritis.

Mononuclear phagocytic system

Tissue macrophages settle at several strategic locations in the body and constitute the tissue-macrophage system. Monocytes and macrophages together constitute the mononuclear phagocytic system. The mononuclear macrophage system was earlier called the reticuloendothelial system (RES), a term that encompassed a large collection of phagocytic cells. Any cell capable of taking up a vital dye like trypan blue and particulate materials like colloidal carbon was called an RES cell. The RES therefore included: (1) reticulum cells of the spleen and the lymph nodes; (2) cells lining the lymphatic and blood sinuses of lymph nodes, spleen, and liver (Kupffer cells); (3) monocytes of blood; and (4) tissue macrophages. The reticuloendothelial system was so named because it included the reticulum cells of spleen and lymph nodes and also the cells lining the splenic and hepatic sinusoids, which were thought to be endothelial cells. It was later discovered that the true endothelial cells and fibroblasts were also capable of taking up the dye and although they were weakly phagocytic, they were for some time included under the RES.

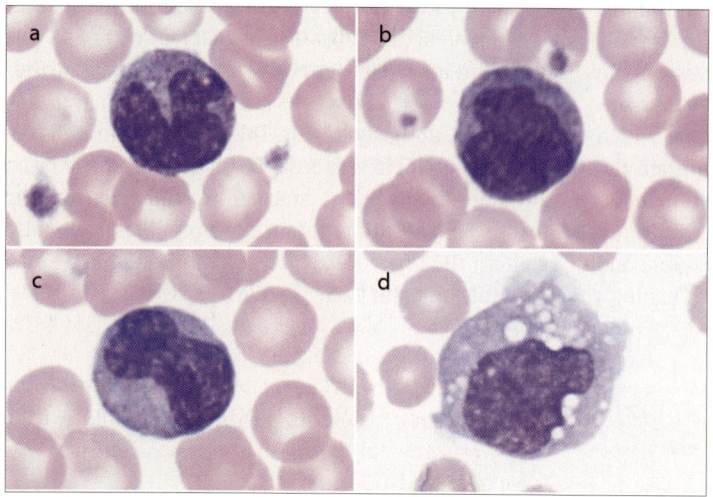

Fig. 30.1 Monocytes. (a-c) Range of appearances of typical monocytes with lobed nucleus, gray-blue stained cytoplasm and fine granulation; (d) Phagocytic monocyte with vacuoles.

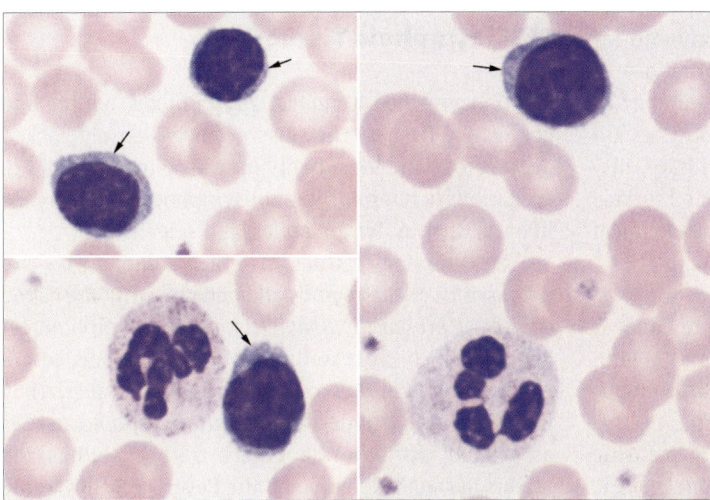

Fig. 30.**2** Lymphocytes (shown with arrows).

The RES was later renamed as the mononuclear phagocytic system (MPS). It includes: (1) precursor cells from the bone marrow; (2) promonocytes from the bone marrow; (3) monocytes from the bone marrow and blood; and (4) macrophages in liver (Kupffer cells), spleen, lymph node, bone marrow, lung (pulmonary alveolar macrophages, also called dust cells), connective tissue (histiocytes), pleura and peritoneum, bones (osteoclasts), and central nervous system (microglial cells). The criteria for identifying a cell of the MPS are vigorous phagocytosis and pinocytosis, and its ability to attach firmly to a glass surface. Reticulum cells of the spleen and lymph nodes, endothelial cells, and fibroblasts are not included in the MPS.

The functions of the MPS are as follows. (1) Macrophages ingest cell debris, broken erythrocytes, fibrin, and bacteria and therefore, are important in inflammation and healing. (2) Bacteria entering through blood or lymph are sequestered and destroyed by the macrophages. (3) Macrophages process antigens and thereby play an important part in the immune response.

Macrophages are also very efficient in phagocytosing antigen-antibody- complement complexes because they have receptors for immunoglobulins and complement. (4) Macrophages, especially those in the spleen, remove aged and effete erythrocytes. The heme moiety is divested of its iron, which is stored in the macrophages. (5) Macrophages store excess lipids and mucoprotein and become swollen to form 'foam cells'. (6) Twenty or more macrophages can fuse to form a multinucleate 'giant cell', up to 50 µm in diameter, that engulfs a bacillus, especially, the tubercle bacillus.

Lymphocyte

Lymphocytes (7–11 µm) have a deeply staining, slightly indented nucleus with a thin rim of clear blue cytoplasm. They contain no specific granules but may have a few small azurophilic granules (Fig. 30.**2**).

Large and small lymphocytes About 10–20% of the lymphocytes in peripheral blood are *large lymphocytes* (11–15 µm). These were earlier believed to be in an early stage of development from the lymphoblast, which is larger than the lymphocyte. However, it is now known that these large lymphocytes are actually the natural killer (NK) cells. They nonspecifically kill any cell that is coated with immunoglobulin IgG. This phenomenon is called antigen-dependent cell-mediated cytotoxicity (ADCC). They neither require any prior sensitization, nor require MHC molecules (p 195) for antigen recognition. They are therefore responsible for innate immunity rather than acquired immunity. They are particularly effective against tumor cells and virus-infected cells.

Table 30.**1** Distribution of B and T lymphocytes in peripheral lymphoid tissues.

Peripheral lymphoid tissue	B–Lymphocyte	T–Lymphocyte
Lymph node	Subcapsular cortex, germinal centers, and medullary cord	Diffuse cortex.
Spleen	Germinal centers and red pulp	Periarteriolar sheath
Alimentary and respiratory tracts	Submucous follicles	Interfollicular areas in submucosa

Small lymphocytes produce acquired immune response. They are classified into the B lymphocytes which mediate humoral immunity, and T lymphocytes which mediate cellular immunity. These subtypes are not morphologically distinguishable, but have distinctive molecules on their cell membranes (surface markers) that are identifiable by immunocytochemical methods.

Lymphocytic count   The normal lymphocyte count is given in Table 22.3. *Lymphocytosis* occurs in: (1) viral infections, e.g., mumps, measles, chicken pox, influenza, and viral hepatitis; (2) bacterial infections, e.g., typhoid, tuberculosis, and whooping cough; (3) parasitic infections, e.g., toxoplasmosis; (4) malignancies, e.g., lymphocytic leukemia and lymphocytic lymphoma; (5) autoimmune diseases, e.g., myasthenia gravis and thyrotoxicosis.

Lymphopenia occurs in (1) acquired immunodeficiency syndrome (AIDS), (2) pancytopenia from any cause (e.g., aplastic anemia, bone marrow infiltration, and hypersplenism), and (3) corticosteroid administration.

Lymphocytic Processing

The committed lymphocytic stem cells soon differentiate into the committed B and T cells. The committed B lymphocytic stem cells are processed (rendered mature and immunocompetent) in the bone marrow (B for 'bursa of Fabricius', the site of B cell processing in birds). The committed T lymphocytic stem cells are processed in the thymus. The processing involves brisk proliferation with frequent mutations during cell divisions. These mutations may have a role in the development of antibody specificity. In the thymus, a factor called thymosin plays an important role in the processing of T lymphocytes.

Following processing, the mature T and B cells enter the circulation where they are present in a 70:30 ratio. Many of them leak out through the postcapillary venules to settle in the peripheral lymphoid tissues (see below). Some of the lymphocytes return again to circulation through the lymphatics draining these lymphoid tissues. At any given time, only about 2% of the lymphocytes are in the peripheral blood while the rest remain in the lymphoid organs. Within the peripheral lymphoid tissues, the T and B lymphocytes are found in distinct zones as shown in Table 30.**1**. After birth, some lymphocytes are still formed in the bone marrow but most are formed in the lymph nodes, thymus, and spleen from precursor cells that originally came from the bone marrow.

Lymphoid Tissues

Lymphocytes occur individually in blood, lymph, and connective tissues. They also occur in densely packed aggregates. The term lymphoid tissue includes all aggregations of lymphocytes and lymphocyte-rich organs like the thymus, lymph nodes, and spleen.

The lymphoid tissues in which lymphocytes are processed, i.e. the thymus and bone marrow, are called *central or primary lymphoid tissues*. The lymphoid organs in which processed lymphocytes are seeded are called *peripheral or secondary lymphoid tissues*. They include lymph nodes, spleen, the lymphoid tissues associated with the alimentary tract (including the tonsils in the oral cavity and the Peyer's patches in the terminal ileum), respiratory, and urinary tracts, and a portion of lymphocytes in the bone marrow.

Thymus

The thymus is located anterior to the heart. In a small child, it may extend up to the neck, almost to the level of the thyroid gland. The thymus reaches its greatest size during puberty. After puberty, the thymus regresses (thymic involution) during which the thymic lymphocytes and epithelial cells are increasingly replaced by adipose tissue though some thymic tissue persists throughout life.

The thymus consists of a cortex and a medulla. The cortex contains lymphocytes and thymocytes (precursors to the lymphocytes). The medulla contains mostly epithelial cells. The thymus is the first organ to become lymphoid in the fetus as it gets seeded by blood-borne lymphocytes from the yolk sac and the fetal liver. The seeding is aided by thymotaxin, a chemotactic factor secreted by the thymus. Lymphocytes processed in the thymus are called T lymphocytes. During their processing in the thymus, the T lymphocytes develop surface molecules (also called surface markers) called CD-4 and CD-8, which form the basis for categorizing T cells into the T4 and T8 cells. The lymphocytes also develop MHC molecules and antigenic receptors. Lymphocytes that develop receptors against self antigens undergo apoptosis which explains why more than 90% of the lymphocytes die in the cortex of the thymus.

Lymph node

The lymph node (Fig. 30.**3**A) has an outer cortex where the lymphocytes are tightly packed, an inner cortex where the lymphocytes are diffusely distributed, and the medullary cords which are dense aggregations of lymphocytes around small blood vessels. Present inside the outer cortex are the germinal centers, which are circular zones with sparsely distributed lymphocytes at their center.

The lymph node is an effective filter located in the path of lymphatics that removes most of the microbes, malignant cells, and macromolecules from the lymph. After entering the lymph node, the lymph vessels open into the larger lymph sinuses. The lumens of these lymph sinuses are criss-crossed by a mesh formed by the modified epithelial cells called reticular cells. Macrophages are attached to this reticular mesh and also line the walls of the sinuses.

As the lymph flows through the lymph sinus, the bacteria are phagocytosed by the macrophages in the sinuses. Most of the bacteria however are filtered out of the lymph vessels and are deposited in the node where they are phagocytosed by the macrophages and initiate an immune response. Viruses enter the lymphocytes but are not destroyed; rather, they persist in the lymphocytes and are distributed throughout the body. The lymph node is not an efficient barrier to malignant cells which tend to accumulate in the lymph node and overflow into other nodes. Hence when removing a malignant tumor, the surgeon removes the draining lymph nodes too for arresting the further metastasis of the malignant cells.

The lymphocytes in the germinal centers of the lymph node proliferate to form more lymphocytes. However most of the lymphocytes present in the lymph node are seeded from blood flowing through the node. The macrophages in the lymph node are derived from monocytes in blood. Granulocytes do not enter the lymph node in significant numbers.

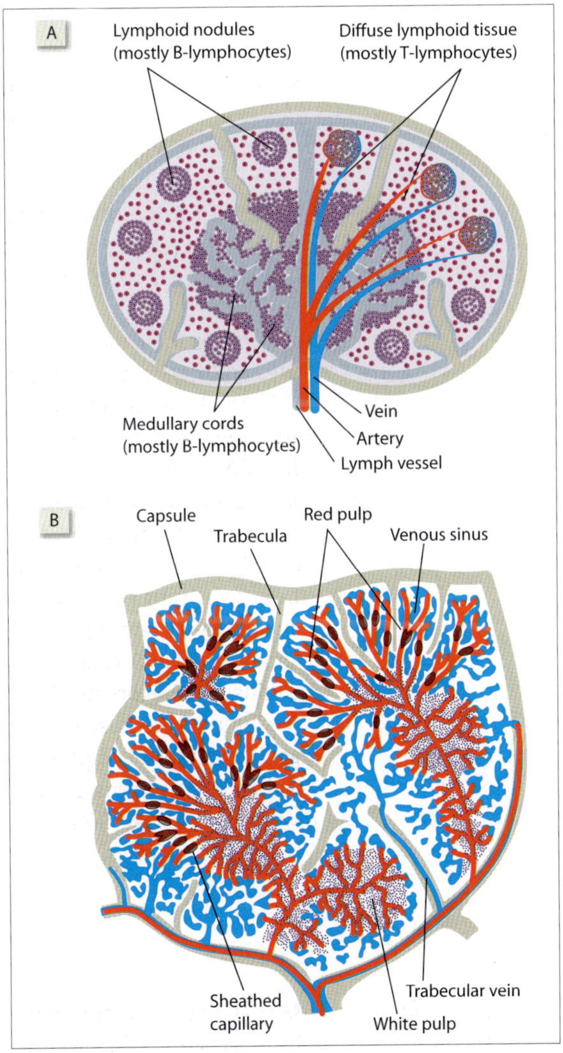

Fig. 30.**3** [A] Structure of a lymph node. [B] Structure of spleen.

Spleen

The spleen (Fig. 30.**3**B) is divided into small compartments by connective tissue trabeculae extending inside from its fibrous capsule. Between the trabeculae is a meshwork of fine reticular connective tissue called the splenic pulp. Splenic arteries and veins travel in the trabeculae before entering the pulp. As they traverse the pulp, the arteries are surrounded by aggregations of immunocompetent lymphocytes, forming a periarterial sheath. At places, the periarterial sheath contains germinal centers, which are circular zones with sparsely distributed lymphocytes at their center. The periarterial sheath and the germinal centers constitute the white pulp. The red pulp consists of splenic sinuses which are engorged with blood and therefore, red in color.

The circulation of the spleen has fast and slow components. The fast component is represented by the nutritive blood supply in which blood stays within blood vessels. The slow component represents the nonnutritive blood supply that leaves arterioles and percolates through large numbers of phagocytes and lymphocytes before entering the splenic sinuses and passing back into the general circulation.

The functions of the spleen are as follows. (1) The spleen produces erythrocytes in the second half of fetal life. It can resume erythrocyte production in adult life if the bone marrow gets destroyed. (2) Spleen is the most effective lymphoid organ for the removal of aged erythrocytes from the circulation. The macrophages lining splenic sinusoids remove mildly-damaged red cells that have reacted with antibodies or drugs. It also removes abnormal red cells like spherocytes, sickle cells and cells containing malarial parasite that are not as flexible as normal red cells and consequently are unable to squeeze through the intercellular gaps between the endothelial cells that line the splenic sinuses. The process is called culling. Hence splenectomy is helpful in elevating the red cell count in conditions like spherocytosis and autoimmune hemolytic anemias that are associated with excessive removal of red cells. (3) The spleen acts as a reservoir for lymphocytes, plasma cells, monocytes, and platelets. It does so by preferentially sequestering newly released cells from the bone marrow, forming a reservoir inside it. In lower animals, it acts as a reservoir for whole blood too. (4) The spleen is important for defense against blood borne infections. Splenectomy renders the body vulnerable to fatal septicemia. (5) The macrophages in spleen perform their usual functions of storage and immunity.

Certain patients of chronic splenomegaly develop neutropenia, anemia, and thrombocytopenia. The blood picture becomes almost normal after splenectomy. The condition is called hypersplenism and occurs probably due to increased sequestration and destruction of blood cells in the splenic pulp.

Immunity

Immunity refers to the body's ability to resist foreign substances (recognized as non-self) like microbes or toxins that are potentially damaging to the body. Immunity may be natural or acquired. Acquired is further classified as actively acquired (active immunity) or passively acquired (passive immunity).

Natural (innate) immunity

Natural (innate) immunity is the nonspecific defense mechanisms of the body which includes the following: (1) Phagocytosis by leukocytes and the monocyte-macrophage system cells; (2) Destruction of ingested microbes by gastric acid; (3) Resistance of the skin to microbial infection; (4) Plasma enzymes like lysozymes that lyse bacteria, basic polypeptides that inactivate certain gram-positive bacteria and complement which destroys foreign substances; and (5) Natural killer (NK) cells which nonspecifically destroy foreign cells, tumor cells, and infected cells.

Inflammation is a form of natural immunity. It is associated with warmth (calor), redness (rubor), pain (dolor), and swelling (tumor). Thus if any external or internal part of the body becomes swollen, red, painful, and warm, it is called an inflammation. The suffix 'itis' is added to the name of the part to signify its inflammation, e.g., gastritis, enteritis, hepatitis, myocarditis, meningitis, encephalitis, etc. The warmth, redness, and swelling of inflammation are due to the vasodilatation associated with inflammation. The vasodilatation, along with the increased capillary permeability associated with inflammation, results in exudation of plasma into interstitial spaces. The exudate contains large numbers of inflammatory cells, mainly neutrophils.

Immune surveillance is another form of natural immunity. It is the continuous recognition and destruction of tumor cells that keep appearing in the body throughout life. Immune cells identify tumor cells by the tumor-specific antigens. Although all types of immune cells are involved in the surveillance, the natural killer (NK) cells are especially dedicated to the surveillance. NK cells do not need to be stimulated by the tumor antigens. The mere contact with NK cells kills the tumor cells.

Tumor cells have several mechanisms of evading the immunological surveillance, some of which are given below. (1) A few isolated tumor cells express too little antigen to stimulate the immune system and by the time the host's tumor immunity is adequate, the tumor burden is overwhelmingly large. The mechanism is called 'sneaking through'. (2) Tumor cells can shed or internalize their antigens and thereby evade detection. (3) Tumor cells decrease the expression of their class I molecules and thereby evade the associated-recognition (p 195) of the tumor antigens. (4) Tumor antigens generally do not express class II molecules and therefore do not stimulate the helper T cells. This makes tumor cells poorly immunogenic. (5) Certain molecules like sialomucin tend to cover and mask the tumor antigens, disabling immune cells from detecting them. This has been called 'antigenic blindfolding'. (6) A subpopulation of T8 cells called *suppressor T cells* (T_s) is tumor-friendly. They suppress the development of the usual immune response to tumor antigens by inhibiting the differentiation of T_c cells and inhibiting the activity of macrophages. (7) Tumor cells can synthesize substances that reduce host immunity that is targeted at them.

Acquired (adaptive) immunity

Acquired (adaptive) immunity is a highly specific defense mechanism of the body that is targeted specifically at foreign materials introduced into the body. The foreign material may be an antigen or a hapten.

An *antigen* is usually a high molecular weight protein, but it may also be a low molecular weight protein (e.g., insulin) or a high molecular weight polysaccharide (e.g., dextran). Immunogenic molecules require some degree of chemical complexity. Large substances lacking chemical complexity, such as nylon, polyacrylamide, and Teflon, are not immunogens. The specificity of an antigen is due to specific areas of its molecule called determinant sites or epitopes. One protein can have several epitopes and these may differ from each other not only in their specificity but also in their antigenic potency.

A *hapten* is usually a nonprotein substance which has little or no antigenic property by itself but which combines with a protein to form a new antigen capable of stimulating production of specific immunoglobulins, the specificity of which depends upon the hapten fraction rather than the carrier protein. A secondary response can be elicited by a subsequent challenge by the same carrier-hapten complex but not by the hapten combined with a different carrier. The hapten however does not require the protein carrier to react with the antibody produced. Lipids and simple carbohydrates that are not antigenic and the more complex polysaccharides that are poorly antigenic are often powerful haptens.

Active immunity involves a direct encounter with the antigen. An essential aspect of active acquired immune response is the development of *immunological memory* of the antigen so that on subsequent encounters, the response to the same antigen is more vigorous.

When an antigen is introduced into an animal for the first time, there is an interval varying from 4 days to 4 weeks before any antibody can be detected in the serum. Then follows a rise in the antibody titer which reaches its maximum by 6 days to 3 months. This response is called the *primary response*. When the same antigen is injected into the body again, there is an immediate drop in the circulating antibody titer due to its neutralization by the injected antigen. However, after 2–3 days, there is a rapid rise in titer that reaches its peak after 7–14 days. The antibody titer is much higher than in the primary response. This response is called the *secondary response*. Even after the antibody titer falls off, it remains higher than the titer achieved during primary response. The main features of the secondary response are therefore threefold: a smaller dose of antigen required to produce it, a shorter lag, and greater antibody production (Fig. 31.1, Table 31.1).

Active immunization against a microbe can be achieved by injecting its antigens into the body. When the microbe invades the body, a secondary response is elicited by its antigens and it is quickly destroyed before it can multiply and infect the body. Subsequent attacks by the same microbe only helps in boosting the secondary response and prolonging the immunity.

Passive immunity is acquired without an encounter with the antigen, as when the mother passes her antibodies to the fetus through the placenta or colostrum, or when antibodies are injected therapeutically. Passively acquired immunity does not confer immunological memory.

Passive immunization can be achieved by injecting preformed antibodies (produced in animals or in another person) against specific microbial antigens. When the microbe subsequently invades the body, it is quickly eliminated by the injected antibodies, which too are consumed in the process. Such prompt elimination however preempts the development of active immunity.

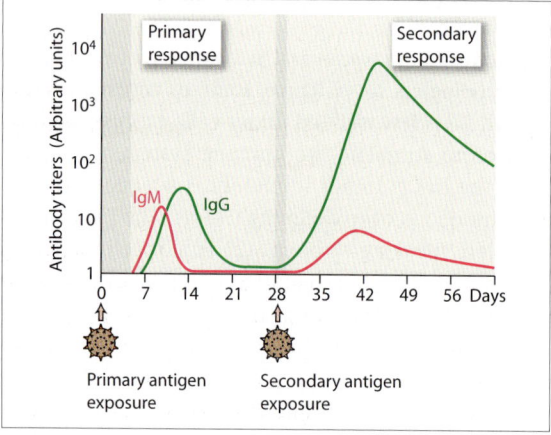

Fig. 31.**1**   Primary and secondary immune responses. Note also the class switch: In the primary response, the rise in IgM occurs before the rise in IgG, and is produced in comparable amount. In the secondary response, IgG appears earlier and its amount far exceeds the amount of IgM produced.

Table 31.**1**   Differences between primary and secondary immune response.

	Primary response	**Secondary response**
Responding B cells	Unsensitized B cells	Memory B cells
Latent Period	5–10 days	1–3 days
Peak antibody titer	Smaller	Larger
Persistence of antibody titer	Short	Long
Predominating antibody class	IgM appears earlier	IgG appears earlier
Induced by	All immunogens including TI antigens	Only T-cell dependent protein antigens
Dose of antigen required	High	Low

Immunotherapy

Immunotherapy is the therapeutic use of immunological methods for eliminating tumor cells.

Nonspecific immunotherapy relies on the activation of immune responses that are not tumor-specific but nonetheless inhibit tumor growth. This is done by injecting cytokines like tumor necrosis factor.

Specific immunotherapy involves injecting the patient with tumor cells or tumor antigens in order to induce immunity against tumor antigens. It is often supplemented by injection of cytokines for stimulating the differentiation of helper T cells which are usually not formed in large numbers in the natural antitumor immune responses. *Monoclonal antibodies* (originating from a single clone of cells and therefore, highly immune specific) directed against tumor antigens have been used in immunotherapy to treat cancer. However, most cancer cells undergo antigenic modulation, rendering the antibodies ineffective. Agents called *immunotoxin* or the 'magic bullet' combine the selectivity of monoclonal antibodies with the killing power of anticancer drugs. It consists of antitumor toxins linked to monoclonal antibodies specific for tumor cells only. Once the antibody binds to the tumor cell, the tumor cell internalizes the antibody along with the toxin that ultimately kills it.

Immunomodulation

Immunoenhancement

Immunoenhancement means enhancing the humoral or cell-mediated immune response against antigens. *Adjuvants* (Latin *adjuvere*, to help) are substances that, when mixed with an antigen before injection, nonspecifically enhance the immune response to the antigen. An adjuvant may be a cell wall constituent (e.g., muramyl dipeptide) from tubercle bacilli or gram-negative bacteria such as diphtheria or whooping couch bacilli. Other adjuvants include alum, mineral oil, lanolin and detergents. The first adjuvant to be discovered was Freund's adjuvant. It contains heat-killed *Mycobacterium tuberculosis*, mineral oil, and lanolin. A similar adjuvant that is in wide use is the Bacillus Calmette-Guerin (BCG). Named after its originators, BCG is an attenuated strain of *M. tuberculosis*.

Adjuvants enhance immunogenicity of soluble protein antigens by converting them into particulate forms. This promotes increased phagocytic uptake and subsequently, a delayed and sustained release of the antigen (the depot effect). Injection of adjuvants often causes granuloma formation at the site of injection after several days. A granuloma is an aggregation of macrophages, lymphocytes, giant cells and fibroblasts, formed within a meshwork of connective tissue fibers.

Immunosuppression

Immunosuppression is used therapeutically in inflammations, organ transplantation, autoimmune disorders, Rh blood antigen sensitization, and in cancers of the hemopoietic system.

Nonspecific immunosuppression can be induced by physical and chemical methods. *Physical methods* of immunosuppression include the removal of lymphoid tissue by surgical means or by irradiation. Surgical removal of the thymus has been performed for the alleviation of myasthenia gravis. Removal of other lymphoid organs changes the immune response only slightly. *Chemical immunosuppressants* include corticosteroids.

Specific immunosuppression is the abolition of the response to a particular antigen, leaving the response to all other antigens intact. *Antigen-induced suppression* is used for allergen desensitization. However, the suppression is transient. The rationale is that if several small doses of an antigen are given over time, the body stops reacting to that antigen. An example of *antibody-mediated suppression* is the prevention of Rh blood antigen sensitization in susceptible pregnant women. *Antilymphocytic serum reacts* selectively with lymphocytes, inactivating their antigenic receptors. Because of several side effects, it has limited use in humans.

Immunodeficiency

Immunodeficiency can be caused by either genetic or environmental factors, or both. Immunodeficiency diseases can affect humoral immunity, cell-mediated immunity, phagocytosis, or the complement system. Most immunodeficiency diseases involve only B cells. Examples are the X-linked agammaglobulinemia and X-linked hyper-IgM syndrome.

Acquired immunodeficiency syndrome (AIDS) was first identified in 1981. It is characterized by dramatic weight loss, night sweats, swollen lymph nodes and high susceptibility to opportunistic infections. It is also associated with neuropsychiatric abnormalities. The definition of acquired immunodeficiency syndrome (AIDS) is explicit in its name: it is 'acquired' because it is not inherited, it is an 'immunodeficiency' because the immune system breaks down, and it is a 'syndrome' because it is associated with several diseases that take advantage of the body's collapsed defenses.

AIDS spreads by intimate homosexual or heterosexual contact, by exposure to infected blood or blood products, or from mother to child during pregnancy. It is caused by an RNA retrovirus that has been named the human immunodeficiency virus type 1 (HIV-1). AIDS patients have lymphopenia involving primarily the T_4 cells because the T_4 molecule acts as a specific receptor that binds with high affinity to the HIV's envelope glycoprotein called gp 120. The number of circulating NK cells in AIDS patients is not significantly reduced but their cytotoxic ability is diminished. The humoral dysfunction associated with AIDS is an inability to produce an adequate IgM response. HIV virus does not kill the infected T_4 cells—clever viruses do not! Rather they activate them to T_H1 cells. This leads to a widespread activation of the immune system which causes immunological destruction.

Monocytes also express the antigens of T_4 cells and therefore get infected with HIV virus. Monocytes are more refractory to the cytopathic effects of the virus—the virus can survive in these cells and get transported to different parts of the body like the brain and lung. Thus, monocytes serve as a major reservoir for HIV in the body. HIV-1 also invades brain cells, causing dementia in over half the patients.

Immune Tolerance

Specific immunological tolerance

Specific immunological tolerance is defined as the acquired inability of a host to express specific humoral or cell-mediated immunity to an antigen to which it would normally respond. The lack of responsiveness is not caused by nonimmunogenicity of an antigen or any impairment of the immune response. To be called specific tolerance, a recipient must have two exposures to the antigen: the first tolerance-eliciting exposure followed by a second challenging exposure of the same antigen that fails to evoke an immune response. Thus, tolerance is the opposite of secondary immune response.

Some of the observations on immune tolerance are as follows. (1) High zone tolerance or immune paralysis is produced by large doses of antigen, probably by depleting the specific clones of cells. (2) Low zone tolerance is induced by repeated subimmunogenic doses

of thymus-dependent antigens. (3) Aggregate-free antigens are tolerogenic, probably because they escape phagocytosis by the macrophages. (4) Tolerance is easier shortly after birth.

Tolerance to fetus

Because of the incorporation of the paternal genes, the fetus differs in genetic make up from its mother. It should therefore induce immune response in the mother in the same way as the fetal Rhesus antigen in a Rh- mother. That however never happens, probably for the following reasons. (1) The placenta is resistant to immune attack because after implantation, the trophoblasts lose much of their immunogenicity due to a decrease in the density of the MHC antigens or the development of an inert coating of mucoprotein on the surface of these cells. (2) The antibodies produced by the mother are absorbed by the placenta, preventing their entry into fetal circulation. (3) Alpha fetoprotein (AFP) and progesterone produced during embryonic development are immunosuppressants.

Tolerance to self

Tolerance to self is probably a high zone tolerance induced during fetal life when large amounts of self-antigens react with and deplete the corresponding clones.

Table 31.**2**   Some autoimmune diseases and their mechanism.

Hashimoto's disease	T cells react with the antigens on the thyroid cells.
Grave's disease	Autoantibodies bind to thyroid cells and stimulate them.
Insulin-dependent diabetes mellitus	Autoantibodies damage the insulin-producing B cells of the pancreas.
Myasthenia gravis	Autoantibodies damage the acetylcholine receptors on the motor end plate.
Rheumatoid arthritis	Autoantibodies damage the joints.
Pernicious anemia	Autoantibodies damage the gastric mucosa.
Autoimmune hemolytic anemia	Autoantibodies damage the erythrocytes.
Thrombocytopenic purpura	Autoantibodies damage the platelets.
Rheumatic fever	Antistrepococcal antibodies cross-react with heart-valve tissues.

Autoimmune diseases result when the body's immune response gets directed toward its own tissues which are normally exempted as self-antigens. Examples of autoimmune diseases are given in Table 31.2. The possible mechanisms of development of autoimmunity are as follows. (1) A sequestered self-antigen which has not come in contact with, and therefore did not deplete, its corresponding clone during fetal life will elicit an immune response in the adult if it comes in contact with immunocompetent cells. This appears to be the case with lens proteins which are enclosed in a capsule and which may leak out in the adult following injury or surgery. (2) Although antibodies are highly specific, sometimes an antibody formed in response to some foreign antigen cross reacts with some tissue of the body itself. This happens in rheumatic heart disease where the heart is damaged by antibodies formed against streptococcal antigens. (3) With age, some tissues of the body undergo an antigenic change which is not recognized as self-antigen. (4) The suppressor T cells may fail to check the immune process adequately.

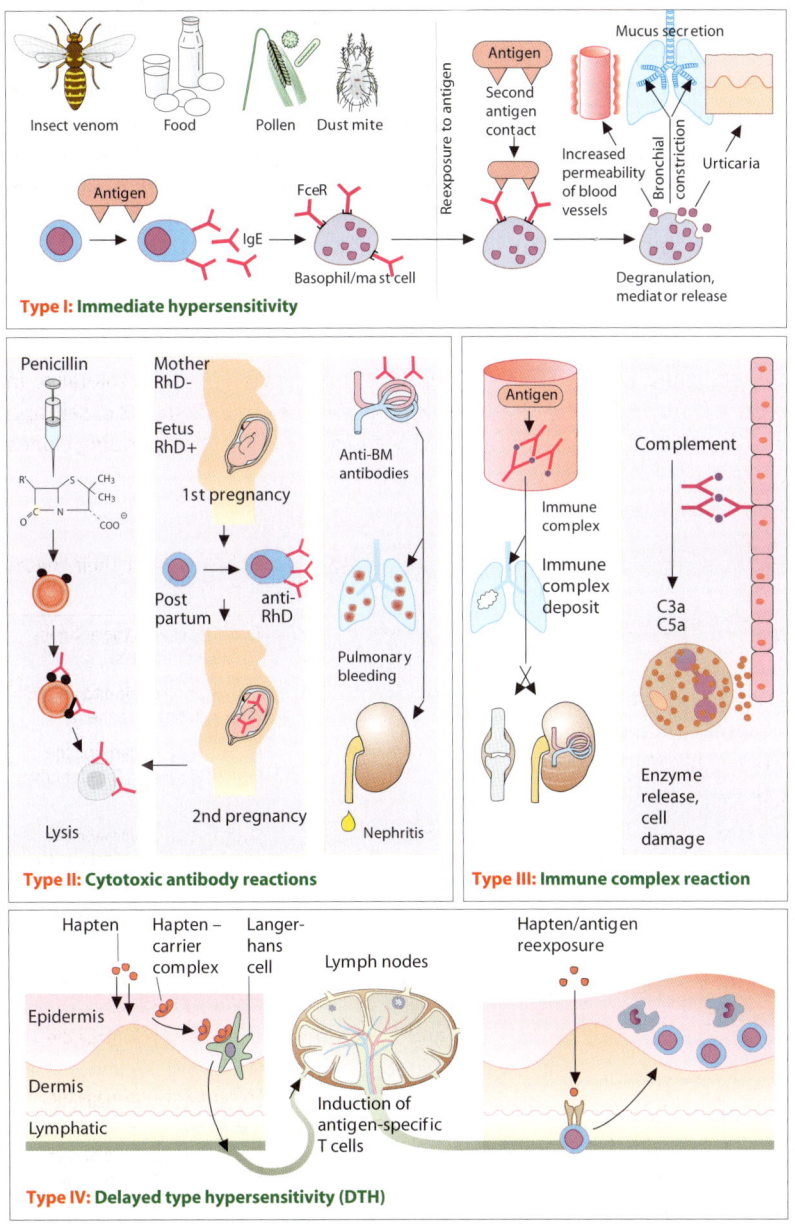

Fig. 31.**2**   Hypersensitivity reactions.

Hypersensitivity

When an immune reaction results in considerable damage to the body it is called hypersensitivity. There are four types of hypersensitivity reactions.

Type-I hypersensitivity (Anaphylaxis) occurs due to mast cell degranulation and is caused by antigens that evoke a strong IgE response. Examples are allergic rhinitis (hay fever), atopic dermatitis (eczema), and acute urticaria (hives). It occurs within minutes after a repeat exposure to the offending allergen (antigen). Mast cell degranulation occurs when the IgE molecules present on them are cross linked by the antigen (Fig. 31.**2**). The mediators include histamine and the slow-reacting substance of anaphylaxis (SRS-A). They cause capillary dilatation and exudation, and sensory nerve stimulation (itching, sneezing, and coughing).

The susceptibility to type-I hypersensitivity is called atopy. Atopic individuals are treated by desensitization, i.e., repeated administration of increasing doses of the offending allergen. The repeated exposure induces formation of IgG rather than IgE. On subsequent challenge with the allergen, the IgG reacts with the allergen before they can react with the IgE or cause mast cell degranulation.

Type-II hypersensitivity (Antibody–mediated cytotoxicity) is an immediate immune reaction that damages antigen-bearing cells. Example of this is seen in incompatible blood transfusion and hemolytic disease of the newborn. When antibodies react with the antigens, they also damage the cells (erythrocytes, leukocytes, or platelets) on which the antigens are located.

Type-III hypersensitivity (Immune complex disorder) occurs when antigen-antibody complexes are deposited in normal tissues of the body where they fix complement. Complement activation damages the tissue cells in the vicinity (damage of 'innocent bystanders'). A form of glomerulonephritis is an example of this type of hypersensitivity.

Type-IV hypersensitivity (delayed-type hypersensitivity, DTH) differs from the preceding types in two ways. (1) It is not mediated by antibodies; rather, it is mediated by macrophages that have been activated by T cells. The cytokines secreted do most of the damage. (2) The hypersensitivity starts after several hours and peaks at 48 to 72 hours. DTH is characteristically associated with granuloma formation. DTH is typically seen in Koch phenomenon—the hypersensitivity to tuberculin, which is present in *Mycobacterium tuberculosis*.

There are three phases of the immune response, viz., the initial phase, the central phase and the effector phase. The phases occur sequentially. The *initial phase* involves innate immunity and includes the events between the entry of antigen and its contact with specific receptors on the lymphocytic membrane. The mono-cyte-macrophage system is essential to this phase of immune response. The *central phase* of the immune response involves cooperation among different subsets of lymphocytes that proliferate and differentiate to form sensitized T and B lymphocytes and memory cells. The *effector phase* involves the inactivation of the antigen by the sensitized B and T cells generated during the central phase.

Initial Phase of Immune Response

T lymphocytes have specialized receptors for antigens. In B lymphocytes, the immunoglobulins present on the cell membrane (IgM and IgD) serve as receptors for antigens. T lymphocytes fail to recognize the antigen in isolation: The antigen has to be closely associated with some major histocompatibility (MHC) molecules in order to be recognized by the T cells. The phenomenon is called *associated recognition*. B lymphocytes do not require MHC molecules for recognizing antigens.

MHC molecules are present on every cell in the body. Just as specific blood group polysaccharides are responsible for transfusion reaction, similarly, specific MHC molecules are responsible for graft rejection following tissue transplantation. In man, the MHC antigens are referred to as the human leukocytic antigens (HLA) and are located on the short arm of chromosome 6. MHC molecules are broadly of two types: class I and class II. *Class I MHC molecules* are present on all cells except macrophages and B cells. *Class II MHC molecules* are present only on macrophages and B cells.

During tissue grafting, the MHC molecules on the donor tissue cells (mostly class I molecules) themselves act as antigen (then called MHC antigens). Antigens other than the MHC antigens have different ways of getting associated with MHC molecules (Fig. 32.1). (1) When a malignant tumor cell develops in the body, its mutant gene codes the synthesis of abnormal antigenic protein molecules on its surface. Thus, the tumor cell gets associated with class I MHC molecules. (2) When a virus invades a cell, it codes the synthesis of protein molecules that present themselves on the cell surface. The viral antigens thereby get associated with class I MHC molecules. (3) When a macrophage phagocytoses an antigen and fragments it, the antigenic epitopes make their way to the surface of the macrophages. Thus, the antigenic epitopes get associated with class II MHC molecules of the macrophage.

The process of engulfing an antigen, fragmenting it, and presenting the epitopes on the cell surface is called *antigen processing* and phagocytes that process antigens are called antigen-processing cells. Macrophages engulf tumor cells too. Thus, a tumor antigen that was originally associated with class I molecules

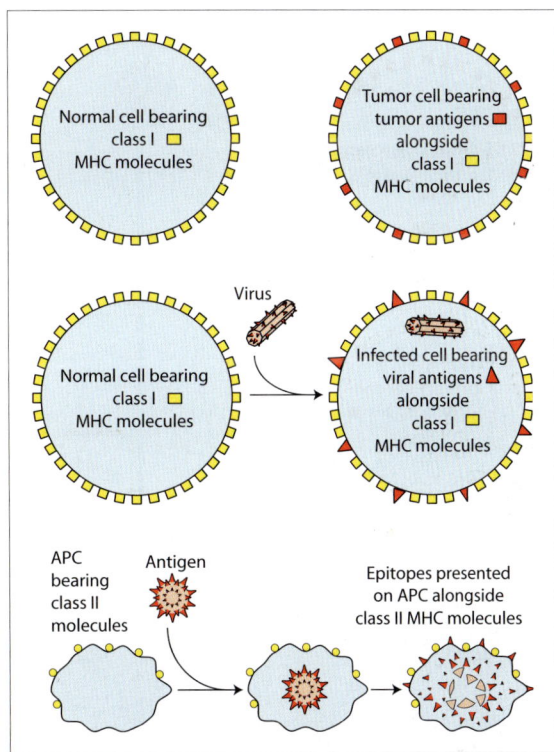

Fig. 32.1   Associated recognition of different types of antigens.

subsequently gets associated with class II molecules. Also, tumor cells are known to shed their antigens (antigen shedding). These antigens are also phagocytosed and processed by macrophages. Therefore, nearly all antigens, including viral and tumor antigens, eventually get associated with class II MHC molecules.

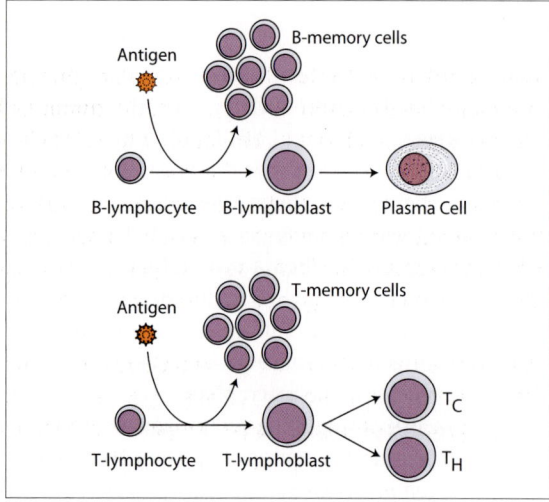

Fig. 32.**2** Differentiation of B and T lymphocytes.

Central Phase of Immune Response

A lymphocyte normally develops from a lymphoblast which is larger in size. However, when stimulated by a suitable antigen, the lymphocyte becomes larger and looks once again like a lymphoblast. This is known as *blast transformation*. The B lymphoblast differentiates into the plasma cell which produces immunoglobulins. The T lymphocyte differentiates into two different effector T cells, the T_H (H for helper) cells and T_C (C for cell mediated cytotoxicity). Some of the B and T lymphocytes (T_4 subtype, see below) do not enlarge or undergo blast transformation when exposed to antigen. Instead, they only proliferate forming a large number of small-sized *memory cells* (Fig. 32.**2**) which are responsible for the secondary immune response.

T-cell differentiation

Depending on the type of their surface antigens, the mature T lymphocytes are further classified into two antigenic subtypes, viz., T_4 cells and T_8 cells. Antigens associated with class II molecules stimulate T_4 cells

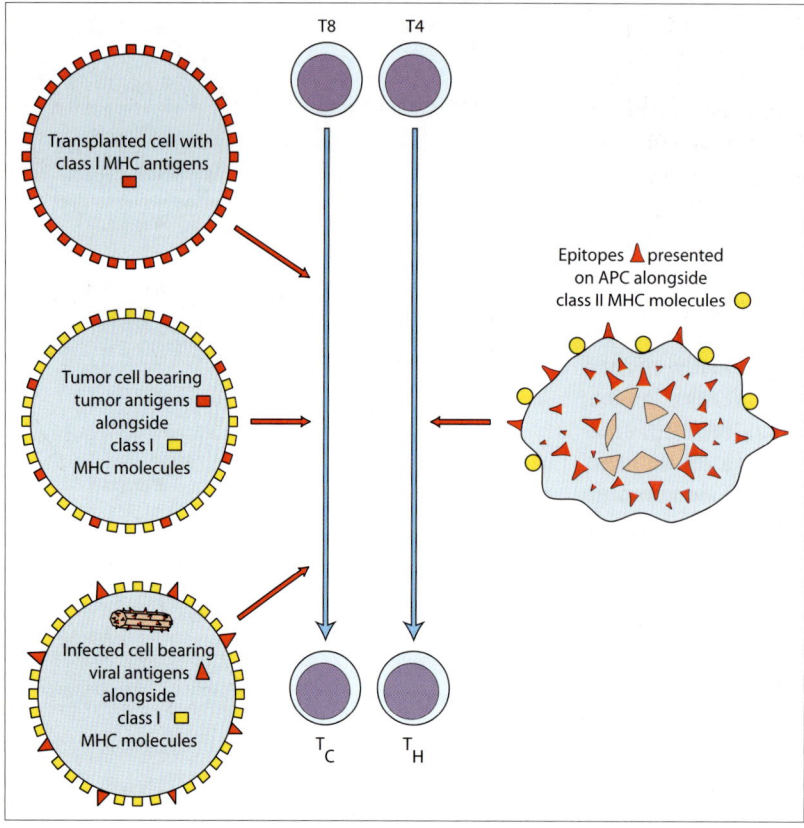

Fig. 32.**3** Role of MHC molecules in T-cell differentiation.

to differentiate into T_H cells. Antigens associated with class I molecules stimulate T_8 cells to differentiate into T_C cells (Fig. 32.**3**). The T_C cell requires the help of the T_H cell for its activity, which is called *T-T cooperation*. Recalling that antigens associated with class I MHC molecules eventually also get associated with class II MHC molecules, it is obvious that most antigens would stimulate both T_4 and T_8 lymphocytes resulting in the production of both T_C and T_H cells.

B cell differentiation

B cells can recognize the antigen but cannot undergo proliferation and transformation into plasma cells until they receive the cooperation of the T_H cells. In other words, the B cells can decipher the 'antigenic signal' but cannot proliferate till they receive the 'proliferative signal' from the T_H cells. This is known as T-B cooperation. Hence, the B cells too are ultimately dependent on MHC molecules without which there would be no T_H cells to enable their differentiation.

Thymus independent (TI) antigens do not require the help of T_H for stimulating the proliferation of B cells. Examples include the ABO agglutinogens and dextran. All TI antigens have an orderly arrangement of repeating antigenic determinants. It seems that about 12 to 16 receptors on the B cell must be cross-linked by identical, properly spaced antigenic determinants to provide the proliferative signal normally provided by the T_H cell. The TI antigens induce only IgM formation and therefore, there is no 'class switch' (Fig. 31.**1**) of antibody. Consequently the secondary response, which is dependent on IgG production, is weak with TI antigens.

Theories of immune specificity

The antibody produced in response to an antigen is highly specific, i.e., it reacts only with the antigen that has evoked its production and not with any other antigen. Such specificity also implies that since there are innumerable antigens to which the body can be possibly exposed, there must be an equal number of antibodies in the body to match them! However the ultimate blueprint for the production of all antibodies resides in the solitary hemopoietic stem cell and somewhere down the line of differentiation, the cells or the antibodies acquire a bewildering range of specificity. The crucial question is: At what stage of differentiation does the specificity develop? Broadly, there are three possibilities, each of which has been shaped into a well-argued theory (Fig. 32.**4**).

The **template (instructive) theory** suggests that antibody specificity does not develop during the course of cell differentiation but develops only following the contact of the antigen with antibody. The antigen acts as a template and the antibody moulds itself to fit with it. Thus it suggests that the antigen 'instructs' the production of its antibody and that immunoglobulins present in the body are nonspecific prior to antigenic exposure.

The **germline (selective) theory** suggests that the complete range of antibodies is present on the surface of each B cell and the antigen has only to 'select' and stimulate the production of the antibody which is specific to it. In other words, the B cell itself is nonspecific but produces the whole range of specific antibodies. It however seems unlikely that so many genes in a single cell are devoted only to antibody production.

The **somatic mutation (clonal selection) theory** differs from the germline theory in suggesting that the complete set of antibodies is produced, not by just one

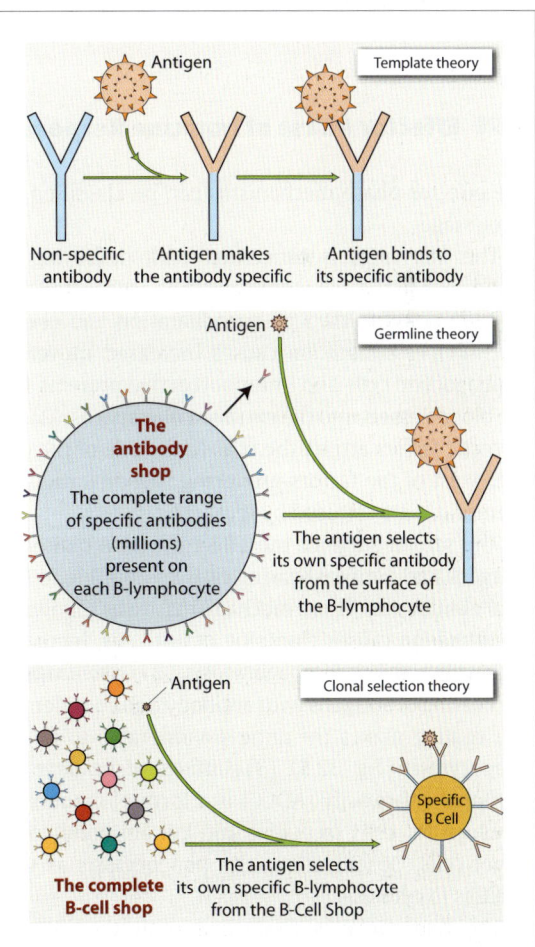

Fig. 32.**4**   Theories of immune specificity.

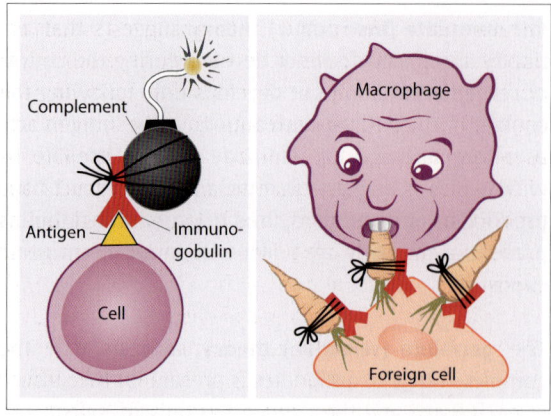

Fig. 32.**5** (*Left*) Cell lysis through complement. The Fc of the antibody is bound to the complement (shown here as a bomb with its fuse lighted). The Fab of the antigen binds to the antigen on the cell surface. The complement destroys the cell. (*Right*) Phagocytosis induced through opsonization. The Fab of the antibody binds to a foreign cell. The Fc of the antibody is bound to some foreign materials (shown here as alluring carrots) that attract macrophages and result in the phagocytosis of the foreign cell.

nonspecific B cell but by a complete set of specific B cells. It proposes that during lymphocytic processing in the central lymphoid tissues, the cells proliferate massively. During proliferation the cells mutate frequently, giving rise to genetically diverse B cells, each capable of producing a different antibody. Each specific B cell multiplies and establishes a population of genetically and immunologically identical B cells called a *clone*. An antigen has only to select and stimulate a specific clone and hence the name 'clonal selection' theory.

Effector Phase of Immune Response

The effector phase mechanisms can be classified into three stages.

The *first stage* is concerned with delivering immune cells to the site where antigens are located. That is made possible through vasodilatation and chemotaxis. (1) *Vasodilatation* causes increased movement of phagocytic cells and immunoreactive proteins from the blood vessels into the extravascular spaces. (2) *Chemotactic factors* attract the cells to the site of the antigen. Most of the factors producing vasodilatation and chemotaxis are released by mast cells.

The *second stage* of the effector phase makes the antigens and antigen-bearing cells susceptible to their final elimination. These mechanisms are as follows. (1) *Agglutination* causes clumping of antigens, increasing their vulnerability to phagocytosis. (2) *Opsonization* is the coating of antigens with antibody and complement. The coating makes the antigen vulnerable to lysis or phagocytosis (Fig. 32.**5**). (3) *Antibody-dependent cell-mediated cytotoxicity* (ADCC) is a form of cytotoxicity in which NK cells recognize and kill antibody-coated target cells. Opsonization is to phagocytosis as what ADCC is to cytolysis.

The *final stage* is concerned with the inactivation and elimination of the antigen, all of which is ultimately dependent on the activation of B and T lymphocytes. These mechanisms are (1) *phagocytosis* of the cell on which it is present, (2) *cytolysis* of the cell on which it is present, (3) *bacteriostasis*, in the case of bacterial antigens, and (4) *neutralization*, in the case of toxic antigens.

Depending on the immune mechanisms employed for eliminating the antigen, immune mechanisms have been classified into humoral and cellular immunity. *Humoral immunity* is a major defense against bacterial infections. It includes effector mechanisms involving antibodies (γ-globulins or immunoglobulins) and complement (enzymatic plasma proteins). *Cell-mediated immunity* is mainly responsible for defense against tumor and transplant cells and infections due to viruses, fungi, and certain bacteria like the tubercle bacilli. It is mediated by T lymphocytes. The distinction between humoral and cellular immunity is blurred by the presence of extensive overlaps between the two systems.

Humoral immunity

An **immunoglobulin** molecule is made of two heavy and two light chains (Fig. 32.**6**). Disulfide bonds anchor the light chain to the heavy chain and also hold the two heavy chains together. The heavy chains are of five subtypes: μ, γ, α, ε, and δ (constituting respectively IgM, IgG, IgA, IgE, and IgD). The light chains are only of two types, κ and λ.

The NH_2 terminal part of each chain (designated as V_L in the light chain and V_H in the heavy chain) has a variable sequence of amino acids and is therefore called the *variable region*. The COOH terminal part has a relatively constant sequence and is called the constant region. The variable regions bind to the antigen.

Disulfide bonds fold each chain into incomplete loops. Each light (L) chain has only two loops, V_L and C_L. The heavy (H) chain has four loops: the V_H, C_H1, C_H2,

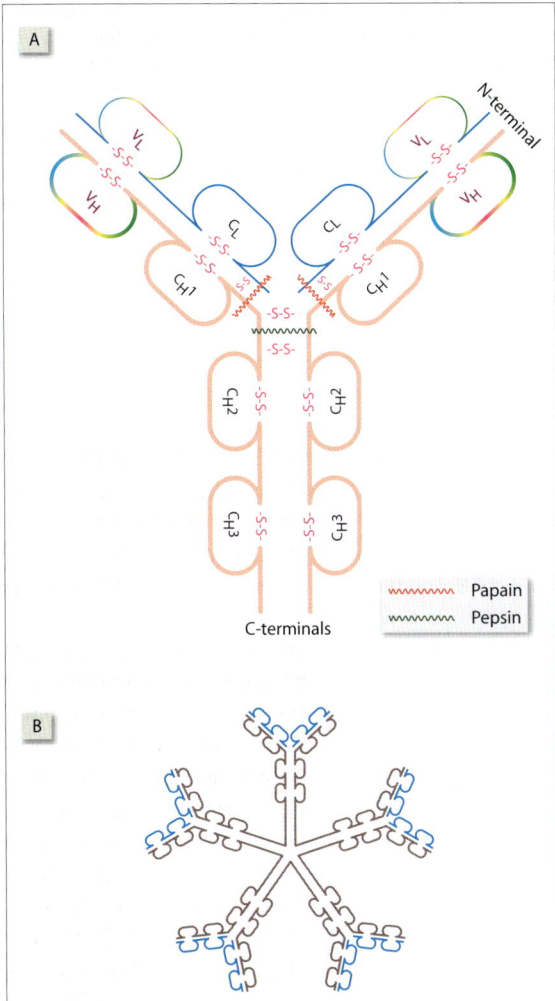

Fig. 32.**6** [A] Structure of an immunoglobulin molecule. [B] The pentamer of immunoglobulin-M.

and C_H3. The V_L and V_H together form the variable (V) region which binds to the antigen. The C_L, C_H1, C_H2, and C_H3 constitute the constant (C) region of the immunoglobulin molecule.

The immunoglobulin molecule can be split enzymatically by papain or pepsin. Papain yields the Fab (antigen binding fraction) which bears all the variable regions and the Fc (crystallisable fraction) which determines such properties of the immunoglobulin as diffusibility, placental transfer, complement-fixation, opsonization, etc. Pepsin splits off a dimer of Fab from the immunoglobulin molecule.

Immunoglobulin-M (IgM) comprises about 10% of the total plasma immunoglobulins. It is the first immunoglobulin produced in the primary response. Its level does not rise significantly in the secondary response. Because of its large size, it is predominantly intravascular. This, coupled with its early production, makes it important in bacteremias. IgM is the only immunoglobulin which is produced before birth. Natural antibodies like anti-A and anti-B agglutinogens are therefore of the IgM type. It has a theoretical valence of 10 which is actually observed only with small haptens. With larger antigens, its effective valence drops to 5 due to steric restriction. However, it has much more affinity and avidity of binding for large antigens with multiple epitopes than for small haptens. IgM produces very effective agglutination. It fixes complement by the classical pathway and the lysis it causes thereby is very effective. Its opsonizing power is rather weak.

Immunoglobulin-G (IgG) is the most abundant immunoglobulin, comprising about 70% of the total immunoglobulins. It is the major immunoglobulin to be synthesized during the secondary response in which the initial IgM production gives way to IgG production (Fig. 31.**1**). This change in the profile of antibody production is called 'class switch'. Unlike IgM, it is able to cross the placenta. Thus it is able to provide passive immunity to the neonate in its first few weeks. The protection is further reinforced by the transfer of colostral IgG across the gut mucosa of the neonates. After birth, the serum IgG concentration steadily falls for 2–3 months and then rises slowly over years. IgG readily diffuses out into the extravascular spaces where its concentration is the same as that in the plasma. It is also found in milk, saliva, nasal, and bronchial secretions. IgG causes opsonization, ADCC, and neutralization of toxins. Although it fixes complement by the classical pathway, the lysis it produces through complement-fixation is weak. The IgG-antigen complex binds to platelets through Fc receptors causing platelet aggregation.

Immunoglobulin-A (IgA) is secreted in colostrum, saliva, nasal and lung secretions, tears, genitourinary, and gastrointestinal fluids, providing 'secretory immunity'. IgA brings about opsonization and neutralization. It also fixes complement by the alternative pathway. The intestinal mucosa secrete IgA bound to a secretory protein which stabilizes it against proteolysis. It coats the microbes and inhibits their adherence to the mucosal cells thereby preventing entry into the body tissues. *Neisseria gonorrhea*, which produces IgA protease, can penetrate the mucosal barrier even in an immune person.

Immunoglobulin-E (IgE) bind to mast cells through their Fc receptors. When an antigen binds to two different IgE molecules, the cross-linking of the IgE molecules triggers the degranulation of the mast cells. Haptens with a single epitope cannot cross-link Fc receptors and therefore cannot degranulate mast cells. Rather, they inhibit IgE by blocking their Fab part. Parasites are particularly efficient in stimulating IgE production.

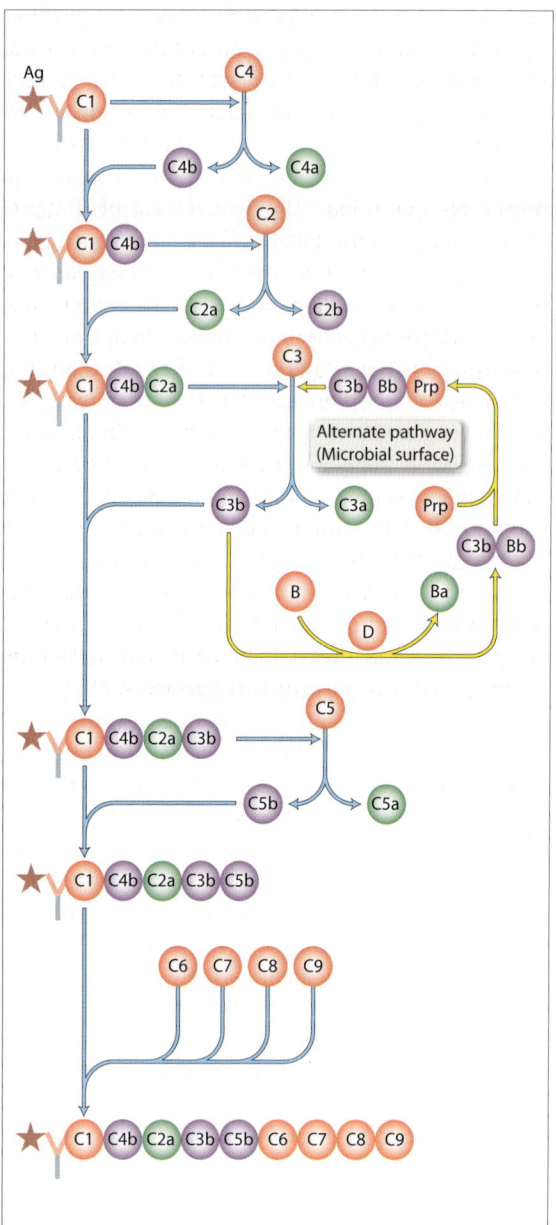

Fig. 32.**7** The pathways of complement fixation. Prp = properdin

vasodilatation, chemotaxis, opsonization, immune adherence, and mast cell degranulation (Table 32.**1**). IgM and IgG fix and activate complement by the classical pathway while IgA fixes complement by the alternative pathway.

In the *classical pathway* of complement activation, C1 cleaves C4 into C4a and C4b. C4b binds to the antigen-Ig-C1 complex. The 'complex' cleaves C2, releasing C2b which is a kinin (a potent vasodilator) while binding C2a to itself. The 'complex' then cleaves C3 into C3a (an opsonin) and C3b which cause mast cell degranulation. C3b adds on to the growing complex. The C4b-C2a-C3b fragment of the complex effects immune adherence. The 'complex' finally cleaves C5, adding on C5b to itself and releasing C5a which effects chemotaxis and degranulates mast cells. C6, C7, C8, and C9 then add on to the complex in successive stages. The C5b-C6-C7 fragment causes chemotaxis. The C5b-C6-C7-C8-C9 fragment is called the 'membrane-attack complex'. It produces cytolysis by enzymatically digesting a part of the membrane and making a hole in it. It may be noted that the activation of C3 to C3a and C3b occupies a central position in complement activation, just as activation of factor IX occupies a central position in the blood coagulation pathway.

The *alternative pathway* of complement activation provides a feedback amplification of C3b production. It involves four plasma proteins: C3, Factor B, Factor D and properdin. All reactions of the alternative pathway occur on microbial surfaces. The membranes of human cells contain sialic acid that destroys factor C3b. The cells of the host tissue are therefore spared the damage caused by complement activation.

The alternative pathway is initiated by factor D which splits factor B into Ba and Bb. Factor Bb combines with C3b to form C3bBb which is called the alternative pathway C3 convertase. The C3bBb complex is unstable and decays rapidly unless it is bound to a factor called properdin. The C3bBb-properdin complex splits more C3 into C3a and C3b.

Immunoglobulin-D molecules are present on the surface of B cells and act as the antigen receptors for the B cells.

The **complement system** is a system of nine enzymatic plasma proteins designated C1 to C9. Normally in the inactive state, their activity is triggered when C1 binds to the antigen-antibody complex. The activated C1 in turn activates the other complement proteins in a series of cascade reactions (Fig. 32.**7**). While the final product of complement activation (C5b-C6-C7-C8-C9) is cytolytic, many of the by-products cause

Table 32.**1**   Effectors of the complement system.

C2b	A potent vasodilator
C3a	An opsonin
C3b	Causes mast cell degranulation
C4b-C2a-C3b	Causes immune adherence
C5b	Causes chemotaxis and mast cell degeneration
C5b-C6-C7	Causes chemotaxis
C5b-C6-C7-C8-C9	The membrane-attack complex.

Cell-mediated immunity

In the following passages, the term 'cell-mediated immunity' has been used in its restricted sense, i.e., when the immune effectors are T lymphocytes or cytokines. In its broadest sense, the term encompasses everything that is related to T lymphocytes, and therefore includes the initial and central phases of immune response too.

Cytotoxic T cells (Tc cells) destroy antigen-bearing cells in two ways: through the induction of apoptosis and by perforating the cell membrane. (1) The Tc cell secretes the tumor necrosis factor-β (TNF-β) which increases the Ca^{2+} permeability of the antigen-bearing cell. The increased intracellular Ca^{2+} concentration activates intracellular enzymes that degrade the nuclear DNA with concomitant nuclear fragmentation. Apoptosis is also induced by a Ca^{2+}-independent mechanism. Activated Tc cells develop receptors for the Fas proteins that are present on the antigen-bearing cells. The binding of Fas proteins with the Fas receptors on T_C cells triggers the onset of apoptosis. (2) Activated TC cells also secrete perforin. In the presence of extracellular Ca^{2+}, perforin polymerizes on the surface of the antigen-bearing cell and dissolves parts of the membrane, forming pores in it. The pores cause cell death by disrupting cell homeostasis.

Helper cell (TH cells)   The T_H cells are of 2 types, T helper 1 (T_H1) and T helper 2 (T_H2).

The T_H1 cells are also called inflammatory T cells. They secrete mainly three cytokines, viz., interleukin-2 (IL-2), γ-interferon (IFN-γ), and TNF-β. These cytokines have three functions: (1) They stimulate macrophages. The stimulated macrophages destroy antigen-bearing cells and are sometimes responsible for the delayed-type hypersensitivity. (2) They help T_8 cells to differentiate into T_C cells (T-T cooperation). (3) INF-γ has the direct ability to kill antigen-bearing cells.

The T_H2 cells are primarily concerned with T-B cooperation, i.e., enabling B cells to produce antibodies. They secrete interleukins 4, 5, and 6. The differentiation of T_H2 is dependent on interleukin-1 secreted by the macrophage and interleukin-2 secreted by T_H1.

33 Hemopoiesis

Hemopoiesis

In the fetus all blood cells originate in the mesenchyme. During the first two months of fetal life blood formation takes place in the yolk sac. Thereafter the liver becomes the main site of hemopoiesis until about the seventh month, with the spleen making a small contribution. Hemopoiesis commences inside the bone marrow in the third month and by the seventh month it is the major site of hemopoiesis (Fig. 33.**1**A).

After birth red cells, granulocytes, and platelets are formed only in the bone marrow. The marrow makes only a small contribution to lymphocyte production, which occurs mainly in other lymphoid tissues. Monocytes are formed mainly in the bone marrow and partly in the spleen and lymphoid tissues. The spleen, liver, and lymph nodes resume their hemopoietic activity only when there is an excess demand for blood cell formation that cannot be met with bone marrow

hyperactivity alone. Such hemopoiesis that occurs outside the bone marrow is called *extramedullary hemopoiesis* and is commonest in infants and young children in whom the entire marrow cavity is occupied by red (hemopoietic) marrow, leaving little space for expansion of the marrow cavity.

Stem cells There are three functionally different stem cells, viz., the *pluripotent stem cell*, which can give rise to any blood cell; the *myeloid stem cell* giving rise to erythrocytes, granulocytes of all types, monocytes, and platelets; and the *lymphocyte stem cell* which gives rise only to lymphocytes. All stem cells possess two fundamental properties, viz., self-renewal, i.e., producing more stem cells through mitosis, and differentiation and commitment, i.e., the ability to differentiate into any of the mature specialized blood cells.

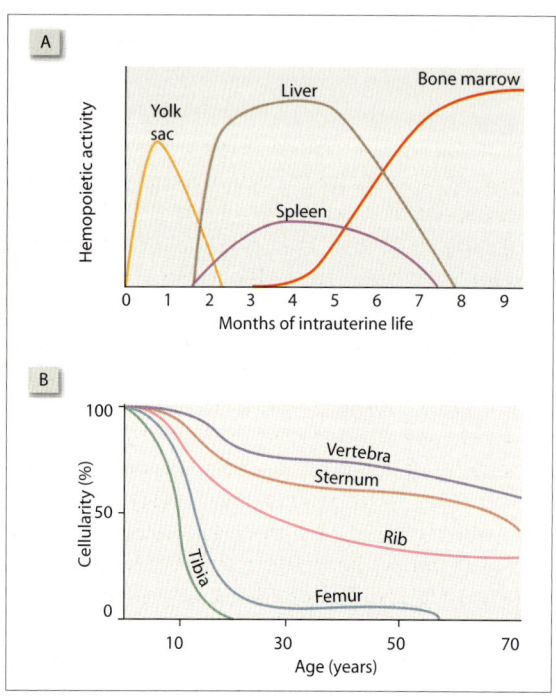

Fig. 33.**1** [A] Sites of extramedullary erythropoiesis. [B] Percentage of red marrow in different bones.

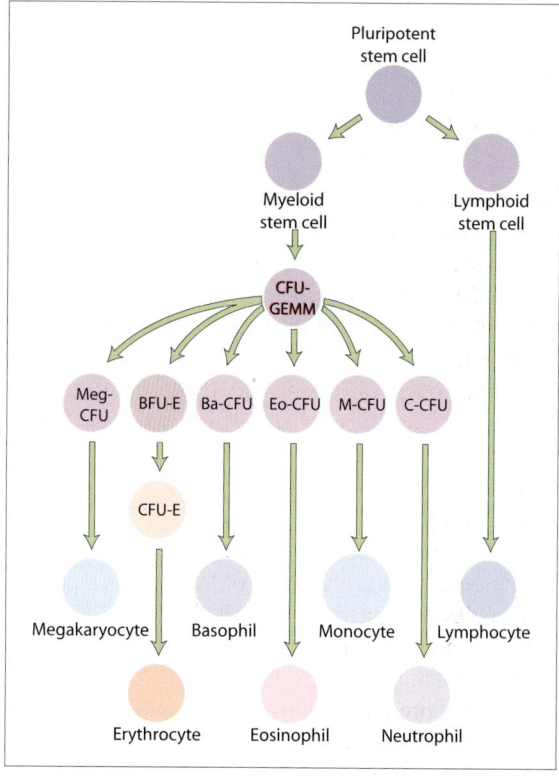

Fig. 33.**2** Hemopoietic pathways.

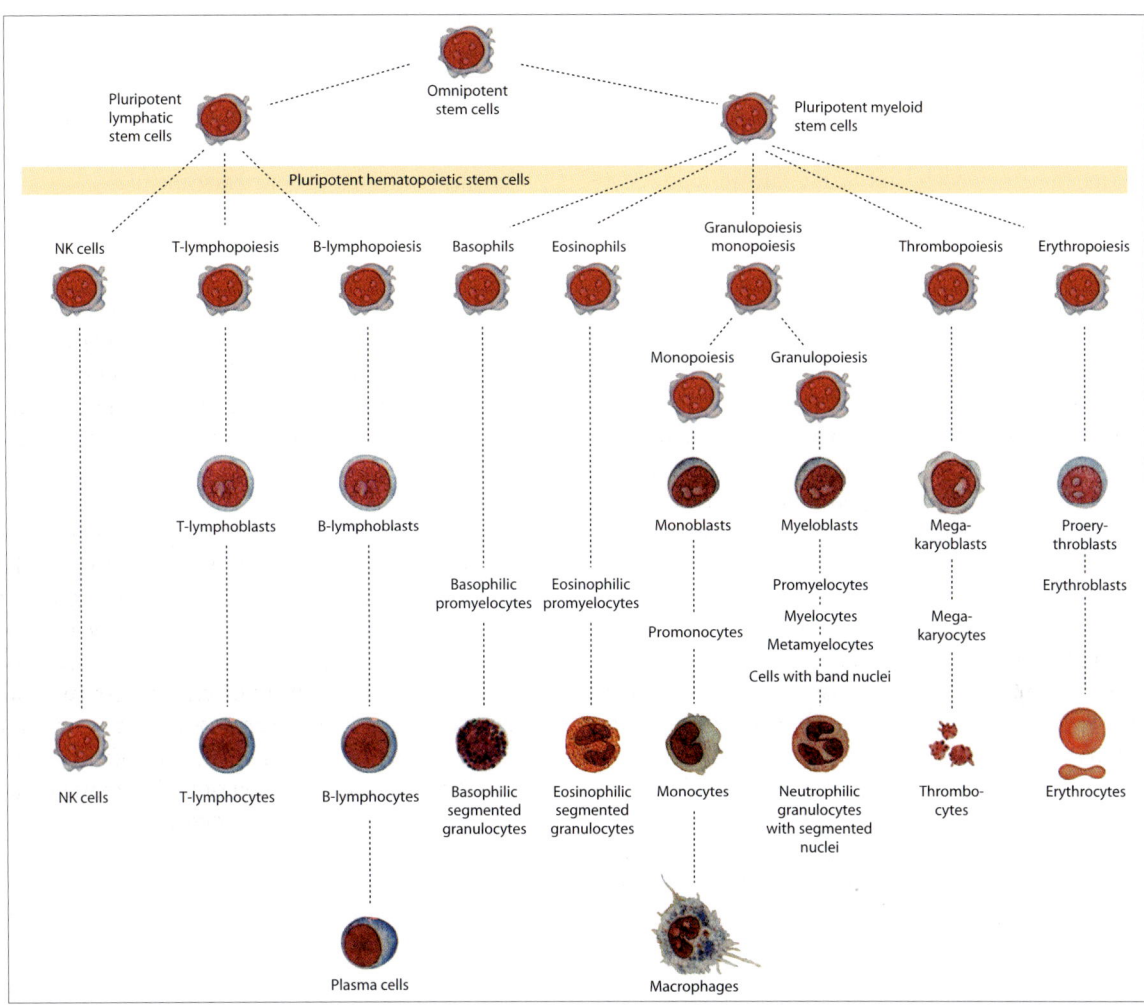

Fig. 33.**3**   Hemopoietic cells.

Progenitor cells   With time, the capability of a stem cell for self-renewal diminishes and its commitment to a particular line of differentiation increases. When the self-renewal capability is lost, the cell is no longer called a stem cell but is termed a progenitor cell. Progenitor cells are committed to one or at the most two lines of development. Progenitor cells are more numerous than stem cells and like stem cells, are present in both bone marrow and circulating blood. Stem cells and progenitor cells cannot be differentiated morphologically. They both look like a large lymphocyte. However, they can be separated by immunological techniques taking advantage of the different types of molecules present on their cell membrane. Progenitor cells are recognized by their ability to give rise to clones (colonies, or a group of cells) of differentiated cells in the presence of growth factors. Hence, progenitor cells are also called colony-forming cells (CFC) or colony-forming units (CFU).

Progenitor cells may be multipotent or unipotent.

An example of multipotent progenitor cells is the CFU-GEMM (colony-forming unit: granulocyte, erythroid, megakaryocyte, macrophage). Examples of unipotent progenitor cells are CFU-E (erythroid) and Meg-CFU (megakaryocytes). The erythroid series has two types of erythroid progenitor cells, viz., burst forming units-erythrocyte (BFU-E) and colony forming units-erythrocyte (CFU-E). BFU-E forms large colonies whereas CFU-E forms much smaller colonies (Fig. 33.**2**, 33.**3**).

Progenitor cells undergo both multiplication and maturation before they are released into the blood stream. The growth and differentiation of hemopoietic cells are controlled by cytokines. Cytokines that control the granulocytic and thrombocytic series are called *colony stimulating factors* (CSF). G-CSF stimulates granulocytic precursors, M-CSF stimulates monocytic precursors, and GM-CSF stimulates both. Cytokines that stimulate the lymphocytic series are called *interleukins*. The cytokine stimulating the erythroid series is called *erythropoietin*.

◾ Erythropoiesis

The regulation of erythropoiesis is discussed on p 143. The first cell of the erythroid series is the pronormoblast which matures successively into the normoblasts, the reticulocyte and finally into an erythrocyte in about 7 days.

The *pronormoblast* (12 to 20 μm) has a large nucleus which occupies most of the cell. The nucleus contains several nucleoli. The chromatin is finely reticular. The cytoplasm is intensely basophilic (dark blue) because of its high RNA content.

The *normoblasts* are classified into the early (basophilic) normoblast, intermediate (polychromatic) normoblast, late (orthochromatic) normoblast, and the reticulocyte. Mitotic division occurs up to the stage of the intermediate normoblast and mitosis is most active at this stage. The late normoblast is not capable of mitotic division. As it proceeds through these stages, the cell and its nucleus become smaller, chromatin becomes thicker and coarser, and the cytoplasm becomes less basophilic. The nucleoli disappear in the early normoblast stage. Hemoglobin appears in the intermediate normoblast stage (Fig. 33.**4**). The acidophilic tint of hemoglobin makes the cytoplasm polychromatic (blue-gray violet). The nucleus becomes pyknotic (a deeply-staining, structureless mass) and is extruded in the late normoblast stage. With the loss of nucleus, formation of new ribosomes in the cytoplasm ceases. The cytoplasm still has a faint polychromatic tint.

The *reticulocyte* is a flat, disk-shaped, non-nucleated cell, of slightly larger volume and diameter than the mature erythrocyte. The hemoglobin content is about the same as that of the mature cell. The cytoplasm still contains small amounts of ribosomic RNA which results in a faint polychromatic tint with Romanowsky stains. With supravital stain such as brilliant cresyl blue, the RNA appears in the form of a fine reticulum and hence the name reticulocyte. As the reticulocyte

matures, the RNA is catabolized and ribosomes disintegrate. The reticulocyte takes about 2 days to mature to an erythrocyte. Reticulocytes are present in circulation too where they comprise 0.2–2.0% of the RBC count. Their number increases in hemolytic anemia and following treatment of deficiency anemias.

◾ Leucopoiesis

Granulocytic (myeloid) series   The first recognizable cell of the granulocytic series is the myeloblast (15 to 20 μm), from which the mature granulocytes develop through a series of cells, viz., the promyelocyte, myelocyte, metamyelocyte, and band cell. Apart from the difference in granules, the myelocyte, metamyelocyte and band form have similar morphological characteristics. The maturation of the granulocytes is characterized by the development of azurophilic granules (in the promyelocyte stage), disappearance of nucleoli and appearance of specific granules (in the myelocyte stage), and the cessation of mitosis (in the metamyelocyte stage). There is progressive loss of cytoplasmic basophilia of the cytoplasm, progressive coarsening of the chromatin and ripening of the nucleus which ultimately becomes segmented, and the development of motility and ability to act as a phagocyte.

Lymphocytic series   The lymphoblast resembles the myeloblast except that its nucleus contains fewer nucleoli and coarser chromatin. Lymphoblasts give rise successively to the large and small lymphocytes, both of which are found in circulation.

Monocytic series   Monocytes are formed mainly in the bone marrow and migrate to the spleen and lymphoid tissues. The earliest cell is the monoblast which gives rise to the promonocyte and the mature monocyte. The monoblast is a large cell that cannot be distinguished from the myeloblast on morphological grounds alone. The promonocyte (20 μm) has a large kidney-shaped nucleus. The chromatin is reticular. The cytoplasm is a dull gray-blue and may contain fine azurophilic granules.

◾ Thrombopoiesis

The megakaryoblast (20 to 30 μm) is the first cell in the platelet series. It develops into a larger promegakaryocyte, and a still larger megakaryocyte (30 to 90 μm). The megakaryocyte has a single multilobed nucleus with coarse clumps of chromatin. The cytoplasm is light-blue and contains fine azurophilic granules. The

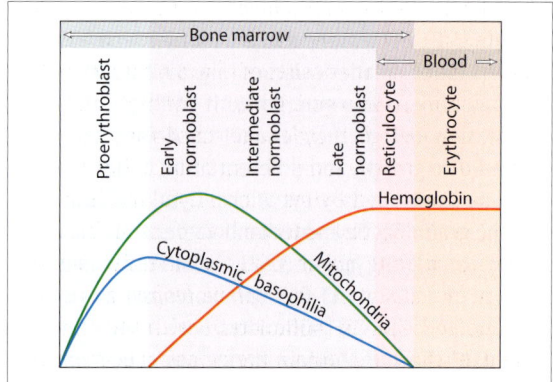

Fig. 33.**4**   Cytoplasmic changes during erythropoiesis.

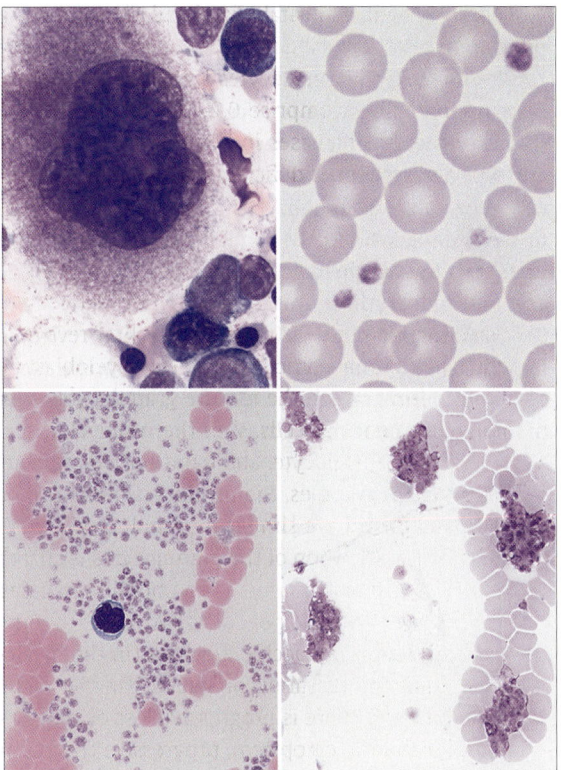

Fig. 33.**5**   Megakaryocytes (*above, left*) and thrombocytes.

cell margin is irregular and may show pseudopodia. Platelets are formed by the breaking off of small fragments of these megakaryocytes (Fig. 33.5).

Bone Marrow

The bone marrow is liquid in consistency. Its color is either red or yellow. Hemopoiesis occurs in the red marrow. The yellow marrow is composed mainly of fat cells. An adult has about 3 to 4 liters of bone marrow, nearly half of which is red marrow. Besides hemopoietic cells, mature blood cells, and fat cells, the bone marrow also contains reticulum cells and reticulin fibers, blood vessels, and nerves. The fine reticulin fibers form a supporting network for the hemopoietic cells that lie outside the blood vessels. The reticulum cells are potentially hemopoietic and form part of the functional reserve of the marrow (see below). The ratio of white cells to nucleated red cells is normally about 3–5:1. Megakaryocytes and lymphoid follicles are also present, though in smaller numbers.

The marrow sinusoids connect the arterial capillaries with the venules. Developing blood cells lie outside the sinusoids. Unlike the spleen, the marrow sinusoids form a closed system with no openings into the extravascular spaces. Hence the newly formed cells have to cross the sinusoid walls before entering circulation. The sinusoids are innervated by sensory nerve fibers. These nerves explain the pain produced when marrow is aspirated.

At birth, the marrow in all the bones of the body appears red and contains no fat. By the age of 5 years, fat cells begin to appear and the red marrow in the long bones gradually is replaced by yellow fatty marrow. The replacement occurs first in the bones of the hands and feet and then in the bones of the arms and legs from the distal to the proximal ends. By the age of 20 years, red marrow is found only in the bones of the thorax (ribs, clavicles, scapula, and sternum), vertebra, skull, pelvis, and the upper one-third of the femur and humerus shafts (Fig. 33.**1**B).

Bone marrow reserve   The bone marrow has a large reserve capacity that enables it to increase its output to as much as 13 times normal. The reserve capacity of the marrow is made up of two components. The *functional reserve* is attributable to stem cells in the marrow that start proliferating and differentiating. Initially, only the intermediate normoblasts and myelocytes start proliferating. Later, even the more primitive early normoblasts and promyelocytes start proliferating. The *anatomical reserve* is in the form of the fat cells that can be readily replaced by active hemopoietic cells. The replacement of fat cells occurs initially in the red marrow itself which becomes redder in color. Later, yellow marrow is transformed into red marrow, either by activation of the dormant stem cells or by migration of stem cells from red marrow. Still later, the bone is eroded by the expanding marrow cavity.

Bone marrow biopsy

Bone marrow biopsy is an important method of investigation in blood disorders. There are two main methods of biopsy. In *aspiration biopsy*, the bone marrow is aspirated through a specially-constructed wide-bore needle. The aspirate consists of marrow diluted with a variable amount of blood. Films of the aspirated marrow are made on glass slides and stained as blood films with a Romanowsky stain. The sternum and the posterior iliac crest are the sites of choice for marrow aspiration (Fig. 33.**6**). The anterior iliac crest and spinous processes of the lumbar vertebrae are also suitable for marrow aspiration. In *trephine biopsy*, a specially constructed trephine (hollow tube with serrated cutting edge) is used to obtain the biopsy specimen. Trephine biopsy gives a histological section in which bony trabeculae, hemopoietic tissue, fat cells, and blood

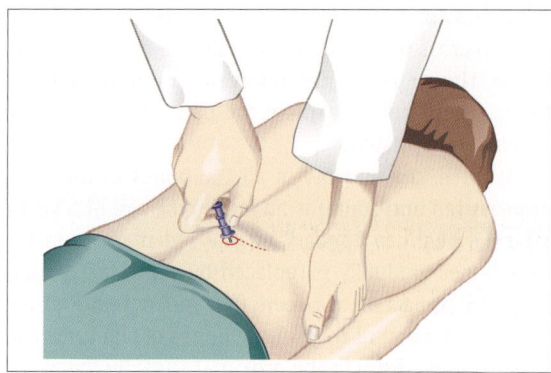

Fig. 33.**6**   Bone marrow biopsy from the superior part of the posterior illiac spine.

vessels are seen, i.e., the architecture of the marrow is preserved.

The following features are systematically noted in bone marrow biopsy: (1) The cellularity of the marrow (indicating whether it is hyperactive or hypoactive); (2) The type and activity of erythropoiesis (which can indicate megaloblastic anemia); (3) The number and type of developing white cells (which can indicate leukemias); (4) The number and type of megakaryocytes; (5) The myeloid–erythroid ratio which is normally about 3–5:1. Reduction of the ratio to 1:1 suggests erythroid hyperplasia; (6) The iron content of the marrow (seen as hemosiderin granules or ferritin); (7) The presence of foreign or tumor cells, parasites or organisms.

Section 4

Cardiovascular System

34.   Cardiac Excitation and the Electrocardiogram

35.   Abnormalities of Cardiac Excitation

36.   Cardiac Cycle

37.   Cardiac Output

38.   Circulatory Pathway and Hemodynamics

39.   Capillary Exchange and Lymphatic Circulation

40.   Chemical Control of Cardiovascular System

41.   Neural Control of Cardiovascular System

42.   Blood Pressure Regulation

43.   Circulatory Shock

44.   Coronary Circulation

45.   Cerebral Circulation

46.   Pulmonary and Pleural Circulation

47.   Cutaneous, Muscle, and Splanchnic Circulation

Cardiac Excitation

The heart as a whole obeys the all-or-none law, i.e., either it contracts completely or it does not contract at all, depending on whether the stimulus applied is of threshold or subthreshold strength: When any part of the heart depolarizes, the depolarization spreads to the entire heart ephaptically and the heart contracts maximally.

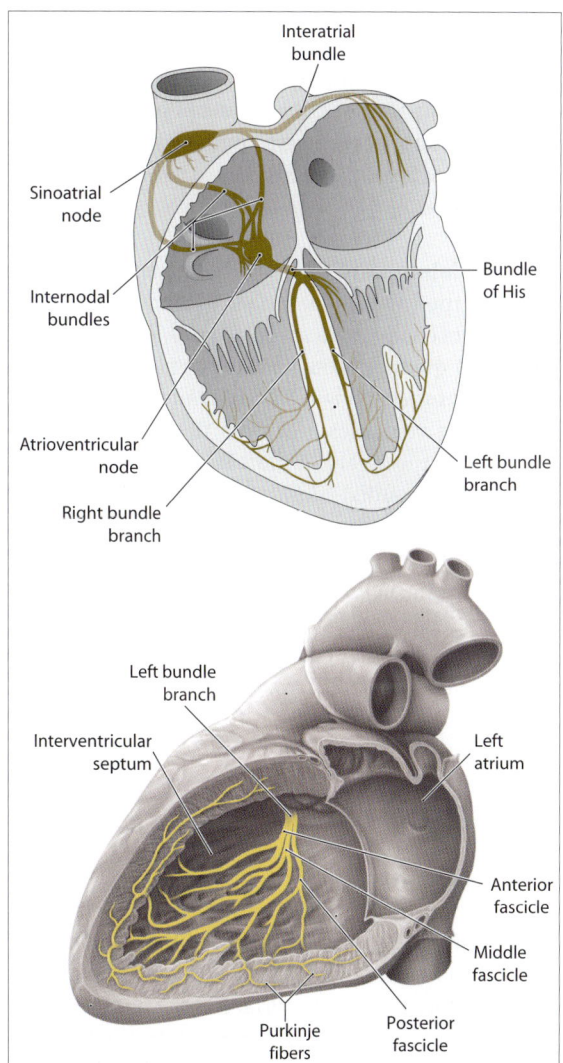

Fig. 34.**1** (*Above*) The conductive system of the heart. (*Below*) The fascicles of the left bundle branch.

Conductive system of the heart

The depolarization of the various parts of the heart proceeds in an orderly and timely way. This is made possible by the presence of the conductive system of the heart (Fig. 34.**1**) which comprises the sinoatrial node (SA node), internodal atrial pathways, atrioventricular node (AV node), bundle of His and its branches, and the Purkinje system. Most cells of the conductive system are specialized for fast conduction though some, like the cells of the AV node, are specialized for slowing down conduction (Table 34.**1**). Cells of the conductive system also possess automaticity, i.e., the capability of spontaneous excitation. The automaticity of the SA node sets the pace of the heart while the automaticity of the other parts of the conductive system normally remains suppressed. The cells of the conductive system are also contractile, though somewhat weakly.

Intrinsic automaticity is present in all cardiac fibers, including the atrial muscle, the AV node, the bundle of His and its branches, the peripheral Purkinje system, and the ventricular muscle. The SA node has the highest automaticity ($72 \ \text{min}^{-1}$) followed by the AV node ($40 \ \text{min}^{-1}$). Other cardiac fibers have even lower intrinsic discharge rates. The automaticity of the SA node, being the highest, paces the excitatory drive to all other cardiac fibers, suppressing their intrinsic automaticity. However if the SA node stops discharging, the AV node, which has the next highest automaticity, takes over the role of cardiac pacemaker (Fig 34.**2**).

The **sinoatrial node** is a small strip of conductive tissue located in the superolateral wall of the right atrium,

Table 34.**1** Conduction velocity of different conductive tissues.

Cardiac tissue	Conduction velocity (m s^{-1})
SA node	0.05
Interatrial pathways	1
Atrial muscle	0.30
AV node	0.05
Bundle of His	1
Purkinje system	4
Ventricular muscle	1

Fig. 34.**2**   The pacemaker! (*Above*) The car, scooter, cycle and roller-skater are racing past at 72 km h^{-1}. Though each have a different 'intrinsic' speed, the fastest vehicle decides their combined speed. (*Below*) With the car gone, the speed slows down to 40 km h^{-1}, which is the intrinsic speed of the scooter.

72 Km h^{-1}

40 Km h^{-1}

immediately below and lateral to the opening of the superior vena cava. The rate of SA node discharge often shows rhythmic changes with the phase of respiration, increasing with inspiration and decreasing with expiration. Such rhythmic variations in the heart rate are called *sinus arrhythmia*.

The **internodal atrial pathways** are the three strips of conductive tissues, the anterior, middle, and posterior, that connect the SA node to the AV node. Impulses from the SA node travel to the AV node along these interatrial pathways. The atrial musculature is excited by impulses spreading out from the interatrial pathways. The interatrial pathways conduct impulses at a speed of 1 m s^{-1}, which is three times faster than the conduction speed in the contractile atrial muscle fibers. Atrial depolarization is complete in about 0.1 s.

The **atrioventricular node** is located in the posterior and inferior part of the interatrial septum. The AV node is the gateway for all action potentials traveling from the atria to the ventricles. This is because the atrium is electrically insulated from the ventricles by a fibrous partition. This partition is bridged only at the AV node and nowhere else can atrial depolarization enter the ventricle.

The conduction speed of AV nodal tissue is only 0.05 m s^{-1} so that impulses take 0.1 s to travel across the AV node. This 0.1 s delay is called the *AV nodal delay* and is critically important for allowing the atrial systole to complete itself before the onset of ventricular systole. The AV nodal delay is shortened by stimulation of the sympathetic nerves to the heart and lengthened by stimulation of the vagi. The increase in AV nodal delay caused by vagal stimulation is put to use in the differential diagnosis of certain types of arrhythmias. Commonly used bedside methods for increasing vagal activity is carotid massage which initiates the baroreceptor reflex (p 269) and pressing on the eyeball which initiates the oculocardiac reflex.

The **bundle of His** originates from the AV node. It penetrates the fibrous barrier separating the atria and ventricle and continues downward in the interventricular septum for about 1 cm. Thereafter, it divides into the left and right bundles that run on the respective side of the septum under the endocardium. The left bundle divides further into the anterior and posterior fascicles. After reaching the apex, the bundle branches turn back toward the base of the heart, coursing through the endocardium where it spreads the excitation to the Purkinje system.

The **Purkinje system** is a network of fast conductive fibers on the endocardial surface of the ventricles. Purkinje fibers conduct at a speed of 4 m s^{-1} and spread the excitation quickly throughout the ventricles, which is important for enabling near-simultaneous contraction of both the ventricles. However, the Purkinje system excites only those cardiac muscle cells that are located near the endocardial surface. Thereafter, the impulses are conducted by the ventricular muscle fibers themselves, and the depolarization spreads outward radially toward the epicardium. Ventricular depolarization is completed in about 0.1 s. The spatial sequence of the spread of ventricular depolarization is described later in the context of the ECG waves.

Electrocardiography

Cardiac vectors

Electrically, the body fluids behave like a volume conductor. The electrocardiogram (ECG) records the potential differences on the surface of the body resulting from the currents set up in body fluids by the electrical activity of the heart. The currents are set up by the array of dipoles formed at the border between the depolarized and repolarized cardiac tissue. The direction of the current is from the depolarized (surface positive) to the repolarized (surface negative) areas of the heart (Fig. 34.**3**). The dipoles are formed when only a part of the heart is depolarized. They disappear when the heart is either completely depolarized or completely polarized. The physics of dipoles is discussed on p 18. The ECG gives an estimate of the magnitude and direction of the body currents set up by cardiac dipoles. However, what it actually measures is the potential drop along the path of the current. The potential drop recorded is taken as a measure of the current, assuming a uniform resistance of body fluids.

Instantaneous electrical axis of the heart

At any moment during the cardiac cycle, the dipoles on the cardiac surface set up currents in all directions. The resultant of these currents is called the *cardiac vector* (or electrical axis). The magnitude and direction of the cardiac vector keep changing through out the cardiac cycle. Both depend on the contour of the border between the depolarized and polarized muscles of the heart. The direction of the resultant vector is perpendicular to the border. The magnitude of the resultant vector depends on the length of the border: the longer the border, the greater is the magnitude of the resultant vector. For example, if the left side of the heart were depolarized and the right side were still polarized, the border between the depolarized and polarized area would be long and a large vector would be directed toward the right. However, if all but the apex of the heart were depolarized, the border would be short and a small vector would be directed toward the apex (Fig. 34.**3**). Thus as the wave of depolarization sweeps across the heart, the instantaneous electrical axis keeps swinging in all directions and its magnitude keeps changing.

The graphic record of the swinging instantaneous electrical axis during the cardiac cycle is called *vectorcardiography* (Fig. 34.**10**). It has never been popular with physicians who are more comfortable with the conventional ECG record. However, it retains its utility in specific situations.

Einthoven's triangle

Einthoven's triangle is an imaginary triangle formed by the two shoulders and the pubis. The electrodes connected to the three corners of the Einthoven triangle are designated as LA (left arm), RA (right arm), and LL (left leg). An electrode connected to all the three corners of Einthoven's triangle will always be at zero

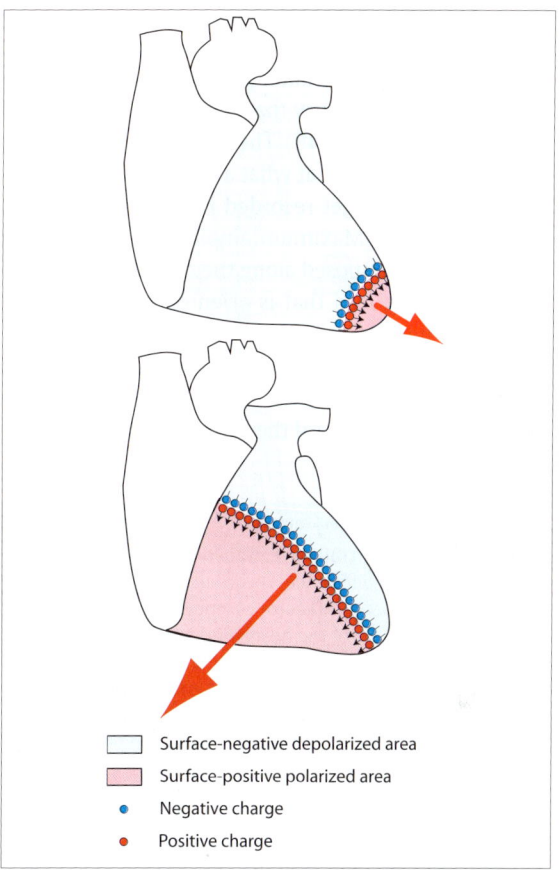

Fig. 34.**3**   The direction of the cardiac vector is perpendicular to the border between the depolarized and repolarized areas. The longer the border, the greater the number of dipoles and therefore the greater is the magnitude of the cardiac vector.

potential regardless of the direction of the cardiac vector. Such an electrode is called the *indifferent electrode* (or the *central terminal of Wilson*) and it serves as a convenient reference for several electrocardiographic measurements. The three limb electrodes needs to be interconnected through high (5000 Ohm) resistors (Fig. 34.**4**A).

A pair of electrodes recording the potential difference between two different sites is called a lead. Leads are classified as unipolar and bipolar, a nomenclature that is fallacious since both types of lead record the potential difference between two electrodes. In the strictest sense, therefore, all recordings are bipolar. However, according to the conventional usage of the terms, bipolar recording involves recording the potential difference between two different sites on the body using two electrodes, one serving as the reference and the other called the exploring or active electrode. In unipolar recording, the indifferent electrode serves as the reference.

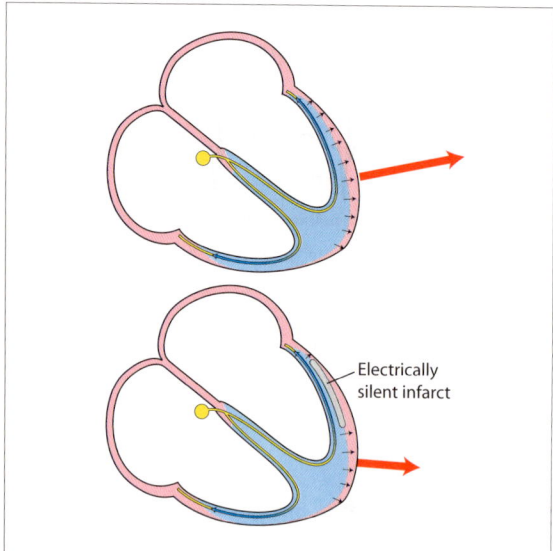

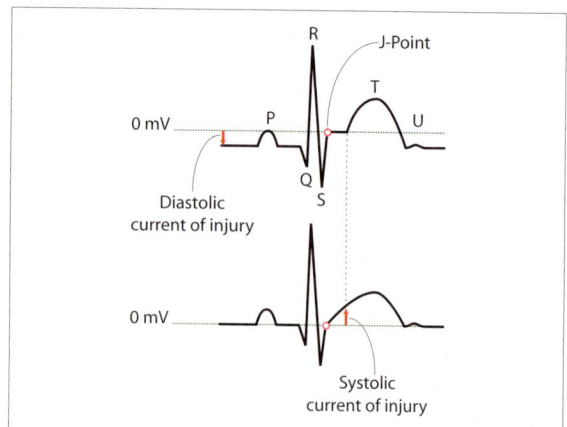

Fig. 34.**15**   The baseline of the ECG is indicated by the J point. A diastolic current of injury affects the potential of the TP segment while a systolic current of injury affects the potential of the ST segment.

Fig. 34.**14**   Change in cardiac vector due to an electrically silent infarcted area. Note the change in both direction and magnitude of the vector.

the surface of the infarcted area positive relative to the surrounding normal area. As a result, the current of injury flows out of the infarcted area into the surrounding areas. This current of injury appears when the cardiac muscle is depolarizing or repolarizing (i.e., during electrical systole) and is therefore called the *systolic current of injury*. It produces an elevation of the ST segment in the ECG recorded from the chest leads overlying the infarcted area. The systolic current of injury disappears when the myocardium is completely repolarized.

Electrical silence   After some days or weeks, the infarcted area becomes electrically silent, i.e., it has no electrical activity whatsoever. The current of injury stops and the ST segment abnormalities subside. However, the presence of an area of electrical silence alters both the direction and magnitude of the instantaneous electrical vectors during the cardiac cycle (Fig. 34.**14**) and results in changes in the Q, R, and T waves. There is increase in the height and width of preexisting Q waves, sometimes resulting in a large QS complex. Q waves may appear in leads in which they are normally absent. R wave amplitude decreases. T waves become very tall within the first few hours, probably as a result of the local rise in extracellular K+ concentration. Later, they gradually become inverted.

Conduction abnormalities   Atrioventricular block or bundle branch block occurs if the septum is infarcted, damaging the conduction system. Arrhythmias occur due to development of microreentrant circuits as well as increased automaticity. Increased automaticity occurs possibly because the infarct damages the autonomic fibers that course through the heart. Epicardial infarcts damage mostly the sympathetic fibers resulting in denervation hypersensitivity to circulating catecholamines. Endocardial infarcts damage mostly the parasympathetic fibers, leaving the sympathetic activity unopposed.

■   ECG isoelectric potential

J point   Normally, both ST segment (when the entire myocardium is depolarized) and the TP segment (when the entire myocardium is repolarized) should be isoelectric, i.e., at zero potential. However, the ST segment does not remain isoelectric when there is a systolic current of injury and the TP segment does not remain isoelectric when there is a diastolic current of injury. Since the systolic current of injury occurs due to delayed depolarization, it manifests itself only after the J point, i.e., after the rapid phase of repolarization. Hence the J point, which is the point where the S wave ends sharply, is a reference point for isoelectric potential in the ECG tracing (Fig. 34.**15**).

Abnormalities of cardiac excitation affect the rhythm of the heart, causing arrhythmias (disruption of the normal cardiac rhythm). Arrhythmias that are associated with bradycardia (decrease in heart rate) are called bradyarrhythmias (brady = slowing). Arrhythmias that are associated with tachycardia (increase in heart rate) are called tachyarrhythmias (tachy = speeding). The electrocardiogram is an essential tool for the precise diagnosis of arrhythmias.

Bradyarrhythmias

Sick-sinus syndrome

The sick-sinus syndrome is a dysfunction of the sinus node that results in marked bradycardia with symptoms like dizziness and repeated episodes of syncope. It can occur due to a reduction of SA node discharge frequency to 60 min^{-1} or less (sinus bradycardia) that does not improve with sympathetic stimulation or vagal inhibition. It can also occur due to a blockade of impulse conduction from the SA node to the atria (sinoatrial block) or a complete stoppage of sinus discharge (sinus arrest).

Atrioventricular block

Any block or slowing down of the conduction of cardiac impulses across the atrioventricular node is called atrioventricular (AV) block or simply heart block (Fig. 35.**1**). There are three degrees of atrioventricular block.

In **first degree A-V block**, there is a slowing of impulse conduction from the atria to the ventricles. It results in the prolongation of the PR interval, which could be due to slow conduction of the impulse anywhere between the SA node and the AV node, in the AV node, or in the bundle of His.

In **second degree A-V block**, there is atrioventricular conduction failure at regular intervals, i.e., some of the impulses fail to reach the ventricle resulting in dropped ventricular beats. Dropped beats at regular intervals result in a 'regularly irregular' pulse.

There are two types of second degree block. In *Mobitz type-I (Wenckebach) block*, there is a progressive delay in the PR interval with eventual failure of AV conduction. Thus every fourth or fifth P wave may be followed by a missing QRS complex, resulting in a 4:3 or a 5:4 ratio of P and QRS waves. In occurs due to block within the AV node. In *Mobitz type-II block*, the PR interval is prolonged but does not show a progressive increase. As with type-I block, the dropped beats occur at regular intervals. This type of block occurs in the bundle of His or its distal branches and is therefore also associated with ECG signs of bundle branch block.

In **third degree A-V block**, the atrioventricular conduction failure is complete, i.e., none of the impulses that originate in the SA node reach the ventricle: Ventricular depolarization is initiated by the AV node and the ventricles beat at a rate determined by the intrinsic AV nodal frequency of 40 s^{-1}. This is called the *idioventricular rhythm*. The atria however continue to beat at the normal sinus rhythm of 72 min^{-1}.

Thus, the atrial and ventricular rhythms become entirely independent, each contracting at its own rate. This is called *atrioventricular dissociation*. The ECG shows the rate and rhythm of the P waves and the QRS complexes to be completely independent of each other.

When a second degree heart block progresses to a third degree heart block, there is a brief period of asystole before the idioventricular rhythm starts. The asystole is associated with syncope. Intermittent third degree block with syncopal attacks is known as *Stokes–Adams syndrome*.

Bundle–branch block

There may be conduction blocks in one or more branches of the bundle of His. Accordingly, there may be different types of blocks whose names are self-explanatory and include the right bundle–branch block (RBBB), left bundle–branch block (LBBB), left anterior hemiblock (LAH), left posterior hemiblock (LPH), bifascicular block (RBBB with LAH or LPH), and the trifascicular block. Conduction blockade of any of the branches results in delayed depolarization of the part of the ventricle that is supplied by the branch. Irrespective of

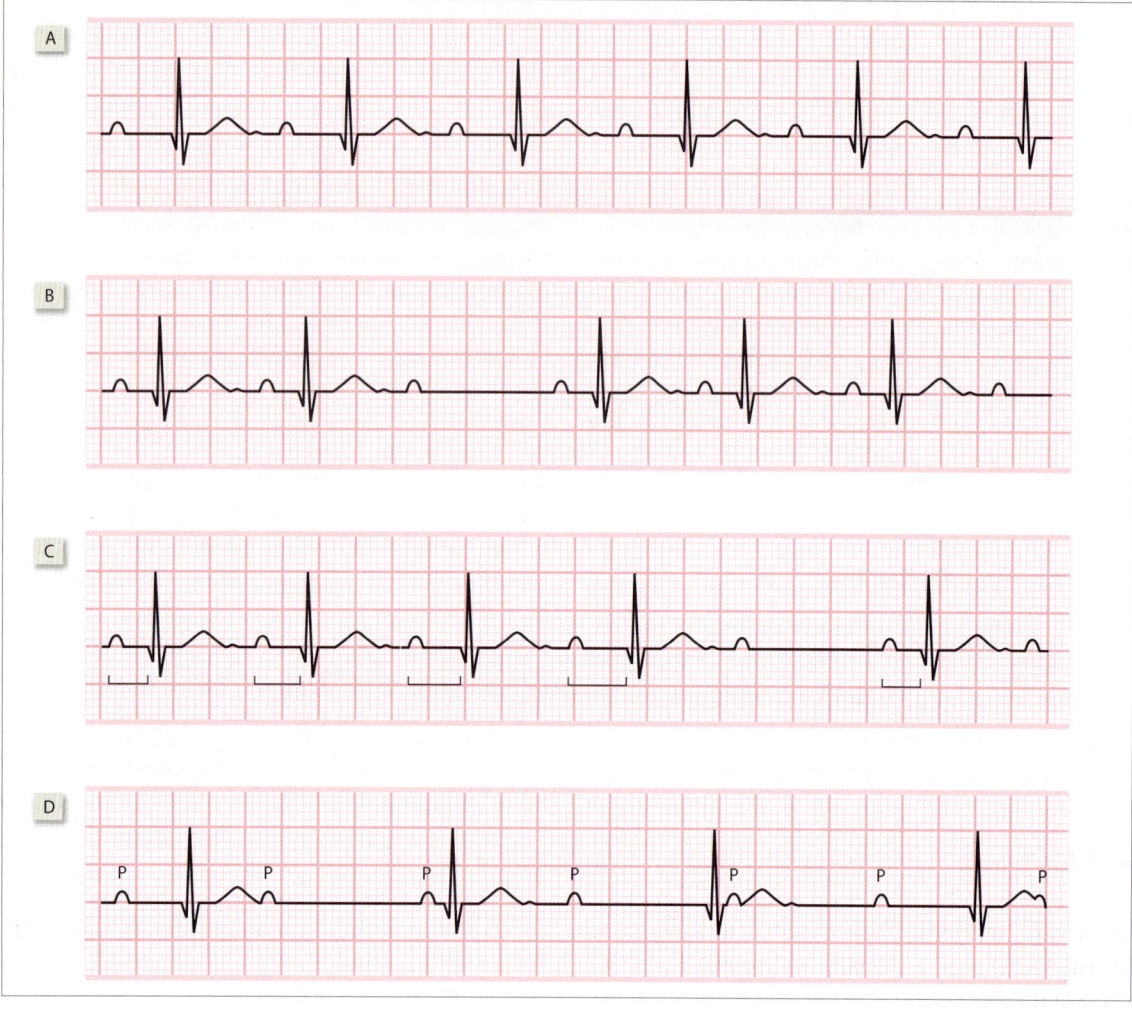

Fig. 35.1.   Heart block. [A] First degree heart block. [B] Second degree (type-1) heart block. [C] Second degree (type-II) heart block showing Wenckebach phenomenon. [D] Third degree atrioventricular block.

the type of block, the total QRS duration is prolonged beyond the normal limit of 0.1 s.

In **left bundle branch block**, ventricular depolarization starts on the right surface of the interventricular septum. Hence the electrical axis is initially deflected to the left, as a result of which the Q wave is prominent in lead V1 and the QRS complex begins with an R wave in lead V6 (Fig. 35.**2**). Recall that in the normal sequence of excitation, ventricular depolarization begins on the left surface of the septum and the axis is initially directed to the right (see Fig. 34.**9**). Hence the Q wave is present in V6 instead of V1.

In **right bundle branch block**, the depolarization of the right ventricle remains incomplete even at the end of 0.1 s, which is the normal time required by the ventricles to depolarize completely. Hence toward the end of ventricular depolarization, the electrical axis remains deviated to the right and front, resulting in a second R wave (denoted by R') in lead V1 and a prominent S wave in lead V6 (Fig. 35.**2**). Recall that when the spread of excitation is normal, the electrical axis deviates to the left toward the end of ventricular depolarization (see Fig. 34.**9**), as a result of which lead V1 shows a prominent S wave while in lead V6, the ventricular complex ends with a prominent R wave.

Tachyarrhythmias

The common types of tachyarrhythmias are premature supraventricular contractions, supraventricular tachycardia, atrial flutter, atrial fibrillation, premature

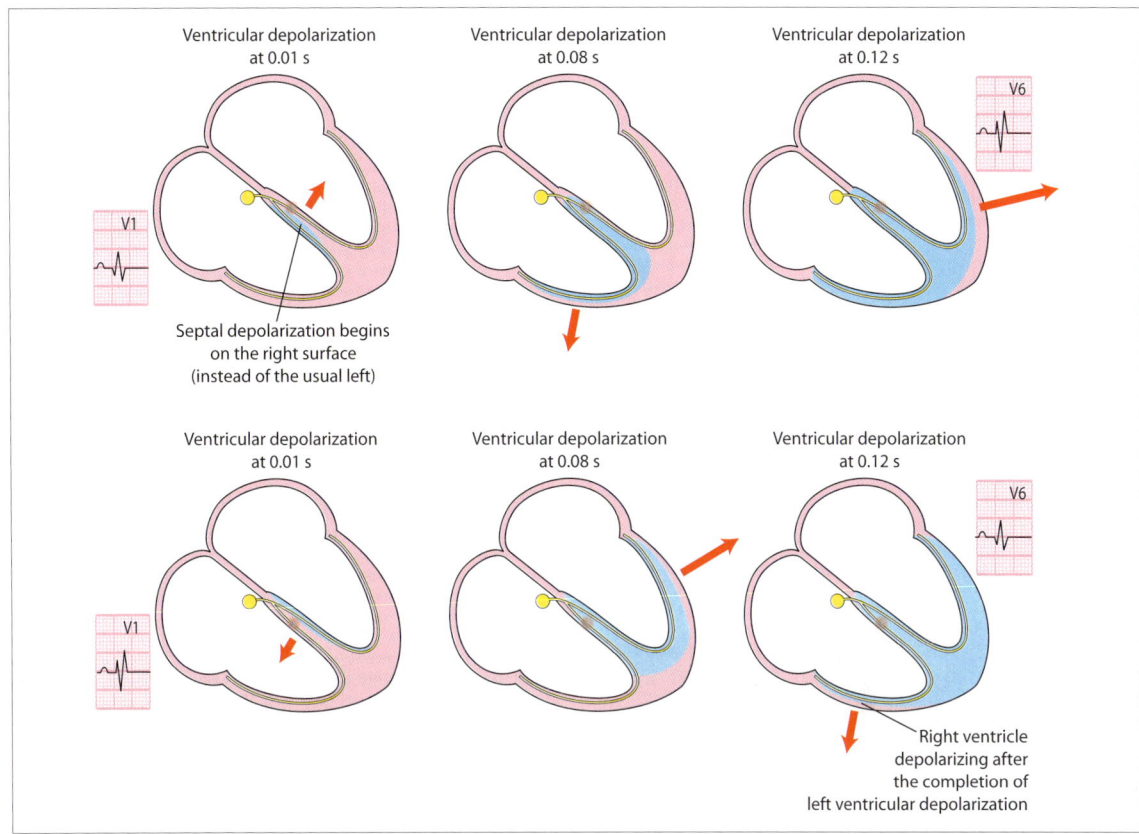

Ventricular depolarization at 0.01 s

Septal depolarization begins on the right surface (instead of the usual left)

Ventricular depolarization at 0.08 s

Ventricular depolarization at 0.12 s

V1

V6

Ventricular depolarization at 0.01 s

Ventricular depolarization at 0.08 s

Ventricular depolarization at 0.12 s

V1

V6

Right ventricle depolarizing after the completion of left ventricular depolarization

Fig. 35.**2**   (*Above*) Left bundle branch block. (*Below*) Right bundle branch block.

ventricular contractions, ventricular tachycardia, ventricular flutter, and ventricular fibrillation. The term 'supraventricular' is used when the site of origin of the arrhythmia could be either the atria or the atrioventricular junction. Tachyarrhythmias can originate in the SA node, in which case it is called *sinus tachycardia*. Most tachyarrhythmias, however, originate in ectopic foci, i.e., sites other than the SA node. They occur either due to increased automaticity or due to re-entry of impulses in closed circuits within the heart. Increased automaticity usually produces premature atrial and ventricular contractions but it can also lead to flutter and fibrillation when ectopics occur simultaneously at multiple sites. Similarly reentrant circuits usually result in paroxysmal tachycardia, flutter, and fibrillation but microreentrant circuits can also produce single, premature contractions.

Increased automaticity is present in all automatic cardiac fibers but usually remains suppressed due to the continuous activity of the SA node. Sometimes, the intrinsic automaticity is abnormally increased and gives rise to ectopic foci of discharge. If the ectopic focus discharges at a critical moment when the cardiac tissue is not refractory, it gets conducted to large areas of the

heart, resulting in depolarization and contraction of the atria or ventricles.

Reentrant circuits are closed-circuit pathways in the heart along which impulses are repeatedly conducted without cessation. Reentrant circuits are broadly of two types: the microreentrant and macroreentrant circuits. The *microreentrant circuits* are confined to a very small area in the atria or ventricles and occur due to local inhomogeneity in cardiac excitability and conductivity. The *macroreentrant circuits* pass through a large part of the heart and involve abnormal anatomical tracts.

A microreentrant circuit, for example, can occur around a postinfarction scar in the atria or the ventricles. Its mechanism is explained in Fig. 35.**3**A. The temporary block referred to in the diagram could be an ectopic focus that is still refractory after having discharged shortly before. More commonly, the microreentrant circuit resides inside the AV node. The intranodal reentrant circuit comprises two conducting pathways inside the node, the α pathway having faster conduction velocity but longer refractory period and the β pathway having slower conduction velocity but shorter refractory period. This results in reentry as shown in Figure 35.**3**B.

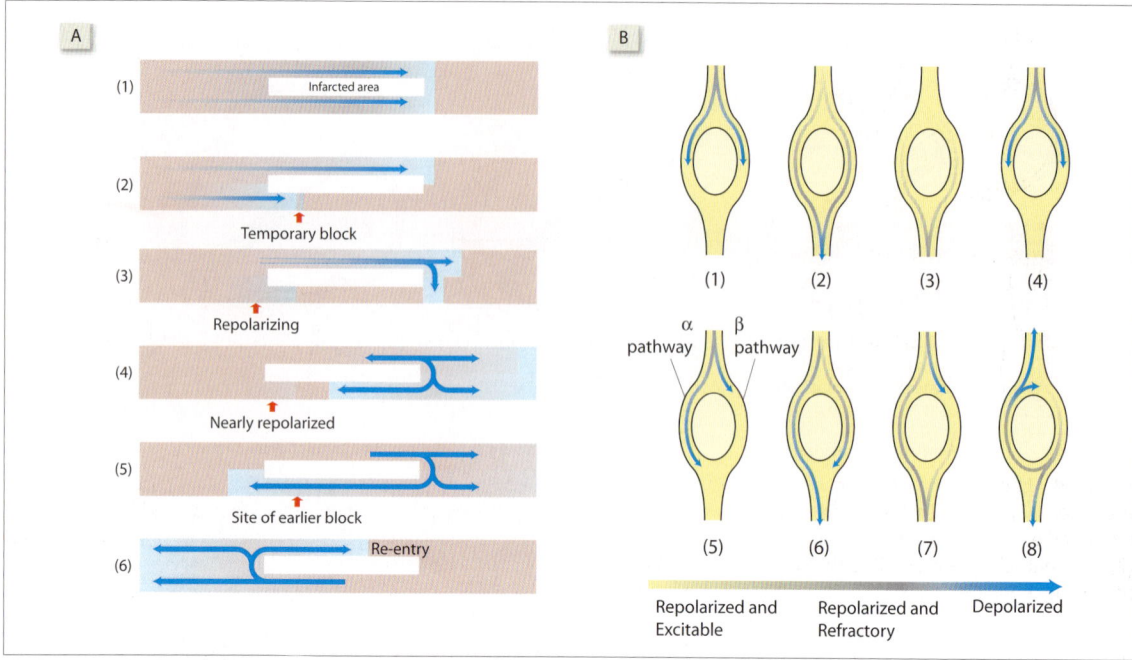

Fig. 35.**3** [A] A microreentrant circuit around an infarct. (1) Impulse travels from left to right. The leading end is depolarized. The trailing end is repolarizing. No reentry occurs. (2) In the lower track, the impulse meets with a temporary block. (3) Impulse in the upper track spreads to the lower track, which has remained polarized due to the block. (4) Impulse in the lower track spreads both forward and backward. Also, the site of the block is now nearly repolarized. (5) The recurrent impulse travels backward in the lower tract to enter the site of the block, which by now has repolarized fully. (6) The recurrent impulse in the lower track reenters the upper track, which by now has repolarized completely. [B] (1–4) Normal propagation of excitation in the AV node. (5–8) A reentrant circuit in the AV node. The excitation travels along two conducting pathways inside the node, the α pathway having faster conduction velocity but longer refractory period and the β pathway having slower conduction velocity but shorter refractory period. The β pathway recovers earlier from its refractory period. An ectopic impulse entering the node at this point will enter the β pathway. The α pathway recovers from its refractory period, allowing a recurrent entry to the impulse in the β pathway.

The best known macroreentrant circuit occurs in *Wolff–Parkinson–White* (*WPW*) *syndrome* that is associated with the *aberrant bundle of Kent*, a congenitally present abnormal band of conductive tissue interconnecting the atria and ventricles, bypassing the AV node. There is no AV nodal delay when the ventricles are excited through this aberrant bundle and therefore the ventricles are excited earlier than usual (i.e., they are preexcited). Such preexcitation shortens the PR interval and results in the appearance of a delta wave in the QRS complex (Fig. 35.**4**).

The aberrant bundle can set up a macroreentrant circuit in two different ways, both resulting in paroxysmal supraventricular tachycardia (see below). One circuit involves passage of impulses from the atria to the ventricles through the bundle of Kent and reentry into the atria through the bundle of His. When this type of circuit is active, the ECG shows delta waves. In the other circuit, there is normal atrioventricular conduction through the AV node and retrograde conduction through the bundle of Kent. When this type of circuit is active, the ECG does not show delta waves or any other evidence of the presence of the aberrant bundle, other than tachycardia. It is therefore called a *concealed bypass*.

Antiarrhythmic drugs that prevent reentry do so in two ways. Some drugs increase the effective refractory period (ERP) of the cardiac muscle so that the site of the temporary block remains refractory for a longer period and does not allow a reentry of the recurrent impulse. Other drugs increase the conduction velocity of cardiac muscle so that the recurrent impulse arrives at the site of temporary block before it has repolarized and thus extinguishes itself.

■ Supraventricular tachyarrhythmias

A **premature supraventricular contraction** is a contraction out of rhythm with the normal ongoing heartbeat. It occurs due to the discharge of an ectopic focus in the atrium (atrial premature beat) or in the AV node (junctional premature beat). The ectopic impulse not only spreads to the ventricle but also travels back to

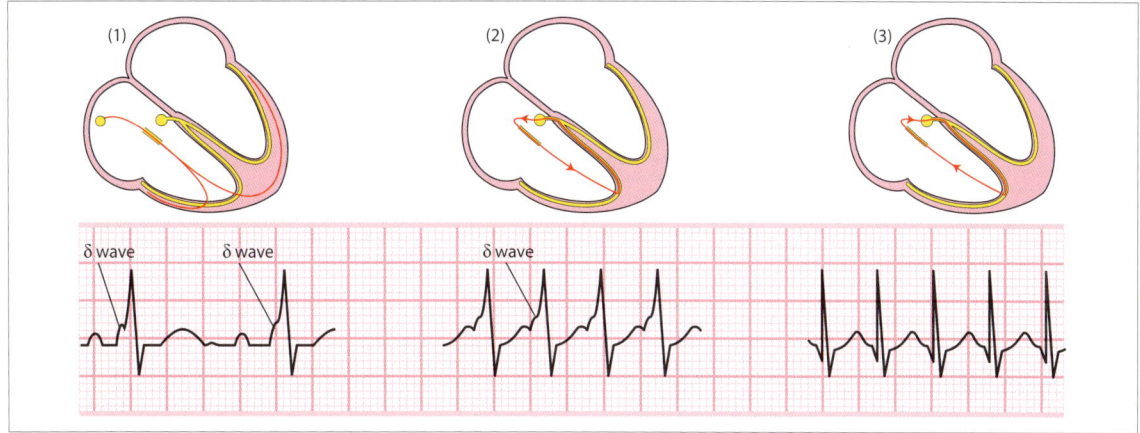

Fig. 35.**4**   Effect of the aberrant bundle of Kent on cardiac excitation. (1) The ventricles are preexcited through the aberrant bundle resulting in a δ wave. No reentrant circuit is set up. (2) The ventricles are preexcited through the aberrant bundle, producing a δ-wave in the ECG. The impulse returns through the AV node, resulting in a reentrant circuit. (3) The ventricles are normally excited through the AV node and therefore there is no δ wave in the ECG. The impulse returns to the atria through the aberrant bundle, setting up a reentrant circuit.

depolarize the SA node and in the process, resets the SA node. After the atrial premature beat, the SA node discharges only after the normal R-R interval has elapsed. Hence, there is no compensatory pause (Fig. 35.**5**A) as in the case of a ventricular extrasystole (see below). Since the atrial activation occurs from below upward, the P wave is inverted in those leads where it is normally upright (e.g., lead II) and upright in those where it is normally inverted (lead aVR).

Supraventricular tachycardia mostly occurs in paroxysms (paroxysm = a sudden outburst) and is therefore called *paroxysmal supraventricular tachycardia*. It often occurs in healthy subjects with abrupt onset and termination, appearing as episodes lasting minutes or hours and separated by periods of days, weeks, months, or years during which no episodes occur. When the tachycardia originates in the AV node, both the atria and the ventricles beat at the same rate of about 120–200 per minute (Fig. 35.**5**B). However, when the tachycardia originates in the atria, the ventricular rate is slower because many of the atrial impulses do not reach the ventricles due to AV delay.

Atrial flutter is similar to paroxysmal atrial tachycardia except that the rate of atrial contractions is much higher at 220–350 beats per minute. It occurs mostly due to a macroreentrant circuit around the tricuspid valve or the vena caval opening. The high rate of atrial contraction is inevitably associated with AV block and only a fraction of all atrial impulses reach the ventricles, resulting in a lower ventricular rate which is usually regular (Fig. 35.**5**C). Atrial flutter causes a reduction in cardiac output because at such high rates, the strength

of atrial contractions reduces, rendering ineffective the atrial pumping that normally contributes about 30% to the cardiac output.

Atrial fibrillation results from random excitation of various parts of the atria. In the absence of a coordinated sequence of excitation, there is no effective contraction of any of the cardiac chambers and the atrium 'quivers' instead of contracting. Fibrillation occurs due to numerous microreentrant circuits. There are several depolarized patches all over the heart and the impulse wanders along the repolarized areas in between the depolarized patches. When the wandering impulse reaches the AV node, it gets conducted to the ventricle. Hence, the ventricular rate is extremely erratic and is said to be 'irregularly irregular' (Fig. 35.**5**D). The ventricular rate usually does not exceed 150 beats per minute.

Ventricular tachyarrhythmias

A **premature ventricular contraction** (ventricular extrasystole) is one that is not preceded by a regular atrial contraction. The ventricular activation does not antidromically activate the atria or depolarize the SA node and therefore the regular SA node rhythm is not disturbed. The regular sinus impulse following the premature beat does not activate the ventricle, which is still refractory from the premature contraction. Rather, the ventricle responds to the next normal sinus impulse. Hence, there is a slight pause, the *compensatory pause*, following the ventricular extrasystole. The interval between the QRS complexes immediately preceding and

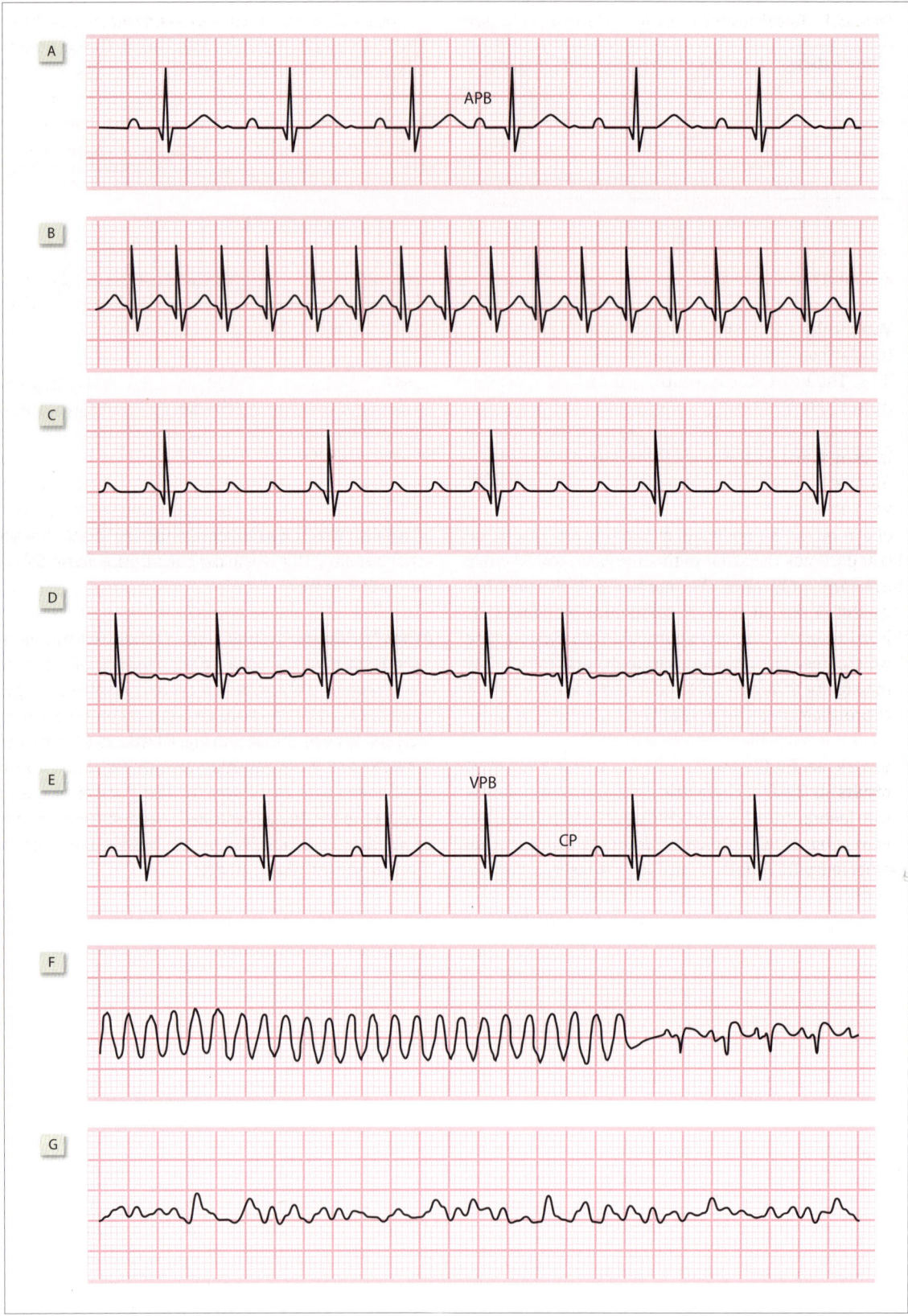

Fig. 35.**5** [A]) Atrial premature beats (APB). [B]) Paroxysmal supraventricular tachycardia. [C] Atrial flutter. [D] Atrial fibrillation. Note the absence of definite P waves and the irregularly irregular pulse. [E] Ventricular premature beat (VPB) and compensatory pause (CP). [F] Ventricular flutter, progressing to fibrillation. [G] Ventricular fibrillation.

Table 35.**1** Three letter codes for different types of pacemakers.

Chamber(s) paced	Chamber(s) sensed	Response
V Ventricle	V Ventricle	T Triggered
A Atrium	A Atrium	I Inhibited
D Dual	D Dual	D Dual
	O None	O None

following the extrasystole is exactly twice the normal R-R interval (Fig. 35.**5**E).

Ventricular tachycardia occurs when premature ventricular contractions occur repetitively for more than 30 s. The heart rate in ventricular tachycardia ranges from 130 to 200 beats per minute.

In **ventricular flutter**, the rate is 200 to 250 beats per minute. It is characterized by wavy QRS complexes with no isoelectric line between them, and the absence of a T wave. The upstroke and downstroke of the QRS complex look the same so that the ECG looks the same even when turned upside down (Fig. 35.**5**F). Ventricular flutter usually leads to ventricular fibrillation and is self-limiting only in exceptional cases. It is a medical emergency that has to be treated immediately. Its management involves DC cardioversion followed by appropriate antiarrhythmic drugs.

Ventricular fibrillation is similar to atrial fibrillation and results in ineffective ventricular contraction and cardiac failure. The ECG is nearly flat with small, *saw-tooth waves* (Fig. 35.**5**G). Like ventricular flutter, it is a medical emergency that needs immediate cardioversion.

DC cardioversion involves passing electricity through the heart by placing electrodes on the thorax (external defibrillation) or directly on the heart (open-chest defibrillation). Defibrillation results in the simultaneous depolarization of the entire heart. Following repolarization, the normal sinus rhythm tends to resume.

Pacemakers

A pacemaker or 'artificial pacemaker', not to be confused with the heart's natural pacemaker, is a device for stimulating the heart when the heart's own pacemaker is not functioning properly. Originally, they were designed to pace the heart in bradyarrhythmias. More sophisticated pacemakers are now used even in tachyarrhythmias.

An artificial pacemaker delivers precisely-timed, preset amounts of electrical energy to the heart, stimulating it to contract. To be able to deliver the stimulus only when required, a pacemaker is enabled with a mechanism to sense the cardiac action potentials.

Pacemakers are classified on the basis of the chamber they stimulate (atrium, ventricle or both), the chamber they sense (atrium, ventricle, both or none), and the response of the pacemaker (triggered, inhibited, both or none). These features have been assigned alphabetic codes that are grouped together to indicate the features of the pacemaker (Table 35.**1**). For example, the simplest pacemaker would be VOO, i.e., the one that stimulates the ventricle, does not sense the cardiac action potential, and delivers repeated stimuli at a fixed rate, i.e., its discharge is neither triggered, not inhibited by cardiac activity.

The heart has two large chambers called the ventricles and two small chambers called the atria. The ventricles pump blood into the aorta and pulmonary artery while the atria are booster pumps that pump blood into the ventricles. The left and right ventricles are separated by the interventricular septum. The left and right atria are separated by the interatrial septum. The pathway of blood through the heart is described on p 243.

The passage of blood from the atrium to the ventricle is guarded by the atrioventricular (AV) valves so that blood cannot flow backward. The left AV valve is called the mitral or bicuspid valve and the right AV valve is called the tricuspid valve. The free margins of the valve cusps are attached to the papillary muscles by tendinous cords called the chordae tendinae (Fig. 36.**1**). During ventricular contraction (systole), the rise in ventricular pressure tends to push open the AV valves back into the atria. This is prevented by the contraction of the papillary muscles at the outset of systole, which restricts the movement of the cusps of the AV valves.

The passage of blood from the left ventricle into the aorta is guarded by the aortic valve. The passage of blood from the right ventricle into the pulmonary artery is guarded by the pulmonary valve. The aortic and pulmonary valves are called the semilunar valves. During ventricular relaxation (diastole), when blood tries to flow back into the ventricle, the cusps of the semilunar valves get distended with blood and thereby occlude the enclosed passage.

■ Phases of the Cardiac cycle

A cardiac cycle refers to the interval between the onset of one heartbeat to the onset of the next heartbeat. It has two main phases: ventricular systole and ventricular diastole, which are often simply called systole and diastole. Although electrical systole begins with the beginning of the Q wave, the onset of mechanical systole begins with the first heart sound ('lub') and roughly coincides with the peak of the R wave. Similarly, although electrical systole ends at the end of the T wave, mechanical diastole begins a little later, with the second heart sound ('dup').

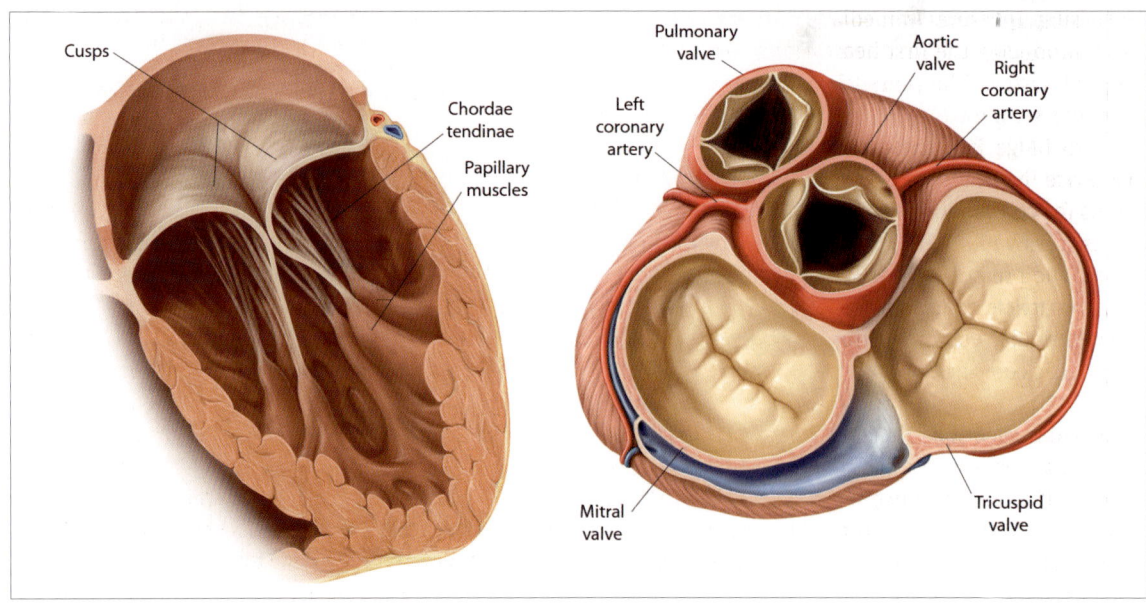

Fig. 36.**1** Heart valves. (*Left*) The papillary muscles with their chordae tendinae anchored to the cusps of the atrioventricular valves. (*Right*) The open semilunar valves and the closed atrioventricular valves.

Table 36.**1**   Duration of cardiac cycle phases.

Systole 0.3 s	
Isovolumetric contraction	0.05 s
Rapid ventricular ejection	0.20 s
Protodiastole	0.05 s
Diastole 0.5 s	
Isovolumetric relaxation	0.05 s
Rapid ventricular filling	0.05 s
Diastasis	0.30 s
Atrial systole	0.10 s

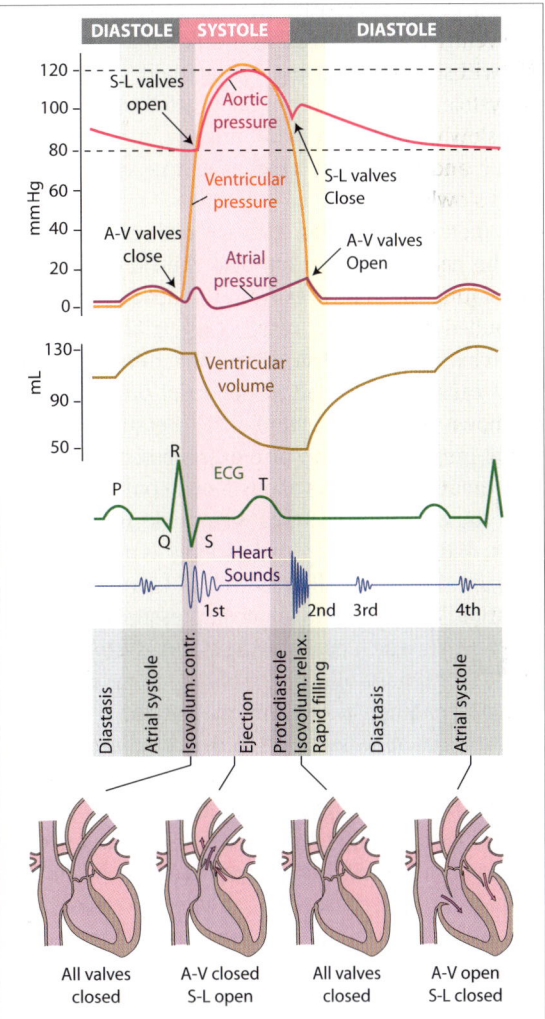

Fig. 36.**2**   (*Above*) Events during the cardiac cycle, showing pressure changes in the left ventricle, left atrium, and the aorta, volume changes in the ventricles, electrocardiogram, and phonocardiogram. (*Below*) The valvular positions in different phases.

Systole and diastole are further subdivided and named after the most prominent event of that phase. The *systolic phase* is divided into: (1) isovolumetric ventricular contraction, (2) rapid ventricular ejection, and (3) protodiastole. The *diastolic phase* is divided into (1) isovolumetric ventricular relaxation, (2) rapid ventricular filling, (3) diastasis, and (4) atrial systole (Table 36.**1**).

Ventricular and Atrial Events

Systole

Isovolumetric ventricular contraction   Before the onset of systole, the intraventricular pressure remains a little above zero mmHg (Fig. 36.**2**). As the ventricle starts contracting, the intraventricular pressure rises. The rising pressure immediately closes the mitral valve, producing the first heart sound that marks the onset of systole. The pressure is not enough to push open the semilunar valves but causes the closed AV valve to bulge into the atrium, causing a small but sharp rise in atrial pressure called the *c-wave*. (The 'c' stands for contraction of the ventricle.) With both the valves (atrioventricular and semilunar) momentarily closed, the pressure inside the ventricle rises steeply as the ventricle contracts. Since there is no change in the ventricular volume in this phase, it is called isovolumetric contraction.

Ventricular ejection   When the steeply rising left ventricular pressure exceeds the aortic pressure (120 mmHg), it is able to push open the aortic valve and eject the blood into the aorta, marking the onset of ventricular ejection. Similarly, the right ventricular pressure exceeds the pulmonary arterial pressure (25 mmHg) and ejects blood into the pulmonary artery after pushing open the pulmonary valve. During

ventricular ejection, the volume of each ventricle reduces from 130 ml (the end-diastolic volume) to 50 ml (the end-systolic volume), pumping out 80 ml of blood (the stroke volume).

Ventricular ejection is rapid at first, slowing down as systole progresses. Concomitantly, the intraventricular pressure rises to a maximum and then declines toward the end of ventricular systole. During rapid ventricular ejection, the ventricular pressure rises because the rate at which the ventricle contracts is greater than the rate at which the blood is ejected. Toward the end of ventricular ejection, however, there is no further contraction of the ventricle but the blood continues to be ejected at a rapid rate due to its momentum set up during rapid ejection. Hence the ventricular pressure drops. The phase toward the end of systole in which the ventricular pressure falls is called the *protodiastole*. Notwithstanding its name, it is actually the last phase of systole. Through most of the protodiastole, the atria get slowly filled with blood flowing in from the systemic and pulmonary veins, and the atrial pressure rises slowly.

When the ventricles contract, not only is the apex of the heart pulled up but also the fibrous partition separating the ventricles from the atria (the AV ring) is pulled down. As a result, the atrial muscles get stretched and the atria dilate. The dilatation of the atria causes a sharp fall in the atrial pressure, which is known as the *x-descent* of atrial pressure. The fall in atrial pressure probably also exerts a suction force on the venous return. As venous blood continues to flow into the atria from the great veins, the atrial pressure rises, and continues to rise as long as the AV valves remain closed, i.e., till the end of isovolumetric relaxation. This results in an atrial pressure wave called the *v-wave*, which peaks at the end of systole. (The 'v' stands for venous filling.)

Another important event that occurs during this phase is the *apex beat*. Due to the spiral arrangement of the ventricular muscle fibers, the heart rotates to the right during ventricular contraction and its apex strikes the anterior wall of the chest. The site of the impact, which is known as the apex beat, can be detected by palpating the chest wall. The apex beat is located at or medial to the left midclavicular line in the fourth or fifth intercostal space. A shift in the location of apex beat is often of diagnostic significance.

▪ Diastole

Isovolumetric ventricular relaxation   When the ventricles begin to relax, the intraventricular pressure falls and the semilunar valves immediately close. With further relaxation, the AV valves open. However, there remains a small intervening period during which the semilunar valves have closed and the atrioventricular valves have not opened. With all valves closed, no change in ventricular volume is possible during this phase and the relaxation is isovolumetric. In the absence of any change in volume, the relaxation is associated with a steep drop in ventricular pressure.

Rapid ventricular filling   As the ventricular pressure continues to drop rapidly during isovolumetric ventricular relaxation, it reaches a point where it falls below the atrial pressure. At this point, the AV valves open and the accumulated blood in the atria rushes into the ventricles. Filling is rapid at first, and hence

the name of this phase. During rapid ventricular filling, there is a sharp fall in atrial pressure, which is known as the *y-descent*.

Diastasis   After the initial rapid ventricular filling, blood flows slowly and smoothly from the superior and inferior vena cava through the right atrium into the right ventricle without any turbulence anywhere along the path. Similarly, blood from the pulmonary veins flows into the left ventricle without any turbulence. This phase of nonturbulent ventricular filling is known as diastasis. The atrial pressure remains slightly greater than the ventricular pressure, since upstream pressure must always be greater. The volume of each ventricle increases to about 105 ml. The pressure in both ventricles increases slowly, the rate of increase being determined by the ventricular compliance.

During the **atrial systole**, the atria contract and pump blood into the ventricles. The onset of atrial systole coincides with the peak of the P wave of the ECG. Atrial systole is associated with a sharp rise in atrial pressure called the *a-wave*. The ventricular volume increases sharply to the end-diastolic volume of 130 ml. The increase in volume of each ventricle during atrial systole is about 25 ml which is approximately 30% of the stroke volume (80 ml). When atrial systole is ineffective, as in atrial fibrillation, the stroke volume decreases by 30% due to ineffective atrial contraction.

▪ Arterial and Venous Events

▪ Aortic pressure

During **ejection phase**, blood flows from the left ventricle into the aorta. Hence, the aortic pressure (downstream pressure) is slightly less than the left ventricular pressure (upstream pressure) during most of the rapid ejection phase. However, toward the end of the ejection phase (protodiastole), left ventricular pressure drops below the aortic pressure because blood flows out of the left ventricle at a faster rate than the rate at which it contracts. On the other hand, blood flows into the aorta faster than it can flow out into the peripheral arteries, leading to a temporary accumulation of blood in the aorta.

During **isovolumetric venticular relaxation**, there is a sharp drop in ventricular pressure. Hence, blood in the aorta tries to rush back into the ventricle, only to collide against the closed aortic valve. The collision causes a small but sharp rise in aortic pressure. This sharp pressure rise is recordable even from peripheral arter-

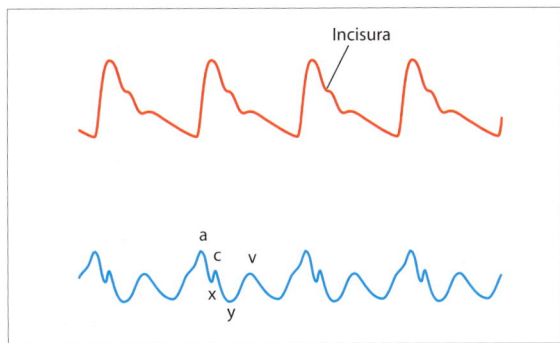

Fig. 36.**3**   (*Above*) The brachial artery pulse. (*Below*) The jugular venous pulse.

ies and is called the *notch* or *incisura*.

During **diastasis and atrial systole** (i.e., the remainder of the diastolic phase), the aortic pressure smoothly declines. If there had not been another systole, the pressure would have declined to about 10 mmHg (the mean systemic filling pressure) which is the pressure of stagnant blood inside the blood vessels. However, by the time the aortic pressure declines to about 80 mmHg, another ventricular systole boosts the aortic pressure again.

■ Jugular venous pressure (JVP)

The variations in right atrial pressure are transmitted to the jugular veins, producing the a-, c-, and v-waves of the venous pressure pulse (Fig. 36.**3**). These waves are low-pressure waves: They are visible but not palpable. The waveform of the jugular venous pulse, when recorded, offers a diagnostic sign. In atrioventricular dissociation, the JVP is characterized by the presence of *cannon waves*—giant c-waves that are produced due to the occasional simultaneous contraction of the right atrium and right ventricle.

■ Heart Sounds

Heart sounds are produced by the heart valves when they close. Normally no sounds are produced when valves open. However the opening of a stenosed AV valve is sometimes associated with an opening snap (Fig. 36.**4**).

The *first heart sound* (S_1) is a low, slightly prolonged 'lub' sound caused by vibrations set up by the sudden closure of the AV valves (mitral and tricuspid) at the onset of ventricular systole. It has a duration of about 0.15 s and a frequency of 25–45 Hz.

The *second heart sound* (S_2) is a shorter, louder high-pitched 'dup' sound caused by vibrations associated with closure of the semilunar valves just at the onset of ventricular diastole. It lasts about 0.12 s and has a frequency of 50 Hz.

The *third heart sound* (S_3) is a soft, low-pitched sound heard during the first-third of the diastole. It is produced due to the turbulence of blood flow during rapid ventricular filling. Generally, S_3 indicates a volume overload (increased preload) of the atria. It is often present in normal young individuals.

A *fourth heart sound* (S_4) is sometimes audible immediately before the first sound when atrial pressure is high or the ventricular compliance is low, as ventricular hypertrophy. It occurs late in the ventricular filling phase. Generally, S_4 indicates a pressure overload (increased afterload) of the atria.

Loud heart sound   Heart sounds become louder when the heart rate is high, the reason for which is as follows. Although the semilunar valves open fully at the onset of systole, they normally start floating back to their closed position toward the end of systole when the ejection of blood from the ventricle slows down. Hence, the sound produced by their closure at the onset of diastole is not very loud. During tachycardia, the

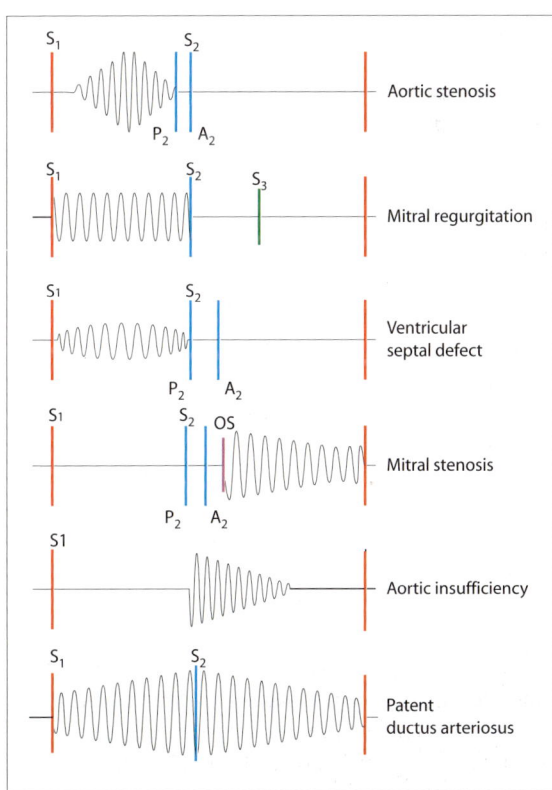

Fig. 36.**4**   Cardiac murmurs. OS=opening snap.

systole is shorter (Table 36.**2**) and the ejection of blood does not slow down much toward the end of systole. The valves therefore remain wide open right till the end of systole and at the onset of the diastole, they shut down from the fully open position to produce a loud S_2. For the same reason, the mitral and tricuspid valves produce a loud S_1 during exercise. Heart sounds are also loud when the diastolic pressure in the aorta or pulmonary artery is elevated (as in systemic or pulmonary hypertension), causing the respective valves to shut forcefully.

Split heart sound   The events on the two sides of the heart are a little asynchronous. The left ventricle starts contracting before the right ventricle. Hence, the mitral (M) valve closes before the tricuspid (T), resulting in the 'splitting' of the first heart sound S_1 into M_1 and T_1, in that order. Although the left ventricle starts contracting earlier, the pulmonary valve opens before the aortic valve because the pulmonary arterial pressure is much lower than the aortic pressure. For the same reason, during diastole, the aortic (A) valve closes before the pulmonary (P) valve, resulting in the splitting of the second heart sound S_2 into A_2 and P_2, in that order. The splitting of S_2 widens during deep inspiration because during inspiration, right ventricular output increases slightly (see Fig. 37.**6**), keeping the pulmonary valve open longer. As a result, P_2 gets delayed further and the split widens.

Murmurs (Fig. 36.**4**) are abnormal heart sounds caused by the turbulence of blood as it passes through defective valves. Stenosed valves open incompletely. Regurgitating valves are leaky valves that do not close properly and allow backflow of blood. Murmurs are produced during systole (systolic murmurs) in stenosed semilunar valves or regurgitant AV valves. Murmurs are produced during diastole (diastolic murmurs) in stenosed AV valves or regurgitant semilunar valves. Murmurs are also heard over ventricular septal defect and patent ductus arteriosus (see Fig. 38.**5**).

Duration of Cardiac Events

The durations of the various phases of the cardiac cycle depend on the heart rate. The durations at cardiac rates of 75 min^{-1} and 200 min^{-1} are given in Table 36.2.

Effect of heart rate   When the heart rate increases, the durations of all the phases decrease. However, the duration of diastole decreases much more than the duration of the systole. The diastole shortens due to greater automaticity of the sinus node. The systole shortens due to quicker repolarization which reduces the duration of cardiac action potential.

The marked reduction in diastolic time during tachycardia has important clinical implications. It is during diastole that most of the ventricular filling occurs. Also, it is during diastole that most of the cardiac muscle, especially the subendocardial portions of the left ventricle, gets adequately perfused by the coronary blood flow. Hence, at heart rates greater than 150 min^{-1}, there is a reduction of both ventricular filling (which tends to reduce cardiac output) and cardiac perfusion (which tends to cause myocardial ischemia and infarction).

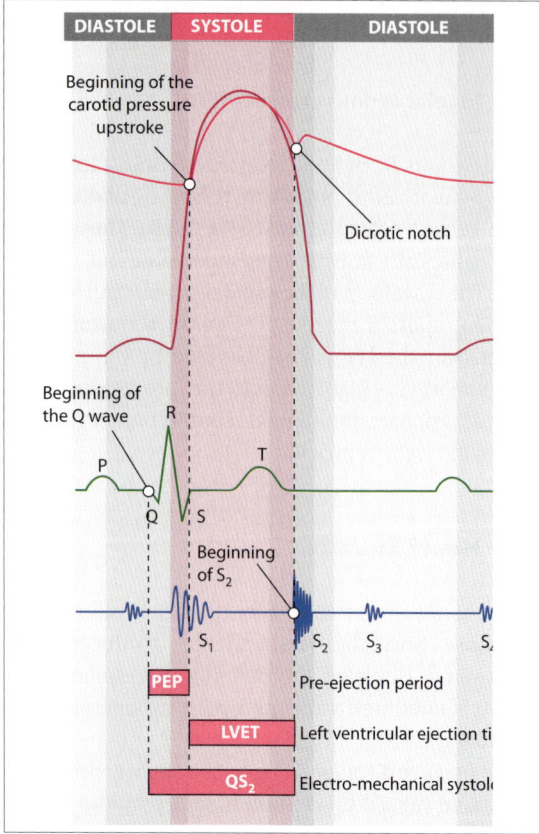

Fig. 36.**5**   Clinical indices of the isovolumetric ventricular contraction.

Table 36.**2**   Duration of cardiac cycle at resting heart rate and in tachycardia.

Duration	HR 75 min^{-1}	HR 200 min^{-1}
Cardiac cycle	0.8 s	0.30 s
Systole	0.3 s	0.15 s
Diastole	0.5 s	0.15 s

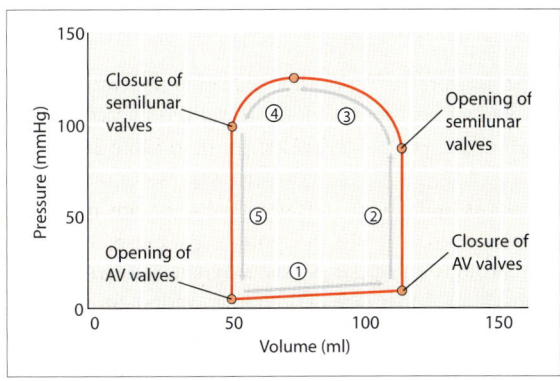

Fig. 36.**6**   Ventricular pressure–volume loop showing (1) Ventricular filling, (2) Isovolumetric contraction, (3) Rapid ejection, (4) Slow ejection, (5) Isovolumetric relaxation.

Clinical indices

The duration of isovolumetric ventricular contraction is of considerable clinical importance. However, it is not easy to measure. Hence certain other indices have been devised as approximations for the duration of isovolumetric contraction. These indices can be calculated by simultaneous recording of the electrocardiogram, phonocardiogram (record of the heart sounds), and carotid pulse (indicating aortic pressure changes).

Electromechanical systole (QS_2) is the time interval between the onset of the ventricular activation (QRS complex) to the closure of the aortic valves (S_2). Left ventricular ejection time (LVET) is the time interval between the beginning of the carotid pressure rise to the incisura. Pre-ejection systole (PEP) is the difference between QS_2 and LVET and gives the duration of electromechanical events preceding systolic ejection (Fig. 36.**5**). The normal PEP/LVET ratio is 0.35. When left ventricular function is impaired, the PEP/LVET ratio increases without a change in QS_2.

Ventricular Pressure–Volume Loop

The ventricular pressure–volume loop (Fig. 36.**6**) is an alternative method of representing the cardiac cycle. In this, the time dimension is eliminated and therefore it is not possible to tell from this loop how fast the events are occurring. However, the advantage is that the work done by the heart is instantly apparent from the area enclosed by the loop. It thus enables the clinician to easily diagnose a failing heart (see Fig. 37.**12**).

37   Cardiac Output

Cardiac output is the amount of blood pumped into the aorta by the left ventricle per minute. The normal cardiac output is 5–6 L min^{-1}. Cardiac index is the cardiac output expressed in relation to the body surface area. The normal cardiac index is about 3.2 L min^{-1} m^{-2}. The left ventricular output is slightly more than right ventricular output due to the presence of a physiological shunt.

Cardiac output is the product of stroke volume and heart rate. The *stroke volume* is the amount of blood pumped out by the left ventricle in each stroke. During a single contraction, each ventricle pumps out 80 ml of blood. Stroke volume is given by the difference between *end-diastolic ventricular volume* (130 ml) and *end-systolic ventricular volume* (50 ml).

Cardiac output (6 L min^{-1})	=	Stroke volume (80 ml)	×	Heart rate (75 min^{-1})

Stroke volume (80 ml)	=	End-diastolic ventricular volume (130 ml)	−	End-systolic volume (50 ml)

The ejection fraction is the percentage of the end-diastolic ventricular volume that is ejected with each stroke. The ejection fraction is a valuable index of ventricular pump function. It decreases in a failing heart.

$$\text{Ejection fraction (61.5\%)} = \frac{\text{Stroke volume (80 ml)}}{\text{End-diastolic ventricular volume (130 ml)}} \times 100$$

Although cardiac output is the product of stroke volume and heart rate, it does not necessarily mean that the cardiac output increases whenever the stroke volume or the heart rate increases. An increase in cardiac output requires simultaneous readjustments in the state of the blood vessels. An increase in heart rate unaccompanied by vascular changes, as occurs during paroxysmal tachycardia, results in a reduction in stroke volume so that the cardiac output remains unchanged or decreases.

■ Cardiac output measurement

Fick method The cardiac output can be estimated by using the Fick principle which states that blood flow (F) through an organ is given by the quotient of the amount of substance (Q) that is added or removed from the blood and the arteriovenous concentration difference ($C_A − C_V$) it produces. This principle is not applicable if the organ exchanges the substance with anything other than the blood that passes through it. Expressed mathematically,

$$F = \frac{Q}{C_A - C_V}$$

For calculating cardiac output using Fick principle, oxygen is used as the indicator substance. Thus Q is the amount of O_2 diffusing every minute from the lungs into blood, C_A is the O_2 concentration in arterial blood and C_V is the O_2 concentration in mixed venous blood (Fig. 37.1). Application of the Fick principle to these parameters gives the pulmonary blood flow (F) which is essentially equal to the cardiac output.

Q can be estimated from the volume difference between volume of inspired air and the volume of expired air that is passed through soda lime to remove CO_2. C_A can be estimated in an arterial blood sample withdrawn from any convenient artery because all systemic arteries have roughly the same P_{O_2} as pulmonary venous blood. C_V is estimated in a sample of mixed venous blood obtained directly from the pulmonary artery through catheterization. This is necessary because almost every vein flowing toward the heart brings in blood with a different level of O_2 desaturation.

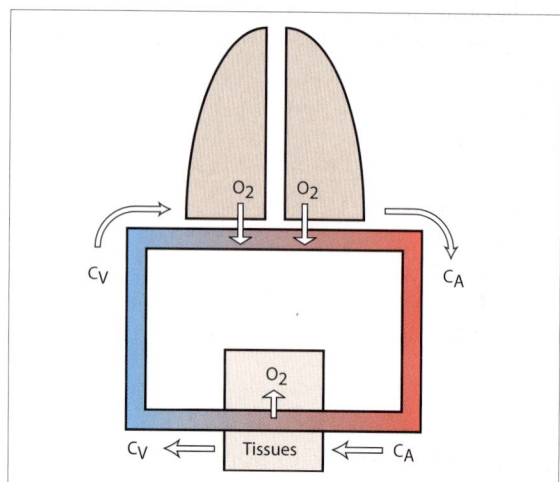

Fig. 37.1   Fick Principle. Explanation in text.

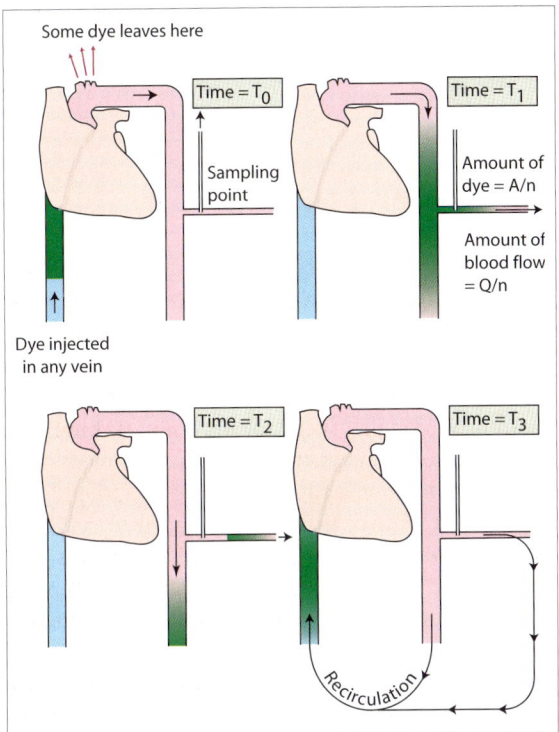

Some dye leaves here

Time = T_0

Sampling point

Dye injected in any vein

Time = T_1

Amount of dye = A/n

Amount of blood flow = Q/n

Time = T_2

Time = T_3

Recirculation

Fig. 37.2 Steps in the circulation of an indicator dye injected into a vein. (Above left) Dye injected into a vein. (Above right) Dye approaches the sampling point in an artery. (Below left) Dye passes the sampling point. (Below right) Dye recirculates and reenters the vein.

Example

If Q = 250 ml min⁻¹, C_A = 195 ml L⁻¹, C_V = 145 ml L⁻¹, then F = 5000 ml min⁻¹.

Dye-dilution technique   A bolus dose (Q) of indicator dye is injected into any superficial vein and blood samples are withdrawn from an artery every few seconds. The indicator must be a substance that stays in the bloodstream during the test and has no harmful or hemodynamic effects. The concentration of the dye in blood (C) is estimated and the time (T) taken by the column of dyed blood to flow across the sampling point is noted.

Suppose, the total amount of dye passing under the sampling point in T seconds = q

Then, the volume of blood passing under the sampling point in T seconds = q / C

∴ The volume of blood (f) passing under the sampling point in 60 seconds is given by the formula:

$$f = \frac{q}{C} \times \frac{60}{T}$$

Two corrections need to be made here. First, the flow rate (f) calculated is that of a single artery (from which blood has been sampled), which is only a fraction (say

1/n) of the cardiac output (F). Second, the amount of dye flowing into the artery (q) is a fraction (also 1/n) of the total amount of dye injected (Q). Applying these corrections,

$$\frac{F}{n} = \frac{Q/n}{C} \times \frac{60}{T}$$

'n' cancels out, and the formula becomes:

$$F = \frac{Q}{C} \times \frac{60}{T}$$

However, the column of blood passing under the sampling point does not have a uniform concentration of dye in it (Fig. 37.2). Moreover, there is recirculation of the dye back to the heart and into the artery from where the dye is sampled. It is therefore difficult to estimate the mean concentration of the dye passing under the sampling point and to estimate the time taken by the dye to pass under it.

The problem is tackled by obtaining multiple blood samples as the dyed blood passes under the sampling point. A graph is plotted between the dye concentration on the Y-axis and time on the X-axis. The time taken (T) is read out from the graph after exponentially extrapolating the dye concentration to a very low value.[1] The integrated value of C × T can be estimated from the area under the curve in Figure 37.3B.

A modification of the dye-dilution technique is the thermodilution technique in which cold saline is used instead of a dye and the dilution of its temperature by blood is measured. This technique has the twin advantages that the saline is completely harmless and the cold is dissipated in the tissues so that recirculation is not a problem.

The **Doppler technique** measures cardiac output by placing an electromagnetic flow meter on the ascending aorta in experimental animals. The Doppler effect is the change in sound velocity when the medium through which it is propagating is itself in motion. Thus, flowing blood changes the velocity of sound through it. The magnitude of the change indicates the velocity of blood flow. In humans, cardiac output is measured by combining Doppler techniques with echocardiography.

The **cineradiographic technique** calculates the ejection fraction by imaging the cardiac blood pool at the end of diastole and the end of systole after injecting radiolabeled red blood cells.

[1] Exponential extrapolation, when attempted directly, is prone to inaccuracies. The accuracy can be improved by plotting the graph on a semilog graph paper (logarithmic y-axis and linear x-axis). Thus plotted, the terminal part of the curve becomes straight and can easily be extended to the base line to find out the value of 'T'.

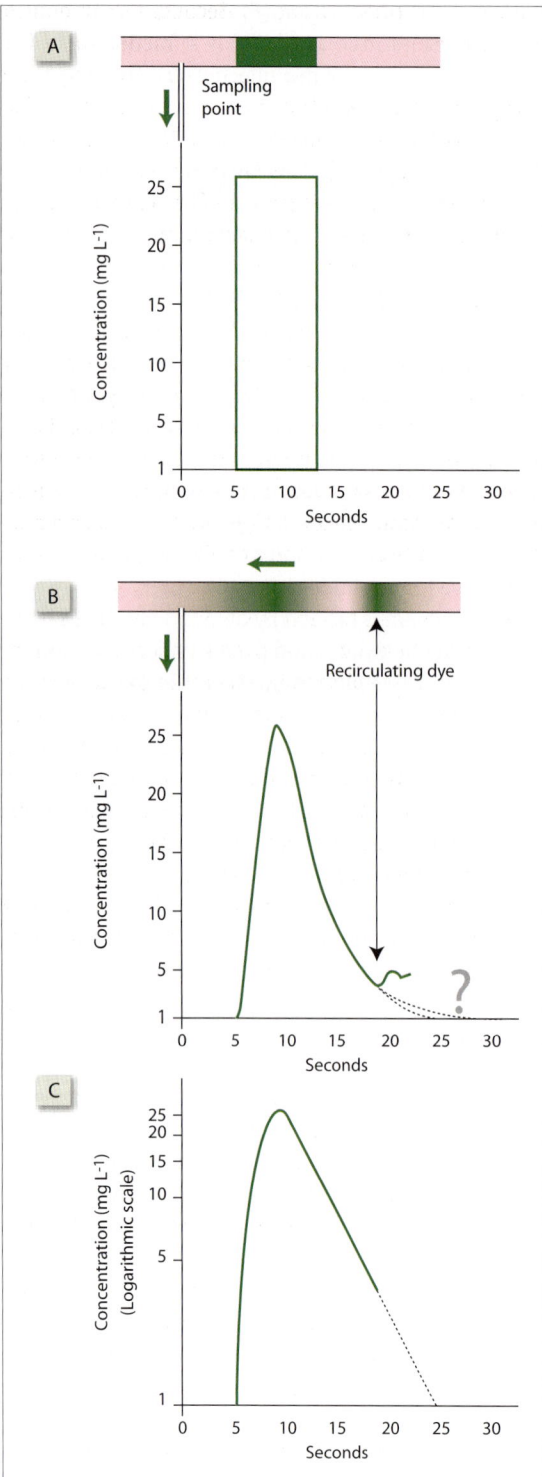

Fig. 37.3 Indicator dilution technique of cardiac output estimation. [A] If the dye moved like a solid block, the graph would be rectangular. [B] The dye concentration fades exponentially at both ends. Due to recirculation of the dye, there is a resurge in dye concentration before the decline is complete. The graph has to be extrapolated exponentially to get the value of T. [C] Extrapolation is easier if the graph in B is plotted on a semilogarithmic scale.

Ballistocardiography measures the recoil force produced by the rapid ejection of a large amount of blood through the aorta. The magnitude of the recoil force gives a measure of the stroke volume.

Control of Cardiac Output

The two main controllers of cardiac output are the cardiac factors and the vascular factors.[2] How well these two factors are performing is indicated by the *right atrial pressure*, also called the *central venous pressure* (CVP), which should remain between 1 to 5 mmHg. A vigorously pumping heart lowers the CVP through a suction force. However, the heart requires an optimum CVP for its pumping action: If the CVP falls, the cardiac output falls with it. If the CVP dropped to subatmospheric levels (below 0 mmHg), the great veins would collapse and no blood would enter the heart. The role of the vascular system is to ensure that the CVP does not fall too low for effective cardiac pumping. The vascular factors include venous pumping, which propels blood toward the heart.

The role of the cardiac and vascular factors can therefore be summarized as follows: The cardiac factors try to prevent the CVP from rising, while the vascular factors try to prevent the CVP from falling. If the CVP rises above 5 mmHg, it indicates that the heart is not pumping adequately and is probably failing. If the CVP falls below 1 mmHg, it indicates that the vascular factors are failing in their role. The CVP can be measured directly by inserting a catheter into the right atrium. CVP is assessed clinically by examining the jugular venous pulse (see Fig. 38.**14**).

Circulatory load on the heart While discussing the function of any pump, it is important to consider the load on it. Talking about circulatory load, it must first be understood that there are two types of load on the heart: the preload and the afterload. A preload stretches the cardiac fibers before they start contracting. Cardiac preload is represented by the volume of venous blood that distends the ventricle. Ventricular filling increases with the CVP. The CVP therefore determines the preload on the heart.

An afterload is the load encountered by the cardiac muscle after it starts contracting. An example of cardiac afterload is the aortic pressure. The left ventricle

---

[2] The concept of cardiac output regulation given here is more in line with Levy's approach rather than with Guyton's. See Physiology by Berne RM and Levy MN, 5th Edition (2004), Mosby St Louis. See also Guyton AC and Hall JE, A Textbook of Medical Physiology, 11th Edition (2006), Saunders WB, Philadelphia.

does not 'feel' the load of the aortic pressure due to the closed aortic valve. However, as soon as the heart contracts, the aortic valve opens and the left ventricle encounters the high aortic pressure. Hence, arterial hypertension imposes a heavy afterload on the heart.

When preload increases, the end-diastolic volume of the ventricle increases. Up to a limit, an increase in preload increases cardiac contractility in accordance with the length–tension relationship of striated muscles. As a result, the cardiac output tends to increase in proportion to the preload imposed on it. This is known as the *Frank–Starling law* of the heart, which states that the ventricular output is proportional to the ventricular end-diastolic volume.

On the other hand, when the afterload on the heart increases, the end-diastolic ventricular volume is unaffected and the cardiac output falls. The fall in the cardiac output occurs despite an increase in myocardial contractility, which is known as the *Anrep effect*. The preload on the heart increases in dynamic exercise, change in posture and aortic regurgitation. The afterload increases in static exercise, hypertension, and aortic stenosis.

Myocardial contractility

There are two mechanisms through which the contractility of cardiac muscle fibers can increase, viz., heterometric and homometric. In the heterometric mechanism, the cardiac muscle fibers are stretched to increase their initial length. In the homometric mechanism, the active tension produced in cardiac muscle fibers is increased without increasing the initial fiber length.

Heterometric mechanism comes into play only when the preload on the heart increases. An increased preload results in a higher end-diastolic volume. In accordance with the Frank–Starling law of the heart, the increase in end-diastolic volume increases the strength of cardiac contraction and the cardiac output increases. However, heterometric regulation is fraught with the risk of excessive increase in the end-diastolic volume with consequent damage to the ventricular muscle fibers. Therefore, a healthy heart avoids heterometric regulation of cardiac output.

Heterometric change in cardiac output occurs, for example, immediately after a change of posture from sitting to lying, which increases the CVP. Heterometric response also occurs continuously for small, momentary adjustments necessary for keeping the outputs of the two ventricles equal. For example, if the output of the right ventricle increases momentarily, the left atrial pressure will increase, which in its turn will increase the output of the left ventricle.

Homometric mechanism increases cardiac performance without changing the initial length of the cardiac fibers. Factors that increase cardiac contractility homometrically include sympathetic discharge, circulating catecholamines, and tachycardia. The relation between heart rate and cardiac contractility is known as the *staircase effect* (see Fig. 21.**4**).

In exercise, regulation of cardiac output is largely (though not entirely) homometric and occurs through the activation of the sympathetic nervous system. Circulating catecholamines and tachycardia also contribute to the increase in contractility. Sympathetic stimulation increases the stroke volume without increasing the end-diastolic volume of the heart. In other words, it increases the ejection fraction. This is made possible by an increase in myocardial contractility and a simultaneous increase in heart rate. The increase in myocardial contractility decreases the end-systolic volume (increases the stroke volume) and reduces the duration of systole. This ensures that with each stroke, the ventricles pump out blood quickly and in larger amounts. The increase in heart rate decreases the end-diastolic volume (the heart has less time for relaxation and venous filling). By increasing both stroke volume and heart rate, sympathetic stimulation increases the cardiac output massively without increasing the cardiac size, i.e., without increasing the end-diastolic volume.

Cardiac pump function

Myocardial contractility should not be confused with the pumping efficacy of the heart. A heart with a valvular defect, for example, will have poor pumping efficacy even if its contractility is normal or high.

Factors resulting in a *hypoeffective heart* are as follows. (1) Decrease in pump power, as occurs in reduced sympathetic discharge or myocardial damage. (2) Inefficient pumping, as in valvular or septal defects of the heart. (3) Reduced pump filling, as in positive intrathoracic pressure (e.g., during expiration or Valsalva maneuver) and pericardial effusion or cardiac tamponade (filling of the pericardial space with blood). These conditions reduce diastolic filling of the heart. (4) Increased afterload on the pump, as in hypertension.

Factors resulting in a *hypereffective heart* are as follows: (1) Sympathetic stimulation of the heart. (2) Inotropic drugs, i.e., drugs like digitalis that increase cardiac contractility. (3) Increased intrathoracic negativity (e.g., during inspiration) which permits greater diastolic filling of the heart.

Effect of cardiac pumping on CVP   The volume of blood that is pumped out by the heart each minute always equals the volume of blood that returns to the right atrium each minute. Why then does the CVP fall as the cardiac output increases? Or, why does CVP rise in cardiac failure when the cardiac output falls?

In answering the question, we shall begin from the stage where the circulation has stopped and the blood pressure in the arteries and veins is 10 mmHg, i.e., at the mean systemic filling pressure. We shall now assume that after every second, 100 ml of blood is quickly pumped by the heart from the venous to the arterial side. (The experiment replicates a cardiac output of 6 L min⁻¹, with a stroke volume of 100 ml and heart rate of 60 s⁻¹.)

Let us consider what happens when for the first time, the heart shifts 100 ml of blood from the venous to the arterial side. The arterial pressure will rise and the venous pressure will fall. Since the arterial compliance is much lower than the venous compliance, the rise in arterial pressure will be higher than the fall in venous pressure. Due to the difference of blood pressure in the arteries and veins, blood starts flowing from the arteries to the veins through the capillaries. The arteriovenous pressure difference starts narrowing. However, not all of the 100 ml of blood that was

pumped into the arteries is able to flow back into the veins in 1 second, and it is time again for the heart to shift another 100 ml of blood from the venous to the arterial side. This time, the arterial pressure rises higher and the venous pressure falls lower than before. The arteriovenous pressure difference is greater and blood flows back into the veins faster.

As the sequence continues, the arterial pressure keeps rising. When the arterial pressure rises to 120 mmHg, the blood flows back into the veins so fast that all of the 100 ml of blood that was pumped into the arteries flows back into the veins in 1 second. At this stage, the arterial and venous pressures stabilize.

If the above steps are repeated, translocating only 50 ml of venous blood every second into the aorta, the result will be similar but at equilibrium, the arterial pressure will be lower and venous pressure will be higher than before (Fig. 37.**4**). The arteriovenous pressure difference will be lower and each second, 50 ml blood will flow back into the veins, stabilizing the blood pressure.

■ Vascular functions

The cardiac pump is not the sole regulator of cardiac output: Even a strong cardiac pump will fail to increase the cardiac output if the afterload on it is very high or the preload on it is very low. The circulatory load is controlled by vascular factors. Two primary vascular factors that control the cardiac output are the peripheral resistance and mean systemic filling pressure. Other vascular factors include venous pumping and gravity.

Peripheral resistance falls when the arterioles are dilated, allowing arterial blood to flow through into the veins with greater ease. The easier arteriovenous flow lowers the arterial pressure and raises the venous pressure. The fall in the arterial pressure reduces the afterload on the heart and therefore increases the cardiac output. The rise in venous pressure and CVP helps the heart to sustain the high cardiac output. Vasodilatation is by far the most important mechanism that increases the venous pressure in exercise. It is through vasodilatation that tissue hypoxia is able to increase the cardiac output. The sequence of events shown in Figure 37.**5** ensures that whenever tissues consume more O_2, the cardiac output increases to meet their O_2 demand.

Peripheral resistance also decreases in arteriovenous (A-V) shunts and fistulae, severe anemia (due to the vasodilation caused by anemic hypoxia), thyrotoxicosis (due to the vasodilatation caused by the increased tissue O_2 consumption), and in wet beriberi (due to

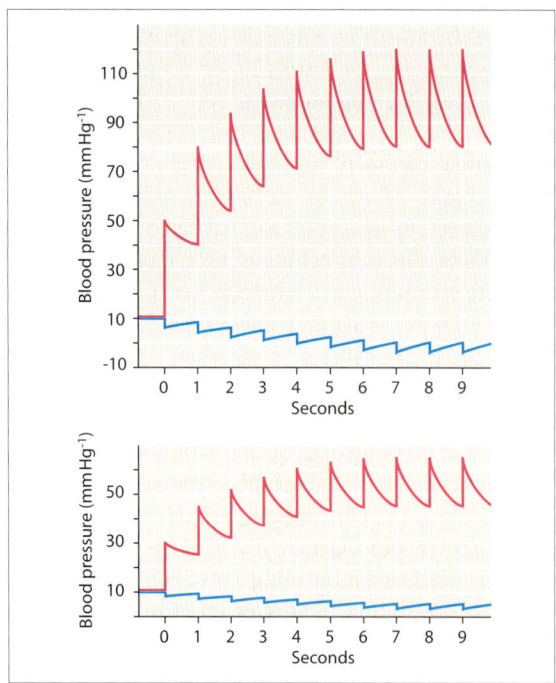

Fig. 37.**4**   The arterial blood pressure (red) and central venous pressure (blue) attained at two different rates of cardiac output, 6 L min⁻¹ (*above*) and 3 L min⁻¹ (*below*).

reduced tissue utilization of nutrients which results in compensatory vasodilatation). All these conditions are associated with high cardiac output.

The **mean systemic filling pressure** is the pressure at which blood is 'filled' into blood vessels. Stated differently, it the pressure inside the circulatory system when circulation is stopped completely. Normally, the filling pressure is about 10 mmHg. A greater filling pressure allows the cardiac output to rise higher before the CVP drops too low. The filling pressure increases if the blood volume increases. It also increases if the venous compliance decreases, as occurs in response to

sympathetic discharge. When the compliance is low, a given volume of blood would produce a greater filling pressure.

Venous pumping   Before discussing venous pumping, it is important to understand the effect of steady venous compression. Application of a steady pressure on a large vein, say the inferior vena cava, results in a biphasic response in cardiac output. There is an *initial increase* in cardiac output for a few seconds. The increase occurs because the blood present in the inferior vena cava gets displaced mostly toward the heart, raising the CVP. (Less blood flows back into the periphery because peripheral veins are smaller and offer more resistance to flow while the central veins are larger and offer less resistance. Moreover, the limb veins are guarded by venous valves.) The initial increase is followed by a *sustained decrease* in cardiac output. The decrease occurs because the pressure on the inferior vena cava increases the resistance to blood flow into the atria, reducing the CVP.

We can now proceed to discuss the effect of the respiratory and muscle pumps that act like small pumps connected in series with the heart and directly contribute to the circulatory flow rate. By propelling blood toward the heart, they increase the CVP and consequently increase the cardiac output.

The *respiratory pump* refers to the effect of rhythmic breathing on circulatory flow rate. During inspiration, the rise in intraabdominal pressure compresses the large abdominal veins. Since the compression is rhythmic and transient, it results in a brief increase in the CVP with each inspiration. The simultaneous inspiratory increase in intrathoracic negativity allows greater diastolic filling of the ventricles and consequently

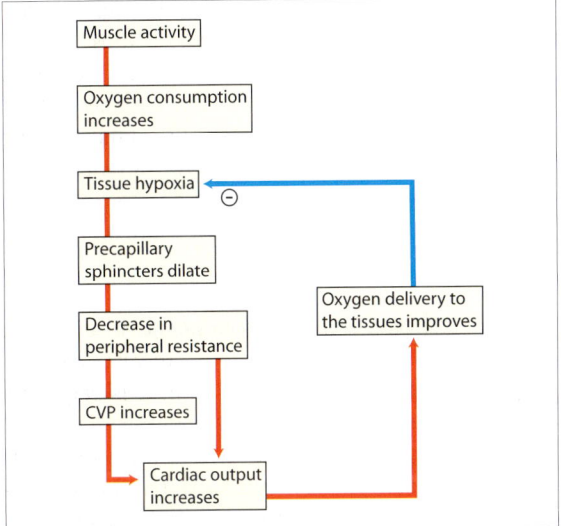

Fig. 37.**5**   Heart as a demand pump. Cardiac output changes according to the oxygen requirement of tissues, which is the ultimate controller of cardiac output.

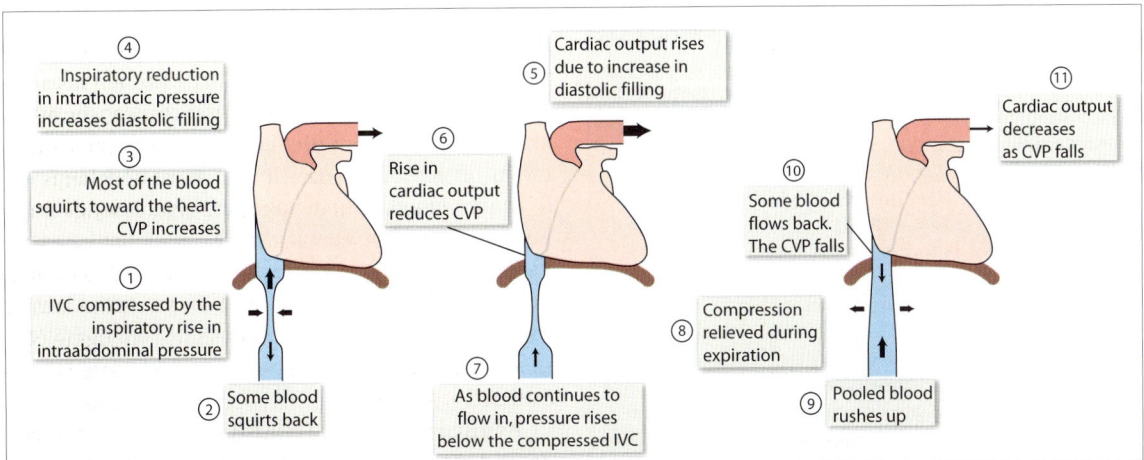

Fig. 37.**6**   The respiratory pump. It is observed that more blood flows from the periphery toward the IVC (in 9) than the amount of blood that flows back from IVC (in 2). Hence although the cardiac output rises and falls during the cardiac cycle, the respiratory pump has an overall propelling effect on the circulation.

increases the cardiac output. During expiration, the compression of the abdominal vein is relieved and blood is sucked in from the periphery (Fig. 37.**6**).

The *muscle pump* refers to the effect of intermittently contracting limb muscles on the circulatory flow rate. When limb veins are squeezed by exercising muscles, the venous blood squirts toward the heart since the venous valves prevent the blood from flowing back. When the muscles relax, the veins suck blood from the capillaries.

Gravity   In the upright posture, venous blood loses pressure energy and gains potential energy as it flows toward the heart (in accordance with Bernoulli's principle). The CVP is therefore low. In the recumbent

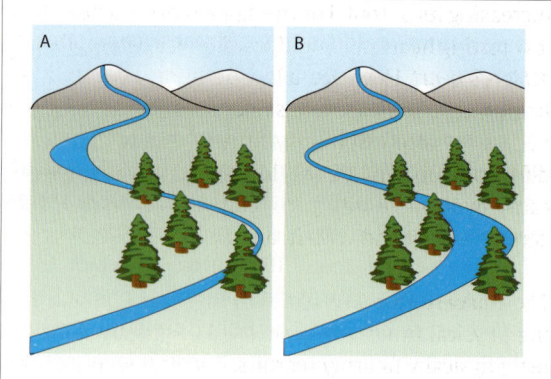

Fig. 37.**7**   Compare pictures A and B. Has the flow of river water through the woods increased in B? Not really! In B, the river has certainly widened where it passes through the woods so that there is more water in the woods. However the amount of river water flowing through the woods has not changed as it remains limited by the amount of water flowing down the hills.

position, the blood flows horizontally toward the heart and therefore there is no conversion of pressure energy into potential energy. The CVP is therefore higher.

Cardiac Output in Exercise

The cardiovascular effects of exercise depend on whether the exercise is static or dynamic (p 274).

Static exercise is associated with a massive increase in peripheral resistance, which leads to a sharp reduction in cardiac output.

Dynamic exercise is associated with a rise in cardiac output. The hypoxia and the metabolites produced during exercise dilate the precapillary sphincter, lowering the peripheral resistance. The fall in peripheral resistance increases the cardiac output by reducing the afterload on the heart. The higher cardiac output would tend to lower the CVP. That however does not happen due to the low peripheral resistance, which also raises the venous pressure. The venous pressure is raised further by the muscle and respiratory pumping of venous blood, both of which are vigorously active in exercise. Since peripheral vasodilatation increases the cardiac output and because the vasodilatation is metabolic in origin, the increase in cardiac output parallels the increase in tissue oxygen consumption. In athletes, the cardiac output can rise to up to 25 L min^{-1} during exercise.

As discussed above, a rise in cardiac output involves an increase in both stroke volume and heart rate. Tachycardia is associated with shorter diastole and reduced diastolic perfusion of coronary arteries. Hence a healthy heart relies more on the stroke volume for

---

What is venous return?

The term *venous return* is usually defined as the amount of blood entering the right atrium per minute. Obviously, it is equal to the cardiac output, i.e., about 5 L min^{-1}. The term is in wide usage and is considered to be an important factor controlling cardiac output through changes in the diastolic filling of the heart. Thus, a fall in venous return, as occurs when the posture changes from supine to standing, reduces the cardiac output. Similarly, in the presence of an arteriovenous fistula, the venous return increases and therefore the cardiac output increases too.

A closer look will reveal the inconsistency between the definition of venous return and the role attributed to it. Venous return cannot be a separate entity from cardiac output, nor can changes in venous return precede the changes in cardiac output since both must change simultaneously. If the venous return decreased to 5.4 L min^{-1} while the cardiac output remained unchanged at 6.0 1 min^{-1}, then 100 ml of fluid would enter the interstitial spaces in 10 seconds, something that never happens. It can be argued, therefore, that cardiac output and venous return are merely two different names for the same quantity, i.e., the total blood flow rate through the circulatory system.

The above-noted inconsistencies disappear if the term venous return is taken to imply the *central venous pressure* (CVP). 'Factors affecting venous return' are essentially the 'vascular factors that affect the CVP'. There is no inconsistency in stating that gravity reduces the CVP and therefore reduces the cardiac output. Similarly, an arteriovenous fistula increases the cardiac output by reducing the peripheral resistance. It also increases the CVP, which enables the heart to sustain the high output.

Is then the term venous return a misnomer? Not quite. The CVP can rise only if there is more blood in the right atrium. More blood can be present in the right atrium only if there is less blood in the lower extremities, assuming that the amount of blood in the arterial circuit does not change much. Thus, the term venous return describes a *translocation of blood* from the extremities to the right atrium (Fig. 37.**7**). To that extent, the term 'return' is justifiable as long as it is not confused with a flow rate. Nonetheless, the term 'venous return' has been studiously avoided in this book, if only to prevent needless confusion.

increasing its output. For the same reason, a heart with low resting heart rate (bradycardia) has been called the *athlete's heart*. However, a low or moderate increase in heart rate following exercise not only indicates cardiac fitness but also respiratory fitness. A reduction in the diffusion capacity of the lungs, for example, would cause hypoxia, resulting in a compensatory increase in cardiac output and with it, an increase in heart rate.

The **Harvard step test** is a simple method for evaluating physical fitness and was originally designed to select physically fit army recruits. The subject is required to step up onto a stool first with one foot and then with the other so that both feet are on the stool and then to step down in the same order so that both feet contact the ground. The height of the stool should be 20 inches. 30 step-ups are performed per minute and continued for 5 minutes. A metronome can be used to assist the subject to maintain the required stepping rate. (The test can also be performed using a stool height of 18 inches and a test duration of 4 minutes.) The ½ minute pulse count is counted from 1–1½, 2–2 ½, and 3–3½ minutes after stopping the test (Fig. 37.**8**). Using the formula below, the *physical efficiency index* (PEI) can be calculated. A gradual decrease in the resting pulse over a period of weeks is indicative of an increase in physical fitness.

$$PEI = \frac{Duration\ of\ exercise\ (sec) \times 100}{2 \times Sum\ of\ pulse\ counts\ in\ recovery\ period}$$

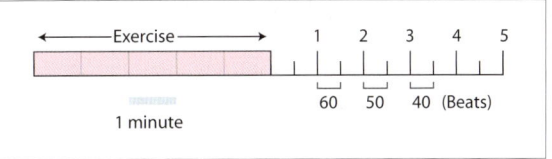

Fig. 37.**8**   Protocol of the Harvard step test.

Table 37.**1**   Interpretation of Harvard step test.

PEI score	Physical conditions
Below 55	Poor
55–64	Low
65–79	High average
80–89	Good
Above 90	Excellent

Based on the PEI score, fitness has been classified as in Table 37.**1**.

Example
After 5 minutes (300 s) of exercise, if the ½ minute pulse counts recorded during 1–1½ , 2–2½, and 3–3½ minutes are 60, 50, and 40, then the PEI would be 100, which is excellent.

Graphic Analysis of Cardiac Output

The graphic analysis of cardiac output is based on the interdependence of cardiac output and CVP. The cardiac

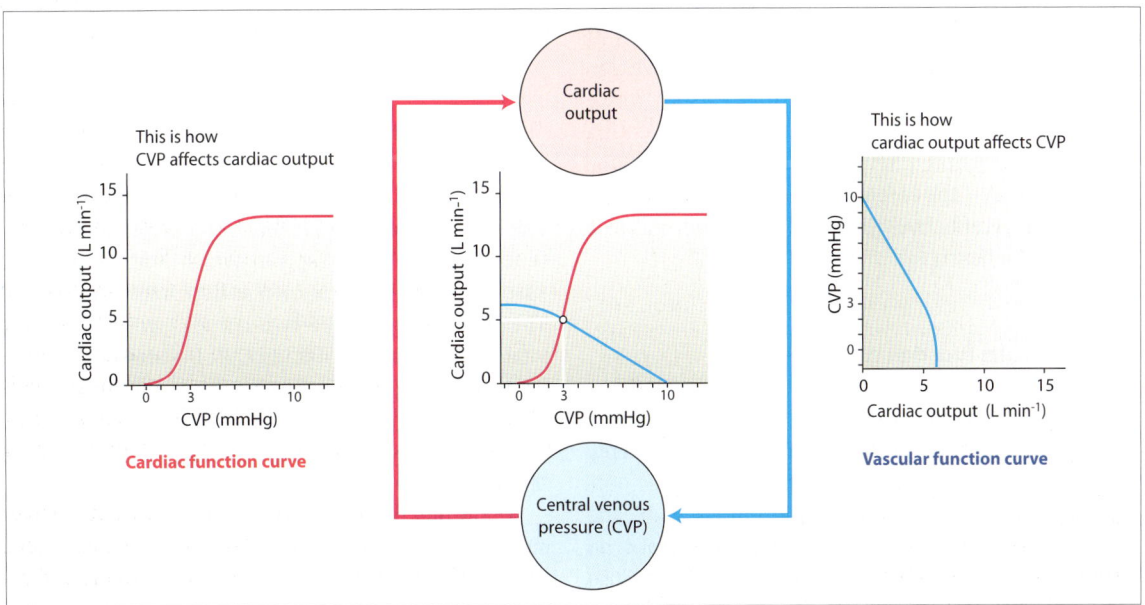

Fig. 37.**9**   The interdependence of cardiac output and central venous pressure.

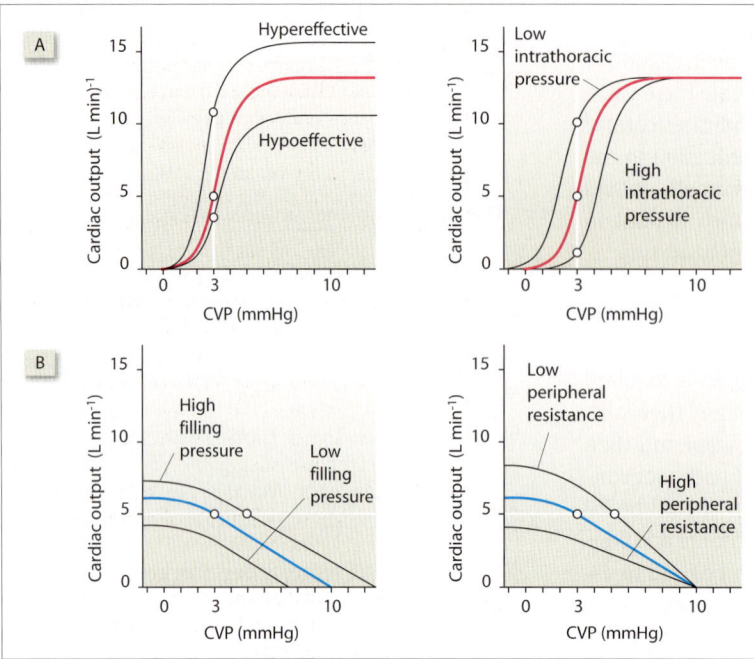

Fig. 37.**10**   [A] Cardiac function curves. (*Left*) Effect of cardiac pump efficiency. (*Right*) Effect of intrathoracic pressure. Note that at a given CVP, the cardiac output increases as pump efficiency increases or intrathoracic pressure decreases. [B] Vascular function curves. (*Left*) Effect of mean systemic filling pressure on CVP. (*Right*) Effect of peripheral resistance on CVP. Note that if the cardiac output is held constant, CVP increases as the filling pressure increases or the peripheral resistance decreases. The graph is unconventional because the dependant variable (CVP) is shown on the x-axis and the independent variable (cardiac output) is shown on the y-axis. (Compare with the vascular function curve in Fig. 37.**9**). This unconventional representation is useful in coupling the cardiac and vascular function curves (Fig. 37.**11**).

function curve shows how the CVP affects cardiac output. The vascular function curve shows how the cardiac output affects the CVP. By combining the two function curves, the operant cardiac output and CVP can be determined (Fig. 37.**9**).

Cardiac function curves

The cardiac function curves depict the heterometric regulation of cardiac output, i.e., when CVP increases, there is increased ventricular filling with consequent increase in cardiac output. The effect plateaus off at higher levels of CVP. The cardiac function curve can be altered by several factors. There thus exists a whole family of cardiac function curves, some representing pump efficiency and others representing pump filling. At any given CVP, the cardiac output will be above normal in a hypereffective heart and below normal in a hypoeffective heart (Fig. 37.**10**A).

Vascular function curves

The vascular function curve shows how changes in the cardiac output affect the CVP. The CVP falls steeply when the cardiac output rises. When the cardiac output exceeds a limit (about 6.0 L min^{-1} in Fig. 37.**9**), the CVP falls steeply to zero or even lower. At this stage, no further increase in cardiac output is possible because the great veins get squeezed by the intrathoracic pressure, effectively stopping the flow of blood through

them. Any further increase in cardiac contractility only makes the CVP more negative without increasing cardiac output.

The vascular factors affecting CVP have been discussed above. A rise in the *mean systemic filling pressure* shifts the vascular function curve to the right, i.e., at a given cardiac output, the greater the filling pressure, the greater will be the CVP (Fig. 37.**10**B).

A fall in *peripheral resistance* rotates the vascular function curve clockwise about the point of systemic filling pressure, as shown in Figure 37.**10**B. Thus at a given cardiac output, a lower peripheral resistance is associated with a greater CVP.

Coupled cardiac and vascular function curves

The cardiac and vascular function curves can be combined into a single graph in which both cardiac and vascular function are shown as dependent on the CVP. The intersection of the two curves gives the operant values of cardiac output and CVP. Besides its other advantages, this graphical approach corroborates that the CVP decreases whenever the cardiac output increases due to an increase in pump efficiency, with the vascular functions remaining unchanged.

The utility of this graph is brought out best by considering the effect of sympathetic discharge on cardiac output (Fig. 37.**11**). Sympathetic discharge causes three changes: (1) It makes the cardiac function curve hypereffective. (2) It shifts the vascular function curve to

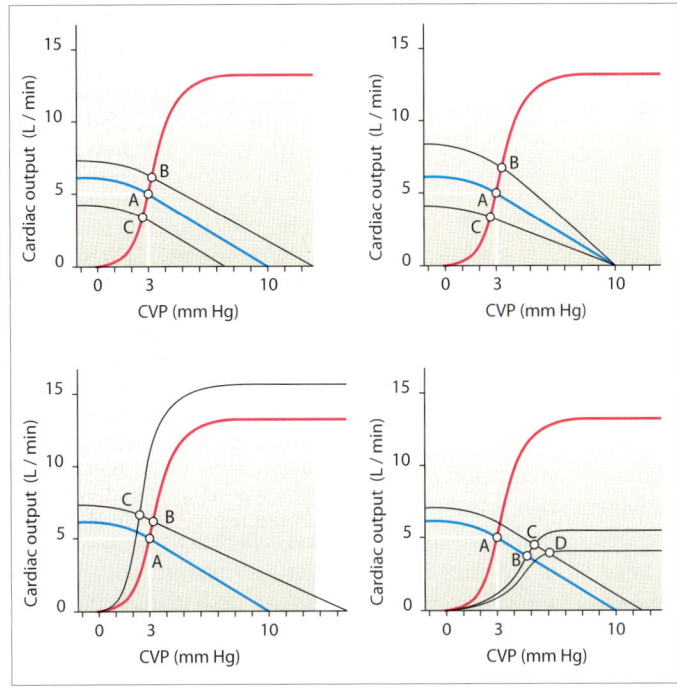

Fig. 37.**11**   The cardiac function curve (red) and vascular function curve (blue) intersect at A, which represents the normal cardiac output. (*Above left*) A rise in blood volume shifts the vascular function curve to the right and thereby, increases the cardiac output from A to B. Conversely, a fall in blood volume shifts the vascular function curve to left and reduces the cardiac output to C. (*Above right*) A fall in peripheral resistance rotates the vascular function curve clockwise and raises the cardiac output to B while an increase in peripheral resistance lowers cardiac output to C. (*Below left*) During sympathetic stimulation, the vascular function curve shows the effect of simultaneous increase in blood volume and peripheral resistance, i.e., the curve shifts as well as rotates. If cardiac function is unchanged, cardiac output increases to B. If cardiac function is augmented, cardiac output rises to C. (*Below right*) When the cardiac function is impaired, the cardiac output falls from A to B. The compensatory increase in blood volume increases the cardiac output to C, which is still inadequate. The retention of fluid and electrolytes impairs myocardial contractility and lowers myocardial function further, and the cardiac output falls to D. The vicious cycle of decompensation continues.

the right due to increase in filling pressure. (3) It also rotates the vascular function curve anticlockwise, due to the increased peripheral resistance caused by vasoconstriction.

Cardiac Failure

Cardiac failure is diagnosed when the atrial pressure is high.[3] It is called left-sided or right-sided failure depending on whether the pressure is elevated in the left or the right atrium. In the long run, failure of one side of the heart leads to the failure of the other side. Cardiac failure is associated with cardiomegaly (cardiac enlargement) because the high atrial pressure results in greater diastolic filling of the ventricle (Fig. 37.**12**).

Left-sided failure is associated with excess blood in pulmonary blood vessels (pulmonary congestion) and transudation of fluid into the alveolar interstitium (pulmonary edema). During physical exertion, there is further increase in pulmonary venous pressure, resulting in dyspnea.

Right-sided failure or *congestive cardiac failure* results in accumulation of blood in the systemic veins. The CVP increases, which is apparent from the raised JVP. Hepatomegaly (hepatic enlargement) occurs as the increased CVP is transmitted backward into the hepatic portal vein. Generalized edema occurs as the raised CVP is transmitted all the way back to the systemic capillaries. It is prominent in the dependent portions of the body where the capillary pressure is higher due to gravity. The high CVP is also transmitted to the coronary veins, reducing the diastolic perfusion of the myocardium and predisposing to myocardial ischemia. The increase in myocardial oxygen demand due to the cardiomegaly also contributes to the risk of myocardial infarction.

Cardiac failure may be associated with a normal cardiac output, a low cardiac output or a high cardiac output. When the cardiac output is normal, it is called a *compensated cardiac failure* to emphasize that the cardiac function is impaired and that the normal cardiac output is being maintained through heterometric mechanism.

When the cardiac output is reduced, it is called a *decompensated cardiac failure*. It is associated with hypotension and poor renal perfusion, leading to renin-degranulation. The activation of the renin-angiotensin-aldosterone system leads to hypervolemia, increasing the CVP and thereby aggravating the failure. If the cardiac output drops very low, it can lead to cardiogenic shock.

---

[3] Cardiac failure was earlier classified as forward failure (low *vis-a-tergo* or the force from behind) and backward failure (high vis-a-fronte or the force from in front). *Vis-a-tergo* is the force with which the heart pumps blood into the circulatory system, and *vis-a-fronte* is the venous pressure which, when high, resists the flow of blood from arteries to the veins. The terms *vis-a-tergo* and *vis-a-fronte* are not only obsolete but ambiguous: *Vis-a-fronte* has also been mentioned as the 'suction force' with which the heart sucks in venous blood.

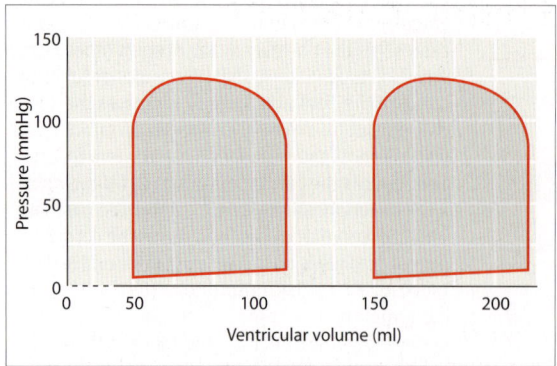

Fig. 37.**12**   Flow-volume loops of cardiac function. The loop to the right depicts a failing heart because it shows that the heart is doing the same amount of work (area of the loop is the same in both) at a higher end-diastolic volume. Note that the failing heart has greater diastolic filling.

High-output cardiac failure is seen in thyrotoxicosis, large A-V fistulae, and wet beriberi. In these conditions, the peripheral resistance is markedly reduced, allowing the high arterial pressure to be transmitted to the venous side and thereby increase the CVP. The high CVP increases the cardiac output. However, the cardiac output does not rise high enough to reduce the CVP to normal level.

The aim of treatment is two-fold: improving cardiac contractility and reducing the load on the heart. Digitalis is used for increasing cardiac contractility. Reduction of the load on heart is achieved by administration of diuretics and ACE inhibitors (inhibitors of angiotensinogen converting enzyme). Both reduce the blood volume.

Circulatory Pathway

Circulatory pathway in the adult

Blood is pumped by the heart through a closed circuit of blood vessels. The right and left ventricles of the heart function as two pumps connected in series, i.e., the entire output of one ventricle enters the other ventricle. From the left ventricle, blood flows through the arteries, arterioles, and systemic capillaries and back through the venules and veins to the right atrium. This circuit constitutes the *systemic circulation*. From the right ventricle, blood flows through pulmonary arteries, pulmonary capillaries, and pulmonary veins to return to the left atrium. This circuit constitutes the *pulmonary circulation.*

As the blood flows through the capillaries, a part of its plasma gets filtered into the interstitial spaces, and then drains out through lymphatic channels as lymph. The lymph flows through the thoracic duct and the right lymphatic duct to drain into the subclavian veins, thereby reentering the systemic circulation. This circuit constitutes the *lymphatic circulation* which is disposed in parallel to the systemic circulation (Fig. 38.1). A similar lymphatic system exists in the lungs also but the lymph is formed only in the pleural and bronchial capillaries and not by the alveolar capillaries. Hence, there is no separate lymphatic system that is disposed in parallel to the pulmonary circuit since bronchial and pleural capillaries belong to the systemic circulation.

Circulatory pathway in the fetus

In the fetus, it is the placenta and not the lungs that oxygenates the blood. Deoxygenated fetal blood is carried to the placenta by the umbilical artery, a branch of the femoral artery, and oxygenated blood is brought back from the placenta by the umbilical vein (Figs. 38.2 and 38.3). The umbilical vein passes through the liver where it joins the portal and hepatic veins which drain into the inferior vena cava. The umbilical vein is also connected directly with the inferior vena cava through a connecting vein called the *ductus venosus.*

Inside the right atrium, the blood from the superior and inferior vena cavae flow in two separate streams without mixing much. Blood from the superior vena cava mostly enters the right ventricle, which pumps the blood into the pulmonary arteries. Since the fetal lungs are collapsed, the pulmonary capillary resistance

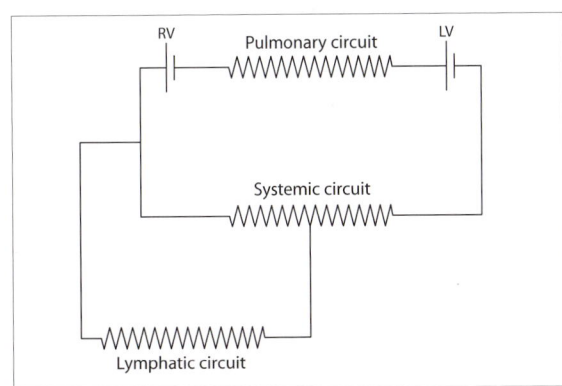

Fig. 38.1   The systemic, pulmonary, and lymphatic circuits. See also Figure 1.17.

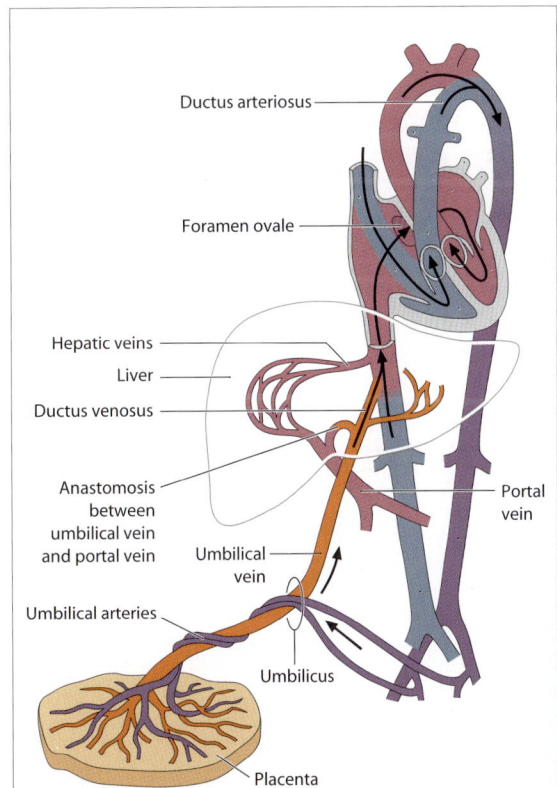

Fig. 38.2   Fetal circulation.

is very high. Hence, most of the blood pumped into the pulmonary artery flows out into the aorta through a connecting artery called the *ductus arteriosus*. A much smaller amount flows through the pulmonary capillaries into the left atrium. Blood from the inferior vena cava mostly enters the left atrium through the *foramen ovale*, a small hole in the interatrial wall guarded by a flap that permits only one-way flow of blood. In the left atrium, it mixes with the small amount of blood brought in by the pulmonary veins and then enters the left ventricle which pumps it into the aorta.

The aorta supplies blood to the brain and upper part of the body through the common carotids and then continues as the descending aorta to supply the lower part of the body. Significantly, the highly deoxygenated blood from the pulmonary vein flowing into the ductus arteriosus enters the aortic arch only after the origin of the common carotids, so that the brain receives relatively better oxygenated blood. It is important to note that in the fetus, both the ventricles are pumping blood into the aorta. Thus the two ventricles operate in parallel and not in series, as in the adult heart.

■ **Circulatory changes at birth**

The umbilical cord constricts when it comes in contact with the relatively colder atmospheric air. Its subsequent clamping completely stops the flow of oxygenated blood to the fetus through the umbilical vein. This asphyxiates the fetus, stimulating it to take its first breath. As air distends the lungs, the resistance of the pulmonary vessels reduces markedly and the left atrial pressure increases. On the other hand, the elimination of the placental circulation increases the resistance of the systemic circulation. The pressure in the pulmonary artery therefore falls below the aortic pressure causing a reversal of blood flow through the ductus arteriosus: For a few days, blood continues to flow through it from the aorta into the pulmonary artery. The rise in Po_2 caused by breathing through the lungs gradually brings about the closure of the ductus arteriosus by causing contraction of the smooth muscles in its walls. $PGE_{2\alpha}$ seems to mediate the closure. Over a period of time, it reduces to a fibrous band called the *ligamentum arteriosum*. Similarly the ductus venosus is

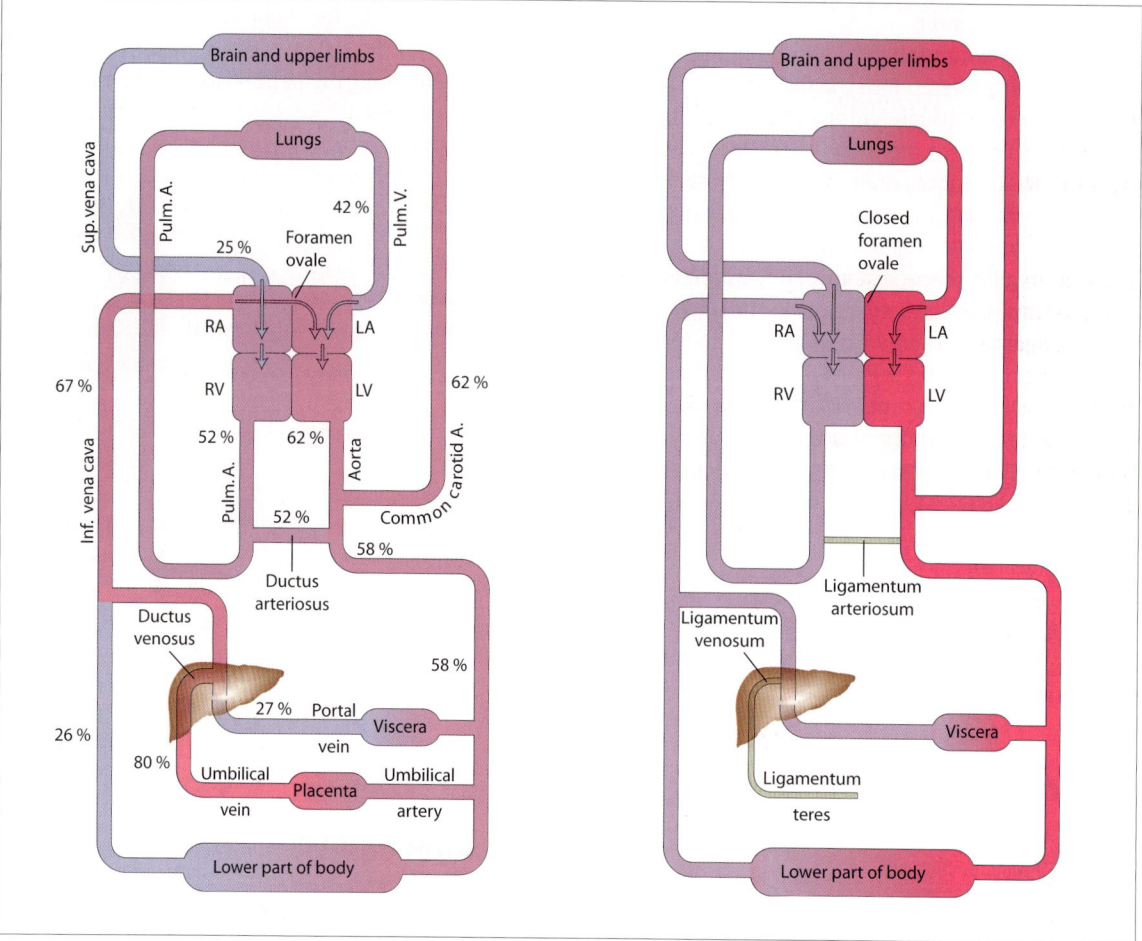

Fig. 38.3   Fetal circulation (*left*) and circulatory changes at birth (*right*).

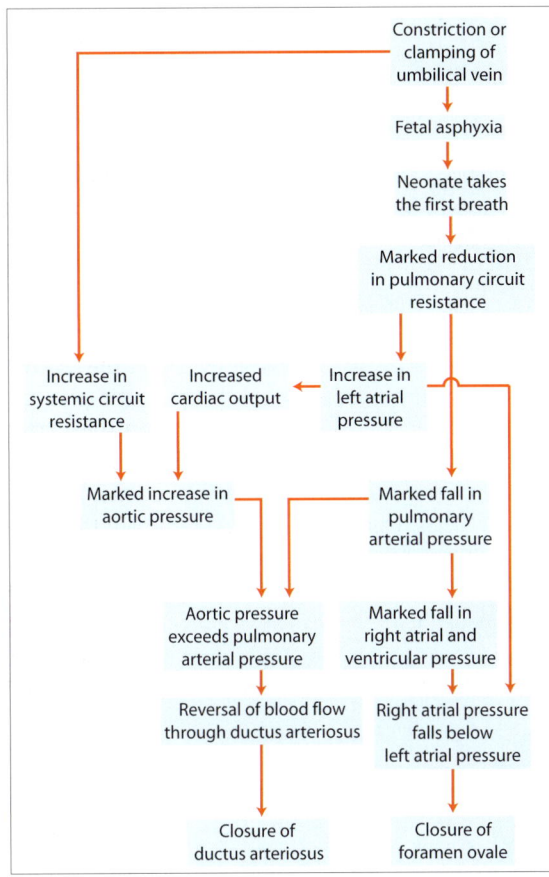

Fig. 38.**4**   Mechanism of circulatory changes at birth.

ventricular septal defect. It is also seen when the ductus arteriosus fails to get obliterated after birth, a condition known as the *patent ductus arteriosus*. As a result of the shunt, a part of the left ventricular output recirculates through the lungs (Fig. 38.**5**). It thus causes volume overload of the pulmonary circulation and right ventricle but does not cause any cyanosis.

Cyanosis occurs in *Fallot's tetralogy.* Two of the four defects that characterize the tetralogy are ventricular septal defect and pulmonary stenosis. The pulmonary stenosis markedly increases the right ventricular pressure, leading to the third defect, i.e., right ventricular hypertrophy. The high right ventricular pressure leads to a *right-to-left shunting* of blood through the ventricular septal defect, i.e., mixing of the deoxygenated blood in the right ventricle with the oxygenated blood in the left ventricle. The oxygenation of arterial blood therefore gets diluted, leading to central cyanosis. The fourth defect in Fallot's tetralogy is the overriding of aorta on the ventricular septum. This allows right ventricular blood to move directly into the aorta, which too constitutes a right-to-left shunt.

■■■ Vascular System

The vascular system begins with the aorta and pulmonary artery which originate from the ventricles. Each ramifies separately and successively into smaller arteries, arterioles, and capillaries. The capillaries join together and drain successively into the venules and veins. The largest veins drain into the atria. The blood vessels of different caliber have different structural and functional characteristics (Fig. 38.**6**).

■ Arteries and arterioles

Arteries   All arteries and arterioles have three concentric layers in their wall: an inner layer called the tunica intima made of squamous endothelial cells, an intermediate layer called the tunica media composed of smooth muscle cells oriented circumferentially, and an outer coat called the tunica adventitia made of fibroblasts and collagen fibers. The tunica intima and tunica media are separated by the internal elastic lamina. The tunica media and tunica adventitia are separated by the external elastic lamina.

Large elastic arteries have a lot of elastic tissues in their wall. The elastic walls prevent abrupt changes in blood pressure, a phenomenon called the *Windkessel effect*. The walls are stretched during systole, preventing a sudden rise in blood pressure. During diastole,

reduced to a fibrous *ligamentum venosum* and the intraabdominal part of the umbilical vein gets converted into the *ligamentum teres* (Fig. 38.**3**).

Another consequence of the fall in pulmonary capillary resistance is that the blood pumped by the right ventricle starts flowing easily through the pulmonary vasculature into the left atrium. This reduces the pressures in the right atrium and right ventricle due to the reduced afterload on them. The pressure in the left atrium now exceeds the right atrial pressure and blood tries to flow 'from left to right' through the foramen ovale. That, however, is not permitted by the flap (septum secundum) guarding the foramen ovale, which subsequently closes permanently.

■ Congenital defects in circulatory pathway

Any abnormal communication between the left-sided and right-sided chambers of the heart usually results in a *left-to-right shunting* of blood due to the higher pressures in the left-sided chambers. This is seen in congenital heart diseases like an *atrial septal defect* or a

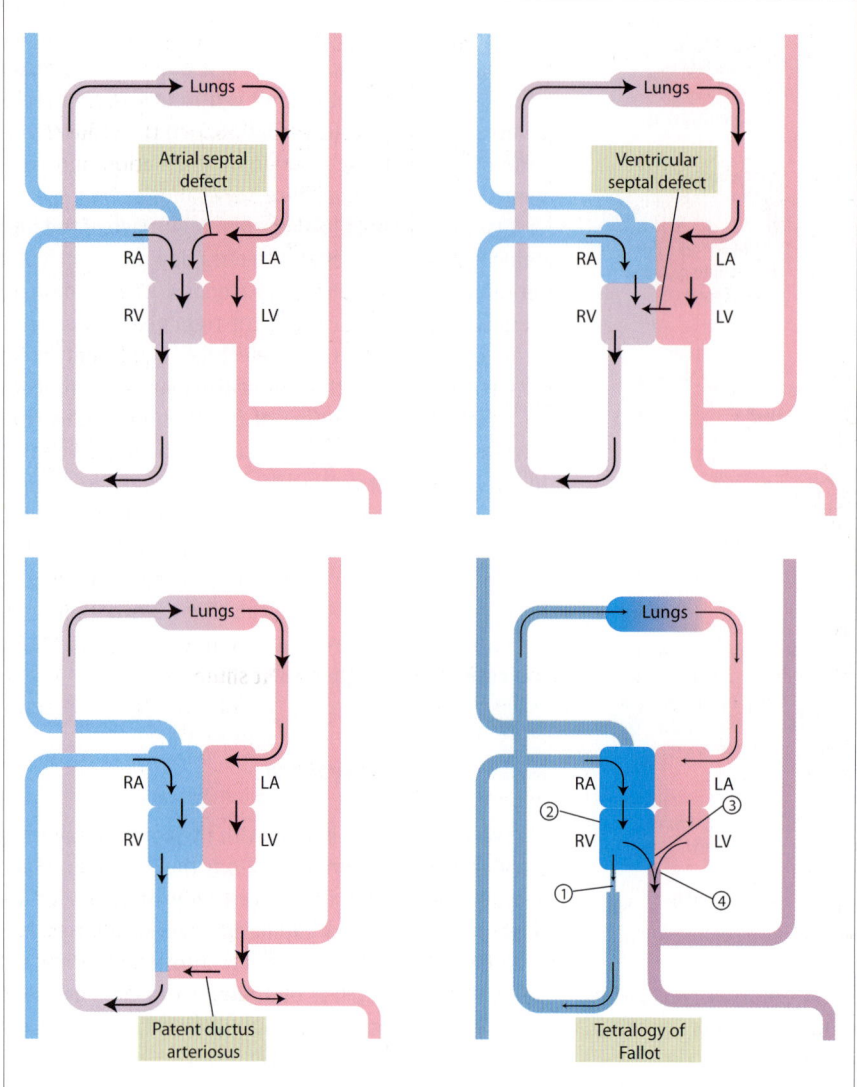

Fig. 38.**5** Congenital heart diseases. The four defects in Fallot's tetralogy are: (1) pulmonary stenosis, (2) right ventricular hypertrophy, (3) ventricular septal defect, and (4) overriding of aorta on the ventricular septum.

they recoil back, preventing a sudden drop in blood pressure.

The **medium and small arteries** contain less elastic tissue and more muscle fibers than the large elastic arteries. The smallest arteries terminate in the muscular arterioles, about 100–50 μm in diameter.

Arterioles act as resistance vessels.   They are the major site of the resistance to blood flow. This is made possible by the presence of considerable amounts of smooth muscle (and less elastic tissue) in their wall, and by the sympathetic (noradrenergic) innervation, which constricts the arterioles by producing contraction of those smooth muscles. The terminal arterioles are the terminal continuation of the muscular arterioles and are about 20 μm in diameter.

Metarterioles branch off almost at right angles from terminal arterioles (Fig. 38.**7**). Its proximal portion is discontinuously coated with smooth muscle cells. Its distal segment, which is also called the *thoroughfare* or *preferential channel*, is devoid of smooth muscle coat. The thoroughfare channels open into the venules.

■ **Capillaries and postcapillary venules**

Capillaries branch off from metarteriole and terminal arterioles. Capillaries are of two types: the true capillaries and the thoroughfare channels (Fig. 38.**7**). The *true capillaries* form an anastomotic network before reuniting and draining into a venule. The origin of a true capillary, whether directly from the terminal arteriole or from the metarteriole, is guarded by a *precapillary*

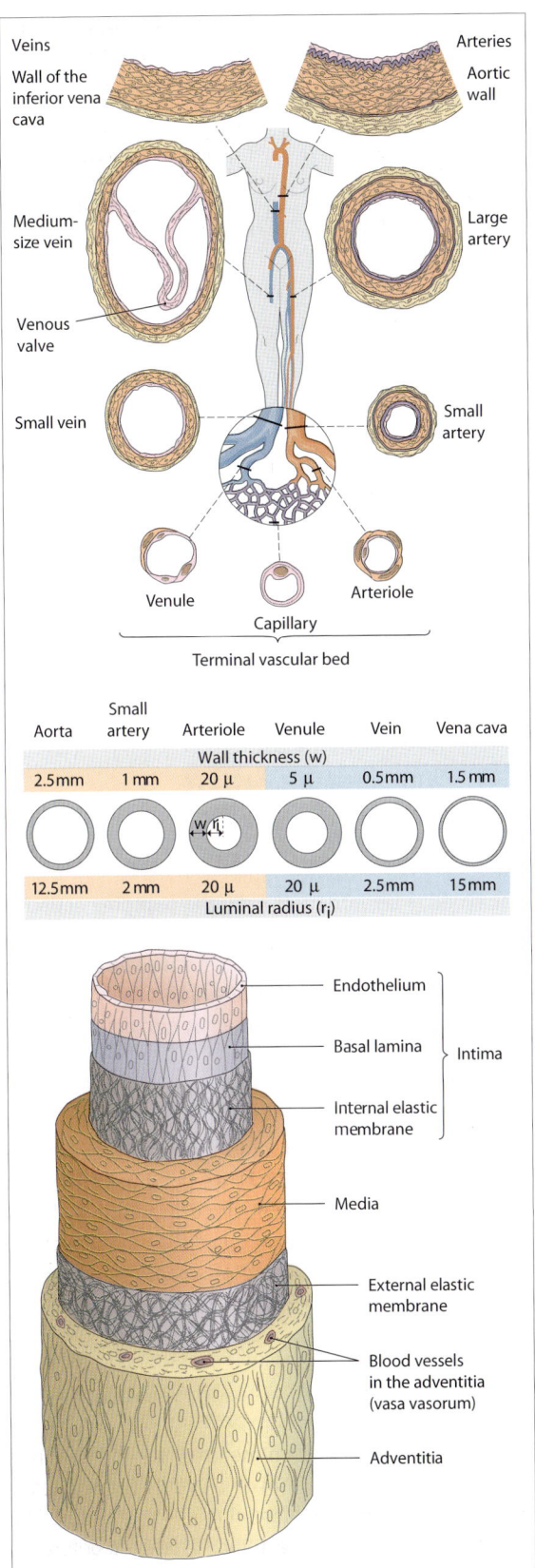

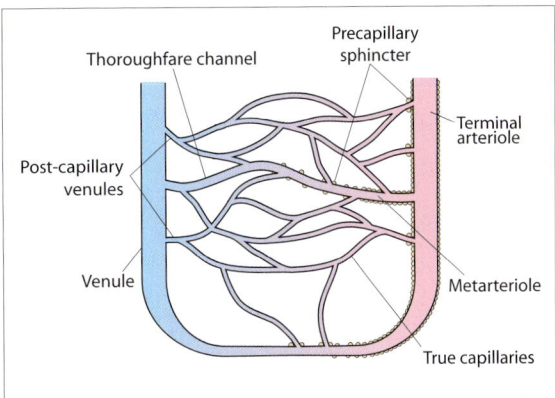

Fig. 38.7 The arteriole and venule are connected through metarteriole and capillaries. The origin of true capillaries, whether directly from the arteriole or from the metarteriole, is guarded by the precapillary sphincter.

sphincter, which is made of a single layer of single unit smooth muscle and is well innervated by sympathetic fibers. True capillaries permit exchange of plasma and solutes through pores in their walls, which is the main function of capillaries and indeed, of the entire circulatory system. *Thoroughfare channels* do not have any nutritive role.

Postcapillary venules are about 20 μm in diameter and are not much different from capillaries. They participate in capillary exchange and are the preferred site for leukocytic emigration. The walls of larger venules (about 50 μm or more) have smooth muscles in them.

Veins

Veins, like arteries, are made of the tunica intima, tunica media, and tunica adventitia. However, the layers and the boundaries between them are less distinct. Veins show considerable venoconstriction in response to sympathetic nerves and circulating vasoconstrictors like endothelins.

Veins are easily distensible, partly because of the sparse smooth muscles in their wall and partly due to their elliptical cross-sectional contour, which becomes circular when distended. Their distensibility makes them excellent capacitance vessels or blood reservoirs (Fig. 38.**8**). They can accommodate large amounts of blood with minimal rise of pressure and are therefore called a *low-pressure system*, unlike the arterial system, which is a *high-pressure system*. The venous system (venules and veins) accounts for more than 50% of the total blood storage capacity of the circulatory system. If 100 ml of blood is transfused into circulation, less than

Fig. 38.6 (*Above*) Different types of blood vessels and their dimensions. (*Below*) Structure of an artery.

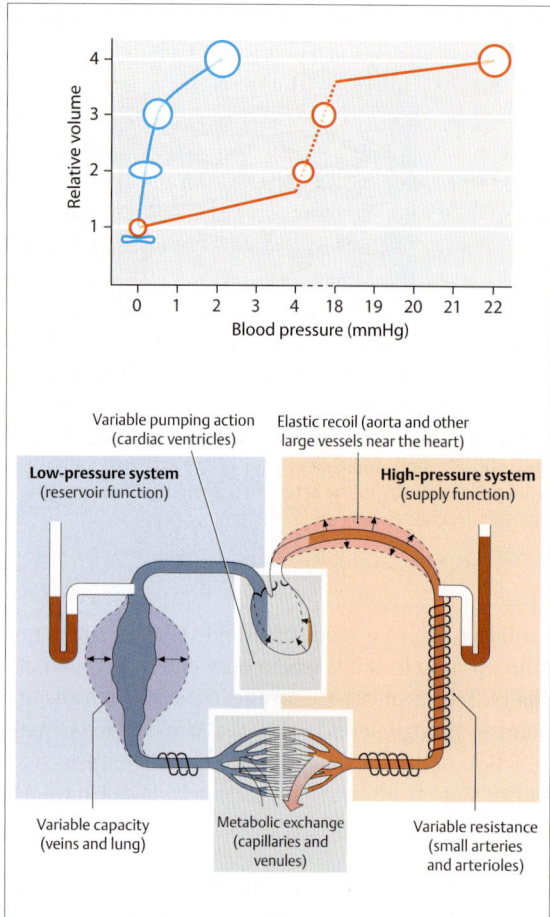

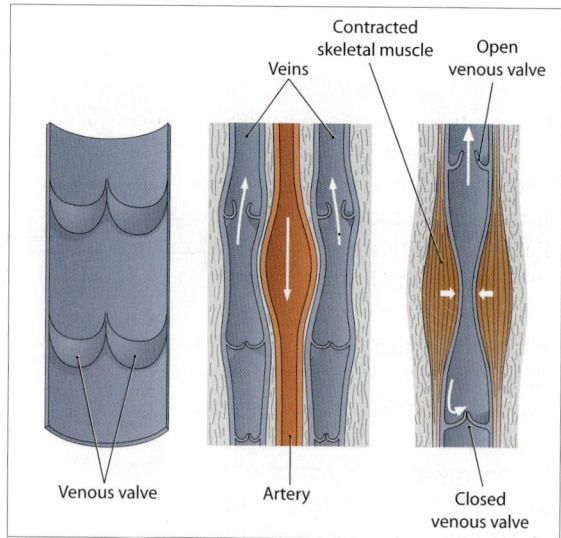

Fig. 38.**9**   Venous valves and muscle pumping.

Fig. 38.**8**   (*Above*) Contour of veins in cross section. When engorged with blood, the contour changes from elliptical to circular. (*Below*) High pressure and low pressure systems.

1 ml of it will enter the high-pressure arterial system and the rest will locate itself in the low-pressure systemic veins, pulmonary circulation, and heart chambers other than the left ventricle.

The intima of the limb veins are folded at intervals to form venous valves that prevent retrograde flow (Fig. 38.**9**). Valves are absent in the great veins and the very small veins.

Arteriovenous anastomoses

The arteriovenous anastomoses are short channels that connect arterioles to venules, bypassing the capillaries. The arterial end of the channel structurally resembles an artery and its venous end structurally resembles a vein. They have muscular walls and are abundantly innervated by both adrenergic and cholinergic nerve fibers. They are present in the skin of fingers, palms, and ear lobes and they have an important thermoregulatory

role. When they are open, they allow a large amount of blood to flow through the skin, dissipating heat to the exterior.

Hemodynamic Factors

Work done

The blood is set into motion by the pumping action of the heart. In the language of physics, work is done *by* the heart *on* blood circulation. Ignoring the wastage of energy in the form of heat, the work done by the heart exactly equals the energy gained by circulating blood.

Work done by the heart   If the ventricles contract against closed valves (see isovolumetric contraction, p 227) there is no change in ventricular volume and therefore no work is done. Work is done by the ventricle only when its volume changes, and is given by the formula:

$$\text{Work done (W)} = \int P dV$$

where, P is the intraventricular pressure and dV is the change in volume. It can be estimated by measuring out the area enclosed in the pressure–volume loop of the cardiac cycle (see Fig. 36.**6**). The rise in intraventricular pressure and the decrease in ventricular volume are determined by the tension developed in the ventricular muscle fibers. Assuming the ventricular cavity to be spherical, the relation between fiber tension (T), ventricular radius (R), and intraventricular pressure (P) is given by Laplace's law.

$$P = \frac{2T}{R}$$

The law tells us that the rise in myocardial fiber tension results in a greater rise in ventricular pressure when the ventricular volume is low. Thus although an increase in the end-diastolic volume increases myocardial contractility and tension (in accordance with Starling's law of the heart), the extra myocardial tension generated fails to produce greater ventricular pressure at higher end-diastolic volume (in accordance with Laplace's law). Hence, in a dilated, failing heart, Starling's law and Laplace's law run at cross-purposes.

The **energy gained by circulating blood** can take various forms. For example, it could lead to the compression of blood. That does not happen because blood, like any other liquid, is incompressible. The efficient circulation of blood is possible only because blood is incompressible. Had it been compressible, cardiac pumping would have compressed the blood instead of making it flow. Gases, however, are highly compressible. *Air embolism*, i.e., entry of air into circulation, has serious consequences. When a large amount of air enters the circulation, it fills the heart and effectively stops the circulation, causing sudden death. When a small amount of air enters the circulation, it flows in blood as bubbles that lodge in the small blood vessels, stopping the flow of blood and causing ischemia. Small air bubbles do not cause any problem in the heart or the large vessels.

Since blood is incompressible, the energy gained by the circulation is converted into other forms of energy: potential energy, kinetic energy, and pressure energy. *Potential energy* is represented by the vertical height through which a column of blood rises against gravity. *Kinetic energy* is represented by the motion of blood. *Pressure energy* is exerted laterally on the walls of the blood vessels. According to Bernoulli's principle, the total of all the three forms of fluid energy will always be constant and in this case, will equal the work done by the ventricle.

Work done by the ventricle =

$$\begin{array}{ccc} \text{Potential energy} & \text{Kinetic energy} & \text{Pressure energy} \\ \text{of blood} & + \quad \text{of blood} & + \quad \text{of blood} \end{array}$$

There are several possibilities for how the total energy is apportioned between its three forms. (1) If the entire circulatory system is horizontal, e.g., in a recumbent subject, there is no possibility of the blood rising through a height. Hence the entire work done by the ventricle will be converted partly into kinetic energy and partly into pressure energy. (2) If the blood is forced up a vertical column, its potential energy increases. The increase in potential energy must necessarily be accompanied by a decrease in pressure energy and kinetic energy. (3) Arteriolar constriction increases the blood flow velocity through the arteriole but decreases blood flow velocity in the arteries. The reduction in kinetic energy of blood in the artery is associated with a rise in arterial blood pressure. The increase in kinetic energy of blood in the arteriole is associated with a fall in arteriolar blood pressure. However only the arterial blood pressure is important clinically. Similarly, when a vessel is narrowed by an atherosclerotic plaque, the velocity of flow in the constricted segment increases. The increase in the kinetic energy results in a reduction in the lateral pressure in the affected segment. The narrowing therefore tends to maintain itself.

It can thus be appreciated from Bernoulli's principle that for a given amount of energy imparted to blood circulation by the contracting heart, the rise in blood pressure depends on the *velocity and rate of blood flow* (which affect the kinetic energy of blood) and the *effect of gravity* (which affects the potential energy of blood). These two factors are discussed below in details.

Velocity and rate of blood flow

Blood flow depends on both the characteristics of the blood vessels and also the viscosity of the blood that flows through them. If we assume that the flow of blood is streamlined and there is no turbulence in the flow, then the effect of vessel characteristics and blood viscosity are explained by the Poiseuille–Hagen formula.

The **effect of vessel characteristics** on blood flow is explained below by taking the example of arterioles (Tables 38.**1** and 38.**2**). The flow velocity has been calculated in two different ways. An arteriole has an internal diameter of 0.02 mm. Its luminal area is therefore 3.142×10^{-4} mm^2 (Area = $\pi/4 \times$ diameter^2). There are 40,000,000 arterioles disposed in parallel. Assuming the cardiac output to be 90,000 mm^3 s^{-1} (5400 cm^3 min^{-1}), it can be calculated that the flow through each arteriole is 0.00225 mm^3 s^{-1}. Dividing this flow rate by the cross-sectional area of an arteriole, we get a flow velocity of 7.16 mm^3 s^{-1} (0.00225 mm^3 s^{-1} ÷ 3.142×10^{-4} mm^2 = 7.16 mm s^{-1}).

There is another method of calculating the flow velocity from the same data. Since there are 40,000,000 arterioles disposed in parallel, each with a cross-sectional area of 3.142×10^{-4} mm^2, the total cross-sectional area of all the arterioles taken together is 12,566.4

Table 38.**1** Comparative dimensions of different types of blood vessels in a dog.

	Diameter (mm)	Length (mm)	Number
Aorta	10	400	1
Large arteries	3	200	40
Main arterial branches	1	100	600
Terminal branches	0.6	10	1800
Arterioles	0.02	2	40,000,000
Capillaries	0.008	1	1,200,000,000
Venules	0.03	2	80,000,000
Terminal veins	1.5	10	1800
Main venous branches	2.4	100	600
Large veins	6	200	40
Vena cava	12.5	400	1

Table 38.**2** Comparison of total cross-sectional area (TCSA), flow rate (FR), flow velocity (FV), and relative resistance (comparison to aortic resistance) of different types of blood vessels.

	TCSA (mm²)	FR (mm³ s⁻¹)	FV (mm s⁻¹)	Relative resistance
Aorta	79	90000	1146	1.00
Large arteries	283	2250	318	1.54
Main arterial branches	471	150	191	4.17
Terminal branches	509	50	177	1.07
Arterioles	12566	0.00225	7	7.81
Capillaries	60319	0.000075	1	5.09
Venules	56549	0.001125	2	0.77
Terminal veins	3181	50	28	0.03
Main venous branches	2714	150	33	0.13
Large veins	1131	2250	80	0.10
Vena cava	123	90000	733	0.41

mm² (3.142 × 10⁻⁴ mm² × 40,000,000 = 12,566.4 mm²). By dividing the cardiac output by the total cross-sectional area, we find the flow velocity through the arterioles to be 7.16 mm s⁻¹ (90,000 mm³ s⁻¹ ÷ 12,566.4 mm² = 7.16 mm s⁻¹).

Note that in one method, the flow rate is calculated first and the flow velocity is derived from it. The other method calculates the flow velocity directly, by introducing the concept of 'total cross-sectional area' of all arterioles. The calculations are illustrative of the following facts: (1) The flow velocity (distance traveled by blood in unit time) is not the same as flow rate (volume of blood flowing in unit time), which is illustrated in Fig. 38.**10**. (2) At a given cardiac output, the flow velocity in a blood vessel is inversely proportional to its total cross sectional area. (3) Smaller blood vessels are larger in number and have a larger total cross sectional area and a lower flow velocity of blood in them.

The resistance of a vessel to blood flow can be calculated by combining Ohm's law with the Poiseuille–Hagen formula. From Ohm's law, we have:

$$\text{Resistance } (\rho) = \frac{\text{Effective perfusion pressure } (\Delta P)}{\text{Flow rate } (Q)}$$

The SI unit of pressure is N m⁻² and that of flow rate is m³ s⁻¹. The SI unit of fluid resistance will therefore be Newton-second meter⁻⁵! A simpler alternative is to express fluid resistance in R units. One R unit is the resistance that will cause a flow of 1 ml s⁻¹ when the pressure is 1 mmHg. Thus if the mean blood pressure is 100 mmHg and the cardiac output is 6 L min⁻¹ (100 ml s⁻¹) the total peripheral resistance is:

$$\frac{100 \text{ mmHg}}{100 \text{ ml s}^{-1}} = 1 \text{ R unit}$$

Most of the major capillary beds of the body are disposed in parallel (see Fig. 1.**17**). When the resistance of any of them decreases, the total resistance falls in accordance with Ohm's law and most of the blood gets directed through the dilated capillary beds. This explains why the total peripheral resistance decreases in

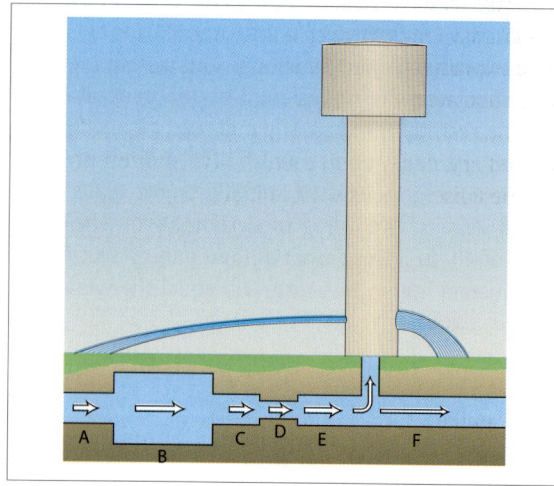

Fig. 38.**10** Flow rate versus flow velocity. There are two holes of different sizes in the water tank at the same level. The leaking water has a higher flow velocity through the smaller hole and a higher flow rate through the larger hole. Both jets have the same kinetic energy. There is also an underground water pipe that feeds the overhead water tank. The flow rate in the pipe is uniform from A to E and decreases at F. The flow velocity is lowest at B and highest at D.

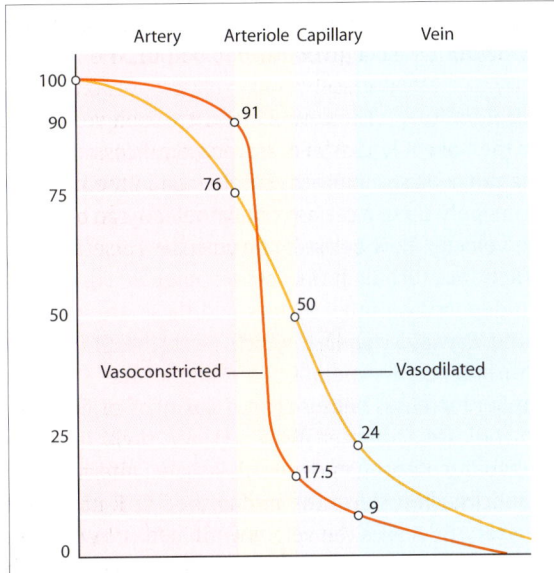

Fig. 38.**11** Pressure drop along arteries, arterioles, and capillaries.

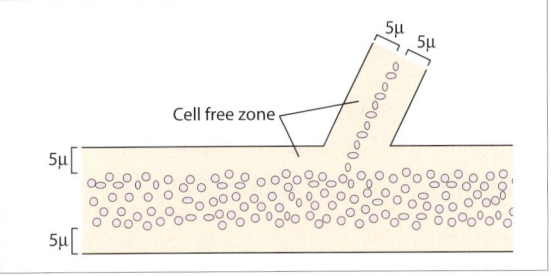

Fig. 38.**12** Axial streaming and plasma skimming.

Dividing,

$$\frac{\rho_{artl}}{\rho_{cap}} = \frac{L_{artl} \times n_{cap} \times R_{cap}^{\,4}}{L_{cap} \times n_{artl} \times R_{artl}^{\,4}}$$

or,

$$\rho_{artl} / \rho_{cap} = L_{artl}/L_{cap} \times n_{cap}/n_{artl} \times (R_{cap}/R_{artl})^4$$

From the data in Table 38.**1**, it can be calculated that L_{artl} / L_{cap} is 2, n_{cap}/n_{artl} is 30 and $(R_{cap} / R_{artl})^4$ is 0.0256. The product of the three ratios is about 1.5. In other words, the arteriolar resistance is about 1.5 times higher than capillary resistance. During vasoconstriction, the pressure-drop across arterioles increases markedly (Fig. 38.**11**) indicating that the vasoconstriction occurs mainly at the arteriolar level. Arterioles are therefore called *resistance vessels*. The relative resistances of the various segments are given in Table 38.**2**. These can be calculated using the data given in Table 38.**1**.

Effect of blood viscosity on blood flow Although the effect of viscosity on fluid flow is given by the Poiseuille–Hagen formula, the formula is valid only for Newtonian fluids and is not precisely applicable to blood which is not a Newtonian fluid. The viscosity of blood decreases with an increase in shear rate. This is especially marked at low shear rates. This anomalous viscosity of blood is attributable to *axial streaming* of blood cells at high shear rates. Axial streaming means that the cells occupy the central axis of the tube through which blood is flowing. This leaves a 5 μm wide cell-free zone immediately adjacent to the vessel wall. This cell free zone produces less friction with the vessel wall and therefore, the viscosity is lower.

Axial streaming also explains the *Fåhreus–Lindqvist effect*, i.e., the reduction in blood viscosity with a decrease in tube diameter. Since the cell-free zone has a constant width of about 5 μm, expressed as a percentage of the tube diameter, it is much greater for smaller tubes and therefore results in a lower viscosity.

Another consequence of axial streaming is *plasma skimming*, i.e., a vessel that branches off from the main

exercise (resistance of muscle capillary bed falls) or in pregnancy (resistance of uterine capillary bed falls). It also explains the phenomena of coronary steal (p 287) and subclavian steal (p 291).

The resistance offered by a single blood vessel, say a capillary, depends on several factors which are given by the Poiseuille–Hagen formula.

$$Q = \Delta P \times \frac{\pi}{8} \times \frac{1}{\eta} \times \frac{R^4}{L}$$

or

$$\Delta P\big/Q \,(\text{Resistance}) = \frac{8\eta L}{\pi R^4}$$

If *n* capillaries are disposed in parallel, then the total resistance (ρ_{cap}) offered by all the *n* capillaries will be:

$$\rho_{cap} = \frac{8\eta L_{cap}^{\,4}}{n_{cap}\pi R_{cap}^{\,4}}$$

where L_{cap}, n_{cap}, and R_{cap} are the length, number, and radius of the capillary. Similarly, for arterioles,

$$\rho_{artl} = \frac{8\eta L_{artl}^{\,4}}{n_{artl}\pi R_{artl}^{\,4}}$$

blood vessel at a large angle carries way more plasma than cells because the cell-free zone lies at the periphery of flowing blood (Fig. 38.**12**). It explains why the hematocrit of capillary blood is about 25% lower than the whole-body hematocrit. In large vessels, a rise in hematocrit causes an appreciable rise in viscosity. However, in vessels smaller than 100 μm in diameter, like the arterioles, capillaries, and venules, the viscosity change is much less than it is in large vessels. This is why hematocrit changes have relatively little effect on the peripheral resistance except when the changes are large.

Critical closing pressure   If one goes by the Poiseuille–Hagen formula, the flow of blood through a blood vessel of finite length can never be zero unless the pressure gradient is zero. It is however observed that the blood flow reduces to zero well before the pressure gradient reduces to zero. The pressure gradient at which blood flow through a vessel reduces to zero is called the *critical closing pressure* or the *zero flow pressure*. This pressure is required to force the slightly oversized red cells through the small blood vessels. The critical closing

pressure increases with rouleaux formation and with sympathetic vasoconstriction of blood vessels.

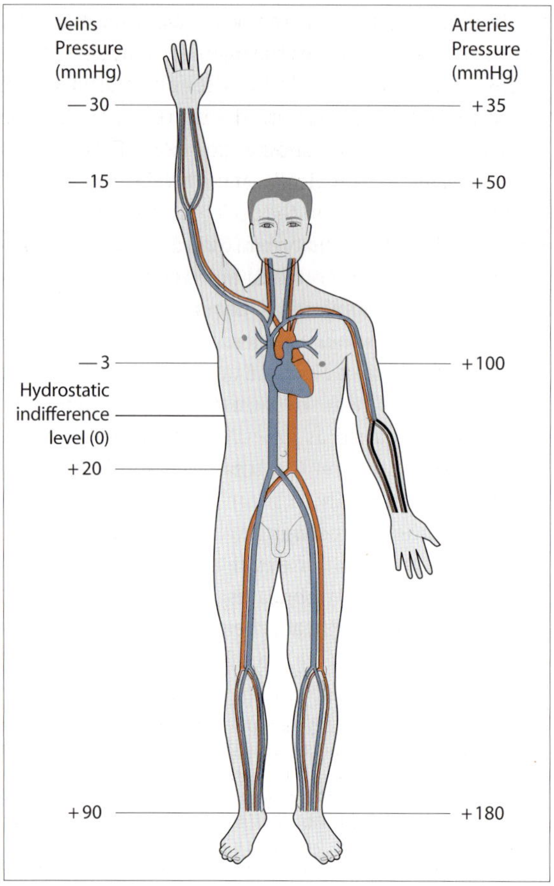

Turbulent flow   The flow of blood in the blood vessels, like the flow of liquids in narrow rigid tubes, is normally laminar or streamlined. The laminar flow however occurs only up to a certain critical velocity. At or above that velocity, flow becomes turbulent. Streamline flow is silent but turbulent flow creates sounds. Turbulence also depends on other factors and these are included in the *Reynolds number* which gives the probability of turbulence. In anemia, for example, the Reynolds number increases because the viscosity of blood is low and the cardiac output is high. Hence, there is higher probability of turbulence, which is associated with soft systolic murmurs over the cardiac area. Constriction of an artery increases the velocity of blood flow through the constriction, producing turbulence as the Reynolds number increases. The turbulence produces sound beyond the constriction, examples of which are the bruits heard over arteries constricted by atherosclerotic plaques, the murmurs over stenosed and leaking valves, and the Korotkoff sounds that are heard during sphygmomanometry.

■ Effect of gravity on circulation

Effect of normal gravity   Both venous and arterial pressures are affected by gravity, increasing or decreasing by an amount that is given by the formula hρg (see Pressure, p 6). Thus the pressure increases by 0.77 mmHg for each centimeter below the right atrium and decreases by the same amount for each centimeter above the right atrium (Fig. 38.**13**).

Venous pressure about 5 mm above the heart is zero. Above 5 mm, all veins are in a collapsed state since the blood pressure inside them is less than zero (i.e., subatmospheric). The jugular vein rises straight up from the right atrium and acts as a manometer for the right atrial pressure, which is also called the central venous pressure (CVP). Since the CVP is greater than zero, blood is pushed up some distance into the collapsed jugular vein, which then gets distended. The upper limit of the distended jugular vein shows pulsations called the *jugular venous pulse* (JVP) which are synchronous with the atrial pressure waves. The vertical height of the distended segment of the jugular vein gives the jugular venous pressure (Fig. 38.**14**).

In the head and neck, all the veins are in a collapsed state. The pressure is subatmospheric (about −10 mmHg) in the dural sinuses, which do not collapse completely as their walls are fixed to the cranium. If a dural sinus is opened up during a neurosurgical

Fig. 38.**13**   Effect of gravity on arterial pressure.

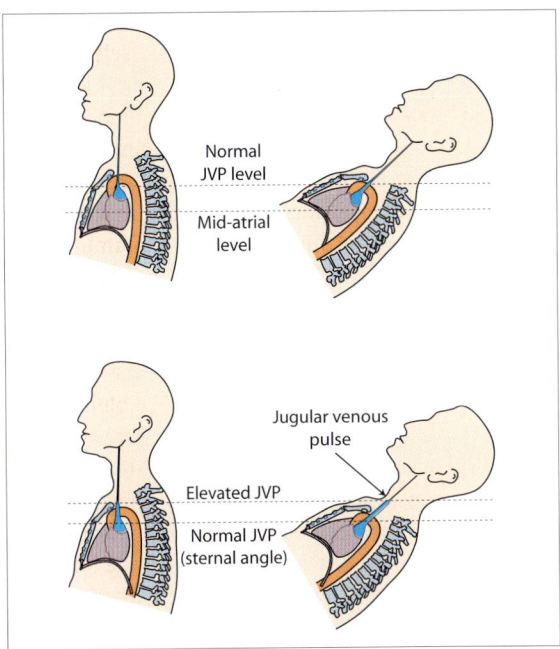

Fig. 38.**14** The jugular venous pressure. Even when slightly elevated, the jugular venous pulse remains hidden behind the manubrium sterni. Tilting the patient makes the pulse visible in the neck. The jugular venous pressure is taken as the vertical height of the pulse from the right atrium. See also Figure 1.**5**A.

Fig. 38.**15** A boy in the middle of a free fall. Will the drink splash out of the glass as the boy has a free fall? Will there be a rush of blood to his head?[1]

procedure with the patient seated, air is sucked into the sinus, resulting in air embolism.

In the lower limbs, blood tends to accumulate in the more dependent parts, especially if the venous valves are incompetent. This is known as *venous pooling* and its cause can be appreciated if one considers that a continuous vertical column of venous blood extending from the right atrium to the foot would exert a downward pressure of 90 mmHg at the foot. The pressure would be transmitted back into capillaries and cause pedal edema due to transudation of fluids. If such a high pressure does not normally exist at the foot, it is because the tall column of venous blood is broken up into shorter columns by the venous valves: Each venous valve bears the downward pressure exerted by the short column of venous blood above it. When the valves are incompetent, the venous pressure at the foot rises and venous pooling occurs.

Effect of G forces There are situations where the body is subjected to high positive Gz, low positive Gz, zero Gz or negative Gz. Any change in Gz changes the downward pressure exerted by a fluid column which is given by hρg. The higher the Gz+, the heavier the fluid becomes and *vice versa*. Thus at a higher Gz+, blood becomes heavier and venous pooling increases in the dependent parts of the body. At very high Gz+

(as experienced by pilots during looping), blood becomes so heavy that arterial blood is unable to rise to the brain, causing cerebral ischemia and unconsciousness. Moments before unconsciousness, there is *gray-out* when everything appears gray due to ischemia of cones and loss of color vision. The gray-out is followed by a *black-out*, i.e., a total loss of vision. The gray-out serves as a warning for pilots who must slow down the aircraft and reduce the Gz+.

At zero Gz, blood becomes weightless and therefore the venous blood reaches the brain as easily as it reaches the lower limbs (Fig. 38.**15**). The circulatory changes associated with zero Gz are the same as the changes associated with the supine position. Gravitational force always acts vertically and its effect on circulation is eliminated when the body is in the supine position and blood flows horizontally. For the same reason, astronauts are able to avoid the effects of high Gz or negative Gz by positioning themselves perpendicular to the direction of the motion of the spacecraft. The tolerance for G forces exerted across the body is much greater in the chest-to-back direction (Gx-) than in the back-to-chest direction (Gx+). Astronauts are therefore positioned to take the G forces of rocket flight in the chest-to-back direction.

In negative Gz, the weight of the blood is directed upward. This causes a rise in cerebral arterial pressure and congestion of cerebral vessels leading to *red-out* and severe throbbing headache. In spite of the great rise

---

[1] It can be experimentally verified, by dropping from a height a paper cup filled with water, that the drink will not splash out of the glass. During free fall, everything becomes weightless: the boy, the glass, and the drink, and the drink remains in the glass because both fall with the same acceleration due to gravity. Yet, as the boy has a free fall, there will be a rush of blood into his head. The reason here is different: the body falls down but the blood is pumped up by the heart. Since blood becomes weightless during free fall, it is pumped with a greater velocity into the head. Even the venous column of blood, which normally rises a few centimeters above the heart, would now reach the head.

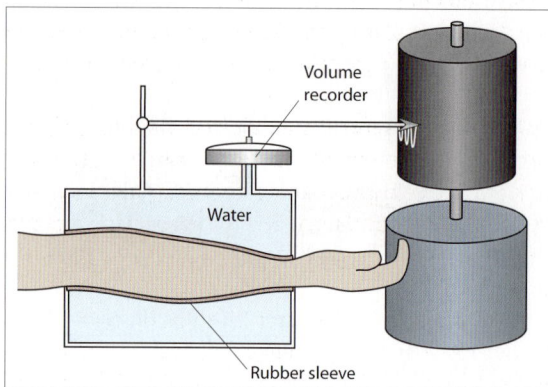

Fig. 38.**16**   Plethysmography of the forearm

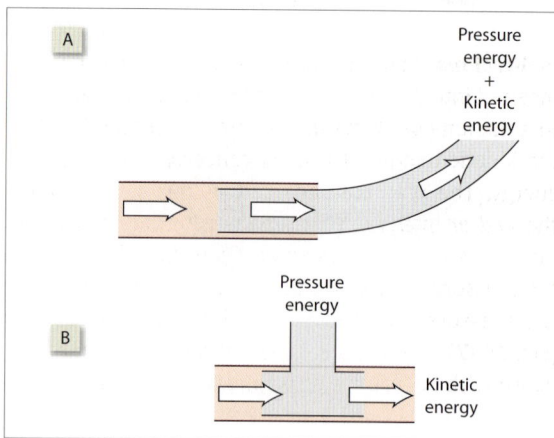

Fig. 38.**17**   Relevance of Bernoulli's principle in direct manometry: Measurement of end-on pressure (*above*) and lateral pressure (*below*).

in cerebral arterial pressure, the vessels in the brain do not rupture because there is a corresponding increase in intracranial pressure which supports the walls of the blood vessels. In other words, the cerebrospinal fluid acts like a G-suit (see below). There is also a general increase in venous return leading to an increase in cardiac output. Most of the cardiac output however moves toward the upper parts of the body.

The *antigravity suit* or the G-suit, in its simplest form, is a water-filled jacket that is worn by astronauts. When there is a high Gz+, there is tendency of venous pooling. Simultaneously, the water in the G-suit also rushes to the lower parts so that there is a proportionate increase in the pressure in the G-suit. Thus, as Gz increases, the pressure in the G-suit parallels the blood pressure in the dependent parts of the body, preventing venous pooling and edema.

Hemodynamic Measurements

Measurement of blood flow velocity

Doppler flow meters   Blood flow velocity can be measured with Doppler flow meters. Ultrasonic waves are sent into a vessel diagonally from one crystal, and the waves reflected from the red and white blood cells are picked up by a second, downstream crystal. The frequency of the reflected waves is higher by an amount that is proportionate to the velocity of flow toward the second crystal because of the Doppler effect.

Electromagnetic flow meters depend on the principle that a voltage is generated in a conductor moving through a magnetic field and that the magnitude of the voltage is proportionate to the speed of movement. Since blood is a conductor, a magnet is placed around the vessel and the voltage, which is proportionate to

the flow velocity, is measured with an appropriately placed electrode on the surface of the vessel.

Clinical estimation   Clinically, an approximate idea of the velocity of the circulation can be obtained by injecting a bile salt preparation into an arm vein and timing the first appearance of the bitter taste it produces. The average normal arm-to-tongue circulation time is 15 seconds. However the normal variations are too much to allow its diagnostic use.

Measurement of blood flow rate

Plethysmography (= volume recording) estimates the blood flow rate to a limb by measuring its change in volume following venous occlusion (Fig. 38.**16**). The part of the body where the blood flow is to be measured (the forearm, for example) is sealed in a watertight chamber (plethysmograph). Changes in the volume of the forearm reflect the changes in the amount of blood and interstitial fluid it contains. When the volume of the forearm increases, it displaces water. The volume of water displaced is measured with a volume recorder. When the venous drainage of the forearm is occluded, the rate of increase in the volume of the forearm is a function of the arterial blood flow (venous occlusion plethysmography).

Fick principle and indicator dilution   Indirect methods like the Fick method (p 232) and dye dilution technique (p 233) have already been described. Methods like the Kety N_2O method for measuring cerebral blood flow or the estimation of renal blood flow by measuring PAH clearance are based on the Fick principle.

■ Measurement of blood pressure

In **direct manometry**, one end of a pressure cannula is inserted into the artery and the other end is connected to a manometer. Flow in the artery is blocked and all the kinetic energy of flow is converted into pressure energy. The pressure recorded in this way is called the *end-on pressure*. If instead, a T tube is inserted into a vessel so that the blood can flow through and the pressure is measured in the side arm of the tube, the pressure recorded is called the *lateral pressure* (Fig. 38.**17**). It is lower than the end pressure, in accordance with Bernoulli's principle. Direct blood pressure

measurement is performed clinically for CVP monitoring during circulatory shock, and for pulmonary wedge pressure measurement in pulmonary hypertension.

Sphygmomanometry is an indirect method of blood pressure measurement. There are three methods of performing sphygmomanometry, viz., palpatory, auscultatory, and oscillatory. For the most part, the procedure for sphygmomanometry is the same in all the three methods.

An inflatable rubber cuff (the Riva-Rocci cuff) attached to a mercury manometer (sphygmomanometer) is wrapped around the arm about an inch above the cubital fossa (Fig. 38.**18**). The cuff is rapidly inflated until the pressure in it is well above the expected systolic pressure so that the brachial artery is occluded. The pressure in the cuff is then lowered slowly at the rate of about 1 cm s^{-1}. When the systolic pressure in the artery just exceeds the cuff pressure, blood spurts through the artery with each heartbeat. This is the point of systolic blood pressure. As the cuff pressure is lowered further, the blood flows through the brachial artery relatively smoothly, though it still passes in spurts, being unable to pass through when the blood pressure is at its lowest. Finally, when the cuff pressure falls to the level of diastolic pressure, the blood flow in the brachial artery becomes entirely free from turbulence. This marks the diastolic blood pressure. The three different methods

> ### The level of the heart in manometry
> An important difference between direct manometry and sphygmomanometry pertains to the level at which the manometer is to be placed. In direct manometry, it is important to place the manometer at the heart level because the conduit conveying the pressure to the manometer is filled with a dense medium (blood or saline). The hydrostatic pressure of the liquid column makes a considerable difference, given by hρg, in the recorded pressure. In sphygmomanometry, the conduit (rubber tube) is filled with air which has a negligible density and therefore, there is no need to place the manometer at the level of the heart. Rather it is important to keep the manometer at eye level so as to avoid parallax while reading out the pressure (Fig 38.**18**).

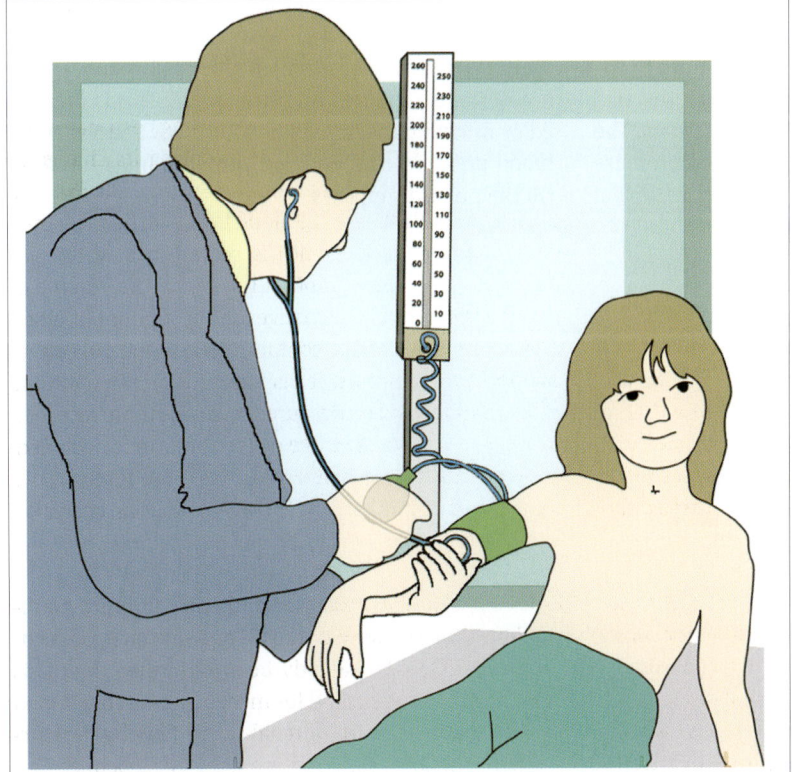

Fig. 38.**18**   Sphygmomanometry by the auscultatory method. See also Figure 1.**5**B.

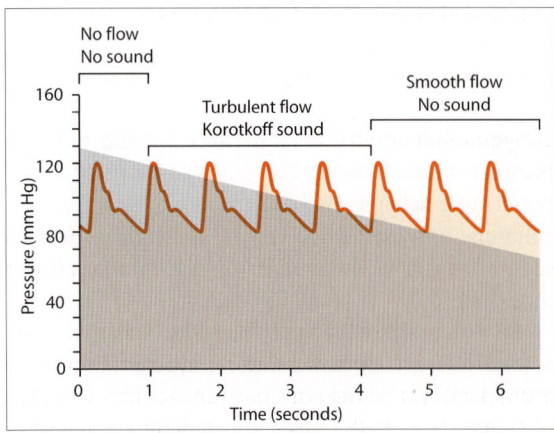

No flow
No sound

Turbulent flow
Korotkoff sound

Smooth flow
No sound

Fig. 38.**19**   Turbulence and Korotkoff sound. The shaded area represents the cuff pressure.

Table 38.**3**   Phases of Korotkoff sounds at different levels of blood pressure.

Phase 1	Tapping sounds, as produced with the fingertips.
Phase 2	Knocking sounds, as produced by the knuckles.
Phase 3	Thumping sounds as produced with the fist.
Phase 4	Patting and wiping sounds as produced by the palm. For best simulation, pat softly and glide the palm over the surface of the table. Technically, it is described as muffled sound.

of sphygmomanometry differ only in the criteria for identifying the systolic and diastolic blood pressures.

In the **oscillatory method**, the mercury column is observed. As the mercury column falls to touch the systolic pressure, it starts showing small oscillations. The oscillations are the largest at the mean blood pressure and abruptly disappear at the diastolic pressure. Visual oscillometry is rarely used. However, the oscillometric method is often used in electronic devices and gives fairly reliable results.

In the **palpatory method**, the radial artery pulse is palpated. As the cuff pressure falls, the pulse appears at the systolic point. It is not an accurate method and the results obtained are about 2–5 mmHg lower than those obtained through auscultatory method. The diastolic point cannot be identified by this method. The palpatory method is routinely used in order to avoid errors due to auscultatory gap (see below).

In the **auscultatory method**, a stethoscope is placed over the brachial artery in the cubital fossa. Sounds called the *Korotkoff sounds* appear at the systolic pressure and disappear at the diastolic pressure. These sounds have been attributed to the turbulence caused by the partial occlusion of the artery (Fig. 38.**19**). There are four phases of Korotkoff sounds and there are authentic descriptions of the sounds. However, the description given in Table 38.**3** should be easier to appreciate and remember. They describe how the sounds can be simulated on a wooden table. These descriptions are unorthodox and should not be quoted in public!

Phase 1 sounds sometimes disappear as the pressure is lowered from the systolic pressure and reappear at a lower level. The interval between the systolic and diastolic pressure when no Korotkoff sound is heard is called the *auscultatory gap*. If the cuff pressure is raised to a pressure in the range of the auscultatory gap and then lowered for measuring the blood pressure, a false-low systolic pressure will be recorded. This is prevented by first palpating the radial pulse while inflating the blood pressure cuff. The cuff pressure must be raised till the pulse disappears.

39 Capillary Exchange and Lymphatic Circulation

Capillary Circulation

The exchange of O_2, nutrients, and waste products between blood and the interstitial tissue takes place across the capillary walls and is known as *capillary exchange*. The diameter of true capillaries is 4–8 μm which is barely enough to permit red cells to squeeze through in single file (see Fig. 23.**1**), and which facilitates gaseous exchange. In resting tissues, most of the capillaries are collapsed (inactive capillaries) and blood bypasses them to flow through the thoroughfare vessels connecting the metarterioles to the venules. In active tissues, the precapillary sphincters dilate and blood starts flowing through the capillaries (active capillaries). The precapillary sphincters are dilated by local metabolic vasodilators and constricted by sympathetic discharge.

Capillary structure

All capillaries are made of a single layer of cells (the endothelial cells) placed on a basement membrane (basal lamina). Often, the entire circumference of the capillary is wrapped around by a single endothelial cell though sometimes there are two or three cells

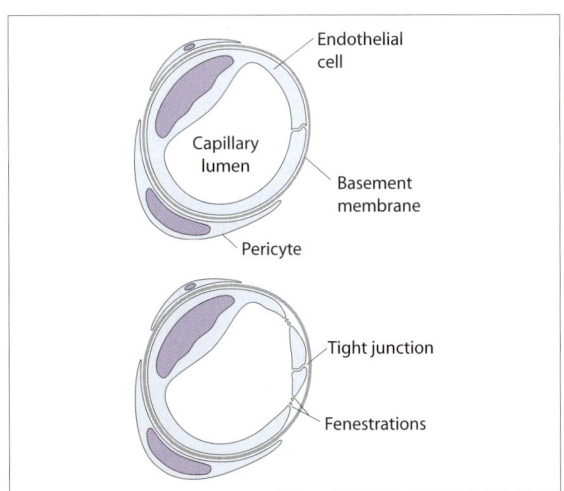

circumferentially. The size of the solutes that diffuse out of the capillaries and the rate at which they diffuse out indicate that the capillary wall has two types of pores: a few large pores with a diameter of about 70 nm and a larger number of small pores with a diameter of 10 nm. The small pores are located at the intercellular junctions between the endothelial cells. The large pores pass through the cell itself and are called fenestrae (windows). The fenestrations permit massive filtration of fluids and the passage of large molecules.

Capillaries that have a large number of fenestrae are called *fenestrated capillaries*. They are found in renal glomeruli and vasa recta, in exocrine and endocrine glands, choroid plexuses, intestinal villi, and in most capillaries that lie close to an epithelium, which includes the skin.

Capillaries that are not densely fenestrated are called *continuous capillaries*. They are found in muscles, brain, and connective tissues. In brain capillaries, tight junctions between endothelial cells constitute a component of the blood brain barrier. The pores in these tight junctions are smaller and they permit only small molecules to pass through (Fig. 39.**1**).

A third type of capillary is found in the sinusoids of the bone marrow, liver, and spleen. In these, there are large intercellular gaps of about 600–3000 nm between the endothelial cells that permit not only the passage of macromolecules but also of erythrocytes. These capillaries are called *discontinuous capillaries*. The basal lamina is much reduced or absent in these capillaries.

Capillary exchange

Capillary exchange occurs in three different ways, viz., through filtration of fluids with bulk flow of solutes, through diffusion of solutes, and through pinocytosis.

Capillary filtration and reabsorption Filtration and reabsorption of fluid across the capillary membrane is determined by the balance of the hydrostatic and oncotic pressures that are called the *Starling forces*.

Figure 39.**2** shows that at the arterial end of the capillary, fluid is filtered out under a filtration pressure of 16 mmHg while at the venous end, fluid is reabsorbed

Fig. 39.**1**   Structure of a capillary, showing how a single endothelial cell is wrapped round its lumen. (*Above*) A continuous capillary. (*Below*) A fenestrated capillary.

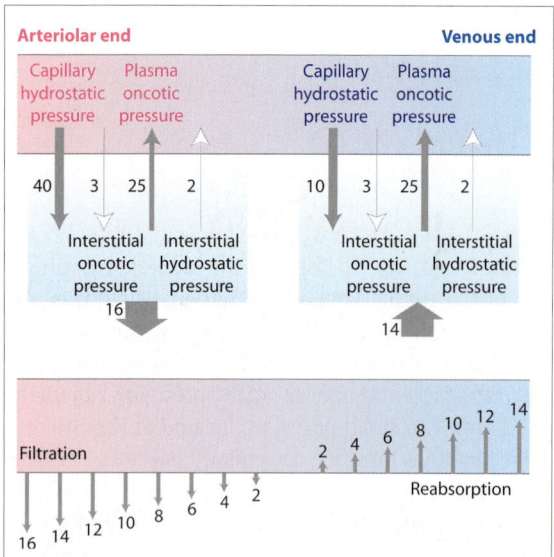

Fig. 39.**2**   (*Above*) Starling forces of capillary exchange at the arterial and venous ends of a capillary. (*Below*) Filtration at the arterial end and reabsorption at the venous end occur under the influence of the Starling forces. Compare with Figure 58.**1**.

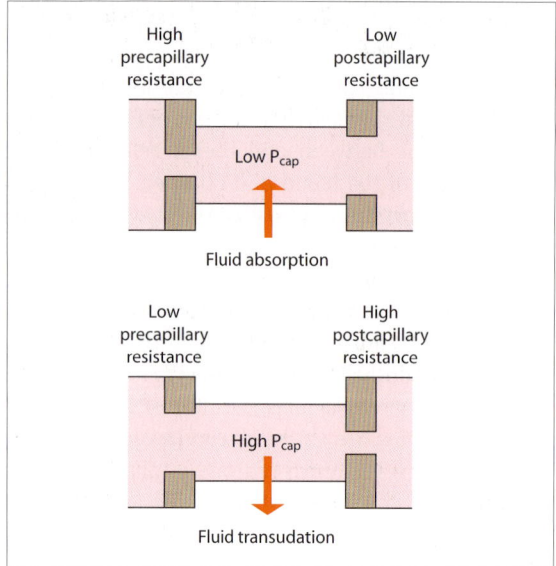

Fig. 39.**3**   Pre:post capillary resistance ratio and the capillary hydrostatic pressure. (*Above*) High pre:post capillary resistance ratio is associated with fluid absorption in capillaries. (*Below*) Low pre:post capillary resistance ratio is associated with fluid transudation in capillaries.

under a reabsorptive pressure of 14 mmHg. Thus, there is a mean filtration pressure of 1 mmHg, i.e., (+16 – 14) ÷ 2. The above dynamics are valid for muscle capillaries. In other capillaries, the balance of Starling forces may be different. For example, fluid moves out of almost the entire length of glomerular capillaries (Fig. 58.**1**B). On the other hand, in intestinal capillaries, fluid

moves into the capillaries through almost their entire length. The calculations of capillary exchange can be simplified by taking the average values of the arterial and venous sides of the capillaries. Thus, the average capillary hydrostatic pressure is taken as P_{Cap} and the average plasma osmotic pressure is represented by π_{Cap}. If P_I and π_I represent the interstitial hydrostatic and oncotic pressure respectively, the equation for the net filtration force (F) can then be written as:

$$F = (P_{Cap} + \pi_I) - (P_I + \pi_{Cap})$$
$$\therefore \quad F = (25 + 3) - (2 + 25)$$
$$= 1$$

The volume of fluid that is filtered under a pressure of 1 mmHg is determined by the capillary filtration coefficient (k_f), which is an index of capillary permeability. Thus, the amount of capillary filtration is given by:

$$V = k_f \times F$$

or

$$V = k_f \times [(P_{Cap} + \pi_I) - (P_I + \pi_{Cap})]$$

Since k_f, π_I, P_I, and π_{Cap} mostly remain unchanged, capillary filtration is controlled largely by P_{Cap}. The capillary hydrostatic pressure increases when the precapillary sphincter dilates or the postcapillary sphincter constricts. In other words, the capillary hydrostatic pressure increases when the *pre:post capillary resistance ratio* decreases (Fig. 39.**3**). Normally, this ratio is **5**:1. Sympathetic vasoconstriction increases this ratio and therefore reduces capillary hydrostatic pressure. Hence, sympathetic discharge favors movement of fluid from the interstitium into the capillary and thereby retains fluid within circulation. Conversely, local metabolites dilate mainly the precapillary sphincter and decrease the pre:post capillary resistance ratio. Hence, the capillary hydrostatic pressure increases, favoring filtration of fluids from the capillary into the interstitium.

The fluid that is filtered out of continuous capillaries into the interstitial space does not have appreciable amounts of proteins in it. This fluid is called a *transudate*. However, if the permeability of the capillaries increases markedly, as happens during inflammation from any cause, large amounts of proteins diffuse out from the capillaries. This protein-rich fluid that comes out of the capillary under conditions of inflammation is called *exudate*. The increase in capillary permeability in inflammation is due to the effect of inflammatory mediators like histamine. Histamine binds to receptors on endothelial cells and stimulates the cells to contract, resulting in the widening of the intercellular junctions between them. The increase in permeability can also be the direct consequence of endothelial injury.

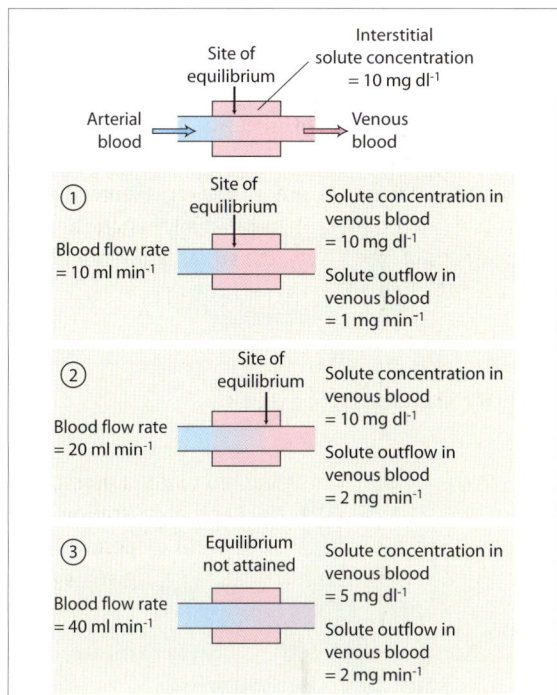

Fig. 39.**4** Flow-limited transport. (1) Capillary blood equilibrates with the tissue at the point marked by the arrow. The solute carried away by the capillary (1 mg min⁻¹) is given by the product of the blood flow rate (10 ml min⁻¹) and the solute concentration in blood (10 mg dl⁻¹). A faster diffusion rate will shift the point of equilibrium to the left but will have no effect on the solute concentration in blood or the solute outflow. (2) The flow rate increases and the point of equilibrium shifts a little to the right. Solute concentration in blood remains unchanged. The solute outflow increases proportionately with the flow-rate. Hence, the solute outflow is called flow-limited. (3) The flow rate increases further, as a result of which the capillary blood fails to equilibrate completely with the tissue. The amount of solute carried away by capillary blood remains unchanged. The solute concentration in blood falls in inverse proportion to the flow rate. The solute outflow is no longer flow-limited because it is not increased further by increase in flow rate. Rather, it has now become diffusion-limited because a faster diffusion rate will allow the capillary blood to equilibrate with the tissue.

Diffusion across the capillary wall Diffusion is quantitatively much more important than filtration in terms of the exchange of nutrients and waste materials between blood and tissue. It may be argued that since plasma is filtered out of the capillaries and then subsequently reabsorbed, dissolved gases and nutrients would move in and out with it, making diffusion unnecessary. However, the amount of filtration is not enough to transport adequate gases and nutrients through bulk flow. O_2 and glucose are in higher concentration in the bloodstream than in the interstitial fluid and diffuse into the interstitial fluid, whereas CO_2 and metabolites diffuse in the opposite direction.

Fig. 39.**5** (*Above*) Flow-limited transport. The trucks are fully loaded with melons. Total transport can increase only if trucks come in more frequently. (*Below*) Diffusion-limited transport. Trucks leaving before they are fully loaded. More trucks will not increase the total transport, which will increase only if the rate of loading increases.

The rate of diffusion of substances from the capillaries to the tissues increases with the concentration gradient of the substance, the permeability of the capillary wall, and the total number of active capillaries. The effect of blood flow rate on diffusion depends on whether the diffusion is flow-limited or diffusion-limited. The concentrations of small molecules like glucose and NaCl in plasma and the interstitial fluid usually equilibrate near the arteriolar end of the capillary. In such situations, the amount of solutes delivered or carried away by the capillary can be increased by increasing blood flow, i.e., the exchange is *flow-limited*. On the other hand, large molecules like sucrose do not reach equilibrium with the tissues during their passage through the capillaries. Their exchange is said to be *diffusion-limited*. An increase in blood flow will not increase the amount of solute exchange. The concept is explained further in Figures 39.**4** and 39.**5**.

Pinocytosis by the endothelial cells is responsible for the transport of large, lipid-insoluble molecules between blood and interstitium. A very small amount of transport across the capillary wall occurs through this mechanism.

Lymphatic Circulation

The extra fluid filtered from the capillaries enters the lymphatics as lymph. The lymph flows through a system of lymphatic vessels and finally drains into the thoracic duct, which opens into the junction of the left subclavian and internal jugular veins. Thus all the lymph is eventually returned to the blood. The normal 24-hour lymph flow is 2–4 L. Agents that increase capillary permeability increase the lymph flow and are called *lymphagogues*. Agents that cause contraction of smooth muscle also increase lymph flow from the intestines.

Lymphatic vessels (Fig. 39.**6**) are of two types: the initial lymphatics and the collecting lymphatics. *Initial lymphatics* lack valves and smooth muscle in their walls. They are found in the intestine and skeletal muscle. Tissue fluid enters them through the loose junctions between their endothelial cells. The fluid is propelled by the massaging effect of muscle contractions and the pulsations of the arterioles nearby. They drain into the collecting lymphatics. Unlike capillaries, lymphatic endothelium does not have any basal lamina and the junctions between the cells have no tight intercellular connections. Lymphatic endothelium does not have visible fenestrations.

Collecting lymphatics have valves and smooth muscle in their walls. They contract in a peristaltic fashion, propelling the lymph along the vessels. Flow in the collecting lymphatics is further aided by contraction of skeletal muscle, the negative intrathoracic pressure during inspiration, and the suction effect of high-velocity flow of blood in the veins in which the lymphatics terminate (Fig. 39.**7**). Of these, muscle contraction is the principal factor propelling the lymph. Collecting lymphatics traverse lymph nodes at regular intervals along their course. Lymphocytes enter the circulation principally through the lymphatics.

The **composition of lymph** is similar to plasma except that its protein content is lower due to the low permeability of the capillary walls to proteins. The walls of the lymphatics are permeable to protein macromolecules and the composition of lymph varies with the capillary permeability in the region it drains. The lymph from liver has the highest protein content (about 6.0 g dl⁻¹). The lymph from the intestine and skeletal muscle has about 4.0 and 2.0 g dL⁻¹ of protein respectively. In the

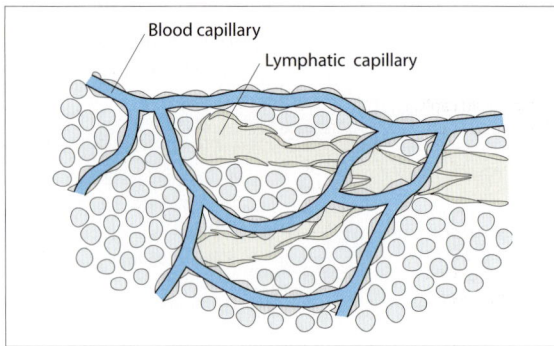

Fig. 39.**6**   Structure of lymph vessels.

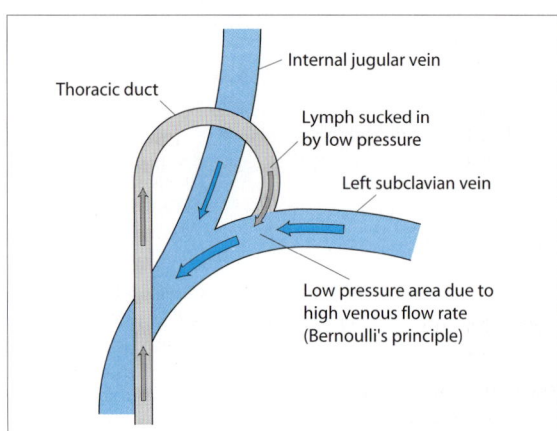

Fig. 39.**7**   Suction of lymph into a large vein.

intestine, fats are absorbed into the lymphatics, and the lymph in the thoracic duct after a meal is milky because of its high fat content.

Edema

Edema refers to an accumulation of excess fluids in the body. The excess fluid may be located outside the cells, or may be confined within the cells, in which case it is called intracellular edema. The term edema, when unqualified as intracellular or extracellular, always implies the latter, that is, a large increase in interstitial fluid volume. The causes of edema are to be found in the factors that control the interstitial fluid volume, viz., (1) increased capillary hydrostatic pressure, (2) reduced plasma oncotic pressure, (3) increased interstitial oncotic pressure, (4) increased capillary permeability, and (5) reduced lymphatic flow (Fig. 39.**8**).

Increased capillary hydrostatic pressure is caused by metabolites and increased venous pressure. Local metabolites cause dilatation of the precapillary

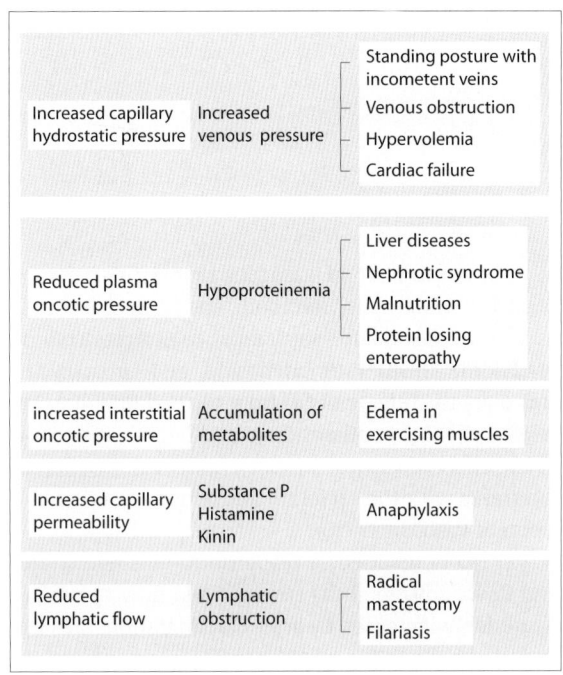

		Standing posture with incometent veins
Increased capillary hydrostatic pressure	Increased venous pressure	Venous obstruction
		Hypervolemia
		Cardiac failure
		Liver diseases
Reduced plasma oncotic pressure	Hypoproteinemia	Nephrotic syndrome
		Malnutrition
		Protein losing enteropathy
increased interstitial oncotic pressure	Accumulation of metabolites	Edema in exercising muscles
Increased capillary permeability	Substance P Histamine Kinin	Anaphylaxis
Reduced lymphatic flow	Lymphatic obstruction	Radical mastectomy
		Filariasis

Fig. 39.**8**   Causes of edema.

sphincter, lowering the pre:post capillary resistance ratio. Increased venous pressure is transmitted back to the capillaries, resulting in elevated capillary hydrostatic pressure. Conditions in which venous pressure is elevated include adoption of the standing posture for long periods, cardiac failure, incompetent venous valves, venous obstruction, and increased blood volume.

Reduced plasma oncotic pressure results from hypoproteinemia, the common causes of which are liver disease, nephrotic syndrome, malnutrition and starvation, and protein-losing enteropathy.

Increase in interstitial oncotic pressure promotes transudation of fluid from the capillary. Osmotically active metabolites in exercising muscles accumulate faster than they can be removed by lymphatics, despite the enhanced lymph drainage during exercise. This accounts for an increase in volume of exercising muscles, which can be as much as 25%. A lowered interstitial oncotic pressure promotes the movement of fluids from the interstitial space into the capillaries. Benzopyrones increase the proteolytic activity of tissue macrophages, which leads to a lowering of interstitial oncotic pressure. They are used in the treatment of lymphedema.

Increased capillary permeability is produced by substance P, histamine, and kinins leading to exudation of large amounts of plasma and causing edema. It is typically seen in anaphylaxis.

Inadequate lymph flow is caused by lymphatic obstruction. Such edema is called lymphedema and is characterized by a high protein content of the edema fluid. Common causes of lymphedema are radical mastectomy and filariasis. *Radical mastectomy* is an operation for breast cancer in which the axillary lymph nodes are also removed. Removal of the axillary lymph nodes on one side reduces lymph drainage and results in edema of the ipsilateral arm. In *filariasis*, the filaria (a helminth) migrate into the lymphatics and obstruct them. Over a period of time, massive edema results, usually of the legs or scrotum, a condition called elephantiasis.

Paracrine Control

Any vasodilator metabolite that accumulates in the tissue during active metabolism can produce autoregulation of blood flow. When blood flow decreases, the vasodilator metabolites accumulate and the vessels dilate. The vasodilatation increases blood flow which washes away the metabolites. The metabolic changes that produce vasodilatation include a decrease in Po_2 or pH and an increase in Pco_2 or osmolarity. These changes cause relaxation of the precapillary sphincters. A rise in temperature exerts a direct vasodilator effect. In exercise, the heat of metabolism raises the temperature in active tissues and contributes to the vasodilatation. K^+ and lactate ions possibly have a role in autoregulation in skeletal muscle. Similarly, adenosine may have an autoregulatory role in cardiac muscle. In injured tissues, histamine released from damaged cells increases capillary permeability and is responsible for some of the swelling in areas of inflammation. Histamine also inhibits the release of norepinephrine; the two are physiological antagonists.

Nitric oxide

Nitric oxide is produced in endothelial cells from L-arginine in the presence of the enzyme nitric oxide synthetase (NOS), which exists in three isoforms. NOS I is present in nervous tissues; NOS II is present in activated macrophages; and NOS III is present in endothelium. NO acts through group IIB hormonal mechanisms, i.e., through cGMP as the second messenger. NO has a very short half-life of less than 5 seconds as it is highly reactive. NO and its products are inactivated through oxidation into nitrite and nitrate which are excreted in urine. The plasma and urine concentrations of nitrate and cGMP are useful indicators for NO production rates. NO has several important physiological functions.

(1) NO relaxes vascular smooth muscle cells. The constant release of NO from endothelial cells produces vasodilatation and therefore NO deficiency causes hypertension. The drug *nitroglycerin*, which is of great value in the treatment of angina, exerts its vasodilator action by being converted to NO.

A large number of vasoconstrictors (e.g., norepinephrine, endothelin, serotonin, and thromboxane A_2) as well as vasodilator substances (e.g., acetylcholine, bradykinin, histamine, adenosine, and prostacyclin) stimulate the release of NO from the endothelium. The release of NO potentiates the vasodilators[1] and reduces the effect of vasoconstrictors. Moreover, NO directly interferes with the secretion and action of endothelin, a potent vasoconstrictor.

(2) NO is released in response to pulsatile stretch and flow-induced shear stress. When flow to a tissue is suddenly increased by arteriolar dilatation, the large arteries to the tissue also dilate. This *flow-induced vasodilatation* (Fig. 40.1A) is due to local release of NO and occurs during physical exercise. Exercise-induced NO formation may explain the beneficial effects of endurance training on the cardiovascular system. Flow-induced vasodilatation is also responsible for *poststenotic vasodilatation* (Fig. 40.2B), i.e., any luminal narrowing of arterial vessels increases the local blood flow velocity, which stimulates NO release and results in vasodilatation beyond the point of narrowing.

(3) NO causes vasodilatation and engorgement of the corpora cavernosa, resulting in penile erection. The drug Viagra® (Sildenafil), a selective inhibitor of cGMP-specific phosphodiesterase, promotes penile erection by inhibiting the inactivation of NO.

(4) NO inhibits platelet adhesion and aggregation. During platelet aggregation, several vasoconstrictors are released. These vasoconstrictors stimulate the release of NO from the endothelium. The NO released inhibits platelet adhesion and aggregation, thereby exerting a negative feedback control on them (Fig. 40.1C). This effect, along with its vasodilatory effect, makes NO important for the maintenance of the normal flow of blood.

(5) NO decreases LDL oxidation and inhibits superoxide anion (O_2^-) production by inhibiting NADPH reductase activity. These actions result in a strong *anti-atherosclerotic effect*.

---

[1] One of the earliest observations that led to the discovery of nitric oxide was that acetylcholine failed to dilate a blood vessel that had been stripped of its endothelium. It was hypothesized that the vasodilatation produced by acetylcholine was mediated by the release of an endothelium-derived relaxation factor (EDRF), which was later identified as nitric oxide.

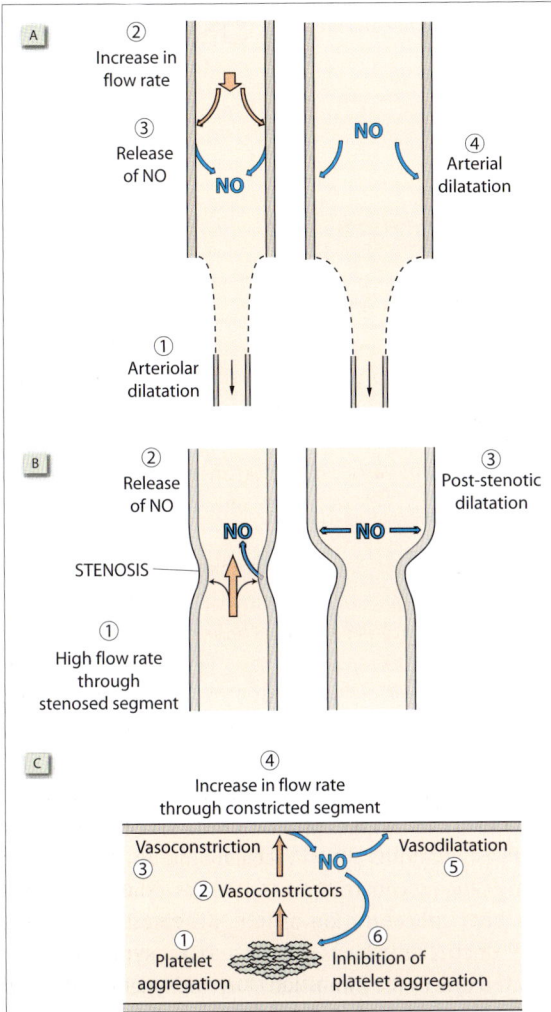

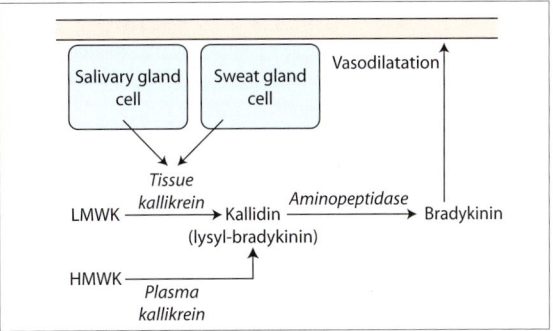

Fig. 40.**2**   Action of bradykinin.

Fig. 40.**1**   Actions of nitric oxide. [A] Flow-induced vasodilatation. [B] Poststenotic vasodilatation. [C] Negative feedback control of platelet aggregation by nitric oxide.

(6) NO is necessary for the cytotoxic activity of macrophages, including their ability to kill cancer cells. It inhibits leukocyte adhesion and migration. Production of NO, which acts as a mediator of the inflammatory response, is elevated in inflammatory diseases.

Kinins

Two forms of kinins with similar actions are found in the body, one is bradykinin (a nonapeptide) and the other is lysyl-bradykinin (a decapeptide), also known as kallidin. They are formed respectively from high-molecular weight and low molecular weight kininogens (HMKW and LMWK) by the action of plasma and tissue kallikreins. Lysyl-bradykinin can be converted to bradykinin by aminopeptidase.

Kinins are present mostly in tissues although small amounts are also found in the circulating blood. Their actions resemble those of histamine. Bradykinin receptors B_1 and B_2 are serpentine receptors coupled to G proteins. They cause contraction of visceral smooth muscle but they relax vascular smooth muscle through NO, lowering the blood pressure. They also increase capillary permeability, attract leukocytes and mediate pain. Tissue kallikrein is formed in sweat glands, salivary glands, and exocrine pancreas during active secretion and mediates the hyperemia in these organs that is normally associated with active secretion (Fig. 40.**2**).

Endothelins

The endothelins (ET) are a family of three highly potent vasoconstrictor peptides (ET-1, ET-2, and ET-3) synthesized and secreted by vascular endothelial cells of the brain, kidneys, and intestine. Endothelins are initially synthesized as preproendothelin, which undergoes preprocessing to proendothelin, also called big-endothelin. Big-endothelin is released and is converted to active endothelin by the endothelin-converting enzyme.

Most of the endothelin-1 is secreted into the scala media of blood vessels where it acts in a paracrine fashion. Endothelin secretion is increased by other vasoconstrictors like catecholamines and angiotensin II, and is inhibited by vasodilators like NO. Endothelin is also secreted when blood flows over the endothelium at high velocity (increased shear stress).

Endothelin is primarily a paracrine regulator of vascular tone. In the brain, endothelins play a role in regulating transport across the blood-brain barrier. In the renal glomerulus, endothelin causes contraction of mesangial cells and thereby decreases the glomerular filtration rate. Endothelins play a role in closing the ductus arteriosus at birth. Endothelin-1 is a potent growth factor for smooth muscle and a chemoattractant for monocytes.

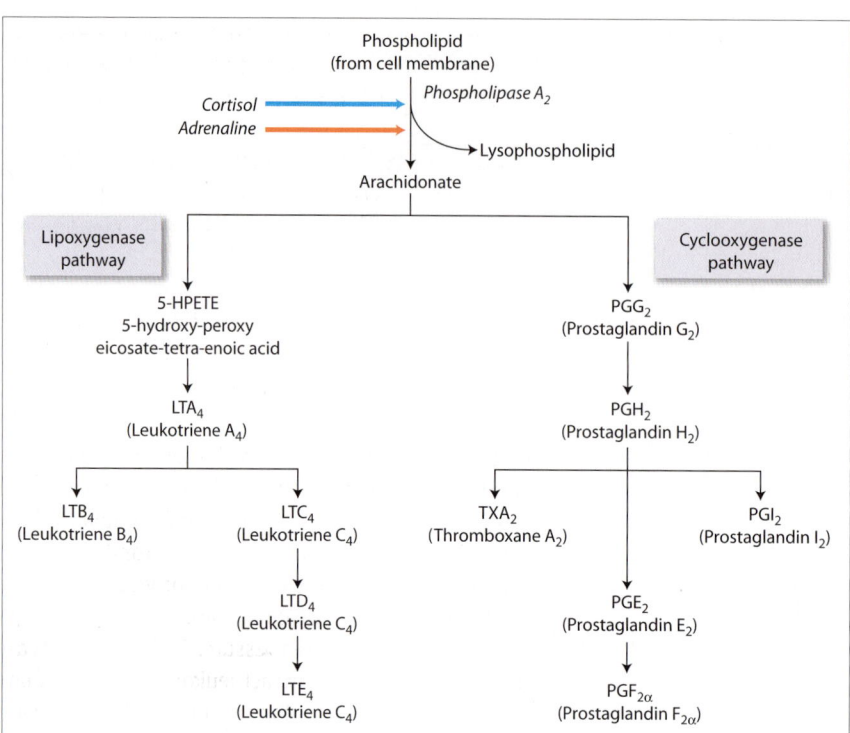

Fig. 40.**3** Prostaglandin synthesis.

The vascular effects of endothelin are mediated by endothelin receptors, of which there are two subtypes (ET-A and ET-B). ET-A is found predominantly on vascular smooth muscles. It is coupled to G_s protein and when activated, stimulates slow and sustained vasoconstriction. ET-B resides on endothelial cells. It is coupled to G_i protein and when activated, stimulates release of NO and thus favors vasodilatation.

Prostaglandins

Prostaglandins[2] are autocrine and paracrine lipid mediators having a wide variety of actions, notably those on smooth muscles and platelets. They are synthesized in the cell from arachidonate, which is derived from membrane phospholipids through the action of phospholipase A_2. Arachidonate passes into the lipoxygenase pathway to form leukotrienes, and into the cyclooxygenase pathway to form thromboxane and prostaglandins (Fig. 40.**3**). Prostaglandins are produced from arachidonic acid by prostaglandin H synthatase. Prostaglandins have a short half-life before being inactivated and excreted.

There are different receptors for different types of prostaglandins: IP receptors for PGI_2, TP receptors for thromboxanes, DP receptors for PGD_2, EP receptors for PGE_2, and FP receptors for PGF_2. Prostaglandins act principally through G-protein-coupled receptors.

Prostaglandins act on vascular smooth muscle cells, some causing constriction and others, dilatation. There is a high concentration of $PGE_{2\alpha}$, a vasodilator, in the fetal ductus arteriosus. At birth, $PGE_{2\alpha}$ synthesis is inhibited due to the inhibition of cyclooxygenase, leading to the closure of the ductus arteriosus. PGF_2, a vasoconstrictor, causes the vasospasm of the uterine arteries that precedes menstrual bleeding, has a role in luteolysis, and increases the intensity of oxytocin-induced contractions of the myometrium. Prostaglandin is also an inhibitor of gastric secretion and a stimulator of renin secretion.

Prostaglandins are important inflammatory mediators, viz., PGE_2 in granulocytes and macrophages, PGD_2 in mast cells, PGI_2 (prostacyclin) in endothelial cells, PGH_2 and thromboxane A_2 in platelets, and leukotrienes in mast cells. The thromboxane A_2 synthesized in platelets causes platelet aggregation while the PGI_2 produced by the endothelial cells inhibits platelet aggregation. Prostaglandins are therefore important for the hemostatic balance. The prostaglandins released from injured tissues act on nociceptors to produce primary hyperalgesia (p 649). PGE_2 is a mediator of fever.

Synthetic prostaglandins are used to induce childbirth, to close a patent ductus arteriosus in newborns, to reduce gastric acid secretion and treat peptic ulcers, as a vasodilator in severe ischemia of a limb and in

[2] The name prostaglandin comes from the prostate gland. When prostaglandin was first isolated from seminal fluid, it was believed to have been added from the prostate.

pulmonary hypertension. Nonsteroidal anti-inflammatory drugs (NSAID) reduce prostaglandin synthesis by inhibiting PGHS.

Endocrine Control

Circulating vasodilators

Vasoactive intestinal peptide (VIP) is a circulating vasodilator which also acts as a gastrointestinal hormone. It is a transmitter in the vagal nonadrenergic, noncholinergic fibers that mediate bronchodilatation. It is also a cotransmitter in the pelvic splanchnic nerve fibers to the penis and mediates penile erection.

Atrial natriuretic peptide (ANP) is a circulating vasodilator with several effects (p 412) but whose exact physiological role is not understood. It reduces blood pressure and in general, its actions are opposite to those of angiotensin II.

Circulating vasoconstrictors

Renin–angiotensin system   Renin is a protease enzyme present in the juxtaglomerular (JG) cells of the kidneys. It is released in response to circulating catecholamines and sympathetic discharge to the kidneys and is eventually converted into angiotensin II which causes vasoconstriction, increases thirst and stimulates

aldosterone secretion. The renin–angiotensin system has important roles in the regulation of blood pressure and ECF volume. It is discussed in detail on p 410.

Vasopressin (ADH) is a potent vasoconstrictor. It also causes renal retention of water. Its actions are discussed in detail on p 410 and its role in blood pressure regulation is discussed on p 275.

Catecholamines include epinephrine, norepinephrine, and dopamine. Of these, epinephrine and norepinephrine act through α and β adrenergic receptors whereas dopamine acts through separate dopaminergic receptors. α receptors mediate excitation of smooth muscles, β1 receptors mediate excitation of cardiac muscle, and β2 receptors mediate inhibition of smooth muscles. The cardiovascular effects of epinephrine and norepinephrine are summarized in Figure 40.**4** and contrasted in Table 40.**1**.

Epinephrine stimulates both α and β receptors in blood vessels. The α-induced vasoconstriction is more than nullified by the β-induced vasodilatation. Hence, the peripheral resistance and diastolic BP remain unchanged or fall slightly. The β-induced increase in stroke volume and heart rate results in higher cardiac output and a rise in systolic BP. The pulse pressure increases.

Norepinephrine has much greater effect on α than on β receptors. Hence, it produces vasoconstriction with a rise in peripheral resistance and diastolic BP. Due to weak β-activation, direct cardiac stimulation is

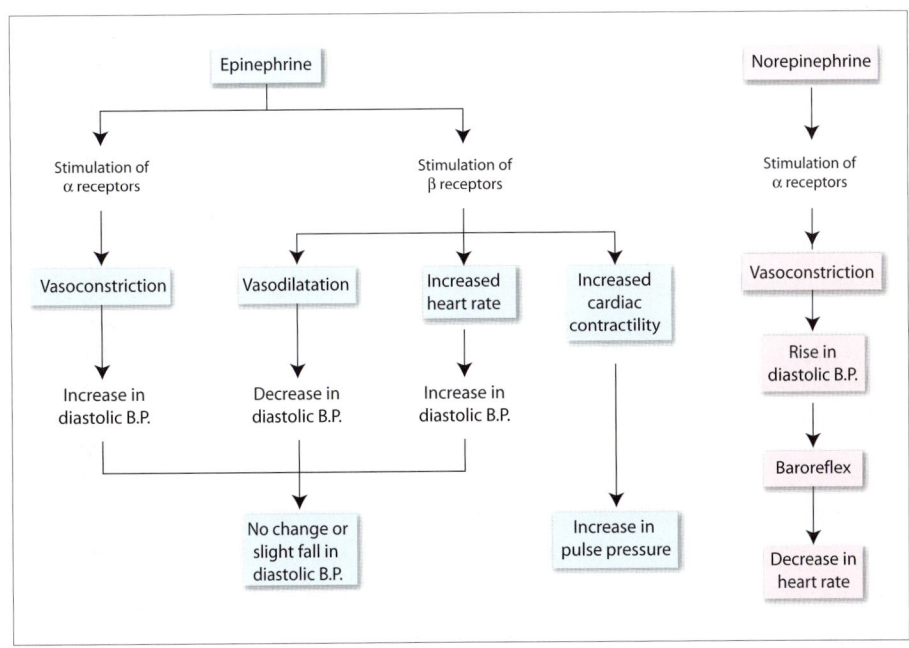

Fig. 40.**4**   Flow chart of the cardiovascular effects of epinephrine and norepinephrine.

Table 40.**1**   Differences in the cardiovascular effects of epinephrine and norepinephrine.

Epinephrine	Norepinephrine
Stimulates the heart through β receptors	Due to weak β-activation, cardiac stimulation is insignificant.
Due to stimulation of both α and β receptors, peripheral resistance and diastolic blood pressure remains unchanged or falls slightly.	Due to predominant α activation, peripheral resistance and diastolic blood pressure increase.
Due to direct stimulation of heart, there is increased stroke volume and heart rate.	The reflex inhibition of sympathetic discharge (due to high diastolic blood pressure) results in reduced heart rate and stroke volume.

insignificant. Rather, there is reflex decrease in heart rate due to the rise in diastolic BP. The cardiac output falls as a result of the low heart rate. The stroke volume and pulse pressure remain nearly the same.

Circulating norepinephrine is of little significance compared to the importance of the norepinephrine released from sympathetic nerves. When α adrenoceptors are blocked, sympathetic stimulation causes vasodilatation (instead of the usual vasoconstriction) because the norepinephrine released from the nerve terminals acts only on the β adrenoceptors. This reversal of the effect of norepinephrine, and therefore of sympathetic stimulation, following the administration of α-blockers is known as *Dale's vasomotor reversal.*

The physiologic function of the dopamine in the circulation is unknown. However, its pharmacological effects make it an important drug in the treatment of circulatory shock because: (1) it produces renal vasodilatation (preventing renal shutdown), acting through dopaminergic receptors; (2) it produces vasoconstriction elsewhere (preventing hypotension), probably by stimulating the release of norepinephrine; and (3) it has a positively inotropic effect on the heart due to the action on β_1-adrenergic receptors.

Cardiovascular reflexes are controlled by centers located in the brainstem reticular formation. These brainstem centers are under the control of higher hypothalamic and cortical centers. The effectors of cardiovascular reflexes are the cardiac muscle (of the heart) and the smooth muscle (of the blood vessels). The receptors initiating cardiovascular reflexes are either mechanoreceptors (sensing blood pressure and volume) or chemoreceptors (sensing gas tension and blood pH). The efferent arc is constituted by autonomic fibers to the heart and blood vessels. The afferent arc is constituted by the glossopharyngeal and vagal fibers.

Efferent control of the cardiovascular system

Sympathetic innervation of heart   The sympathetic nerve cells supplying the heart are located in the intermediolateral horn of spinal segments T1–T5. All parts of the heart (SA node, atria, AV node, and ventricles) receive sympathetic innervation. The sympathetic innervation on the right side is distributed primarily to the SA node, while the sympathetic innervation on the left side is primarily to the AV node. Sympathetic fibers are mostly distributed to the epicardium.

Sympathetic discharge has five effects on the heart: positive inotropic (increase in the force of cardiac contraction), positive chronotropic (increase of the cardiac rate), positive bathmotropic (increase in automaticity), positive dromotropic (increase in conduction velocity), and inhibition of parasympathetic effect, mediated by neuropeptide Y.

Sympathetic innervation of blood vessels   Sympathetic fibers innervating blood vessels originate from the intermediolateral horns in spinal segments T1–L2. They innervate blood vessels of all calibers except the capillaries and postcapillary venules, which do not have smooth muscles in their wall. The arterioles are most densely innervated. The precapillary sphincters too are well innervated. Most sympathetic fibers produce vasoconstriction. They have norepinephrine and sometimes neuropeptide Y as their neurotransmitter. Sympathetic vasoconstrictor fibers show tonic (i.e., continuous) discharge. Hence, sympathectomy produces widespread vasodilatation. Some sympathetic fibers

cause vasodilatation and constitute what is called the *sympathetic vasodilator system*. They supply the arterioles to the skeletal muscles and have acetylcholine and VIP as their neurotransmitters.

Parasympathetic innervation of heart   Parasympathetic fibers to the heart originate from the nucleus ambiguus. They reach the heart through the vagus nerve and relay in ganglia located within the cardiac muscle. The right vagus is distributed mainly to the SA node while the left vagus innervates mainly the AV node. Vagal fibers are mostly endocardial in distribution. Vagal stimulation has a negative chronotropic effect on the

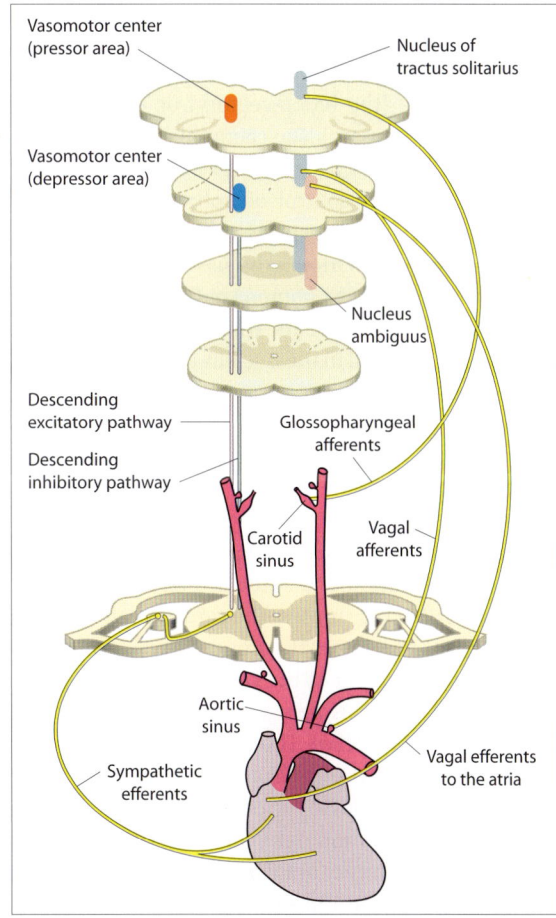

Fig. 41.**1**   The afferents, efferents, and centers of the baroreceptor reflex. The interconnections of the centers are shown in Figure 41.**2**.

heart but it does not have a negative inotropic effect because it does not innervate the contracting myocardial cells of the atria or the ventricles in sufficient numbers. If both noradrenergic and cholinergic systems are blocked, the heart rate is approximately 100 min^{-1}. Since the resting heart rate is approximately 72 min^{-1}, it indicates that at rest, the vagal tone is greater than the sympathetic tone.

Parasympathetic innervation of blood vessels   For the most part, blood vessels do not have any parasympathetic innervation. Some parasympathetic vasodilator fibers run to restricted cranial and sacral areas, supplying vessels in the brain, tongue, salivary glands, external genitalia, bladder, and rectum. They are not tonically active and do not participate in the baroreceptor and chemoreceptor reflexes. In skin, the active vasodilatation associated with these fibers is mediated by bradykinin.

◼ Cardiovascular receptors

The mechanoceptors of the cardiovascular system are stretch receptors that provide information about the blood pressure (the baroreceptors) or blood volume (the volume receptors). The chemoreceptors are located in the carotid and aortic bodies. These are discussed with the chemical control of pulmonary ventilation (see Peripheral chemorecptors, p 352) and their role in blood pressure regulation is discussed in the context of circulatory shock (see Chemoreceptor reflex, p 280).

All peripheral afferents from baroreceptors, and chemoreceptors end in the nucleus of tractus solitarius (NTS). The afferents release the excitatory neurotransmitter glutamate. Cells of the NTS in turn relay the information to other centers that control parasympathetic and sympathetic output.

Baroreceptors are coiled nerve endings located in the adventitia of carotid and aortic arteries at specialized locations called the sinuses. The *carotid sinus* is a small dilatation of the internal carotid artery located just above the bifurcation of the common carotid. It is innervated by the sinus nerve, which is a branch of the glossopharyngeal nerve. The *aortic sinus* is located at the transverse part of the aortic arch, adjacent to the root of the left subclavian artery. It is innervated by the left aortic (mainly) and right aortic nerves, which join the superior laryngeal branch of the vagus. The sinus nerves (from the carotid sinus) and vagal fibers (from the aortic arch) are together called the 'buffer nerves' because they are the afferents of cardiovascular reflexes that buffer abrupt changes in blood pressure.

Location makes the difference

Baroreceptors and volume receptors are structurally similar: it is their location that makes them detect blood pressure or blood volume. An analogy with electricity will be relevant here. Electrical measurements are made with a galvanometer which detects the flow of current through it. The same galvanometer can record potential (voltmeter) or current (ammeter) depending on how it is connected (in parallel or series) and the amount of resistance it offers (high or low). Similarly, the same stretch receptor acts as baroreceptor or volume receptor depending on whether it is located in a high or low pressure area. Another example of location making the critical difference is that of the proprioceptors (see Chapter 104). The muscle spindle detects muscle length because it is disposed in parallel to the muscle fibers while the Golgi tendon organ detects muscle tension because it is disposed in series with the muscle fibers.

Volume receptors are located in low-pressure areas of circulation, e.g., in the walls of the right and left atria at the entrance of the superior and inferior venae cavae and in the pulmonary veins and arteries. Collectively, they are called cardiopulmonary receptors. The immediate effect of an increase in blood volume is an increase in the CVP (see Fig. 42.**3**). The rise in CVP stretches the atrial wall and stimulates the volume receptors present in it.

◼ Cardiovascular control centers

The **spinal sympathetic center** is located in the intermediolateral horn of the spinal segments T1–L2. It has two parts, viz., the *pressor area*, containing the intermediolateral (IML) cells from which the sympathetic fibers originate, and the *depressor area* located a little medially containing the intermediomedial (IMM) cells that inhibit the pressor area.

The **medullary sympathetic center** is also called the *vasomotor center*. It controls the output of the spinal sympathetic center. It has two parts, the *pressor area* located in the rostral ventrolateral medulla (RVLM) which increases the spinal sympathetic output, and the *depressor area*, located in the caudal ventrolateral medulla (CVLM) which reduces the spinal sympathetic output.

The **medullary parasympathetic center** gives origin to the vagal parasympathetic fibers to the heart. It was earlier called the *cardioinhibitory center* and is now called by its specific name, the nucleus ambiguus.

The **hypothalamic autonomic center** controls the lower sympathetic and parasympathetic centers. It

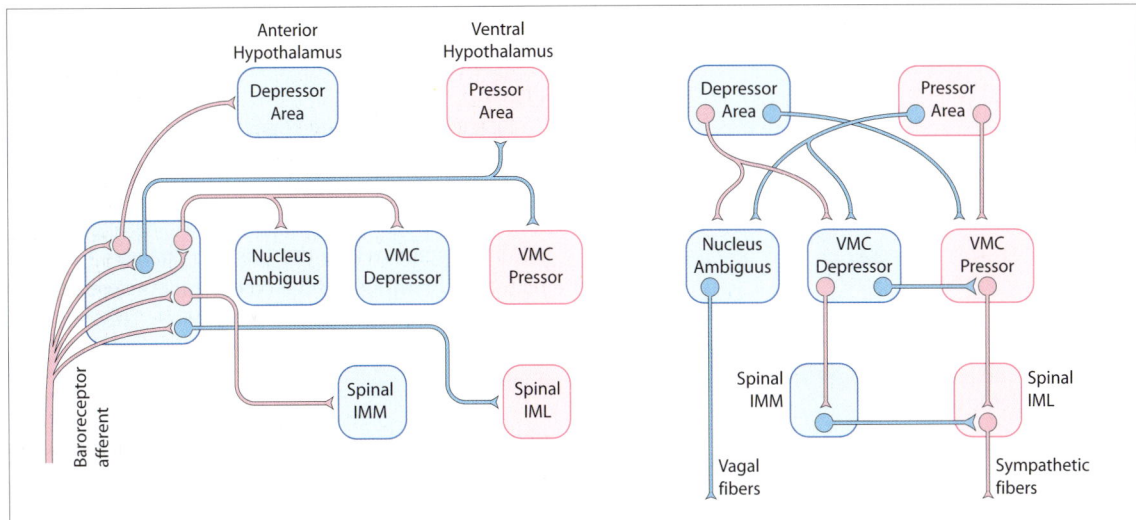

Fig. 41.**2**   (*Left*) Central connections of the baroreceptor afferents. (*Right*) Interconnections of hypothalamic, brainstem, and spinal control centers of the cardiovascular system. Red: excitation, blue: inhibition.

is located in the ventral hypothalamus, ventral to the genu of the corpus callosum. It contains the pressor and depressor areas. The *pressor area*, also called the *defense area*, is connected with the lower autonomic centers such as to increase the sympathetic output and reduce the parasympathetic output. The *depressor area*, which is located in the anterior hypothalamus, is connected with the lower autonomic centers such as to decrease the sympathetic output and increase the parasympathetic output.

The **subthalamic movement center** initiates sympathetic discharge simultaneously with the onset of motor activities. It is probably important for the augmented sympathetic discharge that occurs during exercise.

■ Cardiovascular reflexes

Baroreceptor reflex   At normal blood pressure, the afferent fibers in the buffer nerves discharge at a low rate. When the pressure inside the sinus or aortic arch rises, the discharge rate increases. The discharge rate reaches a plateau at 150 mmHg. When the pressure falls, the discharge rate declines and becomes zero when the blood pressure decreases to 30 mmHg. The carotid receptors respond both to the mean pressure and the pulse pressure. Thus, the baroreceptor discharge increases if the mean pressure rises while the pulse pressure remains unchanged, or if the pulse pressure rises while the mean pressure remains unchanged.

The multiple cardiovascular control centers discharge in a coordinated way. The ultimate output is either sympathetic stimulation with parasympathetic

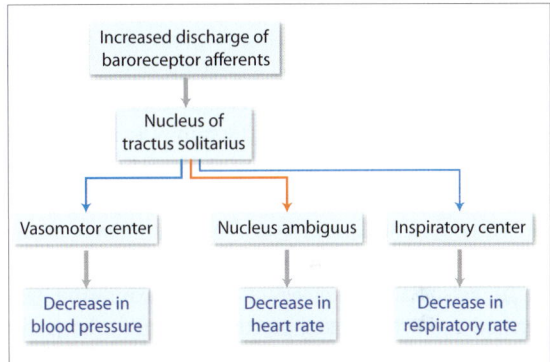

Fig. 41.**3**   The baroreceptor reflex.

inhibition or parasympathetic stimulation with sympathetic inhibition. For example, when the baroreceptors are stimulated, sympathetic outflow decreases. Multiple mechanisms are involved in it (Fig. 41.**2**): The cells of the NTS that receive the baroreceptor input (1) stimulate the IMM cells and inhibit the IML cells, (2) stimulate the VMC depressor area and inhibit the VMC pressor area, (3) stimulate the hypothalamic depressor area and inhibit the defense area, and (4) stimulate the nucleus ambiguus. Baroreceptor stimulation also weakly inhibits respiration (Fig. 41.**3**).

The baroreceptor reflex can be elicited in an experimental animal like the dog by bilateral perfusion of the carotid sinus or by bilateral clamping of the carotid arteries distal to the carotid sinus. As already mentioned, the carotid baroreceptor reflex will cause reduction in blood pressure, heart rate, and respiratory rate. However, the fall in blood pressure will be detected by the aortic baroreceptors, which will partially restore the

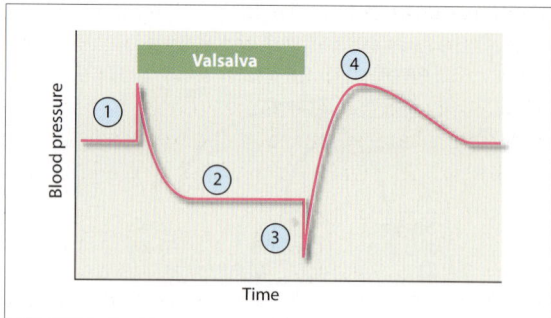

Fig. 41.**4** Blood pressure changes during Valsalva's maneuver. (Explanation in text.)

blood pressure and heart rate. Hence, the reflex effects of carotid baroreceptor stimulation are exaggerated by vagotomy, which abolishes the buffering effects of the aortic baroreceptors.

In humans, the baroreceptor reflex can be elicited by applying pressure on the carotid sinus and through Valsalva's and Müller's maneuvers. The reflex increase in vagal discharge during baroreceptor reflex is of therapeutic value in controlling supraventricular tachycardias and therefore the reflex is sometimes stimulated through *carotid massage*, i.e. by applying pressure on the area of the neck overlying the carotid sinus. The reflex is also activated on wearing a tight collar that presses upon the carotid sinus, causing reflex hypotension and fainting (carotid sinus syncope).

In *Valsalva's maneuver*, a forced expiration performed against a closed glottis results in a large rise in the intrathoracic pressure. The maneuver helps in defecation and parturition as it is also associated with a sharp rise in intraabdominal pressure due to the contraction of abdominal muscles. The rise in intrathoracic pressure during Valsalva's maneuver (+40 mmHg or more) results in a series of changes in cardiac output and arterial blood pressure, which can be described in four phases (Fig. 41.**4**).

In *phase 1*, the arterial pressure increases transiently as the increase in intrathoracic pressure compresses the aorta. *Phase 2* consists of a fall in the diastolic blood pressure and pulse pressure because the high intrathoracic pressure prevents adequate diastolic filling of the heart. The fall in blood pressure initiates the baroreflex, which increases the heart rate slightly but fails to increase the cardiac output in the absence of adequate diastolic filling. The blood pressure therefore remains low. *Phase 3* occurs 1–2 seconds after release of the strain and consists of a transient fall in blood pressure. It is the reversal of phase 1, caused by the release of pressure compressing the aorta. In *phase 4*, the systolic and mean blood pressures rise above the resting level within 10 seconds and return to resting value in 1–2

minutes. The blood pressure overshoot is caused by the continuation of the vasoconstriction initiated in phase 2.

Exactly the opposite changes occur in Müller's maneuver, i.e., forced inspiration against a closed glottis.

A common situation in which the baroreceptor reflex is activated is during venous pooling. Venous pooling occurs when the posture changes from the recumbent (supine) to the erect (standing) posture, or when the body is exposed to high Gz+. This sets off a series of responses (Fig. 41.**5**) that are summarized in Table 41.**1**.

The **Bainbridge reflex** increases the heart rate when the right atrial pressure rises above normal. The increase in CVP excites vagal afferents in the atrial wall and reflexly stimulates sympathetic discharge. The increase in heart rate reduces the diastolic filling time of the ventricles and limits the rise in stroke volume that would otherwise occur due to the increase in right atrial pressure. The reverse, however, is not true: The

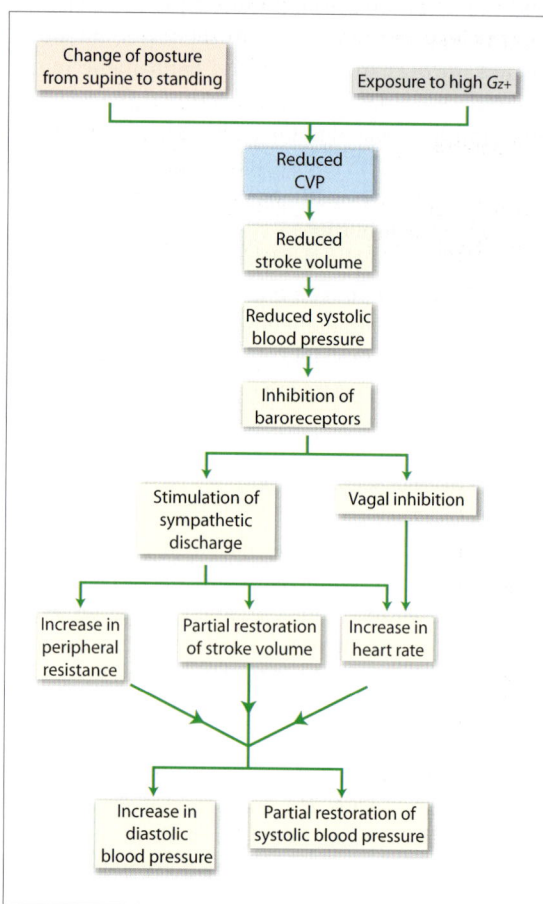

Fig. 41.**5** Flow chart showing changes in blood pressure following a change of posture from supine to standing or an exposure to high Gz+.

heart rate increases even when the CVP falls. At low CVP, the cardiac output and blood pressure fall, triggering the baroreflex which now overrides a weak Bainbridge reflex (Fig. 41.**6**).

The atrial receptors that bring about the Bainbridge reflex are volume receptors. In addition to the Bainbridge reflex, stimulation of the atrial volume receptors also reflexly brings about renal vasodilatation, inhibition of ADH secretion from the posterior pituitary, and the release of atrial natriuretic peptide (ANP).

Table 41.1 Cardiovascular responses to venous pooling. The changes are brought about through the baroreceptor reflex.

Central venous pressure	Reduced
Total peripheral resistance	Increased
Stroke volume	Reduced
Cardiac output	Reduced. Due to a rise in heart rate, the reduction in cardiac output is less marked than the reduction in stroke volume.
Systolic blood pressure	Slightly reduced due to the fall in stroke volume, which remains undercorrected by the reflex sympathetic discharge.
Diastolic blood pressure	Slightly elevated due to the rise in peripheral resistance
Pulse pressure	Reduced
Mean blood pressure	Nearly unchanged

Chemoreceptor reflex Hypoxic stimulation of the chemoreceptors increases blood pressure, heart rate, and respiratory rate. Under experimental conditions in which the respiratory effects are prevented, there is an increase in blood pressure but a decrease in heart rate (see below). The full import of the chemoreflex is discussed in Chapter 54.

The chemoreceptor reflex can be experimentally stimulated by perfusion of the carotid sinus with a perfusate containing 5% oxygen. As already mentioned, the reflex effects would be a rise in blood pressure, heart rate, and respiratory rate. If however the dog is anesthetized to the point where the respiratory center is inhibited, keeping it alive through artificial ventilation, then the heart rate decreases instead of increasing. The reason is as follows. Chemoreceptor afferents (1) stimulate the vasomotor pressor area, resulting in a rise in blood pressure; (2) stimulate the nucleus ambiguus, resulting in a decrease in heart rate; and (3) stimulate the respiratory center, resulting in a rise in the respiratory rate (Fig. 41.**7**). The stimulated respiratory center also inhibits the nucleus ambiguus and therefore increases the heart rate. This increase in heart rate is not seen if the respiratory center is kept inhibited by anesthesia. It is likely that the respiratory center does not directly inhibit the nucleus ambiguus but might be blocking (gating) the excitatory impulses traveling from the NTS to the nucleus ambiguus.

The effects of stimulation of the chemoreceptor afferents (increased blood pressure, heart rate, and respiratory rate) are exactly opposite to the effects of

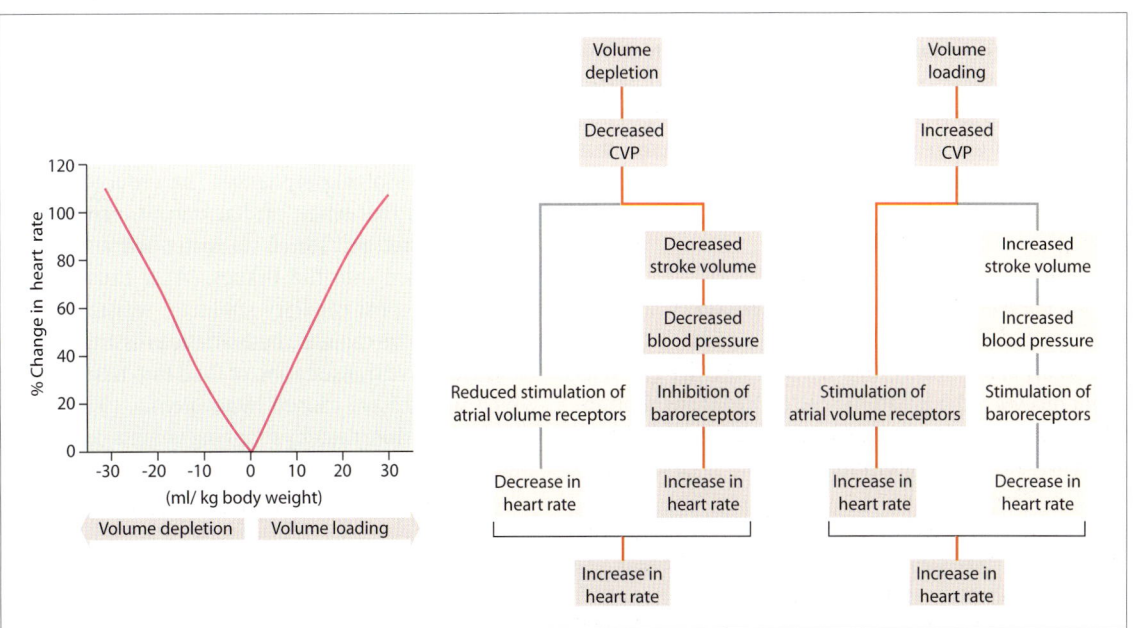

Fig. 41.**6** (*Left*) Changes in heart rate in response to volume depletion and volume loading of circulation. (*Right*) Reflex mechanisms that mediate the biphasic response of heart rate to volume depletion and volume loading.

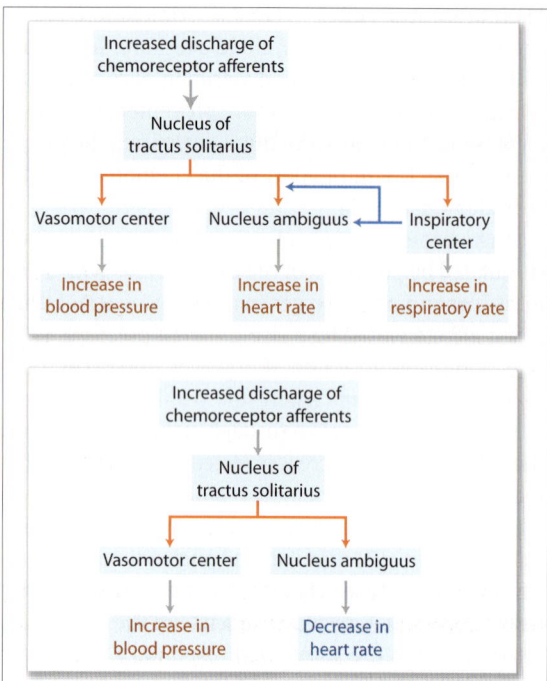

Fig. 41.**7** (*Above*) Chemoreceptor reflex increases the heart rate due to inhibition of the nucleus ambiguus by the stimulated respiratory center. (*Below*) Chemoreceptor reflex decreases the heart rate if its effect on the respiratory center is prevented.

baroreceptor stimulation (decreased blood pressure, heart rate, and respiratory rate). Hence, stimulation of the sinus nerve, which contains both chemoreceptor and baroreceptor afferents, results in a mutual cancellation of their effects. One the other hand, the proximal clamping of the common carotid causes a combination of chemoreceptor stimulation (due to reduced perfusion of carotid bodies) and baroreceptor inhibition (due to reduced pressure in the carotid sinus). The result is a marked stimulation of blood pressure, heart rate, and respiratory rate.

■   **Cardiovascular effects of breathing**

The heart rate shows small oscillations that are synchronous with the breathing cycle. During inspiration, the heart rate increases slightly. During expiration, it decreases. The oscillation in heart rate is called *sinus*

arrhythmia and is partly due to the spilling over of impulses from the inspiratory center to the nucleus ambiguus. The Bainbridge reflex also contributes: During inspiration, the CVP increases due to a rise in the intraabdominal pressure. The increased CVP initiates the Bainbridge reflex and increases the heart rate.

Breathing is also associated with rhythmic changes in the cardiac output (see Fig. 37.**6**) and therefore, in blood pressure too (see Fig. 42.**1**).

■   **Cardiovascular effects of exercise**

The cardiovascular effects of exercise depend on whether the exercise is static or dynamic (p 274). The neural control of the cardiovascular system is important mainly in dynamic exercise in which the sympathetic discharge plays an important role.

The increase in sympathetic discharge that occurs in exercise is not reflex in nature: It originates centrally, probably from the subthalamic movement center, whenever there is an increase in motor activity. Even the thought of exercise stimulates the hypothalamic defense area, bringing about intense sympathetic discharge. The neurons descending from the defense area of the hypothalamus inhibit the baroreceptor afferents as they terminate in the NTS. This explains why the blood pressure remains elevated during exercise and is not reflexly corrected by the baroreceptor reflex.

The sympathetic discharge has numerous important roles in exercise. (1) It increases myocardial contractility which makes the heart less dependent on heterometric mechanisms for increasing the cardiac output. (2) It increases the heart rate which increases the cardiac output and at the same time, prevents an excessive increase in the end-diastolic volume. (3) It causes contraction of the arteriolar smooth muscles, reducing arteriolar compliance. The reduced compliance ensures that when cardiac output increases, the blood does not get 'stored' in distended arteries but moves forward into the tissues. (4) It causes an increase in pre:post capillary sphincter resistance ratio. This reduces the capillary hydrostatic pressure and prevents excessive transudation of fluid into tissue spaces, which would have caused hypovolemia. (5) It causes venoconstriction, thereby reducing venous compliance and increasing the CVP.

When the blood pressure is recorded continuously through direct manometry, it shows three types of waves: the cardiac, respiratory, and Traube–Hering waves (Fig. 42.**1**). The oscillations of the highest frequency are the *cardiac waves*. They are synchronous with the cardiac cycle: the blood pressure rises to its peak with each systole and falls to its lowest during the diastole. The *respiratory waves* are slower oscillations that are synchronous with the breathing cycle. The *Traube-Hering waves* are oscillations in blood pressure with a frequency of about 4 per minute. These are recordable even during apnea and are attributable to rhythmic oscillations in the intensity of sympathetic discharge. Hence, they are also called vasomotor waves.

Cardiac waves The maximum arterial pressure attained during each cardiac cycle is called the *systolic blood pressure* (SBP) and the minimum pressure recorded during diastole is called the *diastolic blood pressure* (DBP). When recorded from the brachial artery of a young adult in the sitting or supine position, the SBP is about 120 mmHg and the DBP is about 80 mmHg. The difference between the systolic and diastolic pressures is called the *pulse pressure*. It is normally about 40 mmHg (120 minus 80 mmHg).

The *mean blood pressure* is the average pressure throughout the cardiac cycle. It can be estimated by measuring out the area under a blood pressure tracing for one cardiac cycle and dividing it by its duration. Because systole is shorter than diastole, greater weight is given to the diastolic blood pressure in the calculation of the mean blood pressure, which is therefore slightly less than the simple average of the systolic and diastolic pressures. As an approximation, the mean pressure is weighted 2:1 in favor of the diastolic blood pressure.

$$\begin{aligned}\text{Mean blood pressure} &= 2/3\ DBP + 1/3\ SBP \\ &= 2/3\ DBP + 1/3\ (DBP + PP) \\ &= 2/3 DBP + 1/3\ DBP + 1/3\ PP \\ &= DBP + 1/3\ PP\end{aligned}$$

For example, if the systolic blood pressure is 120 mmHg and diastolic blood pressure is 90 mmHg, the mean blood pressure is 100 mmHg (and not 105 mmHg).

Determinants of Blood Pressure

Arterial blood pressure is the product of the cardiac output and the total peripheral vascular resistance (peripheral resistance, for short).

$$\begin{matrix}\text{Mean arterial} \\ \text{blood pressure}\end{matrix} = \begin{matrix}\text{Cardiac} \\ \text{output}\end{matrix} \times \begin{matrix}\text{Peripheral} \\ \text{resistance}\end{matrix}$$

If the cardiac output is expressed as a product of stroke volume and heart rate, the formula for blood pressure can be expressed as the product of three variables (the triple product).

$$\begin{matrix}\text{Mean arterial} \\ \text{blood pressure}\end{matrix} = \begin{matrix}\text{Cardiac} \\ \text{output}\end{matrix} \times \begin{matrix}\text{Heart} \\ \text{rate}\end{matrix} \times \begin{matrix}\text{Peripheral} \\ \text{resistance}\end{matrix}$$

Blood pressure is therefore affected by conditions that affect any of these factors. However any inference drawn from this formula must also take into account the interdependence of these three factors. For example, a rise in peripheral resistance reduces the stroke volume (by increasing the afterload on the heart) and the heart rate (through baroreflex). Even then, a rise in peripheral resistance increases the mean blood pressure. It increases the DBP more than the SBP.

An increase in stroke volume increases the pulse pressure and the SBP. The increase in DBP is less marked. SBP is also related to the distensibility of the large arteries. For the same cardiac output, SBP is higher if the

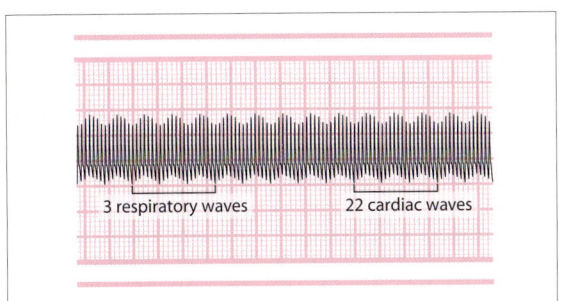

Fig. 42.**1** Cardiac and respiratory waves in a blood pressure tracing. Traube–Hering waves are slower oscillations that are not shown here.

3 respiratory waves   22 cardiac waves

Fig. 42.**2** (*Above*) Rowing, a dynamic exercise. (*Below*) Weight-lifting, a static exercise.

Table 42.1 Mechanisms of blood pressure regulation.

- **Short-term mechanisms**
  Baroreceptor reflex
  Chemoreceptor reflex
  CNS ischemic response
- **Intermediate-term mechanisms**
  Renin–angiotensin mechanism
  Stress-relaxation
  Capillary fluid shift mechanism
- **Long-term mechanisms**
  Pressure diuresis and natriuresis

In **static exercise**, the muscle capillaries get squeezed by the contracting muscles and therefore there is a marked rise in peripheral resistance. The stroke volume decreases due to the combination of an increase in afterload (increased peripheral resistance) and reduced preload (reduced CVP). The diastolic blood pressure rises due to the increased peripheral resistance but the pulse pressure falls due to a reduction in stroke volume. The mean blood pressure rises markedly and can rise to as high as 400 mmHg in heavy weight-lifting! Static exercises cannot be sustained continuously for more than a few seconds since blood supply to the muscles ceases completely due to capillary compression.

arterial distensibility is low, as in aged people with atherosclerosed arteries.

A rise in heart rate, without a concomitant rise in cardiac output, causes a reduction in stroke volume due to reduced diastolic filling of the heart. After the first few seconds of tachycardia, the cardiac output falls and the blood pressure falls with it.

Blood pressure in exercise

Whether blood pressure rises or falls during exercise depends on the type of exercise. If the exercise involves rhythmic contractions and relaxations, it is called dynamic exercise. If the exercise involves intense and sustained isometric contraction, it is called static exercise (Fig. 42.**2**).

In **dynamic exercise**, there is a marked dilatation of the capillary bed due to metabolic autoregulation, leading to an increase in cardiac output (see Fig. 37.**5**). The diastolic blood pressure falls slightly because although the cardiac output rises many fold, there is a marked fall in peripheral resistance. The pulse pressure increases due to increase in stroke volume. The systolic pressure rises and so does the mean blood pressure.

Maintenance of Blood Pressure

The various mechanisms that maintain a near-constant blood pressure in the body can be categorized into the short, intermediate, and long-term mechanisms (Table 42.**1**). The intermediate and long-term mechanisms bring about a change in blood volume, which in turn corrects the blood pressure. The mechanism through which a change in blood volume changes blood pressure is shown in Figure 42.**3**.

It is to be noted that an increase in blood volume does not by itself increase the blood pressure. Any additional volume of blood tends to locate itself in the venous reservoir and not in the arterial system. As shown in Figure 42.**3**, it is the increase in cardiac output caused by hypervolemia that produces the rise in arterial pressure.

Short-term mechanisms

Baroreceptor mechanism Whenever the blood pressure changes rapidly, the baroreceptor reflex quickly brings about a negative feedback, correcting the original change in blood pressure. The baroreceptor reflex prevents erratic fluctuations in blood pressure. In the absence of the reflex, the blood pressure would

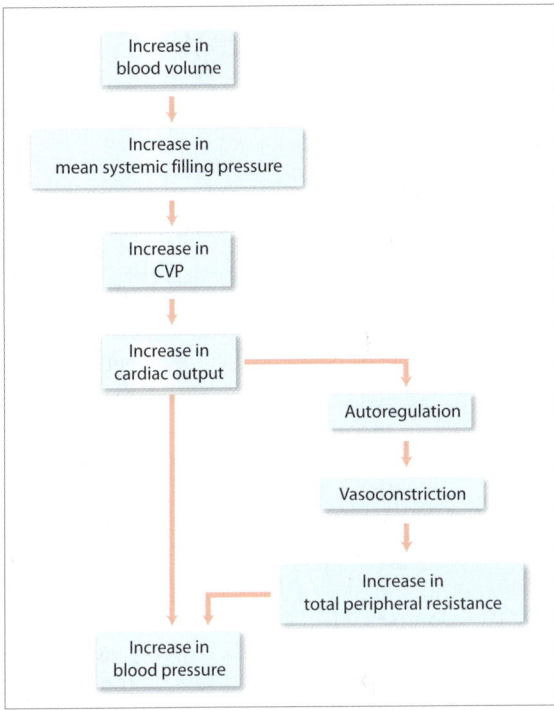

Fig. 42.**3**   The sequence of events through which an increase in blood volume leads to an increase in blood pressure.

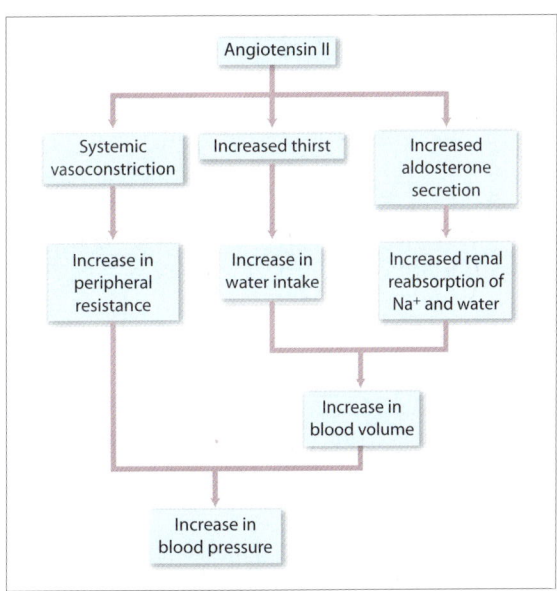

Fig. 42.**4**   The effect of angiotensin on blood pressure.

fluctuate wildly during postural changes, emotional changes or the Valsalva's maneuver associated with defecation and coughing. The baroreceptor reflex fails if the change in pressure is slow and sustained. This is because of baroreceptor resetting, wherein the baroreceptor readjusts itself to a different 'resting' blood pressure. The reset baroreceptor reflex then tries to maintain blood pressure at the new resting blood pressure. The resetting is reversible. Because of baroreceptor resetting, the baroreflex is useless for the long-term regulation of blood pressure. In chronic hypertension, for example, the baroreceptor reflex is reset at a higher blood pressure, say 110 mmHg. Any rapid change in blood pressure from this set point will be quickly restored to 110 mmHg through the baroreceptor reflex.

The **chemoreflex mechanism** and the **CNS ischemic response** are activated in circulatory shock and are discussed with it. These mechanisms have no role in lowering high blood pressure.

■   **Intermediate-term mechanisms**

Renin–angiotensin mechanism is discussed on p 410 and is briefly summarized here. A fall in blood pressure results in a reflex increase in sympathetic discharge. The sympathetic discharge to the kidneys causes renin-degranulation. Renin catalyzes the formation of angiotensin I which is further converted into angiotensin II by angiotensin-converting enzyme (ACE). Angiotensin II is a powerful vasoconstrictor that helps to restore blood pressure by increasing the peripheral resistance. Angiotensin II also brings about an increase in blood volume by increasing thirst and stimulating aldosterone secretion. It thus increases the cardiac output, which helps in restoring the blood pressure (Fig. 42.**4**).

Capillary fluid shift mechanism   When blood pressure rises, it increases the capillary hydrostatic pressure which causes increased capillary transudation. As a result, blood volume decreases and blood pressure falls.

Stress-relaxation mechanism   When there is a sustained increase in arterial blood pressure, the arterial and arteriolar smooth muscles yield to the sustained distending pressure (see smooth muscle plasticity, p 126), leading to a dilatation of these vessels. The vasodilatation decreases the blood pressure.

■   **Long-term mechanism**

The long-term mechanism corrects blood pressure by causing appropriate changes in blood volume through diuresis and natriuresis (Fig. 42.**5**). In fact, it is similar to the capillary fluid shift mechanism except that only the glomerular capillaries are involved in the process.

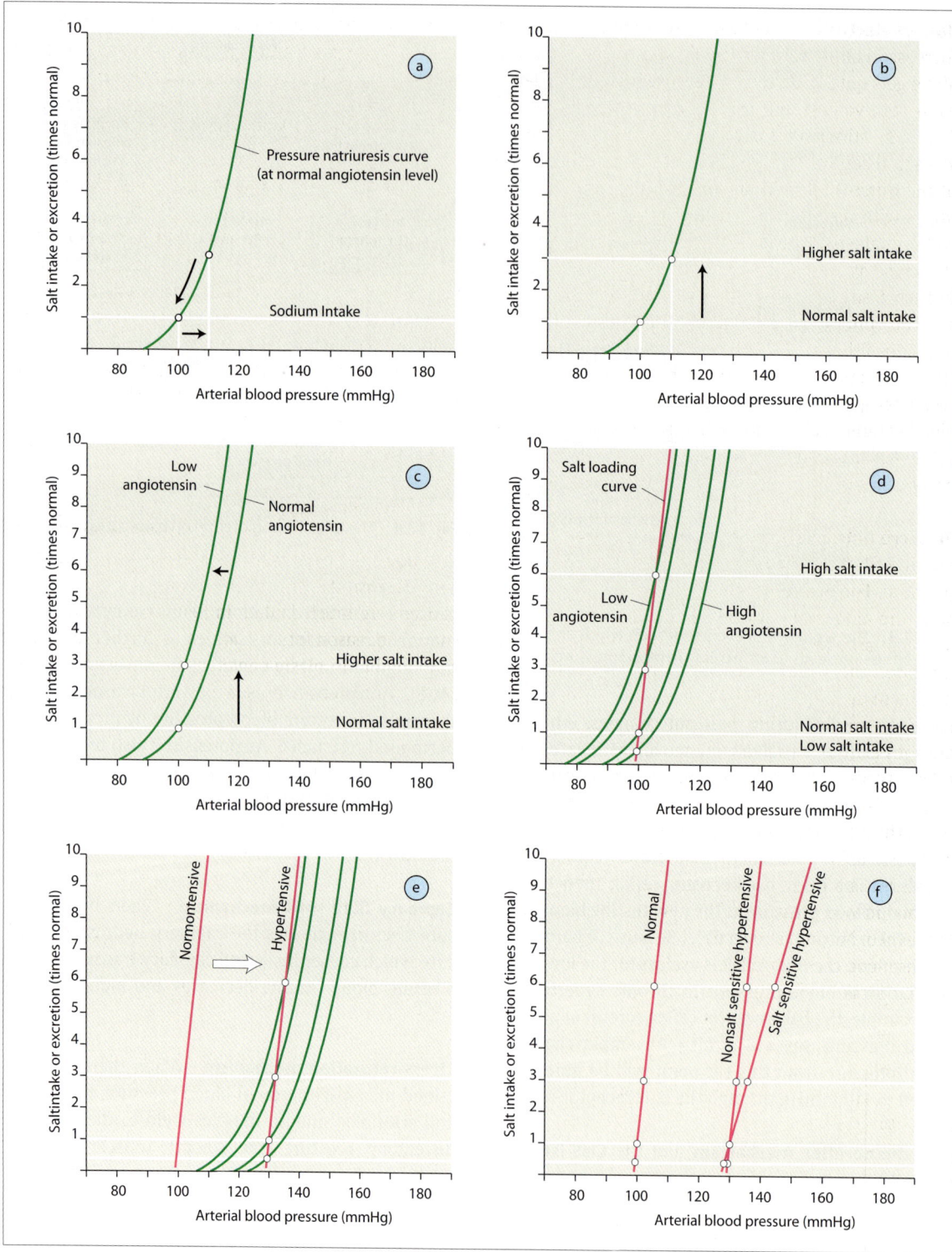

Fig. 42.**5**   The pressure natriuresis curve. (a) If the blood pressure increases, the excretion of sodium increases with it, reducing the blood volume and restoring the blood pressure. (b) If the salt intake increases, blood volume increases with it and raises the blood pressure. In health, this never happens. (c) The increase in salt intake does not increase the blood pressure (unlike in Fig. 42.**5**b) due to the compensatory reduction in angiotensin formation. (d) The salt loading curve. It is constructed by increasing the salt intake and recording the resultant blood pressure. Due to compensatory changes in the amount of angiotensin formation, the salt-loading curve is steep. (e) Hypertension is produced by a right-shift of the salt-loading curve. Compare with Figure 42.**5**d. (f) The salt-loading curve is less steep in salt-sensitive hypertensives. Thus, for an equal increase in salt intake, the rise in blood pressure is greater in salt-sensitive hypertensives.

Role of electrolytes in blood volume regulation   An increase in blood volume produced by water retention alone gets quickly corrected because it leads to changes in both volume and osmolarity of the plasma (see Fig. 61.**3**), stimulating both volume receptors as well as osmoreceptors. This leads to powerful suppression of the thirst-ADH mechanism resulting in prompt diuresis with restoration of blood volume. If the excess water intake is accompanied by excess salt intake too, the blood volume increases but remains isoosmolar. In such a situation osmoreceptors are not stimulated and the volume receptors alone bring about the suppression of the thirst-ADH mechanism. However, being stretch receptors similar to the baroreceptor, volume receptors get reset to a higher level when the rise in blood volume is slow and sustained. Hence, an increase in the Na$^+$ intake or a decrease in the Na$^+$ output can produce a sustained increase in the blood volume.

Pressure natriuresis refers to the changes in urinary output of Na$^+$ in response to changes in arterial blood pressure. For maintenance of a constant blood pressure, the amount of natriuresis must be balanced by an equal amount of Na$^+$ intake. If the blood pressure increases, the urinary output, and with it the Na$^+$ output, increases. The natriuresis and diuresis reduce the blood volume and lower the blood pressure till the Na$^+$ output balances Na$^+$ intake once again (Fig. 42.5a). In other words, the blood pressure cannot increase so long as Na$^+$ intake remains unchanged. If, on the other hand, the Na$^+$ intake increases, blood pressure increases with it and the amount of natriuresis increases till it equals Na$^+$ intake. In the process, a new equilibrium is reached in which the blood pressure is determined by the level of Na$^+$ intake (Fig. 42.**5**b).

The blood pressure however does not rise every time the Na$^+$ intake increases. Hypernatremia is associated with a reduction in plasma angiotensin level. The fall in angiotensin level promotes natriuresis and keeps the blood pressure unchanged despite the higher Na$^+$ intake. As shown in Figure 42.5c, low plasma angiotensin shifts the pressure natriuresis curve to the left.

Salt-loading curve   From the foregoing discussion, it should be clear that for every level of Na$^+$ intake, there exists a corresponding pressure natriuresis curve with which it intersects. The intersection point of the Na$^+$ intake curve with the corresponding pressure natriuresis curve gives the arterial blood pressure. By connecting a series of such intersection points, the salt-loading curve is constructed. The salt-loading curve directly shows how variations in salt intake affect the arterial blood pressure (Fig. 42.**5**d).

Hypertension occurs when the salt-loading curve shifts to the right (Fig. 42.**5**e). Thus, at a given salt intake, the arterial blood pressure is higher in hypertensives than in normal subjects. Hypertensive subjects are known to be of two types: the *salt sensitive* and the *nonsalt sensitive*. For a given salt intake level, the blood pressure is much higher in salt-sensitive hypertensives. The reason for this salt sensitivity is the absence of reduction in plasma angiotensin levels in response to an increase in salt intake. This is apparent from their salt-loading curve (Fig. 42.**5**f).

Hypertension

A sustained elevation of blood pressure above its normal range is called hypertension. The range of normal blood pressure is difficult to define. The present definition of normal blood pressure is the range in which there are no deleterious effects on health. Such a definition however makes it difficult to specify a lower

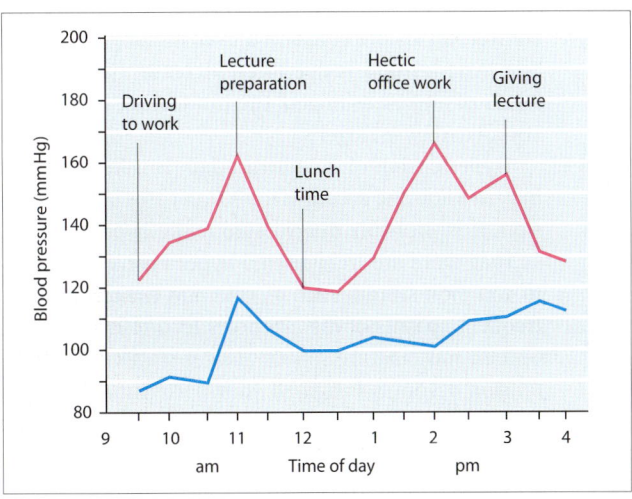

Fig. 42.**6**   Blood pressure fluctuations in a normotensive patient during the daytime and its relation to stresses and activity.

limit of normal blood pressure. The National High Blood Pressure Education Program Coordinating Committee (USA) defines normal blood pressure recorded from the brachial artery as a systolic pressure lower than 130 mmHg and a diastolic pressure lower than 85 mmHg. In healthy humans, both the systolic and the diastolic pressures rise with age.

Hypertension is a sustained elevation of the systemic arterial pressure. Because of nervousness, about 20% of patients have higher blood pressure in the doctor's office than during their normal daytime activity (white coat hypertension). Recordings of blood pressure over prolonged periods in ambulatory normal subjects have shown striking variations in the individual blood pressure response to everyday stresses and activity (Fig. 42.**6**). This emphasizes that a single blood pressure measurement can be misleading.

Experimental hypertension

Sustained hypertension can develop only if the long-term regulation of blood pressure is impaired, i.e., if the amount of pressure natriuresis is reduced. This was demonstrated in a dog by clamping the renal artery of one kidney while removing the other kidney. The hypertension so produced is called *one-clip, one-kidney* Goldblatt hypertension after its pioneer. Following renal artery clamping, the blood pressure increased in two phases, one early and brief and the other late and prolonged. The early rise was due to the systemic vasoconstrictor effect of the angiotensin secreted by the ischemic kidney. The late rise occurred due to the water and electrolyte retention that is associated with the lack of urine formation.

A variation of this experiment is the *one-clip two-kidney* Goldblatt hypertension in which the renal artery is clamped on one side and the other kidney is spared. The hypertension produced is similar and shows a late rise despite the presence of one normally functioning

Table 42.**2** Clinical classification of blood pressure.

Category	Systolic, mmHg	Diastolic, mmHg
Normal	< 130	< 85
High normal	130–139	85–95
Mild hypertension	140–159	90–99
Moderate hypertension	160–179	100–109
Severe hypertension	180–209	110–119
Very severe hypertension	≥ 210	≥ 120

When systolic and diastolic pressures are in different categories, the higher category is selected.

kidney. The late rise in blood pressure is due to water and electrolyte retention caused by the action of angiotensin and aldosterone on the normal kidney. The angiotensin is produced by the ischemic kidney. As explained above, the persistently elevated plasma angiotensin level shifts the pressure natriuresis curve to the right, resulting in salt-sensitive hypertension.

Clinical hypertension

An increase in peripheral resistance is commonly associated with hypertension. This is not merely because mean blood pressure is the product of cardiac output and peripheral resistance. Had it been so, the hypertension would have been corrected through pressure natriuresis. The ultimate reason for hypertension is the impairment of pressure natriuresis in the kidneys. In disease states, increased peripheral resistance is generalized and affects the kidney too. Increased resistance of the renal artery reduces renal blood flow, stimulates angiotensin formation, and impairs pressure natriuresis.

Some of the causes of hypertension are as follows. (1) Preeclampsia and eclampsia cause hypertension due to a pressor polypeptide secreted by the placenta. (2) Hyperaldosteronism and Cushing's syndrome produce hypertension due to salt and water retention. (3) Pheochromocytomas are catecholamine-secreting tumors of the adrenal medulla. Hypertension develops when the catecholamine secreted is norepinephrine. (4) Renal hypertension is due to narrowing of the renal arteries. (5) Coarctation of the aorta, a congenital narrowing of a segment of the thoracic aorta, increases the resistance to blood flow, producing severe hypertension in the upper part of the body. The blood pressure in the lower part of the body is usually normal but may be elevated due to increased renin secretion. (6) Pill hypertension is produced if estrogen-containing oral contraceptives are consumed over prolonged periods. The hypertension is due to an increase in circulating levels of angiotensinogen, the production of which is stimulated by estrogens.

Hypertension of unknown origin is called *essential hypertension* and accounts for 90% of cases of hypertension. It is probably due to autonomic hyperreactivity so that there are exaggerated hypertensive responses to common stimuli such as cold and excitement. The frequent spasms of the arterioles lead to hypertrophy of their musculature so that at later stages, the hypertension becomes sustained. In hypertension of known causes, the pathogenesis can mostly be traced to a right-shifted salt-loading curve.

◼ Adverse effects of hypertension

Clinical hypertension is classified into mild, moderate, severe, and very severe as given in Table 42.**2**. The adverse effects of hypertension are directly related to the severity of the hypertension.

Cardiac effects Hypertension imposes a sustained high afterload on the left ventricle, leading to left ventricular hypertrophy. Ultimately, left ventricular function deteriorates, leading to left ventricular failure. There is also myocardial ischemia due to increased myocardial O_2 demand without a commensurate increase in coronary blood flow: The myocardial O_2 demand rises as the high afterload necessitates development of greater intraventricular pressure and therefore, higher myocardial tension. Most deaths from hypertension are due to myocardial infarction or cardiac failure.

Vascular effects Hypertension is associated with characteristic retinal changes such as narrowing of arterioles, retinal hemorrhages, exudates, and papilledema. Hypertension is commonly associated with occipital headache, particularly in the morning. There may be cerebral infarction due to the increased incidence of atherosclerosis that is seen in hypertensive patients. The high blood pressure also tends to cause cerebral hemorrhage. The atherosclerosis associated with hypertension affects the renal arterioles too and thereby impairs glomerular filtration. About 10% of deaths from hypertension occur due to renal failure.

◼ Treatment of hypertension

Antihypertensive drugs include: (1) α-adrenergic receptor blocking drugs, which reduce the vasoconstrictive effect of sympathetic discharge and circulating catecholamines; (2) β_1-adrenergic receptor blockers, which decreases cardiac contractility; (3) ACE inhibitors, which inhibit the activity of angiotensin-converting enzyme; and (4) calcium channel blockers, which relax vascular smooth muscle. Some patients show marked increase in blood pressure when fed a high-sodium diet, whereas others do not. Since there is no easy test to distinguish salt-responsive from salt-resistant humans, salt-restriction is advised to all patients with hypertension.

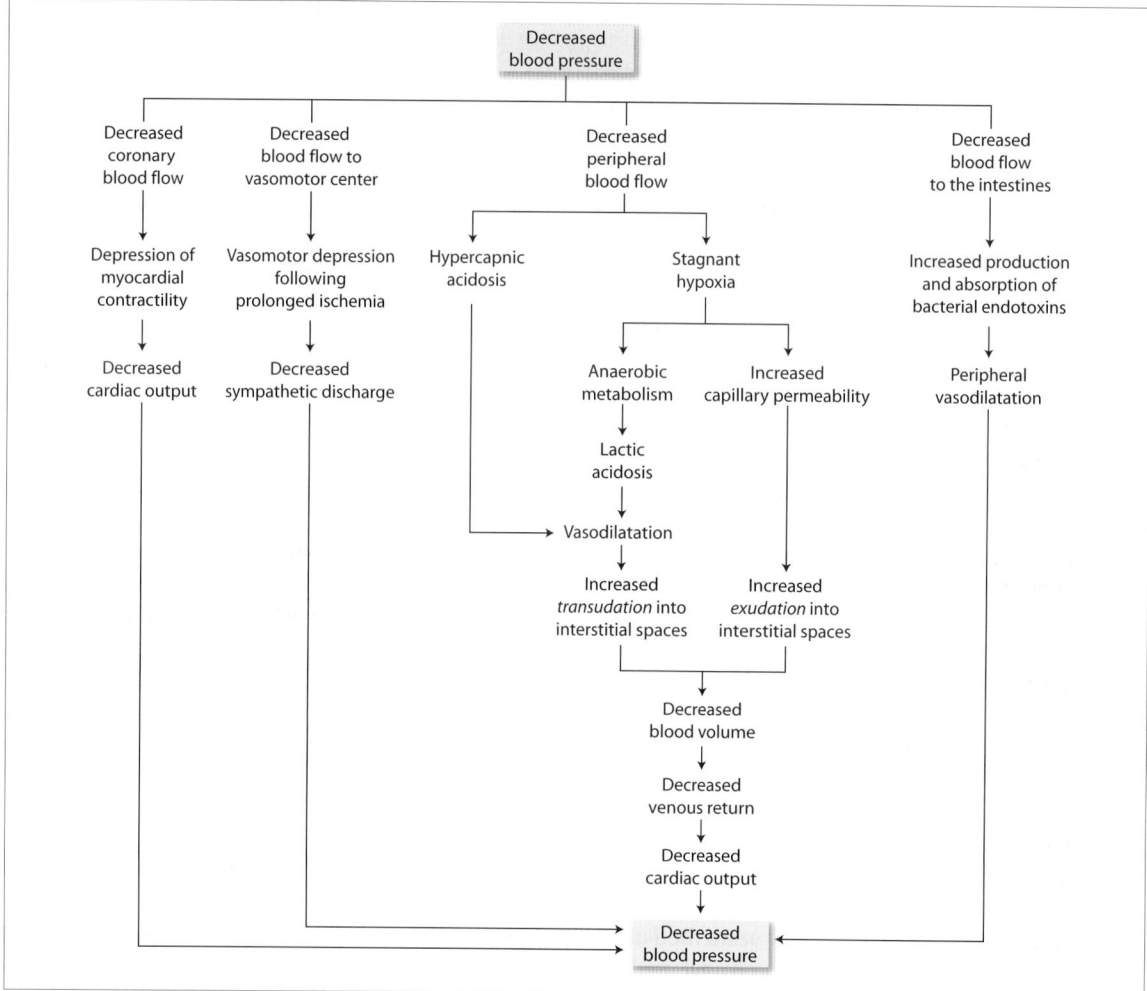

Fig. 43.**2**   Sequence of events leading to the progression of shock.

into the tissues. This further decreases the blood volume, aggravating the shock.

(6) *Release of toxins by ischemic tissues*   Diminished blood flow to the intestines favors the formation of endotoxins by gram negative bacteria and enhances their absorption. These endotoxins cause extensive vasodilatation and depress the myocardium. Endotoxins play a major role in septicemic shock.

(7) *Generalized cellular deterioration*   As shock becomes very severe, there is anoxic damage to the tissues, especially in the liver, lungs, and heart. The heart and liver are especially vulnerable because of their high metabolic rate. The liver is also the first organ to be exposed to toxins from the intestine through the portal vein. The lungs develop pulmonary edema and fail to oxygenate the blood adequately.

(8) *Tissue necrosis* occurs in the cells worst affected by hypoxia. Hence, there is necrosis in zone 3 of hepatic acini (see Fig. 47.**4**), in the tubules of the kidney (acute tubular necrosis), in the lung (shock lung syndrome), and in the heart.

Irreversible stage   Once the shock progresses to the irreversible stage, the cardiac output and blood pressure continue to deteriorate despite transfusion and medication, culminating in death. Probably, the single most important event that marks the point of irreversibility is the depletion of cellular adenosine. The high-energy phosphate reserves, especially in the liver and in the heart, are greatly diminished in severe shock. Nearly all the ATP gets degraded to ADP, AMP, and eventually to adenosine. The adenosine diffuses out of the cells into the circulating blood and is converted into uric acid. New adenosine is synthesized rather slowly so that once depleted, the high-energy phosphate stores of the cells are difficult to replenish.

Causes of shock

Blood pressure is the product of cardiac output and peripheral resistance. Hence, causes of hypotension include conditions causing a reduction in cardiac output and conditions in which peripheral resistance is reduced. Of course, a reduction in peripheral resistance tends to increase cardiac output by reducing cardiac afterload but so long as their product, i.e., the blood pressure is reduced, the possibility of shock remains. If the reduction in cardiac output is due to a hypoeffective heart it is called cardiogenic shock. If the reduction of cardiac output is due to a reduced CVP, the commonest cause of which is hypovolemia, it is called hypovolemic shock. Another cause of shock is vasogenic or low-resistance shock in which there is excessive peripheral vasodilatation. It is also called distributive shock because vasodilatation is associated with a redistribution of blood flow, with more blood accumulating in the venous compartment.

Hypovolemic shock   Common causes of hypovolemic shock are as follows: (1) Loss of blood (external or internal bleeding). (2) Loss of plasma (burns and exudative lesions). In burns, large areas of the skin get denuded. Large volumes of plasma are exuded through these exposed areas so that hypovolemia is a common complication of burns. (3) Loss of fluids (dehydration due to excessive vomiting, diarrhea, sweating). Hypovolemia due to loss of plasma or fluids creates the additional problem of raised blood viscosity, which tends to slow down circulation. (4) *Traumatic shock* is a special type of hypovolemic shock in which there is associated neurogenic shock too, caused by the severe pain that inhibits the vasomotor center.

Cardiogenic shock   Common causes of cardiogenic shock are myocardial infarction, arrhythmias, and valvular disorders.

Obstructive shock occurs due to obstruction to the diastolic filling of the heart, resulting in a decrease in cardiac output. Examples include pulmonary embolism, cardiac tamponade, and tension pneumothorax.

Distributive shock occurs due to massive exudation of plasma from the blood vessels to the interstitial spaces. It may be neurogenic, anaphylactic, septicemic, or endotoxic. (1) *Neurogenic shock* occurs due to a marked reduction in sympathetic vasomotor tone. Its causes include deep general anesthesia, spinal anesthesia affecting the thoracolumbar sympathetic outflow, spinal injury, brain concussion, and antihypertensive drugs. Some cases of neurogenic shock occur due to increase in the vagal tone of the heart, as in vasovagal syncope. (2) *Anaphylactic shock* is associated with reduced peripheral resistance due to the release of histamine, which causes vasodilatation and increased capillary permeability. Hypovolemia occurs due to excessive exudation. (3) *Septicemic shock* is produced by bacterial toxins that produce vasodilatation, not only in the infected tissues but elsewhere too, by stimulating cellular metabolism and producing fever. (4) *Endotoxic shock* occurs in gram-negative septicemia due to release of endotoxin. Endotoxic shock is similar in mechanism to anaphylactic shock. Additionally, endotoxin depresses myocardial contractility.

Clinical features of shock

The patient in shock is typically restless and tachypneic. There may be Cheyne-Stokes breathing. The pulse is rapid and thready (of low volume). The skin is pale, cold, and clammy (moist) and may show peripheral cyanosis. The systolic blood pressure is less than 100 mmHg and the diastolic blood pressure is often not recordable through sphygmomanometry. There is hypothermia due to depressed metabolism. In septic shock, however, there is fever. There is severe muscular weakness.

Impaired renal function   The glomerular filtration rate decreases even in the early stages of shock. It is a part of the compensatory mechanism for restoration of blood volume. In the late stages of shock, the renal tubular epithelial cells get damaged. Certain parts of the tubule are more vulnerable as they have a very high metabolism but receive only a moderate blood supply. The result is acute tubular necrosis with tubular cell death and sloughing, and blockage of the tubules, leading to acute renal failure.

Treatment of shock

Patients in shock are kept in a cold room so that there is no further hypovolemia from sweating. Temperature regulation is a prepotent reflex. If exposed to warmth, there would be sweating which would aggravate the shock further. Shock patients are usually hypothermic and therefore naïve attendants might be inclined to cover them with blankets. This should never be done.

Replacement therapy   Depending on the cause of hypovolemia (loss of blood, plasma or fluids), the patient is transfused respectively with blood, plasma (or plasma expanders like dextran solution), or isotonic saline solutions. Dextran is a large polysaccharide that does not pass through the capillary pores. It therefore promotes osmosis of water from the interstitial to the intravascular spaces, thereby restoring the plasma volume.

Sympathomimetic drugs are especially beneficial in neurogenic shock and anaphylactic shock. The sympathomimetic drug of choice is dopamine because it produces renal vasodilatation and at the same time, produces vasoconstriction elsewhere in the body. It also has a positive inotropic effect on the heart. In neurogenic shock, sympathomimetic drugs fulfill the physiological role of the sympathetic nervous system, which is severely depressed. In anaphylactic shock, histamine plays a prominent role and therefore sympathomimetics are useful in its treatment. Sympathomimetic drugs and histamine are said to be physiological antagonists because the two have opposite effects. For example, sympathomimetic drugs are vasoconstrictors while histamine is a vasodilator. In hemorrhagic shock, the sympathetic system is already maximally active and therefore, sympathomimetic drugs have limited value.

Glucocorticoids are frequently given to patients in severe shock for several reasons. They increase the strength of the heart in the late stages of shock. They

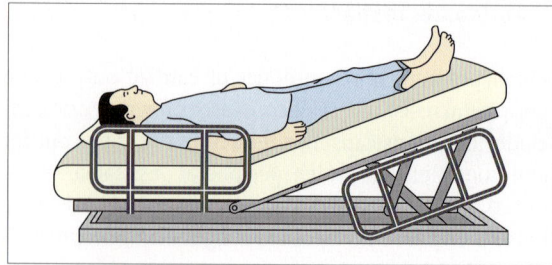

Fig. 43.**3**   A patient in shock kept in leg-up position.

stabilize the lysosomal membranes and prevent release of lysosomal enzymes into the cytoplasm of the cells, thus preventing tissue damage. They also aid in the metabolism of glucose by the severely damaged cells.

Trendelenberg position   Placing the patient with the head lower than the feet helps in increasing the CVP and thereby increasing the cardiac output (Fig. 43.**3**).

Oxygen therapy may be beneficial in some instances. However, the response is not marked which is not unexpected considering that oxygen is beneficial mostly in hypoxic hypoxia. The hypoxia of shock is of the stagnant type.

44 Coronary Circulation

The right and left coronary arteries take their origin from the aorta at the base of the sinuses of Valsalva behind the cusps of the aortic valve (see Fig. 36.**1** and 44.**1**). The right coronary artery supplies the right atrium, the free wall of the right ventricle, parts of the posterior one-third of the ventricular septum, the posterior wall of the left ventricle, the pulmonary conus, the SA node (in roughly half the cases), and the AV node. The left coronary artery further divides into the left circumflex artery and the anterior descending artery. The left circumflex artery supplies the left atrium and the interatrial septum, the lateral and posterior walls of the left ventricle, and the SA node (in roughly half the cases). The anterior descending artery supplies the free wall of the left ventricle, the ventricular septum, and the anterior wall of the right ventricle.

The left ventricular musculature is drained into the coronary sinus by coronary veins. The coronary sinus opens into the right atrium. The anterior surface of the right ventricle is drained by the anterior cardiac veins which empty into the right atrium. The atrial muscles are drained by the Thebesian veins, which empty into the lumen of the right and left atria. In addition to the veins, there are *myocardial sinusoids* that drain the myocardial capillaries directly into the ventricles, and the *arterioluminal vessels* that drain the arterioles directly into the ventricles.

Anastomoses between the coronary vessels are usually quite small and hence sudden occlusion of a coronary artery results in a localized area of myocardial ischemia or infarction (see below). However, if the arterial flow is gradually decreased, as in occlusive coronary artery disease, these anastomotic connections slowly dilate to develop into a collateral channel to the capillary bed of the diseased artery.

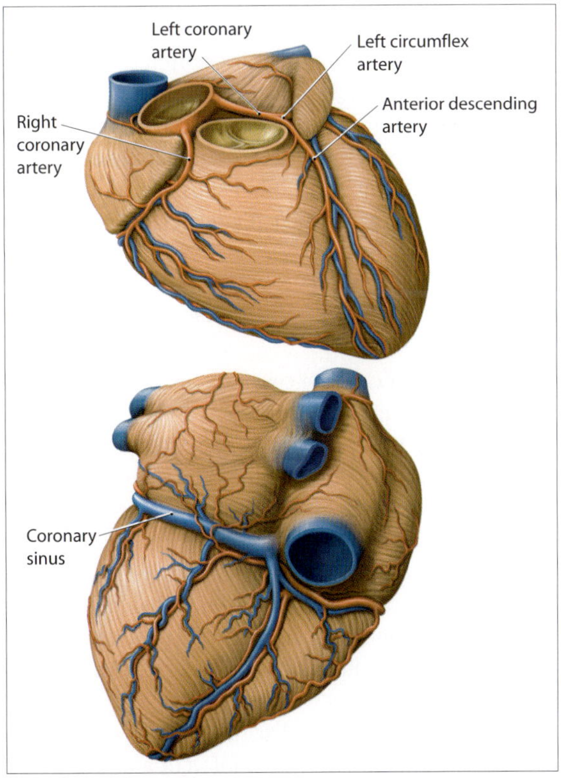

Fig. 44.**1**   Coronary blood vessels on the sternocostal surface (*above*) and diaphragmatic surface (*below*) of the heart.

▪ Regulation of the coronary blood flow

The resting coronary blood flow averages about 225 ml min^{-1} (70–80 ml of blood per min per 100 g of heart) which is 4 to 5% of the total cardiac output.

Myocardial oxygen demand is the main controller of coronary blood flow. The resting heart has an O_2 consumption of 8 ml min^{-1} 100 g^{-1}, which is the highest of all organs. In order to meet this high basal demand for O_2, the myocardium extracts 70% of the O_2 carried by the resting coronary blood supply. As this is the maximum possible extraction ratio, any further increase in O_2 demand can be met only through an increase coronary blood flow. Changes in coronary blood flow and myocardial oxygen consumption therefore parallel each other. During maximal exertion, the myocardial O_2 demand increases five times and to deliver this augmented O_2 supply, the coronary blood flow too must rise five times. The mechanism of this increase is as follows.

Whenever myocardial oxygen demand increases, several metabolites like CO_2, H^+, K^+, and adenosine accumulate in the myocardial tissue. These metabolites produce vasodilatation and thereby increase the coronary blood flow. Though theoretically all vasodilator metabolites can adjust the blood flow to the metabolic

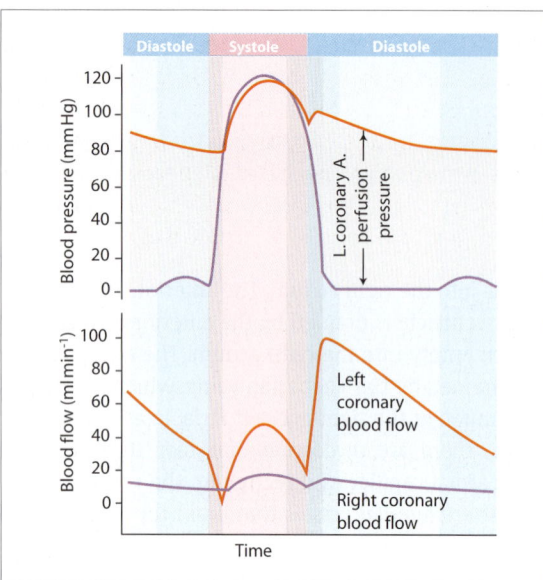

Fig. 44.**2** (*Above*) The aortic pressure and left ventricular pressure. (*Below*) The left and right coronary blood flow. The left coronary blood roughly parallels the difference between the aortic pressure and left ventricular pressure.

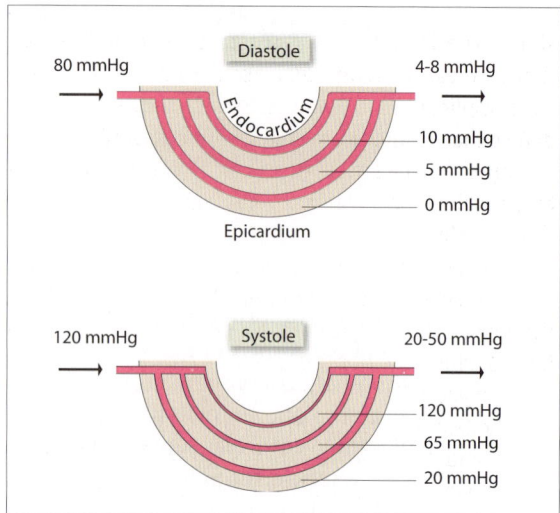

Fig. 44.**3** Myocardial interstitial tissue pressure.

demand, adenosine seems to be the main physiological regulator of coronary blood flow. So long as the myocardial oxygen demand remains unchanged, the coronary blood flow too tends to remain unchanged. This is known as the autoregulation of coronary blood flow and is valid for perfusion pressures between 60 and 180 mmHg. As explained in the context of muscle circulation, the mechanism of autoregulation can be either myogenic or metabolic or both. In case of coronary blood flow, the autoregulation seems to be metabolically mediated with adenosine as its main mediator.

Myocardial interstitial tissue pressure (intramyocardial pressure) is another major determinant of coronary blood flow. In the left ventricle, it varies widely during the cardiac cycle and causes marked changes in coronary blood flow. In the right ventricle, the changes are less marked. During systole, the intramyocardial pressure of the left ventricle is very high, being highest (about 120 mmHg) in the subendocardium and less in the subepicardium. Subendocardial vessels are therefore squeezed much more than the subepicardial ones and the blood flow to the subendocardial capillaries nearly stops during systole. The subepicardial capillaries are better perfused during systole (Figs. 44.**2** and 44.**3**). The subendocardial capillaries however are more numerous and have larger caliber. Hence, they are better perfused during diastole. Generally, the subendocardial and subepicardial capillaries are perfused equally well under normal conditions.

Neural control Coronary blood vessels are innervated by both sympathetic and vagal fibers. Sympathetic fibers constrict and vagal fibers dilate the coronary vessels. In physiological situations however, sympathetic stimulation causes coronary vasodilatation because the direct vasoconstrictor effect of sympathetic discharge is overridden by the marked vasodilatation caused by the increase in myocardial oxygen demand associated with the sympathetic stimulation of the heart.

During coronary angiography, the injection of contrast media into coronary arteries often causes bradycardia and hypotension in the patients. The response is probably due to the *Bezold–Jarisch reflex* (coronary chemoreflex), which can be experimentally elicited by injecting veratradine and other substances that stimulate the nociceptors in the left ventricular wall.

Coronary Insufficiency

The commonest cause of coronary insufficiency is atheromatous coronary artery stenosis (commonly called coronary artery disease). It can also occur due to spasm of coronary artery. Coronary arteries are end-arteries, which perfuse their own exclusive capillary bed. Hence, coronary insufficiency results in myocardial ischemia, especially in the subendocardial layers which are more prone to ischemia than the subepicardial layers. Myocardial infarction (necrosis of myocardial cells) starts occurring after about 20 minutes of continuous myocardial ischemia.

The impairment of cardiac function depends on the extent of myocardial necrosis. Infarction of more than 25% of the left ventricle produces signs of cardiac failure. Infarction of more than 40% produces cardiogenic shock. The infarcted area generates ectopic impulses that can lead to ventricular fibrillation. The mechanical complications include ischemic ventricular septal defect that results from infarction and perforation of the interventricular septum, and mitral regurgitation that results from infarction and rupture of the papillary muscle.

Coronary insufficiency however does not always lead to ischemia or infarction. In some individuals, the myocardium adjusts to the coronary insufficiency by reducing its oxygen demand. This phenomenon is called *myocardial hibernation*. The reduced myocardial metabolism however reduces the power of the cardiac pump and tends to cause cardiac failure. The exact mechanism of myocardial hibernation is not understood though it seems to be associated with reduced entry of Ca^{2+} into the cardiac muscle cell during contraction.

Consequences of myocardial ischemia   During ischemia, myocardial metabolism switches from aerobic to anaerobic, resulting in acidosis. The high concentration of H^+ competes with Ca^{2+} for binding sites on myosin heads, inhibiting cross-bridge cycling and reducing myocardial contractility. Since the ATP yield is much lower in anaerobic metabolism, there is a progressive depletion of ATP. The ATP depletion reduces Ca^{2+} reuptake into sarcoplasmic reticulum and thereby impairs

diastolic relaxation. ATP depletion also impairs ion pumps, causing loss of intracellular K^+ and accumulation of intracellular Na^+ and water. These changes depolarize the muscle, increasing its excitability and triggering arrhythmias. The characteristic anginal pain that is associated with myocardial ischemia is produced by the adenosine formed by degradation of ATP.

Treatment of myocardial ischemia   Nitrates are administered for dilating coronary arteries. However, their use is fraught with the risk of *coronary steal* in which the flow through the stenotic coronary artery decreases instead of increasing following vasodilator administration. It occurs because the vasodilator dilates other coronary arteries that are disposed in parallel to the stenosed artery but is unable to significantly dilate the stenosed artery (Fig. 44.**4**). Most of the blood therefore flows into the dilated artery (which offers less resistance) instead of flowing into the stenosed artery.

The mainstay of medical therapy is therefore to reduce the cardiac workload, thereby reducing the myocardial O_2 demand. This is achieved by controlling heart rate using β-blockers and reducing afterload and preload by using vasodilators like ACE inhibitors, Ca^{2+}-channel blockers, and nitrates. An interesting method of alleviating myocardial ischemia is to provide *counterpulsations* for improving diastolic perfusion. In this method, a balloon is inserted through the femoral artery and positioned below the left subclavian but above the renal arteries. The inflated balloon increases the diastolic pressure in the aorta. As coronary blood supply is mainly diastolic, this 'diastolic augmentation' helps in better perfusion of the coronaries.

Surgical intervention is required only when there is more than 75% reduction in vessel diameter (Fig. 44.**5**). Stenosis of epicardial vessels can be surgically bypassed (coronary artery bypass grafting) using arterial or venous conduits taken from the internal mammary or radial artery or saphenous vein of the patient. Stenosis of coronary artery may also be dilated by passing

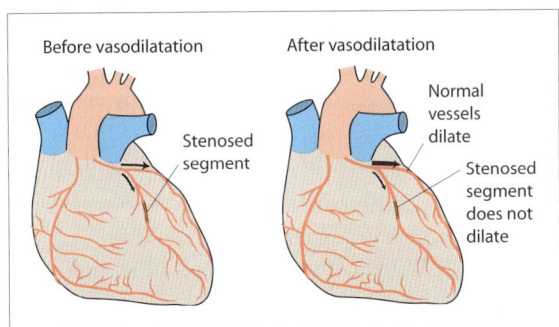

Fig. 44.**4**   Coronary steal.

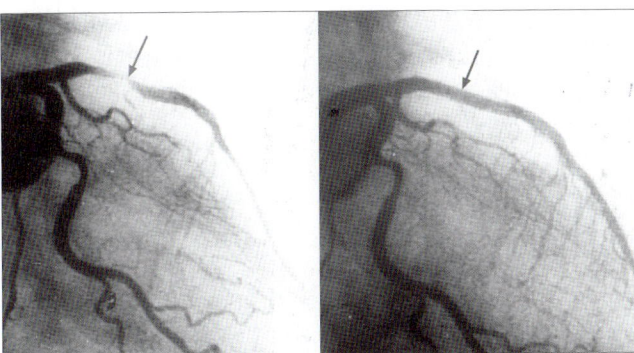

Fig. 44.**5**   Coronary angiogram. (*Left*) Block of the left anterior descending artery, marked with an arrow. (*Right*) Restoration of circulation after angioplasty and stenting.

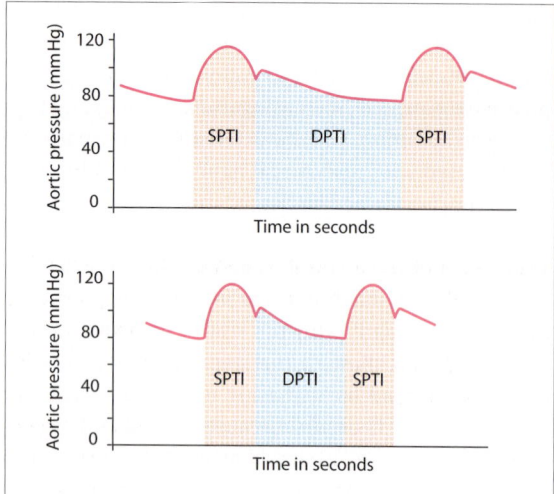

Fig. 44.**6** The aortic pressure curves at normal heart rate (above) and during tachycardia (below). The STPI/DPTI ratios in both cases can be calculated by counting the number of small squares in the shaded areas.

a balloon through femoral or brachial artery and pushing it all the way up into the coronary artery. The balloon is inflated at the stenosis site, causing it to dilate (coronary angioplasty). To avoid recurrence of stenosis, a stent (a tiny tubular mesh of the size similar to coronary vessels) is placed at the stenosed site.

Coronary Vasodilator Reserve

Coronary stenosis may not cause myocardial ischemia at rest, as flow is maintained by compensatory post-stenotic dilatation of the artery. This compensatory dilatation is possible only as long as there is a coronary vasodilator reserve. When the reserve is low, myocardial ischemia appears during physical exertion. When the reserve is exhausted, myocardial ischemia occurs on the slightest exertion. It is therefore important to have an estimate of the coronary vasodilator reserve of an individual. The reserve will be low if the myocardial oxygen consumption is high or if the diastolic perfusion of coronary vessels is low.

Myocardial oxygen consumption (MVO$_2$) In the heart, there is approximately one capillary for each myocardial fiber. With cardiac hypertrophy, the diameter of each muscle fiber increases with concomitant increase in O$_2$ consumption but without a corresponding increase in vascularity. As a result, enlarged hearts have an increased vulnerability to circulatory insufficiency. MVO$_2$ is also high when the myocardial contractility is high.

MVO$_2$ measurement requires determination of coronary blood flow and arteriovenous difference of blood oxygen content (Fick principle), which is not always practicable. However, MVO$_2$ is known to be proportional to the tension developed by myocardial fibers. A rough estimate of MVO$_2$ is given by the tension-time index (TTI), i.e., the product of cardiac muscle tension and the total duration for which the tension is maintained. A clinical index of MVO$_2$ is the systolic pressure-time index (SPTI) which is given by the product of systolic blood pressure and the duration of ventricular systole. Calculation of SPTI would involve graphically measuring out the area under the ventricular pressure curve (Fig. 44.**6**). Clinicians often go by a more convenient index called the *rate-pressure product* (heart rate × systolic blood pressure). The approximation is based on the knowledge that the higher the heart rate, the greater is the relative duration of systole in the cardiac cycle (see Table 36.**2**).

The **diastolic perfusion of myocardium** depends on the diastolic perfusion pressure and the duration of diastole. The product of these two factors gives the diastolic pressure-time index (DPTI), which can be graphically calculated by the area under the diastolic aortic pressure curve. A low DPTI indicates low diastolic perfusion but a normal DPTI does not rule out inadequate diastolic perfusion which can also occur if the coronary arteries are stenosed or the coronary venous pressure is high, as occurs in congestive heart failure.

When SPTI (indicating MVO$_2$) and DPTI (indicating diastolic perfusion of the myocardium) are combined, it gives a fairly reliable clinical index of the risk of myocardial ischemia. Individuals with a high SPTI/DPTI ratio have a greater risk of myocardial ischemia. The SPTI/DPTI ratio gives a simple explanation as to why the risk of myocardial ischemia is higher in conditions like aortic stenosis, severe aortic regurgitation, tachycardia, and increased myocardial contractility.

45 Cerebral Circulation

Cerebral Blood Flow

The blood flow to the brain is about 750 ml min^{-1}, which is 15% of the resting cardiac output. The weight of the brain is about 1500 g and therefore the cerebral blood flow is 50 ml min^{-1} per 100 g of brain tissue. Of all the organs in the body, the brain is most susceptible to ischemia. A reduction of cerebral blood flow below 30 ml min^{-1} 100 g^{-1} for about 5 seconds results in syncope (fainting).

The brain receives blood from two pairs of large arteries: a pair of vertebral arteries and a pair of internal carotid arteries (Fig. 45.1). The vertebral arteries unite into a solitary basilar artery, which divides into a pair of posterior cerebral arteries (the vertebrobasal system). Together, they constitute the posterior circulation of the brain.

The carotid artery divides into the anterior and middle cerebral arteries (the carotid system) and constitutes the anterior circulation of the brain. The anterior and posterior circulations communicate through the posterior communicating artery—a branch of the middle cerebral artery. The circulations of the two hemispheres communicate through the anterior communicating artery—a branch of the anterior cerebral artery. The three cerebral arteries (anterior, middle, and posterior) and the two communicating arteries (anterior and posterior) form an arterial network called the circle of Willis below the hypothalamus.

Blood from each carotid artery is distributed largely to the ipsilateral cerebral hemisphere and very little mixing occurs between the two circulations in spite of their interconnection through the circle of Willis. This separation takes place because the blood pressure is approximately equal in the two carotid arteries so that there is no pressure gradient between them. Moreover, the communicating arteries are too small to permit significant flow when there is a sudden occlusion of one carotid artery. Hence, despite the presence of communicating arteries, occlusion of one internal carotid artery results in ipsilateral cerebral ischemia and brain damage.

The lateral surfaces of the cerebral hemisphere are supplied by the terminal branches of anterior, middle, and posterior cerebral arteries. The border zones between the areas served by these arteries barely receive adequate blood supply and are therefore vulnerable to ischemic damage during sudden hypotension (watershed infarcts).

Measurement of cerebral blood flow

Kety method In this method, the subject breathes a gas mixture of 15% nitrous oxide, 21% oxygen, and 64% nitrogen for 10 minutes, which is sufficient time for equilibration of N$_2$O between brain tissues and the cerebral venous blood. Simultaneous samples of arterial blood (from any artery) and mixed venous blood (from internal jugular vein) are taken at the beginning and at every minute interval for 10 minutes. From these data cerebral blood flow can be calculated by the Fick equation. Since the arterial and venous concentrations are continuously changing with time, it is necessary to get the mean A-V difference over the 10-minute period.

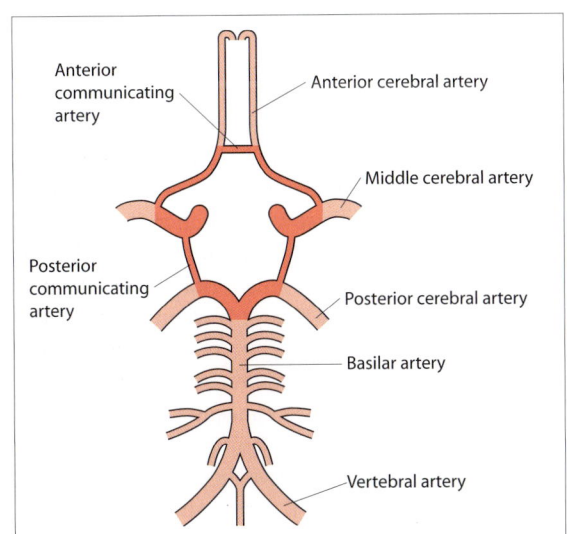

Fig. 45.1 Cerebral arteries (light red) and the circle of Willis (deep red). See also Figure 95.7.

$$\text{Cerebral blood flow} = \frac{\text{N}_2\text{O uptake by brain in 10 min}}{\text{Mean A-V difference of N}_2\text{O over 10 min}}$$

Figure labels:
- Anterior communicating artery
- Anterior cerebral artery
- Middle cerebral artery
- Posterior communicating artery
- Posterior cerebral artery
- Basilar artery
- Vertebral artery

Autoregulation of total CBF

The total cerebral blood flow is held constant in face of considerable changes in the systemic blood pressure (60–140 mmHg). This autoregulation is important since pressures below 60 mmHg cause syncope while pressures above 140 mmHg cause disruption of the blood–brain barrier and cerebral edema. The factors that contribute to the autoregulation of total cerebral blood flow are the Monro–Kellie doctrine, Cushing's reflex, and myogenic autoregulation.

Monro–Kellie doctrine There are three elements, viz., the brain, cerebrospinal fluid (CSF), and blood, that are enclosed in the rigid cranial cavity. If any one of them increases, it has to be at the expense of the other two. For example, when a person stands up from a supine position, or if the body accelerates upward (high Gz+), the blood moves toward the feet and the carotid arterial pressure at the level of the head decreases. However, the cerebral blood flow does not decrease much because the intracranial pressure too falls simultaneously, reducing the pressure on the artery. The reverse happens with a low or negative Gz when the carotid arterial pressure increases. Since the intracranial pressure also rises, the pressure differential across the vessel wall is not increased and the vessels do not rupture.

Cushing's reflex When the intracranial pressure rises, the cerebral blood flow falls, leading to ischemia. The cerebral ischemia causes direct stimulation of the vasomotor area, leading to a rise in the blood pressure and restoration of adequate cerebral blood flow. The rise in blood pressure is proportionate to the rise in intracranial pressure only up to a certain limit beyond which cerebral circulation ceases.

The rise in intracranial pressure is also associated with reflex bradycardia (produced by the rise in blood pressure through baroreflex) and abolition of the pupilloconstrictor response to light (due to the direct compressive effect on the pupilloconstrictor nerves near the aqueduct). Intracranial tension rises in the late stages of extradural hematoma and the three signs of raised intracranial tension (blood pressure, heart rate, and pupil) are therefore of ominous significance to the surgeon.

Myogenic autoregulation As in other circulatory beds, the cerebral precapillary sphincter is sensitive to stretch and responds to an increase in arteriolar pressure with an increase in myogenic tone, thereby limiting the rise in blood flow.

Metabolic regulation of regional CBF

On performing a specific task, the blood flow to the active cerebral area increases (Fig. 45.2) because the activated neurons produce metabolites that cause local vasodilatation. The total cerebral blood flow remains unaltered by localized neuronal activity.

The main metabolic factor responsible for the vasodilatation associated with cerebral activity is the CO_2 produced by the activated neurons. Cerebral vasodilatation is also produced by a rise in blood P_{CO_2} because CO_2 rapidly diffuses into the brain ECF and CSF. Conversely, a decrease in blood P_{CO_2}, as occurs in hyperventilation, causes a reduction in cerebral blood flow.

The vasodilatory effect of CO_2 is indirect and is mediated by the formation of H^+ which has a direct vasodilatory effect on cerebral blood vessels. A fall in the pH of brain ECF or CSF is therefore equally effective in causing vasodilatation. However, a fall in arterial pH fails to cause cerebral vasodilatation if the arterial P_{CO_2} is held constant. Hydrogen ions cannot cross the blood–brain barrier: A rise in arterial H^+ concentration increases cerebral blood flow only when the blood P_{CO_2} rises simultaneously.

Other factors affecting cerebral blood flow are potassium and adenosine. The K^+ concentration in the brain ECF and CSF increases in the initial stages of hypoxia, seizures, and electroconvulsive therapy. The initial rise in cerebral blood flow in these conditions may therefore be due to the rise in K^+. A more likely mediator of the vasodilatation is adenosine, the synthesis of which increases rapidly during tissue hypoxia and remains elevated throughout the period of hypoxia.

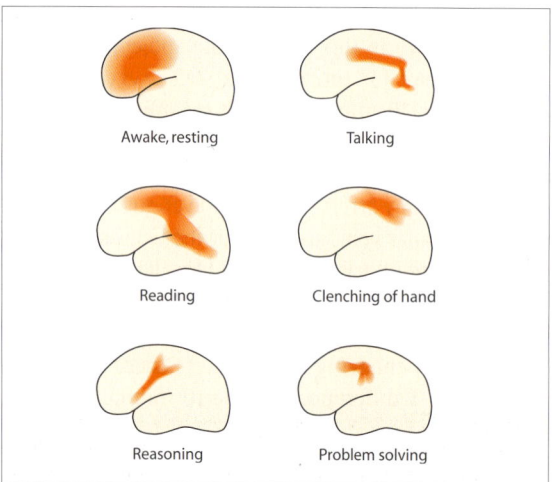

Fig. 45.**2** Increase in regional blood flow during different types of cerebral activities.

Syncope

Syncope (or fainting) is a sudden and transient loss of consciousness due to inadequate cerebral blood flow. Syncope is invariably associated with falling down, which is lifesaving because the change in posture increases cerebral blood flow. Syncope can be due to cardiac or noncardiac causes. Cardiac syncope occurs due to reduction of cardiac output. Noncardiac syncope can be due to a variety of causes, some of which are given below.

Vasovagal syncope (or the common faint) is associated with hypotension and bradycardia. Its cause remains hypothetical and is probably brought about by a reflex similar to the Bezold–Jarisch reflex. *Postural syncope* is the fainting that occurs due to inadequate vasomotor response to a change in posture from sitting/lying to standing. *Carotid sinus syncope* occurs due to excessive sensitivity of the carotid sinus to compression, as by a tight collar. They are common in elderly patients. *Situational syncopes* may be associated with cough, micturition or defecation. All these activities are associated with the Valsalva's maneuver which increases intrathoracic pressure. Micturition and defecation are additionally associated with a sudden reduction in intraabdominal pressure. The two factors (high intrathoracic and low intraabdominal pressures) together bring about a reduction in cardiac output and consequent hypotension and syncope. Other factors, e.g., the orthostatic hypotension associated with getting up from bed before micturating might contribute to situational syncopes. Hypoglycemia or hyperventilation can cause *metabolic syncope*. Hypoglycemia inhibits the vasomotor center while hyperventilation causes hypocapnia, resulting in cerebral vasoconstriction and ischemia.

The *subclavian steal syndrome* is sometimes seen when the subclavian artery is blocked proximal to the origin of the vertebral artery. Exercise of the arm of the affected side may draw blood from the vertebral artery into the arm, reducing the blood flow in the basilar artery and causing cerebral ischemia (Fig. 45.**3**).

Cerebrospinal Fluid

Cerebrospinal fluid (CSF) is a clear, colorless, almost protein free filtrate (transudate) of blood. It is present around the brain and spinal cord (in the subarachnoid space) as well as inside the brain and spinal cord (in its ventricles and the central canal respectively).

The **subarachnoid space** lies between the arachnoid and the pia mater. It is enlarged in a few places to form

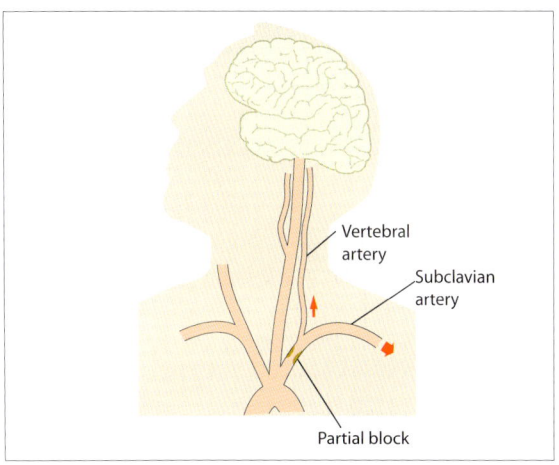

Fig. 45.**3**   Subclavian steal. Exercising muscles draw more blood into the left upper limb through the subclavian artery, reducing the blood flow to the brain through the vertebral artery.

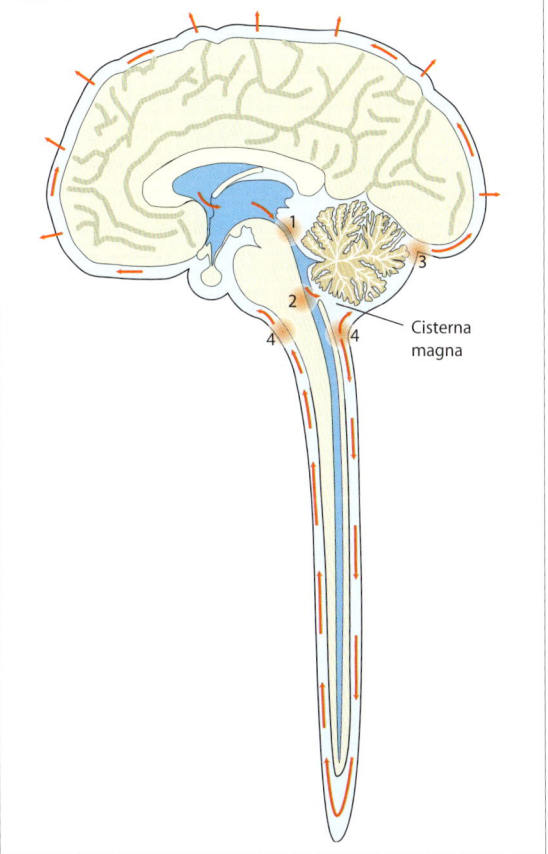

Fig. 45.**4**   CSF circulation within the ventricles and in the subarachnoid space. Obstruction at sites 1 (aqueduct) and 2 (foramen of Magendie and Luschka) results in internal hydrocephalus. Obstruction at sites 3 (tentorial opening) and 4 (foramen magnum) results in external hydrocephalus. Also shown are the subarachnoid cisterns.

subarachnoid cisterns (Fig. 45.**4**). Some of these cisterns (cisterna magna and the spinal cistern) are easily accessible to a needle inserted through skin and are therefore used for obtaining samples of CSF. The cranial nerves pierce the dura but carry with it an investment of arachnoid and pia mater. Thus, the subarachnoid space continues along the cranial nerves. The continuation of the subarachnoid space along the optic nerve leads to the formation of papilledema in conditions of raised intracranial tension. The subarachnoid space also continues along the cerebral blood vessels as perivascular space (also known as Virchow–Robin space).

■ CSF circulation

CSF formation   CSF is secreted by the choroid plexus which is formed by the fusion of the pia mater of the brain with the ventricular ependyma, and which contains within its folds the choroid capillaries. About 500 ml of CSF is formed each day. The rapid production rate keeps up a slight pressure of about 150 mm of water within the subarachnoid space.

The watery part of CSF is secreted by transudation but each of its constituents is actively transported. Na^+ is secreted into the CSF with the help of Na^+-K^+ ATPase. Glucose enters CSF through facilitated diffusion mediated by glucose transporter GLUT-1 (p 407). HCO_3^- is secreted into the CSF with the help of carbonic anhydrase, as in renal tubules.

CSF circulation   From the lateral ventricles, CSF flows through the interventricular foramina (of Monro) into the third ventricle. From there, it flows through the aqueduct (of Sylvius) in the midbrain into the fourth ventricle. Finally, it flows out of the fourth ventricle through the median (Magendie) and two lateral (Luschka) foramina to circulate in the subarachnoid space around the brain and spinal cord.

CSF absorption   The CSF is absorbed into the venous system mostly through small villi called the arachnoid granulations projecting into the dural sinuses (Fig. 45.**5**). About 20% of the CSF is absorbed into the veins through spinal arachnoid granulations while the rest is absorbed through the cerebral arachnoid granulations. The absorption is driven mainly by the difference in colloid oncotic pressure of the CSF (nearly zero) and the plasma (25 mmHg). It is also aided by hydrostatic pressure, which is slightly greater in the CSF than in the dural sinuses.

■ Functions of CSF

The brain weighs about 1500 g but it weighs just 50 g in CSF due to buoyancy. The importance of this buoyancy is apparent from the occasional traction headache that occurs after withdrawal of CSF by lumbar puncture (see below). CSF withdrawal reduces the buoyancy and increases the weight of the brain, which pulls down on the pain-sensitive dura from which it is suspended by the cerebral veins draining into the dural sinuses. Moreover, the brain presses harder upon the floor of the cranial cavity, producing pain due to pressure on the cranial nerves at the base of the brain.

The CSF provides a cushion around the brain and protects it from injury. However, the brain still gets injured sometimes. For example, if there is a severe blow on the head, the skull quickly moves in the direction of the blow but the brain lags behind due to inertia. This results in the *coup injury*. Moments later, the movement of the skull stops but the brain keeps moving due to inertia. As a result, the brain hits the skull at the opposite side to the blow. This is known as *contracoup injury* (Fig. 45.**6**).

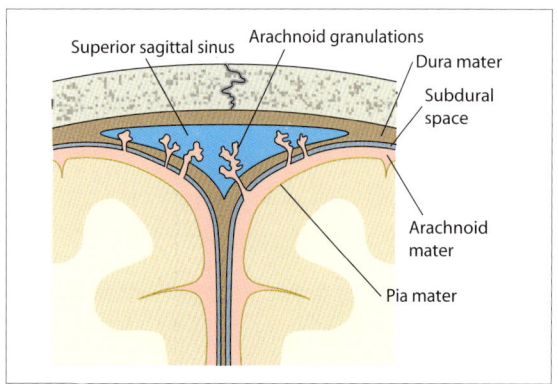

Fig. 45.**5**   The meninges and arachnoid granulations.

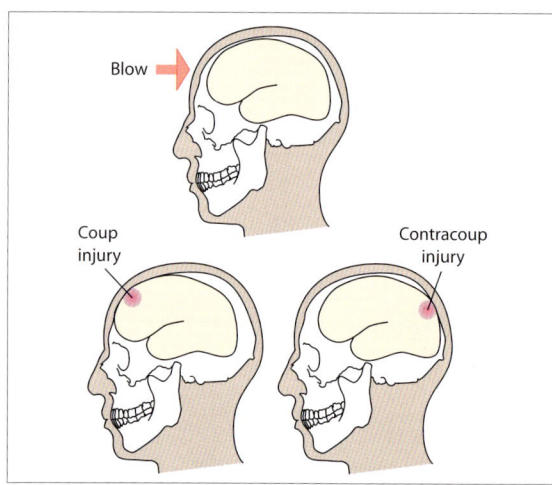

Fig. 45.**6**   Coup and contracoup injuries.

The CSF provides an optimum environment (milieu intérieur) to neurons, which are highly sensitive to changes in their external environment. It acts as a medium for nutrient exchange. Proteins that leak out of capillaries into the brain interstitial fluid are drained by CSF and returned to the bloodstream. Thus, the CSF fulfils the role of lymphatics, which are absent in the brain.

■   CSF analysis

Lumbar puncture   Although samples of CSF can be drawn by lumbar, cisternal or ventricular puncture, the most convenient and commonly used way is the lumbar puncture (Fig. 45.7). It can be done safely between L3 and L4 or L4 and L5 vertebrae because the spinal cord ends at L1 while the dura and arachnoid continues till S2. Besides obtaining a CSF sample, lumbar puncture also allows the estimation of CSF pressure by connecting the puncturing needle to a manometer.

The physical characteristics of CSF are given in Table 45.1 and its chemical composition is given in Table

Table 45.**1**   Normal physical characteristics of CSF.

Amount	125–150 ml
Specific gravity	1.007, a density close to that of brain tissue.
pH	7.33
Pressure	70–200 mmH$_2$O (in the horizontal position)
Potential	+5 mV (CSF Positive)
Cells	< 5 cells mm^{-3} (mostly lymphocytes)

Table 45.**2**   A comparison of the biochemical compositions of CSF and plasma.

	Plasma conc. (mEq kg^{-1})	CSF conc. (mEq kg^{-1})
Cl$^-$	100	130
Pco$_2$ mmHg	40	50
K$^+$	4.5	3.0
Protein	6000	20 (mainly albumin)
PO$_4{}^{3-}$	4.5	3.5
Ca^{2+}	4.5	2.5
Glucose	100	60

45.**2**. Substances present in greater concentration than in plasma are chloride and CO$_2$. Substances present in concentrations equal to that in plasma include Na$^+$ and urea. Substances present in lower concentration than in plasma include K$^+$, Ca^{2+}, phosphates, proteins, and glucose. Most other substances do not pass from plasma to CSF except in minute traces.

Abnormalities in the CSF include the following: (1) Turbidity indicates increased content of protein and cells, indicative of meningitis. (2) CSF glucose is low in meningitis due to utilization of glucose by microbes. (3) Neutrophils in large numbers (reaching up to 1000 to 10,000 cells ml^{-1}) indicate bacterial meningitis or brain abscess. (4) Lymphocytes in large numbers (200-300 cells ml^{-1}) suggest meningeal syphilis. (5) Red cells in large numbers indicate subarachnoid hemorrhage.

■   CSF pressure

The CSF pressure, as measured through a lumbar puncture, is about 150 mm of water when the patient is supine. The pressure recorded is higher in the sitting position. A raised intracranial pressure can cause neuronal damage directly or it can compress cerebral capillaries, leading to ischemic damage. Further damage occurs if a part of the brain herniates through openings in the cranial bones or tentorium.

A characteristic sign of raised CSF pressure is papilledema (swelling of the optic disk) which occurs

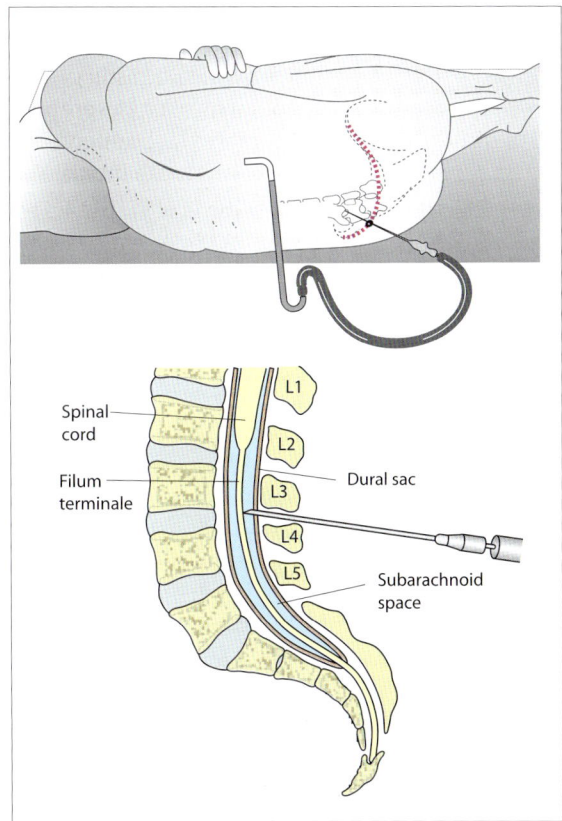

Fig. 45.**7**   Lumbar puncture. (*Above*) Positioning of the patient for lumbar puncture. (*Below*) Positioning of the needle in the spinal subarachnoid space.

when the raised intracranial pressure causes compression of the ophthalmic veins. Blood can still flow along the arteries and reach the optic disk but its return is impeded. Fluid exudes from the capillaries, resulting in the swelling of the optic disk. Other signs associated with raised CSF pressure are those of Cushing's reflex and impaired consciousness like disorientation, stupor or coma.

Hydrocephalus

Hydrocephalus (*Greek:* watery head) occurs due to excessive CSF accumulation within the cranium. Its causes include excessive CSF formation, obstruction to CSF circulation and impaired CSF absorption. There can be other causes too. For example, atrophic (compensatory) hydrocephalus occurs in brain atrophy due to the compensatory enlargement of the ventricular system. It is found in patients of severe head injury and degenerative brain disorders.

Internal hydrocephalus If the fluid accumulates inside the cerebral ventricles, it is called internal hydrocephalus. It could be due to a block at the foramen of Monro (causing dilatation of one or both lateral ventricles), in the aqueduct (causing dilatation of the lateral and third ventricles), or at the foramina of Magendie and Luschka (causing dilatation of the entire ventricular system).

External hydrocephalus If the fluid accumulates both in the ventricles as well as in the subarachnoid space, it is called external hydrocephalus. It is either due to excessive CSF formation (hypersecretory hydrocephalus) or reduced CSF absorption (malabsorptive hydrocephalus). The reduced absorption could be due to a block at the foramen magnum or at the tentorial opening. It could also be due to diseases of the arachnoid mater or thrombosis of the dural sinuses. *Block at the foramen magnum* prevents CSF from entering the spinal cord, thereby cutting off 20% of its absorptive surface. *Block at the tentorial opening* prevents the passage of CSF from the posterior fossa to the supratentorial space, thereby occluding most of the absorptive surface (Fig. 45.**4**).

Two simple clinical tests give clues to the clinician regarding the cause of the hydrocephalus. These tests, which have now been superseded by myeloencephalography, are the phenolphthalein test and Queckenstedt's test. The *phenolphthalein test* involves injecting phenolphthalein into the lateral ventricle and noting the duration after which it appears in the spinal fluid and urine (Fig. 45.**8**). *Queckenstedt's test* is performed while doing a lumbar puncture by connecting the needle to a manometer. Compression of one internal jugular vein, which reduces CSF absorption, causes a rise in CSF pressure followed by a quick fall on release. A similar rise and fall occurs on coughing. When there is a block, the rise and fall is slow or absent.

Appearance of phenolphthalein in urine	Appearance of phenolphthalein in spinal fluid: Normal	Appearance of phenolphthalein in spinal fluid: Delayed/Absent
Normal		Noncommunicating CSF not entering the spinal subarachnoid space
Delayed/Absent	Communicating No obstruction to CSF circulation. CSF not absorbed by arachnoid granulations	Noncommunicating CSF not flowing out of the ventricles

Fig. 45.**8** The phenolphthalein test.

These tests indicate whether or not there is any block to the flow of CSF into the spinal subarachnoid space. If a block is present, it is called a *noncommunicating* hydrocephalus and it could be either due to an internal hydrocephalus or due to an external hydrocephalus in which there is obstruction at the foramen magnum. If there is no block to the flow of CSF into the spinal subarachnoid space, it is called *communicating hydrocephalus* and it rules out an internal hydrocephalus or a block at the foramen magnum.

Blood–Brain Barrier

The blood–brain barrier is a physiological barrier to the movement of several substances in and out of the brain. It maintains a constancy of environment in and around the brain, prevents escape of neurotransmitters, and protects the neuron from harmful substances that may be present in blood. Most of the substances that gain entry into the brain ECF or the CSF are actively transported across the capillary endothelium of the cerebral capillaries. The high density of mitochondria in brain capillary endothelium reflects the higher metabolic activity of these capillaries.

The composition of brain interstitial fluid and CSF are slightly different and hence, the blood–brain barrier (between blood and the interstitial fluid of the

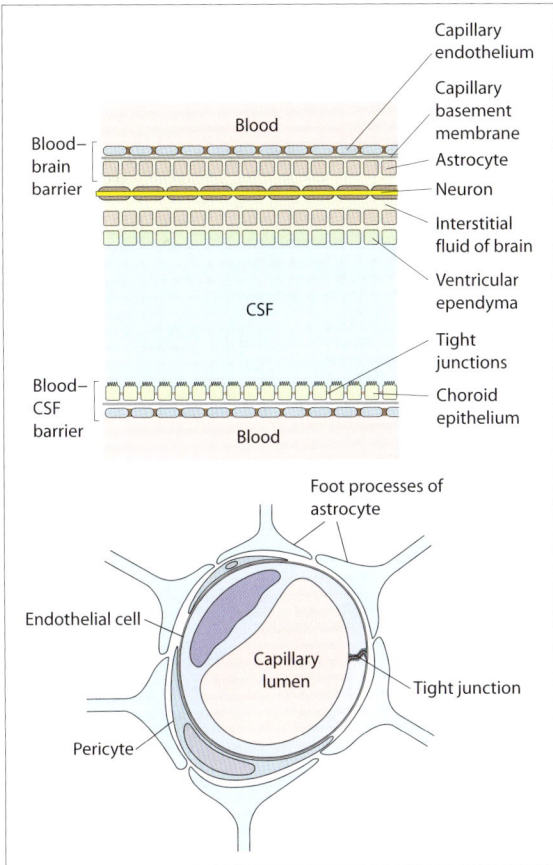

Deficiency of the blood–brain barrier   The blood–brain barrier is not properly developed in infancy and therefore, a high serum bilirubin level leads to accumulation of bilirubin in the brain causing kernicterus. The blood–brain barrier is also deficient at the site of a brain tumor because the new capillaries that proliferate in a brain tumor are not encircled by astrocytes. A radiolabeled amino acid that is injected into circulation gets localized in the tumor and not other parts of the brain. This offers obvious advantage in diagnostic radiography. Inflammation of brain and meninges weakens the blood–brain barrier. Hence antibiotics that normally do not cross blood–brain barrier do so during inflammation. Experimental disruption of the blood–brain barrier can be produced by exposing cerebral capillaries to hyperosmolar solution. It acts by shrinking endothelial cells, thereby making the tight junctions less tight.

The blood–brain barrier is highly permeable to water, CO_2, O_2, lipid-soluble substances like alcohol and most anesthetics. It is slightly permeable to electrolytes, Na^+, Cl^-, K^+, and totally impermeable to plasma proteins and large, lipid-insoluble organic molecules. These permeability characteristics are apparent from the composition of the CSF (Table 45.2).

Structures outside the blood–brain barrier   Certain areas of brain have fenestrated capillaries and because of their permeability, are said to be outside the blood–brain barrier. These areas are located mostly around the third ventricle and are therefore called *circumventricular organs*. Some of them are neurosecretory organs secreting polypeptide neurotransmitters that must be able to enter the circulation freely, while others are chemoreceptor areas for detecting blood-borne substances that must be able to reach the brain. The *neurosecretory areas* are the posterior pituitary and the median eminence of hypothalamus. The *chemoreceptor areas* are the area postrema, subfornical organ, and organ vasculosum of lamina terminalis. The pineal gland and the anterior pituitary also have fenestrated capillaries and are outside blood–brain barrier but both are endocrine organs and not part of brain.

Although the blood–brain barrier is deficient in the circumventricular organs, the blood–CSF barrier in these organs is quite effective due to the presence of tight-junctions between the modified ependymal cells of the third ventricle called the *tancytes*. Hence the substances that leak out of capillaries in the circumventricular organs are unable to enter the CSF.

Fig. 45.9   (Above) The blood–brain and blood–CSF barriers. (Below) The tight junctions and astrocyte foot processes in a cerebral capillary. Compare with Figure 39.1.

brain) is sometimes distinguished from the blood-CSF barrier. Substances diffusing across the *blood–CSF barrier*, i.e., from the choroidal capillary to the CSF have to cross two functionally effective barriers (Fig. 45.9): The tight junctions between capillary endothelial cells and the tight junctions of the choroidal epithelium. The pia mater and the capillary basement membrane do not constitute effective barriers. Substances diffusing across the *blood–brain barrier* from the cerebral capillary to the brain interstitium have to cross only the endothelial barrier, the choroidal epithelium being absent. The foot processes of astrocytes, which encircle the cerebral capillaries and form a complete sheath around them, do not constitute an effective barrier. Yet they contribute to the blood–brain barrier because they induce the formation of the tight junctions between endothelial cells.

46 Pulmonary and Pleural Circulation

Pulmonary Circulation

The pulmonary vasculature is very distensible and so offers a relatively low resistance to the blood flowing from the right to the left ventricle. The distensibility of the pulmonary circulation makes it a low-pressure, high-capacitance system. The afterload on the right ventricle is therefore much less than the afterload on the left ventricle. Compared to the aorta, the pulmonary artery is much shorter, has a larger diameter, and has about one-third its wall thickness. This is also true for pulmonary arterioles in relation to their systemic counterparts.

The bronchi receive oxygenated blood through bronchial arteries, which are systemic arteries (Fig. 46.**1**). Some of the deoxygenated bronchial venous blood mixes with the oxygenated pulmonary venous blood (Fig. 48.**3**). This interconnection between bronchial and pulmonary circulations is called the *physiological shunt*. Besides reducing the O_2 saturation of arterial blood slightly, the shunt makes the left ventricular output slightly greater than the right ventricular output.

Lymphatics are present in the walls of the terminal bronchioles. Particulate matter entering the alveoli is partly removed via these channels. Proteins are also removed from the lung tissues, thereby helping to prevent edema.

Pulmonary blood pressure Pressure measurements in the pulmonary artery, pulmonary vein or in the left atrium are difficult since these areas are not easily accessible to catheters. Pressure in the pulmonary circulation is measured by inserting a catheter through the right side of the heart and the pulmonary artery into one of the small branches of the pulmonary artery and then pushing the catheter until it wedges tightly in the artery (Fig. 46.**2**). The pressure so measured is called the *pulmonary wedge pressure*, and is about 5 mmHg. Since the catheter stops the flow of blood into the artery and faces toward the venous side of the pulmonary circuit, what is measured is closer to the pulmonary venous pressure than the pulmonary arterial pressure. The mean pressure in pulmonary veins and the left atrium is 2 about mmHg in the recumbent position. The wedge pressure is 2 to 3 mmHg greater than the left atrial pressure.

The **pulmonary blood volume** is approximately 600 ml, which is located mostly in the pulmonary arteries and veins. When there is hemorrhage from systemic circulation, it is partly compensated for by translocation of blood from this pulmonary reservoir. The pulmonary blood volume increases when the posture changes from the standing to the lying and decreases when the intrathoracic pressure increases, for example, during a Valsalva maneuver. It also increases in pathological conditions like mitral stenosis and mitral regurgitation.

Effect of gravity In the normal, upright person, the pulmonary arterial pressure varies from the base to the apex of the lung. Compared to the pulmonary arterial pressure at the level of the heart, the apical pressure is 15 mmHg less and the basal pressure is 8 mmHg

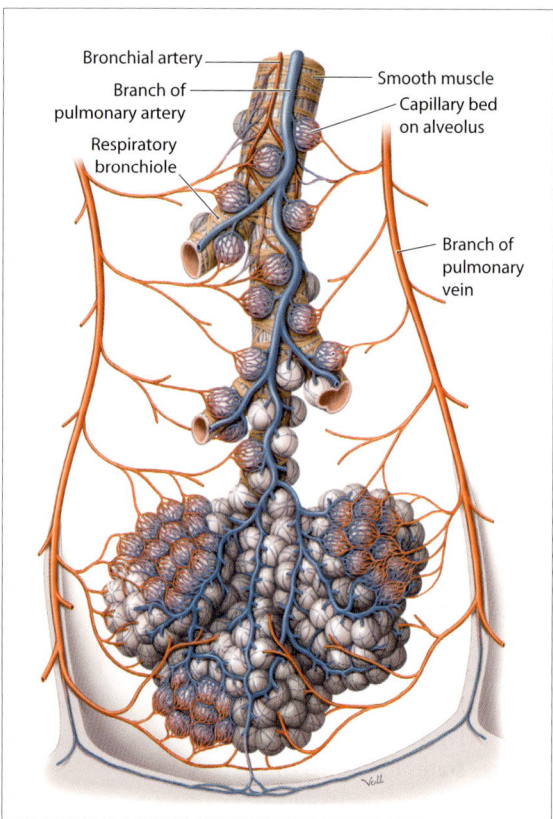

Fig. 46.**1** Pulmonary and bronchial blood vessels.

Table 46.**1** Pressures in pulmonary circulation.

Systolic pressure	25 mmHg
Diastolic pressure	8 mmHg
Mean pressure	15 mmHg
Pulse pressure	17 mmHg

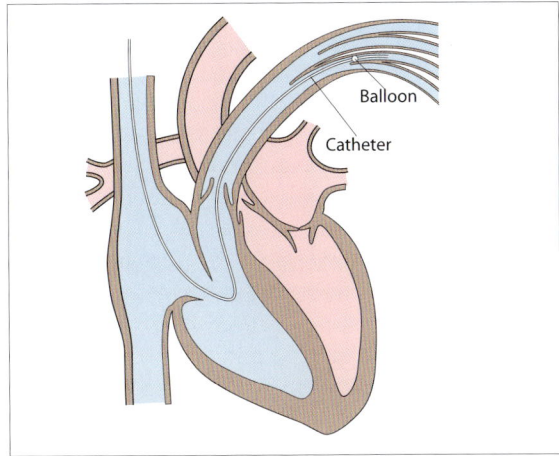

Fig. 46.**2** Pulmonary artery wedge pressure measurement.

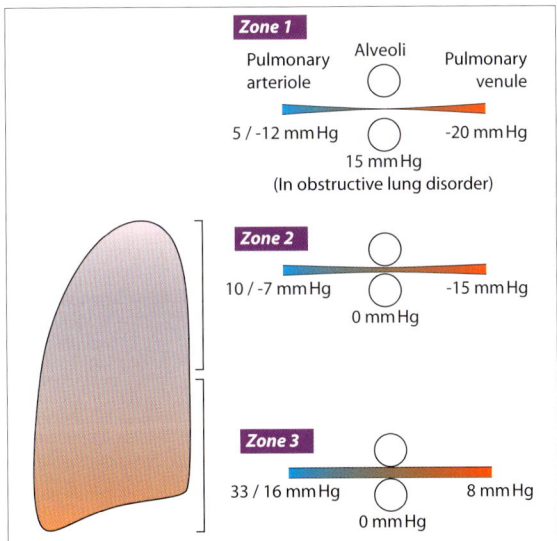

Fig. 46.**3** Zones of pulmonary circulation. Zone 3 is present only in disease states like obstructive airway disorder.

greater. There are corresponding differences in regional blood flow which form the basis of two lung zones: a basal zone called zone 3 extending up to a distance of 10 cm above the heart level, and an apical zone called zone 2 (Fig. 46.**3**). A third zone called zone 1 appears in disease states like COPD (p 365).

Zone 3 has continuous blood flow because the alveolar capillary pressure remains greater than alveolar air pressure throughout the cardiac cycle.

Zone 2 has intermittent blood flow, occurring only during the peaks of pulmonary arterial pressure. The pulmonary systolic and diastolic pressures are 25 mmHg and 8 mmHg when measured at the heart level. The corresponding pressures at the apex, where the pressure is 15 mmHg less, would be 10 mmHg and –7 mmHg respectively. The air pressure in the alveoli is 0 mmHg. Hence, during diastole, the alveolar capillaries collapse and no blood flows through them. However, in the recumbent position no part of the lung is more than a few centimeters above the level of the heart. Therefore the entire lung including the apex becomes equivalent to zone 3, receiving continuous blood flow.

In *zone 1*, no blood flows at all during any phase of the cardiac cycle because the local capillary pressure in zone 1 never rises higher than the alveolar pressure. Zone 1 is not present normally: It is present when the pulmonary systolic arterial pressure is too low (as in hypovolemic states) or the alveolar pressure is too high (as in obstructive lung disorders) to allow blood flow.

Effect of exercise   Blood flow in all parts of the lung increases during exercise, sometimes as much as sevenfold. This extra flow is made possible by the opening up of inactive capillaries and the increase in the flow rate through each capillary. Together, these two factors decrease the pulmonary vascular resistance so much that the pulmonary arterial pressure rises very little even during maximal exercise. However, the rise in pressure is enough to provide continuous blood flow to the apices so that, in effect, the entire lung becomes equivalent to zone 3 during exercise.

The ability of the lung to allow a large amount of blood to flow through it during exercise reduces the load on the right side of the heart. It also prevents a significant rise in pulmonary capillary pressure and therefore prevents development of pulmonary edema during exercise.

◼ Effect of left ventricular failure

The left atrial pressure rises concomitantly with the severity of left ventricular failure. A rise in left atrial pressure up to 7 mmHg does not cause a rise in the pulmonary arterial pressure due to the large compliance of pulmonary venules and the opening up of more pulmonary capillaries. The resultant increase in the pulmonary vascular capacity limits the pressure rise. However, pressures greater than 7 mmHg are transmitted to the pulmonary arteries, imposing a high afterload on the right ventricle. As the pressure is transmitted through the capillaries, the pulmonary capillary hydrostatic pressure rises. When the left atrial pressure exceeds 25 mmHg, the capillary pressure is sufficiently high to produce pulmonary edema.

■ Neural control of pulmonary circulation

Efferent control of pulmonary vessels is exerted only through sympathetic fibers, which cause a slight increase in the pulmonary vascular resistance. However, the effect is insignificant and physiologically unimportant. The main function of sympathetic discharge is to constrict the large pulmonary capacitance vessels, especially the veins, causing translocation of pulmonary blood into systemic circulation.

Afferent control   Pulmonary baroreceptors are present in the adventitia of the pulmonary trunk and the right and left pulmonary arteries. These receptors detect rises in pulmonary arterial pressure and produce reflex inhibition of sympathetic discharge to the pulmonary vasculature. The afferent arc is through the vagus.

■ Chemical control of pulmonary circulation

Hypoxia, hypercapnia, and acidosis induce pulmonary vasoconstriction. This is exactly the opposite of what happens in systemic circulation where hypoxia produces vasodilatation. The physiological significance of this phenomenon is to divert pulmonary blood flow from the alveoli that are poorly ventilated. The smooth muscle of pulmonary arteries contains O_2-sensitive K^+ channels that close down when the P_{O_2} falls. K^+ channel closure is associated with depolarization and contraction of smooth muscle.

The vasoconstriction associated with chronic hypoxia causes a marked increase in pulmonary arterial pressure (pulmonary hypertension) which imposes a heavy afterload on the right ventricle leading to its hypertrophy and eventual failure (cor pulmonale). Pulmonary hypertension is common in high altitude dwellers and in COPD.

■ Pulmonary capillary dynamics

Pulmonary transit time   Blood passes through the pulmonary capillaries in about 0.75 s (the transit time). A rise in cardiac output during exercise shortens the

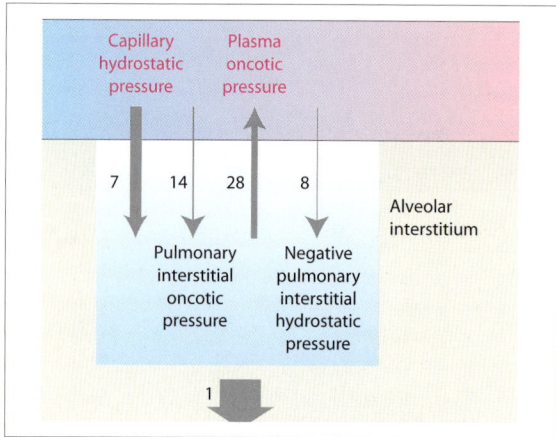

Fig. 46.**4**   Starling forces in pulmonary capillaries.

transit time to about 0.3 s, which is barely adequate for complete oxygenation of blood (see Fig. 51.**8**). The transit time would be shorter if several inactive capillaries did not open up with exercise. The opening up of inactive capillaries increases the total capillary cross sectional area, slowing down the linear velocity of blood and prolonging the transit time.

Pulmonary capillary Starling forces   The balance of Starling forces at the capillary membrane is shown in Fig. 46.**4**. Note that the pulmonary interstitial hydrostatic pressure is negative at –8 mmHg. The net filtration pressure of 1 mmHg (14 + 8 + 7 – 28) causes a slight filtration of fluid from the pulmonary capillaries into the interstitial spaces. Except for a small amount that keeps the alveolar lining moist, this fluid drains back into the circulation through the pulmonary lymphatic system.

Pulmonary edema   A common cause of pulmonary edema is left ventricular failure. It is associated with an increase in the pulmonary capillary hydrostatic pressure that causes transudation (increased capillary filtration) of fluids into the alveolar interstitium. Capillary permeability also increases following exposure to irritant gases. It results in exudation, i.e., leakage of both fluid and plasma proteins out of the capillaries.

In acute cases, pulmonary edema occurs when the pulmonary capillary hydrostatic pressure exceeds 28 mmHg. In chronic cases, there is a compensatory increase in lymphatic drainage and therefore the capillary pressure must rise higher to produce edema. The

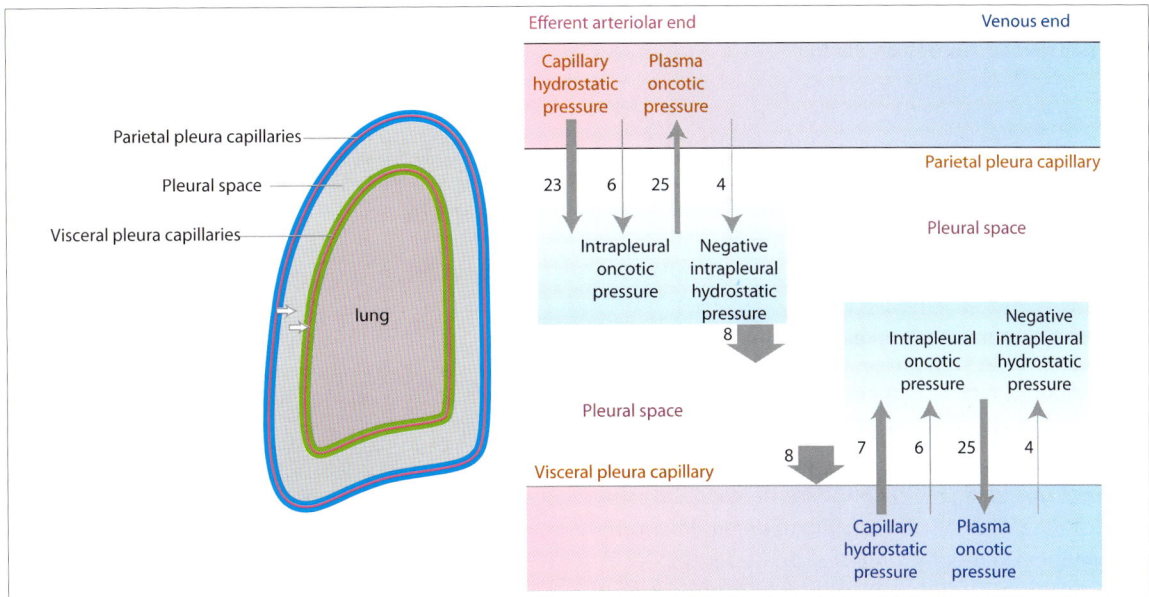

Fig. 46.**5**   (*Left*) The movement of fluid from the capillaries in the parietal pleura (blue) to the capillaries in the visceral pleura (green). (*Right*) Starting forces in pleural capillaries.

edema is initially confined to the interstitial fluid. Once the interstitial fluid pressure becomes positive, the fluid begins to enter the alveoli. The edema reduces the oxygenation of blood in the lungs.

Pleural Circulation

There is a negative intrapleural pressure of about −4 mmHg during normal respiration. The oncotic pressures in both parietal and visceral pleural capillaries are the same, i.e., 25 mmHg. Since there is very little protein in the normal pleural fluid, the pleural oncotic pressure is only about 6 mmHg. The hydrostatic pressure in parietal pleura capillaries fed by systemic arteries is about 23 mmHg while that in visceral pleura capillaries fed by the pulmonary arteries is only about 7 mmHg (Fig. 46.**5**).

Thus there is a net pressure gradient of 8 mmHg pushing the pleural fluid from parietal capillaries to the pleural space. Similarly, there is a net force of 8 mmHg driving the pleural fluid from the pleural space to visceral capillaries and lymphatics. This pressure gradient ensures that the fluid filtered into the pleural space by the parietal pleural capillaries is rapidly absorbed by the visceral pleura capillaries while proteins and particles that enter the pleural space are slowly absorbed by lymphatics opening in the parietal pleura.

Cutaneous Circulation

The oxygen and nutrient requirements of the skin are relatively small. Cutaneous circulation is therefore regulated less by its metabolic activity and more by thermoregulatory requirements. The metabolic control of cutaneous blood flow is revealed by the presence of *reactive hyperemia*, i.e., the increase in blood flow that occurs after a period of occlusion of blood flow. Local metabolites accumulate in the skin during the occlusion and dilate the precapillary sphincters of cutaneous capillaries, producing hyperemia when the occlusion is released. Metabolic control is overridden by thermoregulatory control and therefore it is unable to relieve the ischemia caused by cold-induced vasoconstriction.

In shock, cutaneous vasoconstriction diverts blood to vital tissues. Yet if the ambient temperature is high, the cutaneous blood flow increases at the expense of reducing the blood supply to the vital tissues, i.e., it prevails over other reflexes that tend to reduce cutaneous blood flow. The increase in cutaneous blood flow in response to thermal challenge is thus a *prepotent reflex*, i.e., it overrides all other reflexes.

The resting cutaneous blood flow is the flow when a person is at thermal equilibrium with the environment, i.e., at about 25–27°C (the thermoneutral zone). It is about 13 ml min^{-1} per 100 g of skin tissue (450 ml min^{-1}). At temperatures below the thermoneutral zone, the skin vessels are strongly constricted and blood flow is directed to the deeper tissues. Depending on the amount of sweating, cutaneous blood flow varies from one-tenth to ten times the resting blood flow. Maximal cutaneous blood flow occurring on heat exposure imposes a heavy circulatory load on the heart, which can lead to hypotension and shock.

The thermoregulatory control of cutaneous blood flow may be local or reflex. The local thermoregulatory control is brought about by the direct effect of skin temperature on vascular smooth muscles which contract when cooled and relax when warmed. The center for reflex thermoregulation is located in the hypothalamus), which controls the sympathetic output of the brainstem vasomotor center. This explains why cardiovascular reflexes centered at the vasomotor center are overridden by the thermoregulatory reflexes controlled by the hypothalamus.

Triple response to trauma   A firm strong stroke across the skin using a blunt point evokes three sequential responses: the red reaction, flare, and wheal. The *red reaction* is a red line that develops along the line of the stroke. It occurs due to the dilatation of precapillary sphincters. The dilatation is not neurally mediated but caused by the histamine and bradykinin released from the injured skin. The *flare* is a warm, erythematous (red) area that develops around the red line. It occurs due to a dilatation of the arterioles, terminal arterioles, and precapillary sphincters. The dilatation is mediated by the axon reflex (see Figs. 13.**6** and 103.**6**) in the cutaneous C fibers. The *wheal* is a swelling that develops along and around the line of the stroke. It is caused by the capillary damage, which results in increased capillary permeability and exudation of plasma. Some individuals have a striking triple response reaction so that anything drawn on the skin with a blunt point becomes conspicuous in minutes, a phenomenon called dermographism.

Some of the features of cutaneous circulation are different in the apical and nonapical skin. Apical skin is found in the exposed but poorly insulated areas of the body, like the palm of the hands, sole of the feet, face, and ears. Elsewhere, the skin is called nonapical.

Apical skin arterioles have a high degree of basal sympathetic tone which is under corticohypothalamic control. Heat exposure reduces the sympathetic tone and dilates the arterioles. Conversely, cold exposure constricts the arterioles. However, when the skin temperature falls below 10°C, the arterioles dilate. The cold injury to the tissues causes the liberation of histamine which excites the sensory terminals of cutaneous nerves and produces sustained vasodilatation through the axon reflex. The vasodilatation increases the cutaneous blood flow so as to warm up the skin and prevent frostbite of the highly exposed apical skin. This is known as *cold vasodilatation*.

The apical skin also contains arteriovenous (A-V) anastomoses which are under similar neural control to the arterioles (Fig. 47.**1**). These A-V anastomoses produce quicker and larger increases in the cutaneous blood flow, which makes them especially important in cold vasodilatation. The rapid change in cutaneous blood flow caused by the A-V anastomoses also causes

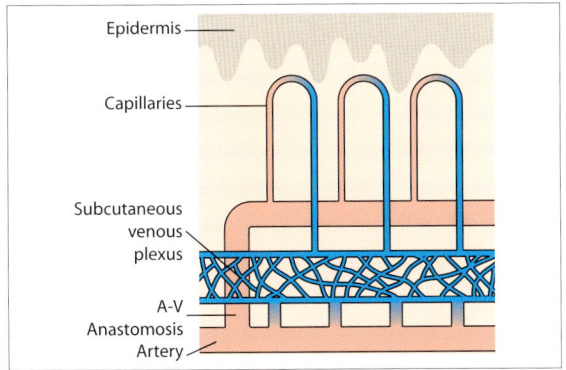

Fig. 47.**1**   Organization of cutaneous blood vessels.

Table 47.**1**   Differences between eccrine and apocrine sweat glands.

Eccrine	Apocrine
Present all over the body but dense on the head and densest on the palms and soles.	Present in the axilla, perianal region, around the nipples, in mons pubis, and labia majora.
Opens on the skin surface.	Opens into the canal of a hair follicle.
Secrete profuse watery, hypotonic sweat.	Secrete a milky fluid that is acted upon by bacteria to release a characteristic odor.
Innervated and stimulated by cholinergic sympathetic fibers.	Not innervated but stimulated by circulating epinephrine.
Important in thermoregulation.	No role in thermoregulation.

quick changes in skin color. Emotions are therefore promptly reflected in the color of the face: Shame and anger cause sympathetic inhibition, resulting in blushing due to dilatation of these vessels, while fear and anxiety cause blanching due to sympathetic vasoconstriction.

Nonapical skin lack A-V anastomoses   The neural control of arterioles in nonapical skin is similar to that in apical skin. The arterioles here have a high degree of intrinsic myogenic tone and low sympathetic tone. Withdrawal of the sympathetic tone therefore does not produce much vasodilatation. Large dilatation of these arterioles is mediated by sympathetic vasodilator fibers to the sweat glands. The stimulated sweat glands release bradykinin which cause vasodilatation (Fig. 40.**2**).

Sweat glands

Sweat glands are of two types, eccrine, and apocrine. Both are present in the dermis and both secrete excessively in emotional states. However they have several differences that are summarized in Table 47.**1**. There are 3–4 million sweat glands in the body which would add up to the size of a kidney and secrete up to 10 liters of sweat per day. They account for the huge increase in cutaneous blood flow that occurs during heat exposure. The blood flow to the actively secreting sweat glands is increased by bradykinin. Since eccrine glands are innervated by cholinergic sympathetic fibers, sweating and the associated increase in cutaneous blood flow are inhibited by atropine and in diseases of the autonomic nervous system.

Muscle Circulation

Blood vessels of skeletal muscle constitute the largest vascular bed in the body. During exercise, the vascular resistance in muscle falls massively, allowing up to 20 times increase in blood flow. Hence, muscle resistance vessels have considerable effect on the arterial blood pressure.

Blood flow to tonic and phasic muscles   Unlike phasic (white) muscle fibers, tonic (red) muscle fibers employ aerobic metabolism and contract all the time. Hence, their O_2 demand is much higher. Their resting blood flow (30 ml per 100 g per min) far exceeds that of phasic muscle (3 ml per 100 g per min). However, since they comprise only a fraction of the total muscle mass, their total resting blood supply is only twice that of white fibers. Red fibers have myoglobin, which allows them to overcome brief periods of ischemia that occur due to capillary compression during muscle contraction. However during sustained contraction of 10 seconds or more, the myoglobin O_2 content is exhausted, resulting in fatigue and ischemic pain.

Myogenic autoregulation   The precapillary resistance vessels in the muscle have a high basal myogenic tone. A rise in arterial blood pressure induces a stretch-induced contraction of the precapillary sphincter (myogenic contraction) which prevents excessive rise in capillary pressure. The response is important in leg muscles where it reduces the rate of capillary filtration in the erect position, preventing pedal edema.

Metabolic control   In exercising muscle, the metabolites accumulate faster than can be carried away by blood flow. The accumulated metabolites like CO_2, K^+, and H^+ cause dilatation of the precapillary sphincters, causing marked reduction of local vascular resistance. This enormously increases the size of the capillary bed and the blood flow increases many fold (exercise hyperemia). The blood flow increases in excess of the O_2

demand of exercising muscles. Hence, the A-V O_2 difference falls during exercise. The capillary hydrostatic pressure rises sharply and there is increased transudation into the interstitium, reducing the plasma volume and causing hemoconcentration. After the cessation of heavy exercise, the blood flow does not subside immediately but falls exponentially from its high level during the exercise to resting values. This is due to the oxygen debt of exercise.

Metabolic control also explains the reactive hyperemia that occurs in muscles following a period of occlusion of blood supply. The occlusion results in the accumulation of metabolites in the capillary bed, which therefore gets maximally dilated. Release of the occlusion brings in a rush of blood through the dilated capillaries, washing out the metabolites.

Neural control   The resistance vessels of muscle possess a high basal tone which is largely attributable to the intrinsic myogenic tone of vascular smooth muscle. The neurogenic tone due to sympathetic discharge is much less. Hence, sympathetic discharge is very effective in causing vasoconstriction (there is a wide margin for increase) but is rather poor in causing vasodilatation (there is little margin for decrease). It is obvious therefore, that a reduction in sympathetic discharge cannot be responsible for the massive exercise hyperemia, which is entirely due to metabolic factors. Rather, sympathetic discharge limits the exercise hyperemia in muscle.

The role of sympathetic discharge to muscle capillaries lies elsewhere. (1) Sympathetic discharge increases the pre:postcapillary resistance ratio and thereby promotes the absorption of interstitial fluid into the capillaries. In exercise, there is transudation of fluids from the capillaries to the interstitial fluid. Sympathetic discharge reduces the transudation though it is not able to stop it completely. (2) During circulatory shock, sympathetic vasoconstriction reduces muscle blood flow profoundly. Since the muscle capillary bed is very large, these help in diverting substantial amounts of blood toward the heart, which helps to raise the blood pressure in circulatory shock.

Intermingled with the sympathetic fibers to the muscle are certain special types of sympathetic fibers that have acetylcholine and not norepinephrine as their neurotransmitter. These *sympathetic vasodilator fibers* are activated by corticohypothalamic-reticulospinal pathways that are quite different from the vasomotor center-thoracolumbar spinal pathway. They are not influenced by the baroreceptor and chemoreceptor afferents and do not participate in the usual cardiovascular reflexes. Rather, they produce muscle vasodilatation in emotional states.

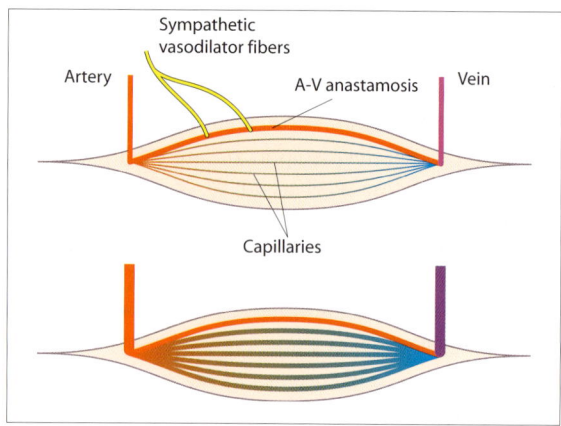

Fig. 47.**2**   Role of sympathetic vasodilator fibers in muscle. (*Above*) Before exercise, the arterioles are dilated by the discharge of sympathetic vasodilator fibers. Blood flow through capillaries is minimal as most of it flows through the thoroughfare channels. (*Below*) Following exercise, capillary blood flow increases markedly due to autoregulation and far exceeds the blood flow through the thoroughfare channels. Note that oxygen desaturation of blood occurs only in the capillaries and not in the thoroughfare channels.

Sympathetic vasodilator fibers do not increase nutritive blood flow to the muscles, either during or before exercise. Their importance lies in preventing an excessive rise in blood pressure before the onset of exercise. Confrontation with a flight-or-flight situation triggers the 'defense reaction' that is associated with considerable sympathetic discharge, causing widespread vasoconstriction in the muscle beds. As a result, the total peripheral resistance and blood pressure tend to rise to dangerous levels. At this time, the discharge of the sympathetic vasodilator nerves lowers the blood pressure to safe levels. It does so by dilating the arterioles in the muscle. Opening of these arterioles does not improve nutritive flow to the muscles since most of the precapillary sphincters are still closed and the extra blood that flows into the muscles through dilated arterioles flows out into the veins through the thoroughfare channels. Once exercise commences, the precapillary sphincters are dilated by local metabolites and most of the blood flows from the arterioles into the muscle capillaries, which now offer less resistance than the thoroughfare channels (Fig. 47.**2**). The importance of sympathetic vasodilatation has not been established in man though it probably underlies vasovagal syncope.

Muscle pump   The rhythmic muscle contraction of dynamic exercise has a propulsive effect on circulation (see venous pumping, p 237). On the other hand, the strong sustained muscle contraction of static exercise compresses the vascular bed and stops the flow of blood through muscles.

Splanchnic Circulation

The vascular beds of the gut, liver, and spleen are together called the splanchnic circulation. The total splanchnic blood flow is approximately 1.5 L min^{-1}. In the gut, the mucosa receives approximately four times the amount of blood as the smooth muscle in its walls. If the entire gastrointestinal tract became simultaneously active, the splanchnic blood flow would have increased to about 4.0 L min^{-1}. However, since during digestion and absorption, different parts of the gastrointestinal tract are sequentially activated, the maximum splanchnic circulation is only about 3.0 L min^{-1}.

Mesenteric circulation

Hormonal control   Food ingestion increases intestinal blood flow. The mediators of this hyperemia include gastrin, cholecystokinin, and products of digestion like glucose and fatty acids. Intestinal blood flow is also increased by the process of absorption of food. Undigested food has no vasoactive influence.

Neural control of the mesenteric circulation is almost exclusively sympathetic. Increased sympathetic activity constricts both the mesenteric arterioles and veins. Both α-adrenergic and β-adrenergic receptors are stimulated but the α-adrenergic (vasoconstrictive) effect predominates.

The sympathetic fibers to the gut are tonically active. When stimulated, these fibers bring about a rise in systemic blood pressure. The main reason for this rise is not an increase in peripheral resistance but venoconstriction, which squeezes out blood from mesenteric circulation and increases the CVP and cardiac output. This translocation of blood is beneficial in fight-or-flight situation where more blood must be available to the heart, muscles, and the brain than to the intestines.

Autoregulation of blood flow in the intestinal circulation is not as well developed. The principal mechanism responsible for autoregulation is metabolic, although a myogenic mechanism probably also contributes. Adenosine is a potent vasodilator of the mesenteric vascular bed and may be the principal metabolic mediator of autoregulation. However, potassium concentration and altered osmolarity of the plasma may also contribute to autoregulation.

Villous countercurrent system   The direction of blood flow in the capillaries and venules in a villus is opposite to that in the main arteriole. This arrangement

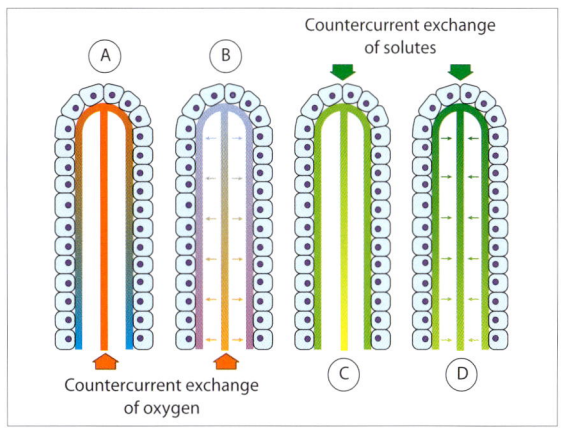

Fig. 47.**3**   Countercurrent exchanges in the villi. (A and B) Due to countercurrent exchange, the arterial blood reaching the tip of the villus is richer in solutes (shown in green) and therefore, is able to absorb less solute from the lumen. (C and D) Due to countercurrent exchange, arterial blood entering the villus is less oxygenated and venous blood is less deoxygenated than it would be without the exchange.

constitutes a countercurrent exchanger that has both advantage and disadvantage (Fig. 47.**3**).

(1) Absorbed lipid-soluble substances that are carried in the venous limbs of the vascular hairpin loop diffuse across into the arterial limb because of a concentration gradient. High concentrations of absorbed substances are thus attained in the outer parts of the villi and these substances leave relatively slowly through the venous drainage. Such a system automatically slows down the entrance of rapidly absorbed solutes into the blood so that the liver can handle them effectively.

(2) The countercurrent exchange also permits diffusion of O_2 from arterioles to venules. At low flow rates, a substantial portion of blood O_2 is shunted from arterioles to venules near the base of the villus. Thus, the supply of O_2 to the mucosal cells at the tip of the villus is reduced. When intestinal blood flow is very low, the shunting of O_2 is exaggerated and may cause extensive necrosis of the intestinal villi.

Hepatic circulation

The definition of the **functional unit of liver** has changed over the years. The earliest concept was a functional unit called the *hepatic lobule*, which was polygonal with a vein (called the central vein) at its center. During the first half of the twentieth century, the concept of a triangular *portal lobule* was preferred. It was centered on the portal triad. Presently, the functional unit of the liver is called the *hepatic acinus*, which is

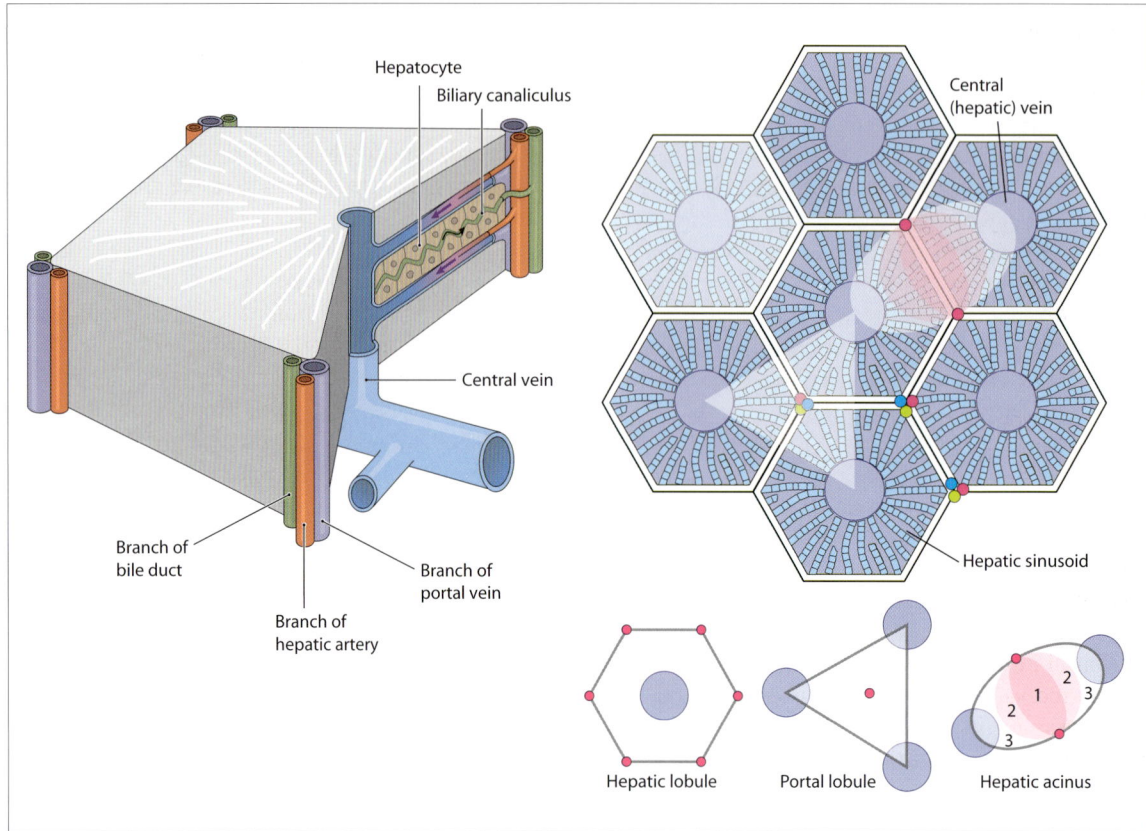

Fig. 47.**4**   Hepatic lobule, hepatic acinus, and portal lobule.

somewhat ellipsoidal in shape. It is symmetrically disposed on either side of the line connecting two central veins. Since the central vein does not occupy a central position in the hepatic acinus, it has been renamed as the terminal hepatic vein (Fig. 47.**4**).

The concept of hepatic acini as the functional unit helps explain the pattern of cell degeneration seen in hypoxic and toxic damage to the liver. The hepatic acinus has three concentric zones. *Zone 1* is an ellipsoid area immediately surrounding the arteriole and terminal portal venule. The cells here are well oxygenated. The area contains mainly those enzymes that are involved in oxidative metabolism and gluconeogenesis. *Zone 2* is intermediate in all respects. It contains a mixed complement of enzymes. *Zone 3* is the peripheral zone of the acinus. It is the least oxygenated zone and therefore, is most susceptible to anoxic injury. It is rich in enzymes involved in glycolysis and lipid and drug metabolism.

The small branches of the portal vein and hepatic artery give rise to terminal portal venules and hepatic arterioles respectively. These terminal vessels enter the hepatic acinus at its center. Blood flows from these terminal vessels into the sinusoids, which constitute the capillary network of the liver. The sinusoids radiate

toward the periphery of the acinus, where they connect with the terminal hepatic venules. Blood from these terminal venules drains into progressively larger branches of the hepatic veins, which are tributaries of the inferior vena cava.

Dual blood flow to liver   Despite having a low O_2 demand, the liver has a relatively high blood supply. This is because only 25% of its blood supply is brought in by the hepatic artery and caters to its metabolic requirements. The remaining 75% is brought in by the portal vein. Both the streams meet in the sinusoids. Blood flows in the portal venous and hepatic arterial systems vary reciprocally; when one reduces, the other compensates partially. The hepatic blood flow is autoregulated. The portal blood flow is not autoregulated; it increases after a meal.

The mean hepatic arterial pressure is about 90 mmHg while the pressure in the hepatic vein is only 5 mmHg. Both the hepatic arterial and portal systems converge on the sinusoids (Fig. 47.**5**). One would therefore expect the sinusoidal pressure to be very high. That however is not so due to the high presinusoidal resistance in the hepatic arterioles. The sinusoidal pressure is only 10 mmHg. On the oth-

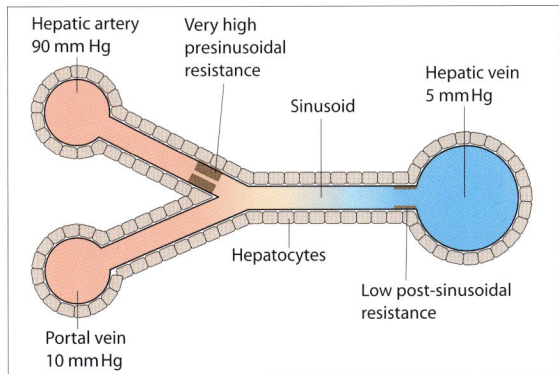

Fig. 47.**5** Hydrostatic pressures in the hepatic artery, portal vein, and hepatic vein.

er hand, the postsinusoidal resistance is low, and any change in the hepatic venous pressure is promptly transmitted to the sinusoids, profoundly affecting the transsinusoidal exchange of fluids. This phenomenon is of clinical importance in the understanding of ascites and portocaval anastomosis, as explained below.

Ascites occurs in congestive heart failure because the elevated CVP is transmitted backward to the hepatic veins, portal veins, and mesenteric capillaries. Rise in hydrostatic pressure in mesenteric circulation results in extensive fluid transudation into the abdominal cavity.

Ascites also occurs in hepatic cirrhosis in which there is extensive fibrosis of the liver. The fibrosis leads to a marked rise in hepatic sinusoidal resistance. The consequent increase in the portal venous pressure is transmitted backward to the mesenteric circulation.

The reverse occurs when a collateral circulation is established between the portal and systemic veins (portocaval anastomosis). In such cases, the pressure rises substantially in systemic veins that anastomose with the portal vein. Such anastomosis can develop, for example, at the lower end of the esophagus, and may enlarge considerably to form esophageal varices. These varices may rupture and lead to severe, frequently fatal internal bleeding.

Oxygen supply to the liver Nearly 70% of the O_2 used by the liver is derived from the portal blood. This is possible because the portal blood entering the liver is three times the volume of hepatic arterial blood entering the liver. Moreover, portal blood is not markedly desaturated because of the low O_2 demands of gastrointestinal tissues. Thus, it contains about 17 ml of O_2 per deciliter of blood compared to the 19 ml of O_2 per deciliter of hepatic arterial blood.

One point of view is that the portal venous blood is more desaturated during digestion and therefore represents an undependable source of oxygen, supplying least during digestion when hepatic activity is greatest. The counter viewpoint is that there is a marked increase of portal blood flow during digestion so that although the P_{O_2} of portal blood is reduced, the total oxygen supply is increased due to hyperemia. Although the oxygen delivery to the liver keeps varying, the liver maintains a constant O_2 consumption and the O_2 extraction varies with it.

Blood storage in liver The liver serves as a blood reservoir, storing about 400 ml of blood in its sinusoids. The sympathetic nerves cause constriction of portal veins, causing in turn a marked reduction in the capacitance of the portal system that helps to divert blood toward the heart.

Section 5

Respiratory System

48. Functional Anatomy of the Respiratory System

49. Mechanics of Pulmonary Ventilation

50. Measurement of Pulmonary Ventilation

51. Alveolar Ventilation, Perfusion, and Gaseous Exchange

52. Transport of Gases

53. Neural Control of Respiratory Rhythm

54. Chemical Control of Pulmonary Ventilation

55. High and Low Pressure Breathing

56. Pulmonary Function Tests and Respiratory Disorders

The process of respiration comprises external respiration, i.e., the intake of O_2 and removal of CO_2 from the body, and internal respiration, i.e., the consumption of O_2 for energy release, production of CO_2 by cells and the gaseous exchanges between the cells and their fluid environment.

Respiratory Passages

The respiratory passages are functionally divided into the upper and lower respiratory tracts. The upper respiratory tract is composed of the nasal and oral cavities, the pharynx and the larynx. The lower respiratory tract is composed of the tracheobronchial tree and the lung parenchyma.

Upper respiratory tract

The nasal cavity warms up the air to the body temperature, humidifies the air to 100% saturation, and cleans and filters the air of its particulate contents by channeling the air through a tortuous path between the turbinates. The particles are deposited at the bends where they adhere to the mucus lining the cavity.

The intrinsic muscles of the larynx are broadly divided into the abductors (which open the glottis) and the adductors (which close the glottis). They are supplied by the recurrent laryngeal branch of the vagus nerve. The abductor muscles (posterior cricoarytenoid) contract early in the inspiratory phase, pulling the vocal cords apart and opening the glottis. When the abductors are paralyzed, there is inspiratory stridor. (Stridor is a harsh, high pitched whistling sound produced during breathing if there is airway obstruction.) In unconscious or anesthetized patients, or when abductors are paralyzed, glottis closure may be incomplete and vomitus may enter the trachea, causing an inflammatory reaction in the lung (aspiration pneumonia). The adductor muscles begin to contract early in expiration but their contraction is not complete. Their main function is protective. During swallowing, there is reflex contraction of the adductor muscles that closes the glottis and prevents aspiration of food, fluid or vomitus into the lungs. Another protective function

of the glottis is its role in the cough reflex (see below). By contracting during the early compressive stage of cough, adductors allow a high intratracheal pressure to be developed.

Maintenance of upper airway patency Keeping the upper airway patent (open, unobstructed) is an important step in any form of resuscitation. The victim is placed in the supine position and the airway is opened by placing a hand under the neck and lifting while keeping pressure with the other hand on the victim's forehead. This extends the neck and lifts the tongue away from the back of the throat (Fig. 48.**1**).

Lower respiratory tract

Tracheobronchial tree After passing through the nasal passages and pharynx where it becomes warm and moist, the inspired air passes down the trachea and through the bronchioles, respiratory bronchioles, and alveolar ducts to the alveoli. Between the trachea and

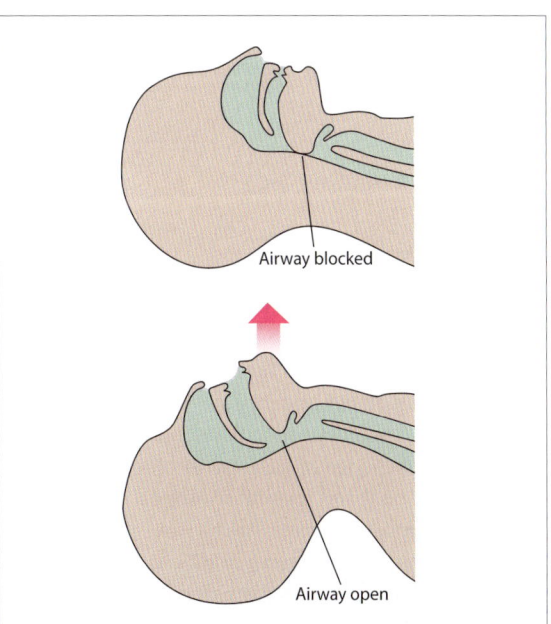

Fig. 48.**1** Upper airway obstruction caused by the soft-tissue (tongue) falling back. Neck extension and chin elevation clears the upper airway obstruction caused by soft tissue.

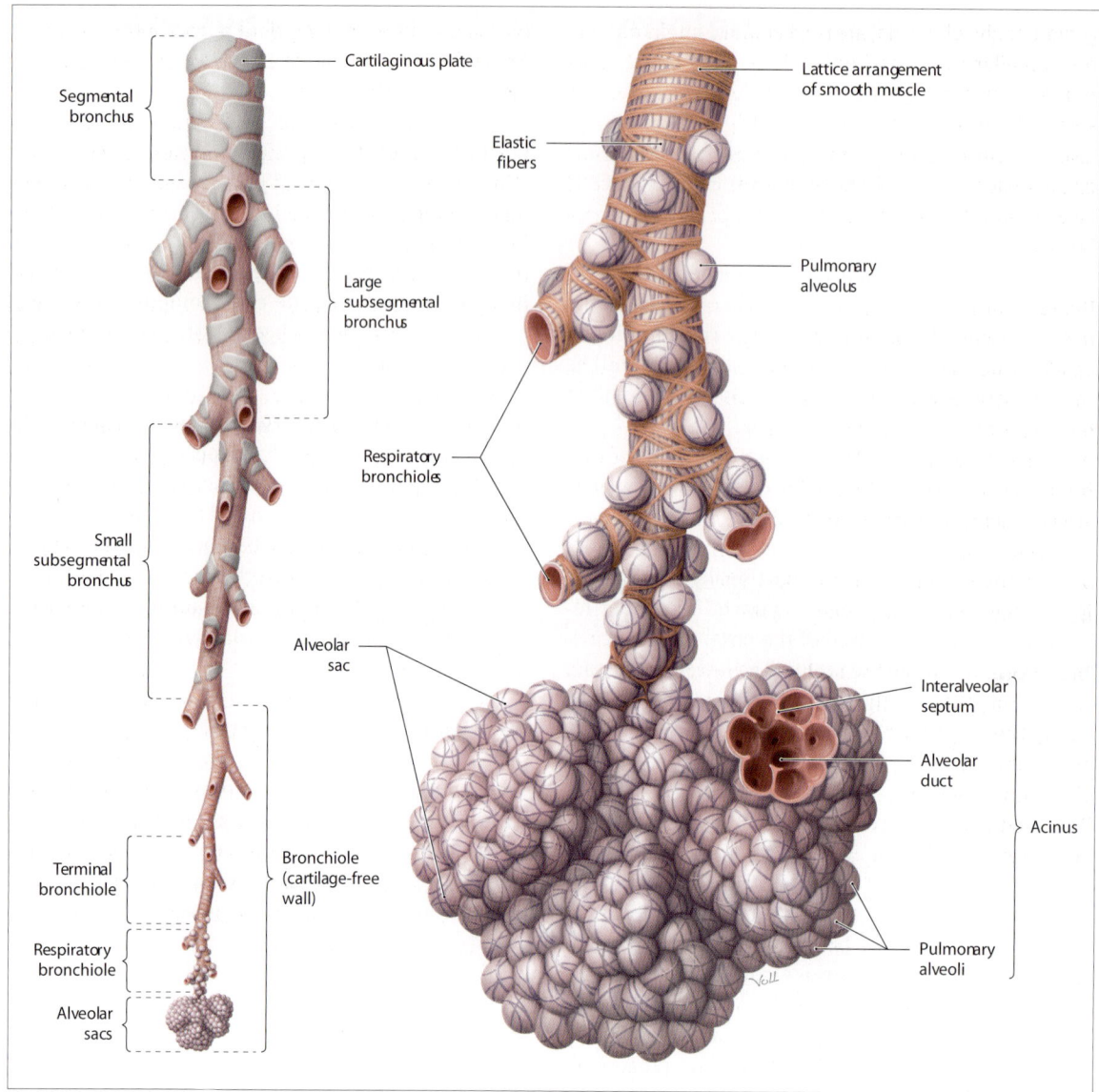

Fig. 48.**2**   The tracheobronchial tree.

the alveolar sacs, the airways divide 23 times. The first 16 generations of passages form the *conducting zone* of the airways. They are made up of bronchi, bronchioles, and terminal bronchioles. The remaining 7 generations form the *transitional* and *respiratory zones*. They are made up of respiratory bronchioles, alveolar ducts, and alveoli. The exchange of gases with blood occurs only in the transitional and respiratory zones.

By definition, *bronchioles* are airways with diameter less than 1 mm, and the term *small airways* is used for airways less than 2 mm in diameter, i.e., small bronchi and bronchioles. Bronchioles and smaller airways have cuboidal epithelium while larger airways have columnar epithelium. Alveolar epithelium is made of flat, squamous cells. The trachea and bronchi have cartilage

in their walls but relatively little smooth muscle. The walls of bronchioles and terminal bronchioles do not contain cartilage; instead, they contain more smooth muscle.

Alveoli   The alveoli are surrounded by pulmonary capillaries. The structures between alveolar air and capillary blood, across which O_2 and CO_2 diffuse, are exceedingly thin and constitute the *respiratory membrane* (see Fig. 51.7). There are 300 million alveoli in a human, and the total area of the alveolar walls in contact with capillaries in both lungs is about 70 m^2! The alveoli are lined by two types of epithelial cells or pneumocytes. *Type I pneumocytes* (squamous alveolar cells) are the primary lining cells. *Type II pneumocytes* (granular alveolar cells) are thicker and contain numerous lamellar inclusion bodies. They are located commonly near the angles between neighboring alveolar septa. These cells secrete surfactant (p 325). The lungs also contain pulmonary alveolar macrophages (PAM) or the dust cells, lymphocytes, plasma cells, Clara cells (see below), APUD (amine precursor uptake and decarboxylation) cells, and mast cells.

Blood supply   The entire right ventricular output passes through the pulmonary artery to the pulmonary capillary bed where it is oxygenated and returned to the left atrium via the pulmonary veins. The smaller bronchial arteries are systemic arteries: They supply bronchial smooth muscles and pleura, and drain into bronchial veins (Fig. 46.**1**, see Fig 48.**3**). The deoxygenated bronchial capillary blood mixes with the oxygenated pulmonary capillary blood, resulting in a *physiological shunt*. Lymphatic channels are more abundant in the lungs than in any other organ.

Nerve supply   Bronchi and bronchioles are kept in a state of slight constriction (the bronchomotor tone) due to the tonic discharge of parasympathetic (cholinergic) fibers in the vagus. The sympathetic (adrenergic) control, which causes bronchodilatation, is feeble. Bronchodilatation is mainly under the control of a *nonadrenergic, noncholinergic* (NANC) system that has VIP

(vasoactive intestinal peptide) as its neurotransmitter. Nerve fibers belonging to the NANC system reach the lungs through the vagus.

The vagus also contains afferent fibers originating in the bronchial epithelium. These afferents are stimulated by a variety of physical and chemical factors. Chemical factors include common air-pollutants like SO_2 and NO_2 that produce bronchoconstriction. Physical factors include foreign bodies in the trachea that trigger cough. Another potent physical stimulus is cold air which causes bronchoconstriction. The bronchoconstriction caused by cold air is not due to a neural reflex as with other physical or chemical factors: Low temperature acts directly on smooth muscle to cause contraction. Exercise too causes bronchoconstriction, possibly due to the excessive ventilation of airways with cool air.

Although the sympathetic innervation of the bronchi and bronchioles is sparse, adrenergic receptors are as abundant in these airways as are the cholinergic receptors. Hence, drugs that stimulate adrenergic receptors (sympathomimetic drugs) are as effective in relieving bronchospasm as are drugs that block cholinergic receptors (anticholinergic drugs). The adrenergic receptors are predominantly of the β_2 subtype (p 632). Also present on the bronchial smooth muscles are histamine receptors. Numerous drugs and allergens cause mast cell degranulation with release of large amounts of histamine. The histamine released binds to the histamine receptors present on bronchial smooth muscles, producing bronchoconstriction.

Nonrespiratory Functions

Physical protection

The respiratory passages humidify and warm or cool the inspired air so that by the time it reaches the alveoli, it is at (or near) body temperature. The respiratory passages also have various mechanisms to prevent particulate matter from reaching the alveoli. Particles larger than 10 μm in diameter are strained out by the hairs in the nostrils. Most of the remaining particles of this size settle on mucous membranes in the nose and pharynx. Because of their inertia, they do not follow the airstream as it curves downward into the lungs but instead, fall on the tonsils and adenoids, which dispose off these particles. Particles 2–10 μm in diameter generally fall on the walls of the bronchi as the airflow slows in the smaller passages. There they are expelled by mucokinesis and coughing. Particles less than 2 μm in diameter generally reach the alveoli where they are ingested by the PAM (Fig. 48.**4**).

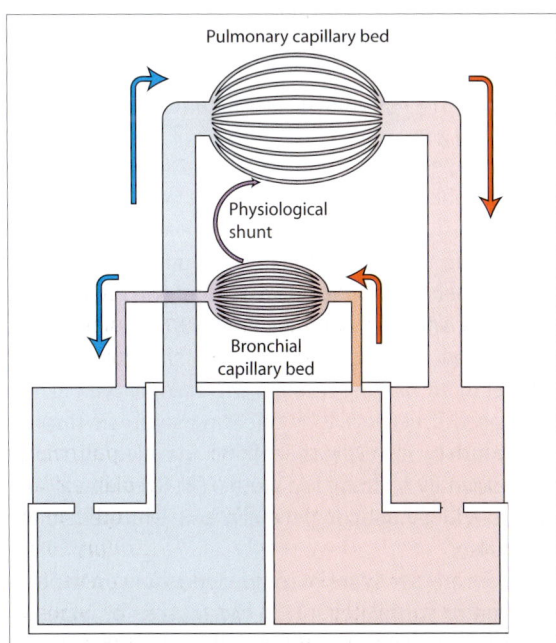

Fig. 48.**3**   The physiological shunt, produced by the mixing of deoxygenated bronchial capillary blood with the oxygenated pulmonary capillary blood.

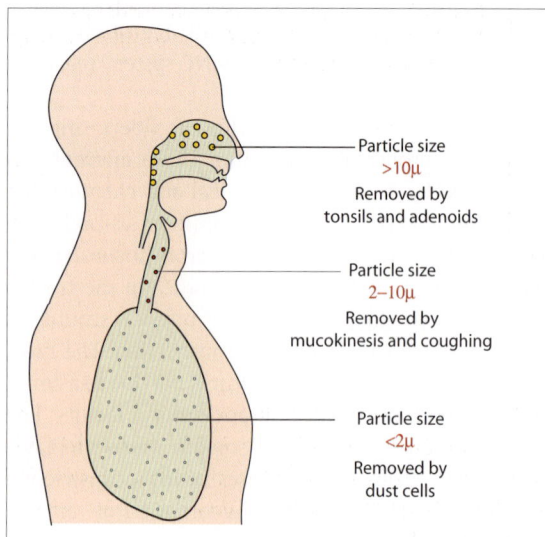

48.**4** Removal of particulate matter from the respiratory tract and lungs.

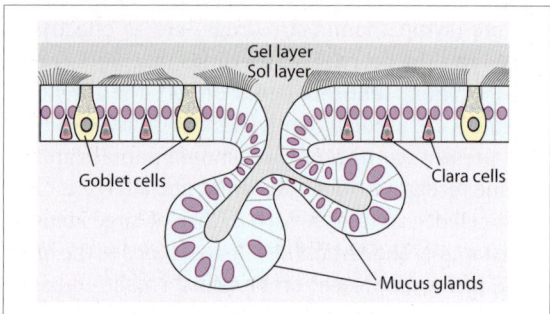

Fig. 48.**5** Bronchial epithelium showing the gel and sol layers, goblet cells, and submucosal mucus glands.

Mucokinesis Inhaled particulate matter gets trapped in the mucus layer which covers the respiratory passage, starting from the nose and covering the tracheobronchial tree epithelium up to the terminal bronchioles. The surfactant layering the alveolar epithelium drains into the bronchiolar mucus layer.

The mucus layer consists of a superficial gel layer that is secreted by the bronchial mucus glands and goblet cells, and a deeper sol layer about 5 μm in thickness (Fig. 48.**5**) that mainly comes from the Clara cells. The cilia, which are covered with mucus, beat in a coordinated fashion (Fig. 48.**6**), moving the gel layer at a rate of 1–2 cm min^{-1} toward the larynx and into the pharynx where it is swallowed. The moving of the mucus layer is called mucokinesis.

When ciliary motility is defective, mucokinesis is almost absent, leading to retained secretions that provide an excellent medium for bacterial growth (stasis pneumonia). The retained secretions also cause plugging

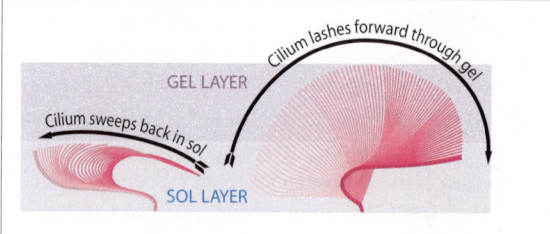

Fig 48.**6** The cilia slowly arch backward, moving through the sol layer. Thereafter, they lash forward, sweeping through the gel layer and propelling it forward.

of bronchioles, leading to collapse (atelectasis) of the connected alveoli.

Coughing and sneezing are concerned with the cleansing of the air passages by removing secretions or inhaled material. Both are reflexes that have their centers in the medulla. Efferent impulses of both the reflexes travel down the vagus to the larynx (or soft palate) and by spinal nerves to diaphragm, abdominal, and pelvic muscle that contract during cough (or sneeze). Both have the following sequence of: (1) a deep inspiration, (2) trapping of air by shutting off its exit, (3) initiation of expiratory effort, raising the intrathoracic pressure, (4) augmentation of the pressure of the trapped air (the compressive stage), and (5) sudden release of the trapped air at high pressure by opening up the exit passages.

The differences between the two reflexes are as follows. (1) Sneezing is stimulated by irritation of the nasal mucosa. The impulses travel up via the trigeminal nerve. Cough occurs due to irritation of sensory receptors in the tracheobronchial tree. Impulses travel up via vagus and glossopharyngeal nerves. (2) In sneezing, the air is trapped by shutting off its exit passages through the nasopharynx (by raising the soft palate to the posterior pharyngeal wall) and the oral cavity (by raising the tongue to the hard palate). In coughing, the air is trapped behind the closed glottis by contraction of the laryngeal adductor muscles (Fig. 48.**7**). (3) In sneezing, the air is released by opening the nasopharynx. The jet of air passing through the nasopharynx can be decreased if necessary by releasing some air through the mouth by lowering the tongue. In coughing, the air is released by opening the glottis. (4) Sneezing always occurs reflexly but coughing can be performed voluntarily also.

There are two types of cough depending on the type of receptor stimulated. (1) In the mucosa of large airways (larynx, trachea, and bronchi), there are rapidly adapting stretch receptors or irritant receptors sensitive to mechanical stimuli. They initiate a forceful expiration without a preceding inspiration. Absence of

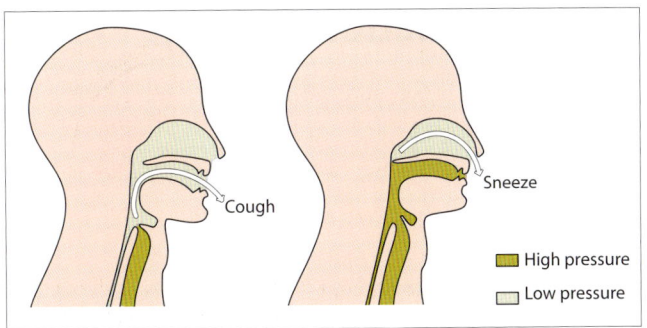

Fig. 48.**7**   In cough, the air pressure builds up below the closed glottis. In sneeze, the air pressure builds up below the nasopharynx.

the preceding inspiration helps prevent aspiration of noxious material. (2) Distal to the large airways up to the acinus, there are chemical receptors. When stimulated, these receptors first initiate a deep inspiration which is followed by a forceful expiration. This type of cough helps in expectoration of mucus containing chemical irritants.

Immunologic protection

Bronchial secretions contain IgA that helps resist infections. In addition, the epithelium of the paranasal sinuses produces nitric oxide that is bacteriostatic.

Pulmonary alveolar macrophages (PAM) or the dust cells are a component of the monocyte-macrophage system. They are actively phagocytic and ingest inhaled bacteria and small particles. They process antigens and secrete cytokines that attract granulocytes and stimulate granulocyte and monocyte formation in the bone marrow. PAM produces α_1-antitrypsin which offers protection against emphysema. When the macrophages ingest large amounts of pollutants like cigarette smoke, silica, and asbestos particles, they release lysosomal enzymes into the extracellular space, causing inflammation.

Metabolic and endocrine functions

The lungs have a number of metabolic functions. (1) Type-II pneumocytes synthesize surfactant. (2) The lungs contain a fibrinolytic system that lyses clots in the pulmonary vessels. (3) The angiotensin-converting enzyme (ACE) responsible for the activation of angiotensin I into angiotensin II is located on the surface of the endothelial cells of the pulmonary capillaries. ACE also inactivates bradykinin. (4) The lungs release several substances (e.g., histamine and kallikrein) into systemic circulation and remove several others (e.g., serotonin and norepinephrine) from circulation. Prostaglandins are both secreted into and removed from the circulation by the lungs.

49　Mechanics of Pulmonary Ventilation

The lungs are elastic, nonmuscular organs that get inflated (by sucking in air through the trachea) or deflated (by expelling air) when the thorax expands or contracts. The lung is separated from the chest wall by a thin layer of pleural fluid that allows the lung and the thoracic wall to glide easily on each other. Yet, the lung resists being pulled off the chest wall in the same way as two wet glass slides are not separated easily. Hence, when the thorax expands or contracts, so do the lungs.

Mechanism of Thoracic Expansion

Changes in chest-wall dimensions

During inspiration, all the three dimensions of the thoracic cavity expand, viz., the vertical (superoinferior) diameter, the anteroposterior diameter and the transverse diameter. There is a different mechanism for expansion in each direction.

The **vertical diameter** increases when the diaphragm descends in the thoracic cavity. The diaphragmatic descent ranges from 1.5 cm (in eupnea) to 7.0 cm (deep breathing) and accounts for nearly 75% of the thoracic expansion during quiet breathing. The extent of

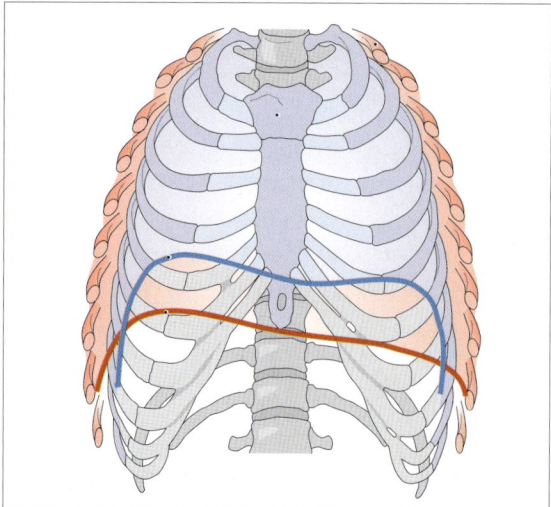

Fig. 49.1　Position of the diaphragm in inspiration (red) and expiration (blue).

diaphragmatic descent (Fig. 49.**1**) can be assessed in a plain radiograph and is of diagnostic importance.

The **anteroposterior diameter** of the thorax increases during inspiration when the upper ribs (2nd to 6th ribs), which normally slope obliquely downward and forward, swing upward to assume a more horizontal position from their joints with the spine. This is called the *pump-handle movement* due to its obvious resemblance to the movements of the handle of a hand pump (Fig. 49.**2**).

The **transverse diameter** also increases during inspiration, but to a lesser degree. This occurs due to the movements of the lower ribs (7th to 10th ribs) that swing outward and upward in inspiration. This is called the *bucket-handle movement* due to its obvious resemblance to the movements of a bucket handle (Fig. 49.**2**).

The relative contribution of thoracic (intercostals) and abdominal (diaphragmatic) breathing differs in the sexes, being predominantly thoracic (thoracoabdominal) in females and predominantly abdominal (abdominothoracic) in males. The lowering of the diaphragm during inspiration is associated with an increase in the intraabdominal pressure, resulting in the bulging out of the abdominal wall. In cases of diaphragmatic paralysis, the inspiratory increase in thoracic volume is brought about entirely by the intercostal muscles and the diaphragm is passively pulled up by the negative intrathoracic pressure. Hence the abdominal wall moves inward during inspiration, which is called the *abdominal paradox*. Sometimes, the reverse occurs in neonates whose chest wall has a high compliance: When the diaphragm contracts, the chest wall caves in, reducing the effectiveness of the inspiratory diaphragmatic movement.

Muscles of inspiration

The diaphragm or the external intercostal muscles alone can maintain adequate ventilation at rest. However, some other muscles called *accessory muscles of inspiration* must contract when greater inspiratory force is required during exercise.

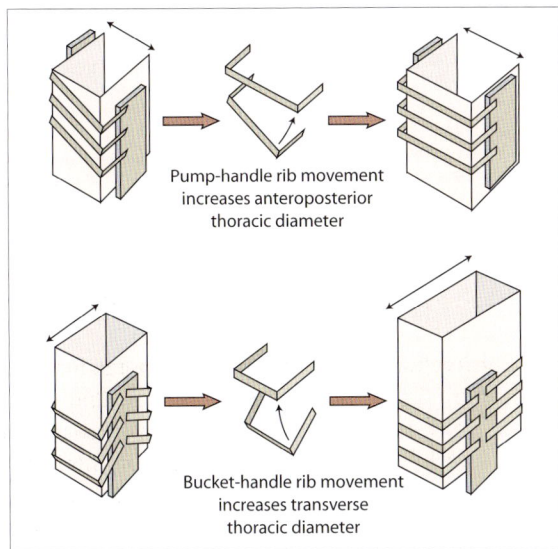

Fig. 49.**2** Pump-handle and bucket-handle movements of the rib cage.

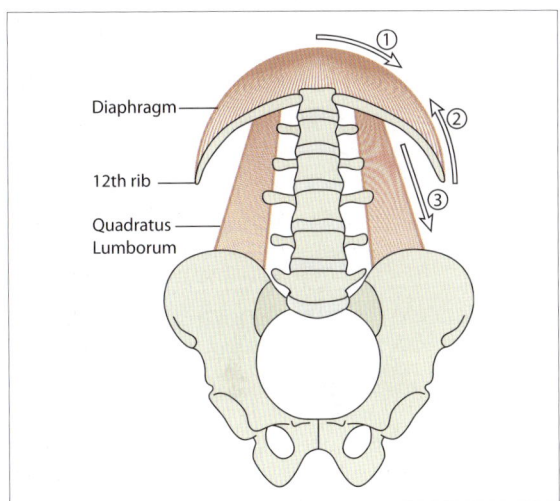

Fig. 49.**3** Action of quadratus lumborum as an inspiratory muscle. Diaphragmatic contraction flattens the dome of the diaphragm, thereby lowering it (1) but simultaneously pulls up the twelfth rib (2), thereby minimizing the effect on the vertical thoracic diameter. The elevation of the twelfth rib is prevented by the quadratus lumborum (3).

The **diaphragm** consists of a central tendinous dome with muscle fibers attached to its periphery. The muscle fibers are attached to the xiphisternum and the inner surfaces of the lower six ribs (costal attachment), and to the lumbar vertebrae and arcuate ligaments (crural attachment). The descent of the diaphragm is brought about by the contraction of its muscular aponeurosis pulling down its central tendinous part. The costal and crural parts of the diaphragmatic muscles

are innervated by different parts of the phrenic nerve (C3,4,5) and can contract separately.

The **quadratus lumborum** is a posterior abdominal muscle that serves a synergistic role with the diaphragm, acting as a stabilizer. When the diaphragmatic muscles contract, the dome of the diaphragm tends to flatten out (action). Simultaneously however, the lower ribs to which the diaphragm is attached tend to get pulled up (reaction), which would elevate the diaphragm instead of lowering it. This is prevented by the quadratus lumborum muscle which anchors the twelfth rib to the iliac crest and prevents it from moving up during inspiration (Fig. 49.**3**).

The **external intercostal muscles** pass obliquely forward and downward from upper rib to the lower rib. When the external intercostals contract, the lower ribs move up. Although the upper ribs would be expected to move down simultaneously, that does not happen for two reasons. The first reason is the mechanical advantage available to the upper rib due to the direction of the muscle fibers that move away from the fulcrum as they reach the lower rib (Figs. 49.**4** and 49.**5**). The second reason is the simultaneous contraction of the scaleneus muscle (see below).

Two prominent **accessory muscles of inspiration** are the scalenei and sternocleidomastoid muscles. The scalenei muscles originate in the cervical transverse process and are inserted into the first rib. They are supplied by C3–C8 spinal nerves. The sternocleidomastoid extends from the mastoid process above to the manubrium sterni and medial part of the clavicle below. It is supplied by the eleventh cranial nerve XI and spinal nerve C2. When the scalenei and sternocleidomastoid contract during inspiration, they tend to elevate the first rib, thereby counteracting the downward pull exerted on the ribs by the external intercostal muscle.

■ **Muscles of expiration**

Normal expiration occurs due to the passive recoil of the lungs and thoracic wall and does not require any muscle action. Muscle action is required only during forced expiration. The expiratory muscles for forced expiration are the internal intercostals and the anterolateral abdominal muscles.

Contraction of the **internal intercostal muscles** pulls the upper ribs down. Although the lower ribs would be expected to move up simultaneously, that does not happen for two reasons. The first reason is the mechanical advantage available to the lower rib due to the direction of the muscle fibers (Figs. 49.**4** and 49.**5**). The second reason is the simultaneous contraction of the abdominal muscles and the quadratus lumborum.

The **anterolateral abdominal muscles** (obliquus external abdominis, obliquus internal abdominis, transversus abdominis, and rectus abdominis) are the most important muscles involved in forced expiration. They have two actions that aid expiration: They pull the ribs downward, reducing the transverse and anteroposterior diameter of the thorax. They also increase the intraabdominal pressure, thereby pushing the diaphragm upward.

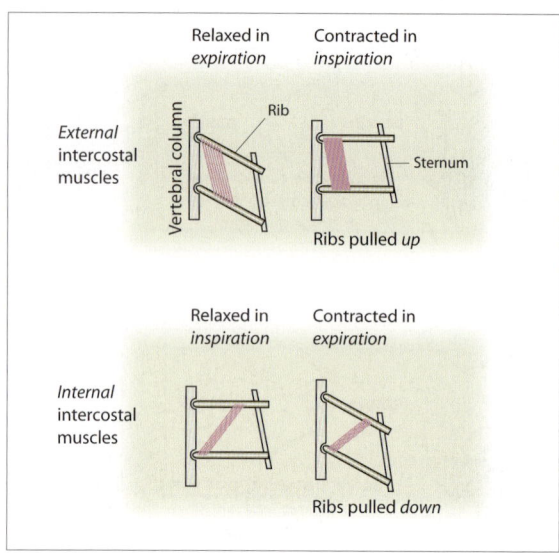

Fig. 49.**4**   Direction of rib movement caused by external and internal intercostals.

Respiratory Resistance

Respiratory resistance is the resistance offered to the inflation or deflation of the lung. The total respiratory resistance is attributable to airway resistance, pulmonary tissue resistance and thoracic resistance.

$$\frac{\text{Total respiratory}}{\text{resistance}} = \frac{\text{Airway}}{\text{resistance}} + \frac{\text{Pulmonary tissue}}{\text{resistance}} + \frac{\text{Thoracic}}{\text{resistance}}$$

The airway resistance and pulmonary tissue resistance are together called pulmonary resistance. Therefore:

$$\frac{\text{Total respiratory}}{\text{resistance}} = \frac{\text{Pulmonary}}{\text{resistance}} + \frac{\text{Thoracic}}{\text{resistance}}$$

Factors affecting respiratory resistance   The *pulmonary tissue resistance* is higher in the recumbent position due to the engorgement of the pulmonary vascular bed with blood, which increases the stiffness of the lung tissue. It also increases when lung surfactant is deficient or in diseases affecting lung tissue, like interstitial fibrosis.

    The *thoracic resistance* comes from the rib cage, diaphragm, and abdominal contents. Thoracic resistance increases in kyphoscoliosis. It also increases in the recumbent position as the abdominal contents press on the diaphragm.

    Of the total *airway resistance* during quiet mouth breathing, 10% is contributed by the peripheral airways (smaller than 2 mm in diameter), 50% by the larger airways, and 40% by the nasal cavity.[1] The resistance offered by the nasal cavity is higher due to the turbulence

Fig. 49.**5**   Leverage at work. While shaking hands, the monkey sitting away from the stem gets pulled up or pulled down due to leverage. It explains the opposite actions of the external and internal intercostal muscles on rib movements (compare with Fig. 49.**4**).

of air in it. The resistance increases markedly during exercise, making it necessary to breathe through the mouth.

    Airway resistance changes with lung volume: It does not decrease much when the lung volume increases but it increases markedly when the lung volume decreases below its functional residual capacity (i.e., about 2 L) due to compression of the small airways. It is also affected by the bronchomotor tone. Fine particles of carbon, chalk, and cigarette smoke, when inhaled, cause bronchoconstriction and increase airway resistance.

---

[1] The resistance of a fluid flowing through a tube varies inversely with the fourth power of the tube radius. Yet, the resistance of the peripheral airways is low because several of them are disposed in parallel. Moreover, the length of the peripheral airways is much smaller than that of the central airways. A more detailed discussion on fluid resistance is given on p 249–251 in reference to vascular resistance.

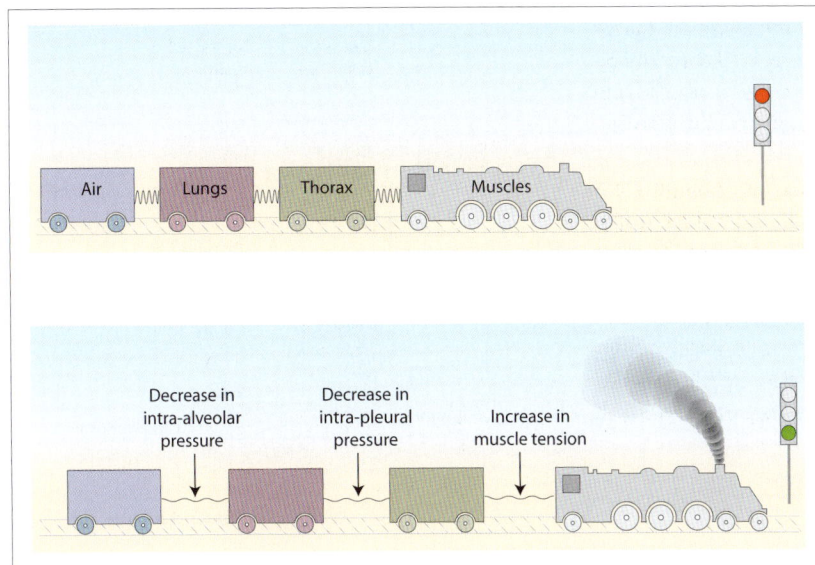

Fig. 49.**6** The 'train' of compliance. When muscles pull, thorax must comply. When thorax expands, lungs must comply or else the intrapleural pressure will fall sharply. When lungs expand, air must move in or else, intrapulmonary pressure will fall sharply.

Finally, the small airways, which do not have rigid walls, are kept open by the traction exerted by the elastic tissue in the lung. When these elastic tissues are destroyed, as in chronic obstructive pulmonary disease (COPD), the smaller airways tend to collapse and the peripheral airway resistance increases. Loss of elastic tissues affects the resistance of the larger airways also, though to a lesser degree.

Compliance

In the living body, the lung and the thorax are attached together by the pleura and they move in unison. For effortless breathing, the lung, thorax, and air must all respond to the forces acting on them. In other words, each must comply (i.e., obey) when asked to move. For example, when the inspiratory muscles pull on the thoracic wall, the thorax must comply and expand. When the thorax expands and pulls the lungs, the lungs must comply and expand. When the lung expands, the air must comply and smoothly flow into the alveoli (Fig. 49.**6**). If the thorax and lungs are stiff or the airways are narrow so that air does not move in and out of the lungs freely, pulmonary ventilation will be impaired.

Lung compliance is defined as the change in lung volume (ΔV) in response to a unit change in the transpulmonary pressure (ΔP), i.e., the pressure difference inside and outside the lungs. The term compliance can be extended to the lung-thorax combine. The term is also applicable to the ventricle, artery, vein, and urinary bladder. Thus compliance of any hollow elastic structure can be defined as the change in its volume in

response to a unit change in the transmural (across the wall) pressure (Fig. 49.**7**).

Compliance is the reciprocal of resistance. Hence the relationship between lung compliance, thoracic compliance, and total (lung-thorax) compliance can be derived as follows:

$$\frac{\text{Total respiratory}}{\text{resistance}} = \frac{\text{Pulmonary}}{\text{resistance}} + \frac{\text{Thoracic}}{\text{resistance}}$$

or,

$$\frac{1}{\text{Total compliance}} = \frac{1}{\text{Lung compliance}} + \frac{1}{\text{Thoracic compliance}}$$

The compliance of lung as well as that of the thorax is approximately 200 ml cmH$_2$O^{-1}. Their combined compliance is therefore 100 ml cmH$_2$O^{-1} (or 147 ml mmHg^{-1}).

$$\frac{1}{100} = \frac{1}{200} + \frac{1}{200}$$

The compliance that is measured is called static or dynamic compliance depending on how it is recorded. If the lung is inflated or deflated in very small steps and the measurements are made only after allowing the lung volume to stabilize completely at each step, then the recorded graph gives the *static compliance*. The graph is curvilinear. On the other hand, if the measurements are made during rhythmic breathing, it gives the *dynamic compliance*. It is calculated from only two sets of pressure–volume measurements, one each at the end of inspiration and expiration. Dynamic compliance varies with the breathing rate.

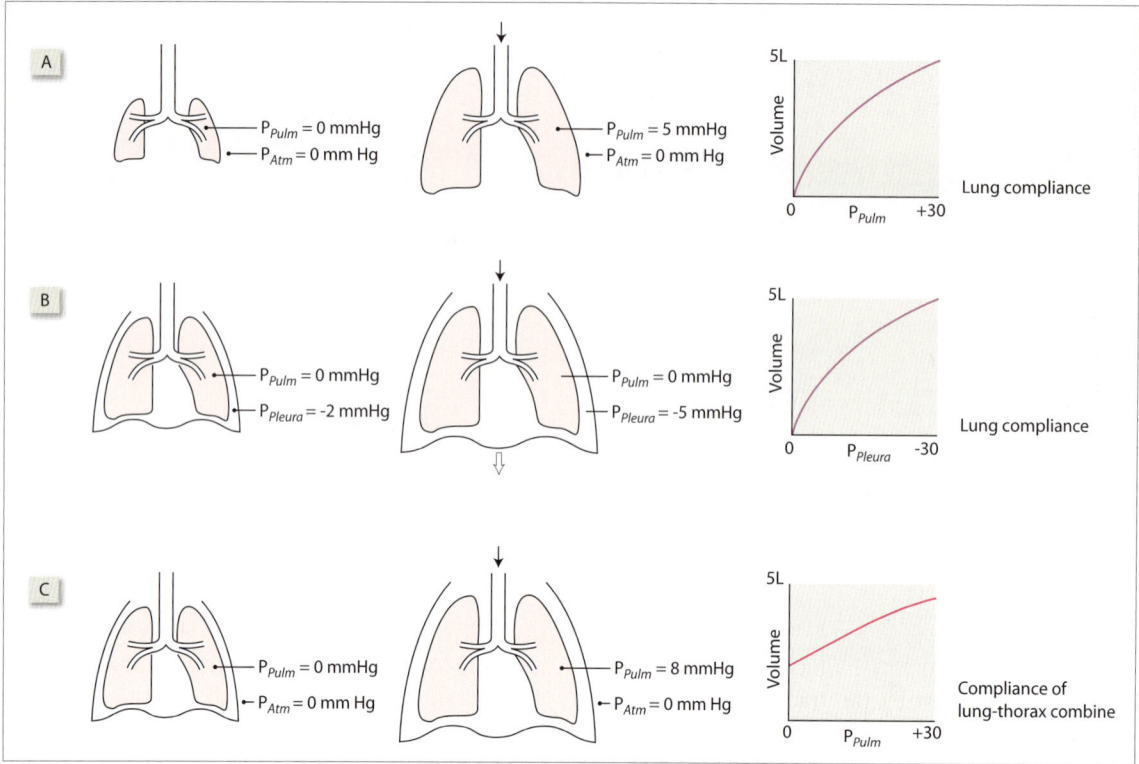

Fig. 49.**7**   Measurement of compliance. [A] The lungs are excised from the thorax and inflated by increasing the intrapulmonary pressure (P_{Pulm}). The distending pressure is given by the difference between P_{Pulm} and atmospheric pressure (P_{Atm}). Since the P_{Atm} is taken as zero, the distending pressure equals P_{Pulm}. [B] The lungs inflate in response to the suction exerted by the negative intrapleural pressure (P_{Pleura}). The suction pressure is given by the difference between P_{Pulm} and P_{Pleura}. Since P_{Pulm} = 0, the suction pressure equals P_{Pleura}. Note that the compliance curve remains identical regardless of whether the volume changes are plotted against P_{Pulm} or P_{Pleura}. [C] The lungs and thorax are distended together by increasing P_{Pulm}, and the distending pressure is given by P_{Pulm} minus P_{Atm}. Since P_{Atm} = 0, the distending pressure equals P_{Pulm}. On plotting the volume changes against P_{Pulm}, what is obtained is the combined compliance of the lung-thorax combine.

■   **Static lung compliance**

Clinically, static lung compliance can be calculated by asking the subject to breathe in a measured volume of air (ΔV) and recording the associated fall in intrapleural pressure (ΔP). Lung compliance is not uniform over a range of pressures. This becomes apparent when the lung is distended in small steps and at each step, the pressure and volume of the lung is noted and plotted on a graph. The line obtained is not straight but flattens out at higher pressures. The flattening of the compliance curve occurs due to the marked increase in pulmonary tissue resistance at high lung volumes.

The compliance at any point on the pressure–volume curve is given by the slope ($\Delta V/\Delta P$) at that point. The steeper the slope, the higher the compliance. The pressure-volume curve can be extended below by forced expiration and noting the volume of expired air and the increase in intrapleural pressure (Fig. 49.**8**).

Specific compliance   A lung compliance of 200 ml cmH_2O^{-1} means that when the intrapleural pressure changes by 1 cmH_2O, the volume of both the lungs taken together changes by 200 ml. For the same pressure change, each lung expands by 100 ml and therefore the compliance of each lung would be 100 ml cmH_2O^{-1}. The compliance of a single lobe of lung would be even less. This difference in compliance does not represent any difference in the lung tissue stiffness, the assessment of which is the primary objective of compliance measurements. Hence the term specific compliance has been introduced.

$$\text{Specific compliance of the lung} = \frac{\text{Compliance of the lungs}}{\text{Functional residual capacity (FRC)}}$$

The specific compliance of the lungs is 200 ÷ 2000 = 0.1 ml cmH_2O^{-1}. The specific compliance of the lungs

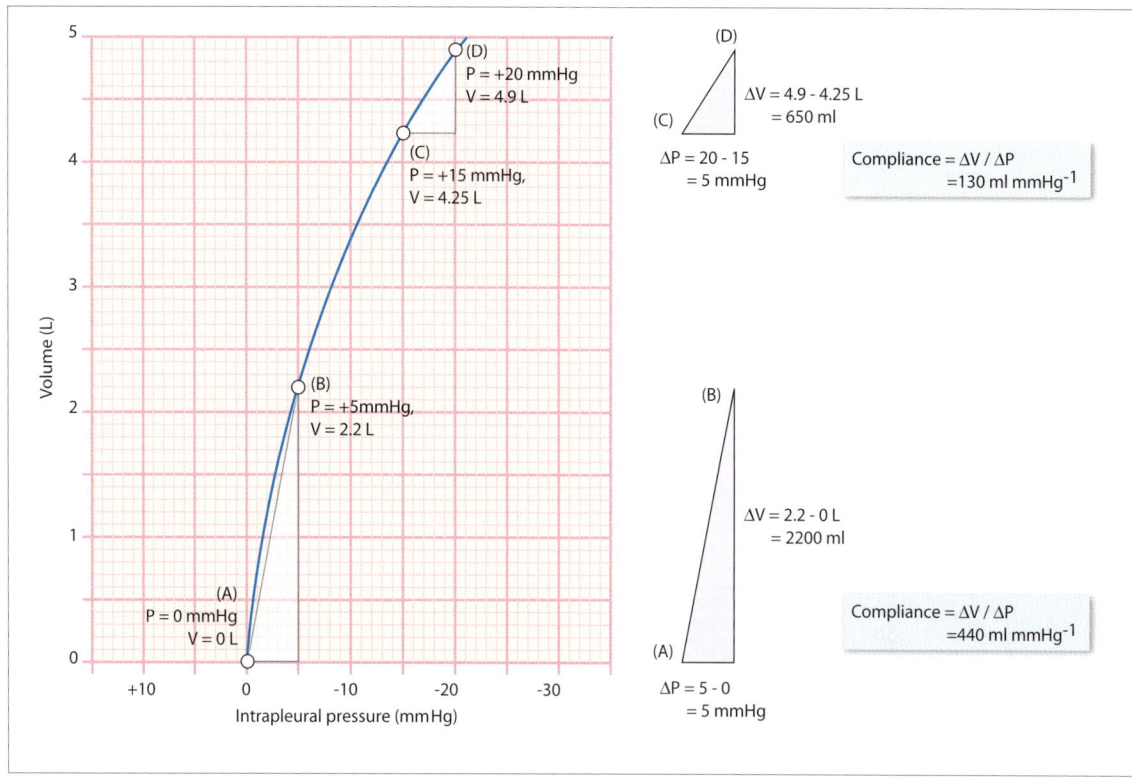

Fig. 49.8 Lung compliance is non-uniform over a range of pressures. Between points A to B, compliance is 440 ml mmHg^{-1} while between points C and D, compliance is 130 ml mmHg^{-1}. The average compliance between points A and D can be calculated to be 245 ml mmHg^{-1}, or 180 ml cmH$_2$O^{-1} (245 ÷ 1.36 = 180).

in neonates is about 0.065 ml cmH$_2$O^{-1}, indicating that their lungs are softer.

▪ Static compliance of lung-thorax combine

Clinically, the lung-thorax static compliance can be measured by asking the subject to breathe in measured volumes of air and then release it, keeping the nose closed, into a manometer that records the pressure. The subject is instructed not to exert force while breathing out. The pressure recorded in the manometer is the *intrapulmonary pressure* and when it is recorded under relaxed conditions (without exerting extra pressure), it is called the *relaxation pressure*. Measuring the relaxation pressure at a given volume is an indirect method of estimating the intrapulmonary pressure required to distend the lung-thorax combine to the same volume.

The compliance curve of the lung-thorax combine (Fig. 49.**9**) can be understood by noting that the distending pressure required to maintain the lung-thorax combine at any volume is the algebraic sum of the distending pressures required to maintain the lung and thorax separately at the same volume. Thus, to maintain

the lung-thorax combine at 2 L, the pressure required is the sum of –4 mmHg (the suction pressure required to contract the thorax to 2 L) and +4 mmHg (the distending pressure required to expand the lung to 2 L). Since the sum is zero, it means that in the absence of any distending pressure, the lung-thorax combine has a volume of 2 L. This volume is known as the functional residual capacity (FRC). In the same way, the pressures required to maintain the volume of lung-thorax combine at 1.5 L and 5 L are shown both graphically and diagrammatically in Figure 49.**9**.

Figure 49.**9** shows that in the absence of any distending pressure (when intrapulmonary pressure = 0 mmHg), the resting volume of the lung-thorax combine is 2 L (point C$_0$ in the graph which is equal to the functional residual capacity). The resting volume of the thorax, if unattached to the lungs, is about 4 L (point T$_0$ in the graph). The resting volume of the lung is zero, i.e., the lungs, unless kept distended, would collapse completely (point L$_0$ in the graph). This means that the lungs are continuously trying to collapse completely while the thorax is continuously trying to spring back to its actual size of 4 L. When the volume of lung-thorax combine is 2 L, the recoil pressure of the lung that is

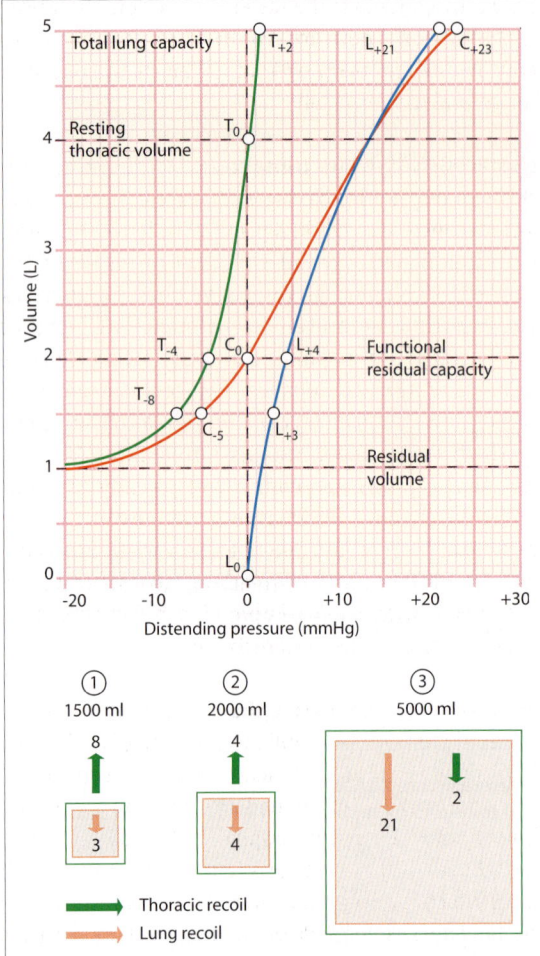

Fig. 49.**9** (*Above*) Static compliances of the thoracic cage, the lungs, and the lung-thorax combine. (*Below*) The recoil pressures (in mmHg) of the thorax and the lungs at different volumes when the two are attached together by the pleura.

trying to collapse it (4 mmHg) is precisely counterbalanced by the recoil pressure of the thorax that is trying to distend it (also 4 mmHg). If the lung elasticity decreases, as in emphysema, the thoracic cage increases in size, resulting in the characteristic barrel-shaped chest.

The compliance curve can be extended below by sucking out air from the lungs and the thorax. When the suction (negative) pressure is sufficiently high, the lung or the thorax reaches its minimum volume below which it cannot be collapsed. The thoracic volume can be reduced to its minimum of 1 L by applying a suction pressure of –20 mmHg, below which the thorax caves in. The lung, as already mentioned above, collapses completely if there is no distending pressure. At high pressures, the curves terminate abruptly because the lung or the thorax ruptures and no further measurements are possible.

Dynamic lung compliance

During quiet breathing, the intrapleural pressure decreases from –2 to –6 mmHg. The 4 mmHg fall in intrapleural pressure should be associated with an increase of 588 ml in lung volume, since the static lung compliance is about 147 ml mmHg^{-1} (147 × 4 = 588). The actual increase in lung volume is only 500 ml. The lung compliance would therefore be 125 ml mmHg^{-1} (500 ÷ 4 = 125), i.e., only 85% of the static compliance. This reduced lung compliance observed during actual breathing is called dynamic lung compliance. The decrease in compliance is attributable to the limitation of time available for inspiration: At a respiratory rate of 15 per minute, each breath lasts for 4 seconds or less. Thus inspiration lasts for only 2 seconds or less, which permits only 500 ml of air to enter the lungs. If the respiratory rate is higher, even less time is available to the lungs for expansion. Moreover, since the air tries to rush in and out in a shorter time, the airway resistance increases, adding further to the reduction in dynamic lung compliance.

The effect of emphysema on compliance is unique in that it results in an increase in static compliance but a decrease in dynamic compliance. Static compliance increases due to destruction of elastic tissues in the alveolar wall. The reduction in elastic tissue makes it easier for the lung to expand in response to a distending pressure. However the loss of elastic tissue in the walls of the smaller airways make them vulnerable to dynamic airway compression (see below). This increases the airway resistance with consequent decrease in dynamic compliance.

Hysteresis of lung compliance

Measurement of compliance requires that the change of volume be recorded for unit change in pressure. The experimenter has the freedom to make the measurements by either increasing or decreasing the pressure. Experimenters, however, soon realized that in the case of the lungs, the compliance measured depended on whether measurements were made during inflation or deflation. Thus, instead of getting a single compliance curve, experimenters obtained two curves, one for inspiration and the other for expiration (Fig. 49.**10**). Together, the curves formed a loop called the hysteresis loop. This phenomenon is called hysteresis (Gr. hysterein = to fall short, to lag behind) because the volume change lags behind the pressure change. Hysteresis occurs due to the presence of viscous resistance to changes in lung volume. Viscosity is the frictional resistance of fluids (liquids or gases). Frictional resistance occurs

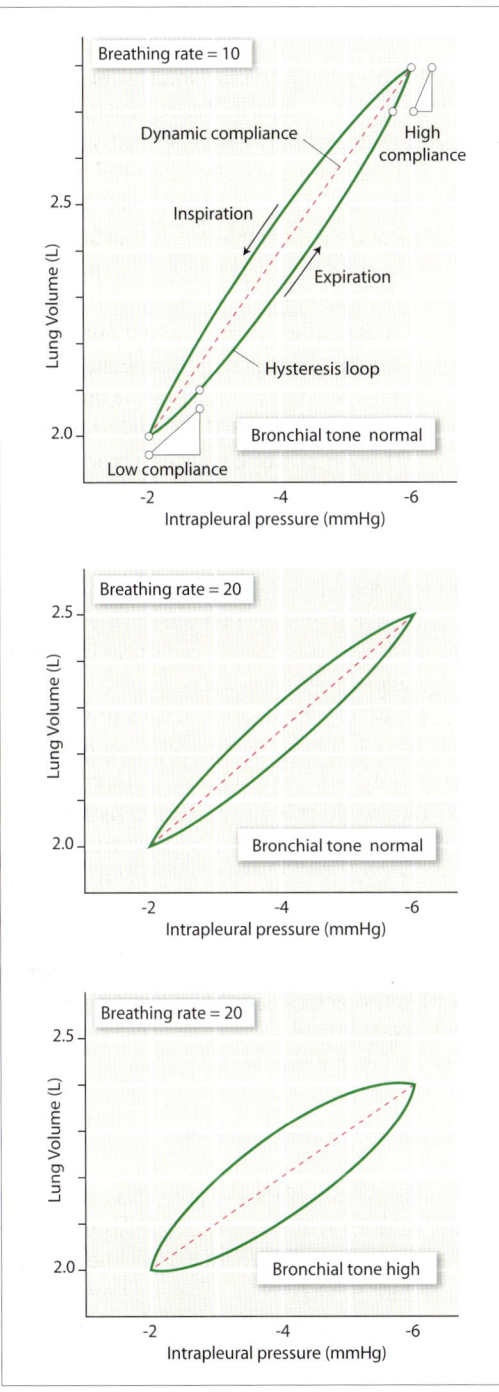

only during motion: the faster the motion, the greater is the friction. There are two main sources of the viscous resistance: the movement of air in and out of the lungs, and the surfactant.

Hysteresis of dynamic compliance When airflow rate through respiratory passages increases, the airway resistance increases markedly and with it, the hysteresis increases. Inspiratory and expiratory forces are highest at the beginning of the respective phases (Fig. 49.**13**). Hence airflow rate and airway resistance are highest at the beginning of inspiration and expiration. Since resistance reduces compliance, the compliance curve is flatter (low compliance) in early inspiration and steeper (high compliance) near its termination. Similarly, the curve is flatter (low compliance) in early expiration and steeper (high compliance) in late expiration. The hysteresis loop occurs due to these variations in compliance with the phase of breathing. In obstructive lung disorders, the dynamic compression of the airways further accentuates the hysteresis loop.

Hysteresis of static compliance During measurement of static lung compliance, the lung is inflated in very small steps and the measurements are made only after the airflow has stopped. Hence airway resistance has no effect on static compliance. Yet, there remains a slight amount of hysteresis in the static lung compliance which is attributable to the presence of surfactant in the lung. Finally, even when the surfactant is washed out with saline solution, some hysteresis still remains. This residual hysteresis is due to the viscous resistance of the lung parenchyma itself, attributable to the intracellular fluid and the fluidity of membranes.[2] The single static compliance curve shown in Figure 49.**9** is the compliance recorded during expiration, which experimenters find to be more consistent than the inspiration curve.

■ **Dynamic airway compression**

Rapid, forceful expiration results in compression, and often closure, of the small airways, especially of the respiratory bronchioles. The closure results in trapping of

Fig. 49.**10** Dynamic compliance and hysteresis. (*Above*) A slow breathing rate of 10 per minute allows adequate time to the lung to inflate and deflate completely. The dashed line connecting the end-inspiratory and end-expiratory points represents the dynamic compliance. (*Middle*) When the breathing rate increases to 20 per minute, less time is available for lung inflation and deflation which remain somewhat incomplete and therefore the dynamic compliance decreases. (*Below*) Bronchoconstriction slows down inspiration and expiration and therefore the dynamic compliance falls further. Also, the hysteresis loop enlarges due to higher airway resistance.

---

[2] The presence of hysteresis in the static compliance curve might appear paradoxical since viscous resistance of any kind should be absent when there is no movement of the lungs. The answer to the paradox lies in the fact that even in static compliance, measurements are made while the lung volume is still changing, albeit extremely slowly. (The experimenter cannot wait indefinitely for the lung volume to stabilize completely!) Since the viscosity of liquid is much higher than that of air, even a slow inflation or deflation of lung is offered considerable resistance by the surfactant, resulting in a slight hysteresis.

air in the alveolus (air trapping) during expiration. In emphysema, the elastic tissues in the walls of the small airways are destroyed. These airways collapse due to dynamic compression in early expiration, resulting in considerable air trapping. The consequent increase in the functional residual capacity (FRC) and reduction in vital capacity are often observed in obstructive airway diseases (see Table 56.1).

Why do the respiratory bronchioles tend to collapse before the alveoli have emptied completely? An explanation to this is offered by the *equal pressure point theory* (Fig. 49.**11**). Suppose the intrapleural pressure at the end of a deep inspiration is –15 mmHg. Although the intraalveolar pressure is negative initially, it soon becomes equal to atmospheric pressure as air rushes into the alveoli. Hence, if the breath is held in inspiration, the airway pressure becomes uniformly atmospheric (Fig. 49.**11**a).

At the end of quiet expiration, the intrapleural pressure increases by 10 mmHg to become –5 mmHg. The intraalveolar pressure too rises by 10 mmHg, only to be restored to atmospheric pressure as the alveolar air rushes out. The flowing alveolar air sets up a pressure gradient along its path, with the elevated alveolar pressure (10 mmHg) at one end and the atmospheric pressure (0 mmHg) at the other (Fig. 49.**11**b).

At the end of a forced expiration, the intrapleural pressure becomes positive, increasing by 35 mmHg (from –15 mmHg) to become +20 mmHg. Intraalveolar pressure also increases to 35 mmHg and air rushes out of the alveoli, setting up a pressure gradient as before (Fig. 49.**11**c). Although the intraalveolar pressure is higher than the intrapleural pressure, somewhere near its exit, the intraairway pressure drops below the intrapleural pressure. The airway collapses distal to this site, which is called the equal pressure point (Fig. 49.**11**d). Airway collapse does not occur so long as the intrapleural pressure remains negative, as during quiet expiration.

■ **Intrapleural pressure**

At the end of a normal expiration, there is no distending pressure inside the lungs. The volume of the lung-thorax combine is therefore 2 L and it is said to be in its resting state. However, as already explained above, at zero distending pressure, the lungs are trying to collapse and the thorax is trying to expand. These opposite pulls result in a negative (i.e., subatmospheric) pleural pressure of about –2.5 mmHg. If the elastic recoil of the lung is reduced, as in older people or in patients with chronic lung disease, the intrapleural pressure may not be negative.

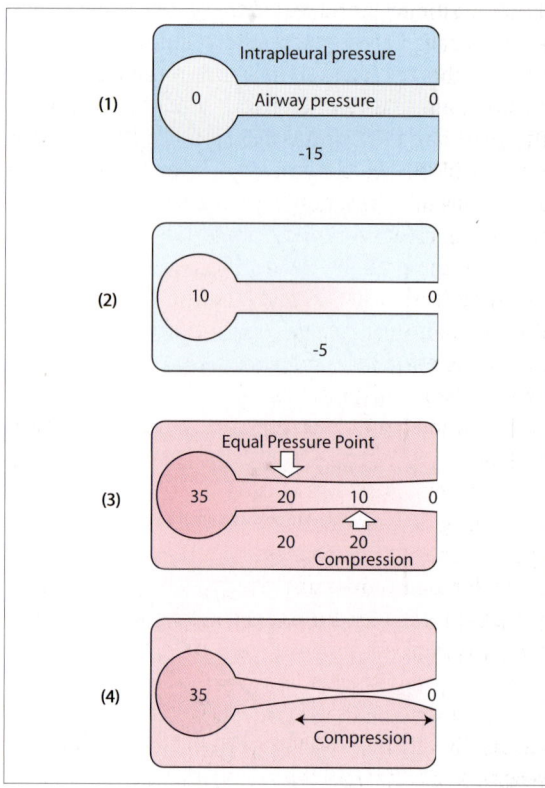

Fig. 49.**11**   Mechanism of airway closure. (1) At end-inspiration, intrapleural pressure is negative while airway pressure is atmospheric. (2) After quiet expiration, intrapleural pressure is still negative and airway pressure is mostly above atmospheric. (3) After forced expiration, intrapleural pressure becomes positive and exceeds the pressure in the distal segment of the airway. (4) The positive intrapleural pressure compresses the distal airway. (Negative pressures are shown in blue while positive pressures are shown in red.)

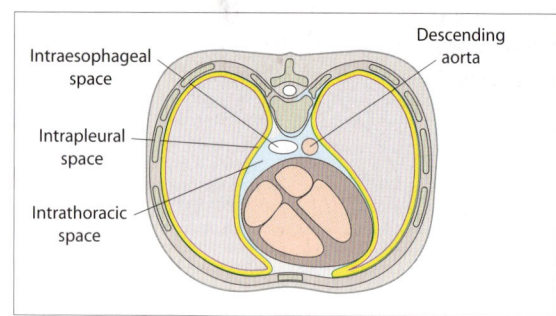

49.**12**   Intrathoracic, intrapleural, and intraesophageal spaces. Since these spaces are separated by thin walls, the pressures in them are nearly equal.

Intrapleural versus intrathoracic pressure   When the thorax and the lungs try to move in opposite directions, the entire space separating the two is subjected to a negative pressure. Not only the pleural space, but everything else that lies in that space (the heart and the mediastinum) is subjected to the negative pressure which is called the *intrathoracic pressure* (Fig. 49.**12**). The intrathoracic pressure is equal to the intrapleural pressure, and the two terms are used synonymously. The heart has a thick muscular wall and therefore the pressure in the heart chambers is relatively unaffected by the negative intrathoracic pressure. However, the pressure in great veins decreases as they enter the thorax. Even the pressure inside the thoracic aorta is affected by large intrathoracic pressure changes during Valsalva maneuver (see below). The intraluminal pressure of the esophagus, which has a soft wall, faithfully reflects the intrathoracic pressure and offers a convenient method for the measurement of the intrathoracic pressure.

Intrapleural pressure is thus measured with an intraesophageal balloon. However, since the esophagus opens to the exterior through the mouth, the pressure in its lumen equilibrates quickly with atmospheric pressure. Hence for measuring intraesophageal pressure, it is important to keep the mouth closed, sealing off the esophageal cavity from the exterior. The intrapleural pressure can also be recorded by inserting a needle directly into the intrapleural space. A small air bubble is injected into the pleural space and the tip of the needle is kept in the bubble. This ensures that the tip of the needle does not get obliterated by the pleural membranes.

The intrapulmonary pressure can be maximized by exerting extra inspiratory or expiratory efforts against an obstruction. A strong expiratory effort against a closed glottis is called the *Valsalva maneuver*. It can raise the intrathoracic and intrapulmonary pressures to +100 mmHg. A strong inspiratory effort against a closed glottis is called the *Müller maneuver*. It can reduce these pressures to –80 mmHg. The maximal pressures that can be achieved by forceful inspiration and expiration indicate the strength of the inspiratory and expiratory muscles respectively, and are called the *forced pressures*. Thus, relaxation pressures indicate lung-thorax compliance while forced pressures indicate the strength of respiratory muscles.

Maximal expiratory pressures can be developed only when the thoracic volume is large. Conversely, maximal inspiratory pressures can be developed only when the thoracic volume is minimum. One reason is that an inflated chest exerts a high relaxation pressure. More importantly, the muscles of expiration are at their 'resting length' (see muscle length–tension relationship, p 106) when the thorax is inflated. Conversely, the inspiratory muscles are at their resting length when the thoracic volume is low (Fig. 49.**13**). When respiratory muscles are severely stretched, they contract with less strength. They can also become fatigued and fail (pump failure), leading to inadequate ventilation.

Intrapulmonary pressure

It has been discussed above that the intrapulmonary pressure recorded during a relaxed expiration is called the relaxation pressure. The relaxation pressures at different lung volumes give the compliance of the lung-thorax combine.

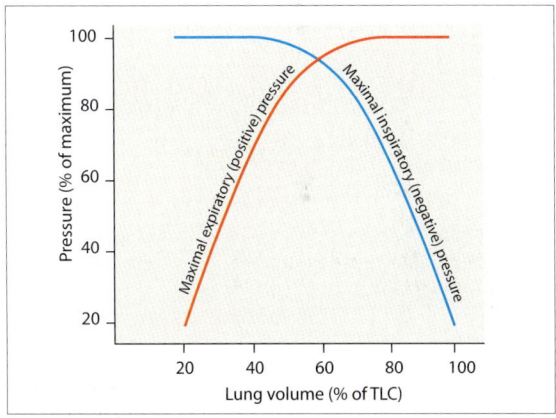

Fig. 49.**13**   Maximal inspiratory and expiratory pressures. Inspiratory force is maximum at low lung volume and declines at higher volumes. The reverse is true for expiratory force.

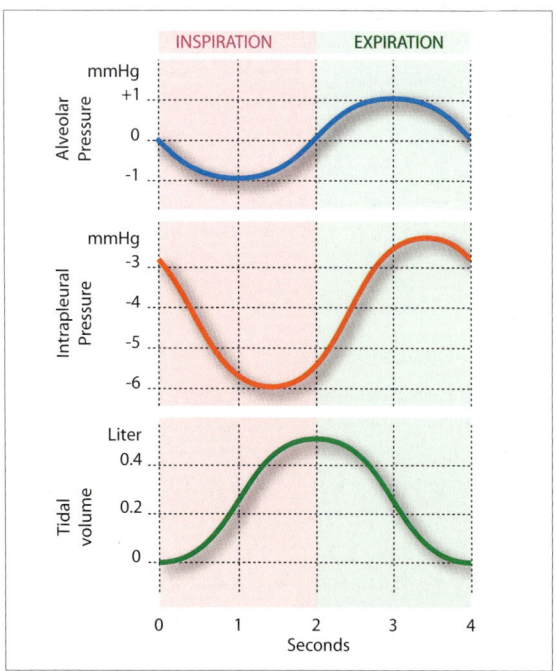

Fig. 49.**14**   Rhythmic changes in alveolar pressure, intrapleural pressure, and tidal volume during normal breathing.

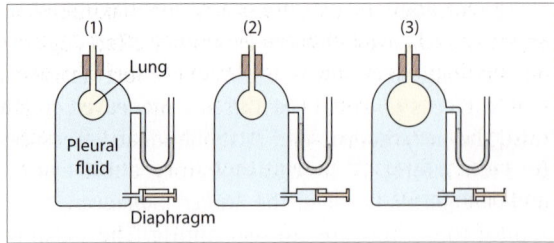

Fig. 49.15 A lung-thorax-diaphragm analog model (after TF Sherman) explaining the changes in pleural pressure during inspiration (1). If the plunger is pulled suddenly (diaphragm contracts), the intrapleural falls sharply before the lung is able to expand (2). As air moves into the lungs, the lung expands and the pleural pressure rises slightly (3).

Breathing Cycle

Pressure-volume changes during breathing

Intrapleural pressure   At the end of quiet expiration, the intrapleural pressure is –2.5 mmHg (Fig. 49.**14**). When the thorax expands, the intrapleural pressure tends to decrease. The lungs comply, i.e., they respond to the fall in intrapleural pressure by expanding. Had the lungs not expanded, there would have been a drastic fall in intrapleural pressure. The intrapleural pressure therefore depends on lung compliance and is affected by the hysteresis of the compliance curve. During forceful breathing, lung compliance decreases due to sharp rise in viscous resistance and therefore the pleural pressure changes are accentuated. It can fall as low as –30 mmHg during mid-inspiration. During normal breathing in a healthy subject, the fall in intrapleural pressure at the end of a maximal inspiration is only 4 mmHg (from –2 to –6 mmHg).

It is observed in Figure 49.**14** that the intrapleural pressure falls to its peak negativity a little before end-inspiration. The reason for this can be easily demonstrated in a lung–thorax–diaphragm analog model (Fig. 49.**15**). Conversely, the intrapleural pressure becomes positive a little before end-expiration.

Intrapulmonary pressure   The intrapulmonary pressure drops to about –3 mmHg during mid-inspiration, but returns to 0 mmHg by end-inspiration due to rapid flow of air into the lungs. During quiet expiration, the elastic recoil of the lung raises the intrapulmonary pressure to about +3 mmHg by mid-expiration. As air flows out of the lungs, the intrapulmonary pressure returns to atmospheric levels by the end of quiet expiration. The changes in intrapulmonary pressure are much greater when airway resistance is high.

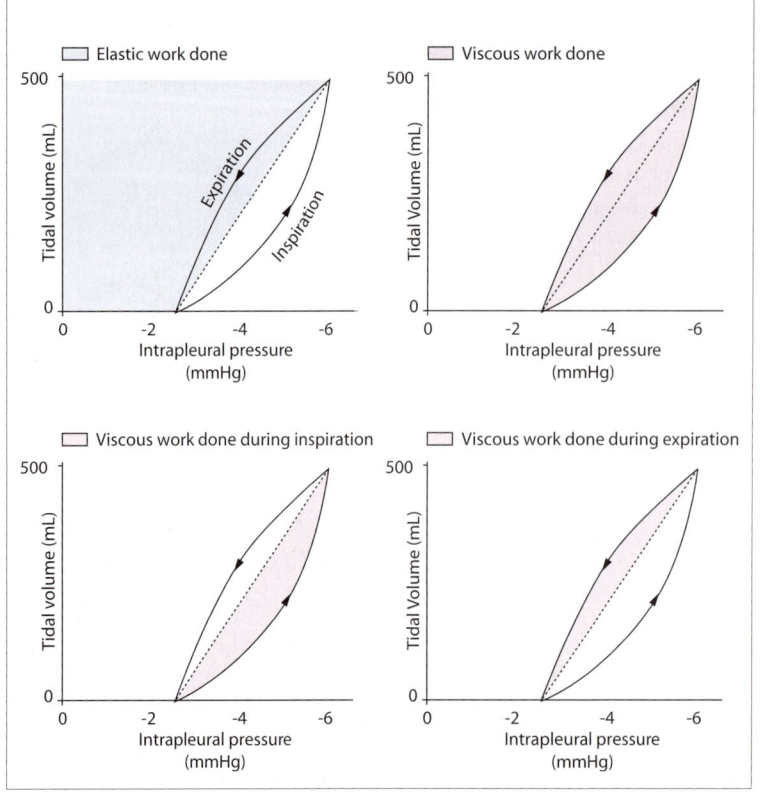

Fig. 49.16   Graphic measurement of work of breathing.

Work of breathing

Work is performed by the respiratory muscles in overcoming the elastic and viscous resistances offered by the chest wall, lungs, and the airways. During heavy breathing, the work of breathing can increase considerably. The work of breathing done in overcoming the viscous resistance can be estimated graphically by measuring the area enclosed by the hysteresis loop (Fig. 49.**16**). The elastic work of breathing, the viscous work during inspiration and the viscous work of expiration can be separately estimated as shown in Figure 49.**16**.

The total work of quiet breathing is 0.3–0.8 kg m min^{-1}. It increases markedly during exercise, but still accounts for <3% of the total energy expenditure during exercise. The work of breathing is greatly increased in diseases such as emphysema, asthma, and congestive heart failure with dyspnea and orthopnea.

Dyspnea or breathlessness is the unpleasant awareness of breathing effort and usually occurs whenever the work of breathing increases. The theory of length–tension inappropriateness holds that there is a constant subconscious comparison between ventilation required and ventilation achieved, and between muscle tension exerted and change in muscle length achieved. Dyspnea occurs when this inappropriateness reaches a certain threshold. Dyspnea may be evoked by stimulation of receptors in the upper airways, in the lungs (see J receptors), respiratory muscles or chest wall. In supine position the abdominal contents tend to press on the diaphragm and cause dyspnea. Such dyspnea, which occurs upon assuming supine position, is called *orthopnea*. Sitting up takes this load off the diaphragm. A severely dyspneic patient therefore tends to sit up and bend forward.

Pulmonary Surfactant

Surfactant is a soap-like substance secreted by the Type-II pneumocytes of the alveolar epithelium. It is a mixture of dipalmitoyl phosphatidylcholine (a phospholipid), other lipids, and proteins. The surfactant molecules form a thin layer on the internal surface of the alveoli (Fig. 49.**17**), with the hydrophilic heads layering the alveolar membrane and the hydrophobic fatty acid tails facing the alveolar lumen.

Surfactant reduces the alveolar surface tension. This has three important physiological consequences. (1) It reduces the collapsing tendency of the lungs. (2) It reduces the natural tendency of the smaller alveoli

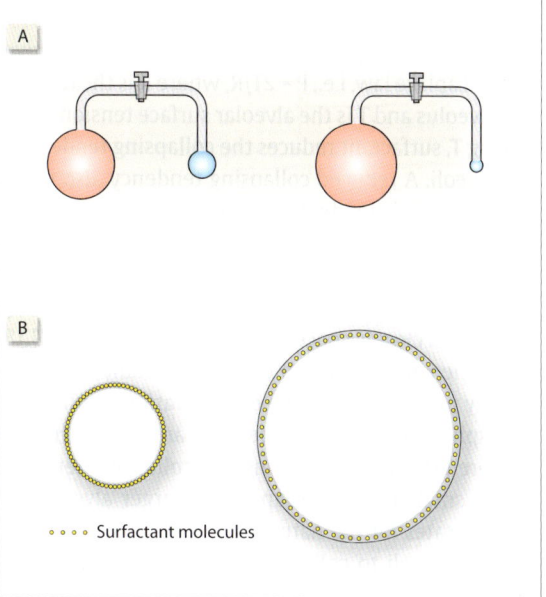

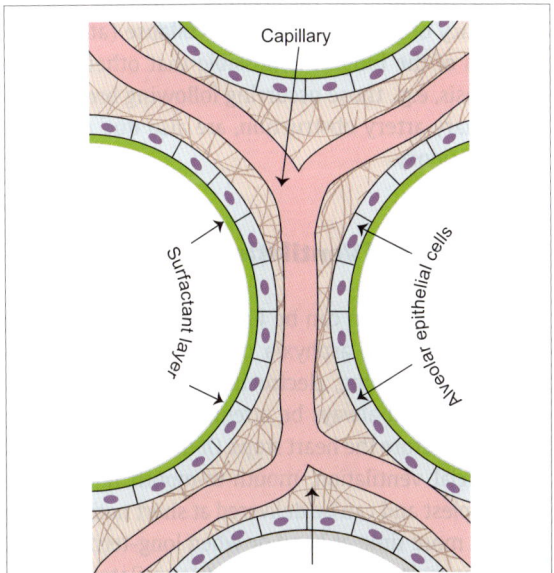

Fig. 49.**17**   Alveoli, internally lined by surfactant and externally separated by interstitial fluid. If alveoli shrink in size, the interstitial fluid pressure becomes more negative.

Fig. 49.**18**   [A] When two inflated balloons of unequal sizes are interconnected, the smaller balloon will collapse and empty completely into the larger one so that at the end, there will be a single large balloon left. It occurs in accordance with the Laplace Law that states that the smaller the radius, the greater is the collapsing tendency. [B] Alveolar stabilization. (*Left*) Surfactant molecules lightly packed in a small alveolus. (*Right*). The same number of surfactant molecules gets spaced apart in a large alveolus and the surfactant concentration falls.

51 Alveolar Ventilation, Perfusion, and Gas Exchange

Dead Spaces

Anatomic dead space

During quiet breathing, about 500 ml of air is inspired and expired (the tidal volume). Not all the air that is breathed in enters the alveoli. About 150 ml of inspired air remains in the upper airways, trachea, and bronchi, and therefore does not participate in gaseous diffusion. This is called the anatomic dead space and can be measured by the *single-breath N_2 method*, which is as follows.

At the end of a quiet expiration, the subject takes in a deep breath of pure O_2 and breathes it out slowly and steadily. The exhaled air is instantaneously monitored for its N_2 concentration. The initial volume of air that is entirely N_2 free gives the anatomic dead space. The air exhaled thereafter, which comes from the alveoli, has a higher concentration of nitrogen (Fig. 51.**1**).

Physiological dead space

Normally, about 350 ml of air enters the alveoli with each breath. Not all the air that enters the alveoli necessarily participates in alveolar gas exchange. This is because the alveolus might not be getting adequate blood supply, or its walls might not permit adequate diffusion. Such spaces in the alveoli constitute the *alveolar dead space*. The sum of anatomical dead space and alveolar dead space is called the physiological dead space. Hence, the physiological dead space is the volume of inspired air, either in the airways or in the alveoli, that does not participate in gaseous exchange. The physiological dead space (PDS) can be calculated using the Bohr equation:

$$P_{CO_2[Exp]} \times TV = P_{CO_2[Art]} \times (TV - PDS) + P_{CO_2[Ins]} \times PDS$$

where:

$P_{CO_2[Exp]}$	is	P_{CO_2} in expired air
$P_{CO_2[Art]}$	is	P_{CO_2} in arterial blood
$P_{CO_2[Ins]}$	is	P_{CO_2} in inspired air
TV	is	tidal volume

Ignoring $P_{CO_2[Ins]}$, the formula reduces to:

$$P_{CO_2[Exp]} \times TV = P_{CO_2[Art]} \times (TV - PDS)$$

Example

$P_{CO_2[Exp]}$	=	28 mmHg
$P_{CO_2[Art]}$	=	40 mmHg
TV	=	500 ml
28×500	=	$40 \times (500 - PDS)$
$\therefore$ PDS	=	150 ml

The derivation of the Bohr equation is based on the knowledge that all the CO_2 in expired tidal air ($P_{CO_2[Exp]} \times TV$) comes from the alveoli. It is also known that the

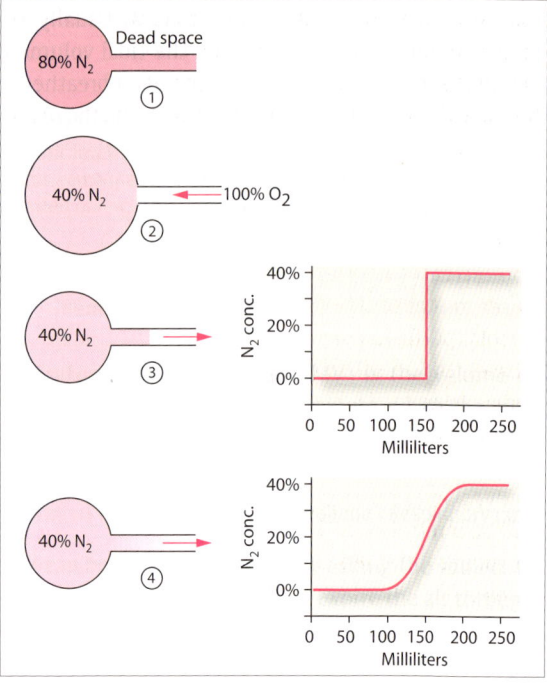

Fig. 51.**1** Measurement of anatomical dead space. (1) Normally, the dead spaces and alveoli remain filled with air containing 80% N_2. (2) Breathing in pure O_2 fills the dead spaces with 100% O_2 and dilutes the alveolar N_2 to 40%. (3) During exhalation, 150 ml of pure O_2 in the dead spaces comes out first followed by 40% N_2. (4) Due to mixing of N_2 with O_2 inside the dead spaces, the transition is not abrupt as shown in 3. The O_2 concentration starts rising from 100 ml and peaks at 200 ml. The mean value of 150 is taken as the dead space volume.

amount of air (nearly CO_2-free) in each breath that equilibrates with blood is TV minus PDS. The P_{CO_2} of this alveolar air equals the P_{CO_2} of pulmonary arterial blood, i.e., $P_{CO_2[Art]}$. Hence, $(TV - PDS) \times P_{CO_2[Art]}$ is equal to the amount of CO_2 entering the alveoli from pulmonary capillaries and leaving with each breath. This must be equal to the total amount of CO_2 exhaled with each breath, which is given by $P_{CO_2[Exp]} \times TV$.

Alveolar Ventilation

Alveolar ventilation is defined as the amount of inspired air entering the alveoli per minute during normal breathing. It is given by:

$$\begin{aligned} \text{Alveolar ventilation} &= (\text{Tidal volume} - \text{Anatomical} \\ &\quad\ \text{dead space}) \times \text{Respiratory rate} \\ &= (500\ ml - 150\ ml) \times 14\ min^{-1} \\ &= 4900\ ml\ min^{-1} \end{aligned}$$

Alveolar air composition can be estimated by alveolar air sampling, i.e., analyzing the last few milliliters of air that issues from the lungs during expiration. The alveolar air composition represents a dynamic equilibrium between the rate at which atmospheric air ventilates the alveoli and the rate at which pulmonary arterial blood perfuses the alveoli and exchanges its gases with them (Fig. 51.**2**). Thus the alveolar air composition depends on the *ventilation–perfusion ratio*.

If alveolar perfusion decreases or alveolar ventilation increases, the alveolar composition will become closer to that of the inspired air. On the other hand, if alveolar perfusion increases or alveolar ventilation decreases, the alveolar partial pressure of gases will become closer to the corresponding partial pressures in arterial blood (Fig. 51.**3**).

Since alveolar P_{O_2} and alveolar P_{CO_2} are both related to alveolar ventilation, it stands to reason that the two will be interrelated. The relationship between alveolar P_{O_2} and P_{CO_2} is given by the alveolar gas equation given below:

$$P_{O_2[Alv]} = P_{O_2[Ins]} - P_{CO_2[Alv]} \times \left[F_{O_2} + \frac{1 - F_{O_2}}{RQ} \right]$$

where:

$P_{O_2[Alv]}$ is the alveolar P_{O_2} (normally 100 mmHg)
$P_{O_2[Ins]}$ is the P_{O_2} in inspired air (normally 150 mmHg)
$P_{CO_2[Alv]}$ is the alveolar P_{CO_2} (normally 40 mmHg)
F_{O_2} is the fraction of O_2 in dry air (normally 0.2)
RQ is the respiratory quotient (normally 0.8). It is the ratio of CO_2 output and O_2 intake by tissues per minute.

Fig. 51.2   The ships are unloading imported cargo (red) in the harbor and taking away the exports (blue). The trucks are taking away the imports from the harbor and unloading on the harbor the cargo for export. The relative frequency with which the ships and trucks come to the harbor is reflected in the type of cargo (imports or exports) that accumulates in the harbor. For example, if the trucks stopped coming in, there would be no blue crates left in the harbor, which would then be full of red crates only.

By substituting the normal values of $P_{O_2[Ins]}$, F_{O_2}, and RQ, the formula simplifies to:

$$P_{O_2[Alv]} = 150 - P_{CO_2[Alv]} \times \left[0.2 + \frac{1 - 0.2}{0.8} \right]$$

or

$$P_{O_2[Alv]} = 150 - (P_{CO_2[Alv]} \times 1.2)$$

The formula can be verified by substituting normal values of $P_{O_2[Alv]}$ and $P_{CO_2[Alv]}$.

$$102 = 150 - (40 \times 1.2)$$

Effect of gravity on alveolar ventilation

Gravity reduces the ventilation of the apical alveoli, i.e., the alveoli that are nearer to the apex of the lungs. The reason is twofold: (1) Gravity reduces the compliance

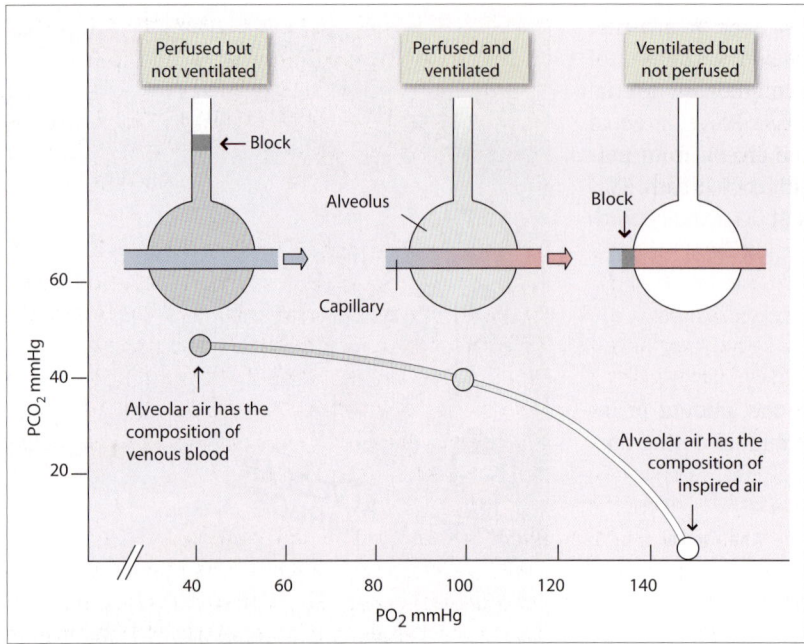

Fig. 51.**3**   Ventilation–perfusion ratio affects alveolar air composition.

of the apical alveoli, and (2) Gravity increases the intrapleural pressure near the base of the lungs.

Effect of gravity on alveolar compliance   In the erect posture, the apical alveoli are larger but poorly ventilated while the basal alveoli are smaller but better ventilated. This is because the weight of the lungs stretches the apical alveoli to nearly their maximum size, leaving little room for further expansion during inspiration (Fig. 51.**4**). In other words, apical alveoli have lower compliance. Although apical alveoli are poorly ventilated, the P_{O_2} of apical alveoli is higher. This is because of the effect of gravity on alveolar perfusion (see below).

A clinical correlate of the effect of gravity on ventilation is that arterial oxygenation improves in unilateral lung diseases when the patient lies on his side so that the good lung is in the dependent position. For reasons not understood, the situation is opposite in infants, who seem to do better with the diseased lung in the dependent position.

Effect of gravity on intrapleural pressure   Gravity also affects the intrapleural pressure. Near the apex, the entire weight of the lungs acts to pull the lung away from the thorax. Near the base, the lungs tend to 'sit' on the thoracic floor and the weight of the lung tends to make the intrapleural pressure positive.

At the end of a forced expiration, intrapleural pressure at the bases of the lungs can exceed the atmospheric pressure in the airways. This high pressure causes the respiratory bronchioles near the base of the lungs to collapse (airway closure). The respiratory

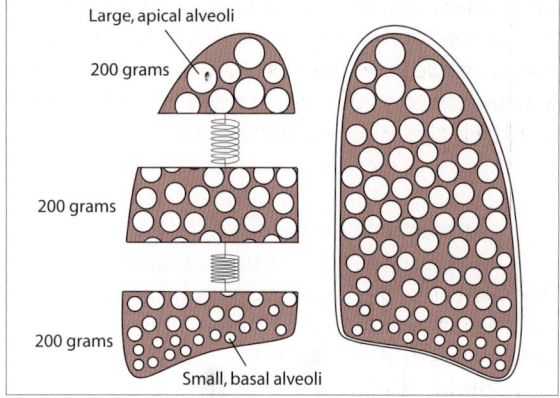

Fig. 51.**4**   (*Left*) Apical alveoli are larger than basal alveoli. Assuming the weight of the lungs to be 600 g, the alveoli in the upper third of the lungs are pulled down by 400 g of lung tissue while alveoli in the middle third are pulled down by only 200 g of lung tissue. (*Right*) The lungs tend to 'sit' on the diaphragmatic pleura. Hence, intrapleural pressure at the apex is more negative than the intrapleural pressure at the base.

bronchioles near the apex however remain open because the intrapleural pressure is more negative near the apex and also because the weight of the lungs keeps the bronchiolar lumen stretched open. In disorders like emphysema that are associated with reduction in lung elasticity, airway closure at the base may occur even at the end of a quiet expiration.

The lung volume at which the basal airways start closing down is called the *closing volume*, and can be estimated by the single-breath O_2 method. The subject takes a deep breath of 100% O_2 and exhales it slowly while the N_2 concentration of the exhaled air is monitored continuously. As explained above, most of the O_2 inspired enters the basal alveoli, greatly diluting the alveolar N_2. Much less O_2 enters the apical alveoli where the air remains rich in N_2. A graph of the N_2 concentration plotted against the volume of expired air therefore shows four distinct phases (Fig. 51.**5**).

In phase I, the pure O_2 present in the dead spaces is exhaled and there is no N_2 in the expired air. In phase II, there is a sharp rise in N_2 as the alveolar air is exhaled. The N_2 concentration reaches a plateau in phase-III. However, there is a slight rise in N_2 concentration during phase III due to the progressive closure of the basal alveoli which have less N_2 so that relatively more air comes from the N_2-rich apical alveoli. In phase IV, the N_2 concentration rises abruptly again as most of the exhaled air now comes from the N_2-rich apical alveoli. By now, the basal alveoli have closed down completely, and the lung volume at the beginning of the fourth phase denotes the closing volume of the lung.

The normal closing volume is about 10% of the vital capacity. It increases when there is a narrowing of the small airways, or there is a reduction in lung elasticity which normally keeps the small airways patent. The elasticity decreases with age and the closing volume can be as high as 25% by the age of 50 years. Closing volume is also high in chronic smokers.

> **The double conundrum**
> 1. The apical alveoli, despite being larger, are poorly ventilated.
> 2. The apical alveoli, despite being poorly ventilated, have higher Po_2.

oxygenated. This blood, which bypasses the alveoli (gets shunted) to supply the bronchi and return to the left atrium, constitutes a physiological shunt (see Fig. 48.**3**). The deoxygenated blood from bronchial circulation drains into the pulmonary veins and mixes with oxygen-rich blood (with a Po_2 of 97 mmHg) from the alveoli. Hence, the Po_2 of mixed pulmonary venous blood is slightly lower, at 95 mmHg.

Effect of gravity on alveolar perfusion   In the erect posture, the basal alveoli are much better perfused than the apical alveoli. The Po_2 and Pco_2 of the well-perfused basal alveoli becomes equal to that of pulmonary arterial blood, i.e., it has low Po_2 and high Pco_2. Conversely, the alveolar air composition of the poorly perfused apical alveoli approximates more to that of inspired air, i.e., it has high Po_2 and low Pco_2. The high Po_2 in apical alveoli is the reason for the vulnerability of apical areas of the lungs to infection by *Mycobacterium tuberculosis*, which is an aerobic bacterium.

Alveolar Perfusion

The normal pulmonary blood flow is the same as the right ventricular output, i.e., about 5.0 L min⁻¹. However, not all the blood that flows into the lungs perfuses the alveoli. Some of the blood supplies the bronchi and gets further deoxygenated instead of getting

Ventilation–perfusion ratio

Considering that the cardiac output is 5.0 L min⁻¹ and alveolar ventilation too is about 5.0 L min⁻¹, the overall ventilation:perfusion (V/Q) ratio is 1:1. Ideally,

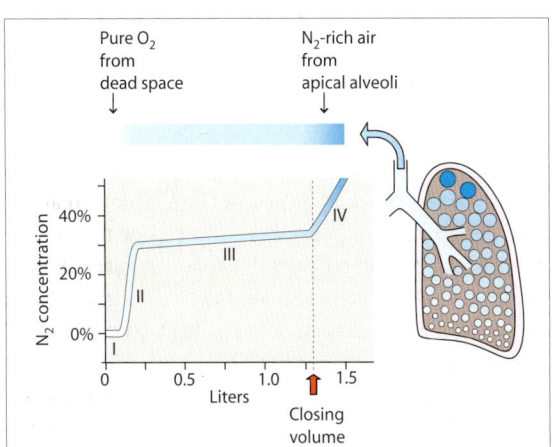

Fig. 51.**5**   Measurement of closing volume.

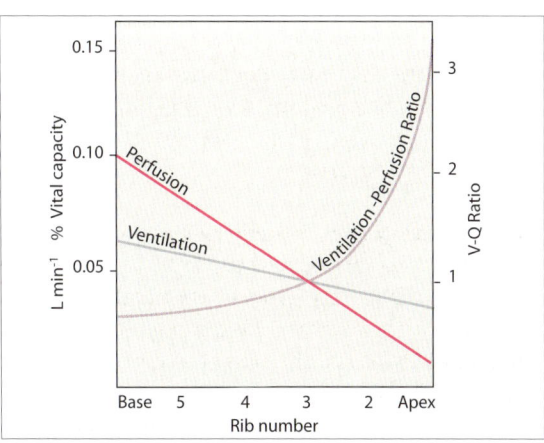

Fig. 51.**6**   Both ventilation and perfusion decrease in the upper part of the lungs, but perfusion decreases more than ventilation. Hence, the ventilation-perfusion ratio increases towards the apex.

therefore, each alveolus should have a V/Q ratio of 1:1. However, that is not so even in normal lungs.

Due to gravity, the apical alveoli are both underventilated and underperfused while the basal alveoli are both overventilated and overperfused (Fig. 51.**6**). However, gravity affects perfusion much more than it affects ventilation. Hence, apical alveoli are more underperfused than underventilated (V/Q = 3) while the basal alveoli are more overperfused than overventilated (V/Q = 0.6). These regional disparities in ventilation and perfusion get exaggerated in diseases, resulting in gross impairment of alveolar gas exchange with consequent hypoxic hypoxia.

Alveolar Gas Exchange

The **respiratory membrane** (or the alveolocapillary membrane) is made up six layers (Fig. 51.**7**). From the alveolar to the capillary side, they are (1) surfactant layer, (2) alveolar epithelium, (3) alveolar basement membrane, (4) alveolar interstitial space, (5) capillary basement membrane, and (6) capillary endothelium. It is about 0.5 μm thick, and its total surface area in the two lungs equals about 100 m². The gases have to pass through these layers while diffusing from the alveoli to pulmonary blood.

Flow-limited vs. diffusion-limited transport Blood spends about 0.75 seconds (the transit time) in the pulmonary capillaries (Fig. 51.**8**). This time is more than double the time taken by O₂ (about 0.3 s) to completely equilibrate with the pulmonary blood. CO₂ takes even less time. Transport of these gases across the respiratory membrane is therefore flow-limited (see Fig. 39.**4**).

In other words, if the pulmonary blood flow increases, the amount of O_2 and CO_2 diffusing across the respiratory membrane will increase with it.

Conversely, carbon monoxide, which does not reach equilibrium in 0.75 seconds, will have a diffusion-limited transport, i.e., its transport across the respiratory membrane will increase only if the rate of diffusion increases.

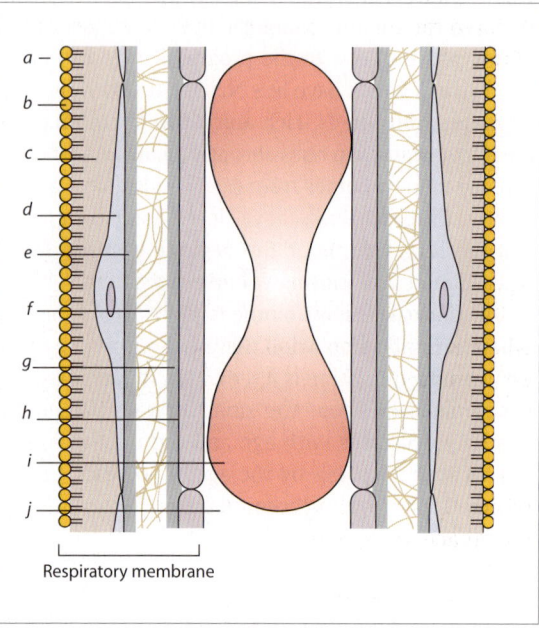

Respiratory membrane

Fig. 51.**7** The respiratory membrane. a. Alveolar space b. Surfactant molecules c. Water layer d. Alveolar epithelial cells e. Alveolar basement membrane f. Alveolar interstitium g. Capillary basement membrane h. Capillary endothelium i. Red blood cell j. Capillary lumen filled with fluid.

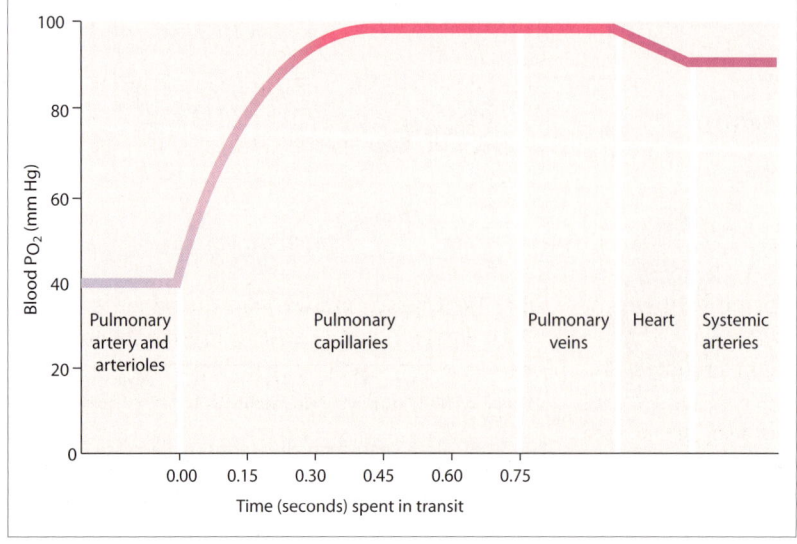

Fig. 51.**8** Time course of gaseous equilibration in pulmonary capillaries. Note that oxygen takes only about 0.3 seconds to equilibrate completely with the pulmonary blood though it remains in the pulmonary capillaries for 0.75 seconds.

■ Diffusion capacity of the lung

The volume V of a gas that diffuses across the respiratory membrane is (1) directly proportional to the pressure gradient ΔP across the respiratory membrane, (2) directly proportional to the duration t of the diffusion, (3) directly proportional to the surface area A of the respiratory membrane, (4) directly proportional to the solubility S_{gas} of the gas in the respiratory membrane, (5) inversely proportional to the thickness d of the respiratory membrane, and (6) inversely proportional to the square root of the molecular weight W_{gas} of the gas.

$$V \propto \frac{\Delta P \times t \times A \times S_{gas}}{d \times \sqrt{W_{gas}}} \quad ...(1)$$

$$V = K \times \frac{\Delta P \times t \times A \times S_{gas}}{d \times \sqrt{W_{gas}}} \quad ...(2)$$

where K is the proportionality constant. The formula can be reduced to:

$$V = \Delta P \times t \times D_L \quad ...(3)$$

where D_L is the *diffusion capacity* of the respiratory membrane, and is given by:

$$D_L = K \times \frac{A \times S_{gas}}{d \times \sqrt{W_{gas}}} \quad ...(4)$$

Equation 3 gives the definition of diffusion capacity, because D_L equals V when t = 1 and ΔP = 1. Hence, the diffusion capacity of the lung for a given gas is defined as the volume of gas diffusing across the respiratory membrane in 1 minute when the pressure gradient is 1 mmHg. The formula for D_L can be reduced to:

$$D_L = K \times \frac{A}{d} \times \text{Diffusion coefficient} \quad ...(5)$$

where:

$$\text{Diffusion coefficient} = \frac{S_{gas}}{\sqrt{W_{gas}}} \quad ...(6)$$

It may be noted that the diffusion coefficient is dependent entirely on the characteristics of the gas. The diffusion coefficients of O_2 and CO_2 are in a ratio of 1:20. On the other hand, diffusion capacity is dependent not only on gaseous characteristics but also on the area and thickness of the respiratory membrane. The diffusion capacity for O_2 = 20 ml min^{-1} mmHg^{-1} while the diffusion capacity for CO_2= 400 ml min^{-1} mmHg^{-1}. The

diffusion capacity for O_2 increases in exercise and is markedly reduced in emphysema and interstitial lung disorder.

Measurement of diffusion capacity for O_2 From equation 3, we have:

$$V = \Delta P \times t \times D_L$$

or

$$D_L = \frac{V/t}{\Delta P} \quad ...(7)$$

For a gas that diffuses from the alveoli into blood, the formula can be written as:

$$D_L = \frac{\text{Increase in the gas content of pulmonary blood in 1 min}}{\begin{array}{c}\text{Partial pressure of} \\ \text{the gas in alveolar air}\end{array} - \begin{array}{c}\text{Partial pressure of the gas} \\ \text{in pulmonary capillary blood}\end{array}} \quad ...(8)$$

The diffusion capacity for O_2 is rarely measured directly. Instead, it is calculated from the diffusion capacity for carbon monoxide which is easier to measure. The reason is twofold: (1) O_2 diffusion across the respiratory membrane is flow-limited and therefore measuring the amount of O_2 transferred to blood will underestimate the true diffusion capacity. CO diffusion on the other hand is diffusion-limited, and the amount of CO transferred to the blood is a correct estimate of the diffusion capacity (Fig. 51.**9**). (2) CO has an added advantage that once in blood, it reacts with hemoglobin to form carboxyhemoglobin. Hence, it has a negligible partial pressure in pulmonary capillary blood, and equation 8 reduces to:

$$D_L = \frac{\text{Increase in the gas content of pulmonary blood in 1 min}}{\text{Partial pressure of carbon monoxide in alveolar air}}$$

The diffusion capacity measured for CO at rest is about 17 ml min^{-1} mmHg^{-1} . Since the diffusion coefficient of O_2 is 1.23 times greater than that of CO, therefore,

$$D_L \text{ for } O_2 = 17 \times 1.2$$
$$= 20 \,\text{ml min}^{-1} \,\text{mmHg}^{-1}$$

Diffusion capacity in exercise The diffusion capacity (D_L) increases about three-fold in exercise due to the opening up of inactive capillaries, resulting in an increase in the total area of the respiratory membrane.

Diffusion rate in exercise Diffusion rate (V/t) is the product of diffusion capacity (D_L) and the pressure gradient (ΔP) of gases between the alveoli and pulmonary

The chemical control of breathing is aimed at maintaining the P_{O_2} and P_{CO_2} of arterial blood at about 95 mmHg and 40 mmHg respectively. This is achieved by readjusting the rate and depth of breathing whenever there is change in the arterial P_{O_2} or P_{CO_2}.

Respiratory Chemoreceptors

Respiratory chemoreceptors are sensory receptors for the detection of P_{O_2}, P_{CO_2}, and pH of blood. Their location may be central (within the central nervous system) or peripheral (in the peripheral nervous system). They bring about reflex changes in the rate and depth of breathing in response to hypoxia, hypercapnia and acidemia. The response to hypoxia is mediated entirely by peripheral chemoreceptors while the response to hypercapnia and acidemia is mediated mainly (75%) by central chemoreceptors and partly (25%) by peripheral chemoreceptors.

Peripheral chemoreceptors

Peripheral chemoreceptors are the carotid and aortic bodies. These bodies are located in the connective tissue associated with the vessel wall, at the bifurcation of the common carotid, and on the arch of aorta, respectively (Fig. 54.**1**). Both are stimulated by a low P_{O_2} and high P_{CO_2} in the arterial blood perfusing them but the response of the aortic bodies is less.

The carotid and aortic bodies are made of two types of cells, type I and type II cells. The type I or glomus cells resemble the chromaffin cells of the adrenal glands (p 534) and release dopamine in response to hypoxia. The cells synapse with the afferent nerve endings bearing dopamine (D_2) receptors on them. The type II cells are glia-like supporting cells.

Afferent neurons from the carotid bodies pass through the carotid sinus nerve and glossopharyngeal nerve to reach the medullary respiratory center. Afferent fibers from the aortic bodies reach the medulla through the vagus nerve. The frequency of discharge in the afferent nerve fibers arising from the glomus cells is proportional to the fall in P_{O_2} or the rise in P_{CO_2} of arterial blood.

Type I glomus cells have oxygen-sensitive K^+ channels. At low arterial P_{O_2}, high P_{CO_2} or low pH, these K^+ channels close down, resulting in the depolarization of the glomus cell. The exact mechanism of how these three different stimuli act on the K^+ channels remains uncertain. Depolarization of the glomus cell opens up the L-type Ca^{2+} channels in the glomus cell membrane, leading to a rise in Ca^{2+} influx. The Ca^{2+} influx triggers the release of dopamine as a neurotransmitter which excites the afferent nerve endings. Like most other cells, glomus cells also depolarize when the plasma K^+ increases. The depolarization initiates increased afferent nerve discharge even in the absence of hypoxia. Since the plasma K^+ level rises during exercise, this may be a contributory mechanism in exercise-induced hyperventilation (see below).

Central chemoreceptors

Central chemoreceptors are located on the ventral surface of the medulla oblongata and are therefore also called medullary chemoreceptors (Fig. 54.**2**). The

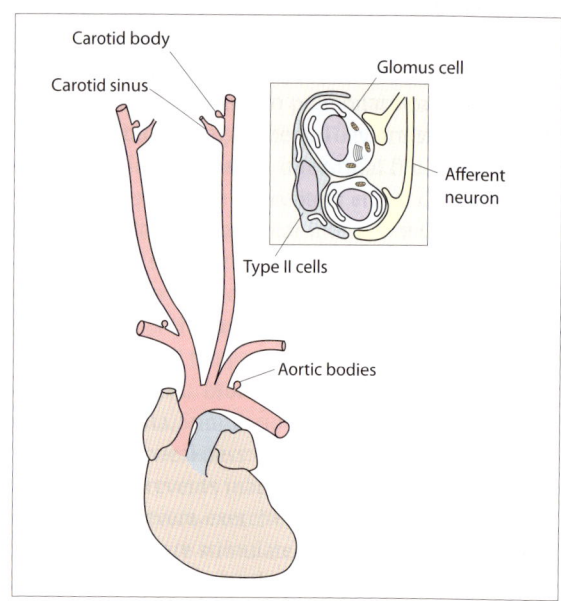

Fig. 54.**1** Location of peripheral chemoreceptors. (*Inset*) Glomus cells present inside the chemoreceptors.

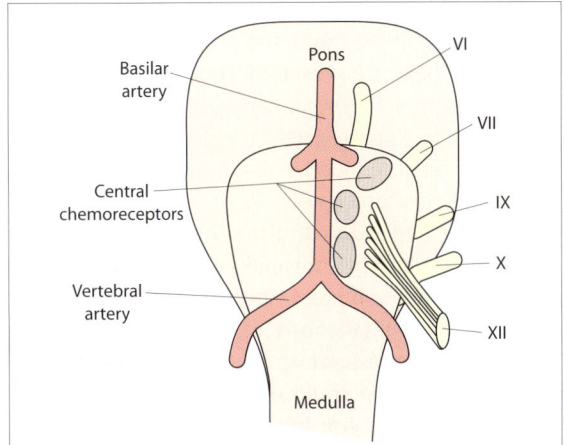

Fig. 54.**2** Central chemoreceptors.

central chemoreceptors are distinct from the respiratory neurons that are located deeper inside the medulla and generate the respiratory rhythm.

Central chemoreceptors are not stimulated by hypoxia; rather, like any other cell, they are depressed by hypoxia. Central chemoreceptors are stimulated by a fall in CSF pH and the pH of brain interstitial fluid. The magnitude of the stimulation is proportionate to the rise in H⁺ concentration. A rise in arterial P_{CO_2} stimulates the chemoreceptors because the CO_2 readily diffuses from arterial blood into the CSF and reacts with water to generate H⁺.

A fall in arterial pH also stimulates the chemoreceptors. However, if the arterial P_{CO_2} is held constant experimentally, a decrease in arterial pH fails to stimulate the chemoreceptors because neither H⁺ ions nor HCO_3^- ions cross the blood-brain barrier readily or in sufficient amounts to affect the CSF pH. Thus, a fall in arterial pH stimulates the central chemoreceptors in three stages: First, the low arterial pH results in a high arterial P_{CO_2}. Next, the CO_2 diffuses into the CSF to lower its pH. Finally, the low CSF pH stimulates the chemoreceptors.

Chemical Regulation of Breathing

Maintenance of respiratory equilibrium

Respiratory disorders are commonly associated with a fall in arterial P_{O_2}, a rise in arterial P_{CO_2}, and a fall in arterial pH. The low arterial P_{O_2} stimulates only the peripheral chemoreceptors while the high arterial P_{CO_2} and low pH stimulate both peripheral and central chemoreceptors. Of the three, hypercapnia provides the strongest respiratory drive. The stimulated chemoreceptors

reflexly increase breathing, as explained above. The hyperventilation produced raises the P_{O_2}, lowers the P_{CO_2}, and raises the arterial pH.

There are situations where the reflex hyperventilation may not have a corrective effect. Take the example of interstitial lung disease (p 367) in which the diffusion capacity of the respiratory membrane is reduced. The low diffusion capacity does not affect the diffusion of CO_2 which is highly diffusible because of its lipid solubility. However, the diffusion of O_2 is impaired, resulting in hypoxia. The hypoxia reflexly stimulates hyperventilation which restores the P_{O_2} to normal but lowers the P_{CO_2} below normal level.

Combined effect of O_2, CO_2, and pH on ventilation Pulmonary ventilation varies linearly with the increase in P_{CO_2}. In hypoxic states, a small rise in arterial P_{CO_2} produces a greater increase in ventilation. In other words, the linear relationship between P_{CO_2} and ventilation becomes steeper at low arterial P_{O_2}. Thus for each value of P_{CO_2}, there is a family of lines with each line representing the P_{CO_2}-ventilation relationship at a different P_{O_2}. A fall in pH shifts the entire 'family of CO_2-response lines' to the left. In other words, if the pH is low, the same amount of respiratory stimulation is produced at a lower arterial P_{CO_2} level (Fig. 54.**3**).

The combined effect of P_{O_2}, P_{CO_2}, and arterial pH explains why ventilation is stimulated only when the arterial P_{O_2} falls below 60 mmHg. Between a P_{O_2} of 100 mmHg and 60 mmHg, there is considerable stimulation of the carotid and aortic chemoreceptors (Fig. 54.**4**). Yet there is little increase in ventilation because any increase in ventilation reduces the hypercapneic ventilatory drive due to CO_2 washout. Respiratory drive is also reduced due to a slight rise in arterial pH that is associated with the increased deoxygenation of

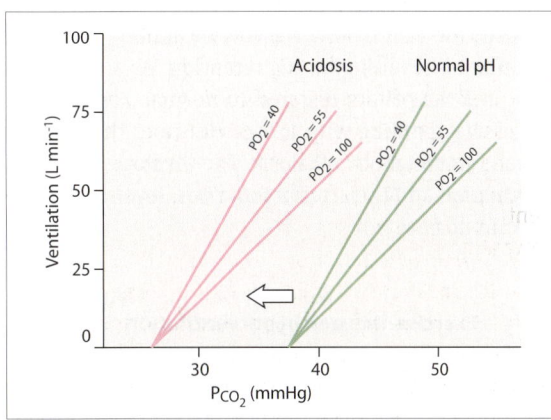

Fig. 54.**3** Combined effects of CO_2, O_2, and pH on pulmonary ventilation.

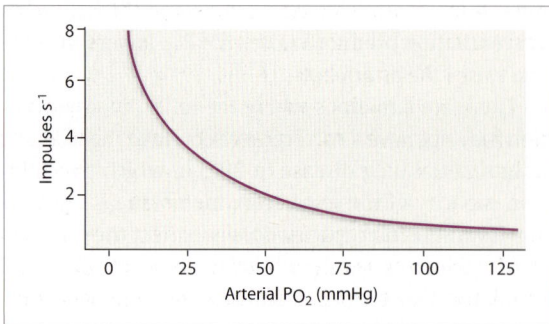

Fig. 54.**4** Effect of O_2 on peripheral chemoreceptor afferents.

hemoglobin in hypoxia. The arterial pH rises because deoxyhemoglobin is a weaker acid than oxyhemoglobin. For hypoxic drive to be effective, it must be strong enough to overcome the respiratory inhibition caused by hypocapnia and a rise in blood pH.

Respiratory correction of acidosis and alkalosis

The lungs can compensate for both acidosis and alkalosis that are of nonrespiratory (metabolic) origin. The related aspects of renal correction of acid-base balance are discussed on p 401.

Respiratory response to acidosis In metabolic acidosis, the P_{CO_2} is high. The high P_{CO_2} stimulates hyperventilation causing blood CO_2 washout. As a result, the P_{CO_2} decreases and the pH is restored to normal. For example, in diabetic ketoacidosis, there is pronounced respiratory stimulation (Kussmaul breathing). The hyperventilation causes CO_2 washout, producing a compensatory rise in pH.

Respiratory response to alkalosis In metabolic alkalosis, the P_{CO_2} is low. The low blood P_{CO_2} depresses ventilation resulting in CO_2 retention. As a result, the P_{CO_2} and the pH are restored to normal. For example, excessive vomiting with loss of HCl from the stomach results in metabolic alkalosis. The alkalosis depresses ventilation and the arterial P_{CO_2} rises, lowering the pH toward normal.

Exercise-induced hyperventilation

Why is exercise associated with hyperventilation? The hyperventilation is unlikely to be due to stimulation of the central or peripheral chemoreceptors since the P_{O_2}, P_{CO_2}, and pH of arterial blood remain almost unchanged in exercise. In the venous blood, however, the P_{O_2} is low, P_{CO_2} is high, and pH is low, and it is possible that these are sensed by chemoreceptors in the large systemic veins or pulmonary artery. It is also possible that a metabolite produced by exercising muscles, e.g., K^+, enters the blood and stimulates the central or peripheral chemoreceptors. The rise in blood temperature during exercise could also stimulate the respiratory center. Yet another possibility is that when the motor cortex sends voluntary commands to the exercising muscles, it simultaneously sends direct stimulatory impulses to the respiratory centers. The hyperventilation during exercise could also be a conditioned reflex: It is possible that one learns to hyperventilate after experiencing the distress caused by not doing so (see avoidance conditioning, p 714). The respiratory center could also be stimulated by afferents from moving joints and the stretch receptors of exercising muscles. Finally, the epinephrine released during exercise could be increasing the sensitivity of the respiratory center so that even small changes in arterial P_{O_2}, P_{CO_2}, and pH cause hyperventilation.

Respiratory Disequilibria

The P_{O_2} and P_{CO_2} of blood are normally maintained within a narrow physiological range that is determined by the equilibrium between the rate of O_2 consumption and CO_2 production on one hand, and the rate and depth of breathing on the other. When this equilibrium is disrupted, it results in respiratory disequilibria, the examples of which are hypocapnia, hypoxia, hypercapnia, and oxygen toxicity.

Hypoxia

Hypoxia is O_2 deficiency at the tissue level. There are four types of hypoxia. In hypoxic hypoxia, the P_{O_2} of the arterial blood is low. In anemic hypoxia, the O_2 content of blood is low despite a normal arterial P_{O_2}. In stagnant hypoxia, the O_2 delivery to the tissues is low despite a normal O_2 content of blood. In histotoxic hypoxia, the tissues suffer from hypoxia despite an adequate delivery of O_2. The principal signs and symptoms of hypoxia are hyperventilation, cyanosis, and cerebral symptoms.

Hyperventilation occurs when peripheral chemoreceptors are stimulated by low arterial P_{O_2}. Hence, hyperventilation occurs in hypoxic hypoxia and also in histotoxic hypoxia if the glomus cells are poisoned. Hyperventilation is not seen in anemic hypoxia so long as the arterial P_{O_2} remains normal. Hyperventilation is not seen in stagnant hypoxia either, because the blood supply to the carotid bodies is extremely high (2000 ml per 100 g of tissue, the highest for any tissue) and the glomus cells rarely if ever suffer from O_2 deficiency so long as the blood Po_2 is normal.

Cyanosis is a bluish discoloration of skin and mucous membrane. It is seen when the deoxyhemoglobin concentration in blood exceeds 5 g dl^{-1}. There are two types of cyanosis: peripheral and central. *Peripheral cyanosis* is seen in the nail beds and is suggestive of stagnant hypoxia. The distal parts of the body are very poorly perfused in hypotensive states and extract large amounts of O_2 from hemoglobin, raising the concentration of deoxyhemoglobin. *Central cyanosis* is seen in the mucous membranes of the tongue-tip and lips and in the earlobes where the skin is thin. These areas normally receive high blood supply and become cyanotic only if the O_2 saturation of blood is low, as in hypoxic hypoxia. Cyanosis does not occur in anemic hypoxia. In fact, it cannot occur in severe anemia if the hemoglobin concentration is 5 g dl^{-1} or less. Cyanosis does not occur in histotoxic hypoxia either, because the O_2 saturation of hemoglobin is normal.

Cerebral symptoms of hypoxia resemble those of alcohol toxicity with impaired judgment, drowsiness or excitement, dulled pain sensibility, disorientation, and headache. Other symptoms include nausea, vomiting, tachycardia, and hypertension.

Hypoxic hypoxia is the commonest form of hypoxia seen clinically. The causes of hypoxic hypoxia are (1) Reduced Po_2 in atmospheric air at high altitudes (high-altitude hypoxia); (2) Reduced ventilation due to restrictive or obstructive lung disorders; (3) Reduced diffusion capacity of O_2 across the respiratory membrane. This occurs when the respiratory membrane becomes thicker, as in interstitial lung disorders or when there is a marked reduction in the area of the respiratory membrane, as in emphysema. (4) Shunting of blood, bypassing the well-ventilated alveoli. This occurs due to alveolar ventilation-perfusion mismatch. It also occurs when there is a right-to-left shunt in the heart, i.e., deoxygenated blood from the right side of the heart bypasses the lung to enter the left ventricle. This is seen in congenital heart diseases like Fallot's tetralogy.

Anemic hypoxia occurs in severe anemia. Mild-to-moderate anemia usually does not produce hypoxia due to compensatory increase in red cell 2,3-DPG. However, such patients usually become hypoxic during exercise. Anemic hypoxia also occurs in carbon-monoxide poisoning and methemoglobinemia.

Carbon monoxide poisoning results in the formation of carboxyhemoglobin (COHb) which reduces the amount of Hb available for O_2 transport. CO has a very high affinity for hemoglobin and once bound, is difficult to dislodge from hemoglobin. Moreover, the COHb shifts the O_2 dissociation curve of the remaining Hb to the left, decreasing the amount of O_2 that can be released. Hence, inactivation of Hb by carbon monoxide poisoning is worse than having a reduced Hb concentration. Both hyperventilation and cyanosis are absent in CO poisoning. The cherry-red color of COHb that is visible through skin and mucous membranes can however be confused with cyanosis. Death results due to hypoxic brain damage. Management of CO poisoning includes termination of exposure to CO and initiation of O_2 therapy, with hyperbaric O_2 if necessary.

Stagnant hypoxia occurs during circulatory shock. The worst affected are the kidneys and heart which have high O_2 demand. Stagnant hypoxia also occurs in congestive heart failure, especially in the liver and brain which are worst affected by venous congestion.

Histotoxic hypoxia occurs when the tissue cells cannot make use of the O_2 supplied to them. It is caused, among other things, by cyanide poisoning, which inhibits cytochrome oxidase. The general treatment for histotoxic hypoxia is hyperbaric oxygen therapy. Specific treatment for cyanide poisoning includes nitrites or methylene blue which act by forming methemoglobin from hemoglobin. Methemoglobin detoxifies cyanide by converting it to cyanmethemoglobin. However, overenthusiastic treatment with nitrites can cause anemic hypoxia by forming too much methemoglobin and reducing the hemoglobin concentration.

■   Oxygen treatment

Oxygen-rich gas mixture is useful when the blood Po_2 is low, either due to reduced alveolar Po_2, or due to inadequate diffusion of O_2 across the respiratory membrane. Hence, it is administered in most types of hypoxic hypoxia and is commonly used in chronic obstructive pulmonary diseases. For the same reason, it is of little use in those types of hypoxia (stagnant, anemic, and histotoxic) in which there is no reduction in blood Po_2. Breathing O_2 is also useless in hypoxic hypoxia if it is due to ventilation-perfusion mismatch, i.e., the deoxygenated blood bypasses the well-ventilated alveoli.

Caution is required while giving O_2 therapy to patients with severe pulmonary failure and high Pco_2. The central chemoreceptors in these patients are inhibited by the high Pco_2 and hence, breathing is driven by hypoxic stimulation of peripheral chemoreceptors. O_2

therapy in such a situation may produce apnea by taking away this hypoxic drive.

Hyperbaric 100% O_2 raises the amount of O_2 dissolved in plasma and therefore, is unaffected by hemoglobin concentration. It is useful in the treatment of carbon monoxide poisoning, decompression sickness and air embolism, severe anemia, and wounds with poor blood supply. For therapeutic use, hyperbaric 100% O_2 should not be administered at pressures higher than 3 bars or for more than 5 hours.

■ Oxygen toxicity

One hundred percent O_2 has toxic effects due to the production of the superoxide anion (a free radical) and H_2O_2. When administered for 8 hours or more, it causes irritation of the airways. Administered chronically, infants may develop bronchopulmonary dysplasia (characterized by lung cysts and opacities) and retrolental fibroplasia, (formation of opaque vascular tissue in the eyes, which can lead to serious visual defects). Normal retinal vascularization is stimulated by mild hypoxia. Oxygen therapy takes away the hypoxic drive and the normal vascular pattern fails to develop. Vitamin E, an antioxidant, has been used for the treatment of retrolental fibroplasia.

Hyperbaric 100% O_2 produces, in addition to airway irritation, nervous symptoms like muscle twitching, tinnitus, convulsions, and coma. The greater the pressure, the faster is the onset of symptoms.

■ Hypercapnia and respiratory acidosis

Hypercapnia is retention of CO_2 in the body. Hypercapnia can occur as a corrective response to metabolic alkalosis but when hypercapnia is the primary problem, it is associated with respiratory acidosis, since any increase in CO_2 promptly generates excess H^+.

An increase in CO_2 production rarely produces hypercapnia because it is promptly washed out by the resulting hyperventilation. The P_{CO_2} does not rise even when there is considerable reduction in the diffusion capacity of the lung because CO_2 is lipid soluble and promptly diffuses out across the respiratory membrane. Hence, hypercapnia occurs essentially in two types of conditions: (1) Hypoventilation, as occurs in restrictive lung disorders and in respiratory depression due to drugs and cerebral diseases. (2) Ventilation-perfusion mismatch, which is commonly present in chronic obstructive pulmonary disease (COPD).

The principal signs of hypercapnia are hyperpnea and CO_2 narcosis. *Hyperpnea* occurs reflexly due to the stimulation of respiration by hypercapnia. Reflex hyperpnea may not be possible in a restrictive lung disorder. In such case, retention of large amounts of CO_2 produces *CO_2 narcosis* which is characterized by symptoms of CNS depression like confusion, diminished sensory acuity, respiratory depression, and coma.

■ Hypocapnia and respiratory alkalosis

Hypocapnia is a reduction of CO_2 in the body. Hypocapnia can also occur as a corrective response to metabolic acidosis but when hypocapnia is the primary problem, it is associated with respiratory alkalosis, since any decrease in CO_2 results in a decrease in H^+ concentration.

Hypocapnia occurs due to hyperventilation. It is seen in the following conditions. (1) Hypoxic conditions that are not associated with restrictive lung disorders, e.g., high altitude hypoxia or the early stages of interstitial lung disease. The latter does not affect diffusion of CO_2, which is highly lipid soluble, but impairs O_2 diffusion, producing hypoxia. The hypoxia stimulates hyperventilation with consequent CO_2 washout and hypocapnia. In restrictive lung disorders, however, the hyperventilation is not possible and therefore hypocapnia is uncommon. (2) Compulsive hyperventilation due to hysteria. (3) Excessive stimulation of the respiratory center due to fever, anxiety, cerebral tumors or in pregnancy. (4) Overenthusiastic artificial ventilation. (5) Excessive exercise.

The respiratory alkalosis associated with hypocapnia causes cerebral vasoconstriction, resulting in faintness. Alkalosis also lowers the ionized Ca^{2+} in plasma, resulting in tetany.

■ Asphyxia

Asphyxia is the simultaneous development of acute hypercapnia and hypoxia. It occurs due to airway obstruction as in choking or drowning. It is associated with violent respiratory efforts, acidosis, and increased catecholamine secretion, causing high blood pressure and heart rate, and predisposing the hypoxic myocardium to ventricular fibrillation. Eventually the respiratory efforts cease, the blood pressure falls and cardiac arrest occurs within 5 minutes.

Drowning Only 10% of deaths in drowning occur due to asphyxia. The asphyxia occurs initially due to breath holding and after breaking breath, due to the laryngospasm triggered by the cold water. The lungs remain dry in these deaths. In others, the lungs are flooded with water. Fresh water drowning causes plasma dilution and intravascular hemolysis. Drowning in hypertonic sea water results in hypovolemia. If rescued and resuscitated, these circulatory effects have to be reversed.

Respiratory Dysrhythmias

Blood gases have an important role in determining the breathing pattern. Given the conditions that predispose to oscillations in a control system (p 27), they can cause respiratory dysrhythmias, some of which are discussed below.

Cheyne–Stokes breathing

Cheyne–Stokes breathing is a periodic breathing characterized by a regular waxing and waning of ventilation at a constant frequency, punctuated by periods of apnea (Fig. 54.5A). The arterial Po_2 and Pco_2 fluctuate during each cycle of Cheyne–Stokes breathing. The Po_2 is lowest and the Pco_2 highest at the end of the apnea. The factors producing Cheyne–Stokes breathing are discussed below.

(1) Hypoxia from any cause predisposes to Cheyne–Stokes breathing if the hypoxic drive is greater than the hypercapneic drive. When driven by hypercapnia, breathing is controlled too promptly to allow any oscillation. (2) A reduction in the functional residual capacity (FRC) or total lung capacity (TLC) results in larger changes in alveolar Po_2 and Pco_2 with any change in breathing, predisposing to oscillations. (3) Cheyne–Stokes breathing occurs when the circulation time is prolonged, as in circulatory shock. (4) Cheyne–Stokes breathing may also appear if there is cerebrovascular disease and the sensitivity of central chemoreceptors to hypercapneic drive is impaired. Together with a prolonged circulation time, it is responsible for the frequent appearance of Cheyne–Stokes breathing in otherwise normal older subjects. Cheyne–Stokes breathing is common in sleep, particularly in stages 1 and 2 NREM sleep (p 692) and in subjects over the age of about 45.

Sedative drugs and any disorder that reduces the level of consciousness may also precipitate Cheyne–Stokes breathing. Premature infants whose cerebral functions are still immature may develop Cheyne–Stokes breathing.

Biot's breathing

Also called ataxic breathing, this is a pattern of slow and irregular breathing with impaired ventilatory response to both CO_2 and O_2 (Fig. 54.5B). It occurs due to the disruption of the normal medullary rhythmicity of breathing and is seen in a wide variety of medullary disorders.

Breathing during sleep

The control of the cerebral cortex over respiration is reduced in sleep and the ventilatory response to hypercapnia changes (Fig. 54.6). During stages 1 and 2 of NREM sleep, there may be periodic breathing but in

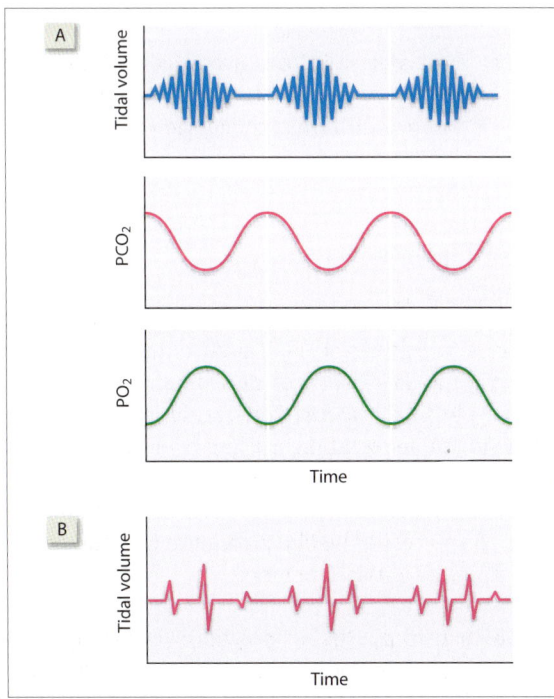

Fig. 54.**5**   [A] Cheyne–Stokes breathing. [B] Biot's breathing.

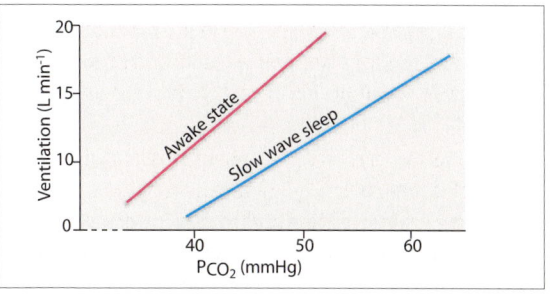

Fig. 54.**6**   Effect of arousal state on hypercapneic respiratory drive.

stages 3 and 4, the rate and rhythm become normal. In REM sleep, both rate and rhythm become erratic, especially when there are rapid eye movements. There may be phases of hyperventilation and apnea. These respiratory disturbances are attributable to reduced central inspiratory drive, reduced sensitivity of peripheral chemoreceptors, and reduced tone of respiratory muscles.

■   Post-hyperventilation periodic breathing

When a subject hyperventilates till there is a feeling of faintness, it is followed by a period of apnea. There is a slow recovery from apnea. However, after a few breaths, apnea follows again. This continues for a few cycles till breathing stabilizes at the normal level. The mechanism of the period breathing is explained as follows.

Hyperventilation causes apnea due to CO_2 washout. During the apneic phase, the P_{CO_2} rises slowly while P_{O_2} falls faster because the RQ is usually less than 1 (p 488). Feeble breathing resumes due to the hypoxic drive even before the P_{CO_2} has been completely restored. However,

> **Apneic because we ought to be!**
>
> After one measures alveolar air at rest, one measures it after hyperventilation. All of us did that, and with Douglas (C.G.Douglas) looking on, we observed the apnea after hyperventilation. The standard figure showing apnea and periodic breathing after hyperventilation, a figure reproduced in every textbook of physiology for many years, was made from results obtained from Douglas as a subject. When I became a teacher and hyperventilated dozens of times, or had students hyperventilate, I never observed apnea. Was Douglas apneic because he thought he ought to be?
>
> Horace W. Davenport

a few breaths again take away the hypoxic drive with resultant apnea. The cycle continues till the arterial P_{CO_2} rises to the normal level and provides sustained respiratory drive.

Although a theoretically sound proposition, whether such periodic breathing actually occurs following hyperventilation is doubtful (see box). Breathing, it should be remembered, is highly susceptible to cortical influences, to our beliefs and expectations.

55 High and Low Pressure Breathing

High-Pressure Breathing

High pressure breathing is commonly employed for under-water breathing and therefore the possible ways of staying under water are discussed below.

Breath-holding is one of the oldest methods of diving into deep waters. Following a deep inspiration, the breath can be held for 4–5 minutes: The duration is shorter if the breath is held at end-expiration. The breath-holding time can be prolonged by hyperventilation or breathing pure O_2 prior to breath-holding. Hypoxia and hypercapnia cannot be the sole factors limiting the breath-holding time because rhythmic respiratory movements against a closed glottis prolongs the breath-holding time by a few seconds beyond the normal breakpoint before breath-holding gives way. For the same reason, breathing a high CO_2-low O_2 gas mixture allows breathing for longer durations than the usual breath-holding time. Encouragement prolongs breath-holding time, implying a role for psychological factors.

Breathing through a snorkel   A snorkel is a breathing tube extending above the surface of the water, used in swimming just below the surface (Fig. 55.**1**). Although it is possible to immerse the head under water and breathe through a snorkel, such a tube represents an extension of the respiratory (anatomical) dead space and the tidal volume must be increased to ensure that adequate air enters the alveoli. Moreover, breathing becomes laborious due to the additional effort required to expand the chest against the pressure of the surrounding water.

Breathing air from a container   It is of course possible to carry under water a container containing air or oxygen. However, it will not be possible to breathe from the container if the air in it is at 1 atm pressure. The water pressure is very high at greater depths.[1] At such high pressure, the thorax gets compressed. The intrathoracic pressure rises and becomes nearly equal to the surrounding water pressure. Hence, a few millimeters of pressure reduction caused by inspiratory expansion of the thorax cannot suck in air from a container that contains air at 1 atm pressure. Hence, for breathing from a container, it is necessary to carry pressurized air or gas mixtures.

For breathing underwater, therefore, divers use the SCUBA equipment (self-contained underwater breathing apparatus) consisting of one or two compressed-air tanks strapped to the back and connected by a hose to a mouthpiece (Fig. 55.**1**). The SCUBA allows the diver to stay under water for long periods.

Problems of high-pressure breathing

In compressed air, say at 4 atm (3000 mmHg), all the gases have partial pressures that are 4 times the normal. Thus, the P_{O_2} in the compressed air will be about 600 mmHg while the P_{N_2} will be about 2400 mmHg. The corresponding figures in the alveoli and arteries will be somewhat less but still very high. The gases

Fig. 55.**1**   (*Above*) Snorkeling, a little below the surface of the water. (*Below*) A diver deep down, equipped with SCUBA.

---

[1] The pressure increases by 1 atmosphere for every 10 meters of depth in sea-water. Thus, at a depth of 30 meters, the pressure will be 4 atm. (3 atm for 30 m of water plus 1 atm).

Pneumothorax

In pneumothorax (Fig. 56.**6**), air enters the pleural space at atmospheric pressure. A primary spontaneous pneumothorax occurs in an otherwise healthy person without underlying lung disease. A secondary spontaneous pneumothorax is a complication of an underlying lung disease. Pneumothorax that occurs following injury to the chest is called traumatic pneumothorax.

Normally, the lung and pleura are held together by the pleural fluid which is inexpansible. When air enters the pleural space, the lung and the thorax are able to move apart because air is expansible. The chest therefore recoils to a larger size and the lung recoils to a smaller size. The pleural pressure becomes less negative. The total lung capacity and the vital capacity decrease, resulting in a restrictive disorder.

Normally, any pneumothorax gets slowly absorbed spontaneously because the partial pressure of gases in the pleural capillaries is about 60 mmHg less than atmospheric pressure (Table 56.**5**). The pressure difference that favors reabsorption of the gas from pneumothorax into the pleural capillary is maximum for O_2 (100 – 40 = 60), marginal for N_2 (573 – 569 = 4), and adverse for CO_2 (40 – 46 = –6). However, as O_2 is absorbed from the pneumothorax, the P_{N_2} and P_{CO_2} in the pneumothorax increase, resulting in a favorable gradient for the absorption of these gases into pleural capillaries. It has been estimated that roughly 5% of the pneumothorax gets absorbed every 24 hours.

In the case of *tension pneumothorax*, a valve-like tear in the visceral pleura allows air to enter the pleural cavity with each inspiration but prevents its exit during expiration. As the intrapleural pressure keeps rising, the mediastinum is pushed to the opposite side

Table 56.**5**   Partial pressures of gases (in mmHg) in pneumothorax (same as in alveolar air) and in pleural capillaries.

	In pneumo-thorax	In pleural capillary	Difference in partial pressures
P_{N_2}	573	569	4
P_{O_2}	100	40	60
P_{CO_2}	40	46	–6
Total	713	655	58

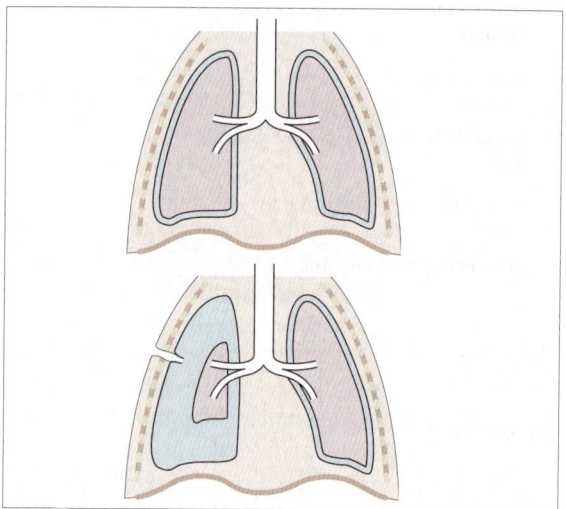

Fig. 56.**6**   Pneumothorax. Note that in comparison to the normal lungs and pleural space (*above*), the lung is collapsed and the pleural space is enlarged in pneumothorax (*below*)

and the lungs of both sides get squeezed, leading to restricted ventilation. The increase in intrapleural (intrathoracic) pressure results in a fall in cardiac output.

Clinical Diagnosis

Clinical diagnosis of respiratory disorders is based mainly on the following observations (Table 56.**6**). (1) Breathing characteristics, (2) shape of thorax, (3) shift of mediastinum (trachea and heart), (4) percussion note, (5) vocal resonance and fremitus, (6) breath sound intensity, (7) breath sound quality, and (8) adventitious sounds.

Breathing characteristics The normal respiratory rate is about 14 to 18 per minute. Higher breathing rate is called *tachypnea* if there is no change in the depth of breathing. An increase in the depth of breathing without a change in the breathing rate is called *hyperpnea*. Neither tachypnea nor hyperpnea is necessarily associated with *hyperventilation*, which refers to an increase

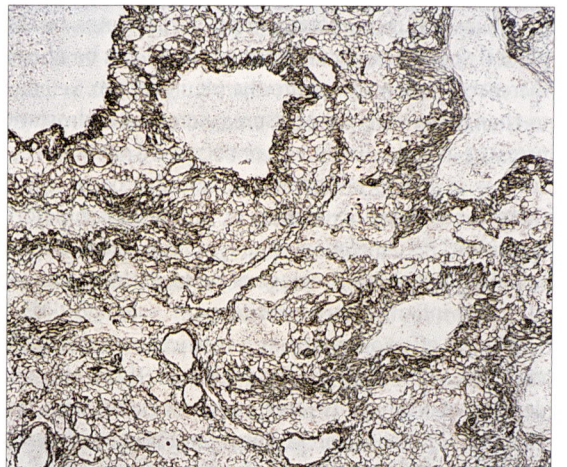

Fig. 56.**5**   Microscopic appearance of a section of the lung in an advanced stage of fibrosis.

Table 56.**6**   Clinical signs in common respiratory disorders.

	Emphysema	Pneumothorax	Atelectasis	Consolidation	Fibrosis	Pleural effusion
Breathing characteristics	Rapid and shallow	Rapid and shallow	Rapid and shallow	Rapid	Rapid and shallow	Rapid and shallow
Shape of thorax	Barrel-chest	Size increased ipsilaterally	Size decreased ipsilaterally	Normal	Normal	Size may be increased ipsilaterally
Tracheal shift	Pushed deeper	Pushed contralaterally	Pulled ipsilaterally	None	None	Pushed contralaterally
Percussion note	Hyper-resonant	Tympanitic	Dull	Dull	Dull ±	Stony dull
Vocal resonance/fremitus	Decreased	Decreased	Decreased	Increased	Normal or increased	Decreased
Breath sound intensity	Decreased	Decreased	Decreased	Normal	Normal	Decreased
Breath sound quality	Vesicular	Vesicular	Vesicular	Bronchial	Vesicular	Vesicular
Adventitious sounds	Wheeze ±	None	Crackles (when partial)	Crackles	Velcro crackles	Pleural rub

± May or may not be present

Table 56.**7**   Difference between vesicular and bronchial breath sounds.

Vesicular	Bronchial
Has a soft, low-pitched, 'rustling' quality.	Harsh and high-pitched. Has a hollow, 'tubular' quality.
The expiratory sound is softer than the inspiratory sound.	The expiratory sound is harsher than the inspiratory sound.
Duration of expiratory sound is twice that of inspiration.	Duration of inspiratory sound is equal to that of expiratory sound.
There is no gap between the inspiratory and expiratory sounds.	There is a definite gap between the inspiratory and expiratory sounds.
The inspiratory sound originates within the lungs while the expiration-phase sound comes partly from the larger airways.	It is generated in the larger airways.
Normally heard on the chest wall over the lungs.	Normally not transmitted to the chest wall. When present, signifies consolidation of lungs. Normally heard over trachea.

in the respiratory minute volume. Breathing movements are normally symmetrical, i.e., they are equal on both sides. Any asymmetry of breathing movements, especially when the subject is asked to breathe deeply, suggests presence of lung or pleural disease on the side moving less.

An excessive inspiratory effort, which occurs in obstructive airway diseases, is often apparent from the prominence of the contracting sternocleidomastoid muscles. Excessive inspiratory effort also increases the negativity of intrapleural pressure, resulting in the retraction of suprasternal and supraclavicular fossae during inspiration. Retraction of intercostal spaces may also be observed.

Shape of thorax   The normal chest has a transverse: anteroposterior diameter of 2:1. When this ratio decreases, the thorax looks barrel-shaped and indicates hyperinflation of the lung. Hyperinflation is typically seen in patients with COPD.

Mediastinal shift   In fibrosis or collapse of the lung, the mediastinum is pulled toward the affected side. In pleural effusion and pneumothorax, the mediastinum is pushed away from the affected side. The most obvious manifestation of mediastinal shift is the shift of trachea from its midline location in the neck, and the shift in the apex beat. Both can be detected through palpation.

Percussion note is the sound produced by tapping the chest with the fingertips. When the finger strikes the chest wall, the chest wall tends to vibrate as a resonant cavity that is partially damped by the thoracic contents. If the underlying lung alveolar air is replaced by fluid in the pleural cavity or by exudative fluid in the alveolar space as in pneumonia, the percussion sound becomes dull, i.e., it has a low amplitude and short duration. Conversely, if the underlying lung is replaced by air as in pneumothorax, the percussion sound is tympanic, i.e., of high amplitude and longer duration.

Vocal resonance is assessed by placing the stethoscope on the chest wall and asking the patient to vocalize. The voice sounds are conducted from the airways to the chest wall. Since solids conduct sound better, vocal resonance is heard better in lung consolidation and is reduced in pneumothorax, pleural effusion, and hyperinflated lungs. If the airway ventilating the auscultated lung region is obstructed, e.g., by a cancerous growth, the vocal resonance is reduced. The sound associated with vocalization may also be felt as the *vocal fremitus* by placing the ulnar border of the hand on the intercostal spaces while the patient is asked to vocalize.

Breath sounds Auscultation over the trachea and major bronchi reveals a hollow sound with 'tubular' quality. It is called *bronchial sound* and is generated in the upper airways between the nasal cavity and principal bronchi. Auscultation over normal lungs over the chest wall reveals a soft sound called the *vesicular sound*. It originates in the smaller airways. The inspiratory phase of this sound is longer than the expiratory phase (normal inspiration to expiration ratio is 2:1), the latter being nearly silent. If bronchial sound is heard over the lung fields, it indicates consolidation of lungs (Table 56.**7**).

Section 6

Renal System

57.   Functional Anatomy of the Kidney

58.   Glomerular Filtration and Tubular Reabsorption

59.   Renal Handling of Sodium

60.   Renal Regulation of Urine Volume and Osmolarity

61.   Body Fluid and Electrolyte Balance

62.   Renal Regulation of Acid-Base Balance

63.   Renal Regulation of Potassium Balance

64.   Renal Handling of Miscellaneous Substances

65.   Hormones Acting on the Kidney

66.   Quantification of Renal Functions

67.   Urine Analysis and Renal Function Tests

68.   Renal Syndromes

69.   Urinary Bladder and Cystometry

The main function of the kidney is homeostasis, i.e., the maintenance of the *milieu interior*. In doing so, the kidney has to deal effectively with the products of protein metabolism, water, and electrolytes. The kidney plays a relatively minor part in the homeostasis of certain other substances like Ca^{2+}, Mg^{2+}, and glucose which are regulated primarily through hormones.

Products of protein metabolism include urea, uric acid, SO_4^{2-}, PO_4^{3-}, creatinine, and several other substances that must be effectively eliminated from the body. Accumulation of these metabolites, mostly nitrogenous compounds, in blood is called *azotemia*. It occurs in kidney dysfunctions and is diagnosed when the serum BUN (blood urea nitrogen) level is elevated. Metabolism of carbohydrates and fats produces only water and CO_2 and therefore, does not load the kidney with metabolites. Hence in kidney dysfunctions, only protein restriction is advised.

Water, Na^+, and K^+ are dealt with according to whether there is a surplus or deficit in the body. Thereby, the kidney regulates water and electrolyte balance, and also the blood and ECF volume. The kidney also contributes to the regulation of blood pH by controlling the excretion of HCO_3^- ions, which are an important blood buffer.

Other functions of the kidney include: (1) the long-term and intermediate-term (through renin secretion) regulation of blood pressure, (2) regulation of erythropoiesis, (3) regulation of Ca^{2+} homeostasis and bone metabolism through the activation of 25-hydroxycholecalciferol to 1,25 dihydroxycholecalciferol, (4) synthesis of metabolic substrate like L-arginine, which is the precursor for nitric oxide (NO), and (5) under exceptional conditions, a significant contribution to the body's gluconeogenetic capability.

A longitudinal section of the kidney (Fig. 57.1) shows two distinct zones: the outer cortex and the inner medulla. The cortex contains most of the glomeruli (see below). The medulla comprises the renal pyramids, four to 14 in number, separated by the cortical columns of Bertin. One renal pyramid and its bounding renal columns constitute a renal lobule. The pyramid shows radial striations which are due to the straight portions of the nephrons. These striations extend some distance upward into the cortex where they are called the medullary rays. The apex of the pyramid is called

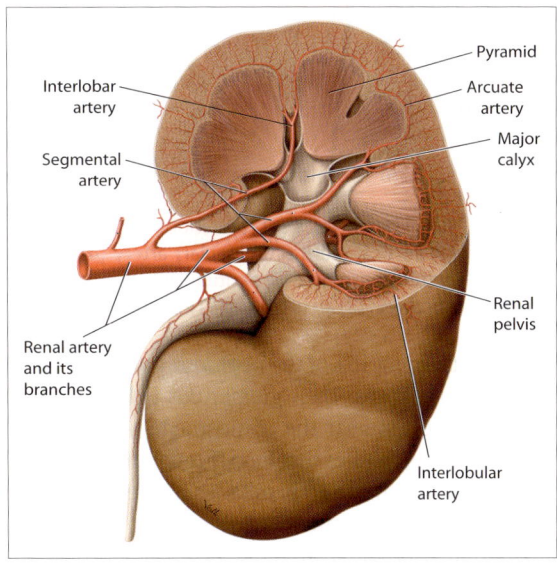

Fig. 57.1   A partial longitudinal section of the kidney, showing the renal pyramids and the arterial supply.

the papilla. The tip of the papilla has visible pores, which are the openings of the collecting ducts into the minor calyx.

The medulla can be subdivided into an outer medulla that is further subdivided into the outer and inner strips, and an inner medulla, also called the papillary zone. The papillae end in the minor calyces. The minor calyces join together to form major calyces which in turn drain into the pelvis. The pelvis narrows down to continue as the ureter which drains into the bladder.

Each kidney is made of about a million nephrons. Each nephron consists of the Bowman's capsule and the uriniferous tubule, a part of which forms a loop called the loop of Henle. Eighty-five percent of the nephrons are located superficially in the cortex and are called *cortical nephrons*. Nephrons filter plasma into the tubule and then reabsorb the plasma solutes in required amounts so that any excess solute present in the blood is eliminated in the urine. The copious reabsorption from these tubules is made possible by a dense peritubular capillary plexus. The cortical tubules do not loop down into the inner medulla.

acellular. It contains hydrated channels approximately 6 nm wide. These channels account for the selectivity of the glomerular membrane. The basement membrane contains negatively charged proteoglycans like chondroitin sulfate proteoglycans and heparin sulfate proteoglycans (HSPG). HSPG is particularly important in imparting selectivity to the glomerular basement membrane. Normally, polyanions like HSPG act as 'anti-clogging' agents that prevent the absorption of plasma proteins so that the pores are not choked. (3) The *Bowman's visceral epithelium* is made of podocytes that are separated by slits approximately 25 nm wide (Fig. 57.**4**). Cells called the *mesangial cells* are present between the capillary endothelial cells and the basement membrane, especially where the basement membrane encloses more than one capillary.

■ Uriniferous tubule

The uriniferous tubule is subdivided into the proximal tubule, the intermediate tubule, the distal tubule and the collecting system (Fig. 57.**5**).

The **proximal tubule** is the longest part of the nephron. It is further subdivided into (1) the proximal convoluted tubule (PCT) located in the cortex and (2) the proximal straight tubule (PST) located in the medullary rays and the outer stripe of the medulla.

The **thin segment** (intermediate tubule) is divided into (1) the descending thin segment (DTS) which traverses the inner stripe and extends deep into the inner medulla, and (2) the ascending thin segment (ATS). In juxtamedullary nephrons, the DTS loops around as the ATS to reach the junction of the outer and inner medulla. In cortical nephrons, the DTS is continuous at the bend of the loop with the distal tubule and therefore, there is no ATS.

The **distal tubule** is divided into the straight and convoluted parts. (1) The distal straight tubule (DST), more commonly known as the thick ascending limb (TAL), extends across the outer medulla. It is subdivided into the medullary thick ascending limb (MTAL) and the cortical thick ascending limb (CTAL). (2) The distal convoluted tubule (DCT) lies in the cortex. It is connected by a short connecting tubule (CNT) to the collecting duct (CD). The pars recta of the proximal tubule, the thin segment, and the thick ascending limb together constitute the loop of Henle.

The **collecting duct** is not, embryologically, a part of the nephron as it is derived from the ureteric bud. It consists of the connecting tubule and collecting duct. (1) The connecting tubule (CNT) lies entirely in the cortex. In this segment, several tubules coalesce to form the collecting duct. (2) The collecting duct (CD) runs through the cortex, medulla, and the papilla, finally opening at the tip of the papilla. Accordingly, it is subdivided into the cortical collecting duct (CCD), the outer medullary collecting duct (OMCD), and the inner medullary collecting duct (IMCD) or the papillary collecting duct. In the IMCD region, several collecting ducts coalesce before finally opening at the tip of the renal papilla.

■ Juxtaglomerular Apparatus

The juxtaglomerular apparatus is located at the angle of the afferent and efferent arterioles where it comes in contact with the CTAL, i.e., the cortical part of the thick ascending limb. It comprises the macula densa, the juxtaglomerular (JG) cells, and the lacis cells (Fig. 57.**3**).

The part of the distal tubule which comes in contact with the afferent arteriole is made of a specialized epithelium called the *macula densa*. The cells of the macula densa have blunt processes at its base, extending toward the juxtaglomerular cells in the afferent arteriole. The Golgi complex is usually located between the nucleus and the cell base. In other tubular cells, the Golgi complex is located near the apical membrane. These structural characteristics suggest the presence of some secretory activity at the base of the cell.

The *juxtaglomerular (granular) cells* are modified smooth muscle cells in the scala media of the terminal part of the afferent arterioles. They contain large

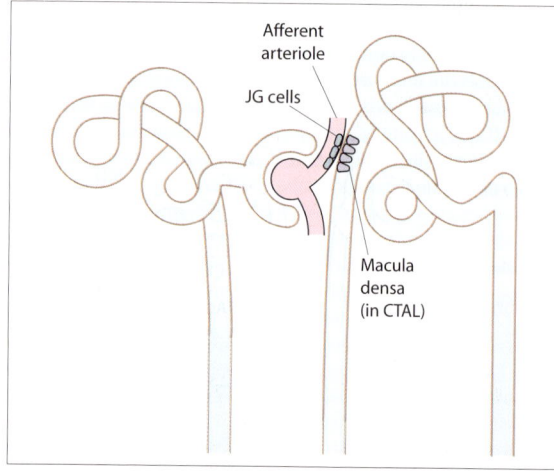

Fig. 57.**6**   Location of the JG apparatus.

granules that secrete renin. In conditions requiring increased renin secretion, additional smooth muscle cells located in the wall of afferent arteriole and even cortical radial artery are transformed into granular cells. Granular cells are densely innervated by sympathetic nerve terminals and they release their renin content in response to sympathetic discharge.

The *lacis cells* (extraglomerular mesangial cells) are derived from smooth muscle cells. They are present in the angular space between the glomerulus and the diverging afferent and efferent arterioles. Structurally, lacis cells may be important as a 'plugging' device at the glomerular entrance which protects the glomerulus against distending forces exerted by the high intra-arteriolar pressure. Functionally, lacis cells possibly relay the signals from the macula densa to granular cells after modulating the signals.

Tubular Cells

The cells of the nephron are mostly cuboidal except in the thin segment where they are flat and squamous. The cells rest on a basement membrane. The apical surfaces of all the cells bear a few microvilli but in the proximal tubule, these microvilli are numerous and dense, giving the apex the appearance of a brush border. The basal cell membrane of all the cells shows fairly extensive invaginations or infoldings except in the thin segment. These infoldings create narrow 'gutters' of extracellular spaces called the *basal space*. The lateral surfaces of the cells bear the lateral cell processes that interdigitate with the corresponding processes of adjacent cells. In between the interdigitations there are small spaces called the *lateral intercellular spaces*. Lateral spaces communicate with each other but they do not freely communicate with the basal extracellular space (Fig. 57.**7**).

The apical surfaces of all cells interdigitate with neighboring cells forming 'tight junctions' or zona occludens. Functionally however, they are not necessarily tight. Hence, they are of two types. The *'leaky' tight junctions* permit water and solutes to diffuse across them. These are present in the proximal tubule. The *'tight' tight junctions* do not permit water and solutes to diffuse across them easily. These are present in the distal tubule.

The different types of tubular cells present in a nephron segment are: (1) the proximal tubular cell, (2) the intermediate tubular cell, (3) the distal tubular cell, and (4) the collecting duct cells which are of two types, the principal cells and the intercalated cells (Fig. 57.**8**).

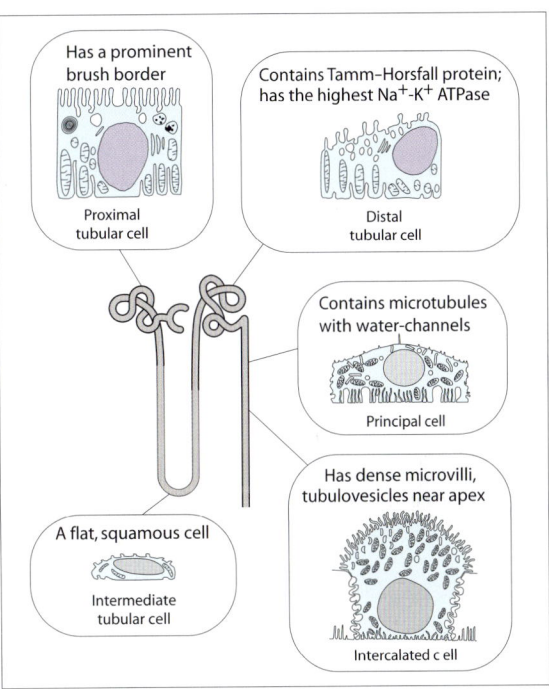

Fig. 57.**8**   The different types of tubular cells present in a nephron segment.

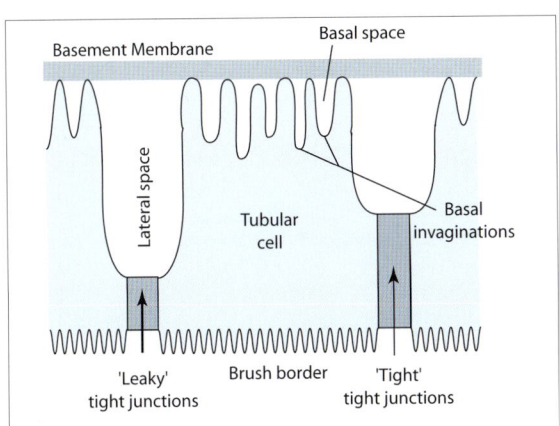

Fig. 57.**7**   A typical renal tubular cell.

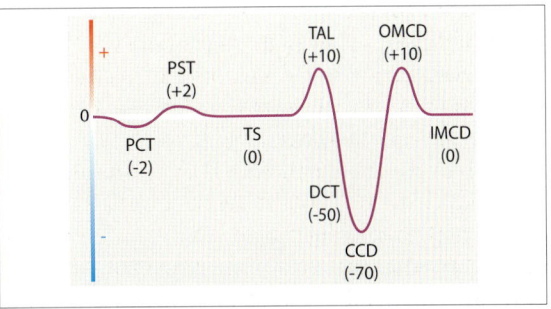

Fig. 57.**9**   Graphic representation (not to scale) of the transepithelial potential difference (TEPD) in different segments of the nephron. The highly negative TEPD in the CCD has several important physiological consequences. The figures are in mV.

Tubular Potentials

An electrical potential difference is recordable across the tubular wall, i.e. between the lumen of the tubule and its exterior. This potential difference, which is called the *transepithelial potential difference* (TEPD), varies in different segments of the tubule (Fig. 57.**9**). The potential in the proximal tubule is very slight (−2 mV in the convoluted part, +2 mV in the straight part). In the thick ascending limb, it is lumen positive (+6 to +10 mV). In the distal tubule, it is highly lumen negative (−70 mV). Tubular potentials, like most other membrane potentials, are produced due to unequal diffusion of anions and cations across the tubular wall. Some of these potentials are important in the tubular transport of certain ions.

In the glomerular capillaries, fluid is filtered out from blood into the uriniferous tubule. The filtered fluid, called the glomerular filtrate, is mostly protein-free. Only low molecular weight proteins whose size is smaller than that of albumin are present in the filtrate. The electrolyte composition of the filtrate is identical to that of the plasma. Most of the glomerular filtrate gets reabsorbed by the peritubular capillaries. Both glomerular filtration and tubular reabsorption of fluids are governed by Starling forces of capillary exchange. While the solutes in the plasma are freely filtered into the tubules, their reabsorption is active and selective.

Glomerular Filtration

The Starling forces that operate at the glomerular capillary level (Fig. 58.**1**) are the same as in systemic capillaries but are different in magnitude. The salient differences are as follows. (1) There is no significant drop in hydrostatic pressure along the glomerular capillaries despite the filtering out of fluids. (2) There is considerable rise in plasma oncotic pressure along the glomerular capillaries. (3) The oncotic pressure of the glomerular filtrate is nearly zero. (4) The capsular hydrostatic pressure is higher than that in the tissue interstitial fluid. (5) The Starling forces equilibrate toward the efferent arteriolar end of the glomerular capillaries. The outcome of the Starling forces is that fluid is filtered out from the arteriolar side of the glomerular capillaries under a net outward pressure of 10 mmHg while there is no reabsorptive or filtrative force at the efferent arteriolar end.

Glomerular filtration of solutes

The concentration ratio of a substance in the Bowman's space and plasma is called its glomerular sieving coefficient. Molecules less than 4 nm in diameter are freely filtered while molecules larger than 8 nm (molecular weight greater than 70 000 Daltons) are excluded from the glomerular filtrate. The glomerular filtration barrier for solutes resides at the glomerular basement membrane which contains negatively charged proteoglycans. Hence, negatively charged molecules

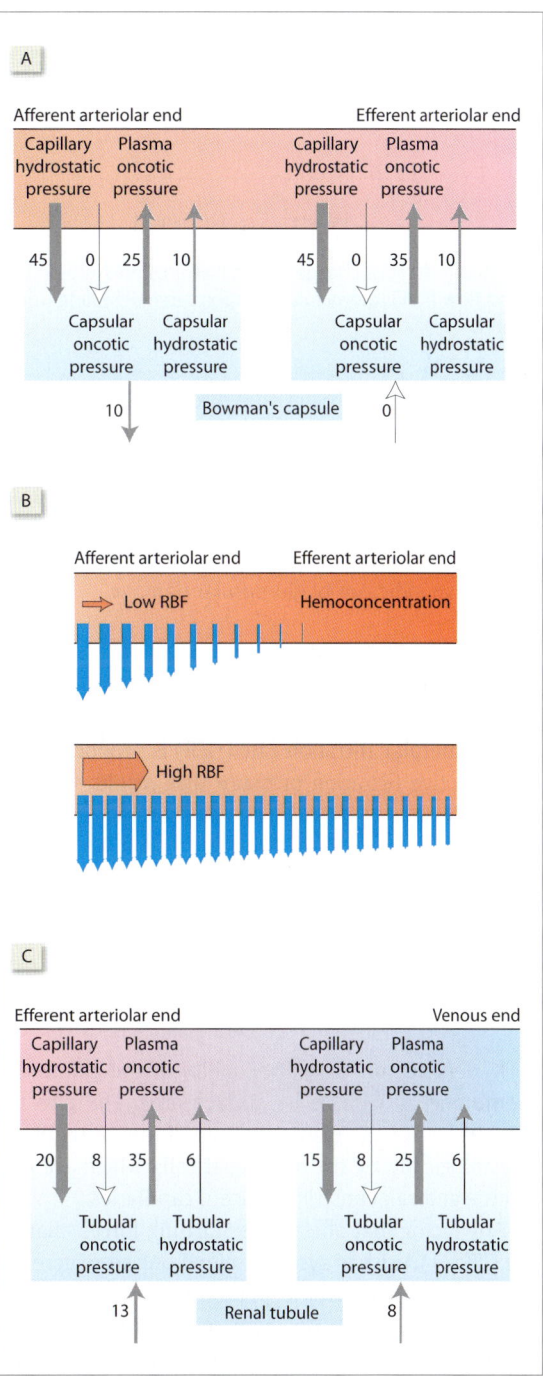

Fig. 58.**1** [A] Starling forces (in mmHg) in the glomerular capillary. [B] Effect of renal blood flow on glomerular filtration. [C] Starling forces (in mmHg) in the peritubular capillary.

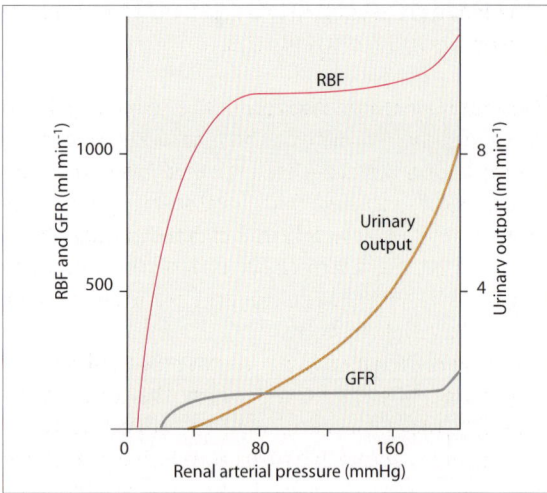

Fig. 58.2   Effect of changes in renal arterial pressure on renal blood flow (RBF), glomerular filtration rate (GFR), and urinary output.

have greater difficulty passing through it. This explains why albumin, which is about 7 nm (69 000 Daltons) is largely excluded from the filtrate. Loss of the negative charge on the basement membrane even without any structural damage to the membrane is enough to produce albuminuria. It explains why albuminuria is such a sensitive indicator of glomerular damage.

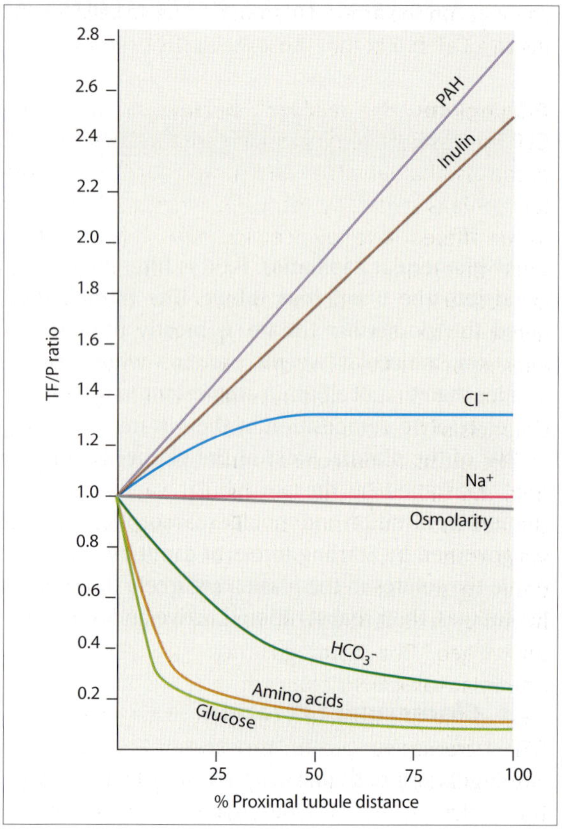

Fig. 58.3   The TF/P ratio for various solutes as a function of the distance it travels in the proximal tubule.

Glomerular filtration of fluids

The glomerular filtration rate (GFR) is proportional to the effective filtration pressure that results from the Starling forces.

$$GFR \propto [(P_{glomerulus} + \pi_{Bowman}) - (P_{Bowman} + \pi_{glomerulus})]$$

$$GFR = K_f \times [(P_{glomerulus} + \pi_{Bowman}) - (P_{Bowman} + \pi_{glomerulus})]$$

where K_f is the *filtration coefficient.*

K_f depends on the permeability of the glomerular membrane as well as its surface area. The effective surface area of the glomerular membrane is regulated physiologically by the mesangial cells which are contractile and can constrict adjacent capillaries.

Any factor that affects the Startling forces changes the GFR. (1) A rise in systemic blood pressure increases the capillary hydrostatic pressure and therefore increases the GFR. (2) An increase in renal blood flow reduces the rise in colloid osmotic pressure along the glomerular capillary (Fig. 58.1C) and therefore increases the GFR. (3) The hydrostatic pressure in the Bowman's capsule increases in ureteric obstruction or renal

edema, resulting in a fall in the GFR. (4) The plasma oncotic pressure increases in dehydration and decreases in hypoproteinemia. Hence GFR decreases in dehydration and increases in hypoproteinemia. (5) GFR is also reduced by renal diseases that reduce the permeability and effective surface area of the glomerular basement membrane. Sympathetic discharge reduces glomerular filtration, the reasons for which are discussed below.

Autoregulation of glomerular filtration

The GFR is normally well autoregulated in the range of 70 to 180 mmHg of systemic pressure (Fig. 58.2). However, this autoregulation can be overridden by renal nerves. There are two plausible hypotheses for explaining the autoregulation of GFR.

The **myogenic hypothesis** of autoregulation suggests that the afferent arterioles constrict in response to augmented blood pressure. Arteriolar constriction restores GFR to normal levels. Possibly, the stretching of arterioles leads to the opening of stretch-sensitive Ca^{2+}

channels on the arteriolar smooth muscle cells, resulting in a Ca^{2+} influx that causes the cells to contract.

Tubuloglomerular feedback hypothesis When the GFR increases, it results in increased delivery of NaCl to the distal tubule. The concentration of Cl^- in the distal tubule is sensed by the macula densa and signaled to the afferent arteriole (see Fig. 57.**6**). Infusion of Na^+ salts other than NaCl does not produce tubuloglomerular feedback. The signal is transmitted from the macula densa to the afferent arteriole probably by some adenosine or eicosanoid compound that causes vasoconstriction of the afferent arteriole by opening up Ca^{2+} channels of the smooth muscles.

The ultimate purpose of renal autoregulation is to hold the GFR constant. This usually necessitates that the RBF too is held constant. That, however, is not always true. Often, the RBF increases or decreases in order to maintain a constant GFR. For example, in hypoproteinemia the GFR increases but RBF remains unchanged. The autoregulatory mechanisms cause arteriolar vasoconstriction which restores GFR but decreases the RBF. Similarly, in ureteric obstruction the GFR decreases but the RBF remains unchanged. Due to autoregulatory mechanisms, there is arteriolar dilatation which restores the GFR but increases the RBF.

When renal perfusion pressure increases, the GFR does not change much due to autoregulation. Yet, the urinary output increases dramatically. Since the kidney is enclosed in a tough capsule that is not easily stretched, any increase in the renal perfusion pressure also increases the renal interstitial hydrostatic pressure. An increase in interstitial hydrostatic pressure decreases the reabsorption of tubular fluids and increases the urinary output.

Tubular Transport

Tubular reabsorption of water

Of the 180 L of fluid filtered into the glomerulus each day, all but 1.5 L is reabsorbed from the tubules. The Starling reabsorptive forces at the peritubular capillaries account only for a small part of this reabsorption, providing only 8 to 13 mmHg of net reabsorptive force (Fig. 58.**1**B). The bulk of the water reabsorption occurs secondary to the active Na^+ reabsorption in the tubules.

Obligatory water reabsorption Eighty-five percent of the water reabsorption occurs irrespective of the body water balance and is called is *obligatory* (must occur). About 65% of the obligatory reabsorption occurs in the

proximal tubules and 20% of the obligatory reabsorption occurs in the distal tubules.

Facultative water reabsorption The remaining 15% of the water may or may not be reabsorbed depending on the body water balance. It is called *facultative reabsorption* (optional). The facultative reabsorption occurs from the collecting tubule as it courses through the renal medulla. It is under the control of the antidiuretic hormone (ADH) which controls the permeability of the collecting tubule to water.

The renal medulla has a very high osmolarity and would normally extract water from the collecting tubules. In the presence of ADH the collecting tubule epithelium is permeable to water, which is reabsorbed in large amounts. However, in the absence of ADH, the collecting tubule is impermeable to water and therefore no water is reabsorbed.

Tubular handling of solutes

Proximal tubular handling of solutes In the proximal tubule, different solutes are handled differently. Most solutes are reabsorbed while some are secreted. About 60% of the tubular load of Na^+, Ca^{2+}, K^+, PO_4^{3-}, Cl^-, and urea are reabsorbed. HCO_3^- reabsorption is more than 60% while glucose and amino acids are reabsorbed nearly completely. SO_4^{2-} is poorly reabsorbed. Reabsorption of almost all solutes is linked directly or indirectly to the active reabsorption of Na^+. Small proteins and peptides are also reabsorbed in the proximal tubules through endocytosis. Substances like creatinine and inulin (exogenous) are not reabsorbed at all.

The amount of water reabsorption can be estimated from the extent to which inulin or creatinine are concentrated in the tubule (Fig. 58.**3**). These solutes, which are not reabsorbed at all, are concentrated 2.5 fold in the proximal tubule, indicating that about 60%[1] of the water load is reabsorbed in the proximal tubules. Sodium ions too are 60% reabsorbed in the proximal tubule, indicating that the reabsorption of Na^+ is isoosmotic, i.e., following the reabsorption of Na^+, its tubular concentration does not fall. Hence, the ratio of Na^+ concentration in the tubular fluid (TF) and plasma (P) remains 1.0 throughout the proximal tubule. Not all solutes are reabsorbed isoosmotically. Glucose, amino acids, and bicarbonates are reabsorbed relatively more than water. The TF/P ratio for these substances drops sharply

---

[1] The percentage of water reabsorption can be calculated by the formula: 100 − (100 ÷ C), where C is the concentration factor. For example, when C = 2.0, the percentage reabsorption is 50, which means that when half the water is reabsorbed, the solute concentration will double.

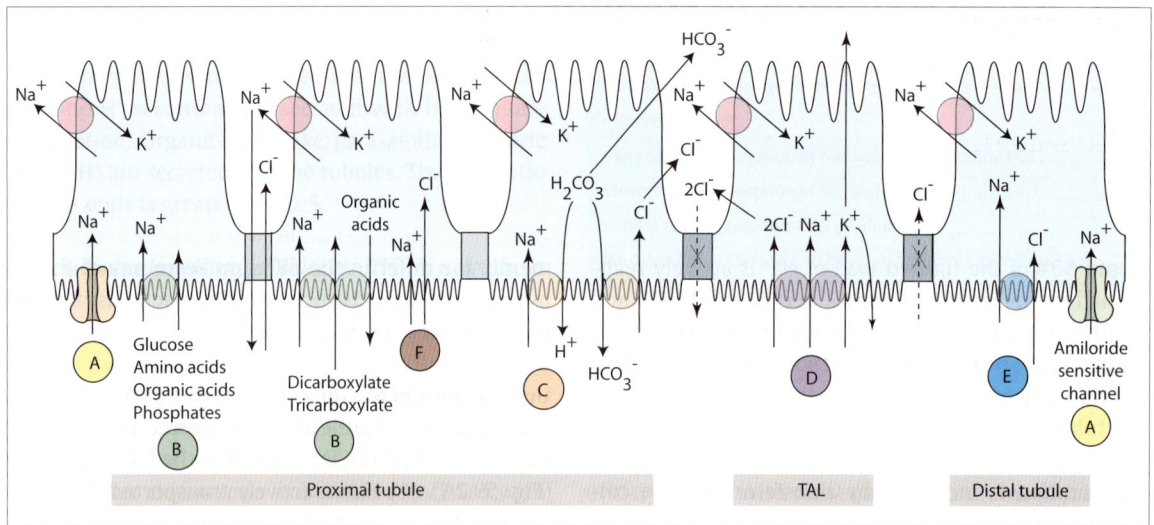

Fig. 59.2 Different mechanisms (A to F) of tubular reabsorption of sodium ions. (A) Unitransport of Na⁺, (B) Na⁺ cotransport with non-Cl⁻, non-H⁺ substrates, (C) Na⁺-H⁺ exchange, usually with a parallel Cl⁻-HCO₃⁻ uniport, (D) Na⁺/K⁺-2Cl⁻ cotransport, (E) Na⁺-Cl⁻ cotransport, and (F) Chloride-driven Na⁺ transport.

Cl⁻ antiport). Cl⁻ diffuses out across the basolateral surface to restore electroneutrality.

Coupled extrusion of H⁺ is limited by the intracellular availability of H⁺. Thus it is increased by a high Pco_2 and decreased by diamox (carbonic anhydrase inhibitor), both of which affect H⁺ production. In alkalosis, H⁺ secretion decreases and the Na⁺-H⁺ antiport is reduced. Hence, Na⁺ reabsorption decreases in alkalosis.

Conversely, since HCO₃⁻ reabsorption in the proximal tubule is linked to Na⁺ reabsorption, bicarbonaturia occurs whenever Na⁺ reabsorption decreases in the proximal tubule, resulting in slight acidosis. Hence, diuretics (including water diuresis) result in not only increased Na⁺ excretion but also slight bicarbonaturia and acidosis.

Sodium cotransport with organic substrates occurs only in the PCT because glucose, amino acids, and organic anions are reabsorbed completely in the PCT. Both glucose and amino acids are cotransported with Na⁺ in the proximal tubule (Fig. 59.**2**B).

The organic anion that is cotransported with Na⁺ is a dicarboxylate or a tricarboxylate organic anion. Other organic ions are coupled to the transport of di/tricarboxylates and thereby, are indirectly coupled to Na⁺ transport. For example, para-amino hippuric acid (PAH) is transported by a PAH-anion antiport working together with a Na⁺-di/tricarboxylate symporter.

Chloride-driven sodium transport occurs in the PST alone and its mechanism is as follows (Fig. 59.**3**). In the glomerular filtrate, the Na⁺ concentration is about 110 mEq L⁻¹ and the HCO₃⁻ concentration is about 24 mEq L⁻¹. The Cl⁻ : HCO₃⁻ concentration in the glomerular filtrate is therefore about 4.5:1. In the PCT, Cl⁻, and HCO₃⁻ are reabsorbed in a ratio of 3:1 because for every four Na⁺ ions, three Cl⁻ and one HCO₃⁻ are reabsorbed. The Cl⁻:HCO₃⁻ concentration in the peritubular fluid is also similar. Therefore, by the time the filtrate reaches the PST, the chloride concentration in the tubule is higher than in the peritubular capillary. Cl⁻ therefore diffuses passively into the peritubular capillaries. The passive diffusion of Cl⁻ across the brush border sets up an electrical gradient along which Na⁺ diffuses in. Cl⁻ ions, which lead the diffusion, tend to produce a lumen-positive TEPD. Na⁺ ions, which follow Cl⁻, nearly nullify the TEPD so that the TEPD in the PST is only +2 mV.

Sodium cotransport with chloride and potassium ions is confined to the TAL only (Fig. 59.**2**D). (1) Na⁺ is actively extruded across the basolateral surface. (2) Na⁺ passively diffuses inside with coupled cotransport of two Cl⁻ ions and one K⁺. Such a cotransport is termed Na⁺/K⁺-2Cl⁻. (3) Two Cl⁻ and one K⁺ ion diffuse out across the basolateral surface to restore electroneutrality. (4) Due to the presence of 'tight' tight junctions, Na⁺ is unable to leak back into the tubule to produce a luminal potential. (5) Some of the K⁺ which enters the cell leaks back across the apical membrane into the tubular lumen, generating a lumen-positive transepithelial potential difference of +6 to +10 mV.

A group of diuretics called the *loop diuretics* inhibit the Na⁺/K⁺-2Cl⁻. Loop diuretics include *frusemide* and *ethacrynic acid*. The Na⁺ reabsorbed from this segment

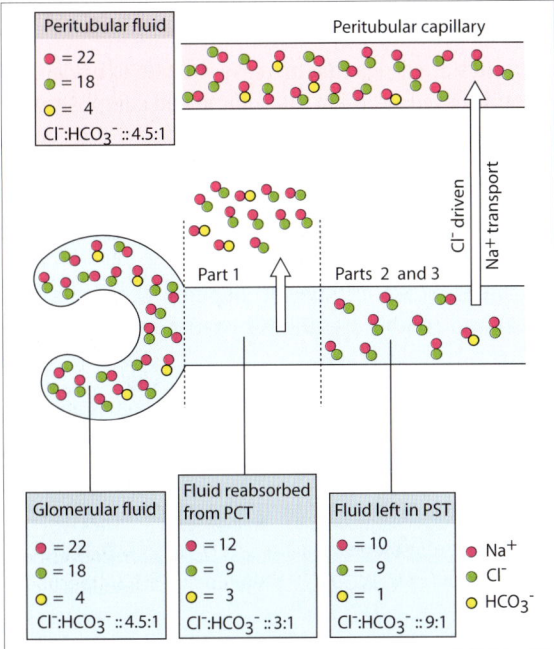

Peritubular fluid
- ● = 22
- ● = 18
- ○ = 4
- Cl⁻:HCO₃⁻ :: 4.5:1

Peritubular capillary

Part 1   Parts 2 and 3

Cl⁻ driven   Na⁺ transport

Glomerular fluid	Fluid reabsorbed from PCT	Fluid left in PST	
● = 22	● = 12	● = 10	● Na⁺
● = 18	● = 9	● = 9	● Cl⁻
○ = 4	○ = 3	○ = 1	○ HCO₃⁻
Cl⁻:HCO₃⁻ :: 4.5:1	Cl⁻:HCO₃⁻ :: 3:1	Cl⁻:HCO₃⁻ :: 9:1	

Fig. 59.**3**   Chloride-driven sodium transport.

is the main driving force behind the countercurrent multiplier system which concentrates Na⁺ and urea in the medullary interstitium. Hence, loop diuretics diminish the medullary hyperosmolarity that is essential for urine concentration. Hence they act as very effective diuretics and are also called high-ceiling diuretics. ADH stimulates Na⁺/K⁺-2Cl⁻ and thereby enhances the urine concentrating ability of the kidneys.

Sodium-chloride cotransport occurs in the DCT alone. It involves the following steps (Fig. 59.**2F**). (1) Na⁺ is actively extruded across basolateral surface of the tubular cell. (2) Na⁺ diffuses across brush border through Na⁺-Cl⁻ cotransport. (3) The chloride diffuses out into the basolateral spaces.

Na⁺-Cl⁻ cotransport is inhibited by a group of compounds called the benzothiazides. Thiazides therefore produce natriuresis with diuresis.

▪ Factors affecting Na⁺ reabsorption

There are three primary factors controlling sodium reabsorption, viz., the apical sodium transport, the activity of the basolateral Na⁺-K⁺ ATPase, and the Starling forces. It might appear unusual that although Na⁺ reabsorption is an active process, it is affected by the passive Starling reabsorptive forces. The reason for it is that the movement of Na⁺ from the basolateral spaces into the peritubular capillary is essentially passive.

The movement of water from the basolateral spaces into the peritubular capillaries 'drags' with it Na⁺ (due to bulk flow). Hence, factors that increase water reabsorption into the peritubular capillaries (e.g., decreased hydrostatic pressure and increased oncotic pressure) also enhance Na⁺ reabsorption. Conversely, the ECF volume expansion that occurs following ingestion of large amounts of water reduces water reabsorption from the proximal tubules due to alteration in the Starling forces. Concomitantly, there is also a reduction in Na⁺ reabsorption from the proximal tubules.

Glomerulotubular balance Na⁺ reabsorption in *proximal tubule* is load-dependent, i.e., when the amount of Na⁺ filtered increases, the reabsorption of Na⁺ increases proportionately. This is known as glomerulotubular balance and it occurs because the tubular Na⁺ reabsorption is flow-limited. Because of glomerulotubular balance, the urinary Na⁺ output does not increase massively when the GFR increases. The following example will illustrate the point.

Example
When the GFR is 125 ml min⁻¹, the amount of Na⁺ entering the tubule each minute is about 20 mOsmoles, of which 19.25 mOsm get reabsorbed and 0.75 mOsmoles get excreted in urine. The Na⁺ reabsorption is associated with 124 ml of water reabsorption and the urinary output is 1 ml min⁻¹, i.e., 1.44 L per day. Now, suppose the GFR increases by 20%, to 150 ml min⁻¹. If the Na⁺ and water reabsorption remains unchanged, the urinary output increases to 26 ml min⁻¹, i.e., 37 L per day! However, due to glomerulotubular balance, the Na⁺ reabsorption increases proportionately to 23.1 mOsm, and the water reabsorption increases to 148.8 ml. Thus both urinary Na⁺ excretion and urinary output increase only by 20%, to 0.9 mOsm and 1.2 ml min⁻¹ respectively.

Na⁺ reabsorption in *distal nephron* too is load-dependent and therefore is proportional to the Na⁺ load. Thus, when proximal tubular Na⁺ reabsorption decreases, distal tubular Na⁺ load increases, and the Na⁺ reabsorption increases to compensate. Hence, Na⁺ delivery to the distal tubules is an important controller of distal tubular Na⁺ reabsorption.

▪ Transepithelial potential difference

Transepithelial potential difference (TEPD) is the electrical potential recorded between the tubular lumen and its exterior. It is recordable in all segments of the nephron that are involved in active Na⁺ transport. *Lumen negative TEPD* is produced by active Na⁺ reabsorption that is unaccompanied by transport of equivalent

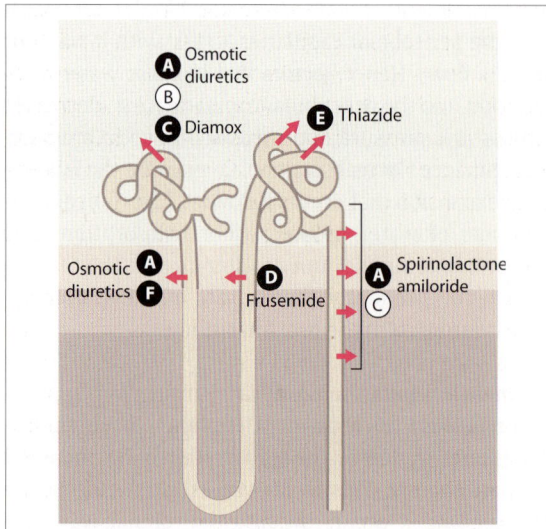

Fig. 59.**4** Sodium-chloride cotransport. Related A to F to Fig. 59.**2**.

Diuretics

Diuretics are drugs that increase the rate of urine flow. They are used to adjust the volume and composition of body fluids in conditions like hypertension and edema. The diuresis produced is almost always secondary to natriuresis (increased Na$^+$ losses in urine). Most diuretics have undesirable side effects like hypokalemia and pH disturbances (Table 59.1).

Diuretics act on specific nephron segments (Fig. 59.**4**) and inhibit specific Na$^+$ transport mechanisms

Table 59.1 The mechanism of action of diuretics and their side effects. The numbers 1 to 5 indicate the site of diuretic action. 1 = PCT, 2 = PST, 3 = TAL, 4 = DCT, 5 = CD.

Diuretic	Segment and mechanism	K$^+$ depletion	Acidosis/alkalosis
Osmotic diuretics	1-(A), 2-(A), 2-(F)	Yes	Acidosis
CA inhibitors	1-(C)	Yes	Acidosis
Loop diuretics	3-(D)	Yes	Alkalosis
Thiazides	4-(E)	Yes	Alkalosis
Aldosterone antagonists	5-(A)	No*	Acidosis
Na$^+$ channel inhibitors	5-(A)	No*	Acidosis

* These are called potassium-sparing diuretics.

amounts of anions. The high TEPD in the CCD is due to Na$^+$ unitransport which is not associated with any paracellular Cl$^-$ reabsorption. *Lumen-positive TEPD* is produced in the PST (by the Cl$^-$ reabsorption which exceeds Na$^+$ reabsorption), in the TAL (by the back diffusion of K$^+$ into the tubule), and in the OMCD (by active, electrogenic secretion of H$^+$ into the tubule).

The high-negative TEPD in the CCD is of considerable physiological importance. As already mentioned, an increased Na$^+$ delivery to the distal nephron increases Na$^+$ unitransport and thereby, increases the TEPD. (1) Factors that increase distal Na$^+$ delivery and thereby increase the TEPD include *dehydration* and *diuretics* acting on the proximal tubule. (2) *Aldosterone* increases the activity of the Na$^+$-K$^+$ pump in the distal nephron, thereby promoting Na$^+$ unitransport and with it, the TEPD. (3) *Impermeable anions* like SO$_4^{2-}$ and NO$_3^-$, which are not reabsorbed with Na$^+$, increase the TEPD when present in large amounts in the tubular fluid. (4) The secretion of K$^+$ into the lumen decreases the lumen-negative TEPD. Hence, *K$^+$ depletion*, which is associated with reduced tubular secretion of K$^+$, increases TEPD. (5) The secretion of the positively charged H$^+$ into the lumen decreases the lumen-negative TEPD. *Alkalosis* makes the TEPD more negative because H$^+$ secretion decreases in alkalosis. H$^+$ and K$^+$ secretion are reciprocally related since both are facilitated by the TEPD and on secretion, both tend to neutralize the potential difference. For example, when H$^+$ secretion increases, the H$^+$ ions tend to neutralize the luminal negativity. The reduction in TEPD reduces K$^+$ secretion. In the same way, when K$^+$ secretion is high, the H$^+$ secretion decreases.

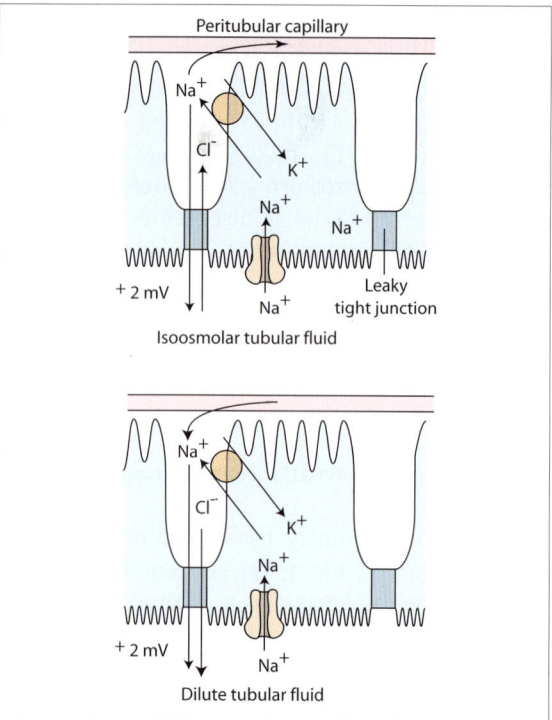

Fig. 59.**5** Mechanism of action of osmotic diuretics.

(Table 59.**1**). Diuretics acting on the proximal tubule have limited efficacy since the TAL, which has a great reabsorptive capacity, compensates for any decrease in Na^+ and H_2O reabsorption that might occur in the proximal tubule. Diuretics acting on sites distal to the TAL also have limited efficacy because only a small part of the filtered solute load and fluid reaches the distal tubule. Diuretics acting on the TAL are called loop diuretics or high-ceiling diuretics. They are the most efficacious of all diuretics. They abolish the urine concentrating ability of the nephron.

Most of the diuretics (amiloride, spirinolactone, thiazide, and frusemide) have been discussed above with Na^+ reabsorption. Only osmotic diuretics are discussed here.

Osmotic diuretics   Commonly used osmotic diuretics are glycerin, mannitol, and urea. Being nonreabsorbable, they hold water in the tubule. They are most effective in the proximal tubule where the maximum amount of water is reabsorbed. The water retained in the tubule dilutes the tubular concentration of Na^+ and other electrolytes, thereby reducing their reabsorption too. Hence, osmotic diuretics increase the urinary excretion of nearly all electrolytes including Na^+, K^+ (kaliuresis), Cl^-, HCO_3^- (resulting in acidosis), Ca^{2+}, Mg^{2+}, and PO_4^{3-}.

As already explained above, the movement of Na^+ from the tubules into the tubular cells and from the lateral spaces to the peritubular capillaries occurs through passive diffusion. Dilution of the tubular fluid not only slows down the movement of Na^+ but also tends to reverse their direction: Na^+ from the lateral spaces diffuses back into the tubules and Na^+ from the peritubular capillaries diffuses back into the lateral spaces (Fig. 59.**5**).

60 Renal Regulation of Urine Volume and Osmolarity

Osmolarity Changes in Tubule

In the proximal tubules, solutes and water are reabsorbed in isoosmolar proportions. Hence, the tubular fluid remains isoosmotic to body fluids (290 mOsm L^{-1}, often rounded off to 300 mOsm L^{-1} for convenience) till the end of the proximal tubule (Fig. 60.**1**).

As the thin segment descends into the deeper parts of the renal medulla, the tubular fluid becomes progressively hyperosmolar (up to 1200 mOsm L^{-1}). This happens because the interstitium of the renal medulla is extremely hyperosmolar. As a result, water moves out of the tubular fluid into the hyperosmolar medulla. As the thin segment ascends back from the hyperosmotic medulla, the tubular fluid again becomes nearly isoosmolar. The thin ascending limb is impermeable to water. Hence, the osmolarity changes occur due to the diffusion of Na$^+$ from the tubular fluid into the interstitium of the outer medulla and not by the diffusion of water in the reverse direction. The entire sequence of osmolar changes in the thin segment (first becoming hyperosmolar and then isoosmolar again) might appear purposeless. On the contrary, the sequence of osmolar changes is a part of a larger mechanism called the countercurrent multiplier: It is the countercurrent multiplier that makes the medulla hyperosmotic.

In the thick ascending limb (TAL), the urine becomes hypoosmolar because TAL cells actively pump out Na$^+$ from the tubular fluid into the interstitium. Water cannot follow since the TAL is impermeable to water.

The permeability of the collecting duct to water is variable. In the presence of ADH, it is highly permeable to water. In the absence of ADH, it is impermeable to water. Accordingly, as the collecting tubule descends through the hyperosmolar medulla, the osmolarity of the tubular fluid can change in one of the two following ways. *In the absence of ADH*, the osmolarity of the tubular fluid remains unchanged. The urine formed is hypoosmolar. *In the presence of ADH*, the osmolarity of the tubular fluid tends to equal that of the inner medulla. The urine formed is therefore hyperosmolar. It is in the collecting tubule that the urine becomes hyperosmolar a second time. (It becomes hyperosmolar for the first time in the thin segment). The maximum possible osmolarity of the tubular fluid is 1200 mOsm L^{-1}, i.e., equal to the osmolarity in the innermost medulla. Thus, while urine invariably gets diluted in the distal tubule, it may or may not get reconcentrated in the collecting tubule.

Renal medullary hyperosmolarity

The interstitial fluid in the renal cortex has the same osmolarity as that of the plasma, i.e., 300 mOsm L^{-1}. In the renal medulla however, it is much higher, and more so in the inner medulla where it is 1200 mOsm L^{-1}. This hyperosmolarity is generated by a mechanism called the countercurrent multiplier system that operates in the loop of Henle.

The solutes concentrated in the medulla diffuse into the blood that flows through the medulla in the vasa recta. The flowing blood therefore carries away extra solutes with it and reduces the medullary osmolarity. This dissipation of medullary hyperosmolarity is minimized by the countercurrent exchanger system that operates between the ascending and descending limbs of the vasa recta.

The main driving force behind the countercurrent multiplier system is called the *single effect*, i.e., the

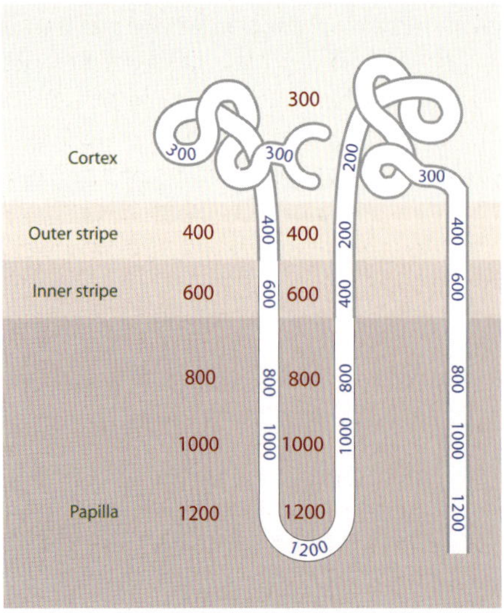

Fig. 60.**1** Osmolarity of the filtrate at different sites in the tubule.

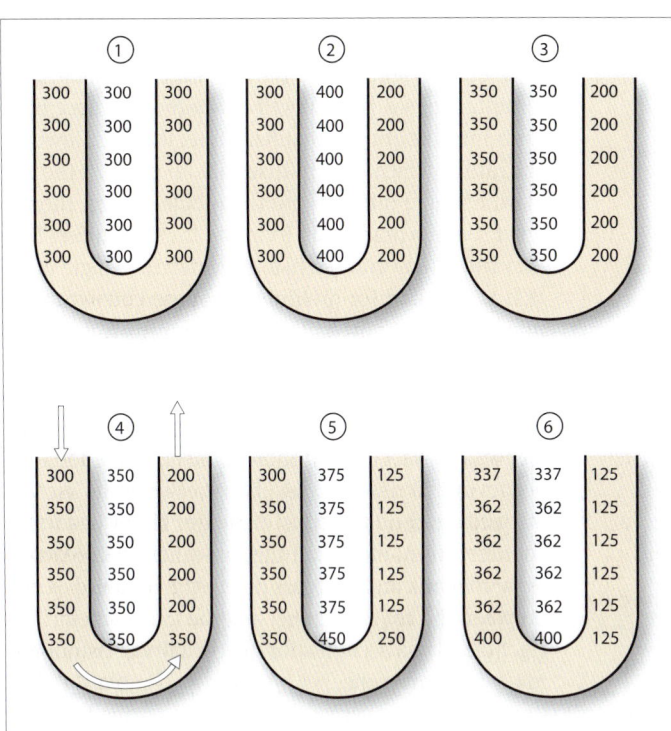

Fig. 60.**2** Step-by-step changes in the osmolarity of tubular fluid due to countercurrent multiplier effect.

osmotic gradient of approximately 200 mOsm L^{-1} that exists between the tubular fluid in the ascending limb of the loop of Henle and the adjacent interstitium. The mechanisms of the single effect in the outer medulla and the inner medulla are different. *In the outer medulla*, the single effect is produced by the active reabsorption of Na$^+$ from the TAL. *In the inner medulla*, the thin segment cannot reabsorb Na$^+$ actively. Hence, the mechanism of the single effect remains hypothetical.

■ Countercurrent multiplier in the tubule

The countercurrent multiplier system can be appreciated through a step-by-step approach (Fig. 60.**2**). (1) Initially, the tubular fluid and the renal interstitium have uniform osmolarity of 300 mOsm L^{-1}. (2) The TAL pumps out Na$^+$ ions (along with Cl$^-$ and HCO$_3^-$) into the adjacent areas of medullary interstitium.

Since the TAL is impermeable to water, the osmolarity of the medullary interstitium rises and that of the tubular fluid in the TAL decreases. The difference in osmolarity between the interstitium and the tubular fluid is around 200 mOsm L^{-1}, which is limited by the power of the Na$^+$-K$^+$ pump. Establishment of this osmotic gradient is called the *single effect*. (3) The portion of the DTS that is located in the outer medulla is moderately permeable to Na$^+$. The Na$^+$ in the adjacent interstitium therefore diffuses into the tubular fluid in the DTS, which equilibrates with the hyperosmolar medullary interstitium around it. (4) Fresh isoosmolar filtrate (300 mOsm L^{-1}) flows down into the descending loop and pushes some of the hyperosmolar fluid into the ascending limb round the bend. (5) As the cycle of above steps are repeated several times over, the tubular fluid and the medullary interstitium in the deeper regions of the medulla become more and more hyperosmolar.

Fig. 60.**3** Two pairs of workers are building two mounds A and B. To the left, the workers are dumping the earth close to the site where they are digging. As a result, mound-A soon becomes very steep and the workers have to stop work as they are unable to negotiate the incline. To the right, the workers are dumping the earth at a distant site. As a result, mound-B does not become steep even as it becomes much higher than mound-A. The workers can continue to dig more earth and make the mound taller!

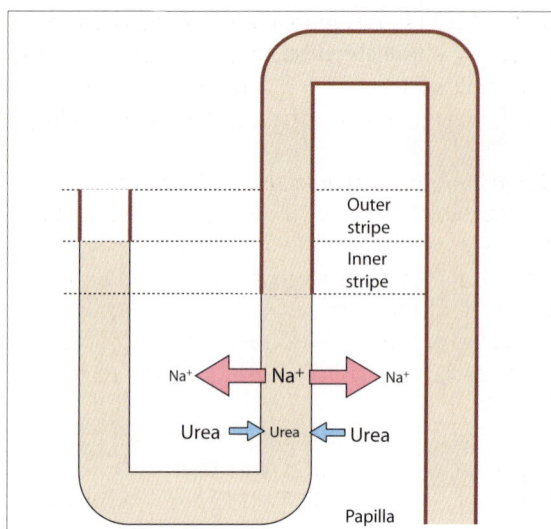

Fig. 60.4 Passive countercurrent multiplier in inner medulla due to urea recycling.

There is a simpler way of understanding it. Consider that Na⁺ is continuously flowing in from the glomerulus into the tubule. However, most of the Na⁺, instead of flowing out, gets trapped in a circular path and recycles between the ascending and descending limbs of the loop of Henle. Consequently, the Na⁺ concentration becomes very high at the tip of the loop.

The above approach to understanding the countercurrent system cannot explain why a longer loop of Henle produces a higher medullary osmolarity, as observed in desert rats. An alternative approach explains the importance of the length of the loop of Henle. Consider the analogy illustrated in Figure 60.3 which shows that it is possible to build a tall mound only if the earth dug out is dumped at a distance. In the same way, if all the NaCl pumped out at the TAL accumulated locally, the Na⁺-K⁺ pump would stop after creating a concentration gradient of 200 mOsm. However, since the countercurrent system removes the NaCl to a distant site, the Na⁺-K⁺ pump can continue to pump while the NaCl concentration at the tip of the loop of Henle continues to rise.

Single effect in the inner medulla   Unlike the TAL, the inner medullary ATS does not have the ability to actively reabsorb NaCl. The single effect in the inner medulla is therefore produced passively (Fig. 60.4). In the inner medulla, the osmolarity of the tubular fluid and the interstitium are initially the same. Although their osmolarities are the same, the composition of the tubular fluid is different from the composition of the interstitium. The tubular fluid in the thin ascending limb is rich in NaCl but has less urea in it.

The adjacent interstitium is rich in urea but has less NaCl in it. Consequently, urea moves into the tubular fluid. Simultaneously, NaCl moves out of the thin ascending limb into the inner medullary interstitium. However, the rate at which NaCl diffuses out far exceeds the rate at which urea diffuses into the tubule. Hence, there is a *net solute reabsorption* from the thin ascending limb into the interstitium, which provides the single effect that is required for driving the countercurrent multiplier in the inner medulla. This hypothesis of the single effect in the inner medulla is called the *passive equilibration model*.

The countercurrent multiplier discussed above increases the Na⁺ concentration in the renal medulla and occurs due to the *recycling of Na⁺* between the ascending and descending loops of Henle. There is another countercurrent multiplier that concentrates urea in the renal medulla and it occurs due to the *recycling of urea* between the collecting duct and the loop of Henle (see Fig. 64.2). Urea accounts for nearly half the osmolarity in the inner medulla.

■   **Countercurrent exchanger in the vasa recta**

Countercurrent exchange is essentially a physical principle that has several laboratory and industrial applications. In the human body too, exchangers are found in the scrotum, skin, and intestinal villi. The basic principle of countercurrent exchange in the renal medulla is as follows.

If blood flowed through the hyperosmolar medulla as shown in Figure 60.5A, it would equilibrate with medullary interstitium and carry away the solutes concentrated in the medulla. However, what actually happens is shown in Figure 60.5B. As the vasa recta dips into the hyperosmolar medulla, the blood equilibrates with the surrounding interstitium and becomes hyperosmolar. Thereafter, as the vasa recta loops around and ascends toward the cortex, the osmolarity of the blood keeps decreasing as it equilibrates with the surrounding interstitium. By the time the vasa recta leaves the medulla, the blood in it is only slightly more hyperosmolar than when it entered the medulla. In other words, the solutes concentrated in the medulla are not washed away by the blood flowing through the medulla.

Since the blood flowing through the medulla equilibrates completely with the medullary interstitium, the amount of solutes carried away by the blood is flow-limited. Hence, the slow rate of blood flow through the vasa recta contributes to the conservation of medullary hyperosmolarity. Conversely, a high blood flow rate through the medulla reduces medullary hyperosmolarity.

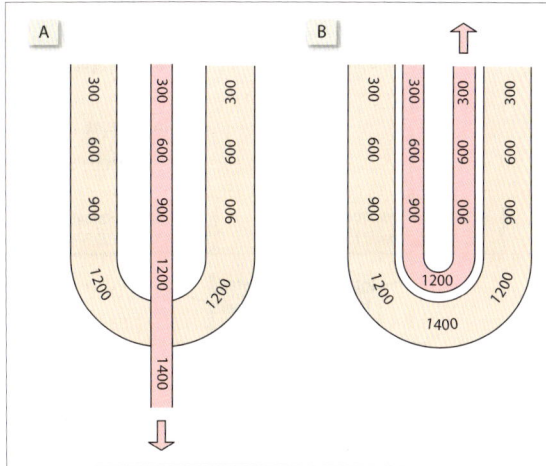

Fig. 60.**5**   Countercurrent exchanger system in the vasa recta. [A] A blood vessel flowing past the loop of Henle. [B] A blood vessel looping with the loop of Henle

■ Factors affecting medullary hyperosmolarity

Urea concentration in urine   Urea reabsorption from the IMCD is dependent on a urea-transport protein. Since the urea reabsorbed from IMCD is a major contributor to medullary hyperosmolarity, a genetic lack of urea-transport protein impairs urine-concentrating ability. Factors that reduce plasma urea concentration, e.g., a low protein diet, also affect the urine concentrating ability of the kidneys. Conversely, a high protein diet increases the concentrating ability of the kidney.

ECF volume expansion increases the RBF and GFR and suppresses ADH secretion. Each of these factors contributes to the reduction of medullary hyperosmolarity with consequent impairment of the urine concentrating ability of the kidney.

An *increase in RBF* increases the blood flow through vasa recta and washes out solutes from the renal medulla.

An *increase in GFR* increases the tubular flow rate. When tubular flow is high, the active transport of Na^+ is not able to sufficiently dilute the tubular fluid in the TAL, and the single effect is reduced in magnitude. Similarly, the reabsorption of water from the collecting tubule is not able to sufficiently concentrate urea in the IMCD if the tubular flow is high. Thus, the medullary hyperosmolarity decreases.

In the *absence of ADH*, the CCD, OMCD, and the initial part of the IMCD become impermeable to water. Hence, urea does not get concentrated in the collecting duct. As a result, urea reabsorption from the terminal part of collecting duct is markedly reduced (to ~20%) and urea concentration in the papillary interstitium is lowered.

while phosphoric acid is formed in the metabolism of phospholipids, nucleic acids, phosphoproteins, and phosphoglycerides. Organic acids like lactic acid, acetoacetic acid, and β-hydroxy butyric acid are normally oxidized to CO_2 and water but they appear in blood following incomplete combustion of carbohydrates and fats. Uric acid is formed through the metabolism of nucleoproteins.

The body pH is also affected by consumption of acids (vinegar, curd) and alkalis (slaked lime) or potentially acidic and alkaline substances in the diet like lysine (acidic) and citrates (alkaline). Acids are lost from the body in large amounts in vomiting, causing alkalosis. In diarrhea, alkalis are lost from the body, causing acidosis.

■ Blood buffers

The major blood buffers are hemoglobin, plasma proteins, bicarbonates, and phosphates.

Hemoglobin is the largest contributor to the buffering capacity of blood. The NH_2 groups on intermediate histidine residues of the hemoglobin provide most of the buffering capacity.

Plasma proteins are the next largest contributors to the buffering capacity. They also owe their buffering action to their NH_2 residues.

Bicarbonates are the most important buffers in blood. This is because the components of this buffer system can be adjusted by the body. The HCO_3^- concentration is controlled by the kidney while the Pco_2 is controlled through pulmonary ventilation.

Phosphates are the main intracellular buffer in the body. In blood, their concentration is low and therefore, they are not very important as blood buffers.

■ Metabolic causes of pH disturbances

The common causes of metabolic acidosis can be grouped into five types (Table 62.1). The rise in H^+ concentration has to be accompanied by a rise in an anion, which can be either Cl^- or organic anions like lactate, acetoacetate etc. Accordingly, metabolic acidosis is classified into the hyperchloremic variety (types III, IV, and IV) and the high-anion gap variety (types I and II).

The common causes of metabolic alkalosis are given in Table 62.2. The milk–alkali syndrome occurs when patients (e.g., with ulcer) consume, over a prolonged

Table 62.1 Causes of metabolic acidosis.

I.	Increased acid production	Ketoacidosis, lactic acidosis, poisoning with methanol or salicylates (aspirin)
II.	Decreased filtration of acids into tubules	Renal failure
III.	Decreased H^+ secretion by tubules	Renal tubular acidosis
IV.	Excess alkali loss from body	Diarrhea
V.	Increased acid consumption	Intake of NH_4Cl, lysine

Table 62.2 Causes of metabolic alkalosis.

I.	Loss of acids, NaCl	Vomiting, diuretic therapy
II.	Increased tubular H^+ secretion	Hyperaldosteronism, Cushing's syndrome, severe K^+ depletion
III.	Increased intake of alkali	Milk–alkali syndrome

period of time, large amounts of milk and antacids (which are alkaline) for soothing their pain. Besides alkalosis, these patients also suffer from hypercalcemia and deposition of $Ca_3(PO_4)_2$ in the kidney.

■ Effects of pH disturbances

Ventilation Acidosis produces *Kussmaul breathing*, a form of hyperventilation associated with increased tidal volume and a lesser increase in frequency. Alkalosis depresses ventilation. These effects are mediated largely through the central chemoreceptors.

Oxygen delivery Acute and chronic acidosis have opposite effects on oxygen delivery. Acute acidosis shifts the oxygen dissociation curve to the right (the Bohr effect). Chronic acidosis inhibits 2,3-DPG synthesis which in turn shifts the oxygen dissociation curve to the left. Alkalosis has the opposite effects.

Cardiac contractility Acidosis depresses cardiac contractility. However, acidosis also stimulates the release of catecholamines which increase cardiac contractility and partly nullify the direct effect. Acidosis results in enhanced cardiac response to vagal stimulation. Hence, vagal reflexes are accentuated.

Vascular tone Acidosis causes vasodilatation of peripheral vessels due to the direct effect on smooth muscles. Acidosis also causes vasoconstriction of larger vessels. The vasoconstriction is mediated by the catecholamine release produced by acidosis.

Serum electrolytes   Serum potassium levels rise in acidosis, but only in the hyperchloremic type of metabolic acidosis. In metabolic alkalosis, there is hypokalemia. Acidosis also increases the plasma Ca^{2+} levels. Conversely, alkalosis depresses plasma Ca^{2+}, producing symptoms of neuronal hyperexcitability.

Nervous system   In acidosis, CNS functions are depressed, resulting in lethargy, stupor, and even coma. In alkalosis, there is increased neuronal excitability due to the decrease in ionized Ca^{2+}. The hyperexcitability is manifested peripherally as muscle spasms and tetany. Centrally, the hyperexcitability results in mental confusion, parasthesias, and seizures.

◼ Renal correction of acid–base balance

Renal correction of acidosis *Metabolic acidosis* occurs due to an excess of H^+ ions which react with HCO_3^- to form CO_2 and water. Consequently, the HCO_3^- decreases (HCO_3^- is consumed) and Pco_2 increases (CO_2 is generated) in plasma.

$$H^+ + HCO_3^- = H_2O + CO_2$$

Respiratory acidosis occurs due to a high Pco_2. Consequently there is production of excess H^+ (which results in the acidosis) and excess HCO_3^- in the plasma.

$$H_2O + CO_2 = H^+ + HCO_3^-$$

It may be noted, therefore, that the Pco_2 rises in both metabolic acidosis (high Pco_2 is the cause of acidosis) as well as in respiratory acidosis (high Pco_2 is the result of acidosis). Regardless of whether it is the cause or the effect of acidosis, a rise in Pco_2 results in increased tubular secretion of H^+. The H^+ secreted binds to tubular HCO_3^-, resulting in increased HCO_3^- reclamation. In metabolic acidosis, the plasma HCO_3^- is low and hence, the filtered load of HCO_3^- in the tubules too is also low. Hence, there is not much tubular HCO_3^- available to be reclaimed. However, acidosis is also associated with increased renal ammoniagenesis, which increases the amount of ammonia available in the tubule for buffering the augmented H^+ secretion in acidosis. Hence there is new HCO_3^- generation. As a result, plasma HCO_3^- increases and the pH rises back to normal. Without increased renal ammoniagenesis, correction of metabolic acidosis is not possible.

Renal correction of alkalosis *Metabolic alkalosis* occurs due to an excess of hydroxyl ions, which react with CO_2 to produce HCO_3^-. Consequently, the plasma

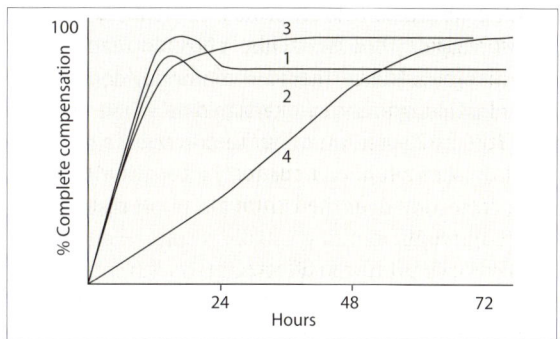

Fig. 62.**3**   (1) Respiratory compensation to acidosis. (2) Respiratory compensation to alkalosis. (3) Metabolic compensation to alkalosis. (4) Metabolic compensation to acidosis.

HCO_3^- increases (HCO_3^- is generated) and the Pco_2 falls (CO_2 is consumed).

$$OH^- + CO_2 = HCO_3^-$$

Respiratory alkalosis occurs due to hyperventilation (CO_2 washout) which reduces the Pco_2. Consequently, the above reaction is driven in the reverse direction in accordance with principles of chemical equilibrium. Therefore, the plasma HCO_3^- decreases (HCO_3^- decomposes) and the pH rises (OH^- is generated).

It may be noted, therefore, that the Pco_2 falls both in metabolic alkalosis (low Pco_2 is the result of alkalosis) as well as in respiratory alkalosis (low Pco_2 is the cause of acidosis). Regardless of whether it is the cause or the effect of alkalosis, a fall in Pco_2 decreases the tubular secretion of H^+ and the amount of HCO_3^- reclaimed. The reduced tubular secretion of H^+ decreases the blood pH. If the plasma HCO_3^- is high (as in metabolic alkalosis), large amounts of it are filtered into the tubules and go unabsorbed due to reduced reclamation. As a result, plasma HCO_3^- decreases and the pH falls further, which contributes to the restoration of normal pH.

Respiratory versus renal compensation   Respiratory compensation to acidosis or alkalosis occurs quickly. Metabolic compensation to alkalosis is slower. Metabolic compensation to acidosis is much slower still. Respiratory compensation is nearly complete in acute cases but is not sustained in chronic cases. Metabolic compensation is nearly complete in chronic cases, but since it is quite slow, it remains somewhat incomplete in acute cases (Fig. 62.**3**).

Conflicting compensatory responses   Metabolic alkalosis would normally be corrected through a reduced reabsorption (and therefore, a greater loss) of HCO_3^- in the filtrate. However, if there is associated volume contraction, there is a compensatory increase in Na⁺ reabsorption and with it, HCO₃⁻ reabsorption also increases. Hence, there are conflicting demands on the renal compensatory mechanisms when alkalosis and volume depletion are to be corrected simultaneously. In such situations, the body gives priority to volume correction over the correction of pH disturbance. Thus, although vomiting usually results in alkalosis, a persistent vomiting that is associated with dehydration results in paradoxical acidemia. Recalling the response to conflicting demands of volume and tonicity (p 396), the complete axiom would be 'volume overrides tonicity, which overrides acid–base balance'.

◼ Anion gap

The **plasma anion gap** is the total concentration of anions, excluding HCO₃⁻ and Cl⁻, that are present in the plasma. It therefore represents the total plasma concentration of anions like albumin, phosphates, sulfates, and other organic acids. Since the molar concentrations of plasma anions and cations must be equal, the plasma anion gap can be calculated by subtracting the concentrations of HCO₃⁻ and Cl⁻ from the total concentration of cations. Since Na⁺ is the predominant cation in plasma (concentrations of K⁺, Ca²⁺, and Mg²⁺ being negligible in comparison), the formula for anion gap can be written as:

Plasma anion gap = plasma Na⁺ − (plasma HCO₃⁻ + plasma Cl⁻)

The normal value of the anion gap is 12 ± 2 mEq L⁻¹. Plasma albumin accounts for most of the anion gap. The estimation of the plasma anion gap provides a quick laboratory method of distinguishing types I and II of metabolic acidosis (high anion gap variety) from types III, IV, and V of metabolic acidosis (hyperchloremic variety). In type I metabolic acidosis, the anion gap increases due the accumulation in plasma of anions of acetoacetic acid, lactic acid or other organic acids. In type II metabolic acidosis (renal failure), the anion gap increases because sulfates, phosphates, and organic acid anions are not excreted efficiently.

The **urine anion gap** provides an indirect way of estimating the urinary NH₄⁺. It is given by the formula:

Urine anion gap = urinary Na⁺ + urinary K⁺ − urinary Cl⁻

It assumes that the major cations in urine are Na⁺, K⁺, and NH₄⁺ while the major anion is only Cl⁻ since urinary HCO₃⁻ is zero at pH more than 6.5. Normally, the urinary anion gap is negative and gives the urinary NH₄⁺ concentration. It becomes zero if renal ammonium production is defective and becomes positive in bicarbonaturia.

◼ Graphic analysis of pH disturbances

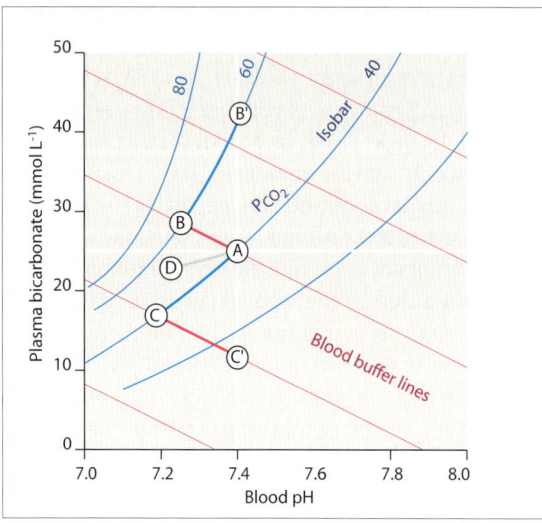

The **Davenport diagram** (Fig. 62.4) gives a grid of Pco_2 isobars and blood buffer lines. Any point on this grid gives a set of three values, viz., the blood pH, Pco_2, and plasma HCO₃⁻. The Pco_2 isobars give the plasma HCO₃⁻ concentration at different blood pH, the Pco_2, remaining constant. The blood buffer lines give the blood pH and plasma HCO₃⁻ concentrations at different Pco_2 values.

Point A on the graph gives the normal pH, Pco_2, and plasma HCO₃⁻. When pH changes are due to purely respiratory causes, Point A will slide up or down the blood buffer line, while when they are due to purely metabolic causes, the point will slide up or down Pco_2 isobars. Point A will fall off the grid when the pH disturbance is a mixed one.

In respiratory acidosis, the HCO₃⁻ concentration, Pco_2, and pH change from Point A to point B. Although all points to the left of Point A are at a lower pH and

Fig. 62.**4** Davenport diagram. Point A represents the normal blood pH, Pco_2, and plasma HCO₃⁻. A shift of point A to point B along the blood buffer line indicates that it is a respiratory acidosis. Subsequent metabolic compensation shifts the point up the Pco_2 isobar to B'. Similarly point A moves to point C due to metabolic acidosis and from there to point C' due to respiratory compensation. For moving to point D, point A has to cut across the grid, indicating that point D represents a mixed metabolic and respiratory acidosis. Note that the slope of the blood buffer lines depends on the buffering capacity of blood, which is determined largely by the hemoglobin concentration. The slope is less when the buffering capacity is lower.

In metabolic acidosis, the values of HCO_3^-, Pco_2, and pH initially change from Point A to point C, sliding down the Pco_2 isobar. During respiratory compensation, point C slides down along the blood buffer line to point C' with a pH of 7.4.

The **Siggaard–Anderson nomogram** (Fig. 62.5) gives a readout of parameters like the standard bicarbonate, buffer base and the base excess, enabling quick calculations of the requirements for pH correction. The *standard bicarbonate* is the HCO_3^- concentration of a blood sample when its Pco_2 is adjusted to 40 mmHg. *Buffer base* refers to the total amount of buffer anions present in blood. It includes hemoglobin, plasma proteins, and plasma HCO_3^-. The *base excess* is a measure of the degree of acidosis or alkalosis present. It is positive in alkalosis and negative in acidosis.

The protocol for using the Siggaard–Anderson nomogram is as follows. (1) The pH of blood sample is measured after incubating it at two different CO_2 partial pressures (e.g., 80 and 25 mmHg). (2) The paired values of pH and Pco_2 measured after incubation, e.g., (7.12 , 80) and (7.41 , 25) are plotted on the nomogram and joined by a line. This line is called the *blood buffer line*. (3) The values of buffer base (41 mEq L^{-1}), standard bicarbonate (19 mEq L^{-1}), and base excess (–8 mEq L^{-1}) are read out from the points at which the blood-buffer line intersects the buffer base line, the standard bicarbonate line, and the base excess line respectively. If the pH of the blood sample before incubation with CO_2 is known (7.23), then its actual Pco_2 (52 mmHg) can also be read out from the blood buffer line.

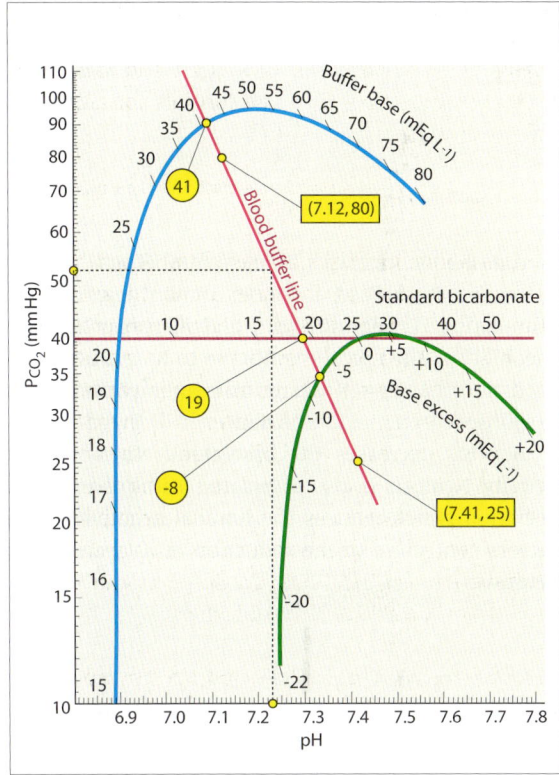

Fig. 62.5   The Siggaard–Anderson nomogram.

higher Pco_2, Point A slides only along the blood buffer line. With metabolic compensation, point B slides up along the Pco_2 isobar till it reaches point C at pH 7.4.

63 Renal Regulation of Potassium Balance

Renal Handling of K⁺

The renal handling of K⁺ is summarized in Figure 63.1A. The proximal convoluted tubule (PCT) has the largest reabsorptive capacity for K⁺ while the connecting tubule (CNT) and the initial part of the cortical collecting duct (CCD) have the greatest capacity for K⁺ secretion. The overall result may be net K⁺ reabsorption or net K⁺ secretion. The distal convoluted tubule (DCT) cells can tilt the balance between net reabsorption and net secretion because these cells can reabsorb as well as secrete K⁺ depending on the body's potassium balance.

The **cellular mechanisms** of K⁺ transport are summarized in Figure 63.1B. Unlike the transport mechanisms of Na⁺ which have important implications in diuretic mechanisms, the different K⁺ transport mechanisms have no such clinical importance and therefore, are not discussed further.

All renal tubular cells have, on their basolateral membrane, a Na⁺-K⁺ ATPase that accumulates K⁺ inside the cell. A cell secretes K⁺ if the K⁺ leaves the cell through the apical membrane. Conversely, a cell reabsorbs K⁺ if the K⁺ enters through the apical membrane using K⁺-H⁺ ATPase and exits through the basolateral membrane. A cell may neither secrete nor reabsorb K⁺ if all the K⁺ that enters the cell through the basolateral membrane exits through basolateral K⁺ channels. In other words, the K⁺ merely recycles across the basolateral membrane.

Factors affecting potassium secretion In the distal nephron, there may be net reabsorption or secretion depending upon the K⁺ balance in the body. K⁺ secretion increases when the activity of the apical K⁺-H⁺ ATPase is low. K⁺ secretion also increases if the intracellular K⁺ concentration is high (due to increased activity of the basolateral Na⁺-K⁺ ATPase), the tubular K⁺ concentration is low (due to increase in the tubular flow of filtrate), and the TEPD is highly negative (due to high Na⁺ unitransport).

A *fall in plasma* K⁺ stimulates K⁺ reabsorption by stimulating the apical K⁺-H⁺ ATPase in the intercalated cells of the collecting duct. This change in K⁺ secretion in response to plasma K⁺ is an important chronic response to external K⁺ imbalance.

Aldosterone increases K⁺ secretion because it stimulates Na⁺-K⁺ ATPase, increases lumen negativity by stimulating Na⁺ unitransport, and increases the permeability of the apical membrane to K⁺. Aldosterone-mediated change in K⁺ secretion is an important acute response to external K⁺ imbalances.

Acidosis decreases the basolateral Na⁺-K⁺ ATPase activity. Acidosis is also associated with increased H⁺ secretion which reduces the luminal negativity. Both factors contribute to the reduction in K⁺ secretion in acidosis.

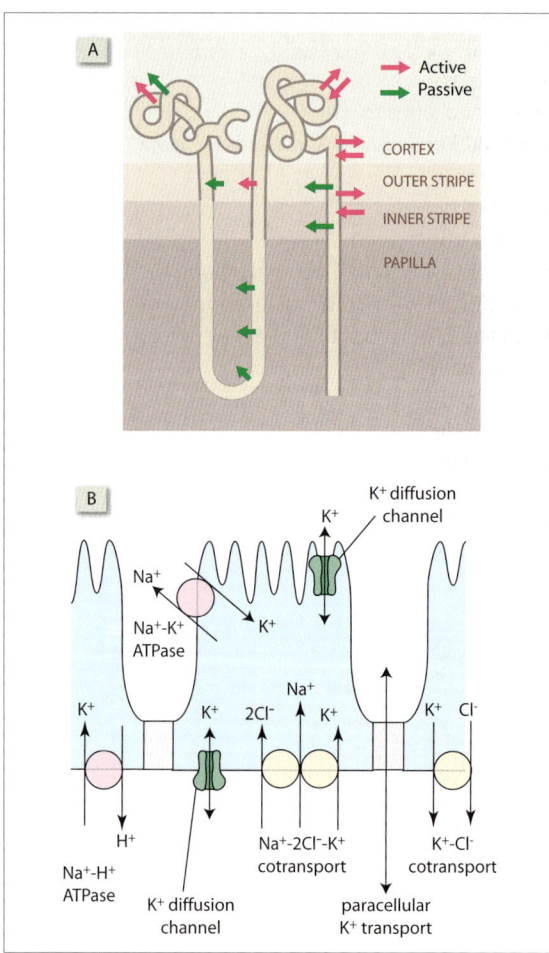

Fig. 63.1 [A] Sites of potassium ion reabsorption and secretion in the nephron. [B] Cellular mechanism of potassium ion reabsorption and secretion.

Body Potassium Balance

The total body K⁺ content is about 4500 mmol. More than 95% of K⁺ is located within cells, mostly muscle cells, with smaller quantities in liver cells and blood cells (Fig. 63.**2**). The normal serum K⁺ level ranges from 3.5 to 5.3 mmol L⁻¹.

External potassium balance

External K⁺ balance refers to the constancy of total body K⁺ and is achieved when the daily intestinal absorption of K⁺ equals its daily urinary excretion (Fig. 63.**2**). A normal external balance ensures long-term, chronic constancy of body K⁺ content. However, rapid losses of large amounts of K⁺ can cause serious hypokalemia.

The average diet contains approximately 100 mmol d⁻¹ of potassium (5 g d⁻¹). 90% of ingested potassium is absorbed and the remaining 10% appears in the stools. Most of the gastrointestinal tract absorbs as well as secretes K⁺. In the colon, secretion is highly developed. Diarrheal states in which colonic K⁺ secretion is stimulated are associated with severe hypokalemia. Gastrointestinal absorption of K⁺ is not physiologically regulated and hence, the external K⁺ balance is maintained entirely through renal regulation of K⁺ excretion.

K⁺ reabsorption in the proximal tubule is load-dependent, always absorbing nearly 90% of the filtered K⁺ load. Hence, the proximal tubules are ineffective in the regulation of external K⁺ balance. The appropriate renal response to external K⁺ imbalance occurs in the DCT and the collecting duct. The renal response has short-term and long-term mechanisms.

In the *short-term mechanism*, an increase in plasma K⁺ stimulates aldosterone secretion. Aldosterone acts on the distal nephron and increases K⁺ excretion within 2 hours.

The *long-term mechanism* takes several hours and is aldosterone-independent. An increase in plasma K⁺ inhibits the apical K⁺-H⁺ ATPase in the intercalated cells of the collecting duct and thereby increases K⁺ secretion, restoring normal plasma K⁺ levels. Conversely, K⁺ deprivation stimulates K⁺-H⁺ ATPase.

Internal potassium balance

Internal K⁺ balance refers to the constancy of the distribution of K⁺ in the intracellular and extracellular compartments. What makes internal balance important is that even a small leakage of intracellular K⁺ drastically raises the plasma K⁺ levels, with lethal consequences. This does not happen because any acute change in plasma K⁺ results in a rapid (occurring within minutes) redistribution of K⁺ between the ECF and ICF, acting as a buffer against extreme changes in plasma K⁺ concentration. As in the case of external balance, the internal K⁺ balance also has short-term and long-term mechanisms.

The *short-term mechanism* of restoration of the internal K⁺ balance is initiated by the membrane depolarization that occurs when the plasma K⁺ increases (see Fig. 10.**6**A). The sequence of events is depicted in Figure 63.**3**.

In the *long-term mechanism*, chronic hypokalemia decreases the number of Na⁺-K⁺ ATPase molecules in the membrane. This reduces the intracellular K⁺ concentration and increases the extracellular K⁺ to normal level.

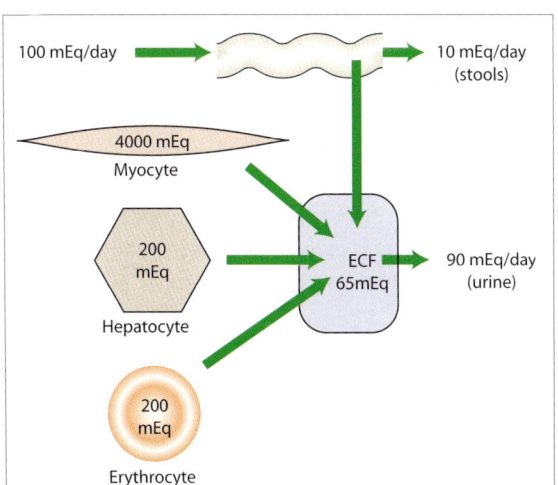

Fig. 63.**2**   External potassium balance.

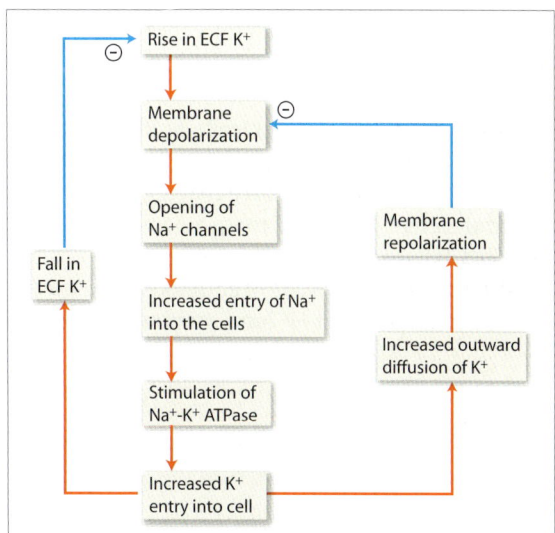

Fig. 63.**3**   Short-term mechanism of internal K⁺ balance.

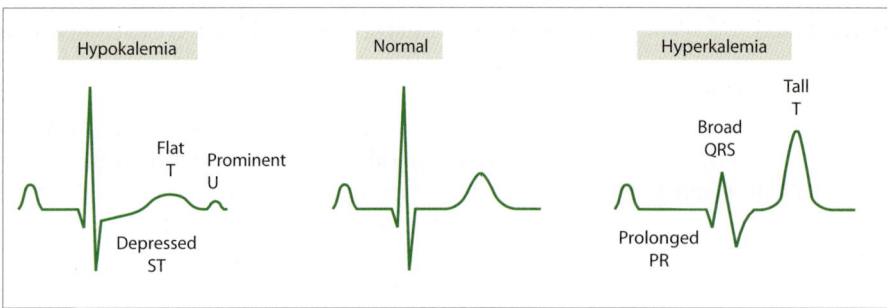

Fig. 63.**4** Electrocardiographic changes associated with changes in serum potassium ion concentration. See also Figure 34.**8**.

Table 63.**1**   Causes of hypokalemia and hyperkalemia.

Causes of hypokalemia	Causes of hyperkalemia
Decreased dietary intake	Excessive dietary intake
	Tissue damage (muscle crush, hemolysis, internal bleeding)
Metabolic alkalosis	Metabolic acidosis
Renal tubular acidosis	
Hyperaldosteronism	Hypoaldosteronism
Cushing's disease	Addison's disease
Diuretics	Potassium-sparing diuretics
Renal failure (ARF, CRF)	

■ Disorders of potassium homeostasis

Body K$^+$ balance is disturbed in several clinical situations (Table 63.**1**). The pathophysiology of only some of them are discussed below.

Pathophysiology   *Diabetic ketoacidosis* results in hyperkalemia with reduced intracellular stores of K$^+$. The reason is twofold. (1) Insulin promotes the entry of K$^+$ into cells. Hence, in diabetes mellitus, K$^+$ moves out of the cells into the ECF. (2) The loss of insulin's growth factor activity diminishes intracellular RNA. Normally, intracellular anions like organic phosphates, DNA, and RNA counterbalance the positive charge of intracellular K$^+$ and therefore, the loss of these anions is associated with loss of K$^+$ from the cells.

Metabolic acidosis results in entry of H$^+$ into the ICF. The rise in intracellular H$^+$ concentration may or may not be associated with a fall in intracellular K$^+$. In the case of mineral acids, the conjugate bases like Cl$^-$, SO$_4^{2-}$ and NO$_3^-$ are unable to enter the cell simultaneously with H$^+$. The rise of intracellular H$^+$ therefore depolarizes the membrane, allowing intracellular K$^+$ to diffuse outward along a more favorable electrochemical gradient. However, organic acids like lactate or acetoacetate enter cells readily along with H$^+$ so that the membrane potential remains unchanged and there is no redistribution of K$^+$. *Respiratory acidosis* has very little effect on serum K$^+$ levels because the membrane potential does not change when CO$_2$ (which is electroneutral) enters the cells or when it reacts with water to form H$^+$ and HCO$_3^-$ (cation-anion pair) inside the cell.

The internal K$^+$ balance is also disturbed during *exercise*. During the depolarization that accompanies muscle contraction, K$^+$ exits from cells. Most of the K$^+$ released enters the T tubules, from which K$^+$ is rapidly taken back into the cell on repolarization. Very little K$^+$ enters the interstitial fluid and plasma. With vigorous exercise, however, there is a rise in plasma K$^+$ levels. Patients with disturbed T tubule architecture may develop hyperkalemia with the slightest exercise, like fist-clenching. The local hyperkalemia in muscle causes vasodilatation and activation of glycogenolysis, both of which are useful in exercise.

Clinical features   *Neuromuscular symptoms* are prominent in hypokalemia. It ranges from muscle weakness to total paralysis. *Cardiac symptoms* like arryhthmias are more prominent in hyperkalemia. ECG changes are present in both hypokalemia and hyperkalemia (Fig. 63.**4**). *GIT symptoms* of hypokalemia include abdominal distension due to paralytic ileus.

Glucose

Glucose is freely filtered into the glomerular filtrate and is completely reabsorbed in the proximal tubule. At the apical membrane of the proximal tubular cell, there is a carrier-mediated Na^+-glucose cotransport. The carrier is called the Sodium-dependent GLucose Transporter (SGLT). The cotransport derives its energy from the Na^+ concentration gradient that exists between the high tubular Na^+ concentration and the low intracellular Na^+ concentration produced by the pumping out of Na^+ through the basolateral surface.

The glucose that enters the cell through the apical membrane diffuses out of the basolateral membrane through facilitated diffusion. The carrier for facilitated diffusion across the apical membrane is called the GLUcose Transporter (GLUT) which belongs to a different family of glucose transporters than the SGLT.

In the PCT, where the tubular concentration of glucose is high, the apical glucose carrier is SGLT-2, which is a high-capacity, low-affinity SGLT. The basolateral glucose carrier is GLUT-2, which is high-capacity, low-affinity GLUT.

In the PST, the tubular glucose concentration is low. Appropriately, the glucose carrier is SGLT-1, which is a high-affinity and low-capacity $2Na^+$-glucose-galactose cotransporter (Fig. 64.1). The glucose reabsorbed here is mostly used for cellular nutrition. Some of it exits the basolateral membrane using GLUT-1.

Proteins

Amino acids are normally completely reabsorbed in the proximal tubule. They are transported across the apical border of proximal cells through secondary active transport. There are different carrier proteins for neutral, basic, and acidic amino acids and imino acids. Cystine and methionine have specific carriers.

Peptides and small proteins that are filtered into the tubules are endocytosed by the proximal tubular cells. Inside the cell, they are cleaved into the constituent amino acids which diffuse out into peritubular capillaries.

Urea

Urea is filtered freely into the glomerular filtrate. About half of the filtered urea is reabsorbed passively in the proximal tubule. In the proximal straight tubule (PST) and the thin segment, urea diffuses into the tubular lumen from the medullary interstitium where it is present in high concentration. The remaining part of the tubule, except its terminal part, is impermeable to urea and therefore loses water to the hyperosmolar medullary interstitium. As water is reabsorbed from the cortical collecting duct (CCD), outer medullary collecting duct (OMCD), and the initial part of inner medullary

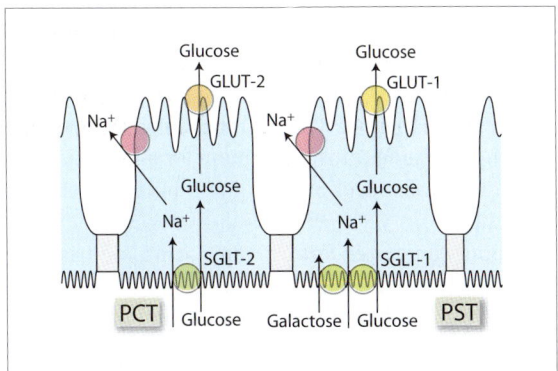

Fig. 64.**1** Tubular reabsorptive mechanisms for glucose.

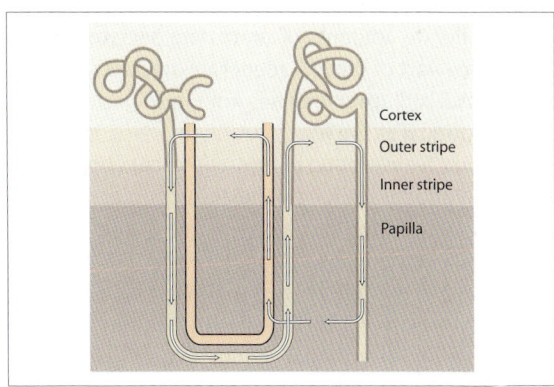

Fig. 64.**2** Urea recycling.

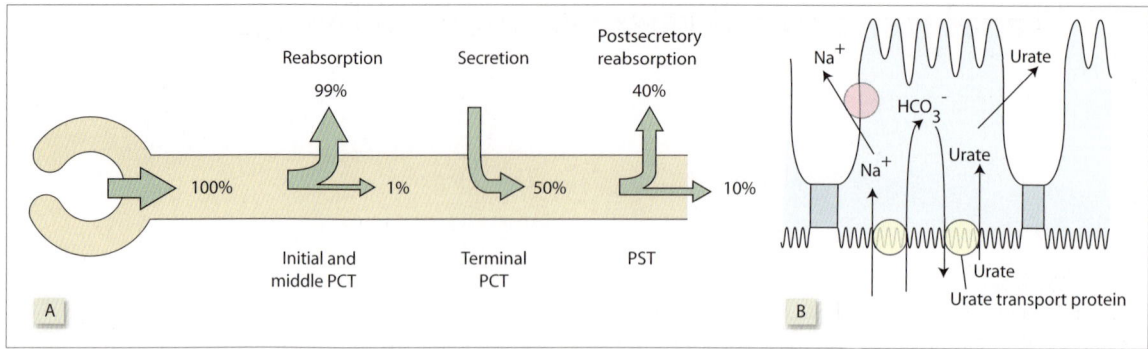

Fig. 64.**3** [A] Postsecretory reabsorption of urates in the proximal tubule. [B] Tubular reabsorptive mechanism for urates.

collecting duct (IMCD), the urea in the collecting duct becomes more and more concentrated. When the urea-rich tubular fluid reaches the urea-permeable terminal IMCD, a large amount of urea is reabsorbed into the interstitium using a special *urea transport protein*. The synthesis of this protein is stimulated by ADH.

From the medullary interstitium, most of the urea enters the vasa recta and is carried up toward the renal cortex by the ascending vasa recta. From there, urea diffuses out to enter the PSTs of cortical nephrons. The urea is carried back to the OMCD from where it diffuses out again, resulting in a constant recycling (Fig. 64.2). This *recycling of urea* concentrates it in the inner medulla and explains why nearly 50% of the medullary hyperosmolarity is attributable to urea.

Uric acid

Urate is freely filtered into the glomerular filtrate. The tubular transport of uric acid is confined almost exclusively to the proximal tubule and involves both reabsorption and secretion. In the initial and middle part of the PCT, the reabsorption far exceeds secretion and the urates are reabsorbed almost completely. In the distal portion of the PCT and the beginning of the PST, moderate amounts of urates are secreted. In the remaining part of the PST, moderate amounts of urates are reabsorbed again which is known as postsecretory reabsorption (Fig. 64.**3**A).

Tubular reabsorption of uric acid occurs both transcellularly and paracellularly. The transcellular reabsorption occurs through secondary active transport while the paracellular reabsorption is passive. The apical membrane of the cells of the proximal tubule contain a *urate transport protein* involved in the countertransport of urate with intracellular anions like Cl^-, HCO_3^-, and also organic anions like lactate. The urates move out through the basolateral membrane using another an-

ion-exchanger. The same anion-exchangers are employed for urate secretion too (Fig. 64.**3**B).

Since urate reabsorption in the proximal tubules is linked to Na^+ reabsorption, *diuretics* that decrease the reabsorption of Na^+ and water in the proximal tubule decrease urate reabsorption too. On the other hand, diuretics acting on the distal tubules decrease the ECF volume, which in turn increases the Na^+ reabsorption in the proximal tubule. Hence, these diuretics increase urate reabsorption and tend to cause hyperuricemia.

Analgesics like salicylates and phenylbutazone, when administered in low dose, reduce urate excretion by inhibiting urate secretion. When administered in higher doses, these analgesics also inhibit urate reabsorption and therefore increase urate excretion.

Organic acids and bases

In the ionized form, organic acids and bases are secreted through secondary active transport. For example, para-amino hippurate (PAH) is secreted across the apical membrane by PAH-anion antiport coupled with Na^+-anion symporter. The anion is a dicarboxylate or a tricarboxylate. The entire transport mechanism is nonelectrogenic (see Fig. 59.**2**B). These transport systems are nonspecific, e.g., one anion transporter can transport several endogenous (bile salts, hippurate, urate) and exogenous (PAH, penicillin, probenecid, aspirin) substances. There is competition among substances sharing the same transport system for elimination. For example, probenecid inhibits the tubular secretion of penicillin.

In the *nonionized form*, organic acids and bases are sufficiently lipid-soluble to diffuse passively across the tubular epithelium. Passive diffusion is faster than transport of the ionized forms. Many drugs and their metabolites are weak acids or bases that are predominantly nonionized and therefore diffuse out into the

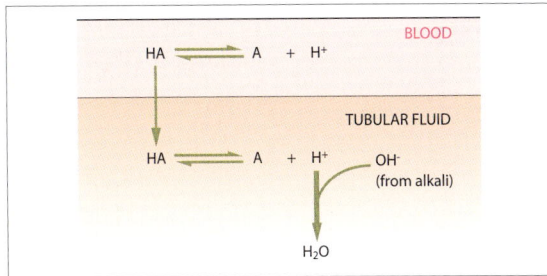

Fig. 64.**4** Nonionic diffusion and diffusion trapping produced by urine alkalinization.

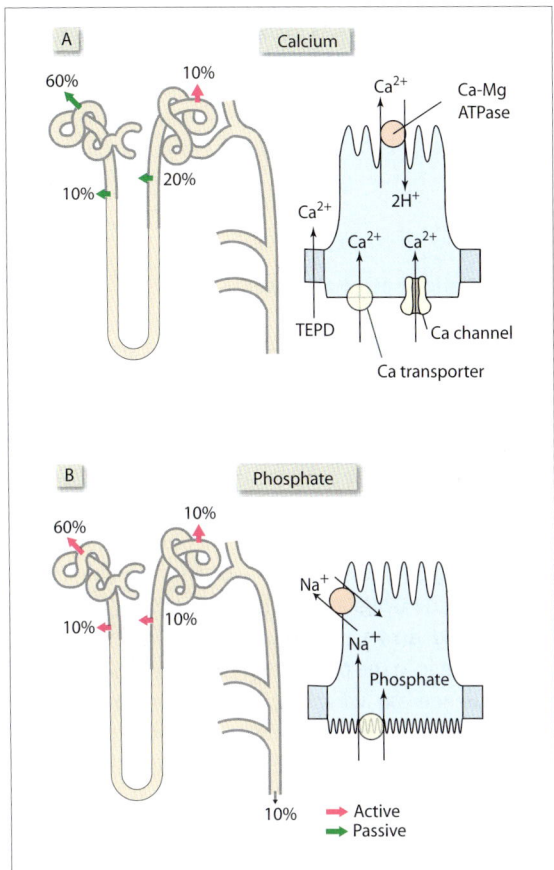

Fig. 64.**5** Tubular reabsorption of calcium and phosphates.

tubules quickly. Alkalinization of urine ionizes the drug in the tubular fluid, resulting in a reduction of its non-ionized form in tubular fluid. This increases the concentration gradient of the nonionized drug between blood and tubular fluid and therefore favors its passive diffusion from blood into the tubule.[1] This process is known as *nonionic diffusion* with *diffusion trapping* (Fig. 64.**4**).

■ Calcium and phosphates

About 45% of the plasma calcium is bound to plasma proteins and therefore does not get filtered into the tubules. The free Ca^{2+} that is filtered into the tubules is almost completely reabsorbed.

The cellular mechanisms of calcium and phosphate handling are summarized in Figure 64.**5**. The reabsorption of Ca^{2+} is both paracellular and transcellular. In transcellular transport, the apical transport is mostly through facilitated diffusion and partly through Ca^{2+}-channels that are responsive to PTH and calcitonin. The facilitated diffusion employs a carrier protein called calbindin. Ca^{2+} is actively extruded across the basolateral membrane through primary active transport using Ca^{2+}-Mg^{2+} ATPase, a Ca^{2+}-stimulated, Mg^{2+}-dependent ATPase. Phosphate is reabsorbed through a Na^+-phosphate symport.

PTH stimulates Ca^{2+} reabsorption in the distal tubules. However, since PTH increases the plasma Ca^{2+}

concentration through its actions on bones and the gastrointestinal tract, large amounts of Ca^{2+} are filtered into the tubules and the net excretion in the urine might actually increase. PTH also increases the excretion of phosphates.

Vitamin D promotes the reabsorption of both Ca^{2+} and phosphates. Vitamin D increases the synthesis of calbindin and thereby increases Ca^{2+} reabsorption. It also stimulates the activity of Ca^{2+}-Mg^{2+} ATPase.

Loop diuretics produce a large increase in Ca^{2+} excretion by reducing the TEPD in the TAL. Thiazide diuretics markedly reduce Ca^{2+} excretion by increasing Ca^{2+} reabsorption in the DCT and have been used in the treatment of hypercalcuria.

---

[1] It can be argued that consumption of alkali mixture would first alkalinize the blood and therefore, the excretion of the drug should decrease. This may be true initially but as the alkali is excreted in urine, the blood pH becomes normal while the urine pH becomes alkaline.

65 Hormones Acting on the Kidney

■ Antidiuretic hormone

Antidiuretic hormone (ADH), also called *arginine vasopressin* (AVP), is a neurohormone, i.e., a hormone secreted into circulation by nerve cells. The precursor of ADH is *prepropressophysin* which is synthesized in the cell bodies of the neurons in the *supraoptic* and *paraventricular nuclei* (see Fig. 82.**3**). Prepropressophysin is cleaved in the endoplasmic reticulum to form ADH and packaged into secretory granules (called *Herring bodies*) in the Golgi apparatus. The Herring bodies are transported down the axons by axoplasmic flow to their endings in the posterior pituitary. ADH is released from the neurons when stimulated by afferent signals from the osmoreceptors and volume receptors.

There are two types of ADH (vasopressin) receptors. V_1 receptors act through the group II-C hormonal mechanism while V_2 receptors act through the group II-A mechanism (see Chapter 81).

V_1 receptors mediate smooth muscle contraction. Activated *V_2 receptors* promote antidiuresis: (1) They increase the permeability of the collecting duct to water (by causing insertion of water channels[1] into the luminal membrane of the collecting duct cells) and urea (by activating urea transporter proteins). (2) They stimulate Na^+/K^+-2Cl^- cotransport in the thick ascending limb (TAL).

In the **Syndrome of Inappropriate ADH secretion (SIADH)**, secretion of excess ADH results in the retention of large amounts of water in the body with consequent dilutional hyponatremia. The ECF expansion increases Na^+ excretion through the Starling mechanism (see Fig. 61.**2**) and aggravates the hyponatremia. The normal plasma level of ADH is 0–10 pg ml^{-1}. In SIADH, the plasma osmolarity is low and the plasma level of ADH is at its maximum, i.e., 10 pg ml^{-1}. Although in the normal range, it actually signifies a high rate of ADH secretion because it occurs despite a low plasma osmolarity which normally should inhibit the ADH secretion completely. SIADH occurs in cerebral diseases and ADH-secreting lung tumors. SIADH can be treated with demeclocycline which reduces the renal response to ADH.

[1] The water channels are protein molecules called aquaporin-2. They are stored in endosomes inside the cells. When stimulated by V_2 receptors, they are rapidly translocated to the cell membrane.

Diabetes insipidus is associated with the formation of large volumes of urine with a low specific gravity. Diabetes insipidus may be neurogenic, nephrogenic, or gestational.

The commonest cause of *neurogenic diabetes insipidus* is surgical hypophysectomy. It occurs 4 to 6 weeks after surgery due to retrograde degeneration of the ADH-secreting neurons. Since most ADH-secreting neurons terminate in or above the infundibulum, sectioning of the pituitary *below* the infundibulum does not usually cause diabetes insipidus. Neurogenic diabetes insipidus can also occur as an autosomal dominant disease. It is treated by administering ADH.

Nephrogenic diabetes insipidus is usually genetic and may be a X-linked recessive disorder in which V_2 receptors are not responsive to ADH, or an autosomal recessive disorder in which aquaporin-2 formation is impaired.

Gestational diabetes insipidus occurs in pregnancy due to an abnormal increase in plasma vasopressinase levels. It is treated by giving desmopressin, an analogue of vasopressin which is resistant to inactivation by vasopressinase.

■ Renin-angiotensin system

Renin is a protease enzyme. Its primary source is the JG cells of the kidneys. Renin production also occurs in unspecialized smooth muscle cells in the afferent arteriole up to the interlobular arteries in fetal kidneys, and also in adults when there is increased demand for renin secretion.

The principal regulators of renin degranulation are as follows. (1) The juxtaglomerular (JG) cells themselves are probably the sensors of the afferent arteriolar pressure. They degranulate when the pressure rises. (2) Increase in NaCl (Cl^- in particular) in the distal convoluted tubule (DCT) decreases renin secretion. The amount of NaCl in the tubular fluid is sensed by the macula densa. The information is signaled to the JG cells. Adenosine is a probable mediator of the signal. (3) The JG cells are innervated by sympathetic fibers. They release renin in response to sympathetic discharge and circulating catecholamines. (4) Prostacyclin (PGI_2) stimulates renin secretion through a direct action on the JG cells.

The renin released from the JG cells enters circulation and acts on an α_2 globulin plasma protein called *angiotensinogen* and splits off from its N-terminal a decapeptide called *angiotensin-I*. An enzyme called *angiotensin converting enzyme* (ACE) then acts on angiotensin-I to split off from it an octapeptide called *angiotensin-II*. The enzyme is found on the surface of capillary endothelium in the lungs and kidneys.

Angiotensinogen $\xrightarrow{\text{Renin from JG cells}}$ Angiotensin-I
(α_2 globulin)   (decapeptide)

Angiotensin-I $\xrightarrow{\substack{\text{ACE on the capillary} \\ \text{endothelium of the} \\ \text{lungs and kidneys}}}$ Angiotensin-II
(decapeptide)   (octapeptide)

Central effects   Angiotensin-II acts through angiotensin receptors present in the subfornical organ (SFO) and the organ vasculosum of the lamina terminalis (OVLT) to increase thirst and stimulate ADH release. Angiotensin-II also acts through angiotensin receptors located in the area postrema of the brain to bring about an increased sympathetic discharge.

Renal effects   Angiotensin receptors are present on afferent and efferent arterioles, proximal tubules, and mesangial cells. Through these receptors, angiotensin-II exerts the following effects.

Effect on renal arterioles   Angiotensin-II constricts both afferent and efferent arterioles, thereby reducing both GFR and RBF. The efferent arterioles, however, are more sensitive to angiotensin-II than are the afferent arterioles. Hence, angiotensin-II causes greater constriction of the efferent arterioles, thereby increasing the filtration fraction.

Effect on glomerular capillaries   Angiotensin receptors are present in the glomerular capillaries. Angiotensin-II decreases the pore-size of the glomerular basement membrane. ACE-inhibitors have been used in nephrotic syndrome to decrease the proteinuria.

Effect on glomerular mesangial cells   Angiotensin receptors are present on the mesangial cells. Angiotensin-II results in mesangial cell hypertrophy and an increase in the extracellular matrix production by the mesangial cells. Since both mesangial cell hypertrophy and extracellular matrix expansion are observed in diabetes mellitus, angiotensin-II has been implicated in the pathophysiology of diabetes mellitus. ACE inhibitors tend to retard these changes in diabetic nephropathy.

Effect on adrenals   Angiotensin receptors are present on the zona glomerulosa of the adrenal glands. Through these receptors, angiotensin-II causes stimulation of aldosterone secretion.

Vascular effect   Angiotensin-II is a powerful vasoconstrictor. Its vasoconstrictor effect along with its other effects like thirst and stimulation of aldosterone secretion leads to a rise in blood pressure (Fig. 65.**1**). The role of angiotensin in the regulation of blood pressure in discussed in Chapter 42.

■   **Aldosterone**

The secretion of aldosterone and the factors affecting it are discussed on p 531. Only the renal effects of aldosterone are discussed here.

The actions of aldosterone are localized to the distal tubular and collecting duct cells. (1) It stimulates the basolateral Na^+-K^+ ATPase, (2) increases the permeability of the apical membrane to Na^+ unitransport channels, and (3) increases the permeability of the apical membrane to K^+. As a result of these actions, aldosterone increases the reabsorption of Na^+ and with it, of water too. The increased Na^+ reabsorption makes the transepithelial potential difference (TEPD) highly negative. The increase in K^+ secretion produced by aldosterone is due to the combined effects of the highly negative TEPD together with the increased apical permeability to K^+.

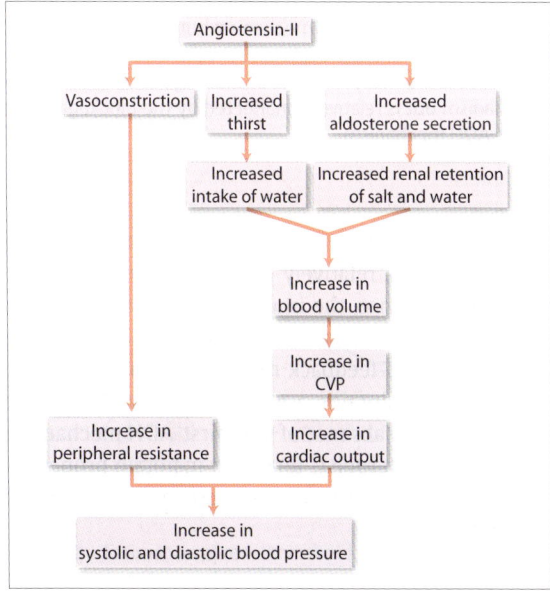

Fig. 65.**1**   Effect of angiotensin-II on blood pressure.

When the plasma concentration (P_G) equals the renal threshold, the tubular reabsorption (Tr) reaches its maximum (i.e., the TmG) and exactly balances the filtered load of glucose. At that stage:

P_G = renal threshold for glucose

Tr = TmG

UGV = 0

∴ TmG = GFR × renal threshold

or Renal threshold = TmG / GFR

There is however a discrepancy between the calculated and the measured values of the renal threshold. The renal threshold calculated from TmG would be:

$$\frac{375 \text{ mg min}^{-1}}{125 \text{ ml min}^{-1}}$$

$$= 3 \text{ mg ml}^{-1}$$
$$= 300 \text{ mg dl}^{-1}$$

Thus glucose would be expected to appear in urine when the arterial concentration of glucose exceeds 300 mg dl^{-1}, which corresponds to a venous glucose concentration of about 200 mg dl^{-1}. However, the measured value is lower, at about 180 mg dl^{-1}. This is because the calculated value represents the average value of 2 million nephrons. A nephron can have a TmG that is either higher or lower than the average of 375 mg min^{-1}. Nephrons with a lower TmG leak glucose into urine at plasma glucose levels below the calculated threshold. This discrepancy shows up graphically as the splay (see Fig. 66.**3**).

Renal Clearance

The clearance of a substance is defined as the volume of plasma that is cleared of that substance completely in 1 minute. Clearance is a 'virtual' volume since blood is never cleared completely of any of its constituents. The unit of clearance is ml min^{-1} and is calculated by the formula:

$$\frac{UV}{P}$$

where:

U = urinary concentration of the substance

V = rate of flow of urine

P = plasma concentration of the substance

Osmolar clearance is the volume of plasma that is cleared of all its electrolytes in 1 minute. If the urine excreted is isoosmotic to the plasma, then the osmolar clearance is equal to the urinary output in 1 minute.

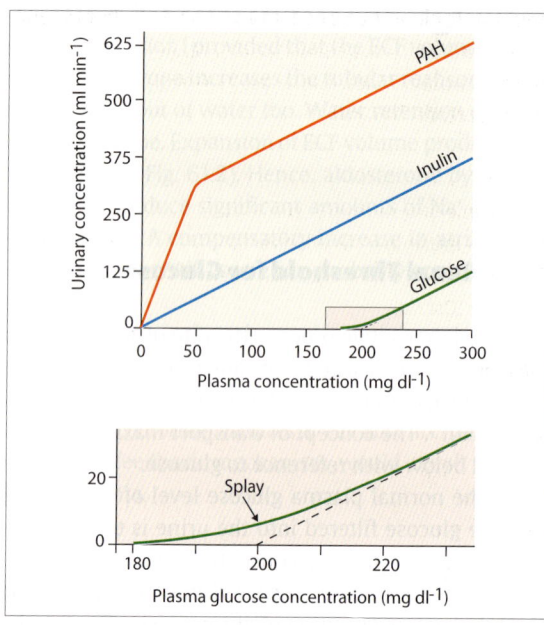

Fig. 66.**3**   (*Above*) Changes in the urinary concentration of para-amino hippuric acid (secreted into tubule), glucose (reabsorbed from tubule), and inulin (neither secreted nor reabsorbed) as a function of plasma concentration. (*Below*) The splay, showing the divergence between the calculated renal threshold for glucose (200 mg dl^{-1}) and the measured renal threshold for glucose (180 mg dl^{-1})

Osmolar clearance is given by the formula:

$$C_{Osm} = \frac{U_{Osm} V}{P_{Osm}}$$

where

U_{Osm} = osmolarity of urine

P_{Osm} = osmolarity of plasma

If $U_{Osm} = P_{Osm}$, then $C_{Osm} = V$

Thus, defined in another way, osmolar clearance is the volume of isotonic urine that must be excreted per minute for eliminating the excretory solute load.

Free-water clearance   (C_{H_2O}) is defined as the amount of water that must be removed from the urine volume excreted to render it isotonic with plasma. It is given by the formula:

$$C_{Osm} = V - C_{H_2O}$$

or

$$V = C_{Osm} + C_{H_2O}$$

In other words, the urine volume can be viewed as the sum of the volume of electrolyte-free urine (C_{H_2O}) and the volume of isoosmolar urine (C_{Osm}). A positive free-water clearance occurs when urine is hypoosmolar and a negative free-water clearance when urine is hyperosmolar. When the urine is isoosmolar, C_{Osm} = V, and therefore C_{H_2O} = 0.

Example

The volume of urine excreted in 24 h is 2 L. Plasma osmolarity is 300 mOsm L^{-1}. Calculate the free-water clearance when the urine osmolarity is (i) 1200 mOsm L^{-1} (ii) 150 mOsm L^{-1} (iii) 300 mOsm L^{-1}.

Solution

$$C_{H_2O} = V - C_{Osm}$$

or

$$C_{H_2O} = V - \frac{U_{Osm}V}{P_{Osm}}$$

(i)

$$C_{H_2O} = 2.0 - \frac{1200 \times 2.0}{300}$$

$$= -6.0 \text{ L d}^{-1}$$

(ii)

$$C_{H_2O} = 2.0 - \frac{150 \times 2.0}{300}$$

$$= 1.0 \text{ L d}^{-1}$$

(iii)

$$C_{H_2O} = 2.0 - \frac{300 \times 2.0}{300}$$

$$= 0 \text{ L d}^{-1}$$

◼ Plasma concentration and clearance

The general formula for urinary excretion of solutes is:

$$UV = (GFR \times P) - Tr + Ts$$

where:
   Tr = amount reabsorbed from the tubule
   Ts = amount secreted into the tubule
   GFR × P = amount filtered into the tubule per minute
   U × V = amount of the solute excreted in urine per minute.

Dividing throughout by P:

$$UV/P = GFR - Tr/P + Ts/P$$

If a substance is neither secreted into the tubule nor reabsorbed from the tubule, the formula reduces to:

$$UV/P = GFR$$

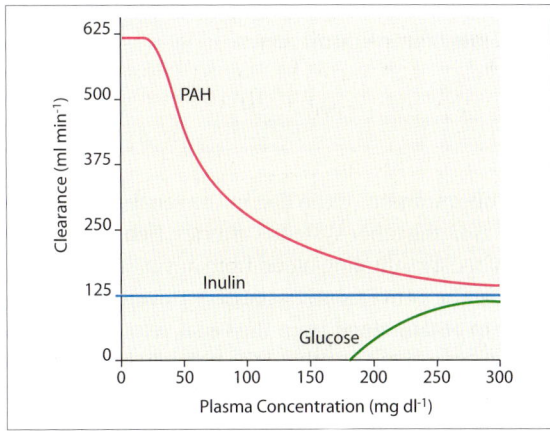

Fig. 66.4   Changes in renal clearance of PAH (secreted into tubule), glucose (reabsorbed from tubule) and inulin (neither secreted nor reabsorbed) as a function of plasma concentration.

Inulin, and to a lesser extent creatinine, meet the above criteria. Hence, GFR can be estimated by measuring inulin clearance or creatinine clearance. Since the values of Tr or Ts cannot increase indefinitely (they are limited by the Tm), Tr/P and Ts/P approach zero at very high values of P. Hence, the clearance of all substances approaches the value of GFR (inulin clearance) at higher values of P (Fig. 66.4). The clearance of glucose is normally zero since the normal urinary concentration of glucose (U) is zero. However as the plasma level of glucose exceeds the threshold, its clearance exceeds zero. At higher plasma levels the clearance of glucose too, like that of other substances, approaches the value of inulin clearance. For the same reason, the clearance of para-amino hippuric acid (PAH) only at low plasma levels gives the renal plasma flow (see below) while at higher values, it will approach the value of GFR.

◼ PAH clearance and renal plasma flow

If a dye is infused into the blood and its arterial and venous concentrations are estimated, it is possible to calculate the renal plasma flow (RPF) using the Fick principle:

$$RPF = \frac{Q}{(P_a - P_v)} \qquad ...(1)$$

where Q is the amount of dye removed from the kidney, P_a is the dye concentration in the renal artery and P_v is the dye concentration in the renal vein.

   The amount of dye (Q) removed from the kidney is equal to the amount of the dye excreted in the urine.

$$\therefore Q = UV \qquad ...(2)$$

where U is the urinary concentration of the dye, and V is the rate of urine formation.

$$RPF = \frac{UV}{(P_a - P_v)} \qquad ...(3)$$

The concentration of the dye in the renal artery is the same as in the other systemic arteries. Hence, P_a can be estimated by sampling blood from any systemic artery that can be conveniently accessed. However, it is difficult to obtain blood from the renal vein. Hence, it is expedient to use a dye that is completely excreted into the urine in a single passage through the kidney so that the concentration of the dye in renal venous blood is zero. If $P_v = 0$, then equation 3 reduces to:

$$RPF = \frac{UV}{P_a} \qquad ...(4)$$

The dye that comes nearest to satisfying the condition of complete extraction in a single passage through kidney is para-amino hippuric acid (PAH). Hence, RPF equals the renal clearance of PAH. However, PAH is only 90% excreted in the urine in a single passage. In other words, its extraction ratio is 90%. Hence, its concentration in renal venous blood is about 10% of its concentration in the renal arterial blood:

$$P_v = P_a \times (1 - \text{extraction ratio})$$
$$= P_a - (P_a \times \text{extraction ratio}) \qquad ...(5)$$

Substituting equation (5) in equation (3), we have:

$$RPF = \frac{UV}{(P_a \times \text{extraction ratio})} \qquad ...(6)$$

The RBF (renal blood flow) can be calculated from RPF using the formula,

$$RBF = \frac{RPF}{(1 - \text{hematocrit})} \qquad ...(7)$$

Inulin clearance and the GFR

Inulin is a polysaccharide. Inulin clearance gives an estimate of the GFR because it is freely filtered into the glomerular filtrate and is neither secreted nor reabsorbed in the kidneys. It also satisfies the basic requirements of a good indicator dye, i.e., it is neither metabolized nor stored in the kidneys, is neither toxic nor has any effect on the filtration rate itself, and can be easily estimated in the laboratory.

Inulin clearance (GFR) is given by the formula:

$$\text{Inulin clearance (GFR)} = \frac{UV}{P_a}$$

where U is the urinary concentration of inulin, V is the rate of urine flow, and Pa is the arterial concentration of inulin.

During estimation, it is important to maintain a stable plasma concentration of inulin. This is achieved by giving a single bolus dose of inulin solution followed by a continuous intravenous infusion of the same. GFR is more easily estimated by using creatinine as an indicator. Creatinine is an endogenous substance having a fairly constant plasma value (P) of about 0.6–1.5 mg dl^{-1} and therefore does not need continuous intravenous infusion. Creatinine, however, is slightly secreted into the tubules, increasing its urinary concentration (U). The value of GFR obtained, although not accurate, is still acceptable to clinicians. All clearance techniques depend on accurately timed collections of urine, which is the major source of error.

The **filtration fraction** (FF) is given by the ratio of GFR and RBF. It serves as an approximate index of glomerular filtration coefficient (Kf).

$$FF = \frac{GFR}{RPF}$$

Urea clearance

The concept of clearance originally developed around efforts to use urea excretion as an index for renal function. Part of the problem in such efforts was that urea excretion even by a normal kidney varies greatly depending on the rate of urinary outflow. Two empirical indices were therefore developed, one for urinary flow less than 2 ml min^{-1}, (called the standard clearance) and the other, for urinary flow more than 2 ml min^{-1}, (called the maximal clearance). These indices are fairly constant as long as there is no impairment of renal function.

$$\text{Standard urea clearance} = \frac{U\sqrt{V}}{B}$$

$$\text{Maximal urea clearance} = \frac{UV}{B}$$

It was realized later that UV/B actually signified the 'volume of blood' that was cleared of urea per minute. The blood concentration (B) was later replaced by plasma concentration (P) and the term clearance was extended to other solutes, some of which more accurately indicated the GFR or RBF. Maximal urea clearance underestimates GFR because urea is reabsorbed from the collecting duct. Standard urea clearance does not denote a volume of blood and therefore, cannot be called clearance according to its present definition.

Urine Analysis

Urine examination is always performed during routine health check-up and is the first test performed on patients with suspected renal disorders. Many of the tests performed in urine examination may not be indicative of renal functions. Radiographic tests like retrograde pyelography, ultrasonography or arteriography cannot be called renal function tests. The same is true for renal biopsy, which is done only when a renal disease is strongly suspected. However, *intravenous pyelography* and *renal scan* are indicative of renal functions and are employed by surgeons to assess the renal functions of the two kidneys separately before deciding which of them to operate upon.

Physical examination

The normal volume of urine passed per day ranges from 500 to 2500 ml. It increases after meals, after drinks, and on exposure to cold, and decreases if water intake is low and after excessive sweating. Polyuria occurs in diabetes insipidus, diabetes mellitus, and in the diuretic phase of acute renal failure. Oliguria occurs in acute glomerulonephritis, hypotension, and dehydration. Anuria occurs in lower urinary tract obstruction.

Color and turbidity   Urates precipitate in acidic urine on standing, making the urine cloudy. Urinary urate excretion increases when purine metabolism in the body increases, as in gout. Strongly alkaline urine appears cloudy due to the precipitation of $Ca_3(PO_4)_2$. The cloudy appearance increases on warming the urine which makes it more alkaline as the CO_2 bubbles off from it. Urinary tract infection increases the pus cells and bacteria in urine, giving it a cloudy appearance (Table 67.**1**).

   Hematuria means the presence of red blood cells in the urine. Hematuria does not necessarily indicate a renal abnormality: The blood can come from the urinary tract. Red cells that enter urine through the damaged glomerulus are usually distorted, which helps in differentiating glomerular from nonglomerular bleeding. *Hemoglobinuria* refers to the presence of free hemoglobin in the urine.

The normal specific gravity of urine is 1.003–1.030 and the normal urine osmolarity is 100–1000 mOsm kg^{-1}. If the early morning urine sample after an overnight fast has an osmolarity of >600 mOsm kg^{-1} (specific gravity >1.018), it indicates that the patient has a normal urine concentrating ability. If the osmolarity remains constant at 300 mOsm kg^{-1} (specific gravity 1.009) it is called *isosthenuria*. It occurs in chronic renal failure (CRF). Urine is normally slightly acidic except shortly after a meal due to the postprandial alkaline tide. Abnormal causes of alkaline urine include alkali consumption, impairment of tubular acidification, and urinary tract infection (UTI) with urea-splitting organisms.

Chemical analysis

Proteins   Up to 150 mg of protein is excreted daily in urine. Of this, 15 mg is albumin. The rest comprises low molecular weight proteins (LMWP). About 25 mg of the LMWP is Tamm–Horsfall protein derived from the cells of the thick ascending limb (TAL). The remainder of the LMWP is derived from plasma proteins, e.g., β_2 microproteins, retinal binding protein, and light chains of immunoglobulin. Excretion of >150 mg per day of proteins is called proteinuria. *Transient proteinuria can*

Table 67.**1**   Causes of urine color and turbidity.

Normal tint	Urochrome, uroerythrine urobilin (formed on standing)
Cloudy	Strongly alkaline urine, excessive urates, infection
Smoky	Hematuria
Frothy	Proteinuria
Milky	Chyluria
Orange	Excess urobilin
Brown	Bilirubinuria
Red-dark brown	Porphyrins (on standing), frank hematuria
Red-dark brown-black	Hemoglobinuria, melanin (on standing)

occur in fever and after exercise. *Orthostatic proteinuria* occurs only on standing and is not associated with any renal damage. The clinically important proteinurias are glomerular proteinuria, tubular proteinuria, overflow proteinuria, and nephrogenic proteinuria.

In *glomerular proteinuria* large amounts of albumin and other high molecular weight proteins (HMWP) are filtered into the tubule. Hence, glomerular proteinuria is also called high molecular weight proteinuria. Massive glomerular proteinuria is called *nephrotic syndrome*.

In *tubular proteinuria*, the LMWP that are normally filtered into the tubule in fairly large amounts are not reabsorbed. Hence, this type of proteinuria is also called low molecular weight proteinuria.

Overflow proteinuria is the name given to the low molecular weight proteinuria that occurs when the plasma concentration of LMWP rises markedly, as in multiple myeloma, rhabdomyolysis, and intravascular hemolysis. The proteins appear in urine when the protein reabsorptive capacity of the tubules is exceeded.

Nephrogenic proteinuria is the name given to the appearance of tubular enzymes like N-acetyl β–glucosaminidase (NAG), and γ-glutamyl transferase (γGT) in the urine. These enzymatic proteins are released when the proximal tubular cells are damaged.

Sugars are normally absent in the urine. *Glycosuria* may be due to diabetes mellitus, renal glycosuria, or alimentary glycosuria. *Galactosuria* and *fructosuria* occur due to inborn errors of metabolism. Lactosuria occurs in late pregnancy and lactation. Pentosuria is caused by consuming large quantities of plums, cherries, or grapes.

Ketones are normally absent in urine. *Ketonuria* occurs in diabetic ketoacidosis, starvation, and prolonged diarrhea and vomiting.

Bilirubin is normally absent in urine but appears in conjugated hyperbilirubinemia. The daily urobilinogen excretion is normally 1.0–3.5 mg.

Heme pigments are normally absent in urine. They appear following intravascular hemolysis and hemolysis in renal tubules. They also appear following crush injury to muscles and in myopathies which release heme from myoglobin.

Nitrites Urine normally contains nitrates (NO_3^-) but not nitrites (NO_2^-). Urinary nitrites suggest the presence of urinary tract infection because organisms infecting the urinary tract produce nitrites from urinary nitrates.

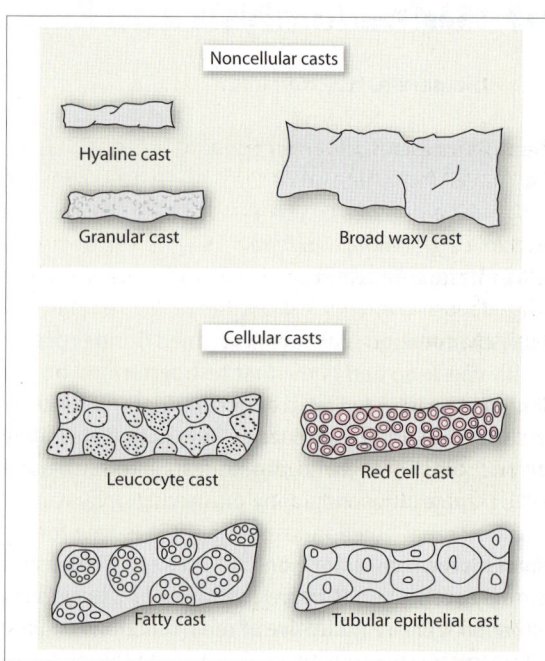

Fig. 67.1 The different types of noncellular and cellular casts found in the urine.

Microscopic examination

The common types of *cells* found in the urine are leucocytes, tubular epithelial cells, and squamous epithelial cells. *Crystals* are usually of no pathological significance. However, uric acid crystals and cystine crystals, when present in large amounts, have diagnostic importance. Casts are formed within the nephron and washed out by the flow of tubular fluid. Casts may be cellular or noncellular (Fig. 67.**1**).

Cellular casts *Leukocyte casts* are typically found in acute bacterial pyelonephritis. Fatty casts are found in nephrotic syndrome. They contain epithelial cells laden with fat droplets. *Red cell casts* are pathognomonic of glomerular bleeding and are almost diagnostic of acute glomerulonephritis. *Tubular epithelial cell casts* are most commonly present in acute tubular necrosis.

Noncellular casts *Hyaline casts* are structureless, transparent proteinaceous plugs, made largely of Tamm–Horsfall protein. Up to 1 hyaline cast per low power field (×10) is normal. They increase in proteinuria. *Granular casts* are similar to hyaline casts except that granular aggregates of proteins are embedded in them. Fewer than 1 granular cast per low power field of the microscope is considered normal. *Broad waxy casts* (*renal failure casts*) are formed in chronic renal failure in the dilated nephrons.

Renal Function Tests

Glomerular function tests

Electrophoresis of urinary proteins helps to distinguish the HMWP from the LMWP that are normally present in the urine in fairly large amounts. When present in excess of 2 g/24 hours, high molecular weight proteinuria indicates the presence of glomerular disease. If the urine deposit contains red cell casts, then glomerular inflammation is virtually certain.

GFR measurement   Clearance of inulin or creatinine can be measured for estimating the GFR. However, they are not routinely estimated in the clinics. The plasma levels of urea and creatinine are used instead.

Plasma urea and creatinine   Neither blood urea nor plasma creatinine increases much until the GFR falls below 30 ml min^{-1}. Thereafter, the plasma creatinine is an accurate and sensitive guide to further changes in renal function.

Plasma urea may give a false high in conditions of low urine flow rates like dehydration, heart failure, and nephrotic syndrome. In these conditions, the urea diffuses out of the tubular lumen to reenter the blood. Plasma urea is also abnormally high following consumption of a high protein diet, gastrointestinal bleeding (blood proteins are digested), and catabolic states like infection and steroid therapy.

Plasma creatinine levels are less affected by extrarenal causes though the urinary excretion of creatinine is proportional to the whole body muscle mass.

Proximal tubular functions tests

Electrophoresis of urinary proteins helps to detect low molecular weight proteins in the urine. Damage to the proximal tubule is associated with low molecular weight proteinuria. Large amounts of light chains of immunoglobulin can also appear in the urine in multiple myeloma and lymphoma. One differentiating feature is that in tubular dysfunction, all classes of light chains appear in the urine but in immunological disorders, they are monoclonal, i.e., they all belong to the same class (κ or λ). This can be easily be ascertained through immunoelectrophoresis.

Urine chromatography for amino acids helps in detecting the presence of abnormal amounts of amino acids. The presence of N-acetyl β-glucosaminidase (NAG) and γ-glutamyl transferase (γGT) in the urine indicates damage to the proximal tubular epithelial cells.

The **renal threshold for glucose** is a fairly good index of proximal tubular function. The renal threshold is an easily measurable index of glucose reabsorption in tubules. TmG is a more stable index than the renal threshold: the threshold is affected by the 'splay' but the TmG is not (p 413).

Distal tubular function test

Tests for renal water handling   The *specific gravity* and *free-water clearance* give preliminary information about the osmolarity of the urine. More diagnostic clues are obtained on studying the urine-diluting and urine-concentrating ability of the kidney.

The *oral water-loading* test gives accurate information regarding the urine-diluting ability of the kidneys. After voiding, the subject drinks 20 ml kg^{-1} of body weight of water over 15 minutes. Half-hourly samples are collected over the next 5 hours. Normally, more than 75% of the administered water load is excreted by 4 hours. Urine osmolarity falls to <100 mOsm kg^{-1}. The ability to produce a dilute urine is lost in chronic renal failure, which is associated with isosthenuria. Urine-diluting ability is also lost in cardiac failure because the impaired renal perfusion causes excessive proximal reabsorption of Na$^+$. In chronic liver diseases, there is secondary hyperaldosteronism, leading to an inability to concentrate urine.

The *water-deprivation test* gives accurate information regarding the urine-concentrating ability of the kidneys. Oral fluids are withheld for a period of 12 hours after which a urine sample is collected and analyzed for its volume and osmolarity. To prevent serious and life-threatening dehydration, the test should be abandoned if body weight falls by 3% at any time during the test. Normally, urine osmolarity is >900 mOsm L^{-1}. In diabetes insipidus, urine osmolarity is <300 mOsm L^{-1}. If the diabetes insipidus is nephrogenic, subsequent ADH administration will not improve it. If the polyuria is due to compulsive water drinking, the osmolarity would be normal following water-deprivation test.

Renal acidification tests indicate the H$^+$-concentrating ability of the kidney. They are unnecessary if the early morning sample has a pH of <5.3, which is sufficient proof of the renal concentrating ability.

In the *oral ammonium chloride test*, ammonium chloride is administered in gelatin capsules to maximally acidify the plasma. Hourly urine samples are collected and tested for pH. Normal subjects can reduce the urine pH to <5.0 in 2 hours. The nausea and vomiting due to ammonium chloride can be troublesome.

Liver patients should not be given ammonium salts. Alternative acidifying agents include arginine hydrochloride and calcium chloride.

The *sodium sulfate infusion test* helps to differentiate whether the impaired renal acidification is due to reduced H^+ secretion as in renal tubular acidosis or due to increased back-leakage from the tubules. Sodium sulfate solution is infused slowly. Hourly urine volumes are collected and tested for pH. Normal subjects can reduce urine pH to <5.0. If the distal tubular defect is due to the back diffusion of hydrogen ions, then urine can be normally acidified. This is because sulfates, being nonreabsorbable ions, trap hydrogen ions in the tubular lumen. If the problem is failure to secrete hydrogen ions, then the urine is not acidified even in the presence of sulfate ions.

Renal Syndromes

Nephritic syndrome

Nephritic syndrome is characterized by hematuria, red cell casts in urine, moderate proteinuria, edema, hypertension, and occasionally, oliguria. Nephritic syndrome occurs due to immunological damage to the glomerular basement membrane. The initiating cause of this immunological damage is mostly unknown except for *acute poststreptococcal glomerulonephritis* that follows an infection with group-A β-hemolytic streptococci. It occurs due to cross-reaction of anti-streptococcal antibodies with the glomerular basement membrane.

Nephritic syndrome occurs in three forms that differ in their mode of onset and clinical course. *Acute glomerulonephritis* (AGN) is diagnosed when the onset is abrupt and is not due to an acute exacerbation of a previously existing disease. AGN is mostly reversible and self-limiting. Less commonly, it progresses to RPGN, CGN or nephrotic syndrome. *Rapidly progressive glomerulonephritis* (RPGN) is diagnosed when the symptoms deteriorate rapidly, leading to chronic renal failure in a matter of weeks or months. *Chronic glomerulonephritis* (CGN) is diagnosed when the symptoms persist and progress over years or decades to chronic renal failure.

The above terms are clinical syndromes and are not to be confused with the actual disease processes (glomerulopathies), the names of which are based on the pathological changes in the glomeruli. For example, the commonest form of glomerulopathy associated with AGN, especially acute poststreptococcal glomerulonephritis, is *diffuse endocapillary glomerulonephritis*. The commonest form of glomerulopathy associated with RPGN is *crescentic glomerulonephritis*. The glomerulopathies associated with CGN are not identifiable except in the early stages. In late stages of CGN, all glomerulopathies look alike as the glomeruli become hyalinized.

Nephrotic syndrome

Nephrotic syndrome is diagnosed when proteinuria exceeds 4 g per day. The differences between nephritic

Table 68.1 Differences between nephrotic and nephritic syndromes.

	Nephrotic syndrome	Nephritic syndrome
Proteinuria	Gross	Moderate
Plasma albumin	Markedly reduced	Slightly reduced
Hematuria	Absent or traces	Marked
Blood pressure	Normal or low	Raised
Edema	Marked	Moderate
Urine volume	Normal / reduced	Reduced
Plasma lipid	Grossly elevated	Minimal increase

syndrome and nephrotic syndrome are summarized in Table 68.1. Most cases of nephrotic syndrome are of immunological origin due to unknown causes. The commonest glomerulopathy associated with nephrotic syndrome is *minimal change* GN or *lipoid nephrosis*. Some cases of nephrotic syndrome are secondary to known causes like diabetes mellitus, amyloidosis, exposure to allergens (e.g., bee stings) and toxins (e.g., mercury), infections, and drugs.

Nephrotic syndrome is characterized by *hypoalbuminemia* (due to heavy proteinuria), *edema* (because plasma oncotic pressure decreases due to hypoalbuminemia), *hyperlipidemia* (probably because the reduction in plasma oncotic pressure stimulates hepatic lipoprotein synthesis), *microcytic hypochromic anemia* (due to urinary loss of transferrin), *increased tendency to thrombosis*, especially of renal veins (due to urinary loss of antithrombin III), *vitamin D deficiency* (due to urinary loss of cholecalciferol-binding protein), *thyroid abnormalities* (due to urinary loss of thyroid-binding globulin), and *susceptibility to infections* (due to urinary loss of IgG).

Defects in tubular transport

Renal glycosuria is an autosomal-recessive disorder in which there is impaired reabsorption of glucose from the tubules. As a result, glycosuria occurs in the presence of normal blood glucose and the measured renal threshold for glucose is low. Renal glycosuria is of two types: *type A* is characterized by a low TmG and *type B* is characterized by an increased splay (Fig. 68.1).

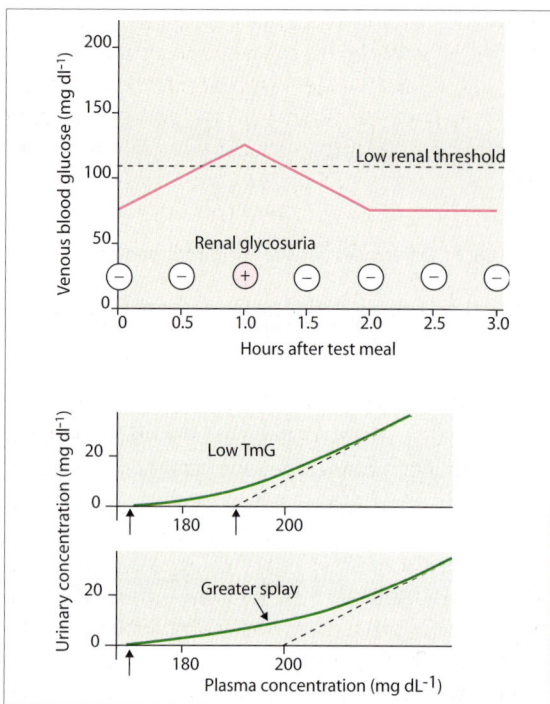

Fig. 68.1   Glucose tolerance test graph in a case of renal glycosuria (*above*), the cause of which could be a low TmG or a greater splay (*below*).

Renal tubular acidosis is a condition in which systemic acidosis occurs either due to inadequate H^+ secretion by the tubules or due to excessive back-diffusion of the secreted H^+ from lumen to blood (see sodium sulfate infusion test, p 420). Impaired H^+ secretion reduces bicarbonate reclamation. There is therefore a compensatory increase in Cl^- reabsorption that accompanies Na^+ reabsorption. Hence, the acidosis is hyperchloremic and the anion gap is normal.

In **nephrogenic diabetes insipidus**, the tubules respond poorly to circulating ADH, resulting in polyuria.

Fanconi syndrome results from a defect in proximal tubular transport of several ions (Na^+, K^+, HCO_3^-, Ca^{2+}, PO_4^{3-}) and organic substances (glucose, uric acid, proteins, and amino acids). It usually occurs as an autosomal recessive disorder.

■   Acute renal failure

Acute renal failure (ARF) is defined as an abrupt but reversible impairment of renal function which is always associated with an increase in blood urea nitrogen (BUN) and an increase in serum creatinine, and is usually associated with oliguria.

In a broad sense, acute renal failure includes prerenal azotemia, intrinsic renal failure, and postrenal azotemia. In *prerenal azotemia*, there is inadequate renal perfusion resulting in a rise in nitrogenous waste products in the blood, but without causing ischemic necrosis of renal tubules. Intrinsic renal failure is due to some renal disease, the commonest being acute tubular necrosis (see below). Other causes include rapidly progressive glomerulonephritis. *Postrenal azotemia*, i.e., obstruction to the flow of urine causes renal dysfunction and azotemia.

In a more restricted sense, ARF is used synonymously with acute tubular necrosis, which is by far the commonest cause of ARF.

Clinically, **acute tubular necrosis** goes through the following phases. The *initiating phase* (*stage of onset*) lasts for about 36 hours. The signs of the underlying cause, e.g., shock or the toxin, are prominent in this phase. The *maintenance phase* (*oliguric stage*) lasts from a few days to up to 3 weeks. It is characterized by oliguria less than 400 ml per day (anuria is rare), salt and water overload with dilutional hyponatremia, hyperkalemia, uremia, and acidosis. The *recovery phase* (*diuretic stage*) is characterized by polyuria and natriuresis, hypokalemia, and decrease in BUN.

The mainstay of the management, besides treating the underlying cause, is to overcome the oliguric phase so as to allow spontaneous recovery. Hemodialysis may be required during the crisis period. However, for the most part, the management comprises: (1) restricting water intake to match urinary output and eliminating electrolytes from administered fluids, and (2) restricting dietary proteins to reduce the load on the kidneys while allowing plenty of carbohydrates and fats to provide energy.

■   Chronic renal failure

Chronic renal failure is defined as a slowly progressive impairment of renal function associated with a reduction in the functioning renal mass. The decreased functioning renal mass has two broad effects: decreased excretion and decreased biosynthesis. The *decreased excretory capacity* results in clinical features like azotemia, fluid imbalance and/or hypertension, hyperkalemia, and metabolic acidosis. The *decreased biosynthetic capacity* results in normocytic normochromic anemia (due to reduced erythropoietin secretion) and renal osteodystrophy (due to decreased activation of vitamin D). Management of CRF includes hemodialysis and kidney transplantation.

The ions and metabolites that are affected by the decreased excretory capacity of the kidney can be classified into three groups. (1) Substances like *creatinine* and *urea* depend largely on glomerular filtration for their excretion into urine. Therefore, as the GFR falls, the plasma levels of urea and creatinine rise progressively and serve as useful indicators of GFR impairment. Even so, the plasma levels of creatinine and urea do not start rising till the GFR falls to about 30% of the normal. (2) Substances like PO_4^{3-}, K^+, H^+, and urate are excreted into urine at least partly through tubular secretion. Hence, the plasma levels of these substances rise only when the GFR falls to very low levels. (3) Substances like Na^+, Cl^- do not show any change in their plasma level even when the GFR drops to very low levels. The net urinary excretion of these ions does not change due to compensation by surviving nephrons.

The **fluid and electrolyte imbalances** associated with chronic renal failure are due largely to the reduction in the total nephron mass in the kidneys. With decrease in nephron mass the GFR decreases and with it, the filtration fraction decreases too since the RBF does not decrease as much as the GFR. The GFR in the individual surviving nephrons (the *single nephron GFR*) increases and therefore there is a rapid flow of tubular fluid in these nephrons, which reduces the tubular reabsorption of water. Hence, despite the reduced GFR, the volume of urine excreted might actually increase, resulting in polyuria. Another consequence of the rapid tubular flow is that the composition of the tubular fluid is less affected by tubular reabsorption of fluids and electrolytes. Hence, urine osmolarity is limited to a narrow range of 250 to 350 mOsm L^{-1} (specific gravity of about 1.008). The excretion of urine within a narrow range of osmolarity is called *isosthenuria*.

The surviving nephrons also show a compensatory *natriuresis* for two reasons: (1) The peritubular capillary hydrostatic pressure rises due to the arterial hypertension that is usually present in CRF. (2) The peritubular capillary oncotic pressure decreases due to a decrease in the filtration fraction and due to the hypoalbuminemia that is commonly present in CRF. Due to compensatory natriuresis, the total Na^+ concentration is unaffected in CRF.

The natriuresis is somewhat exaggerated in certain forms of CRF associated with *salt-wasting nephropathies*. In these nephropathies, the renal medulla (which contains most of the tubules) is affected more than the cortex (which contains most of the glomeruli). Hence, the GFR is relatively unaffected but the Na^+ reabsorption is greatly impaired, resulting in natriuresis.

Acidosis   The reduction of renal mass results in reduced ammoniagenesis which is a major source of urinary buffer. Reduction in urinary ammonia buffers in the tubules decreases new HCO_3^- generation and results in acidosis. Since the Na^+ concentration is unaffected in CRF, so also must be the total concentration of anions. However, since the plasma HCO_3^- level is decreased, other anions must increase in plasma. In advanced CRF, the drop in plasma HCO_3^- level is compensated by an increase in sulfates and phosphates that are retained in large amounts. Hence, a large anion gap develops. In less severe CRF, the drop in plasma HCO_3^- levels is compensated by an increase in Cl^- reabsorption and the anion gap remains normal.

Potassium balance   The plasma K^+ level is fairly well maintained in CRF. A decrease in renal mass would be expected to decrease the tubular secretion of K^+. However, there is a compensatory increase in the distal tubular secretion of K^+ for three reasons: (1) The rapid flow of tubular fluid keeps the tubular concentration of K^+ low. This facilitates tubular secretion of K^+. (2) The presence of high amounts of sulfates and phosphates in the tubular fluid increases the luminal negativity of the tubule, which promotes potassium secretion. (3) Any increase in plasma K^+ level increases aldosterone secretion. Aldosterone increases the tubular secretion of K^+.

Toxic metabolites   Although a high concentration of blood urea is characteristic of CRF, the urea is relatively nontoxic. However, urea does cause anorexia, nausea, vomiting, and hiccups, which in turn contribute to protein-calorie malnutrition in CRF. The actual toxic substances are the products of protein, amino acid, and nucleic acid catabolism. The bleeding tendency in CRF is caused mainly by guanidosuccinic acid which interferes with the activation of platelet factor-III.

Calcium and phosphate imbalances   The Ca^{2+} level decreases for two reasons. (1) The kidney is normally a major site for the activation of vitamin D (p 522). The decrease in renal mass is associated with a decrease in the circulating levels of calcitriol and a consequent decrease in the Ca^{2+} absorption from the gastrointestinal tract. (2) Decreased GFR also results in decreased PO_4^{3-} excretion and thereby, causes a slight hyperphosphatemia. The increase in plasma PO_4^{3-} causes reciprocal decrease in the ionized Ca^{2+} levels in the plasma due to deposition of $Ca_3(PO_4)_2$ in the bone and other tissues. The hypocalcemia stimulates high levels of parathyroid hormone secretion, resulting in the bone changes described below.

Bone changes   CRF results in hypocalcemia and hyperphosphatemia, which lead to two types of pathological changes: *renal osteodystrophy* and metastatic calcification (see Bone Chemistry, p 517). Renal osteodystrophy is the collective name given to the several bone changes characteristically associated with CRF. It is caused by parathyroid hormone, the secretion of which is persistently elevated in CRF due to the decrease in plasma Ca^{2+}. Renal osteodystrophy can take the form of osteomalacia or rickets, osteitis fibrosa cystica, or osteosclerosis (see Bone disorders, p 519).

Dialysis

Dialysis is a term of physical chemistry. It is the process of separation of colloids from crystalloids in a complex solution. In medical diction, it refers to the process by which some of the excess crystalloids and other toxic waste products that accumulate in blood in renal failure are eliminated from the plasma while colloids like plasma proteins and the cellular elements are retained in the plasma.

Hemodialysis is done by interposing a semipermeable membrane between the patient's blood and the dialysate, an electrolyte solution with a composition similar to that of normal plasma. As the blood equilibrates with the dialysate, its crystalloid composition becomes similar to that of the dialysate. By varying the dialysate composition, the plasma composition can be modified as desired by the clinician.

In *peritoneal dialysis*, the dialysate is infused into the peritoneal cavity and allowed to equilibrate with blood across the peritoneal membrane, which thereby acts as the dialyzing membrane.

Dialysis involves transport of both solute and water. Hence, by appropriately altering the colloid composition of the dialysate, it is also possible to use dialysis for removal of extra water from the body. Concepts of osmolar clearance, free-water clearance, and clearance of specific substances are applicable to dialysis too for quantifying its effectiveness.

Urinary Bladder

The urinary bladder stores urine at a low pressure. It has a serous, a muscular, and a mucous coat. The muscle coat is made of the detrusor muscle and consists of the outer, middle, and inner layers of smooth muscle fibers.

The *mucous membrane* epithelium is of the transitional variety. There is no muscularis mucosae in the bladder wall. The transitional cell epithelium stretches as the bladder distends, prevents the loss of fluids and electrolytes, and secretes a glycosaminoglycan barrier that prevents bacterial adherence.

Urinary sphincters

There are four urinary sphincters.[1] Two of the sphincters, viz., sphincter vesicae and intrinsic leiomyosphincter, are made of smooth muscle and are called the *internal sphincters*. The other two are skeletal muscle sphincters or rhabdomyosphincters, and are called the *external sphincters* (Fig. 69.**1** and Table 69.**1**).

The **sphincter vesicae** encircles the neck of the bladder. It is composed of smooth muscle derived from the middle layer of the detrusor. Its main function is to prevent retrograde ejaculation and therefore it is not prominent in females. Neurally mediated contraction of the sphincter vesicae occurs simultaneously with seminal emission, just prior to ejaculation. It is not essential for continence as evidenced by patients who undergo prostatectomy. During voiding, the sphincter vesicae is 'pulled open' by the contraction of the intrinsic leiomyosphincter.

The **intrinsic leiomyosphincter** is the inner longitudinal muscle of the urethra. It is an extension of the inner longitudinal muscle layer of the bladder. It has little role in continence. During voiding, its contraction 'pulls open' the bladder neck and proximal urethra.

The **intrinsic rhabdomyosphincter**, also called the sphincter urethrae, is a striated muscle sphincter. It is located inside the urogenital diaphragm and surrounds the membranous urethra. It is the tightest of all the sphincters. It is the main sphincter that maintains continence at rest (tonic continence).

The **extrinsic rhabdomyosphincter** is formed by the periurethral muscles of the levator ani (pelvic diaphragm). It has mainly fast-twitch, easily fatigable fibers. It is not essential for tonic continence. However, it causes urethral closure quickly in response to sharp increases in intravesical pressure such as those that accompany coughing or straining (phasic continence). Contraction of this sphincter causes inhibition of detrusor activity through a spinal reflex. Hence, it is especially suitable for stopping micturition in midstream. After radical prostatectomy, patients are able to voluntarily interrupt urination in midstream although they have tonic incompetence due to lack of the intrinsic rhabdomyosphincter.

Sensory innervation of the urinary tract

Sensory afferents from the bladder wall reach the spinal cord (T10–L2) through the pelvic splanchnic nerve

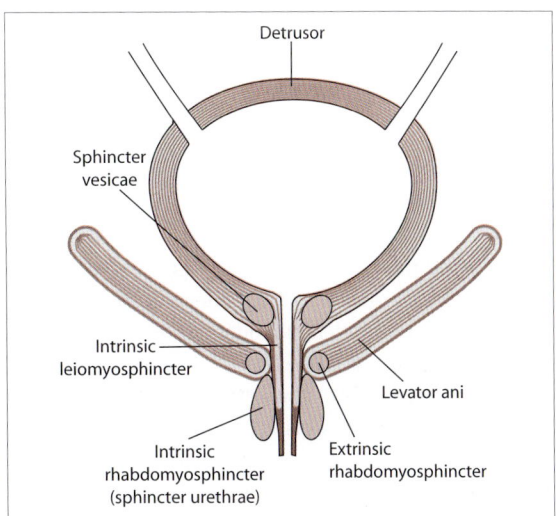

Fig. 69.**1**   Urinary sphincters.

---

[1] Anatomists describe only two sphincters, viz., the sphincter vesicae and the intrinsic rhabdomyosphincter, sometimes simply called the internal and external sphincter respectively. Surgeons describe two more sphincters that are described here.

Table 69.**1** Comparison of the urinary sphincters.

Urinary sphincter	Function
Sphincter vesicae	Prevents retrograde ejaculation.
Intrinsic leiomyosphincter	When contracted, pulls open the bladder neck.
Intrinsic rhabdomyosphincter	Main provider of tonic continence.
Extrinsic rhabdomyosphincter	Main provider of phasic continence.

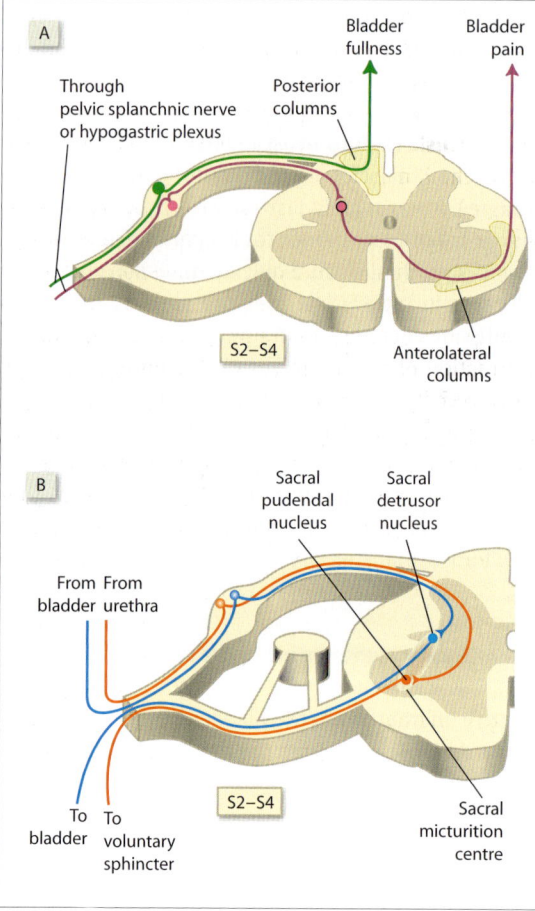

Fig. 69.**2** [A] Afferent pathway for conscious bladder sensations. [B] The reflex arc for the micturition reflex.

(nervi erigentes) and the hypogastric plexus (Fig. 69.**2**A). Sensory afferents from the urethra reach the spinal cord through the pudendal nerve.

The sensation of bladder distension arises from the stretch receptors of the detrusor. The sensation of imminent voiding associated with maximal bladder filling originates in the periurethral striated muscle. These sensations ascend in the dorsal columns of the spinal cord to reach the pontine and suprapontine micturition centers.

The pain fibers from the bladder ascend in the anterolateral columns of the spinal cord. Bladder pain is therefore relieved by cutting the anterolateral columns of the spinal cord. After the anterolateral cordotomy, the patient is still aware of bladder filling and of the desire to micturate.

Motor innervation of the urinary tract

The **sympathetic innervation** of the bladder originates from the intermediolateral gray horn of the spinal cord at the level of T10–L2 segments of the cord. They travel via the hypogastric nerves to the bladder, bladder neck, and urethra. Sympathetic fibers are inhibitory to the detrusor and the intrinsic leiomyosphincter but excitatory to the sphincter vesicae. Damage to sympathetic nerves predisposes to retrograde ejaculation.

The **parasympathetic innervation** originates from the sacral detrusor nucleus located in the intermediolateral gray horn of the sacral spinal cord (S2–4). The nucleus is formed by the soma of preganglionic parasympathetic neurons. Parasympathetic efferents from the sacral detrusor nucleus leave the cord through its ventral root, pass through the pelvic splanchnic nerve, and relay in the ganglia near or within the bladder and urethra. The postganglionic fibers innervate the muscles of the bladder and urethra. The parasympathetic fibers are excitatory to the detrusor muscle and the intrinsic leiomyosphincter and inhibitory to the sphincter vesicae.

Somatic motor innervation The somatic motor nerves originate from the *sacral pudendal nucleus*, also known as the nucleus of Onuf, located in the ventral horn of the S2 segment of the spinal cord. They reach the extrinsic rhabdomyosphincter through the perineal branch of the pudendal nerve (S2–4), and the intrinsic rhabdomyosphincter through the inferior hypogastric plexus and the pelvic splanchnic nerve.

The efferent bladder control is summarized in Table 69.**2**.

Micturition Centers

The **sacral micturition center** (S2–4) is the spinal center for the micturition reflex (Fig. 69.**2**B). It consists of the sacral detrusor nucleus and the sacral pudendal nucleus. Afferent impulses from the detrusor and urethra reach the *sacral micturition center* (SMC) through the dorsal root of the sacral cord. In the SMC, the afferents excite the *sacral detrusor nucleus*, causing detrusor

The **cerebral cortex** has a detrusor motor area that is located in the medial frontal lobe (superior-frontal gyrus). The limbic area of the cerebral cortex (anterior cingulate gyrus and anterior genu of the corpus callosum) excites the SMC. It provides the anatomical basis for the effect of emotions on voiding.

The **basal ganglia** inhibits the SMC and thereby, the detrusor activity. Parkinson's disease (p 618) is often associated with poor detrusor function.

◼ Initiation of micturition

Common experience suggests that micturition is a voluntary action. Yet in certain situations, the voluntary initiation of micturition may be difficult, e.g., if the bladder is not full or there is an awareness of being watched in the act! Thus, micturition is not entirely voluntary but is dependent on certain involuntary reflexes.

Micturition is voluntarily initiated by relaxing the pelvic diaphragm and the external urethral sphincters. Relaxation of the pelvic diaphragm reduces the support under the bladder, which therefore sags under its own weight. The sagging pulls open the bladder neck. The sagging also stretches the intrinsic leiomyosphincter, triggering its contraction and further pulling open the bladder neck. As the bladder neck opens, urine trickles into the urethra through the relaxed external sphincters. The presence of urine in the urethra triggers reflex contraction of the detrusor (Barrington's reflex), which forces more urine down the urethra. This sets up a positive feedback cycle which ends with the complete emptying of the bladder.

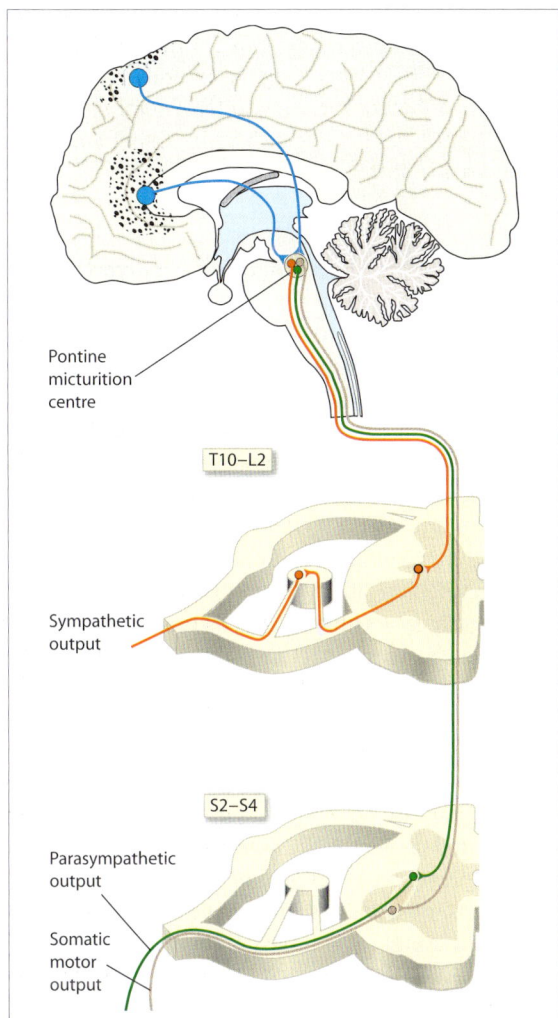

Fig. 69.3 Supraspinal control centers for micturition.

contraction. They also inhibit the *sacral pudendal nucleus*, thereby relaxing the intrinsic and extrinsic rhabdomyosphincters.

The **pontine micturition center** (also called Barrington's center) corresponds to the locus ceruleus of the rostral pons (Fig. 69.**3**). Neurons from the pontine micturition center (PMC) descend in the reticulospinal tracts and exert control over the sacral micturition center and thoracolumbar sympathetic outflow. The PMC coordinates the activity of the bladder and urinary sphincter and relays inputs from suprapontine centers. Any lesion involving the PMC or its descending pathways to the SMC causes detrusor-sphincter dyssynergia, i.e., loss of coordination between the bladder and distal sphincteric mechanism. The associated disruption of suprapontine influences results in decreased storage and incomplete voiding.

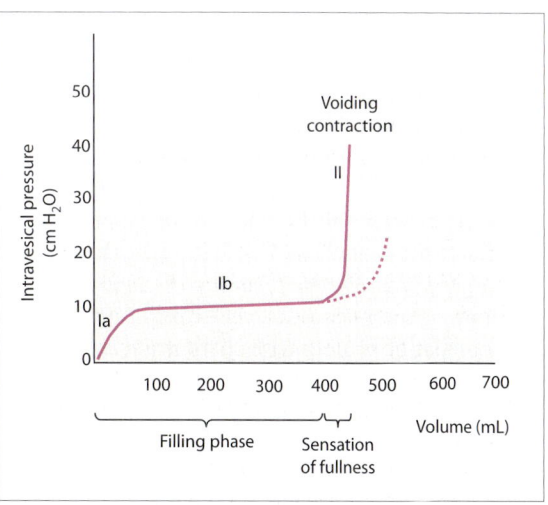

Fig. 69.4 A normal cystometrogram.

Table 69.**2** Efferent nervous controls of the bladder.

Nerve supply	Origin	Passes through	Contracts	Relaxes
Sympathetic fibers	Intermediolateral spinal gray horn (T10–L2)	Hypogastric plexus	Sphincter vesicae	Detrusor and intrinsic leiomyosphincter
Parasympathetic fibers	Sacral detrusor nucleus (S–4)	Pelvic splanchnic nerve	Detrusor and intrinsic leiomyosphincter	Sphincter vesicae
Somatic fibers	Sacral pudendal nucleus (S2–4)	Hypogastric plexus and pelvic splanchnic nerve	Rhabdomyosphincters	—

Table 69.**3** Comparison of neuropathic bladders.

Bladder disorder	Reflexes	Overflow	Bladder size
Flaccid neuropathic	Absent	Present	Large
Hypertrophic areflexic	Absent	Present	Small
Spastic neuropathic	Exaggerated	Absent	Small

Neuropathic Bladders

A **flaccid neuropathic bladder** results from deafferentation of the bladder. In the absence of the micturition reflex, the bladder is flaccid and becomes overfilled till urine starts leaking out (overflow incontinence). The bladder always remains full and the residual urine is high, but the intravesical pressure remains low. Due to overfilling, the bladder wall is thin and distended.

A **hypertrophic areflexic bladder** results when there is damage to both the afferents and efferents of the bladder. In other words, it occurs in a decentralized (autonomous) bladder. Due to the abolition of the micturition reflex, decentralization initially results in a flaccid neuropathic bladder with overflow incontinence, as occurs in a deafferented bladder. Subsequently, the detrusor develops denervation hypersensitivity and results in a small, hypertrophic areflexic bladder. Due to decreased bladder distensibility, bladder filling is associated with a steep rise in bladder pressure with consequent overflow incontinence. The bladder never voids reflexly and always remains full. Residual urine is low.

A **spastic neuropathic bladder** occurs when the SMC is isolated from the brain due to suprasacral spinal cord injury. *Immediately following the injury*, there is abolition of the micturition reflex resulting in a flaccid neuropathic bladder with all its attendant features, e.g., overflow incontinence etc. *In the initial stages of recovery*, the reflex excitability of the striated muscle of the sphincters are restored. The incontinence disappears and there is urinary retention. *In the later stages of recovery*, the reflex excitability of the smooth muscle returns and the bladder capacity is reduced. In the absence of supraspinal inhibition, the micturition reflex is exaggerated, resulting in spasticity. The micturition reflex is triggered whenever the bladder is distended, resulting in incontinence. Sphincter spasticity and detrusor-sphincter dyssynergia leads to detrusor hypertrophy and high voiding pressures (i.e., intravesical pressure at which voiding is initiated). A few patients develop the ability to empty the bladder reflexly by using trigger techniques, i.e., by tapping or scratching the pubic area or external genitalia.

Autonomic dysreflexia occurs in patients with spinal cord lesions above the level of sympathetic outflow, i.e., above T1. In these patients, afferent impulses to the sacral spinal cord can trigger a series of sympathetically mediated responses, including hypertension, headache, piloerection, and sweating. This phenomenon is known as *autonomic dysreflexia*. The triggering stimuli could be visceral (overdistension of bowel or bladder, penile erection) or somatic (spasm of lower extremities, insertion of a catheter, dilatation of the external urethral sphincter, ejaculation) in origin. Treatment includes bladder catheterization (for immediate relief), sphincterotomy, and bladder deafferentation.

Cystometry

Cystometry is one of a battery of tests called the *urodynamic studies* that record the urinary bladder activity, urethral sphincter activity, and urinary flow rate. The urodynamic studies relating to the bladder function are cystometry and radiographic (cinefluoroscopy) studies. The latter is now done only infrequently. There are two methods of doing cystometrography, viz., voiding cystometry and static cystometry.

Voiding cystometry allows physiologic filling of the bladder with urine. Intravesical pressure is recorded through a urethral catheter, starting when the patient's bladder is empty and continuing until the bladder is full. The patient is then asked to urinate. The disadvantage with this method is that the bladder volume is inferred from the amount of urine voided, assuming that there is no residual urine.

In **static cystometry**, the bladder is progressively filled with water through a urethral catheter, and the intravesical pressure is recorded either through the catheter itself or through a suprapubic cannula. This method permits accurate determination of the bladder volume and pressure at each level of filling. The disadvantage with this method is that the fluid is introduced at a more rapid rate than occurs naturally through urine, which could affect bladder function.

In both methods, the pressure recorded from the bladder is actually the sum of both intraabdominal and intravesical pressure. Hence, true detrusor pressure is the intravesical pressure minus the intraabdominal pressure. Intraabdominal pressure is recorded simultaneously by a small balloon catheter inserted high in the rectum and connected to a separate transducer. In clinical practice, however, it is usually sufficient to make a note of when the patient is straining (by observing the abdominal contractions) and take them into consideration while interpreting the cystometrogram.

A normal cystometrogram shows three phases of filling (Fig. 69.**4**). *Phase Ia* is the initial phase of filling up to 50 ml. It is associated with a slight increase in pressure to about 10 cm of H_2O. *Phase Ib* is the next phase of filling that lasts till the bladder volume is about 400 ml. As the bladder fills up, the bladder smooth muscles relax (due to smooth muscle plasticity), permitting an increase in the bladder volume. Thus the pressure remains unchanged in accordance with Laplace's law. *Phase II* is denoted by a sharp rise in pressure as voiding is initiated. Normally, the voiding contractions raise the intravesical pressure by about 30 cm of water which is called the voiding pressure. The dotted lines beyond phase II denotes the intravesical pressure changes with further filling if voiding is not initiated.

Cystometry is done to ascertain the following bladder functions: (1) The *bladder capacity* is the bladder volume at which voiding is irresistible. It is normally 400–500 ml. It is higher in women who have trained themselves to retain large volumes of urine. (2) *Accommodation* is the ability of the bladder to accommodate large volumes of urine without significant rise in the intravesical pressure. Normally, the bladder pressure remains almost constant during filling up to the point of voiding. (3) *Bladder sensations* enable the perception of bladder fullness. The sensation of fullness is first perceived when the volume reaches 150–250 ml. There is a definite sense of fullness at 350–450 ml. The desire to void occurs when the bladder feels full. (4) *Bladder contractions* are assessed from the ability of the bladder to contract when full, to sustain a contraction till it is empty, and to contract in response to parasympathomimetic drugs. Normally, there is no residual urine in the bladder after voiding. Bladder contractions are not present during the filling phase, and the presence of premature bladder contractions during filling is called unstable bladder activity. (5) *Voluntary bladder control* is the ability to initiate voiding even before filling is complete, to inhibit voiding till the bladder capacity is maximal, and the ability to stop voiding in midstream.

Gastrointestinal System

70.   Events in the Mouth and Esophagus

71.   Events in the Stomach

72.   Events in the Duodenum

73.   Events in the Small Intestine

74.   Events in the Colon

75.   Gastrointestinal Hormones

76.   Gastrointestinal Disorders

Mechanical Events

Mastication

Mastication involves movements of the jaws, action of the teeth, and coordinated movements of the tongue and other muscles of the oral cavity. The masticatory muscles (masseter, temporalis, and pterygoids) and the tongue show automatic coordinated movements that are controlled by a brainstem center. Mastication is aided by the secretion of saliva which acts as a lubricant.

Deglutition

The coordination center of swallowing (deglutition) is located in the reticular formation between the IX and X cranial nerve nuclei. The afferent arc comprises the trigeminal, glossopharyngeal, and vagus nerves. The efferent arc reaches the pharyngeal musculature and the tongue through the trigeminal, facial, and hypoglossal nerves. Swallowing has three phases: the oral phase, the pharyngeal phase, and the esophageal phase.

The **oral phase** is the voluntary phase in which the bolus is squeezed out of the oral cavity into the pharynx. This squeezing is made possible through the following steps: (1) The jaws are shut and the lips are closed. (2) The tongue brings the bolus into the midline between the anterior portion of the tongue and the hard palate. The tip of the tongue then presses firmly against the roof of the hard palate and limits the bolus anteriorly. (3) The voluntary contraction of mylohyoid muscle pushes the bolus toward the posterior pharyngeal wall.

In the **pharyngeal phase**, the bolus enters the pharynx and stimulates the sensory nerve endings of the glossopharyngeal nerve in the posterior pharyngeal wall, soft palate, and epiglottis to start off the deglutition (swallowing) reflex which has its center, the *deglutition center*, in the medulla. The deglutition reflex coordinates the sequential occurrence of the following steps (Fig. 70.**2**).

(1) The oral cavity is shut off from the pharynx by the approximation of the palatopharyngeal arch. (2) The nasopharynx is shut off from the pharynx by the elevation of the soft palate. (3) The glottis is shut off from the pharynx by the approximation of the vocal cords. (4) The hyoid is raised by the contraction of the digastric and the geniohyoid. The larynx rises with the hyoid and brings the epiglottis in the path of the bolus. (5) The bolus tilts the epiglottis backward over the closed glottis. The lumen of the esophagus is pulled open by the forward movement of the larynx and trachea, the posterior walls of which are attached to the anterior walls of the pharynx and the esophagus respectively. (6) Respiration is arrested. (7) The upper esophageal (hypopharyngeal) sphincter (formed by the cricopharyngeus muscle, which guards the upper end of the esophagus, briefly relaxes and the bolus enters the upper esophagus. (8) The cricopharyngeus contracts, glottis reopen, and breathing resumes.

Esophageal phase Once the bolus enters the esophagus, it is propelled by peristalsis at a velocity of 2 to 3 cm s^{-1}. The movement of the bolus is aided by gravity and therefore food travels through the esophagus faster in the standing than in the supine position. However, peristaltic waves are strong enough to propel food against gravity and hence, swallowing is possible even when a person is standing on his head.

There are two types of peristaltic waves in the esophagus. *Primary peristaltic waves* originate in the pharynx during the pharyngeal phase of swallowing and travel down the esophagus. The contraction of the esophageal wall begins at the upper esophageal sphincter when the bolus enters the esophagus and travels down the lower esophageal sphincter. Primary peristalsis thus ensures that the lower sphincter relaxes in synchrony with the relaxation of the upper esophageal sphincter.

Secondary peristaltic waves originate in the esophagus itself when the esophageal wall is stretched by the bolus. Secondary waves continue to be produced till the bolus is dislodged from the esophagus into the stomach. They begin above the bolus and push the bolus down. The propulsive force produced is proportional to the bolus size.

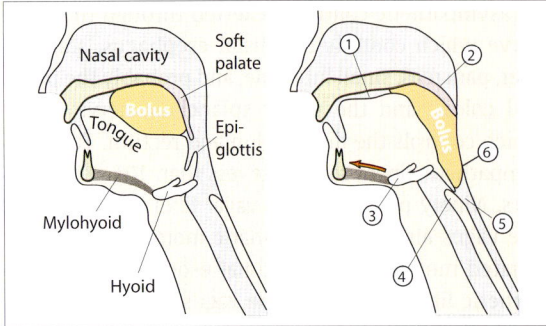

Fig. 70.**2** Stages of swallowing. (1) The tongue rises, obliterating the oropharynx. (2) The epiglottis rises, shutting off the nasopharynx. (3) The hyoid rises and moves forward. (4) The anterior wall of the esophagus is drawn forward by the hyoid. (5) The esophageal lumen is pulled open. (6) The epiglottis tilts backward to shut off the glottis.

Prevention of esophageal reflux

Anatomically, the *lower esophageal sphincter* is only vaguely identifiable as a slight thickening of muscle coat of the esophageal wall. It is about 5 cm in length of which 3 cm lie below the diaphragm and 2 cm above it. Part of the ability of the esophagus to resist reflux depends upon the 3 cm segment that lies below the diaphragm. Any rise in the intraabdominal pressure squeezes not only the stomach but also the intraabdominal part of the esophagus, occluding its lumen so that the reflux is prevented. In pregnancy, the enlarged uterus pushes the stomach up and pushes the intraabdominal portion of the esophagus into the thorax. Hence, reflux is common in pregnancy. Another cause of the reflux is the high progesterone level in pregnancy, which reduces the lower esophageal sphincteric tone.

Reflux is also prevented by the oblique angle of entry of the esophagus into the stomach, which is called the *angle of His* (Fig. 70.**3**). Any increase in intragastric pressure tends to push the stomach upward and to the right, and thereby compress and close the end of the esophagus. The angle of His is almost nonexistent in many infants and the esophagus tends to form a straight line with the stomach. Hence, reflux is quite common in infants.

The tone of the lower esophageal sphincter is mainly under vagal (cholinergic) control. Hence, parasympatholytic agents like atropine should not be given to a patient with reflux esophagitis. When reflux does occur, it stimulates secondary peristaltic waves that sweep the offending material back down into the stomach.

Secretory Events

The principal secretory event in the mouth is the secretion of the saliva. About 1 liter of saliva is secreted each day by the three salivary glands: 250 ml is secreted by the parotid, 700 ml by the submandibular, and 50 ml by the sublingual glands. Depending on the type of secretion, salivary acini are categorized into two types, viz., serous and mucous. Serous acini secrete the watery saliva containing more than 90% water and mucous acini secrete a more viscous fluid containing the glycoprotein mucin, which gives the saliva its sticky and viscous texture. The parotid gland is purely serous, the sublingual gland is largely mucous while the submandibular is a mixed gland (Fig. 70.**4**).

Saliva has the following functions: (a) It keeps the mouth moist and clean, and protects the tooth enamel. (b) It acts as a lubricant and thereby aids in speech, chewing, and swallowing. (c) It helps in bolus formation by acting as a glue. (d) It dissolves food particles and is therefore necessary for taste. (Taste receptors respond only to dissolved substances.) (e) It is alkaline and therefore helps to neutralize the gastric juices that might regurgitate into the esophagus. (f) It contains amylase which initiates digestion of carbohydrates and lipase which contributes to fat digestion. (g) It contains lysozyme and lactoferrin which have antibacterial action.

Composition of saliva   Saliva is composed of water, electrolytes, enzymes, glycoproteins, and growth factors. The composition of saliva depends on the particular salivary gland, the stimulus, and the rate of flow. The composition of saliva at basal flow rates is given in Table 70.**1**.

The organic components of saliva are secreted by the duct cells of the salivary gland. The salivary fluid is formed by transudation of plasma and therefore

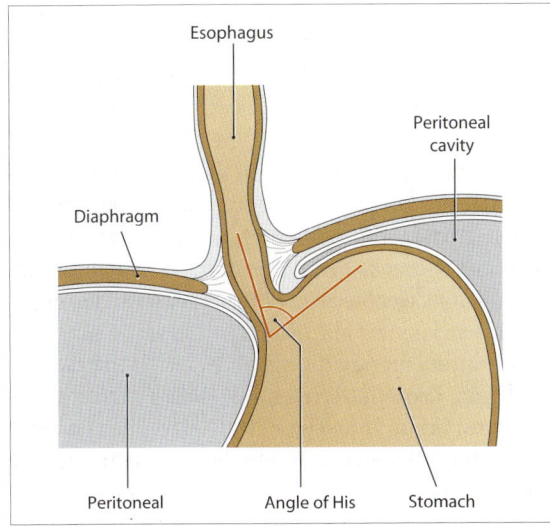

Fig. 70.**3**   The angle of His.

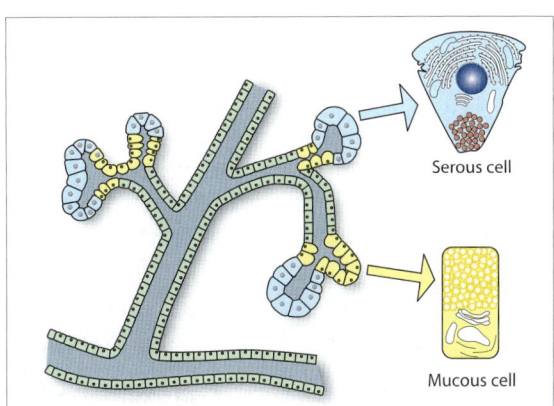

Fig. 70.**4**   Salivary acini and ducts. Insets show the serous and mucous cells.

Table 70.**1**   Characteristics of saliva.

Tonicity	Hypotonic
pH	7.0 – 8.0
Na$^+$ and Cl$^-$	Lower than in plasma.
K$^+$ and HCO$_3^-$	Higher than in plasma.
Other ions	Ca^{2+}, Mg^{2+}, PO$_4^{3-}$
Organic contents	Proteins (e.g., amylase) and glycoproteins (e.g., mucin)

is isotonic when freshly formed. During its transit through salivary ducts, Na$^+$ and Cl$^-$ are reabsorbed from it while K$^+$ and HCO$_3^-$ are secreted into it. Since the absorption of NaCl is more rapid than the secretion of KHCO$_3$, and since the duct epithelium is not freely permeable to water, the saliva becomes hypotonic. Aldosterone promotes Na$^+$ reabsorption and K$^+$ secretion in the salivary ducts. The slower the salivary flow rate, the greater is the change in the ionic composition of saliva in the ducts. Despite the secretion of bicarbonates into the saliva, the salivary pH at low flow rates is about 7.0 because some amount of Na$^+$ is reabsorbed from the ducts through Na$^+$-H$^+$ antiport.

Regulation of salivary secretion

Salivary secretion is exclusively under neural control (Fig. 70.**5**). The salivary glands are innervated by both parasympathetic cholinergic nerves and sympathetic adrenergic nerves.

The *preganglionic parasympathetic fibers* to the salivary gland originate in the salivatory nuclei of the brain stem (see Fig. 96.**4** and Table 96.**3**). They reach the submandibular and lingual glands through the facial nerve and the parotid gland through the glossopharyngeal nerve. Parasympathetic postganglionic cholinergic nerve fibers secrete acetylcholine and vasoactive intestinal peptide (VIP). These nerves may be stimulated (a) through *unconditioned reflex*, e.g., by the presence of food (especially, dry or sour food) in the mouth, or (b) through *conditioned reflex*, e.g., the smell or even the thought of good food.

Parasympathetic stimulation causes profuse secretion of watery saliva. The secretion of amylase is also increased. However, due to the large increase in watery secretion, the amylase concentration in the saliva decreases. The local vasodilatation required for increased salivary secretion is caused by VIP, which is a cotransmitter with acetylcholine in some of the postganglionic parasympathetic neurons. The tissue kallikrein secreted by the actively secreting salivary cells contributes to the hyperemia (see Fig. 40.**2**).

Sympathetic discharge causes vasoconstriction and inhibits the secretion of serous saliva but increases the enzyme concentration in saliva. The sympathetic discharge associated with fear and excitement makes the mouth dry. Cholinergic and α-adrenergic agonists act through the group-IIA hormonal mechanism while β-adrenergic agonists act through group-IIC hormonal mechanism (see Chapter 81).

Reflex salivary secretion occurs when mechanoreceptors and chemoreceptors in the oral cavity respond to the texture of food and its chemical composition. They are also stimulated by dryness of the mouth. The afferent impulses are integrated in the *salivation center* (distinct from the salivary nucleus) in the medulla. This center is near the centers regulating respiration and vomiting. The salivation center also receives inputs from the cerebral cortex, amygdala, and hypothalamus, which explains the conditioned reflex described by Pavlov in which dogs were conditioned to salivate at the sound of a bell.

Digestive Events

Digestive events in the mouth include the actions of two enzymes, an amylase and a lipase. *Salivary amylase* is an α-amylase (also called ptyalin) with an optimum pH of 6.7. It breaks down the starches by hydrolyzing the 1:4 α linkages (see Fig. 72.**6**). Its action is very short-lasting because it is inactivated by the acidic gastric pH shortly after entering the stomach.

Lingual lipase is secreted by Ebner's glands on the dorsal surface of the tongue. The lipase remains active in the stomach and contributes substantially to fat digestion in the stomach.

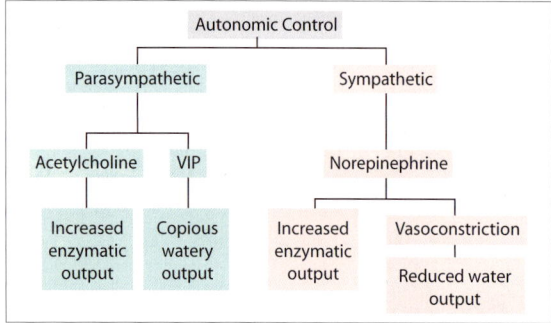

Fig. 70.**5**   Effect of autonomic nerve stimulation on salivary secretion.

The stomach is divided into the cardia, fundus, corpus, antrum, and pylorus (Fig. 71.**1**). Movement of food in and out of the stomach is guarded by two sphincters, the cardiac and the pyloric. The gastric mucosa forms prominent folds, called rugal folds, that increase its surface area.

The stomach has motor, secretory, and digestive functions. (1) The stomach acts as a reservoir: it relaxes to accommodate large volumes of food. (2) It grinds food to optimum-sized particles. (3) It mixes the bolus with the gastric juice and converts the bolus into a chyme. (4) It partitions the food, retaining the solid portion of the meal until most of the liquid has emptied. (5) It sieves the food, retaining larger particles, permitting more time for their further breakdown. (6) It regulates the amount of chyme delivered to the intestine. (7) It secretes hydrochloric acid which disinfects the food. (8) It secretes pepsin which begins the digestion of proteins.

Mechanical Events

Motility of the empty stomach

Migrating motor complexes (MMCs) are the contraction waves produced by the electrical activity of the single-unit smooth muscle of the gastrointestinal tract. The electrical activity is called the *basal electrical rhythm* (BER). It originates in pacemaker cells located in the outer circular muscle layer near the myenteric plexus. The pacemaker cells are star-shaped mesenchymal cells (the interstitial cells of Cajal) whose processes make synaptic contacts (electrical synapses) with the smooth muscle cells. When there are no MMCs, the BER consists of rhythmic oscillation of the resting potential between -65 and -45 mV. The oscillations occur due to rhythmic changes in Ca^{2+} and K^+ permeabilities. The MMCs occur when the electrical oscillations are superimposed with spikes.

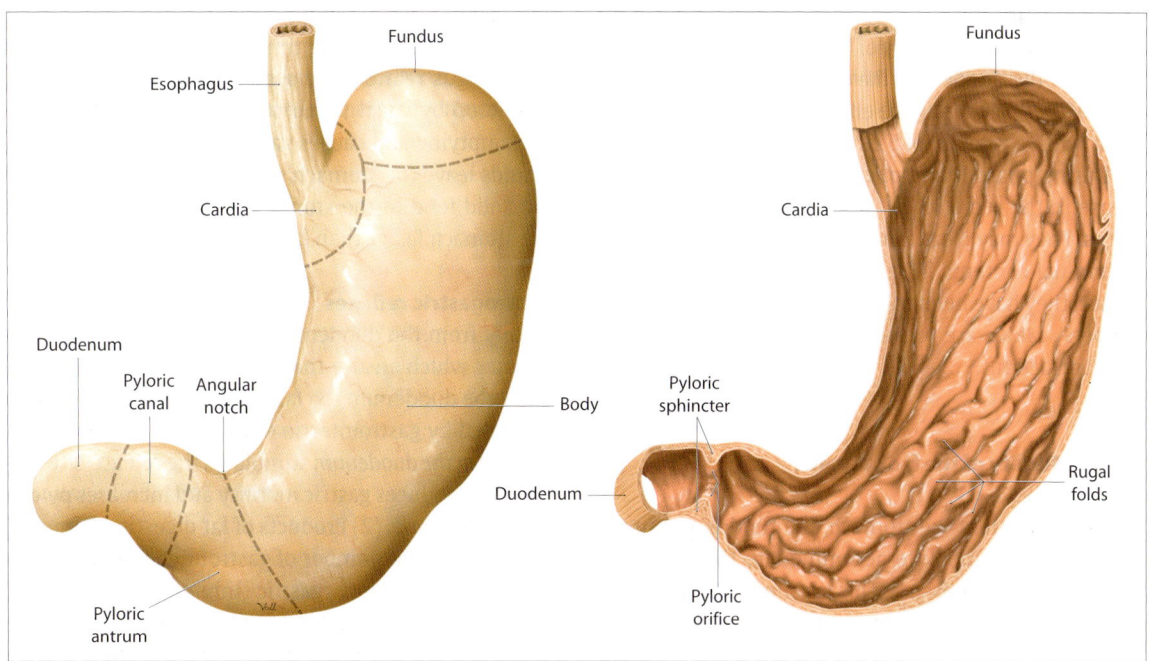

Fig. 71.**1** (*Left*) Parts of the stomach. (*Right*) Interior of the stomach.

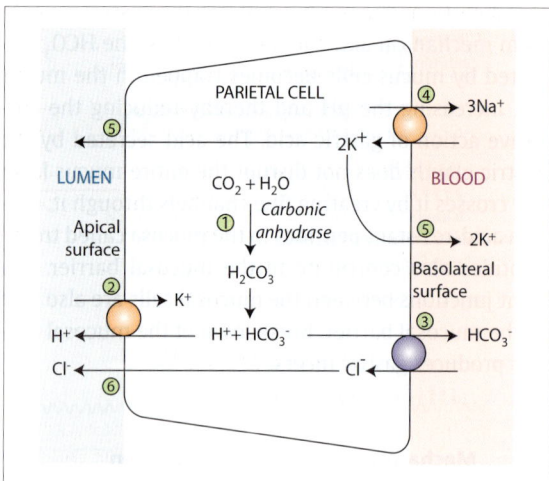

Fig. 71.**4**   Steps (1–6) in the secretion of HCl by the parietal cell of the gastric gland.

Table 71.**1**   Neurohormonal controllers of gastric acid secretion.

Neurohormone	Mechanism*	Action
Acetylcholine (muscarinic)	Gp IIC	Stimulatory
Gastrin	Gp IIC	Stimulatory
Histamine	Gp IIA (Gs)	Stimulatory
Prostaglandin E$_2$	Gp IIA (Gi)	Inhibitory
Somatostatin	Gp IIA (Gi)	Inhibitory

* See Chapter 81 for details.

a meal, there is a rise in blood pH which is called the post-prandial *alkaline tide*.

Secretion of chloride ions   (1) The Na$^+$-K$^+$ pump located on the basolateral membrane of the parietal cell pumps out three Na$^+$ ions for every two K$^+$ ions pumped in. The interior of the parietal cell therefore becomes negative. (2) The K$^+$ ions that are pumped in diffuse out through the K$^+$ channels present on the basolateral as well as apical membranes. This diffusion further increases the intracellular negativity of the parietal cell. (3) The high intracellular negativity forces out Cl$^-$ ions through the Cl$^-$ channels located on the apical membrane.

■   **Regulation of gastric acid secretion**

In an unstimulated parietal cell, considerable amounts of inactive H$^+$-K$^+$ ATPase are stored inside the cells on the membranes of the endoplasmic reticulum. When activated, these H$^+$-K$^+$ ATPase-bearing membranes fuse with the apical membrane of the cell, greatly increasing its surface area and delivering the H$^+$-K$^+$ ATPase to the site where it can pump out H$^+$. On removal of the stimulus, the apical cell membrane is internalized into the cell to reform the cytoplasmic tubulovesicular membrane system and storing the H$^+$-K$^+$ ATPase. The fusion of the cytoplasmic tubulovesicular membranes with the apical cell membrane requires high concentrations of Ca^{2+}, which is brought about by neurohormonal mediators.

There are five well-known neurohormonal mediators (Table 71.**1** and Fig. 71.**5**) of gastric acid secretion. Three of these are stimulatory and two are inhibitory.

Receptors for each of these mediators are present on the membrane of the parietal cell. These mediators are secreted mainly by three types of paracrine cells.

G cells are located at the base of the gastric glands and are especially abundant in the pyloric gastric glands. They secrete gastrin which stimulates HCl secretion. Gastrin secretion is stimulated by GRP (gastrin-releasing peptide) and inhibited by somatostatin.

D cells are located adjacent to the G cells and parietal cells. They secrete somatostatin which inhibits HCl secretion in two ways; directly, by inhibiting the parietal cells and indirectly, by inhibiting the G cells. GRP stimulates acid secretion by inhibiting somatostatin release (see Fig. 87.**7**). On the other hand, secretin, enteroglucagon, GIP, and VIP inhibit gastric secretion by stimulating somatostatin release.

Enterochromaffin-like (ECL) cells are found in the corpus of the stomach, in the base of the gastric gland. They secrete histamine which stimulates HCl secretion from parietal cells. ECL cells bear both gastrin receptors and ACh receptors. They release histamine in response to both circulating gastrin as well as the ACh released by vagal fibers. Stimulation of ECL cells seems to be an important mechanism through which gastrin stimulates acid secretion.

■   **Effect of vagus on gastric secretion**

Vagal fibers to the stomach have two neurotransmitters, gastrin releasing peptide (GRP) and acetylcholine. The GRP increases gastrin secretion from G cells with consequent increase in acid secretion. The GRP also inhibits somatostatin secretion from D cells and thereby disinhibits HCl secretion from parietal cells. Acetylcholine increases the secretion of gastric acid, pepsin, and mucus. Part of the acid secretion is mediated by ECL cells that secrete histamine.

■ Phases of gastric acid secretion

Interdigestive phase (basal acid secretion) Acid is continuously secreted by the stomach even between meals and during sleep. A circadian rhythm is seen, with basal secretion reaching its peak around midnight and its trough around 7.00 am. Since most of the basal secretion is abolished by vagotomy, it is believed that the interdigestive phase of gastric acid secretion is vagally mediated. Like the cephalic phase, the basal acid secretion is influenced by psychic factors (Fig. 71.**6**).

The **cephalic phase** of gastric secretion accounts for up to 50% of the acid secreted in response to a normal meal. It is vagally mediated and is easily conditioned. The unconditioned stimulus is the presence of food in the mouth. Conditioned stimuli increasing gastric secretion include the sight, smell, and thought of food. The conditioned reflex involves activation of the anterior hypothalamus and parts of the adjacent orbital frontal cortex. The cephalic phase of gastric secretion is influenced by psychic states: it is increased with anger and hostility and is reduced in fear and depression.

The **gastric phase** of acid secretion comes into play when food makes contact with the gastric mucosa. It accounts for up to 50% of the acid secretion in response to a meal. Acid secretion in this phase is brought about by two factors: (1) Gastrin secretion, which in turn is brought about by two factors, a reduction in antral acidity due to the buffering effect of the meal, and the stimulatory effect of small peptides and amino acids. (2) Stretch of the stomach wall, which activates a vago-vagal reflex as well as a local intragastric reflex.

Intestinal phase When food enters the intestine, gastric secretion is inhibited by the same intestinal factors that reduce gastric motility through the enterogastric reflex (see above). Briefly, they are: (1) acid in the duodenum, (2) products of fat digestion, (3) osmolarity of the duodenal chyme, and (4) mechanical distension of the duodenum. Products of protein digestion, however, have a slight stimulatory effect on gastric acid secretion. It accounts for about five percent of the total gastric acid secretion that occurs following a meal. The hormone enterogastrone was thought to mediate the inhibition of gastric secretion in the intestinal phase.

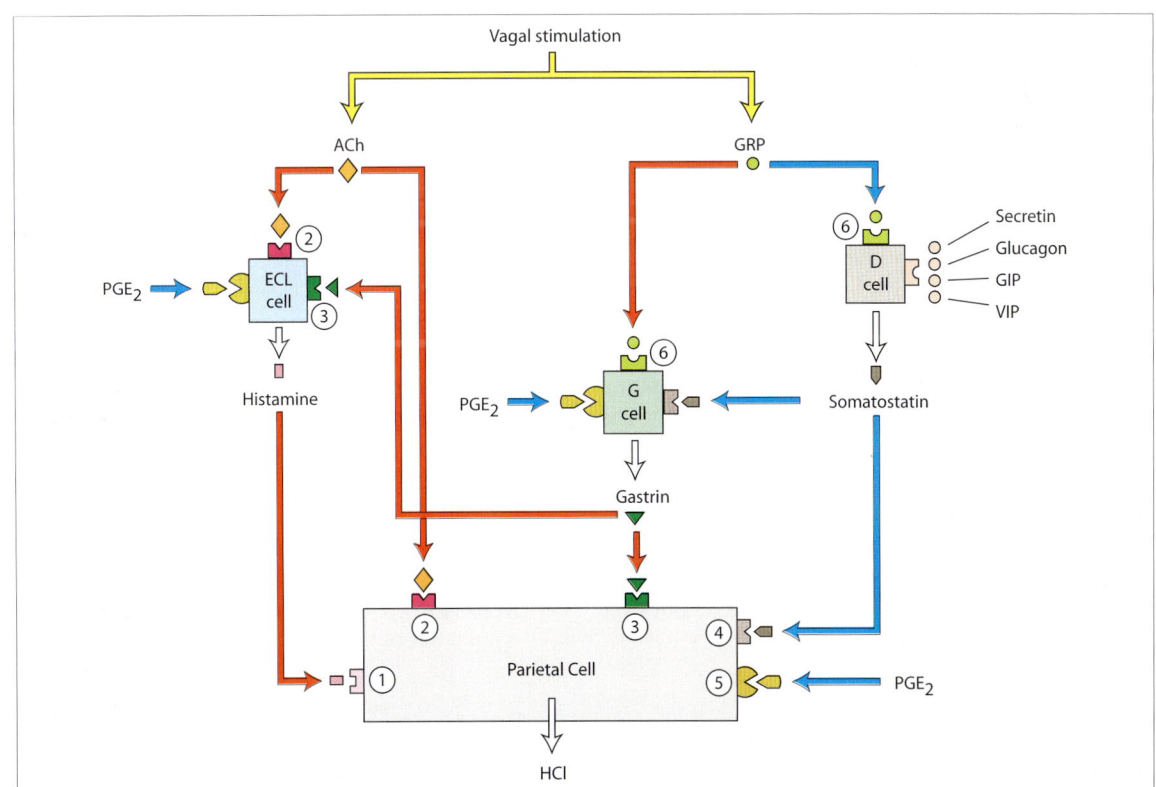

Fig. 71.**5** Regulators of gastric acid secretion. (1) Histamine receptor, (2) Acetylcholine receptor, (3) Gastrin receptor, (4) Somatostatin receptor, (5) Prostaglandin E$_2$ receptor, (6) GRP receptor.

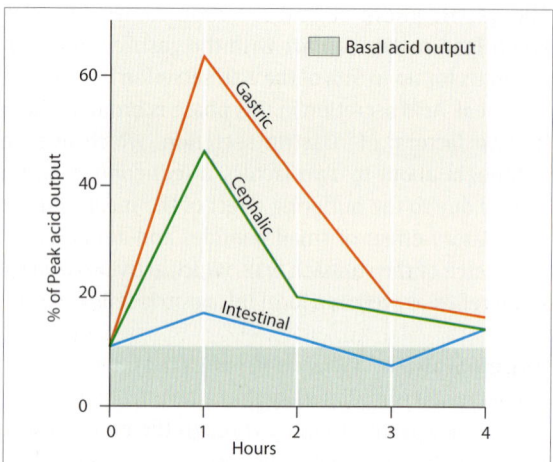

Fig. 71.**6**   Phases of gastric acid secretion.

Other factors   Hypoglycemia increases acid and pepsin secretion by stimulating central vagal discharge. Other stimulants include alcohol and caffeine, both of which act directly on the mucosa.

■ Gastric acid output

Measurement of gastric acid output has considerable clinical importance. Gastric acid secretion is increased in duodenal ulcer, gastric ulcer, and Zollinger–Ellison syndrome while it is reduced after vagotomy and in pernicious anemia.

Basal acid output (BAO) is the rate of acid secretion in the absence of all avoidable stimulations. In the night fasting secreting test, continuous gastric suction is carried out through an indwelling nasogastric catheter from 9.00 pm to 9.00 am. The room is kept devoid of food and even its odor. Normally, about 400 ml of gastric juice is collected overnight. The HCl concentration in the juice collected gives the basal acid output. The normal BAO is <10 mmol h^{-1}.

Maximal acid output (MAO) is the gastric acid secretion produced by stimulants like pentagastrin or histamine. The acid output is measured every 15 minutes for 1 hour. MAO represents the total acid output over

1 hour and is normally <50 mmol h^{-1}. The *peak acid output* (PAO) is calculated by totalling the two highest consecutive 15-minute outputs and multiplying by 2.

The **Hollander's insulin test** is performed after surgically interrupting the vagal nerve supply to the stomach (an operation performed as a surgical remedy of peptic ulcer) in order to test the completeness of the vagotomy performed. The vagus is stimulated by insulin-induced hypoglycemia. The gastric juice is collected and analyzed for its acid content.

■ Regulation of pepsinogen secretion

The regulation of pepsinogen secretion is not well understood. In general, acid secretion and pepsinogen secretion are similarly affected by most stimuli. However, pepsinogen secretion does not show distinct phases like cephalic, gastric, and intestinal.

■ Digestive Events

Gastric (hydrochloric) acid has a pH of approximately 1.0 and serves the following functions: (1) It acts as a good solvent that dissolves foodstuffs insoluble in water. (2) Its acidic pH is required for activation of pepsin. (3) It is a strong disinfectant, killing bacteria and other microorganisms in the ingested food. (4) It stimulates the duodenum to secrete secretin.

The **gastric enzymes** are pepsin, which digests proteins and gastric lipase, which digests fats. Gastric lipase is of little importance in fat digestion except in pancreatic insufficiency. Pepsin cleaves food proteins, forming small peptides. When secreted by the chief cells, pepsin is in its inactive form, a larger protein called pepsinogen. Acid in the lumen converts pepsinogen to pepsin. Pepsin once formed also attacks pepsinogen, producing more pepsin molecules (*autocatalysis*).

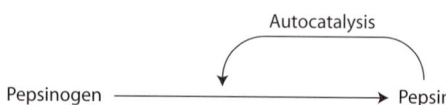

Secretory Events

Duodenal secretions are derived from two main sources, viz., the bile juice and the pancreatic juice. In addition to the above, the Brunner's glands in the duodenal mucosa secrete a thick alkaline mucus that protects the duodenal mucosa from the gastric acid. There is also an appreciable secretion of HCO_3^- that is independent of Brunner's glands.

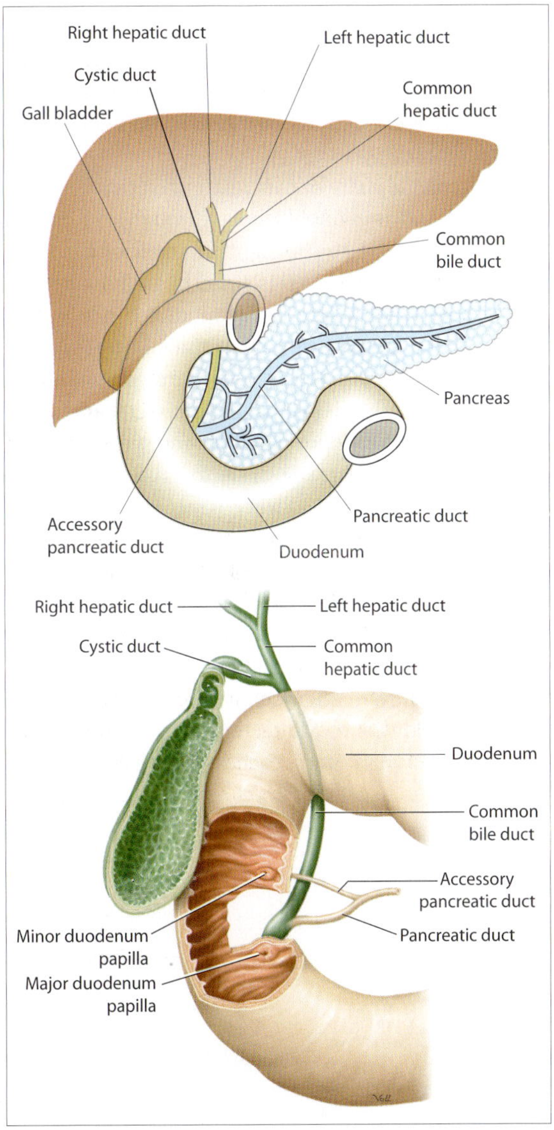

Bile juice

The bile juice is secreted continuously by the liver into the bile canaliculi that drain into the right and left hepatic ducts (Fig. 72.**1**). The two hepatic ducts join to form the common hepatic duct. Shortly after its formation, the common hepatic duct joins with the cystic duct of the gall bladder to form the common bile duct. Shortly before draining into the duodenum through the duodenal papilla, the common bile duct joins the pancreatic duct to form the ampulla of Vater. The duodenal papilla is guarded by a sphincter called the sphincter of Oddi that opens only when fatty food enters the duodenum. At other times it remains closed so that the bile secreted by the liver is diverted through the cystic duct to the gall bladder where up to 50 ml of bile is stored. When stimulated by fatty food, the gall bladder contracts with simultaneous relaxation of the sphincter of Oddi, emptying its stored bile into the duodenum.

The major constituents of the bile juice are bile salts (mainly, taurocholates and glycocholates), bile pigments (bilirubin and biliverdin), lipids (cholesterol, lecithin, fatty acids, and triglycerides), and electrolytes. Cholesterol is present in a concentration of 60–170 mg L^{-1}. Higher concentrations of cholesterol in bile predispose to gallstones. The cations Na^+, K^+, and Ca^{2+} are all present in concentrations about 20% greater than in the plasma. The two major anions are Cl^- and HCO_3^-. Cl^- is present in concentrations less than in plasma while HCO_3^- concentration is much more than in plasma, which makes the bile juice alkaline (pH = 7.0–7.4).

Of all the constituents of bile, only the bile salts are of importance to the digestive system. Bile salts, like phospholipids, are amphipathic molecules because the steroid nucleus (hydrophobic) lies in a single plane while the polar (hydrophilic) groups, e.g., the hydroxyl and carboxyl groups as well as the peptide bond, project on one side (Fig 72.**2**). The amphipathic property makes bile salts important for fat emulsification and micelle formation. They also have a choleretic action.

Fig. 72.**1** (*Above*) The bile and pancreatic ducts and the associated organs. (*Below*) The opening of the ducts inside the duodenum

Right hepatic duct
Left hepatic duct
Cystic duct
Gall bladder
Common hepatic duct
Common bile duct
Pancreas
Accessory pancreatic duct
Pancreatic duct
Duodenum

Right hepatic duct
Left hepatic duct
Cystic duct
Common hepatic duct
Duodenum
Common bile duct
Accessory pancreatic duct
Pancreatic duct
Minor duodenum papilla
Major duodenum papilla

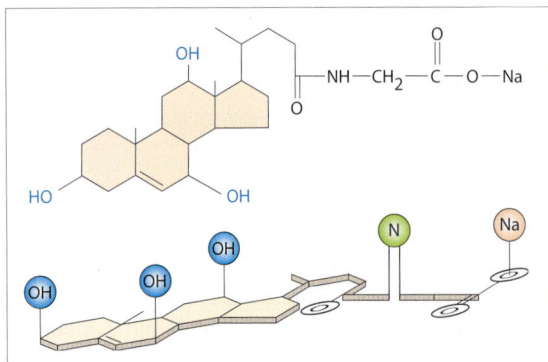

Fig. 72.**2** (*Above*) Chemical structure of bile salt sodium gly-cocholate. (*Below*) Chemical structure of bile salt redrawn to show that the hydrophilic radicals project outside the plane of the steroid ring.

Regulation of biliary secretion   The entry of bile into the duodenum increases either when the liver cells increase their bile secretion or when the gall bladder pours out its stored bile.

(1) Substances that increase hepatic bile secretion are called *choleretics*. Examples of choleretics are secretin and bile salts. Vagal stimulation also increases bile secretion. Bile salts increase the hepatic secretion of bile but inhibit the synthesis of new bile salts. Despite the inhibition of fresh bile salt synthesis, the amount of bile salts in the bile does not decrease. This is because the bile salts secreted into duodenum are reabsorbed from the intestine and resecreted into the bile juice (enterohepatic circulation). Drugs that stimulate the liver to increase the output of bile of low specific gravity are called *hydrocholeretics*.

(2) Substances that cause gall bladder contraction are called cholagogues. A well-known cholagogue is cholecystokinin (CCK). Fatty acids and amino acids in the duodenum stimulate the release of CCK which causes gall bladder contraction.

■   **Pancreatic Juice**

The portion of the pancreas that secretes the pancreatic juice is called the exocrine pancreas. Structurally, it is quite similar to the salivary glands. The smaller pancreatic canaliculi coalesce into a single duct called the duct of Wirsung and occasionally, another accessory duct called the duct of Santorini. The duct of Wirsung joins the common bile duct to form the ampulla of Vater. The duct of Santorini, when present, either opens into the duct of Wirsung or directly into the duodenum through a separate opening (Fig. 72.**1**).

The pancreatic juice is the major source of digestive enzymes that digest all components of the food

—proteins, carbohydrates, fats, and nucleic acid. Its highly alkaline pH neutralizes the gastric HCl in the chyme that enters the duodenum. The volume of pancreatic juice secreted per day is about 500–1500 ml. It is highly alkaline (pH 8.4) due to its high HCO_3^- concentration, which is 2–5 times higher than the plasma concentration. As the flow rate of pancreatic juice increases, its HCO_3^- concentration increases and Cl^- concentration decreases. The reciprocal relationship with flow rate occurs because HCO_3^- is secreted in the small ducts but is reabsorbed in the large ducts in exchange for Cl^- (Fig. 72.**3**). The magnitude of the exchange is inversely proportional to the flow rate. The Na^+ and K^+ concentrations are similar to those of plasma and do not change with flow rate.

Regulation of pancreatic secretion   Pancreatic juice secretion has a cephalic phase that is neurally mediated and an intestinal phase that is hormonally mediated. There is no well-defined gastric phase of pancreatic secretion.

The *cephalic phase* of pancreatic secretion is vagally mediated and is brought about through conditioned reflex in response to the sight or smell of food. The pancreatic juice produced is small in amount but rich in enzymes. The vagal effect is mediated not only by acetylcholine but by other neurotransmitters like VIP and GRP that are also released by the vagal endings.

The *intestinal phase* of pancreatic secretion is controlled by two hormones, viz., CCK and secretin. Secretin acts on the pancreatic ducts to cause secretion of large amounts of a very alkaline pancreatic juice that is rich in bicarbonates and poor in enzymes. CCK acts on the acinar cells to cause the release of zymogen granules and production of a pancreatic juice that is rich in enzymes.

Cells secreting secretin and CCK are found at several sites but notably, in the mucosa of the upper intestine. Both these hormones are secreted just when they are needed, i.e., when food reaches the duodenum. The

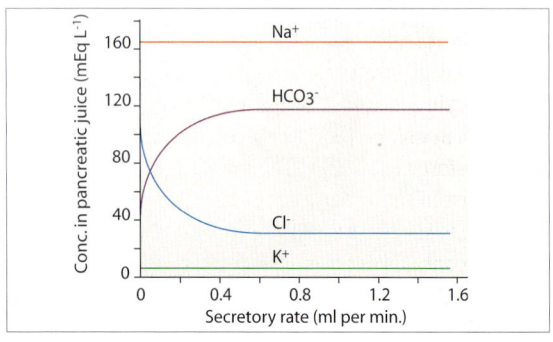

Fig. 72.**3**   Effect of secretory rate on the ionic concentration of pancreatic juice.

secretions of both are stimulated by the contact of the intestinal mucosa with the acidity of the chyme and by the peptides and amino acids present in the chyme. CCK secretion is additionally stimulated by the presence of long chain fatty acids in the duodenum. Thus, both CCK and secretin mediate a physiological reflex wherein the food itself stimulates the secretion of the digestive juice required to digest it.

Insulin, it is now believed, has long-term effects on the regulation of pancreatic enzyme synthesis. The venous blood from the islets of Langerhans passes through the pancreatic acini before returning to the systemic circulation. This exposes the pancreatic acini to high concentrations of insulin, which influences the synthesis of pancreatic enzymes.

Digestive Events

Action of bile salts

The role of bile salts in fat digestion is threefold, namely, (1) emulsification of fats, (2) formation of micelles, and (3) activation of an enzyme called bile salt-activated lipase which is present in milk.

Emulsification is the division of large lipid droplets into smaller droplets about 1 millimeter in diameter (Fig. 72.**4**). Emulsification increases the surface to volume ratio of the lipid droplets, facilitating the action of lipases. The process of emulsification requires churning of the fat (done by the pyloric antrum) and the presence of detergents in the form of bile salts (in bile juice) and phospholipids (in bile juice as well as in the food). Churning breaks up large lipid droplets into smaller ones while the detergents prevent the reaggregation of the smaller droplets into larger ones. Detergents form a coating on the small lipid droplets with their polar residues facing outward. These polar residues on small lipid droplets repel each other, preventing the reaggregation into larger droplets.

Formation of micelles   Micelles are much smaller than emulsified droplets, about 5 nanometers in diameter, and are cylindrical in shape. Most of the lipid content is in the form of fats that have already been digested and are therefore absorbable. The main function of micelles is to assist in the absorption of fats. The composition of micelles is similar to that of emulsified droplets. Each micelle contains detergents (bile salts and phospholipids) and absorbable fats (fatty acids, monoglycerides, and cholesterol). The detergents are located on the micellar surface while the absorbable fats are present in the hydrophobic center of the micelle.

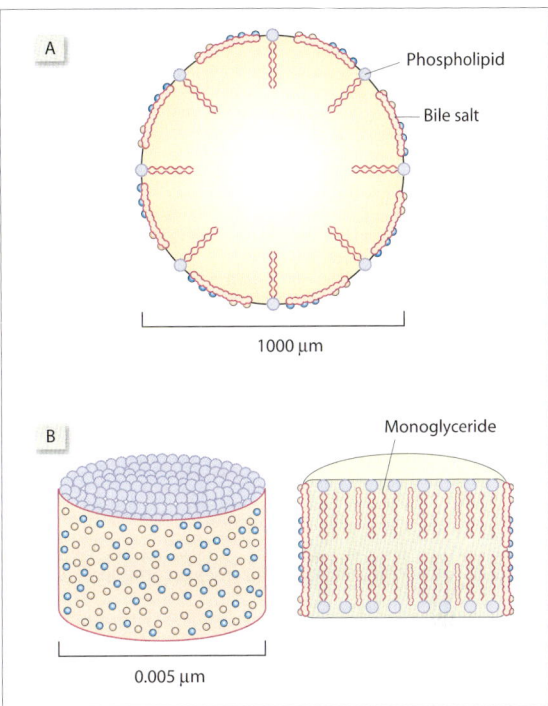

Fig. 72.**4**   (*Above*) An emulsified lipid droplet. (*Below, left*) A micelle. (*Below, right*) Structure of a micelle sectioned along its axis. Note that the lipid droplets are about a hundred thousand times larger than the micelles.

Actions of pancreatic enzymes

Digestion of proteins   The pancreatic juice contains three endopeptidases (trypsin, chymotrypsin, and elastase) and two exopeptidases (carboxypeptidase A and B) that cleave proteins at different sites (Fig. 72.**5**). Endopeptidases break the peptides somewhere in the middle. Exopeptidases break the peptide chain near its end, releasing single amino acids.

All the above enzymes are secreted from the pancreas as inactive precursors (zymogens). Chymotrypsinogen, proelastase, and procarboxypeptidases are converted into their active forms by the action of trypsin. Trypsin itself is activated from trypsinogen by the action of enteropeptidase (enterokinase) secreted by the intestinal mucosa. Once activated, trypsin activates a large number of proenzymes, viz., chymotrypsinogen, proelastase, procarboxypeptidase, prophospholipase A$_2$, and procolipase. Trypsin also autoactivates itself by acting on trypsinogen.

Digestion of carbohydrates   The pancreatic juice contains pancreatic α-amylase which digests starch, hydrolyzing the 1:4α linkages but sparing the 1:6α linkages, terminal 1:4α linkages, and the 1:4α linkages

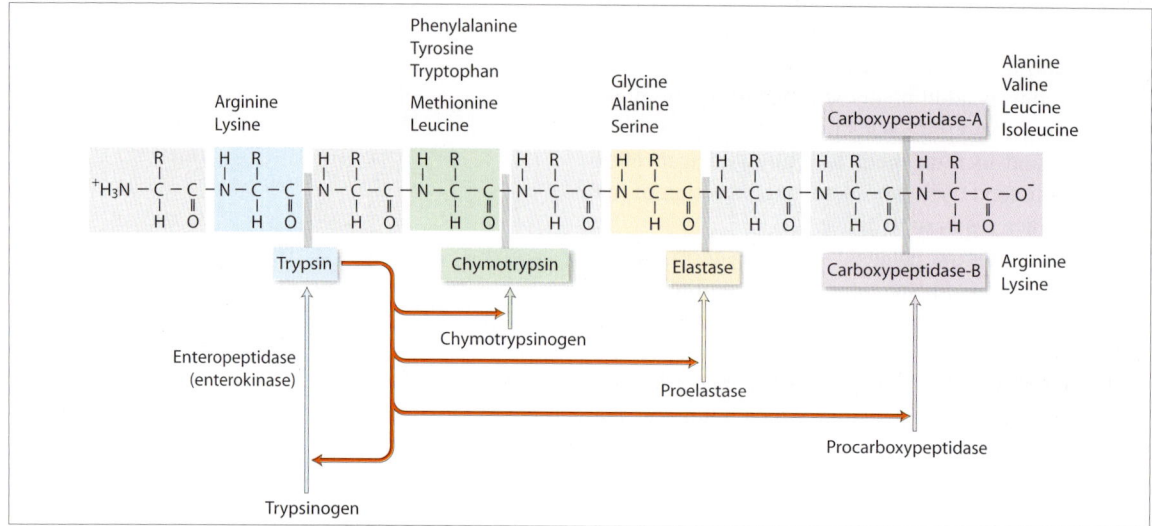

Fig. 72.**5**   Action of pancreatic enzymes on proteins. Each enzyme cleaves peptide bonds that are adjacent to specific amino acids.

next to branching points (Fig. 72.**6**). Consequently, the end products of α-amylase digestion are mostly the disaccharide maltose (two α-glucose residues linked by 1:4α bonds), the trisaccharide maltotriose (three α-glucose residues linked by 1:4α bonds), oligosaccharides (several glucose residues linked by 1:4α bonds), and α-limit dextrins (polymers of glucose containing an average of about eight glucose molecules with 1:6α linkages).

Digestion of fats is brought about by the following pancreatic enzymes (Fig. 72.**7**): (1) Pancreatic lipase hydrolyzes primary ester linkages at positions 1 and 3 of triacylglycerols, yielding mostly free fatty acids and 2-monoglycerides. Pancreatic lipase is inhibited by bile salts. (2) Colipase helps overcome the inhibition of lipase by bile salts. (3) Bile salt-activated lipase breaks down triacylglycerol completely into glycerol and fatty acids. It also catalyzes the hydrolysis of cholesteryl esters, esters of fat soluble vitamins, and phospholipids. Human milk contains an enzyme that is very similar to bile salt-activated lipase. It ensures complete digestion of milk fat, and is of particular importance in premature infants whose pancreatic secretions are not fully operational. (4) Cholesteryl ester hydrolase acts only on cholesteryl esters, releasing cholesterol in nonesterified free form. (5) Phospholipase A$_2$ hydrolyzes the ester bond in the 2- position of glycerophospholipids to form lysophospholipids which, being detergents, aid emulsification and digestion of lipids.

Digestion of nucleic acids is brought about by ribonuclease and deoxyribonuclease.

Absorptive Events

Significant amounts of absorption occur in the duodenum. In general, the absorptive pattern in the duodenum resembles that in the jejunum. For example, the duodenum and upper jejunum have the highest capacity to absorb sugars, dipeptides and tripeptides, and fats. Calcium and phosphate absorptions are especially high in the duodenum and ileum. The duodenum is also the principal site of absorption of nonheme iron.

Water may either move into or out of the duodenum depending on the tonicity of the chyme across the duodenal mucosa. Usually, the semisolid chyme extracts water into the duodenum lumen. However, when the chyme is watery, water may be reabsorbed from the duodenum. Sodium ion reabsorption occurs throughout the intestine, beginning in the duodenum. Bicarbonates are secreted into the duodenal lumen. Lower down in the ileum, bicarbonates are reabsorbed.

Pancreatic Function Tests

Analysis of pancreatic juices

The pancreatic secretions are collected by passing a double-lumen radioopaque (Dreiling) tube into the alimentary canal in such a way that one lumen drains the stomach and the other drains the duodenum. In this way, the duodenal contents are collected free from gastric contamination. The pancreatic juice secretion can be stimulated either directly by injecting secretin or cholecystokinin or by consuming a standardized test meal.

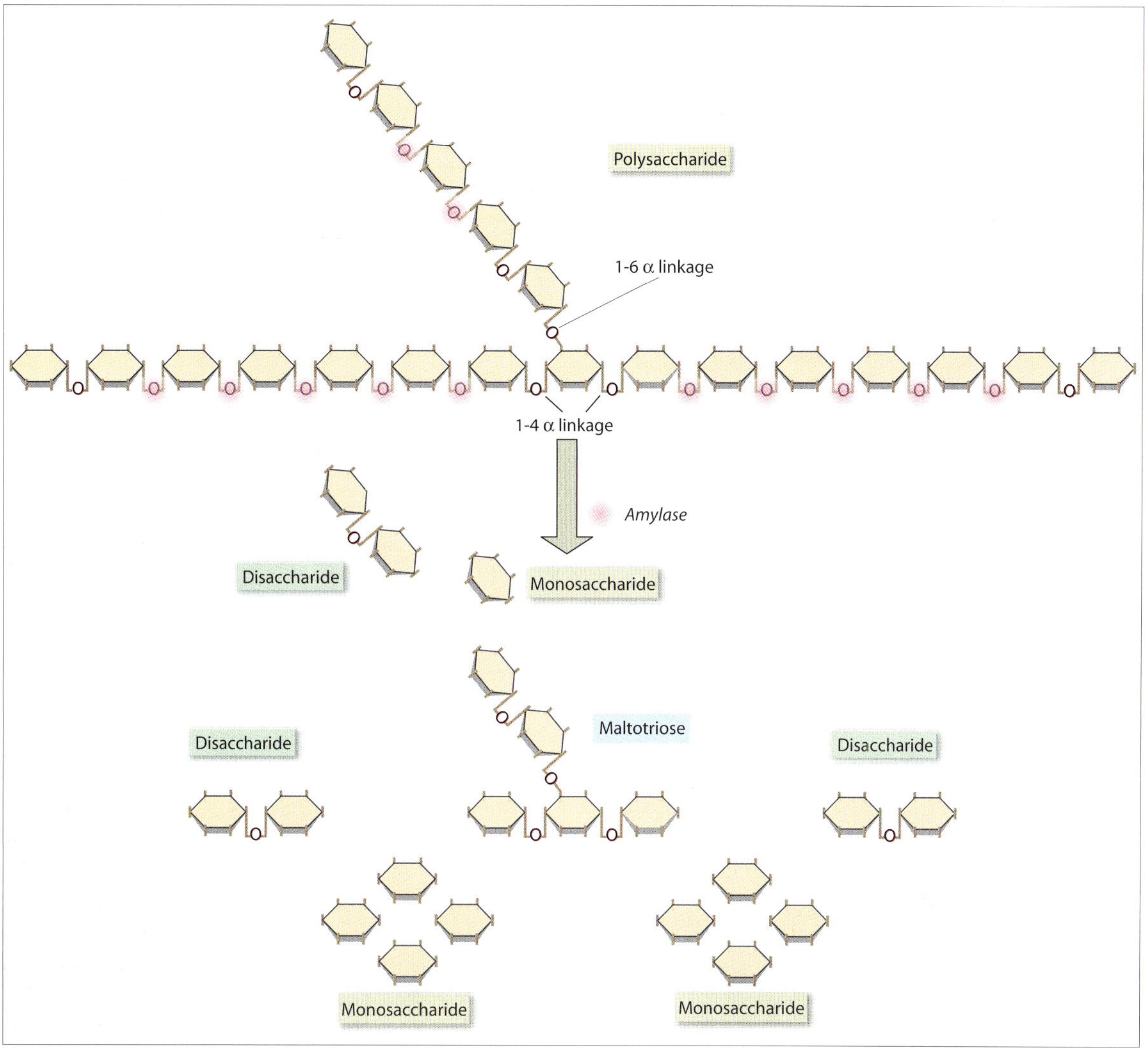

Fig. 72.**6** Digestion of carbohydrates.

The secretin test measures the secretory capacity of the pancreatic duct. It is decreased in chronic pancreatitis. In the combined secretin-CCK test or following a test meal, both bicarbonate and enzyme output are stimulated in normal subjects. With mild pancreatic damage, only the bicarbonate output is affected. With advanced damage, both are affected. Test meals can however give false positive results. For example, the enzyme activity may be low, not due to pancreatic insufficiency but due to some disease of the intestinal mucosa that results in inadequate release of cholecystokinin.

Analysis of digestion products

Stool examination The stool is examined microscopically for undigested meat fibers and fat, the presence of which indicates a lack of proteolytic and lipolytic enzymes. The test is reliable and simple but not sensitive enough to detect milder cases of pancreatic insufficiency.

Fecal fat test The subject is fed a test diet containing adequate amounts of fats. A red carmine dye indicator is ingested before and after the test meal for identifying the stool resulting from the test meal. Normally, fecal fat is < 7% of the dietary intake.

The **triolein breath test** involves oral administration of radiolabeled triolein. Metabolism of the triolein releases radiolabeled CO_2 which is exhaled. The amount of radiolabeled CO_2 exhaled is less if the triolein is maldigested or malabsorbed.

The **tripeptide hydrolysis test** utilizes a synthetic peptide N-benzoyl-l-tyrosyl-p-aminobenzoic acid (Bz-Ty-PABA) for testing chymotrypsin activity. The peptide is cleaved by chymotrypsin into Bz-Ty and PABA. The

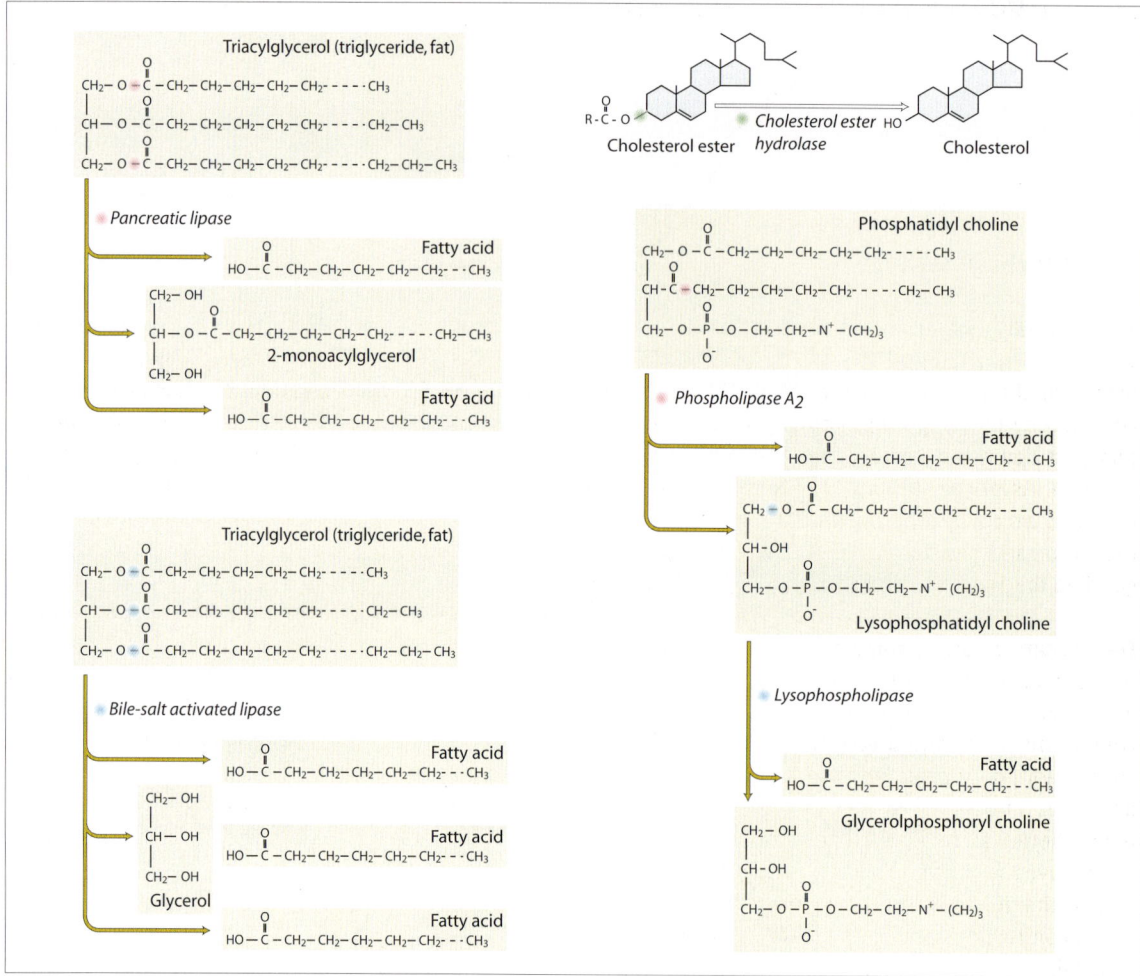

Fig. 72.**7**   Digestion of fats.

PABA is rapidly absorbed. If chymotrypsin activity is low, PABA excretion in urine decreases.

The **dual-label Schilling test** is based on the knowledge that trypsin plays a role in vitamin B_{12} absorption. It is known that cobalamin requires intrinsic factor for its absorption. However, before it can bind to the intrinsic factor, the ingested cobalamin is attached to a protein present in the gastric juice called the R-protein. In the duodenum, trypsin degrades the R-protein, releasing the cobalamin which can then bind to the intrinsic factor. Hence, in pancreatic insufficiency, there is malabsorption of vitamin B_{12}.

The patient is given to ingest a mixture of [57]Co-cobalamin complexed with intrinsic factor, and [58]Co-cobalamin complexed with R-Protein. Normally, both [57]Co-cobalamin and [58]Co-cobalamin are absorbed and excreted in urine in equal amounts. In pancreatic insufficiency, the urinary excretion of [58]Co-cobalamin (which was complexed with R-protein) is reduced.

73 Events in the Small Intestine

Mechanical Events

Motility patterns

In the unfed state, the small intestinal motility is characterized by the migrating motor complexes (MMCs) that pass down the stomach and intestines at regular intervals. As in the stomach, the MMCs are replaced following a meal by different motility patterns like segmentation and peristalsis. The MMCs are much stronger than the peristaltic waves.

Segmentation contractions are ring-like contractions of the circular smooth muscle of the gut that appear at regular intervals. The intestine becomes transiently compartmentalized into several short segments. The contractions disappear after a few seconds, only to reappear as another set of ring contractions in the segments between the previous contractions (Fig. 73.**1**). They move the chyme to and fro and increase its exposure to the mucosal surface. A variant of the segmentation contractions are the *tonic contractions* that last somewhat longer, isolating one segment of the intestine from another. Segmentation and tonic contractions slow down the transit of chyme in the small intestine. This permits longer contact of the chyme with the enterocytes (cells of the intestinal mucous membrane) and thereby improves absorption.

Peristaltic waves propel the intestinal chyme toward the large intestines. Peristaltic waves are contraction rings of the intestine that travel short distances along the intestine at velocities of 2–3 cm s^{-1}. Very intense peristaltic waves called peristaltic rushes are not seen normally but occur when the intestine is obstructed. Most waves pass regularly in an oral to caudal direction.

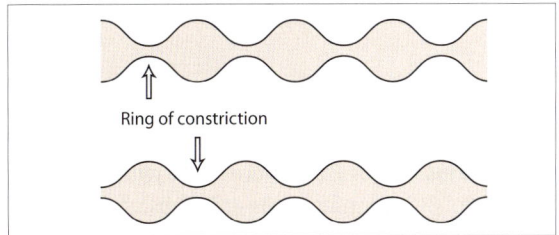

Fig. 73.**1**   Segmentation contractions of the intestine.

Reflexes

Peristaltic reflex is induced by a localized distension of the intestine. It results in peristaltic contraction proximal to the distension and inhibition distal to it. It is unlikely that the peristaltic reflex is of any physiological significance because the intestine is rarely distended to the degree required to induce the reflex. The peristaltic reflex is not a spinal reflex: It is coordinated by the enteric nervous system of the intestine.

In the **gastroileal reflex**, excessive secretory and motor activity of the stomach brings about a reflex increase in the motility of the terminal part of the ileum and accelerates the movement of material through the ileocecal sphincter. The reflex is vagally mediated.

In the **intestinointestinal reflex**, overdistension of one segment of the intestine relaxes the smooth muscle in the rest of the intestine.

Secretory Events

The tubular intestinal glands, also known as the crypts of Lieberkuhn, secrete an isotonic fluid called the succus entericus. Most of the enzymes usually found in this secretion come from the desquamated mucosal cells. Cell-free succus entericus hardly contains any enzymes. The secretion of succus entericus is stimulated by gastrointestinal hormones such as vasoactive intestinal peptide (VIP) but is unaffected by vagal stimulation. The mucus secreted by the intestine comes from the enterocytes and the goblet cells in the epithelium.

Digestive Events

Digestion of all constituents of food is completed in the small intestine. Intestinal digestion occurs at three locations, viz., the intestinal lumen, the brush border, and inside the mucosal cells. Relatively little digestion occurs in the lumen itself. Most of the digestion occurs at the intestinal brush border to which the digestive enzymes are attached. Some di- and tripeptides are actively transported into the intestinal cells and hydrolyzed by intracellular peptidases. The amino acids so formed diffuse out into the bloodstream. The same is true for some lipids.

Carbohydrate digestion   There are five enzymes responsible for carbohydrate digestion at the intestinal brush border. (1) Sucrase breaks down sucrose into glucose and fructose. (2) Maltase (α-glucosidase) breaks the 1:4α linkages and releases glucose. (3) Isomaltase (α-dextrinase) breaks down the 1:6α linkages and releases glucose. (4) Lactase (β glucosidase) breaks down lactose into glucose and galactose. (5) Trehalase hydrolyzes trehalose, a 1:1α-linked dimer of glucose, into two glucose molecules. Trehalose is found in mushrooms.

Protein digestion   There are five enzymes responsible for protein digestion at the intestinal brush border. (1) Enteropeptidase (enterokinase) is meant only for activating trypsinogen to trypsin. (2) Aminopeptidase is an exopeptidase that breaks the peptide bonds next to N-terminal amino acids of peptides. (3) Carboxypeptidase is an exopeptidase that breaks the last peptide bond toward the C-terminal. (4) Endopeptidases break peptide bonds somewhere in the middle of the polypeptide. (5) Dipeptidase splits dipeptides into amino acids.

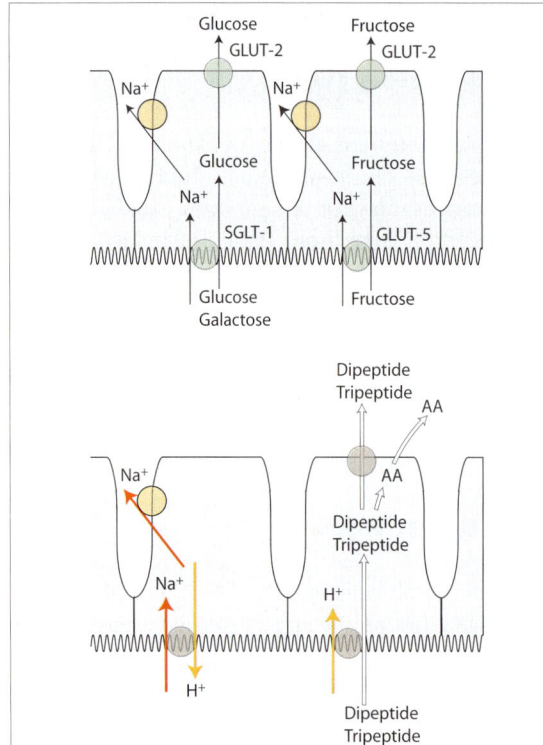

Fig. 73.**2**   Transport across intestinal mucosa. (*Above*) Transport of monosaccharides. (*Below*) transport of small peptides.

About half the protein in the intestine comes from ingested food and the rest comes from digestive juices and desquamated mucosal cells. Most of the protein is digested in the small intestine. Any undigested protein is digested in the colon by bacterial action. The protein in the stools is not of dietary origin but comes from bacteria and cellular debris.

Nucleic acid digestion   Three intestinal enzymes help in digestion of nucleic acids. (1) Polynucleotidases split nucleic acids into nucleotides. (2) Nucleosidases catalyze the phosphorylation of nucleosides to give the free nitrogen base plus a pentose phosphate. (3) Phosphatase removes the phosphate from pentose phosphate formed as a result of nucleic acid digestion. It also removes the phosphates from organic phosphates in the diet.

Fat digestion in the intestine is brought about mostly by the pancreatic juice. However, the intestinal mucosal cells have on them a phospholipase that attacks phospholipids to produce glycerol, fatty acids, phosphoric acid, and bases such as choline.

Absorptive Events

Absorption of Carbohydrates

Monosaccharides are rapidly absorbed from the intestine before the meal reaches the terminal part of the ileum. Pentoses are absorbed by simple diffusion. Glucose and galactose are absorbed by facilitated diffusion employing sodium-dependent glucose transporter (SGLT) and glucose transporter (GLUT-2). Fructose employs GLUT-5 and GLUT-2 for diffusion across the luminal and basolateral membranes respectively (Fig. 73.**2**). When SGLT is congenitally defective, the resulting glucose or galactose malabsorption causes severe diarrhea that is often fatal if glucose and galactose are not promptly removed from the diet.

Insulin has little effect on intestinal transport of sugars. In this respect, intestinal absorption resembles glucose reabsorption in the proximal convoluted tubules of the kidneys: Neither of the processes requires phosphorylation and both are essentially normal in diabetes but are depressed by the drug phlorhizin.

Absorption of Proteins

Amino acids are absorbed more in the jejunum than in the ileum. Some amino acids are reabsorbed through simple diffusion. However, most amino acids employ transporters, both for entering the enterocytes across the brush border as well as for diffusing out across the basolateral membrane. There are several transporters specific to individual or groups of amino acids. Some of these involve cotransport with Na⁺ while others are Na⁺-independent.

Congenital defects in amino acid transport usually affect both intestinal absorption and renal tubular reabsorption, e.g., cystinuria and Hartnup disease which occur due to impaired transport of basic and neutral amino acids respectively. The purine and pyrimidine bases formed by the digestion of nucleic acids are absorbed through active transport.

Small peptides   Significant amounts of small peptides also enter the portal blood, passing through the enterocytes. Absorption of small peptides is more in the ileum but less in the jejunum. Dipeptides and tripeptides are transported across the brush border through secondary active transport (Fig. 73.**2**). Most of the peptides that enter the enterocyte are cleaved into amino acids inside the cell. The amino acids diffuse out of the basolateral membrane into the portal blood. However, some small peptides are not broken down inside the enterocyte: they diffuse out into the bloodstream unchanged, possibly using a transporter.

Proteins   A small amount of protein passes unaltered from the intestine into the blood: The protein is endocytosed by the intestinal epithelium and exocytosed into the blood stream. The amount of protein absorbed in this way is substantial in infants but declines with age. In infants, the IgA present in maternal colostrum is thus able to enter the circulation and provide passive immunity.

In adults, certain food proteins may produce allergy after being absorbed without digestion. Several bacterial and viral proteins are absorbed by the large microfold (M) cells present on Peyer's patches. The M cells pass the antigens to the lymphoblasts in Peyer's patches. The activated lymphoblasts enter the circulation, only to return to the intestinal mucosa and other epithelia where they secrete IgA in response to subsequent antigenic exposures.

Absorption of Fats

A barrier to the absorption of fat molecules is the unstirred layer, a stationary layer of luminal fluid that is in contact with the mucosal surface of the intestine. Digested fats must therefore be converted to micellar form for passage across the unstirred layer. The thickness of the unstirred layer increases in certain disease states like celiac sprue, thereby contributing to malabsorption.

More than 95% of the ingested fat is absorbed in the intestine. The micelles move down their concentration gradient through the unstirred layer to the brush border of the mucosal cells. The lipids diffuse out of the micelles and a saturated aqueous solution of the lipids is maintained in contact with the brush border of the mucosal cells.

All fatty acids enter the enterocytes by facilitated diffusion (Fig. 73.**3**). Once inside the mucosal cell, short and long chain fatty acids are dealt with differently. Short-chain fatty acids (containing <10 carbon atoms) pass from the mucosal cells directly into the portal blood. Long-chain fatty acids (>10 carbon atoms) are rapidly reesterified to triglycerides in the mucosal cells, maintaining a favorable diffusion gradient of lipids. Most of the triglyceride is formed by the acylation of the absorbed 2-monoglycerides. The triglycerides are then coated with a layer of protein and phospholipid to form *chylomicrons*. The cholesterol that enters the enterocyte is esterified with fatty acids and incorporated into chylomicrons. The chylomicrons leave the cell and enter the lymphatics. During fat absorption, the lymph in the villi becomes milky due to suspended chylomicrons. The lymphatic channels in the villi are therefore called lacteals and the lymph they carry is called *chyle*.

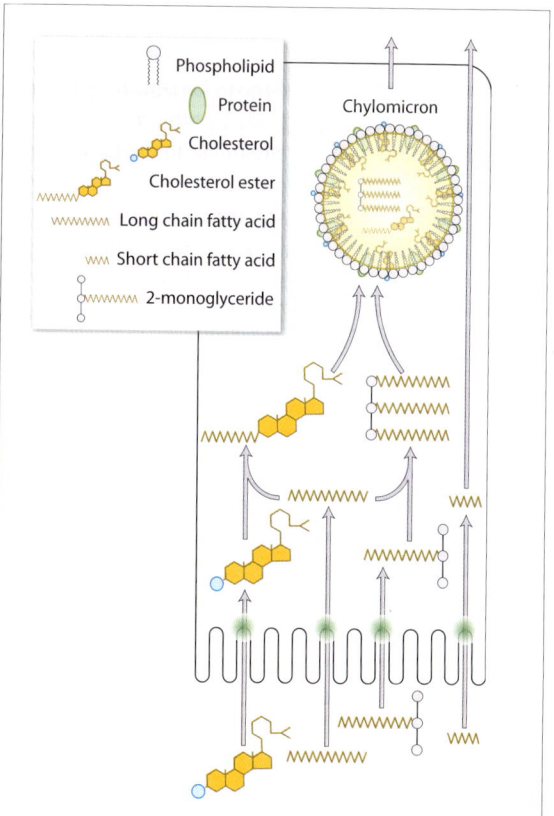

Fig. 73.**3**   Absorption of fats.

Metabolism and Nutrition

77. Dietary Nutrients and Fibers

78. Nutritional Assessment and Dietary Planning

79. Metabolic Pathways

80. Metabolic States and the Liver

Macronutrients

The bulk of a diet is constituted by proteins, carbohydrates, and fats which are called *macronutrients*. Macronutrients provide energy and are consumed in much larger quantities than other nutrients like vitamins and minerals, which are called *micronutrients*.

Fat

Fat makes food rich and tasty and is physiologically important for producing satiety. Fat is required for the absorption of fat-soluble vitamins. More than 90% of dietary fat is in the form of triacylglycerol (triglycerides), which supply the essential fatty acids required by the body. Triacylglycerols obtained from plants are generally richer in unsaturated fatty acids than those from animals. Salient exceptions are coconut and palm oils which are rich in saturated fat, and fish which is rich in unsaturated fats. The unsaturated fatty acids may be monounsaturated (olive oil) or polyunsaturated (corn and soybean oils).

Essential fatty acids are required for fluidity of membrane structure and for synthesis of prostaglandins, prostacyclins, thromboxanes, and leukotrienes. Dietary fat provides the essential fatty acids, viz., linoleic and linolenic acids. Arachidonic acid becomes essential if linolenic acid, from which it is produced, is absent in the diet. A deficiency of essential fatty acids is characterized by scaly dermatitis, hair loss, and poor wound healing. Because they are widely distributed in nature, a deficiency is rare.

Cholesterol is an essential component of cell membranes and serves as a precursor of bile acids, steroid hormones, and vitamin D. However, elevated levels of cholesterol increase the risk of coronary heart disease due to the formation of atherosclerotic plaque in coronary vessels. Plasma cholesterol level is moderately decreased when low-cholesterol diets are consumed. Egg yolk and meat are very rich in cholesterol, which is found only in animal products. Plant products, including margarine and vegetable oils, contain no cholesterol.

Saturated fatty acids are present mainly in dairy and meat products and some vegetable oils, such as coconut and palm oils. They tend to raise the total plasma cholesterol and the cholesterol in low-density lipoproteins (LDL), resulting in increased risk of coronary artery disease.

n-6 (ω-6) polyunsaturated fatty acids like linoleic acid are found in corn, safflower, soybean, and sunflower oils. Consumption of these fatty acids (with a double bond located six carbons from the methyl end) lowers plasma cholesterol. They lower plasma LDL but also lower the high-density lipoprotein (HDL) which is considered healthier.

n-3 (ω-3) polyunsaturated fatty acids are found in oily fish and soybean. Diets rich in these fatty acids (with a double bond three carbons from the methyl group end) reduce plasma triacylglycerols and decrease the risk of heart disease. They inhibit the conversion of arachidonic acid to thromboxane A$_2$ (TXA$_2$) by platelets and are themselves converted to TXA$_3$, which is less thrombogenic than TXA$_2$. Thus, they decrease platelet aggregation and are antithrombogenic.

Monounsaturated fats are as effective as polyunsaturated fats in lowering blood cholesterol when substituted for saturated fatty acids. In contrast to the effect of n-6 polyunsaturated fats, monounsaturated fats do not lower HDL.

Trans fatty acids do not occur naturally in significant amounts but are formed during the hydrogenation of liquid vegetable oils into ghee and margarine. These fatty acids, unlike the naturally occurring cis isomers, raise plasma cholesterol levels.

Carbohydrates

Carbohydrate is not an essential nutrient because the carbon skeletons formed by deamination of amino acids can release energy or can be converted into glucose. However the breakdown of dietary proteins for energy release is wasteful since high-protein food is much more expensive than high-carbohydrate food.

Carbohydrate is 'protein-sparing' because it allows amino acids to be used for repair and maintenance of tissue protein rather than for gluconeogenesis. Fats too can release energy though they cannot be converted to glucose. However, if carbohydrates are totally excluded from the diet, the energy released by proteins and fats decreases drastically and they are converted mostly into ketone bodies. Even the glucose formed from the carbon skeletons of amino acids is not oxidized completely and forms ketone bodies. These facts are summed up by the adage, "fats burn in the flame of carbohydrates" and explain not only the ketogenic effect of the Atkins diet (a zero-carbohydrate diet with no restrictions on proteins and fats) but also its weight-reducing effect despite the liberal intake of fats and proteins (Fig. 77.1). Medical opinion is however overwhelmingly against the Atkins diet for its long-term complications, like atherosclerosis.

Carbohydrates in the normal diet are either monosaccharides and disaccharides (simple sugars) or polysaccharides (complex sugars). Glucose and fructose are the principal monosaccharides found in food and are abundant in fruits and honey. The most abundant disaccharides are sucrose (glucose + fructose), lactose (glucose + galactose), and maltose (glucose + glucose). Ordinary table sugar is sucrose, which is abundant in sugar cane. Lactose is the principal sugar found in milk. Maltose is a product of enzymatic digestion of polysaccharides. It is found in significant quantities in beer and liquors produced from barley. Polysaccharides are mostly polymers of glucose and do not have a sweet taste. Starch is a polysaccharide that is abundant in plants. Common sources of starch include wheat and other grains, potatoes, peas, beans, and vegetables. Animal starch is the common name for the glycogen present in meat and liver.

Fig. 77.1   The athletic boy envying his obese cousins who are on the Atkins diet for weight reduction. The diet recommends zero carbohydrates, and allows liberal consumption of proteins and fats.

Protein

Protein is required for repair and maintenance of tissue proteins and for growth in general. Protein consumed in excess of the body's needs is deaminated into carbon skeletons which are metabolized to provide energy or to synthesize fatty acids. Proteins are made of different combinations of up to 20 different types of amino acids. Like the amino acids that constitute them, one end of the protein molecule has an NH_2 group (the N-terminal) and the other end has a COOH group (the C-terminal). Ten of the amino acids are nutritionally essential since they cannot be synthesized in the body in adequate amounts. The essential amino acids are valine, leucine, isoleucine, threonine, methionine, arginine, lysine, histidine, phenylalanine, and tryptophan. Of these, arginine and histidine are required only during periods of rapid tissue growth characteristic of childhood or recovery from illness.

The ability to provide the essential amino acids required for tissue maintenance determines the biologic value of a dietary protein. The nutritive quality of a protein, i.e., its biologic value, is measured on a relative scale, with whole egg protein or egg albumin scored at 100. Proteins from animal sources (meat, poultry, milk, fish) have a high biologic value because they contain all the essential amino acids in proportions similar to those required for synthesis of human tissue proteins. Proteins from wheat, corn, rice, and beans have a lower biologic value than animal proteins. In the average Indian diet, cereals are the main source of proteins.

Protein-energy malnutrition (PEM)

Insufficient intake of protein or energy-yielding food causes loss of both body mass and adipose tissue. There are two syndromes of PEM, viz., marasmus and kwashiorkor, that result from a reduction in the consumption of protein as well as energy (Fig. 77.2).

Marasmus results from catabolism and depletion of the somatic protein compartment represented by the skeletal muscles. The visceral protein compartment is depleted only marginally. It is for this reason that serum albumin levels remain normal or only slightly decreased. Subcutaneous fat is mobilized resulting in emaciated extremities and monkey-like facies. The skin appears dry and inelastic. In comparison, the head appears too large for the body. The main brunt of the disease process falls on weight gain.

In **kwashiorkor**, protein deprivation is relatively greater than reduction in total calories. Unlike

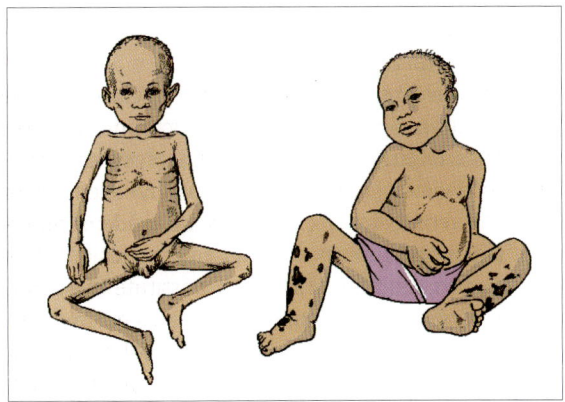

Fig. 77.**2**   Protein-energy malnutrition. (*Left*) Marasmus. (*Right*) Kwashiorkor. (Reproduced with permission from WHO/FCH/CAH/Integrated management of childhood illness.)

marasmus is the most frequent presentation of severe PEM. Cases of pure kwashiorkor are seldom seen in India. What is seen mostly has been called marasmic kwashiorkor. It is not clear which factors decide whether PEM will lead to kwashiorkor or marasmus. The classical theory holds that a predominantly protein deficiency leads to kwashiorkor while a predominantly energy deficiency leads to marasmus. Alternatively, it has been suggested that the outcome of PEM is determined not by the quality of diet but by the child's response to deficient nutrients. According to this theory, marasmus occurs as a result of adaptation to chronic nutritional deficiency through excess cortisol secretion while kwashiorkor is an acute condition wherein the body fails to adapt to the nutritional stress.

marasmus, marked protein deprivation is associated with severe loss of the visceral protein compartment represented mainly by protein stores in the liver. This leads to hypoalbuminemia and edema. Compromised weight-for-age may be masked by fluid retention and relative sparing of subcutaneous fat and muscle mass. Fatty liver (p 495) is often present. Hairs are sparse, brittle, and depigmented. Many of these changes are due to damage by free radicals, which is excessive in kwashiorkor.

The commonest form of PEM is borderline undernutrition or mild PEM where the classical picture of marasmus or kwashiorkor is not seen. In India,

Table 77.**1**   Comparison of kwashiorkor and marasmus.

Similarities	
* Both occur due to insufficient intake of protein and/or energy	
* Both set of patients show apathy, inactivity, and irritability	
* Both set of patients show growth retardation	
Dissimilarities	
Kwashiorkor	**Marasmus**
Usually seen in recently weaned infants	Usually seen in children who have been weaned early or have never been breast-fed
Visceral protein compartment (liver) is depleted	Somatic protein compartment (muscle) is depleted
Hypoalbuminemia present	Serum albumin normal or slightly decreased
Edema is characteristically present	Edema not present
Body weight remains normal	Severe wasting is present
Face puffy and moon-shaped	Face shriveled and monkey-like
Impaired appetite	Voracious appetite.

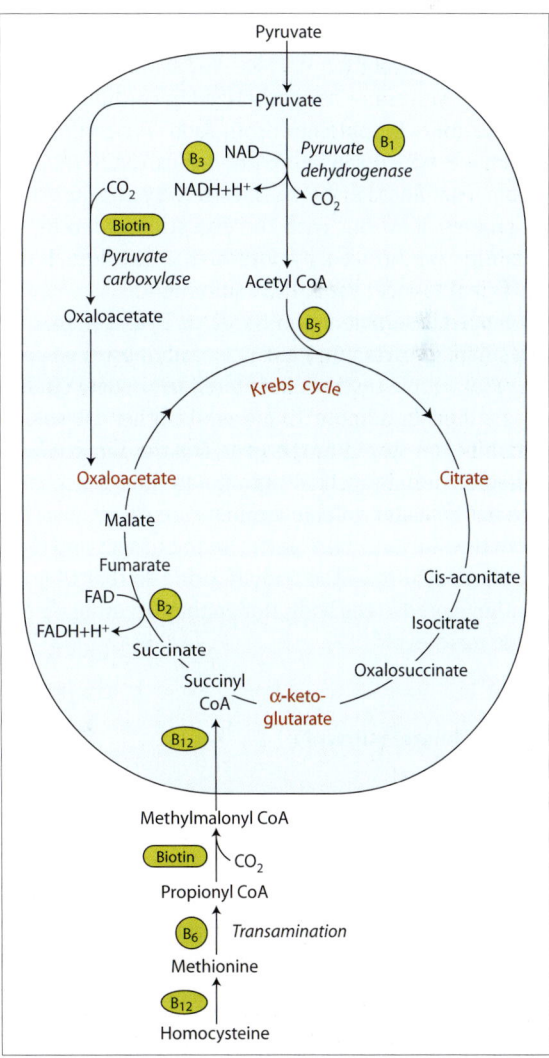

Fig. 77.**3**   Vitamins and their role in metabolic reactions.

Micronutrients

Vitamins

Vitamins are physiologically important dietary components that are required in small quantities in the diet. Unlike the major nutrients, they do not generate metabolic energy but serve specialized functions in the body. Their deficiency can result in serious symptoms and even death (Table 77.**2**). Vitamins have been categorized on the basis of their solubility as the fat-soluble vitamins (vitamin A, D, E, and K) and the water-soluble vitamins (vitamin B_1, B_2, B_3, B_5, B_6, B_{12}, folate, and vitamin C).

Water-soluble vitamins are precursors of coenzymes (Fig. 77.**3**). The fat-soluble vitamins A and D act more like hormones. For example, vitamin A is important in visual transduction while vitamin D is important in calcium homeostasis. Vitamin K is a coenzyme. Vitamins C and E and β-carotenes have antioxidant properties. They inactivate oxygen radicals which are harmful to the body.

Vitamins cannot be synthesized in the human body and so must be obtained from food. An exception is vitamin D which can be photosynthesized in the skin. Some vitamins can be synthesized by the intestinal microorganisms but with the exception of biotin, the quantity synthesized by intestinal microflora is not sufficient to meet the daily requirement.

Since fat-soluble vitamins (A, D, E, and K, or their precursor provitamins) are not easily absorbed from the diet, ample reserves are stored in tissues. Vitamin E is stored in adipose tissue while other fat-soluble vitamins are stored in the liver. For the same reason, fat-soluble vitamins can accumulate and reach toxic levels. The water-soluble vitamins are easily absorbed from the intestine, transported to the tissues and readily eliminated from the body in urine so that they are neither stored in the body, nor do they accumulate easily to toxic levels.

Mineral nutrients

Mineral nutrients are divided into *macrominerals* (ions of sodium, calcium, potassium, chlorine, sulfur, and magnesium) that are required in quantities of more than 100 mg day^{-1}, and *microminerals* or *trace minerals* (ions of chromium, cobalt, copper, iodine, iron, manganese, molybdenum, selenium, zinc, and several others) that are required in much smaller quantities in the diet. Macrominerals are major components of the body fluids and the inorganic matrix of bone. The microminerals are mostly important components of enzymes.

Deficiencies of trace metals (except for iron) are relatively uncommon.

Intestinal absorption of dietary minerals requires specific carrier proteins (see iron absorption, p 154). Synthesis of these carrier proteins serves as an important mechanism for control of mineral levels in the body. Excretion of minerals occurs through the renal or the hepatobiliary route. The minerals are transported in blood as complexes with carrier proteins. For example, iron is bound to transferrin in plasma. Apoceruloplasmin, a glycoprotein, is synthesized in the liver, and binds with six copper atoms to form ceruloplasmin. Ninety percent of the copper in the body is tightly bound in ceruloplasmin and the rest is loosely bound to albumin. Since ceruloplasmin binds very tightly to copper, it refuses to deliver copper to tissues easily and is of less use as a transport protein than albumin.

Metallothioneins The tissue levels of certain metals are partly regulated by metallothioneins, a group of small proteins found in the cytosol of liver, kidney, and intestine. Metallothioneins have a high content of cysteine, and the SH group of cysteine binds to metals like copper, zinc, cadmium, and mercury. A sudden rise in the concentration of these minerals induces the synthesis of metallothioneins, which suggests that metallothioneins have a protective role against the toxic effects of these minerals. The fact that an excessive amount of dietary zinc causes copper deficiency is explained by the higher binding affinity of metallothioneins for copper. Excess dietary zinc induces synthesis of metallothioneins, which trap copper within the intestinal mucosal cell. Copper is subsequently lost with mucosal cell exfoliation.

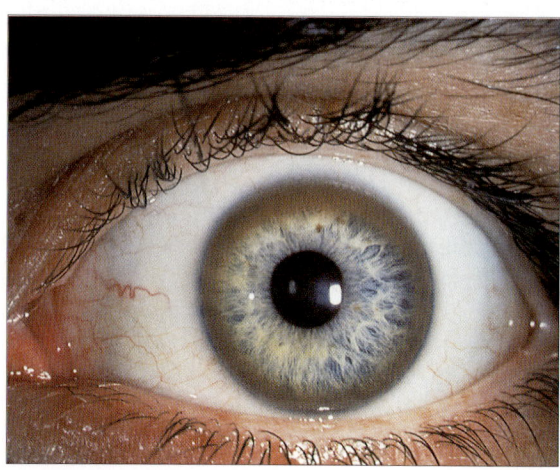

Fig. 77.**4** Kayser-Fleischer ring.

Table 77.**2**   Vitamins, their physiological role, dietary sources and deficiency signs and symptoms.

Vitamin	Functions	Dietary sources	Causes and features of deficiency disease
Water-soluble vitamins			
Thiamine Vitamin B$_1$	Thiamine is converted in tissues to thiamine pyrophosphate (TPP), which serves as a coenzyme for the oxidative decarboxylation of ketoacids and transketolase reaction in the hexose monophosphate shunt pathway.	Legumes, pork, liver, nuts, the germ of cereals, yeast, and outer layers of seeds.	Deficiency is uncommon except in chronic alcoholics because of poor intestinal absorption and inadequate dietary intake. Thiamine deficiency causes beriberi, which has two forms. In *dry beriberi*, neuromuscular symptoms predominate, with demyelination of somatic nerves and wasting of muscles. *Wet beriberi* occurs in severe thiamine deficiency. It is associated with edema due to cardiac insufficiency, which in turn is probably due to inadequate metabolism and accumulation of pyruvic acid and lactic acid. In alcoholics, thiamine deficiency results in Wernicke–Korsakoff syndrome. The acute stage of this disease is known as Wernicke's encephalopathy, which is characterized by delirium and ataxia. In the chronic stage, there is Korsakoff psychosis.
Riboflavin Vitamin B$_2$	Riboflavin is converted to flavin mononucleotide (FMN) in intestinal mucosa and to flavin adenine dinucleotide (FAD) in liver. The flavin nucleotides serve as coenzymes for the flavin dehydrogenases in the redox reactions of the electron transport chain.	Milk, eggs, liver, and green leafy vegetables.	Ariboflavinosis is rare. It is associated with glossitis, stomatitis, cheilosis (reddening of the mucous membranes of the lips), and dermatitis. The dermal lesions occur especially around the nasolabial and scrotal areas.
Niacin Vitamin B$_3$	Niacin is converted to the coenzymes NAD$^+$ (nicotinamide adenine dinucleotide) and NADP$^+$ (nicotinamide adenine dinucleotide phosphate). NADH+H$^+$ and NADPH+H$^+$ are essential for anabolic redox reactions like cholesterogenesis and lipogenesis in the extramitochondrial compartments of the cell while NAD$^+$ and NADP$^+$ are used in the catabolic redox reactions in the mitochondrial matrix.	Unrefined grains, yeast, liver, legumes, and lean meats.	Limited amount of niacin can be synthesized in the body from tryptophan. Niacin deficiency is endemic among poor people who subsist chiefly on maize, which is deficient in tryptophan. Pellagra (meaning rough skin) affects the gastrointestinal tract, skin, and central nervous system. It causes the three Ds: diarrhea, dermatitis, and dementia.
Pantothenic acid Vitamin B$_5$	Pantothenic acid is converted into coenzyme A, which plays a key role in several reactions. An example is conversion of succinate to succinyl CoA, which is a precursor for heme.	Yeast, liver, and eggs.	Rare.
Pyridoxine Vitamin B$_6$	Pyridoxine is converted into pyridoxal phosphate, which acts as a coenzyme for aminotransferases.	Whole-grain cereals, wheat, corn, nuts, muscle meats, liver, and fish.	Rare, except in alcoholics, in women taking oral contraceptives, in infants fed formula diet low in this vitamin, and in patients receiving isoniazid for the treatment of tuberculosis. Isoniazid binds with pyridoxal and the pyridoxal-hydrazone complex is rapidly excreted in urine. Clinical symptoms include neuronal dysfunctions, which may be due to impaired synthesis of neurotransmitters like norepinephrine and serotonin, and anemia due to impaired heme biosynthesis. Sideroblastic anemia and personality changes are seen in severe deficiency.
Biotin	Biotin serves as the prosthetic group for the carboxylases, transferring CO$_2$ to acceptor molecules (carboxylation reactions).	Liver, kidney, milk, egg yolk, corn, and soya milk.	Intestinal microorganisms synthesize biotin. Hence, biotin deficiency is caused by (1) antibiotics that inhibit growth of intestinal microorganisms, and (2) consumption of excess raw egg. Egg white contains the protein avidin, which binds biotin tightly and prevents its utilization in the diet. Biotin deficiency causes seborrheic dermatitis, anorexia, nausea, and muscular pain. Children with biotin deficiency have immunodeficiency.

Contd...

Folic acid	Folic acid is converted into tetrahydrofolic acid (THF) which is essential for transmethylation reactions in DNA synthesis.	Leafy green vegetables, yeast, pulses, and liver.	Folate deficiency occurs due to (1) inadequate intake which is not uncommon in alcoholics, teenagers, and infants, (2) increased requirements, as in pregnancy, infancy, malignancy, and increased hematopoiesis (chronic hemolytic anemias), and (3) malabsorption. Folate deficiency affects rapidly multiplying cells like the hemopoietic cells, resulting in megaloblastic anemia.
Cobalamin Vitamin B$_{12}$	Vitamin B$_{12}$ is required for recovery of THF from methyl trap and formation of SAM (S-adenyl methionine), which is important for myelination of neurons.	Animal products (meat, liver, kidney, fish, egg) and milk.	Vitamin B$_{12}$ deficiency occurs mostly due to a reduction in its intestinal absorption, the causes of which could be (1) inadequate production of intrinsic factor (IF), as occurs in Pernicious anemia and gastrectomy, or (2) diseases or resection of terminal ileum. Deficiency can also occur if (3) the vitamin is consumed in the intestine by bacteria, as in the blind-loop syndrome or by fish tapeworm. Vitamin B$_{12}$ deficiency results in megaloblastic anemia and neurological signs and symptoms.
Ascorbic acid Vitamin C	Vitamin C is a reducing agent and functions as a coenzyme in hydroxylation reactions. It is important for the synthesis of collagen, epinephrine, bile acid, and steroids. It is also required for absorption of iron, bone mineral formation, and degradation of tyrosine. It is one of the three nutrients with antioxidant properties; the other two are vitamin E and β-carotene. Hence, it provides protection against the free radicals and is thought to prevent atherosclerosis, coronary artery disease, and cancer.	Citrus fruits, potatoes (particularly their skins), green vegetables, and tomatoes. *Emblica officinalis* (amla, Indian gooseberry) is the richest source.	Vitamin C deficiency occurs due to decreased dietary intake, which is uncommon. It can occur in infants on processed milk formulas unsupplemented by citrus fruits and vegetables. Vitamin C deficiency causes scurvy characterized by sore, spongy gums, fragile blood vessels, swollen joints, and anemia. There is impairment of wound healing and bone formation, which may lead to osteoporosis. Integrity of the blood vessels is decreased because of decreased collagen strength of the vessel walls. It leads to frequent rupture of blood vessels. There is reduction in immunity.

Fat-soluble vitamins

Vitamin A Retinol	Retinol derivatives (retinoids) are essential for vision, which is mediated by retinal. Another retinol derivative, retinoic acid, is essential for growth, reproduction, and maintenance of epithelial tissues, serving as a transcriptional regulator like other group-I hormones.	Liver, kidney, butter fat, oils, egg yolk, green leafy vegetables, fruits.	Causes of vitamin A deficiency include (1) decreased dietary intake of the vitamin or the carotene provitamin, (2) malabsorption of fats, as in obstructive jaundice, and (3) increased excretion, as in proteinuria. In blood, vitamin A is bound to plasma proteins. Night blindness is the earliest symptom, followed by retinal degeneration. The bulbar conjunctiva becomes dry (xerosis) and small gray plaques with foamy surfaces develop (Bitot's spots). There is ulceration of the cornea (keratomalacia), and dryness and hyperkeratosis of the skin.
Vitamin D Calciferol	The active form of Vitamin D is DHCC (dihydrocholecalciferol). On bone, DHCC has an antirachitic effect. In the kidney and intestine, DHCC stimulates the absorption of Ca^{2+} and phosphates by inducing the synthesis of carrier proteins.	Liver, egg yolk, and butter.	Causes of vitamin D deficiency include decreased dietary intake of the vitamin or the carotene provitamin, malabsorption of fats as in obstructive jaundice, and increased losses as in nephrotic syndrome. In blood, vitamin D is bound to plasma proteins. Vitamin deficiency results in rickets in children and osteomalacia in adults.
Vitamin E Tocopherol	Vitamin E is the most potent fat soluble antioxidant and scavenger of free radicals. Vitamin E delays aging and cataract formation and improves athletic performance. There is an inverse relationship between vitamin E intake and risk of coronary heart diseases.	Vegetable seed oils, liver, and eggs.	Absorption of vitamin E is dependent on appropriate intestinal fat absorption, which requires bile salts. Vitamin E deficiency is rare but is associated with liver atrophy, neurological disorders, and red blood cell hemolysis due to decreased protection for RBCs against peroxides.
Vitamin K Quinone	Reduced hydroquinone, which is the active form of vitamin KH$_2$, acts as a cofactor for the carboxylation reactions in the synthesis of coagulation factors 2, 7, 9, and 10.	Green leafy vegetables, egg yolk, and liver.	The main causes of vitamin K deficiency are: (1) inadequate dietary intake, (2) malabsorption of fats, as in obstructive jaundice, (3) hepatocellular disease, and (4) antibiotics that kill gut microflora. Vitamin K deficiency impairs blood coagulation resulting in prolonged bleeding time.

Wilson's disease (hepatolenticular degeneration) is a rare autosomal, recessively inherited disorder in which the biliary excretion of copper is impaired. It leads to copper accumulation, initially in the liver and subsequently, in other organs including the brain, especially in the basal ganglia. This results in hemolytic anemia, liver damage, and neurological disorder. A common finding is the Kayser–Fleischer ring (Fig. 77.**4**), a green or golden pigment ring around the cornea due to deposition of copper in Descemet's membrane.

The increase of copper in liver cells inhibits the coupling of copper to apoceruloplasmin and leads to low levels of plasma ceruloplasmin. The disease is treated by administration of D-penicillamine, a copper chelator, which permits urinary excretion by forming a water-soluble copper complex.

Dietary Fibers

Dietary fibers are plant carbohydrates (with the exception of lignin) in our diet that are resistant to digestion in the gut. They consist of insoluble fibers, e.g., celluloses and lignin and soluble fibers, e.g., hemicelluloses, gums, mucilages, and pectin. Whole-grain cereals, fruits, vegetables, and pulses provide a good natural source of both soluble and insoluble fibers. Wheat bran and other whole grains are rich in insoluble fiber while oat bran, peas, and beans are rich in soluble fiber.

Physiological effects Dietary fiber provides no energy but has several beneficial effects. Fiber absorbs up to 15 times its own weight in water and swells up, increasing the bulk of fecal matter. Increased bulk of stool increases colonic motility and hastens the intestinal transit of chyme. Anaerobic bacterial fermentation of fiber in the colon results in the formation of water, carbon dioxide, hydrogen, and short-chain fatty acids such as acetic, propionic, and butyric acids. The short-chain fatty acids are well absorbed by the colon and have a trophic effect on the colonic mucosa. They also promote the absorption of water, sodium, and chloride.

If the amount of dietary fiber is small, the diet is said to lack bulk. When the bulk of fecal matter in the colon is small, the colon is inactive and bowel movements are infrequent. Starvation and parenteral nutrition lead to atrophy of the mucosa of the colon. The atrophy is reversed if fibers like pectin are present in the colon.

Therapeutic role The recommended intake of fiber is 25 to 35 g day^{-1}. It should be obtained from natural dietary sources. In constipation, fibers act as *bulk-laxatives*, providing a larger volume of indigestible material to the colon and making the stool softer. In diarrhea, restricting the amount of dietary fiber slows intestinal transit and therefore decreases the frequency and volume of stool. Fibers bind to toxic compounds including certain carcinogens, decreasing their absorption. Fibers also bind to carcinomatous foci on the colonic mucosa and wash them away. Consumption of a high-fiber diet is said to lower the serum cholesterol and result in a low incidence of constipation, hemorrhoids, diverticulosis, cancer of the colon, diabetes mellitus, and coronary artery disease.

High-fiber supplements available are mostly either methyl cellulose preparations or varying combinations of the seeds and husks of the plant *Plantago lanata*, rich in hemicellulose and used since ages past in Indian medicine (wherein it is called *isoghul*). Patients should drink large amounts of water along with the fiber to enable it to swell and retain water.

Overenthusiastic consumption of high-fiber dietary supplements should be avoided. Large amounts of fiber can impair the absorption of iron, calcium, and fat-soluble vitamins. Swollen fibers can form a solid mass inside the esophagus called *bezoar*; these supplements are best avoided in people with disorders of esophageal motility.

Carbohydrate Metabolism

Glycolysis

Embden–Meyerhof pathway Glycolysis is the major pathway for glucose utilization in all cells. It can work under both aerobic and anaerobic conditions. In glycolysis, one molecule of glucose is oxidized to two molecules of pyruvate. In the process, there is a net gain of two ATP molecules (two are consumed while four are produced). In addition, two molecules of NADH+H⁺ are generated. These are oxidized through the electron transport chain, generating six molecules of ATP. The total yield of glycolysis under aerobic conditions is therefore eight molecules of ATP.

$$\text{Glucose} + 2\text{ATP} \rightarrow 2\text{Pyruvate} + 2\text{NADH+H}^+ \,(= 6\text{ATP}) + 4\text{ATP}$$

The pyruvate molecule has two possible courses. *In aerobic conditions*, it mostly diffuses into the mitochondria, is decarboxylated into acetyl CoA and enters the Krebs cycle.

$$\text{Pyruvate} + \text{NAD}^+ + \text{CoA} \rightarrow \text{Acetyl-CoA} + \text{NADH+H}^+ + CO_2$$

In anaerobic conditions, the NADH+H⁺ produced by glycolysis reduces the pyruvate to lactic acid. The reduction of pyruvate regenerates the NAD⁺ from NADH+H⁺, allowing the glycolysis to continue. However, since the two molecules of NADH+H⁺ are spent in reduction of two molecules of pyruvate and not in ATP production,

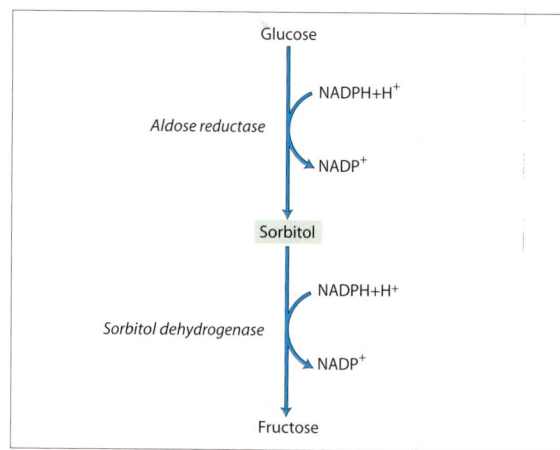

Fig. 79.**5**   Glucose-sorbitol-fructose pathway.

the net yield of glycolysis per molecule of glucose falls to two ATP molecules only.

$$\text{Pyruvate} + \text{NADH+H}^+ \rightarrow \text{Lactate} + \text{NAD}^+$$

HMP shunt pathway Between glucose-6-phosphate and glyceraldehyde-3-phosphate, there is an alternate pathway for glycolysis called the *pentose phosphate pathway* or the *hexose monophosphate shunt pathway* (Fig. 79.**4**). No ATPs are consumed or generated in this pathway. Its importance lies in the fact that (1) it produces ribose sugar that is essential for nucleic acid synthesis and (2) it generates NADPH+H⁺ (instead of NADH+H⁺) which is essential for fatty acid and steroid biosynthesis. NADPH+H⁺ is also required by the liver microsomal enzyme system that hydroxylates steroids, alcohols, and several drugs. NADPH+H⁺ is required for antioxidant reactions and for the respiratory burst in phagocytes.

Glucose-sorbitol-fructose pathway Glucose is also converted into fructose by way of sorbitol (Fig. 79.5). The pathway is present in sperm where fructose is the preferred source of energy. The pathway is also present in liver cells, providing the cells with a mechanism through which dietary sorbitol can be fed into the glycolytic or gluconeogenic pathway.

The cells of the retina, lens, kidney, and nerves contain aldose reductase but not sorbitol dehydrogenase. Hence these cells convert glucose into sorbitol but do not convert it further to fructose. When there is hyperglycemia, a large amount of glucose enters these cells and is converted into sorbitol. The accumulated sorbitol holds water and the cell swells up due to osmosis. The retinopathy, cataract, nephropathy, and neuropathy of diabetes mellitus have been attributed partially to this osmotic swelling of the cells.

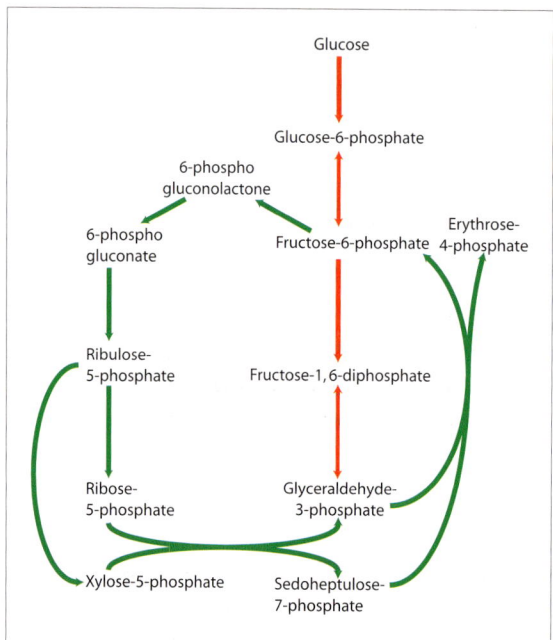

Fig. 79.**4**   Hexose monophosphate (HMP) shunt pathway.

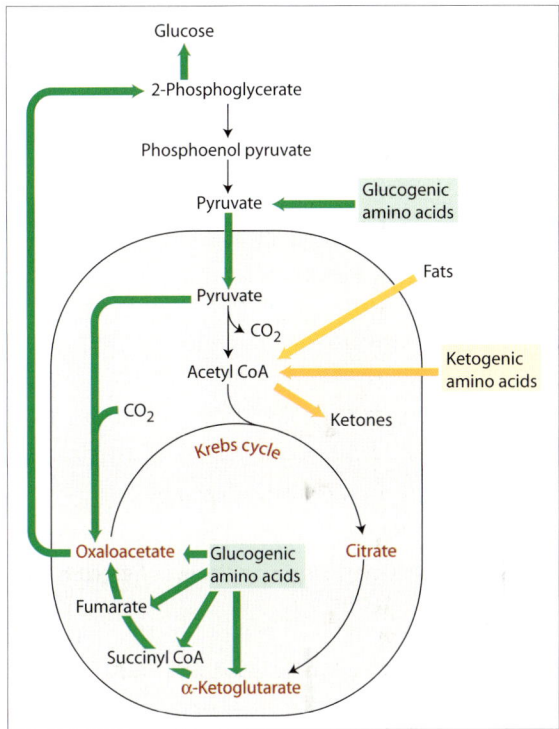

Fig. 79.**6**   Glucogenic and ketogenic amino acids.

Gluconeogenesis

Gluconeogenesis (Fig. 79.**3**) is the synthesis of glucose from noncarbohydrate sources. It ensures a continuous supply of glucose (necessary for organs like the brain, exercising muscles, erythrocytes, cornea, and testes) even during starvation when there is no carbohydrate intake and body glycogen stores are depleted. Ninety percent of the gluconeogenesis occurs in the liver and 10% occurs in the kidneys. Gluconeogenesis is made possible by three key gluconeogenic enzymes that drive the glycolytic pathway in reverse gear, resulting in the formation of glucose from acetyl CoA. These three enzymes are *phosphoenol pyruvate kinase, fructose-1,5-biphosphatase,* and *glucose-6 phosphatase.*

Gluconeogenesis makes it possible to convert pyruvic acid to glucose. However, it cannot convert acetyl CoA to glucose due to the irreversibility of the reaction in which pyruvate is converted to acetyl CoA.[2]

The irreversibility of the pyruvate-to-acetyl CoA conversion explains why fat, which yields acetyl CoA

---

[2] It can be argued that since acetyl CoA is converted to oxaloacetate through the Krebs cycle, it should be convertible to glucose as well. That, however, is not so because when a molecule of acetyl CoA enters the Krebs cycle, it actually consumes a molecule of oxaloacetate rather than producing one. Hence, the molecule of oxaloacetate formed later in the cycle is only a regeneration of the molecule consumed earlier and therefore, cannot result in a net glucose gain.

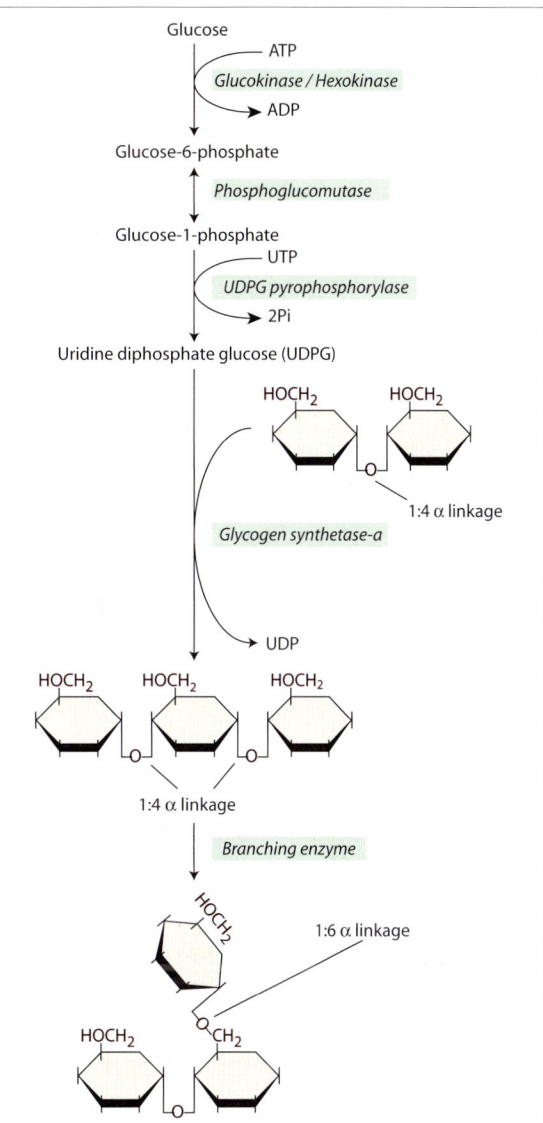

Fig. 79.**7**   Glycogenesis.

on hydrolysis, results in ketogenesis and not gluconeogenesis. For the same reason, leucine and lysine, which yield acetyl CoA on deamination, are called *ketogenic amino acids.* On the other hand, amino acids that yield pyruvate are called *glucogenic amino acids.* Amino acids that yield oxaloacetate or any of its precursors in the Krebs cycle are also glucogenic (Fig. 79.**6**).

Glycogenesis

Glycogenesis, or the synthesis of glycogen from glucose, occurs mainly in the organs that store glycogen, viz., muscle and liver. Glycogenesis begins with the phosphorylation of glucose to glucose-6-phosphate.

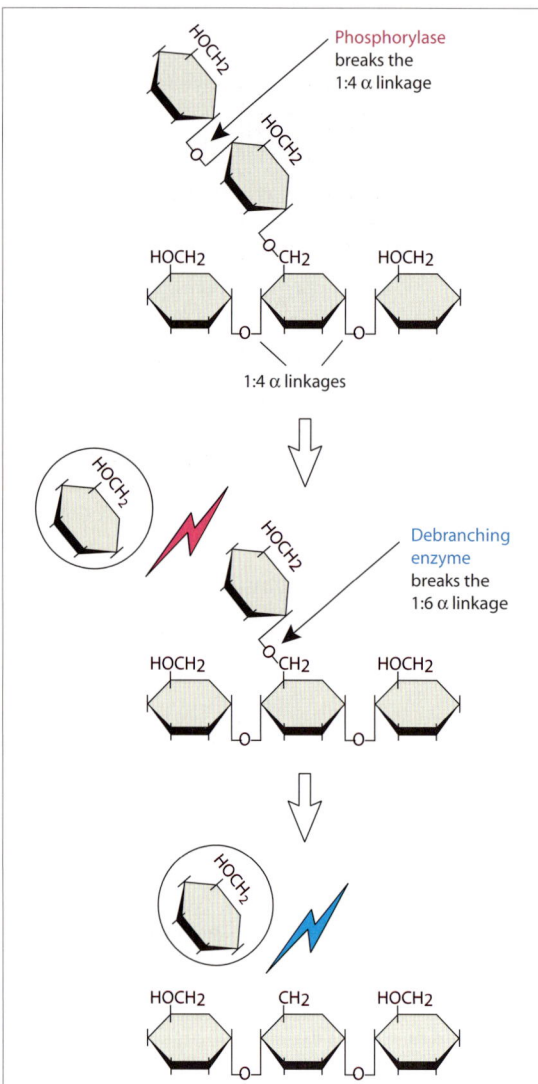

Fig. 79.**8**   Glycogenolysis.

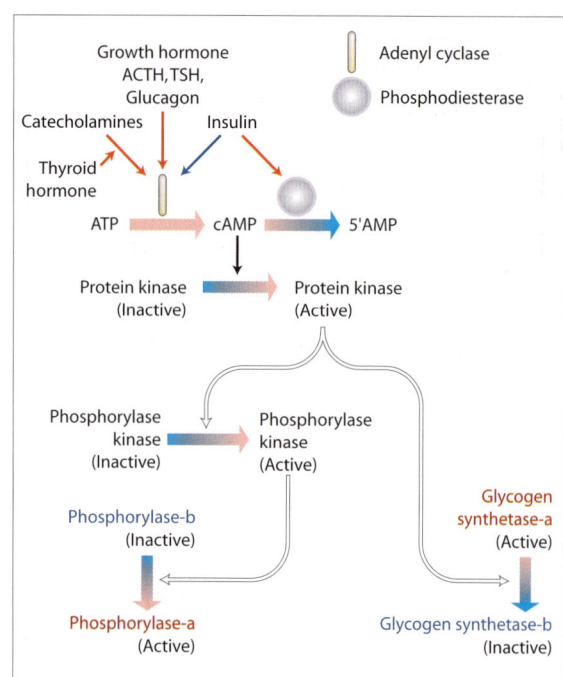

Fig. 79.**9**   Hormonal control of glycogenesis and glycogenolysis.

The reaction is catalyzed by *hexokinase* in muscle and *glucokinase* in liver. The glucose-6-phosphate isomerizes to glucose-1-phosphate and reacts with uridine triphosphate (UTP) to form uridine diphosphoglucose (UDPG). The UDPG molecules are the building blocks of glycogen. They bind to each other through 1:4α linkages in the presence of the enzyme *glycogen synthetase-a* to form long straight chains of glycogen. A *branching enzyme* brings about 1:6α linkages between UDPG molecules, resulting in the formation of branched glycogen molecules (Fig. 79.**7**).

Glycogenolysis

Glycogenolysis (Fig. 79.**8**) is the breaking down of glycogen into glucose molecules. It involves delinking of single glucose molecules from glycogen by the enzyme *phosphorylase*, which breaks the 1:4α linkages and the *debranching enzyme*, which breaks the 1:6α linkages. Since the linkages are phosphorylated before they are broken, the product of glycogenolysis is glucose-6-phosphate (and not glucose). In the liver, glucose-6-phosphate is dephosphorylated to glucose and released in blood. In the muscle, the glucose-6-phosphate produced is consumed by the muscle itself through glycolysis.

Glycogenesis and glycogenolysis are under hormonal control. The control, which is exerted through the enzymes glycogen synthetase and phosphorylase, is reciprocal: when one is stimulated, the other is inhibited (Fig. 79.**9**).

Fat Metabolism

Fatty acid synthesis

Fatty acid synthesis (Fig. 79.10) occurs primarily in liver and lactating mammary glands, and to a lesser extent, in adipose tissue and kidney. Fatty acids that occur in natural fats are straight chain derivatives and contain an even number of carbon atoms because the building units from which they are synthesized are the molecules of acetyl CoA, which have two carbon atoms each. Acetyl CoA is produced inside the mitochondria

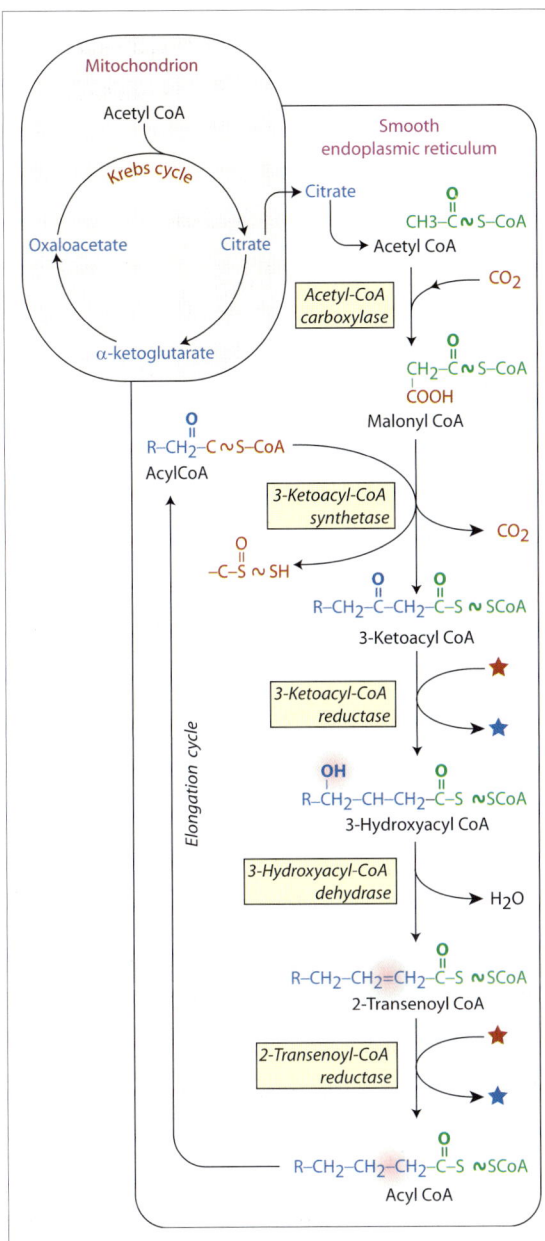

Fig. 79.**10**   Fatty acid synthesis.

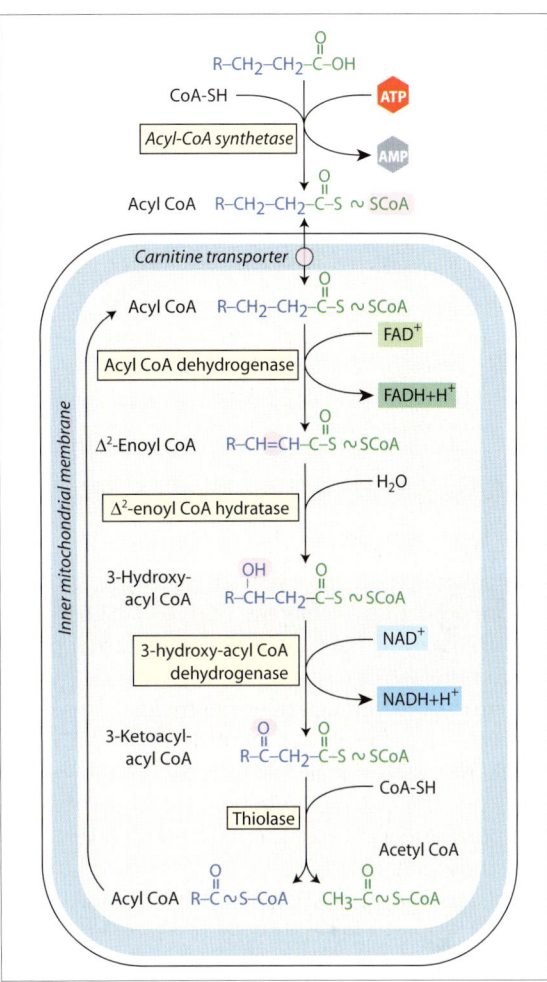

Fig. 79.**11**   β-oxidation of fatty acids.

acetyl CoA. Fatty acid biosynthesis requires NADPH+H⁺ (instead of NADH+H⁺), which comes from the HMP shunt pathway.

▪ β-oxidation of fatty acids

The β-oxidation of fatty acids occurs in all cells except neurons, erythrocytes, and adrenal medulla. In β-oxidation of fatty acids (Fig. 79.**11**), two carbon atoms are cleaved at a time from the acyl CoA molecule, starting at the carboxyl end. The chain is broken between the C-2 (earlier called α-carbon) and C-3 (earlier called β-carbon) and hence the name. Much smaller amounts of α-oxidation, i.e., removal of one carbon atom at a time, occur in the brain.

In β-oxidation, the fatty acid is first activated with a molecule of ATP. Further oxidation occurs in the mitochondria. The smaller acyl CoA molecules (like acetyl CoA) enter the mitochondria easily but the larger ones

from various metabolic pathways. From there, it must enter the endoplasmic reticulum where the biosynthesis of fatty acids takes place. Acetyl CoA cannot diffuse across the mitochondrial membrane. Hence, it is converted into citric acid and then reconverted into acetyl CoA after crossing the mitochondrial membrane and entering the cytosol. Once inside the endoplasmic reticulum, the enzyme *fatty acid elongase* condenses several molecules of acetyl CoA together into a growing chain of fatty acid. One by one, molecules of acetyl CoA are transferred from malonyl CoA to the growing chain. Malonyl CoA itself is formed by the carboxylation of

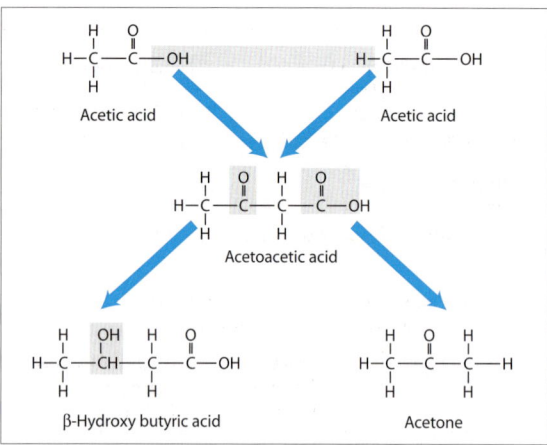

Fig. 79.**12** Structure and formation of ketone bodies.

(like palmitoyl CoA) must be transported across the inner mitochondrial membrane through facilitated diffusion employing the membrane-bound carrier palmitoyl transferase. Once inside, the acyl CoA is converted to ketoacyl CoA through three intermediate steps that generate one molecule each of $FADH+H^+$ and $NADH+H^+$. In the final step, a molecule of acetyl CoA is split off from the acyl CoA molecule by the enzyme thiolase. The process repeats itself till the chain is broken down into acetyl CoA molecules.

A large number of ATP molecules are formed in the process. For example, in the case of a molecule of palmitate, five molecules of ATP are produced for each of the seven molecules of acetyl CoA that are split off, giving a total of 35 ATP molecules. Each of the eight molecules of acetyl CoA formed yields 12 molecules of ATP when fed into the Krebs cycle, giving a total of 96 ATP molecules. Subtracting the two molecules of ATP that are consumed for regenerating from AMP the single molecule of ATP consumed during the initial activation of fatty acid, the net gain from β-oxidation of one molecule of palmitate is 129 molecules of ATP.

Ketogenesis

Ketone bodies, i.e., acetone and β-hydroxy butyric acid (Fig. 79.**12**), are formed in the mitochondria of hepatocytes by the condensation of two molecules of acetyl CoA. The liver constantly produces ketone bodies at a low rate. However, during starvation, large amounts of ketone bodies are formed by the liver. These are released into blood and transported to peripheral tissues for energy release. The heart prefers ketone bodies to glucose as a fuel. The brain can utilize ketone bodies only when their concentration in the blood is high. Erythrocytes cannot utilize ketone bodies at all.

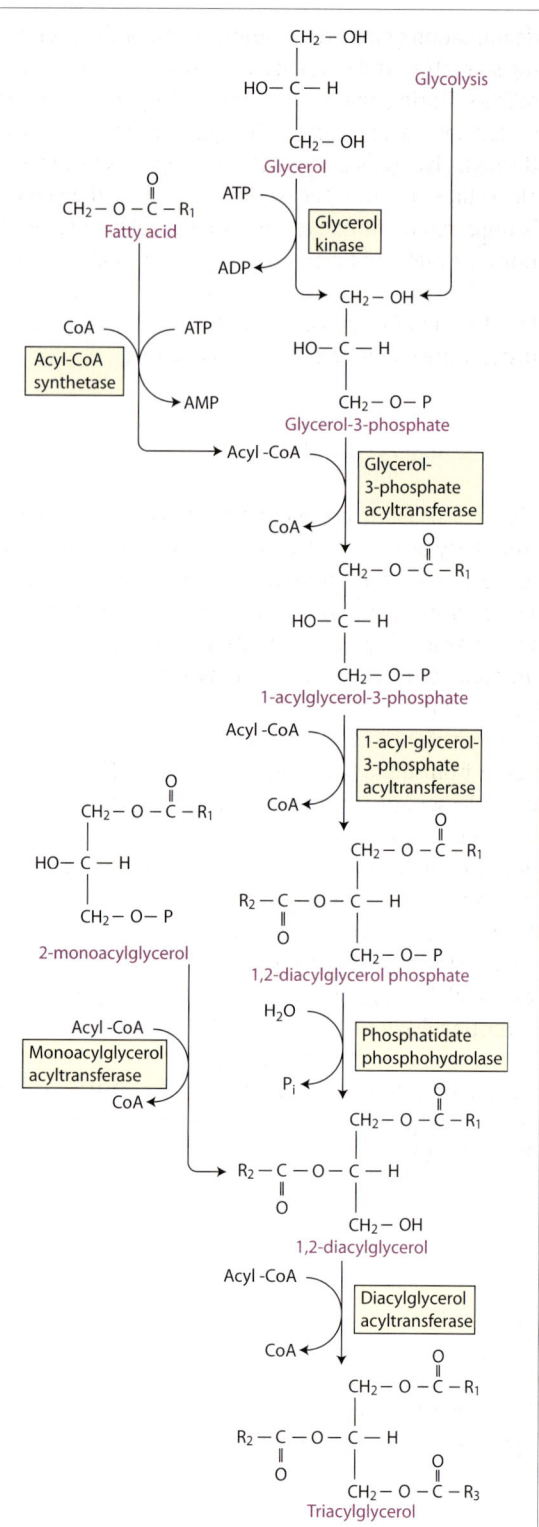

Fig. 79.**13** Lipogenesis.

The two prerequisites for ketone body formation are: (1) reduction in the amount of glycolysis, and (2) production of large amounts of acetyl CoA from other sources, like the oxidation of fatty acids and deamination of ketogenic amino acids. Both occur during starvation and in diabetes mellitus. The reason is as follows. During glycolysis, some of the pyruvate is converted into oxaloacetate. The oxaloacetate produced through glycolysis is important for replenishing the oxaloacetate lost from the Krebs cycle. Thus, if glycolysis is impaired or carbohydrates are completely removed from the diet, the Krebs cycle comes to a halt. The acetyl CoA molecules that are formed from proteins and fats are therefore unable to enter the Krebs cycle and instead, are converted into ketone bodies.

Lipogenesis

Lipogenesis is the production of lipids (triacylglycerol) from fatty acids and glycerol. It occurs in the liver and adipose tissues. Adipose tissue stores the synthesized lipids within itself while the liver releases it into blood in the form of VLDL. As mentioned earlier, only small amounts of *de novo* fatty acid synthesis occur in adipose tissue.

Most of the fatty acids for lipogenesis in adipocytes come from dietary fat (in the form of chylomicrons) and a lesser amount from the liver as VLDL. For the reaction to occur, fatty acid must first be activated into acyl CoA by the enzyme acyl CoA synthetase and glycerol must be phosphorylated by the enzyme glycerol kinase (Fig. 79.**13**). Glycerol kinase is absent in adipose tissue. However, the presence of glycerol kinase is not essential for lipogenesis because glycerol phosphate (phosphoglycerate, see Fig. 79.**3**) is also produced, in the liver as well as in adipose tissue, during glycolysis. Formation of triacylglycerol from acyl CoA and glycerol phosphate proceeds through four steps as shown in Figure 79.**13**.

Lipolysis

Lipids are stored only in adipose tissue and hence, lipolysis occurs only in adipose tissue. The triacylglycerol is hydrolyzed by the action of hormone sensitive lipase into fatty acids and glycerol (Fig. 79.**14**). The enzyme is activated by circulating catecholamines.

Protein Metabolism

About 1–2% of the total body protein, mostly muscle protein, is broken down daily into amino acids. Nearly 25% of these amino acids are deaminated while the rest are reutilized for protein synthesis. The breakdown of amino acids occurs in two steps.

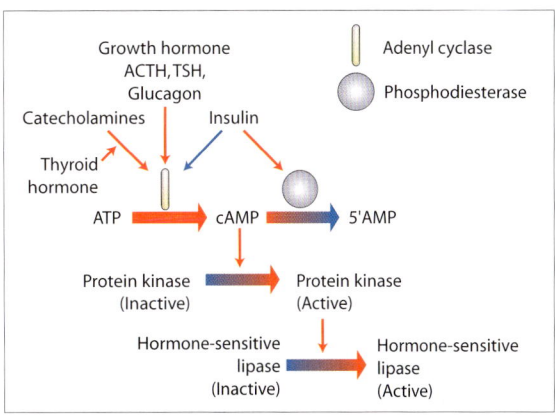

Fig. 79.**14**   Activation of hormone-sensitive lipase.

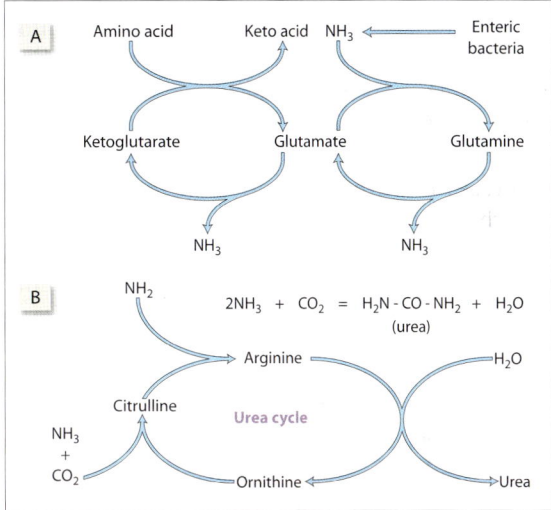

Fig. 79.**15**   [A] Amino acid catabolism and detoxification of ammonia. [B] Urea cycle.

In the first step, the amino acid that is to be degraded undergoes transamination, i.e., it transfers its NH$_2$ group to a keto acid called ketoglutarate, resulting in the formation of glutamate, an amino acid (Fig. 79.**15**A). The degraded amino acid itself changes to the corresponding keto acid and enters the Krebs cycle. Glutamate therefore represents the byproduct of deamination of various amino acids.

In the second step, the NH$_2$ group of glutamate reacts with CO$_2$ to form urea, which is excreted as a nontoxic waste product. The formation of urea involves intermediates like ornithine, citrulline, arginosuccinate, and arginine, all of which are regenerated and form the urea cycle (Fig. 79.**15**B).

Glutamate also rapidly mops up the NH$_3$ that is formed by enteric bacteria and which enters the liver through portal blood. The product of glutamate and NH$_3$ is glutamine. The glutamate is later regenerated when the NH$_3$ is removed from glutamine to form urea.

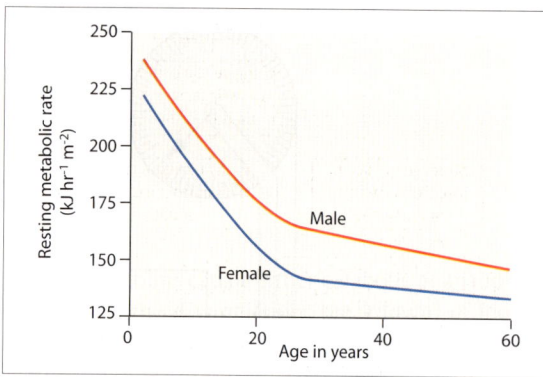

Fig. 80.**3**   BMR changes with age.

metabolic rate falls lower during sleep and is called the basal metabolic rate (BMR).

It is important to record BMR during the postabsorptive period so that it is unaffected by the *thermic effect* (or the *specific dynamic action*) of food. The stimulating effect of food on metabolic rate is least with carbohydrate and fat (5–10% increase) and greatest with protein (10–35% increase). The stimulating action of carbohydrates is attributed to the metabolism of extra carbohydrate for providing the energy for glycogenesis. The greater effect of protein on metabolic rate occurs probably because the carbon skeletons of the deaminated amino acids have a stimulating effect on metabolism.

It is also important to record BMR in a thermoneutral ambient temperature because metabolic rate increases with fall in ambient temperature. The rise in

metabolism helps in greater heat production for maintaining the body temperature. If the exposure to cold is prolonged, a gradual fall in the metabolic rate takes place.

Since metabolism correlates much better with the surface area than with height or weight, BMR is usually expressed per unit body surface area. Similarly, energy consumption is expressed in kilojoules, which is the SI unit of energy (1 kcal = 4.2 kJ). The normal BMR is about 165 kJ hr^{-1} m^{-2} in the male, and about 155 kJ hr^{-1} m^{-2} in the female. Clinicians however like to express it as 1 kcal hr^{-1} kg^{-1} and call it 1 MET (METabolic equivalent). One MET is associated with oxygen consumption at the rate of approximately 3 ml min^{-1} kg^{-1}. Higher metabolic rates are expressed in METs, i.e., in multiples of the BMR.

Factors influencing BMR   (1) Starvation or prolonged undernutrition decreases the metabolic rate. The fall in the metabolic rate explains why, when an individual tries to shed weight, the weight loss is initially rapid but slows down thereafter. (2) BMR varies with age: it is considerably greater per square meter of surface in children than in adults. There is a further gradual fall in the metabolism during adult life as age advances (Fig. 80.**3**). (3) Like any other chemical reaction, the biochemical reactions of the body are speeded up by a rise in body temperature. For every rise of 0.5 °C in the internal temperature of the body, the resting metabolism increases by 7%. (4) Thyroid hormones speed up the metabolic activities of the tissues and hence, the BMR increases in thyrotoxicosis up to +100. In myxedema,

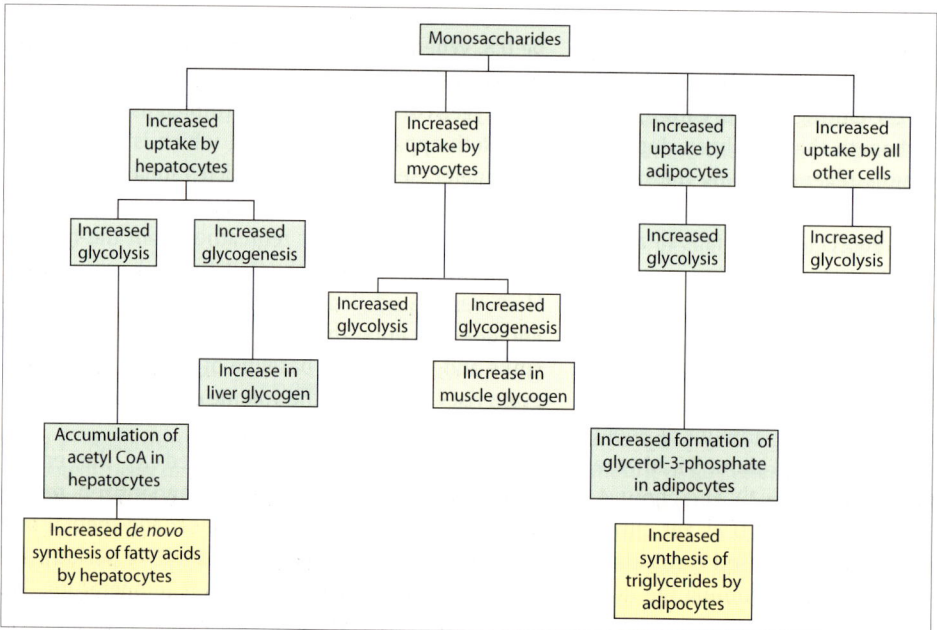

Fig. 80.**4**   Carbohydrate metabolism in the postabsorptive state.

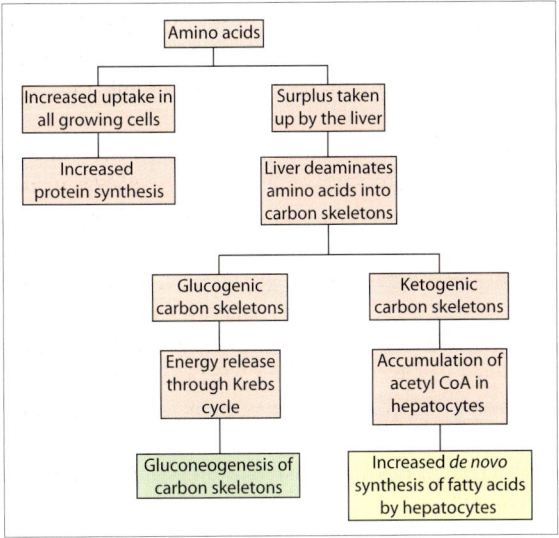

Fig. 80.**5**   Protein metabolism in the postabsorptive state.

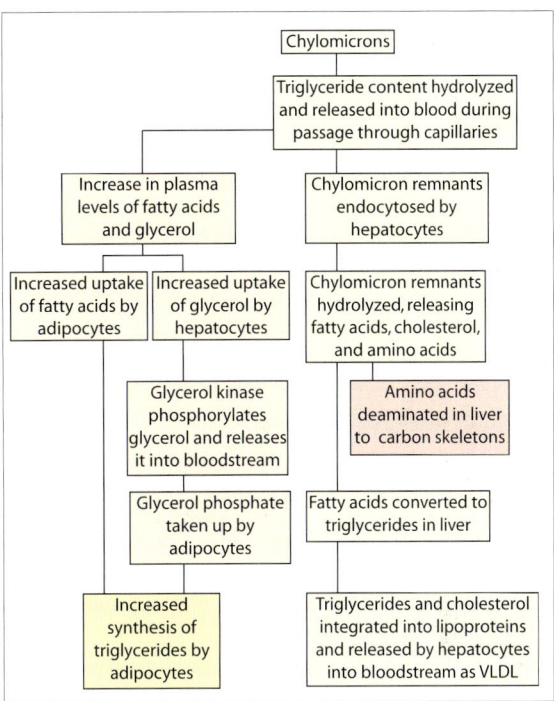

Fig. 80.**7**   Fat metabolism in the postabsorptive state.

the BMR can fall to as low as –40, which is its lowest recorded value. Adrenalin increases the metabolic rate, but to a lesser extent than thyroxine. (5) Anxiety and tension elevate BMR because they cause epinephrine secretion and tensing of the muscles, even when the individual is quiet. On the other hand, depressed patients may have a low BMR.

Anabolic States

Postabsorptive state

The postabsorptive state is the 2–4 hour period following ingestion of food during which digested food is absorbed, and the monosaccharides, amino acids, and

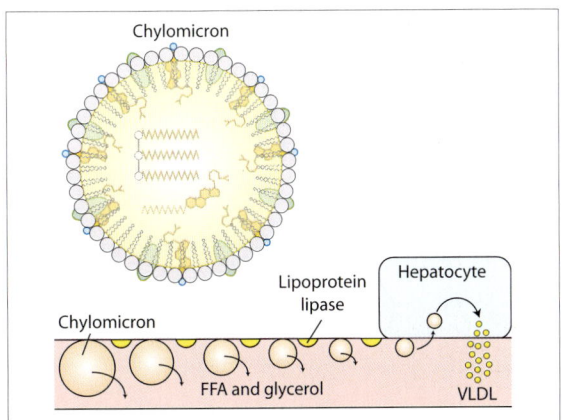

Fig. 80.**6**   Removal of triglycerides from a chylomicron during its passage through a capillary and the uptake of a chylomicron remnant by the liver.

chylomicrons enter the bloodstream in large amounts. The high levels of glucose and amino acids in the blood stimulate insulin secretion. Many of the subsequent metabolic events are under the control of insulin.

Carbohydrates   Most of the surplus glucose is taken up by the liver and muscles where it is trapped as glucose-6-phosphate, converted to glycogen, and stored. The increased glucose uptake by muscle and liver cells occurs under the influence of plasma insulin, which increases in the postabsorptive state. In the liver, the stimulation of glycolysis produces acetyl CoA that is used for fatty acid synthesis (Fig. 80.**4**).

Proteins   The surge of amino acids in the blood during the postabsorptive period briefly stimulates protein synthesis in all cells, resulting in the replacement of the proteins that were degraded after the preceding postabsorptive period. In liver too, the synthesis of plasma proteins increases. Surplus amino acids are taken up by the liver and the kidney, are deaminated, and converted either to glucose (for storage as glycogen) or to fatty acids (Fig. 80.**5**).

Fats   The chylomicrons released from the intestine (nascent chylomicrons) rapidly lose their triacylglycerol content due to the action of lipoprotein lipase present on the surface of capillary endothelial cells in various tissues of the body (especially, adipose tissue

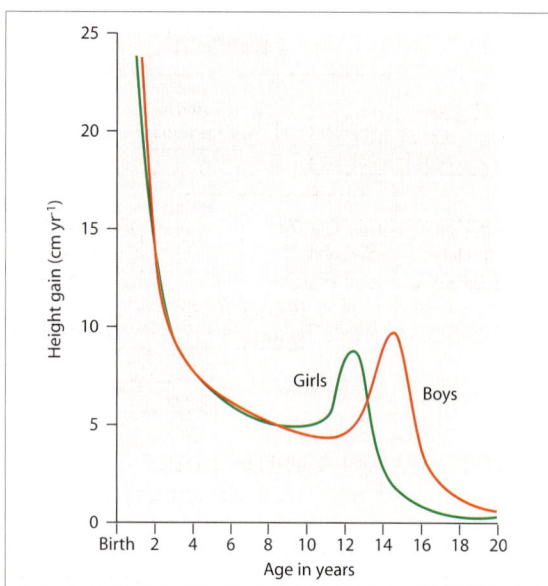

Fig. 80.**8**   Age of growth spurts in boys and girls.

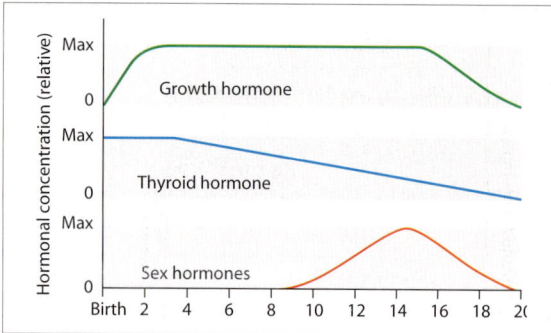

Fig. 80.**9**   Secretion of three anabolic hormones decline after the growing age.

and striated muscle), with the notable exception of the liver. None of these organs takes up chylomicrons: they hydrolyze the triglyceride content of chylomicrons passing through them (Fig. 80.**6**). After its triglyceride content has been hydrolyzed, whatever remains of the chylomicron is called the chylomicron remnant.

The *chylomicron remnants* are taken up by the liver and degraded, releasing cholesterol, fatty acids, and amino acids. The fatty acids derived from chylomicron remnants as well as those synthesized *de novo* are converted into triacylglycerol, incorporated into lipoproteins and released into the bloodstream as VLDL (*very low density lipoproteins*) for transport to all tissues. The adipose tissues synthesize triacylglycerol from glycerol-3-phosphate (derived from glycolysis) and fatty acids (derived mainly from chylomicrons and partly VLDL). The high levels of insulin (which stimulates lipoprotein lipase) and glucose in the postabsorptive state favor lipogenesis in adipocytes. Insulin also inhibits hormone sensitive lipase and thereby inhibits lipolysis (Fig. 80.**7**).

Growth

Growth involves increase in length and size of the body due to protein accretion and is associated with maturational changes. In humans, there are two growth spurts, one during infancy and the other during puberty (Fig. 80.**8**). The growth during infancy is not continuous but episodic, i.e., it occurs in brief bursts of 0.5 to 2.5 cm growth in a few days separated by up to two months of little or no growth. The pubertal growth spurt is largely

under the control of the anabolic hormones, viz., growth hormone, thyroid hormone, and the sex hormones. These hormones are secreted in larger quantities during the growing age (Fig. 80.**9**). The fourth anabolic hormone, viz., insulin is also important for growth though it is not secreted in higher amounts during the growing age. The growth-promoting effects of growth hormone are mediated by somatomedins which are also called insulin-like growth factors (IGF) due to their structural and functional resemblance to insulin.

Proper growth requires adequate intake of proteins, vitamins, and minerals. Injuries and diseases retard growth because they increase protein catabolism. Following recovery from the illness, there is a growth spurt called *catch-up growth* which continues till the lag in growth is compensated. Growth also requires adequate caloric intake so that proteins are not diverted for meeting the body's energy requirements. Growth

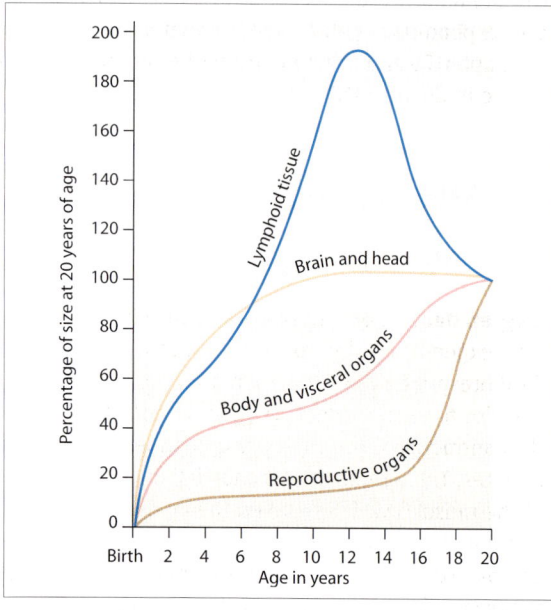

Fig. 80.**10**   Growth of different organs.

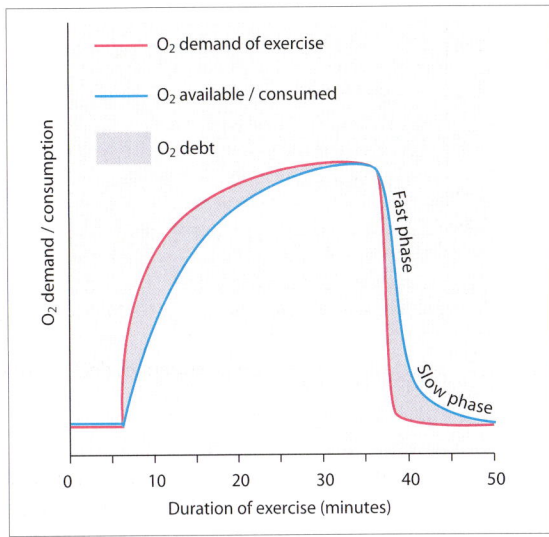

Fig. 80.**11**   Oxygen debt and its recovery.

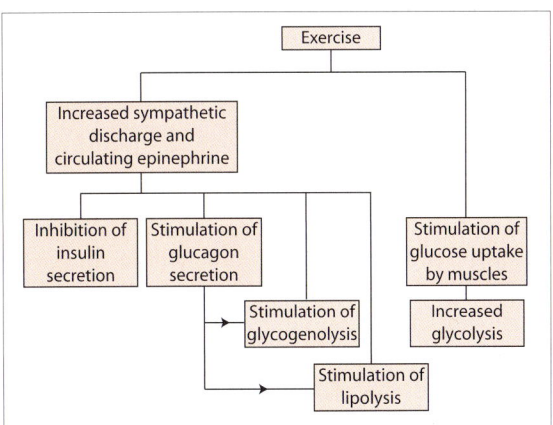

Fig. 80.**12**   Metabolic events during exercise.

continues despite caloric inadequacy if it coincides with a growth spurt. Conversely, unless it coincides with a growth spurt, excess protein intake does not cause growth and is either used for energy release or causes obesity.

Not all organs of the body grow at equal rates. The brain and skull complete their growth by puberty while others show a growth spurt during puberty. Lymphoid tissues show a growth spurt before the onset of puberty (Fig. 80.**10**).

Obesity

Obesity tends to occur if the caloric intake exceeds calorie expenditure. Caloric requirements are discussed in Chapter 78 and the regulation of food intake is discussed in Chapter 110.

Catabolic States

Exercise

Oxygen debt   Exercise is associated with excess energy expenditure due to increased muscular activity. Cardiorespiratory readjustments are required during exercise to ensure adequate glucose and O_2 supply to the contracting muscles. These adjustments take a few minutes. Till then, there is a shortfall of oxygen supply to the muscles which is called the O_2 *debt* (Fig. 80.**11**). The initial muscle contraction is therefore always anaerobic, i.e., its energy requirements are met through anaerobic glycolysis. In long-distance races, it is important not to incur excessive O_2 debt early in the race lest

there should be too much anaerobic metabolism and accumulation of lactic acid. Hence, long-distance runners begin the race very slowly to allow the cardiorespiratory system to gear up to the energy demands of muscular activity. Gradually, the O_2 supply to the muscle improves due to cardiorespiratory readjustments and the muscle switches to oxidative metabolism. A steady state is then attained in which the O_2 supply balances the O_2 requirements of the muscle. This state can continue for hours, as evident in marathon runners. After the muscle stops contracting, it continues to take up extra O_2 called the recovery O_2 which is equal to the O_2 debt incurred at the onset of contraction. The recovery oxygen has two components: (1) The *fast component* is expended quickly and immediately after the end of exercise and is utilized for replenishing the muscle stores of ATP and phosphocreatine that were exhausted in the

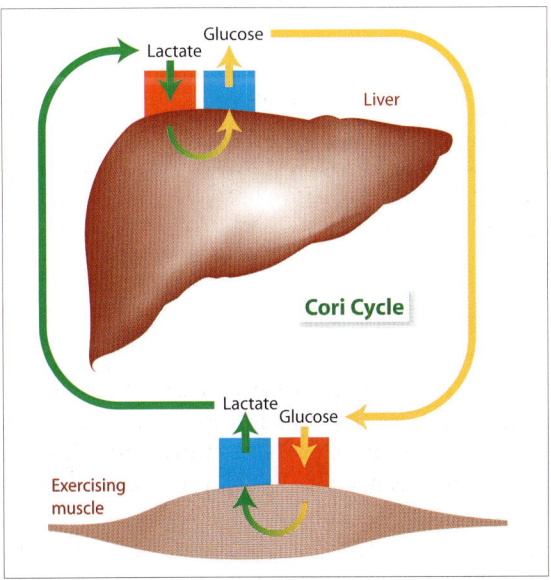

Fig. 80.**13**   Cori cycle.

initial phase of contraction. (2) The *slow component* of recovery O_2 is expended slowly and after the completion of fast oxygen recovery. It is utilized for oxidizing the lactic acid that accumulated during glycolysis.

Catabolic reactions Exercise is associated with increased sympathetic discharge and increased plasma levels of circulating epinephrine, which act directly on the pancreatic islets to inhibit insulin secretion and stimulate glucagon secretion. Although muscle cells require insulin for glucose uptake, the inhibition of insulin secretion in exercise does not affect glucose uptake by the contracting muscle cells, which take up glucose independently of insulin. Both epinephrine and glucagon stimulate glycogenolysis and lipolysis. Lipolysis in adipose tissue releases fatty acids and glycerol which are taken up by skeletal muscles for oxidation (Fig. 80.**12**).

Glycogenolysis in the liver results in release of glucose into the blood. Glycogenolysis in muscles produces glucose-6-phosphate, which is consumed by the muscle itself through glycolysis. Glycolysis converts glucose to pyruvate. Under aerobic conditions, the pyruvate is completely oxidized through the Krebs cycle with the release of large amounts of energy. Under anaerobic conditions, as is commonly seen in severely exercising muscles, the pyruvate is reduced to lactate and released into the bloodstream. The liver takes up the lactate and converts it back to glucose or glycogen. This exchange of lactate between muscle and liver is called the Cori cycle (Fig. 80.**13**).

Grading of exercise Based on the rate of its energy expenditure, exercise is categorized into very light, light, moderate, heavy, and very heavy (Table 80.**1**). The exercise is called severe if it can be kept up only for a very short time, for example, a 100 meter sprint at top speed.

Table 80.**1** Examples of different grades of exercise.

Grade	METS	Examples
Very light	1–3	Sedentary work, walking at 3 km hr^{-1}, driving a car.
Light	3–5	Carrying objects weighing 10–15 kg, walking at 6 km hr^{-1}, cycling at 10–12 km hr^{-1}.
Moderate	5–7	Carrying objects weighing 15–25 kg, walking at 7 km hr^{-1}, cycling at 15 km hr^{-1}, swimming, climbing stairs slowly.
Heavy	7–9	Carrying objects weighing 25–40 kg, running at 8 km hr^{-1}, cycling at 20 km hr^{-1}, climbing stairs quickly.
Very heavy	9–15	Carrying objects weighing more than 40 kg, running at 10 km hr^{-1}, cycling at more than 20 km hr^{-1} or up a steep slope, running upstairs or carrying a load upstairs.

Can heart rate reliably grade exercise?

The fact that heart rate increases linearly with exercise in an individual does not make it suitable for grading exercise *per se*. For a given amount of exercise, the heart rate will be lower in an individual with a higher cardiorespiratory fitness, not to mention athlete's bradycardia. The heart rate is higher in cardiac as well as respiratory disorders, and is susceptible to emotions, autonomic imbalances, and medications. It is therefore important to appreciate that the term exercise refers to the exercise of skeletal muscles, which does not necessarily parallel cardiac exercise.

Muscle metabolism is completely anaerobic in such severe exercise.

The maximum amount of O_2 uptake by an individual during exercise is called the Vo_{max} and is a measure of the cardiorespiratory fitness of the individual. The maximal exercise capacity for most individuals rarely exceeds 15 METs, which is equivalent to a Vo_{max} of 52.5 ml min^{-1} kg^{-1}. Champion athletes can go up to 20 METs.

Starvation

Starvation is a form of stress (see Fig. 85.**7**) and therefore it is associated with increased sympathetic discharge and epinephrine secretion. As in the case of exercise,

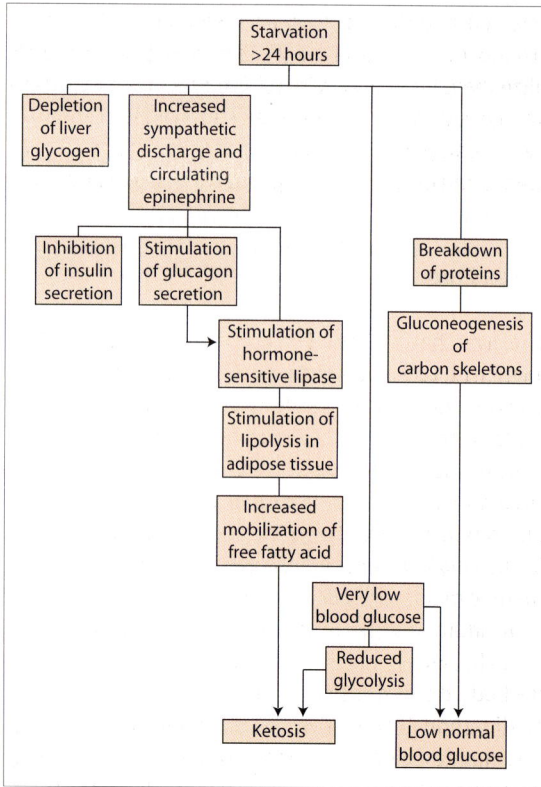

Fig. 80.**14** Metabolic events during starvation.

some of the metabolic consequences of starvation are brought about by sympathetic activity and circulating epinephrine which stimulate glucagon secretion.

Epinephrine and glucagon stimulate glycogenolysis and within 24 hours of starvation, the liver is depleted of its stored glycogen, which is only about 100 g. Thereafter, energy is obtained mostly by the oxidation of fatty acids and amino acids. The dissimilation of proteins and fats occur in a ratio of roughly 1:2. Amino acids are mobilized by breakdown of proteins. Proteins of the brain and heart are largely spared and most of the protein that is dissimilated comes from the spleen, liver, and muscles, in decreasing order. The excess protein breakdown is reflected in increased urinary urea excretion. Adequate blood glucose is necessary for the brain, nerves, erythrocytes, and adrenal medulla that cannot metabolize fatty acids. Till the very end, the liver maintains normal blood glucose levels through gluconeogenesis, using the carbon skeletons of deaminated glucogenic amino acids and the glycerol obtained from lipolysis (Fig. 80.**14**).

The sympathetic discharge as well as the circulating epinephrine and glucagon stimulate hormone sensitive lipase which hydrolyzes neutral fats present in adipose tissue. In the absence of adequate glycolysis, the fatty acids mobilized from adipose fats are converted into ketone bodies, resulting in ketoacidosis. The formation of ketone bodies by the liver is beneficial because they are released into the blood and transported to tissues like striated muscles and the renal cortex, which can metabolize them for energy release. Even the brain can utilize ketone bodies when their level is sufficiently high in the blood. Complex lipids that form part of the cell membranes and organelles are spared till the very end.

Liver

The liver has a central role in metabolism, both in anabolism and catabolism. *In the anabolic state*, it synthesizes glycogen from glucose and lipids from fatty acids. It stores the glycogen in hepatocytes. It incorporates the lipids into lipoproteins and releases them into circulation. It deaminates amino acids that are not used for growth and uses them instead for gluconeogenesis. *In the catabolic state*, it releases glycogen into blood, thereby maintaining the blood sugar at normal levels.

In addition to glycogen, the liver also stores vitamins A and B_{12}, which makes it an important storage organ of the body. It synthesizes and secretes bile which helps in fat digestion. It detoxifies several endogenous (porphyrins, amides from deaminated amino acids) and exogenous substances (ammonia produced by enteric bacteria and drugs) and excretes them in bile. Detoxified porphyrins are excreted as bile pigments, and detoxified amides and ammonia are excreted as urea. The liver also synthesizes plasma proteins and coagulation factors II, VII, IX, and X. The mast cells in the liver synthesize heparin. In the fetus, the liver forms erythrocytes. In adults, the Kupfer cells destroy aged red cells. Kupfer cells adhere to the walls of hepatic sinusoids. They belong to the monocyte-macrophage system.

Extirpation of the liver causes hypoglycemia since the glucose that is consumed by the body is not replenished by glycogenolysis which occurs only in the liver. On the other hand, intravenous administration of glucose, even when combined with insulin administration, results in marked hyperglycemia because the glucose is not rapidly removed through glycogenesis, which occurs only in the liver. These effects demonstrate the homeostatic role of liver on blood glucose level.

Removal of liver results in unconjugated hyperbilirubinemia because the bilirubin that is produced in extrahepatic sites (bone marrow and spleen) is not excreted in the bile. The urea level falls while amino acid levels rise in blood indicating that the liver is the only site of urea synthesis. For the same reason, administered galactose (unlike other sugars) does not change to glucose since the liver is the only site where the conversion occurs. Reduced synthesis of coagulation factors results in hypocoagulability of blood.

Fatty liver occurs when the amount of fat synthesized in the liver exceeds the amount of triglyceride secreted by it as VLDL. Common causes of fatty liver include ethanol ingestion and uncontrolled diabetes mellitus in which there is excessive hepatic fat synthesis. A deficiency of choline also causes fatty liver by impairing lipoprotein synthesis and therefore, VLDL output. Conversely, choline aids in recovery from fatty liver by increasing VLDL secretion and hence, it is called a *lipotropic factor*. Substances that contain choline (like lecithin, present in egg yolk) as well as substances that lead to increased synthesis of choline (like methionine, an amino acid) act as lipotropic factors. A fatty liver ultimately leads to cirrhosis (fibrosis) of the liver and impairs liver functions.

Liver function tests

Hepatic failure is associated with impaired hepatic conjugation and excretion, reduced hepatic synthesis and impaired metabolic reactions. Damage to liver cells is associated with leakage of hepatocellular enzymes into the blood. Accordingly, liver function tests can be categorized into four types.

Mechanism of Hormonal Action

Hormones are chemical mediators controlling cellular functions. They are secreted into the bloodstream by ductless endocrine glands and thereby exert widespread actions.

Several hormones act in a *paracrine way*, i.e., they exert their effect on neighboring targets, e.g., the effect of gastrin on D cells secreting somatostatin (see Fig. 87.**7**) or the effect of testosterone on spermatogenesis.

Sometimes, a hormone acts in an *autocrine way*, i.e., it acts on the cell from which it is secreted. The term cytokine is a general term for small proteins that act in an autocrine or paracrine manner. Some cytokines

	Group-I	Group-IIA	Group-IIB	Group-IIC	Group-IID
				TRH	
Hypothalamic		CRH			
				GnRH	
		TSH			
		ACTH			
		MSH			
		FSH			
		LH			
		HCG			
Pituitary					HCS
					GH
					IGF
					Prolactin
				Oxytocin	
		ADH			
Thyroid	TH				
		PTH			
Calcitropic		Calcitonin			
	DHCC				
	Cortisol				
Adrenal	Aldosterone				
		Adrenaline α_2		Adrenaline α_1	
					Insulin
Pancreatic		Glucagon			
		Somatostatin			
	Estrogen				
Reproductive	Progesterone				
	Testosterone				
Gastro-intestinal				Gastrin	
		Secretin			
				CCK	
			ANP		
			NO		
Others	Retinoic acid				
					Erythropoietin
	Angiotensin-II				
				Acetylcholine	

Table 81.**1** Examples of group-I and group-II hormones.

have systemic effects as well. For example, IL-1 and IL-3 produce fever. Cytokines are secreted mostly by lymphocytes and macrophages and also by endothelial cells, neurons, and glial cells.

Cellular functions are also controlled by neurotransmitters released by neurons. Neural effects are much more localized and much faster than hormonal effects. The distinction between hormones and neurotransmitters is however becoming blurred as evident from terms like neurohormone and neurosecretion that are applicable, for example, to posterior pituitary hormones. Another example is cholecystokinin which is secreted by S cells in the mucosal glands and also by neurons in the brain.

Hormones control cellular activity by modifying the flow of substrates within the cell and the turnover of enzyme proteins. All hormones act through specific receptors present on hormone-sensitive target cells. Hormones bind to the receptors with high specificity and affinity. Receptors have two functional domains: a *recognition domain* which binds the hormone and a *coupling domain* which generates a signal that couples hormone recognition to some intracellular function. Based on their mechanism of action, hormones have been classified as shown in Figure 81.**1**.

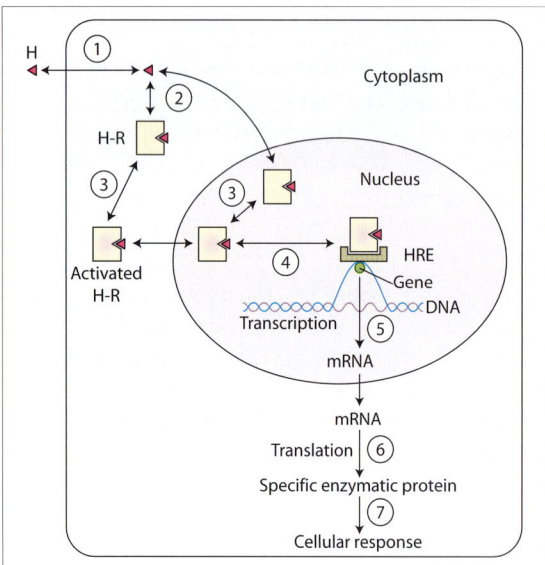

Fig. 81.**1** Mechanism of group-I hormonal action. Hormone (H) diffuses across the plasma membrane (1) and binds to its specific receptor (R) protein (2), either in the cytoplasm or in the nucleus to form the hormone-receptor complex (H-R). The binding brings about a conformational change in the receptor which now becomes activated (3). The activated H-R complex binds to specific regulatory regions of the DNA called the hormone response element or HRE (4). This binding facilitates transcription of the adjacent gene(s) by RNA polymerase, thereby increasing rate of messenger RNA (mRNA) formation (5). The newly synthesized protein (6) induces cellular responses (7).

Group-I Hormones

Group-I hormones (steroid/retinoid/thyroid hormones), examples of which are listed in Table 81.1, are lipid soluble and easily diffuse across the cell membrane. Inside the cell, they bind to intracellular receptors and affect gene expression, the broad mechanism of which is explained diagrammatically in Figure 81.**1**.

Group-II Hormones

Group-II hormones are water soluble peptide hormones that cannot enter the cell. They act by binding to the receptors on target cells and initiating a chain of reactions within the cell membrane and inside the cell. In doing so, they activate a series of enzymes or messengers. Depending upon the type of second messenger (the hormone itself is the first messenger), the group-II hormones are classified into four subgroups called IIA, IIB, IIC and IID (Tables 81.**1** and 81.**2**). In general, all group-II hormones employ a specific receptor, a G protein (GTP-dependent protein), a GTP molecule, a membrane-located enzyme or ion channel, a second messenger, and an effector enzyme (Fig. 81.**2**).

Hormonal receptors are integral proteins present in the cell membrane (Fig. 81.**3**). Receptors coupled to G proteins have seven membrane-spanning domains. The G protein is a complex of three subunits (α, β, and γ) that is anchored to the plasma membrane. In the absence of hormone, an inactive molecule of GDP is bound to its α subunit which has intrinsic GTPase activity. When the hormone binds to the receptor, the receptor undergoes a conformational change and activates the G protein. Activation of G protein is associated with the replacement of the GDP molecule on its α subunit by GTP. The GDP–GTP exchange on the G protein leads to the separation of its α subunit bound to GTP from its β and γ subunits. The α-GTP complex binds to a membrane-located enzyme and activates it. The activation is terminated only when the GTP is split by the intrinsic GTPase activity of the α subunit.

The membrane-located enzyme may be adenylyl cyclase, guanylyl cyclase or phospholipase-C. Instead of an enzyme, there can be a membrane-located ionic channel that becomes activated through phosphorylation (Fig. 81.**4**). The second messenger may be cAMP, cGMP or DAG and IP_3. Group-IID hormones do not have a separate membrane-located enzyme or a second messenger. Rather, the receptor itself has the enzymatic activity for activating target enzymes (Table 81.**2**).

The effector enzymes are mostly the different types of kinases (e.g., cAMP-dependent protein kinase, cGMP-dependent protein kinase, protein kinase-C, etc.) that

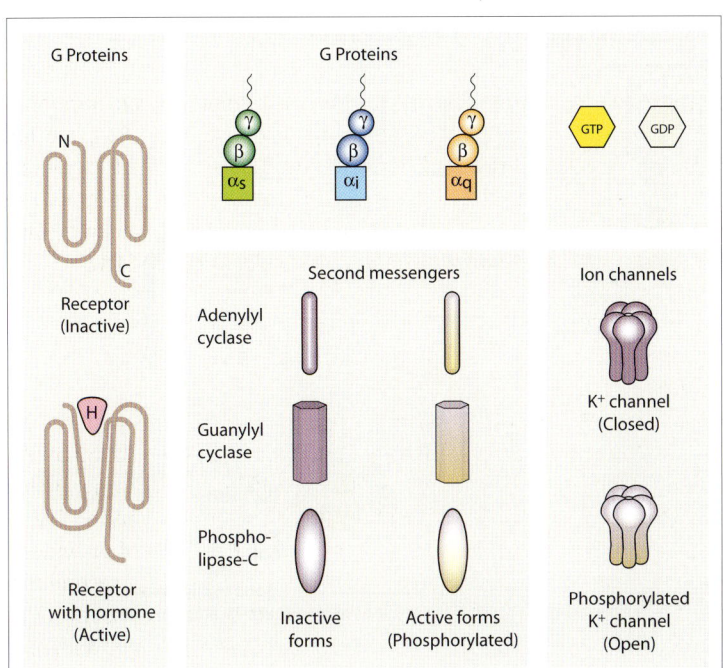

Fig. 81.**2** The hormone action 'kit'. The shapes and colors used here are the same in all figures of this chapter.

activate various protein enzymes by phosphorylating them. Thus glycogenolysis and glycogenesis are regulated by controlling the activity of glycogen synthetase (see Fig. 79.**9**) and lipolysis is regulated by controlling the activity of hormone-sensitive lipase (see Fig. 79.**14**).

Group-IIA hormones

The second messenger for group-IIA hormones is cyclic AMP and the enzyme that catalyzes its formation is adenylyl cyclase. The protein kinase involved is called cAMP-dependent protein kinase. The G protein may either stimulate or inhibit adenylyl cyclase and is accordingly called stimulatory G protein (G_s) or inhibitory G protein (G_i). The actions of G_s or G_i are attributable to their α_s and α_i fractions respectively. α_s and α_i also

have other actions in addition to their effect on adenylyl cyclase. For example, α_i stimulates K$^+$ channels and inhibits Ca^{2+} channels while α_s does the opposite. The activation of cAMP protein kinase by a group-IIA hormone is shown in Figure 81.**5**A.

The cAMP is hydrolyzed to 5'AMP by the enzyme phosphodiesterase. Inhibitors of phosphodiesterases such as methylated xanthine derivatives (caffeine) increase intracellular cAMP and thereby prolong the action of hormones. In intestinal epithelial cells, cholera toxin irreversibly inactivates the GTPase activity of G_s. The adenylyl cyclase therefore remains in a perpetual state of activation. This results in continuous formation of cAMP. The cAMP activates protein kinase and phosphorylates various membrane transport proteins, resulting in the active transport of electrolytes into the intestinal lumen. Water follows passively, causing life-threatening diarrhea.

Group-IIB hormone

In group-IIB hormones (Fig. 81.**5**B), cyclic GMP (guanosine 3'5'-cyclic monophosphate) acts as a second messenger and the enzyme that catalyzes its formation is guanylyl cyclase. Guanylyl cyclase has two isomers, one of which is present in the membrane as an integral protein. The other is present in the cytosol. The membrane isozyme of guanylyl cyclase is activated by the hormone atrial natriuretic peptide (ANP). The cytosolic guanylyl cyclase is activated by nitric oxide (NO).

Table 81.**2** Classification of hormones.

Class	Membrane-located enzyme	Second messenger	Kinase
IIA	Adenylyl cyclase	cAMP	cAMP-dependent protein kinase
IIB	Guanylyl cyclase	cGMP	cGMP-dependent protein kinase
IIC	Phospholipase-C	DAG and IP$_3$	Protein kinase-C, calmodulin-dependent protein kinase.
IID	-	-	Serine or threonine kinase.

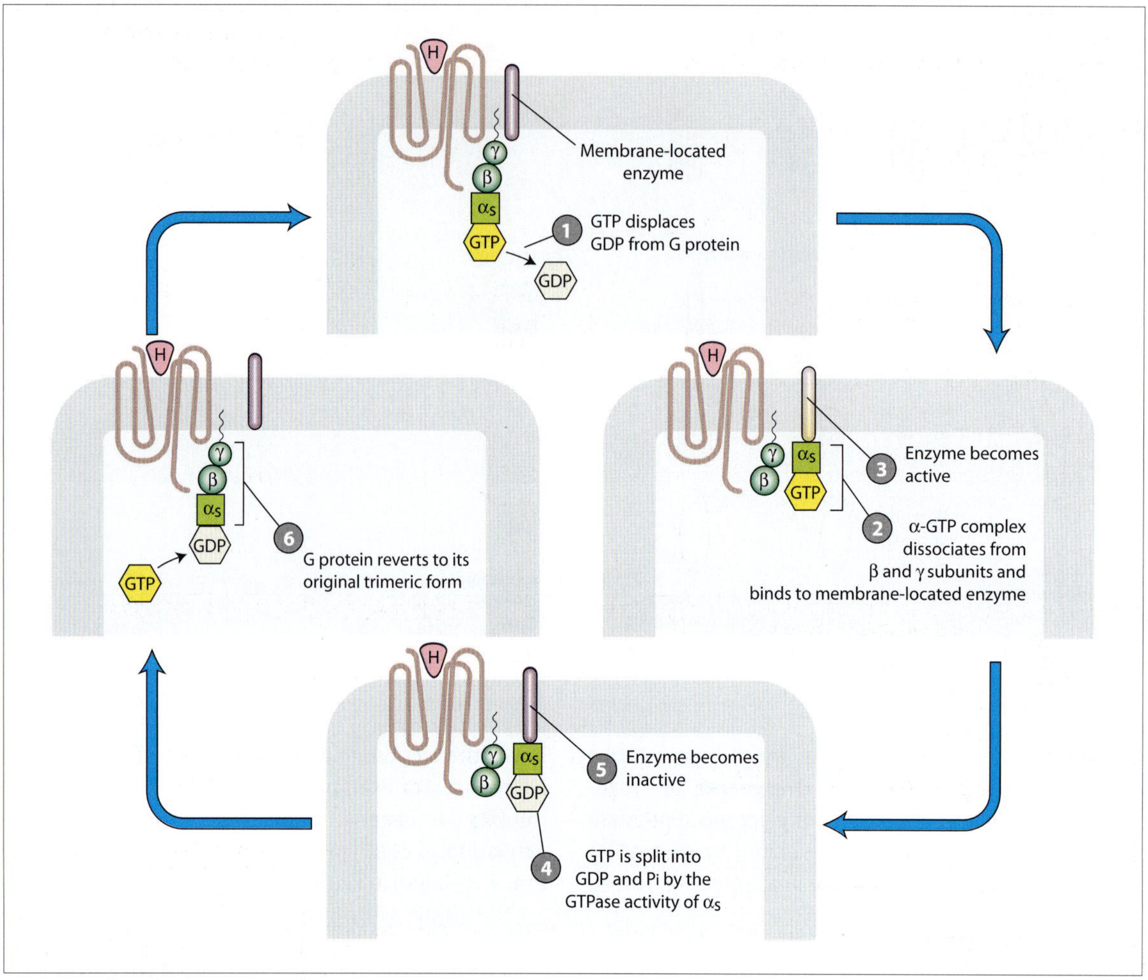

Fig. 81.3   Activation of membrane-located enzymes by group-II hormones.

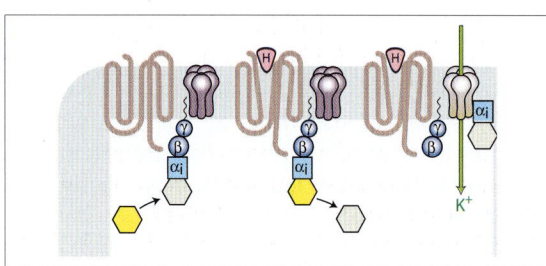

Fig. 81.4   Opening of a potassium channel by G_i.

cGMP activates cGMP-dependent protein kinase (also called protein kinase-G). The kinase phosphorylates the myosin light chain of smooth muscle, thereby causing its relaxation. Thus, ANP acts by relaxing afferent arterioles and mesangial cells. NO mediates the hypotensive action of nitroprusside. Inhibitors of cGMP phosphodiesterase like sildenafil (Viagra®) enhance and prolong these responses.

Group-IIC hormones

The second messengers in group-IIC hormones are diacylglycerol (DAG) and inositol-1,4,5-triphosphate (IP_3) which are produced from phosphatidylinositol-4,5-biphosphate (PIP_2) by the action of phospholipase-C (PLC) (Fig. 81.6A). PIP_2 is a membrane phospholipid. DAG activates the enzyme protein kinase-C. IP_3 acts on the endoplasmic reticulum, releasing Ca^{2+} from it. The increase in cytosolic Ca^{2+} activates calmodulin-dependent protein kinase. The protein kinases phosphorylate enzymatic proteins into their physiologically active forms and thereby mediate hormonal actions.

Group-IID hormones

These hormones (insulin, GH, prolactin, and IGF) act through a protein kinase cascade. Receptors for these hormones have intrinsic tyrosine kinase activity. The

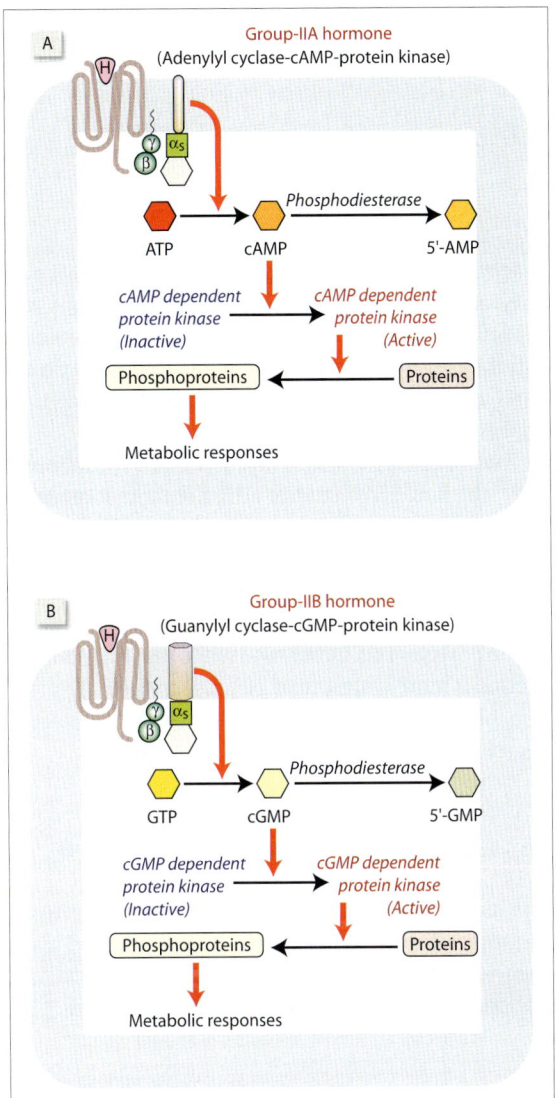

Fig. 81.**5** [A] Mechanism of a group-IIA hormone that activates G_s. Activation of G_s results in the formation of cAMP as the second messenger. The cAMP in turn activates a specific type of protein kinase. [B] Mechanism of action of group-IIB hormones. The second messenger is cGMP.

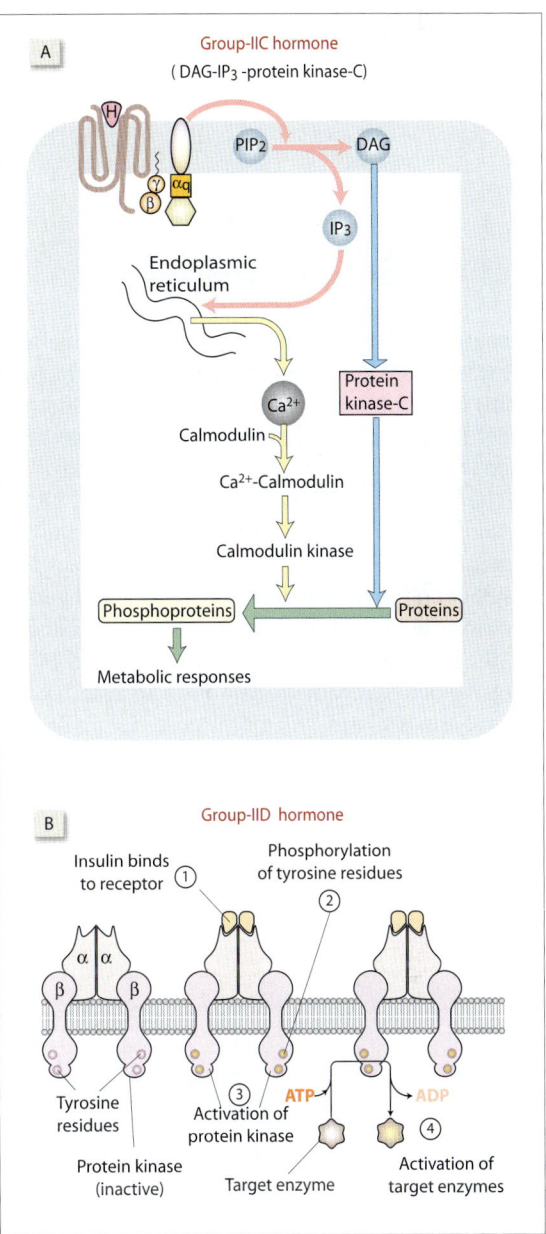

Fig. 81.**6** [A] Mechanism of group-IIC hormones. Note that the hormone produces two second messengers, DAG and IP_3, which activate protein kinase-C and calmodulin-kinase respectively. [B] Mechanism of group-IID hormones. Note that the enzyme protein kinase forms a part of the receptor molecule.

insulin receptor consists of two α chains located on the outer surface of the plasma membrane and two β chains that span the entire membrane thickness and protrude into the cytosol. The α chains contain the insulin-binding domain and the β chains have the tyrosine kinase domain. Binding of insulin to the α-chain activates the tyrosine kinase of the β chains. Tyrosine kinase autophosphorylates itself at its tyrosine residues in the α chain (Fig. 81.**6**B). The phosphorylated β

chain acquires an enzymatic property and activates a second protein kinase, which may then activate a third serine or threonine kinase. Eventually, phosphorylation of serine or threonine residues alters the activity of enzymes crucial to certain cellular functions.

Hypothalamic Hormones

A major function of the endocrine system is to maintain homeostasis of the internal environment, which explains why much of the endocrine system is ultimately under the control of the hypothalamus, the part of the brain that controls visceral functions of the body.

The control arm of the hypothalamus (Fig. 82.1) is the longest in the case of insulin secretion by the pancreatic islets. The islets are innervated by postganglionic autonomic fibers that are ultimately under hypothalamic control. The hypothalamic control arm is shorter for the adrenal medulla, which is innervated by the preganglionic sympathetic neurons. The hypothalamic control arm is shorter still for the posterior pituitary in which the axons of the hypothalamic neurons directly reach the posterior pituitary and secrete the hormones (oxytocin and ADH). The shortest control arm is represented by the hypothalamic control of the anterior pituitary in which the hypothalamic neurons do not send axons far out but simply release their secretions (releasing hormones) into the bloodstream.

Two facts are obvious from the above examples. First, the longer the control arm, the more localized is the domain of control; and second, the difference between a neurotransmitter and a neurosecretion lies in the extent of their effect. Though both are secreted by a neuron, the effect of a neurotransmitter is localized to the postsynaptic membrane while that of a neurosecretion is more widespread. Neurosecretions are mostly peptides, except dopamine which controls prolactin secretion (p 581).

Magnocellular and parvocellular systems

The magnocellular neurosecretory system refers to the neurosecretory neurons of the supraoptic and paraventricular nuclei, which secrete the neurohormones antidiuretic hormone (ADH) and oxytocin.

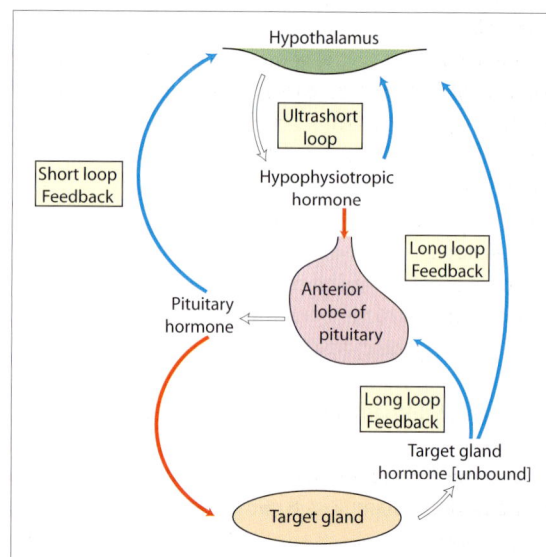

Fig. 82.**1**   The 'length' of the hypothalamic control 'arm'. The arm is longest for control of insulin secretion and shortest for anterior pituitary hormones.

Fig. 82.**2**   The hypothalamo-hypophysial control of a target gland. Several negative feedback loops are built into the control system.

The *parvocellular neurosecretory system* refers to the neurosecretory neurons originating in the arcuate nucleus (see Fig. 98.**2**) and terminating directly on the capillaries in the median eminence. These neurons secrete the hypophysiotropic hormones (releasing or inhibiting hormones), which reach the anterior pituitary through the hypothalamo-hypophysial portal system and stimulate or inhibit its hormonal secretion. Examples of hypophysiotropic hormones are the somatotropin releasing hormone (SRH), prolactin inhibiting hormone (PIH, now known to be dopamine), thyrotropin releasing hormone (TRH), corticotropin releasing hormone (CRH), and gonadotropin releasing hormone (GnRH). These hypophysiotropic hormones are discussed separately with the hormones they control.

Hypothalamo-hypophysial axis   The secretory activity of the anterior lobe of the pituitary gland is controlled by hypothalamic hormones which reach the pituitary (hypophysis) through the hypothalamic-hypophysial portal system. Under the influence of the hypothalamic releasing hormones, the anterior lobe of the pituitary releases a set of hormones called the *hypophysiotropic hormones* which increase the secretory activity of the target glands like the thyroid gland, adrenal cortex, and gonads. The hormones produced by target endocrine organs inhibit the hypothalamus and the pituitary, causing a decrease in the secretion of their tropic hormones. This is called *negative feedback control* and is an important mechanism regulating hormone synthesis and secretion (Fig. 82.**2**). Many hormones, e.g., parathyroid hormone, calcitonin, insulin, and glucagon are not under hypothalamo-hypophysial control though they do have their own regulatory mechanisms.

The negative feedback control occurs in three ways. In *long loop feedback*, the target organ hormones (or the substrates produced through their action) inhibit both the hypothalamus and the anterior lobe of the pituitary gland. *Short loop feedback* is a negative feedback exerted by the anterior pituitary tropic hormones on the hypothalamus, decreasing its secretion of hypophysiotropic hormones. In *ultrashort loop feedback*, the hypophysiotropic hormones inhibit their own secretion.

Pituitary Gland

The pituitary gland lies in a bony walled cavity, the sella turcica of the sphenoid bone at the base of the skull, and is closely associated with the hypothalamus of the brain: The infundibular stem (neural stalk) of the posterior lobe arises in the median eminence of the hypothalamus.

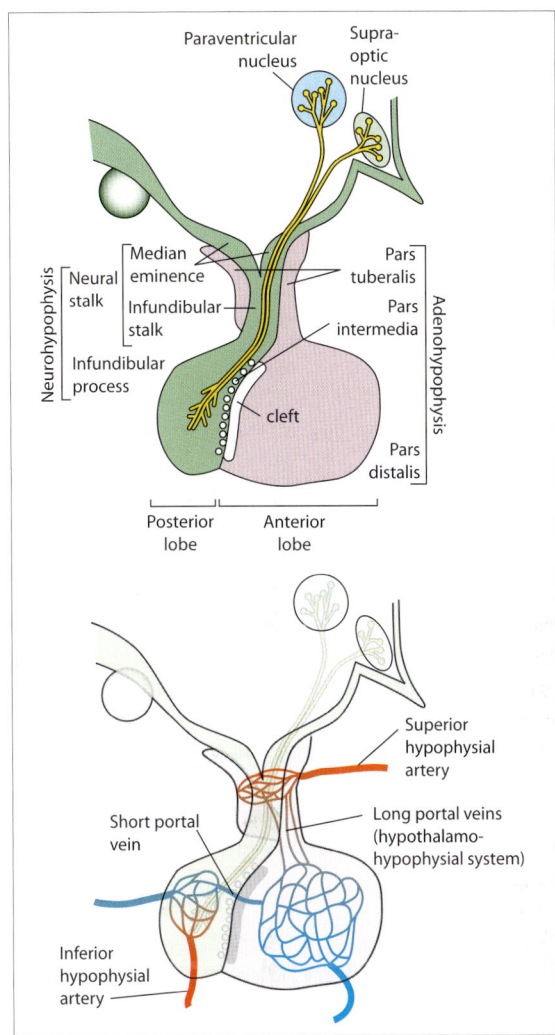

Fig. 82.**3**   (*Above*) Parts of the pituitary gland. Also shown are the connections of the posterior pituitary with the hypothalamic nuclei. (*Below*) Hypothalamo-hypophysial portal system.

The pituitary gland (hypophysis) originates in early embryonic life from two sources: The adenohypophysis (anterior lobe of the pituitary or the anterior pituitary) is formed by an upward evagination of Rathke's pouch (ectoderm) and grows dorsally toward the infundibulum where it meets the *neurohypophysis* (posterior lobe of the pituitary or the posterior pituitary), which is derived from a downward outgrowth of the infundibular process from the diencephalon. The posterior pituitary, therefore, represents an extension of the brain.

Hypothalamo-hypophysial portal system   The adenohypophysis receives 90% of its blood supply from the long portal veins and the remaining 10% from the short portal vein (Fig. 82.**3**). The *long portal veins* drain the capillary bed of the superior hypophysial artery which

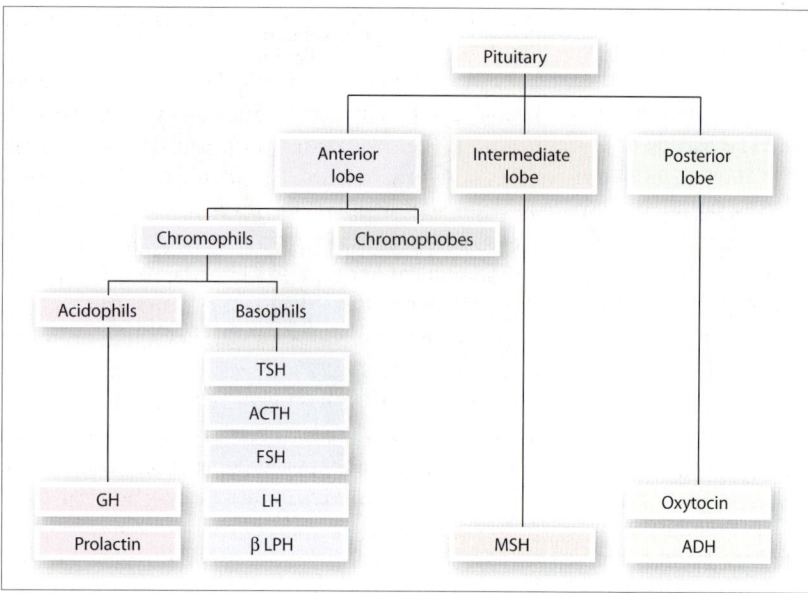

Fig. 82.**4**   Hormones secreted by the anterior, intermediate, and posterior lobes of the pituitary gland.

is located in the median eminence and infundibular stalk. This hypothalamo-hypophysial portal system is important for the transport of the hypophysiotropic hormones from the hypothalamus to the pituitary. The *short portal vein* drains the capillary bed of the inferior hypophysial artery which is located in the neurohypophysis.

Anterior pituitary (Adenohypophysis)

The anterior pituitary has three parts (Figs. 82.**3** and 82.**4**). The *pars distalis* represents the bulk of the anterior lobe in humans and is the source of the pituitary tropic hormones. The *pars tuberalis* surrounds the infundibular stem. It does not secrete any hormone. Between these two parts is the *pars intermedia* which is almost nonexistent in humans.

The pars distalis of the anterior lobe contains two types of cells, viz., the chromophobes and the chromophils. The *chromophobes* do not have any physiological significance. The *chromophils* exist in two forms: the acidophils (80%) and basophils (20%). *Acidophils* secrete prolactin (p 581) and growth hormone (GH).[1] *Basophils* secrete tropic hormones that stimulate other endocrine glands, viz., thyroid-stimulating hormone (TSH), adrenocorticotropic hormone (ACTH), luteinizing hormone

(LH), follicle-stimulating hormone (FSH), and β-lipotropic hormone (β-LPH).

Intermediate lobe of the pituitary

The intermediate lobe of the pituitary secretes the melanocyte stimulating hormone (MSH), a peptide hormone that is structurally similar to ACTH. Both MSH and ACTH are derived from a larger molecule called the proopiomelanocortin (POMC). MSH, ACTH, and other POMC derivatives (Fig. 82.**5**) have similar actions: They increase skin pigmentation (due to increased melanin synthesis), stimulate adrenal glucocorticoid production, and reduce food intake.

Melanin is the pigment that lends color to the hair and skin. The organelles containing melanin are called melanosomes which are present in cells called melanocytes. Melanocytes are different from the melanophores present in fish, reptiles, and amphibians. Melanophores can quickly change color through a redistribution of their colored and refractile granules that is controlled by MSH. Melanocytes cannot change their color quickly and therefore the role of MSH in humans remains unknown.

Albinism is a congenital condition in which there is a genetic inability to synthesize melanin. In *piebaldism* there is patchy depigmentation of the skin due to impaired migration of pigment cell precursors from the neural crest during embryonic development. The condition is congenital and even the pattern of depigmentation is inherited. In *vitiligo* the patchy depigmentation develops after birth and is progressive.

---

[1] GH-secreting acidophils are called somatotropes while prolactin-secreting acidophils are called mammotropes. Some acidophils secrete both GH and prolactin. These cells are called somatomammotropes.

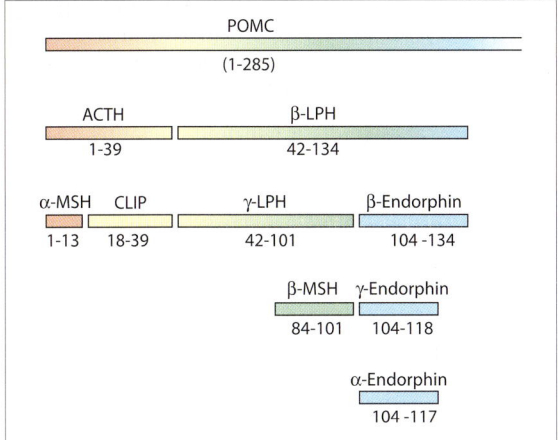

Fig. 82.**5**   Formation of ACTH and β-MSH from POMC.

Posterior pituitary (Neurohypophysis)

The posterior pituitary is made of neurosecretory neurons originating from the magnocellular neurosecretory system of the hypothalamus. These unmyelinated nerve fibers arise from the supraoptic and paraventricular nuclei and descend through the infundibulum to terminate in the posterior lobe. The oxytocin (p 581) and antidiuretic hormone released from the posterior lobe are the neurosecretions of these nerve fibers. The posterior pituitary does not have secretory cells of its own.

Growth Hormone

Growth hormone (GH) is also known as somatotropin. It is a polypeptide synthesized and secreted by somatotropes which are a subpopulation of the acidophils present in the adenohypophysis. GH is stored in very large amounts in the pituitary and is secreted episodically at 2-hour intervals. About half of the plasma GH is bound to a GH-binding protein.

Control of growth hormone secretion

GH secretion is controlled by two hypothalamic hormones, viz., somatotropin releasing hormone (SRH) which increases GH secretion, and somatostatin which decreases GH secretion. GH exerts a negative feedback (short loop) on the secretion of SRH. Also, SRH inhibits its own release via an ultrashort feedback loop (Fig. 82.**6**).

GH increases the synthesis of somatomedin in liver. Somatomedin reduces GH secretion by inhibiting SRH

secretion (long loop negative feedback) and stimulating somatostatin secretion.

GH secretion is increased in hypoglycemia or when hypoglycemia is imminent, as in fasting, and decreases in obesity. It also increases in stressful situations like fever or emotional trauma. The physiological significance of this is explained below. There are however several other stimuli for GH secretion whose physiological significance is not understood. For example, GH secretion increases during deep sleep and decreases during REM sleep. Dopamine agonists like bromocriptine stimulate GH secretion and have been used in the treatment of GH deficiency. Certain amino acids, like arginine and lysine, also increase GH secretion. Estrogens increase but progesterone decreases GH secretion, which is the reason why a decline in GH secretion is observed in late pregnancy. Glucocorticoids decrease GH secretion but their predominant effect is the interference with the metabolic actions of GH.

Physiologic effects of growth hormone

Skeletal growth   The effect of GH on skeletal growth is mediated by somatomedins, a family of polypeptides also called insulin-like growth factors (IGFs). They are

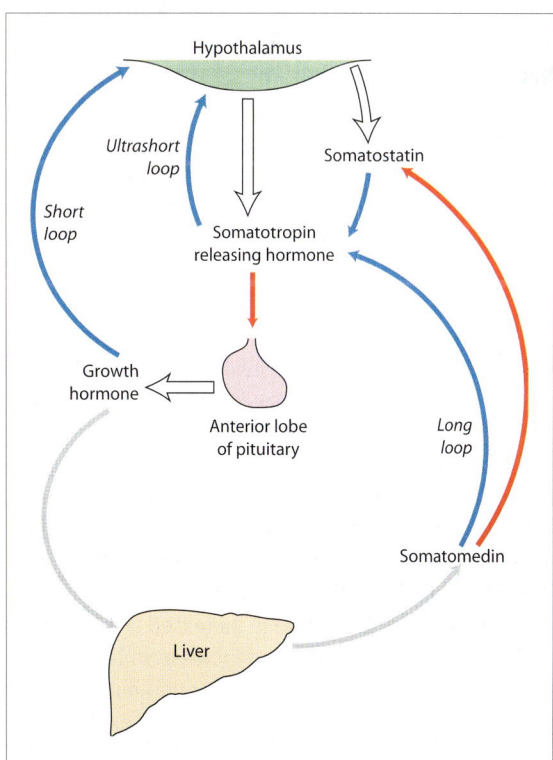

Fig. 82.**6**   Feedback regulation of growth hormone secretion.

Starvation

Well-fed state

Fig. 82.**7** Interrelationship between growth hormone, IGF, and insulin. IGF is formed from growth hormone (hence, they look similar in this cartoon) but its actions are similar to those of insulin (hence, they are seated on the same side of the seesaw). (*Above*) In starvation, growth hormone level increases and prevents hypoglycemia through its diabetogenic effects while the reduction of IGF suppresses growth. Insulin suppression prevents hypoglycemia. (*Below*) In the well-fed state, the high level of insulin promotes assimilation of food and growth is stimulated by the rise of IGF level.

synthesized mainly in the liver. The growth-promoting action of somatomedins is helped by their insulin-like actions (see below).

GH, through somatomedin, stimulates proliferation of chondrocytes and osteocytes resulting in increased deposition of cartilage and increased ossification of the newly-formed cartilage. Before the closure of epiphyseal plates, the increase in chondrogenesis exceeds its ossification, resulting in the widening of the cartilaginous epiphyseal plate. The bones grow longer, resulting in rapid increase in height. After epiphyseal closure, chondrogenesis does not occur and only subperiosteal bone deposition occurs due to increased activity of the osteocytes. Hence, there is no increase in bone length but bone thickening continues through subperiosteal bone deposition. Osteocytic activity, as explained in Chapter 84, is associated with both deposition as well as resorption of bone and therefore, is associated with increased urinary excretion of hydroxyproline. It is this

growth that accounts for the changes seen in acromegaly (see below). GH promotes renal reabsorption of Ca^{2+} and PO_4^{3-} which are important for bone growth.

Protein metabolism GH has predominantly anabolic effects on skeletal and cardiac muscle where it promotes amino acid transport into cells and increases protein synthesis. GH causes positive nitrogen balance, i.e., it promotes protein anabolism and reduces the plasma concentration and urinary excretion of nitrogenous products of protein catabolism like amino acids and urea. GH also promotes renal reabsorption of Na^+, K^+, and Cl^-.

Carbohydrate and fat metabolism The effects of GH on carbohydrate and fat metabolism are complicated by the fact that while GH itself has anti-insulin effects, the somatomedins it produces have insulin-like effects.

The *anti-insulin effect* of GH is primarily its lipolytic effect on the adipose tissues, which results in the mobilization of large amounts of FFA and glycerol. The FFA is oxidized to acetyl CoA. There is suppression of glycolysis and stimulation of gluconeogenesis, converting the large amounts of acetyl CoA into glucose which accumulates intracellularly. Excess acetyl CoA is converted into ketone bodies. The entire sequence takes about 2 hours to manifest. The large amounts of glucose formed through gluconeogenesis and the inhibition of glycolysis lead to hyperglycemia, which has been called *pituitary hyperglycemia* to differentiate if from diabetic hyperglycemia. In the long run however, pituitary hyperglycemia stimulates excessive insulin secretion from the pancreas, leading to B cell exhaustion and frank diabetes mellitus.

The *insulin-like effects* of GH are attributable to the formation of somatomedins and are relatively quicker, appearing in about 30 minutes. Somatomedins bind to insulin receptors and induce most of the metabolic effects of insulin, although to a lesser degree.

Another twist to the metabolic effects of GH is lent by the fact that the induction of somatomedins by GH requires a high insulin level along with adequate nutrients. When the insulin level is high, the GH level is low but the amount of somatomedin formed from it increases (Fig. 82.**7**). Thus, in the *well-fed state* when the insulin level is high, somatomedin level is also high and it brings about increased growth. Conversely, in the *fasting state*, the low insulin level reduces somatomedin formation and therefore growth is suppressed, while the high GH level prevents hypoglycemia. In diabetes mellitus, reduced somatomedin formation retards growth and the high GH level aggravates the diabetes through its antiinsulin metabolic effects.

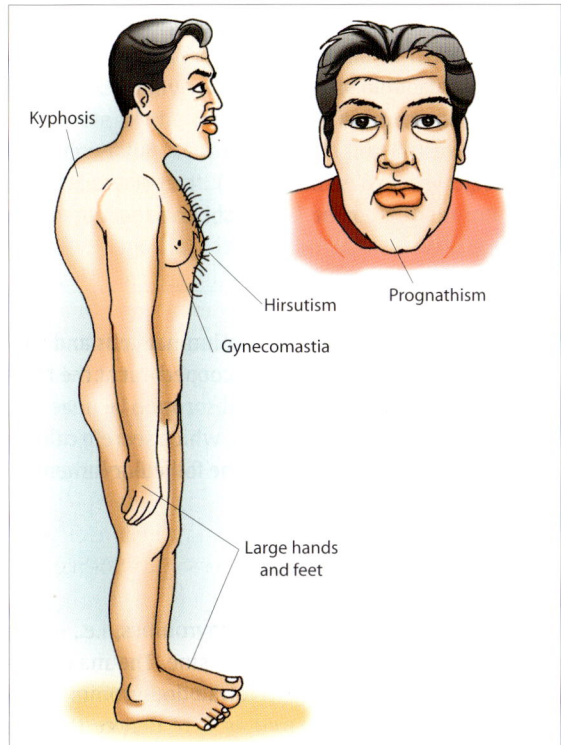

Fig. 82.**8**   Characteristics of an acromegalic patient.

Growth Hormone Disorders

Hyposecretion of GH causes dwarfism while hypersecretion causes gigantism and acromegaly. In some cases of GH hypersecretion, growth retardation can occur if the somatomedin levels are depressed, e.g., in kwashiorkor. Both GH and somatomedin levels are normal in the African pygmies.

Gigantism and acromegaly

Gigantism and acromegaly occur due to hypersecretion of GH. Tumors of somatotropes secrete large amounts of GH. Tumors of somatomammotropes secrete both GH and prolactin.

Overproduction of GH during adolescence results in gigantism, which is characterized by excessive growth of the long bones. Patients may grow to heights of as much as 8 feet. Excessive GH secretion during adulthood, i.e., after the epiphyseal (growth) plates of long bones have fused, causes growth in those areas where cartilage persists. This leads to acromegaly.

In acromegaly, increased growth hormone secretion results in enlargement of the hands and feet (acro: extremities) and soft tissue hypertrophy (e.g., cardiomegaly, hepatosplenomegaly, and renomegaly). The protrusion of the lower jaw (mandibular prognathism) together with the prominent brow, cheek bone, and other facial bones produce the coarse facial features called acromegalic facies (Fig. 82.**8**). Body hair is increased in amount. About a quarter of the patients have abnormal glucose tolerance tests and a few develop lactation in the absence of pregnancy.

Signs of acromegaly that are related to the local effects of the tumor include enlargement of the sella turcica, headache, and visual disturbances like bitemporal hemianopia.

Acromegaly is treated by selective surgical excision of the pituitary adenoma. Bromocriptine, a stimulator of GH secretion in normal individuals, is effective in suppressing GH levels in many acromegalic patients.

Pituitary dwarfism

Decreased GH secretion in immature persons leads to stunted growth or dwarfism. It is characterized by maxillary prognathism in contrast to the mandibular prognathism that characterizes acromegaly. Hair growth is impaired and hypoglycemia may be present. Unlike cretins (see Fig. 83.**8**), the body proportions of the pituitary dwarf are not infantile but like those of an adult.

GH deficiency may be part of an overall lack of anterior pituitary hormones (panhypopituitarism) or from an isolated genetic deficiency, which is rare in adults. Dwarfism due to GH deficiency can be treated with human GH. Genetic defects usually affect the GH receptors rather than GH secretion. In *Laron type dwarfs*, the hepatic synthesis of somatomedin is impaired because the GH receptors in the liver are resistant to the action of GH. In *pygmies*, even the GH receptors are normal suggesting that the defect lies in the mechanisms subsequent to the receptor binding of the hormone.

the energy released in the mitochondria through oxidative phosphorylation is captured into ATP and 32% is wasted as heat. Hence increase in oxidative phosphorylation not only increases the ATP yield but also increases the heat produced (thermogenesis). Earlier it was believed that thyroid hormone caused uncoupling of oxidative phosphorylation, i.e., it reduced the ATP yield so that more energy was released as heat. It is now known that this is true only for pharmacological doses (supraphysiological levels) of thyroid hormones. At physiological doses of thyroid hormone, the ATP:heat output remains unaltered.

Growth and maturation   Thyroid hormone is essential for normal ossification of cartilage and bone growth, normal erythropoiesis, and normal onset of puberty and lactation. TH is also essential for the normal myelination and synaptic development in the CNS.

Metabolic rate   Thyroid hormone seems to adjust the set point for the metabolic rate of the body. Thyroid hormone increases the BMR and also increases the body temperature. Calorigenesis is the most striking effect of thyroid hormones.

Carbohydrate metabolism   Thyroid hormone has both hypoglycemic and hyperglycemic effects. The hypoglycemic effect is the increased glycolysis. The hyperglycemic effects are stimulation of glycogenolysis and gluconeogenesis and enhanced intestinal glucose absorption due to increased activity of glucose transporters and increase in gastrointestinal motility. The net effect is hyperglycemia and depletion of liver glycogen. There is increase in food intake in response to the increased glucose utilization.

Protein metabolism   Thyroid hormone has a potent protein anabolic effect but in large doses, it has a protein catabolic effect. Thyroid hormone inhibits synthesis of glycosaminoglycans and fibronectin in fibroblasts.

Fat metabolism   Thyroid hormone stimulates both lipogenesis and lipolysis. The lipolysis exceeds lipogenesis. Hormone-sensitive lipase mobilizes fat from adipose tissues, increasing the plasma concentrations of FFA and glycerol. The elevated levels of FFA and glycerol promote hepatic triglyceride synthesis. However when the triglycerides synthesized in the liver are released into circulation, they are again broken down into FFA and glycerol by lipoprotein lipase. Thyroid hormone also enhances *de novo* cholesterol synthesis as well as bile acid synthesis. Yet, plasma cholesterol decreases due to increased LDL receptor formation in the liver and the consequent increase in the removal of cholesterol from circulation. Overall, there is an increase in the plasma free fatty acids and glycerol and a decrease in the plasma triglycerides, phospholipids, and cholesterol.

Vitamin metabolism   Thyroid hormone is required for the hepatic conversion of carotene to vitamin A. In hypothyroid states, the serum carotene is elevated and the skin becomes yellow. The color of the sclera of the eye is not affected in carotenemia, which distinguishes it from jaundice in which the sclera too is stained yellow.

Cardiovascular effects   The increase in metabolic rate and body temperature imposes greater demands on the cardiorespiratory system. The rise in body temperature is associated with increased cutaneous vasodilatation and sweating. Also, the increase in metabolic rate imposes a greater oxygen demand which produces autoregulatory vasodilatation in the muscle bed. Vasodilatation lowers the diastolic blood pressure. The fall in blood pressure results in a compensatory increase in blood volume through renal mechanisms. The vasodilatation and hypervolemia result in a rise in CVP and with it, a rise in cardiac output. The increase in cardiac output tends to lower the CVP. However, when the CVP is very high, it remains elevated despite the increase in cardiac output, resulting in high-output cardiac failure.

The increase in cardiac output is also attributable to a direct effect of thyroid hormone on cardiac contractility, partly because it increases the synthesis of β-adrenergic receptors which potentiates the cardiac effect of catecholamines, and partly because it increases the synthesis of the α isoform of the myosin heavy chain which has greater ATPase activity than the β isoform.

Thyroid Disorders

Goiter

A prominent enlargement of the thyroid gland is called a goiter. Thyroid enlargement does not necessarily mean that the thyroid is functionally overactive. A goiter may be associated with a hypothyroid, hyperthyroid, or euthyroid state. A euthyroid goiter, for example, can occur if thyroid hormone synthesis is impaired by goitrogens and is subsequently restored to normal by increased TSH secretion through hypothalamo-hypophysial feedback.

The recommended minimum intake of iodine is 150 μg per day, which is equal to the amount of iodine taken up daily by a normal thyroid gland. During pregnancy the recommended intake is 200 μg per day. Goiter occurs when the intake decreases to less than half of the recommended amount. Iodide is added to commercial preparations of salt and bread to prevent the occurrence of goiter.

Hypothyroidism

Although mild iodine deficiency causes a euthyroid goiter as explained above, severe iodine deficiency is the commonest cause of hypothyroid goiter. Less common causes of hypothyroidism are diseased or maldeveloped thyroid, genetic enzyme deficiency, antithyroid therapy, and goitrogens in the diet.

The signs and symptoms of hypothyroidism include fatigue, lethargy and sleepiness, muscular weakness, bradycardia, decreased cardiac output, hypovolemia, weight gain, constipation, mental sluggishness, sparse and coarse hair, characteristic facies (Fig. 83.**7**), scaly skin, husky voice, and in severe cases, an edematous appearance throughout the body called *myxedema*. Myxedema is not due to fluid retention but rather, to the deposition in the interstitial fluid of large quantities of gel-like substances (proteins mixed with hyaluronic acid and chondroitin sulfate). Hence instead of the usual pitting-edema that occurs due to increase in interstitial volume, myxedema results in non-pitting edema. The huskiness of voice is common enough in hypothyroidism to justify the clinical gag that myxedema can be diagnosed over the telephone! Finally, the mental sluggishness can progress to myxedema madness in severe cases of hypothyroidism. Hypothyroidism is associated with a rise in blood cholesterol, predisposing to atherosclerosis with all its complications: peripheral vascular disease, coronary artery disease, and deafness.

The laboratory diagnosis of hypothyroidism includes the estimation of plasma concentration of thyroid hormones (lowered), the BMR (reduced), and the systolic ejection time (prolonged).

Cretinism   When thyroid deficiency occurs during fetal life, infancy, or childhood, the result is cretinism. Skeletal growth in the cretin is inhibited more than the soft tissue growth. Therefore soft tissues enlarge excessively, producing the characteristic appearance of cretins (Fig. 83.**8**), which includes pot belly and large protruding tongue. The ratio of the upper part (iliac crest to the crown of the head) to the lower part (iliac crest to the heel) of the body is about 1.7:1 at birth and reduces to 1:1 by the age of 7–9 years. In cretins however, the infantile proportions persist so that the upper part of the body is taller than the lower part.

Development of the fetus is dependent on the maternal T_4 that reaches it through the placenta and

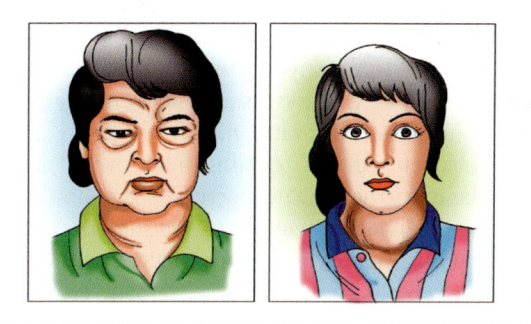

Fig. 83.**7**   Contrasting facies of hyperthyroidism (*left*) and hypothyroidism (*right*).

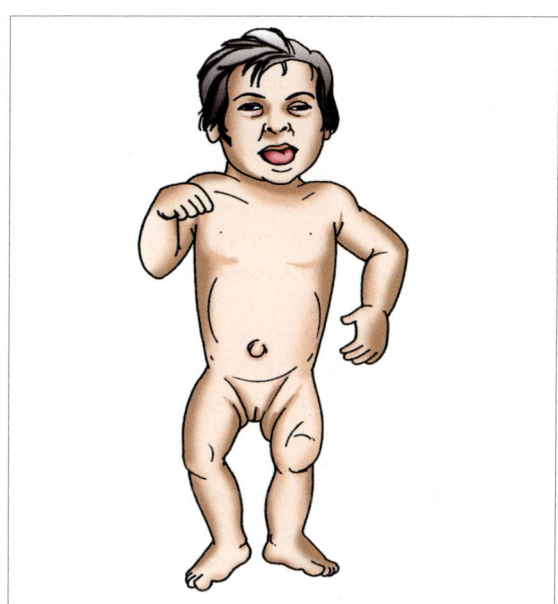

Fig. 83.**8**   Characteristic appearance of a cretin.

therefore, maternal hypothyroidism results in neonatal hypothyroidism. It is important to treat neonatal hypothyroidism immediately at birth. If left untreated till two years of age, there is a marked decrease in the myelination and arborization of neurons in the brain, which is dependent on thyroid hormone for its early development. These changes are irreversible after two years of age and result in mental retardation, as observed in cretins.

Hyperthyroidism

The most common cause of hyperthyroidism is Graves' disease, an autoimmune disease in which autoantibodies are formed against TSH receptors. The TSH receptors are activated by the autoantibodies, resulting in hypersecretion of thyroid hormone. Hyperthyroidism can also occur due to a TSH-secreting tumor of the anterior pituitary.

Clinical features of hyperthyroidism include intolerance to heat, excessive sweating and weight loss (due to the high metabolic rate), muscle weakness (thyrotoxic myopathy, due to increased breakdown of muscle proteins), diarrhea (due to increased gastrointestinal motility), nervousness and psychic disorders, inability to sleep, and tremors of the hands. Palpitations and arrhythmias occur due to increased cardiac response to circulating catecholamines, and the pulse pressure is high. High output cardiac failure may develop in severe cases. The BMR rises and the systolic ejection time shortens.

Patients of Graves' disease develop *exophthalmos*, i.e., proptosis of the eyeballs (Fig. 83.**7**). In this condition, the eyelids do not close completely when the person blinks or is asleep. It is caused by the accumulation of fluid and cells in retrobulbar tissues, and a varying degree of spasm of the upper eyelid.

Severe exacerbation of hyperthyroidism is called *thyroid storm* or *thyrotoxic crisis*. In the past, it was seen postoperatively in patients poorly prepared for surgery. Its incidence has decreased now with preoperative use of antithyroid drugs and iodide.

The treatment of hyperthyroidism includes surgical removal of the thyroid and use of antithyroid drugs.

Thyroid Function Tests

Thyroid dysfunction should be suspected if the BMR, serum cholesterol, and systolic ejection time are outside their respective normal ranges. The following tests are necessary for confirming a thyroid dysfunction.

Tests of thyroid activity   The normal serum T_4 is about 8 µg dl^{-1} and normal serum T_3 is 0.15 µg dl^{-1}. They are measured by radioimmunoassay. Higher or lower values are suggestive of hyperthyroidism or hypothyroidism respectively.

Radioactive iodide uptake by the thyroid gland indicates the functional status of the thyroid gland. A 24-hour uptake normally ranges between 5% and 35% of the administered dose of I^{131}, a beta emitter. The uptake increases in hyperthyroidism and decreases in hypothyroidism.

Scintiscanning localizes the sites of accumulation of the radionuclides and thereby detects localized areas of thyroid hyperactivity or hypoactivity. Pertechnetate-99m, a gamma emitter, is used for radioimaging of the thyroid because it is transported into the thyroid, but unlike iodide, it is not organified and therefore diffuses back into the circulation.

Tests for hormonal feedback control   The basal concentration of serum TSH is 7 µU ml^{-1}. It increases on TRH administration and decreases on TH administration.

The *TRH stimulation test* is helpful in diagnosing mild abnormalities of thyroid function. The thyroid hormone feedback at the level of the pituitary reduces the TRH receptors on the pituitary and also inhibits the transcription of TSH. In hyperthyroidism of thyroid origin, the high level of circulating TH renders the pituitary insensitive to TRH. Stimulation by TRH therefore elicits little TSH secretion from the pituitary. Conversely, there is an exaggerated TSH secretion in response to TRH administration in hypothyroidism of thyroid origin.

In the *thyroid suppression test*, TH is administered to suppress the pituitary secretion of TSH. Suppression of TSH secretion reduces the radioactive iodine uptake to 50% of normal. Lack of suppression indicates autonomous production of TH.

Tests of thyroid damage   These tests detect actual or potential damage to the thyroid gland. In carcinoma of the thyroid, thyroglobulin is released into the bloodstream. The presence of autoantibodies to the thyroid gland indicates thyroid disorders. For example, antimicrosomal antibody suggests Hashimoto's disease and antithyroglobulin antibody suggests Graves' disease.

Antithyroid Drugs

Drugs that inhibit thyroid hormone synthesis are used in the treatment of hyperthyroidism. Antithyroid drugs are grouped into two classes: agents that block iodide transport and agents that inhibit the coupling of iodotyrosyl residues in thyroglobulin.

Drugs that block iodide uptake include pertechnetate (TcO_4^-), perchlorate (ClO_4^-), thiocyanate (SCN^-), and nitrate (NO_3^-). These monovalent anions are competitive inhibitors of iodide transport. ClO_4^- and SCN^- are no longer used because of their toxicity.

Drugs preventing iodotyrosine formation are the thionamides (propylthiouracil, methimazole, and carbimazole), which are competitive inhibitors of thyroxine peroxidase. Thionamides also inhibit the coupling reactions mediated by thyroxine peroxidase. Propylthiouracil additionally inhibits 5'-deiodinase, leading to a reduction in the extrathyroidal synthesis of T_3.

Some other drugs that inhibit thyroid function are radioiodine I^{131}, which is used in the treatment of Graves' disease, and lithium carbonate, which inhibits the iodination of thyroglobulin. The radiation from I^{131} destroys the thyroid tissues.

There are three calcitropic hormones: parathyroid hormone, calcitonin, and calcitriol. These hormones control the calcium balance of the body by acting on three tissues: bone, intestine, and kidney. Before discussing these hormones, it is necessary to understand the physiology of bone tissue and its role in body calcium balance.

Bone Tissue

Bone chemistry

Bone consists of bone cells and an extracellular matrix. One-third of the matrix is made of organic components called the osteoid and two-thirds is made of inorganic mineral crystals. The osteoid is made of collagen and glycosaminoglycans and is deposited by the bone cells around themselves. The inorganic minerals are the hydroxyapatite and fluorapatite crystals, the chemical formulae for which are $[(Ca_3(PO_4)_2]_3 \cdot Ca(OH)_2$ and $[(Ca_3(PO_4)_2]_3 \cdot CaF_2$ respectively. The Ca^{2+}: P ratio in bone is about 1.7:1. Amorphous $Ca_3(PO_4)_2$ is deposited in the matrix when a certain critical value[1] called the *solubility product* is exceeded. It is slowly converted to hydroxyapatite and fluorapatite crystals through the addition of hydroxides and fluorides. When the solubility product exceeds 60, there is *metastatic calcification*, i.e., deposition of $Ca_3(PO_4)_2$ in tissues other than the bone.

Bone histophysiology

New bone is always formed in thin layers. These layers may form concentric lamellae as in cortical (compact) bone, or a meshwork of bone spicules, as in trabecular (spongy) bone. The concentric lamellae of cortical bone are arranged around a central channel (Haversian canal) that contains the capillary blood supply (Fig. 84.**1**). The entire structure is called the osteon and constitutes the basic unit of cortical bone. Present in the bone tissue are three types of cells: the osteoblasts, the

osteocytes, and the osteoclasts. The osteoblasts and osteocytes develop from the osteoprogenitor stem cells while the osteoclasts belong to the monocyte-macrophage system.

Osteoblasts are bone-forming cells present on bone surfaces. They secrete matrix constituents, i.e., collagen, other non-collagenous proteins like osteocalcin and osteonectin, and ground substance. They also contain abundant alkaline phosphatase which hydrolyzes phosphate esters. The phosphate liberated by active osteoblasts raises the local phosphate concentration to the point where the solubility product is exceeded and mineral crystals precipitate. Although primarily concerned with bone accretion, osteoblasts also facilitate bone resorption, ensuring that there is continuous bone remodeling (see below).

Osteocytes are osteoblasts that have been buried in the bone matrix. Each cell is surrounded by its own lacuna but an extensive canalicular system connects the osteocytes to the osteoblasts present on the bone surface. The long cell processes of osteocytes are connected through gap junctions to other osteocytes and to the surface osteoblasts, forming a functional syncytium. To survive, osteocytes must ensure that the canalicular system is not obliterated completely. Hence, osteocytes promptly break down any freshly formed mineral crystals (*osteocytic osteolysis*) and transport the calcium released to the exterior. Osteocytic osteolysis should not be confused with bone resorption (see below) which involves the complete breakdown of the bone matrix as well.

Osteoclasts are giant, multinucleated cells formed by the fusion of several precursor cells. Like osteoblasts, they are present on the bone surfaces. They attach to bone surface through integrins, sealing off a small enclosed area (Fig. 84.**2**). The part of the cell membrane that faces the bone surface becomes ruffled and is called the ruffled border. Collagenase, acid phosphatase, lysosomal enzymes, and H^+ are secreted across the ruffled border into the enclosed space, causing bone resorption in the underlying bone surface. H^+ dissolves the bone minerals while the enzymes digest the organic bone matrix.

---

[1] There are several practical difficulties in calculating the exact value of the solubility product in bone tissue.

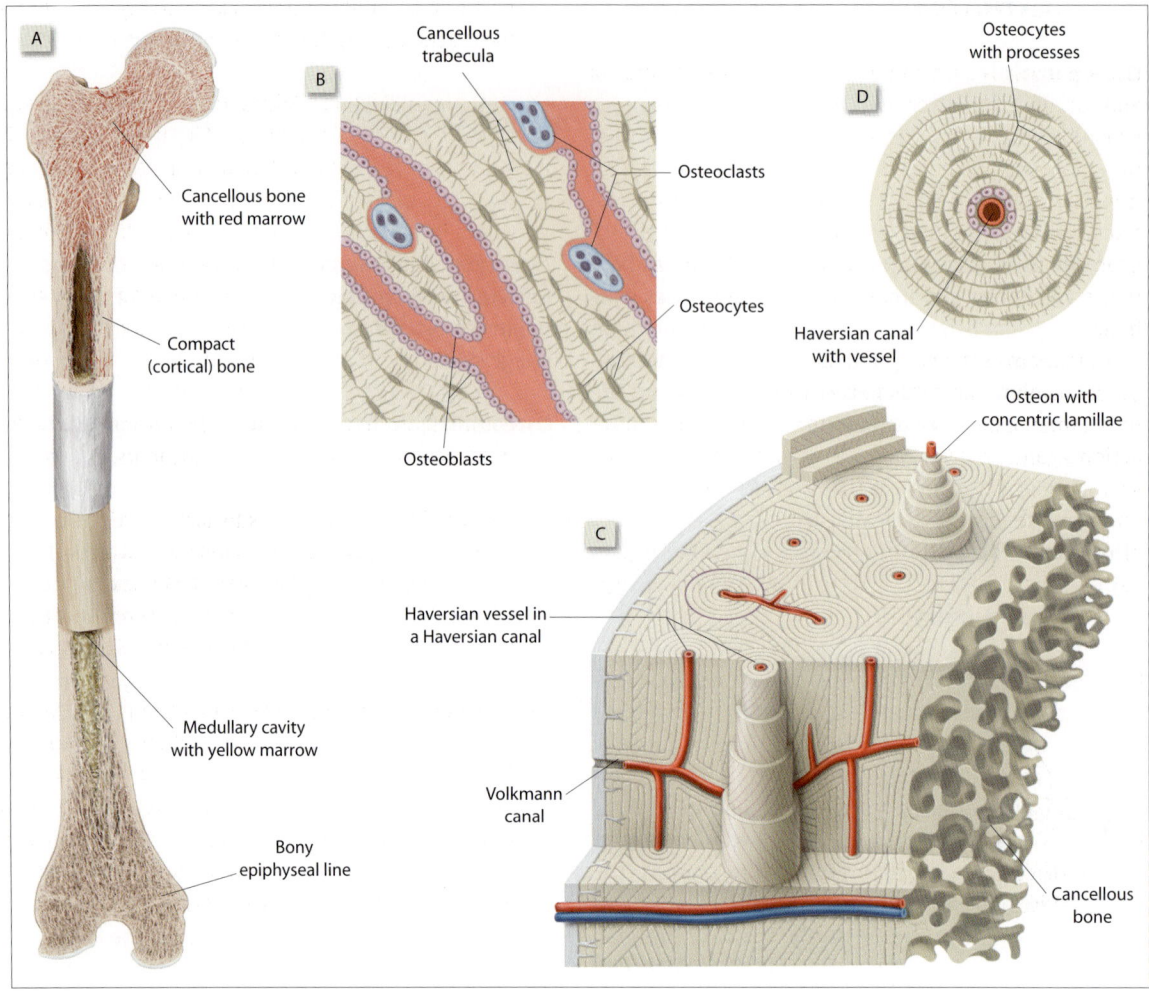

Fig. 84.**1**   [A] A coronal section through parts of a femur, showing compact and cancellous bone. [B] Cancellous trabecula and bone cells. [C] 3-D representation of the structure of compact bone, showing the Haversian canals and Volkmann canals. [D] Microstructure of an osteon.

Bone turnover

Throughout life, a significant fraction of bone is continuously replaced by new bone. This process is called bone turnover (or bone remodeling) and involves both bone accretion and bone resorption. Most of the bone turnover occurs at bone surfaces. In compact bone, bone turnover occurs both at the endosteal surface adjoining the marrow cavity and at the subperiosteal surface. Trabecular bone presents with a very large surface area and therefore has a much greater bone turnover.

Bone turnover is brought about by osteoblasts (causing bone accretion) and osteoclasts (causing bone resorption) working in tandem. Activated osteoblasts secrete an osteoclast-stimulating factor that activates osteoclasts. Osteoblasts also secrete the enzyme procollagenase and plasminogen activator. Plasminogen activator catalyzes the conversion of serum plasminogen to plasmin. This in turn releases collagenase from its proenzyme procollagenase. Collagenase depolymerizes collagen, softening the matrix and making it accessible to the activated osteoclasts (Fig. 84.**2**).

The extent of bone accretion, bone resorption, and bone turnover (both bone accretion and resorption) are indicated by the serum level of certain enzymes. *Alkaline phosphatase* is secreted by active osteoblasts and therefore its serum concentration increases during bone accretion. *Hydroxyproline*, the major metabolite of collagen, is produced during bone resorption and therefore its urinary level provides an index of bone resorption. *Osteocalcin* is released from osteoblasts during bone accretion and from the bone matrix during bone resorption. Hence, its serum level is an index of bone turnover.

Bone Disorders

Osteoporosis is associated with a loss of bone matrix and occurs whenever osteoclastic activity exceeds osteoblastic activity. It is associated with a reduction in bone mass per unit volume with a normal ratio of mineral-to-organic matrix (Fig. 84.**3**). Osteoporosis and fractures are commoner in bones containing a higher proportion of trabecular bone, e.g., distal forearm (Colle's fracture), vertebral bodies (kyphosis), and hip bone.

Osteoporosis is observed normally after the age of 35 years (involutional osteoporosis) and is more marked in postmenopausal women. Estrogen has a protective action against osteoporosis because it stimulates the secretion of cytokines that inhibit the development and increase the apoptosis of osteoclasts. Osteoporosis also occurs following prolonged immobilization (disuse osteoporosis). Calcitonin and biphosphonates (e.g.,

etidronate) inhibit osteoclastic activity and therefore have been used in the treatment of osteoporosis.

Osteomalacia and **rickets** occur due to inadequate mineralization of the bones (Fig. 84.**4**). Failure of the organic matrix (osteoid) to mineralize normally results in excess of unmineralized bone. Osteomalacia occurs in adults while rickets occurs in children prior to the closure of the epiphyses. Hence, it affects the mineralization of not only the osteoid but also of the epiphyseal cartilages which increase in thickness. Moreover, rickets is associated with bony deformities like bow legs. In osteomalacia, gross bony deformities are uncommon. Both rickets and osteomalacia are associated with hypocalcemia.

Osteitis fibrosa cystica is characterized by osteoclastic bone resorption which results in subperiosteal erosions, especially in phalanges, long bones, and distal ends of clavicles. The endosteal surface shows fewer trabeculae, and the bone marrow becomes fibrous.

In **osteosclerosis** there is enhanced bone density in the upper and lower margins of vertebrae. It is seen in calcium deficiency and occurs mainly due to bone remodeling and redistribution of bone minerals. *Osteopetrosis* is a congenital form of osteosclerosis in which the osteoclasts are defective. The unopposed action of osteoblasts results in increase in bone density (Fig. 84.**4**).

Body Calcium Pools

The total body calcium content of about 1 kilogram is distributed in two major pools: the extracellular fluid (ECF) pool and the bone pool. The ECF calcium pool contains only about 1.2 g of calcium while the rest is present in the bone pool. There is continuous exchange of calcium between the two pools. Within the bone calcium pool, there is a smaller *rapidly exchangeable* bone calcium pool of about 4 g. The calcium in this

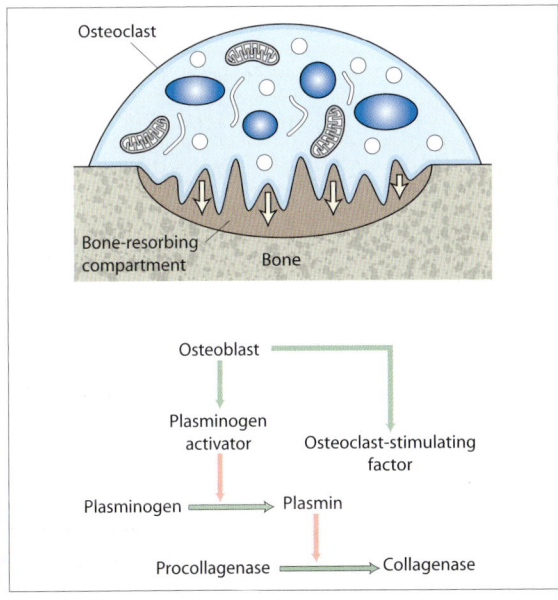

Fig. 84.**2** (*Above*) Bone resorption by an osteoclast. (*Below*) Bone resorptive mechanism of an osteoblast.

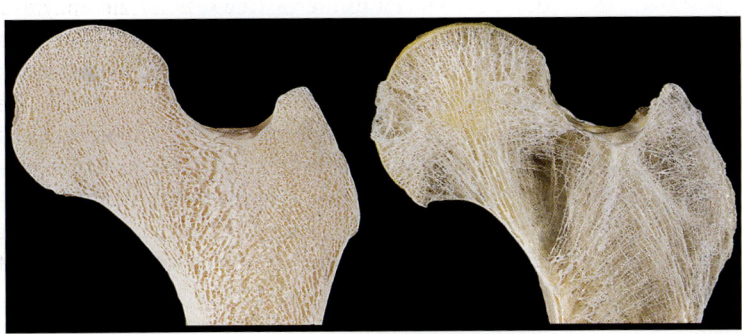

Fig. 84.**3** (*Left*) Normal femur (*Right*) Osteoporotic femur.

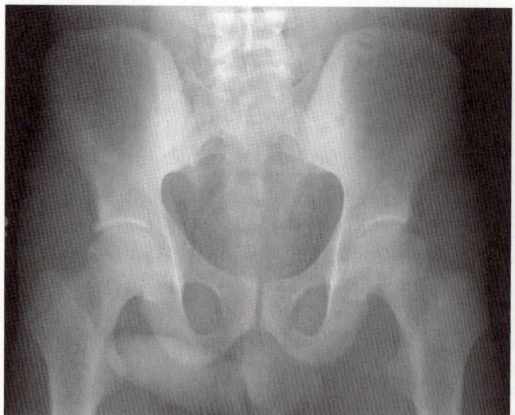
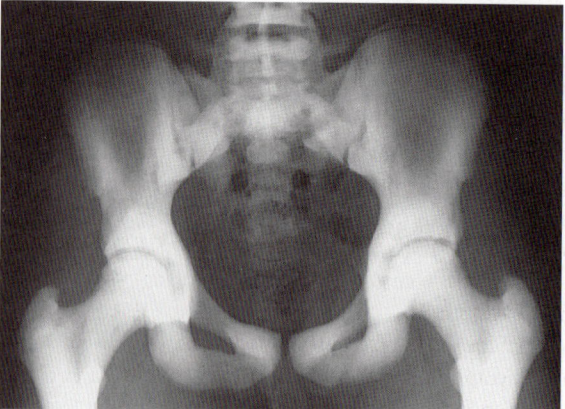

Fig. 84.**4**   (*Left*) Reduced bone density in osteomalacia (*Right*) Increased bone density in osteopetrosis.

pool rapidly exchanges with the calcium in the ECF. The remaining bone calcium pool is called the *slowly exchangeable* bone calcium pool that requires parathyroid hormone for its mobilization. The calcium in this pool enters the ECF only when the bone tissue is broken down (bone resorption). Conversely, ECF Ca²⁺ enters this pool only when new bone is deposited (bone accretion). The total size of the body calcium pool remains constant since the amount lost daily in urine (0.1 g) is replenished by an equal amount of dietary Ca²⁺ absorbed from the gastrointestinal tract (Fig. 84.**5**).

The plasma concentration[2] of total Ca²⁺ is about 10 mg dl⁻¹. About 45% is *ionized* or *free Ca²⁺* which is freely diffusible and brings about the physiological effects of Ca²⁺. Another 10% is present as *poorly ionizable Ca²⁺* salts of phosphate, bicarbonate, and citrate which are freely diffusible and provide an immediate reserve of Ca²⁺ during sudden hypocalcemia. The remaining 45% of the Ca²⁺ is *nondiffusible*, being bound to albumin. The albumin releases Ca²⁺ slowly when plasma Ca²⁺ concentration decreases.

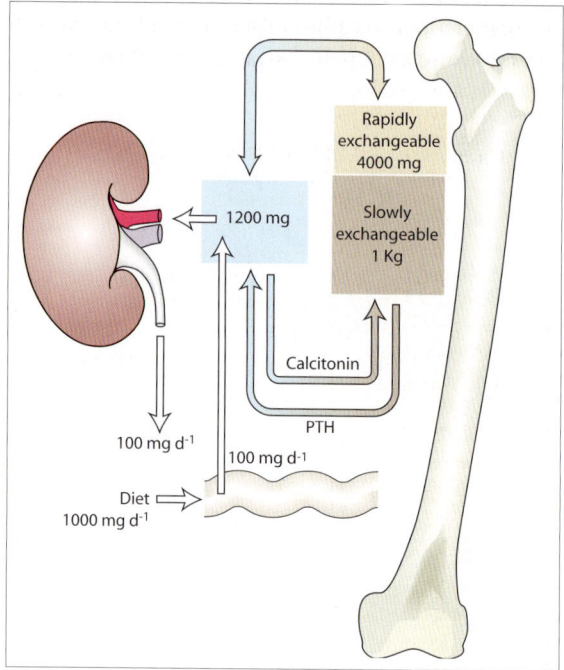

Fig. 84.**5**   Body calcium compartments.

Calcitropic Hormones

Parathyroid hormone

Parathyroid hormone (PTH) is a group-IIA polypeptide hormone secreted by the chief cells of the parathyroid glands (Fig. 84.**6**). It increases bone remodeling by directly stimulating osteoblasts and indirectly stimulating osteoclasts through the osteoclast-stimulating factor. However, the bone resorption exceeds bone accretion and therefore PTH causes a net bone resorption.

Calcium homeostasis   When plasma Ca²⁺ falls, PTH secretion increases. PTH raises plasma Ca²⁺ in three ways: (1) It mobilizes bone Ca²⁺ by increasing bone resorption. (2) It increases the gastrointestinal absorption of Ca²⁺ by increasing calcitriol synthesis: PTH increases calcitriol formation by stimulating 1α-hydroxylase activity in the proximal tubular cells of the kidney. (3) It increases Ca²⁺ reabsorption in the distal nephron. However, since it also produces hypercalcemia, the large amounts of Ca²⁺ filtered into the tubules result in hypercalcuria despite the enhanced tubular reabsorption. Interestingly, a parathyroid hormone-related protein (PTHrP) can bind to PTH receptor and mimic its actions. PTHrP is the major cause of hypercalcemia in cancer.

---

[2]   Ca²⁺ concentration expressed in mg dl⁻¹ can be converted to mmol L⁻¹ (mM) by dividing by 4.

The PTH-induced dissolution of the hydroxyapatite crystals in the bone is also associated with bicarbonate release, which tends to produce metabolic alkalosis. However, the simultaneous *bicarbonaturia* produced by the tubular action of PTH minimizes the alkalosis produced by the dissolution of hydroxyapatite crystals.

Bone resorption releases both Ca^{2+} and phosphates into plasma. A simultaneous elevation of phosphates along with Ca^{2+} would cause precipitation of $Ca_3(PO_4)_2$ in tissues and a lowering of plasma ionized calcium. PTH prevents such precipitation of $Ca_3(PO_4)_2$ through its *phosphaturic effect* which lowers the plasma phosphate levels. The phophaturia occurs because PTH decreases the proximal tubular reabsorption of phosphate.

Phosphate homeostasis In an indirect way, PTH also maintains constancy of plasma phosphate level. Increase in plasma phosphate is associated with increased precipitation of $Ca_3(PO_4)_2$ as the solubility product is exceeded. The resultant decrease in plasma Ca^{2+} stimulates PTH secretion, which in turn increases the urinary excretion of phosphates (Fig. 84.**7**).

Calcitonin

Calcitonin is a polypeptide hormone secreted by the parafollicular (C) cells of the thyroid gland. Its secretion is stimulated by hypercalcemia (Fig. 84.**8**). Calcitonin secretion is also influenced by gastrointestinal hormones such as gastrin and glucagon (stimulators) and somatostatin (inhibitor). Calcitonin produces hypocalcemia and hypophosphatemia by inhibiting osteoclastic bone resorption and also by increasing the urinary excretion of calcium and phosphate ions. Osteoclasts have receptors for calcitonin on them.

Calcitonin deficiency causes little problem so long as the parathyroid gland functions normally. Calcitonin might be important during pregnancy and lactation when the calcium demand of the body rises considerably. The excess calcium demand can be met either by higher dietary intake or by mobilizing calcium through bone resorption. Calcitonin might be important in preventing excessive bone resorption in such situations.

Calcitriol

Calcitriol (dihydrocholecalciferol) is a lipid-soluble hormone. It acts like a group-I (steroid) hormone that interacts with nuclear receptors in target cells. However, it is also hydrophilic as it contains three hydroxyl groups. Hence it also acts like a group-IIB hormone, acting through membrane receptors with cGMP as the second messenger.

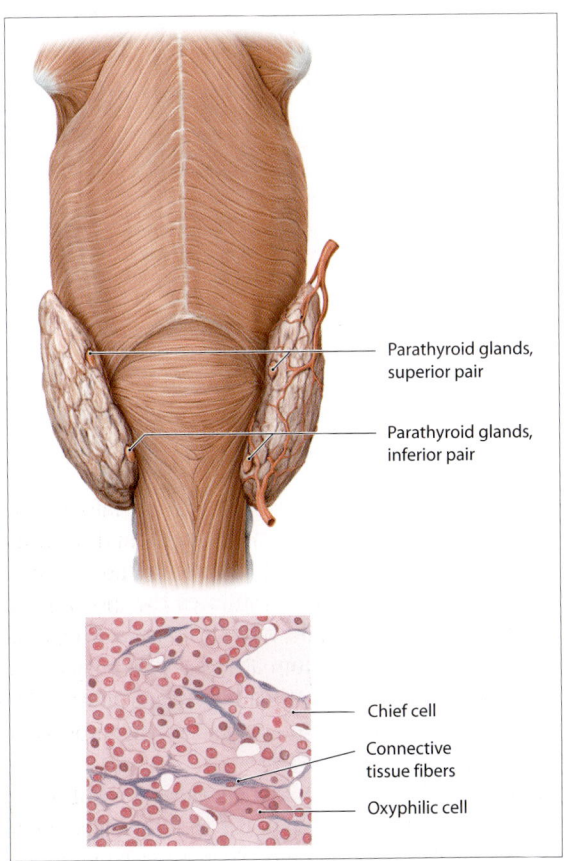

Fig. 84.**6** (*Above*) The parathyroid glands, posterior view. (*Below*) Microscopic structure of parathyroid gland.

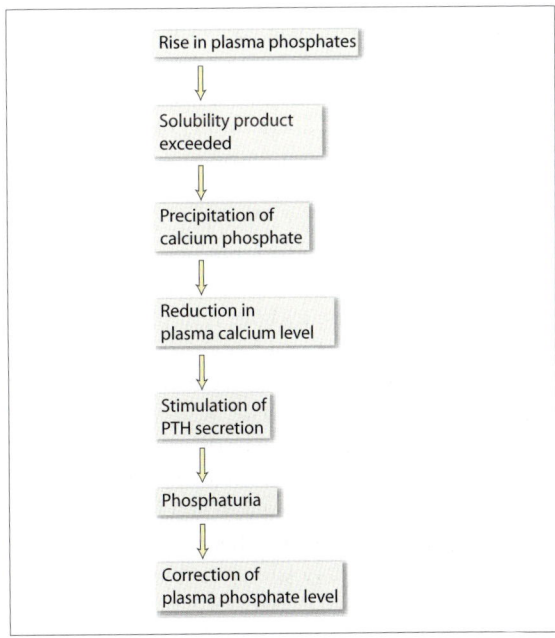

Fig. 84.**7** Role of PTH in phosphate homeostasis.

Calcitriol is synthesized from vitamin D$_3$ (cholecalciferol) in two steps (Fig. 84.**9**). Vitamin D$_3$ is first converted in the liver to calcidiol (25-hydroxycholecalciferol) by 25-hydroxylase. Next, it is metabolized in the kidney to calcitriol (1,25 dihydroxycholecalciferol) by 1α-hydroxylase. Calcidiol is the major circulating form of vitamin D$_3$ while calcitriol is the final active form of vitamin D$_3$. Mole for mole, calcitriol is 100 times more potent than calcidiol and 300 times more potent than vitamin D$_3$.

Vitamin D$_3$ is synthesized in skin and is also present in the diet. The keratinocytes of the stratum corneum of the epidermis contain 7-dehydrocholesterol which is converted into previtamin D$_3$ through photoactivation by solar ultraviolet radiation in the frequency range of 290 to 315 nm. Previtamin D$_3$ is then slowly converted into vitamin D$_3$ (cholecalciferol). Fatty fish and cod liver oil are rich dietary sources of vitamin D$_3$. Absorption of dietary vitamin D$_3$ occurs mainly in the ileum and requires bile salts.

Calcitriol production is increased by PTH and hypophosphatemia (Fig. 84.**10**A). PTH activates renal 1α-hydroxylase, which catalyzes the conversion of 25-hydroxycholecalciferol to 1,25 dihydroxycholecalciferol. In its turn, calcitriol exerts a negative feedback on PTH secretion. Hypophosphatemia increases calcitriol by directly activating renal 1α-hydroxylase. Hypocalcemia increases calcitriol synthesis indirectly through its stimulation of PTH secretion.

Effect on bone   Calcitriol, like PTH, has the opposite effects on bone, increasing both bone deposition and bone resorption. However, its net effect is increased mineralization of the bone. Calcitriol stimulates osteoblasts to lay down osteoid (osteoblasts and their precursors have nuclear receptors for calcitriol) and further promotes the mineralization of the osteoid by maintaining adequate concentrations of extracellular Ca^{2+} and phosphates. Activated osteoblasts secrete osteoclast-stimulating factor which induces the development of osteoclasts from their precursors. (Osteoclasts themselves do not have any calcitriol receptor.) Osteoclastic activation promotes bone resorption.

Effect on intestine   Calcitriol promotes intestinal absorption of Ca^{2+} and phosphate, which contributes to increased bone mineralization. Ca^{2+} absorption occurs principally in the duodenum while phosphates are absorbed mostly from the jejunum and ileum. Both are absorbed through secondary active transport. Ca^{2+} is extruded across the basolateral membrane of duodenal enterocytes by a Ca^{2+} ATPase pump (Fig. 84.**10**B). Calcitriol induces the synthesis of this pump protein. Increased activity of the Ca^{2+} ATPase pump lowers the

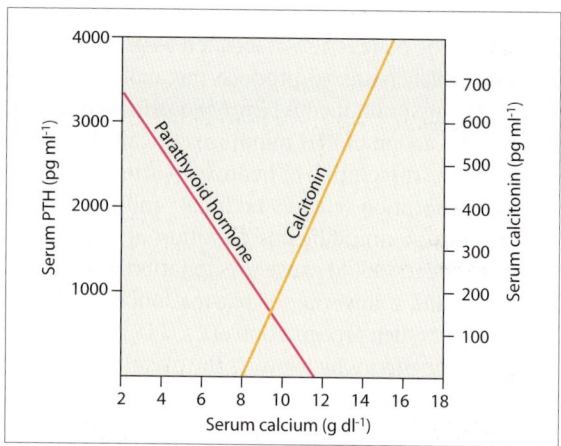

Fig. 84.**8**   Reciprocal relationship between PTH and calcitonin secretion.

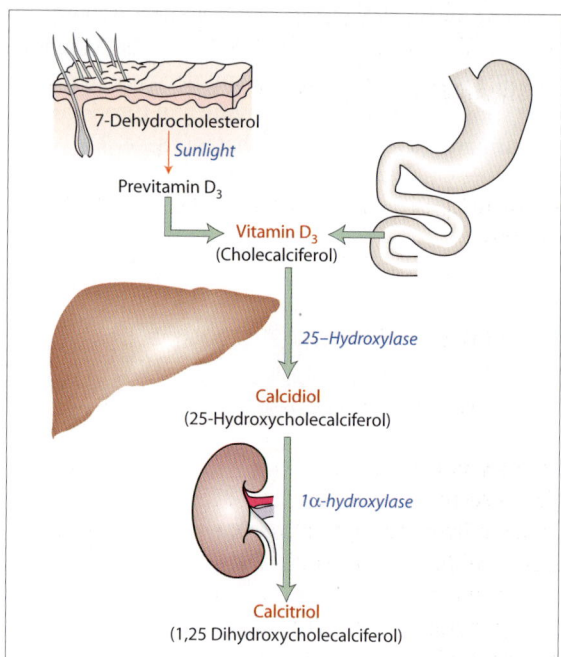

Fig. 84.**9**   Calcitriol formation.

intracellular Ca^{2+}, which promotes the facilitated diffusion of Ca^{2+} at the brush border membrane of intestinal microvilli. Calcitriol also induces the synthesis of a calcium-binding protein called calbindin. The exact role of calbindin in calcium absorption is not known. Calcitriol stimulates intestinal phosphate uptake via a Na$^+$-phosphate symporter.

Effect on kidney   In the kidney, calcitriol stimulates reabsorption of phosphate in the proximal tubule and reabsorption of Ca^{2+} in the distal tubule. As in enterocytes, calbindin is present in renal tubular cells too and probably has a role in increasing Ca^{2+} absorption.

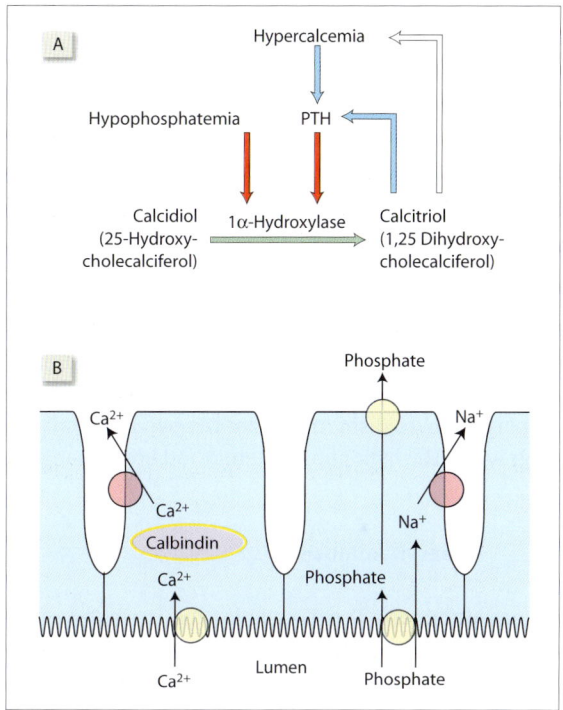

Fig. 84.**10**   [A] Control of calcitriol production. [B] Effect of calcitriol.

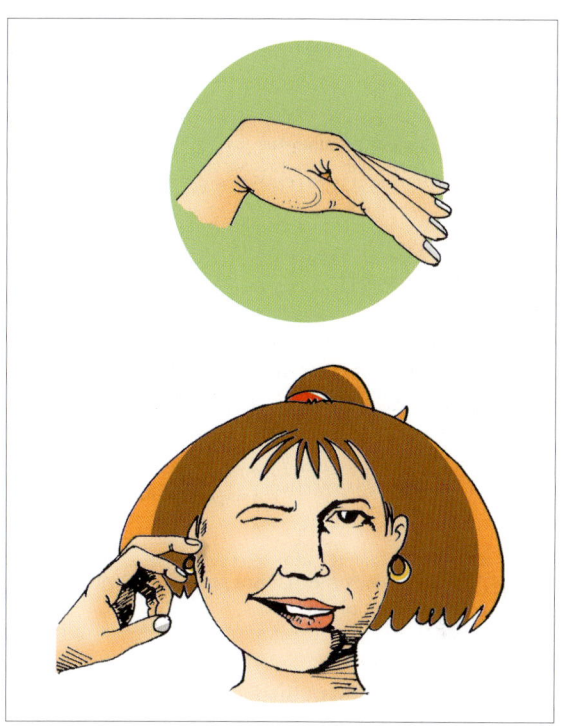

Fig. 84.**11**   (*Above*) Carpopedal spasm. (*Below*) Chvostek's sign.

Calcium Balance Disorders

Hypoparathyroidism

Primary hypoparathyroidism   When the parathyroid glands secrete inadequate amounts of PTH, it is called primary hypoparathyroidism. The reduced secretion may be either due to a deficiency in PTH secretion or due to the accidental removal of parathyroid tissue during thyroid surgery. It results in hypocalcemia and hyperphosphatemia. Soft tissue calcification is also common, especially in the basal ganglia, and is probably due to the hyperphosphatemia.

If the hypocalcemia is severe enough, it results in tetany, a condition associated with spontaneous excitability of nerves leading to muscle spasms. The spasms mostly involve the extremities and the larynx. Laryngeal spasm can be fatal. Latent tetany can be diagnosed by eliciting the Chvostek's and Trousseau's signs. *Chvostek's sign* is the ipsilateral contraction of the facial muscles elicited by the percussion of the facial nerve just anterior to the ear lobe. *Trousseau's sign* is the carpopedal spasm (Fig. 84.**11**) elicited by application of occlusive pressure (with a blood pressure cuff) on the arm. The carpopedal spasm appears as thumb adduction, metacarpophalangeal joint flexion, and interphalangeal joint extension.

Table 84.**1**   Comparison of different types of hypoparathyroidism.

	PTH secretion	Plasma calcium
Primary hypoparathyroidism	Reduced	Reduced
Secondary hypoparathyroidism	Reduced	Increased
Pseudohypoparathyroidism	Increased	Reduced

Secondary hypoparathyroidism is caused by the feedback suppression of PTH secretion by increased plasma Ca^{2+} concentration, e.g., by excessive intake of vitamin D.

In **pseudohypoparathyroidism**, PTH receptors are resistant to the action of PTH. Hence hypocalcemia and hyperphosphatemia are present but PTH secretion is elevated due to feedback stimulation of the parathyroid by hypocalcemia (Table 84.**1**).

Hyperparathyroidism

Primary hyperparathyroidism   When the excessive secretion of PTH is caused by a tumor of the parathyroid gland or by ectopic parathyroid tissue, it is called

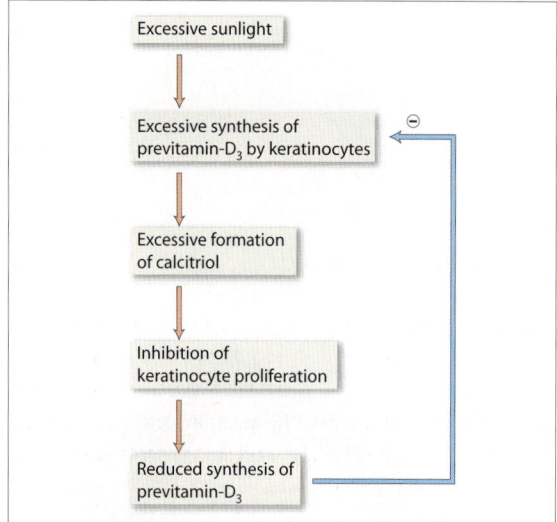

Fig. 84.**12**   Regulatory mechanism preventing hypervitaminosis on exposure to sunlight.

primary hyperparathyroidism. Excess PTH leads to hypercalcemia and hypophosphatemia with increased urinary excretion of Ca^{2+}, phosphates, and hydroxyproline. Nephrolithiasis (kidney stones) is common. The stones are composed of calcium oxalate or calcium phosphate The bones show osteopenia (reduction in bone mass) and may develop osteitis fibrosa cystica .

Secondary hyperparathyroidism   If the excessive PTH secretion is secondary to hypocalcemia, it is called secondary hyperparathyroidism. Such hypocalcemia can occur if vitamin D is deficient in diet, is poorly absorbed due to malabsorption of fats, or is not converted to calcitriol due to renal disease. Secondary hyperparathyroidism can also occur when there is an increased demand for Ca^{2+} as during pregnancy and lactation.

■ Hypovitaminosis D

Hypovitaminosis D results in rickets or osteomalacia, moderate hypocalcemia, severe hypophosphatemia, and secondary hyperparathyroidism. The secondary hyperparathyroidism minimizes the hypocalcemia but aggravates the hypophosphatemia and bone changes.

■ Hypervitaminosis D

Hypervitaminosis D occurs following ingestion of excess vitamin D. Hypervitaminosis D results in hypercalcemia and hypercalcuria. In severe cases, there is widespread ectopic calcification.

Although sunlight stimulates vitamin D production, excessive exposure to sunlight does not cause vitamin D toxicity because of the slow release of the freshly synthesized vitamin D_3 from the skin. Moreover, calcitriol inhibits the proliferation of the keratinocytes, which constitutes another regulatory mechanism for preventing excessive synthesis of vitamin D_3 by the skin (Fig. 84.**12**).

85 Adrenocortical Hormones

The adrenal glands are paired structures situated above the kidneys (Fig. 85.1). The adrenal gland consists of an outer cortex and an inner medulla. The hormones secreted by the adrenal cortex are called corticosteroids, which are classified into glucocorticoids, mineralocorticoids, and androgens.

The fetal adrenal cortex is much larger. By the third year of life, three distinct zones of cells have formed in the adrenal cortex. The zona glomerulosa is the outermost layer and is the site of mineralocorticoid synthesis. The wider middle zone is the zona fasciculata and the innermost layer is the zona reticularis. The middle and inner zones of the adrenal cortex synthesize and secrete glucocorticoids and androgens. The middle zone secretes more glucocorticoids and the inner zone secrete more androgens. The adrenal cortex is essential for life; the adrenal medulla is not. Following adrenalectomy, the mineralocorticoid deficiency results in hypotension, circulatory insufficiency, and eventually fatal shock.

Corticosteroids

Corticosteroids with predominantly glucocorticoid activity are cortisol, corticosterone, and 11-dehydrocorticosterone. Corticosteroids with predominantly mineralocorticoid activity are aldosterone and 11-deoxycorticosterone. Corticosteroids with androgenic activity are dehydroepiandrosterone (DHEA), dehydroepiandrosterone sulfate (DHEAS), and androstenedione.

The adrenal cortex also secretes testosterone and estrogens but the amounts secreted are too small to be effective physiologically. On the other hand, the androgenic activities of androstenedione, DHEA and DHEAS are too weak to be physiologically important. However, DHEA is converted in adipose tissues to potent androgens like testosterone and dihydrotestosterone as well as estrogens, mainly estrone. The peripheral conversion of DHEA to potent androgens is of little significance in males but is of significance in females as they cause

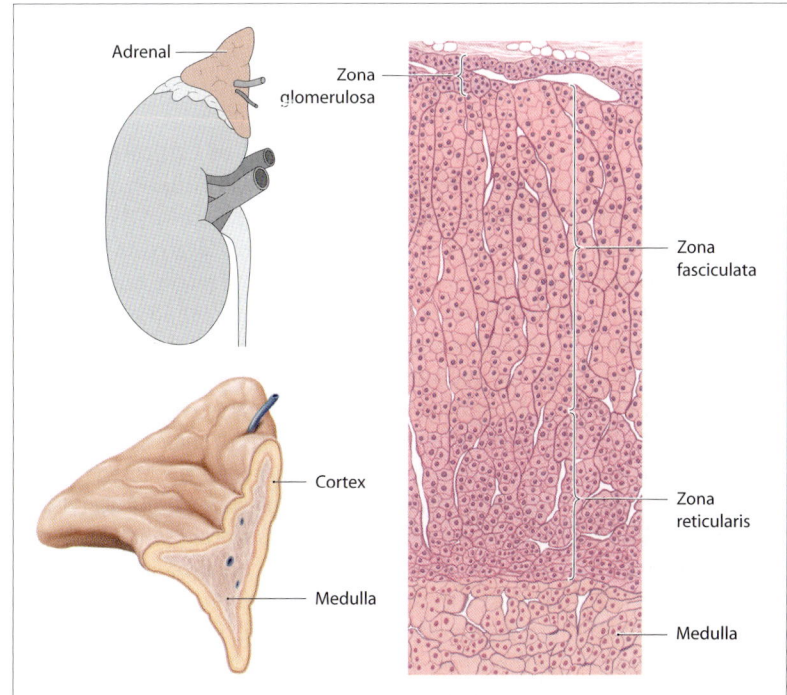

Fig. 85.1 (*Above, left*) Location of the adrenal gland. (*Below, left*) Right adrenal gland cut open. (*Right*) Histological section from an adrenal gland.

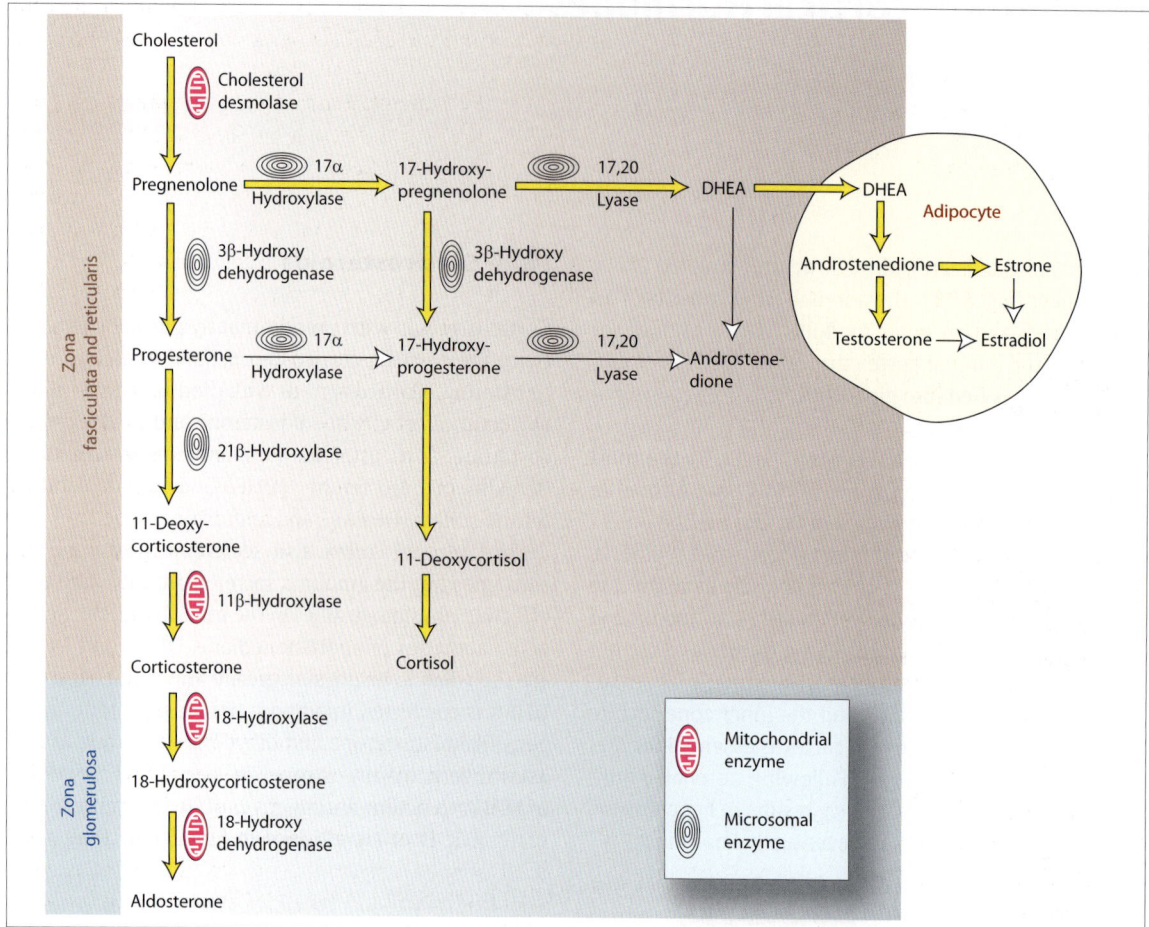

Fig. 85.2 Biosynthesis of steroid hormones. The thick yellow arrows indicate heavy flow of substrates. Details of androgen and estrogen production are shown in Figure 88.**1**.

axillary and pubic hair growth and stimulate libido. The peripheral conversion of DHEA to estrogens is important in both males and females, as explained below.

In males, 85% of the estrogen comes from the peripheral conversion of adrenal DHEA and only 15% comes from the testes. Estrogens hasten the closure of epiphyses, enhance the secretion of growth hormone at puberty, mediate the inhibition of gonadotropin secretion in the pituitary by testosterone, and may regulate the plasma level of high-density lipoproteins. They may be responsible for gynecomastia in the aged due to an imbalance of the testosterone:estrogen ratio.

In premenopausal females, 60% of the estrogen is estradiol from ovaries and 40% of the estrogen is estrone formed by peripheral conversion of adrenal DHEA. *In postmenopausal females*, the peripheral formation of estrogen predominates. Because adipose tissue is the major site of estrone synthesis from DHEA, the total estrogen in a massively obese postmenopausal woman may be even greater than in a premenopausal woman.

■ Biosynthesis

The steps in the biosynthesis of corticosteroids are shown in Figure 85.**2** and the structures of the hormones are shown in Figure 85.**3**. Some of the enzymes in the biosynthetic pathway are located in the mitochondria while others are located on the smooth endoplasmic reticulum. Hence, during steroid synthesis, the substrates must move in and out of the mitochondria for the appropriate reactions to occur.

Corticosteroid synthesis begins with the uptake of cholesterol, side-chain cleavage of cholesterol, and formation of pregnenolone. Although the adrenal cortex can synthesize its own cholesterol from acetyl-CoA, most of the cholesterol for its steroid synthesis is provided by plasma low-density lipoproteins (LDLs). The adrenal cortex stores cholesterol in the form of esters and deesterifies it during steroid synthesis. Cholesterol is converted to pregnenolone by 20,22 desmolase, a mitochondrial enzyme. It is the rate-limiting step in

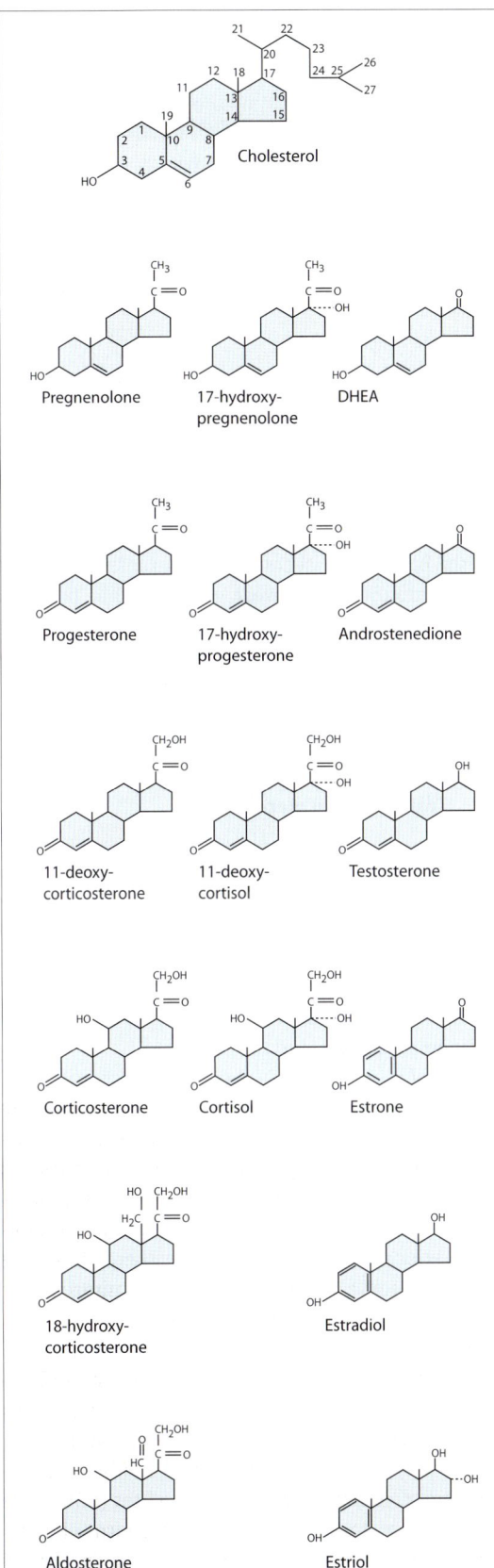

Fig. 85.**3**   Chemical structure of steroid hormones

corticosteroid biosynthesis. Pregnenolone is the common precursor of all steroid hormones, both mineralocorticoids and glucocorticoids.

In the *mineralocorticoid pathway*, pregnenolone is converted to progesterone by the enzyme 3β-hydroxy dehydrogenase. Progesterone is then sequentially hydroxylated at the C-21, C-11, and C-18 positions to form 18 hydroxycorticosterone (Fig. 85.**3**). 18-hydroxycorticosterone is converted to aldosterone by the mitochondrial enzyme 18-hydroxy dehydrogenase, which is found only in the zona glomerulosa.

In the *glucocorticoid pathway*, pregnenolone is sequentially hydroxylated at the C-17, C-21, and C-11 positions to form cortisol. The zona glomerulosa cannot synthesize cortisol as it lacks 17α-hydroxylase.

In the *androgenic pathway*, DHEA and androstenedione are produced respectively from 17α-hydroxypregnenolone and 17α-hydroxyprogesterone by the action of 17,20 lyase.

■   **Biosynthetic defects**

Genetic deficiency of the enzymes necessary for corticosteroid biosynthesis results in reduction in cortisol synthesis (Fig. 85.**4**). Low plasma cortisol causes feedback stimulation of ACTH secretion, leading to hyperplasia of the adrenal cortex, giving it the name *congenital adrenal hyperplasia* (CAH). Hypersecretion of ACTH results in excessive production of those substrates that are proximal to the deficient enzymes, leading to various clinical manifestations (Table 85.**1**). For example, the production of sex hormones is affected, resulting in genital abnormalities. Hence the condition is also known as the *adrenogenital syndrome*.

21β-hydroxylase deficiency results in deficient production of both glucocorticoids and mineralocorticoids. Mineralocorticoid deficiency results in excessive loss of body salt and water and the condition is called the *salt-losing form of CAH*. Most of the substrates are diverted to the androgenic pathway, resulting in the production of large amounts of DHEA, which is a weak androgen and contributes to the increased urinary level of 17-ketosteroids. Outside the adrenal gland, DHEA is metabolized to testosterone, estrone, and estradiol. In girls, this enzyme deficiency results in female pseudohermaphroditism and virilization.

11β-hydroxylase deficiency results in impaired production of corticosterone and aldosterone but excessive production of 11-deoxycorticosterone, which has about 3% of the activity of aldosterone. Hence, there is Na^+ and water retention, hypertension, and hypokalemia

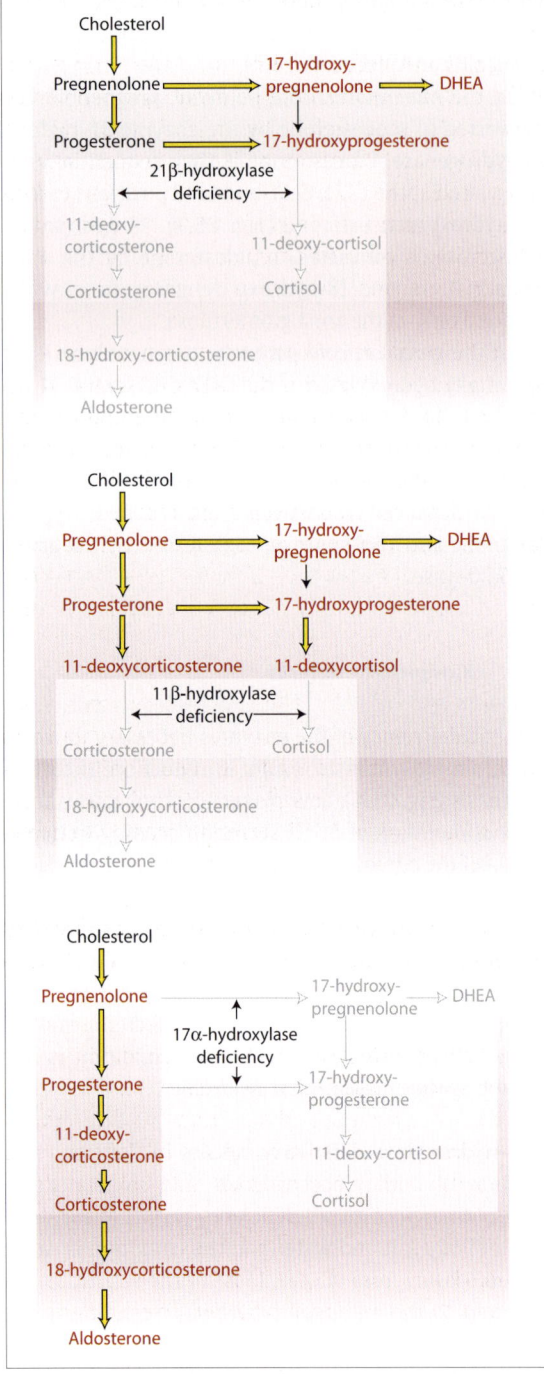

Fig. 85.**4**   Steroid biosynthesis in 21β-hydroxylase deficiency (*above*), 11β-hydroxylase deficiency (*middle*) and in 17α-hydroxylase deficiency (*below*).

Table 85.**1**   Effects of different enzyme defects in congenital adrenal hyperplasia.

	Glucocorticoid effect	Mineralocorticoid effect	Androgenic effect
21β-hydroxylase	Low	Low: salt-losing	High: virilizing
11β-hydroxylase	Low	Moderate: hypertensive	High: virilizing
17α-hydroxylase	Low	Excess: hypertensive	Low: feminizing

17α-hydroxylase deficiency occurs in both adrenal cortex and gonads. A deficiency of 17α-hydroxylase results in impaired synthesis of cortisol, androgens, and estrogens. Most of the substrates enter the mineralocorticoid pathway causing hypertension and disturbances of water, electrolyte, and pH balance. Although cortisol production decreases, the excessive production of corticosterone, a weak glucocorticoid, prevents the symptoms of glucocorticoid deficiency. In females, the low estradiol levels result in sexual infantilism. In males, the low testosterone level is associated with feminization and male pseudohermaphroditism.

Transport and metabolism

About 90% of the plasma cortisol is bound to cortisol-binding globulin (CBG or transcortin), which is an α-globulin, and about 6% of the plasma cortisol is bound to plasma albumin. About 4% is unbound and represents the physiologically active steroid. The bound cortisol represents a reservoir from which the hormone can be released as required.

Cortisol is inactivated in the liver to dihydrocortisol and tetrahydrocortisol and conjugated with glucuronic acid and sulfates to form water-soluble metabolites that are readily excreted by the kidney. Aldosterone is excreted in urine as a glucuronide conjugate. Androgens are excreted in urine after conversion to androsterone, etiocholanolone, and several other metabolites that are collectively called urinary 17-ketosteroids.

Glucocorticoids

Control of glucocorticoid secretion

The adrenal cortex is under control of the hypothalamo-hypophysial axis (Fig. 85.**5**). The corticotropin-releasing hormone (CRH) released from the hypothalamus stimulates adrenocorticotropic hormone (ACTH) secretion from the anterior pituitary. ACTH stimulates

and the condition is called the *hypertensive form of CAH*. There is reduced cortisol production but the production of 11-deoxycortisol, which is not bioactive, is increased. There is also an excessive secretion of DHEA with all its consequences.

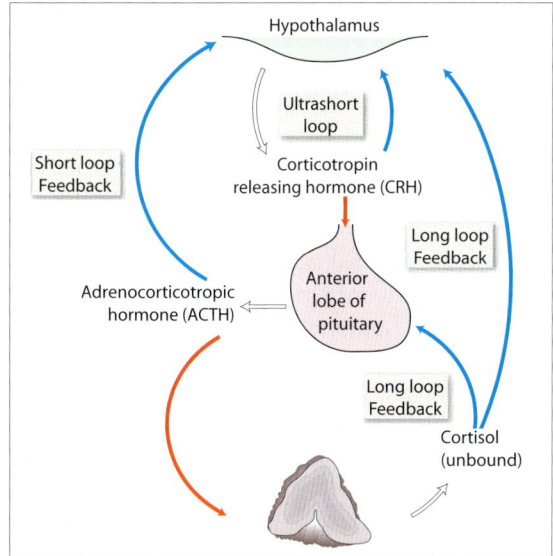

Fig. 85.**5** Hypothalamic-hypophysial control of adrenocortical secretion.

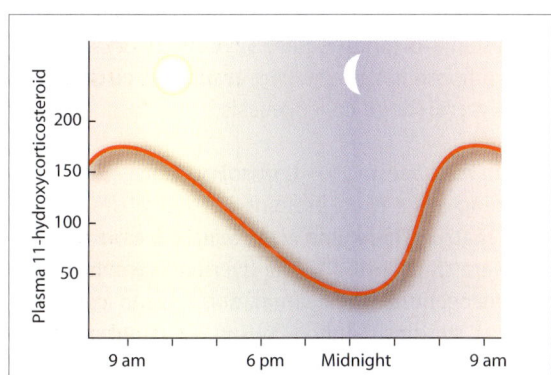

Fig. 85.**6** Diurnal variation in glucocorticoid secretion.

the output of glucocorticoids as well as aldosterone and DHEA.

The hypothalamo-hypophysial axis is susceptible to feedback control by cortisol. Thus a fall in cortisol level stimulates ACTH and CRH secretions (long loop feedback) resulting in the restoration of cortisol levels. Short loop feedback (ACTH inhibiting CRH) and ultrashort loop feedback (CRH regulating its own output) also occur. The integrity of this feedback control system can be tested by the dexamethasone suppression test. Dexamethasone, a synthetic glucocorticoid, is a potent inhibitor of ACTH and CRH secretion. If administration of dexamethasone reduces cortisol secretion, it indicates that the pituitary is responsive to CRH and the adrenal cortex is responsive to ACTH.

Glucocorticoid secretion shows a diurnal variation (Fig. 85.**6**) and increases sharply in response to stress. These responses are mediated by CRH.

Physiologic actions of glucocorticoids

Effect on blood cells Cortisol stimulates hemopoiesis and increases the number of circulating erythrocytes, neutrophils, and platelets. Neutrophilia occurs due to excess release from bone marrow, reduced margination, and reduced diapedesis into tissues. Cortisol reduces the number of circulating lymphocytes, monocytes, eosinophils, and basophils by promoting their migration from blood into tissues.

Anti-inflammatory effects Glucocorticoids inhibit inflammatory and allergic reactions in several ways. (1) They stabilize lysosomal membranes and thereby inhibit the release of proteolytic enzymes. (2) They decrease capillary permeability, thereby inhibiting leukocytic diapedesis and reducing inflammatory exudations. (3) Cortisol decreases the release of inflammatory mediators like serotonin, histamine, and hydrolases from granulocytes, mast cells, and macrophages. It also inhibits prostaglandin synthesis.

Anti-immunity effects Glucocorticoids cause involution of the lymph nodes, thymus, and spleen. In high doses, they suppress both humoral immunity (by reducing B-cell proliferation) and cell-mediated immunity (by inhibiting T-cell proliferation and cytokine release).

Antiallergic effects The basophilopenia caused by cortisol reduces allergic response due to a decrease in histamine release.

Renal effects Glucocorticoids facilitate rapid excretion of a water load (increases free-water clearance) and also enhance uric acid excretion. A fall in plasma cortisol is associated with reduction in GFR and renal plasma flow. Mineralocorticoids do not have these effects.

Gastric effects Cortisol increases gastric acid secretion and decreases gastric mucosal cell proliferation. Hence, prolonged cortisol treatment predisposes to peptic ulceration. Stress, which is invariably associated with excessive glucocorticoid secretion, often results in gastric ulcers (*stress ulcers*).

Psychoneural effects High cortisol levels can cause irritability, depression, insomnia, amnesia, and lower seizure threshold.

Antigrowth effects Large doses of cortisol reduce Ca^{2+} absorption from the gut (by antagonizing calcitriol), inhibit mitosis of fibroblasts, and cause degradation of collagen. These effects lead to osteoporosis. The breakdown of collagen leads to an increase in urinary hydroxyproline excretion. Cortisol delays wound healing because of the reduction of fibroblast proliferation. Connective tissue is reduced in quantity and strength. Cortisol inhibits the anabolic actions of growth hormone and insulin-like growth factor-1, particularly in bone. Excess cortisol suppresses growth hormone secretion and inhibit somatic growth. In large doses, cortisol causes muscle atrophy and weakness (*steroid myopathy*).

Vascular effect Cortisol enhances catecholamine synthesis by activating PNMT. In pharmacological doses, cortisol enhances the vascular responsiveness to norepinephrine (the *permissive action*) and thereby helps in maintaining normal blood pressure and volume. Absence of cortisol results in vasodilatation and hypotension.

Stress adaptation Any condition that disrupts or threatens to disrupt homeostasis is called a stress. Stress increases ACTH secretion, resulting in increased glucocorticoid secretion. The physiological significance of this is however poorly understood because resistance to stress is not increased by the administration of glucocorticoids.

Stressful situations that increase ACTH secretion include severe physical trauma, pain, surgery, circulatory shock, fever, hypothermia, hypoglycemia, infections, emotional trauma, and severe exercise.

ADH secretion Cortisol has a negative feedback effect on the ADH secretion. The water intoxication caused by cortisol deficiency is largely related to increased release of ADH. Hence, ADH antagonists are able to relieve the water intoxication caused by cortisol deficiency.

The **metabolic effects** of cortisol (Fig. 85.7) can be summarized as an overall catabolic effect with an anabolic effect on the liver where both glycogenesis and gluconeogenesis are stimulated. Cortisol helps the body during starvation by maintaining normal blood glucose and storing up liver glycogen at the expense of breaking down body fats and expendable proteins.

Carbohydrate metabolism (1) Cortisol causes hyperglycemia due to increased gluconeogenesis. It also blocks glucose transport into muscle and adipose tissue, which contribute to the hyperglycemia. These effects result in glucose intolerance and eventual *steroid diabetes*. The hyperglycemia causes a compensatory hyperinsulinemia. (2) Cortisol also promotes hepatic glycogenesis and increases the glycogen content of the liver.

The increase in gluconeogenesis occurs mainly due to two reasons: (1) Cortisol mobilizes amino acids from muscles and bones, making available amino acids for gluconeogenesis. Cortisol also augments the synthesis and activity of key gluconeogenic enzymes, especially, fructose-1,6-bisphosphatase. (2) Cortisol causes lipolysis in adipose tissues, making available free fatty acids and glycerol for gluconeogenesis.

Protein metabolism (1) Cortisol enhances the release of amino acids from proteins in skeletal muscle and bone matrix. The amino acids released, especially alanine, are transported to the liver and deaminated for gluconeogenesis. The deamination leads to an increase in urea synthesis and excretion. (2) Glucocorticoids also have an antianabolic effect as they inhibit protein synthesis, probably at the translational level.

Fat metabolism (1) Cortisol causes lipolysis in adipose tissue by stimulating hormone-sensitive lipase. The

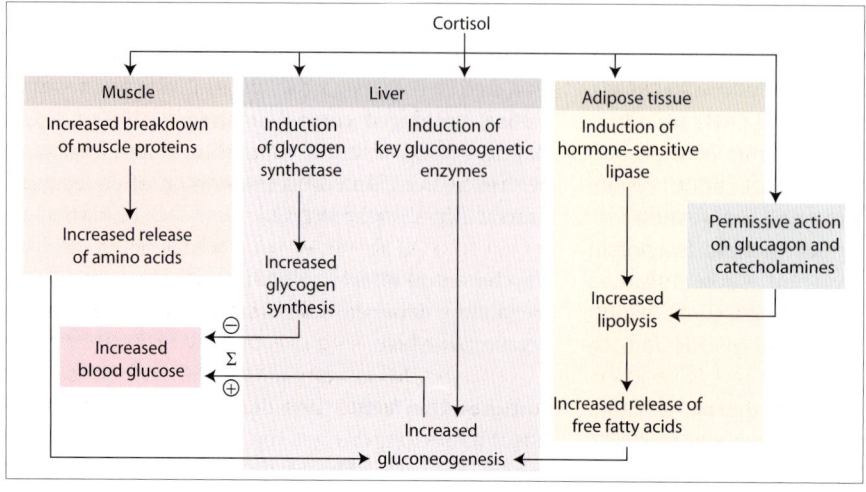

Fig. 85.**7** Metabolic effects of glucocorticoids.

fatty acids mobilized from adipose tissue are used for gluconeogenesis. Cortisol also has a permissive (potentiating) action on the lipolytic effects of catecholamines and glucagon. (2) Despite the lipolysis, cortisol increases total body fat and causes a redistribution of fat with characteristic centripetal distribution that is seen in Cushing's syndrome. The lipogenic effect is probably due to the hyperinsulinemia produced by cortisol. Other contributory factors could be the stimulation of leptin synthesis in adipocytes and the stimulation of differentiation of preadipocytes to adipocytes.

Mineralocorticoids

Aldosterone is the principal mineralocorticoid secreted by the adrenal, although corticosterone is secreted in sufficient amounts to exert a minor mineralocorticoid effect. Deoxycorticosterone too has a weak mineralocorticoid activity but is not secreted in appreciable amounts normally. A large amount of progesterone also has some mineralocorticoid activity, but it does not have any physiological role in the control of Na^+ excretion.

Control of mineralocorticoid secretion

Mineralocorticoid secretion is stimulated by hyperkalemia, angiotensin-II, ACTH, and hyponatremia, in reducing order of efficacy. (1) Aldosterone secretion in response to hyperkalemia forms the basis for renal regulation of body potassium balance. (2) Stimulation of aldosterone secretion by angiotensin-II is important for the correction of hypovolemia and hypotension. It also explains why aldosterone secretion increases on prolonged standing. (3) Stimulation of aldosterone secretion by ACTH results in the diurnal variation of aldosterone secretion. It also explains why aldosterone secretion increases in response to stress. ACTH is not an important physiological regulator of aldosterone secretion. When ACTH tends to increase aldosterone secretion, angiotensin-II and potassium ions promptly oppose any change in aldosterone level. However, in an interesting syndrome called *glucocorticoid-remediable aldosteronism*, ACTH exerts a stronger control of aldosterone secretion than angiotensin-II or hyperkalemia and therefore results in hyperaldosteronism. It is treated by cortisol administration which decreases ACTH secretion through negative feedback.

A sharp fall of about 20 mEq L^{-1} in plasma Na^+ stimulates aldosterone secretion but such changes are rare. Physiologically, hyponatremia is a weak stimulator of aldosterone secretion, and fortunately so! We know that excessive water intake results in dilutional hyponatremia and water diuresis. The diuresis would not occur if hyponatremia were a potent stimulus for aldosterone secretion.

Physiologic actions of mineralocorticoids

Aldosterone acts mainly on the cortical collecting ducts to increase the reabsorption of Na^+ and the secretion of K^+ and H^+ ions. Aldosterone is a major controller of K^+ homeostasis. It is of secondary importance in regulation of fluid and electrolyte balance despite its effect on Na^+ and water reabsorption. Mineralocorticoids also increase the reabsorption of Na^+ from the sweat and digestive juices as they flow out through the glandular ducts.

Adrenocortical Dysfunctions

Cushing's syndrome

Cushing's syndrome results from hypersecretion of cortisol. It can be caused by ACTH-secreting pituitary tumors (ACTH-dependent Cushing's syndrome) or cortisol-secreting adrenal or ectopic tumors (ACTH-independent Cushing's syndrome). ACTH-independent Cushing's syndrome is associated with feedback suppression of ACTH secretion. (Earlier, ACTH-dependent Cushing's syndrome was called Cushing's 'disease'—a practice that causes needless confusion and is better avoided.)

Cushing's syndrome has all the features that would be expected from the excessive actions of cortisol and only the clinically important ones are recapitulated here. (1) The patient has a characteristic appearance (Fig. 85.**8**) due to redistribution of body fat; a moon face, truncal obesity with pendulous abdomen and buffalo (or dowager's) hump. (2) Due to a general increase in catabolism, there is osteoporosis of bones, atrophy of muscles and thinning of skin. The abdomen shows reddish-purple striae due to stretching of its skin by excess fat deposition. Tell-tale signs of easy bruisability (due to skin thinning) and poor wound healing are present. (3) Sex-related problems include amenorrhea and hirsutism (in women), impotency (in men), and decreased libido (in both). Psychological problems range from irritability to severe depression and even psychosis. (4) Investigations reveal hypertension (due to Na^+ and water retention), glucose intolerance, hyperlipidemia, hypercholesterolemia, and excessive urinary 17-ketosteroids.

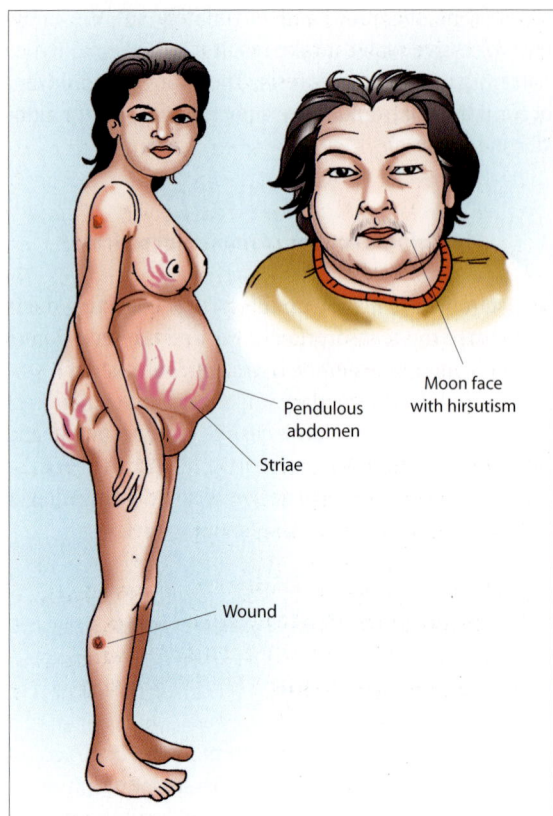

Pendulous
abdomen

Striae

Moon face
with hirsutism

Wound

Fig. 85.**8**   Cushing's syndrome

Adrenal insufficiency

Adrenal insufficiency is called *primary* when the reduced adrenocortical secretion occurs due to a disorder of the adrenal gland itself. It is called *secondary* when it is due to inadequate stimulation of the adrenals by ACTH.

In **primary adrenal insufficiency (Addison's disease)**, there is deficiency of cortisol, aldosterone, and DHEA. Deficiency of cortisol results in hypoglycemia and hyperpigmentation of the skin. The hyperpigmentation is caused by the feedback *hypersecretion of ACTH* which has β-MSH-like structure and activity. In fact, a part of the ACTH peptide is identical in structure to β-MSH (see Fig. 82.**5**). *Deficiency of aldosterone* causes hyponatremia with hypotension, and hyperkalemia with metabolic acidosis. These electrolyte abnormalities cause gastrointestinal symptoms like nausea, vomiting, diarrhea, and abdominal cramps. *Deficiency of androgens* leads to the loss of axillary and pubic hair, reduced muscle mass, and loss of libido in men and women.

Secondary adrenal insufficiency is similar to primary adrenal insufficiency except for two major differences. (1) The secretion of aldosterone remains nearly unaffected and therefore fluid and electrolyte disturbances do not occur. The hypotension that occurs occasionally is attributable to cortisol deficiency. (2) There is absence of hyperpigmentation since there is no hypersecretion of ACTH.

Hyperaldosteronism

Primary hyperaldosteronism (Conn's syndrome) is seen in patients with aldosterone-producing adenoma. Benign adenomas typically exhibit the normal diurnal pattern of aldosterone secretion, which suggests that the aldosterone synthesis is still under the control of ACTH. However, there is absence of the increase in aldosterone secretion that normally occurs on standing. The absence is attributable to the marked suppression of the renin–angiotensin system by the hypokalemia, which is prominent in Conn's syndrome. The hypokalemia also impairs insulin secretion, causing glucose intolerance, depolarizes the muscle membrane causing muscle weakness, and impairs the urine concentrating ability of the kidney (hypokalemic nephropathy) causing polyuria.

Hyperaldosteronism also causes *hypernatremia* with reduced Na^+ excretion in urine and reduced Na^+ secretion in sweat, saliva, and gastrointestinal secretions. The hypernatremia is not severe because of the *mineralocorticoid escape*—an escape from the Na^+-retaining effects of chronic aldosteronism. For the same reason and also due to the polyuria associated with hypokalemic nephropathy, edema almost never occurs in Conn's syndrome.

Secondary hyperaldosteronism is prominently associated with edema. The movement of Na^+ and water out of circulation into interstitial spaces results in hypovolemia, which stimulates aldosterone secretion. The rise in aldosterone level is unable to correct the hypovolemia since the retained water and electrolytes promptly move out again into the interstitium, aggravating the edema. Hypokalemia is not prominent because the hypovolemia reduces the urinary flow rate through renal tubules, inhibiting tubular K^+ secretion.

Corticosteroid Therapy

Use Adrenocorticosteroid therapy is often life-saving. (1) Corticosteroids are used for substitution therapy in primary or secondary adrenal insufficiency and also in nonendocrine conditions. (2) Corticosteroids have both antiinflammatory and immunosuppressive effects and therefore are used in the treatment of chronic inflammatory disorders like rheumatoid arthritis and collagen disorders. For the same reason, they are administered to organ transplant recipients for reducing the chances of graft rejection. However, treatment with large doses of glucocorticoids predisposes to infections, making antibiotics a necessary adjunct to the steroid therapy. (3) Due to their antiallergic effects, they are used in bronchial asthma and skin diseases. (4) Due to their antilymphocytic effect, they are used in malignancies like lymphocytic leukemias and lymphomas. In breast cancers that are aggravated by estrogens, corticosteroids are administered for indirectly suppressing estrogen secretion. Corticosteroids suppress adrenocortical activity by feedback inhibition of ACTH secretion so that the adrenal cortex produces less of the androgenic precursors of estrogens. (5) Since corticosteroids reduce edema, they are used in cerebral edema. Patients of cerebral stroke are often administered corticosteroids. The stress associated with the cerebral stroke tends to cause stress ulcers, and the administration of corticosteroids aggravates them. Hence, simultaneously with steroid therapy, these patients are administered antacids. (6) Since glucocorticoids enhance the vascular responsiveness to norepinephrine, they are frequently administered in circulatory shock along with other drugs. However, their efficacy in shock remains debatable.

Since long-term corticosteroid therapy is associated with suppression of the axis, the therapy should not be stopped suddenly. Abrupt cessation of steroid therapy is associated with life-threatening adrenal insufficiency. Hence, the dose of the steroid must be slowly decreased, i.e., tapered. Full recovery from hypothalamic-hypophysial-adrenocortical suppression may require as long as one year following cessation of all steroid therapy.

Misuse Steroids are commonly used as antiinflammatory agents but they only suppress inflammation without eliminating its cause. Application of a cortisone cream alleviates an itchy rash but the sensitivity to the allergen persists: the next exposure will produce the rash again. Unscrupulous quacks use steroids indiscriminately to make their patients symptom-free overnight, impressing the patients but actually harming them by making them immunosuppressed.

Reproductive System

88.   Testicular and Ovarian Hormones

89.   Puberty and Gametogenesis

90.   Menstrual Cycle

91.   Sperm Transport and Fertilization

92.   Sexual Differentiation of the Fetus

93.   Pregnancy

94.   Parturition and Lactation

Hormones of the Testes

The hormone-secreting cells in the testes are the Leydig cells and Sertoli cells. The Leydig cell secretes all the androgens, viz., dihydrotestosterone (DHT), testosterone, androstenedione and dehydroepiandrosterone (DHEA). All of them have 19 carbon atoms, and their androgenic potencies are in a ratio of 60:20:2:1 respectively. The amount of testosterone secreted is 100 times more than the other hormones. The Leydig cell also secretes some androgenic precursors like pregnenolone and progesterone. The Sertoli cell secretes small amounts of estradiol.

Testosterone is synthesized in the Leydig cell through the same biosynthetic pathway as the adrenocortical steroids (see Fig. 85.**2**). Cholesterol esters, the major precursors for testosterone biosynthesis, are stored in the lipid droplets in the Leydig cells. Androgen biosynthesis in the human testes proceeds mostly from pregnenolone to DHEA and then to androstenedione, testosterone, and dihydrotestosterone, i.e., the Δ^4 pathway (Fig. 88.**1**). Only 20% of dihydrotestosterone is synthesized by the testes. The rest is derived from the peripheral conversion of testosterone in the skin and male reproductive tract (epididymis, prostate gland, and seminal vesicles).

Two-thirds of the total plasma testosterone is bound to albumin, and one-third is bound to testosterone-binding globulin which is also called sex hormone-binding globulin, because it binds estradiol as well. Less than 2% of the plasma testosterone is in the free form.

Testosterone and androstenedione are converted to estradiol and estrone respectively through the action of aromatase, which is a microsomal enzyme found in the brain, skin, liver, mammary tissues, adipose tissue, and placenta. These active metabolites (having hormonal activity) are further converted into inactive metabolites like androsterone and etiocholanolone. The inactive metabolites are conjugated in the liver and excreted into the urine as 17-ketosteroids. Of the total urinary ketosteroids, only one-third is of testicular origin while the remainder comes from the DHEA produced by the adrenal cortex. Hence, the urinary 17-ketosteroid reflects mainly adrenocortical activity and is not a good index of testicular function.

Hormonal control of testicular function   Luteinizing hormone (LH) increases testosterone synthesis in the Leydig cells by activating cholesterol desmolase, which is the rate-limiting enzyme for the conversion of cholesterol to pregnenolone. In the developing male fetus,

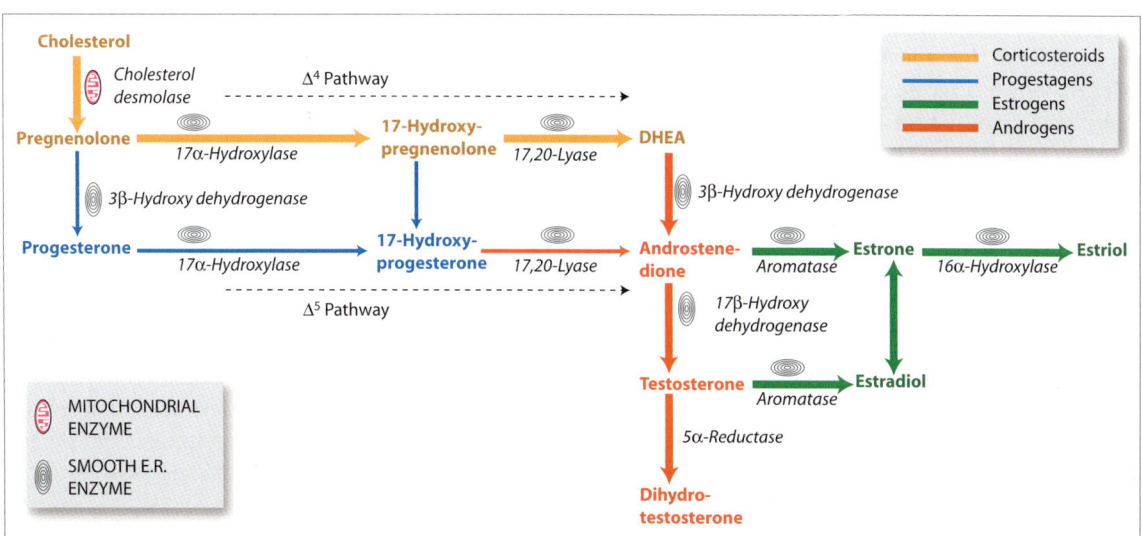

Fig. 88.**1**   Biosynthesis of sex steroids. Estrogens are shown in green, progestagens in blue, and androgens in red. Note that estrone and estradiol are interconvertible.

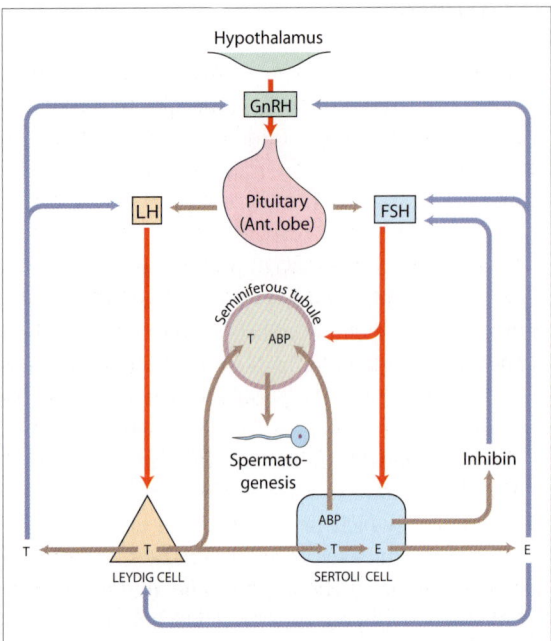

Fig. 88.**2** Hormonal control of the testes. The inhibition of testosterone secretion by estradiol is a paracrine effect. E = estradiol, T = testosterone.

the stimulus for testosterone synthesis is HCG (human chorionic gonadotropin), which is structurally and functionally similar to LH.

FSH stimulates the Sertoli cells to synthesize androgen-binding protein (ABP), which binds testosterone, increasing its local concentration in the testes and thereby stimulating spermatogenesis. FSH also indirectly increases testosterone synthesis by increasing the number of LH receptors on the Leydig cell. The secretion of FSH and LH from the anterior pituitary is stimulated by gonadotropin-releasing hormone (GnRH), which is secreted by the parvocellular neurosecretory neurons in the hypothalamus. The GnRH secretion is pulsatile and occurs at a frequency of 8 to 14 pulses per day.

Consistent with the phenomenon that the secretion of the target hormone inhibits its trophic hormone, testosterone inhibits LH secretion while estradiol and inhibin inhibit FSH secretion. Inhibin is secreted by Sertoli cells and acts only at the pituitary level while testosterone and estradiol act at both pituitary and hypothalamic levels (Fig. 88.**2**).

Physiological actions of androgens (1) In the fetus, testosterone stimulates the differentiation of the male internal genitalia while dihydrotestosterone stimulates the differentiation of the male external genitalia. (2) During puberty, androgens stimulate the development

Table 88.**1**   Differences between the actions of testosterone and dihydrotestosterone.

Testosterone	Dihydrotestosterone
Necessary for the differentiation of the internal genitalia	Necessary for the differentiation of the external genitalia
During puberty, promotes the growth of penis, seminal vesicles, larynx, muscles, and skeleton	During puberty, promotes the growth of facial, body, and pubic hair, scrotum, sebaceous glands with increased sebum production and development of acne, and prostate with stimulation of prostatic secretions
Promotes spermatogenesis, increases libido, promotes erythropoiesis and brings about feedback inhibition of LH	Has none of these effects

of the secondary sexual characteristics, libido, and potency (erectile functions). Most of these are brought about by testosterone. Dihydrotestosterone promotes growth of pubic hair, increased sebum production by sebaceous glands with consequent development of acne, growth of scrotum and growth of prostate, and stimulation of prostatic secretion (Table 88.**1**). (3) Testosterone is essential for normal spermatogenesis. (4) Testosterone is a protein anabolic hormone. It stimulates cell division and maturation. In the adolescent, testosterone produces linear skeletal growth and broadens the shoulders; it also causes muscular development (myotrophic effect) and retention of potassium, nitrogen, and phosphorus. Testosterone has a dual effect on skeletal growth because it also accelerates epiphysial fusion of the long bones and thereby limits linear growth. Treatment with testosterone reduces bone loss and osteoporosis. (5) Testosterone stimulates erythropoiesis directly as well as by stimulating erythropoietin secretion.

Hormones of the Ovary

Endocrine cells of the ovary

Three types of ovarian cells (Fig. 88.**3**) are involved in hormone secretion, namely, the granulosa cells, luteal cells, and thecal cells (cells of theca interna). Together they secrete estrogens (estrone, estradiol, and estriol), progestagens (17α-hydroxy progesterone and progesterone), androgens (androstenedione and testosterone), and relaxin.

Thecal cells produce androstenedione and testosterone that are converted in the granulosa cells to estrone and

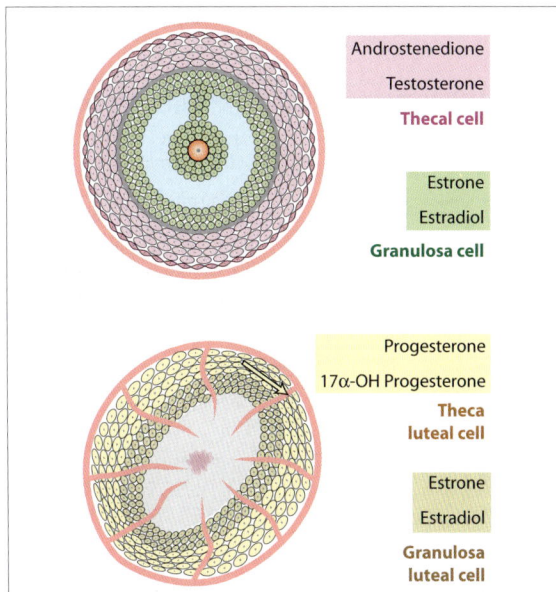

Fig. 88.**3**   Hormones secreted by the granulosa cells, thecal cells, and luteal cells of the ovary. Note that the theca luteal cells and granulosa luteal cells secrete different hormones. Also note that the granulosa cells secrete estrogens into the follicular fluid but the granulosa luteal cells secrete estrogens into blood.

estradiol respectively by the enzyme aromatase (Fig. 88.**4**). Thecal cells do not produce appreciable amounts of progesterone or 17-hydroxy progesterone because most of the androgens are formed via the Δ^4 and not the Δ^5 pathway (Fig. 88.**1**) and therefore progestagen levels remain low during the follicular phase of the menstrual cycle (see Fig. 90.**2**).

Granulosa cells cannot produce estrogens or progestagens directly from cholesterol because they lack 17α-hydroxylase and 17,20-lyase. Estrogens are synthesized in granulosa cells from the androstenedione obtained from thecal cells. Most of the estrogen produced by granulosa cells is secreted into the follicular fluid and therefore estrogen levels remain low during the early follicular phase.

Luteal cells are of two types: the theca luteal cells and the granulosa luteal cells formed by the luteinization of granulosa and thecal cells respectively. Theca luteal cells produce mainly progesterone and 17-hydroxy progesterone due to the activation of the Δ^5 pathway of steroid biosynthesis. The granulosa luteal cells secrete estrone and estradiol. These estrogens are secreted directly into the blood capillaries that grow into the corpus luteum and therefore estrogen levels rise during the luteal phase (see Fig. 90.**2**).

Estrogens

Estradiol (E$_2$) is the main estrogen secreted by the ovary and is also the most biologically active. Estrone (E$_1$) is a weak ovarian estrogen. In postmenopausal women, estrone is the dominant plasma estrogen and is mostly formed by the conversion of adrenocortical androstenedione in peripheral tissues, mainly liver and adipose tissue. Estriol (E$_3$) is not secreted by the ovary. It is formed in small amounts in the liver from estradiol and estrone. Large amounts are secreted only during pregnancy, when it is secreted by the placenta. Note that estr(one) has one hydroxyl group, estra(di)ol has two hydroxyl groups, and es(tri)ol has three hydroxyl groups (see Fig. 85.**3**). Estrone and estradiol are interconvertible (Fig. 88.**1**). The synthesis of estriol by the placenta is described on p 579.

Over 70% of circulating estrogens are bound to sex steroid-binding globulin and 25% are bound to plasma albumin. Estrogens are metabolized in the liver. Estradiol and estrone are hydroxylated to form estriol and catecholestrogens, which are excreted primarily as glucuronides. Estrone is excreted primarily as a sulfate.

The **physiological actions** of estrogens can be summarized as follows: (1) *Metabolism*. Estrogens have important protein anabolic effects. They mediate the growth and development of the female reproductive organs, especially, that of the gravid uterus. They also promote cellular proliferation in the mucosal linings

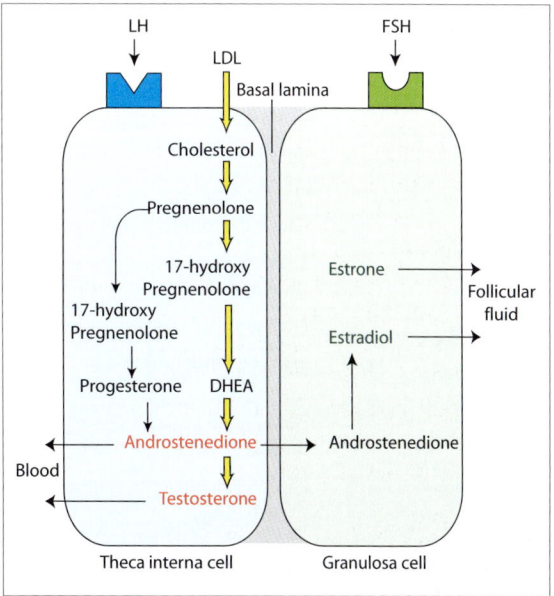

Fig. 88.**4**   Granulosa cells synthesize estrogens from the androgens provided by the thecal cells.

of these structures. (2) *Endometrium*. Estrogens stimulate the regeneration of the stratum functionalis during the proliferative phase of the endometrial cycle by increasing mitosis. The spiral arterioles of the stratum functionalis grow rapidly under estrogenic influence. (3) *Myometrium*. Estrogens cause hypertrophy of the myometrium and sensitize it to the action of oxytocin, which promotes uterine contractility. (4) *Cervix*. Under the influence of estrogens, the uterine cervix secretes an abundance of copious thin, watery mucus during the preovulatory phase. (5) *Vagina*. The vaginal epithelium cells mature under the effect of estrogen, resulting in the thickening and cornification of the vaginal lining. (6) *Breast*. Estrogens promote the development of the duct and lobuloalveolar system of the mammary gland. (7) *Bone*. Estrogens, like androgens, exert a dual effect on skeletal growth in that they cause an increase in osteoblastic activity, resulting in a growth spurt at puberty. However, estrogens also hasten bone maturation and promote the closure of the epiphyseal (cartilaginous) plates in the long bones more effectively

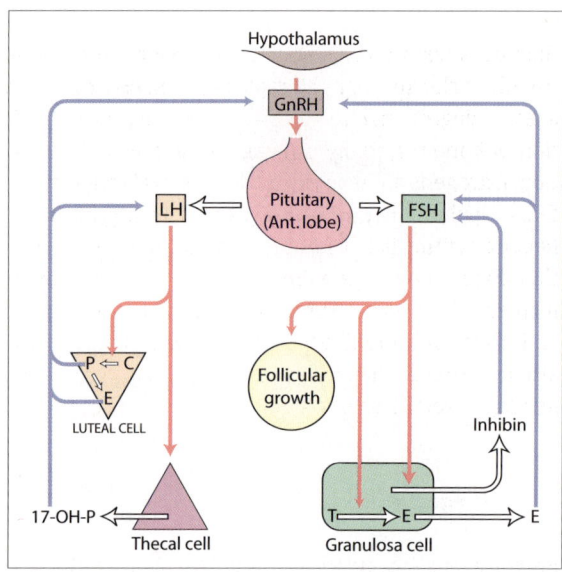

Fig. 88.**5** Hormonal control of the ovary. E = estradiol, P = progesterone, 17-OH-P = 17α-hydroxy progesterone, T = testosterone.

Table 88.**2** Differences in the physiological actions of estradiol and progesterone.

Site of action	Estradiol	Progesterone
Pituitary-hypothalamus	Inhibits FSH secretion by negative feedback	Inhibits LH secretion by negative feedback
	Increases prolactin secretion	
	Inhibits LH secretion by negative feedback (in high concentration)	
	Stimulates midcycle surge of LH and FSH by positive feedback on pituitary	
Uterus	Promotes hypertrophy of myometrium	
	Promotes hyperplasia of endometrium	Arrests endometrial mitosis and induces secretory activity
	Promotes uterine motility	Inhibits uterine motility
		Maintains the decidua and assists in the implantation of blastocyst
Cervix	Causes thinning of cervical mucus	Causes thickening of cervical mucus
Vagina	Causes maturation of vaginal epithelial cells and the thickening and cornification of vaginal mucosa.	
Breast	Promotes ductal and lobuloalveolar growth	Promotes lobuloalveolar growth
Kidney	Promotes renal Na⁺ retention	Antagonizes the action of aldosterone on the kidney and promotes the renal excretion of Na⁺
Bone	Enhances bone growth, density and maturation and causes early epiphyseal closure	
Miscellaneous	Promotes development of female secondary sexual characteristics	
	Makes sebaceous secretion more fluid and thus inhibits formation of acne and comedones	Serves as a precursor for steroid hormones
		Increases basal body temperature
		Stimulates breathing

than does testosterone. Therefore, the female skeleton usually is shorter than the male skeleton. Estrogens, to a lesser degree than testosterone, promote the deposition of bone matrix by causing Ca^{2+} and HPO_4^{2-} retention. Estrogens are responsible for the oval shape of the female pelvic inlet. (8) *Liver*. Estrogens stimulate the hepatic synthesis of the transport globulins, including thyroxine-binding globulin and transcortin. Estrogens increase the hepatic synthesis of coagulation factors and therefore predispose to thrombosis. Estrogens also increase the hepatic synthesis of angiotensinogen, leading to retention of Na^+ and water in the body.

■ Progestagens

Progestagens (progesterone and 17α-hydroxy progesterone) are secreted mainly by theca luteal cells. Progesterone is not bound to sex steroid-binding globulin. The progesterone is bound primarily to cortisol binding globulin (transcortin) and albumin. The liver metabolizes progesterone to pregnanediol, and 17α-hydroxy-progesterone to pregnanetriol.

The **physiological effects** of progestagens are attributable entirely to progesterone since 17α-hydroxy progesterone has little if any biological effect. The physiological actions of progesterone are as follows: (1) *Endometrium*. Progesterone promotes secretory changes in the stratum functionalis of the endometrium. The endometrial glands become elongated and coiled and secrete a glycogen-rich fluid. Progesterone is required for the implantation of the blastocyst and the maintenance of decidua. Antiprogesterone agents like mifepristone are used for medical termination of pregnancy. (2) *Myometrium*. Progesterone inhibits uterine motility by hyperpolarizing the uterine smooth muscles. (3) *Cervix*. Under the influence of progesterone, the mucus secreted by the cervical glands is reduced in volume and becomes thick and viscid. (4) *Breast*. Progesterone promotes lobuloalveolar growth in the mammary gland. (5) *Kidney*. Progesterone promotes renal excretion of Na^+ (antialdosterone effect). This antagonizes the effects of the elevated aldosterone levels found in pregnancy. (6) *Fetus*. Progesterone contributes to the growth and development of the fetus by acting as a precursor for corticosteroid synthesis by the fetal adrenal cortex. (7) *Brain*. Progesterone stimulates breathing through a direct effect on the brainstem respiratory center.

■ Hormonal control of ovarian function

The development of the ovarian follicle is largely under the control of FSH. Ovulation is caused by LH which also stimulates secretion of progesterone from the corpus luteum. Consistent with the phenomenon of negative feedback in which the secretion of the target hormone inhibits its trophic hormone, progesterone inhibits LH and inhibin inhibits FSH secretion. Estradiol, whose secretion is stimulated by both LH and FSH, inhibits both LH and FSH. Progesterone and estrogen act at both hypothalamic and pituitary levels while inhibin secreted by granulosa cells acts only on the pituitary (Fig. 88.**5**). Under certain conditions, estrogen causes stimulation rather than inhibition of LH, resulting in a positive feedback cycle (see Ovulation, p 559).

Puberty

Puberty or adolescence is that stage of development when the endocrine and gametogenic functions of the gonads develop for the first time to the point where reproduction is possible. Puberty generally occurs between the ages of 8 and 13 in girls and 9 and 14 in boys. Spermatogenesis in boys and folliculogenesis in girls begin at puberty. It coincides with a surge of sex hormone secretion resulting in the development of the secondary sexual characteristics. In girls, the most striking events associated with the onset of puberty are thelarche (the development of breasts) followed by pubarche (development of axillary and pubic hair) and menarche (the first menstrual period). The initial periods are generally anovulatory: regular ovulation appears about a year later. A less striking event associated with puberty is adrenarche, an increase in the secretion of adrenal androgens that occurs probably because an adrenal androgen-stimulating hormone secreted from the pituitary gland stimulates the enzyme systems in the adrenal glands to channel more pregnenolone to the androgenic pathway.

Secondary sexual characteristics

Male secondary sexual characteristics At puberty, the penis and scrotum increase in size and become pigmented. Rugal folds appear in the scrotal skin. Facial hair develops. The scalp line undergoes temporal recession. Pubic hair develops as a triangle with apex up. Axillary and body hair appear. The pubertal growth spurt occurs. Shoulders broaden. The prostate and seminal vesicles enlarge and start secreting seminal fluid. Sebaceous gland secretion thickens and increases, predisposing to acne. The pitch of the voice becomes lower due to the enlargement of the larynx and the thickening of the vocal cords. The attitude becomes more aggressive and interest develops in females. The muscle bulk and strength increases and there is a positive nitrogen balance.

Female secondary sexual characteristics At puberty, the external genitalia develop further and menstruation begins. Pubic hair develops as a triangle with apex below. Axillary hair appears. The breasts develop. A growth spurt occurs with a characteristic feminine fat deposition. Interest develops in males.

Control of pubertal onset

A plausible mechanism of the onset of puberty is that till puberty, the GnRH secretion from the hypothalamus is highly sensitive to feedback inhibition by testosterone and estrogens. As a result, the levels of testosterone and estrogen are never able to rise sufficiently high to induce puberty. From birth to puberty, there is a decrease in the sensitivity of the hypothalamus to feedback inhibition by testosterone or estrogens. By puberty, the hypothalamus is no longer sufficiently sensitive to feedback inhibition by testosterone and estrogen and therefore it secretes adequate amounts of GnRH in its usual pulsatile pattern. In the presence of the normal pulsatile release of GnRH, testosterone and estrogens are secreted in adequate amounts.

Role of leptin Over the past couple of centuries, the age of pubertal onset has been decreasing at the rate of 1–3 months per decade. Such changes are called secular changes and have been attributed to a general improvement in health and nourishment. The link between nourishment and earlier pubertal onset is provided by leptin, a satiety-producing hormone secreted by fat cells. Leptin facilitates the release of GnRH, thereby helping in pubertal onset. It is observed that young women often stop menstruating when they lose weight and resume menstruation once they regain weight. It therefore seems that a critical body weight is required for leptin release and pubertal onset.

Abnormalities of pubertal onset

Precocious pseudopuberty Exposure of immature males to androgens or females to estrogens causes early development of secondary sexual characteristics without gametogenesis. This syndrome is called precocious pseudopuberty to distinguish it from true precocious puberty that occurs due to early secretion of pituitary gonadotropins. Hypothalamic diseases are frequently associated with precocious puberty. Lesions of the ventral hypothalamus near the infundibulum cause precocious puberty, either due to interruption of neural pathways that inhibit GnRH secretion or due to chronic stimulation of GnRH secretion originating in irritative foci around the lesion. Sometimes LH secretion is high despite inadequate GnRH secretion, resulting in *gonadotropin-independent precocity*.

Delayed puberty Puberty is considered to be pathologically delayed if menarche fails to occur by the age of 17 or testicular development fails to occur by the age of 20. Failure of maturation can occur in panhypopituitarism and Turner's syndrome as well as in some otherwise normal individuals. In males, this clinical picture is called *eunuchoidism*. In females, it is called *primary amenorrhea*.

Male Gametogenesis

Structure of the testes

Seminiferous tubules The testes are made of convoluted loops of seminiferous tubules (see Fig. 91.**2**) that drain at both ends into the ducts in the epididymis. Between the coils of seminiferous tubules are the *interstitial cells of Leydig* which secrete testosterone. The seminiferous tubules contain the male germ cells. The tubule walls are made of Sertoli cells that rest on a basement membrane. The cell membranes of adjacent Sertoli cells are attached through tight junctions, forming a barrier that divides the lumen of the seminiferous tubule into two compartments: a *basal compartment* containing the spermatogonia and the early spermatocytes, and an *adluminal compartment* containing spermatocytes, spermatids, and spermatozoa (Fig. 89.**1**).

The Sertoli cell serves numerous functions, which are as follows: (1) It provides nourishment to the developing spermatozoa. (2) The intercellular junctions between adjacent Sertoli cells constitute the blood–testes barrier. (3) It produces inhibin which suppresses FSH secretion. (4) It secretes androgen-binding globulin (ABP) which binds testosterone with high affinity and is responsible for the high testosterone concentration in the tubular lumen. (5) It secretes transferrin for transporting iron to tubular cells, ceruloplasmin for transporting copper to the tubular cells, and plasminogen activator, which may mediate proteolytic reactions important for migration of maturing germ cells from the basal compartment to the adluminal compartment. (6) It produces Mullerian duct inhibiting substance (p 572). (7) It converts the androstenedione and testosterone produced by the Leydig cells to estrone and estradiol, respectively. The small amounts of estrogens secreted inhibit testosterone secretion through a paracrine effect. (8) It absorbs the unnecessary cellular organelles that are cast off from the spermatozoa.

Blood–testes barrier The blood–testes barrier is formed by tight intracellular junctions between adjacent Sertoli cells in the seminiferous tubule. The barrier separates the basal and adluminal compartments of the seminiferous tubule. Thus, the germ cells too have to cross the barrier as they pass from the basal compartment to the adluminal compartment. The tight junctions between Sertoli cells loosen up to enable the passage of maturing germ cells through the junction, only to tighten up again after they have passed. The blood–testes barrier protects the spermatocytes, spermatids, and spermatozoa from blood-borne toxic substances and circulating antibodies. It also prevents byproducts of gametogenesis from entering circulation lest they should stimulate an autoimmune reaction. Not unexpectedly, the breakdown of the barrier sometimes leads to autoimmune response against the germ cells. Steroids penetrate the barrier with ease. The blood–testes barrier enables the seminiferous tubule to maintain a somewhat different composition of fluid inside its lumen. The fluid in the lumen of the

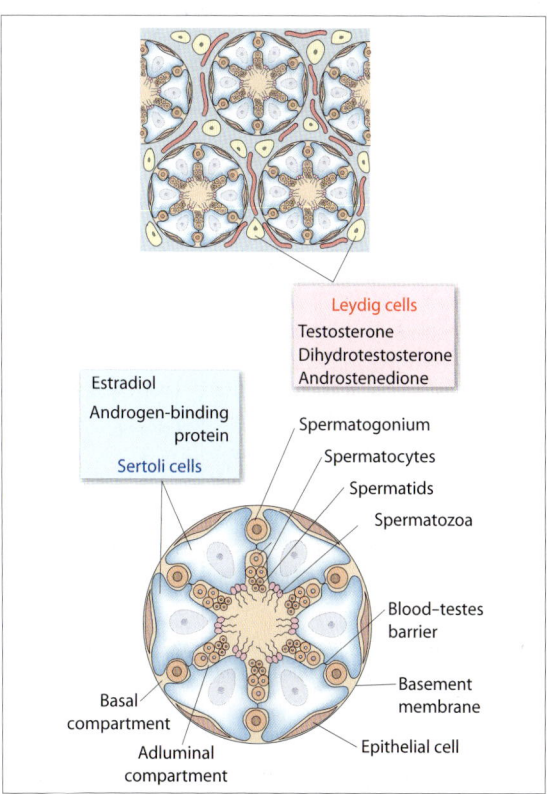

Fig. 89.**1** Leydig and Sertoli cells.

seminiferous tubules contains very little protein and glucose but is rich in androgens, estrogens, and potassium ions.

Sperm (spermatozoon)   Each sperm (Fig. 89.2) has a head and a long tail. The head contains an elongated nucleus with highly compact chromatin. Covering the head like a cap is the acrosome, a lysosome-like organelle rich in enzymes involved in sperm penetration of the ovum. The sperm tail is divided into the middle piece, principal piece, and the end piece. Through its entire length, the tail has a central axoneme made of two central microtubules surrounded by nine microtubule doublets (see Fig. 4.5B). Except in the end piece, the axoneme is surrounded by seven to nine outer dense fibers. The middle piece is surrounded by the spiral mitochondrial sheath made of several mitochondria tightly aligned end-to-end.

Spermatogenesis

The male gametes or the sperms are formed inside the seminiferous tubules. Male gametogenesis or spermatogenesis is the formation of mature sperms from the primitive germ cells or spermatogonia. Spermatogenesis has three phases (Fig. 89.3), viz., the *proliferative phase* in which there is mitotic multiplication of spermatogonia (stem cells) to form primary spermatocytes, the *meiotic phase* which leads to the formation of

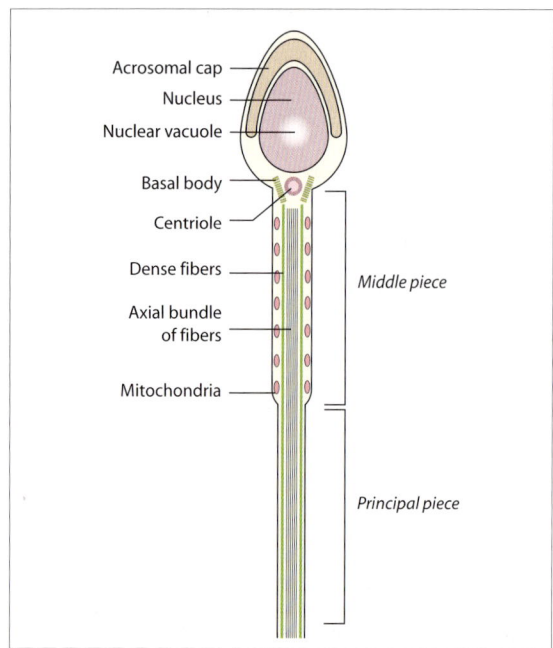

Fig. 89.**2**   Structure of a sperm.

primary spermatids, and the *cytodifferentiative phase* in which the spermatids differentiate into the primary spermatozoa (spermiogenesis).

The spermatogonia divide and multiply in the basal compartment of the tubules. The spermatocytes formed enter the adluminal compartment where the remainder of the maturation occurs. Each spermatogonium divides 7 times to form 128 primary spermatocytes. The 128 primary spermatocytes undergo the first meiotic division to form 256 secondary spermatocytes, which then complete the second meiotic division to form 512 spermatids. Finally, each spermatid develops into a mature sperm in a span of 74 days.

Spermiogenesis is the transformation of the spermatids into mature spermatocytes or sperms. The spermatids mature into spermatozoa inside the membrane recesses of the Sertoli cells. Mature spermatozoa are released from the Sertoli cells and become free in the lumens of the tubules. The transformation of spermatids into spermatozoa is associated with the following changes. (1) The Golgi apparatus, containing hyaluronidase and other proteases, is transformed to the acrosome—a cap-like structure covering the anterior two-thirds of the sperm head. By releasing its enzyme-rich contents during fertilization, the acrosome penetrates the ovum. (2) The centrioles and mitochondria are transformed into the flagellae (sperm tails). The mitochondria provide energy for the flagellar movement. One centriole is converted into the basal body from which the dense fibers originate. (3) The nuclear protein histone is replaced by protamine, which surrounds the inactive and highly condensed chromatin in the spermatozoa. The nuclear condensation in the spermatozoa protects the genome from the deleterious effects of mutagens. (4) Unnecessary cellular organelles like ribosomes, lipids, degenerating mitochondria, and Golgi apparatus are cast off from the spermatozoa as residual bodies. The Sertoli cells absorb these bodies. (5) The spermatozoa do not grow or divide any further.

Factors affecting spermatogenesis   The maturation of spermatids to spermatozoa depends on the action of testosterone on the Sertoli cells in which the developing spermatozoa are embedded. Proper spermatogenesis requires that the local concentration of testosterone inside the seminiferous tubule remains high. The testosterone concentration is increased by the action of FSH on the Leydig cells. FSH also increases the number of testosterone receptors on the Sertoli cells.

The spermatic arteries to the testes run parallel but in the opposite direction to the spermatic veins. This permits a countercurrent exchanger mechanism for conservation of the high local testosterone

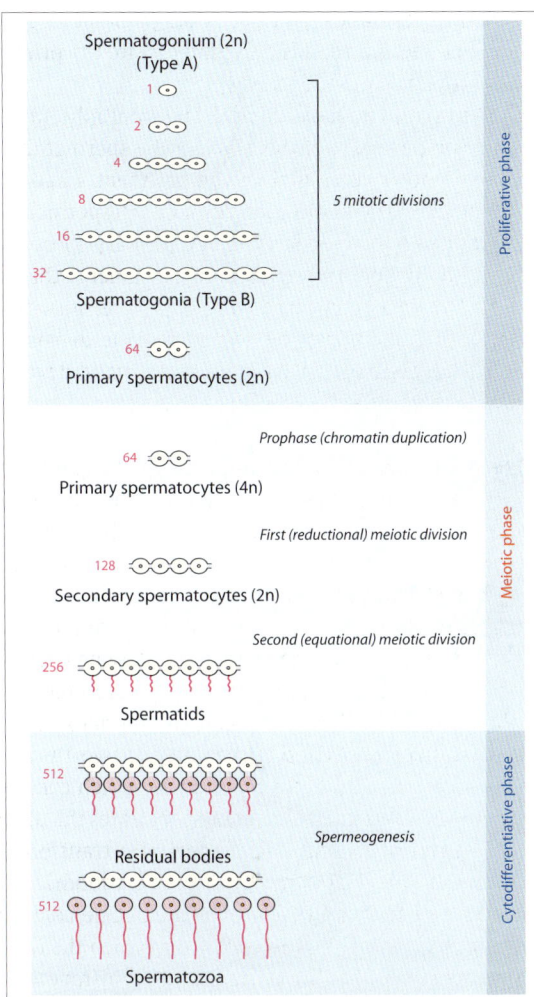

Fig. 89.**3**   Spermatogenesis. The proliferative phase occurs in the basal compartment of the seminiferous tubule. Due to incomplete cytokinesis, all cells derived from a single spermatogonium remain connected through cytoplasmic bridges. The connections persist till the formation of the spermatozoa.

concentration in the testes (Fig. 89.**4**A). It is to be noted that systemic administration of testosterone inhibits FSH secretion and therefore lowers the local testosterone concentration in the testes.

Spermatogenesis requires a temperature of about 32°C, which is lower than the normal body temperature. This is possible because the testes are located in the scrotum, which is outside the body cavity. Failure of testicular descent (cryptorchidism) results in sterility because of thermal damage to the spermatogonia.

The temperature inside the scrotum is kept low by evaporative cooling. Warm arterial blood at 37°C flowing into the testes tends to increase the testicular temperature. However, this tendency is minimized by the countercurrent heat-exchanger mechanism operating

between the spermatic arteries and veins (Fig. 89.**4**B). Varicocele (dilatation of veins of the pampiniform plexus) is associated with incompetence of the venous valves, thus resulting in retrograde blood flow. This interferes with the countercurrent heat exchanger mechanism, causing impaired sperm production.

There is also a seasonal effect on spermatogenesis, with sperm counts increasing in winter regardless of the temperature to which the scrotum is exposed.

Sperm production continues till the age of 80 or 90 although the production rate slows down after the age of 40.

Female Gametogenesis

Oogenesis

The female germ cell is called the oogonium, which develops into the primary oocyte and then into the secondary oocyte. At all stages of development, it is called by the general name ovum or egg (pleural: ova). At birth itself, the oogonium enters the prophase of meiosis I, forming the primary oocyte. However immediately after it completes prophase, the meiosis I is arrested for several years. The remaining phases of meiosis I (metaphase, anaphase, and telophase) do not occur until puberty (Fig. 6.**5**).

When meiosis I resumes at puberty, it is completed quickly and results in the formation of two daughter cells. One of the daughter cells, the secondary oocyte, receives most of the cytoplasm, while the other, the first polar body, fragments and disappears. The

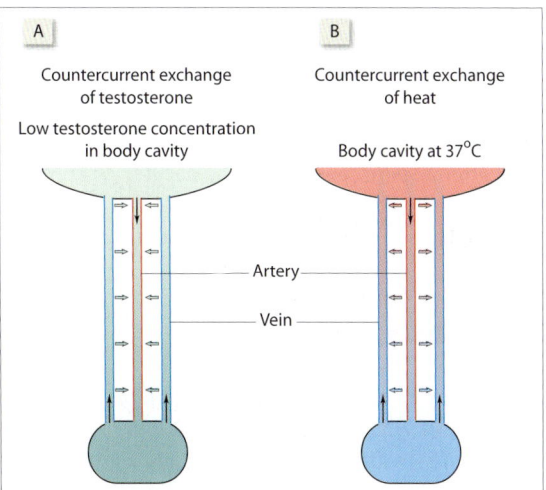

Fig. 89.**4**   Countercurrent exchanger mechanism for conservation of local temperature and testosterone concentration in the testes.

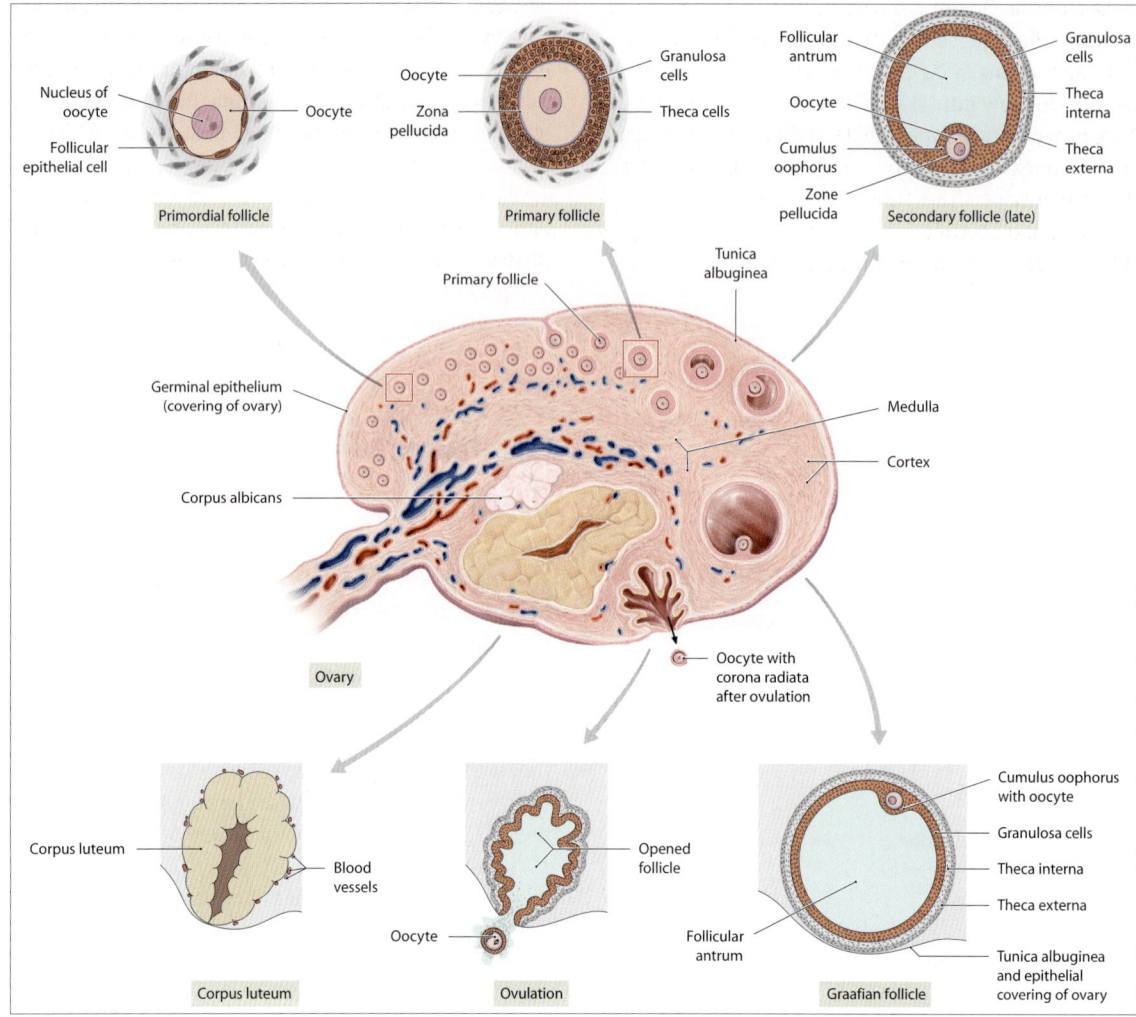

Fig. 89.**5** Ovarian follicles. Note that the oocyte released from the ovary retains its coat of the granulosa cells, called the cumulus oophorus or the corona radiata.

secondary oocyte immediately goes into meiosis II but the division stops again, this time at metaphase. It is at this stage that ovulation occurs. Meiosis II is completed only after a sperm penetrates the oocyte. When meiosis II resumes, the second polar body is cast off and the fertilized ovum proceeds to form the embryo.

■ Folliculogenesis

At the time of birth, there are two million oogonia in the ovary of which fewer than 300,000 ova are left by puberty. Thereafter, once every 28 days, a group of ova start maturing of which only one ovum reaches maturity while the rest degenerate. A maturing primordial follicle grows successively into the primary, secondary, tertiary, and antral follicle (Fig. 89.**5**).

Primordial follicles are composed of an outer single layer of flat epithelial cells and a small immature oocyte arrested in the prophase of meiosis I. The granulosa and the oocyte are enveloped in a thin basal lamina. When the flat epithelial cells become cuboidal, the follicle is called a *primary follicle*.

Secondary follicles form when the cells inside the basal lamina divide mitotically to form a multilayered stratum granulosum. The oocyte enlarges and forms a coating of mucoid substance called the zona pellucida which separates the granulosa cells from the oocyte.

Tertiary follicles are characterized by the formation of a fluid-filled cavity called the antrum. The stromal cells outside the basal lamina differentiate into two concentric layers of thecal cells called the theca interna and theca externa. One of the several tertiary follicles out-grows all the rest to form an antral (Graafian) follicle.

It takes about 14 days for the primary follicle (about 0.4 mm) to grow into a Graafian follicle (about 30 mm). Once every 28 days, one follicle, with the mature ovum in it, is released from the ovary (ovulation) into the abdominal cavity. The other follicles gradually regress through apoptosis and form the corpus albicans. Some of these events are discussed again in Chapter 90.

A primordial follicle can differentiate into the primary, secondary, and early tertiary follicle in the absence of FSH. It is only after reaching the early tertiary stage that its continued development depends on FSH and its own ability to secrete estrogen inside it. FSH receptors are located exclusively on the granulosa cells and LH receptors are located on the thecal cells. FSH regulates the growth and maturation of the follicle. LH regulates the function of the corpus luteum.

The plasma levels of ovarian hormones are not constant but show cyclical fluctuations. These fluctuations occur due to cyclical changes in the ovary (ovarian cycle). The hormones bring about cyclical changes in the uterine endometrium (uterine or the menstrual cycle), the uterine cervix (cervical cycle), and in the vagina (vaginal cycle). Cyclical changes also occur in the breast and body temperature.

Menstrual cycles mark the beginning of estrogen and progesterone secretion in substantial amounts. The onset of menstruation at puberty is called *menarche*. In early adolescence the cycles are anovulatory and the menses are irregular. The cessation of menses is called *menopause* which typically occurs at about the age of 50 years, and is associated with the cessation of estrogen and progesterone secretion.

The menstrual cycle is counted from the first day of menstruation (menstrual bleeding). A typical cycle has a length of 28 days with ovulation occurring at about the middle of the cycle (Fig. 90.**1**). All the cycles, ovarian, uterine, cervical, and vaginal, can be divided into preovulatory and postovulatory periods. The length of the postovulatory period is fairly constant at approximately 14 days. When the length of the menstrual cycle is not exactly 28 days, the time of ovulation is estimated by subtracting 14 days from the duration of the menstrual cycle. Thus in a 30-day cycle, ovulation occurs on the 16th day.

Preovulatory period

Hormonal changes The estrogen and progesterone levels are at their lowest at the beginning of the menstrual cycle (Fig. 90.**2**). Thereafter till ovulation, there is progressive rise in the plasma estrogens due to the secretion of estradiol by the granulosa cells of the growing ovarian follicle. The progesterone level remains very low and nearly unchanged during this period. The preovulatory changes in uterus, cervix, and vagina are largely attributable to the rising estrogen levels. The preovulatory phase of the ovarian cycle is therefore referred to as the *estrogenic phase*.

The early growth of the ovarian follicle in the preovulatory period occurs under the combined influence of FSH and LH on the granulosa and theca cells respectively. However, FSH secretion gradually declines in the preovulatory period due to feedback inhibition by the estrogens and inhibin secreted by the proliferating granulosa cells. The LH level rises slowly since there is not enough progesterone secretion to cause its feedback suppression.

Ovarian changes The preovulatory period of the ovarian cycle is called the *follicular phase*. It is marked by follicular growth and secretion of estradiol by the developing follicles in response to FSH and LH from the pituitary. By the end of this phase, one follicle reaches the final stage of growth (see Fig. 89.**5**). The development of the ovarian follicle is shown in Figure 89.**5**.

Uterine changes During the preovulatory period, the uterus goes through two phases, viz., the menstrual phase and the proliferative phase. Although it occurs in the preovulatory period, the *menstrual phase* is described in this chapter in the context of postovulatory uterine changes, where it will be better understood.

During the *proliferative phase*, estrogens stimulate mitosis of the stratum basale (endometrial proliferation) which regenerates the stratum functionale. The

	1	2	3	4	5	6	7	8	9	10	11	12	13	14	15	16	17	18	19	20	21	22	23	24	25	26	27	28
Ovary								Follicular phase							O	Luteal phase												
Uterus	Menstruation						Proliferative endometrium								Secretory endometrium													
Cervix							Watery mucus secretion								Thick mucus secretion													
Vagina							High glycogen content of epithelium								Low glycogen content of epithelium													

Fig. 90.**1** Phases of the menstrual cycle. Ovulation (O) occurs 14 days before the end of the cycle.

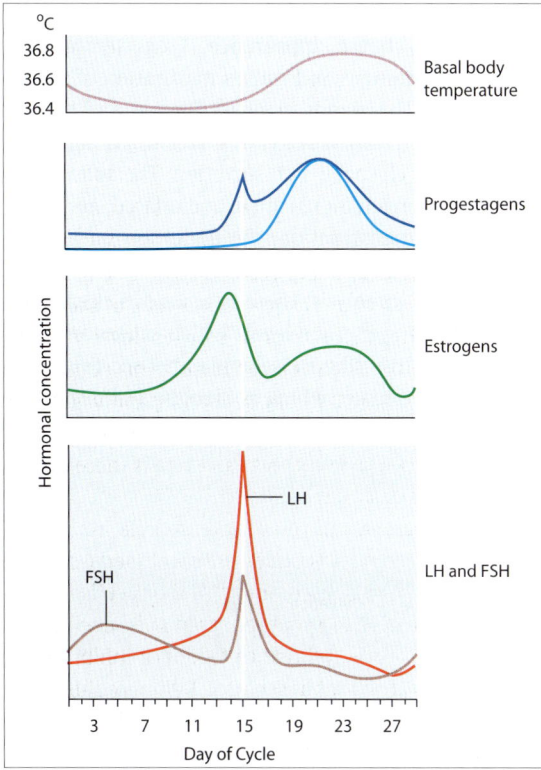

Fig. 90.**2**   Basal body temperature and hormonal concentrations through the menstrual cycle. The two progestagens shown are progesterone (thick line), which comes only from the corpus luteum, and 17α-hydroxy progesterone (thin line), which comes from the thecal cells as well. Hence, progesterone concentration in the preovulatory phase is nearly zero, while 17α-hydroxy progesterone is secreted in small quantities in the preovulatory phase and shows a surge during ovulation.

endometrium grows to about 4 mm in thickness. The blood vessels in the stratum functionale become long and coiled and are called spiral arteries. The uterine glands increase in length. The cells lining the glands accumulate glycogen but remain nonsecretory in this phase.

Cervical changes   During the preovulatory phase, estrogen makes the cervical mucus watery and more alkaline. These changes promote the survival and transport of sperms. The mucus is thinnest at the time of ovulation and its elasticity, or *spinnbarkeit*, increases so much that by midcycle, a drop can be stretched into an 8–12 cm-long thin thread. In addition, it dries in an arborizing fern-like pattern when a thin layer is spread on a slide (Fig. 90.**3**). The characteristic ferning pattern is due to crystallization of sodium chloride.

Vaginal changes   The vaginal epithelium cells mature under the effect of estrogen, resulting in the thickening and cornification of the vaginal lining. By

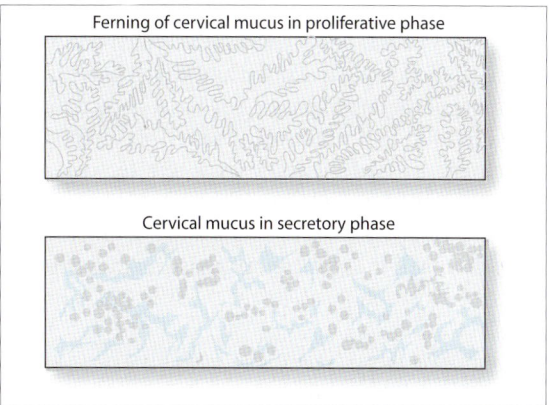

Fig. 90.**3**   Ferning of cervical mucus. The persistence of a fern-like pattern in the secretory phase indicates anovulation.

evaluating the ratio of mature to immature cells microscopically in vaginal smears, this maturation can be indicated by what is known as the karyopyknotic index (KPI). KPI peak coincides with the estrogen peak near ovulation. Estrogen also increases the glycogen content of the epithelial cells, which are exfoliated throughout the menstrual cycle. The glycogen in the exfoliated cells is a rich substrate for bacterial flora which breaks down glycogen into lactic acid and lowers the vaginal pH.

Breast changes   During menstruation, the decrease in estradiol and progesterone is associated with a decrease in the size of the mammary duct and acini. In the preovulatory phase, there is proliferation of mammary ducts under the influence of estrogens. A gradual increase of epithelial tissue occurs with each successive cycle.

■   **Ovulation**

Ovulation refers to the extrusion of the secondary oocyte from the Graafian follicle into the peritoneal cavity. Ovulation is triggered by a sharp rise in the LH level, called the LH surge, which occurs 24 hours before ovulation. The LH surge occurs because when the plasma estrogen exceeds a critical concentration, it exerts a positive feedback (instead of the usual negative feedback) on the hypothalamic-hypophyseal axis, causing a reflex release of GnRH and a concomitant surge in pituitary LH secretion 24 hours later. A lesser FSH surge also occurs simultaneously. The midcycle LH surge requires a plasma estradiol concentration of about 150 pg ml⁻¹ for at least 36 hours, a condition which is normally achieved by day 13. The onset of the LH surge is a fairly precise indicator of ovulation.

Tests for ovulation Progesterone is associated with a 0.5°C rise in *basal body temperature* which occurs immediately following ovulation and which persists during most of the luteal phase. The basal body temperature dips during the follicular phase. This temperature increment is used clinically as a test for ovulation. The other test for ovulation is the *thinning of cervical mucus*: The persistence of spinnbarkeit and ferning of cervical mucus in the secretory phase indicates the absence of ovulation.

Postovulatory period

Hormonal changes The LH surge not only triggers ovulation but also initiates the process of luteinization of the thecal and granulosa cells. FSH and LH stimulate the corpus luteum to secrete progesterone and estradiol. The plasma concentration of both progesterone and estradiol peak in the midluteal phase (day 21). Since progesterone is secreted in greater amounts (Fig. 90.**2**), the postovulatory phase is also called the *progestational phase*. FSH and LH levels continue to decline during the luteal phase due to negative feedback of estrogens and progesterone.

Ovarian changes The ruptured follicle promptly fills up with blood forming the corpus hemorrhagicum. Minor bleeding from the follicle into the abdominal cavity may cause peritoneal irritation and fleeting lower abdominal pain called *mittelschmerz*. This midcycle pain may be unilateral, indicating which of the two ovaries has ovulated. Thereafter, the clotted blood is replaced by yellowish, lipid-rich luteal cells (derived from the granulosa and thecal cells) forming the corpus luteum. The postovulatory phase of the ovarian cycle is therefore called the *luteal phase*.

The oocyte that is released from the ovary is picked up by the fimbriated ends of the uterine tube and is transported to the uterus. Fertilization occurs in the ampulla of the tube. *If fertilization does not occur*, the corpus luteum begins to degenerate by day 24 (luteolysis) and is eventually replaced by fibrous tissue, forming the corpus albicans. Luteolysis seems to be mediated by PGF$_2$. *If fertilization occurs*, the functional lifespan of the corpus luteum is extended under the trophic influence of HCG from the placenta and it continues to secrete estradiol and progesterone till the third month of pregnancy when the placenta takes over its endocrine function (the luteal-placental shift, see Fig. 93.**3**).

Uterine changes The endometrium enters its secretory phase and is prepared for a possible implantation of the fertilized ovum. The endometrium thickens to about 5 mm, becomes hyperemic and develops a scalloped Swiss cheese appearance. Progesterone halts endometrial mitosis and causes maturation of the endometrium. The glands elongate further and become tortuous. Their lumen becomes wider and is filled with mucus and a glycogen-rich secretion. The spiral arteries become longer, more coiled, and dilated. About two days before menstruation, the endometrium is infiltrated with neutrophils (Fig. 90.**4**).

If there is luteolysis, there is a sudden drop in estrogen and progesterone levels which causes arteriolar constriction, resulting in ischemia and necrosis of the stratum functionale, which is sloughed off. The stratum basale remains intact and the total endometrial

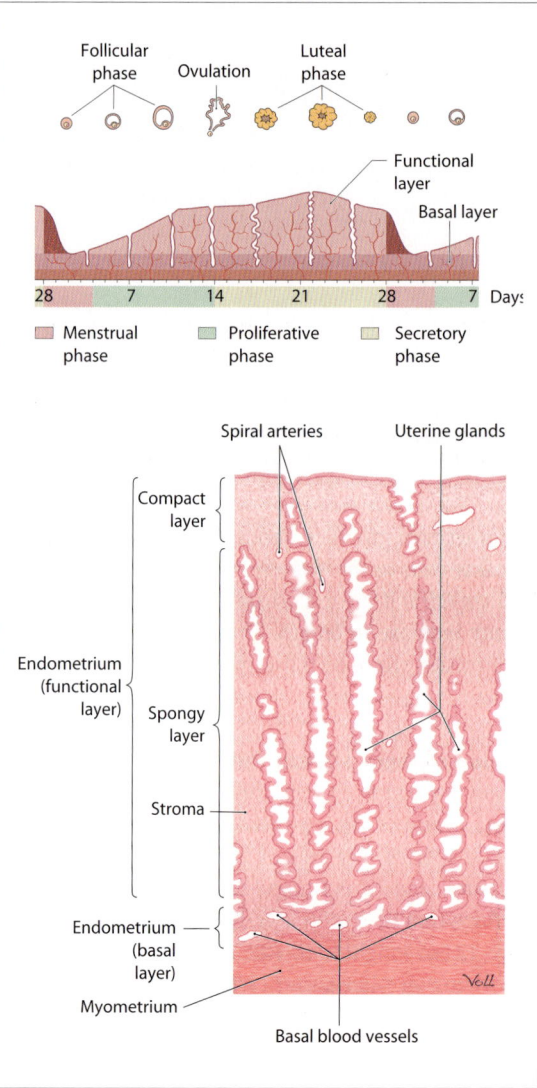

Fig. 90.**4** (*Above*) Ovarian and endometrial changes during menstrual cycle. (*Below*) Structure of the endometrium at the peak of the secretory phase.

thickness reduces to about 2 mm. The vasospasm seems to be mediated by PGF_2. Bleeding starts when the arterioles relax. Since menstrual bleeding follows a sharp fall in the plasma concentration of ovarian hormones, it is also called *withdrawal bleeding*. Menstrual blood clots in the uterus but liquefies again in the vagina due to fibrinolysis. The average blood loss during menstruation is 40 ml (10–200 ml) over a period of 4 days.

Cervical changes Progesterone makes the cervical mucus thick, tenacious, and cellular. It fails to form the fern pattern on drying and the spinnbarkeit is no longer possible.

Vaginal changes The glycogen content of the epithelial cells decreases during this phase. Consequently, less lactic acid is produced in the vagina by the bacterial flora. The rise in pH makes the vagina prone to infection by *Trichomonas vaginalis*. The epithelium becomes infiltrated with leukocytes.

Breast changes In the luteal phase of the cycle, progesterone stimulates the proliferation of the terminal duct and alveoli, distension of the ducts, and hyperemia and edema of the interstitial tissue of the breast. All these changes, which cause a sense of breast fullness, regress during menstruation.

Sperm Transport in the Male Tract

Sperms mature as they pass through the male ducts (Figs. 91.**1** and 91.**2**). Transport of sperms through the male ducts does not require much sperm motility. Spermatozoa show significant motility only after they are ejaculated.

Sperm passage through the rete testis

Sperms are concentrated as they pass through the rete testis. If the concentration does not occur, the sperms entering the epididymis become diluted in a large volume of fluid and infertility results. The concentration occurs due to the reabsorption of Na^+ and water under the influence of estrogen. Spermatozoa leaving the rete testis are not fully motile and are therefore incapable of fertilizing the ovum.

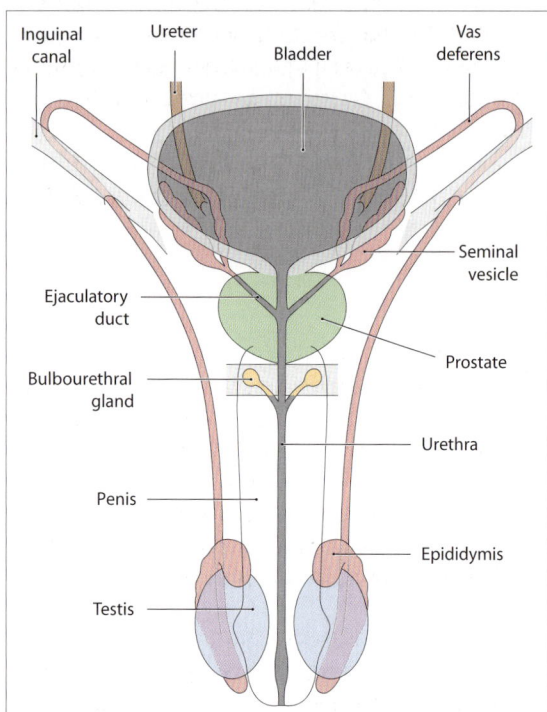

Fig. 91.**1** Male reproductive tract. The epididymal duct continues into the vas deferens, which has thicker walls. Near its termination, the vas deferens joins the duct of the seminal vesicles to form the ejaculatory duct, which opens in the urethra.

Sperm passage through epididymis

The sperms continue to mature and acquire motility during their passage through the epididymis. Sperms take 2 to 11 days to pass through the epididymis. During this transit, the epididymis serves four major functions related to sperm: its maturation, storage, decapacitation, and protection from immunological damage.

Sperm maturation occurs in the proximal epididymis (caput and corpus epididymis) whereby sperms gain their ability to fertilize eggs. It involves (1) development of sperm motility with a higher velocity of forward progression and a straighter path, (2) development of zona-binding capacity due to development of ZP3 receptors on the sperm, (3) development of the ability to undergo acrosome reaction, (4) development of the potential to fuse with the egg, and (5) condensation of the sperm chromatin which is later decondensed by the egg ooplasm.

Sperm storage The storage capacity of the human epididymis is small and is exceeded after two weeks of abstinence whereafter sperms start overflowing into urine. Since these aged sperms still retain their viability, a couple of weeks of abstinence is recommended for increasing the sperm output of oligozoospermic patients. For the same reason, sperms continue to be released in semen for a few weeks after vasectomy, often causing unwanted pregnancy unless additional contraceptive measures are taken.

During their storage in the epididymis, sperms come in contact with antioxidants like superoxide dismutase and glutathione peroxidase which protect sperms from lipid peroxidation, and with acrosin inhibitor which protect sperms from damage by leaking acrosomal enzymes.

Decapacitation of spermatozoa Spermatozoa show motility only after they are ejaculated. The motility of the sperm, which is acquired in the head and body of the epididymis, is suppressed (decapacitated) again in the tail of the epididymis. The decapacitation keeps the sperms quiescent till the right moment and place of fertilization. Sperm motility is inhibited when the intracellular pH is acidic and stimulated when its

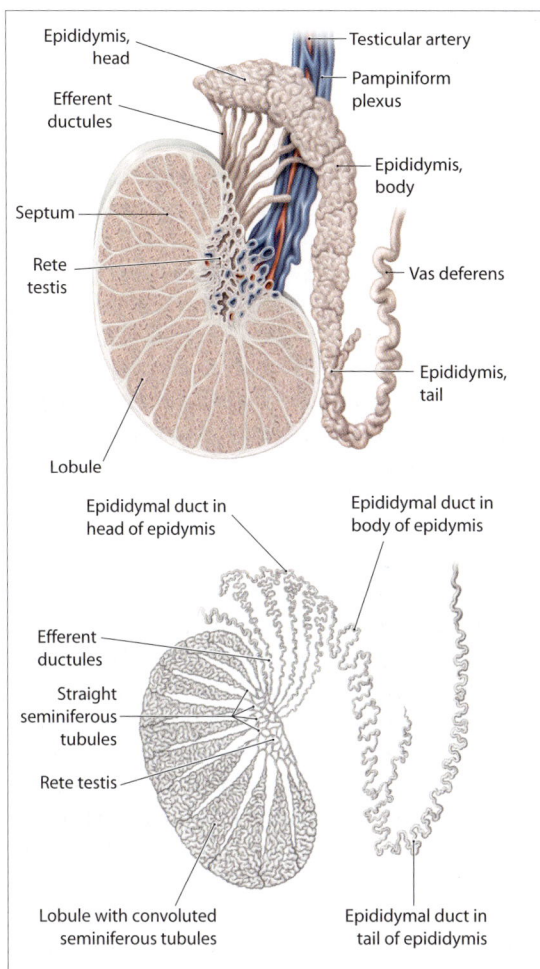

Fig. 91.2 Seminiferous tubules, rete testis and epididymis. The seminiferous tubules in the testes drain into a plexus of epithelium-lined spaces called the rete testis. About a dozen efferent ducts from the rete testes drain into the epididymis. The epididymis is divided into the caput (head), the corpus (body) and the cauda (tail). It is made of over 6 meters of coiled duct (the epididymal duct).

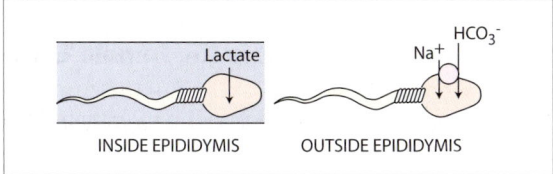

Fig. 91.3 Changes in pH as the sperm moves out of the epididymis.

intracellular pH is alkaline. Epididymal fluid contains lactates which diffuse into the sperm, lowering the pH. Once outside the epididymis, the sperm finds itself in a Na⁺-rich environment similar to plasma. The Na⁺ is transported into the cell with HCO_3^-, raising the intracellular pH and stimulating spermatic motility (Fig. 91.3). This HCO_3^- cotransport with Na⁺ cannot occur inside the epididymis because the epididymal fluid has a low Na⁺ and high K⁺ concentration as compared to plasma.

Immune protection of spermatozoa   Immune tolerance to self develops in fetal life. As spermatozoa are formed several years after the development of immune tolerance, they are considered 'foreign' if they are encountered by the immune system. To prevent immunological damage, sperms are separated within the epididymis from circulating immune cells. There are tight junctions in the epididymis that prevent paracellular transport. If the epididymal tubule is ruptured, as may occur after vasectomy, sperm antigens can encounter immune cells. Indeed, antisperm antibodies are present in the serum of men with epididymal occlusion.

Sperm passage through urethra   During their passage through the urethra, the sperms come in contact with the secretions of seminal vesicles, prostate, and bulbourethral glands. The seminal plasma comes mostly from these accessory glands. During ejaculation, the accessory glands sequentially contribute their secretions to the seminal plasma. First, the bulbourethral glands secrete an alkaline solution with glycoproteins (the preejaculatory fluid) to neutralize the urinary tract acidity and lubricate the tract before ejaculation. Next, the epididymis and prostate contract together, discharging spermatozoa and prostatic secretions. Finally the seminal vesicles contract and expel the spermatozoa to the exterior with their secretions.

■   Semen

Seminal fluid analysis is performed on samples obtained by masturbation after 24 to 36 hours of abstinence. Analysis should be performed within an hour. Normal semen is white or opalescent in color with a specific gravity of 1.028 and a pH of 7.35–7.50. The normal ejaculate volume is greater than 2 ml. Immediately after ejaculation, coagulation of the seminal fluid occurs, followed within 30 minutes by liquefaction.

Sperms   The normal sperm count is 40–100 million ml⁻¹ with fewer than 20% having abnormal morphology. For normal fertility, the sperm count should be at least 20 million ml⁻¹ and at least 60% of the sperm should be motile with normal morphology. However many fertile men have lower counts and many infertile men have higher counts.

Seminal plasma The physiological role of several constituents of seminal plasma is not known, but these constituents are estimated to assess the functioning of the accessory glands.

The *epididymis* contributes carnitine, glycerophosphocholine, and neutral α-glucosidase to the seminal fluid.

The *prostate* contributes about 20% to the volume of the semen. The prostatic fluid is slightly acidic (pH 6.5) due to the presence of citric acid. It contains substances important for sperm motility (notably albumin), fibrinolysin, acid phosphatase, and an antibacterial substance. It also contributes zinc and prostate-specific antigen to the seminal plasma.

The *seminal vesicle* contributes 70% to the volume of the semen. It contains fructose, citrate, ascorbic acid, prostaglandins, and various enzymes. The fructose is a source of energy for the spermatozoa. It is broken down into lactate through anaerobic glycolysis. The function of prostaglandins is not understood. Seminal fluid contains hyaluronidase, which breaks down hyaluronic acid found in cervical mucus and thereby allows the sperms to pass easily through the cervix.

Sperm Transport Across the Sexes

Coitus allows the transfer of sperms from the male to the female reproductive tract. Successful coitus is the culmination of sexual arousal, which in both sexes has four stages, viz., excitement, plateau, orgasm, and resolution.

During **sexual excitement**, parasympathetic activity is increased, resulting in vasocongestion in the genitalia. *In the male*, there is dilatation of penile arteries and accumulation of blood in the corpora cavernosum and spongiosum, resulting in erection (penile tumescence). VIP and NO are important for penile erection. Increased blood flow to the testes causes an increase in testicular size. Contraction of the longitudinal muscles of the vas deferens lifts the testes closer to the body.

In the female, vasocongestion of the vagina causes transudation of fluids from the vaginal epithelium, resulting in vaginal lubrication, which facilitates insertion of the penis without discomfort. Vasocongestion also occurs in the external genitalia and breast. There is a general increase in muscle tone which results in nipple erection, sometimes in men too.

The **plateau phase** is characterized by the intensification of the changes seen during the phase of excitement. *In the male*, blood pooling causes the glans to enlarge and become darker in color. Preejaculatory fluid

from the bulbourethral glands trickles out of the penis. The fluid contains a few sperms and accounts for the poor safety rate of the withdrawal method of contraception.

In the female, increased blood flow to the labia produces a deepening in the color of the labia minora and labia majora. The clitoris retracts. Swelling of the vaginal walls narrows the vaginal orifice. The uterus is elevated away from the vagina (tenting). There is increased size of the breast areola and reddening of the skin (sexual flush).

Orgasm is the climaxing of sexual excitement during intercourse. *In the male*, it culminates in ejaculation, i.e., a forceful ejection of sperm from the urethra. Only sperms from the distal cauda epididymis enter the ejaculate. Ejaculation involves two processes, viz., emission and expulsion of ejaculatory fluid. *Emission* is the entry of the ejaculatory fluid into the urethra. It is produced by rhythmic muscular contractions of the vas deferens, seminal vesicles, and prostate. *Expulsion* is the ejection of ejaculatory fluid out of the urethra. It is brought about by rhythmic muscular contractions of the urethra, aided by the contraction of bulbocavernous muscles. The sphincter vesicae contracts during ejaculation, preventing the ejaculatory fluid from entering the bladder.

In the female, orgasm is characterized by rhythmic muscular contractions of the uterus and vagina. Unlike males, females are multiorgasmic, i.e., they are capable of experiencing more than one orgasm.

In the **resolution phase**, all the changes associated with sexual arousal revert to the prearousal state. *In the male*, penile detumescence is caused by increased sympathetic nerve activity that constricts the penile arteries. Ejaculation is followed by a refractory period during which orgasm is not possible again. Females generally do not experience the refractory period.

Sperm Transport in the Female Tract

Sperm motility is essential for sperm migration out of cervical mucus. It helps in sperm transport to the site of fertilization, and is essential for penetrating the egg's investments. During intercourse, spermatozoa are deposited high up in the vagina. Soon after its deposition, the semen coagulates due to the presence of thrombin-like enzymes in the prostatic fluid and fibrinogen-like substrates (seminogelin) in the seminal vesicle fluid. The coagulum helps to retain the sperms in the vagina and protect them against the acidic vaginal pH. Within an hour of deposition, the coagulum is dissolved by the

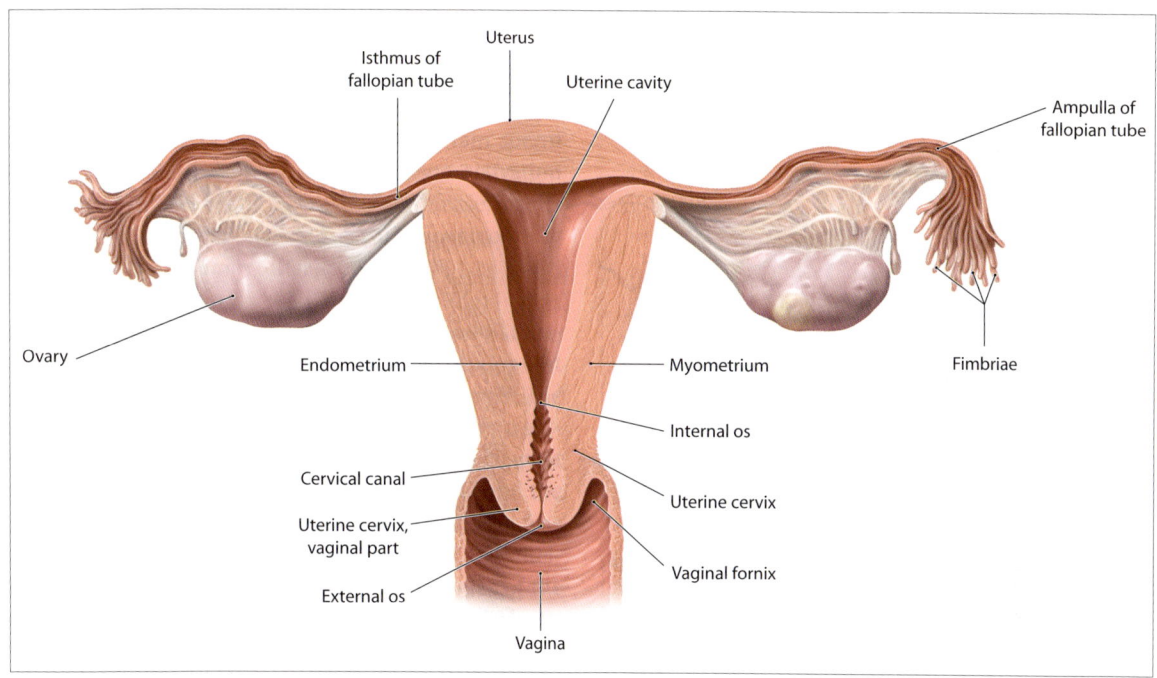

Fig. 91.4 Female reproductive tract. The uterus consists of the corpus (which includes the fundus) and the cervix, separated by a narrow isthmus. The cervical cavity communicates with the uterine cavity through the internal os, and opens into the vagina through the external os. The upper part of the vaginal cavity extends around the cervix as the vaginal fornices. The Fallopian tube consists of the infundibulum, ampulla, isthmus, and uterine part. The opening of the infundibulum into the peritoneal cavity is surrounded by fimbriae.

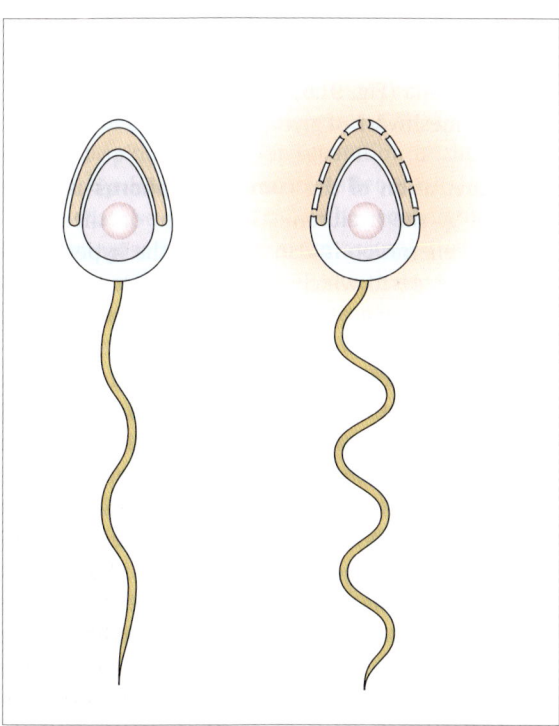

Fig. 91.5 Capacitation of spermatozoon. (*Left*) Sperm before capacitation. (*Right*) Capacitated sperm showing hyperflagellation and acrosomal reaction.

fibrinolysin present in the prostatic secretions and the sperms migrate out of the coagulum.

During the secretory phase, the cervical mucus is watery and favors the passage of sperms through it. The same mucus however prevents the passage of sperms with abnormal head or subnormal flagellation and antibody-coated spermatozoa. This observation is put to diagnostic use in the *in vitro* cervical mucus contact test.

Sperms reach the uterine tubes within an hour of intercourse. Ejaculated spermatozoa have to reach the ampulla of the tube to fertilize the ovum (Fig. 91.4). The movement of spermatozoa is further assisted by contractions of the female tract, which is stimulated by prostaglandins present in the semen. The tubal fluid arrests migration of spermatozoa until ovulation. After ovulation, the progesterone present in follicular fluid, which is released with ovulation, stimulates sperm motility.

Capacitation   Ejaculated sperms are ordinarily unable to fertilize the ovum in vitro. Sperms undergo capacitation, i.e., gain the ability to fertilize eggs, during their passage through the female tract. For capacitation to occur, the decapacitation factors acquired in the epididymis and seminal vesicle have to be removed. The

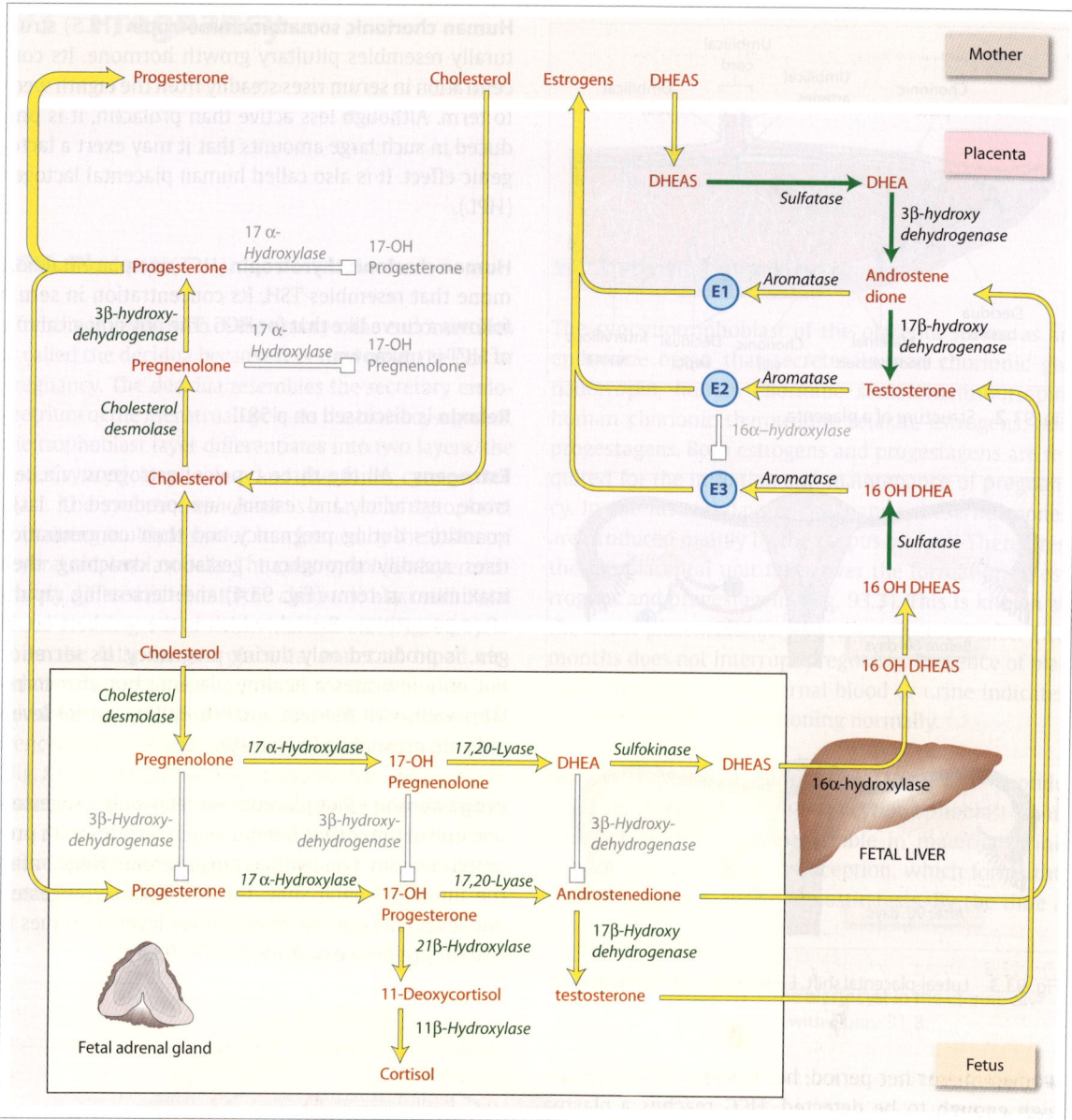

Fig. 93.**5**   Hormone synthesis in the fetoplacental unit.

Fetoplacental unit

The fetus, placenta, and mother are interdependent and constitute a functional unit called the fetoplacento-maternal unit or simply, the fetoplacental unit. This interdependence is especially apparent in the placental synthesis of progesterone and estrogens and the fetal synthesis of cortisol, as explained below (Fig. 93.**5**).

Placental synthesis of progesterone   The placenta synthesizes progesterone from cholesterol. Since it cannot synthesize its own cholesterol from acetate, it obtains the cholesterol from the maternal and fetal circulations.

Ninety percent of the progesterone synthesized in the placenta diffuses back into the maternal circulation, resulting in the characteristic maternal changes in pregnancy. Ten percent of the progesterone synthesized in the placenta enters the fetal circulation where it is converted to cortisol. Up to the tenth week of gestation, the fetus is dependent on placental progesterone for synthesizing corticosteroids, because its own adrenal cortex is deficient in 3β-hydroxy dehydrogenase. After 10 weeks, the fetus is able to synthesize its own progesterone. It is also able to synthesize androstenedione and testosterone. These androgens diffuse into the placenta where they are converted into estrogens.

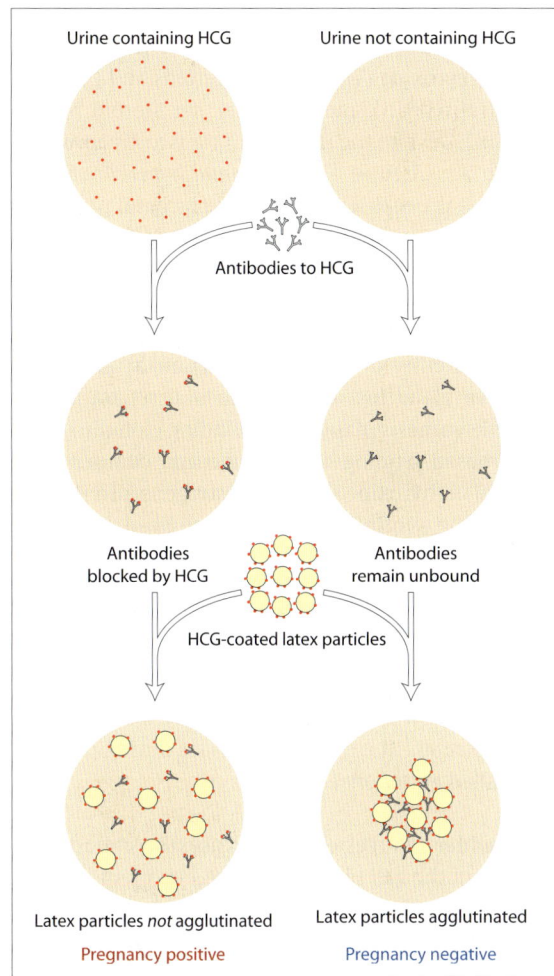

Fig. 93.**6**   The immunological basis of the Gravindex test.

Placental synthesis of estrone and estradiol   The placenta cannot synthesize estrone and estradiol from cholesterol because it lacks 17α-hydroxylase and 17,20-lyase. The placenta synthesizes estrone and estradiol from DHEAS (dehydroepiandrosterone sulfate) which it obtains from the maternal circulation. As with progestagens, 90% of the placental estrogens enter the maternal circulation and the remainder enters the fetus.

Placental synthesis of estriol   The placenta cannot synthesize estriol from estrone because it lacks 16α-hydroxylase. Hence, it obtains 16-hydroxy DHEA sulfate (16-OH DHEAS) from fetal circulation, removes the sulfate and aromatizes it to estriol. Since the placental synthesis of estriol is entirely dependent on the 16-OH DHEAS produced by the fetal adrenal gland, the urinary excretion of estriol in the mother is an index of fetal health.

Pregnancy diagnostic tests

All pregnancy diagnostic tests are based on the presence of HCG in urine, which can be detected as early as 14 days after conception, i.e., within a day or two of a missed period. The accuracy of these tests is about 99%.

In **biological tests**, laboratory animals (like frog, mouse, and rabbit) are injected with the patient's urine and observed for physiological responses like ovulation or release of sperms. Although very sensitive, they have been abandoned as they are cumbersome.

Immunological tests are quick (give results in minutes), cheap, and accurate enough to replace biological methods. The most commonly used immunological test is the Gravindex test, the kit for which consists of Gravindex antigen (latex particles coated with HCG), Gravindex antibody (serum containing HCG antibodies), dark slide, sticks, and a pipette. If serum containing antibodies to HCG is allowed to react with HCG-coated red cells or latex particles, agglutination of the red cells or latex particles occurs. But if the serum is allowed to react with urine containing HCG prior to the reaction with red cells or latex particles, the agglutination of red cells or latex particles is inhibited, since the antibodies are blocked by HCG from the urine (Fig. 93.**6**). Hence, inhibition of agglutination confirms pregnancy.

Maternal Changes in Pregnancy

Most of the physiological changes associated with pregnancy are directly or indirectly related to the high concentrations of progesterone and estrogens in maternal blood, the dilatation of uterine blood vessels, and the elevation of the diaphragm. These are discussed below.

Effects of progesterone and estrogens

Breathing   Progesterone directly stimulates the respiratory center, resulting in hyperventilation. The hyperventilation is more than that required to wash out the extra CO_2 produced by the fetus and therefore causes hypocapnia and respiratory alkalosis with bicarbonaturia.

Circulation   Estrogens increase uterine blood flow, helping to provide nourishment to the fetus through the placenta. Progesterone causes fluid and electrolyte retention, resulting in hypervolemia, increase in renal

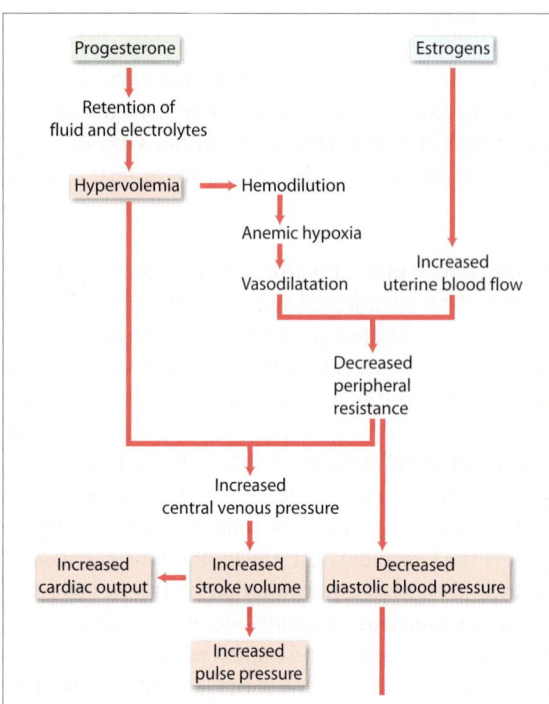

Fig. 93.**7**   Cardiovascular changes in pregnancy.

blood flow and increase in GFR. Water retention exceeds sodium retention, resulting in hypoosmolarity of the ECF. Water retention also results in hemodilutional anemia which is called the *physiological anemia of pregnancy*. The resulting anemic hypoxia causes vasodilatation, reducing the peripheral resistance and increasing the CVP. The CVP also increases as a direct consequence of the increase in blood volume. The increase in CVP increases the cardiac output. Despite the increase in cardiac output, the mean blood pressure and even the systolic blood pressure remain normal because of the large fall in peripheral resistance. The peripheral resistance also decreases as a direct consequence of the increased uterine blood flow. These events are summarized in Figure 93.**7**.

Plasma proteins   Estrogens stimulate liver enzymes to increase the synthesis of transport globulins that bind to thyroxine (TBG), corticosteroids and progestagens (transcortin), and estrogens (sex steroid binding globulin). Despite the increase in transcortin, pregnant women are often in a state of *mild hyperadrenocorticism*. This is because the elevated placental progesterone competes with cortisol for binding sites on transcortin, thus increasing plasma free cortisol. Estrogens also increase fibrinogen synthesis, resulting in *hypercoagulability of blood* and an increase of ESR. The hypercoagulability of blood is important in preventing excessive blood loss during placental separation. Increased synthesis of thyroxine-binding globulin results in increased binding of thyroxine and compensatory increase in thyroxine secretion. Estrogens also increase the hepatic synthesis of angiotensinogen leading to increased angiotensin II synthesis and aldosterone secretion.

Glucose tolerance   Glucose tolerance is reduced in pregnant women due to the antiinsulin effects of progesterone, estrogen, and human chorionic somatomammotropin. The transient diabetes during pregnancy is called *gestational diabetes*.

Gall bladder   Progesterone decreases the smooth muscle tone of the gall bladder and thereby predisposes to gall bladder stasis. Hence, gallstones are commoner in females who have been pregnant several times.

■ Effects of diaphragmatic elevation

As the gravid uterus enlarges, the diaphragm is elevated. Elevation of the diaphragm affects the lung volumes and capacities of pregnant woman (Fig. 50.**5**). It also rotates the apex of the heart upward and to the left, resulting in an outward shift of the apex beat, a left axis deviation in the ECG and an increase in the cardiac shadow in the radiograph.

Hormones of Parturition and Lactation

Oxytocin

Oxytocin is a nonapeptide that is synthesized in the paraventricular and supraoptic nuclei of the hypothalamus and is stored in the posterior lobe of the pituitary gland. It causes (1) galactokinesis (see below) by stimulating the myoepithelial cells of the mammary gland, and (2) stimulates uterine contractions. It plays a role in labor and is a useful therapeutic agent in the induction of labor. Along with estrogen, oxytocin is used therapeutically in arresting uterine bleeding. Oxytocin injections are given to dairy animals before milking them.

Oxytocin secretion is stimulated by (1) suckling and (2) genital stimulation, as occurs during coitus and parturition. Oxytocin is released in men too during genital stimulation but its role is not clear. Oxytocin secretion is inhibited by sympathetic discharge and circulating catecholamines, by pain and enkephalins, by emotional stress, especially fear, and by alcohol.

Relaxin

Relaxin is a polypeptide hormone secreted by the corpus luteum and placenta in women and by the prostate in men. During pregnancy, it relaxes the pubic symphysis and other pelvic joints and softens the cervix, facilitating parturition. In men, it probably has a role in sperm motility and sperm penetration of the ovum.

Prolactin

Prolactin is a polypeptide hormone secreted by the pituitary acidophils. Its physiological actions are (1) mammogenesis, lactogenesis, and galactopoiesis, and (2) suppression of GnRH secretion which results in lactation amenorrhea.

Normally, the control of prolactin secretion is under constant inhibition by the prolactin-inhibiting factor (PIF) secreted by the arcuate (infundibular) nucleus of the hypothalamus (see Fig. 98.**2**). PIF is now known to be dopamine which acts through D_2 receptors. Hence, prolactin secretion is disinhibited by dopamine antagonists like bromocriptine. Prolactin secretion is stimulated by serotonin which probably mediates the increase in prolactin secretion during sleep, and by VIP which mediates the suckling reflex. Prolactin secretion is also stimulated by estrogens and thyrotropin-releasing hormone (TRH).

Plasma prolactin levels start increasing by the eighth week of pregnancy and reach their peak two weeks before term (see Fig. 93.**4**). The increase is brought about by estrogens, which directly stimulate the pituitary lactotrophs to synthesize and secrete more prolactin. After parturition, prolactin secretion falls but the fall is gradual since prolactin secretion is stimulated during each session of suckling. Prolactin secretion is also stimulated by other factors of unknown physiological significance, e.g., it increases during sleep, exercise, and stress.

In **hyperprolactinemia** galactorrhea occurs in only about 30% of cases. In women, it causes infertility and amenorrhea, while in men, it causes impotence and decreased libido. Treatment for prolactin hypersecretion includes administration of a dopamine agonist like bromocriptine.

Parturition

Phases of parturition

Parturition is the process of giving birth, which involves the preparation for childbirth, the process of childbirth and the recovery from childbirth. During parturition, the myometrium and the cervix undergo a series of events that are divided into four phases (Fig. 94.**1**).

Uterine phase 0 of parturition is the prelude to initiation of parturition and extends from before implantation until late gestation. During this phase, the myometrium remains relaxed and the cervix remains rigid and unyielding under the effect of progesterone. The conduction of action potentials by myometrial cells is slow and myometrial contractility is low. There is increased degradation of endogenously produced *uterotonins* like prostaglandins, oxytocin, and histamine by

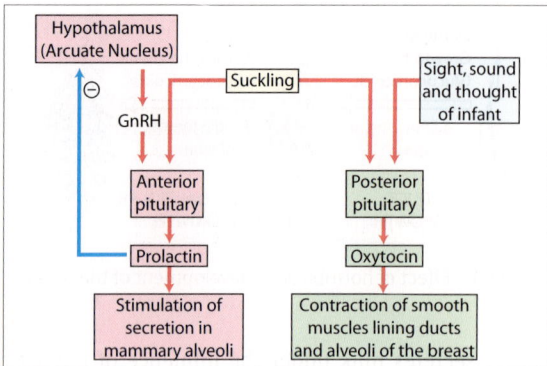

Fig. 94.**4**   The control of milk ejection.

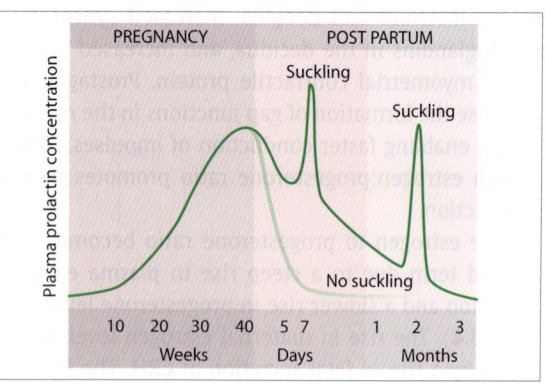

Fig. 94.**5**   Prolactin levels during pregnancy and suckling.

offer themselves for hire for suckling another's baby. The withdrawal of prolactin is associated with involution of the breast. The alveolar epithelium undergoes apoptosis and remodeling, and the gland reverts to its prepregnant state.

The high level of prolactin inhibits GnRH secretion, resulting in low levels of FSH and LH. Hence suckling, which stimulates prolactin secretion, is associated with anovulation and amenorrhea. Such amenorrhea is commonly called *lactation amenorrhea*.

Section 11

Central Nervous System

95.   Anatomy of the Central Nervous System

96.   Spinal Cord and Brainstem

97.   Cerebellum

98.   Diencephalon

99.   Basal Ganglia

100.   Cerebral Cortex

101.   Autonomic Nervous System

102.   Synaptic Mechanisms and Neurotransmitters

103.   Sensory Mechanisms

104.   Regulation of Muscle Length and Tone

105.   Motor Planning, Programming, and Execution

106.   Thought, Emotion, and Conation

107.   Electroencephalogram and Epilepsies

108.   Sleep

109.   Regulation of Body Temperature

110.   Regulation of Food Intake

111.   Memory and Learning

112.   Language and Speech

The central nervous system consists of the brain (Fig. 95.**1**) and the spinal cord (Fig. 95.**10**). The further subdivisions of the brain are shown in Figure 95.**2**. The medulla oblongata, pons, and the midbrain together constitute the brainstem.

The functional anatomy of the different parts of the brain is discussed in the next few chapters. However the study of the parts does not necessarily lead to a clear understanding of the anatomy of the brain as a whole or the relationship between its different parts. An attempt has been made in this chapter to give a holistic, three-dimensional concept of the brain through schematic diagrams. Such a concept helps in visualizing the neural connections between different parts of the brain and brainstem, many of which are of considerable physiological importance. For the most part, the brain has been schematically 'dismantled' in steps so that we are able to see all the structures inside the cerebrum. The brief account given below should be combined with a more comprehensive text in a standard anatomy textbook.

Brain

Cerebral hemispheres

The cerebrum is made of two cerebral hemispheres. Each hemisphere has a superolateral surface, a medial surface, and an inferior surface. The surface of the cerebrum is thrown into folds. The surface elevations created by these folds are called gyri and the grooves separating them are called sulci. Each cerebral hemisphere is also divided into the frontal, parietal, occipital, and temporal lobes. The *frontal lobe* is the part of the hemisphere anterior to the central sulcus and above the lateral sulcus. The *temporal lobe* is inferior to the lateral sulcus. Posteriorly it is limited by an imaginary line joining the preoccipital notch to the parietooccipital sulcus. The parietal lobe is bounded anteriorly by the central sulcus and posteriorly by the line joining the preoccipital notch to the parietooccipital sulcus. The occipital lobe lies behind the line joining the preoccipital incisure to the parietooccipital sulcus (Fig. 95.**3**A).

The two cerebral hemispheres are interconnected by the corpus callosum which must be cut to separate the two hemispheres. The medial surface of the left hemisphere (Fig. 95.**3**B) shows the cut surface of the corpus callosum, third ventricle, septum pellucidum, and a part of the fornix. The prominent sulci on the medial surface are the cingulate, calcarine, and parietooccipital sulci.

The surface of the cerebrum has a thin layer of gray matter, less than 4 mm thick. The gray matter is made of cell bodies of neurons. The interior of the cerebrum contains mainly the white matter, which is made of myelinated axons. Also present in the interior is a large

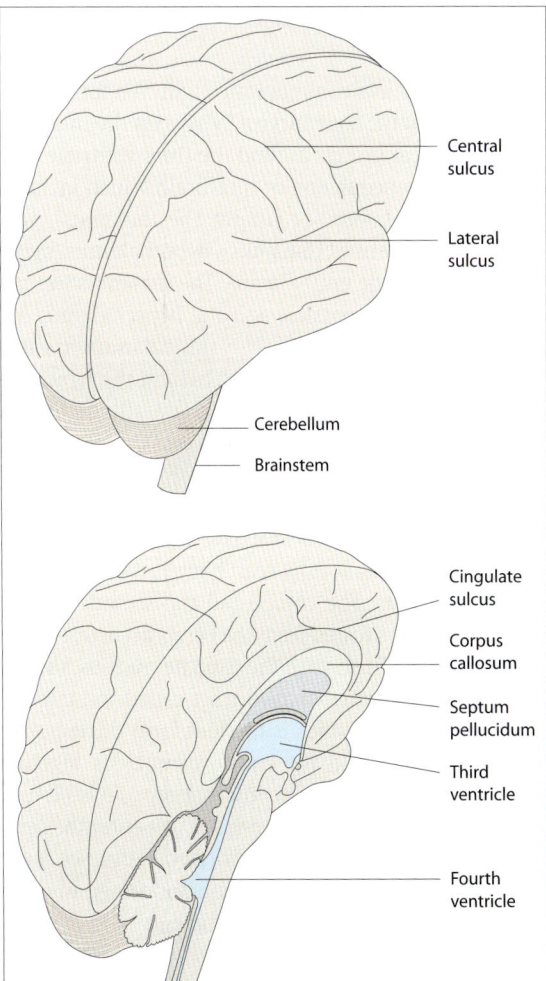

Central
sulcus

Lateral
sulcus

Cerebellum

Brainstem

Cingulate
sulcus

Corpus
callosum

Septum
pellucidum

Third
ventricle

Fourth
ventricle

Fig. 95.**1** (*Above*) The brain, viewed from above, behind and right. (*Below*) The left cerebral hemisphere.

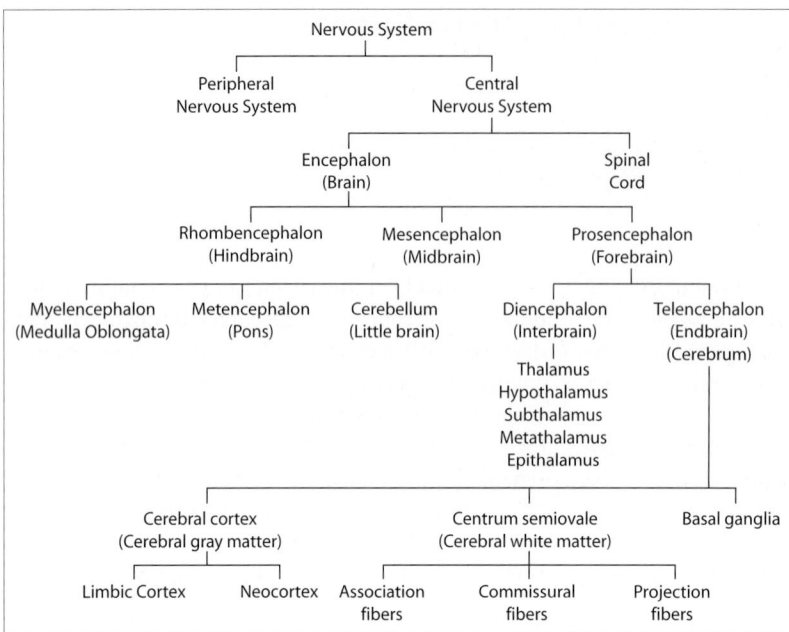

Fig. 95.**2**   Subdivisions of the nervous system.

aggregation of cell bodies called the basal ganglia[1] and large fluid-filled spaces called the cerebral ventricles. The ventricles are lined by the membranous ependyma which is shown in blue in all the diagrams. To understand the anatomy of these structures, let us imagine that the thin layer of the cerebral cortex on the medial surface has been removed and some of the white matter has been scooped out. The brain would then look as shown in Figure 95.**4**A.

Figure 95.**4** shows the corpus callosum in its entirety. The corpus callosum has a body, rostrum, splenium, forceps minor and major, and tapetum. The corpus callosum contains commissural fibers, i.e., fibers that interconnect the two hemispheres. Also seen is the optic radiation, running lateral to the tapetum and coursing toward the calcarine sulcus. On removing the corpus callosum, the structures that come into view are the ventricles and the corona radiata (Fig. 95.**4**B).

There are four cerebral ventricles, viz., the two lateral ventricles, the third ventricle, and the fourth ventricle (Fig. 95.**4**C). Each lateral ventricle has a body and three horns or cornu: an anterior cornu extending into the frontal lobe, a posterior cornu extending into the occipital lobe, and an inferior cornu extending into the temporal lobe.

The third ventricle is a midline structure. It communicates with the two lateral ventricles at the Y-shaped interventricular foramen (of Monro). Posteroinferiorly,

the third ventricle communicates with the fourth ventricle through the aqueduct (of Sylvius) in the midbrain.

The nuclear groups around the third ventricle constitute the diencephalon or interbrain which includes the dorsal thalamus (or simply, the thalamus), subthalamus, (or ventral thalamus), hypothalamus, metathalamus, and the epithalamus. The thalamus lies lateral to the third ventricle and the hypothalamus lies below the third ventricle. The subthalamus has two parts, a small midline part lying behind the hypothalamus and a thin sheet extending laterally below the dorsal thalamus (Fig. 95.**4**D).

Removal of the ventricles exposes the underlying structures which include the dorsal thalamus, fornix, stria terminalis, and the caudate nucleus (Fig. 95.**5**). The corona radiata is now seen lying just lateral to the caudate nucleus. The corona radiata is made of long projection fibers to and from the cerebral cortex. When it is transected flush with the caudate nucleus, the cut edge presents a knee-shaped tract called the internal capsule, lateral to which is the lentiform nucleus. The caudate nucleus and the lentiform nucleus together form the basal ganglia. The internal capsule is pierced from below by the striate arteries which enter the cerebrum through the anterior perforated substance. Rupture of the striate artery causes hemorrhage into the internal capsule, resulting in hemiplegia.

The medial side of the temporal lobe contains the hippocampus, dentate gyrus, fimbria of fornix, amygdala, the origin of the stria terminalis, and the inferior cornu of the lateral ventricle. These structures have been shown in Figure 95.**6**A by removing the cerebral

---

[1]   The term basal ganglia is a misnomer. An aggregation of cells inside the central nervous system is called a nucleus and therefore the correct name should be basal nuclei.

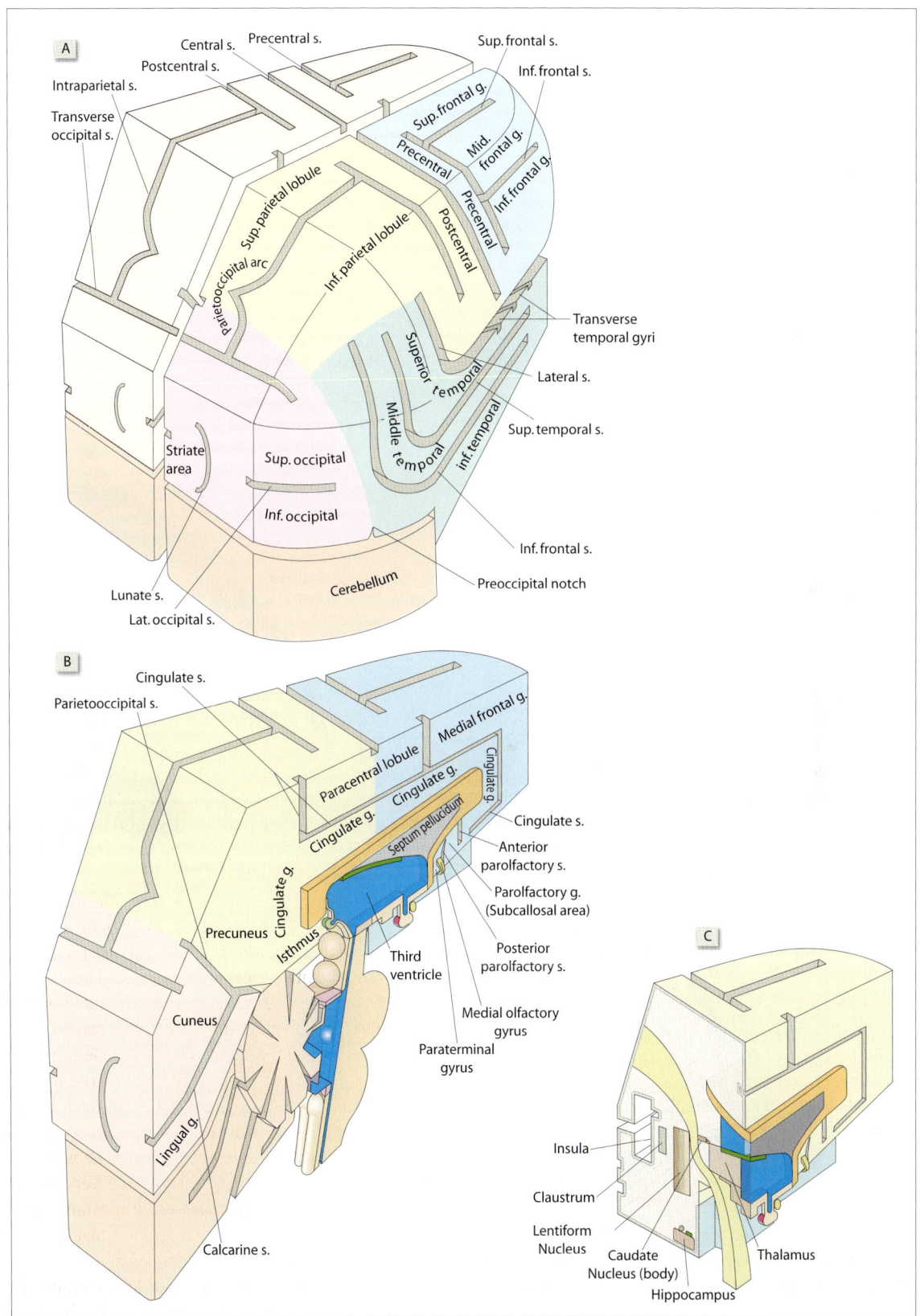

Fig. 95.**3** [A] The 'schematic' brain viewed from above, behind and right, showing the sulci and gyri on the superolateral surface of the right hemisphere. Also shown are the demarcations between the frontal, parietal, temporal, and occipital lobes. [B] The left hemisphere showing the sulci and gyri on its medial surface. [C] A coronal section through the left hemisphere.

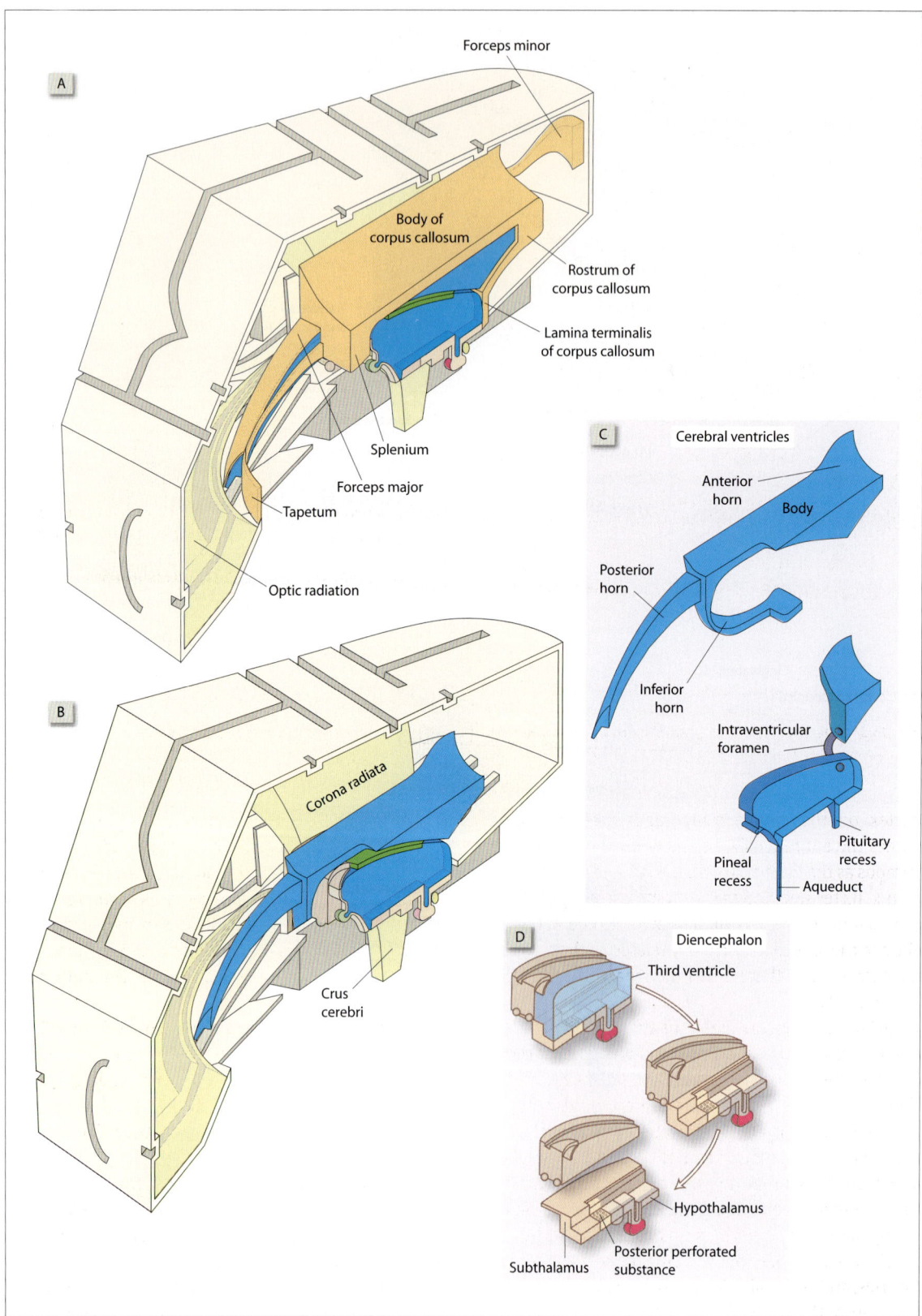

Fig. 95.**4** [A] Parts of the corpus callosum. Also shown is the optic radiation ending near the calcarine sulcus. [B] The corona radiata, and the cerebral ventricles. [C] Parts of the lateral and third ventricles. [D] The diencephalon, showing the relation between its parts.

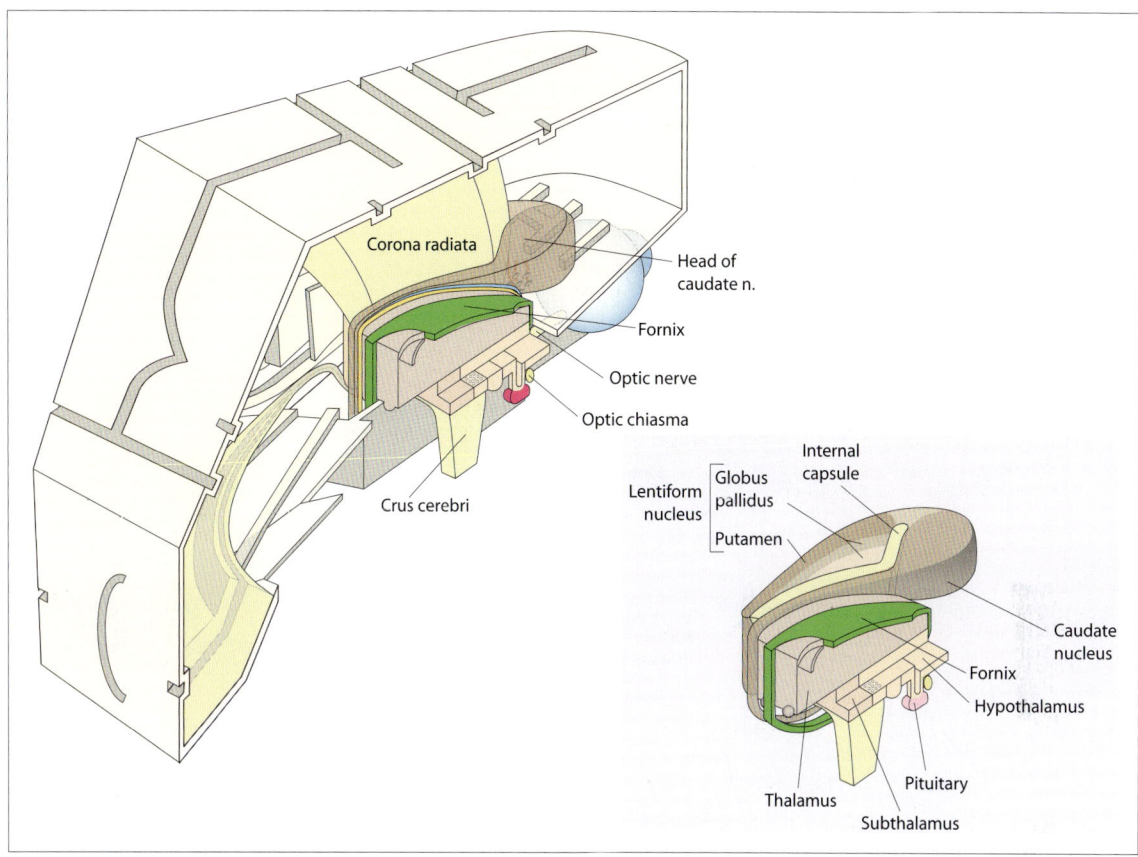

Fig. 95.**5** The caudate nucleus and the fornix. Inset shows the basal ganglia, thalamus, and internal capsule.

cortex on the medial temporal surface. The figure shows the entire fornix, i.e., its origin from the hippocampus as the fimbria, its crus, body, and column of the fornix. Its termination in the mamillary body of the hypothalamus is not shown. It should also be noted that though the amygdala is closely related to the tail of the caudate nucleus, there is no functional relationship between the two.

In Figure 95.**6**B, the thalamus and basal ganglia have been removed so that all the sulci and gyri on the inferior surface of the left cerebral hemisphere are visible. The sulci are shown as elevations because they are viewed from the inside. The inset shows the posterolateral view of the anterior half of the right hemisphere. This view shows that the insula is essentially a sulcus with a greatly expanded floor. The expanded floor is triangular in shape, with its tip directed medially.

The origin of the twelve cranial nerves, the cerebral arteries, and cerebral veins are shown in an anteroinferior view of the brain (Fig. 95.**7**). It shows that the olfactory tract occupies the olfactory sulcus. It also shows the optic chiasma anterior to the pituitary and the optic tracts as they wind round the cerebral peduncles.

Figure 95.**8** shows the brainstem and the cerebellum from the same perspective as in Figure 95.**3**. It shows that the fourth ventricle is bounded posteriorly by the cerebellar vermis and anteriorly by the pons and upper part of the medulla oblongata. It also shows the three cerebellar peduncles connecting the cerebellum to the brainstem. The four deep nuclei of the cerebellum are shown in Figure 95.**8**B.

The fourth ventricle is shown fully in Figure 95.**8**C by removing the cerebellum. The recesses and the foramina of the fourth ventricle are also shown. The foramina are important for the circulation of cerebrospinal fluid. Also shown in the figure is the left half of the visual pathway.

Removal of the fourth ventricle exposes the floor of the fourth ventricle which is shown in Figure 95.**9**. The locus ceruleus and the area postrema are located on the floor of the fourth ventricle.

The nuclei and tracts of the brainstem are shown by sections made at conventional sites (Fig. 95.**9**). Each section shows certain characteristic features. For example, the midbrain is sectioned at the levels of the upper and lower colliculi. Note the origin of the third

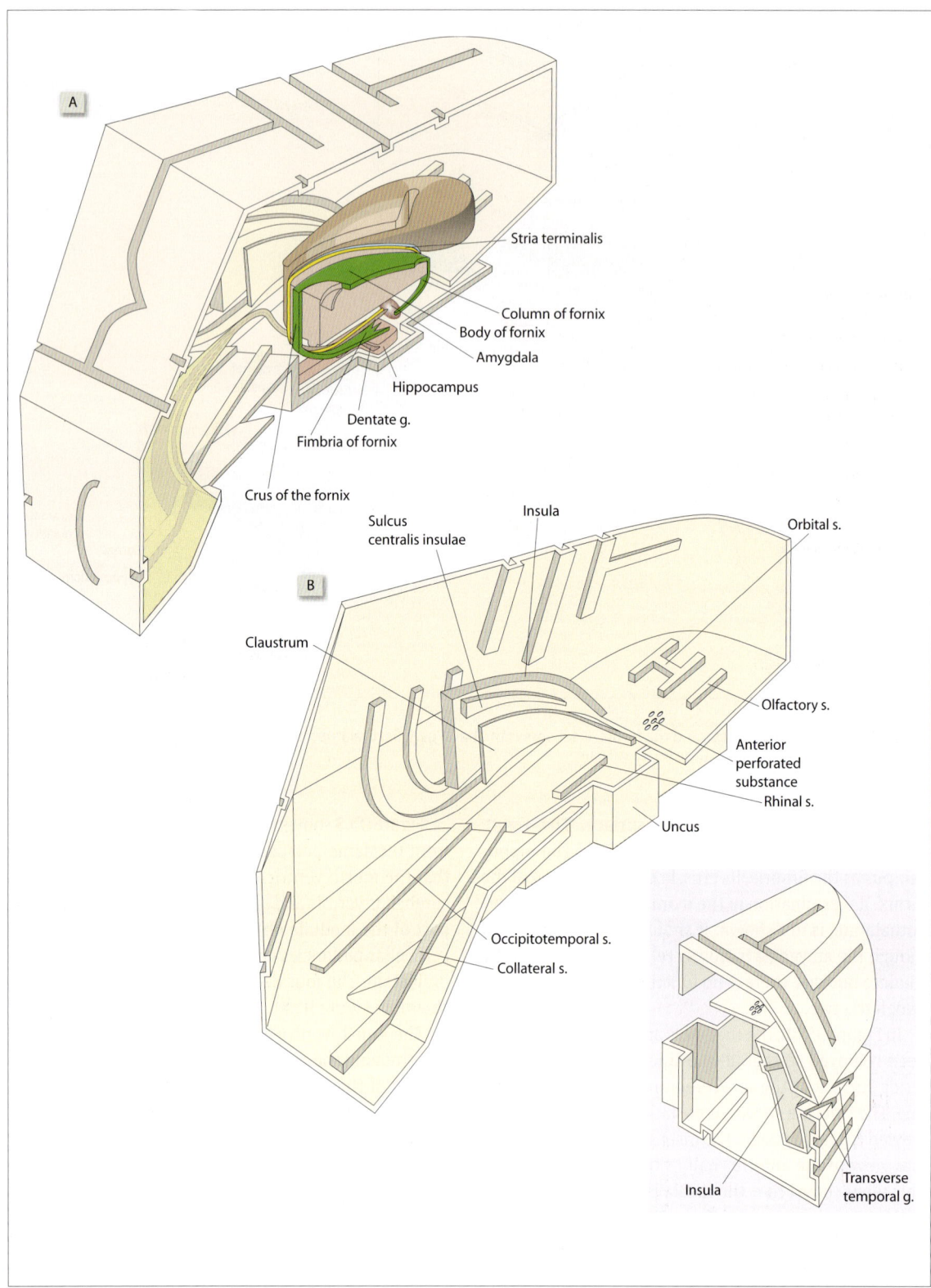

Fig. 95.**6**   [A] The hippocampus and the fimbria of the fornix, located inside the temporal lobe. Also shown are the amygdala and the stria terminalis originating from it, the tail of the caudate nucleus, and the termination of the column of fornix in the region where the mamillary body would have been. [B] The sulci and gyri on the inferior surface of the left cerebral hemisphere viewed from the inside. Inset shows the continuity of the insula with the lateral sulcus. Also shown are the transverse temporal gyri on the floor of the lateral sulcus.

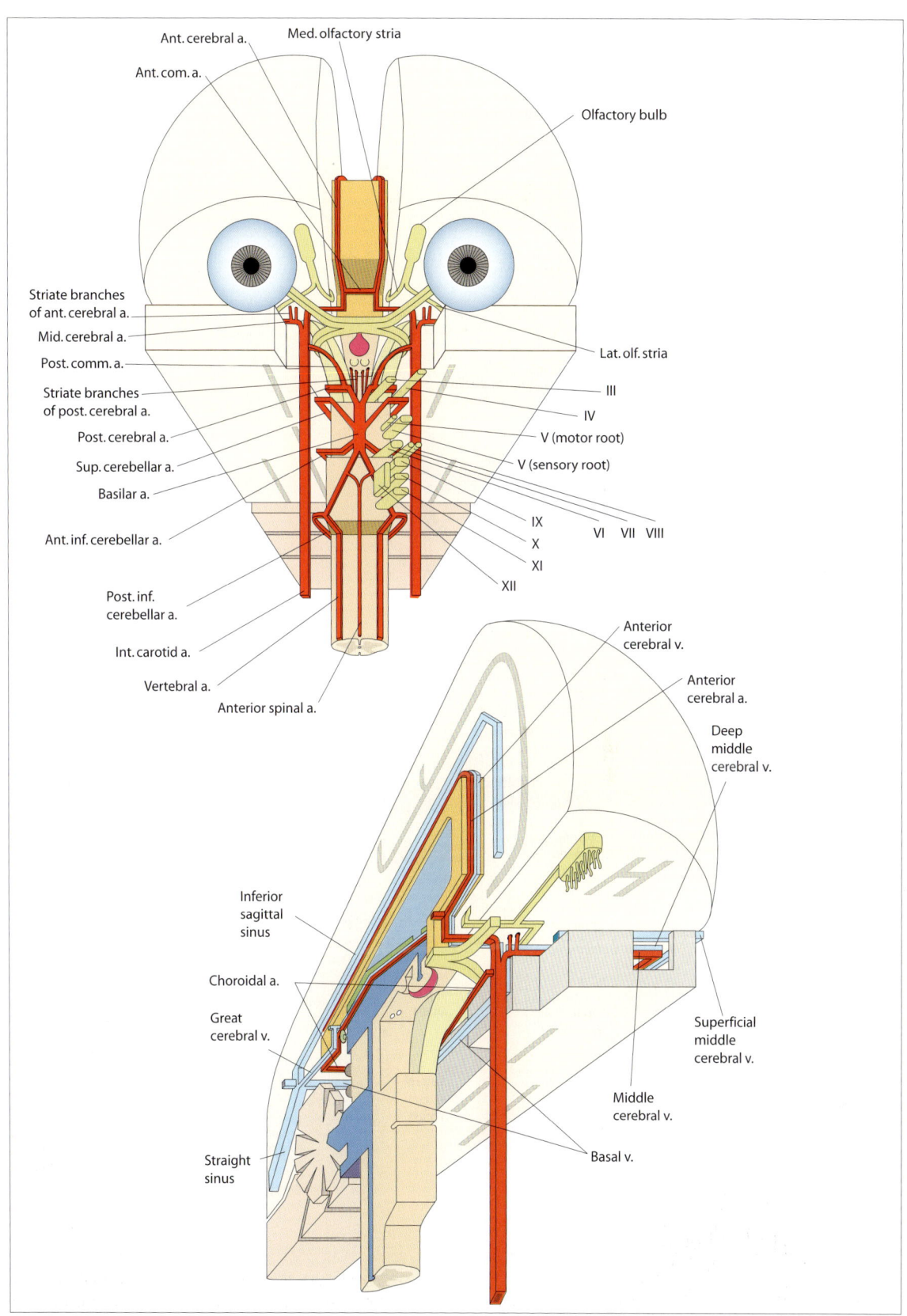

Fig. 95.**7**   (*Above*) The anteroinferior view of the brain showing the cerebral arteries and cranial nerves. (*Below*) The arteries and veins on the medial surface of left hemisphere.

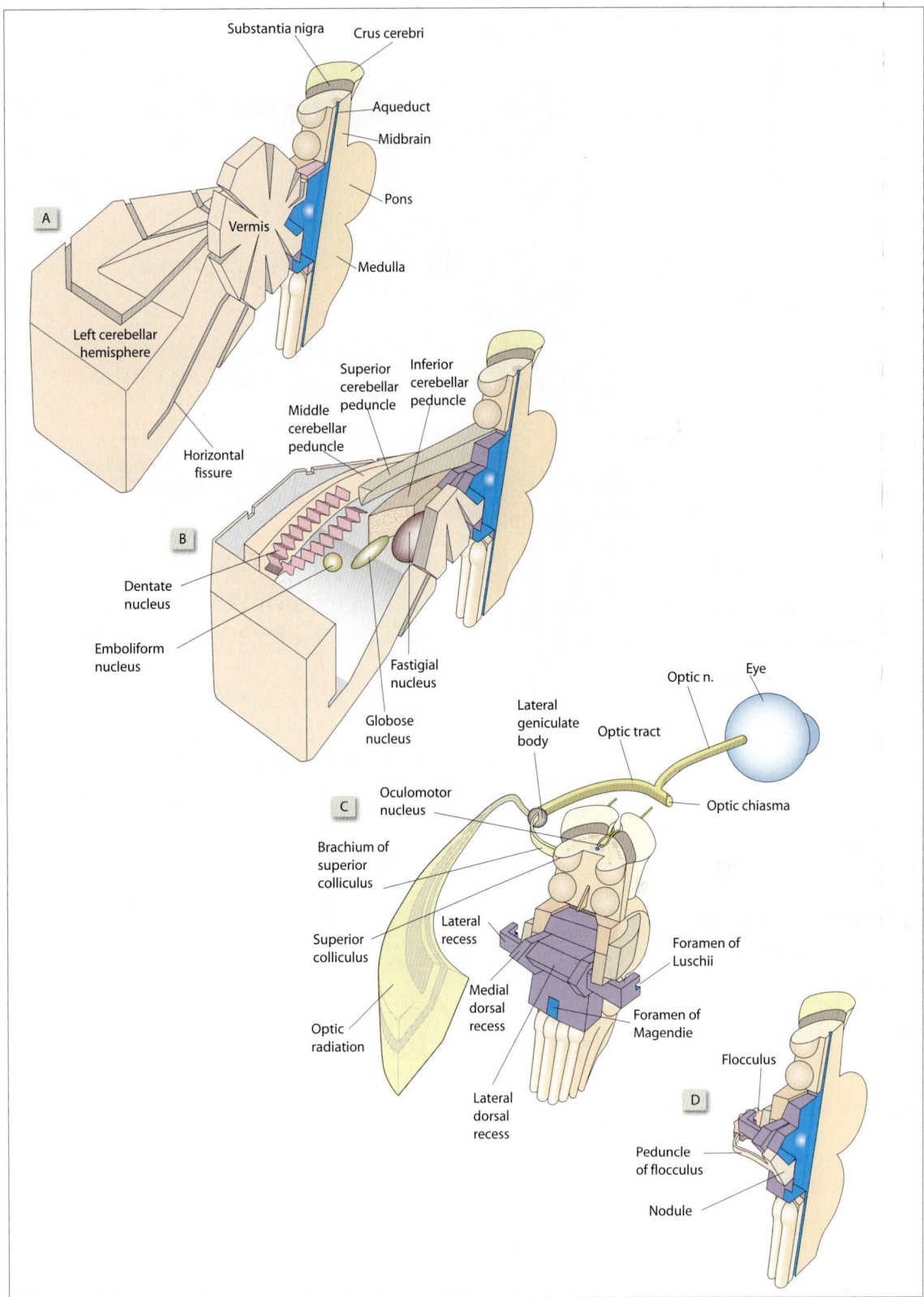

Fig. 95.**8** The cerebellum and the fourth ventricle. [A] Cerebellar vermis and left cerebellar hemisphere. It shows that the cerebellum forms the roof of the fourth ventricle. [B] Left cerebellar hemisphere cut open to show the deep cerebellar nuclei. [C] Cerebellum removed to show the fourth ventricle and its recesses. Also shown is the optic pathway, showing its relation to the midbrain. [D] Left half of the brainstem and the flocculonodular lobe of the cerebellum.

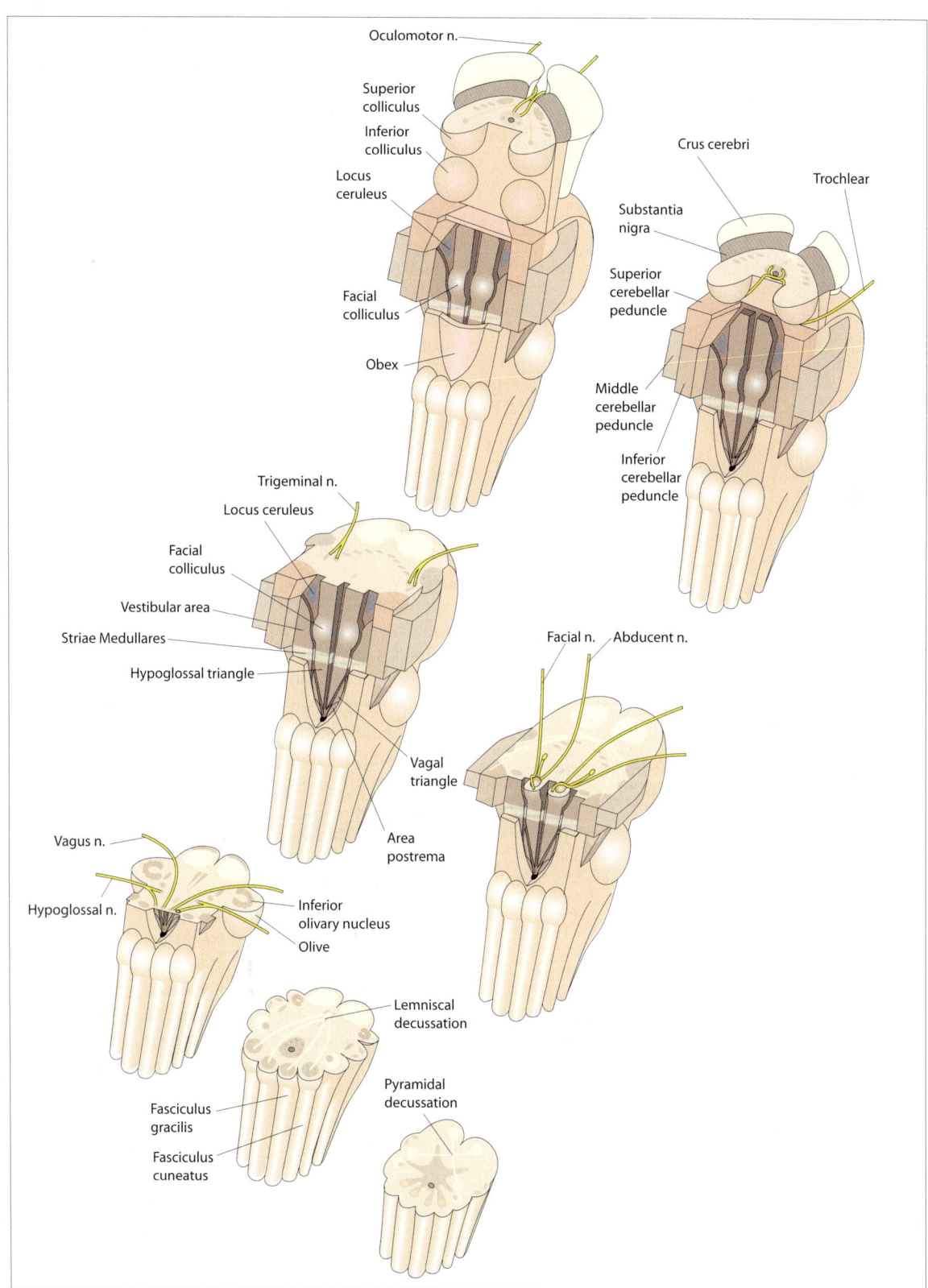

Fig. 95.**9**   Posterior aspect of the brainstem that has been sectioned at various levels. (*Above*) Section through the midbrain at the level of superior colliculus (left) and inferior colliculus (right). (*Middle*) Section through the pons at the level of trigeminal nerve (left) and facial colliculus (right). (*Below*) Section through the medulla oblongata at the level of inferior olive (left), lemniscal decussation (middle) and pyramidal decussation (right).

The medulla oblongata is conventionally sectioned at the levels of the inferior olive, lemniscal decussation, and pyramidal decussation. Some of the important structures shown in the figure are the inferior olivary nucleus and the gracile and cuneate fasciculi.

Spinal Cord

The spinal cord is the caudal continuation of the medulla oblongata. It is enclosed in the vertebral canal and extends below to the border between first and second sacral vertebra. The caudal tip of the spinal cord is called the conus medullaris. The conus is attached to the coccygeal bone with a connective tissue cord called the filum terminale. The central canal of the medulla continues into the spinal cord.

Spinal nerves originate in pairs from both sides of the spinal cord and emerge through the intervertebral foramina. Each spinal nerve consists of a ventral motor root and a dorsal sensory root. Both roots unite to form the mixed spinal nerve. The dorsal root bears a ganglion.

There are 31 pairs of spinal nerves: 8 pairs of cervical (C), 12 pairs of thoracic (T), 5 pairs of lumbar (L), 5 pairs of sacral (S), and 1 pair of coccygeal (Co) nerves. The segment of spinal cord giving rise to a pair of spinal nerves is called a spinal segment. The roots of all nerves below spinal segment L1 form a bunch known as the cauda equina. The spinal cord is shorter than the vertebral column and therefore a spinal segment is higher than the corresponding vertebra (Fig. 95.**10**). The distance between corresponding spinal and vertebral segments increases progressively toward the lower end. This knowledge is necessary to the surgeon during laminectomy for relieving spinal cord compression.

The spinal cord presents two fusiform enlargements, a cervical enlargement extending from C4 to T2 segments and a lumbosacral enlargement extending from L2 to S3 segments. The cervical and lumbosacral enlargements accommodate extra motor neurons to supply the muscles of the upper and lower limbs respectively.

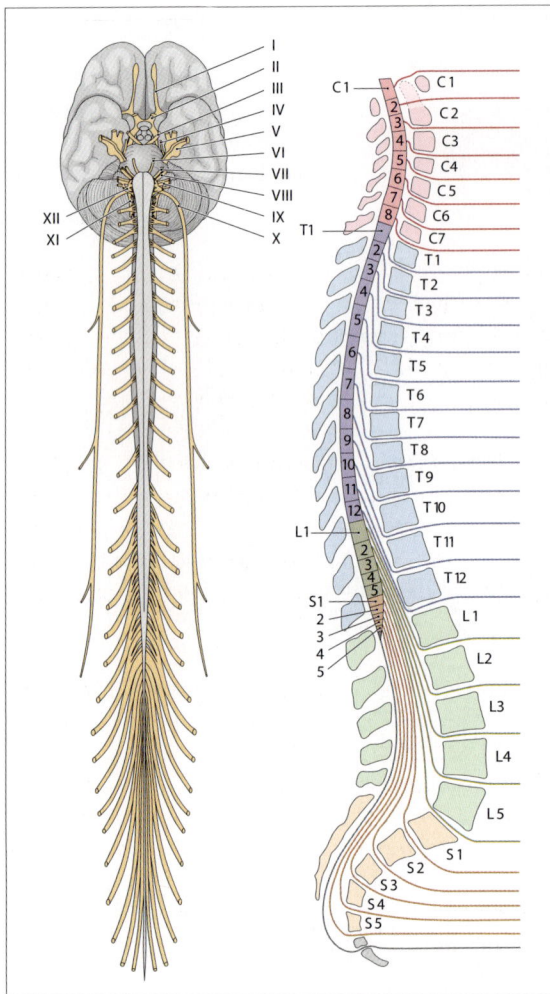

Fig. 95.**10**   (*Left*) The brain, cranial nerves, spinal cord, and spinal nerves. (*Right*) Spinal nerves emerging from different segments of the spinal cord.

and fourth nerves and also the characteristic shape of a midbrain section.

Pontine sections show the origin of the trigeminal and abducent nerves. The figure shows how the fibers of the facial nerve loop round the nucleus of the abducent nerve, forming an elevated ridge on the floor of the fourth ventricle, called the facial colliculus.

Spinal Cord

In cross-section, the spinal cord presents gray matter (aggregation of cell bodies of neurons) in the interior and white matter (aggregation of myelinated axons) at the periphery. Each half of the spinal gray matter shows three pointed extensions named as the posterior (dorsal), lateral (intermediate), and anterior (ventral) horns. The intermediate horn is present only in the thoracic and upper lumbar segments. These horns separate the white matter into three pairs of funiculi: the anterior, lateral, and posterior. A gray commissure connects the symmetrical halves of gray matter across the middle line. The anterior column is broader in cervical and lumbosacral enlargements for accommodation of the numerous motor neurons supplying the muscles of the upper and lower limbs.

Spinal gray matter

Cytoarchitecturally, the spinal gray matter has been divided into ten Rexed's laminae (Fig. 96.1). These laminae are useful when referring to specific areas of the spinal gray matter. At some places in the spinal gray matter, the cell bodies of neurons are closely aggregated together to form nuclei. Unlike the globular nuclei present in most parts of the central nervous system, the spinal nuclei are columns of cells extending vertically through several spinal segments.

The **dorsal (posterior) horn** contains the cells of the spinothalamic tract and other ascending tracts. It also contains the axonal endings of first order sensory neurons. Together, lamina II and parts of lamina III form a gelatinous mass of nerve cells called the *substantia gelatinosa*.

The **lateral (intermediate) horn** is present in the T1 to L2 segments of the cord, and also from S2 to S4. In the T1 to L2 segments, the lateral horn contains the cells of the preganglionic sympathetic nervous system and constitutes the thoracolumbar sympathetic outflow. In the S2–S4 segments, the lateral horn contains the soma of the preganglionic parasympathetic neurons and constitutes the sacral component of the craniosacral parasympathetic outflow (p 631).

The **ventral (anterior) horn** contains the soma of the lower motor neurons, which directly innervate skeletal

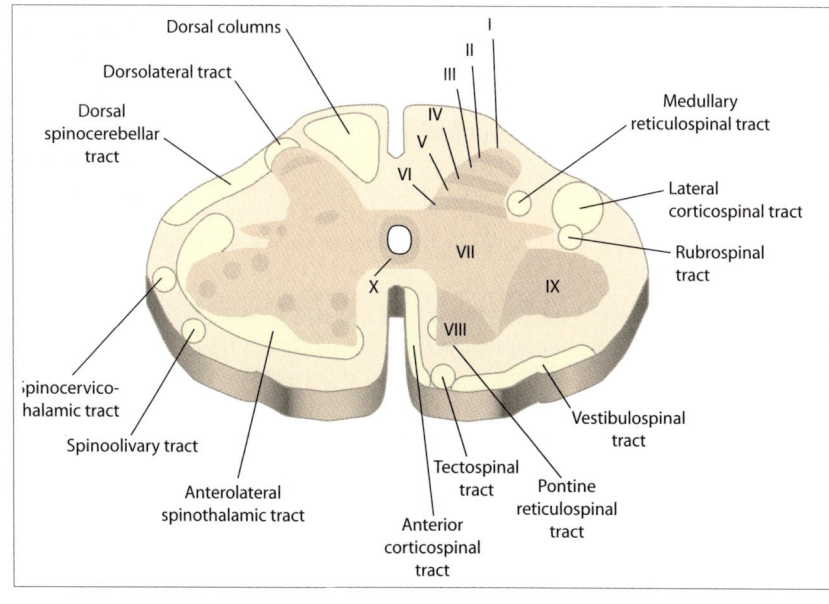

Fig. 96.**1** Physiologically well-understood ascending and descending tracts of the spinal cord. The spinoreticular fibers are not organized into a well-defined tract and therefore, are not shown here. Also shown are Rexed's laminae.

muscles. The soma of the lower motor neurons are organized into three groups of nuclei. The lateral group is present only in the cervical and lumbar spinal segments and supplies the limbs. The *medial group* is present only in the thoracic segments and supplies the trunk. The central group is present only in the C3 to C5 segments and innervates the diaphragm.

Spinal white matter

The spinal white matter is composed largely of the long ascending and descending spinal tracts that are listed in Table 96.**1** and illustrated in Figure 96.**1**.

Autonomic fibers descending from the hypothalamus do not form a compact tract but descend as scattered nerve fibers. Hence, they cannot be seen in a cross-section of the spinal cord but their presence in the spinal cord can be inferred from the effects they produce. Descending autonomic fibers carry the information from the hypothalamus and brain stem visceral nuclei (respiratory and cardiovascular) to the preganglionic autonomic neurons of intermediate horn of the spinal gray matter.

The spinal cord also contains *intersegmental tracts* that run between different segments of the spinal cord. An important group of intersegmental fibers are the C3/C4 propriospinal neurons that have their cell bodies in spinal segments C3 and C4. These neurons allow selection of the correct muscle synergies that are required for visually-guided reaching movements of

the upper limbs, and also contribute to the associated postural readjustments. Lesions of these neurons affect the precise aiming of the hand towards the target even as the major descending inputs remain intact.

Spinal nerve

As the spinal nerve approaches the spinal cord, it divides into the dorsal (sensory) and ventral (motor) roots (Fig. 96.**2**). All sensory fibers reach the spinal cord through dorsal nerve roots and all motor nerves exit the spinal cord through the ventral nerve root. This is known as the *Bell–Magendie law*.

Each ventral root carries motor axons supplying a group of anatomically and functionally related skeletal muscles called the *myotome*. In the thoracolumbar segments, the ventral root also contains axons of the preganglionic and postganglionic sympathetic neurons. (The dorsal root is described on p 650.)

Spinal lesions

Depending on its exact site, a lesion in the spinal cord can cause motor deficits (p 677), sensory deficits (p 658) or both. Lesions causing both motor and sensory deficits are discussed here. For understanding the neurological deficits, the full course of the corticospinal tracts (Fig. 96.**5**) and the spinothalamic tracts (see Fig. 13.**9**) have to be read before proceeding further.

Brown–Sequard syndrome is a rare clinical condition in which there is segmental damage on one side of the spinal cord due to traumatic injury or compression by extramedullary tumors. It results in several neurological deficits *below the level of the lesion*. These are summarized in Table 96.**2** and explained diagrammatically in Figure 96.**3**. *At the level of the lesion*, there is flaccid paralysis (due to lower motor neuron lesion) and

Table 96.**1**   Physiologically well-understood long spinal tracts.

Ascending
1. Pathways to the thalamus and cortex
(a) Fasciculus gracilis and cuneatus (dorsal columns)
(b) Spinothalamic tracts, (anterior and lateral)
2. Pathways to the cerebellum
(a) Spinocerebellar tracts
(b) Spinoreticulocerebellar tract
(c) Spinoolivocerebellar tract
Descending
1. Lateral pathways
(a) Lateral corticospinal tract
(b) Rubrospinal tract
2. Medial pathways
(a) Anterior corticospinal tract
(b) Vestibulospinal tract
(c) Reticulospinal tract
(c) Tectospinal tract

Table 96.**2**   Neurological deficits in Brown–Sequard syndrome.

Site of lesion	Neurological signs
Dorsal columns	Ipsilateral loss of position and vibratory senses; disturbances of stereognosis and tactile discrimination
Lateral spinothalamic tract	Contralateral and segmental loss of pain and temperature—one or two segments below the level of the lesion
Corticospinal tract	Ipsilateral spastic paralysis with exaggerated tendon reflexes and positive Babinski sign

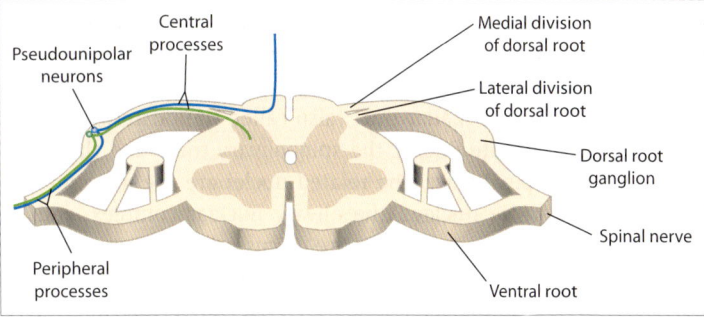

Fig. 96.**2**   A transverse section of the spinal cord showing the gray and white matter. Also shown are the dorsal and ventral roots of the spinal nerves. Note that the dorsal root is further subdivided into the medial and lateral divisions.

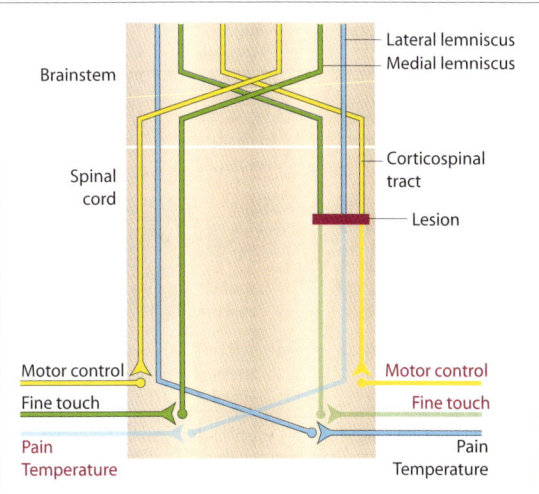

Fig. 96.**3**   Brown–Sequard syndrome. The neurological deficits are summarized in Table 96.2.

sensory deficits that are confined to the affected myotome and dermatome respectively.

Complete transection of the cord occurs following spinal fracture or dislocation, and results in loss of all sensations and voluntary movements below the level of the lesion. Immediately after injury, there is a period of *spinal shock* that may last from a few days to several weeks. During this period, all somatic and visceral reflex activities are abolished. Thereafter, the reflex activities return and the muscles become spastic with exaggerated tendon reflexes. Voluntary control over the bladder and bowel functions are lost if the lesion takes place above S2. Such dysfunction of the urinary bladder is called automatic bladder or cord bladder.

If the transection occurs in the cervical cord, the patient becomes quadriplegic. If the site of lesion is between C1 and C4, it is called *high quadriplegia* and causes respiratory paralysis due to the involvement of the phrenic nerve. These patients require respiratory support. If the site of lesion is between C5 and C8, it is

called low quadriplegia. If the transection occurs at T1 or below, the patient becomes paraplegic. An isolated spinal cord favors flexor muscles and therefore a complete spinal transection results in *paraplegia in flexion*. If the paraplegia is associated with predominantly extensor spasms, it is known as *paraplegia in extension* and it indicates that the transection is not complete.

Subacute combined degeneration of spinal cord is most often seen in pernicious anemia. The dorsal columns and the lateral corticospinal tract undergo bilateral degeneration, especially involving the lumbosacral segments. The disease results in loss of position and vibratory senses of the lower extremities, with signs of upper motor neuron lesions.

Brainstem

The internal structure of the brainstem is conventionally studied by making sections at levels with salient landmarks. The *medulla* is thus sectioned at four levels, viz., at the level of pyramidal decussation, lemniscal decussation, the midolivary level, and at the pontomedullary junction. The pons is sectioned at two levels, one above and the other below the level of the facial colliculus. The midbrain too is sectioned at two levels, one at the level of the inferior colliculus and the other at the level of the superior colliculus. The salient features seen in these sections are shown in Figure 95.**9**.

Cranial nerves nuclei

All cranial nerves except the olfactory and optic nerves originate from the brainstem. The midbrain gives origin to the III and IV nerves, the pons gives origin to the V to VIII cranial nerves, and the medulla gives origin to the IX to XII cranial nerves. All cranial nerves are associated with one more nucleus in the brainstem. These nuclei are shown in Figure 96.**4** and are listed in Table 96.**3**.

Spinocerebellum

Somatotopy The spinocerebellum has two somatotopic maps, one in the anterior lobe and the other in the posterior lobe (Fig. 97.**5**). The trunk of the body is represented in the vermis and the extremities are represented in the intermediate zone. The two maps are inverted with respect to each other with the head oriented toward the primary fissure in both the maps. The maps are both motor and sensory in nature, i.e., stimulation of the cerebellar cortex at any part of the map results in movements of the corresponding parts of the body, and stimulation of any part of the body triggers electrical activity in the corresponding part of the map. Visual and auditory information reach the areas that represent the head.

Afferents Sensory information from the spinal cord reaches the spinocerebellum mainly through three tracts, viz., the spinocerebellar, spinoolivocerebellar, and spinoreticulocerebellar tracts. All of them enter the cerebellum through the inferior peduncle.

There are two spinocerebellar tracts. The *ventral spinocerebellar tract* is important in locomotion. The *dorsal spinocerebellar tract* originates from the nucleus dorsalis or Clarke's column located in the dorsal horn of the spinal gray matter (Fig. 97.**6**). Clarke's column is located in spinal segments T1 to L2 and receives touch, pressure, and proprioceptive information from the trunk and lower limbs. The axons of Clarke's column ascend to enter the ipsilateral cerebellar hemisphere through the inferior cerebellar peduncle. Since Clarke's column does not extend above the thoracic segments, the same sensations from the forearms and upper part of the body ascend in the dorsal columns to terminate in the accessory cuneate nucleus. Axons of the accessory cuneate nucleus form the

cuneocerebellar tract (posterior external arcuate fiber) and reach the spinocerebellum through the inferior cerebellar peduncle.

Efferents The vermal part of the spinocerebellar cortex projects to the fastigial nucleus while its paravermal part projects to the interpositus nucleus (Fig. 97.**7**). The vermal part (B zone) of the anterior lobe of the cerebellar cortex also has a direct projection to the ipsilateral Dieter's (lateral vestibular) nucleus.

The fastigial nucleus projects bilaterally to all the vestibular nuclei (except Dieter's nucleus) and the descending facilitatory reticular formation. The fastigial projection to the facilitatory reticular formation is excitatory in nature and therefore tends to increase the α-motor discharge to the muscles (p 663). On the other hand, the projections of the cerebellar cortex (B zone of anterior lobe) to the fastigial nucleus as well as to Dieter's nucleus, are inhibitory. Hence, if one considers

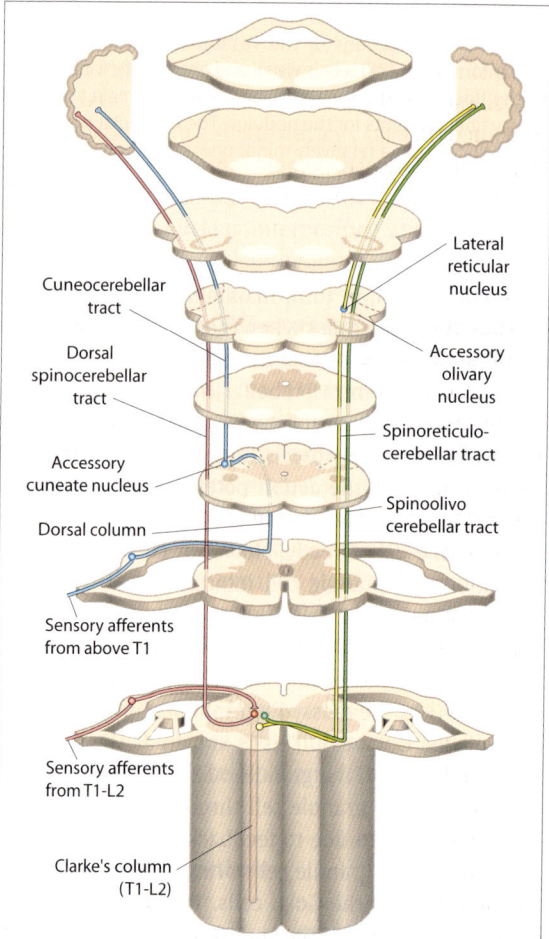

Fig. 97.**6** Spinal tracts ascending to the spinocerebellum include the spinocerebellar and cuneocerebellar tracts, the spinoolivocerebellar tract, and spinoreticuloolivary tract.

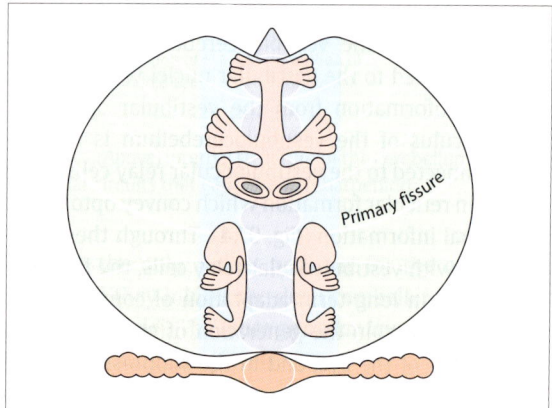

Fig. 97.**5** Somatotopy in spinocerebellum.

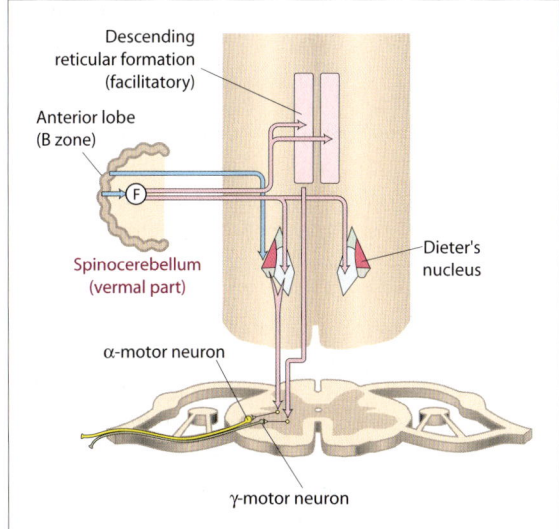

Fig. 97.**7** Connections of the vermal part of the spino-cerebellum. The magnocellular part and the rubrospinal tract issuing from it are prominent in cat, a commonly used animal for cerebellar research, but are poorly developed in man.

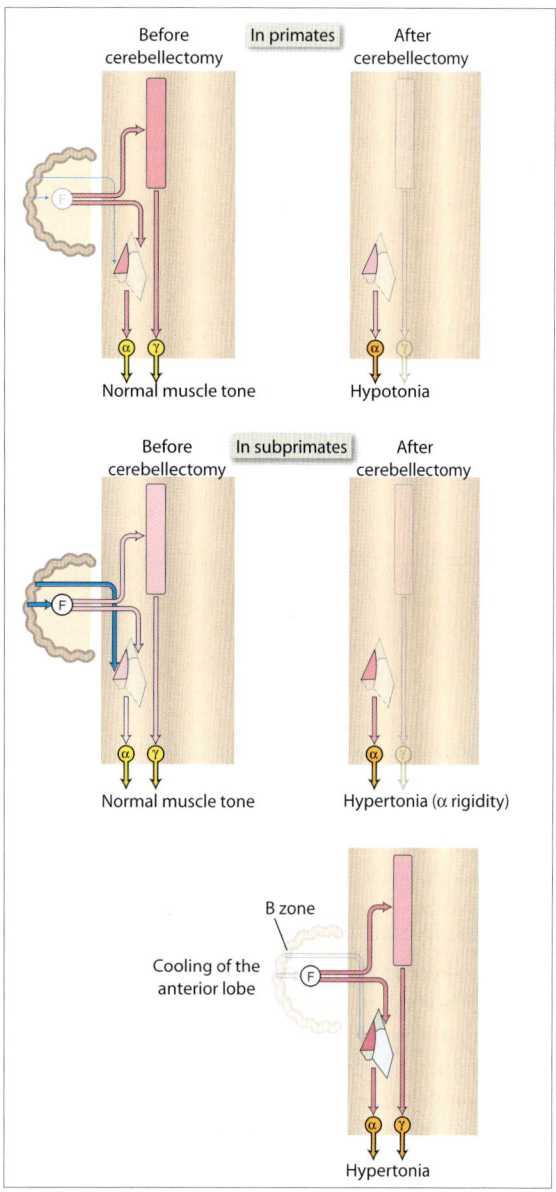

Fig. 97.**8** Experiments with the cerebellum. Cerebellectomy causes hypertonia in subprimates but hypotonia in primates. However, selective inactivation (by cooling) of the B zone of the anterior lobe vermis always results in hypertonia.

the net output of the spinocerebellum to the spinal motor neurons, it depends on the balance between the inhibitory output of the cerebellar cortex and the excitatory output of the fastigial nucleus. In subprimates, the net output is inhibitory and therefore cerebellectomy causes hypertonia (Fig. 97.**8**). In primates, the inhibitory output of the cerebellar cortex is weak and the net output is excitatory. Hence, cerebellectomy in primates, including humans, produces hypotonia. However, even in primates, the selective inactivation of only the B zone (by cooling) results in hypertonia. The role of cerebellum in the control of muscle tone is discussed again in Chapter 104.

The interpositus nucleus of the cerebellum projects ipsilaterally to the spinal cord through the interstitio-rubrospinal tract (Fig. 97.**9**). The interstitiorubral fibers cross the midline to the magnocellular (made of large cells) caudal part of the red nucleus. Fibers from the magnocellular red nucleus cross the midline again before descending to the spinal cord. In humans, the rubrospinal tract contains very few fibers. They intermix with the fibers of the lateral corticospinal tract and terminate on the spinal α-motor neurons.

The interpositus nucleus is essential for the conditioned eye blink reflex. The blinking of the eye occurs as an unconditioned reflex in response to a puff of air into the eye. However, when the air puff is regularly preceded by a sound, the blink starts occurring in response to the sound. This conditioned reflex is abolished by a lesion of the interpositus nucleus.

Cerebrocerebellum

The cerebrocerebellum (also called pontocerebellum) does not have any direct connection with the spinal cord as it is concerned with motor programming and not motor execution. The principal afferents to the cerebrocerebellum are the corticopontocerebellar and olivocerebellar tracts. The cerebrocerebellum projects mostly to the dentate nucleus (Fig. 97.**10**), which in turn sends efferents back to the cortex via thalamus

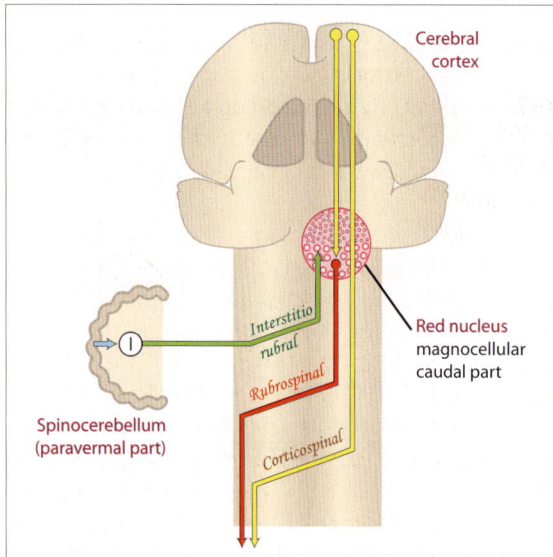

Fig. 97.**9**   Connections of the paravermal part of the spinocerebellum.

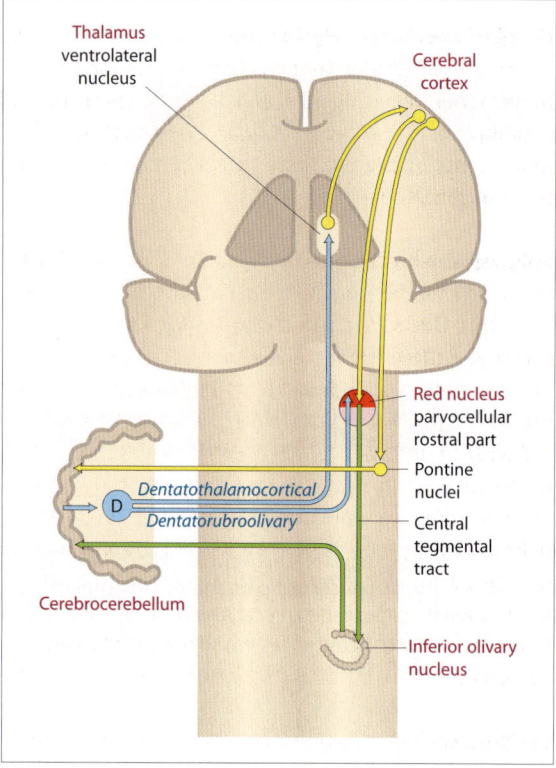

Fig. 97.**10**   Connections of the cerebrocerebellum.

(dentatothalamocortical fibers) and to the inferior olivary complex via the red nucleus (dentatorubroolivary fibers).

The dentatothalamocortical fibers relay in the ventrolateral nucleus (VLc) of the thalamus. The dentatorubroolivary fibers relay in the rostral parvocellular part of the red nucleus. It seems that the dentato-rubro-olivo-cerebello-dentate loop is meant for subconscious rehearsal of a ballistic movement (p 669) prior to its actual commencement.

An important function of the cerebrocerebellum is to enable simultaneous, well-coordinated motor actions of the limbs. Dysfunction of the cerebrocerebellum results in decomposition of movement (see below). The patient complains of inability to perform movements subconsciously and that every step of a movement has be to thought out. The cerebrocerebellum is also important for adaptive motor learning. Patients with damaged inferior olive do not show adaptive motor learning.

Contrary to the general notion that the cerebellum is a purely motor organ, it is now known that the cerebrocerebellum has some purely cognitive functions. The dentate nucleus enables a judgment of time and is important for the sensory processing that is required for performing complex spatial tasks. The cerebrocerebellum has a role in speech too, being involved in verb-generation tasks (p 717).

Cerebellar Dysfunctions

Cerebellar dysfunctions usually result from trauma and ischemia. Sometimes, a selective degeneration of the anterior cerebellum (vermis and leg representation area) occurs in chronic alcoholics resulting in cerebellar symptoms only in the lower limbs. The cerebellum is also affected by a number of hereditary disorders that are collectively called hereditary cerebellar ataxias.

Unilateral lesions of the cerebellum produce ipsilateral lesions. This is due to double crossing of cerebellar circuits. Examples of the double crossing exist in spinocerebellum as well as cerebrocerebellum. *In the spinocerebellum*, efferents of the interpositus nucleus terminate on the contralateral red nucleus (magnocellular part), which in turn gives rise to descending (rubrospinal) fibers to the contralateral spinal cord (Fig. 97.**9**). *In the cerebrocerebellum*, efferents of the dentate nucleus relay in the contralateral thalamus before reaching the contralateral motor cortex. The corticospinal fibers originating from the motor cortex cross back again to the original side.

Vestibulocerebellar dysfunction disrupts the ability to use vestibular information for coordination of movements, resulting in disturbances in balance and equilibrium. It also causes nystagmus (rapid oscillating movements of the eye) due to disruption of the vestibuloocular reflex.

Spinocerebellar dysfunction produces *hypotonia* (decrease in muscle tone) and ataxia (lack of motor coordination). Due to the somatotopic projection of spinocerebellar afferents, vermal lesions affect axial (trunk) musculature and produce *ataxia of gait* (often called a drunken gait because the patient walks like a drunkard with an unsteady balance and a wide stance) while paravermal lesions cause *ataxia of limbs*, leading to intention tremor and decomposition of movements (see below). Since vermal lesions affect facial musculature too, they also cause *dysarthria* (defect in speech articulation), leading to slurring and slowing of speech. Patients tend to exhibit a speech rhythm called scanning speech in which successive syllables emerge slowly.

Cerebrocerebellar dysfunction results in defective motor programming, delays in initiation and termination of movements, and errors in the range and force of movements (dysmetria), resulting in the clinical signs like overshoot, dysdiadokokinesia, intention tremor, and decomposition of movement. The intention tremor and decomposition of movement are also seen in spinocerebellar disorders.

Overshoot or *past-pointing* is a failure to stop a movement at the appropriate time. *Dysdiadokokinesia* is the inability to perform rhythmic, rapidly alternating movements like pronation and supination. *Intention tremor* is a coarse tremor (oscillations with low frequency and high amplitude) of the hand that occurs during reaching movement and which increases as a target is approached. The tremor probably results from repeated overshoots and their corrections. The tremor does not occur at rest, which distinguishes it from the fine tremors (oscillations with high frequency and low amplitude) of Parkinson's disease that occur at rest. *Decomposition of movement* is observed in a motor task that involves simultaneous movements at multiple joints. The movement is performed awkwardly, with the movements at the different joints performed sequentially rather than simultaneously.

The diencephalon (interbrain) has five parts, viz., dorsal thalamus (or simply, the thalamus), epithalamus (the pineal body), metathalamus (medial and lateral geniculate bodies), hypothalamus, and subthalamus (also called ventral thalamus). Functionally, the medial geniculate body belongs to the auditory system, the lateral geniculate body belongs to the visual system, and the subthalamus belongs to the basal ganglia.

The **pineal body** secretes melatonin (N-acetyl-5-methoxytryptamine). Melatonin secretion peaks between 2:00 am and 6:00 am, and probably serves to control the circadian rhythm of other physiological events. Melatonin secretion is controlled by sympathetic nerves originating from the superior cervical ganglion. The sympathetic fibers to the pineal gland are under the control of the suprachiasmatic nucleus of the hypothalamus which is linked to the day–night cycle through the retinohypothalamic fibers.

Thalamus

A vertical sheet of white matter, the internal medullary lamina, divides the thalamus into medial and lateral nuclear masses (Fig. 98.**1**). Rostrally, the lamina splits in a Y-shaped manner to enclose the anterior group of nuclei. A group of midline nuclei intervenes between the ependymal lining of the third ventricle and the medial nucleus. Another thin sheet of white matter, the external medullary lamina, covers the lateral surface of the thalamus.

Thalamic nuclear groups

The thalamus contains six groups of nuclear masses, viz., the ventral, medial, lateral, intralaminar, midline, and reticular groups of nuclei.

The **ventral group** is further subdivided into the ventral anterior, ventral lateral, and ventral posterior nuclei. The *ventral anterior* (VA) nucleus receives input from the basal ganglia-frontal cortex loop. The *ventral lateral* (VL) nucleus receives input from two major sources. Its anterior and medial parts (VL_o and VL_m) receive afferents from the GP_i/SN_{pr} (see Chapter 99) and project to

the premotor and supplementary motor cortices (Figs. 99.**2** and 100.**6**). Its posterior part (VL_c) receives afferents from the deep cerebellar nuclei and projects to the primary motor cortex. The *ventral posterior* (VP) nucleus acts as the specific somatosensory relay nucleus of the thalamus. Also called the ventrobasal complex, it is subdivided into the ventral posterolateral (VPL) nucleus which receives the medial and spinal lemnisci, and the ventral posteromedial (VPM) nuclei which receive the trigeminothalamic tract and solitariothalamic tracts.

The **anterior group of nuclei** is a relay station in the Papez circuit (see Fig. 16.**3**) and may have a role in attention and memory.

The **medial group of nuclei** includes the mediodorsal nucleus, which is a part of the anterior cingulate circuit (p 617) that is important for procedural memory (p 710).

The **lateral group** consists of the lateral dorsal nucleus, lateral posterior nucleus, and the pulvinar. All of them receive afferents from the superior colliculus and establish two-way connections with the sensory association cortex. The lateral dorsal nucleus is also well connected with the limbic cortex. The physiological role of the lateral group is not well understood. The pulvinar is particularly large in man compared to other primates. It projects to area 7 of the parietal cortex (see Fig. 103.**15**) and is important for visual-attention and visually-guided movements.

The **intralaminar nuclei** include the centromedian nucleus which is particularly large in man. The intralaminar nuclei receive ascending fibers from the brainstem reticular formation and are relay stations for the spinoreticulothalamic pain fibers.

The **midline nuclei** are the major thalamic target of ascending noradrenergic and serotonergic axons from the locus ceruleus and raphe nuclei respectively, and a cholinergic input from the midbrain. Their efferents pass mainly to the hippocampus, amygdala, and nucleus accumbens. They appear to be involved in memory and arousal, and may be important in the regulation of seizure activity.

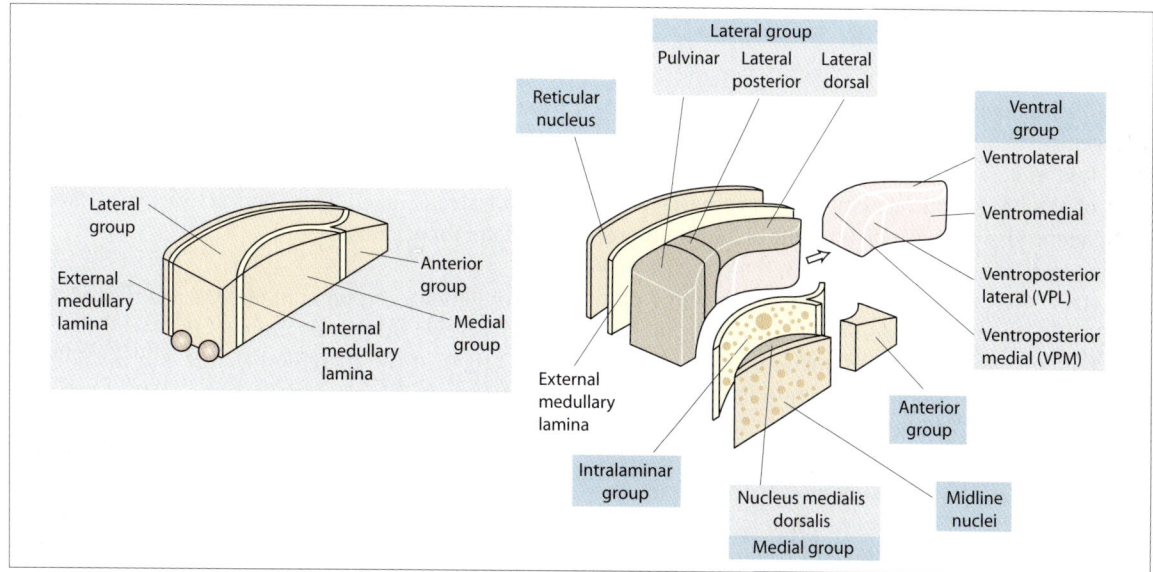

Fig. 98.**1**   Schematic representation of the nuclear groups in dorsal thalamus. See also Figure 95.**4**D.

The **reticular thalamic nucleus** forms a thin shell of nerve cells around the thalamus and intervenes between the external medullary lamina and the posterior limb of the internal capsule. The nucleus does *not*, as the name suggests, belong to the reticular activating system. The nucleus contains GABAergic neurons and receive collaterals from both thalamocortical and corticothalamic axons that pass through it (see Fig. 107.**6**). The reticular nucleus has bidirectional connections with the thalamus relay nucleus which is topographically organized. They are important in EEG synchronization (p 688).

Specific and nonspecific nuclei

The thalamic nuclei are categorized into the specific and nonspecific nuclei. Specific nuclei are further categorized into the relay nuclei and the association nuclei. The *relay nuclei* belong to the long ascending pathways to the sensory cortex and include the ventroposterior nucleus and the medial and lateral geniculate bodies. The *association nuclei* have reciprocal connections with the association areas of the cerebral cortex and include the lateral group of nuclei and part of the medial dorsal nucleus. The *nonspecific nuclei* do not have direct connections with the cerebral cortex and they include the intralaminar nuclei, midline nuclei, and reticular nucleus of thalamus.

Stimulation of specific nuclei produces a sharply localized ipsilateral cerebral response with a short latency while stimulation of nonspecific nuclei produces a delayed, widespread, and bilateral cerebral response. It was assumed that the nonspecific nuclei provided the background level of cortical preparedness necessary for the processing of information from the specific nuclei. Although the terms specific and nonspecific nuclei are still in use, the distinction between them is becoming blurred as extensive interactions between the two types of nuclei have been found.

Functions of thalamus

The broad functions of the thalamus are as follows: (1) The thalamus is a relay station in all sensory pathways, including gustatory and olfactory, that reach the cerebral cortex. It modulates the sensory information that is relayed through it and determines whether or not it reaches conscious awareness. (2) The thalamus is concerned with the conscious interpretation of crude touch, pain, and temperature. The final discrimination of sensory modalities however takes place in the sensory cortex after integrating the information provided by the thalamus. (3) The VA and VL nuclei of the thalamus integrate the activities of the motor information from the cerebellum and globus pallidus with the premotor cortex. (4) The pulvinar of the thalamus is important for visual-attention and visually-guided movements. (5) The reticular nucleus of the thalamus modulates the synchronization and desynchronization of the brain waves and influences attention and consciousness. (6) Being a relay station in the Papez circuit, the thalamus has a role in emotions and memory.

Thalamic syndrome

The thalamic syndrome (of Dejerine-Roussy) occurs due to a lesion of the thalamus, usually after the thalamogeniculate artery is blocked. It results in a temporary loss of cutaneous sensation on the contralateral side of the body (contralateral hemianesthesia) with permanent loss of sense of position of the limbs resulting in ataxia. After a few weeks, the patients may complain of agonizing pain, called the thalamic pain (p 655), on the affected side. Stimuli such as a pinprick, which do not give rise to anything more than discomfort on the normal side, may cause quite severe pain on the affected side. The threshold for pain is raised but the reaction to pain is exaggerated. The reason for the overreaction to pain is unknown. Sometimes the patient with eyes closed is unable to locate the position of a limb or may develop an illusion that the limb is lost. This is known as the thalamic phantom limb. Thalamic lesion may be associated with abnormal involuntary movements in the form of choreoathetosis or intention tremor when the projection fibers from the basal ganglia or cerebellum to the VA and VL nuclei of the thalamus are involved.

Hypothalamus

The hypothalamus acts as the head ganglion of the autonomic nervous system. It regulates autonomic, endocrine, and visceral functions and is central to the maintenance of homeostasis. The basic drives of life, i.e., hunger, thirst, and sex, originate in the hypothalamus.

The hypothalamus is subdivided into the lateral and medial zones. The most lateral part of the lateral zone is sometimes called the periventricular zone. The nuclear masses of these zones are arranged rostrocaudally into four regions, viz., preoptic, supraoptic, tuberal, and mamillary (Fig. 98.**2**).

The **preoptic region** is important for the regulation of body temperature.

The **supraoptic region** contains the supraoptic, suprachiasmatic, and paraventricular nuclei. Together, the supraoptic and paraventricular nuclei secrete oxytocin, ADH, CRH, and somatostatin. The suprachiasmatic nucleus regulates circadian rhythms.

The **tuberal region** contains several nuclei. The ventromedial and the lateral hypothalamic nuclei regulate food intake and serve as the center for satiety and feeding respectively. The arcuate (infundibular) nucleus secretes GnRH, GHRH, TRH, and PIH. The dorsomedial

nucleus is involved, together with the ventromedial nucleus, in the regulation of complex integrative functions like growth, feeding, maturation, and reproduction. The posterior hypothalamic nucleus contains groups of large histaminergic neurons that are concerned with stimulating sympathetic activity, regulation of sleep-wakefulness, and activation of thermoregulatory mechanisms.

The **mamillary region** contains the mamillary nuclei which receive the terminations of the fornix and provide origin to the mamillothalamic and mamillotegmental tracts. The mamillothalamic tract is a component of the Papez circuit.

Connections of the hypothalamus

The hypothalamus has bidirectional connections with the brainstem, other parts of the diencephalon, and the cerebral cortex. These connections enable the hypothalamus to receive somatosensory, visual, gustatory, and olfactory information, and to influence the autonomic and the limbic systems. Some of the connections are organized into anatomically prominent tracts while others are not. One of the most prominent tracts associated with the hypothalamus is the *medial forebrain bundle*, which consists of ascending and descending

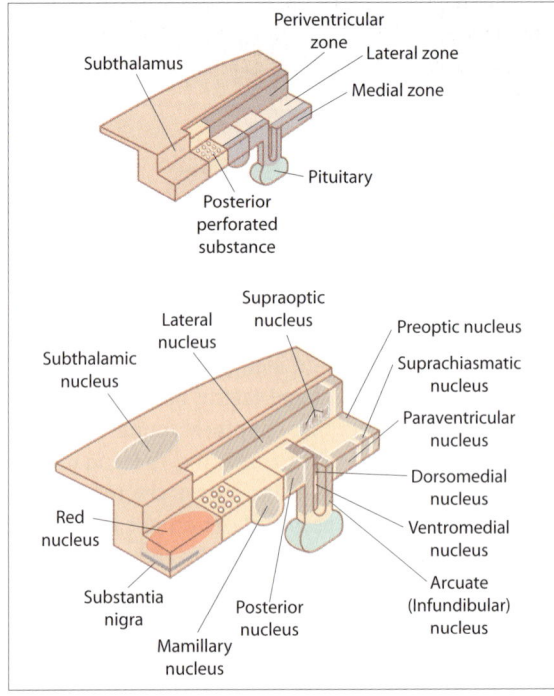

Fig. 98.**2**   Schematic representation of hypothalamic nuclei. See also Figure 95.**4**D.

fibers that run through the lateral hypothalamus to connect the cerebral cortex with the brainstem.

Connections with brainstem   The hypothalamus receives somatosensory information (including gustatory information) through collaterals of the lemniscal pathways and through the *dorsal longitudinal fasciculus* which connects the hypothalamus to several cranial nerve nuclei. The mamillary body of the hypothalamus is also connected to the brainstem reticular formation through the mamillotegmental tract and mamillary peduncle. The hypothalamic efferents to the brainstem terminate on preganglionic autonomic neurons present in the cranial nerve nuclei in the brainstem and in the lateral horn of the spinal cord. Descending fibers also reach somatic motor neurons.

Connections with other parts of diencephalon   The hypothalamus sends efferents to the anterior nucleus of the thalamus through the mammillothalamic tract which forms a part of the Papez circuit (see Fig. 106.**3**). It also has reciprocal connections with the medial dorsal nucleus and the midline nuclei of the thalamus. These fibers run beneath the ependymal lining of the third ventricle and form the anatomically distinct periventricular system of fibers. The suprachiasmatic nucleus receives retinohypothalamic fibers that relay in the lateral geniculate bodies of the thalamus. The hypothalamus probably receives auditory afferents too though none has been demonstrated.

The hypothalamus has neurovascular links with the posterior pituitary and the infundibulum (see Figs. 82.**3** and 82.**4**). The axons of the supraoptic and paraventricular nuclei of the hypothalamus reach the neurohypophysis to form the hypothalamo-hypophyseal tract. The axons end in dilated terminals called Herring bodies which contain ADH and oxytocin. Axons of the arcuate nuclei of the tuberal region of the hypothalamus reach the median eminence and infundibulum to form the tuberoinfundibular tract. The axon terminals end on the capillary plexuses of the hypothalamo-hypophyseal portal system into which they secrete the hypothalamic releasing or inhibiting hormones.

Connections with the limbic cortex   Anatomically, the most prominent connection between the limbic system and the hypothalamus is represented by the fornix, which originates from the hippocampus and ends in the mamillary body. The fornix forms a part of the Papez circuit. Other anatomically prominent connections between the limbic system and the hypothalamus are the stria terminalis and the ventral amygdalofugal fibers that arise from different parts of the amygdala and terminate in different areas of the hypothalamus.

Functions of the hypothalamus

The functions of the hypothalamus can be summarized as: (1) visceral control through autonomic nervous system and endocrine system, (2) thermoregulation, (3) control of emotions and instinctive behavior like food intake, water and salt intake, and sexual behavior, and (4) control of sleep-wakefulness and circadian rhythms.

subthalamic nucleus (STN), and the third from GP_i/SN_{pr} to thalamus. Thus the striatum disinhibits the STN. The STN in its turn stimulates GP_i/SN_{pr} which then increases its inhibitory output to the thalamus. The overall effect of the circuit is therefore inhibitory. It seems that the indirect pathway assists in braking voluntary movements while the direct pathway simultaneously facilitates the movement. Hence in conjunction, the two pathways are able to bring about the required degree of scaling of movements. It could also result in focusing, i.e., allowing only the voluntary movement to occur while preventing any change in the background posture.

A third component of intricacy of the basal ganglia circuitry is introduced by the SN_{pc} which is inhibited by the striatum through the striatonigral pathway. The SN_{pc} in turn exerts both a facilitatory as well as an inhibitory effect on the cholinergic neurons in the striatum. Through these cholinergic neurons, a facilitatory effect is exerted on the direct pathway to the GP_i/SN_{pr}, pr while an inhibitory effect is exerted on the indirect pathway to the GP_i/SN_{pr}.

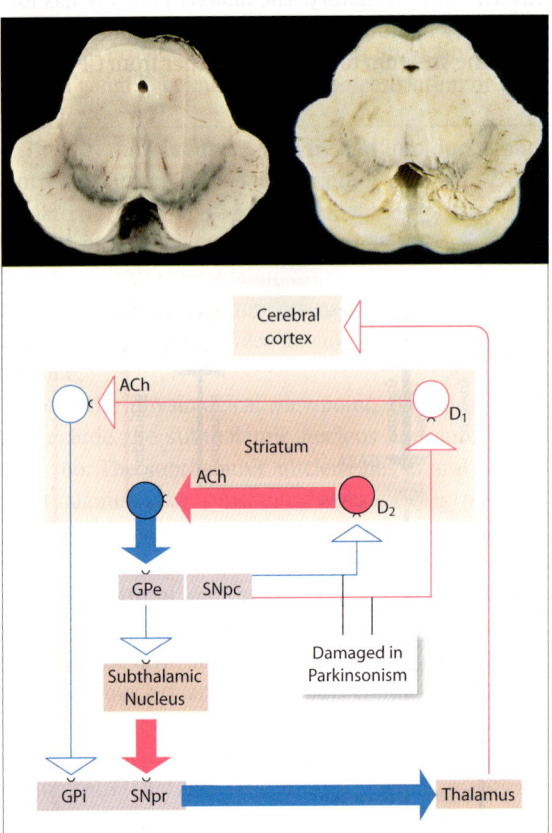

Fig. 99.**4** Parkinson's disease. (*Above, left*) Normal substantia nigra in a brain section. (*Above right*) Substantia nigra in Parkinson's disease. (*Below*) Consequences of damage to the nigrostriatal pathway.

▪ Neurotransmitters in basal ganglia

The neurotransmitter secreted by the nigrostriatal pathway is dopamine. Dopamine exerts an excitatory effect when it acts through D_1 dopamine receptors while it exerts an inhibitory effect when it acts through D_2 receptors. These receptors are present on excitatory cholinergic interneurons in the striatum. In other areas of the basal ganglia, excitatory neurotransmission is through glutamate while inhibitory neurotransmission is mostly through *GABA*. The other inhibitory neurotransmitters involved are enkephalins and substance P.

▪ Disorders of Basal Ganglia

Dysfunction of the basal ganglia result in both hyperkinetic and hypokinetic features. Hyperkinetic abnormalities occur in choreoathetosis and hemiballismus. Parkinson's disease is characterized by both hyperkinetic (rigidity and tremor) and hypokinetic features.

▪ Parkinson's disease (Paralysis agitans)

Although dopamine and dopamine receptors decrease normally with age, in Parkinson's disease aging is associated with an accelerated loss of dopamine and dopamine receptors. Degenerative changes are observed in the substantia nigra with marked reduction of dopamine in the striatum and substantia nigra. In the absence of dopamine, there is a decrease in the excitatory output from the direct striatothalamic circuit while there is an increase in the net inhibitory output from the indirect striatothalamic circuit via the subthalamic nucleus (Fig. 99.**4**). Either way, there is an increased inhibitory output from the globus pallidus internus which explains why pallidectomy is effective in the treatment of Parkinsonism.

The term Parkinsonism refers to conditions that result in symptoms similar to those in Parkinson's disease. These include certain viral infections and treatment with certain drugs (phenothiazines) which block dopamine D_2 receptors. Experimental Parkinsonism can be produced by the drug MPTP (methylphenyl-tetrahydro-pyridine).

Hypokinesia Dysfunction of the putamen loop hampers the smooth transition from one motor program to another. It results in akinesia, bradykinesia, hypokinesia, inability to execute simultaneous actions, and defective kinetic automatism. The patient has a characteristic festinating gait (festinate = hurry). He bends slightly forward and walks with short, quick steps as if to catch up with his own center of gravity to prevent

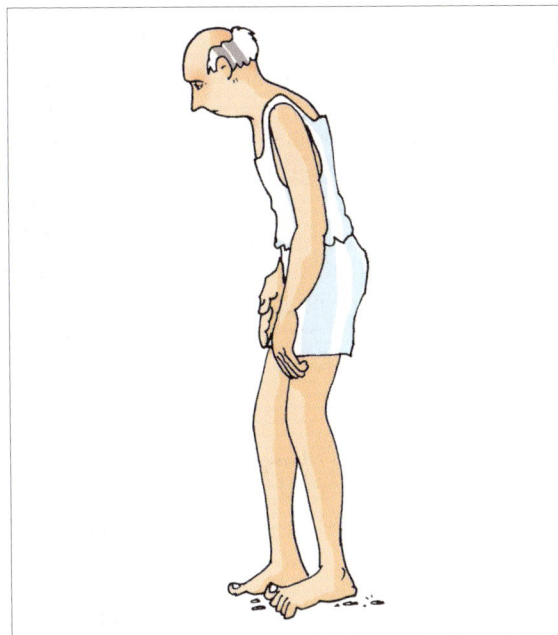

Fig. 99.**5**   Typical posture of a patient with Parkinson's disease while walking.

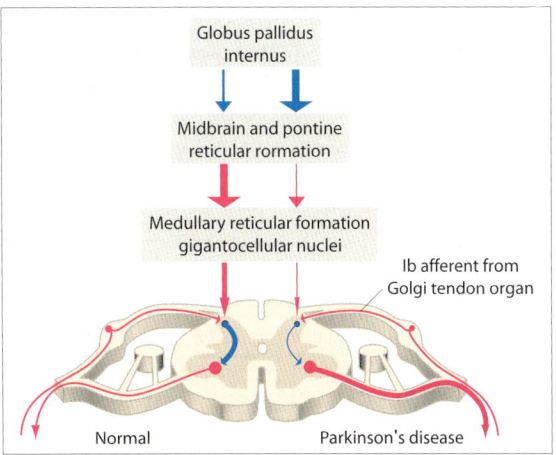

Fig. 99.**6**   Mechanism of Parkinsonian rigidity.

himself from falling (Fig. 99.**5**). He experiences difficulty in taking initial steps and in stopping.

(1) Akinesia is difficulty in initiating movements and a decrease in spontaneous movements. The delayed motor initiative becomes evident from the prolongation of visual or auditory response time. For example, if a subject is asked to press a button as soon as a bulb is lighted, the usual time lag is less than 0.5 seconds. This response time is prolonged in Parkinson's disease. (2) Bradykinesia means slow performance of voluntary movement. (3) Hypokinesia means difficulty in reaching a target with a single continuous movement: the movement must stop and resume a few times before completing itself. (4) Inability to execute simultaneous actions is apparent when the patient, for example, can salute when seated but not when walking. (5) Inability to execute sequential actions is apparent when the patient becomes stuck in the middle of an otherwise automatic motor action, and after performing one motor act, he finds it difficult to initiate the next motor act. (6) Defective kinetic automatism is the loss of associated movements like reduced facial expression and hand gestures during speech or absence of arm swinging during walking. As a result, the patient has a masked face appearance with no emotional response.

Rigidity occurs due to increased muscle tone of both agonist and antagonist muscles. During passive flexion or extension of a limb, the muscular resistance increases and decreases alternately as if it is overcoming a series

of catches (cog-wheel rigidity). Sometimes, there may be a more uniform resistance to passive flexion (lead pipe rigidity). The cause of rigidity remains a subject of debate. One of the theories suggests that it results from withdrawal of facilitation of inverse-stretch (Golgi tendon) reflex. In Parkinson's disease, the increased inhibitory output from the globus pallidus internus reduces the descending facilitatory influences on the inhibitory interneuron in the Golgi tendon reflex pathway (Fig. 99.**6**). Since the Golgi tendon reflex is inhibitory to muscle tone, reduction of supraspinal facilitatory inputs to the reflex increases the muscle tone. The resulting rigidity is different from the spasticity that results from excessive facilitation of the stretch reflex.

Tremor at rest is a prominent sign in Parkinson's disease. The frequency of tremors ranges from 3 to 6 Hz and they are described as fine tremors to distinguish them from cerebellar tremors. The tremor activity is usually suppressed during voluntary movements and sleep and is exaggerated by stress and anxiety. The tremors often take the form of pin-rolling movement of the hand. The tremor seems to occur due to a pacemaker activity in the nucleus ventralis intermedius of the thalamus. Thalamic neurons have an intrinsic autorhythmicity and it is possible that this automaticity is unmasked by the increase in the inhibitory input from the globus pallidus. The unmasked thalamic pacemaker activity induces oscillations in the long-loop reflex pathways originating from muscle spindle, resulting in tremors.

Parkinsonism is treated by administration of L-dopa. Direct administration of dopamine is useless since it does not cross the blood brain barrier. Anticholinergic drugs too are effective. Surgical destruction of the globus pallidus (pallidectomy) or the subthalamus

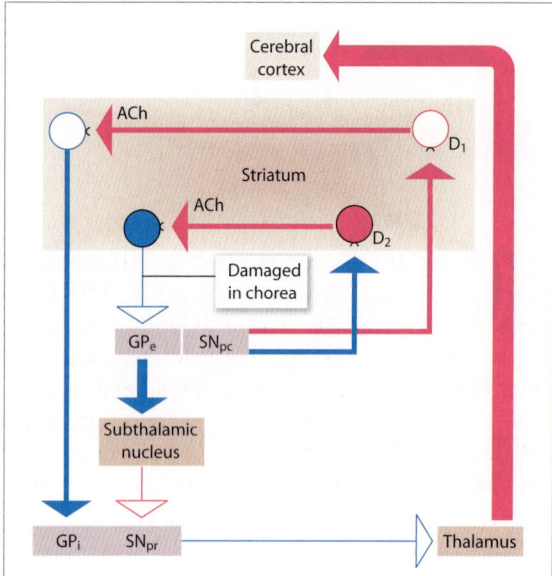

Fig. 99.**7**   Site of neuronal damage in chorea.

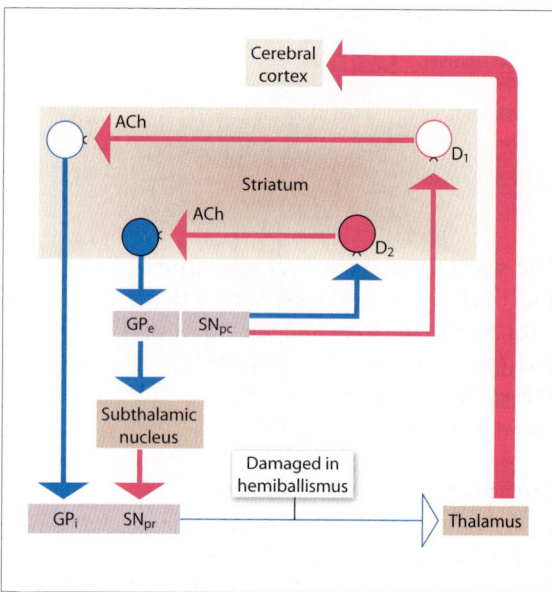

Fig. 99.**8**   Site of neuronal damage in hemiballismus.

is effective in patients poorly controlled by medication. Improvement has also occurred following surgical implantation of dopaminergic cells from the patient's own adrenal medulla or carotid body or better still, from fetal striatal tissue.

Choreoathetosis

Chorea is characterized by brisk, jerky, and purposeless movements of the distal parts of the extremities, and is usually associated with twitching of the face. A slower version of chorea is athetosis which is characterized by slow, worm-like writhing movements of the extremities, affecting chiefly the fingers and the wrists.

Choreoathetosis is associated with the degeneration of the indirect striatothalamic pathway via the subthalamic nucleus (Fig. 99.**7**). Since the overall effect of this pathway is inhibitory, damage to this pathway results in hyperkinesia. Athetosis is frequently seen in damage of putamen as a result of birth-injury while chorea is mostly due to damage to the caudate nucleus. Examples of chorea that are genetically inherited are Huntington's chorea and Wilson's disease.

Huntington's chorea is associated with degeneration of caudate nucleus (most affected), putamen (moderately affected), and globus pallidus (least affected). It is associated with increased glutamate neurosecretion that results in glutamate excitotoxicity.

Wilson's disease (earlier called hepatolenticular disease) is the widespread manifestation of copper toxicity which occurs due to impaired biliary excretion of dietary copper. The toxic effects are most pronounced in the liver and the brain. In the brain, the lesions are widespread but most marked in the putamen and to a lesser degree, in the globus pallidus and caudate nucleus. Wilson's disease is characteristically associated with low plasma levels of ceruloplasmin. This is because the increase in copper in liver cells inhibits the binding of copper to apoceruloplasmin; the two must be in an optimum proportion for the hepatic synthesis of ceruloplasmin.

Quite a few immunological disorders are associated with chorea. Of these, the best known is *Sydenham's chorea* which occurs following streptococcal infection: antibodies to β-hemolytic streptococci cross-react with neurons of the corpus striatum and damage them.

Hemiballismus

Hemiballismus is a rare disease caused by degeneration of subthalamic nucleus on the opposite side. Damage to the subthalamus reduces the inhibitory output from the GP_i/SN_{pr}, resulting in the disinhibition of the thalamic output (Fig. 99.**8**). The condition is characterized by wild, flail-like movements of the contralateral arm.

The cerebral cortex is a layer of gray matter, less than 4 mm thick, present on the surface of the cerebral hemispheres. It is thrown into folds which increase its surface area many times. The surface elevations created by these folds are called gyri and the grooves separating them are called sulci (Fig. 100.1).

Cortical cells Although there are several types of cells in the cortex, most of them are either pyramidal cells or granule (stellate) cells. Pyramidal cells are projection cells (Golgi type-I) that leave the cortex while the granule cells are interneurons (Golgi type II cells) that are confined to the cortical laminae. Pyramidal cells are so named because they have a pyramidal-shaped cell body. Pyramidal and granule cells mostly contain excitatory neurotransmitters like glutamate and aspartate. The rest contain the inhibitory neurotransmitter GABA.

Cortical areas *Phylogenetically*, the cortex is subdivided into two parts; the allocortex (old cortex) which forms about 10% of the entire cortex, and the neocortex which comprises the remaining 90%. Since most of the allocortex is located around the peripheral margin of the diencephalon, it is also called the limbic cortex (limbus: periphery).

Cytoarchitecturally, six laminae are discernible in the cerebral cortex. These laminae are numbered I to VI from the outside in, as shown in Figure 100.2. Lamina I contains a dense network of dendrites of pyramidal cells and the axons of the granule cells. Lamina II contains mainly granule cells and some small pyramidal cells. Lamina III contains predominantly medium-sized pyramidal cells. Lamina IV contains predominantly granule cells. Lamina V contains predominantly large-sized pyramidal cells. Lamina VI contains all types of cells.

Based on the variations in the laminar structure, five variants of cortex have been described, viz., granular, agranular, frontal, parietal, and polar. In frontal, parietal, and polar cortices, all the laminae are well developed. Their names are misleading since the so-called frontal cortex can be found in the parietal area too and *vice-versa*. The granular cortex, which is present characteristically in the somesthetic areas, contains numerous granular cells but few pyramidal cells. The agranular cortex, which is characteristically present in the motor cortex, contains numerous and some of the largest pyramidal cells but has few granular cells.

Cortical areas are also divided into several Brodmann areas (Fig. 100.3) based on a more detailed histology.

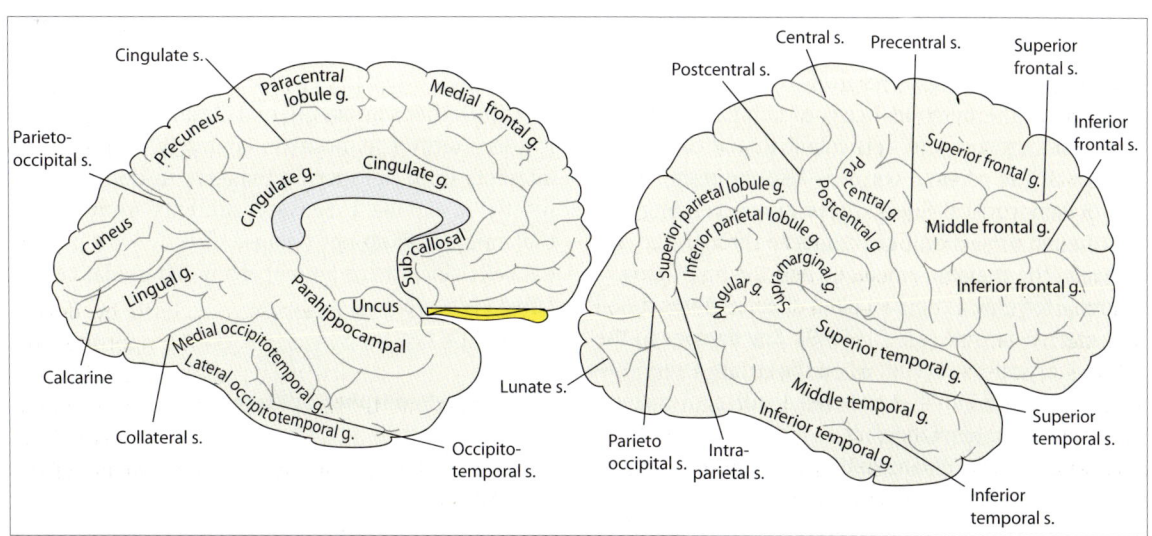

Fig. 100.**1** Sulci and gyri on the medial surface of the left cerebral hemisphere (*left*) and the superolateral surface of the right cerebral hemisphere (*right*). Compare with Figures 95.**1** and 95.**3**.

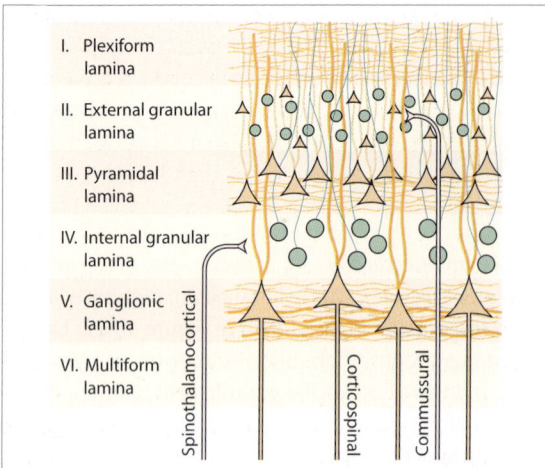

Fig. 100.**2** Cerebral laminae and the predominant cells present in them.

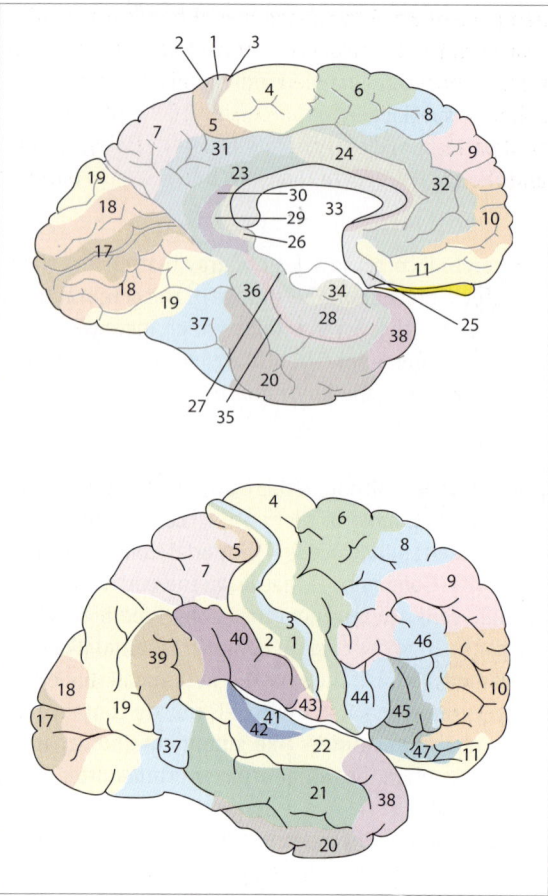

Fig. 100.**3** Brodmann areas on the medial surface of the left cerebral hemisphere (*above*) and the superolateral surface of the right cerebral hemisphere (*below*).

The Brodmann areas do not reflect much functional difference as was previously assumed but still retain their usefulness as a convenient grid for referring to different cortical areas.

Functionally, the cortex is divided into the motor and sensory areas. The sensory areas are more extensive than the motor areas, and are further subdivided into the unimodal (visual, auditory, olfactory, and vestibular) and the multimodal (or polysensory) association areas. There are three polysensory association areas, namely, the posterior (parietooccipitotemporal) association area, the anterior (prefrontal) association area, and the limbic association area (Fig. 100.**4**). These areas are discussed in detail below.

Cortical projection fibers Underneath the cortical gray matter lies the white matter which is made of fibers that originate or terminate in the cortex. Projection fibers that descend from the cerebral cortex originate in the pyramidal cells of lamina V. They include the corticostriate, corticobulbar, and corticospinal fibers. The cerebral cortex is also the terminal for the thalamocortical fibers. These long ascending and descending projection fibers squeeze through the gap between the thalamus medially and the basal ganglia laterally, forming a compact band of nerve fibers called the internal capsule (see Fig. 95.**5**). Any damage to the internal capsule, as occurs when the striate artery (see Fig. 95.**7**) bleeds into it, is associated with contralateral hemiplegia and hemianesthesia.

Pyramidal cells of lamina II and III give rise to association fibers (axons interconnecting different cortical areas of the same hemisphere) and commissural fibers (fibers interconnecting the two hemispheres). Prominent examples of association fibers are the neurons connecting Wernicke's area with Broca's area or those connecting the visual cortex with the frontal eye field. The major commissural pathway is the corpus callosum, the function of which is discussed on p 709 (see Fig. 111.**4**).

Third-order sensory afferents from specific thalamic nuclei terminate mostly in lamina IV, while sensory afferents from nonspecific thalamic nuclei terminate mostly in lamina I. Sensory afferents from specific thalamic nuclei do not synapse directly with pyramidal cells since there are not many pyramidal cells in lamina IV.

Interhemispheric differences

In most individuals, the left hemisphere is specialized for generating complex voluntary movements. Hence, most individuals are right-handed and the left hemisphere, which controls the right hand, has been called the dominant hemisphere. The term dominant hemisphere was later abandoned when it was realized that both the hemispheres showed certain specialization. In most right-handed individuals, the right hemisphere is specialized for processing of complex geometrical patterns and of faces, auditory processing of music and the tactile recognition of complex spatial patterns,

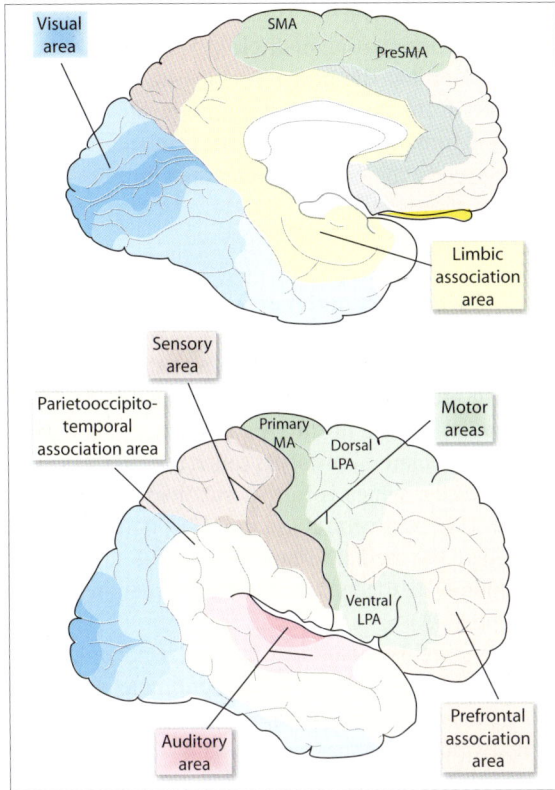

Fig. 100.**4** Functional areas on the medial surface of the left cerebral hemisphere (*above*) and the superolateral surface of the right cerebral hemisphere (*below*).

and mental rotation of complex shapes. This hemisphere is called the *representational hemisphere*. On the other hand, visual processing of written language, auditory processing of spoken language and language-related motor functions like speech, reading, and writing are all performed better by the left hemisphere. This hemisphere is also called the *categorical hemisphere*. By way of generalizing, therefore, it can be stated that the representational hemisphere is involved in visuospatial processing while the categorical hemisphere is specialized for *sequential-analytical processing*. Thus, the representational hemisphere does geometry better while the categorical hemisphere does arithmetic better!

There are several anatomic differences between the two cerebral hemispheres. A major difference is attributable to the specialization of the left hemisphere (in right-handed persons) in language functions and the right for music functions. Thus, the planum temporale (a part of Wernicke's area, p 716), which is involved in language-related auditory processing, is larger on the left than on the right whereas there are two Heschl's gyri on the right but only one on the left.

▪ Sex differences

Sex differences in cerebral functions are observed in motor skill, spatial analysis, mathematical aptitude, perception, and verbal abilities. In general, women have more fluent language whereas men tend to perform better in spatial analysis. In mathematics, women are better in computations while men are better in mathematical reasoning. Women are better in perception of faces and body postures while men are better in perception of mechanical devices and geometrical designs. Women have better spatial memory, e.g., of landmarks

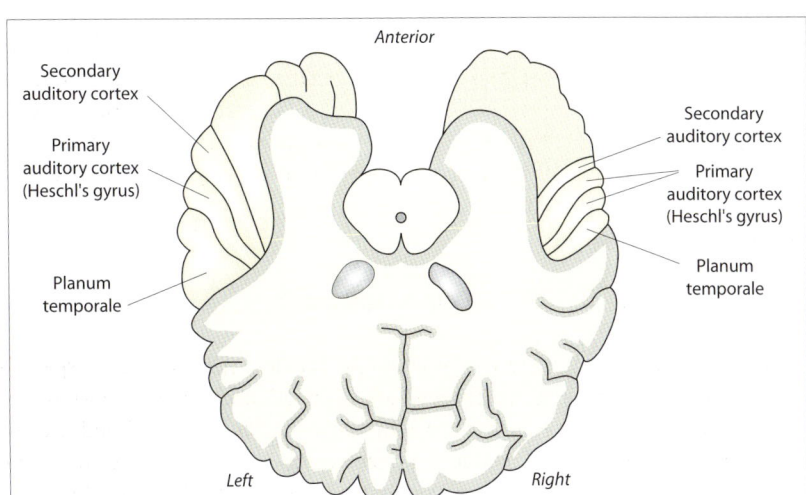

Fig. 100.**5** The superior view of a horizontal section of the brain through the lateral sulcus, showing interhemispheric differences in the temporal lobe. Note that the planum temporale is larger on the left side. The primary auditory cortex has two gyri on the right side and only one gyrus on the left.

along a route, while men are better in spatial analysis as required for navigational tasks or mental rotation. Women show better fine motor control in delicate tasks while men perform better in throwing objects at targets or in intercepting flying objects. Anatomically, men have larger brains with more neurons than women of comparable body size.

Motor Areas

The motor areas of the cerebrum comprise the primary motor area and the non-primary motor areas that are also called Ms-I and Ms-II respectively, indicating that although these areas are primarily motor areas, they have sensory functions too. Most of these motor areas are anterior to the central sulcus and are called precentral areas. All motor areas, whether primary or nonprimary, have an agranular cortex (i.e., the lamina IV is either deficient or absent). They give rise to the pyramidal (corticospinal) neurons which terminate in the brainstem and spinal cord.

Primary motor area

The primary motor area occupies most of area 4. It includes the precentral gyrus on the superolateral surface and the anterior part of paracentral lobule on the medial surface of the hemisphere. The primary motor area receives somatotopically organized inputs directly from the primary somatosensory area (Fig. 100.**6**). It also receives some somatosensory neurons relayed through the VPL_o nucleus of the thalamus by spinothalamic afferents. Hence each neuron in the primary motor area has a receptor field (p 648) in the periphery. For the same reason, a sensory stimulus can travel all the way up to the primary motor area and produce muscle contraction reflexly. Such a reflex is called a transcortical reflex, an example of which is the long-loop stretch reflex (p 662). The primary motor area also receives afferents from area 5 in the parietal cortex, which is the sensory association area involved in integrating multiple sensory modalities for motor planning. The primary motor area also receives afferents from the nonprimary motor areas (see below) and from the cerebellum. The cerebellar afferents are relayed to the primary motor area through the VL_c nucleus of the thalamus.

Nonprimary motor areas

The nonprimary motor areas are located in areas 6 and 8 which are spread over the superolateral as well as the medial surface of the cerebrum. The part located on the superolateral surface is called the lateral premotor area (LPA) which is further subdivided into dorsal (dorsal LPA) and ventral (ventral LPA) parts. The part located on the medial surface includes the supplementary motor area (SMA) and presupplementary motor area (preSMA).

A nonprimary motor area, in general, not only sends pyramidal neurons to the spinal cord directly but also has corticocortical connections with the region of the primary motor area that sends pyramidal neurons to the same spinal segment. The premotor area receives major input from areas 5 and 7 of the parietal cortex as well as

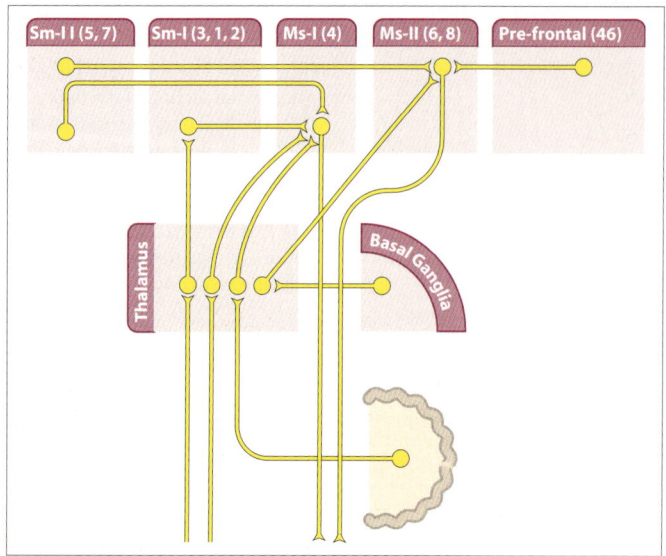

Fig. 100.**6**   Connections of the different areas of the cerebral cortex.

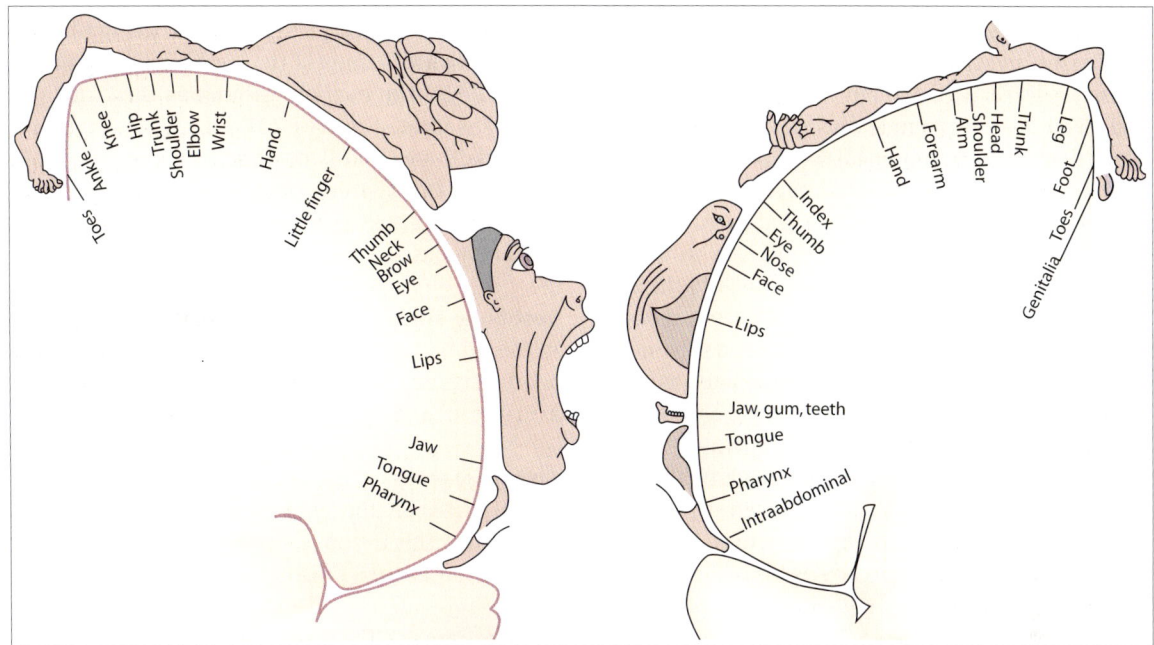

Fig. 100.**7**   (*Left*) Motor homunculus. (*Right*) Sensory homunculus.

from area 46 of the prefrontal cortex, which is important for working memory. The premotor cortex also receives afferents from the basal ganglia that are relayed through the VL$_o$ nucleus of the thalamus. The functions of the subdivisions of the nonprimary areas (see Fig. 105.**3**) are discussed in the context of motor programming.

Electrical stimulation of all parts of the motor area (primary or nonprimary) elicits gross movements of some part of the body. However, the primary motor area has the lowest threshold of electrical stimulation for eliciting gross movements. Centers for movements are represented somatotopically, with the head end below and the leg end up (the motor homunculus). The extent of the cortical representation of a muscle is related to the precision requirements of its movements and not on its size. On the lateral surface of the hemisphere the centers of movements from the bottom up are as shown in Figure 100.**7**. Centers for movements of contralateral lower extremities below the knee, and the control of micturition and defecation, are located on the medial surface in the anterior part of the paracentral lobule.[1]

---

[1] Artificial stimulation of the cortex producing simple movement does not necessarily give the correct picture of cortical motor control. The reason can be appreciated if one considers the analogy of the sensory system. The physiological stimulus for a sensory fiber is not an electrical stimulus. The physiological stimulus for touch fibers is touch, and for temperature fibers, it is warmth or cold. In the case of motor fibers, the physiological stimulus is a complex pattern of inputs from sensory neurons. It is impossible to replicate this pattern of stimulation through artificial electrical stimulation. This is the reason why the motor output differs with Galvanic stimulation and Faradic stimulation of the cortex.

The inverted motor homunculus is evident clinically in extradural hematoma which causes descending paralysis as the blood clot spreads upward over the motor cortex: The paralysis occurs contralaterally, first in the face, then the arm, and finally in the leg.

The *frontal eye field* which regulates the voluntary conjugate movements of the eyes is located in area 8 and is included in the premotor area. Stimulation of this area produces conjugate deviation of the eyes to the opposite side. A *supplementary eye field* is present in the rostral-most area of the SMA. The writing center (of Exner) which coordinates writing movements is located in the upper part of area 6 and is included in the premotor area. The Broca's area for speech (areas 44 and 45) is considered an extension of the lower part of area 6 and can be considered to be a part of premotor area.

Unimodal Sensory Areas

General sensory areas

The somesthetic areas occupy most of the parietal lobe of the cerebral cortex. The functions of the various sensory areas are discussed on p 655 in the context of sensory processing and only the functional subdivisions of the somesthetic area are discussed below.

The **primary somesthetic area** (Sm-I) is located in the postcentral gyrus and extends onto the medial surface in the posterior part of the paracentral lobule (Brodmann areas 3, 1, and 2). It has a granular cortex densely packed with stellate cells with a few small and medium-sized pyramidal cells. It receives afferent projections from posteromedial (VPM) and postero-lateral (VPL) parts of the ventral nucleus of the thalamus, which convey exteroceptive and proprioceptive impulses from the contralateral side. The pyramidal cells of the sensory area contribute fibers to the corticospinal and corticobulbar tracts. However, areas 3, 1, and 2 are primarily sensory and secondarily motor in function. Hence these areas are designated as Sm-I. Sensations from various parts of the body are somato-topically represented in Sm-I upside down with head below and leg up (the sensory homunculus). The sensory area in the paracentral lobule receives the sense of distension from the bladder and rectum. The lower part of the postcentral gyrus acts as the taste receptive center (Fig. 100.**7**).

The **secondary (supplementary) somesthetic area** (Sm-II) is situated along the upper lip of the posterior ramus of the lateral sulcus. Most of the sensory inputs are contralateral. The area for the face is represented in front and that for the leg, behind. The most rostral part of this region produces movements when stimulated. Hence this region is designated as Sm-II.

The **somesthetic association area** is located in the posterior parietal cortex behind the postcentral gyrus in the superior parietal lobule. Unilateral damage to areas 5 and 7 results in contralateral sensory neglect (see Fig. 103.**20**). Bilateral damage results in Balint syndrome, i.e. the inability to create an internal representation of the external world (amorphosynthesis). There is neglect of everything other than a single object, which is reflected in the paucity of saccadic eye movements.

Special sensory areas

Visual areas   The *primary visual area* (visuostriate area or visual area 1) is located in area 17. The visual association areas are located in areas 18, 19, and 20. Area 20 is the inferior temporal gyrus (inferotemporal cortex) which is important for the recognition of faces and facial expressions. It sends a heavy projection to the amygdala, which has an important role in emotions. The projection underlines the importance of facial expressions in emotions.

Auditory areas   The *primary auditory area* (auditory area I) is located in the anterior transverse temporal gyrus (Heschl's gyrus) which is situated on the upper surface of the superior temporal gyrus along the floor of the posterior ramus of the lateral sulcus (areas 41, 42, and 52). The *auditory association area* (auditory area II) is located in area 22 which is also known as Wernicke's area and is involved in higher, language-related auditory processing. Damage to Wernicke's area, where spoken words are analyzed, produces sensory aphasia.

Vestibular areas   A cortical vestibular area is present in the primary somatosensory areas 2 and 3. There is another vestibular area in the secondary somesthetic area (Sm-II) and another in parietal association area 7.

Gustatory areas   The gustatory area is located at the border between the anterior insula and its frontal operculum. This region is rostral to the area where the somatic senses from the tongue are represented.

Olfactory areas   The cortical areas of olfaction include the piriform cortex, corticomedial amygdala, olfactory tubercle, and a part of the entorhinal cortex.

Multimodal Sensory Areas

Parietooccipitotemporal association area

The parietooccipitotemporal area represents the area of confluence of the parietal (sensory), occipital (visual), and superior temporal (auditory) areas of the cortex. It includes the posterior parietal cortex (area 7), the supramarginal gyrus (area 40), and the angular gyrus (area 39).

Our extrapersonal space, with all its somesthetic, visual, and auditory cues, is represented in area 7. Thus, a patient with a lesion in this area may be able to feel and see objects in his extrapersonal space but is unable to synthesize the cues to generate a holistic and meaningful idea of his extrapersonal space. For the same reason, the patient is unable to perform any purposeful movement directed at objects in his extrapersonal space.

Areas 39 and 40 are involved in the higher level of visual processing. The angular gyrus is concerned with language and receives both visual input (reading) and somatosensory input (Braille). Injury to this area produces alexia (inability to read).

Prefrontal association area

The part of the frontal lobe rostral to the premotor area is known as the *prefrontal cortex* (areas 9 to 12). The prefrontal area is highly developed in the human brain. It is connected reciprocally with the dorsomedial nucleus of the thalamus, hypothalamus, and limbic system. It receives long association fibers from almost all areas of the cerebral cortex. Several aspects of prefrontal functions are exemplified by the consequences of damage to the area, as in the case of Pheneas Gage (see Fig. 106.**2**). Broadly, the prefrontal functions may be summarized as follows:

(1) The prefrontal cortex is the site of the *attention allocation system* of working memory and therefore, is important for behavioral continuity. Neurons in the prefrontal cortex fire if visual attention is directed to any object placed at a particular spot in the contralateral visual hemifield. Even if the object is removed, the neurons continue to fire so long as the object's location is remembered. Thus, these prefrontal neurons are said to possess a 'memory field', i.e., a delimited area of the extrapersonal space whose memory it carries. Lesions of these neurons produce 'blind spots' in visuospatial memory.

(2) The prefrontal cortex is important for the central, evaluative component of emotions (see Arnold theory of emotions, p 680). Lesions in the subgenual prefrontal cortex severely compromise reasoning ability and rationality. Emotionally-arousing stimuli therefore are either ineffective or produce abnormal autonomic responses. Schizophrenia, which is associated with a flat affect, is associated with prefrontal hypofunction. Conversely, irritative lesions that stimulate the prefrontal cortex are associated with anger and aggressive behavior.

(3) The prefrontal cortex is the storage site for episodic memory and therefore prefrontal lesions cause source amnesia (p 707).

(4) The prefrontal cortex is important for motor planning (p 667). Two important neuronal circuits, the dorsolateral prefrontal circuit and the lateral orbitofrontal circuit, interconnect the prefrontal cortex to the basal ganglia. Damage to the dorsolateral prefrontal circuit causes cognitive disorder while damage to the lateral orbitofrontal circuit produces obsessive-compulsive disorder.

The operation of bilateral *prefrontal leucotomy* or lobotomy is sometimes practiced in patients with symptoms of mental illness and distressing somatic pain, by severing the connections between the prefrontal area and the dorsomedial nucleus of thalamus. The resulting changes include a loss of anxiety and relief from intractable pain. Surgical lobotomies make the individual docile and hyporesponsive to changing surroundings. It is sometimes argued, therefore, that surgical intervention does not alleviate the mental illness: rather, it demotes the patients to subhuman forms.

■ Limbic association area

The limbic cortex includes large parts of the cortex on the medial wall of the hemisphere, principally the parolfactory gyrus (subcallosal area 25), cingulate gyrus (areas 23, 24, and 32), parahippocampal gyri (area 36), and perirhinal area (area 35). It also includes the hippocampal formation, which consists of the hippocampus, the dentate gyrus, the subiculum, and the entorhinal cortex (area 28). The limbic cortex is functionally related to certain subcortical nuclei like the amygdala and the nucleus accumbens. Together, they constitute the limbic system.

The **hippocampus**, also known as Ammon's horn or cornu Ammonis (CA), is a trilaminar archicortex. It consists of a single pyramidal cell layer with plexiform layers above and below. It is divided into three distinct fields: CA-1, CA-2, and CA-3 (Fig 100.**8**). Neurons in the multimodal association areas of the neocortex first relay in the entorhinal cortex before entering the dentate gyrus (Fig. 100.**9**). The dentate gyrus is the 'gatekeeper' of hippocampal excitability. Neurons originating in the entorhinal cortex terminate on the dendrites of the granule cells in the dentate gyrus, forming the *perforant pathway*. The axons of the granule cells of the dentate gyrus constitute the *mossy pathway* and terminate on

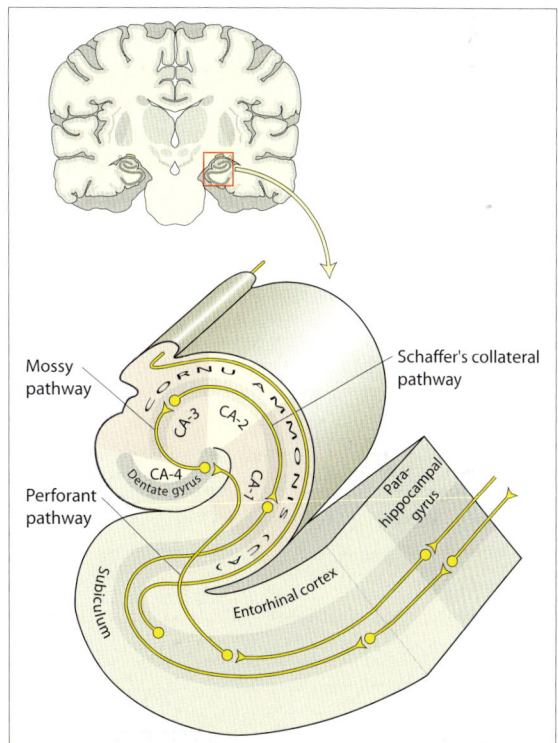

Fig. 100.**8**   The hippocampus, showing the perforant pathway, mossy fiber pathway, and Schaffer's collateral pathway.

the dendrites of CA-3 pyramidal cells. The projections from the pyramidal cells of fields CA-3 and CA-2 to CA-1 constitute *Schaffer's collateral pathway*. The CA-1 field projects heavily to the subiculum which projects to the entorhinal cortex. The subiculum of the hippocampal formation also projects to the mamillary body through the fornix, forming part of the Papez circuit.

The hippocampus slowly transfers newly acquired information present in the short-term memory to the long term memory storage system in the neocortex, ensuring that the new information does not disrupt the existing information. Hence hippocampal damage abolishes neither the long term memory, nor the acquisition of new short-term memory but impairs the transfer of new short-term memory into long-term memory. The hippocampus is concerned with explicit memory only. Hippocampal damage does not affect implicit memory or reflexive learning and new motor skills can still be learnt at the normal pace. For example, the patient with hippocampal damage can learn to ride a bicycle (implicit memory) and yet fail to recall subsequently that he ever rode a bicycle before (explicit memory)!

In explicit memory, the hippocampus is particularly important for spatial representation while the memory for objects is mainly processed in the entorhinal cortex. The hippocampus contains a cognitive map of the spatial environment in which we move. Our location in a particular space is encoded in the firing pattern of the hippocampal pyramidal cells. These cells are therefore called *place cells* that encode a particular position in space. For example, in a house which we are thoroughly familiar with, we can move about even in perfect darkness and reach a particular spot in it with considerable ease. This is possible because of our 'spatial memory' of the house. Spatial learning involves long-term synaptic potentiation in the hippocampal circuits described above.

The hippocampus is one of the sites where seizures are known to originate. The dentate gyrus of the hippocampus is considered to be the gateway to hippocampal excitability. When the granule cells in the dentate gyrus are excited, they produce recurrent inhibition of the afferents through the mossy cells and basket cells. The mossy cells are excitatory interneurons and the basket cells are GABAergic inhibitory interneurons (Fig. 100.**10**). Degeneration of the mossy cells is known to be associated with temporal lobe seizures.

The **amygdala** has two main groups of nuclei: the corticomedial and the basolateral groups. The corticomedial group receives afferents from the olfactory bulb. It is small in humans. The large basolateral group receives afferents from the entorhinal cortex. Amygdalar efferents mainly reach the hypothalamus and limbic areas. The amygdala is an important relay station in the Papez circuit.

The connections of the amygdala with the cingulate gyrus, parahippocampal gyrus, and prefrontal cortex make it important in the central evaluative component of emotions. The amygdala also has strong connections with the autonomic centers of the hypothalamus and brainstem, which makes it important in the peripheral component of emotions. The basolateral nucleus of the amygdala receives a robust input from the inferior temporal cortex (important for the explicit memory of faces). The amygdala therefore is important for 'social cognition'. Amygdalar damage impairs emotional response and social behavior and abolishes fear-conditioning.

The amygdala is also important in instinctive behavior and its lesion causes psychic blindness. Olfactory and gustatory fibers project to those regions of the amygdala that project to the hypothalamus. These olfactory fibers to not mediate conscious olfactory sensation but are involved in reproductive behavior. The gustatory fibers are important for discriminative appetite.

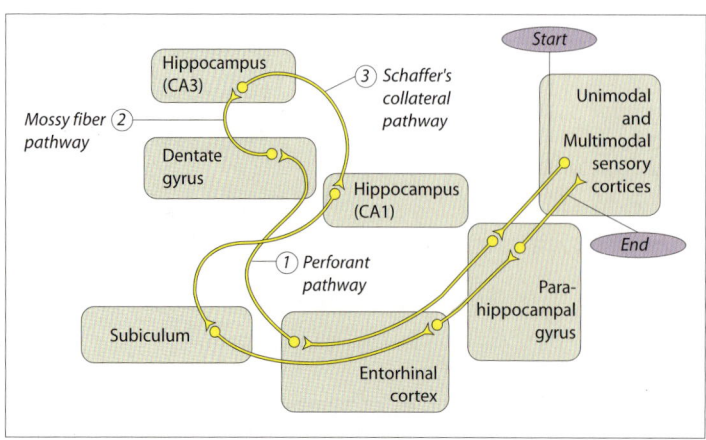

Fig. 100.**9** Schematic representation of the hippocampal circuit. Compare with Figure 100.**8**.

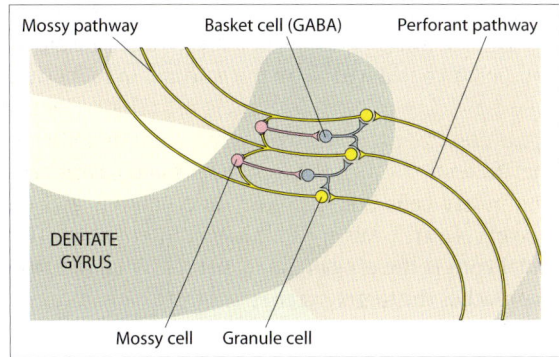

Fig. 100.**10** Dentate gyrus acting as the gateway to hippocampal excitability.

The **nucleus accumbens**, together with the olfactory tubercle, constitutes the ventral system. It receives dopaminergic innervation from the midbrain ventral tegmental area and provides the locomotor drive associated with rewarding stimuli. Psychotropic drugs like amphetamine and cocaine accentuate the locomotor drive by stimulating the mesolimbic dopamine system. It controls the body language in emotional states, a function that is made possible by the afferents it receives from the amygdala through the stria terminalis.

Bilateral ablation of the parahippocampal gyrus and hippocampus causes *Korsakoff's psychosis* which is characterized by severe anterograde amnesia, i.e., inability to establish new memories. Ablation of the piriform cortex produces hypersexuality. Cats and monkeys, after the ablation, even approach animals of other species, e.g., hen. Destruction of the amygdalar cortex abolishes rage and induces docility, hyperphagia, and oral tendencies. Bilateral temporal lobectomy removes a sizeable part of the limbic areas and results in *Kluver–Bucy syndrome* characterized by docility, hyperphagia, hypersexuality, visual agnosia, increased oral tendency, loss of memory, and hypermetamorphosis (failure to ignore peripheral stimuli and therefore easily distracted).

The autonomic nervous system (ANS) is essentially an *efferent visceral nervous system* that controls (excites or inhibits) the contraction of cardiac and smooth muscles as well as certain secretory and metabolic processes. Visceral afferents are sometimes called autonomic afferents: However, many disagree and consider the autonomic system to be essentially an efferent system. In this chapter, the term autonomic nervous system will be considered strictly as an efferent nervous system.

The hypothalamus plays an important role in the regulation of autonomic activity and has been called the head ganglion of the ANS. However, it is now known that the limbic cortex is equally important in the regulation of the ANS. Efferent fibers from the hypothalamus descend through the brainstem to terminate on the preganglionic autonomic neurons of intermediate regions of the spinal gray matter. Descending autonomic fibers do not form well-defined tracts but are intermixed with fibers of other tracts. Their presence in the brainstem can be inferred through physiological experiments. It was earlier believed that the ANS is not under voluntary control. However, persons skilled in the arts of yoga, meditation, and relaxation have demonstrated voluntary control over their blood pressure and heart rate that are normally regulated by the ANS.

The autonomic system has two components: the sympathetic and the parasympathetic (Fig. 101.**1**). The two components have both anatomic as well as functional characteristics. *Anatomically*, the sympathetic fibers originate from the thoracic and lumbar segments of the spinal cord (the thoracolumbar outflow), while the parasympathetic fibers originate from the brainstem and the sacral segments of spinal cord (the craniosacral outflow). *Functionally*, the sympathetic system is triggered during 'fight-or-flight' situations and is associated with catabolic processes while the parasympathetic system is concerned with the vegetative aspects of day-to-day living and supports the anabolic processes of the body.

Unlike in motor nerves where a single motor neuron travels all the way from the spinal cord to the muscle, the autonomic pathway from the spinal cord to the target organ is made of two neurons that synapse in an autonomic ganglion. The neuron originating from the spinal cord is the preganglionic neuron while the one that reaches the target organ is the postganglionic neuron. In general, sympathetic ganglia are located close to the spinal cord whereas parasympathetic ganglia are located close to the target organ. However, there are exceptions to this rule, as discussed below. It is important to stress here that no nerve is entirely made of autonomic fibers. Even the vagus nerve, which carries most of the parasympathetic fibers to the viscera, also carries sensory fibers and nerve fibers supplying skeletal muscles.

■ Sympathetic nervous system

Preganglionic sympathetic neurons originate from cells located in the intermediolateral horn of the thoracolumbar (T1–L3) spinal gray matter. The preganglionic fibers leave the spinal cord through its ventral root along with the somatic nerves (Fig. 101.**2**). However, they soon exit the ventral root through the *white rami communicantes* to enter the paravertebral sympathetic ganglion. The white ramus communicantes is white because it is formed entirely of preganglionic sympathetic fibers that are thinly myelinated B fibers. Postganglionic neurons originating in the sympathetic chain leave the ganglion through the *gray rami communicantes* and reenter the ventral root to enter the spinal nerve. The gray ramus communicantes is gray in color because the post-ganglionic sympathetic fibers are unmyelinated C fibers. Some fibers ascend or descend along the sympathetic trunk to a variable extent and make synapses with the cells of the upper or lower sympathetic ganglia.

Some preganglionic sympathetic fibers pass uninterrupted through the paravertebral sympathetic ganglia. They emerge from the paravertebral ganglia as splanchnic nerves and synapse in the collateral ganglia (e.g., celiac, otic, mesenteric etc.). Some preganglionic fibers pass uninterrupted even through the collateral ganglion to synapse in ganglia in or near the target organ (e.g., in the genital tract). The postganglionic neurons innervating these organs are very short. Finally, some preganglionic fibers synapse directly with the chromaffin cells of the adrenal medulla. The chromaffin cells secrete adrenaline when stimulated.

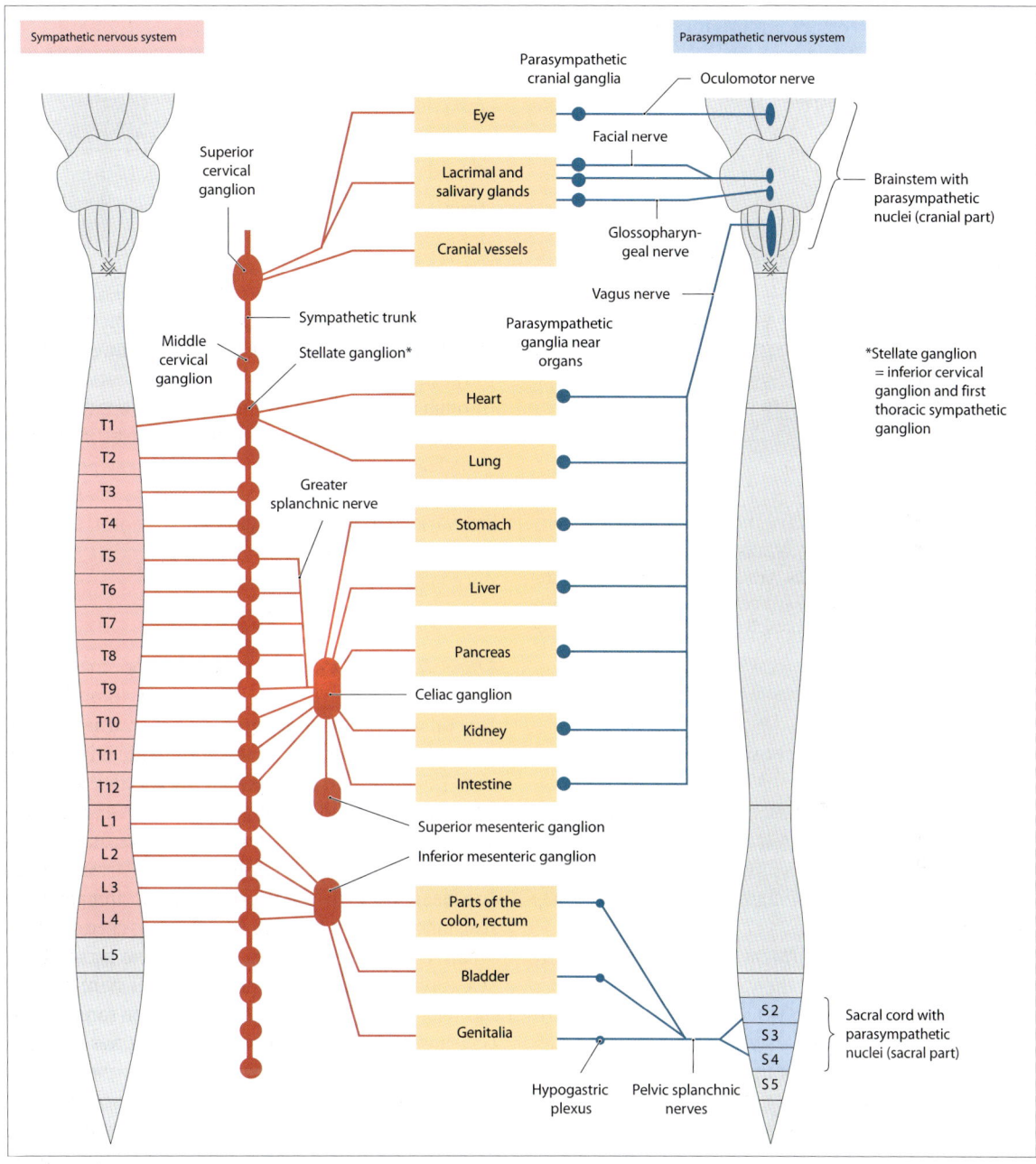

Fig. 101.**1** Organization of the sympathetic and parasympathetic nervous system.

Parasympathetic nervous system

Preganglionic parasympathetic neurons are present in the craniosacral outflow from the central nervous system. *In the cranial nerves*, the parasympathetic nerves are present in the III, VII, IX, and X cranial nerves and originate in the nuclei of the corresponding cranial nerves (see Table 96.**3**). *In the sacral part* of the spinal cord, the parasympathetic fibers originate in the intermediolateral gray horn of S1–S4 and pass out through the ventral spinal root of the corresponding nerves. In either case, the parasympathetic preganglionic fibers travel all the way to the target organ and synapse with the cells of the postganglionic parasympathetic neurons present in the parasympathetic ganglia located near or in the organ.

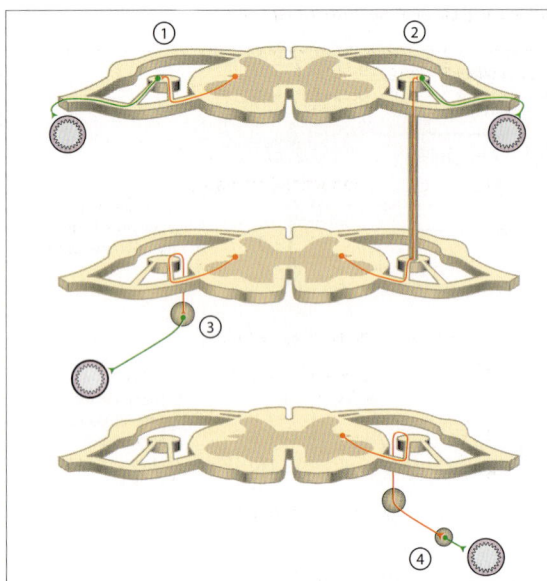

Fig. 101.2 Sympathetic preganglionic and postganglionic neurons. The preganglionic fiber may relay in (1) the segmental sympathetic ganglion, (2) the sympathetic ganglion of a higher or lower spinal segment, (3) a collateral ganglion, or (4) a ganglion near the target organ.

Autonomic Neurotransmission

Postganglionic parasympathetic fibers release mostly acetylcholine as their neurotransmitter. Acetylcholine has two types of receptors: muscarinic and nicotinic. Muscarinic receptors are found on target organs. They are blocked by atropine.

Postganglionic sympathetic fibers release mostly noradrenaline. However, there are exceptions to this general rule: sympathetic vasodilator fibers and sympathetic fibers innervating sweat glands have acetylcholine as their neurotransmitter.

The adrenergic receptors on target organs are of three types, viz., α_1, β_1, and β_2. (See also α_2 inhibitory autoreceptors, p 641.) Norepinephrine acts mainly on α receptors and to a lesser degree, on β receptors. In general, α receptors mediate excitation of smooth muscles, β_1 receptors mediate excitation of cardiac muscle, and β_2 receptors mediate inhibition of smooth muscles. The autonomic effects on different organs are summarized in Table 101.**2**.

Ganglionic transmission At the synapse between the preganglionic and postganglionic neurons, the neurotransmitter is mostly acetylcholine which acts on nicotinic cholinergic receptors located on the postganglionic neuron. These nicotinic receptors are blocked by ganglion blockers like hexamethonium. Some

Table 101.**1** Major neurotransmitters of the autonomic nervous system.

Neurotransmitter	Receptor	Postsynaptic potential
Acetylcholine	Nicotinic	Fast EPSP (30 ms)
Acetylcholine	Muscarinic	Slow EPSP (30 s)
Dopamine	D_2	Slow IPSP (2 s)
GnRH	GnRH receptor	Late slow EPSP (4 s)

cholinergic receptors in the ganglia are of the muscarinic type. Moreover, there are other types of ganglionic neurotransmitters too, like dopamine and GnRH. They act through D_2 receptors and GnRH receptors respectively.

Activation of the postsynaptic receptors in the ganglion leads to the generation of EPSP in the postganglionic cell, and occasionally, produces an IPSP (p 642). The various neurotransmitters, the receptor, they activate and the type of postsynaptic potentials they produce are summarized in Table 101.**1**.

Autonomic Function Tests

Several autonomic function tests are available for diagnosing underactivity of the autonomic nervous system. The heart rate responses to the Valsalva maneuver (Valsalva ratio) and a change of posture from standing to lying (S/L ratio) have been used to assess parasympathetic function. The galvanic skin resistance (GSR) has been used to assess sympathetic function. The resistance offered by skin to a galvanic current is low when it is moist and therefore, GSR gives a measure of sweating. Since sweating increases during anxiety, GSR has been used in the lie detector. For the same reason, GSR has been used in anxiety patients for providing biofeedback on the amount of nervous sweating.

Autonomic dysfunctions

Sympathetic adrenergic failure causes orthostatic (postural) hypotension and ejaculatory failure, while sympathetic cholinergic failure results in anhidrosis (lack of sweating). Parasympathetic failure causes dilated pupils, a fixed heart rate, a sluggish urinary bladder, an atonic large bowel, and erectile failure. With autonomic hyperactivity, the reverse occurs.

Autonomic dysfunction can also be due to localized disorders. Thus, Horner's syndrome is not only an early sign of generalized autonomic failure but also occurs due to involvement of the stellate ganglion in apical lung neoplasm. Horner's syndrome comprises ptosis

Table 101.**2**   Responses of effector organs to autonomic nerve impulses and circulating catecholamines.

Effector organs	Cholinergic response	Noradrenergic impulses	
		Receptor	Response
Eyes			
Radial muscle of iris		α	Contraction (mydriasis)
Sphincter muscle of iris	Contraction (miosis)		
Ciliary muscle	Contraction for near vision	β	Relaxation for far vision
Heart			
SA node	Decrease in heart rate, vagal arrest	β	Increase in heart rate
Atria	Decrease in contractility and increase in conduction velocity	β	Increase in contractility and conduction velocity
AV node	Decrease in conduction velocity	β	Increase in conduction velocity
His-Purkinje system	Decrease in conduction velocity	β	Increase in conduction velocity
Ventricles	Decrease in contractility	β	Increase in contractility
Arterioles			
Coronary	Constriction	α	Constriction
		β	Dilation
Skin and mucosa	Dilation	α	Constriction
Skeletal muscle	Dilation	α	Constriction
		β	Dilation
Cerebral	Dilation	α	Constriction
Pulmonary	Dilation	α	Constriction
		β	Dilation
Abdominal viscera		α	Constriction
		β	Dilation
Salivary glands	Dilation	α	Constriction
Renal		α	Constriction
		β	Dilation
Systemic veins		α	Constriction
		β	Dilation
Lungs			
Bronchial muscle	Contraction	β	Relaxation
Bronchial glands	Stimulation	α	Inhibition
		β	Stimulation
Stomach			
Motility and tone	Increase	α, β	Decrease
Sphincters	Relaxation	α	Contraction
Secretion	Stimulation	α	Inhibition
Intestine			
Motility and tone	Increase	α, β	Decrease
Sphincters	Relaxation	α	Contraction
Secretion	Stimulation	α	Inhibition
Gallbladder and ducts	Contraction	β	Relaxation

Contd...

Contd...

Urinary bladder			
Detrusor	Contraction	β	Relaxation
Trigone and sphincter	Relaxation	α	Contraction
Ureters			
Motility and tone	Increase	α	Increase
Uterus	Variable	α	Contraction
		β	Relaxation
Male sex organs	Erection	α	Ejaculation
Skin			
Pilomotor muscles		α	Contraction
Sweat glands	Generalized secretion	α	Slight, localized secretion
Spleen capsule		α	Contraction
		β	Relaxation
Adrenal medulla	Secretion of epinephrine and norepinephrine		
Liver		α, β	Glycogenolysis
Pancreas			
Acini	Increased secretion	α	Decreased secretion
Islets	Increased insulin and glucagon secretion	α	Decreased insulin and glucagon secretion
		β	Increased insulin and glucagon secretion
Salivary glands	Profuse, watery secretion	α	Thick, viscous secretion
		β	Amylase secretion
Lacrimal glands	Secretion	α	Secretion
Nasopharyngeal glands	Secretion		
Adipose tissue		α, β	Lipolysis
Juxtaglomerular cells			Increased renin secretion
Pineal gland			Increased melatonin synthesis and secretion

(drooping of the eyelid), miosis (pupillary constriction), and hypohidrosis (reduced sweating of the face) on the affected side.

Similarly, gustatory sweating may follow surgery to the parotid gland (Frey's syndrome) or be the result of diabetic autonomic neuropathy. Some gustatory sweating is normal after eating hot, spicy foods. However, excessive gustatory sweating is most commonly a result of damage to the auriculotemporal nerve. One theory holds that the parasympathetic fibers in the auriculotemporal nerve that normally innervate the parotid are misdirected to the overlying sweat glands.

A synapse is the site where an action potential travels across the 20 to 40-nm-wide gap (called the synaptic cleft) separating one neuron (the presynaptic neuron) from another (the postsynaptic neuron). Most commonly, the axon terminal forms the presynaptic element of the synapse while the dendrites form the postsynaptic element. Accordingly, the commonest type of synapse is the axodendritic synapse though axosomatic and axoaxonic synapses are also quite common. The transmission at the synapse is mediated by chemicals called neurotransmitters released from the presynaptic neuron terminal.

The brief sequence of events at a synapse (Fig. 102.1) is as follows: (1) An action potential arrives at the presynaptic nerve terminal and depolarizes the presynaptic membrane. (2) When the membrane depolarizes to –40 to –20 mV, the voltage-gated Ca^{2+} channels in the presynaptic terminal open up, resulting in an influx of free Ca^{2+} into the nerve terminal. (3) The rise in intracellular Ca^{2+} stimulates the exocytosis of synaptic vesicles from the presynaptic terminal. (4) The neurotransmitters present in the synaptic vesicles are released into the synaptic cleft. (5) The neurotransmitter binds to ionic channels present on the postsynaptic neuron. (6) The binding of neurotransmitter to the ion channels opens up the channel, allowing ions to flow across the postsynaptic membrane. If the ionic current is a depolarizing current, the postsynaptic neuron depolarizes, triggering off an action potential in it and thereby completing the process of synaptic transmission. (8) The neurotransmitter released is cleared from the synaptic cleft quickly through reuptake into the presynaptic terminal or nearby glial cells (the commonest mechanism), rapid enzymatic degradation, or quick diffusion away from the synaptic cleft. The neurotransmitter glutamate has its own specific transporter for reuptake while other small-molecule transmitters share a common transporter.

The process of synaptic transmission involves several steps and therefore it takes about 0.5 milliseconds to cross the 20-nm-wide synaptic cleft. In comparison, an impulse traveling the same distance along an uninterrupted nerve membrane takes only 0.2 to 40 nanoseconds. Synaptic transmission is thus a much slower process than nerve conduction and the time taken to cross the synapse is called the *synaptic delay*. The greater the number of synapses in a neural path, the slower is the speed with which an impulse travels along it. The extent to which a nerve impulse slows down gives an estimate of the number of synapses in its path. It was this type of analysis that led to the conclusion that the reflex arc for the stretch reflex (p 661) has only one synapse in it.

The synaptic transmission described above depends on a neurotransmitter. Hence it is called chemically mediated transmission and the synapse is called a chemical synapse to distinguish it from an electrical synapse, which is essentially a gap junction that allows ephaptic conduction between two neurons. The differences between chemical and electrical synapses are summarized in Table 102.1.

Table 102.1 Differences between chemical and electrical synapse.

Chemical synapse	Electrical synapse
Is mediated by the release of neurotransmitters.	Is ephaptic transmission of electrical impulses through gap junctions.
Can occur only in one direction: from the presynaptic terminal containing neurotransmitter to the postsynaptic membrane bearing receptors for the neurotransmitter.	Can occur in both directions.
Speed of synaptic transmission is much slower than the speed of nerve conduction, which results in a synaptic delay of about 0.5 ms.	Speed of synaptic transmission equals the speed of nerve conduction.
Vulnerable to synaptic fatigue on repeated stimulation.	Much less susceptible to fatigue.

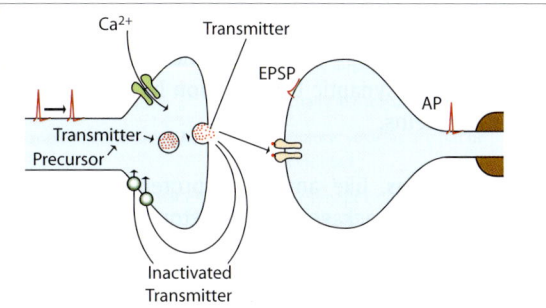

Fig. 102.1 Steps in synaptic transmission.

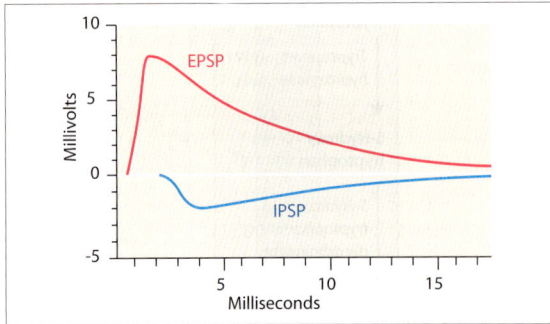

Fig. 102.**8**   Excitatory postsynaptic potential (EPSP) and inhibitory postsynaptic potential (IPSP).

Neuropeptides

The three main endogenous opioid peptides are enkephalins (leucine enkephalin and methionine enkephalin), β-endorphin, and dynorphin. The three major classes of opioid receptors are μ, δ, and κ. Morphine exerts its actions through μ. Naloxone blocks it. The μ receptors are highly concentrated in the periaqueductal gray matter, ventral medulla, and substantia gelatinosa of the dorsal horn, all of which are important in the regulation of pain. Opioid receptors are also located on peripheral terminals in skin, joints, and muscle.

Postsynaptic Potentials

Excitatory postsynaptic potential (EPSP) The binding of the neurotransmitter to receptors on the postsynaptic membrane triggers the opening of the ligand-gated Na^+ channels on the postsynaptic membrane. The resultant rise in the Na^+ conductance depolarizes the postsynaptic membrane. The depolarization thus produced is called the excitatory postsynaptic potential or EPSP (Fig. 102.**8**). The magnitude of the EPSP is 8 mV. The depolarization starts with a latency of 0.5 ms, rises to its peak in 2.0 ms, and then declines with a half-life of 4.0 ms.

Inhibitory postsynaptic potential The soma and dendrites of a neuron usually have synaptic connections with a large number of axon terminals. Not all of them produce an EPSP. Any number of them can produce an inhibitory postsynaptic potential (IPSP). The IPSP is produced when the neurotransmitter increases the conductance of the postsynaptic membrane to K^+ ions, producing hyperpolarization of the membrane. Its magnitude is 2 mV. The hyperpolarization has a latency of 2.0 ms, attaining its maximum at 4 ms and then returning toward the resting potential with a half-life of 3 ms (Fig. 102.**8**).

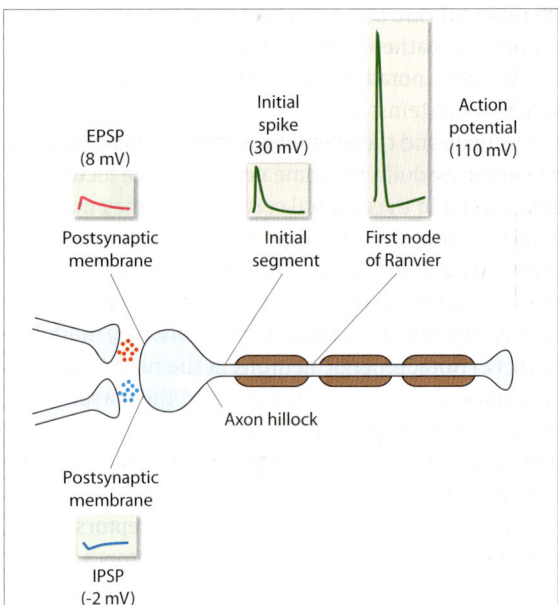

Fig. 102.**9**   Spatial summation of an EPSP and an IPSP, and the generation of initial spike and action potential.

Table 102.**3**   A comparison of EPSP and IPSP.

EPSP	IPSP
Produced when excitatory neurotransmitters (like glutamate) are released	Produced when inhibitory neurotransmitters (like GABA and glycine) are released.
Associated with the opening of ligand-gated Na^+ channels.	Associated with the opening of ligand-gated K^+ channels.
Has a potential of +8 mV	Has a potential of –2 mV.

Synaptic integration   The soma-dendritic tree acts as an integrator that summates all the EPSPs and IPSPs. Depending on the summated postsynaptic potential, synaptic transmission may or may not occur. Suppose there are five presynaptic neurons synapsing on one postsynaptic neuron. If two of the presynaptic neurons produce an EPSP (8 mV each) and the other three produce IPSPs (–2 mV each), the summated potential will be +10 mV and synaptic transmission will occur. On the other hand, if only one of them produces an EPSP and the other four produce IPSPs, then the summated potential will equal zero and no synaptic transmission will occur.

The type of summation described above is called *spatial summation* because the postsynaptic potentials are 'separated in space', i.e., they are produced simultaneously at different sites on the soma or dendrite of the postsynaptic neuron (Fig. 102.**9**). Another type of summation called *temporal summation* occurs when the postsynaptic potentials are 'separated in time'. For example, if an inhibitory presynaptic neuron fires off thrice in quick succession before the decay of the previous ones, each time producing an IPSP of –2 mV at the same site, then the summated potential will be about –6 mV.

Summation of postsynaptic potentials is made easier by the membrane characteristics of the dendrites: The EPSPs and IPSPs produced on the dendrites do not fade quickly, a property called the *holding capacity* of dendrites. Temporal summation is more complete when the time constant of the membrane is high. Spatial summation is more complete when the space constant of the membrane is high. (Time and space constants are explained on p 93.) The holding capacity is more in the axodendritic synapses on the dendritic spines. The thin neck of the spine (see Fig. 8.**1**) prevents the high Ca^{2+} concentration in the spine (during synaptic transmission) from spreading to the main shaft, increasing the holding capacity of the dendrites.

Initial spike and action potential   If the summated postsynaptic potential is an excitatory potential, it depolarizes the *initial segment* (the axon hillock and the proximal unmyelinated part of the unmyelinated axon) by about 30 mV. This depolarization, which is called the *initial spike*, is more than adequate for triggering a fully-fledged action potential at the first node of Ranvier. The initial segment has the lowest threshold of excitation as compared to the other parts of the soma because it has a higher density of voltage-dependent Na^+ channels.

The action potential triggered by the initial spike travels not only anterogradely along the axon but also retrogradely to the soma and dendrites. This retrogradely conducted potential, although identical to the action potential, is called by a different name, the *soma-dendritic (SD) spike*. It serves to restore the soma and dendrites to the resting potential, allowing fresh summation of subsequently generated EPSPs and IPSPs. It is like making calculations on a slate and then wiping the slate clean for a fresh set of calculations.

Synaptic Plasticity

The effectiveness of chemical synapses can be modified for both short and long periods. This modifiability is called synaptic plasticity and is controlled by two types of processes: homosynaptic and heterosynaptic (Fig. 102.**10**). Homosynaptic processes, like postsynaptic facilitation and low frequency depression, occur within the neuron and result from changes in the resting membrane potential or the firing of action potentials. Heterosynaptic processes, like presynaptic inhibition and presynaptic facilitation, involve synaptic input from other neurons.

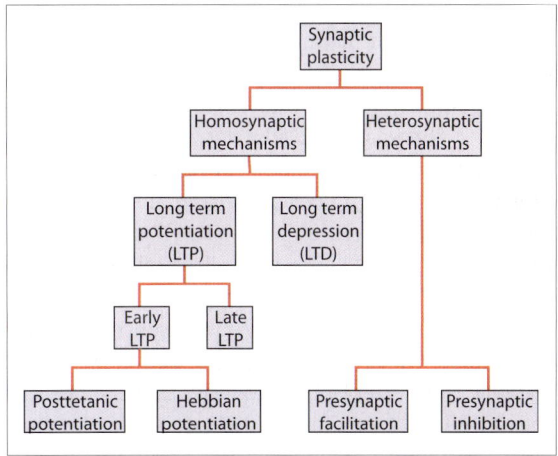

Fig. 102.**10**   Classification of synaptic plasticity mechanisms.

Homosynaptic plasticity

Long-term potentiation (LTP) is the increase in EPSP that occurs following several trains of tetanic stimulation. It persists for hours to days. Long-term potentiation has an early and a late phase. *Early LTP* sets in following a single train of tetanic stimulation, lasts 1 to 3 hours and does not require new protein synthesis. *Late LTP* requires 4 or more trains of stimulation to set in, lasts for at least 24 hours and occurs due to growth of new presynaptic release sites as well as new clusters of postsynaptic receptors.

There are two mechanisms of early LTP phase, viz., posttetanic potentiation and Hebbian potentiation. *Posttetanic potentiation* is typically seen in the neurons of the mossy fiber pathway of the hippocampus (see Figs. 100.**9** and 100.**10**). Since each action potential triggers the entry of Ca^{2+} into the axon terminal, tetanic stimulation results in a high calcium concentration in the axon terminal. The high intracellular calcium level activates Ca^{2+}/calmodulin-dependant protein kinase. The activated kinase phosphorylates synapsin, allowing synaptic vesicles to be freed from their cytoskeletal restraints so that they can be released. Hence, after posttetanic potentiation, each action potential is associated with greater release of the neurotransmitter.

Hebbian potentiation is typically seen in the hippocampus at the synapse between the CA3 neurons (presynaptic) and the CA1 neurons of the Schaffer collateral pathway (postsynaptic). Unlike posttetanic potentiation which requires the repeated stimulation of the presynaptic neuron alone, Hebbian potentiation requires the simultaneous stimulation of both presynaptic and postsynaptic neurons. Activation of the postsynaptic NMDA and AMPA receptors results in a large Ca^{2+} influx into the postsynaptic terminal. The high concentration of cytoplasmic Ca²⁺ activates calcium-dependent protein kinases (like Ca²⁺/calmodulin protein kinases and protein kinase C) which phosphorylate the non-NMDA (AMPA) channels, increasing their sensitivity to glutamate as well as activating some inactive channels. Simultaneously, a retrograde messenger, probably nitric oxide, is released from the postsynaptic terminal and diffuses into the presynaptic terminal, resulting in greater neurotransmitter release through some yet unidentified mechanism.

Late LTP too is initiated by the high concentration of cytoplasmic Ca²⁺. The Ca²⁺ activates membrane-bound adenylyl cyclase, leading to the formation of cAMP and the cascading activation of cAMP-dependent protein kinase and mitogen-activated protein kinase (MAPK). These two kinases diffuse into the nucleus where they activate a transcription protein called CREB-1 (cAMP response element binding protein) and inhibit a transcription-repressing protein called CREB-2, initiating the synthesis of new proteins that lead to structural changes in the synapse.

Long-term depression (LTD) is typically seen in the cerebellum at the synapses between the parallel fibers and the Purkinje cells. It occurs when there is repeated synchronous stimulation of the parallel fiber terminal (through mossy fibers) and the Purkinje cell. It occurs when a prolonged depolarization causes a large influx of Ca²⁺ into the nerve terminal, activating cGMP-dependent kinases and phosphorylating the AMPA channel.

◼ Heterosynaptic plasticity

In heterosynaptic plasticity, the facilitatory or inhibitory neuron (neuron C in Fig. 102.**11**) does not make direct contact with the postsynaptic neuron (neuron B). Instead, it makes an axoaxonic synapse with the presynaptic axon terminal (neuron A) and increases or inhibits the release of the excitatory neurotransmitters from it. There are two ways in which the presynaptic release of neurotransmitters is altered.

In **presynaptic facilitation**, the facilitatory neuron stimulates the release of excitatory neurotransmitters. The facilitatory neuron usually releases serotonin which acts through serotonergic receptors to activate

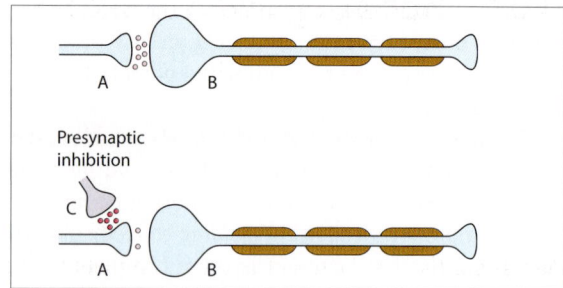

Fig. 102.**11**   Presynaptic inhibition.

cAMP-dependent phosphokinase A (as in group-IIA hormones) and phosphokinase C (as in group-IIC hormones). Together, phosphokinase A and phosphokinase C open up the L-type Ca²⁺ channels, thereby enhancing neurotransmitter release. Phosphokinase A additionally shuts down K⁺ channels, resulting in prolongation of the depolarization which keeps the Ca²⁺ channels open for a longer period.

In **presynaptic inhibition**, the inhibitory neuron releases GABA, which binds to GABA$_A$ or GABA$_B$ receptors on the presynaptic neuron terminals. Stimulation of the ionotropic GABA$_A$ receptors increases the membrane permeability to Cl⁻ ions. Since the Nernst potential of Cl⁻ is close to the resting membrane potential, the rise in Cl⁻ permeability does not change the membrane potential. However, the high Cl⁻ conductance tends to stabilize the membrane potential near E$_{Cl}$ (i.e., the membrane potential does not change easily) in accordance with the general rule that the membrane potential tends to stay near the Nernst potential of the highly permeable ion. Thus, when an action potential arrives at the presynaptic axon terminal, the depolarization of the presynaptic membrane is less than usual and the presynaptic action potential is dwarfed. The smaller action potential is associated with the release of less neurotransmitter from the presynaptic axon terminal.

Stimulation of the metabotropic GABA$_B$ receptors increases the membrane permeability to K⁺ ions and decreases the membrane permeability to Ca²⁺ ions. The rise in K⁺ permeability hyperpolarizes the membrane, dwarfing the presynaptic action potentials and reducing Ca²⁺ influx. The reduction in Ca²⁺ permeability further reduces the Ca²⁺ influx and with it, the neurotransmitter release too.

Sensations that can be discerned at the conscious level are called *conscious sensations*. They include the senses of touch, position of body parts (sensed by proprioceptors), temperature, pain, gravity and acceleration (sensed by the labyrinth), vision, hearing, taste, and smell. Senses whose receptors are present only in the head, i.e., vision, audition, vestibular sense, olfaction, and gustation are called the special senses. There are other *unconscious senses* that never reach our conscious appraisal but nevertheless effect crucial reflex responses. These senses are mostly visceral, e.g., baroreception, but may be somatic, e.g., the position of joints as sensed by the muscle spindles and Golgi tendon organs. Unconscious senses also include the touch and pressure senses that reach the cerebellum through the spinocerebellar tracts and help in postural regulation.

Sensory Receptors

Sensory receptors act as transducers that transform different types of stimuli (mechanical, light, sound, chemical) into electrical signals. Most often, it is the axonal ending of a neuron that is modified into a sensory receptor. In the olfactory epithelium, it is the cell body of the neuron that is modified into the olfactory receptor. The sensory receptor can also be a nonneural cell. In the retina, for example, the photoreceptors are modified epithelial cells.

Classification of sensory receptors

Classification based on stimulus source   Sensory receptors are broadly classified on the basis of the stimulus source, according to which there are three types of sensory receptors: (1) *Exteroceptors* provide information about the external environment, like touch, pressure, temperature, light, sound, taste, smell etc. Exteroceptors are further subdivided into general exteroceptors that are present in the skin and the special exteroceptors present in the head, represented by the receptors for vision, hearing, taste, and smell. (2) *Proprioceptors* provide information about the position and posture of our body in space. They are present in the muscles, tendons, and joints, and also in the vestibular apparatus. Proprioceptors are further subdivided into general proprioceptors present in the locomotor system (muscle spindles, Golgi tendon organ, and Pacinian corpuscles of the joints) and the special proprioceptors present in the head (receptors of the vestibular apparatus). (3) *Interoceptors* or visceroceptors sense the state of the viscera or the internal environment, e.g., receptors sensing blood pressure, plasma osmolarity, blood glucose concentration or the degree of stretching of the urinary bladder.

Classification based on stimulus energy   Further classification of sensory receptors is based on the stimulus energy. According to this classification, sensory receptors are of five types: (1) *Photoreceptors* sense light, e.g., the retinal rods and cones. (2) *Chemoreceptors* sense chemicals. These are used for the sensations of taste and smell as well as for sensing the internal milieu, e.g., the glucoreceptors in the hypothalamus. (3) *Thermoreceptors* sense temperature. Thermoreceptors present in the skin sense the external temperature while those present inside the body sense the core temperature. (4) *Mechanoreceptors* sense force and pressure. For example, the inner hair cells of the ear sense sound pressure, the otolith organs sense the direction of gravitational force, Pacinian corpuscles sense pressure applied on the skin, the muscle spindle responds to stretching forces applied to muscle, and baroreceptors sense blood pressure. (5) *Nociceptors* respond to high concentrations of any chemical, intense heat or cold, or intense pressures. Accordingly, they are called thermonociceptors, chemonociceptors, etc. (Fig. 103.**1**).

General exteroceptors

General exteroceptors (cutaneous and subcutaneous receptors) detect cutaneous sensations that have been grouped into the epicritic and protopathic sensations.

Epicritic sensations involve the finer aspects of touch, pressure, and proprioception and are mediated by encapsulated receptors. These sensations include the ability to: (1) detect a brief, light touch to the skin (as when tested with a wisp of cotton or the finger) and localize its position (tactile localization), (2) resolve by

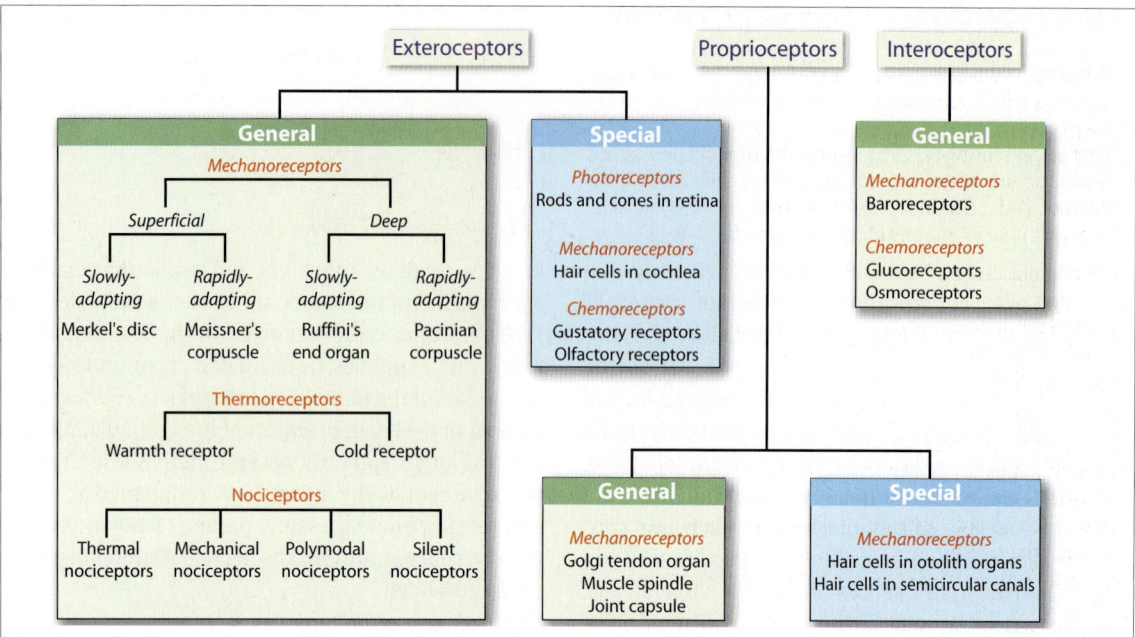

Fig. 103.**1** Classification of sensory receptors based both on stimulus source and stimulus energy.

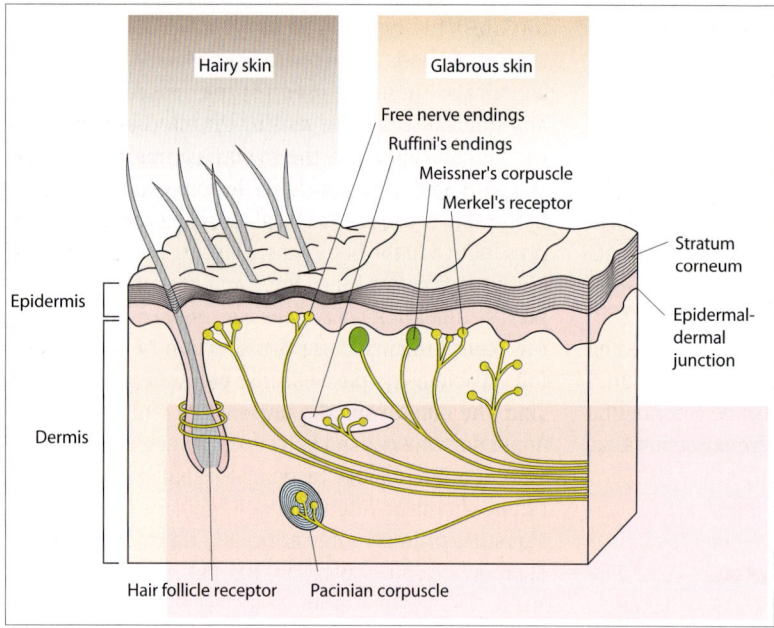

Fig. 103.**2** Location of cutaneous receptors in the skin.

touch spatial detail, such as texture of surfaces (topognosis), and the spacing of two points touched simultaneously (two-point discrimination), (3) recognize the shape of objects grasped in the hand (stereognosis), and (4) detect vibration and determine its frequency and amplitude (pallesthesia).

Protopathic sensations involve pain and temperature senses and are mediated by receptors with bare nerve endings. Protopathic sensations are cruder than epicrit-ic sensations and have a protective function of warning of injury. The pain that occurs due to the activation of nociceptors is called nociceptive pain to distinguish it from neuropathic pain that occurs due to direct injury to neurons. Inflammatory pain is an example of nociceptive pain. Examples of neuropathic pain are the neuralgias, phantom limb pain, and anesthesia dolorosa. Anesthesia dolorosa is the occurrence of pain in the absence of sensation. It sometimes follows therapeutic

Cutaneous receptors

Meissner's corpuscles are rapidly-adapting superficial receptors which are coupled mechanically to the edge of the papillary ridge, a relationship that confers to them fine mechanical sensitivity. The receptor is a pear-shaped, fluid-filled structure that encloses a stack of flattened cells which are modified Schwann cells. The axon terminal zigzags through the various layers of the corpuscle (Fig. 103.**3**).

Merkel's disk receptor is a slowly-adapting superficial is a slowly adapting superficial receptor. It is a small epithelial cell (Merkel cell) in contact with a disk-like expansion of the axon terminal. Merkel's receptors are found in clusters (called Iggo's touch domes) at the center of the papillary ridge.

Ruffini's endings are slowly-adapting deep receptors. They are fusiform structures that link the subcutaneous tissue to folds in the skin at the joints and in the palm or to the fingernails and contribute to our perception of the shape of grasped objects. Parallel bundles of collagen fibers run through the core of the corpuscle and intermingle with the collagen bundles of the surrounding collagen tissues. The tension on the collagen bundles stimulates the axon endings running through the corpuscle.

Pacinian corpuscles are rapidly-adapting deep receptors. They are ovoid structures about 1 mm in length. The corpuscle contains the axon terminal which is surrounded by several concentric lamellae made of very thin, flat cells separated by narrow, gel-filled spaces. The entire structure is enclosed in a thin connective tissue capsule.

There are several other types of sensory receptors, many of which structurally resemble the Pacinian corpuscle and are therefore called Paciniform corpuscles. However, they are much smaller than Pacinian corpuscles and their exact function remains speculative. An example of a Paciniform corpuscle is the end bulbs of Krause.

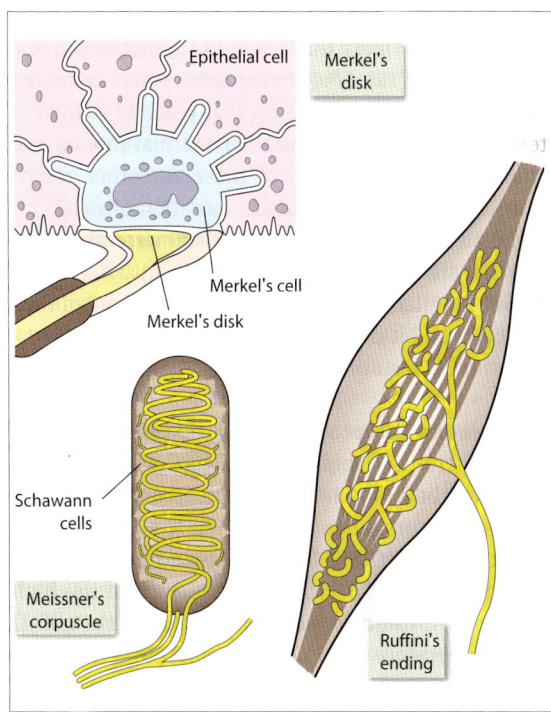

Fig. 103.**3** Morphology of cutaneous receptors.

transection of sensory nerves (e.g., dorsal root nerves) performed in an attempt to block neuralgic pain.

Tactile receptors, i.e., the general exteroceptors for epicritic senses (Figs. 103.**2** and 103.**3**), are subdivided into superficial and deep receptors. *Superficial receptors* are present in the epidermis or the papillary layer of the dermis. In glabrous (nonhairy) skin, the superficial receptors are crowded under the epidermal ridges, especially under the fingerprint ridges, and include Meissner's corpuscles and Merkel's disks. Superficial receptors in hairy skin include the hair follicle receptors that respond to hair displacement and the field receptors that respond to skin stretch. The *deep receptors* are present deeper in the dermis or in the subcutaneous tissues. The deep receptors are the same in both glabrous and hairy skin and include Pacinian corpuscles and Ruffini's end organs.

Touch, pressure, and vibration are different forms of the same sensation. Pressure is felt when the force applied on the skin is sufficient to reach the deep recep-

tors whereas touch is felt when the force is insufficient to reach the deep receptors. Vibrations are rhythmic variations in pressure. Whether a tactile receptor senses pressure or vibration depends on whether the receptor is rapidly-adapting or slowly-adapting. *Slowly-adapting deep receptors* are meant for signaling sustained pressure: they are useless for signaling vibrations. Conversely, *rapidly-adapting deep receptors* stop discharging in response to sustained pressure: They are useful only when the pressure fluctuates rapidly, i.e., during vibration. The higher the adaptation rate of the receptor, the greater is the vibration frequency it can detect.

When the external force reaches only the superficial receptors, the feeling of touch is produced. The superficially located Meissner's rapidly-adapting receptors sense the texture of a surface, i.e., detect whether a surface is rough or smooth. If the skin is rubbed on a rough surface, these receptors are stimulated repeatedly in rapid succession by the irregularities present on the surface. On the other hand, the slowly-adapting Merkel's disk has the smallest receptor field (see below) and provides the sharpest spatial resolution enabling the finest two-point discrimination.

Pain receptors The naked nerve endings of some of the primary sensory neurons (see below) function as nociceptors. The nociceptors are of four types. The *thermal nociceptors* are the naked nerve endings of Aδ fibers that respond to extreme temperatures (>45°C

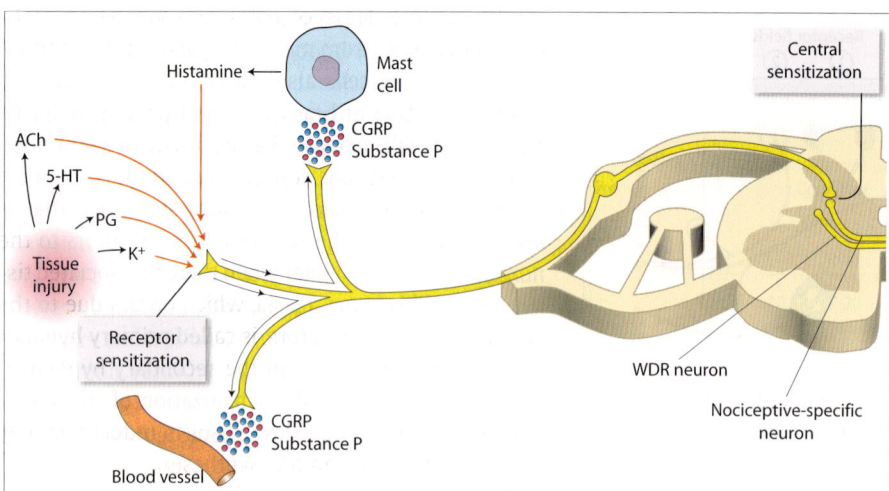

Fig. 103.**6** Sensitization of nociceptors through axon reflex. WDR=Wide dynamic range.

receptors, notably nociceptors, vestibular receptors, and muscle spindle, do not show adaptation.

Adaptation may be slow or rapid. *Slow adaptation* is brought about by the inactivation of Na^+ and Ca^{2+} channels or through the activation of Ca^+-dependent K^+ channels. *Rapid adaptation* occurs when, as in the case of the Pacinian corpuscle, the receptor structure filters out the steady component. When pressure is applied to a Pacinian corpuscle, it initially causes deformation of all its concentric layers and deforms the nerve ending in its core. However, the inner layers slowly regain their original position, relieving the pressure on the nerve ending although the outer layers remain deformed (Fig. 103.**7**).

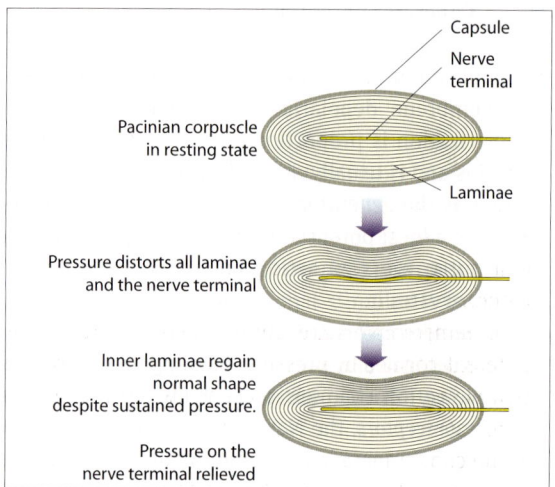

Fig. 103.**7**   Adaptation of Pacinian corpuscle.

▬▬ Sensory Pathways

The various sensations from all parts of the body (except the face) are carried to the spinal cord through the spinal nerves. As the spinal nerve approaches the spinal cord, it divides into the dorsal (sensory) and ventral (motor) roots (see Fig. 96.**2**). The dorsal root ganglion contains the soma of sensory neurons that are the first in a chain of three neurons that convey sensory information to the cortex via the thalamus. It is called the *primary (first order) sensory neuron* and it carries sensory information from the sensory receptor to the spinal cord or the brainstem. The *secondary (second order) sensory neuron* carries the sensations from the spinal cord or brainstem to the specific relay nuclei of the thalamus (the ventroposterior nuclei). The *tertiary (third order) sensory neuron* carries the sensations from the specific thalamic nuclei to the sensory cortex, mostly to its fourth layer.

▪ Dorsal spinal nerve root

The primary sensory neuron is a T-shaped pseudounipolar neuron with peripheral and central processes. Its cell body is located in the dorsal root ganglion. The *peripheral processes* reach the sensory receptors, and the area of the skin supplied by it is known as a *dermatome* (Fig. 103.**8**). The central processes constitute the dorsal nerve root of the spinal nerve. The primary sensory neurons in the dorsal nerve root comprise Aα, Aβ, Aδ, and C group fibers and are often referred to as group, I, II, III, and IV respectively by sensory physiologists. Aγ and B fibers are not present in sensory pathways (Table 103.**1**).

At root entry, each rootlet presents the medial and lateral divisions. The *medial division* contains neurons

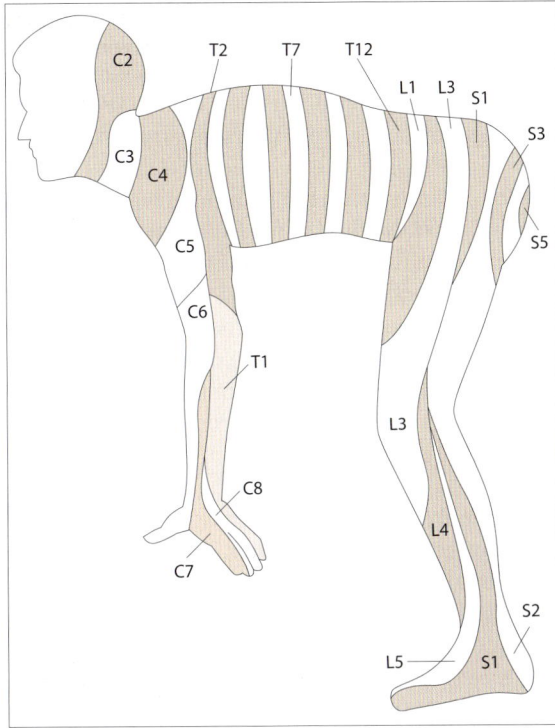

Fig. 103.**8** Dermatomes.

Table 103.**1** Classification of sensory nerve fibers.

Fiber type	Sensory group	Origin
Aα	Ia	Annulospiral endings on intrafusal muscle fibers
	Ib	Golgi tendon organ
Aβ	II	Flower-spray endings on intrafusal muscle fibers
		Touch and pressure receptors
Aγ		Absent in the sensory system
Aδ	III	Receptors for pain (fast), cold, and crude touch
B		Absent exist in the sensory system
C	IV	Pain (slow) and temperature

that carry fine touch, pressure, and proprioception while the *lateral division* contains neurons that carry pain, temperature, and crude touch. The two follow different pathways up to the thalamus. The continuation of the medial division constitutes the dorsal column pathway while the continuation of the lateral division constitutes the anterolateral pathway. These are discussed below in detail. The two pathways converge on the thalamus, from where both travel to the sensory cortex by the thalamocortical pathway. The anatomic separation of the two different tracts has important clinical implications. First, it permits the surgical abolition of pain by cutting the pain-carrying fibers in the anterolateral spinothalamic tract. Second, since touch travels through both pathways it is often spared in lesions of the central nervous system.

The terms fine and crude touch need further explanation. A subject is said to possess the sensation of *fine touch* if the abilities of two-point discrimination and topognosis are intact. A lesion in the dorsal column pathway abolishes these abilities and what remains is only the sensation of *crude touch*, i.e., the ability to detect and localize touch. The spinothalamic fibers carrying crude touch originate only in Meissner's corpuscles and hair follicle receptors. The dorsal column neurons carrying fine touch originate from all types of receptors.

Sensory tracts in spinal cord

Dorsal column-medial lemniscal pathway The medial division of the dorsal root contains the myelinated group-I (Aα) and -II (Aβ) fibers that carry fine touch, pressure, and proprioceptive information and ascend through the spinal cord ipsilaterally. The fibers from the lower part of the body ascend as the *fasciculus gracilis* while the fibers from the upper part of the body ascend as the *fasciculus cuneatus*. The two fasciculi are together called the dorsal columns and they terminate respectively on the nucleus gracilis and the nucleus cuneatus in the medulla. The secondary sensory neurons originating in these dorsal column nuclei decussate in the medulla as the internal arcuate fibers and ascend in the brainstem as the medial lemniscus, terminating finally in the ventroposterior nucleus of the thalamus (Fig. 103.**9**).

Anterolateral-spinothalamic pathway The lateral division of the dorsal root consists of thinly myelinated group-III (Aδ) fibers and unmyelinated group-IV (C) fibers that carry pain and temperature. After entering the spinal cord, these fibers divide into short ascending and descending branches that constitute the dorsolateral tract of Lissauer. They finally terminate in the dorsal horn of the spinal gray matter, one or two segments rostral or caudal to the level of entry of the dorsal root. In the dorsal horn, they synapse with the cell bodies of the spinothalamic fibers.

The secondary sensory neurons carrying *pain and temperature* originate in the dorsal gray horn, mostly from its laminae I and V. These secondary neurons are of two functional subtypes: the *nociceptive-specific neurons* that respond only to nociceptive stimuli, and

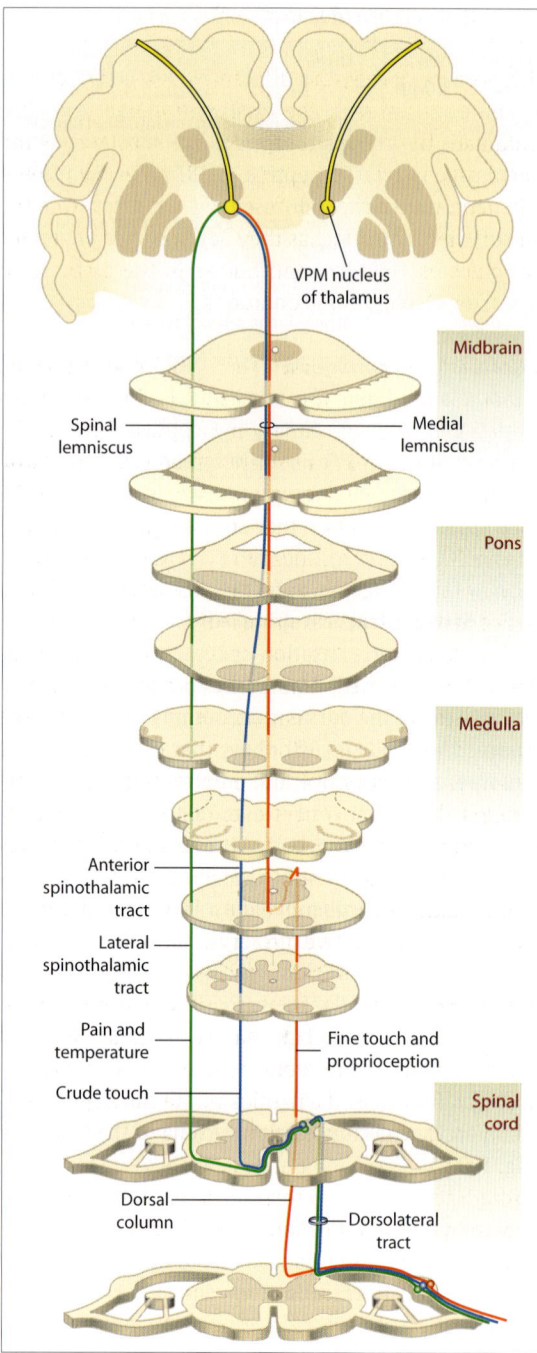

Fig. 103.**9**   Somatosensory pathways.

touch cross even more obliquely.[2] After crossing, the fibers for pain and temperature ascend contralaterally in the lateral spinothalamic tract which, higher up in the brainstem, is called the spinal lemniscus.

The secondary sensory neurons carrying the sensation of *crude touch* ascend in the contralateral anterior spinothalamic tract. In the brainstem, the anterior spinothalamic tract joins the medial lemniscus.

Since pain is conducted by both Aδ and C fibers that differ considerably in conduction velocities, it results in two types of pain: the fast and the slow pain. Following an injury, there is an immediate sharp pain (conducted by Aδ fibers) called the *fast pain* followed after a few seconds by a dull aching pain (conducted by C fibers) called the *slow pain*.

Trigeminal lemniscus   The sensations from the face travel directly to the brainstem through the trigeminal nerve. Fibers carrying fine touch terminate on the *principal sensory nucleus* of the trigeminal in the pons. Fibers carrying proprioceptive information end on the *mesencephalic nucleus* of the trigeminal. Fibers carrying pain, temperature, and crude touch end on the *spinal nucleus* of the trigeminal (Fig. 103.**10**, and see Table 96.**3**). Fibers from these three nuclei of the trigeminal nerve ascend to the thalamus as the trigeminal lemniscus. The fourth nucleus of the trigeminal is its motor nucleus that sends fibers to the muscles of mastication.

Spinal tracts for pain only   There are three other spinal tracts that conduct pain only. The *spinoreticulothalamic tract* conveys pain sensation to the intralaminar nuclei of the thalamus after relaying in the medullary and pontine reticular formation. Pain fibers to the reticular formation produce diffuse pain and bring about a general arousal. The *spinomesencephalic tract* conveys pain to the reticular formation and periaqueductal gray matter in the midbrain. From there, the pain fibers project to the amygdala via the parabrachial nucleus. The amygdala mediates the emotional component of pain. The *spinohypothalamic tract* conveys pain fibers to the hypothalamus, which activates neuroendocrine and cardiovascular responses to pain.

■   **Thalamocortical pathway**

Fibers of both spinal lemniscus, and medial lemniscus, terminate in the same thalamic nuclei, i.e., mainly the ventroposterior (VP) nuclei of the thalamus. Pain and

the *wide dynamic range (WDR) neurons* that respond to both noxious and nonnoxious stimuli (Fig. 103.**6**). The axons of these cells cross the midline in front of the central canal in the anterior white commissure. The pain fibers cross horizontally in the same cord segment. The temperature fibers cross obliquely, i.e., they ascend slightly while crossing. Fibers carrying crude

---

[2]   It is important to bear in mind the obliquity of the crossing fibers while diagnosing the level of a spinal lesion.

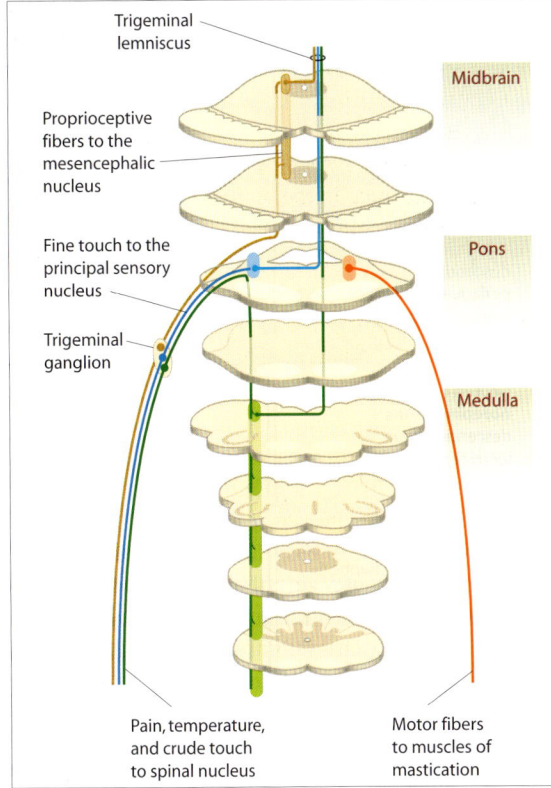

Fig. 103.**10** Trigeminal pathways.

temperature are primarily appreciated by the thalamus. For other forms of sensation, the thalamus is unable to analyze the details of sensations. Fibers carrying tactile and proprioceptive information ascend through the thalamocortical pathway to area 3 of the primary somatic sensory cortex through the thalamic radiations. The sensory cortex receives general sensation from all parts of the body. The various parts of the body have a cortical representation proportionate not to their size but to their innervation density. When the cortical areas are charted out for their peripheral connections and depicted diagrammatically, the distorted shape of a 'little man' emerges. This is called the sensory homunculus (see Fig. 100.**7**).

Subcortical Sensory Processing

Sensory processing occurs as early as in the secondary sensory neurons. This is true for both the dorsal column and the anterolateral pathways. *Processing in the dorsal columns* occurs in the nucleus gracilis et cuneatus. Processing in the anterolateral pathway occurs mostly in the dorsal horn of the spinal cord. Further processing occurs higher up in the thalamus.

Dorsal horn of spinal cord

The primary nociceptive afferents release two classes of neurotransmitters: the *excitatory amino acids* like glutamate and the *neuropeptides* like substance P and calcium-gene-related peptide (CGRP). The neuropeptide neurotransmitters diffuse long distances from the sensory nerve terminal as they have no specific uptake mechanisms, thereby contributing to the unlocalized character of many pain conditions.

Secondary hyperalgesia The nociceptive-specific spinothalamic fibers are sensitized by long-term potentiation when the primary nociceptive C fibers discharge continuously. This sensitization is called *central sensitization* or *winding up* to distinguish it from the nociceptor sensitization that occurs in primary hyperalgesia. Sensitized secondary nociceptive fibers show an exaggerated response to the discharge of primary nociceptive neurons, resulting in secondary hyperalgesia. The central sensitization underlies a clinical symptom called phantom limb pain, an excruciating pain in a nonexistent limb that has been amputated. The sensitization occurs due to the excessive stimulation of the primary nociceptive afferents during amputation. *Phantom limb pain* occurs even if the amputation is performed under general anesthesia and the only way of preventing it is to apply local anesthetics to the site of amputation so as to prevent excessive stimulation of the primary nociceptive afferents.

Dorsal horn gating Lamina II of the dorsal horn of spinal cord (substantia gelatinosa, SG) contains enkephalinergic inhibitory interneurons that inhibit, presynaptically as well as postsynaptically, the nociceptive spinothalamic neurons present in lamina V. The tactile Aβ fibers of the dorsal root stimulate these interneurons through collaterals and thereby inhibit the conduction of pain by the spinothalamic tract. The nociceptive Aδ and C fibers themselves inhibit the SG inhibitory interneurons through collaterals (Fig. 103.**11**). In other words, the SG inhibitory interneurons serve as a gate in the pain pathway. The gate remains open when only Aδ and C fibers are stimulated but tends to close down when Aβ fibers are stimulated simultaneously.

The gate-control theory provides the basis for using vibratory stimulation for reducing pain. In transcutaneous electrical stimulation (TENS), surface electrodes are used to stimulate Aβ fibers that overlap the area of injury. Stimulation of dorsal columns via surface electrodes has also been used. Even the simple massage owes its pain-relieving efficacy to the gate control of pain.

The SG inhibitory enkephalinergic interneurons are also stimulated by the serotonergic neurons descending

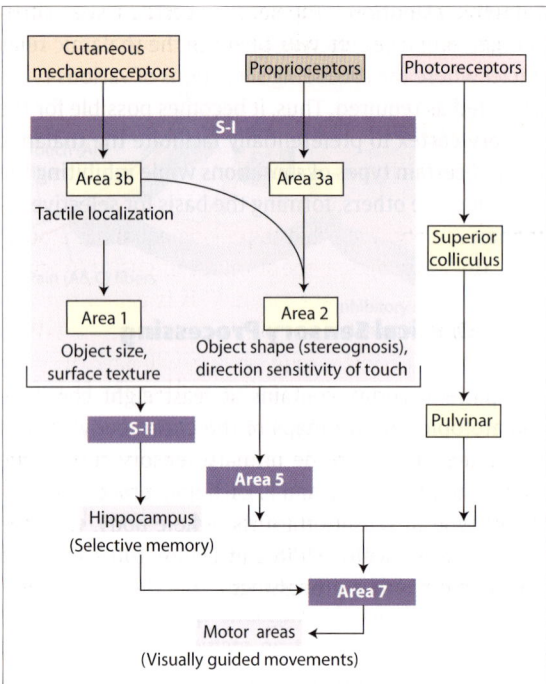

Fig. 103.**15**   Cortical sensory processing.

Secondary sensory cortex (S-II)

The secondary somatic sensory cortex (S-II) receives information from all areas of S-I and determines whether a piece of tactile information is remembered or not. This selective memory is made possible by the projection of S-II to the hippocampus via the insular cortex.

Posterior parietal cortex

The posterior parietal cortex (areas 5 and 7) receives input from S-I as well as input from the pulvinar. The posterior parietal cortices of the two hemispheres are connected through the corpus callosum and are therefore able to integrate information from both sides. *Area 5* integrates tactile information from mechanoreceptors in the skin with proprioceptive inputs from the underlying muscles and joints. *Area 7* belongs to the multimodal association cortex. It integrates the tactile and proprioceptive inputs with visual information, enabling visually-guided hand movements. The posterior parietal cortex enables the perception of extrapersonal space. It projects to the motor areas of the frontal lobe and plays an important role in the sensory initiation and guidance of reaching and grasping movements.

Plasticity of sensory cortex   The afferent connections of the sensory cortex can change with extensive use

or total disuse of the senses. Rigorous sensory use of a particular part of the body leads to an increase in the area of representation of that part in the sensory cortex. Similarly, if a part of the body is ablated, sensory afferents from the neighboring parts establish connections with the deafferented area of the sensory cortex. The same is true for different modalities of sensations. Thus, auditory and tactile afferents have been observed to invade the visual cortex in the blind. Such plastic changes occur not only during development but also in adults.

Cortical pain areas

The cingulate gyrus processes the emotional component of pain. The anterior cingulate gyrus is critical to the perception of thermal pain. The insular cortex receives afferents from the thalamic nuclei and integrates the sensory, affective, and cognitive components of pain. Lesions of the insular cortex results in *asymbolia for pain*: These patients perceive noxious stimuli as painful and can distinguish sharp from dull pain but do not display appropriate emotional response to the pain.

Severe pain can be dissociated from its unpleasant feeling by cutting the deep connections between the frontal lobes and the rest of the brain (prefrontal lobectomy). Patients operated in this way feel the pain but are 'not bothered by it' (p 681).

Sensory Perception

Sensory perception is constructed internally by the brain and is not an exact reflection of the real world. For example, the colors of the rainbow reside only in our brain because electromagnetic radiations, which include light, do not have any color of their own. The relationship between the physical characteristics of a stimulus and the attributes of its sensory perception forms the subject of *psychophysics*. Mostly however, our perception gives a reasonably accurate picture of the reality. When it does not, it is called an *illusion* (Fig. 103.**16**). When we perceive nonexistent things, it is called *hallucination*. For example, cocaine consumption causes tactile hallucinations that give the feeling of ants crawling under the skin although there are none! Hallucinations are common in schizophrenia.

Sensory conflict   In certain situations, different sensory organs can send conflicting information to the sensory cortex. It is believed that motion-sickness results at least partly from the conflicting visual and vestibular signals regarding the inclination of the horizon.

Doctrine of specific nerve energies

The type of sensation perceived depends on the area of the sensory cortex stimulated, which in turn depends upon the class of receptors stimulated. Hence, a neuron connected to thermal receptors will always evoke thermal sensation, whether it is stimulated naturally through its receptors or whether the nerve is stimulated directly. Even if a sensory nerve is denuded of its receptors, it will still evoke the sensation characteristic of the severed receptors. The phenomenon is called the doctrine of specific nerve energies. If a cortical touch area becomes connected with a skin temperature receptor, then the temperature stimulus will either not be recognized or will be misinterpreted as touch stimulus.

The Weber–Fechner law

It has been observed that to provide equal increments in sensory perception, the stimulus intensity has to be increased by a constant factor. This is called the Weber–Fechner law. The law helps explain why a 1-Watt and a 10-Watt bulb (or loudspeaker) will have the same difference in brightness (or loudness) as between a 10-Watt and a 100-Watt bulb (or loudspeaker). This law forms the basis of the decibel scale of loudness, as depicted comically in Figure 103.**17**. Similarly, the law explains why a 2-kg weight feels distinctly heavier than a 1-kg weight but it is difficult to distinguish a 12-kg weight from a 11-kg weight although the difference in weight in both cases is 1 kg.

Sensory Projection

The sensation evoked by a sensory unit will always be projected to (appear to originate from) the area innervated by its peripheral endings. This remains true even if the peripheral end of the nerve is severed and the central stump is stimulated, or even when the area innervated by the nerve is ablated. Thus, in a person with an amputated leg, if a sensory nerve originally carrying pain from the toe is stimulated in the thigh region, pain will be felt in (referred to) the nonexistent toe! The phenomenon is known as the *phantom limb* (nonexistent limb).

Tactile localization becomes inaccurate when abnormal connections are established between the periphery and the cerebral cortex. Suppose an area X

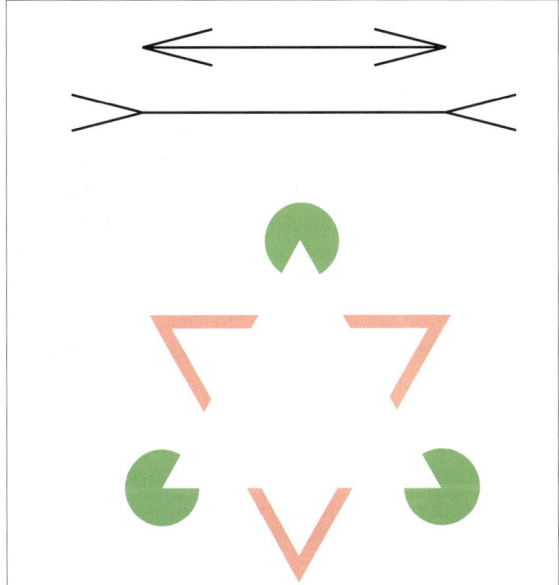

Fig. 103.**16** (*Above*) Müller–Lyer illusion. The line above appears longer than the line below. Measurement will show them to be of equal length. (*Below*) Kanizsa triangle. The white triangle with apex up is visible to most observers although its contours are illusory.

Fig. 103.**17** The Weber–Fechner law at work. Ten babies, not two, must cry together to produce twice the noise produced by a single crying baby.

of the sensory cortex that was originally connected with the base of the thumb becomes connected to the tip of the thumb following peripheral nerve injury and regeneration. Experience has taught the brain that cortical activity at area X represents stimulation of the thumb base. Hence, after aberrant peripheral connections are established, sensory impulses from the thumb tip reach area X and are falsely localized to the thumb base. The false localization can only be overcome, if at all, by prolonged practice.

Referred pain   A sensation originating from the viscera is often not referred to the appropriate visceral area. Instead, it is referred to a dermatome corresponding to the spinal segment through which the visceral afferents relay. The classical example of this is referred pain.

There are two theories of referred pain (Fig. 103.**17**). One theory holds that the visceral and somatic afferents belonging to the same spinal segment converge on the same secondary sensory neuron (in lamina V of the spinal gray matter) and any sensory signal reaching the cortex through this common path is interpreted as coming from the somatic area rather than its visceral counterpart. Another possibility is that any visceral sensory traffic facilitates the adjacent synapses

relaying somatic sensations in the same spinal segment. The subthreshold somatic sensations from the corresponding dermatome are consequently perceived with augmented intensity.

Sensory Disorders

Tabes dorsalis, a consequence of tertiary syphilis, destroys the central processes of the large diameter neurons in the dorsal root ganglia, resulting in the degeneration of the dorsal columns (Fig. 103.**19**). It affects mostly the lumbar and lower thoracic segments of the spinal cord and is associated with loss of fine touch, pressure, and proprioception on the side of the lesion. Loss of proprioception results in *sensory ataxia*: The patient walks on a broad base with the legs apart, eyes fixed to the ground for correcting the steps, raising the legs excessively high and slapping the feet on the ground. Since these patients are heavily dependent on visual cues for maintaining posture, the sensory ataxia is aggravated by closing the eye (positive Romberg's sign), which distinguishes it from cerebellar ataxia.

Syringomyelia is a rare condition that is characterized by overgrowth of neuroglial tissue and cavitation of the gray matter around the central canal of the spinal cord. Fibers that decussate in the gray commissure are destroyed (Fig. 103.**19**), resulting in loss of pain and temperature sensitivity. However, fibers carrying the sensation of fine touch that ascend in the dorsal column escape damage. Hence, syringomyelia is characterized by *dissociated anesthesia*, i.e., loss of pain and temperature sensitivity with retention of touch sensation. The deficiency is bilateral and usually occurs in the hands and arms due to the predilection of syringomyelia for the cervical enlargement of the cord. The gliosis and cavitation may spread, resulting in motor deficits.

Somatic pain fiber
Visceral pain fiber
Spinal lemniscus

Convergence theory of referred pain

Somatic pain fiber
Visceral pain fiber
Spinal lemniscus

Facilitation theory of referred pain

Fig. 103.**18**   Two theories of referred pain.

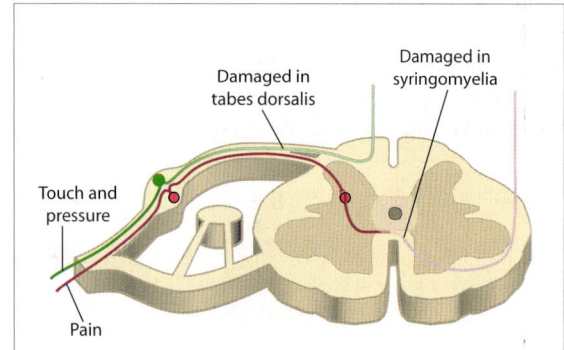

Touch and pressure

Damaged in tabes dorsalis

Damaged in syringomyelia

Pain

Fig. 103.**19**   Site of lesion in tabes dorsalis and syringomyelia.

Fig. 103.**20**   (*Above*) Spatial neglect. The patient was asked to draw a clockface and set it to quarter past twelve. (*Below*) Personal neglect. The food is eaten from one side of the plate.

Fig. 103.**21**   Example of representational neglect. (*Above*) A view of the office desk of a patient. (*Below*) The patient draws the diagram of the desk imagining himself to be seated, first on his own chair (*below left*) and then, on the opposite side (*below right*).

Cortical lesions   Lesions in the primary somatosensory cortex produce *astereognosis*, i.e., loss of the ability for stereognosis. Damage to the right parietal association area results in left-sided sensory neglect as well as left-sided motor deficits like poor eye–hand coordination during reaching, grasping, and hand-orientation. Sensory neglect can take three forms.

In *personal neglect*, the patient disowns the left half of his body or its parts: He is either unaware that the left half of his body exists or believes it to belong to someone else (Fig. 103.**20**).

In *spatial neglect*, the patient is unaware of the left side of his extrapersonal space. Thus when asked to sketch out a clock placed in front of him, he will ignore the parts of the clock in the left half (including numbers 7 to 11) on the clock and draw out the rest.

In *representational neglect*, the patient ignores the left half of mental images that he recalls from memory. For example, if he is asked to sketch out his office desk imagining himself to be seated in his own chair, he forgets to sketch out things to the left. On the other hand, if he is asked to sketch out the same desk imagining himself to be seated on the opposite side of the desk, the sketch once again reveals his left-sided neglect (Fig. 103.**21**). Thus although the patient is evidently aware of the complete spatial organization of his desk, the representational neglect depends on the perspective from which he recalls the mental image. It indicates that spatial memory recalled from the hippocampus must be routed through the sensory association areas.

There are two reflex mechanisms through which a muscle is maintained at a constant length. One is the monosynaptic reflex which is a spinal reflex, and the other is the long-loop stretch reflex whose center is located in the cerebral cortex. The sensory receptor involved in these reflexes is the muscle spindle, an encapsulated proprioceptor located inside the muscle.

Muscles also have a reflex mechanism through which an excessive rise in muscle tension is prevented. The reflex is called the negative stretch reflex or the Golgi-tendon reflex. The proprioceptor involved in the reflex is the Golgi tendon organ which, as the name suggests, is located in the muscle tendon.

Muscle Spindle

Muscle spindles are located within the muscle, intermingled with the muscle fibers. The number of spindles in a muscle is proportionate to the degree of precision required of the muscle. Thus, the gastrocnemius muscles contain as few as 5 to 10 spindles per gram of muscle while the interossei muscles of the hand contain as many as 100 spindles per gram of muscle. The ends of the muscle spindle are fixed to an adjacent muscle fiber.

Within the spindle, a few specialized muscle fibers are present. These fibers are called *intrafusal fibers* to distinguish them from the fibers of the muscle itself which are called *extrafusal fibers* (fusus: spindle). Intrafusal fibers are of two subtypes: the nuclear bag fibers and the nuclear chain fibers. The *nuclear bag fibers* have a bulge in the equatorial region due to the presence of a bunch of nuclei. The *nuclear chain fibers* have no such bulge since the nuclei in its equatorial region are arranged in single file (Fig. 104.**1**).

Sensory innervation of muscle spindle   Two types of sensory nerve fibers originate from the intrafusal fibers. The *annulospiral (primary) endings* are wound around the equatorial regions of intrafusal fibers. They are present on both nuclear bag and chain fibers. Annulospiral endings are Aα nerve fibers and are called type-Ia fibers by sensory physiologists. The *flower-spray (secondary) endings* innervate the peripheral parts of the intrafusal fibers, mostly of the nuclear chain fibers. Flower spray endings are Aβ fibers and are called type-II fibers by sensory physiologists.

Motor innervation of muscle spindle   The central nuclear region of the intrafusal fiber does not contain any contractile protein. It is only the peripheral parts of the intrafusal fibers that contain the contractile proteins as obvious from the cross striations present in the periphery. These peripheral parts are innervated by motor fibers which are of the Aγ type and are called γ-*motor neurons or fusimotor neurons*. These γ-motor neurons have their cell bodies in the anterior gray horn of the spinal cord where they are interspersed among the cells of α-motor neurons (Aα) which supply the extrafusal fibers. As with sensory innervation, motor innervation of the spindle is also of two types, the γ_1 and γ_2 fibers, which innervate the polar regions of the bag and chain fibers respectively. Aγ fibers alter spindle sensitivity to stretch. γ_1 fibers alter the dynamic sensitivity while γ_2 fibers alter the static sensitivity of the stretch reflex.

Spindle loading and unloading

When the extrafusal fiber contracts, it pulls its ends nearer and tries to move any external load attached to it. On the other hand, when the intrafusal fiber contracts, its ends do not come nearer because they are fixed to the adjacent extrafusal fibers; instead, its own noncontractile equatorial region becomes stretched. The stretching of the equatorial region of the intrafusal fiber is called *spindle loading*. The spindle becomes loaded when the muscle is stretched or when the γ-motor neurons are stimulated by descending motor influences from the brain. The spindle is unloaded when the α-motor neurons are stimulated and the muscle contracts (Fig. 104.**2**).

Spindle loading through muscle stretch   When muscle is stretched, the muscle fibers as well as the spindles are stretched. In the spindle, the stretch is unequally shared by its peripheral and equatorial parts. The stretching occurs more at the center (where the compliance is high) and less at the periphery (where the compliance is low). Stretching of the equatorial region stimulates the sensory afferents originating there.

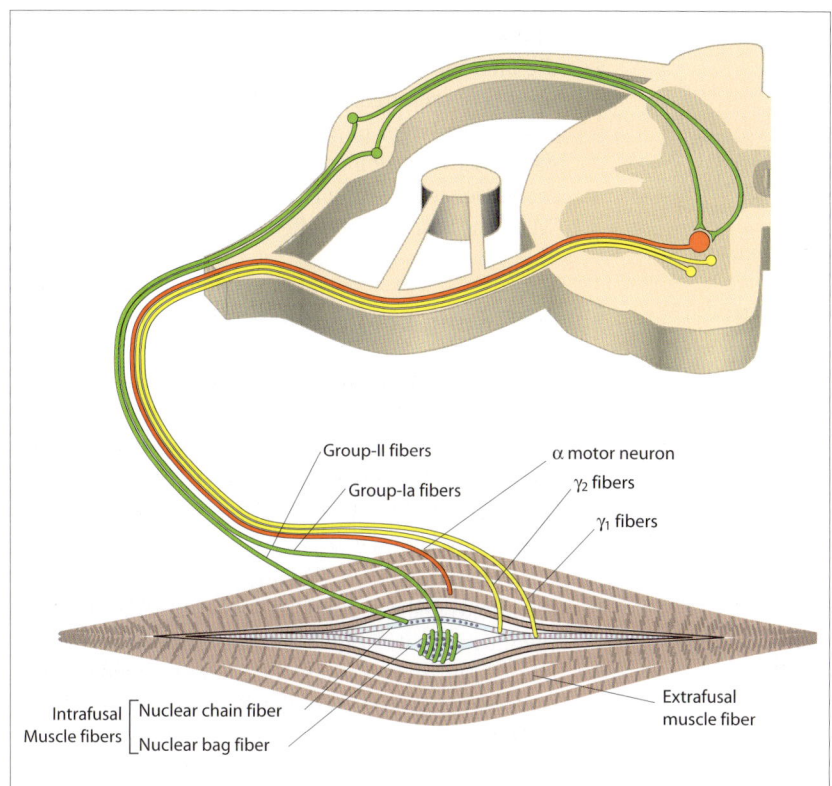

Fig. 104.**1**   Muscle spindle and the monosynaptic reflex arc.

Spindle loading through γ-motor discharge   The intrafusal muscle fibers are more contractile at the periphery (which contains numerous striations) than at the center (which is sparsely striated). Moreover, it is only the peripheral part that receives the motor innervation (Aγ fibers). When the Aγ fibers are stimulated, the peripheral parts of the intrafusal fibers contract. As explained above, contraction of the peripheral part of the spindle results in the stretching of its central part and stimulates the spindle afferents.

Spindle unloading through α-motor discharge   α-motor neurons produce contraction of the extrafusal fibers. Since the ends of the spindle are attached to extrafusal fibers, the spindle folds up when the extrafusal fibers contract. To stimulate the spindle, it must first be stretched to its original length before it can produce afferent discharge of the spindle. Hence, α-motor discharge decreases the spindle afferent discharge.

■   Static and dynamic spindle responses

When the intrafusal fiber is stretched, the primary and secondary afferents respond differently. Afferents originating in the chain fibers discharge with a frequency that is proportional to the degree of stretch

(static response). Afferents originating in the bag fibers discharge with a frequency proportional to the rate at which the fiber is stretched (dynamic response).[1] Since the group-Ia afferents originate from both bag and chain fibers, they show both static as well as dynamic responses. Group-II afferents originate only from the chain fibers and therefore show only static response. The significance of having separate static and dynamic responses is explained on p 27.

■   Monosynaptic Reflex

The monosynaptic reflex (MSR) is triggered when the muscle spindle, or more specifically, the equatorial region of the intrafusal fiber is stretched. Stretching of the equatorial region of the intrafusal fiber stimulates the afferent nerve fibers (Ia and II) of the spindle. The afferents in turn stimulate the α-motor neuron in the ventral horn of the spinal cord. The α-motor neurons stimulate contraction of the extrafusal muscle fibers and bring about a shortening of the muscle. Contraction of

---

[1]   The mnemonic "(B)right (D)ay, (C)lear (S)ky" should help in remembering that (B)ag fibers show (D)ynamic response while (C)hain fibers show (S)tatic response.

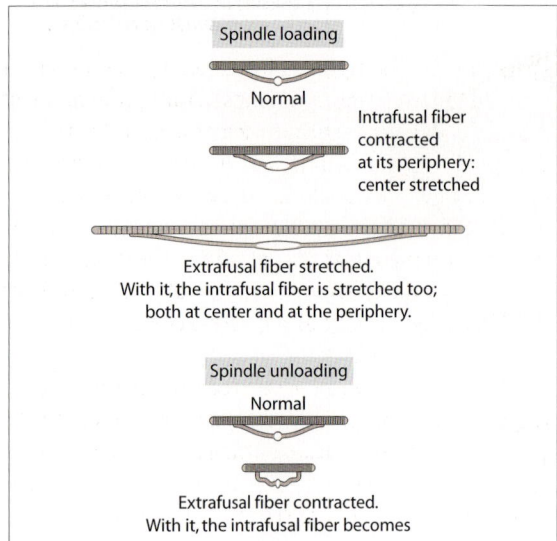

Fig. 104.**2** Spindle loading and unloading.

the extrafusal fibers releases (i.e., reduces) the stretch on the intrafusal fibers and the reflex stops.

The MSR pathway has only a single synapse, that is located in the anterior horn of the spinal cord. The MSR is the only reflex of its kind in the entire body: All other reflexes are either bisynaptic or polysynaptic.

The gain of the MSR

The gain of the monosynaptic reflex refers to the magnitude of response evoked by a given amount of stimulus. When the gain is high, a small stimulus evokes a huge response. When gain is low, even a large stimulus evokes a small response. When gain is zero, the reflex is absent. The gain of the monosynaptic reflex depends on two factors, viz., spindle sensitivity to stimulus and motor recruitment by group-Ia afferent discharge.

Spindle sensitivity to stimulus The sensitivity of the reflex depends on the extent to which a given amount of stretch is able to distort the equatorial region of the muscle spindle. When there is a high fusimotor (γ-efferent) discharge, the peripheral parts of the intrafusal fiber contract. When the peripheral parts are in a contracted state, its compliance to stretch decreases. Hence when the muscle is stretched, the periphery of the intrafusal fibers allows very little stretching and most of the stretching now occurs at the center of the intrafusal fiber. This results in greater afferent discharge in response to stretch. Thus when fusimotor (γ-efferent) discharge is high, a given amount of stretch causes greater spindle loading.

Motor recruitment by group-Ia afferents When spindle afferent discharge is low, only the smaller motor units are recruited. At higher levels of spindle discharge, the larger units are recruited. This is in accordance with the Henneman principle of motor recruitment. Hence, the motor response to stretch is greater when the basal spindle afferent discharge is high.

Long-Loop Stretch Reflex

Imagine a subject stretching out his hand with the forearm semiflexed. If a load is suddenly dropped on his hand, the hand along with the forearm is displaced down. Soon thereafter however, the limb regains its original position. The restoration to the original position occurs in three steps.

When a muscle is suddenly stretched, it first elicits the monosynaptic stretch reflex which tends to restore the muscle to its original length. However, the MSR is usually not strong enough to achieve its purpose. Next, the long-loop stretch reflex (also called functional stretch reflex) is elicited which nearly restores the muscle to its original length and sometimes, even overcorrects it. The final precise restoration of the muscle length is brought about through voluntary muscle contraction. The long-loop stretch reflex is polysynaptic and its reflex arc is centered in the cerebral cortex. This reflex is continuously active in the erect posture and brings about a continuous correction of the 'sways' that occur continuously during standing.

Golgi Tendon Reflex

The Golgi tendon reflex is a bisynaptic reflex (Fig. 104.**3**), initiated by the Golgi tendon organ located in muscle tendons. Like muscle spindles, Golgi tendon organs are also stretch receptors. However, unlike the muscle spindle which acts as a length-detector, the Golgi tendon organ acts as a muscle tension-detector. This difference in sensory function occurs because muscle spindle is disposed in parallel to the extrafusal fibers while the Golgi tendon organ is disposed in series with the extrafusal fibers.

The Golgi tendon organ is innervated by Aα sensory fibers which are called type-Ib fibers by sensory physiologists. These afferent fibers terminate on Ia inhibitory interneurons (Golgi bottle neurons) in the dorsal gray horn of the spinal cord. The interneurons terminate on the α-motor neurons in the ventral gray horn of the spinal cord.

The Golgi tendon reflex is a protective reflex that prevents excessive rise in muscle tension. When the

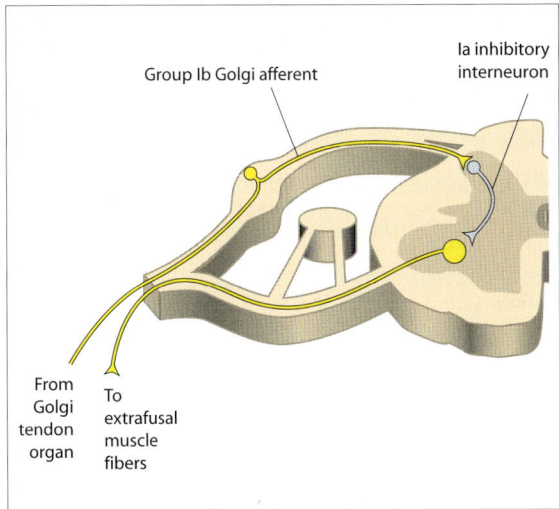

Fig. 104.**3**   The reflex arc of the Golgi tendon reflex.

muscle contracts isometrically, the tendon is stretched and the tension in the tendon rises markedly. This rise in tension is sensed by the Golgi tendon organ which stimulates the group-Ib afferents. These afferents stimulate the Ia inhibitory interneurons in the spinal cord and thereby, inhibit the α-motor neuron discharge to the muscle, which consequently relaxes. This reflex relaxation of the extrafusal muscle fibers in response to a rise in muscle tension is also called the *negative (inverse) stretch reflex*.

Regulation of Muscle Length

The importance of regulation of muscle length can be appreciated from the following example. Consider the case of a person holding a cup in his hand (Fig. 104.**4**). If tea is poured into the cup, the forearm will instantly sag in response to the increased load. If the stretch reflex is absent, it requires fresh initiative by the brain to institute a feedback correction. Thus, the brain has to bother about every small perturbation in muscle load. If however the stretch reflex is present, the spindle will be stretched in response to the increase in load and the MSR will be triggered to correct the displacement. The correction occurs without the intervention of the brain, since the MSR is a spinal reflex. The brain only has to send out the initial 'length signal' through the γ-motor neurons, specifying the amount of flexor muscle shortening required to semiflex the forearm. Through the stretch reflex, the spinal cord not only contracts the muscle to the desired length but also maintains it constant against external perturbations.

Maintenance of a constant muscle length

When the MSR is triggered by stretching the muscle, it is called the *stretch reflex*. It helps in the maintenance of a constant muscle length in face of external perturbations in load. When a muscle is stretched, its extrafusal fibers elongate and so do the muscle spindles attached to them. In the spindle, the stretch is borne mainly by the equatorial zone of the intrafusal fibers and therefore, the spindle afferents are stimulated. This initiates the MSR which brings about reflex contraction of the extrafusal muscle fibers.

The stretch reflex can be elicited as a tendon jerk (e.g., the knee jerk, biceps jerk) through a sharp tap on the muscle tendon. The nature of the jerk gives clues in the clinical diagnosis of neurological disorders: The tendon jerk is exaggerated in upper motor neuron lesion and in anxiety. It is absent in lower motor neuron lesion.

Contracting a muscle to the desired length

The muscle can be contracted to the desired length by two possible mechanisms: the α-mechanism (direct) and the γ mechanism (reflex).

Alpha-led activation   In this type of activation of muscles, the brain evaluates the load to be moved and based on its previous experience, sends down a calculated dose of stimulation to the α-motor neurons in order to contract the muscle to the desired length. This method has the advantage that muscle contraction

Fig. 104.**4**   A girl pours tea into the cup held in the out-stretched hand of her father who is reading the newspaper. Will his hand sag as the cup fills up? (After RHS Carpenter)

Fig. 104.**5**   Jack and Jill went up the hill to fetch a pail of water. The wise man suggested to them three different methods for doing so. If Jack is an α-motor neuron, Jill is a γ-motor neuron, and the wise man is brain, then each method of fetching the pail of water exemplifies a mechanism for regulation of muscle length, i.e., alpha-led, gamma-led, and alpha-gamma coactivation, respectively.

is quick and powerful but has the disadvantage that there is no guarantee against a miscalculation by the brain, not to mention that any unexpected changes in the load upsets all calculations made by the brain (see open loop control, p 26). Moreover, when there is contraction of extrafusal muscle fibers without simultaneous contraction of intrafusal fibers, it results in the unloading of the muscle spindle. An unloaded spindle will be unable to produce stretch reflex and therefore, unable to maintain a constant muscle length.

Gamma-led activation   In this type of activation of skeletal muscle, the brain makes the extrafusal fibers contract reflexly through the MSR instead of stimulating the Aα directly. Contraction of intrafusal fibers is not affected by external load. Hence, to make the intrafusal fibers contract to the desired length, the brain does not have to evaluate the external load. The intrafusal fibers contract to the desired length regardless of the uncertainties of the external load. The contracted intrafusal fibers trigger the MSR and this results in contraction of the extrafusal fibers. The MSR does not stop till the extrafusal fibers contract through exactly the same distance as the intrafusal fibers, thereby unloading the spindle to its original level. The advantage of this mechanism, which is called the length-servo mechanism, is that it is unaffected by perturbations of external load. Also, at the end of contraction, the spindle sensitivity is retained. It is now known that this type of reflex contraction is too weak to move heavy loads but it is through this mechanism that all the muscles of the body are kept in a slightly contracted state called the muscle tone. The muscle tone and its control are discussed below in detail.

Alpha-gamma coactivation   It seems that the brain takes advantage of both the direct as well as indirect mechanisms by sending its length-signal simultaneously to both extrafusal and intrafusal fibers through α-motor neurons and γ-motor neurons respectively (α-γ coactivation). The α-mediated stimulation makes the muscle contract powerfully and quickly. If the extrafusal fibers fail to contract to the desired length due to unexpected increase in the external load, it results in a mismatch between the degree of extrafusal and intrafusal fiber shortening. This causes spindle loading and triggers off the MSR till the disparity is eliminated and the extrafusal fibers contract to the desired length. According to this hypothesis, the MSR provides only a *follow-up servo mechanism* to ensure that the extrafusal fibers contract to the desired length. In the follow-up servo mechanism, the spindle is not unloaded and therefore retains its sensitivity to stretch. Hence, the stretch reflex is elicited whenever external perturbations tend to change the muscle length from the desired position, and the muscle remains at the desired length (Fig. 104.**5**).

α-γ coactivation is rarely so perfect that the shortening of the extrafusal and intrafusal fibers are exactly equal. Such exact matching is symbolically referred to as α = γ. More commonly, the two are unequal. If the shortening of the extrafusal fibers is greater than that of intrafusal fibers (α > γ), then the contractions are called α-led, as occurs in ballistic movements (see Fig. 105.**4**). At the end of contraction, the spindle remains unloaded. Alternatively, if the shortening of the extrafusal fibers is less than that of intrafusal fibers (α < γ), then the contractions are called γ-led, as occurs in ramp movements. The spindle sensitivity is retained even at the end of the contraction.

When a task is being learnt (say typewriting), the movements are γ-led. The movements are error-prone and require constant feedback correction. The muscles are stiff and the movements are slow. As the facility with the task increases, the contraction changes to the α-led mode. The brain, which by now is thoroughly acquainted with the task, can accurately calculate the Aα discharge required and is less dependant on feedback corrections through the stretch reflex. The switchover to the α mode is attended with increased speed and the disappearance of the background muscle stiffness.

Regulation of Muscle Tone

All muscles in the living body are maintained at a slightly contracted state which is called the muscle tone. This muscle tone is normally brought about through γ-led contraction of the muscle fibers. Muscle tone is assessed by observing the resistance offered by the MSR to passive stretch.

Hypotonic muscles offer little resistance to passive stretching. Conversely, a hypertonic muscle offers unusually high resistance to passive stretching. Sometimes, when a hypertonic muscle is stretched passively, the intense initial resistance suddenly gives way, producing the feel of a clasp knife. Hypertonia of this kind is known as *spasticity*. The initial resistance is due to the exaggerated stretch reflex. The resistance increases muscle tension, resulting in the activation of the Golgi tendon reflex which causes the sudden yielding to stretch. Spasticity is observed mostly in extensor group of muscles and occurs due to increased fusimotor discharge to extensor muscles. A hypertonic muscle is not necessarily a spastic muscle: Spasticity is absent in α-rigidity and in Parkinsonian rigidity.

A spastic muscle shows an exaggerated tendon jerk. Sometimes the exaggerated contraction of the stimulated muscle triggers a stretch reflex in the antagonistic muscle. This results in *clonus*, i.e., the rhythmic oscillations of a limb due to the alternating contractions of an agonist-antagonist muscle pair.

Supraspinal control

Muscle tone is normally under elaborate supraspinal controls. The supraspinal control is effected mainly through the vestibulospinal tracts and the reticulospinal tracts. The *medial reticulospinal tract* descends ipsilaterally from the pontine reticular formation to the spinal cord. The *lateral reticulospinal tract* descends bilaterally (but mainly ipsilaterally) to the spinal cord (see Fig. 96.**1**).

The γ-motor neurons in the ventral horn of the spinal cord are stimulated by descending fibers from the facilitatory reticular formation which extends from the upper medulla to the midbrain. They also receive descending inhibitory fibers from the inhibitory reticular formation in the pons and medulla. However, the facilitatory effect predominates, resulting in a tonic discharge of γ-motor neurons to the muscles, which is the main cause of muscle tone (Fig. 104.**6**).

As already mentioned, muscle tone is normally not under tonic control of α-motor neurons. (Tonic control of α-motor neurons should not be confused with the stimulation of α-motor neurons through corticospinal

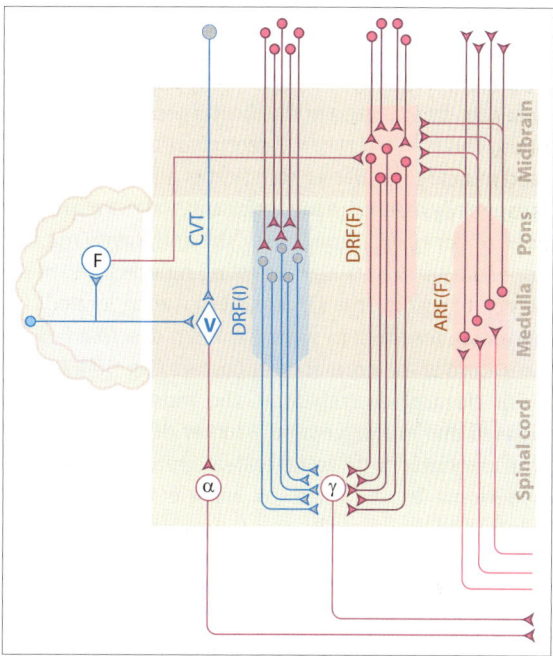

Fig.104.**6**   Supraspinal control of muscle tone. CVT=Corticovestibular tract, DRF(I)=Descending reticular formation (inhibitory), DRF(F)=Descending reticular formation (facilitatory), ARF(F)=Ascending reticular formation (facilitatory).

fibers during voluntary contractions.) Tonic control of α-motor neurons is exerted almost entirely through vestibulospinal pathways. However, the vestibular nucleus (especially, Dieter's nucleus) is itself constantly inhibited by corticospinal fibers and the fastigiovestibular fibers from the cerebellum. It is only under certain abnormal or experimental situations that the vestibular nucleus is disinhibited, resulting in an exaggerated muscle tone that is α-led rather than γ-led.

Spinal shock

The net supraspinal influence descending on the γ-motor neuron is excitatory. When this facilitatory supraspinal influence is suddenly removed, the γ-motor neurons become quiescent for a period of days to weeks. This is called *spinal shock* (p 599). The duration of the spinal shock is much less in lower animals, possibly because their supraspinal facilitation is normally much less. Since the γ-motor discharge determines the gain of spinal reflex, all spinal reflexes are abolished in the absence of γ-motor discharge. After a period of weeks, the γ-motor neurons start discharging excessively, possibly due to denervation hypersensitivity and due to the sprouting of collaterals from interneurons which excite the γ-motor neurons. The excessive γ discharge is associated with muscle spasticity.

Decerebrate rigidity

Decerebrate rigidity is produced experimentally, mostly in cats, by making a midcollicular section of the midbrain. This is called *classical decerebration*. The exaggerated muscle tone affects both flexors and extensors but is especially marked in the antigravity extensor muscles. Even under normal conditions, the antigravity muscles are endowed with a higher resting muscle tone. In decerebrate rigidity, the same pattern is exaggerated (hypertonia), resulting in what has been described as the caricature of the normal erect posture (Fig. 104.**7**).

In *classical decerebration*, the muscle tone is increased due to exaggerated γ-motor discharge. Hence, it is associated with exaggerated stretch reflexes. The hypertonia is abolished by muscle deafferentation (cutting of the group-Ia and II proprioceptive muscle afferents) which interrupts the stretch reflex. The rigidity observed following classical decerebration is actually a form of spasticity. Spasticity is not observed if the increase in muscle tone is due to α-overdrive.

Since classical decerebration was frequently associated with death of the experimental animal, the *ischemic decerebration* was attempted in which the cerebral cortex was rendered ischemic and nonfunctional by tying off the basilar artery. Soon it was discovered that *ischemic decerebration* did not result in spasticity: It resulted in a different form of rigidity that was due to exaggerated α-motor neuron discharge (α-rigidity). The α-overdrive resulted in direct stimulation of extrafusal muscle fibers. Since the rigidity is not reflex in nature, it is not abolished by muscle deafferentation.

Mechanism of decerebrate rigidity   As already mentioned, the γ-drive to muscles is maintained by descending fibers from the descending reticular formation. The descending influences are both facilitatory (from facilitatory reticular formation) as well as inhibitory (from the inhibitory reticular formation). The facilitatory influence predominates. The facilitatory and inhibitory descending reticular formations have no intrinsic activity of their own: They have to be kept activated by neurons from other areas. The inhibitory reticular formation is kept activated by descending supraspinal fibers, mostly from the basal ganglia. Hence, following decerebration, it is totally inactivated and its inhibitory effect on γ-motor neurons vanishes. The facilitatory reticular formation, in contrast, remains activated by ascending sensory stimuli which relay to it through the ascending reticular formation. Hence even following decerebration, the facilitatory reticular formation remains quite active and continues to facilitate the γ-motor neurons. The balance of descending influence on γ-motor neurons therefore tilts heavily toward facilitation.

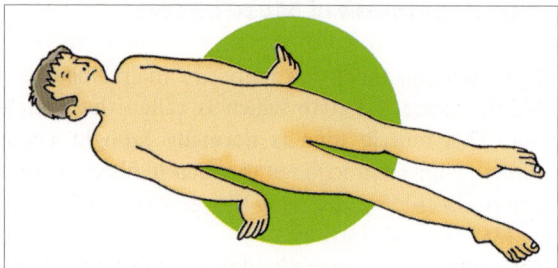

Fig. 104.**7**   Decerebrate posturing.

Cerebellectomy in subprimates results in hypertonia (see Fig. 97.**8**) due to α-overdrive. Cerebellectomy in a decerebrate preparation changes the γ-rigidity into α-rigidity and increases the muscle tone further. Although cerebellectomy decreases the γ-motor discharge, it cannot lead to a reduction of the muscle tone when the direct α-motor discharge to the muscle fibers is high.

The α-rigidity of ischemic decerebration occurs for a similar reason. Ischemic decerebration is unavoidably associated with ischemia and consequent necrosis of a large part of the cerebellum. As already explained above, a reduction in cerebellar activity leads to a disinhibition of α-motor neurons and a reduction of γ-facilitation. This leads to α-rigidity.

Decorticate rigidity

A decorticate preparation is made by removing the whole cerebral cortex but leaving the basal ganglia intact. The decorticate animal does not have such intense hypertonia as a decerebrate preparation. This is because the basal ganglia, which are intact in the decorticate animal, activate the descending inhibitory reticular formation, and thereby prevent excessive hypertonia. The tonic neck reflexes become exaggerated and override the tonic labyrinthine reflexes. Therefore, the tonic neck reflexes can be elicited in patients of decorticate rigidity (see Fig. 105.**17**).

Experimental decerebration or decortication are somewhat different from their clinical counterparts. Clinical *decerebrate posturing* is usually caused by bilateral midbrain or pontine lesions. It is associated with bilateral extension of the lower extremities, adduction and internal rotation of the shoulders, and extension at the elbows and wrist (see Fig. 104.**7**).

Clinically, *decorticate posturing* is a much poorer localizing sign as it results from lesions in several locations, usually above the brainstem. It is associated with bilateral flexion at the elbows and wrists, with shoulder adduction and extension of the lower extremities.

The motor control of the body has a three-tier system of planning, programming, and execution. The highest level is concerned with the generation of the motor plan, i.e., the framing of the right ideas for a motor act. The middle level of motor control is concerned with developing and perfecting the motor program and sub-programs for bringing about a motor act. The middle level also supervises the implementation of the motor program. The lowest level of control is vested with the spinal cord which implements the motor program designed and perfected by the middle level.

Motor Planning

Motor planning, as already mentioned, is the framing of the right motor ideas. The EEG activity (see Chapter 107) associated with motor planning has been studied in instructed-delay tasks (Fig. 105.1). An example of an instructed-delay task would be when a sprinter is instructed to start running (the task) on hearing the gun

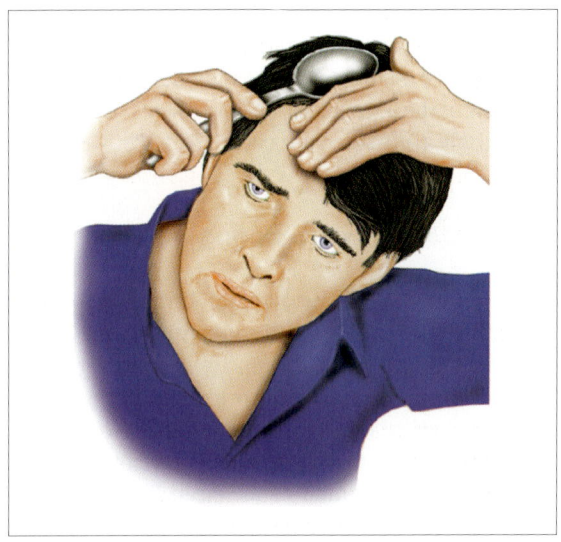

Fig. 105.2   A patient tries to comb his hair with a spoon: An example of ideational apraxia.

shot (the cue) after a while (the delay). As the sprinter waits for the cue, an EEG potential called the *preparatory-set activity* or the *set-related activity* is recordable from his scalp. It reflects his motor planning for getting off to a good start.

The importance of a definite motor plan is best appreciated from the consequences of its absence. The absence of a motor plan results in *ideomotor apraxia*,[1] which usually affects movements of the face or limbs. A patient of facial apraxia, when asked to blow out a burning matchstick, may be unable to do so but may instead put it out by shaking the matchstick in his hand. However, the same patient may be able to whistle through his lips when in a happy mood, indicating the absence of any muscle paralysis. This type of motor error indicates that the patient has the right motor idea but is unable to transform the motor idea into a motor plan.

While ideomotor apraxia refers to the inability to develop a motor plan corresponding to a motor idea, a

Fig. 105.1   (*Above*) Sprinters waiting for the start of the race is an example of an instructed-delay task. (*Below*) The crowd bursting into applause is an example of a self-initiated task.

---

[1]   Apraxia is an inability to carry out learned, purposeful movement that cannot be accounted for either by weakness, incoordination, or sensory loss, or by incomprehension of, or inattention to commands. Geschwind identified three types of apraxia: ideomotor, ideational, and akinetic.

less common condition called *ideational apraxia* is the result of a faulty motor idea (Fig. 105.**2**). A patient of ideational apraxia who is provided with a cigarette and a matchbox might try to light the cigarette by striking it on the side of the matchbox.

Lesions of the lower parts of the posterior parietal areas 5 and 7 (sensory association areas) in the left hemisphere can lead to pronounced apraxia of the limbs because these are important in motor planning.

The anterior part of the corpus callosum probably conveys the motor plan to the right hemisphere. Hence, lesions of the anterior part of the corpus callosum produce only a left-sided apraxia.

Motor Programming

The motor program is a set of simultaneous or sequential commands issued to the muscles. It involves formulation of the program, its sequencing, scaling, and correction. The motor cortex and the cerebellum are involved in motor programming. Motor programming is associated with the appearance of a characteristic negative potential in the scalp EEG, about a full second before the commencement of a self-initiated movement (Fig. 105.**1**). This negative potential is called the *Bereitschaft (readiness) potential* and its activity increases before difficult movements like a complex sequence of finger movements. The Bereitschaft potential is recordable even when the complex sequence is imagined mentally.

■ Role of primary motor area

An individual neuron in the primary motor cortex does not control an individual muscle. Rather, they simultaneously stimulate a set of synergistic muscles such that a single contralateral joint moves in a particular direction. The greater the discharge of this neuron, the greater the force of movement of the joint in that particular direction. The contributions of a population of neurons in the primary motor cortex are vectorially summed to produce a population vector, moving the contralateral joint in the desired direction with the desired force.

■ Role of premotor area

In contrast to the primary motor area, the premotor area is important for movements that involve multiple joints. It is also important for motor learning. The functions of the various parts of the premotor area are discussed below and are summarized in Figure 105.**3**. A lesion in the premotor cortex causes *motor (kinetic) apraxia*.

The **medial premotor area** is important for the learning and performance of a memorized sequence of movements in the absence of visual cues. This area has two parts: the supplementary motor area (SMA) and the presupplementary motor area (preSMA). The preSMA is involved in learning novel motor sequences and is

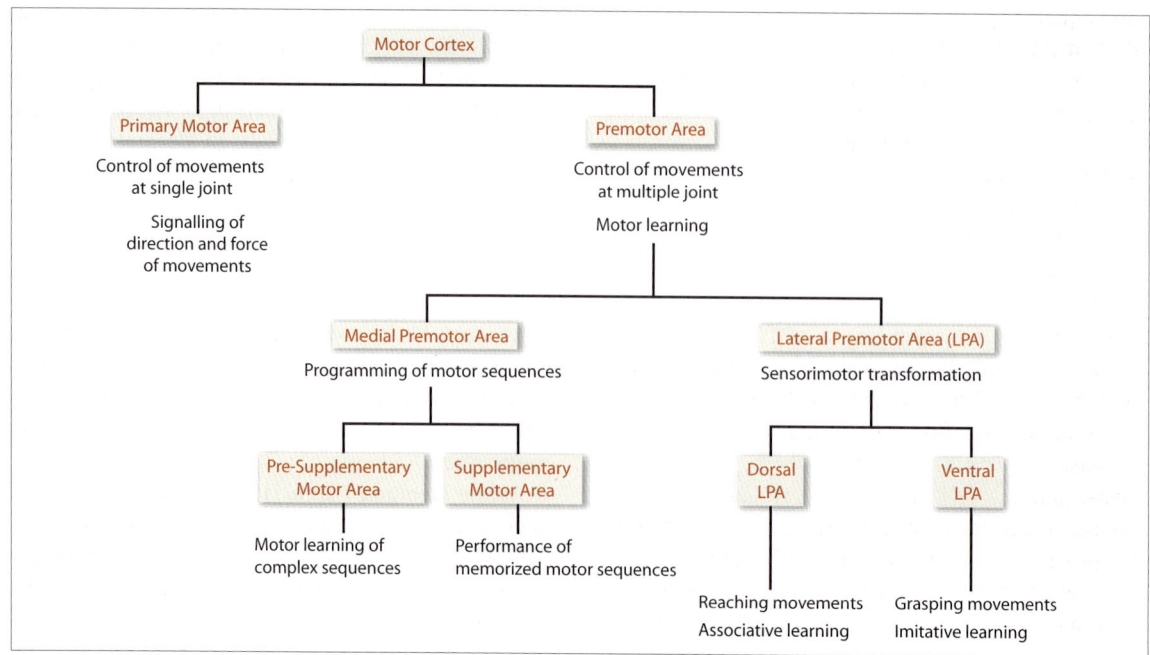

Fig. 105.**3**   Different areas of the motor cortex and their functions. See also Figure 100.4.

active only as long as a subject learns the sequence. The motor learning of the preSMA involves a continuous interchange of information with the prefrontal cortex. The SMA designs the motor program for the execution of the memorized sequence and is the site where the readiness potentials are generated. The activity in the SMA gradually shifts to the primary motor area as the newly learnt task becomes automatic.

The **lateral premotor area** (LPA) is important for the programming of visually-guided movements. It designs the motor program through a process called *sensorimotor transformation*. Hence, the LPA requires continuous sensory cues from the extrapersonal space. Visually-guided movements are either reaching movements or grasping movements. Reaching movements are programmed by the dorsal-LPA while grasping movements are processed by the ventral-LPA. Moreover, the dorsal-LPA is also involved in associative learning, i.e., how to associate a specific sensory cue with a specific movement. The ventral-LPA probably helps in imitating a movement: Certain neurons in this area, called mirror neurons, are activated when one observes a grasping movement being performed by others!

Role of basal ganglia

The role of basal ganglia in the middle level motor control is attributable to the caudate nucleus through its caudate loop and to the putamen, through its putamen loop. The caudate loop plays an important role in motor sequencing. When there is dysfunction of the caudate nucleus, the transition from one motor program to another appears extremely difficult and the patient becomes stuck in the middle of an otherwise automatic motor action. This difficulty results in akinesia, an apparent inability to initiate voluntary actions.

The putamen loop scales the intensity of movements. For example, a person signing his name on a paper uses his finger and wrist muscles for the motor act. When the person is asked to sign his name on a large blackboard, he uses his shoulder, elbow, and wrist joints and associated muscles. He signs on the board perfectly even if he is doing it for the first time, provided he has enough practice of signing on paper. This is because the same motor plan is used for both the motor acts–signing on the paper and on the blackboard. The putamen loop formulates for them two different motor programs differing both in quality (the joints and muscles used) and in intensity (the magnitude of the movements). The scaling ability of the putamen seems to be affected in patients of Parkinsonism who retain the ability to execute motor patterns but make the constituent movements too small. Hence, their writing is small and their speech is slurred and rapid, just as their stride is a shuffling sequence of short steps.

Role of cerebellum

The cerebrocerebellum is concerned with higher motor planning, a view that is suggested by two observations. First, it has no direct connections with the spinal cord. Second, the dentate nucleus fires well before the onset of movement and even before the activation of the motor cortex. The dentato-rubro-olivo-cerebello-dentate loop of the cerebrocerebellum is probably involved in the mental rehearsal of ballistic movements before initiating them. Lesion of the dentate nucleus does not abolish the movement but delays its initiation.

In contrast to the basal ganglia which enable a smooth transition from one motor subprogram to the other, the cerebrocerebellum coordinates the simultaneous performance of multiple motor subprograms. For example, reaching out to pick up an object involves two simultaneous actions: one, the forward movement of the hand, and second, a gradual opening up of the grip. Both occur simultaneously. However, in disorders of the cerebrocerebellum, the two occur sequentially instead of simultaneously: first the hand reaches the object and then the grip opens. This abnormality is known as decomposition of movement due to the breaking up of complex motor programs into sequential subroutines.

The function of the cerebellum as a comparator enables it to continuously refine the motor programs after obtaining sensory feedback on the accuracy of the motor actions. Long-term modification of the motor program for achieving the desired result is called *adaptive motor learning* and is made possible by the plasticity of cerebellar neurons.

Role of sensory feedback

Sensory feedback may be proprioceptive, somatosensory or vestibular. Development of the motor program may or may not be dependent on sensory feedback from the muscles, depending on whether the movements are of the ramp or ballistic type. Threading a needle is a *ramp movement*: every time there is an overshoot, there is an immediate correction of the motor activity. Ramp activity is precise but slow, and usually lacks much power. Throwing a dart is a ballistic movement: once thrown, the direction of the dart cannot be altered or guided to the bull's eye. *Ballistic movement* is an example of open loop control system (p 26): It

Fig. 105.**4** The throwing of a boxing punch is an example of ballistic movement. A missed punch cannot be corrected.

may not be precise but it allows great speed and power (Fig.105.**4**). Before the onset of the ballistic movement, the brain makes all possible calculations to formulate the motor subprogram, i.e., decides on the direction and the force of movements. If a breeze causes the dart to drift, then the brain also tries to estimate the wind velocity and incorporates it as a corrective factor into the subprogram formula. Hence, before throwing a dart or a ball, it is a usual practice to swing the arm a few times to get a feel of the weight of the dart as well as any wind velocity. Finally, if the dart does not hit the bull's eye, the brain makes further corrections to its subprogram formula and the second throw is usually better than the first. In general, ballistic movements are α-led while ramp movements employ a follow-up servo mechanism.

The importance of sensory feedback is also underlined by the various postural reflexes that help in steadying the posture. A striking example is Romberg's sign (see sensory ataxia, p 658) which demonstrates the heightened dependence on sensory (visual) feedback for standing straight in the absence of proprioceptive control.

Motor Execution

Upper motor neurons are the neurons that descend from the cerebral cortex, cerebellum, and brainstem to terminate on the neurons that directly innervate the skeletal muscles. The earlier categorization of descending fibers into pyramidal fibers and extrapyramidal fibers (descending fibers other than the pyramidal fibers) is an ill-conceived one that still somehow lingers

on in clinical parlance. Grouping descending fibers in lateral and medial pathways is more meaningful.

The *lateral pathways* terminate either directly on motor neurons (as in the case of fingers which require very fine movements) or on interneurons in the lateral parts of the spinal gray matter to control the movements of the distal limbs as well as those that activate supporting musculature in the proximal limbs. It includes the lateral corticospinal tract and the rubrospinal tract, which is largely vestigial in man. Fewer than 200 rubrospinal fibers are present in man, and their function has been completely taken over by the corticospinal tract. Also included in the lateral pathways are the corticobulbar fibers supplying the lower part of the facial nerve nucleus and the hypoglossal nucleus.

The *medial pathways end* in the medial ventral horn on the medial group of interneurons. These interneurons connect with motor neurons that control the axial musculature bilaterally and thereby contribute to balance and posture. Medial pathways include the anterior corticospinal tract, most of the corticobulbar tracts, the vestibulospinal tract, the reticulospinal tract, and the tectospinal tract.

Lower motor neurons are the neurons that directly innervate the muscle fibers. All motor commands must ultimately be routed through this *final common pathway* in the spinal cord. The cell body of these neurons is located in the ventral horn of the spinal cord. These neurons, which are of the Aα type (hence called the α-motor neurons) receive and integrate inputs from various parts of the brain as well as from sensory receptors. In general, two types of inputs are funneled through the final common pathway: (1) segmental inputs originating from receptors in the muscle (the muscle spindle), the tendon (Golgi tendon organ), and skin (nociceptors), and (2) supraspinal inputs descending from various parts of the brain like the cerebral cortex, basal ganglia, cerebellum, and descending reticular formation.

Spinal interneurons Some segmental and suprasegmental inputs terminate directly on the α-motor neuron. More often, there are intervening neurons called interneurons. Two important spinal interneurons are the Ia interneuron and the Renshaw interneuron. Whenever a muscle contracts, the Ia interneurons (Golgi bottle neurons) ensure the relaxation of their antagonistic muscle. This is known as *reciprocal inhibition* (Fig. 105.**5**) and occurs regardless of whether the muscle contracts in response to a supraspinal input or a segmental input.

The Renshaw cell is present in Rexed lamina VII of the spinal cord (Fig. 105.**6**). It is an inhibitory neuron

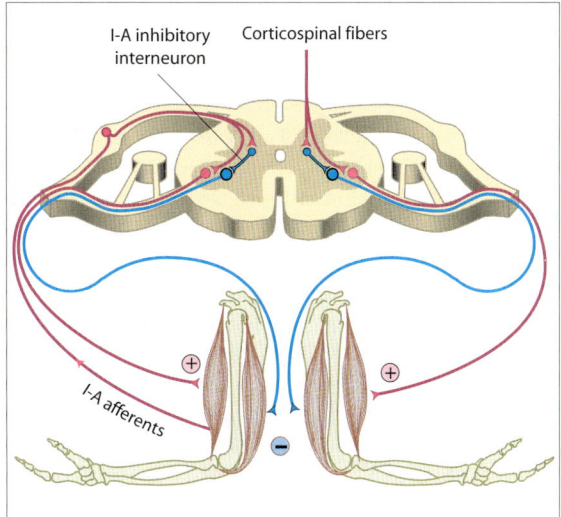

Fig. 105.**5**   Reciprocal inhibition.

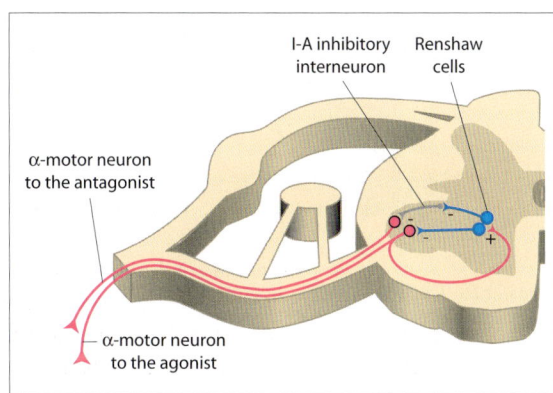

Fig. 105.**6**   Renshaw cell inhibition.

that is stimulated by the axonal branch of the α-motor neuron. The stimulated Renshaw cell inhibits the α-motor neuron and also inhibits the Ia interneuron. Inhibition of the Ia interneuron results in the disinhibition of the antagonist muscles. The functions of the Renshaw cells are debatable but they probably bring lateral inhibition as seen in the sensory system, resulting in motor focusing so that only the intended muscles contract. By inhibiting the Ia interneuron, they prevent the reciprocal inhibition of antagonist muscles, thereby promoting the simultaneous contraction of the agonist and antagonist muscles (Fig. 105.**7**). They probably also influence the balance of α-γ coactivation.

■   **Spinal reflexes**

When a sensory input relays, directly or indirectly, to the α-motor neurons, it constitutes a reflex arc. Depending on the number of synapses in the reflex arc, the reflex

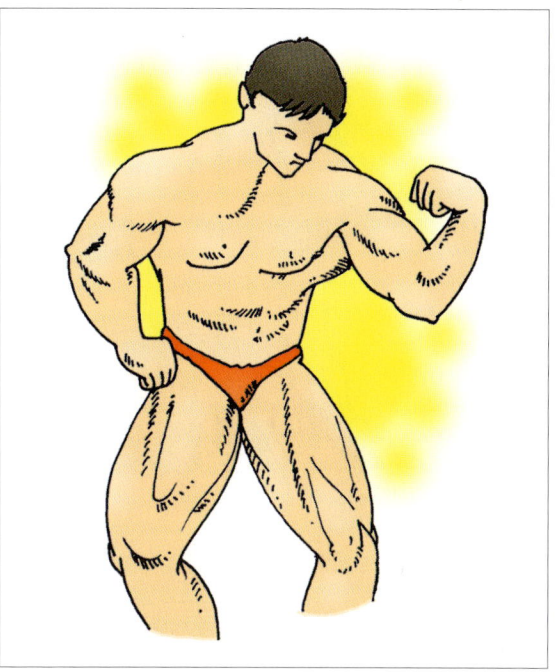

Fig. 105.**7**   Whither reciprocal inhibition? Showing off the muscles involves simultaneous contraction of agonist-antagonist muscle pairs, overriding reciprocal inhibition.

may be monosynaptic, bisynaptic, or polysynaptic. Monosynaptic and bisynaptic reflexes are described in Chapter 104. An example a of polysynaptic reflex is the *withdrawal (flexor) reflex*, which is a protective reflex triggered by painful stimuli. It results in flexion[2] of the stimulated limb. Withdrawal reflex is associated with a crossed-extensor reflex, the physiological significance of which is discussed in the context of postural control.

Clinicians classify reflexes as superficial and deep. Withdrawal reflexes are called superficial reflexes because they are elicited by stroking the skin, which is superficial. The stretch reflex and the inverse stretch reflex are called deep reflexes as they are elicited by striking the tendon, which lies deep beneath the skin.

In reflexes, the sensory afferents and the α-motor neurons do not have a one-to-one correspondence. Rather, a single sensory afferent makes contact with several motor neurons. As a result, some reflexes might show the phenomenon of subliminal fringe while others might indicate the presence of occlusion.

---

[2]   The term flexion can have different meanings to the anatomist and the physiologist. The gastrocnemius muscle, for example, is a flexor when considered anatomically but is physiologically an extensor. Physiological flexion can be defined as the position of the fetus in utero. Flexor muscles help in withdrawing a limb from a noxious stimulus. Thus physiologically, the so-called 'plantar flexion' is an extension of the foot since it takes the foot closer to a painful stimulus.

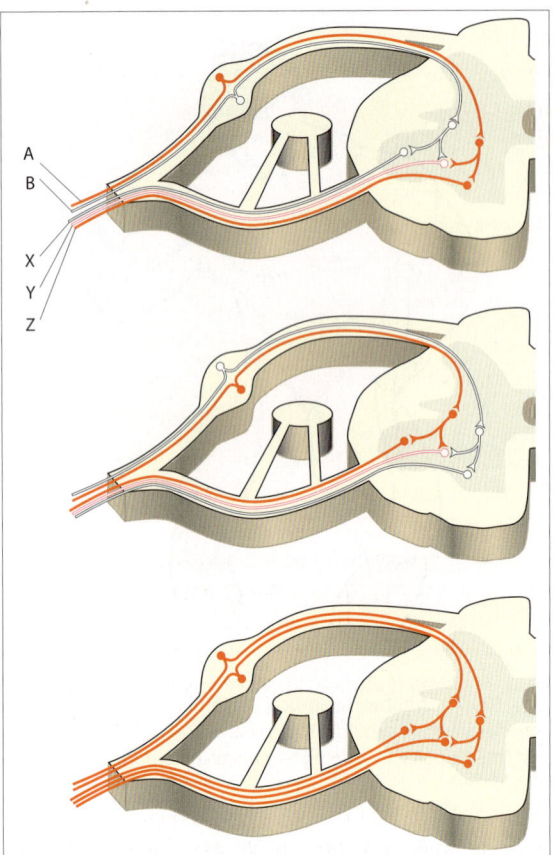

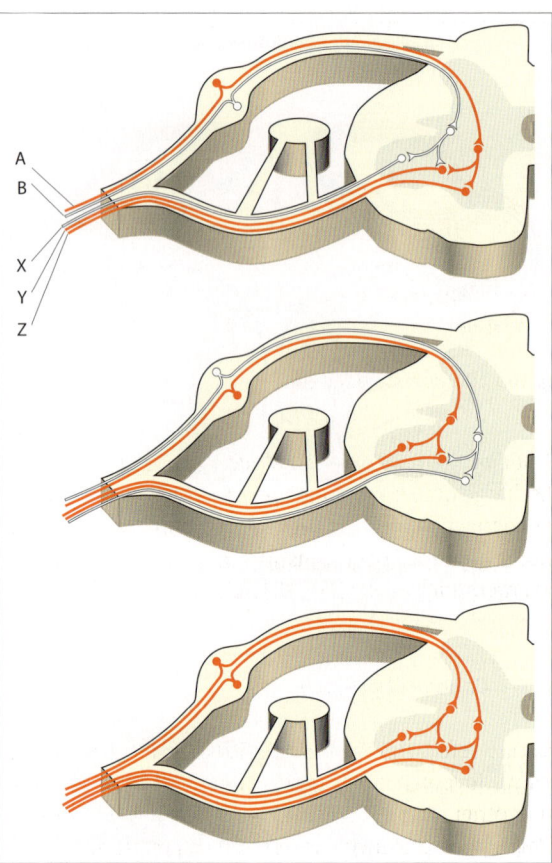

Fig. 105.**8** Subliminal fringe. Threshold stimulation of the afferent neuron A causes excitation (shown in red) of neuron Z. Similarly, stimulation of B excites neuron X. In both cases, neuron Y, which has a low excitability, is subliminally stimulated (shown in pale magenta). When A and B together, neuron Y is excited due to spatial summation. Neuron Y is therefore said to be in the subliminal fringe of neurons A and B.

Fig. 105.**9** Occlusion. When neuron A is stimulated with a train of suprathreshold stimuli, neurons Y and Z are stimulated. When neuron B is stimulated with a train of suprathreshold stimuli, neurons X and Y are stimulated. Neuron Y, although of a lower excitability, is stimulated in both cases due to temporal summation. Hence, when neurons A and B are stimulated together, the total number of efferents fibers excited (three) is less than the number of fibers excited when neurons A and B are stimulated separately (four).

Subliminal fringe and occlusion are quite marked in the withdrawal reflex.

Subliminal fringe is evident when the afferent neurons of a reflex arc are stimulated using single, threshold stimuli. Due to the subliminal fringe, the strength of the motor output obtained by stimulating all the afferent neurons together is more than the sum of outputs obtained by stimulating the afferent neurons singly. As explained diagrammatically in Figure 105.**8**, it occurs when some of the neurons remain subliminally excited, i.e., slightly depolarized but not depolarized to the firing level.

Occlusion is evident when the afferent neurons are stimulated by a train of suprathreshold stimuli. Due to occlusion, the strength of the motor output obtained by stimulating all the input neurons together is less than the sum of outputs obtained by stimulating the input neurons singly (Fig. 105.**9**).

The occurrence of subliminal fringe in reflexes

explains the phenomenon of reinforcement of tendon jerks through Jendrassik's maneuver.[3] When one muscle is stretched, it not only results in its own contraction through the monosynaptic reflex but also places a substantial number of α-motor neurons in the subliminal fringe. When the state of activity in the α-motor neuron pool (also called the *central excitatory state*) is high, other motor reflexes too are exaggerated.

In the withdrawal reflex, and especially in the crossed extensor reflex, the tension in the contracting muscle rises slowly and also declines slowly. The slow rise in muscle tension is due to the phenomenon of *motor recruitment*. The slow decline in the muscle tension

---

[3] It is a method of enhancing the patellar reflex. The subject hooks his hands together by the flexed fingers and pulls against them with all his strength.

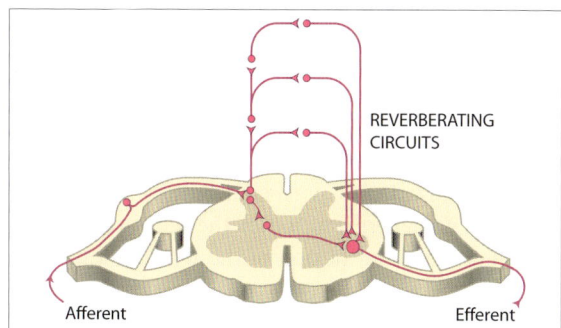

REVERBERATING CIRCUITS

Afferent

Efferent

Fig. 105.**10** Reverberating circuits.

Table 105.**1** Postural reflexes.

Postural reflex	Integrating center
Muscle tone	Spinal cord
Positive supporting reaction	Spinal cord
Crossed extensor reflexes	Spinal cord
Tonic labyrinthine reflexes	Medulla oblongata
Tonic neck reflexes	Medulla oblongata
Righting reflexes	Midbrain
Long-loop stretch reflexes	Cerebral cortex
Hopping and placing reactions	Cerebral cortex

is due to the phenomenon of *after discharge*. The after discharge occurs due to the presence of multiple parallel connections between the sensory afferent and the motor efferent fibers, constituting what are called the *reverberating circuits* (Fig. 105.**10**).

■ Control of posture

Posture refers to the static position of any part of the body. Movements are the transition from one posture to another. The motor system is endowed with wide-ranging postural mechanisms, which are as follows.

(1) *The antigravity muscle* tone ensures that there is a good amount of basal contraction (i.e., muscle tone) in those extensor muscles that keep the body upright (the antigravity muscles). This ensures that the weight-bearing joints like the knee joint do not give way under the effect of gravity.

(2) The *positive supporting reaction* ensures that once in the standing position, the ankle joint is steadied in such a way that it can neither flex nor extend under the body weight. This is made possible by simultaneous contraction of the ankle flexors and extensors.

(3) The *crossed extensor reflexes* ensure that the body is not thrown off balance when one limb is flexed. Sudden flexion of a limb in a quadruped, say the right forelimb, can occur if a painful stimulus is inflicted on it. This will throw the quadruped off balance unless the contralateral forelimb and ipsilateral hindlimb compensate by increasing their extensor tone.

(4) The *tonic labyrinthine reflexes* and the *tonic neck reflexes* ensure that the body is not thrown off balance even when standing on an inclined plane.

(5) The *rescue reactions* like the long-loop stretch reflexes, and the hopping and placing reactions ensure that the body is not thrown off balance when tipped over its center of gravity.

(6) The *righting reflexes* ensure that the body regains its upright stance even after it is thrown off-balance.

The centers for these reflexes are at different levels of the central nervous system (Table 105.**1**).

Antigravity muscle tone Antigravity muscles are those extensor muscles which keep the weight-bearing joints of the body extended, e.g., the extensors of the neck (erector capitis), spine (erector spinae), hip (glutei), and knee (quadriceps). The antigravity muscles of the body are endowed with higher muscle tone than the other muscles of the body. This higher muscle tone is provided by elaborate supraspinal mechanisms that are strikingly apparent in decerebrate rigidity. The hypertonia in decerebrate preparations has a characteristic extensor distribution, resulting in a caricature (i.e., exaggerated mimicry) of the normal posture. Interestingly, in bats, it is the flexor muscles that act as antigravity muscles because bats hang from the tree. Accordingly in bats, the exaggerated muscle tone is present in the flexor muscles and not the extensor muscles.

Positive supporting reaction While standing, steadying the ankle joint is much more difficult than steadying other weight-bearing joints like the knee joint. The knee joint can be steadied by simply contracting the knee extensors because the knee joint does not permit hyperextension. At the ankle joint, however, both dorsiflexion and plantar flexion are possible (Fig. 105.**11**). During standing, neither of them is desirable. Dorsiflexion of the foot would tip the body forward, while plantar flexion would throw the body backward. What is required therefore is that the ankle is steadied at an intermediate position. This is possible only by the simultaneous contraction of the flexors and extensors of the foot. This in turn is brought about by the positive supporting reaction, which is stimulated by pressure on the sole of the foot. Afferent impulses from the stimulated skin and the muscles (specially, the interossei) cause reflex contraction of both the agonist and antagonist muscles acting on the ankle joint, fixing the ankle joint at a right angle.

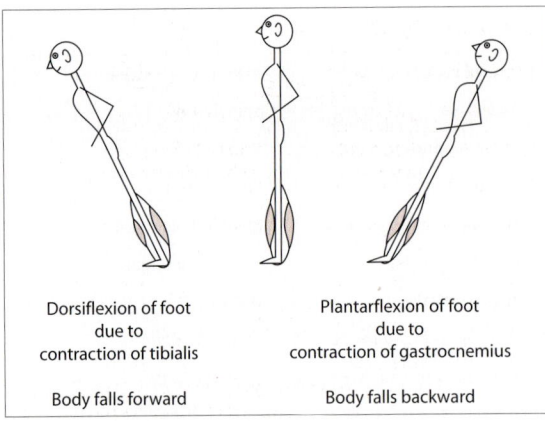

Dorsiflexion of foot due to contraction of tibialis

Plantarflexion of foot due to contraction of gastrocnemius

Body falls forward

Body falls backward

Fig. 105.**11** Positive supportive reaction is necessary for steadying the ankle joint.

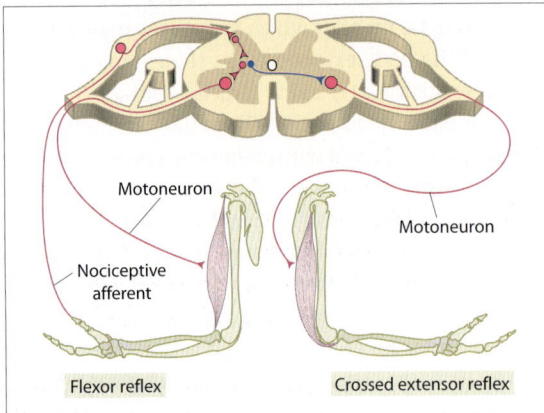

Motoneuron

Motoneuron

Nociceptive afferent

Flexor reflex

Crossed extensor reflex

Fig. 105.**12** Crossed extensor reflex.

Fig. 105.**13** A man standing erect has greater extensor tone in the lower limbs than in the upper limbs. This distribution of extensor muscle tone is attributable to the tonic labyrinthine reflex when the head is held straight (*left*), and to the tonic neck reflex when the head is bent forward slightly (*middle*). In the crawling infant (*right*), the neck is extended and therefore the forelimbs have higher extensor tone than the lower limbs.

Crossed extensor reflex The importance of this reflex in postural regulation is best understood with reference to quadrupeds. If a cat inadvertently places its right front paw on a sharp object and is hurt, it reflexly flexes the limb (withdrawal reflex). The flexion of the right forelimb will tend to destabilize its erect posture. The destabilization is prevented by other associated reflexes, i.e., hyperextension of the left forelimb and right hindlimb, and flexion of the left hindlimb (Fig. 105.**12**).

Tonic neck and labyrinthine reflexes In a person standing straight, the neck is neither flexed nor extended and the tonic neck reflexes are not active. However, the vestibular apparatus is tilted 30° backward (see Fig. 119.**1**). The backward tilt of the labyrinth reflexly increases the extensor tone in the lower limbs and reduces the extensor tone in the upper limbs. If the person flexes his neck forward, the vestibular apparatus becomes horizontal and therefore the labyrinthine reflex ceases. However, flexion of the neck activates the tonic neck reflex which increases the extensor tone in the lower limbs and reduces the tone in the upper limbs. Thus regardless of the position of the head, the extensor tone is high in the lower limbs and reduced in the upper limbs so long as the trunk remains upright.

The distribution of muscle tone in the extremities changes when the trunk is tilted. In the crawling infant (Fig. 105.**13**), the trunk is horizontal and therefore the neck has to be extended to hold the head up. The extension of the neck increases the extensor tone in the forelimbs and reduces the extensor tone in the hindlimbs. Similar reversal of muscle tone distribution is seen in a falling man (Fig. 105.**14**). The head of the falling man is tilted slightly forward so that the tonic labyrinthine reflex would be absent. However his neck is fully extended, which increases the extensor tone in the upper limbs and reduces the extensor tone in the lower limbs. However, as with all other postural reflexes, these reflexes too can be suppressed voluntarily.

Tonic neck and labyrinthine reflexes are also elicited when the head is tilted sideways (Fig. 105.**15**). If the head is tilted to the left, the extensor tone increases in the left upper and lower limbs and decreases on the right side due to the tonic labyrinthine reflexes. If the neck is flexed to the left, the extensor tone decreases in the left upper and lower limbs and increases on the right side due to the tonic neck reflexes. Hence when both the reflexes are active, their effects are mutually cancelled. However, whenever the trunk inclines, only one of them is active and therefore its postural effects are apparent.

It should be clear from the above examples that the tonic neck and labyrinthine reflexes work in concert,

Fig. 105.**14**   While falling with head down, the tonic neck and labyrinthine reflexes come into play. However, the reflexes can be voluntary suppressed, as during diving.

Fig. 105.**15**   Tonic neck and labyrinthine reflexes are a common occurrence in daily activities. Here they are seen in a game of soccer, in both the striker and the goalkeeper.

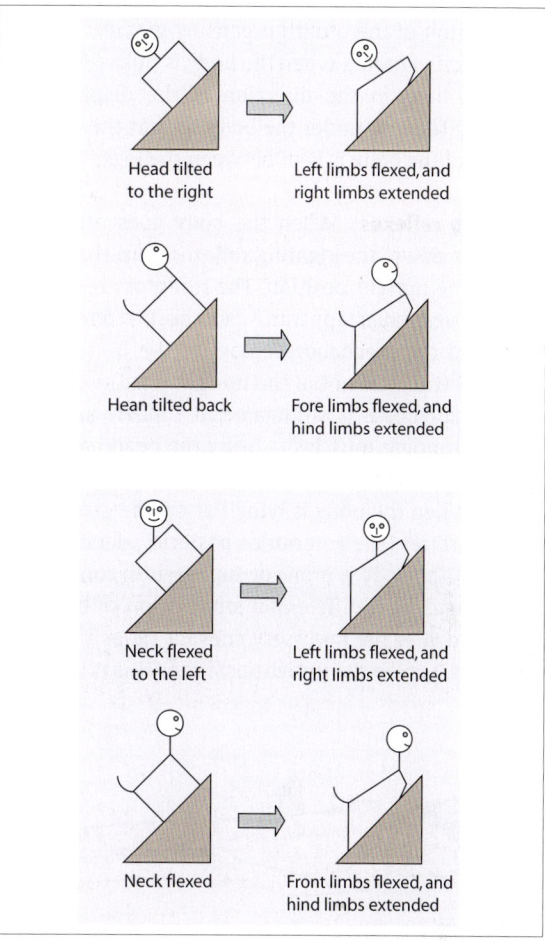

Fig. 105.**16**   Tonic labyrinthine reflexes (*above*) and tonic neck reflexes (*below*) in quadrupeds. The redistribution of muscle tone in the limbs gives postural stability on inclined planes.

bringing about a redistribution of extensor tone in the limbs depending on the inclination of the trunk. These reflexes are important because while standing on an inclined plane, it is important to redistribute the muscle tone in the limbs for maintaining a stable posture (Fig. 105.**16**). They do not occur while standing erect on a level plane because the effects of the two reflexes cancel each other. The exaggerated tonic neck reflexes can be seen clinically in the decorticate posture (Fig. 105.**17**).

Rescue reactions   *Long-loop stretch reflexes* are continuously active in the erect posture and bring about a continuous correction of our 'sways' that occur from moment to moment during standing. The body behaves like an inverted pendulum hinged at the ankle joint. When the body sways forward, there is stretching of the gastrocnemius muscle. This triggers not only the monosynaptic stretch reflex but also the long-loop stretch reflex which brings about reflex contraction in the gastrocnemius muscle with consequent correction of the sway.

Long-loop postural reflexes are also triggered by visual inputs that suggest that the body is swaying. The importance of these two long-loop reflexes (one proprioceptive, and the other visual) is obvious in patients with lesions of the dorsal columns as in tabes dorsalis. Such patients suffer from *sensory ataxia*. The patient walks on a broad base with the legs apart, raising the legs excessively high and slapping the feet on the ground. The eyes are fixed to the ground for correcting the steps visually. Closing of the eye accentuates the sensory ataxia (Romberg's sign). Romberg's sign is characteristic of sensory ataxia and helps to distinguish it from cerebellar ataxia in which the sign is absent.

Hopping and placing reactions are programmed reflexes that depend on the motor cortex. When the foot comes in contact with any firm surface, the foot is reflexly placed on the surface and the leg muscles are adjusted so as to support the body. This is the *placing reaction*. The placing reaction is also triggered during a fall under gravity, preparing the animal for landing by causing extension of the forelimbs. It is called the vestibular placing reaction and occurs due to the stimulation of the otolith organs by gravity. The *hopping reaction* occurs when the body is pushed forward. The leg hops in the direction of the displacement, bringing the foot under the body so that the center of gravity of the body is kept between the legs.

Righting reflexes   When the body goes off balance and falls down, the righting reflexes help the body to regain the upright position. The receptors responsible are the vestibular apparatus, the neck stretch receptors, and the mechanoreceptors of the body surface. When the head is not in the upright position, the vestibular apparatus is stimulated. It reflexly stimulates the appropriate muscles to bring the head back to the upright position. This is known as the *head righting reflex*. When the body is lying flat on the ground, one of its surfaces (the anterior or posterior, depending on whether the body is prone or supine) is in contact with the ground. This differential stimulation of body surfaces provides the necessary cues for reflexly bringing the head back to the upright position. This is known as

the *body-on-head righting reflex*. Righting of the head (through the head righting the reflex) while the body continues to be in the horizontal position results in the flexion or extension of the neck and activation of the neck-on-body righting reflex which stimulates the appropriate postural muscles to bring the whole body back to the upright position. When the body is lying flat on the ground, the differential stimulation of its anterior and posterior surfaces provides the necessary cues for bringing the body back to the erect position. This is known as the *body-on-body righting reflex*.

Control of locomotion

Phases of locomotion   The normal rhythm for locomotion consists of alternating swing and stance phases of the lower limbs (Fig. 105.**18**). The *swing phase* of locomotion refers to the forward stepping movement of a leg due to flexion at the hip joint. During the *stance phase*, the leg (that has just swung forward) remains firmly rooted on the ground while the trunk moves forward due to the extension at the hip joint. The phases alternate in both the legs: When one leg is in the swing phase, the other is in the stance phase.

Central pattern generator   The basic pattern for locomotion is generated in the spinal cord by a group of aminergic neurons called the *central pattern generator*. It contains two half-centers, one for swing and the other for stance, with each half-center inhibiting the other. Thus, the swing begins only at the end of the stance and the stance begins at the end of the swing phase. A transection disconnecting the spinal cord from the cerebrum does not abolish walking in cats and dogs. Although the same is generally not true for humans, cases have been reported where patients with spinal transection have shown rhythmic stepping movements of the lower limbs that were triggered by the extension at the hip joint. Human infants produce stepping movements

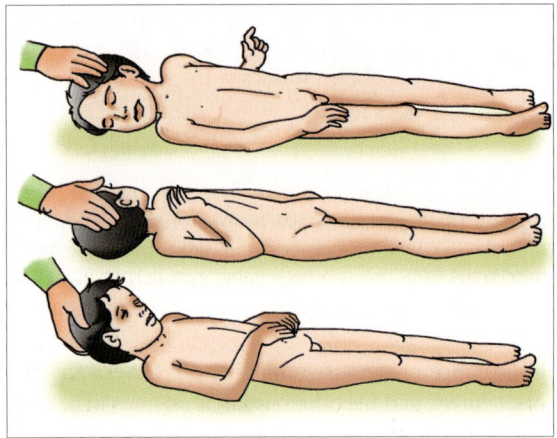

Fig. 105.**17**   Decorticate posture. It is not certain whether the clinically observed decorticate posture is caused by a lesion similar to the experimental lesion that produces the decorticate posture in animals.

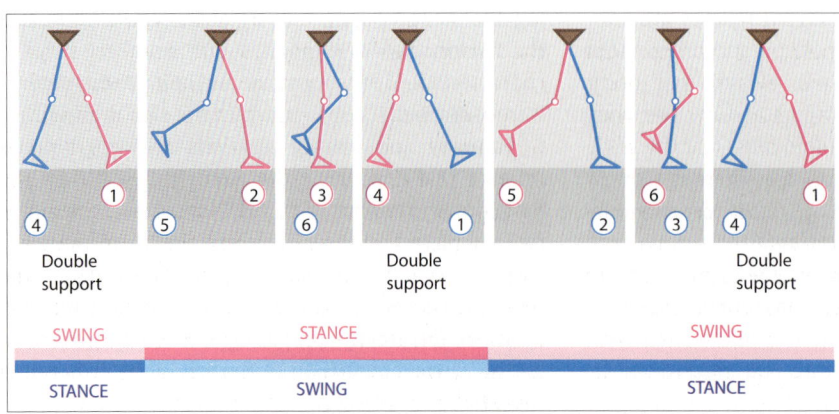

Fig. 105.**18**   Rhythm of locomotion in humans. (Red) right leg events. (Blue) left leg events. (1) heel strike (2) foot flat (3) mid-stance (4) heel raise (5) toe off (6) mid-swing.

immediately at birth if held erect and moved on a horizontal surface. Such stepping movements are known to occur even in anencephalic infants. These stepping movements however are soon suppressed and reemerge at the end of the first year in a more mature, well-coordinated form that has greater cerebral dependence.

The basic spinal rhythm of locomotion can be modified by higher control centers in the medulla, midbrain, and the motor cortex. The speed and initiation of the stepping reflex is controlled by the *mesencephalic locomotor region*. The region has no direct connections with the spinal cord but exerts its influence through the medullary reticular formation, which controls the spinal center through the reticulospinal tract.

Role of sensory feedback The importance of sensory feedback in locomotion is demonstrated by the fact that when a spinal or decerebrate cat is placed on a motorized treadmill, the rate of its stepping matches the speed of the treadmill belt. One of the most important sensory stimuli that influences stepping movements comes from the *muscle spindles of the hip flexors*. Stretching of hip flexors occurs during normal locomotion at the end of the stance phase and it stimulates the onset of swing phase. The response can be obtained in a spinal preparation by stretching the hip flexors.

Cutaneous exteroceptors also play an important role in locomotion. They detect external obstacles and adjust the stepping movements to avoid them. For example, when the plantar surface of the foot is stroked when the leg is in its swing phase, the leg flexes. The reflex is called the *stumbling-corrective reaction*, and the flexion it produces is meant for clearing any obstacle. Interestingly, when the same stimulus is applied to the plantar surface when the leg is in the stance phase, the leg extends. This is an example of *phase-dependent reflex reversal*. The physiological significance of the reflex reversal is obvious: when in the stance phase, the leg supports the entire weight of the body and its flexion during the stance phase would have sent the body tumbling down.

Role of cerebellum The cerebellum has an important role in locomotion and cerebellar disorders result in ataxia. The cerebellum compares the actual movements of the legs with the intended movements. Information regarding the former originates from proprioceptors and reaches the cerebellum through the *dorsal spinocerebellar tract*. Information about the latter originates in the central pattern generator in the spinal cord and reaches the cerebellum through the *ventral spinocerebellar tract*. The cerebellum then computes the error and sends corrective signals to the spinal cord via the *medullary reticular formation*.

Role of motor cortex The motor cortex enables skilled, visually-guided walking. Lesions of the motor cortex do not impair walking on a smooth surface but they seriously impair the ability to walk past a series of obstacles.

Motor Deficits

Lower motor neuron lesion (flaccid paralysis)

Lower motor neuron lesion (LMNL) is typically observed in poliomyelitis, in which the polio virus selectively affects the lower motor neurons of the spinal cord and brainstem. The damage to the lower motor neuron results in segmental paralysis of all voluntary movements and all reflexes, both superficial and deep. The abolition of the deep tendon reflex results in loss of muscle tone in the affected muscle (resulting in muscle flaccidity) and therefore, the condition is called *flaccid paralysis*. The paralyzed muscles often contract spontaneously due to denervation hypersensitivity, resulting in fibrillation and fasciculation. Prolonged absence of any strong muscle contraction, voluntary or reflex, results in atrophy (wasting) of affected muscles.

Upper motor neuron lesion (spastic paralysis)

Upper motor neuron lesion (UMNL) refers to damage to the corticospinal tracts and other descending motor pathways. It abolishes the voluntary contraction of the affected muscles (paralysis, usually in the form of contralateral hemiplegia) but increases their muscle tone (resulting in muscle spasticity) due to the exaggeration of the stretch reflex. Hence, the condition is called *spastic paralysis*. The spasticity is characteristically associated with clasp-knife rigidity and occasionally clonus. The immediate effect of UMNL is flaccid paralysis due to spinal shock. The spasticity develops a few days or weeks later. Pure pyramidal lesions (see below) are associated with hypotonia rather than spasticity. Hence, the spasticity of UMNL must be due to the interruption of the descending extrapyramidal fibers.

Superficial reflexes (abdominal or cremasteric reflex) are lost or diminished as their reflex pathways are transcortical and the efferent path is formed by the corticospinal tract. The *Babinski sign* is positive, i.e., the great toe is reflexly dorsiflexed, and the other toes fan out when the lateral aspect of the sole of the foot is scratched with a blunt point. The Babinski sign indicates the involvement of the corticospinal tract. In normal infants the Babinski sign is positive prior to the myelination of the corticospinal tract.

A comparison of the clinical features of upper and lower motor lesions is given in Table 105.**2**.

Pure pyramidal lesion   Pyramidal fibers do not originate from motor areas alone. A fairly large proportion of pyramidal neurons also originate from postcentral areas, viz., the primary somatosensory cortex and the sensory association (parietal) cortex. Pyramidal neurons owe their name, not to the pyramidal cells from which they originate, but to the medullary pyramids through which they all pass. In the human brain, only 3% of pyramidal neurons originate from giant-sized pyramidal cells called Betz cells. These Betz cells are confined to the primary motor area only.

Although it is tempting to believe that the robust corticospinal tract is primarily responsible for voluntary contraction of muscles, it is not true. There are surprisingly few motor deficits after a pure pyramidal lesion, which interrupts only the corticospinal neurons. A pure pyramidal lesion can be produced experimentally by sectioning the medullary pyramids. Elsewhere, the corticospinal tract is interspersed with several descending motor fibers other than the corticospinal fibers. The only deficit consistently observed after a pure pyramidal lesion is the inability to produce fine, independent (fractionated) finger movements.[4]

Table 105.**2**   Differences between upper and lower motor paralysis.

Upper motor neuron paralysis	Lower motor neuron paralysis
Muscles affected in groups, never individually.	Individual muscles may be affected.
Muscle atrophy can occur but is not marked.	Muscle atrophy is marked.
Muscles are spastic and deep tendon reflexes are exaggerated. Clasp-knife effect is present.	Muscles are flaccid and deep tendon reflexes are absent. Clasp-knife effect is absent.
Babinski sign (extensor plantar reflex) is present.	Plantar reflex absent only if L5–S1 are affected.
Muscle fasciculations are not present.	Muscle fasciculations may be present.
Electromyogram (EMG) normal.	EMG shows fibrillations and evidence of reduced numbers of motor units.

Lesions of the primary motor area produce the same deficits. These observations have led to the conclusion that corticospinal fibers, especially those originating from the primary motor area, only superimpose speed and fractionation upon the movements produced by other descending systems.

[4] Fractionated movement means controlling the digits independently of each other, a difficult proposition considering that digits are not controlled by separate muscles but by muscles that supply them together. For example, the flexor digitalis longus, when contracted, would flex all the digits together. To flex only one digit, the flexion of the other digits has to be opposed through the contraction of the extensors.

Sensory perception is a product of the brain's analysis of sensory input. The term cognition has a much broader meaning as it refers to all the processes by which the sensory input is transformed, reduced, elaborated, stored, recovered, and used. Cognition leads to thoughts, emotions, and conation (motor activity), in that order. Thus, on hearing a loud sound (cognition), one might think it to be a dangerous explosion (thought), feel afraid (emotion), and look for a shelter (conation). The terms cognitive disorder and thought disorder are often used synonymously.

Thoughts

Thinking is the ability to have ideas and to infer new ideas from old ones. Thought processes have been divided into primary and secondary processes. *Primary thought process*, like a dream, disregards logic, is based on visual images and is dominated by wish and fantasies. *Secondary thought process* is based on both visual images and words. It is logical and goal-directed.

Thought disorders

Thoughts can be retarded in depression. *Thought blocking*, seen in schizophrenia, is a sudden break in the train of thought. *Flight of ideas*, seen in mania, is an increased flow of thought resulting in frequent switching from one train of thought to another. Crowding of thoughts occurs in schizophrenic patients who complain that a large number of thoughts have been compressed into their head. *Delusions* are fixed, false beliefs that are strongly held and will not change even with evidence to the contrary. *Obsessional thinking* is stereotyped, repetitive, persistent thinking which one knows to be irrational but cannot resist. Obsessional thoughts are usually seen in conjunction with compulsive behavior.

Emotions

The term emotion must be distinguished from related terms like mood and affect. Emotion is a complex feeling state with psychic, somatic, and behavioral components, that is related to affect and mood. *Affect* is the observed expression of emotion over a short span of time, usually in response to an external event. *Mood*[1] is a sustained emotion experienced by a person. Mood is to affect as climate is to weather.

An emotional state has two components: a peripheral component and an affective component. For example, the feeling of fear (affective component) is associated with the bodily events of palpitation and heavy breathing, sweating of palms, dryness of mouth, and tense muscles (peripheral components). The peripheral components of emotion prepare the body for action (preparatory function) and communicate the emotional states to other people (communicative function). In humans, social communication of emotion is mediated primarily by the facial muscles and posture, both of which involve the skeletomotor system. The preparatory function involves both general arousal, which prepares the organism as a whole for action, and specific arousal, which prepares the organism for a particular behavior. For example, sexual arousal involves general arousal as indicated by an increase of heart rate, a change that prepares for physical exertion. In addition, it involves features of specific arousal such as penile tumescence. Arousal usually enhances both intellectual and physical performance.

Theories of emotion

James–Lange theory   According to this view, the perception of emotion results from the perception of one's own physiological changes. The theory suggests that "we feel sorry because we cry, angry because we strike, afraid because we tremble". In other words, we do not cry when we feel sorry, but it is because we find ourselves crying in a certain situation that we realize that we must be feeling sorry! In support of this theory, a very common situation is cited: A person

---

[1] Hippocrates believed that mood was determined by the balance of four factors or humors, viz., blood, phlegm, yellow bile, and black bile. According to him, an individual becomes sanguine (cheerful and optimistic) due to excess blood, phlegmatic (calm even in difficult situations) due to excess phlegm, choleric (ill-tempered) due to yellow bile, and melancholic (sad and depressed) due to black bile.

sitting in pin-drop silence is startled by a sudden harsh sound. Instantly, his heart starts racing and his mouth goes dry even before he realizes the significance of the sound. However, he takes notice of his racing heart and dry mouth and realizes that he was scared! Thus, according to the James–Lange theory, emotions are only physical sensations and therefore, originate in the peripheral organs.

The theory is partly borne out by the observation that distinct emotions tend to have distinct patterns of autonomic, endocrine, and voluntary responses. Also, patients of spinal injury who have sensory deficits appear to experience a reduction in the intensity of their emotions. However, the theory fails to explain why one often continues to be emotionally aroused even after the physiological changes have subsided. Conversely, some feelings arise faster than the changes in bodily state normally associated with those feelings.

The *Schachter theory* refines the James–Lange theory saying that the cortex takes into consideration the individual's expectations while creating the affective response to peripheral information. Schachter injected volunteers with epinephrine: Some subjects were informed of the side effects (e.g., pounding heart) while others were not. All were then exposed either to annoying or amusing conditions. The subjects who had been warned about the side effects of epinephrine exhibited less anger or less pleasurable feelings: they attributed their arousal to the drug. The other group perceived their arousal as an intense anger or joy. In support of this theory, it may be recalled that the general arousal

produced by exercise is known to result in specific arousal, such as sexual arousal.

The **Cannon–Bard theory** points out that both fight and flight (rage and fear) lead to similar autonomic (sympathetic) responses. It suggests that when a person encounters an emotional event, nerve impulses travel to the thalamus. From there they travel to the cerebral cortex to produce the affective component of fear and to the hypothalamus, where they produce simultaneous physiological changes. The theory therefore locates emotion centrally and not peripherally. Autonomic responses are however not as uniform as Cannon believed: Rather, different emotional states are typically accompanied by different patterns of autonomic responses, such as changes in blood flow or heart rate.

The Arnold theory refines the Cannon–Bard theory by saying that the peripheral component of emotion results from the unconscious evaluation of a situation as potentially harmful or beneficial, while the affect is the conscious reflection of the unconscious appraisal. In other words, emotions have their own logic (Fig. 106.**1**). Thus, the sight of the New York Twin Towers crumbling to dust would have evoked feelings as varied as fear, rage, grief or joy in the millions watching around the world!

◼ Neural substrates of emotion

The **hypothalamus** coordinates the peripheral, autonomic component of emotions. Stimulation of different hypothalamic regions evokes different patterns of autonomic reactions that are characteristic of different emotions. For example, animals with lesions in the lateral hypothalamus become placid, whereas animals with lesions of the medial hypothalamus are highly excitable and aggressive.

Cats in which the whole cerebral cortex has been removed retain fully integrated emotional responses, termed *sham rage* because the responses lack the conscious experience that is characteristic of genuine rage (sham = false). Sham rage also differs from genuine rage because responses can be triggered by very mild stimuli such as a weak touch, or can even occur spontaneously without provocation. It is undirected, and the animals sometimes even bite themselves. No matter how it is elicited, sham rage subsides very quickly once the stimulus is removed. Sham rage largely disappears when the hypothalamus is ablated.

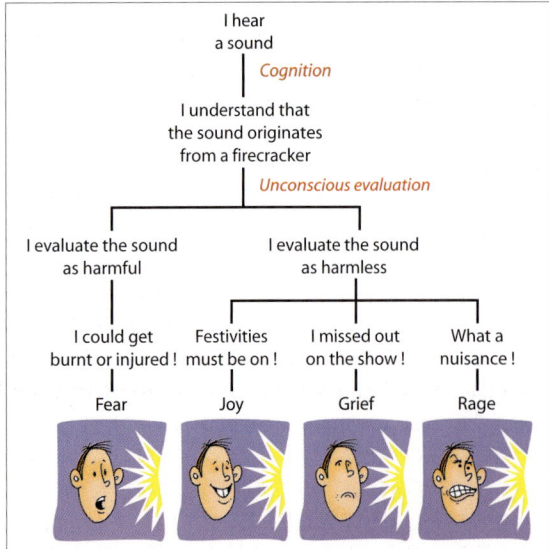

Fig. 106.**1** An example of the Arnold theory of emotions, emphasizing the importance of the unconscious evaluation of a situation as potentially harmful or beneficial.

The **cerebral cortex** is the site of the evaluative, central component of emotions. It provides the means by which memory and imagination too can evoke emotional feelings. A cortical area that can evoke autonomic responses and a general arousal when stimulated is the *orbitofrontal cortex*. Lesions of the orbitofrontal cortex reduce the normal aggressiveness and emotional responsiveness.

Cortical mechanisms also provide the means by which conscious thought can suppress reflex emotional responses. For example, once we know that an explosive sound came from only a firecracker (Fig. 106.**2**), the fear subsides. One source of this cognitive control of emotional responses is the *ventromedial frontal cortex*. Patients with frontal lesions have normal galvanic skin responses (sweating measured electrically, an index of sympathetic activity) to startle stimuli such as unexpected loud noises or bright lights, indicating a normal autonomic response mechanism. However, when they are shown emotionally disturbing pictures, they fail to show the expected autonomic responses to these emotionally charged stimuli. Some patients said that they knew they should have been disturbed by the pictures but found themselves unmoved! It was this area that was probably damaged in Phineas Gage (see box).

Another cortical area for control of emotional responses is the *anterior cingulate cortex*. Lesions of this area reduce the emotional response to chronic intractable pain. Patients in whom the cingulate gyrus has been removed are no longer bothered by pain. They experience pain as a sensation and exhibit appropriate autonomic reactions, but the sensation is not perceived as intensely unpleasant.

The fact that the cognitive information and the emotional information contained in a picture are processed independently is brought out by the existence of two contrasting clinical syndromes, viz., prosopagnosia and Capgras syndrome. In *prosopagnosia* (p 747), the patients cannot consciously identify faces, even those of familiar associates and relatives. Yet they exhibit autonomic responses (e.g., skin conductance change) to familiar faces but not to unfamiliar faces. In *Capgras syndrome*, the patients can readily recognize familiar faces but apparently do not have emotional responses to them.

The **amygdala** plays a pivotal role in coordinating the peripheral and central components of emotional processing. The autonomic responses to emotion involve the hypothalamus while the affect involves the cerebral cortex, especially the cingulate, parahippocampal, and the prefrontal cortices. These connections of the amygdala form the extended Papez circuit (Fig. 106.**3**).

It is different from the Papez circuit which was earlier thought to be important for emotions and in which the hippocampus was assigned a central role.

Phineas Gage

In September 1848, Phineas Gage had a nasty accident while working on the Rutland and Burlington Railroad. While tamping down the blasting powder for a dynamite charge, Gage inadvertently sparked an explosion. The inch-thick tamping rod flew through his cheek and brain (Fig. 106.**2**). Gage collapsed but a moment later, he stood up and spoke! His fellow workers took him to a local doctor who dressed his wounds. Within two months, Phineas Gage recovered physically but his personality changed completely. Honest and diligent Gage had become a foulmouthed liar given to extravagant schemes that he never followed through. The tamping rod had damaged the left ventromedial region of his brain, making him antisocial. His battered skull is preserved in the Warren Medical Museum at Harvard.

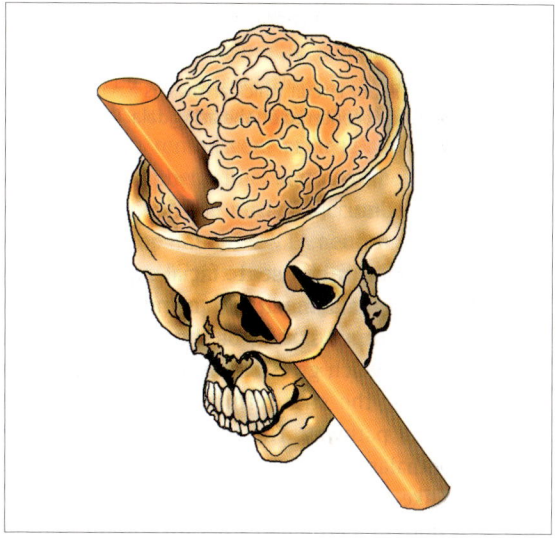

Fig. 106.**2** The path of the rod that accidentally pierced the skull and brain of Phineas Gage.

The role of the Papez circuit in emotions is now largely discarded. Parts of the circuit may be important to the hippocampus for its role in explicit (declarative) memory for facts and personal events (p 707) that are associated with emotions. The unimportance of the hippocampus in emotion is apparent in fear-conditioning, e.g., a flash of light (conditioned stimulus) followed by a loud explosion (unconditioned stimulus). Hippocampal damage will interfere with remembering the features of the conditioned stimulus like the color of light or when it occurred, etc. However, the affective as well as the peripheral response to the loud explosion persists.

The *basolateral nucleus* of the amygdala receives important afferent information from all sensory modalities. A major amygdalar input comes from the *inferotemporal cortex*, which is involved in the explicit memory of facial identity only. Since the amygdala receives input from the inferior temporal cortex and has strong connections to the autonomic nervous system, it can

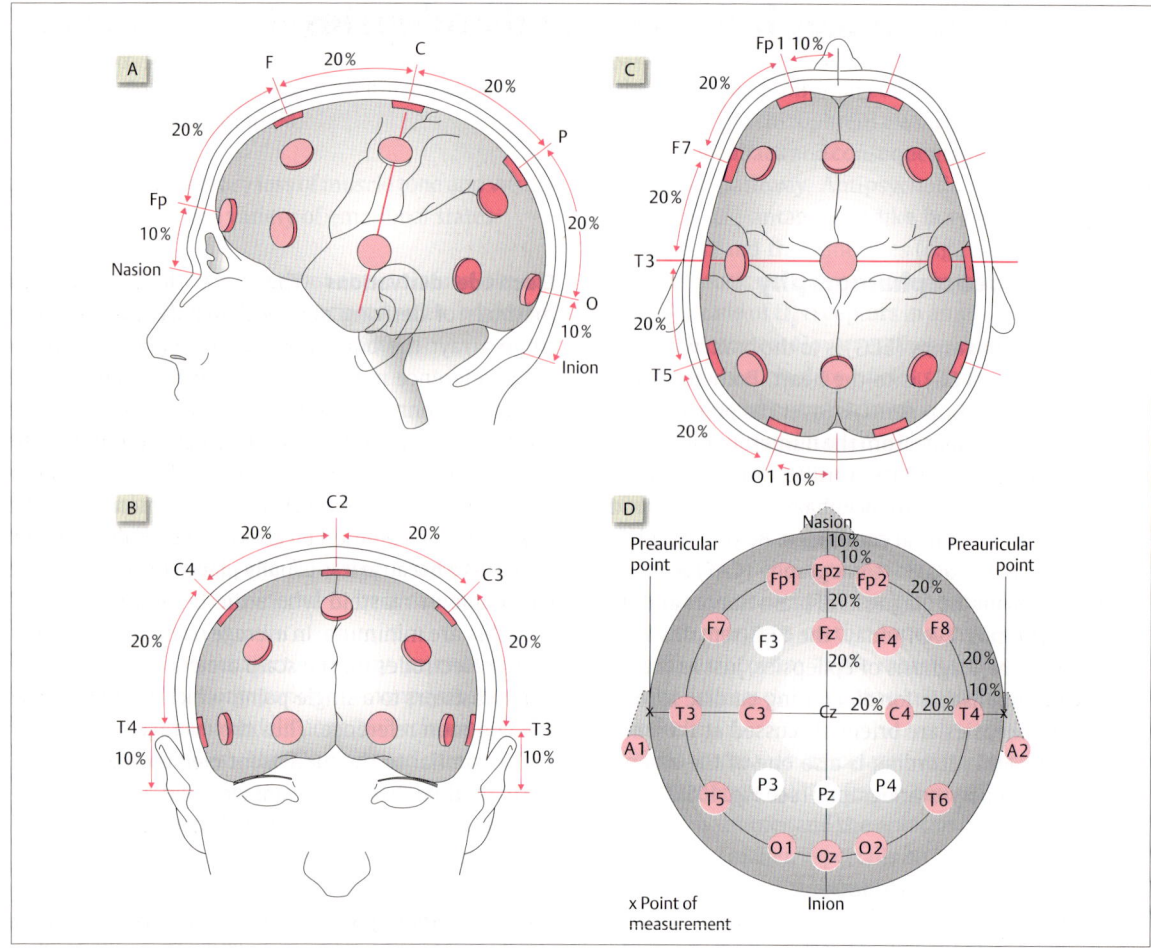

Fig. 107.**1**   The 10–20 international system of electrode placement. [A] Lateral view. The electrodes are placed at fixed percentage intervals between the nasion and the inion. [B] Frontal view. The preauricular points serve as reference points for the placement of the central transverse row of electrodes. C2 is the intersection of the central transverse and longitudinal rows. [C] Superior view. [D] Names of the electrodes in the 10–20 system.

the parietooccipital area of the scalp when the person is awake and relaxed with eyes closed. The α rhythm disappears on opening the eyes and being attentive. This transformation is known as *EEG desynchronization* (Fig. 107.**4**). This desynchronization occurs due to attention being paid to a stimulus and not its perception since even trying to see in a dark room brings about EEG arousal. The α rhythm disappears entirely during deep sleep. The mean peak alpha frequency is 10.2 Hz which decreases in hypoglycemia, hypothermia, hypercapnia, and hypocortisolemia.

Beta waves have frequency more than 13 cycles per second and may be as high as 25 Hz. They have lower voltage than alpha waves. They are frequently recorded from the parietal and frontal region. Barbiturates induce beta (18–24 Hz) activity.

Theta waves have a frequency between 4 and 8 Hz and have larger amplitude than alpha waves. They are seen in the parietal and temporal region in children. They are seen in emotional stress in adults, and also occur in many brain disorders. The incidence of transient theta rhythm is about 30% in normal adult subjects in the alert stage. It is recordable directly from the hippocampus in experimental animals.

Delta waves have a frequency of less than 3 Hz. They are seen in deep sleep and in infancy. When delta activity occurs in an awake adult, it indicates serious organic brain disease. Delta waves also occur in the cortex of animals that have had subcortical transections separating the cerebral cortex from the thalamus, indicating that delta rhythm occurs in the cortex independent of activities in the lower regions of the brain.

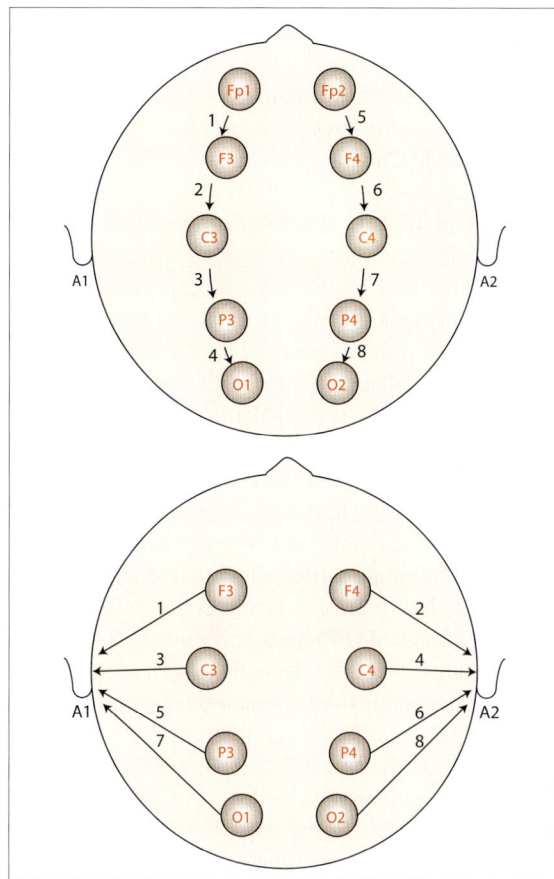

Fig. 107.**2**   EEG montages. (*Above*) Longitudinal bipolar montage. (*Below*) Common referential montage referred to the ipsilateral ear lobule.

	Designation	Morphology	Definition
1	β rhythm		Regular sequence of waves at 14–30 Hz
2	Spindles		Regularly waxing and waning waves at 14–30 Hz
3	α rhythm		Regular sequence of waves at 8–13.3 Hz
4	ϑ rhythm		Regular sequence of waves at 4–7 Hz
5	δ rhythm		Regular sequence of waves at 1–3.5 Hz
6	δ activity		Irregular sequence of polymorphic waves at 1–3.5 Hz
7	Subdelta wave		Wave with duration > 1 s
8	Steep waves (steep potential)		Conspicuous, blunt, steep individual waves
9	Sharp waves (sharp potential)		Sharp and steep waves of 80–250 ms duration ascending phase usually steeper than descending phase
10	Spike		Sharp and steep wave of duration < 80 ms
11	Polyspikes		Compact series of spikes
12	Spike–wave complex		Complex consisting of a spike and slow wave
13	Rhythmic spikes and waves		Sequence of regular spike–wave complexes at about 3 Hz
14	Sharp and slow waves	2s	Sequence of complexes of sharp waves and slow waves of 500–1000 ms duration, often rhythmic

Fig. 107.**3**   Some important EEG waves.

Epileptiform EEG discharges

While there are several types of abnormal EEG waves, the epileptiform waves are of particular importance because a major use of EEG lies in the diagnosis of epilepsy. In general, the epileptiform wave may be: (1) a spike (duration of 20–70 ms), (2) a sharp wave (duration of 70–200 ms) or (3) a spike and wave complex, in which a spike or a sharp wave is followed by a slower wave. In grand mal epilepsy, there are polyspikes. In petit mal, there are spikes and waves at a rate of 3 Hz.

Activation procedures   In a suspected case of epilepsy, an EEG record showing epileptiform waves will confirm the diagnosis. The epileptiform waves, however, usually appear during an epileptic attack. In between attacks, i.e., in the *interictal period*, epileptiform waves may or may not occur. Since EEG records are mostly obtained during interictal periods, attempts are made to trigger the appearance of epileptiform waves and thereby clinch the diagnosis of epilepsy. The procedures for triggering epileptiform waves in the EEG are called activation procedures. The most widely used activation procedure is *hyperventilation* for 3 to 5 minutes. However, the most effective activation procedure is *intermittent photic stimulation* in which flashes of bright light (frequency of 1 to 30 per second) are delivered for a period of 10 seconds.

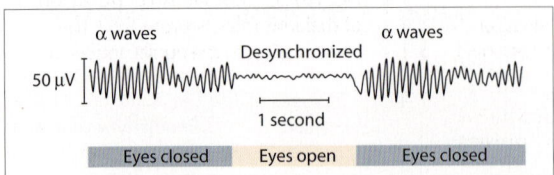

Fig. 107.**4**   Alpha block. Alpha waves appear when the eyes are closed and the mind is relaxed. They disappear as soon as the eyes are opened.

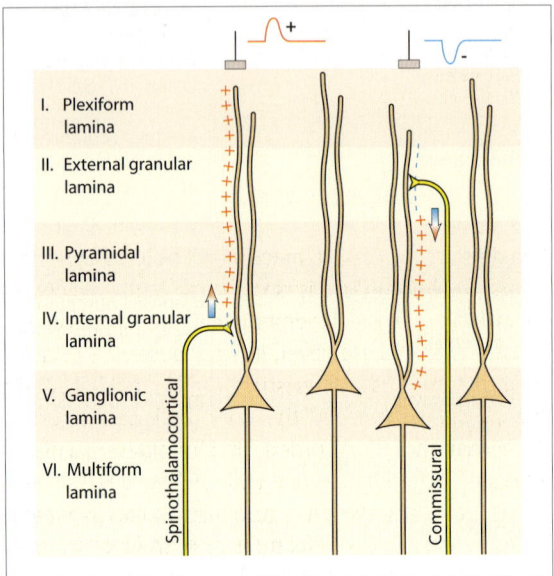

Fig. 107.**5**   Dipoles formed on the apical dendrite of a pyramidal cell (see also Fig. 100.**2**). Commissural fibers produce an EPSP on the dendrite in layers 2 and 3. The dipole is therefore negative near the surface and the EEG wave is negative. Spinothalamocortical fibers produce an EPSP on the dendrite in layer 4. The dipole is therefore positive near the surface and the EEG wave is positive. The polarity of the EEG waves reverses if an IPSP is produced instead of an EPSP.

Origin of EEG waves

The scalp potentials recorded in the EEG are produced by the currents set up by the dipoles formed in the cerebral cortex. Layer 5 of the cerebral cortex contains the pyramidal cells. At its apex, the pyramidal cell gives off a long dendrite (the apical dendrite) that is directed perpendicular to the cortical surface. The dipoles that produce the EEG are formed on these apical dendrites of pyramidal cells.

As explained in the context of the ECG, a dipole is formed when different parts of an excitable tissue are in different states of polarization. In case of the cortical dendrites, the polarity of the dipole depends on whether the postsynaptic potential is excitatory (EPSP) or inhibitory (IPSP), and whether the postsynaptic potential develops in layers 2 and 3 (i.e., near the cortical

surface) or in layer 4 (i.e., a deeper layer). Axons from the contralateral cortex terminate in layers 2 and 3 while thalamocortical neurons terminate in layer 4. Figure 107.**5** shows how these factors influence the polarity of the EEG wave.

Mechanism of EEG synchronization   The thalamic relay nuclei have two different physiological states: the transmission mode and the burst mode. When the relay nucleus is in the *transmission mode*, its resting potential remains close to its threshold potential so that it fires off a single action potential each time it is stimulated by the spinothalamic neurons. Thus each impulse traveling up the spinothalamic neuron is faithfully transmitted by the relay cell to the sensory cortex. When the relay nucleus is in the burst mode, it is in a hyperpolarized state and when depolarized to the threshold potential, it fires off a burst of action potentials. These bursts are conducted to the sensory cortex and are associated with high amplitude EEG waves.

The transition of the thalamic relay nuclei from the transmission mode to the burst mode is brought about by the GABAergic inhibitory neurons of the reticular nucleus of the thalamus (Fig 107.**6**) which hyperpolarize the thalamic relay cells through $GABA_B$ receptors, switching the cells to the burst mode. Since the thalamic relay cells and the reticular cells are reciprocally connected, bursts of action potentials in the relay cells produce similar bursts in the reticular nucleus too, so that the reticular nucleus remains stimulated and the thalamic relay cells remain hyperpolarized. The burst cycles end when the reticular nucleus is inhibited by the neurons of the mesopontine and basal forebrain

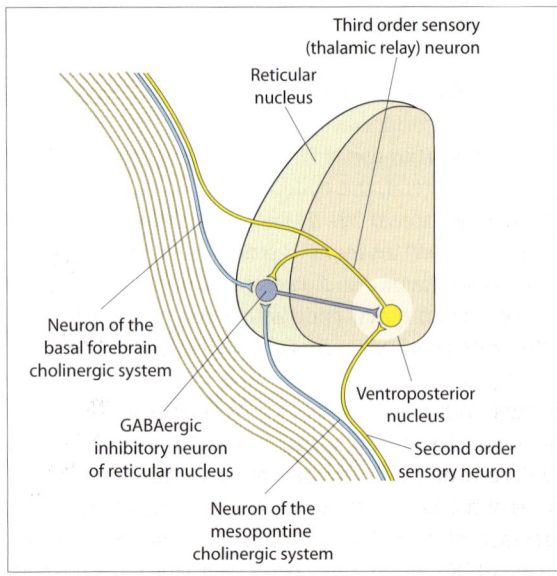

Fig. 107.**6**   Role of the reticular nucleus in EEG synchronization.

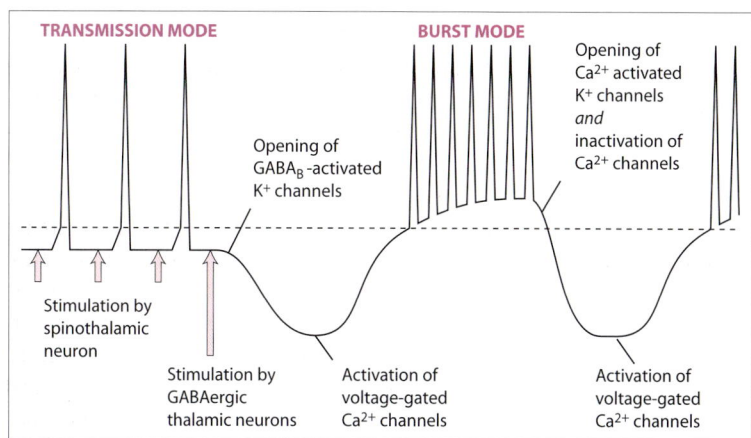

Fig. 107.**7**   Mechanism of transition of thalamic relay neurons from the transmission to the burst mode.

cholinergic systems. Similar bursts of the thalamic relay cells and reticular nucleus cells may underlie seizure discharges too.

The ionic mechanism of the alternations between the transmission and burst modes is as follows. The thalamic relay cell has a population of voltage-gated Ca^{2+} channels that remain inactivated when the membrane potential is near the firing level, and become activated when the membrane hyperpolarizes. In the hyperpolarized relay cell, therefore, an inward Ca^{2+} current flows through these Ca^{2+} channels, depolarizing the cell to the threshold potential and triggering a burst of action potentials. Following membrane depolarization, the Ca^{2+} channels close down but not before enough Ca^{2+} has entered the cell to activate a population of calcium-activated K^+ channels. The opening of these K^+ channels hyperpolarizes the cell and the burst of action potentials stops. However, hyperpolarization also restores the activity of the Ca^{2+} channel, so that the burst cycles repeat over and over again (Fig. 107.**7**).

EEG Analysis

Visual analysis

Certain features of the EEG that are of diagnostic importance can be appreciated merely by looking at the record carefully. These changes include the epilepsy-seizure potentials, the irregular slowing associated with cerebral trauma or metabolic coma, the local changes associated with tumors, and the changes associated with psychotropic drugs. Brain death results in an isoelectric or flat EEG. It is important that artifacts of EEG recording are not interpreted as clinical abnormalities of the EEG. Some common EEG artifacts are sweat artifact, blink artifact, EMG artifact, and ECG artifact (Fig. 107.**8**). An experienced electroencephalographer will readily identify these and other artifacts.

Quantitative analysis

Looking at the EEG, it may be possible to comment whether the voltage is somewhat high or low, or whether it shows a predominance of a particular frequency of waves. However, it is not possible to exactly quantify these characteristics just by looking at the EEG, if only because of the sheer bulk of the record. A 16-channel EEG recorded for 5 minutes at a chart-speed of 30 mm per second will produce 9 × 16 meters of EEG record! However, a computer makes it possible to summarize the key features of such large volumes of EEG record (data reduction) in terms of voltage and frequency. The common methods of data reduction are frequency analysis and spatial mapping. Frequency analysis gives the relative power of a particular EEG wave that is present in the EEG. When this information is depicted graphically for all the electrode sites on the scalp, it is called a spatial map. The spatial mapping of the EEG activity is called brain electrical activity mapping (BEAM).

Evoked potentials   If a stimulus is delivered to a subject while his EEG is being recorded, it results in the appearance of several electrical waves in the EEG after a brief latency. These small but consistent EEG waves produced in response to the stimulus are called the evoked potentials. However, the evoked potentials (the signal) that are recorded are barely discernible against the dense backdrop of the spontaneous EEG waves (the noise). In order to identify the evoked potential, the signal-to-noise ratio is enhanced using the technique of averaging (Fig. 107.**9**). The evoked potential is called the signal because it is produced consistently and interests us, while the background EEG is called noise because it is produced randomly and does not interest us.

Depending on the stimulus modality, evoked potentials are called auditory, visual, or somatosensory

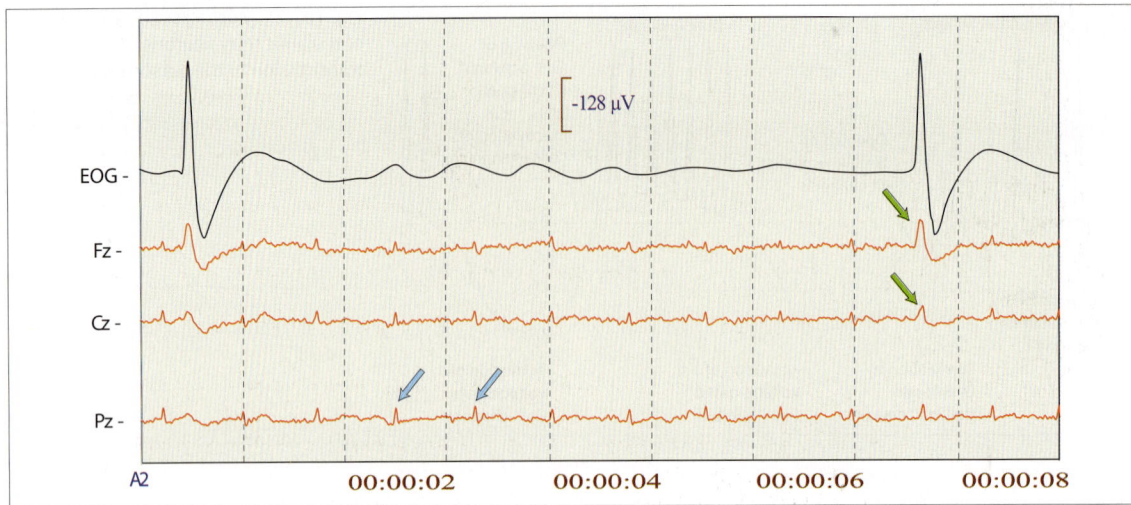

Fig. 107.**8**   Blink artifact (green arrows) and ECG artifact (blue arrows) in the EEG. Note that the blink artifact is most promi- nent in the frontal (Fz) EEG lead and coincides with the blink recorded in the electrooculogram (EOG).

evoked potential. Also, depending on their latencies, these waves are called short-latency (<10 ms), mid-latency (10–50 ms) or long-latency (>50 ms) evoked potentials (see Fig. 118.**14**). Evoked potentials with latencies less than 250 ms are called *stimulus-related potentials* because they are strongly related to the intensity of the stimulus and are recordable even in comatose patients. They are of interest to the neurologists as they give considerable information about the conduction speed of auditory, visual, and somatosensory impulses through the brain. The commonest type of evoked potential in clinical use is the short-latency auditory evoked potential, popularly known as the *brainstem auditory evoked potential* as it originates in the brainstem. The mid-latency (10–50 ms) response is thought to reflect activity arising in the thalamocortical radiations and the cerebral cortex. The neural generators of the long-latency (>50 ms) response are debatable.

Event-related potentials   Evoked potentials with latencies greater than 250 ms are of more interest to the psychiatrists and psychologists who have hailed it as a brain wave that reflects cognitive processes in the brain. These waves, which are called event related potentials (ERP), may or may not appear prominently in response to a stimulus depending on the attention devoted to it. When they do appear, they are independent of the stimulus intensity. The most easily recordable event-related potential is the P300 (or P3 for short), which is a large, positive potential recorded about 300 ms after the stimulus, provided the subject pays special attention to it. It is generally agreed that the amplitude of P3 is related to the amount of working memory allocation demanded by the task. The P3 wave has a

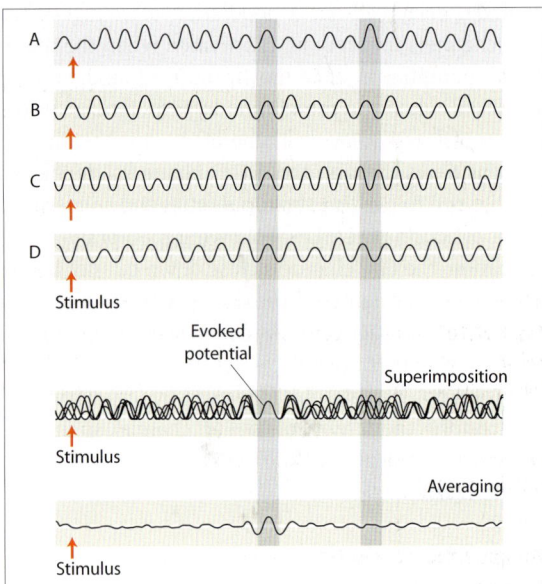

Fig. 107.**9**   The principle of averaging. Strips A to D shows the EEG record for a certain duration after delivering a stimulus (auditory or visual). The stimulus results in a consistent EEG wave (called evoked potential) after a fixed time interval. It is difficult to identify the wave in strips A to D but when all the strips are superimposed or averaged, the evoked potential stands out prominently.

longer latency and smaller amplitude in dementia.[1] P3 has also been used in the evaluation of attention-deficit disorders.

---

[1]   Dementia is an abnormal deterioration of intellect affecting several areas of cognitive function, such as abstraction, orientation, judgment, and memory.

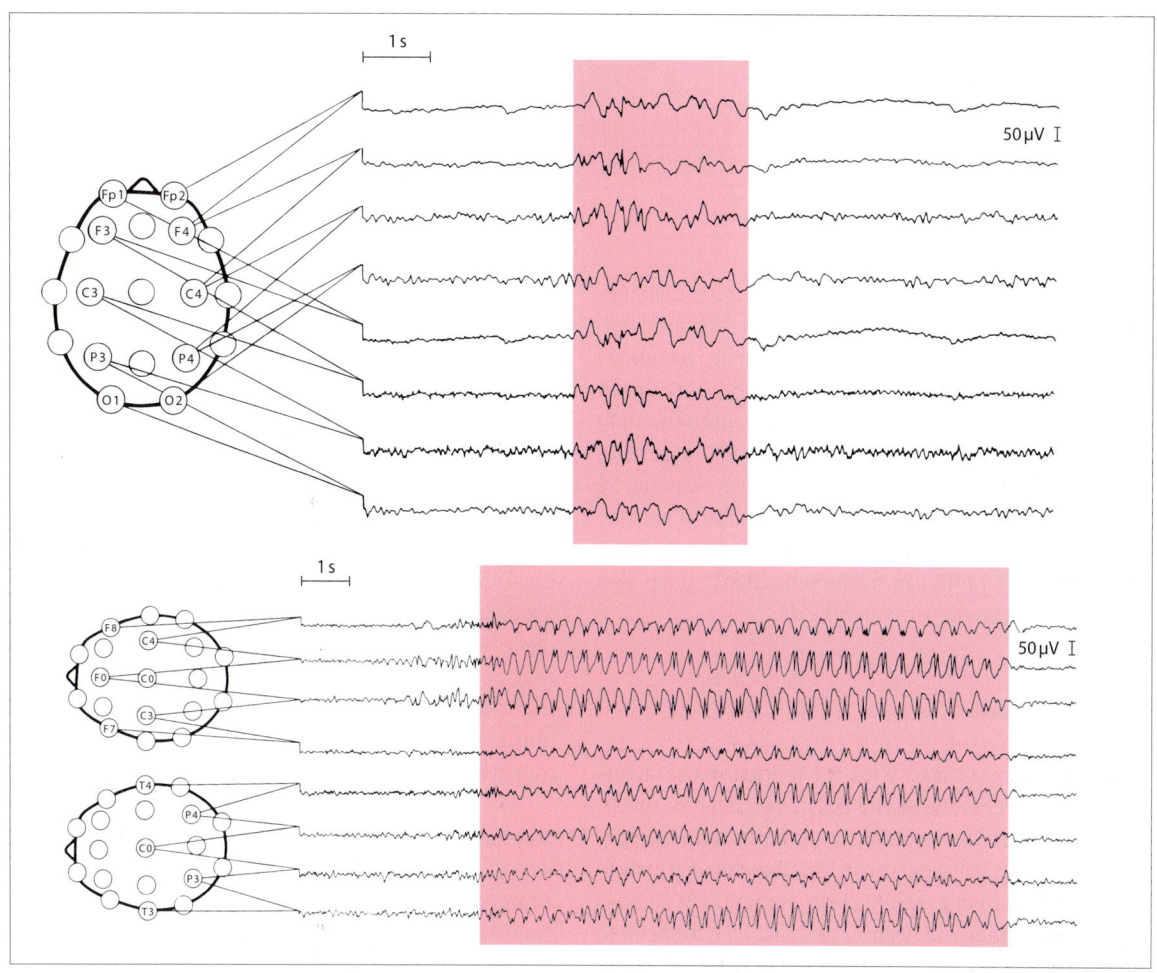

Fig. 107.10 (*Above*) Interictal EEG in a patient with grand mal seizures, showing a synchronous paroxysm of generalized, partly atypical spikes and waves. (*Below*) EEG in a patient with absence seizures, showing generalized spikes and waves at 3–4 Hz, induced by hyperventilation.

Epilepsies

An epileptic fit or seizure is a brief disorder of cerebral function, associated with disturbance of consciousness and accompanied by a sudden, excessive discharge of cerebral neurons. They are triggered by a population of abnormally hyperexcitable neuron cells. Seizures are of two types: generalized and focal.

In **generalized seizures**, the excitation originates in the midline, diencephalic areas and spreads to large areas of the brain. Clinically, there are two common subtypes of generalized seizures, the grand mal and the petit mal (Fig. 107.**10**). In *grand mal epilepsy*, there are tonic-clonic seizures lasting about a minute. In the first 30 seconds, there is loss of consciousness and a sustained, tonic spasm of all the muscles. Spasm of respiratory muscles results in inadequate ventilation, leading to cyanosis. In the next 30 seconds, there is jerky, 'clonic'

contraction of muscles. Contraction of the jaw and tongue results in frothing at the mouth and biting of the tongue. In *petit mal epilepsy*, there are 'absence' seizures, lasting about 15 seconds. During these absences, there is a transient loss of consciousness and the patient may stare blankly ahead.

In **focal seizures**, the clinical features depend on the areas of the brain that are affected. However, the focal discharge may spread to become generalized. The temporal lobe is the commonest site of focal epilepsy. Temporal lobe seizures result in hallucination and disturbances of memory, including déjà vu phenomenon, i.e., a sensation of reliving an experience or a feeling of great familiarity with the surroundings. A less common type of focal epilepsy is Jacksonian epilepsy. It originates in a small part of the body as muscle twitches or jerky movements, and slowly spreads to the other parts.

108 Sleep

Sleep is a reversible behavioral state of unresponsiveness and perceptual dissociation from the environment. Sleep is usually, but not necessarily, accompanied by postural recumbency, quiescence, and closed eyes. Most young adults report sleeping approximately 8 hours a night. Sleep length depends upon genetic factors, volitional factors (staying up late, waking by alarm, etc.), and circadian rhythms (the time of the day when one sleeps). Within sleep, there are two separate states, viz., the rapid eye movement (REM) sleep and the nonrapid eye movement (NREM).

REM sleep

REM sleep is associated with the following characteristics: (1) The EEG shows the desynchronized or wakefulness pattern. Hence, REM sleep is also called *desynchronized sleep*. Yet, a person is usually more difficult to awake in REM than in NREM sleep. Hence, REM sleep is also called *paradoxical sleep*. (2) REM sleep is associated with longer episodes of dreaming than NREM sleep. The earlier belief that dreaming occurs only in REM sleep is incorrect. (3) Inhibition of spinal motor neurons via brainstem mechanisms causes atonia of antigravity muscles in REM sleep. (4) During REM sleep, there are bursts of rapid, saccadic eye movements, twitches in other phasic muscles, and changes in the breathing pattern. (5) During NREM sleep, there is sweating or shivering due to impaired thermoregulation. These cease during REM sleep. (6) In cats, rapid eye movements are associated with bursts of pontogeniculooccipital (PGO) waves. In humans, PGO waves are not seen in scalp EEG but are recordable by depth EEG recordings.

NREM sleep

NREM sleep is subdivided into four stages based on the EEG (Fig. 108.1). Sleep is lightest in stage 1 and deepest in stage 4. A series of body movements usually signals a drift to the lighter NREM sleep stages. NREM sleep is usually associated with a lack of mental activity. The EEG pattern in NREM sleep is synchronous, and is associated with characteristic waveforms like sleep spindles, K complexes, and high-voltage slow waves.

In **stage 1 NREM sleep**, the EEG changes from a rhythmic alpha (8 to 13 Hz) activity, particularly in the occipital region, to a relatively low-voltage, mixed-frequency pattern. Bursts of high-voltage, theta (3 to 7 Hz) activity are common during the onset of stage 1 sleep in children and young adolescents.

There is considerable sensitivity to sensory stimuli during stage 1 sleep. However, mild or moderate stimuli are often unable to produce a full arousal; instead, they produce a sharp electronegative wave at the vertex called the *vertex sharp wave* or *V wave*. They can be regarded as the rudimentary form of the K complexes that appear in stage 2. They look similar too, except that the V waves are sharper and have shorter duration. Slow eye movements commonly precede the EEG transition from wakefulness to stage 1 sleep.

Stage 2 NREM sleep is signaled by the appearance of sleep spindles in the EEG. These are up to 100 μV in amplitude and have a frequency of about 14 Hz. Each burst lasts about 0.5 to 1.5 seconds. Auditory stimuli during this phase readily evoke the *K complexes* in the EEG. They also occur spontaneously during this stage. The K complex consists of one or two high voltage waves followed by a brief 14 Hz activity. As stage 2 sleep progresses, there is a gradual appearance of high-voltage slow wave activity in the EEG till the onset of stage 3 sleep.

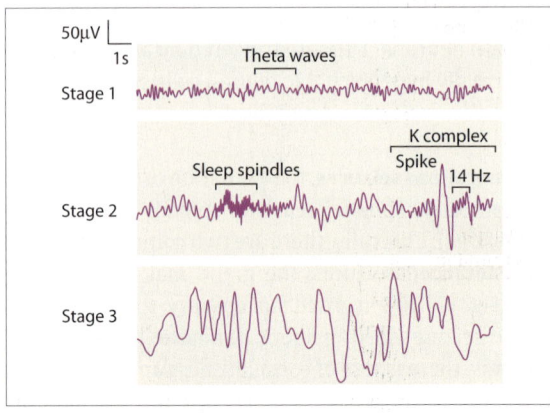

Fig. 108.**1** Electroencephalographic characteristics of different stages of NREM sleep.

Stage 3 NREM sleep is characterized by high-voltage and slow frequency waves called the delta waves, which account for 20 to 50% of the EEG activity.

Stage 4 NREM sleep is characterized by delta activity accounting for more than 50% of the EEG activity. The stage 3 and stage 4 sleep together are referred to as slow wave sleep, delta sleep, or deep sleep.

Sleep onset

Attempts have been made to define the exact point of sleep onset in terms of polysomnographic features, behavioral responses, and intellectual functions. Polysomnography refers to the record of electroencephalogram (EEG), electromyogram (EMG), electrooculogram (EOG), and other physiological parameters that indicate the onset and depth of sleep (Fig. 108.**2**).

Electrophysiological responses In the EMG, the motor unit potential (MUP) amplitudes show a gradual diminution as stage 1 sleep approaches. The EOG indicates the presence of slow rolling movements of the eye called the slow eye movements (SEMs) as sleep approaches. However, the onset of stage 1 does not

necessarily coincide with the perceived onset of sleep. Hence, the presence of specific EEG patterns like the K complex or sleep spindle (i.e., stage 2 sleep), has been taken as the time of sleep onset.

Behavioral responses In stages 1 and 2, the subject responds to visual stimuli infrequently. Auditory reaction time becomes longer as stage 1 sleep approaches. Auditory response is absent at sleep onset. There is some response to meaningful stimuli even in sleep, which indicates that sensory processing continues even after the onset of sleep. This is apparent from the discriminant responses during sleep to meaningful versus nonmeaningful stimuli (see box). The response to a meaningful stimulus may be indicated by the evoked K complexes instead of a full arousal. The likelihood of a response to a neutral stimulus during sleep can be

Examples of Discriminant Responses
- A person tends to have a lower arousal threshold for his/her own name versus someone else's name.
- A sleeping mother is more likely to hear her own baby's cry than the cry of an unrelated infant.
- A captain wakes up to the cry of 'iceberg' in the midst of the din and bustle of a ship.

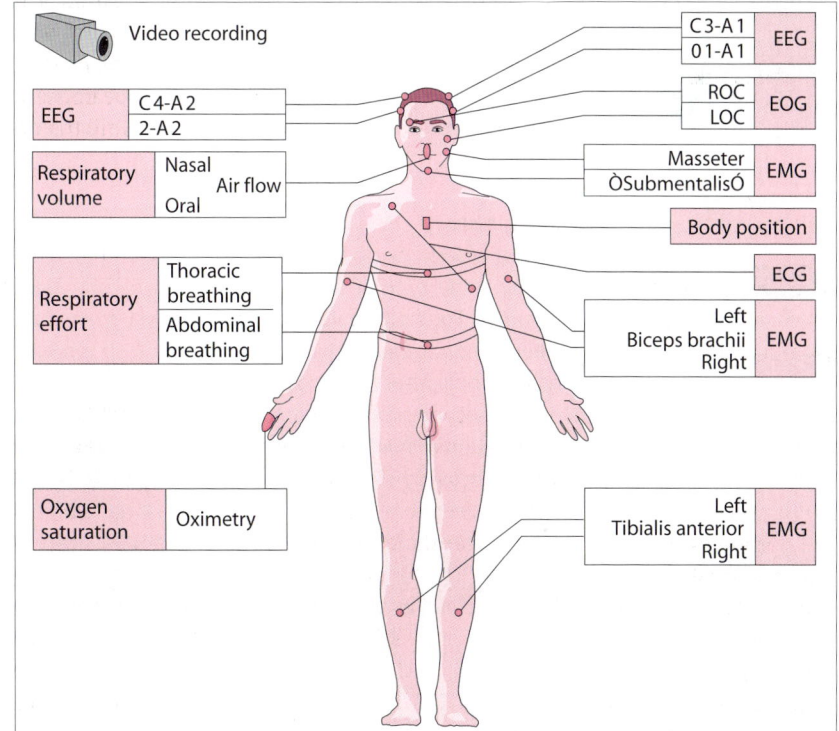

Fig. 108.**2** Recording scheme for polysomnography.

increased through instrumental conditioning, e.g., by linking the absence of response to punishment (a loud siren or an electric shock).

Intellectual functions Toward the onset of sleep, thoughts become illogical and incoherent. The transition from wakefulness to sleep tends to produce a retrograde amnesia (see box). This is because sleep inactivates the consolidation of short-term into long-term memory. At least 10 minutes of sleep is required for retrograde amnesia to occur. Patients suffering from excessive sleepiness may experience similar memory problems in the daytime if there are frequent catnaps.

Sleep cycles

In normal adults, sleep mostly begins with NREM sleep. Thereafter, NREM sleep and REM sleep alternate cyclically through the night. In infants, entry into sleep occurs through REM sleep. In adults, REM-onset sleep occurs in jet lag, chronic sleep deprivation, narcolepsy, acute withdrawal of REM-suppressing drugs and endogenous depression.

A night's sleep usually progresses through 4 or 5 sleep cycles (Fig. 108.**3**). The average duration of a sleep cycle is 90 min, with the first cycle a little shorter. A typical sleep cycle progresses through stages 1 to 4 and

> **Examples of Retrograde Amnesia**
> - Inability to grasp the instant of sleep onset in memory.
> - Not remembering the ringing of the alarm clock.
> - Morning amnesia for sleep-talking.
> - Poor recall of midnight dreams.

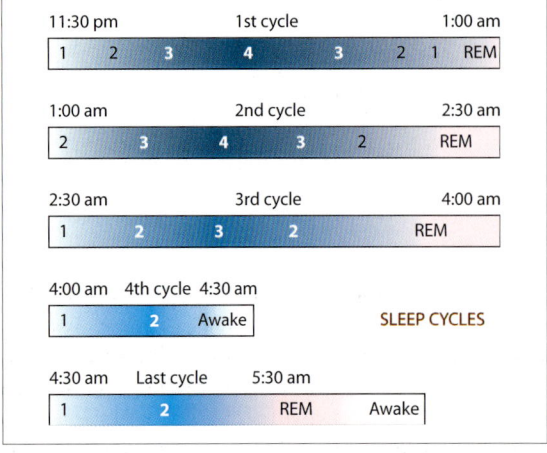

Fig. 108.**3** Sleep cycles during a typical night's sleep. The fourth cycle was interrupted prematurely by wakefulness.

then back through stages 3 to 1. The cycle ends with the REM sleep. Toward the end of the night, the cycles are often separated by brief intervals of awakening. These interruptions in sleep are not remembered in the morning due to retrograde amnesia. The transition from wakefulness to sleep is always through stage 1. The transition from sleep to the awake state occurs either at the end of REM sleep or at the end of stage 2.

Sleep stage distribution

The duration of sleep stages are different in the first cycle and in the later cycles. In the first cycle, stages 2, 3, and 4 constitute 20%, 30%, and 40% of the duration respectively, with stage 1 and REM sleep accounting for 5% each. In the later cycles, the duration of stage 2 increases greatly to occupy most of the NREM portion. The later cycles also show an increase in REM sleep. Stages 3 and 4 (slow wave sleep) occupy less time in the second cycle and may disappear altogether from later cycles.

Effect of age Sleep cycles are present from birth. Neonates have REM-entry sleep cycles of 60 minutes with equal durations of REM and NREM sleep in each cycle. Slow wave sleep is maximal in young children and it is nearly impossible to wake them in the slow wave sleep of the night's first sleep cycle. After adolescence, the slow wave sleep keeps decreasing and by the age of 60, slow wave sleep may no longer be present, particularly in men. The age-related decline in nocturnal slow wave sleep seems to parallel the reduction in cortical synaptic density. REM sleep duration does not decrease much while arousals during sleep increase markedly with age.

Sleep history When an individual is selectively deprived of REM or slow wave sleep, a rebound increase in the same type of sleep occurs when natural sleep resumes. Following total sleep deprivation, slow wave sleep tends to be recovered first. REM sleep tends to recover only after the recovery of slow wave sleep, e.g., on the second or subsequent recovery nights. Chronic deprivation of nocturnal sleep can result in REM-onset sleep. Recovery sleep, whether following REM, NREM or total sleep loss, is usually deeper, i.e., it has a higher arousal threshold throughout NREM and REM sleep.

Drugs Slow wave sleep is suppressed by benzodiazepines (e.g., diazepam). REM sleep is suppressed by tricyclic antidepressants and monoamine oxidase inhibitors (MAOI). Withdrawal from drugs that selectively suppress a particular stage of sleep tends to be associated with a rebound increase in the same sleep stage. Thus, acute withdrawal from benzodiazepine produces an increase of slow wave sleep while acute withdrawal from a tricyclic antidepressant or MAOI produces an increase of REM sleep. Alcohol intake immediately before sleep produces REM suppression early in the night and a rebound increase in REM sleep later in the night as the alcohol is metabolized.

Neuronal systems of sleep-wakefulness

The sleep spindles of NREM sleep originate in the same thalamic circuit that produces EEG synchronization. The rhythmic discharge of this circuit (between thalamic relay nuclei and the thalamic reticular nuclei) is terminated by the neuronal discharge originating in the cells of the *mesopontine cholinergic system*. These cholinergic neurons also promote REM sleep and arousal by stimulating cells of the nucleus reticularis pontis oralis, which is present in the lower midbrain and upper pons.

The *nucleus reticularis pontis oralis* is made of four types of cells that are responsible for different aspects of REM sleep (Fig. 108.**4**). (1) The *cholinergic PGO-on cells* project to the lateral geniculate body and onward to the occipital cortex. They are responsible for the generation of the pontogeniculooccipital (PGO) spikes that are characteristic of REM sleep. The firing of these cells is blocked by the noradrenergic neurons originating in the locus ceruleus and serotonergic neurons of the REM-off cells. (2) The *locus ceruleus* and *REM-off cells* are kept inhibited by the GABAergic REM-on cells, which in turn are kept activated by the mesopontine cholinergic system. (3) The mesopontine cholinergic system also keeps the glutamatergic *REM-on cells* activated. Descending axons from these REM-on cells terminate on glycinergic neurons in the medulla that inhibit stretch reflex and thereby result in the atonia that is characteristic of REM sleep. (4) The *REM-waking-on cells* descend to the spinal cord to stimulate muscles, especially the extraocular muscles, causing muscle twitches and the rapid eye movements that are characteristic of REM sleep.

The hypothalamus also has a role in the sleep-arousal mechanism. The posterior hypothalamus contains histaminergic cells that stimulate the ascending reticular activating system. These cells are, however, inhibited by a group of GABAergic cells in the anterior hypothalamus called the *NREM-on cells*. These cells also suppress REM sleep by inhibiting the nucleus reticularis pontis oralis.

The above theories explain some of the early observations in sleep research. For example, following a

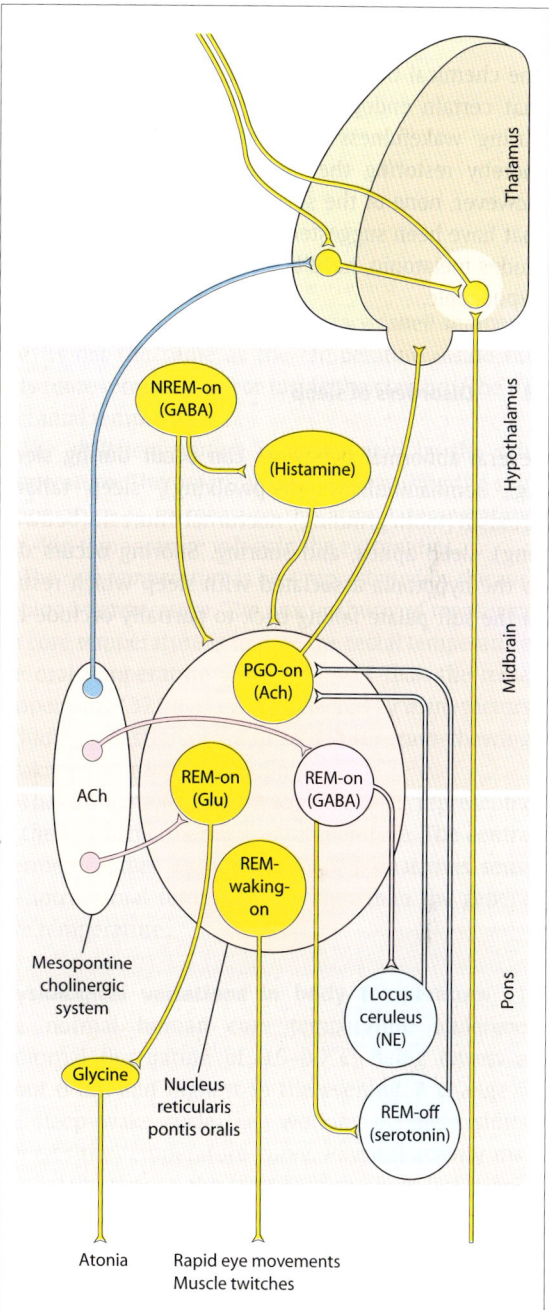

Fig. 108.**4**   Neural centers involved in sleep-wake mechanisms.

transection separating the brain from the spinal cord, the isolated brain (*encéphale isolé*) continued to show EEG characteristics of wakefulness. However, when transected just cranial to the midbrain, the isolated forebrain (*cerveau isolé*) showed EEG characteristics of sleep because the thalamic circuits for EEG synchronization are not inhibited by ascending brainstem influences.

■ Body heat exchange with exterior

The heat exchange between the body and its environment can be viewed as a two-step process: The first step is the transfer of heat between the core of the body and its shell (the skin). The second step is the heat transfer between the body shell and the exterior (Fig. 109.**1**).

The transfer of heat from the core to the shell occurs almost entirely through convection and is regulated by the cutaneous blood flow. When the cutaneous vessels are dilated, blood flows into the skin carrying the heat from the core to the shell. When the cutaneous vessels are constricted, blood does not flow into the skin and the heat remains trapped in the core of the body. The arteries supplying the limbs have accompanying veins (venae comites) that serve as countercurrent heat exchangers. These exchangers minimize the convective heat transfer from the core to the shell.

The transfer of heat from the shell to the exterior occurs through conduction, convection, and radiation. The conductive heat loss is partly regulated by the layer of hair on the skin. Heat is conducted from the skin to the air trapped between hairs and from the trapped air to the exterior. Convective (evaporative) heat loss is regulated by the amount of sweating. Radiative (nonevaporative) heat loss is proportional to the temperature difference with the exterior.

A large amount of heat is lost from the scalp due to its high vascularity. At subzero temperatures, up to 50% of the body heat is lost from the scalp of bald persons. It explains why old people, who have a relatively less efficient thermoregulation, should wear a cap in winter, especially if bald.

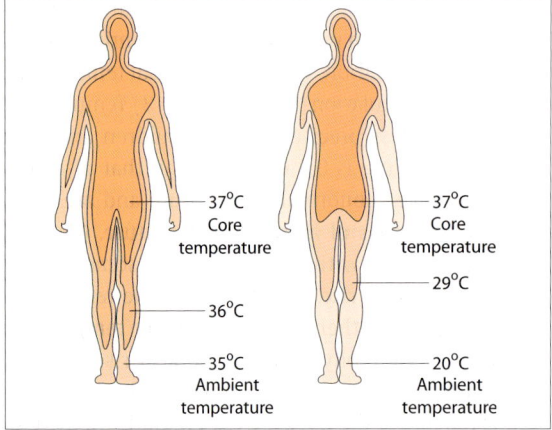

Fig. 109.**1**   Core and shell temperatures. Regardless of the ambient temperature, the core temperature is maintained at 37°C.

■ Thermoregulatory Mechanisms

The temperature of the body is determined by a 'thermostat' located in the hypothalamus. The set point of the hypothalamic thermostat is 37°C but it can be reset. When the body temperature rises above or falls below the set point, the thermostat activates appropriate effector mechanisms for restoration of the body temperature.

The various effector mechanisms activated by the hypothalamus (Table 109.**1**) broadly fall into three categories: the sympathetically mediated mechanisms (vasoconstriction or vasodilatation, sweating, chemical thermogenesis), the somatic mechanism (shivering), and behavioral mechanisms (hyperactivity or lethargy, hyperphagia or hypophagia).

Following transverse damage to the spinal cord above the thoracolumbar sympathetic outflow, the sympathetic and somatic thermoregulatory mechanisms do not remain under hypothalamic control. Some local vasodilatation and sweating are still possible through spinal reflexes but these are too weak for effective control of body temperature. Such patients have to depend largely on conscious behavioral thermoregulatory mechanisms.

■ Body heating mechanisms

Cutaneous vasoconstriction is sympathetically mediated and occurs reflexly in response to cold. It reduces the cutaneous blood flow. Thus, the heat loss from the body core to the body shell is minimized. Cutaneous vasoconstriction occurs through a direct effect on smooth muscles and local reflexes, as well as in response to descending hypothalamic command. When cutaneous blood vessels are cooled, they become more sensitive to catecholamines and the arterioles and venules constrict. This local effect of cold directs blood away from the skin.

Chemical (nonshivering) thermogenesis Sympathetic discharge and adrenaline cause an immediate increase in the rate of metabolism due to uncoupling of oxidative phosphorylation, i.e., the oxidation of foodstuffs is associated with release of heat rather than generation of ATP. Heat produced in this way is called chemical thermogenesis.

The amount of chemical thermogenesis is proportional to the amount of brown fat in the tissues. Brown fat contains large numbers of special mitochondria where the uncoupled oxidation occurs. These cells have sympathetic innervation. Adults do not have brown fat and therefore chemical thermogenesis contributes less than 15% to the total heat production in them. Infants have some brown fat in the interscapular region and are therefore able to double its heat production through chemical thermogenesis. Chemical thermogenesis is promoted by thyroxine. Cold temperature stimulates hypothalamic release of TRH which in turn stimulates the secretion of TSH and thyroxine. However, several weeks of cold-exposure are required before the thyroid gland secretes more thyroxine.

Shivering is a cortical reflex in which stimulation of cold-receptors in the skin reflexly increases the muscle tone of the body. When the muscle tone increases above a certain threshold, it results in muscle clonus, which is commonly known as shivering. Shivering, or even the increase of muscle tone, is associated with greater metabolic activity and greater heat generation. A primary motor center for shivering is located in the posterior hypothalamus.

Horripilation is the fluffing of the feathers or erection of the hairs seen in animals in response to cold: It traps air in its layers, reducing heat loss from the skin. Horripilation is not seen in man: What is seen instead is piloerection, i.e., the cold-induced contraction of the piloerector muscles attached to the hairs on the skin, resulting in 'goose pimples'.

Behavioral mechanisms Cold environmental temperature also stimulates several behavioral mechanisms like hyperactivity and hyperphagia. Hyperphagia helps because the thermic effect of food increases the body metabolism. Hyperactivity generates excess body heat. A particular type of hyperactivity is the rubbing of palms: It generates frictional heat in the palms and keeps them warm. A behavioral mechanism for minimizing heat losses is curling up while sleeping. Curling up reduces the body surface area in contact with the environment. Behavioral responses in winter also include seeking out heat sources and putting on warm clothes. Heat-seeking behavior includes standing out in the sun or by a fire and consumption of hot food and drinks.

Body cooling mechanisms

Nonevaporative heat loss is radiative heat loss. It is proportional to the difference between body temperature and the ambient temperature. As the ambient temperature decreases, there is a linear rise in heat loss.

Insensible perspiration About 50 ml of water evaporates every hour from the skin at all times. This is called insensible perspiration and results in an obligatory evaporative heat loss.

Vasodilatation and sweating A rise in the ambient temperature produces cutaneous vasodilatation and sweating. The sweating causes evaporative cooling of the skin. Vasodilatation increases the cutaneous blood flow and promotes heat transfer from the core to the cooler surface. Like vasoconstriction, vasodilatation and sweating too are produced by a local response, a spinal reflex, and a hypothalamic reflex. Persons acclimatized to hot climates can sweat much more than others. Their sweat has a low Na^+ concentration due to increased aldosterone secretion and therefore they can sweat without much depletion of body sodium.

Behavioral mechanisms A rise in ambient temperature also activates behavioral responses like lethargy and hypophagia. There is also a search for shade and a preference for cold food and drinks. Some mammals lose heat by panting. This rapid, shallow breathing greatly increases the amount of water that evaporates in the mouth and respiratory passages and therefore the amount of heat lost also increases. Because the breathing is shallow, there is not much change in the alveolar air composition.

Table 109.**1** Thermoregulatory effector mechanisms.

Mechanisms activated by cold
Increase heat production
Shivering
Hyperphagia
Hyperactivity
Increased secretion of
Norepinephrine and epinephrine
Decrease heat loss
Cutaneous vasoconstriction
Curling up
Horripilation
Mechanisms activated by heat
Increase heat loss
Cutaneous vasodilatation
Sweating
Increased respiration
Decrease heat production
Anorexia
Lethargy

Thermoneutrality and thermal comfort

The thermoneutral zone (TNZ), also called the zone of least thermoregulatory effort, is defined as the range of ambient temperature (normally 25°–27°C) within which the heat produced in the body is balanced by the nonevaporative and obligatory evaporative heat losses, without stimulating any reflex heating or cooling mechanisms of the body. The lower limit of the TNZ is called the *critical temperature* below which the metabolic heat production increases to maintain thermal balance.

The zone of thermoneutrality is not the same as the *preferred ambient temperature* which is the range of ambient temperature associated with thermal comfort. When the humidity is high, the preferred temperature is lower than the TNZ, while if there is brisk air movement, the preferred ambient temperature is higher than the TNZ. Thermal comfort also depends on the level of activity and the amount of clothing. Thermal comfort is maximum when the skin temperature is about 33°C.

Figure 109.2 shows the thermal balance at increasing environmental temperature. Several points are to be appreciated in the graph. (1) The deep body temperature is maintained remarkably constant through a large range of environmental temperatures. It changes only at extremes of environmental temperature, resulting in hypothermia or hyperthermia. (2) Any fall in the environmental temperature below the critical temperature is associated with progressive increase in metabolic heat production. Above the critical temperature, the metabolic heat production remains constant at a minimum level necessitated by the basal metabolic rate. The heat production increases again at high environmental temperature if there is hyperthermia, leading to an unregulated increase in the metabolic rate of the body. (3) The nonevaporative heat loss is zero at 37°C. It rises linearly as the environmental temperature decreases. The rise is less in the thermoneutral zone due to cutaneous vasoconstriction, which minimizes the heat flow from the body core to body surface. (4) At environmental temperatures below the thermoneutral zone, the evaporative heat loss is minimal and attributable to the evaporation of insensible perspiration. Above the thermoneutral zone, the evaporative heat loss rises linearly due to sweating and its evaporation.

To summarize, there is a steep rise in metabolic heat production when the ambient temperature falls below the thermoneutral zone, and there is a steep rise in evaporative heat loss when the temperature rises above the thermoneutral zone. Both the processes are at their minimum within the thermoneutral zone. The thermoneutral zone can therefore be defined as the ambient temperature range in which a person neither shivers not sweats.

Temperature receptors

Temperature regulation is dependent on information originating in both central and peripheral receptors. *Central thermoreceptors* are present in the hypothalamus and spinal cord. They sense the temperature of blood flowing through them. Peripheral *thermoreceptors* are of two types; the cutaneous thermoreceptors that sense the ambient temperature and the visceral thermoreceptors present in the abdominal viscera and in or around the great veins. Both cutaneous as well as visceral thermoreceptors detect cold rather than warmth.

The central and peripheral thermoreceptors rarely provide identical information about body temperature. In dealing with body temperature, therefore, the posterior hypothalamus calculates an integrated temperature from the information obtained from both the central and peripheral thermoreceptors.

Hypothalamic thermoregulatory mechanisms

The hypothalamus has two thermoregulatory mechanisms built into it; one for raising, and the other for lowering the body temperature (Fig. 109.3). Both the

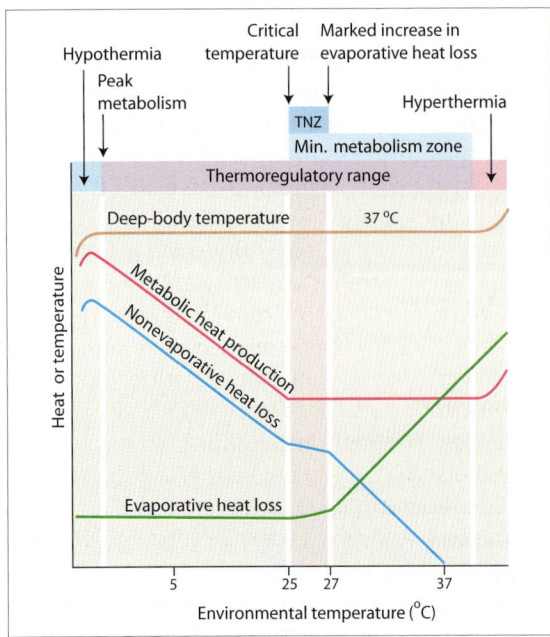

Fig. 109.2 Changes in the intensity of thermoregulatory mechanisms with increase in environmental temperature.

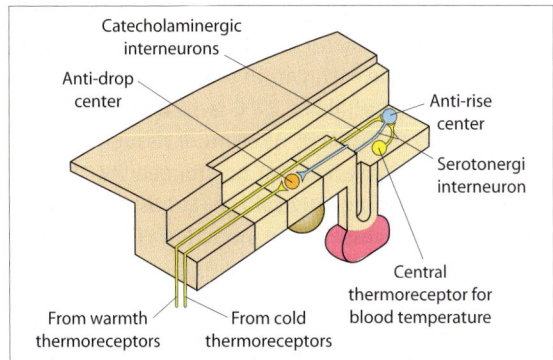

Fig. 109.3   Hypothalamic thermoregulatory centers.

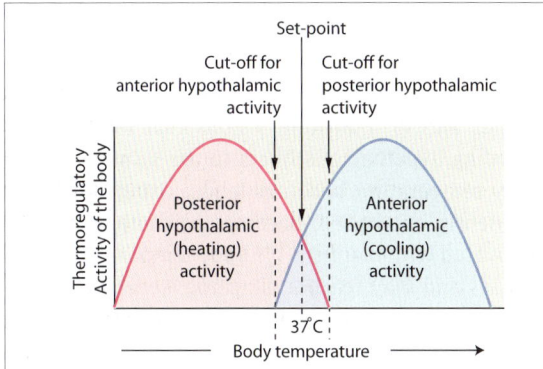

Fig. 109.4   Graph showing the hypothalamic set point and the cut-off temperatures for posterior and anterior hypothalamic activity.

systems have their own cut-off temperature at which the heating or cooling mechanisms are activated or deactivated (Fig. 109.**4**).

Antirise center   The anterior or rostral hypothalamus acts as a heat-loss (heat-dissipating or antirise) center and opposes any rise of body temperature. Stimulation of this region, especially the preoptic area, produces heat loss by cutaneous vasodilatation, sweating, panting, and probably by reducing, heat production. The antirise center receives catecholaminergic afferents from peripheral and central thermoreceptors. Destruction of the rostral hypothalamus results in hyperthermia (neurogenic fever).

Antidrop center   The posterior or caudal hypothalamus, near the mamillary body, is concerned with heat production (antidrop center). Stimulation of this region activates heat production through shivering and increased TSH secretion. It also reduces heat loss by causing cutaneous vasoconstriction. The antidrop center receives serotonergic afferents from the peripheral cold thermoreceptors. It does not have any direct input

from the central thermoreceptors but it is inhibited by impulses from the anterior center that arise when the blood temperature rises. It is also stimulated by blood-borne pyrogenic substances like viruses and toxins.

Disorders of Thermoregulation

Hyperthermia

Hyperthermia is an elevation of the body temperature above the normal range. It is also called fever or pyrexia. Fever is presumably beneficial because many microbes grow best at a specific temperature, and a rise in temperature inhibits their growth. Moreover, antibody production increases when body temperature is elevated. Before the advent of antibiotics, fever was artificially induced for the treatment of neurosyphilis and proved to be beneficial. Hyperthermia also slows the growth of some tumors.

Body temperatures above 41°C (hyperpyrexia) are harmful. Hyperpyrexia is accompanied by tachycardia, tachypnea, weakness, headache, mental confusion, and finally brain damage with loss of consciousness. A persistent temperature of over 43°C is not compatible with life.

In malignant hyperthermia, a defective ryanodine receptor leads to excess Ca^{2+} release during muscle contraction. The resulting muscle contracture greatly increases heat production, leading to hyperthermia which is fatal if not treated.

Fever is produced if the body produces or gains heat at rates greater than can be balanced by heat loss. This can happen in severe exercise and hyperthyroidism. However, fever most commonly occurs when the hypothalamic set point is reset to a higher temperature, resulting in the activation of heat-producing and heat-conserving mechanisms of the body till the body temperature equals the hypothalamic set point. The onset of fever is signaled by the 'chills' and its termination, by the 'crisis'. *Chills* are felt in fever when the heat generating and heat conserving mechanisms of the body are active. These mechanisms continue to be active until the body temperature rises to the elevated hypothalamic set point of fever. *Crisis* is characterized by sudden sweating. It occurs when the hypothalamic set point suddenly drops to normal and the heat-dissipating mechanisms are activated.

The hypothalamic set point is elevated by fever-producing substances called pyrogens. Lipopolysaccharide endotoxins derived from bacterial cell membranes are called *exogenous pyrogens*. The cytokines (1L-IB, IL-6, β-IFN, γ-IFN, and TNF-α) released from actively phagocytosing monocytes and macrophages are called *endogenous (leukocytic) pyrogens*. The cytokines act by inducing the synthesis and release of prostaglandin E$_2$. Antipyretics like aspirin prevent the formation of prostaglandin E$_2$ from arachidonic acid. The cytokines act on the OVLT (organ vasculosum of lamina terminalis) which is outside the blood brain barrier. The activated OVLT, in its turn, raises the hypothalamic set point.

The septal nuclei located anterior to the preoptic area serve as the antipyretic area. The area contains vasopressin-secreting neurons that reset the thermostat. Injection of vasopressin into this area reduces fever directly.

Heat exhaustion and heatstroke *Heat exhaustion* occurs due to excessive sweating leading to circulatory failure. *Heatstroke* occurs at very high ambient temperatures that upset the normal thermal balance of the body, leading to a rise in body temperature. The hyperthermia impairs the thermoregulatory capability of the hypothalamus, leading to a further rise in body temperature. A vicious cycle is thus set off, culminating in loss of consciousness.

■ Hypothermia

Thermoregulatory mechanisms fail when the body temperature falls below 32°C. Death usually occurs when the core temperature falls below 25°C. The temperature-regulating mechanism of old people is often defective, which is why in cold countries many elderly people die each winter of hypothermia.

Hypothermia of about 27°C is sometimes induced artificially so that the O$_2$ demand of tissues is greatly reduced and circulation can be stopped for 15 to 20 minutes, permitting surgical operations on the heart and large blood vessels. The induction of hypothermia during surgery is made easier with the use of anesthesia and muscle relaxants, both of which abolish shivering. The patient is warmed up slowly during recovery. Ventricular fibrillation is a common complication of hypothermia.

■ Poikilothermia

Thermoregulation can be impaired in hypothalamic lesions and in brainstem lesions that interrupt descending hypothalamic fibers to the spinal cord. The body temperature becomes labile, rising and falling frequently. The patient becomes poikilothermic: If he is covered with blankets, his body temperature quickly rises and may reach a dangerous level. An experimental lesion of the caudal hypothalamus produces poikilothermia because efferents from both antirise and antidrop centers descend through the caudal hypothalamus.

The primitive goal of feeding behavior is to achieve homeostasis of the nutrient concentration in the body. Food intake depends on both internal and external factors. *Internal stimuli* like hunger and satiety regulate food intake by producing a conscious sensation of the adequacy of the food consumed. *External factors* like the taste and smell of food are necessary for discriminative appetite which helps in the consumption of a balanced diet. Thus, salt intake is known to be regulated through salt craving and a similar mechanism might exist for sweet carbohydrates.

Appetite is an emotional desire to eat, which may or may not be associated with the need for food. *Hunger* is a conscious sensation that stimulates feeding behavior for fulfilling the caloric requirements of the body. The conscious sensation that causes the cessation of eating is called *satiety*. The mechanisms of hunger and satiety are inborn while those of appetite are partly acquired and dependent upon past hedonic experiences. When hungry, an individual will eat almost any wholesome food as long as it is reasonably palatable. After he has reached a state of satiety, he might still be left with the appetite for some dessert, which is entirely unnecessary so far as caloric requirements are concerned.

Central Regulatory Mechanisms

Neural regulation of feeding behavior

Hypothalamic control   The hypothalamus has a feeding center and a satiety center. The *feeding center* is located in the lateral nuclei of the hypothalamus. It stimulates feeding behavior. The *satiety center* is located in the ventromedial nucleus of the hypothalamus (VMH) (see Fig. 98.**2**). It inhibits feeding behavior. Lesions of the feeding center produce marked reduction in food intake and body weight. Lesions of the satiety center cause overeating, leading to obesity. A lesion of the VMH produces overeating only if the lateral nuclei are intact, suggesting that the VMH acts by inhibiting the lateral feeding center.

The feeding center is composed of two parts: the midlateral and the farlateral. The *midlateral hypothalamus* controls the basic feeding behavior but does not provide the motivation to cross a difficult barrier to obtain the food. The *farlateral hypothalamus*, through which the medial forebrain bundle passes, is important in 'hunger motivation' that is necessary for overcoming a barrier to obtain food. A lesion in the medial forebrain bundle fails to abolish the basic feeding responses but the animal will not 'work' for this food. Only the farlateral hypothalamus is inhibited by the VMH.

The amount of weight gained following a VMH lesion depends on the starting weight. If the experimental rats are already obese, they do not overeat or gain much weight after the satiety center is lesioned. Thus, it seems that the feeding and satiety centers together determine the 'set point' of food intake and body weight. The feeding center raises the set point while the satiety center lowers the set point. A lesion of the satiety center therefore raises the set point. However, if the animal is made obese by forced feeding before the experiment, its subsequent self-motivated food intake actually reduces even if its satiety center is lesioned. It is also observed that an obese person who has reduced to normal weight by strict dietary measures usually develops intense hunger that is far greater than that of the normal person. This indicates that the set point of the obese person's feeding control system is at a higher level of nutrient storage than that of the normal person.

The VMH is stimulated by the entry of glucose into it. The glucose uptake by the VMH cells is facilitated by insulin. The stimulated VMH inhibits insulin secretion and thereby completes a negative feedback loop (Fig. 110.**1**). In diabetes mellitus, the deficiency of insulin

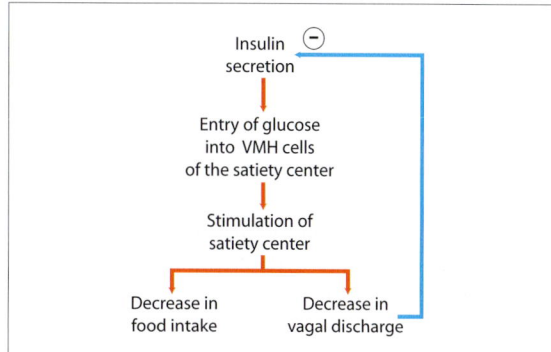

Fig. 110.**1**   Feedback control of the satiety center.

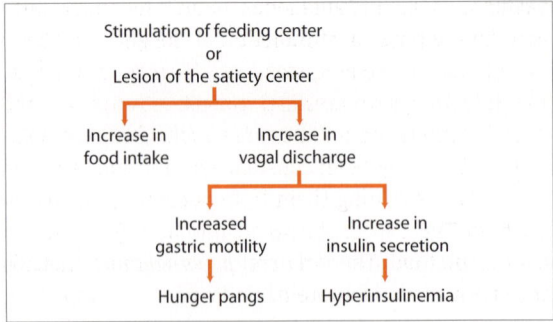

Fig. 110.**2**   Hypothalamic obesity.

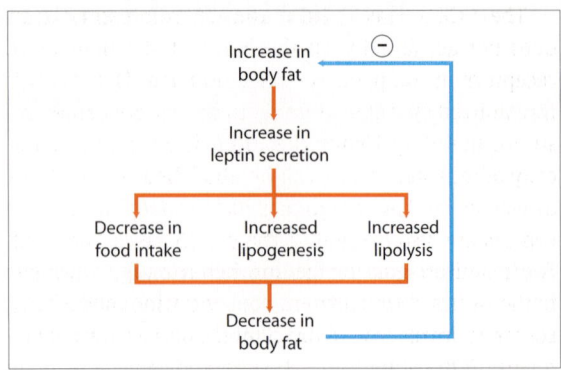

Fig. 110.**3**   Leptin feedback control of body fat stores.

reduces the glucose uptake in VMH cells and therefore the satiety produced is less, resulting in polyphagia.

Obesity of hypothalamic origin is also seen in humans and is called *hypothalamic obesity*. It is attributable not only to a defect in the VMH but also in the paraventricular nucleus (PVN). A lesion in the PVN causes excess consumption of sweet carbohydrates. Hypothalamic obesity is reduced by vagotomy, indicating that the vagus plays an important role in obesity. Vagal discharge is stimulated by the feeding center and inhibited by the satiety center. The hunger contractions of the stomach that occur before meals are due to excessive vagal discharge triggered by the hypothalamic feeding center. The vagal discharge also produces hyperinsulinemia, which is known to coexist with hyperphagia (Fig. 110.**2**). Hence, a lesion of the VMH causes both hyperphagia (behavioral overfuelling) and hyperinsulinemia (cellular overfuelling).

Other brain areas affecting feeding behavior include the brainstem, limbic cortex, and the neocortex. The basic mechanical features of feeding like salivation, licking of lips, chewing of food, and swallowing are controlled by brainstem centers. The amygdala and the prefrontal cortex are important for *discriminative appetite*. Bilateral lesions of the amygdala produce *psychic blindness* in the choice of foods, i.e., indifference to the type and quality of food. Psychic blindness is a feature of the Kluver–Bucy syndrome. The neocortex modifies feeding behavior through learning.

■   Hormonal regulation of feeding behavior

Leptin (Greek: thin) is a protein hormone produced by fat cells. It acts on the hypothalamus to reduce food intake, decrease lipogenesis, and increase lipolysis, thereby reducing the body fat stores. Leptin thus regulates the size of the body's fat deposits through a negative feedback mechanism (Fig. 110.**3**).

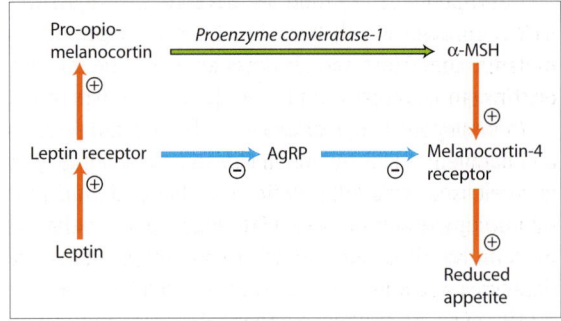

Fig. 110.**4**   Mechanism of leptin action on appetite.

The effect of leptin on appetite is mediated through α-MSH (α-melanocyte stimulating hormone), a pro-opiomelanocortin (POMC) derivative (see Fig. 82.**5**). Leptin acts through a group-I hormonal mechanism to increase the synthesis of POMC which is converted into α-MSH. The α-MSH acts through melanocortin-4 receptors to depress appetite (Fig. 110.**4**). Leptin also inhibits the secretion of AgRP (see below).

Plasma leptin levels are proportional to the amount of body fat and are therefore higher in women and obese individuals. A decrease in plasma leptin is interpreted by the body as a signal for caloric deficiency and is therefore associated with adaptive responses like inhibition of the pubertal onset and thyroid function, and stimulation of glucocorticoid secretion.

Neuropeptides, many of which are also present in the gut as gastrointestinal hormones, have an effect on food intake. Neuropeptides inhibiting food intake include CCK (cholecystokinin), GRP (gastrin-releasing peptide), glucagon, somatostatin, CART (cocaine and amphetamine-regulated transcript), α-melanocyte stimulating hormone (α-MSH), and corticotropin releasing hormone (CRH). Food entering the gastrointestinal tract triggers the release of these hormones, which produce the brief periods of satiety between meals.

The CCK released from the gastrointestinal tract does not act directly on the brain: It stimulates CCK receptors in the pylorus of the stomach. The signal is transmitted by vagal afferents to the satiety center, resulting in satiety. Hence, the anorexic effect of systemically administered CCK can be abolished by vagotomy as well as by CCK antagonists like devazepide. CCK receptors are also present in the satiety center and CCK is the neurotransmitter in some of the vagal afferents to the satiety center. Hence, the satiety induced by CCK seems to involve both peripheral and central mechanisms. CCK seems to work synergistically with serotonin, which explains why antiserotonergic drugs like fenfluramine are effective stimulants of appetite.

Neuropeptides that increase food intake include NPY (neuropeptide Y), Agouti-related peptide (AgRP), melanin-concentrating hormone (MCH), galanin, orexins, and ghrelin.

Neuropeptide Y is produced in the arcuate nucleus and paraventricular nuclei. In the arcuate nucleus, NPY is coreleased with AgRP which stimulates appetite by inhibiting the action of α-MSH. In the paraventricular nucleus, NPY is coreleased with norepinephrine which depresses appetite. Amphetamine, which increases the release of norepinephrine in the brain, suppresses appetite. NPY induces a strong appetite for carbohydrates.

Melanin-concentrating hormone is a polypeptide found in the lateral hypothalamus. *Galanin* seems to stimulate fat intake in particular. *Orexins* are synthesized in the lateral hypothalamus. A defect in orexin receptor genes causes narcolepsy in dogs. This is interesting since narcolepsy and hyperphagia often coexist—recall Dickens' fat sleepy boy Joe in Pickwick Papers! *Ghrelin* is released from the stomach in the fasting state. It increases hunger by inhibiting the VMH. It also stimulates GH secretion.

Hunger contractions contribute to, but are not solely responsible for, the hunger sensation. Cutting the splanchnic and vagus nerves abolishes hunger contractions but does not abolish hunger. They cease immediately on gastric filling. However, emptiness of stomach by itself is not the cause of hunger contractions, which intensify as the fast is prolonged. Hunger contractions are induced and accentuated by insulin-induced hypoglycemia due to the direct action of low blood glucose on the hypothalamic feeding center.

Peripheral Feedback Mechanisms

Peripheral feedback mechanisms that regulate food intake are broadly categorized into gastrointestinal feedback and the metabolic feedback. *Gastrointestinal feedback* is an instantaneous neural feedback initiated by the physical stimulation of the gut. It makes a person eat in frequent, small meals so as to optimize the rate of gastrointestinal transit, absorption, and postabsorptive assimilation. *Metabolic feedback* helps to maintain long-term constancy of nutrient stores in the body, preventing them from becoming too low or too high. These mechanisms are initiated by the chemical stimulation of sensory receptors within or outside the central nervous system.

Gastrointestinal feedback

Gastrointestinal filling Distension of the stomach stimulates stretch receptors located in the muscular wall of the stomach. Impulses from these receptors reach the brain through the vagus and inhibit hunger and appetite. Vagotomy abolishes the effect, leading to hyperphagia.

Oral metering of food intake Oral receptors related to feeding, such as chewing, salivation, swallowing, and tasting, inhibit the hypothalamic feeding center after a certain amount of food has passed through the mouth. This is demonstrable by sham feeding of an esophagostomized animal so that the food it ingests is drained out through an opening in the esophagus (Fig. 110.**5**). Although no food enters its stomach, the animal still stops eating after a time. If some of the food is allowed to enter the stomach, the sham feeding stops earlier.

Metabolic feedback

Metabolic feedback inhibition of food intake is initiated by both glucose (glucostatic feedback) and adipose tissues (lipostatic feedback). The roles of both are explained above. Glucose directly acts on the satiety center while the effect of adipose tissue is mediated by leptin.

Fig. 110.**5**　Sham feeding.

A *thermostatic hypothesis* holds that a fall in body temperature below a given set point stimulates appetite and a rise above the set point inhibits appetite. Considering the thermic effect of the food, it is possible that the rise in body temperature caused by the food may be a factor limiting further food intake.

Energy Balance and Obesity

Obesity is caused by an excess of energy input over energy output. Muscular activity is by far the most important means by which energy is expended in the body, accounting for more than a third of the daily energy expenditure. Obesity results from a high ratio of food intake to daily exercise.

For each 9 kcal of excess energy consumed, 1 gram of fat is stored in the body. Excess energy input is required only during the developing phase of obesity. Once a person becomes obese, all that is required to remain obese is that the energy input equals the energy output. For the person to reduce weight, caloric input must be less than caloric output. Indeed, the food intake of most obese people whose weight has stabilized is about the same as that for normal people.

Obesity is usually caused by abnormalities of the feeding regulatory mechanism. Excess eating results in excess deposition of fats in the fat cells. Overeating in childhood is particularly harmful because it also increases the rate of formation of new fat cells. The number of fat cells is about three times more in obese children.

Psychogenic obesity A large proportion of obesity results from psychogenic factors like the commonplace belief that three meals a day is a physiological necessity. Eating also helps in the release of tension and therefore some people gain considerable weight during stressful situations such as bereavement, a severe illness or melancholic depression.

Neurogenic and endocrine obesity Frohlich's syndrome causes obesity with hypogonadism due to hypothalamic dysfunction. Sometimes, it occurs due to a hypophysial tumor that presses on the hypothalamus. Endocrine disorders leading to obesity are Cushing's syndrome and hypothyroidism.

Genetic obesity Obesity definitely runs in families. Identical twins usually maintain weight levels within 1 kilogram of each other throughout life, if their lifestyles do not differ markedly. This might result partly from eating habits acquired during childhood, but it is likely that this close similarity between twins is genetically controlled. Genetic defects in leptin, leptin receptor, POMC, melanocortin-4 receptor, and pro-enzyme convertase-1 are all known causes of human obesity.

Memory is the acquisition, storage, and retrieval of sensory information. *Learning* is an enduring change in behavior that results from sensory information. The change in behavior must satisfy three criteria: First, it must last at least a few seconds to distinguish it from transient reactions to sensory stimuli. Second, it must be based on individual experience as opposed to species experience, e.g., an infant learning to walk. Third, it must have a teleonomic functions, i.e., it must serve a purpose.

The premise that learning is the process of acquisition of sensory information while memory is the storage and retrieval of the same information is fallacious. There can be no acquisition without some form of temporary storage and therefore it is impossible to draw a line between acquisition and memory, both of which must occur simultaneously. Hence, it is important to incorporate *a change in behavior* as the critical element of learning (Fig. 111.**1**).

Memory is of two types: explicit and implicit. *Explicit memory* is also called declarative or reflective memory. It is the memory of knowledge, i.e., of facts and events that we are aware of. *Implicit memory* is also called nondeclarative or reflexive memory. These are memories of skills that we are never aware of and do not have to make any conscious effort to memorize

Fig. 111.**1**   A schoolteacher chides the students, alluding to behavior as the essential difference between memory and learning.

or recall. Explicit memory often progresses to implicit memory. For example, learning to drive a car involves explicit memory of a series of coordinated movements. However, over a period of time, the memory of these movements become implicit and the driving behavior becomes reflexive.

Explicit Memory

Explicit memory is of two types: *semantic memory*, which is the memory of objects, facts, and concepts as well as words and their meaning, and *episodic memory*, which is the memory of events and factual knowledge of people, places, and things and also the feelings they aroused.

Semantic (factual) memory is distributed at multiple sites in the brain. For example, the word 'flute' conjures up in our mind several of its attributes: a visual memory of a long narrow cylinder, an auditory memory of its tonal quality, and a somatosensory memory that reminds us that it is light and has a smooth surface. Memories of each of these attributes that are perceived by our visual, auditory, and somatic senses are stored in different areas of the brain. Each time the knowledge about flute has to be recalled, the recall is built up from distinct bits of information, each stored in specialized and dedicated memory stores. As a result, damage to a specific cortical area can lead to loss of specific information and therefore, to a fragmentation of knowledge. Thus, damage to the posterior parietal cortex can result in *associative visual agnosia*, i.e., patients cannot name objects (e.g., flute) but they can identify objects by selecting the correct drawing and can faithfully reproduce detailed drawings of the object. In contrast, damage to the occipital lobes and surrounding region can result in an *apperceptive visual agnosia*, i.e., patients are unable to draw objects (e.g., flute) but they can name them if appropriate perceptual cues (its sound or its feel) are available.

Episodic (autobiographical) memory is stored in the prefrontal cortex. Patients with frontal lobe damage have a tendency to forget how a piece of information was acquired, a deficit called *source amnesia*. However, patients with loss of episodic memory still have the ability to recall vast stores of factual (semantic) knowledge. Such a patient may be able to name an old friend just by looking at his face but may fail to recall their past meetings and therefore, fail to recall his identity.

■ Processing of explicit memory

The storage of episodic memory proceeds through four stages: encoding, consolidation, storage, and retrieval of memory. *Encoding* is the process by which newly learned information is attended to and processed when first encountered. *Consolidation* is the process that alters the newly stored and still labile information so as to make it more stable for long-term storage. *Storage* refers to the retention of memory at specific sites in the brain. It involves the expression of genes and synthesis of new proteins. *Retrieval* refers to the recall and use of stored information.

Encoding of memory The briefest form of memory, lasting less than a second, is the sensory memory. Things we see, hear or feel continuously in our environment register only in our *sensory memory*, only to be forgotten in less than a second. A slightly longer form of memory is the primary or *short-term memory* which lasts several seconds. A special type of short-term memory that is required for maintaining behavioral continuity is called the *working memory*. The importance of working memory in behavioral continuity becomes apparent when we consider that

to do anything properly, it is important to remember what has just been done and what remains to be done. Similarly, to speak fluently, it is important to remember what has just been spoken and what remains to be spoken. The deficiency in working memory in monkeys whose prefrontal lobe has been ablated has been demonstrated as follows. A piece of food was hidden randomly under one of the two containers placed in front of the monkey. Although the monkey was allowed to observe where it was hidden, it was unable to discover it barely 5 seconds later.

Both the initial encoding and the subsequent recall of memory are routed through the working memory store. Working memory has an attentional control system and two temporary rehearsal systems, one for language and the other for vision and action (Fig. 111.**2**). The *attentional allocation system* is located in the prefrontal cortex and has a very limited capacity of less than 12 items. It regulates the apportioning of attention between the two *rehearsal systems*. The information processed in these rehearsal systems can then enter the primary memory.

The *rehearsal system* for language is called the *articulatory loop*. The articulatory loop is a temporary storage circuit where memory for words and numbers can be held in mind only through repeated mental speech, as is often done with a new telephone number as one prepares to dial it. The rehearsal system for the visual characteristics and the spatial location of objects to be remembered is the *visuospatial sketchpad*. For example, this system allows one to store the image of the face of a person one meets for the first time. Visual characteristics (details of what it is) are stored in the inferior temporal cortex while the spatial characteristics (details of where it is) are stored in the posterior parietal cortex.

Consolidation and storage of memory A critical amount of time, about 5 to 10 minutes, is required for the transfer of short-term memory to long-term memory through a process called consolidation of memory. If this time is not allowed for the consolidation to occur, the data in short-term memory is completely forgotten. This is seen in patients of concussion and electroconvulsive therapy who are unable to recall the events immediately preceding the concussion or convulsion. This phenomenon is called *retrograde amnesia*. A similar retrograde amnesia occurs before the onset of sleep, which is the reason why one is unable to remember the precise time of one's own sleep onset. For the same reason, damage to brain areas responsible for memory consolidation results in *anterograde amnesia* in which new memory formation is impaired without affecting old memories.

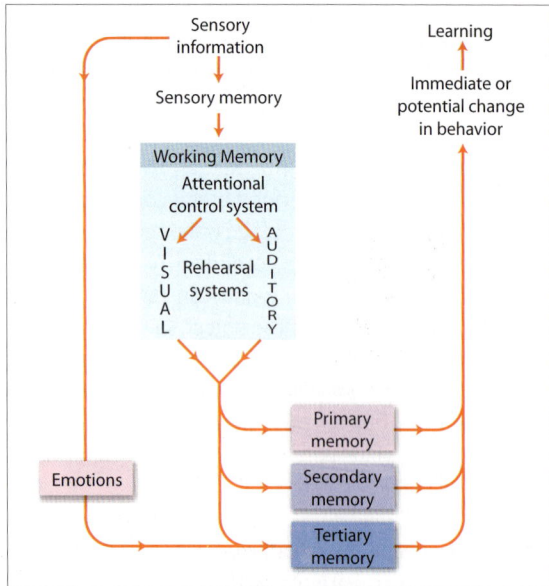

Fig. 111.**2** Levels of memory and their relation with learning and emotions.

The area of brain responsible for the consolidation of memory is the medial temporal lobe (parahippocampal gyrus, entorhinal cortex, and hippocampus). The right hippocampus processes spatial memory and the left hemisphere processes verbal memory (see cerebral asymmetry, p 622). The parahippocampal and entorhinal cortices process memory of object attributes. In processing information for explicit memory storage in the hippocampus, the entorhinal cortex is a major relay station both for hippocampal input as well for hippocampal output (see Figs. 100.**9** and 100.**10**). Polymodal information from the association cortices reaches the hippocampus through the entorhinal cortex which projects to the dentate gyrus via the perforant pathway. Also, the hippocampal output relays in the entorhinal cortex before reaching the polymodal cortices. Hence, memory impairments associated with damage to the entorhinal cortex are particularly severe and affect all sensory modalities. In fact, the earliest pathological changes in Alzheimer disease, which affects mainly explicit memory storage, occur in the entorhinal cortex.

Memory that lasts for minutes to years is called *secondary* or *intermediate-term memory*. Memory that is permanent and lasts a lifetime is called *tertiary* or *long-term memory*. Our own name resides in our tertiary memory. The first word recalled by a patient recovering from coma is his own name. Consolidation of memory into long-term memory occurs when sensory experiences are associated with profound emotions. Hence, when we reminisce about our childhood days, the events recalled are only those that were associated with deep emotions. For the same reason, we remember even the trivial things we were doing moments before we received shocking news. This is the brain's ways of deciding what is to be remembered and what is to be forgotten. Experiences with no emotional adjunct are better forgotten so that they do not take up memory space in the brain. The amygdala stores the components of memory concerned with emotions.

Interhemispheric memory transfer If an experimental animal (cat or monkey) is conditioned to a visual stimulus presented only to the left eye, will the conditioned response occur subsequently if the visual stimulus is presented only to the other eye? Normally, yes. The response occurs even after the optic chiasma has been sectioned in the middle so that the visual stimulus is transmitted only to the ipsilateral visual cortex (Fig. 111.**3**). It suggests that the memory and associated learnt behavior are somehow transferred to the contralateral cerebral hemisphere. However, if the training is imparted after sectioning the corpus callosum too (in addition to the optic chiasma), the intercortical

transfer fails to occur. Such failure of intercortical transfer of learning and memory is also seen in human subjects in whom the corpus callosum is congenitally absent or in those in whom the corpus callosum has been surgically sectioned in an effort to control epileptic seizures. Studies in these subjects indicate that the transfer of visual memory occurs in the posterior part of the corpus callosum while transfer of auditory and somesthetic memory occurs in the anterior part of the corpus callosum.

■ Learning associated with explicit memory

Almost any fact or event that is consigned to explicit memory has the potential to alter behavior at a later date. Such learning occurs in the absence of an identifiable instructive situation and is known as *incidental learning* because the individual acquires information while attending 'incidentally' to sensory inputs and thereby develops the potential to behave differently. Incidental learning is often not apparent immediately.

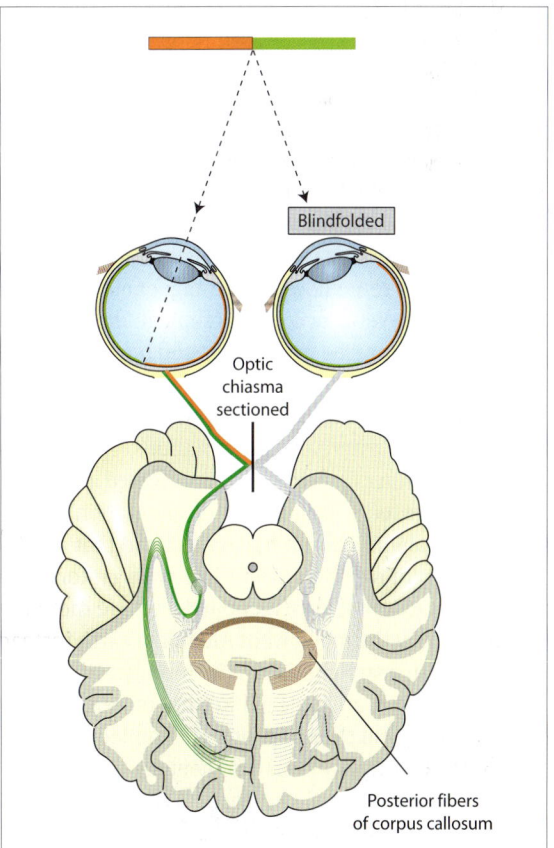

Fig. 111.**3** Interhemispheric transfer of visual memory occurs even with sectioning of the optic chiasma.

Fig. 111.**4**   The child learns to make friends! An example of explicit memory with incidental learning.

For example, consider a child who fears a burly stranger (Fig. 111.**4**). She remembers the stranger and his benign manners on several occasions (explicit memory). In the course of time, she tries to befriend him (incidental learning).

Incidental learning may also contribute to the reorganization of internal representations, resulting in sudden solutions to previously unresolved problems (insight). Newton's formulation of gravitation theory is a case in point. Newton had enunciated that a body continues to move in a straight line with uniform velocity till an external force is applied on it. He therefore wondered why the moon went round the earth instead of flying off in a straight line. As he brooded sitting under an apple tree, an apple fell in front of him. He suddenly realized that the same force that pulled the apple to the ground (gravity) was also keeping the moon in a circular orbit around the earth. That was insight! Newton had seen both earlier: the moon going round the earth and apples falling to the ground. But all of a sudden, this knowledge was 'reorganized' to give him a new insight.

Implicit Memory

Unlike explicit memory which is not immediately associated with behavioral changes, implicit memory is associated with an immediate behavioral change that is called *reflex learning*. Implicit memory and reflex learning are therefore used synonymously. Unlike incidental learning where no clear instructive situation is identifiable, reflex learning is associated with well-defined instructive paradigms and helps in achieving the desired outcome in a novel situation. While explicit memory stores the feelings associated with events, implicit memory stores the autonomic and somatic responses associated with events. The various types of reflex learning can be classified as shown in Figure 111.**5**.

Different parts of the brain are involved in different types of reflex learning. Classical conditioning involves the activity of the cerebellum which mediates the motor responses (as in adaptive motor responses and the eye-blink reflex) and the amygdala which mediates the autonomic responses (as in fear conditioning). Operant conditioning involves the striatum and the cerebellum. Habituation and sensitization involve the reflex pathways in the spinal cord and brainstem.

Priming

Priming means having a greater ability to identify an object or a word after a recent exposure. It is common experience that the more often we see something, the easier it is for us to identify it. This phenomenon is known as *perceptual priming* and it explains why we are able to quickly identify known faces, voices, dialects, scripts, handwriting, currencies, gadgets, streets, and landmarks in comparison to the new and unknown ones. Similarly, the more often we read a word, the better we recall its meaning. This is known as *semantic priming*.

Procedural learning

Procedural learning or memory refers to the learning of cognitive skills or motor skills. Learning to tell the time by looking at a clock is an example of cognitive skill. Learning to cycle, swim or even to handle a computer mouse are examples of motor skills. The pre-SMA (presupplementary motor area) is involved in the motor learning of complex sequences. A special kind of procedural learning is *motor adaptation*, which involves learning to adjust to a changing stimulus. It occurs when the earlier

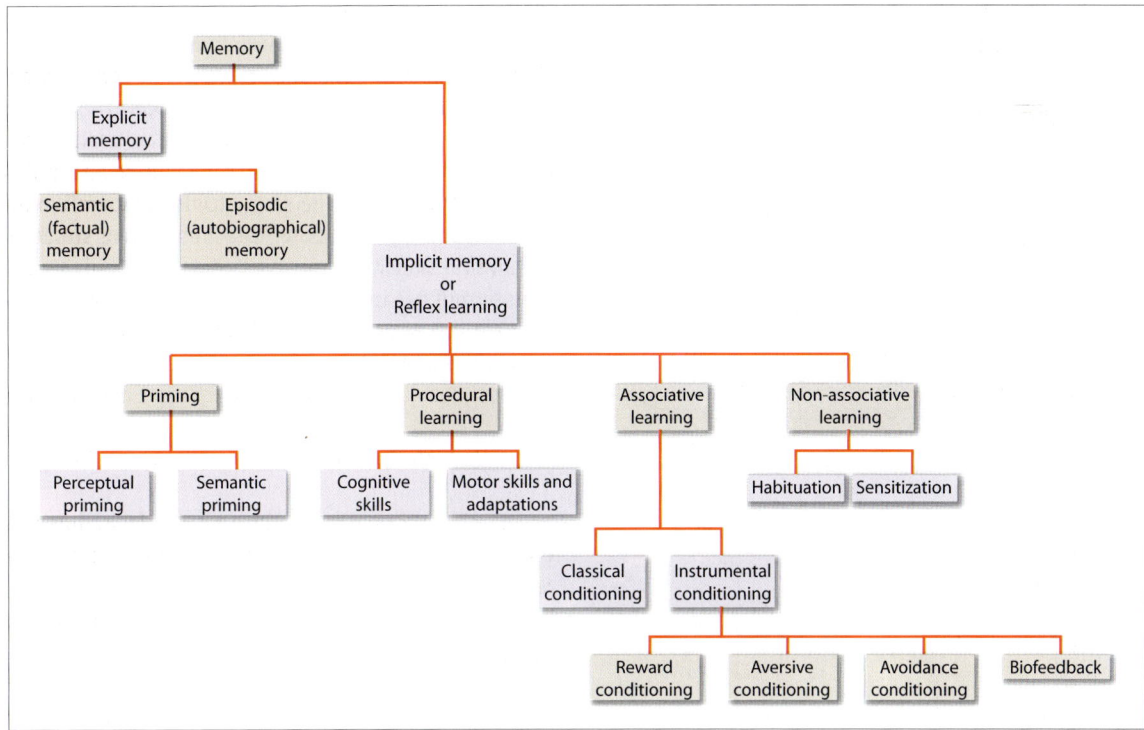

Fig. 111.**5**   Classification of memory and learning.

behavior stops having the desired effects. A common example of motor adaptation is observed in our handling of the computer mouse. Each time the speed and acceleration of the cursor produced by the mouse is changed, there is an adaptive motor response in which the hand movements adapt to the changes appropriately. Adaptive motor responses are abolished if the cerebrocerebellum or the inferior olivary nucleus is lesioned.

A well studied adaptive motor response is the abolition of the vestibuloocular reflex (VOR) on wearing prisms. The VOR ensures that any abrupt rotation of the head rotates the eye to the opposite side, thereby helping to keep the gaze fixated on an object. If the subject wears prismatic glasses so that the rotation of the eye fails to bring back the object to the fixation point, the VOR is eventually abolished.

Nonassociative learning

Nonassociative learning is the most simple and elementary form of learning. It takes the form of either habituation or sensitization. Sensitization occurs through a heterosynaptic mechanism in contrast to the homosynaptic mechanism of habituation (see synaptic plasticity, p 643). The synaptic mechanisms of habituation and sensitization have been best studied in lower animals like the aplysia (sea slug) with a simple nervous system. The reflex mostly studied is the gill-withdrawal reflex in which the slug withdraws its gill when its siphon is touched or when a painful stimulus is applied to its tail or its mantle shelf. The neurotransmitter released at the synapse between the sensory and motor neuron is glutamate, which acts on the non-NMDA receptors present on the motor neuron.

Habituation is learning *not* to respond to repetitive, low intensity stimuli and ignore them as unimportant. It is a gradual diminution of response to a stimulus, following repeated presentation of the same (or a very similar) stimulus. Habituation is not due to fatigue; rather, habituation helps in avoiding fatigue. Habituation involves decline of existing responses but the acquisition of novel responses remains unaffected. Habituation thus helps in adaptation to the environment and prevents unnecessary defense responses, making it possible to devote more attention to novel stimuli than to familiar ones.

Following habituation by repeated application of a stimulus, if this stimulus is withheld, the response recovers over time. If repeated series of habituation training and spontaneous recovery are given, habituation becomes successively more rapid. The weaker the

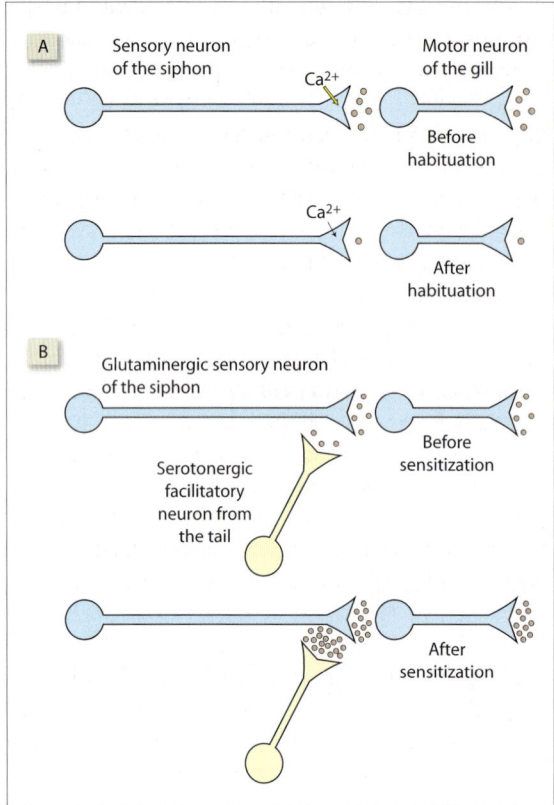

Fig. 111.**6**   Synaptic mechanism of [A] habituation and [B] sensitization.

stimulus and more rapid the frequency of stimulation, the more rapid and pronounced is habituation. Strong stimuli may fail to produce habituation. Presentation of another, usually noxious stimulus results in recovery of the habituated response, i.e., *dishabituation*.

Habituation of the gill-withdrawal reflex occurs when the siphon of the slug is touched repeatedly, and the slug does not withdraw its gill anymore. Habituation is associated with a decrease in neurotransmitter released at the synapses, which in turn is due to the inactivation of Ca²⁺ influx at the axon endings (Fig. 111.**6**A). However, what causes the inactivation of the Ca²⁺ channels is not known.

Sensitization is the augmentation of response to nonspecific stimuli. More precisely, it is learning to intensify the response to any stimulus following an exposure to a strong or noxious (sensitizing) stimulus. Sensitization is termed nonassociative because it does not result from specific associations between particular stimuli, as opposed to associative learning. Rather, a sensitizing stimulus increases the responsiveness to a wide variety of stimuli. Thus, a man bitten by a snake might react sharply to just about anything serpentine, like a piece of rope. Sensitization increases arousal and attention and lowers the threshold of defensive responses. Repetitive exposure to noxious stimuli may lead to long-term sensitization, lasting weeks or more.

Sensitization of the gill-withdrawal reflex occurs if a painful stimulus is applied to the tail repeatedly. Following sensitization, the slug withdraws its gill much more vigorously in response to an innocuous touch on the siphon. Sensitization occurs due to presynaptic facilitation of synaptic transmission by a facilitator neuron that synapses presynaptically on the sensory nerve terminal (Fig. 111.**6**B).

Associative learning

Associative learning is the learning of relations among events, either among stimuli or among stimuli and effects. It is a more discriminatory form of learning than nonassociative learning. Conditioning can be either classical or instrumental. In *classical conditioning*, a stimulus that is physiologically unimportant becomes important and starts eliciting a specific behavior due to its close association with a physiologically important stimulus. In *instrumental conditioning*, a behavior that is feeble and inconsequential becomes intense when it starts producing an effect that is physiologically important.

Conditioning of the gill-withdrawal reflex requires Hebbian potentiation (p 643) involving the activation of NMDA receptors on motor neurons, as opposed to the non-NMDA receptors that are activated during sensitization.

In the **classical (Pavlovian) conditioning**, the organism is first presented with a stimulus that evokes an instinctive behavioral response. The stimulus is called the unconditioned stimulus (US) and the response to the US is called the unconditioned response (UR). Another stimulus, neutral with respect to the UR, is then presented in association with the US. This stimulus is called the conditioned stimulus (CS). The association between the CS and US alters the response to the CS, resulting in a conditioned response (CR), which is identical to the UR. Successful conditioning is critically dependent on the duration of the US and CS (both must be of adequate duration), their rate (should be frequent), the interval between them (should be minimum), and the order of their presentation (CS must precede US).

In Pavlov's classical conditioning (Fig. 111.**7**A), when the dog was presented with food (US), there was salivation (UR). After the dog had been repeatedly presented with food immediately after ringing a bell (CS), it started salivating on hearing the bell (CR).

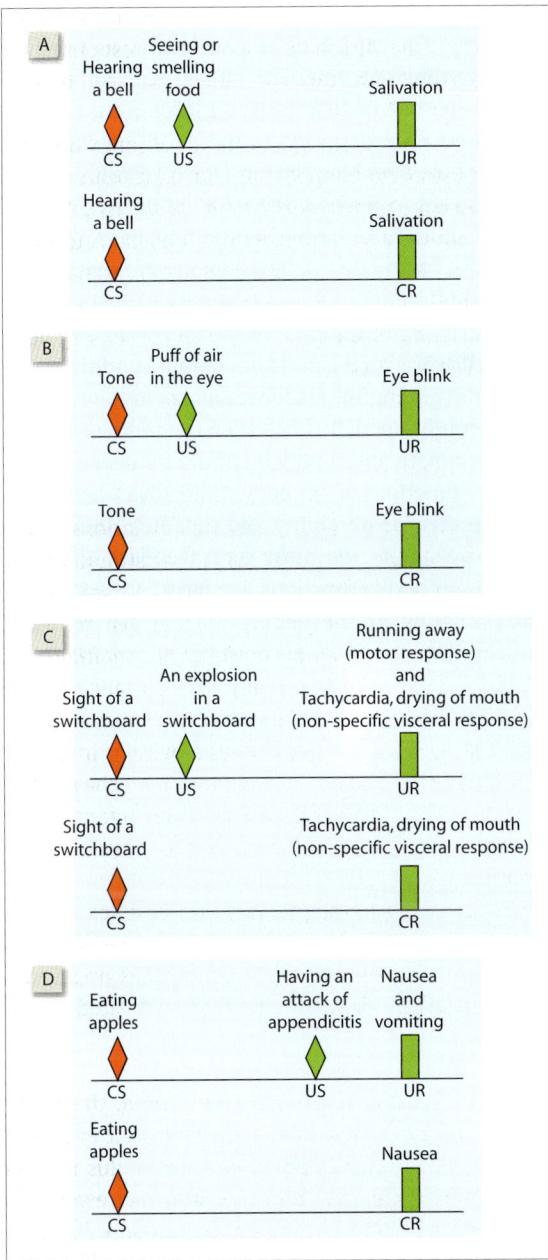

Fig. 111.7 (A) Classical (Pavlovian) conditioning of a visceral response. (B) Conditioning of a skeletomotor response: the nictitating membrane reflex. (C) Conditioned fear response showing several nonspecific responses. (D) Visceral conditioning with a long interval between conditioned and unconditioned stimuli.

The response in the above example of conditioning is a *visceromotor response*. An example of classical conditioning in which there is *skeletomotor response* is the eye-blink reflex. In this reflex, the unconditioned stimulus is a puff of air in the eye and the unconditioned response is the eye blink. The conditioned stimulus is a tone (Fig. 111.7B).

Specific CRs are frequently associated with nonspecific responses. For example, when an innocuous CS is paired with an aversive US, specific motor CRs are accompanied with nonspecific visceral reactions signifying arousal and fear, e.g., altered heart rate, blood pressure, and respiration (Fig. 111.7C). These visceral reactions are called *conditioned emotional response* or conditioned fear and are an important factor in the development of anxiety neurosis (p 683).

Conditioning is sometimes possible even when there is an extensive delay between the CS and the US. These cases involve gustatory reflexes, in which a meal is followed hours later by vomiting and cramps. Nature seemingly has evolved a mechanism to ensure that when survival is at stake, conditioning tolerates a long delay between cause and effect. For example, a person falling ill shortly after consuming a food item frequently develops an aversion for that food even if it was not responsible for the illness (Fig. 111.7D).

If the CS is presented repeatedly without the US, the conditioned reflex eventually dies out. This process is called extinction or *internal inhibition*. On the other hand, if the animal is disturbed by an external stimulus immediately after the CS, the conditioned response may not occur, which is called *external inhibition*.

In **instrumental (operant) conditioning**, an organism learns from the impact of its actions on its environment. In this type of learning, the presence or absence of a particular behavior is reinforced by the consequence of the behavior. Hence in instrumental conditioning, it is the consequence of a spontaneous behavior (or its absence) that acts as the *reinforcer*. The term 'instrumental' denotes that the organism's behavior is instrumental (i.e., necessary) in the delivery of the US. In classical conditioning the organism learns that an external event (US) predicts the occurrence of a reinforcer (CS). In instrumental conditioning the organism learns which of its own actions presages the occurrence of a reinforcer.

There are three major categories of instrumental conditioning. In *reward conditioning*, a positive (rewarding) reinforcer is provided in response to a particular behavior, which consequently becomes intensified. Addiction to drugs occurs as a result of reward conditioning: The consumption of the drug becomes a compulsive behavior due to the pleasure it gives.

In *aversive conditioning*, a negative (punishing) reinforcer is delivered in response to a particular behavior, which consequently becomes infrequent. The use of disulfiram for alcohol deaddiction is an example of aversive conditioning. When an alcoholic on disulfiram therapy consumes alcohol, he develops severe unpleasant reactions that serve as a punishing reinforcer, discouraging alcohol consumption. Aversion conditioning has also been attempted for stopping bedwetting in children. When the bed is soaked with urine, a sensor is activated and triggers a loud sound that wakes up the child. The loud sound in the middle of sleep acts as a punishing reinforcer to the child and discourages bedwetting.

In *avoidance conditioning*, a punishing reinforcer is delivered in the absence of a particular behavior, which consequently becomes intensified. Drug dependence is the consequence of avoidance conditioning. Whenever attempts are made to give up drugs, it results in severe unpleasant symptoms called withdrawal symptoms that serve as negative reinforcers and encourage the continuance of drugs.

Instrumental conditioning can be demonstrated in experimental animals too. In Thorndike's paradigm, a hungry cat is placed inside a puzzle box and a piece of fish is kept outside in sight of the cat. The box has a complicated system of pulleys, strings, and levers designed to open its door. The cat initially manages to open the door by chance but its performance improves with time. The cat learns to associate its actions with the rewarding outcome. The cat can even be manipulated to scratch or lick itself to escape from the box! Thorndike's paradigm led to the development of the Skinner box, a problem box in which an animal learns to press a key to get food (reward conditioning) or avoid electric shock (avoidance conditioning) or to avoid the pressing of a key that delivers an electric shock (aversion conditioning).

Biofeedback is an example of operant conditioning in which it is autonomic response and not motor response that is conditioned. It is based on the premise that we have an innate ability to voluntarily influence the autonomic functions of our body. Biofeedback helps us to develop this innate ability and puts it to therapeutic use. For example, we must get a feedback about the state of our bodily functions like blood pressure, heart rate, sweating or the resting tone of our voluntary muscles if we are to try to control them voluntarily.

Language

The collection of all the words in a given language is called the lexicon. The meanings of the words and sentences are called semantics. Memorizing words and their meanings is necessary for speaking a language. However, a language can never be spoken fluently till its structure is understood.

The smallest unit of a language is the phoneme. Phonemes are fundamental sounds that are combined to form morphemes, the smallest meaningful units of words. A morpheme may be a base ('do' in 'undo'), an affix ('un' in 'undo' and 'unfold' or 'er' in 'painter' and 'washer'), or an inflectional form ('ing' in 'doing' or 's' in 'girls'). Morphemes may be words themselves or they may be combined to form words. Several words are strung together using certain rules called the syntax. The syntax can change the semantics, e.g., 'speed thrills but kills' versus 'speed kills but thrills'. The semantics can also be modified by vocal intonation or prosody. The importance of prosody is illustrated by the following example. When a child is asked "What is this?" he replies "This is my *toy*." When asked "Whose is this?" he replies "This is *my* toy." When asked "Which one is yours?" he replies "*This* is my toy." Each time, the answer remains the same but the prosody changes its meaning. Finally, the stringing together of sentences to form a meaningful narrative is called pragmatics. The term grammar mainly encompasses phonology (rules of phonemes), morphology (rules of morphemes), and syntax. These components of language are applicable not only to sound-based language but also to sign languages. For example, a morpheme in sign language is the smallest meaningful movement.

Table 112.**1**   Structural components of a language.

Phonemes	Sound units whose sequence produces morphemes
Morphemes	Smallest meaningful unit of a word
Syntax	Grammatical combination of words in phrases and sentences
Prosody	Vocal intonation that can modify the meaning of words and sentences
Pragmatics	Linking of sentences into a narrative.

Aphasia

Aphasia refers to a disorder of language apparent in speech, writing, or reading. Aphasias must be distinguished from dysarthria, a disorder of speech due to motor defects in the muscles of articulation. Aphasias are categorized into nonfluent and fluent. In *fluent aphasias*, the speech is fluent but there is difficulty in the comprehension or repetition of words, phrases, or sentences spoken by others. In *nonfluent aphasias*, the speech is limited and effortful but there is fairly good comprehension of spoken speech. Discussed below are some of the commoner forms of aphasia.

Wernicke's aphasia or *sensory aphasia* is the inability to comprehend words or to arrange sounds into coherent speech. There is no dysarthria and the speech is fluent and indeed, the patient may speak a lot. Auditory comprehension is defective due to the inability in distinguishing the phonemes of the speech, a phenomenon called *word deafness*. An example will illustrate the point. In the Japanese language, the sounds *l* and *r* are not distinguished. Therefore, a Japanese-speaking person hearing English cannot distinguish the sounds *l* and *r* although this distinction is perfectly clear to English-speaking persons. A person with Wernicke's aphasia has a similar problem in his or her own language.

Visual comprehension or reading is affected too, i.e., there is *word blindness*, because a person who has problems in distinguishing phonemes has problems with graphemes too. Graphemes are the pictorial or written representations of a phoneme that combine to form a word. The word blindness is associated with *agraphia*, i.e., impairment of writing.

Speech is fluent but is often associated with anomia, neologism, and paraphasias. *Anomia* is the difficulty in finding an appropriate word to express a thought. *Neologism* is using or creating new words or new meanings for established words. *Paraphasias* are syllables, words or phrases that are spoken completely unintentionally (for example, "my mother" instead of "my wife") or that represent a distortion of the intended word (for example, "lime" instead of "line"). Confusion associated with phonemes while speaking results in what has been called a *word salad*.

Just as the epidermis grows by the proliferation of the basal cells, with shedding of the old superficial cells, in the same way, the lens grows by the proliferation of the cells on its surface. The older cells, however, cannot be cast off but undergo sclerosis and accumulate in the center as the nucleus of the lens. The newly formed cells elongate into fibers.

The lens is surrounded by a hyaline membrane called the lens capsule, which is attached to the zonule. The zonule is composed of connective tissue fibers. It is split medially into anterior and posterior laminae, which enfold the lens capsule and blend with it. Laterally the zonule is attached to the inner surface of the ciliary body.

The main source of energy of the lens is glucose, most of which is metabolized through glycolysis with the HMP shunt pathway and citric acid cycle making smaller contributions. The sorbitol pathway, which is normally negligible, metabolizes large amounts of glucose in hyperglycemia. The sorbitol produced becomes trapped inside the lens because the lens capsule is impermeable to sorbitol. The sorbitol retains water osmotically, causing swelling of the lens and resulting in cataract formation.

Uveal tract

The uveal tract is made of highly vascular connective tissue and is innervated by sensory nerve fibers of the trigeminal as well as vasomotor fibers. Due to the presence of nociceptive fibers, the inflammation of the uveal tract (uveitis) is intensely painful. The uveal tract consists of three parts: the choroid, ciliary body, and the iris.

The **iris** encloses a central aperture called the pupil and contains two muscles that control pupillary movements, viz., the *sphincter pupillae*, a circular bundle running round the pupillary margin, and the *dilatator pupillae*, arranged radially near the root of the iris. The sphincter pupillae is supplied by parasympathetic fibers from the oculomotor nerve. The dilatator pupillae is supplied by sympathetic fibers from the cervical sympathetic ganglia. The iris forms a circular diaphragm which, together with the lens, partitions the eye into the anterior and posterior chambers.

The **ciliary body** has an outer and an inner part. The outer part of the ciliary body is composed of smooth muscle called the *ciliary muscle*. The muscle consists of the longitudinal and circular (sphincteric) parts with a common origin in the ciliary tendon, a structure that runs circumferentially round the globe blending with

the scleral spur. The inner part has tufts of blood capillaries. The anterior part of the inner surface of the ciliary body is thrown into folds called the ciliary processes, which secrete the aqueous humor.

The **choroid** is an extremely vascular membrane in contact with the sclera. On the inner side, the choroid is covered by a thin elastic membrane, the membrane of Bruch. Immediately beneath the membrane of Bruch is a plexus of the choriocapillaries. The outer choroid layers contain the medium-sized and large vessels.

Aqueous humor

Aqueous humor is a clear transparent fluid, filling the anterior and posterior chambers of the eye. It provides nutrition and oxygen to the lens and the cornea, and also serves as a vehicle for elimination of metabolic waste products.

Aqueous circulation   The aqueous humor is secreted into the posterior chamber by the ciliary body (Fig. 113.**2**). The aqueous humor is formed mostly through active secretion though ultrafiltration also contributes. The aqueous humor flows from the posterior to the anterior chamber through the pupil. It then flows into the canal of Schlemm, a venous channel at the limbus (junction between the iris and the cornea) and thence, into the episcleral veins. The main resistance to the outflow of the aqueous humor is offered by the trabecular meshwork that forms the medial wall of the canal of Schlemm.

The pressure of the aqueous humor in the anterior chamber of the eye is called the *intraocular pressure* (IOP). The normal IOP is between 10 and 21 mmHg and rises by up to 6 mmHg when reclining from a sitting to

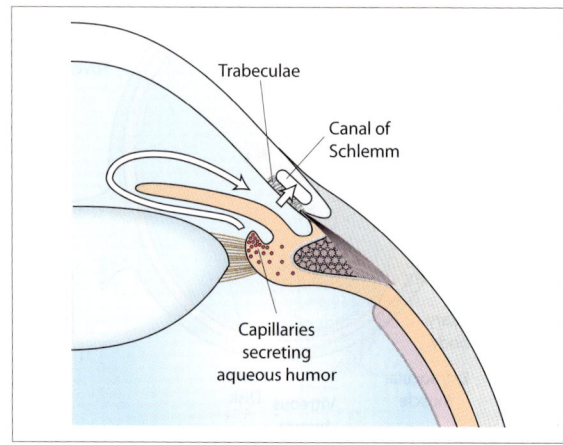

Fig. 113.**2**   Aqueous circulation

supine position. It increases with age, especially after 40 years. Myopes and individuals with a family history of glaucoma tend to have a higher IOP. Glucocorticoids increase the IOP, and the diurnal changes in IOP is related to the diurnal fluctuations in glucocorticoid secretion.

Vitreous humor

The vitreous humor is a transparent, hydrated gel containing over 90% water. A vitreoretinal barrier prevents the vitreous constituents from equilibrating with blood and surrounding fluids. The vitreous humor also contains hyalocytes which secrete hyaluronic acid and fibrocytes which secrete collagen fibrils.

Lacrimal Apparatus

The lacrimal apparatus comprises the lacrimal glands which secrete tears and the lacrimal passages which drain the tears into the nasal cavity. The lacrimal glands of each eye consist of the superior (or orbital) gland, the inferior (or palpebral) gland, and the accessory

lacrimal glands (also called Krause's glands). All are serous acinous glands. Tear secretion is increased by parasympathetic nerves and decreased by sympathetic nerves.

A *tear film* has three layers (Fig. 113.**3**). The middle layer is the aqueous layer secreted by the lacrimal glands. It lubricates the ocular surface, removes debris, oxygenates the cornea, and has an antimicrobial action. The outermost lipid layer is secreted by the meibomian glands in the eyelids. It prevents evaporation of the aqueous layer. The innermost mucin layer is secreted by the goblet cells of the conjunctiva. The mucinous coating on the corneal epithelium is necessary for soaking the tears. In its absence, the lipid membrane of the underlying epithelial cells makes the corneal epithelium hydrophobic and the eye becomes dry. The condition is called *xerophthalmia* and is seen in vitamin A deficiency which causes malfunctioning of the goblet cells. Xerophthalmia is associated with corneal damage and repeated infections. Application of *artificial tear fluids* alleviates the condition. Methylcellulose solution and polyvinyl alcohol solution are the two main types of artificial tear supplements. The eyes can also be kept moist by occluding the lacrimal puncta to decrease the drainage of tears.

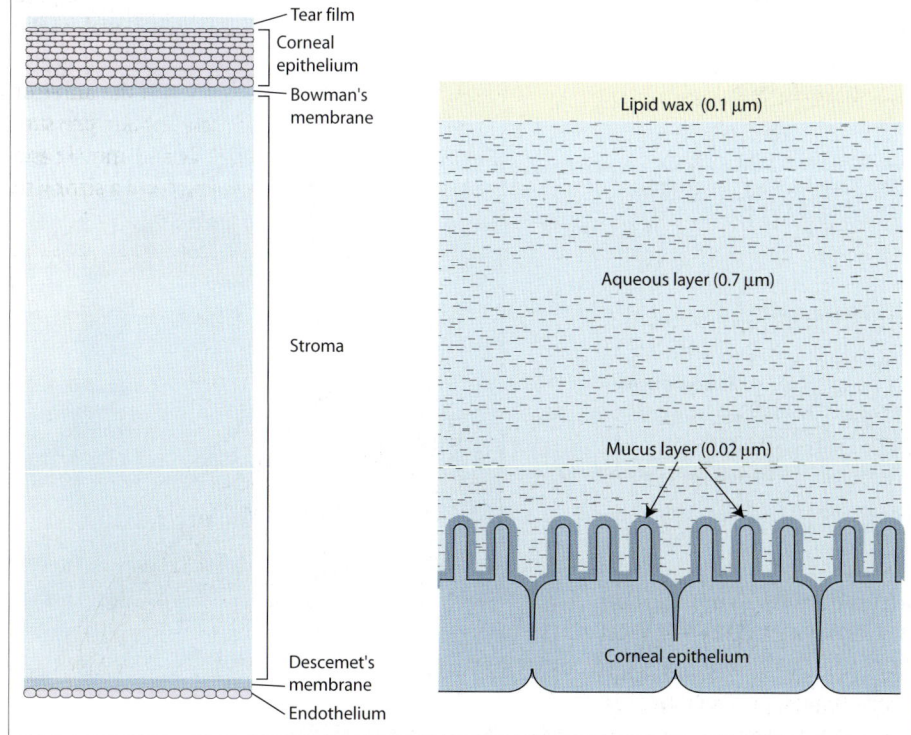

Fig. 113.**3** (*Left*) Layers of the cornea. (*Right*) Layers of the tear film.

Glaucoma

Glaucoma is an elevation of IOP with characteristic degenerative changes in the optic nerve head. Patients with normal IOP can also show glaucomatous optic nerve changes: They are classified as normal tension glaucomas. On the other hand, patients with increased IOP may not show any glaucomatous changes in their optic nerve head: They are classified as ocular hypertensives. Therefore, the diagnosis of glaucoma is based, not on the elevated IOP but on certain characteristic (glaucomatous) changes in the optic nerve head. There are two distinct types of glaucoma, which differ considerably in their pathophysiology and treatment. They are called angle-closure glaucoma and open-angle glaucoma. The differentiation is based, as the names suggest, on the magnitude of the *sclerocorneal angle* (Fig. 113.4A), which can be assessed with an instrument called the gonioscope.

In **angle-closure glaucoma** (Fig. 113.4B), the sclerocorneal angle (which is also equal to the angle of anterior chamber) is narrow and therefore easily occludable by the peripheral part of the iris. The occlusion, which may be slight initially, causes pooling of aqueous humor in the posterior chamber and further accentuates the occlusion. This is called the *appositional closure* at the angle and results in a sudden rise in the IOP. Appositional closure is often relieved by punching a small hole in the iris (*peripheral iridectomy*). The hole allows aqueous flow from the posterior to the anterior chamber and thereby relieves the IOP (Fig. 113.4C). Repeated attacks of appositional closure may be followed by irreversible closure of the angle due to adhesions (synechiae). This type of synechial closure is not relieved by peripheral iridectomy and may require trabeculectomy. The rise in IOP in this type of glaucoma is sudden: It is extremely painful and associated with red congested eye and corneal edema, thus accounting for its previous name of *acute congestive glaucoma*.

Fig. 113.**4**   (A) Normal (left) and narrow (right) sclerocorneal angle. (B) Obliteration of angle by pupillary dilatation in angle closure glaucoma. (C) Facilitation of aqueous drainage by peripheral iridectomy in angle closure glaucoma. (D) Facilitation of aqueous drainage by trabeculectomy in open angle glaucoma.

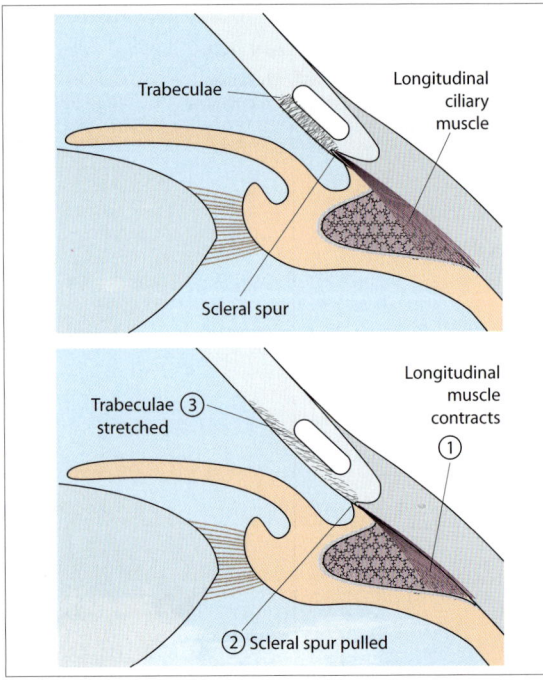

Fig. 113.**5**   Effect of miotics on open-angle glaucoma.

In **open-angle glaucoma** (Fig 113.4D), the anterior chamber angle is not reduced, and the increased IOP is due to the increased resistance of the trabeculae to aqueous outflow. Increased trabecular resistance is often due to deposition of pigments and mucopolysaccharides in the trabecular meshwork. What causes such deposition is not known but some predisposing factors include myopes above 50 years of age, patients with diabetes mellitus and thyroid disorders, and patients with a family history of glaucoma. Such cases are diagnosed as having primary open angle glaucoma. Cases where the cause can be discerned are called secondary open angle glaucoma. This can be due to either an increase in the episcleral venous pressure or due to deposition of pigment and exfoliative materials from the iris and lens capsule. Surgical treatment of open angle glaucoma involves *trabeculectomy*, i.e., the creation of a fistula between the anterior chamber and the subconjunctival space.

The treatment in both types of glaucoma comprises an immediate reduction in the IOP with the help of hyperosmotic agents like mannitol, glycerol, sorbitrate or urea which causes shrinkage of the vitreous humor, and intravenous acetazolamide which reduces aqueous humor formation. Subsequent reduction in pressure can be maintained with drugs that reduce aqueous humor formation like acetazolamide, β and α_2 adrenergic blockers, cholinergic agonists, and $PGF_{2\alpha}$ agonists.

Medical management is the mainstay of treatment in open angle glaucoma. Miotics like pilocarpine are useful in both types of glaucoma. In angle-closure glaucoma, pupillary constriction flattens the iris and draws it away from the cornea, thus increasing the outflow of aqueous humor. In open angle glaucoma, the ciliary muscle contraction induced by miotics exerts traction on the scleral spur, thereby opening out the sclerosed trabecular meshwork and improving the outflow (Fig. 113.5). The patient requires surgery if medical therapy fails.

Physiological Optics

The passage of light through the eye involves refraction at multiple interfaces. Thus, there is bending of light when it passes from air to cornea, cornea to aqueous humor, and aqueous humor to vitreous humor. The degree of bending at an interface is proportional to the ratio of the refractive indices of the two media it separates. When one considers the refractive indices of the various media shown in Figure 114.**1**, it is obvious that the maximum bending of light occurs at the air–cornea interface (for air, $\mu = 1$). With light bending at multiple interfaces, any mathematical treatment of visual optics becomes complex. This is avoided by confining the mathematical approach to a simplified model of the eye, called the reduced eye of Listing.

In the **reduced eye of Listing**, the multiple media, viz., cornea, aqueous humor, and vitreous humor, are all replaced with a single medium with a refractive index of 1.336. The surface of this hypothetical ocular medium (air–medium interface) is spherical (radius of curvature = 5.6 mm) and is located 1.4 mm behind the corneal surface. The nodal point or optical center of the medium is therefore located 5.6 mm behind the air–medium interface. Rays of light passing through the nodal point do not undergo any refraction.

Optical axis versus visual axis The line passing through the nodal point and the center of the pupil is called the optical axis. On the other hand, the line passing through the nodal point and the fovea centralis is called the visual axis. The two axes subtend a small angle between them, which is called the angle γ (Fig. 114.**2**). When the optical axis intersects the retina medial to the fovea centralis, the angle γ is said to be positive. When the optical axis intersects the retina lateral to the fovea centralis, the angle γ is said to be negative.

In emmetropes, the γ angle is 5°. The angle is greater in hypermetropes. In myopes, the angle is 0° or even negative. The line of vision is the same as the visual axis. However, an external observer judges the direction of the line of vision by the position of the pupil. Hence, the greater the size of the positive angle γ, the more the eyes will appear to look outward, and therefore squinted (Fig. 114.**3**). Since the normal angle γ is 5°, we are accustomed to regard a 10° divergence of the eye (5° for each eye) as normal.

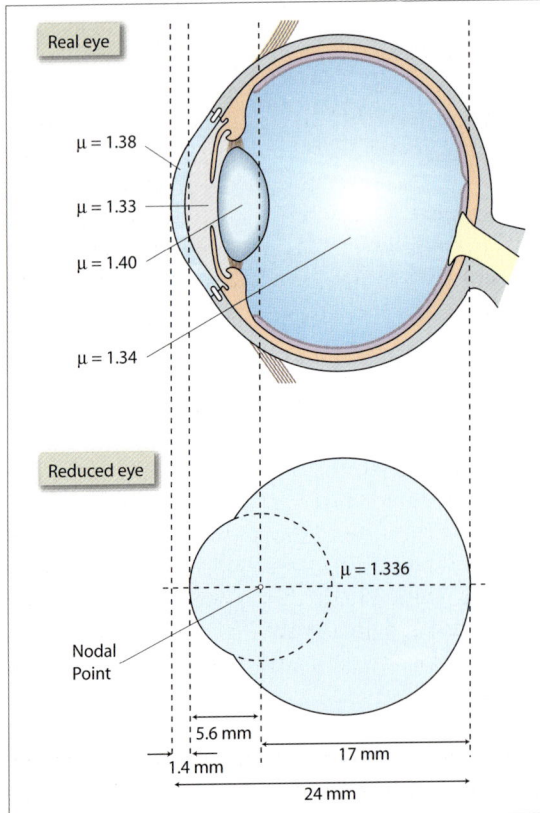

Fig. 114.**1**   Reduced eye of Listing.

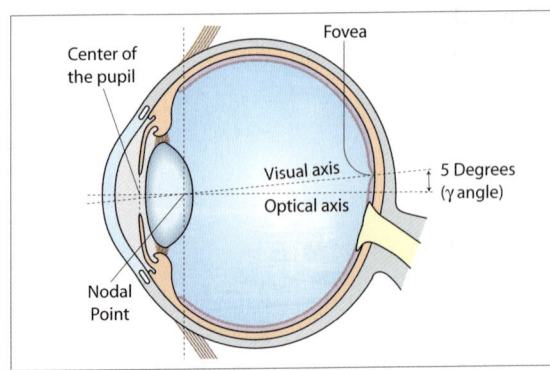

Fig. 114.**2**   Visual and optical axes of the eye.

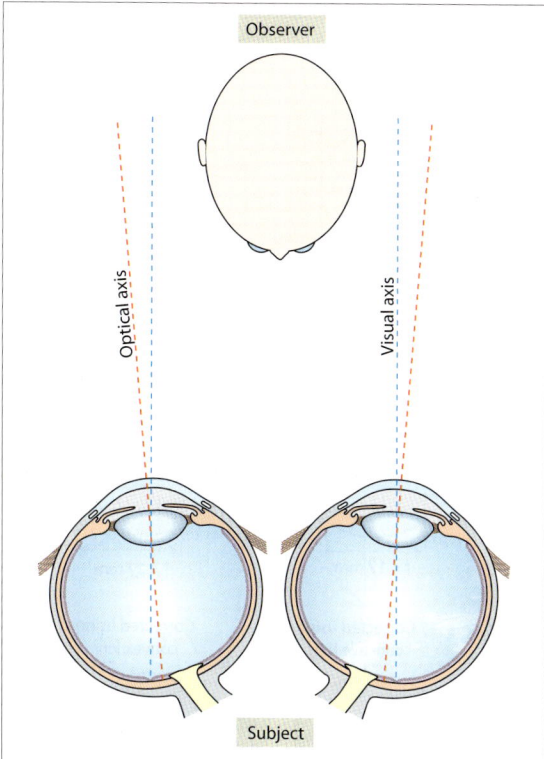

Fig. 114.**3**   Apparent squint. The direction of gaze of a subject is determined by his visual axes. Hence, when he looks straight ahead, his visual axes are parallel. Due to the positive γ angle, his optical axes are divergent. The observer judges the direction of gaze by the optical axes. Therefore the eyes of the subject appear divergent to the observer.

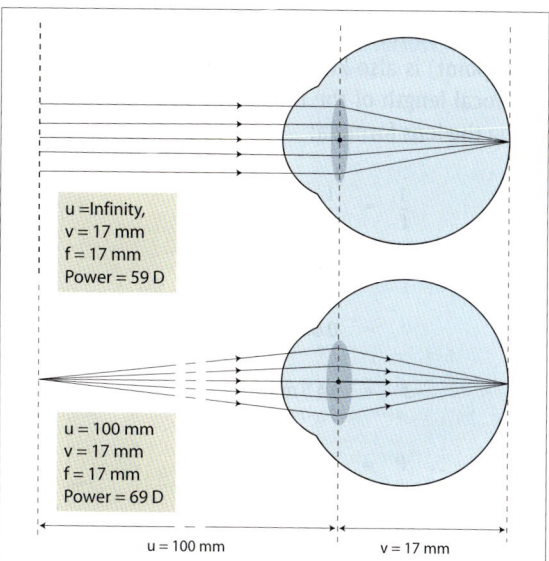

Fig. 114.**4**   Focal length of the eye, before (*above*) and after maximum accommodation (*below*).

Diopteric power of the eye   Although the refraction of light first occurs at the air–medium interface, for all practical purposes, the entire refraction is assumed to occur at the plane of the nodal point as shown in Figure 114.**4**. In an eye with normal dimensions and refractive power, all parallel rays come to focus on the retina. Thus, the focal length of the eye equals the distance between the nodal plane and the retina, which is equal to 17 mm. The relation between focal length and diopteric power is given by the formula:

$$\text{Power in diopters} = \frac{1}{\text{Focal length in meters}}$$

Substituting:

$$\text{Power in diopters} = \frac{1}{0.017}$$

$$= 59 \text{ D}$$

Although the calculations are based on the schematic eye, the diopteric power of the real eye too is 59 D. This power represents the combined refractory power of the cornea, aqueous humor, lens, and vitreous humor. The contribution of the lens in it is 16 D.

Accommodation

Far point   Only rays that originate at infinity are parallel. Rays originating from finite distances are divergent: The nearer the object, the greater is the divergence of rays reaching the eye from it. Since, the eye is able to focus parallel rays on the retina, it means that it is able to see objects at infinity effortlessly. In other words, the far point of the normal eye is infinity (∞).

Near point   The eye can bring to focus on the retina divergent rays. This is made possible by an increase in the diopteric power of the eye. By increasing its diopteric power, the eye can see clearly objects as close as 10 cm. This minimum distance at which the eye can see objects clearly is called the near point. The near point is not the same as the least distance of distinct vision which is about 25 cm. This is the distance at which near objects should be placed for viewing without unduly straining the eyes, i.e., without excessive accommodation.

Amplitude of accommodation   At maximal accommodation, the focal length (f) of the eye becomes less than 17 mm. However, the distance of the retinal image from the nodal plane (v) remains 17 mm. The distance

Astigmatism occurs when the eye shows two different powers in two different axes. It is corrected with cylindrical lenses, which have two different focal lengths in the two different planes (Fig. 114.**6**).

Aberrations of image formation

Spherical aberration The refractive index of the lens of the eye is not uniform throughout: Its central part has a higher refractive index than its periphery. Thus, rays passing through the periphery of the lens and the rays passing through the central part of the lens come to focus at different planes. This is called the spherical aberration and would normally cause blurring of the retinal image, were it not for the iris, which keeps the periphery of the lens covered most of the time. In fact, the smaller the pupil, the less is the spherical aberration and the sharper the image (Fig. 114.**7**).

Chromatic aberration When white light passes through the lens, light of different wavelengths comes to focus at different planes. This is called chromatic aberration. The eye tackles this problem by keeping the green-red wavelengths in focus while allowing the blue light to become blurred. Since blue cones are entirely absent from the fovea and comprise less than 10% of all cones, vision is not much affected by the blurring of blue light (Fig. 114.**7**).

Visual acuity

When two point-objects are brought very close together, they are no longer seen as separate. To be seen as discrete points, they must be so separated that between them, they subtend a visual angle of at least 1' (1 minute). It can be calculated that a 1' visual angle corresponds to a 4.95 μm separation of point-images on the retina[1] (Fig. 114.**8**). Since a cone is about 3 μm wide, it can be argued that two different cones must be stimulated if the images falling on them are to be perceived as discrete (see two-point discrimination, p 648). The smaller the visual angle subtended between two distinctly visible points, the higher is the visual acuity. The tangent of the minimum visual angle is called the visual acuity.

The visual acuity read out from the Snellen chart is a comparison with normal vision, e.g., 6/9 or 6/12, etc. These fractions should not be reduced. This avoids confusion with the direct method of expressing acuity that gives the tangent of the minimum visual angle, i.e., the minimum object size visible divided by the distance it is kept at. According to this direct system, an acuity of 6/6 would correspond to 8.75/6000. In the readout of the Snellen chart, a small fraction denotes poor visual acuity whereas in the direct system, a small fraction denotes high acuity.

The Snellen chart For routine evaluation of visual acuity, the Snellen chart is used. In the Snellen chart, the types consist of a series of letters arranged in lines of diminishing size. When the letters are read from the distance indicated alongside, the stroke width of the letters subtend an angle of 1 minute at the nodal point of the eye. Each letter is of such a shape that it can be placed in a square, the sides of which are five times the stroke width. Hence, the whole letter will subtend an angle of 5 minutes at the nodal point of the eye at the given distance (Fig. 114.**9**).

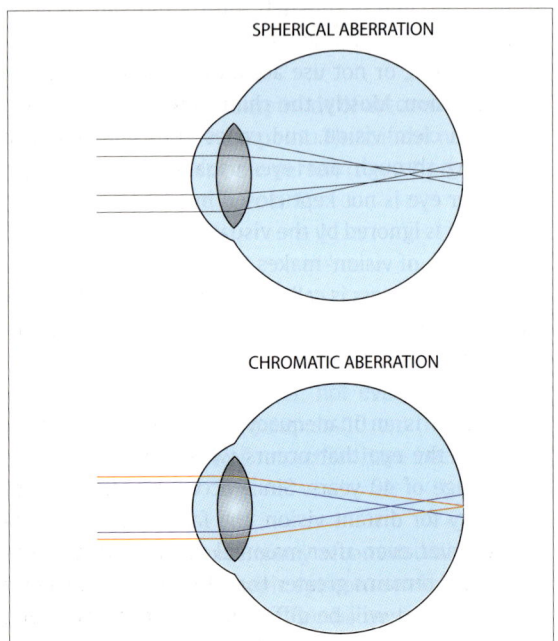

Fig. 114.**7** Spherical and chromatic aberrations.

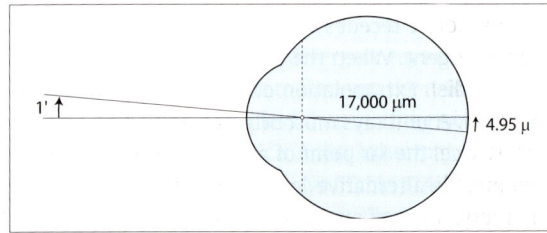

Fig. 114.**8** A visual angle of 1' corresponds to a distance of about 5 μm on the retina.

---

[1] The distance between the nodal point and retina is 17,000 μm (17 mm). 1 minute is equal to 1/60 degrees, which is equal to 0.0167°. The height of the retinal image can be calculated to be $17,000 \times \tan(0.0167) = 4.95$ μm.

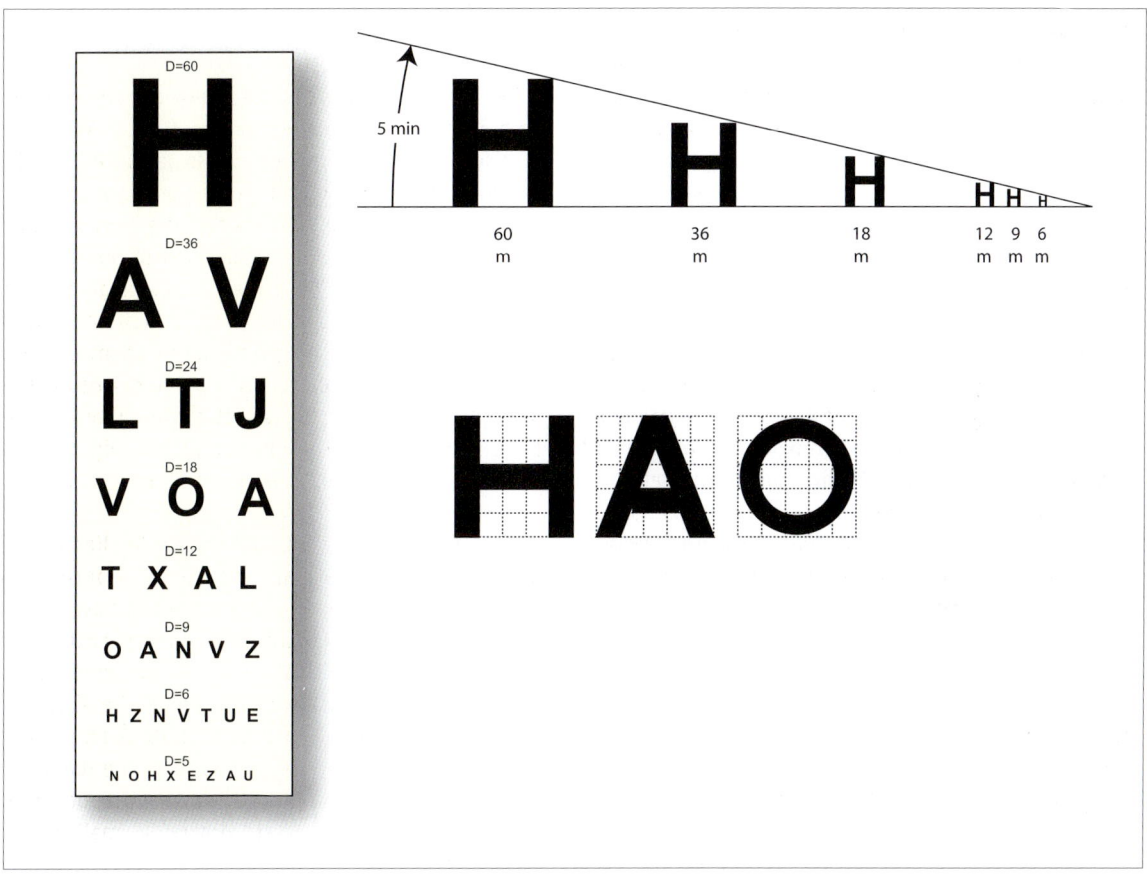

Fig. 114.9   The Snellen chart. The size of each letter in the chart is such that when placed at the distance mentioned alongside, it will subtend an angle of 5 minutes at the eye.

In the Snellen chart, the largest letter will subtend 5 minutes at the nodal point if it is placed 60 meters from the eye. Those in the subsequent lines will subtend 5 minutes if they are 36, 24, 18, 12, 9, and 6 meters from the eye. Sometimes, smaller letters corresponding to 5 and 4 meters are used. A person with average acuity of vision is able to read the top letter at 60 meters, the second at 36 meters, the third at 24 meters, and so on.

For convenience, the patient is kept at a fixed distance of 6 m from the types. Rays incident on the eye from a distance of 6 meters or more can be considered parallel. A normal subject standing 6 meters from the chart ought to be able to read every letter from the topmost line (marked 60) to the last line (marked 6). A numerical fraction is used to express the visual acuity. Thus a patient who can read only up to the line marked 18 has a visual acuity of 6/18. The numerator denotes the distance (in meters) of the patient from the types. The denominator is the distance (in meters) from which a person with normal vision can read the line. Normal vision is recorded as 6/6. Many people can read more in a good light and have a visual acuity of 6/5.

Visual Field

Although it is rarely realized, the eye can see clearly only an extremely small area at a time. This point-sized area, which the eye focuses and looks directly at, is called the *fixation point*. There is however no difficulty in seeing a large part of the world around because of the extremely efficient mobility of the eyes. The total sector of the external world visible to the eye that is fixated straight ahead is called the *field of vision*. The measure of the visual field in a given direction is denoted by the angle its boundary subtends with the visual axis at the center of the cornea.

Using an instrument called the perimeter, the entire field of vision can be charted out, and the technique is called *perimetry*. The visual field is divided into the nasal and temporal halves by a vertical line passing through the fixation point in the visual field. The field is further charted out by meridians and isopters (Fig. 114.**10**). The normal field of vision is 50° superiorly (restricted by eyebrow), 60° nasally (restricted by nose), 70° inferiorly (restricted by zygoma), and 90° temporally (unrestricted).

Visual acuity of peripheral retina

Perimetry   The Snellen chart measures the visual acuity only at the fovea centralis of the retina, and not of the rest of the retina. The acuity of the peripheral parts of the retina can however be indirectly assessed. For example, the field of vision shown in Figure 114.**10** is obtained in good daylight by using a 5 mm object kept at a distance of 330 mm. The area charted out therefore represents the field with a visual acuity of 5/330 or more. When a 1 mm object is used, the visual field is constricted. The constricted field denotes the areas of the retina with an acuity of 1/330 or more. The field is constricted further when the 1 mm object is placed at a distance of 1200 mm. This highly constricted field has an acuity of 1/1200 or higher.

The minimum visual angle that is required for vision in retinal areas with an acuity of 5/330 is given by $\tan^{-1}[5/330]$, which is equal to 55 minutes. The corresponding angles for 1/330 and 1/1200 are 11 minutes and 3 minutes respectively. The visual acuity is thus greater near the center of the visual field. Since the photoreceptors are distributed more sparsely nearer the periphery, the visual acuity declines toward the periphery.

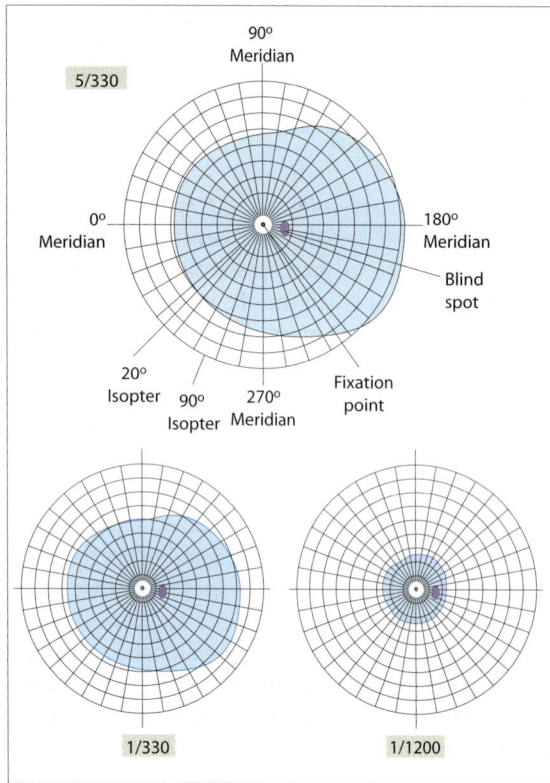

Fig. 114.**10**   The field of vision shrinks as the size of the object with which it is tested is made smaller.

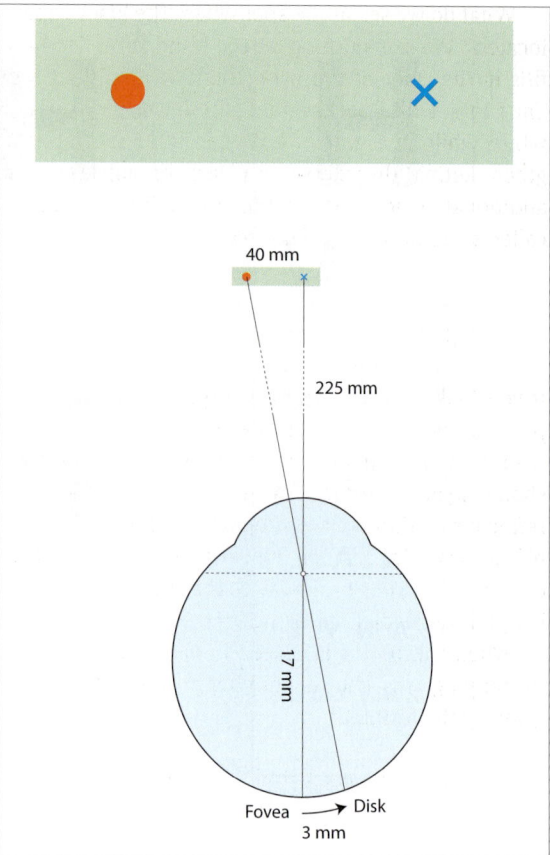

Fig. 114.**11**   Appreciation of the blind spot (See text).

Blind spot   The part of the visual field that is focused on the optic disk remains invisible to the eye and it is called the blind spot. Normally, the blind spot is located between the 12° and 18° isopters, a little below the 180° meridian, i.e., on the temporal side.

The presence of the blind spot can be appreciated by following the procedure outlined here: Close your right eye and hold Figure 114.**11** at a distance of about 30 cm in front of your eye. Fixate the cross sign with your left eye. You will still be able to see the circle to your left, though less clearly. Slowly draw the figure nearer to you, keeping the cross sign fixated (and resisting the instinctive urge to fixate the black circle). When the figure is about 22 cm from your eye,[2] the black circle suddenly disappears from your visual field, only to reappear when you draw the figure still nearer.

---

[2] Note the distance (say 22 cm) from your eye at which the blind spot is observed. Given that the distance between the cross and the circle is 4 cm, the distance between the optic disk and the fovea centralis can be calculated to be about 3 mm by applying the principle of similar triangles:

$$\frac{\text{Retinal distance between disk and fovea}}{\text{Distance between lens and fovea (17 mm)}} = \frac{4\,\text{cm}}{22\,\text{cm}}$$

What do we see at the spot where the black circle is located? We see 'nothing' at the blind spot. The brain fills in the color of the background in the blind spot. Thus in a white background, the blind spot appears white while in Figure 114.**11**, the blind spot appears green. Retinal disorders can cause the appearance of additional blind spots in the visual field. These are called scotomata (singular = scotoma).

Ophthalmoscopy

If you look into the eye of a subject through his pupil, what do you see? Considering that you are seeing the retina through 59 diopters of refractory power, you should see a magnified image of the retina. Here, the refractive medium of the subject's eye serves as a magnifying glass for his/her own retina. A simplified ray diagram (Fig. 114.**12**) illustrates the situation. The focal length is taken as 17 mm (1/59 = 0.017 m).

When the object is placed at the focal plane of a magnifying lens, the magnifying power of the lens is given by the formula:

$$\frac{\text{Least distance of distinct vision (D)}}{\text{Focal length of the lens (f)}}$$

Substituting, we get:

$$\frac{30 \text{ cm}}{1.7 \text{ cm}} = 17.5 \text{ (approx)}$$

Thus, if we look through the pupil, we will see a retina that is about 17 times magnified. No external lenses are required for obtaining this magnification. This is the underlying principle of direct ophthalmoscopy.

Although a sound proposition theoretically, it is not practically possible to see the retina as suggested above. The pupil is a very small window. If we look in through the pupil, we will block all the light that enters the eye. On the other hand, if we shine light through the pupil, we are left with no space to bring our eye in front of the pupil. Hence, for seeing the retina, an ophthalmoscope is required. An ophthalmoscope is a simple device that allows both the light source as well as the observer's eye to come in front of the pupil of the subject. Essentially, the ophthalmoscope is a concave mirror with a central hole in it that the observer can look through. Usually, it is equipped with its own source of light—an electric bulb (Fig. 114.**13**).

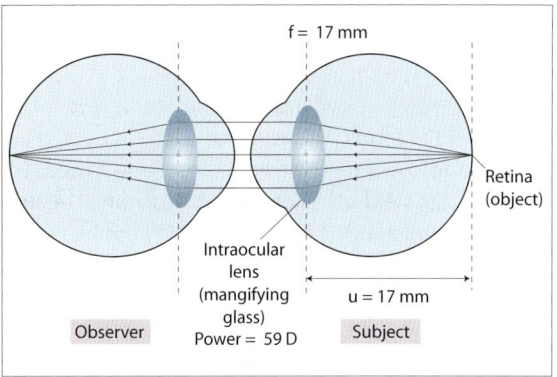

Fig. 114.12   An observer looks into the eye of a subject. This creates an optical system in which the retina of the subject serves as the object, and his intraocular lens serves as a magnifying glass for the observer.

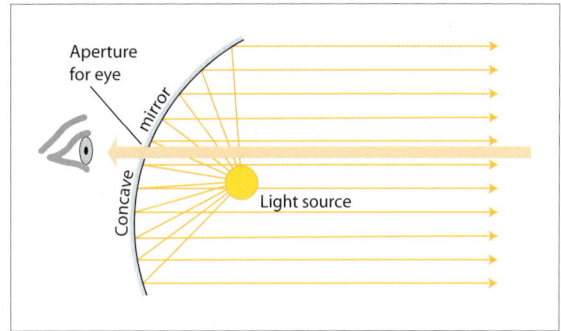

Fig. 114.13   An ophthalmoscope projects a parallel beam of light while helping in bringing the visual axis in line with a beam of light.

The ophthalmoscope is used to examine the fundus (interior) of the eye. During ophthalmoscopy, both the examiner and the patient have to take off their glasses. Hence, most ophthalmoscopes are equipped with an array of convex and concave lenses. These lenses are only corrective lenses meant for compensating the refractory errors of the subject and the observer.

Retinoscopy

Retinoscopy (or more correctly, skiascopy, which means observing the shadow) makes it possible to objectively prescribe lenses of appropriate power to myopic and hypermetropic patients. In essence, the retinoscope is similar to the ophthalmoscope in that both are devices that help in bringing the axis of vision in line with a source of light. The procedure for retinoscopy can be summarized as follows:

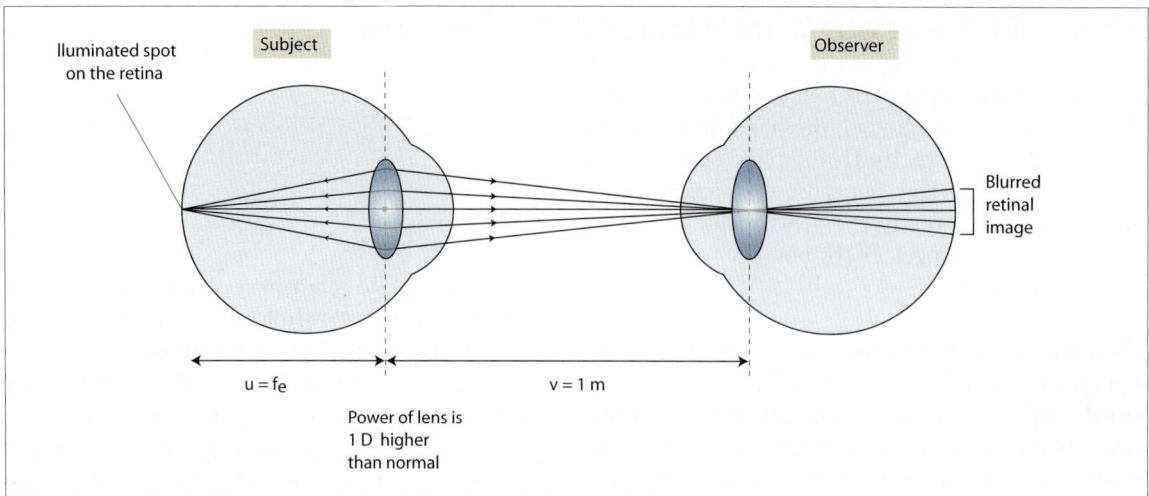

Fig. 114.**14**   Principle of retinoscopy. If the observer sees a completely blurred image of the subject's retina, the subject must be 1 diopter myopic. The illuminated spot on the retina acts as the object.

The observer places himself 1 m away from the subject and views through the retinoscope a small circular spot on the subject's retina that is illuminated by the light from the retinoscope. He then adds lenses in front of his eyes till the image of the illuminated retina becomes completed blurred. At this stage, the subject with all the lenses he is wearing, is exactly 1 diopter myopic. Adding a concave lens of −1 diopter would now make him emmetropic.

Example

If the subject is wearing +3 diopters of lenses when the retinal image becomes completely blurred (endpoint), it means that the subject is 1 diopter myopic when wearing a +3 diopters (convex) lens. To make the subject emmetropic, another −1 diopter (concave) lens has to be added. Overall, the subject requires corrective lenses of +2 diopters (i.e., +3 plus −1).

Retinoscopy is based on the analysis of the optical system in which the subject's retina serves as the object. The refractive medium of the subject's eye acts as the converging lens of the optical system. Additional lenses that are placed in front of the subject eye help in forming a real image of his retina exactly at the pupil of the observer seated 1 m away (Fig. 114.**14**). When an image forms exactly at the pupil of the observer, the observer sees nothing but diffuse light. In this optical system, the distance of the image (v) from the lens is 1 m. The distance of the object (u) from the lens is given by f_e, i.e., the focal length of the lens in the emmetropic eye. This is because in an emmetropic eye, the retina is always at the focus of the lens. Suppose, at endpoint, the eye power of the subject with lenses on is $P_{subject}$.

Using the lens formula,

$$P_{subject} = \frac{1}{v} - \frac{1}{u}$$

$$= \frac{1}{1} - \frac{1}{-f_e}$$

$$= 1 + P_e$$

In other words, the power P of the refractive medium of the subject is 1 diopter greater than the power of an emmetropic eye.

By tilting the retinoscope sideways, the observer can move around the illuminated spot on the retina. By doing so, the observer can have a preliminary idea as to whether the refractive error is hypermetropia or myopia. In hypermetropia, the retina is located within the focal length of the lens and therefore would form an erect, virtual image. Hence, if a spot of light moved across the retina from left to right, it would appear to move in the same direction to the observer too. In emmetropia, the retina is located at the focus of the lens and therefore would form an erect, virtual image. Hence, in emmetropia too, if a spot of light moved across the retina from left to right, it would appear to move in the same direction to the observer. In subjects with myopia greater than 1 diopter, the retina is located beyond the focal length of the lens and therefore would form an inverted, real image. Hence, if a spot of light moved across the retina from left to right, it would appear to move in the opposite direction to the observer. If the myopia is 1 diopter or less, the image of the light is highly blurred as it is formed near the observer's pupil.

The Retina

The retina lies on the inner surface of the choroid. It has a purple tint because the rod cells of the retina contain visual purple or rhodopsin. Light rapidly bleaches the visual purple; in darkness, the color gradually reappears. The retina contains six types of cells, viz., photoreceptor cells, bipolar cells, ganglion cells, horizontal cells, amacrine cells, and Müller cells. These cells are organized into ten layers (Fig. 115.**1**).

Photoreceptors

There are two types of visual receptors, the rods and the cones, located in the retina. In the retinal periphery, the photoreceptors are fewer, larger, and less evenly distributed. Photoreceptors are absent from the optic disk, which is the area where the optic nerve fibers exit the eye.

Each photoreceptor (rod or cone) has an outer segment, an inner segment, and an innermost segment. The outer segment contains stacks of membranous disks. The double-membrane disks are formed by invagination of the cell membrane. The disk membrane contains photopigment molecules 'stitched' into it. The outer segment contains Na$^+$ channels. The inner segment contains the nucleus, mitochondria, and Na$^+$-K$^+$ pumps. The outer and inner segments are connected by an eccentrically placed modified cilium. The innermost segment is the synaptic zone. In darkness, it releases transmitters continuously.

The photoreceptor synapses with the bipolar cell, which in its turn synapses with the ganglion cell. The axons of the ganglion cells form the optic nerve, which leaves the eye to reach the brain. The horizontal and amacrine cells have important roles in image processing.

Fovea centralis   About 2.5 mm lateral to the border of the optic disk, the inner surface of the retina shows a shallow round depression called the fovea, which contains a dense and orderly aggregation of cones (Fig. 115.**3**). The area around the fovea contains yellow pigment and is called the macula lutea or the 'yellow spot'. Vision at the fovea is extremely clear (high

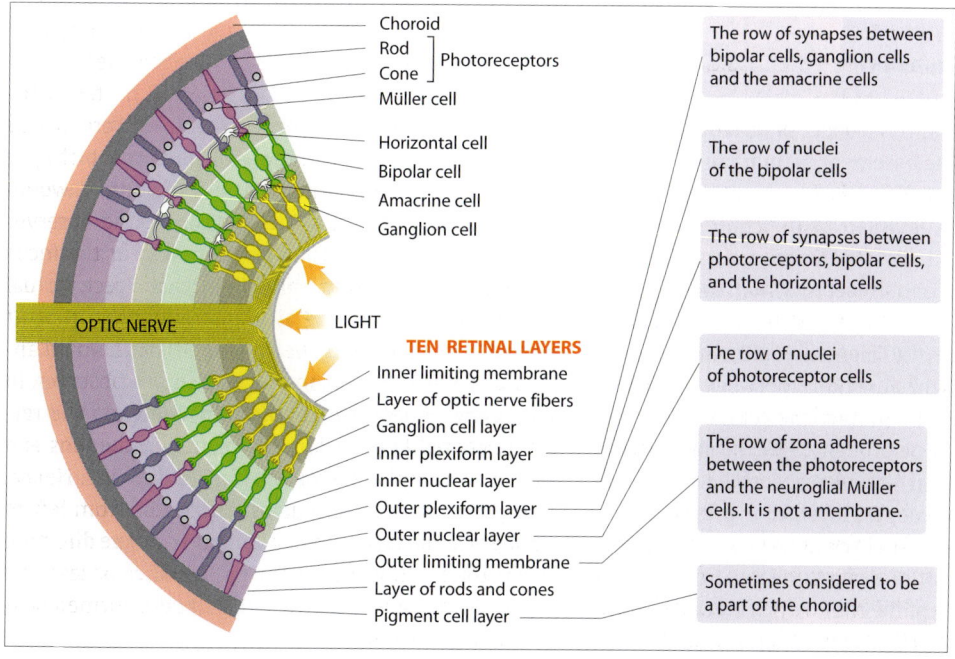

Choroid
Rod ⎤
Cone ⎦ Photoreceptors
Müller cell

Horizontal cell
Bipolar cell
Amacrine cell
Ganglion cell

OPTIC NERVE

LIGHT

TEN RETINAL LAYERS
Inner limiting membrane
Layer of optic nerve fibers
Ganglion cell layer
Inner plexiform layer
Inner nuclear layer
Outer plexiform layer
Outer nuclear layer
Outer limiting membrane
Layer of rods and cones
Pigment cell layer

The row of synapses between bipolar cells, ganglion cells and the amacrine cells

The row of nuclei of the bipolar cells

The row of synapses between photoreceptors, bipolar cells, and the horizontal cells

The row of nuclei of photoreceptor cells

The row of zona adherens between the photoreceptors and the neuroglial Müller cells. It is not a membrane.

Sometimes considered to be a part of the choroid

Fig. 115.**1**   Retinal layers.

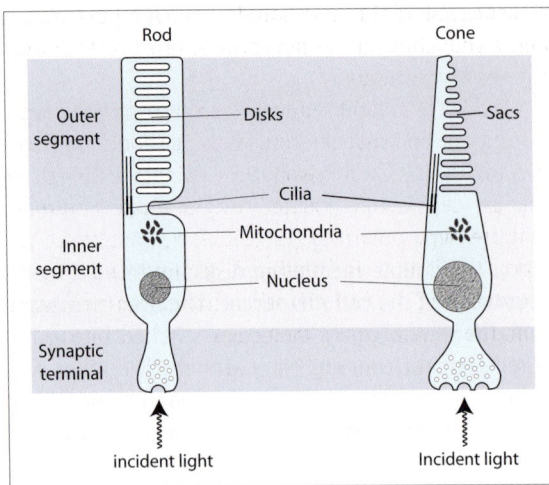

Fig. 115.**2**   Rods and cones.

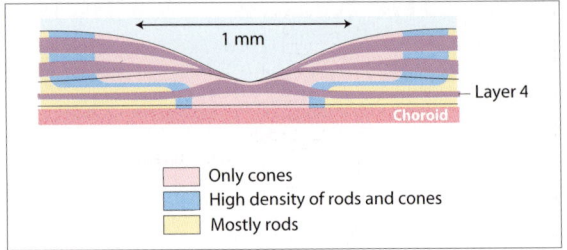

Fig. 115.**3**   Fovea centralis.

visual acuity) for several reasons: (1) There is a high concentration of photoreceptors at the fovea. The foveal photoreceptors are all cones: There are no rods in the fovea. (2) The inner retinal layers are thinned out at the fovea exposing the cones better to light. In other areas of the retina, light has to pass through all the layers before it can strike the photoreceptors. (3) The fovea contains only fine arteries, veins, and capillaries. At the very center of the fovea, even the capillaries are absent, greatly increasing its transparency. (4) There is no convergence of the efferents of the foveal cones. Each foveal cone relays to a single ganglion cell. Hence, there is a disproportionately large representation of the fovea in the visual cortex.

Table 115.**2**   A comparison of rods and cones.

Characteristics of rods	Characteristics of cones
The outer segment of a rod is a slender cylindrical structure.	The outer segment of the cone is a long conical structure.
In rods, the invaginated cell membrane becomes separated from the cell surface.	In cones, the invaginated cell membrane retains continuity with the cell surface.
The photopigment is rhodopsin, which mediates scotopic vision (vision in dim light). Can detect even a single photon of light!	The photopigment is photopsin, which mediates photopic vision (vision in bright light) and color vision.
Membrane proteins synthesized in the inner segments are assembled into new disks at the inner end of the outer segment. The newly formed disks slowly move toward the tip of the rod as additional disks are formed behind them.	Membrane proteins synthesized in the inner segment are inserted into the disk membranes throughout the outer segment. Disks do not move toward the pigment epithelium continuously as in rods.
The nucleus of the rod is smaller and its chromatin is more condensed.	The nucleus is larger and relatively pale staining.
The terminal part of the inner segment is narrower than that of cones.	The terminal part of the inner segment is expanded into a triangular cone-pedicle.

Pigment epithelium

The pigment epithelium is a sheet of heavily pigmented epithelial cells. The basement membrane of the pigment epithelium is called Bruch's membrane and is considered by some as a part of the choroid because in retinal detachment, the separation occurs along the plane between the photoreceptors and the pigment epithelium.

The functions of the pigment epithelium are as follows: (1) The tight junctions between adjoining pigment epithelial cells protect the retina from toxic metabolites that may be brought in by the choriocapillaries. (2) The pigment granules absorb light that has traversed the photoreceptor layer, thus preventing its reflection from the external ocular layers. (3) The membranous disks are continuously exfoliated from the tips of the outer segment of the rods, which are surrounded by processes of the pigment epithelium. These disks are phagocytosed by the pigment epithelium. In mice and rats, certain genetic disorders of photoreceptors are associated with impairment of this process of continuous destruction of disks. These disorders bear resemblance to a disorder called *retinitis pigmentosa* in humans. (4) The pigment epithelium participates in the visual cycle by reducing all-trans-retinal to all-trans-retinol (vitamin A), followed by isomerization of the vitamin to 11-cis-retinol and regeneration of 11-cis-retinal. In the visual cycle, the vitamin A derivatives move to the pigment epithelium during light adaptation and return to the photoreceptors during dark adaptation. Because vitamin A derivatives are fat-soluble, they can pass from cell to cell when bound to a special transport protein. In the pigment epithelial cells, they are stored in the membranes of the smooth endoplasmic reticulum.

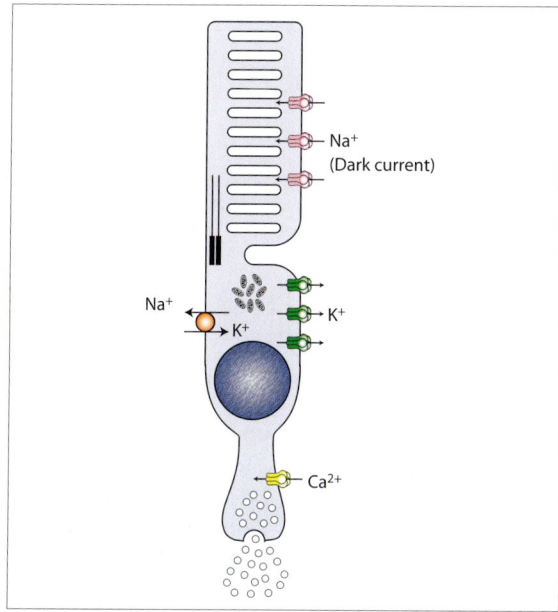

Fig. 115.**4**   The dark current in a photoreceptor.

Phototransduction

The photoreceptor cell has a membrane potential of −40 mV. The K⁺ channels are located in the inner segment of the photoreceptors while the Na⁺ channels are located in the outer segments. The Na⁺-K⁺ pump is located in the inner segment. The presence of the Na⁺ and K⁺ channels at different locations on the cell does not make any difference to the basic mechanism of generation of membrane potential. Thus, the membrane potential is attributable largely to a greater outward diffusion of K⁺ than the inward diffusion of Na⁺.

The Na⁺ channels are open when no light falls on the photoreceptor. Hence in darkness, Na⁺ continuously flows into the outer segment of the photoreceptor. This is called the dark current (Fig. 115.4). Due to the dark current, the photoreceptor remains depolarized in the dark and a steady stream of neurotransmitter is released from its terminal. As is usual, the neurotransmitter release is mediated by Ca²⁺ influx into the synaptic terminal through voltage-gated Ca²⁺ channels. The neurotransmitter is glutamate.

When light falls on the retina, the light energy is trapped in the photosensitive pigments present in the photoreceptors and through a series of steps called the visual cycle, results in the closure of the Na⁺ channels with consequent reduction in the dark current. Reduction in Na⁺ current hyperpolarizes the photoreceptor to −70 mV and reduces its neurotransmitter output. Phototransduction is thus a unique example of sensory

transduction that is associated with hyperpolarization and a reduction in the neurotransmitter release from the sensory receptor.

The fact that light reduces neurotransmitter output should not be cause for confusion: It should be borne in mind that there are two types of bipolar cells—those that are depolarized (excited) by glutamate, and those that are hyperpolarized (inhibited) by glutamate. The bipolar cells that are inhibited by glutamate become disinhibited (i.e., excited) when the glutamate output from the photoreceptor decreases.

The visual cycle

The photosensitive pigments (rhodopsin and iodopsin) are composed of a protein moiety called opsin (scotopsin in rods, photopsin in cones) and 11-cis-retinal (retinal is also called retinene), an aldehyde of vitamin A. When light falls on the receptors, the 11-cis-retinal in the photopigment is converted into all-trans-retinal, forming prelumirhodopsin. This cis–trans transformation triggers a chain reaction (proceeding through lumirhodopsin, meta-rhodopsin I and meta-rhodopsin II), which culminates in the separation of retinal from meta-rhodopsin II, leaving behind opsin (Fig. 115.**5**).

Meta-rhodopsin II closes the Na⁺ channels in the outer segment, thereby hyperpolarizing the membrane and reducing the neurotransmitter output from the synaptic zone of the receptor. The action of meta-rhodopsin II is mediated through group-IIb hormonal mechanism, which employs a G-protein called *transducin*. The Na⁺ channels are kept open in darkness by the action of cGMP. In the presence of light, the lowering of cGMP concentration due to its conversion to 5'GMP results in the closure of Na⁺ channels.

The 11-cis-retinal is regenerated from all-trans-retinal by the enzyme retinal isomerase. Some of the all-trans retinal is converted into all-trans-retinol (vitamin A) by retinal reductase in the presence of NADH + H⁺. The 11-cis-retinol thus produced, as well as that obtained as vitamin A from the diet, is converted to 11-cis-retinal, which is required for refurbishing the rhodopsin stock of the retina.

Since vitamin A is important for replenishing the retinal stores in photoreceptors, a dietary deficiency of vitamin A produces visual abnormalities. The earliest to appear is nyctalopia (night blindness) due to impaired function of rods. Impairment of cone function occurs later. Chronic avitaminosis A is associated with structural abnormalities in rods and cones followed by degeneration of other layers of the retina. Treatment with vitamin A restores retinal function if administered before the receptors are destroyed.

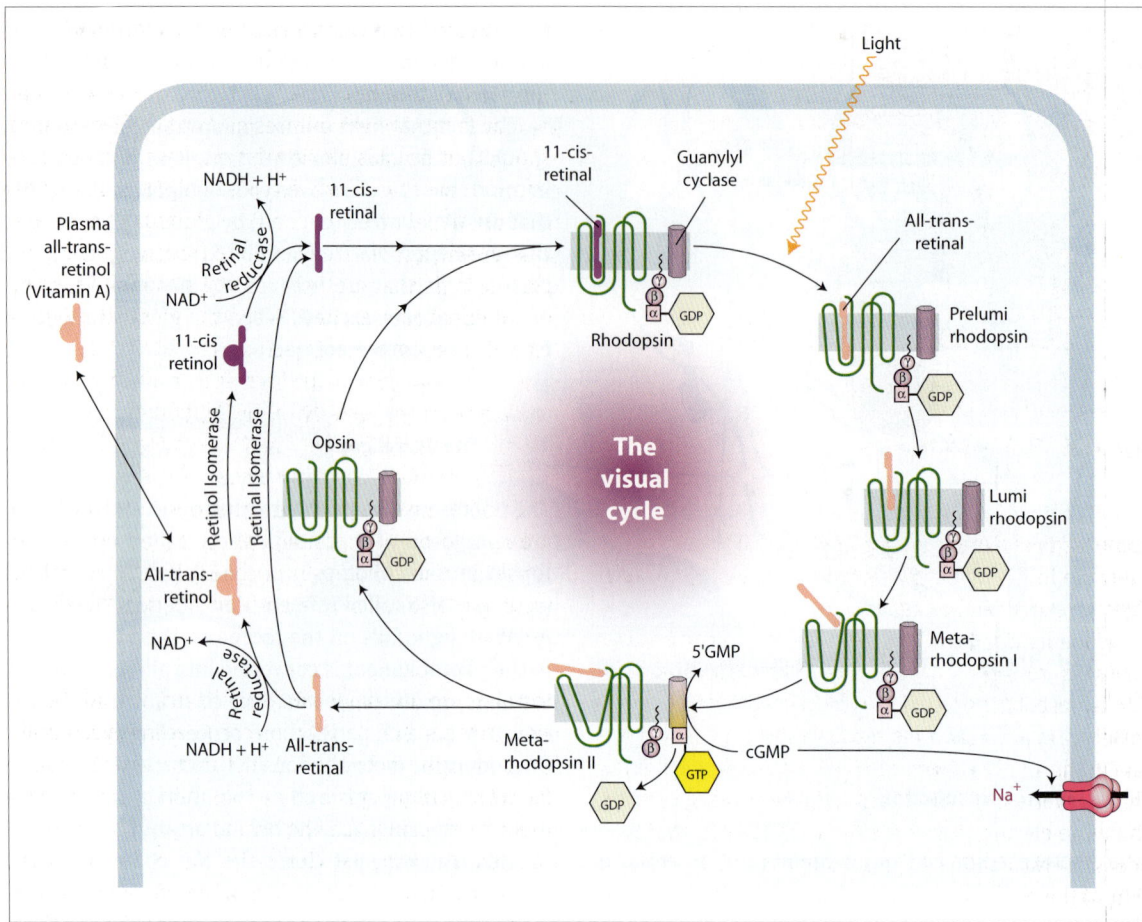

Fig. 115.**5** The visual cycle. See also Figs. 81.3 and 81.5B.

Photopic versus scotopic vision

The human eye responds to a remarkable range of luminance. A white surface lit by a moonless night sky has a luminance of 10^{-6} Cd m^{-2} whereas in bright sunlight, the same surface has a luminance of 10^4 Cd m^{-2}. Neither rods nor cones have the ability to provide effective vision over such a wide range of luminance. Rods function effectively in the range of 10^{-7} to 10^{-3} Cd m^{-2}. Vision in this range of luminance is called *scotopic vision*. Cones function effectively in the range of 1 to 10^8 Cd m^{-2}. Vision in this range of luminance is called *photopic vision*. Prolonged exposure to higher luminance damages the retina. Comfortable reading requires 30 Cd m^{-2}.

Rods and cones also differ in their spectral sensitivities. The spectral sensitivity of rods is maximum at 500 nm, i.e., the blue-green wavelength. The spectral sensitivity of cones is maximum at 560 nm, i.e., the green-yellow wavelength. This difference in peak spectral sensitivity of rods and cones accounts for the *Purkinje shift*, a difference in the luminosity of colors in lights of different intensities. A red and a blue flower may have the same luminosity during daytime when vision is photopic. This is because the peak spectral sensitivity of photopic vision (λ = 560 nm) is roughly equidistant from the wavelengths of red (λ = 650 nm) and blue (λ = 450 nm) light. When vision becomes scotopic at dusk, the blue flower appears to glow whereas the red flower turns nearly black. This is because the peak spectral sensitivity of scotopic vision (λ = 500 nm) is nearer to blue light than red light.

Light adaptation When one passes suddenly from a dim to a brightly-lit environment, the light seems intense until the eyes adapt to the increased illumination. This adaptation occurs over a period of about 5 minutes and is called light adaptation. As already explained above, photoreceptors signal light by decreasing their neurotransmitter output. Intense light reduces their neurotransmitter output to zero so that no further decrease is possible. This renders rods useless for signaling bright light, but the neurotransmitter output of cones gradually increases again even in the

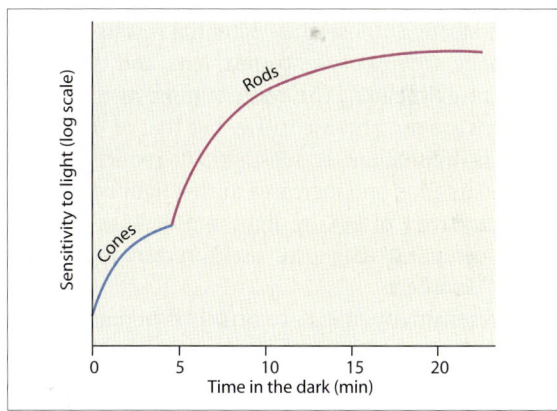

Fig. 115.**6**   Dark adaptation.

continued presence of bright light. Ca^{2+} plays an important role in this recovery of conal sensitivity to bright light, as explained below.

When light falls on the cone, there is closure of Na^+ channels resulting in membrane hyperpolarization. The hyperpolarization causes closure of Ca^{2+} channels, resulting in a reduction in intracellular Ca^{2+}. Intracellular Ca^{2+} normally inhibits guanylyl cyclase—the enzyme that converts GTP into cGMP. A low intracellular Ca^{2+} therefore elevates intracellular cGMP levels. The cGMP keeps the Na^+ channels open, leading to the depolarization of the outer segment and the resumption of neurotransmitter output.

Dark adaptation   If a person spends a considerable length of time in brightly-lit surroundings and then moves to a dimly-lit environment, the retina slowly become more sensitive to light as the individual becomes accustomed to the dark. This rise in visual sensitivity is known as dark adaptation (Fig. 115.**6**). It is nearly maximal in about 20 minutes. The time required for dark adaptation is the time required to build up the photopigment stores, which are constantly depleted in bright light. The dark adaptation response has two components. The *initial rapid but small rise* in visual sensitivity is due to the dark adaptation of cones. A *late, slow but large rise* in visual sensitivity occurs due to the dark adaptation of rods. It occurs only in the peripheral portions of the retina and not in the fovea, which contains only cones.

Aircraft pilots taking off into the night sky need to become dark adapted immediately on take-off. Hence they wear red goggles when on the ground in bright light. Red light ($\lambda = 650$ nm) is a poor stimulator of rods which have peak spectral sensitivity at $\lambda = 500$ nm. However red light stimulates cones reasonably well. Therefore, pilots wearing red glasses can see in bright light and yet their rods remain dark-adapted.

Electroretinogram and electrooculogram

The **electroretinogram** (ERG) is the electrical potential generated by the retina in response to a brief flash of light. It is recordable outside the eye, usually at the cornea, with a contact lens electrode. The ERG is produced by the radially oriented processes of the photoreceptors, bipolar, and Müller cells. The processes of the ganglion cells have a relatively tangential orientation and do not contribute to the ERG waves. Hence, the ERG is normal in optic atrophy despite the loss of ganglion cells. The ERG is useful in the diagnosis of certain retinal disorders, particularly retinitis pigmentosa.

The typical ERG wave has three components (Fig. 115.**7**): The photoreceptors generate the a-wave of the ERG, which is a cornea-negative (downward) wave. The bipolar and Müller cells together produce the b-wave which is cornea-positive (upward). Processes of the inner plexiform layer generate low voltage oscillatory waves.

The **electrooculogram** (EOG) is the steady voltage recorded between electrodes placed near the eye. It is affected by eye movements. The source of the voltage is the corneoretinal potential, which renders the cornea up to 10 mV positive with respect to the back of the eye. For recording the EOG, electrodes are placed at the inner and outer canthi. When the eyes move to the right, the positive cornea becomes closer to one of the electrodes, which is accordingly more positive than the other electrode. The opposite happens when the eyes move to the left (Fig. 115.**8**).

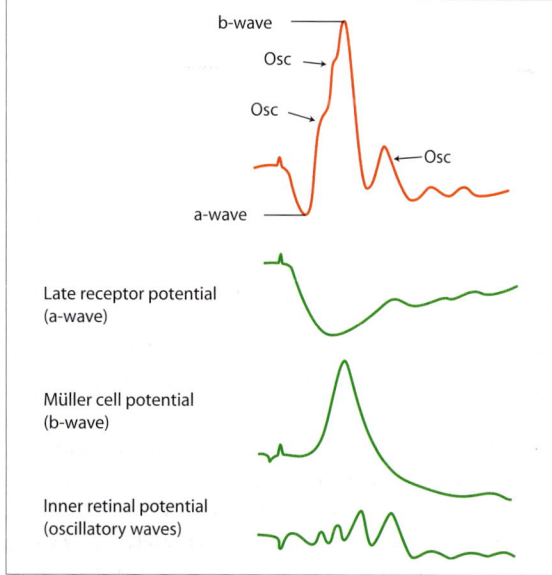

Fig. 115.**7**   The electroretinogram (red) and its component waves (green).

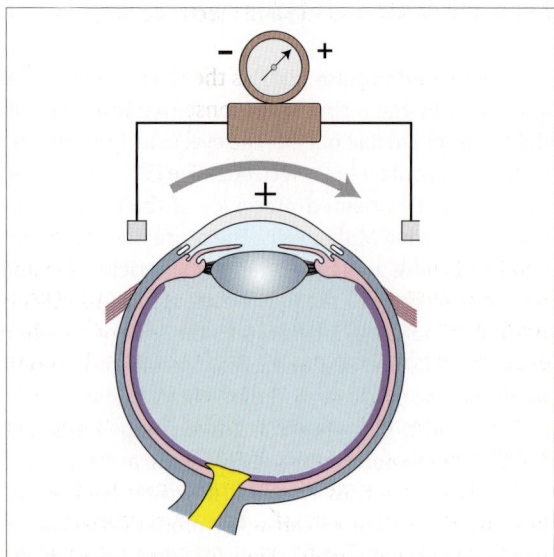

Fig. 115.**8**   The principle of electrooculogram. See also Figure 107.**8**.

The steady corneoretinal potential results from the metabolic activity of the cornea, lens, and the retinal pigment epithelium. The contributions of the cornea and lens are not photosensitive, but that of the retinal pigment epithelium is substantially reduced during dark adaptation and increases during light adaptation. The sensitivity of EOG to light and dark adaptations have been put to diagnostic use in certain uncommon retinal disorders.

The sensitivity of EOG to ocular movements is utilized in polysomnography for identifying REM sleep and for recording nystagmus. It is also used for identifying the blink artifacts in the EEG (see Fig. 107.**8**). Whenever the eye blinks, the eyes roll upward (Bell's phenomenon) and therefore the EEG potentials, especially in the frontal leads, become more positive. This positive drift of the EEG potential caused by the blink is called the blink artifact. The simultaneous recording of the EOG helps in identifying the blink artifact in the EEG.

Visual Pathways

The axons of the retinal ganglion cells run in four different pathways. Ninety percent of the optic fibers reach the striate cortex via the lateral geniculate body, forming the geniculostriate pathway. The other pathways are the colliculopulvinar pathway, the accessory optic pathway, and the retinohypothalamic pathway.

Geniculostriate pathway

The visual field is divided into temporal and nasal halves by a vertical line passing through the fixation point. Accordingly, the retina too is divided into nasal and temporal halves by a vertical line passing through the fovea centralis. The axons of retinal ganglion cells relay in the lateral geniculate nucleus (LGN) of the thalamus. Fibers from the nasal half of the retina decussate in the optic chiasma before relaying in the contralateral LGN. Fibers from the temporal half of the retina relay in the ipsilateral LGN without decussating.

The LGN has a lamellar structure comprising six layers numbered ventrodorsally. Each layer receives input from one eye only. The fibers from the temporal half of the retina relay to the ipsilateral LGN in layers 2, 3, and 5 while those from the nasal half relay to the contralateral LGN in layers 1, 4, and 6.

Efferents from the LGN fan out as the optic radiation to terminate in the ipsilateral primary visual cortex,

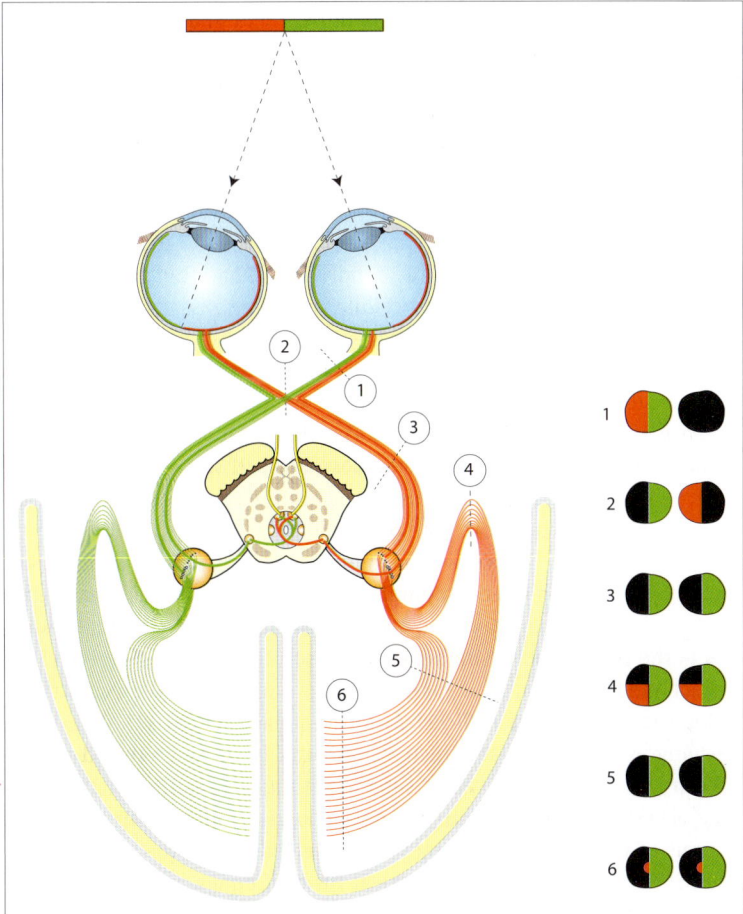

Fig. 116.**1**   Lesions of the geniculostriate pathway. Note that the image formed on the left half of the retina represents the right half of the visual field, and vice versa. 1. Lesion of the optic nerve causes ipsilateral blindness. 2. Midsagittal lesion of the optic chiasma causes bitemporal hemianopia. 3. Lesion of the optic tracts causes contralateral homonymous hemianopia. 4. Lesion of the temporal lobe affects the fibers of Meyer's loop coming from the lower half of the retina and causes contralateral homonymous upper quadrantanopia. 5. Lesion of the optic radiation causes contralateral homonymous hemianopia. 6. Lesion of the visual cortex causes contralateral homonymous hemianopia with macular sparing.

which is also known as area 17, striate cortex or visual area 1 (V1). Fibers from the inferior half of the retina take a detour of the temporal lobe as Meyer's loop before terminating in the cortex (Fig. 116.**1**).

Lesions of the geniculostriate pathway All lesions of the geniculostriate pathway, other than those in the optic nerve, tend to cause hemianopia, i.e., blindness in only one half of the visual field. The term homonymous is used when the hemianopia affects only the left halves or right halves of the visual field, while the term heteronymous is used when the hemianopia affects only the temporal halves (bitemporal) or nasal halves (binasal) of vision. The visual defects associated with lesions at different sites in the geniculostriate pathway are shown in Figure 116.**1**. The fovea projects to a relatively large part of the striate cortex, and lesions of the striate cortex are rarely large enough to abolish macular (foveal) vision completely. Hence, hemianopia with macular sparing is characteristically seen in lesions of the striate cortex.

Colliculopulvinar pathway

The second strongest retinal projection is to the superior colliculus of the midbrain. The superior colliculus is made of superficial and deep layers of cells. The retinal fibers terminate in the superficial layer where they have a retinotopic representation. From here, fibers project to the pulvinar of the thalamus, which in turn projects to the extrastriate cortex (Fig. 116.**2**). The extrastriate visual cortex projects back to the cells of the deep layer of the superior colliculus which are important for saccadic movements of the eye and its coordination with the grasping reflex. When the geniculocalcarine pathway is damaged, this colliculopulvinoextrastriate pathway is responsible for the persistence of some very limited stimulus detection and eye movement toward objects in the visual field. This residual vision is called *blindsight*.

Accessory optic pathway

Fibers of the accessory optic pathway relay in the pretectal nuclei, viz., the nucleus of the optic tract and the terminal nucleus. Efferents from these nuclei reach the contralateral vestibulocerebellum after relaying in the ipsilateral medial vestibular and inferior olivary nuclei (Fig. 116.**2**). Thus, the visual signals in the accessory optic pathway interact with vestibular signals at two sites: in the medial vestibular nucleus and in the cerebellum. These sites are important for the synergistic actions of the optokinetic reflex and the vestibuloocular reflex. The relative effects of the two reflexes can be readjusted by the cerebellum, resulting in a form of adaptive motor learning.

Retinohypothalamic pathway

A very small number of axons terminate immediately above the optic chiasma in the suprachiasmatic nucleus of the hypothalamus. This retinohypothalamic pathway provides information about light–dark cycles to the suprachiasmatic nucleus which generates circadian rhythms (1-day-long cyclic rhythms) in sleep and wakefulness, eating and drinking, physical activity, and melatonin and adrenocortical secretions (see Fig. 85.**5**).

Parallel Processing of Image

Starting from the ganglion cells, all the cells in the geniculostriate pathway up to the striate cortex and probably beyond, are of two types: the large (M) cells with large dendritic field and thick axon, and the small (P) cells with small dendritic field and thin axon. M cells constitute the magnocellular pathway and the P cells constitute the parvocellular pathway. The magnocellular and parvocellular pathways are involved in parallel processing of the image, i.e., analysis of different aspects of the image. Cells of the colliculopulvinar pathway or the accessory optic pathway do not conform to either M or P cells: They have small to medium-sized cells with large dendritic field and thin axons.

The magnocellular pathway extends to the middle temporal area (posterior parietal cortex) as the *dorsal cortical pathway* while the parvocellular pathway extends to the inferior temporal cortex as the *ventral cortical pathway*. The two pathways meet at several sites in the geniculostriate pathway. For example, the simple and complex cells in V1 (see below) receive input from both pathways.

M cells have higher sensitivity to light contrast and higher temporal resolution, while P cells have higher sensitivity to color contrast and higher spatial resolution. A higher spatial resolution results in higher visual acuity. A higher temporal resolution increases the of *critical flicker-fusion* frequency, which is the maximum rate at which stimuli can be presented and still be perceived as separate stimuli. In motion pictures, the perception of motion occurs only when individual picture frames are projected at a frequency exceeding the normal critical fusion frequency, which is about 16 per second.

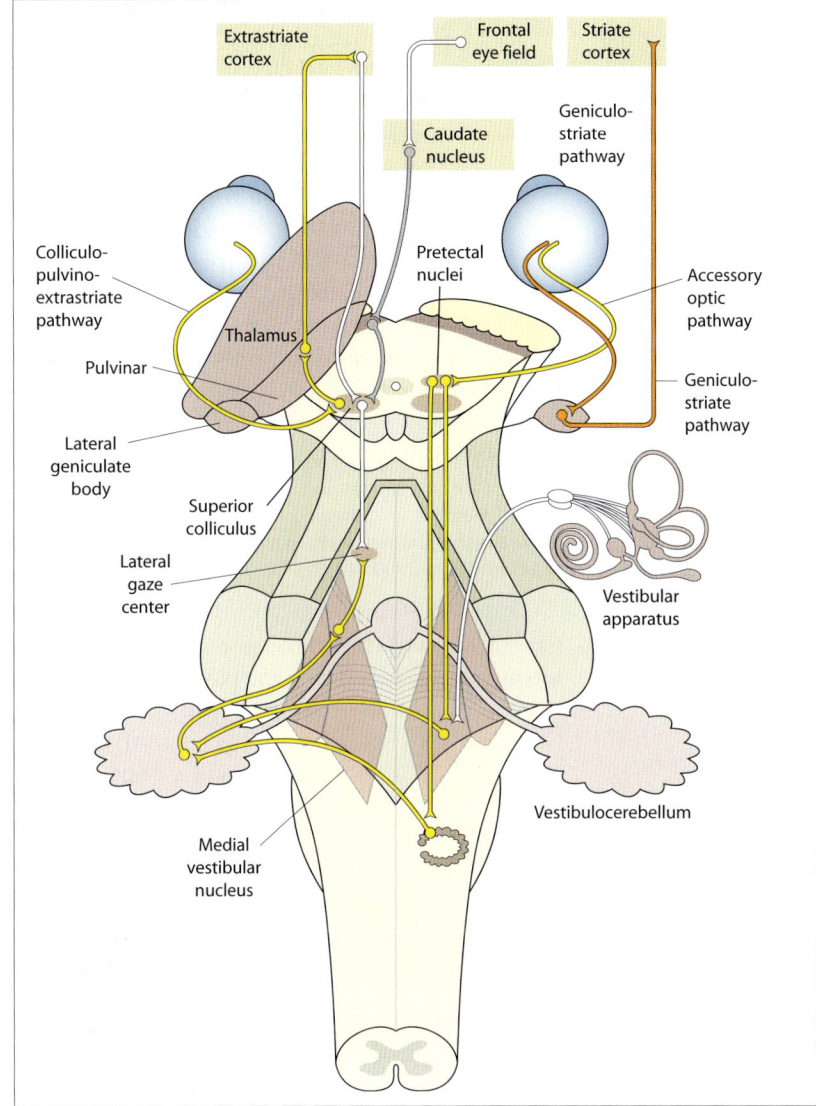

Fig. 116.**2**   Colliculopulvinar and accessory optic pathways. Note that the vestibular afferents meet the accessory optic pathway in the vestibular nucleus. Also shown are the geniculostriate pathway and the oculomotor circuit of the basal ganglia.

Serial Processing of Image

Successive cells in the geniculostriate pathway are involved in increasingly complex analysis of the image. This is called sequential or serial processing.

Photoreceptors and bipolar cells

The photoreceptors break up the image into small spots of light or darkness, quite like a digital camera that breaks down a picture into small pixels. Photoreceptors reduce their neurotransmitter (glutamate) output in response to light. As already mentioned above, there are two types of bipolar cells: those that are depolarized (excited) by the glutamate, and those that are

hyperpolarized (inhibited) by glutamate. Those bipolar cells that are inhibited by glutamate are disinhibited (i.e., excited) when the glutamate output from the photoreceptor decreases.

Ganglion cells and LGN cells

The image processing in ganglion cells and LGN cells results in the sharpening of the image contrast through lateral inhibition (see Fig. 103.**13**). The image is thus analyzed mostly in terms of the contours of the light–darkness boundaries, with areas of uniform light or uniform darkness eliciting very little neural response. Thus image processing in ganglion cells and LGN cells yields a high-contrast line diagram of the original image

(Fig. 116.**3**). The advantage in such a conversion is that relatively fewer neurons are required for transmitting the distinctive features of the image to the brain. The accentuation of the image contrast is made possible by the center-surround organization of the receptive fields of the ganglion cells and LGN cells.

Center-surround organization of receptive fields The bipolar cells as well as the ganglion cells have concentric-antagonistic receptor fields. Depending on the stimulation mode of their receptor fields, the ganglion cells are called on-center or off-center cells. An *on-center cell* is stimulated when light falls on the center of its receptor field and inhibited when light falls on the periphery of the field. Conversely, an *off-center cell* is stimulated when light falls on the periphery of the field and inhibited by light on its center. Similar concentric-antagonistic receptor fields are present in the LGN cells too.

The mechanism of an on-center response is as follows: The bipolar cells of the on-center receptor field are inhibited by glutamate. Hence, they are disinhibited

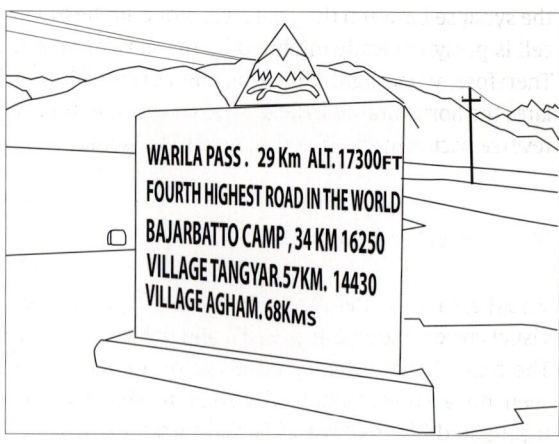

Fig. 116.**3** This black-and-white sketch has fewer details and shades than photograph in Figure 55.**2**. Yet, both are nearly equally effective in conveying the scenery. This is because in the brain, the photograph is converted to a line sketch anyway (through lateral inhibition).

(stimulated) when the glutamate output of photoreceptors decreases, as occurs in the presence of light. Also,

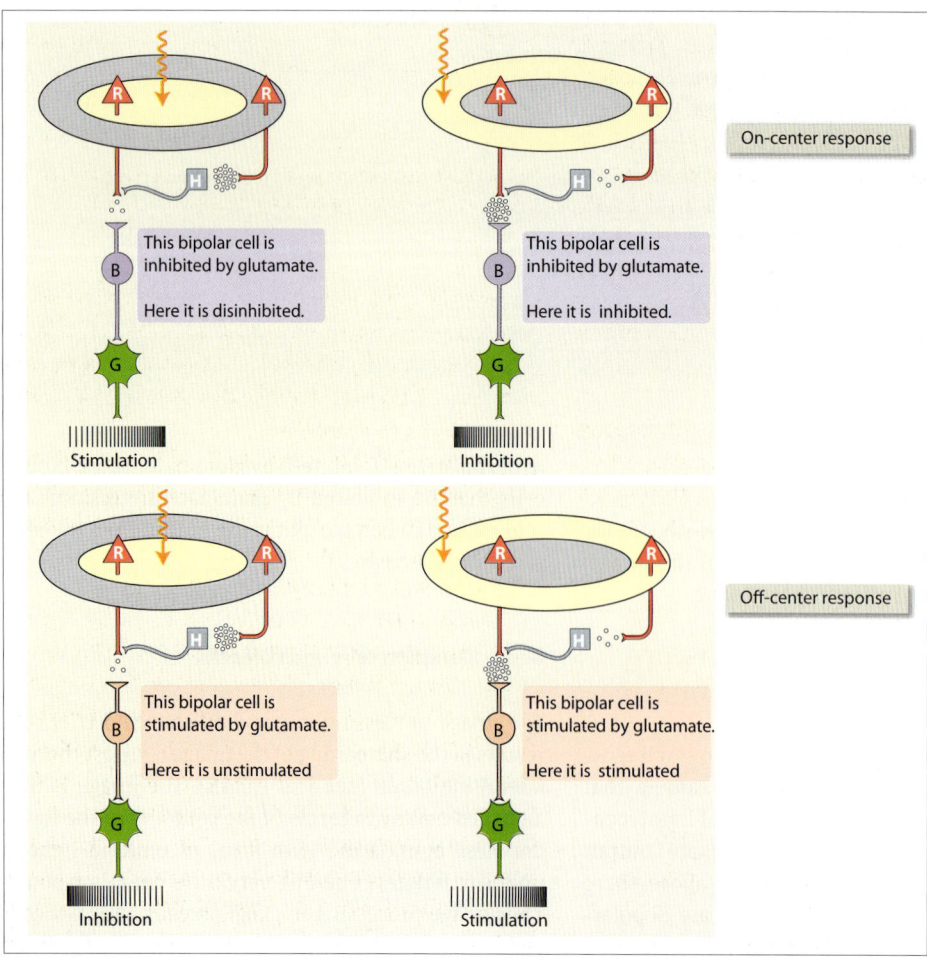

Fig. 116.**4** The on-center and off-center responses of the ganglion cell.

the synapse between the photoreceptor and the bipolar cell is presynaptically inhibited by the horizontal cells. Therefore, when light falling on the periphery stimulates the horizontal cell, the bipolar cell is inhibited. The reverse occurs in an off-center field (Fig. 116.**4**).

■ Visual area 1 (V1)

Visual area 1 (V1) is variously known as the primary visual cortex, Brodmann area 17, and the striate cortex. The name 'striate' comes from the prominent stripes seen on a cross section of cortex produced by the myelinated fibers of the LGN that terminate there. V1 is situated along the lips and walls of the posterior part of the calcarine sulcus. The peripheral part of the retina is represented in the anterior part of area 17. The upper quadrants project to the upper wall of the calcarine sulcus and the lower quadrants, to the lower wall of the sulcus. The macular part of the retina projects mainly to the posterior part of area 17 and anteriorly, to a thin strip along the calcarine sulcus. The macular area occupies nearly one-third of area 17 (Fig. 116.**5**).

The striate cortex is made of six layers of cells numbered 1 to 6 (Fig. 116.**6**). Layer 4 is further subdivided into four sublayers, viz., 4A, 4B, 4Cα, and 4Cβ. Afferents from the LGN enter layer 4. The magnocellular afferents from the LGN enter layer 4Cα. The parvocellular afferents enter layer 4Cβ from where they project to the blobs and interblobs (see below). Layers above 4C project to higher visual cortices like V2, V4, and V5 and, through the corpus callosum, to the contralateral visual areas. Layers below 4C project to subcortical areas: Thus layer 5 projects to the superior colliculus, pons, and pulvinar, and layer 6 projects back to the LGN and to the claustrum.

Simple and complex cells Most striate cells respond optimally only to stimuli that have linear properties, such as a line or bar, and generally do not respond to small spots of light. These cells belong to two major

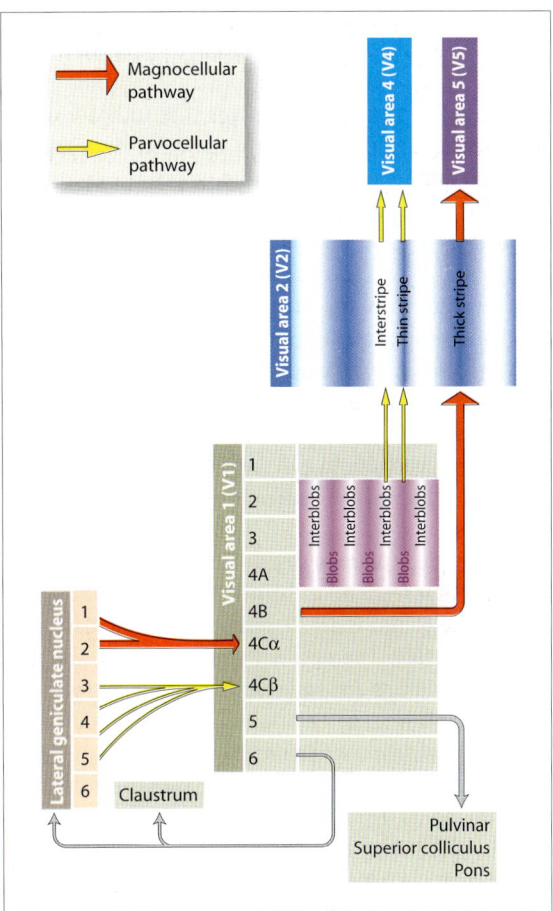

Fig. 116.**6** Magnocellular and parvocellular pathways from the lateral geniculate ganglion to the visual cortices.

groups, the simple and the complex cells. The simple and complex cells receive input from both the M and P pathways, both pathways contributing to the *primal sketch*, i.e., the initial two-dimensional approximation of the shape of a stimulus.

Simple cells have a receptor field that is oblong in shape. They are best stimulated by lines and edges of light that have the same orientation as the long axis of their oblong receptive field. Simple cells also have excitatory and inhibitory zones in their field.

Complex cells have larger receptive fields. They respond to linear stimuli with specific orientation but the precise position of the stimulus within the receptive field is less crucial because there are no clearly defined 'on' or 'off' zones. Thus, movement of light across the receptive field is a particularly effective stimulus for certain complex cells. *M-type complex cells* respond better to moving edges than stationary edges of light. Some of them are directionally sensitive. These cells are stimulated when an edge of light moves across their receptor field in one direction but not in the

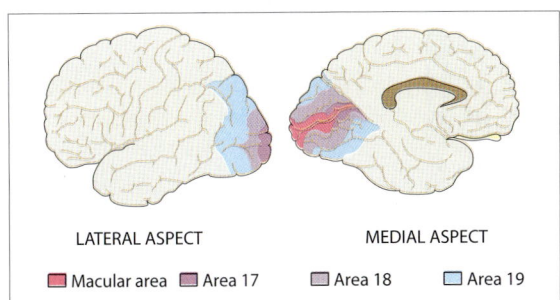

LATERAL ASPECT   MEDIAL ASPECT

■ Macular area ■ Area 17   ☐ Area 18   ☐ Area 19

Fig. 116.**5** Visual area 1 (V1) medial and lateral surfaces.

reverse direction. *P-type complex cells* respond not only to lines of specific orientation but also to lines of specific length. This is made possible by their ability to respond to right-angled edges that indicate where a line ends. Given their ability to detect both the orientation and length of lines and edges, the complex P cells are able to analyze the detailed form and texture of an object.

Hypercolumns in the striate cortex   Each receptor field of a retinal ganglion cell is represented in the striate cortex by a hypercolumn, which occupies an area of 1 mm² and processes multiple aspects of the image, viz., orientation, binocular interaction, color, and motion. The hypercolumn has three components: the orientation columns, the ocular dominance columns, and the blobs.

Orientation columns (Fig. 116.**7**) are columns of simple cells that respond to stimuli with a specific orientation. The columns are disposed perpendicular to the cortical surface. The optimal stimulus orientations of adjacent columns are progressively tilted, and within a width of 1 mm, all orientations are represented.

Ocular dominance columns (Fig. 116.**7**) are important for stereopsis (binocular vision). Cells of these columns receive inputs from corresponding points (see Fig. 116.**10**) of both retinas. The inputs from the two eyes in most cases are not equal, and the effect from one of the eyes usually dominates. Columns of cells that are dominated by the same eye form the ocular dominance columns. Dominance alternates between output from the left and right eye at intervals of about 0.5 mm.

Columns of blobs are apparent in V1 when it is stained with cytochrome oxidase. Blobs and the intervening interblobs are sites of color processing and are prominent in layers 2, 3, and 4A of the striate cortex. The blobs contain color-opponent neurons (see below).

■   **Visual area 2 (V2)**

Visual area 2 (V2) is also called the secondary visual cortex. It occupies much of area 18 but is not coextensive with it. Cells in V2 are sensitive to the orientation, color, and binocular disparity of the stimulus. V2 carries out more complex analysis of contours than V1 and can respond even to illusory contours, like the Kanizsa triangle (see Fig. 103.**16**).

Staining with cytochrome oxidase gives a striped appearance to the V2 cortex. The stripes are of two types: the thick stripes and the thin stripes. The interstripe regions are sometimes called pale stripes. The thick stripes receive their input from layer 4B of V1. The thin stripes receive their input from the blobs in V1 while the pale stripes (interstripes) receive their input from the interblobs in V1. The thick stripes project to V5. The thin and pale stripes project to V4 and onward, into the inferior temporal cortex.

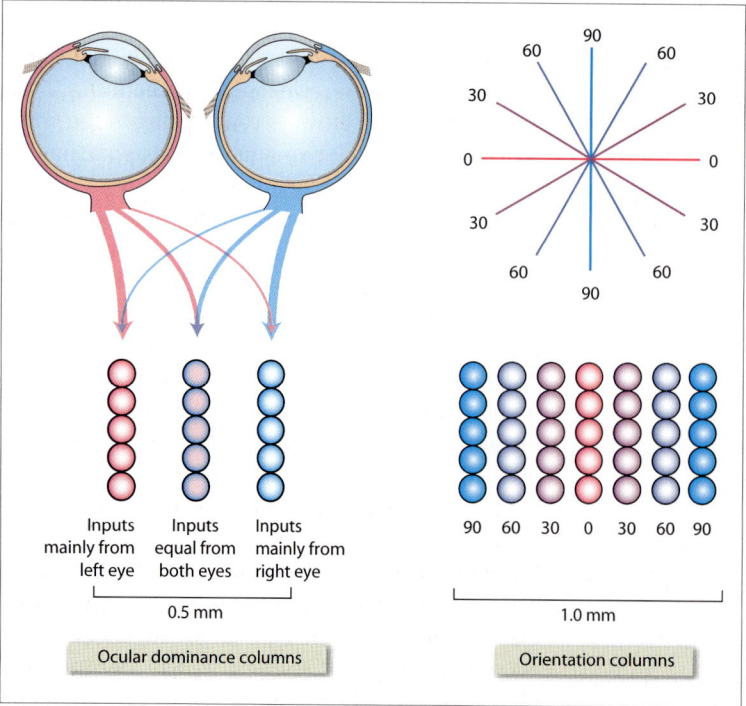

Fig. 116.**7**   (*Left*) Ocular dominance columns. (*Right*) Orientation columns.

Visual area 4 (V4)

Visual area 4 (V4) lies within area 19 anterior to visual area 3[1] (V3) and may include parts of area 20. Its major input is from the thin and pale stripes of V2 but not from the thick stripes. V4 cells are responsive to both color and form. V4 cells are also responsible for color constancy, i.e., the ability of the visual system to identify different colors under different illumination conditions. V4 is a key relay station for the ventral (parvocellular) pathway to the inferior temporal cortex.

Visual area 5 (V5): middle temporal (MT) area

The cortical area that is dedicated to detection of motion across the visual field is designated as V5. In monkeys, it is located toward the posterior end of the superior temporal sulcus and has been called the middle temporal (MT) area. In the human brain, the area corresponds to the junction of the parietal, temporal, and occipital cortices, lying in Brodmann area 19, but no separate name has been assigned to it. Since most of the experimental work is done in monkeys, the name MT area is widely used.

The MT area receives a strong input from the directionally sensitive M-type complex cells in V1. MT cells show a more complex directional sensitivity than the V1 cells. The latter respond only to moving lines and edges and are therefore called *component direction-selective*. MT cells respond to moving objects and forms and are therefore called *pattern direction-sensitive*. The receptor field of the MT cell is ten times larger than that of the V1 M cell. This enables the MT cell to detect entire objects that move across its large receptor field.

A few MT cells are sensitive to form and color. They help in detecting motion of colored spots and shapes. A cortical area adjacent to MT, called the *medial superior temporal area* (MST), contains neurons that respond to visual motion. These neurons may process a type of global motion in the visual field called *optic flow*, which is important for detecting a person's own movements through the environment.

Inferior temporal cortex (Area 20)

The inferior temporal cortex is located in inferior temporal gyrus, the temporal pole and the lateral and medial occipitotemporal gyri. The most prominent visual input to the inferior temporal cortex is from V4. The

cells of this area have a very large receptor field, sometimes including the entire visual field. Importantly, the fovea is always represented in every receptor field. Such large fields enable *position invariance*, i.e., the ability to recognize the same object regardless of its location in the visual field.

Some inferotemporal cells respond only to specific types of complex stimuli, such as the hand or face. For cells that respond to a hand, the individual fingers are especially critical since these cells do not respond when there are no spaces separating the fingers. However, all orientations of the hand elicit similar responses. Among neurons selective for faces, the frontal view of the face is the most effective stimulus for some, while for others it is the side view. Moreover, whereas some neurons respond preferentially to faces, others respond preferentially to specific facial expressions. Some cells are activated even by two dots and a line appropriately positioned, e.g., ☺. Some cells respond to facial dimensions (distance between the eyes) while others respond to the familiarity of the face. Cells responding to similar facial features are probably organized in columns. Although the proportion of cells in the inferior temporal cortex responsive to hands or faces is small, lesions of this region lead to *prosopagnosia*, i.e., deficits in face recognition.

Color Vision

The three primary colors, viz., red, blue, and green, when added in appropriate proportions can create all other colors (Fig. 116.**8**). Hence they are also called *additive colors*. This can be demonstrated by projecting overlapping primary colors on a white screen. The area where all the three colors overlap looks white. The screen appears black when no light falls on it. Lights of different colors have different wavelengths.

Colored pigments do not follow the same rules as colored lights and are called *subtractive colors*. A red pigment looks red because it absorbs (subtracts) all other colors from the white light that falls on it. Similarly a blue pigment reflects only blue light and absorbs all other colors. When red and blue pigments are smeared together, they subtract all colors from white light and therefore look black. Cyan, magenta, and yellow are three subtractive colors which when mixed in appropriate proportions produce pigment colors of all hues.

Trichromatic coding   The simplest explanation of the existence of primary colors is the trichromatic coding theory of Young and Helmholtz. There are three types of cones, namely, the L cone for red (λ = 650 nm) color, M cone for green (λ = 530 nm) color, and S cone for blue

---

[1] V3 is a narrow strip adjoining the anterior margin of V2, still within Brodmann area 18. Its functions are not well defined.

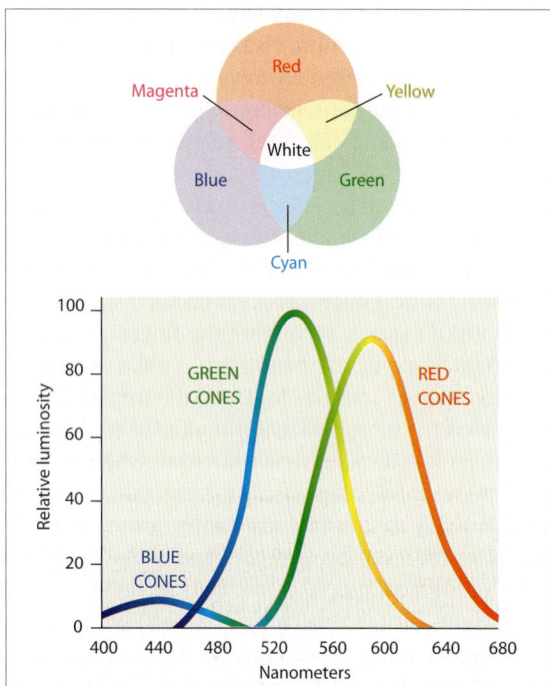

Fig. 116.**8**   (*Above*) The primary colors red, blue, and green produce magenta, cyan, and yellow when added in pairs. Addition of all the three primary colors produces white light. (*Below*) Wavelength sensitivity of retinal cones according to the Young–Helmholtz theory.

(λ = 440 nm) color. The letters L, M, and S denote the long, middle, and short wavelength of light absorbed by the cones. All colors are coded at the receptor by differential stimulation of these three types of cones. This is known as trichromatic coding.

Opponent color coding   Simple trichromatic color coding occurs only at the receptor level. Beyond receptors, there is opponent color coding. There are three distinct opponent mechanisms. One captures red-green variation in the image: It is excited by red light and inhibited by green light. Another captures blue-green variation in the image: It is excited by blue light and inhibited by yellow light. A third opponent mechanism captures the achromatic (light-dark) variations in the image: It is excited by light and inhibited by dark. All the three opponent mechanisms are formed by summation or subtraction of signals from the three types of cones. Summation occurs when the photoreceptors facilitate each other. Subtraction of signals occurs when one photoreceptor inhibits the other. The achromatic opponent mechanism occurs by the summation of signals from all the three cones, i.e., S + M + L. The red-green opponent mechanism occurs by the subtraction of signals of the M and L cones, i.e., L – M. The yellow-blue opponent mechanism occurs by the subtraction

of S cone signals from the summated signals of M and L cones, i.e., S – (M + L). All signals from cones, color signals as well as the achromatic signals about image brightness, are transmitted through parvocellular pathways. The P cells are very densely distributed, especially in and around the fovea.

The activation of the opponent mechanism requires the stimulation of several cones in the receptor field and therefore the mechanism is not activated when light falls on a small point on the retina and stimulates a single cone. The cones themselves respond to a wide range of wavelengths and, on their own, signal the brightness of the light rather than its color. Hence, the P cell detects the brightness of the finest dots in the image and detects shades of color in larger fragments of the image. In the striate cortex, color-opponent neurons are found in the blobs. Further processing of color occurs in V4. In these neurons, maximum response occurs to those hues for which the human languages have separate words, i.e., the seven colors of the rainbow.

■ Color blindness

Color blindness is common in men (8%) and uncommon in women (0.4%). It is inherited as an X-linked recessive character. Though hereditary color blindness is much more common, defects of color vision can also be acquired as a result of diabetes mellitus or disease of the retina, optic nerve or visual cortex. The detection of color blindness is important in selecting candidates for jobs in which it is necessary to distinguish colored markings or colored light signals. A convenient quick test is the

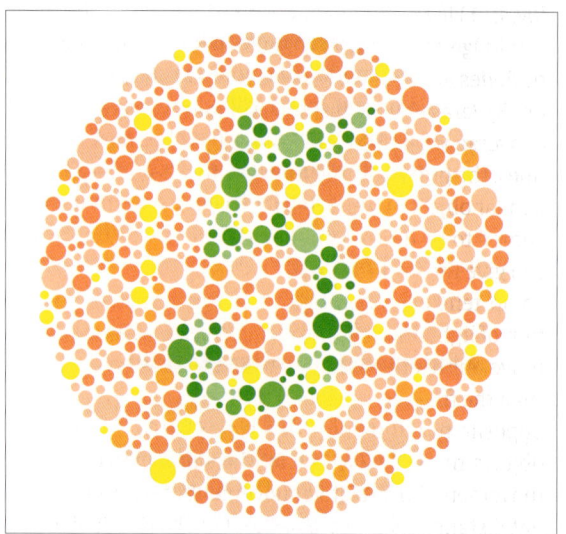

Fig. 116.**9**   The Ishihara chart. A color blind person may not be able to see the green figure of five hidden in the red and yellow dots.

Ishihara's chart (Fig. 116.**9**). Color blind individuals may be monochromats, dichromats or trichromats.

Monochromats are unable to distinguish any color. There are two kinds of monochromats. *Rod monochromats* lack normally functioning cones. Like dark-adapted subjects, they see very poorly in bright light. *Cone monochromats* lack functioning rods. Like light-adapted subjects, they have poor vision in dim light. Monochromats, especially cone monochromats, are rare.

Dichromats, who can see two colors, are of three kinds. The *protanopes* (red-blind) and *deuteranopes* (green-blind) are often grouped together as 'red-green' blind. They confuse red and green objects, though they can usually distinguish yellow objects. *Tritanopes*, sometimes called 'blue-blind', are rare. They have little ability to distinguish blue from green.

Anomalous trichromats can see all the colors but still have defective color vision. The cause of anomalous trichromacy is unknown, though they might have abnormal pigments.

Depth Processing

Monoocular mechanisms

At distances greater than about 30 meters the retinal images seen by each eye are almost identical, so that looking at a distance, we are practically one-eyed. Nevertheless we can perceive depth with one eye by relying on a variety of monocular depth cues, which are as follows: (1) If we are familiar about the size of a person, we can judge the person's distance. (2) If one person partly occludes another, we assume the person in front is closer. (3) Parallel lines, like those of railway track, appear to converge with distance. The greater the convergence of lines, the greater is the impression of distance. (4) If two similar objects appear different in size, the smaller is assumed to be more distant. (5) Patterns of light and dark can give the impression of depth, e.g., brighter shades of colors tend to be seen as nearer. In painting this distribution of light and shadow is called chiaroscuro. (6) As we move our head or body from side to side, objects nearer than the fixated plane move quickly and in the direction opposite to our own movement, whereas more distant objects move slowly and in the same direction as our movement. This relative movement of objects at different distances is called parallax. (7) The depth of accommodation required for focusing an object gives an idea of its distance from the eye.

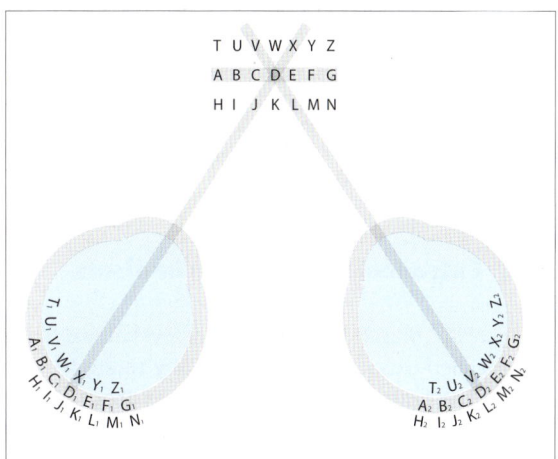

Fig. 116.**10** Corresponding points and depth perception. The corresponding points on the retina are A_1 and A_2, B_1 and B_2, C_1 and C_2, till G_1 and G_2. When identical images form on corresponding points, they are perceived as single. When the images on corresponding points are nonidentical, they are interpreted as two different objects. When the eyes are fixed on the letters ABCDEFG (the middle plane), the images of all the letters in that plane fall on corresponding points. The letters in the rear and front planes do not cast their images on corresponding points and therefore appear doubled, providing depth cues.

Binocular mechanisms (Stereopsis)

When the eyes fixate an object, its image is formed on the retina of each eye. The two retinal points on which the images of a fixated point-object are formed are called corresponding points. Conversely, when two identical images are formed on the corresponding points, the brain interprets them as originating from the same object. However, not all images formed on corresponding points are identical. This occurs with objects that are located either behind or in front of the fixation plane. The displacement of the images from corresponding points, called *binocular disparity*, is proportionate to the distance of the object from the fixation plane (Fig. 116.**10**). Cells sensitive to binocular disparity are found in V1, V2, and V3. These cells are organized into ocular dominance columns. MT cells are even able to judge whether an object is nearer or farther away from the plane of fixation. There are specific cells that respond to objects in front of the fixation plane (near cells) or behind the plane (far cells).

The importance of stereopsis in depth perception can be easily appreciated through a simple experiment. Hold both your arms straight out in front, keeping them wide apart. Now, bring a finger of each hand together so that their tips meet. First, try to do it with one eye closed and then, with both eyes open.

Extraocular Motor Mechanisms

The skeletal muscles attached to the eyeball external-ly are called extraocular muscles. They rotate the eye around the center of rotation, which lies about 12 mm behind the center of the cornea. Three types of rotation are possible around this center of rotation (Fig 117.**1**): *Horizontal rotation* around the vertical axis can be ei-ther abduction (temporal rotation) or adduction (nasal rotation). *Vertical rotation* around the horizontal axis can be either elevation (upward rotation) or depres-sion (downward rotation). Cyclic rotation around the anteroposterior axis can be either intorsion (inward rotation) or extorsion (outward rotation).

Each pair of muscles that help in directing the eye toward any of the six cardinal directions of gaze are known as yoke muscles. Each pair of yoke muscles are an agonist–antagonist pair that are reciprocally innervated.

Versions and vergences

A visual target is normally foveated with both eyes and therefore requires synergistic binocular movements. Such movements may be conjugate (version) or dis-juate (vergence). In conjugate movements the eyes move in the same direction. In disjugate movements the eyes move in opposite directions.

Conjugate movements (versions) may be horizontal or vertical. The *lateral gaze center* is present in the para-median pontine reticular formation. It produces only lateral versions. Stimulation of the lateral gaze center produces conjugate version to the same side and there-fore pontine lesions result in the conjugate deviation of the eye to the opposite side. The *vertical gaze center* is present in the mesencephalic reticular formation, or more precisely, in the rostral interstitial nucleus of the medial longitudinal fasciculus. It produces vertical ver-sions. Pure vertical versions require bilateral activity of the gaze centers. Oblique versions are produced by combining the activities of the lateral and vertical gaze centers (Fig. 117.**2**).

Disjugate movements (vergences) make take the form of convergence and divergence. They are observed when the eyes are fixed on a target that moves nearer to or farther away from the eye and are mediated by the near reflex (see below).

Fixation of visual target

Eye movements are meant for 'foveating' or 'fixating' a visual target, i.e., ensuring that a steady image of the visual target is formed on the fovea of the retina. The image tends to slip away from the fovea whenever there is a relative movement between the visual target

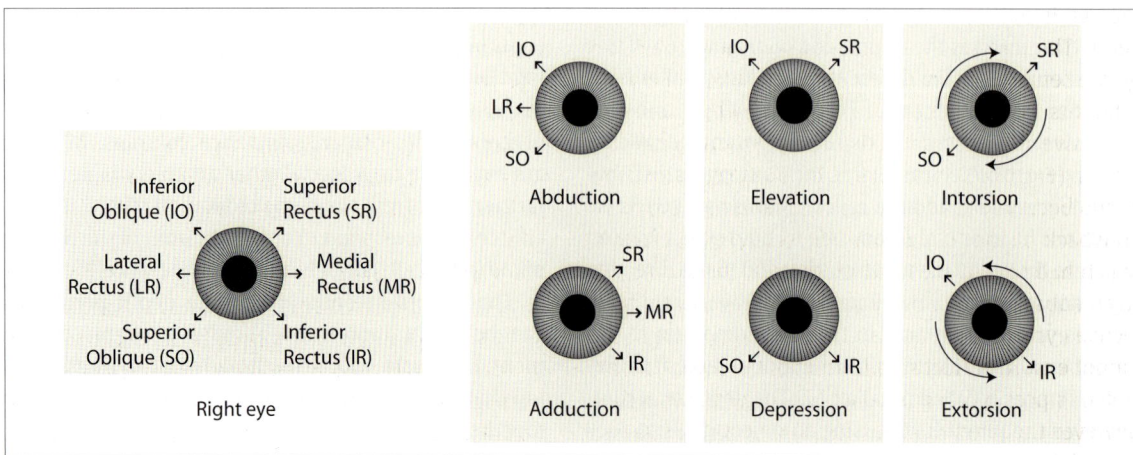

Fig. 117.**1**   Monocular movements.

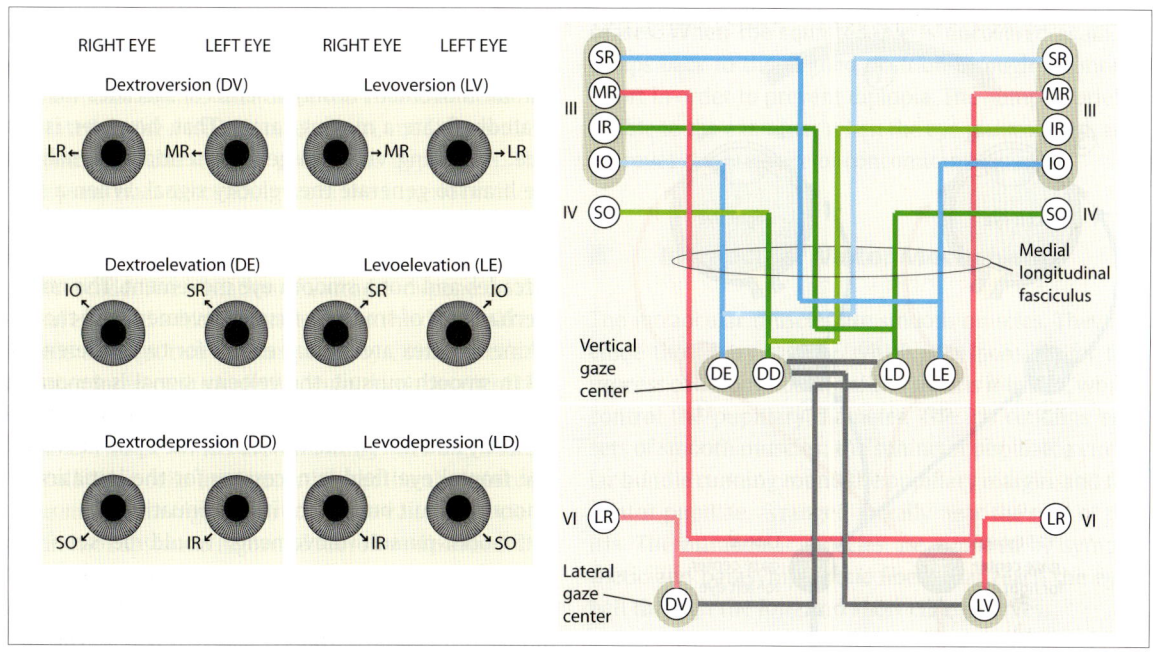

Fig. 117.**2**   Versions, elevations, and depressions. (*Left*) Conjugated movements produced by pairs of yoke muscles. (*Right*) The lateral and vertical gaze centers and their connections with the nuclei of the yoke muscles.

and the observer. This slippage of the foveated image is prevented by different types of reflex eye movements: The vestibuloocular and optokinetic reflexes keep the eye stationary in space when the head is moving. The smooth pursuit movement keeps a moving visual target foveated. When the visual target approaches or recedes from the eye, it is kept foveated by vergence movements. A different kind of movement, called saccadic movement, occurs when the eye fixation shifts from one object to another. Saccadic movements are quick and brief, and employ a neural mechanism that is quite different from the other reflex eye movements.

The **vestibuloocular reflex** (VOR) is triggered when the eye is fixed on a target and the head moves suddenly. The direction of the head movement is sensed by the semicircular canals of the vestibular apparatus. Afferents from the semicircular canal reflexly activate corrective movements of the eye in the opposite direction (Fig. 117.**3**). Were it not for the VOR, our vision would become blurred during a jerky ride in a car. The drawback of the VOR is that the semicircular canals, which had sensed the head movement, do not receive any feedback information as to whether or not the corrective eye movements are adequate. Hence, the reflex cannot ensure perfect stabilization of the foveal image, which is possible only through the optokinetic reflex. However the latency of the optokinetic reflex (80 ms) is much longer than that of the VOR (12 ms) and therefore during rapid jerks, the stabilization of the foveal

image is maintained almost entirely by the VOR. Also, the VOR can operate even in the dark because it is not dependent on vision. Finally, VOR is only meant for stabilization of the foveated image. It is suppressed by the vestibulocerebellum during shifting of gaze, be it a saccade or a smooth pursuit.

The VOR can be demonstrated easily in a subject seated on a rotating stool. When the stool rotates fast, the subject is unable to fix the eyes on any particular target. Nonetheless, if the subject rotates clockwise, the eyes will rotate anticlockwise due to the VOR. When the eyes reach about 60° version, they jump back to the opposite end in a saccadic motion resulting in oscillatory eye movements called *nystagmus*. The nystagmus thus has two components: a slow vestibular component and a fast saccadic component. A post-rotatory nystagmus occurs for a few seconds after rotation ceases. It occurs as the endolymph in the semicircular canals continues to flow due to inertia even after rotation has stopped. The VOR forms the basis of the vestibular function tests (p 769).

The **optokinetic reflex** is triggered by the slippage of the foveated image when the head moves. The velocity of the slippage is calculated by the cells of the nucleus of the optic tract (see accessory optic pathway, p 742) in the pretectum. The information is passed on to the medial vestibular nucleus which cannot distinguish between the velocity signals conveyed from the retina and those coming from the vestibular apparatus. The

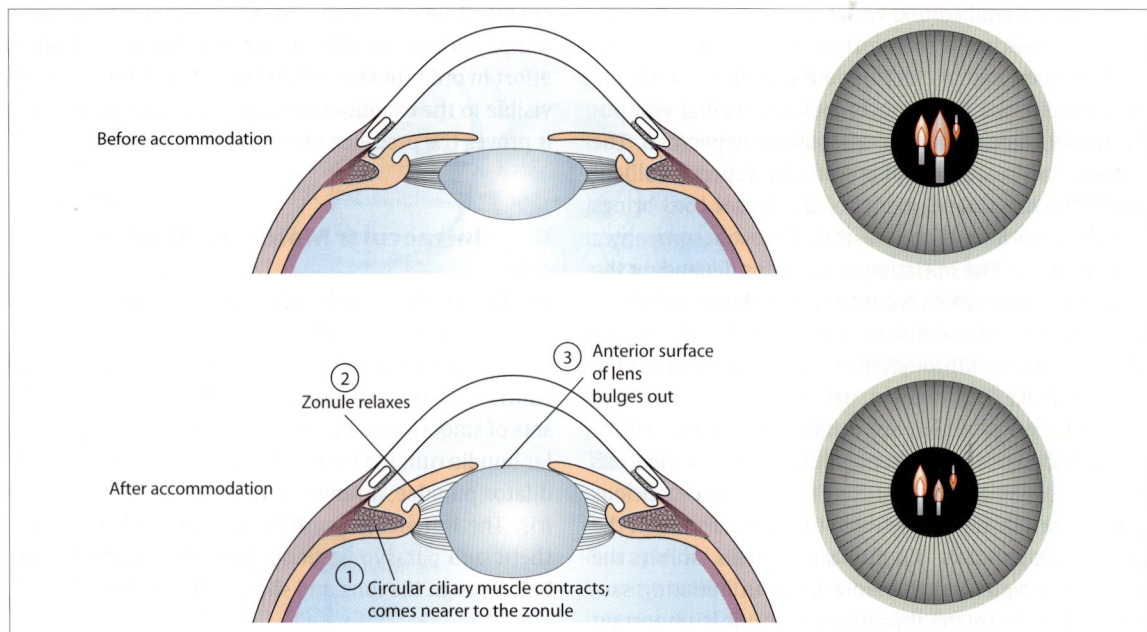

Fig. 117.**4**   Accommodation reflex and the Purkinje–Sanson images.

two images do not show any change with accommodation. These images, which demonstrate that it is the anterior surface of the lens that changes its curvature during accommodation, are called the Purkinje–Sanson images.

Pupillary reflexes

Light reflex   When light is shone in one eye, there is simultaneous constriction of the ipsilateral pupil (direct light reflex) and the contralateral pupil (consensual light reflex). The afferent pathway of the reflex is through the optic nerve and the efferent pathway is through the oculomotor nerve (Fig. 117.**5**).

The afferents from the retinal ganglion cells run in the optic nerve and partially decussate to enter the optic tract. The fibers pass through the lateral geniculate body uninterrupted and enter the superior colliculus to relay in the pretectal nucleus. Fibers originating in the pretectal nucleus partly decussate in the midbrain and enter the Edinger–Westphal (EW) nucleus on both sides. The EW nucleus is one of the several cell groups in the third nerve nucleus. It gives rise to parasympathetic fibers to the eye (see Table 96.**3**). Parasympathetic fibers originating in the EW nucleus leave the midbrain through the third cranial nerve and relay in the ciliary ganglion. Postganglionic parasympathetic fibers leave the ciliary ganglion through short ciliary nerves to innervate the pupillary sphincter.

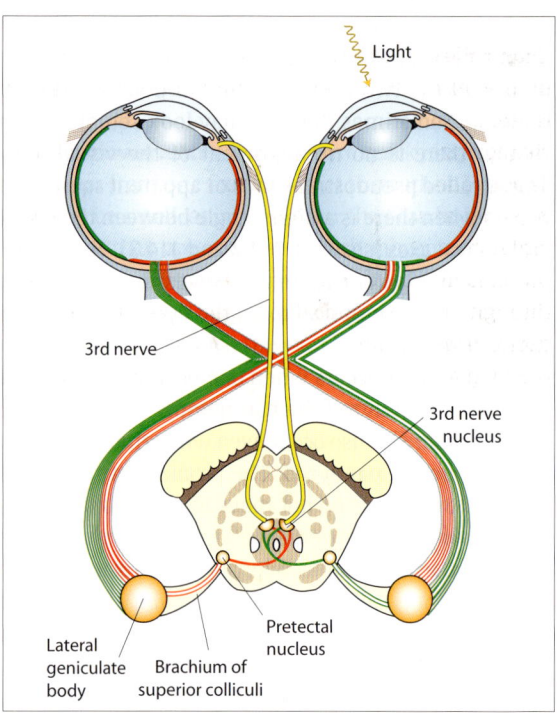

Fig. 117.**5**   Light reflex.

The *consensual light reflex* occurs because the optic fibers from each eye reach the pretectal nuclei of both sides due to incomplete decussation at the chiasma. Also, the pretectal nucleus of each side innervates the EW nuclei bilaterally.

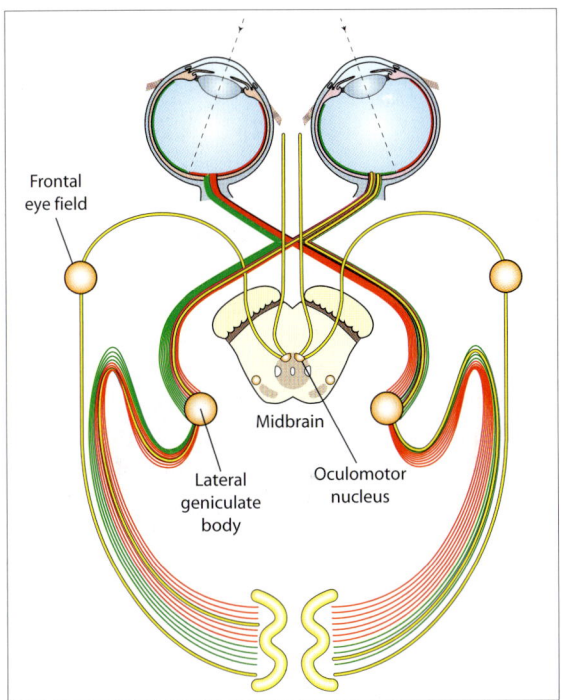

Fig. 117.**6**   Near reflex.

Near reflex   The near reflex (or the accommodation reflex) occurs when an object, on which the eye is fixed and focused, is moved nearer to the eye. It is characterized by the triad of convergence, accommodation, and pupillary constriction. The reflex is triggered by any of the several monocular and binocular visual cues of depth perception that suggest that the object is moving closer to the eye. The afferents travel through the geniculostriate pathway to the visual cortex from where they reach the frontal eye field through the long association fibers. Projection fibers from the frontal eye field descend through the internal capsule and terminate on the oculomotor nuclear complex where they stimulate the nucleus of medial rectus and the EW nucleus (Fig. 117.**6**).

Psychosensory reflex   If, while the eye is maximally constricted, the examiner observes the pupillary size when there is a sudden stimulus like pain or a loud noise, the pupils dilate due to stimulation of the sympathetic fibers and inhibition of the EW nucleus.

Abnormal pupillary reflexes

Argyll–Robertson pupil (ARP)   A lesion of the pretectal nucleus, which is almost always due to tertiary syphilis, abolishes the pupillary light reflex. However, the visual pathway to the cortex is unaffected. Also not affected is the pathway for the accommodation reflex. Thus, there is pupillary constriction in response to the accommodation reflex but not in response to the light reflex (ARP = Accommodation Reflex Present).

Marcus–Gunn pupil   If there is a lesion in the optic nerve on one side, it diminishes the direct pupillary reflex on the same side and the consensual reflex on the opposite side. This is called the Marcus–Gunn pupil and it results in an interesting phenomenon when light is shone alternately in both eyes every 2 to 3 seconds (the swinging-flashlight test). It is observed that when light is shone into the eye with optic nerve lesion, the direct pupillary reflex is not seen; instead, the pupil dilates. The dilatation actually signifies the cessation of the consensual light reflex triggered from the opposite eye a few seconds before.

118 Auditory Mechanisms

▰ The Ear

The ear consists of external, middle, and internal parts. The *outer ear* is made of the pinna and the external acoustic meatus, which is 2.5 cm in length. The *middle ear* or tympanic cavity contains three auditory ossicles, viz., the malleus (hammer), incus (anvil), and stapes (stirrup). The *internal ear* comprises the cochlea (concerned with hearing) and the vestibular apparatus (concerned with equilibrium). Except the pinna, all other parts of the ear are located inside the petrous part of the temporal bone. The external auditory meatus and the middle ear are separated by the tympanic membrane which is about 1 cm in diameter. The middle ear communicates with the internal ear through the round and the oval windows (Fig. 118.**1**).

Sound waves are deflected by the pinna into the external auditory canal which channels the waves to the tympanic membrane. The vibrations of the tympanic membrane are transmitted by the malleus and incus to the stapes, the base of which is attached to the circumference of the oval (vestibular) window. When the stapes vibrates, it moves rhythmically in and out of the oval window, setting up pressure waves in the cochlear fluid. The waves travel to the secondary tympanic membrane covering the round window.

The sound waves in air can set up pressure waves in the cochlear fluid as described above only if there is proper impedance matching. The acoustic impedance of water is much higher than that of air and without impedance matching, most of the sound reaching the cochlea is reflected back instead of being transmitted into the cochlear fluid. Impedance matching is brought about in two ways. (1) Since the force exerted by sound on the tympanic membrane is concentrated on a much smaller area of the stapes footplate, it results in pressure gain. (2) The mechanical advantage of the ossicle chain is higher than 1, which produces further pressure gain. The pressure gains ensure that more than half the sound energy striking the tympanic membrane is transmitted to the cochlea. If the function of the middle ear is impaired, the threshold of hearing is elevated by up to 20 dB.

The **auditory (Eustachian) tube** connects the middle ear to the nasopharynx. It usually remains collapsed so that debris and infectious agents cannot pass from the oral cavity to the middle ear. Children suffer from middle ear infections more than adults because the direction of the auditory tube is nearly horizontal in children. When open, the auditory tube equalizes the air pressure in the middle ear with the atmospheric pressure. The auditory tube opens up during swallowing, chewing, yawning, and sneezing. When driving up a mountain, the tympanic membrane tends to bulge out due to the low atmospheric pressure. Swallowing relieves it with a popping sensation in the ears as it opens up the auditory canal, allowing air to flow out from the higher pressure in the middle ear to the lower pressure in the nasopharynx.

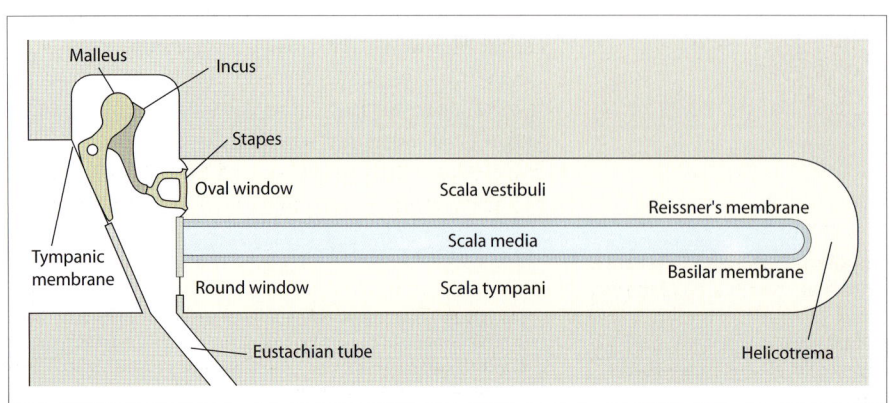

Fig. 118.**1** Schematic anatomy of the middle and inner ear.

Acoustic reflex   The two muscles present inside the middle ear are the tensor tympani and the stapedius. These muscles are attached to the malleus and stapes respectively. Contraction of the *tensor tympani* pulls the malleus medially and decreases the vibrations of the tympanic membrane by making it taut. Contraction of the *stapedius* pulls the footplate of the stapes out of the oval window so that the sound waves are not conducted to the internal ear.

Loud sounds initiate reflex contraction of these muscles, dampening the movements of the ossicular chain. The reflex is called the tympanic or acoustic reflex and is useful in preventing ear damage due to continuous loud sound. However, a sudden explosion, unless it is anticipated, will still damage the middle and inner ears because the reflex does not occur quickly enough.

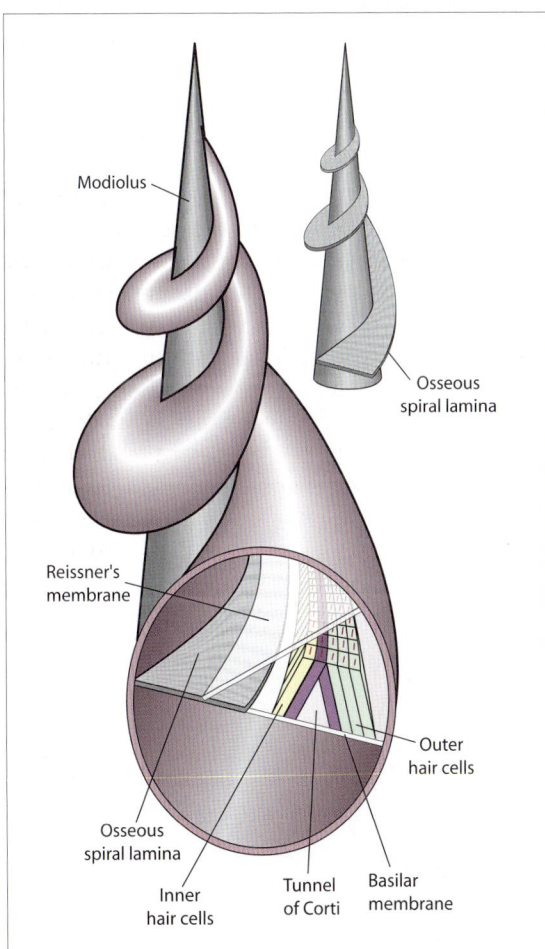

Fig. 118.**2**   Schematic anatomy of the cochlea and the tunnel of Corti. The cochlear tube is shown as it winds around the modiolus. The osseous spiral lamina, the basilar membrane and the organ of Corti are partly visible through its open proximal end. The entire osseous spiral lamina is also shown separately, as it winds around the modiolus.

■ Inner ear

Cochlea   The membranous cochlea is about 35 mm long in humans and takes approximately $2^2/_3$ turns from base to apex. It is compartmented by the vestibular (Reissner's) and basilar membranes into three spiraling tunnels: the scala vestibuli, scala media (or the cochlear duct), and scala tympani (Figs. 118.**1** to 118.**3**). The cochlear duct ends blindly. It does not extend till the apex of the cochlea and therefore leaves a small gap between its tip and the cochlear wall. The scala media and the scala tympani communicate through this gap, which is called the *helicotrema*. The inner edge of the basilar membrane is attached internally to the spiral lamina that arises from the modiolus, the bony core around which the cochlea is wound. Externally, the basilar membrane is anchored to the wall of the cochlea by the spiral ligament. Contained within the spiral ligament is a vascular structure called the *stria vascularis* which secretes the endolymph into the scala media.

Organ of Corti   The sensory hair cells are located on the basilar membrane, surrounded by supporting cells with a small amount of perilymph intervening between them. The apex of the hair cells bears stereocilia which are bathed in endolymph. Near the apex, the hair cells have tight junctions with the supporting cells. These junctions prevent the mixing of the perilymph that bathes the base of the hair cells with the endolymph into which their stereocilia project (Fig. 118.**8**).

The hair cells are arranged in rows. There is a single row of inner hair cells and three rows of outer hair cells. Between the rows of inner and outer hair cells are the inverted Y-shaped rods of Corti which enclose the tunnel of Corti. The stereocilia of the hair cells are embedded within a gelatinous tectorial membrane that overhangs the hair cells.[1] The entire assembly of the basilar membrane, the tectorial membrane, and the hair cells with their sensory afferents is called the spiral organ or the organ of Corti (Fig. 118.**3**).

The scala media is filled with the endolymph secreted by the stria vascularis. The composition of the endolymph resembles intracellular fluid (high K^+ and low Na^+ concentration). The fluid in the scala vestibuli and scala tympani has a composition similar to that of extracellular fluid (high Na^+ and low K^+) and is called the perilymph. The perilymph of the scala vestibuli

---

[1]   The stereocilia of the outer hair cells are definitely embedded in the tectorial membrane while the stereocilia of the inner hair cells seem only to touch the membrane. The difference does not seem to have any physiological significance.

and scala tympani is continuous at the apex of the co-chlea through the helicotrema. The tunnel of Corti is filled with perilymph because the basilar membrane is relatively permeable to perilymph in the scala tym-pani. The bases of the hair cells are also bathed in peri-lymph but the apices of the hair cells are bathed in en-dolymph. The scala media is electrically positive (+80 mV) relative to the scala vestibuli and scala tympani. This positive endolymphatic potential occurs because the amount of K^+ secreted by the stria vascularis ex-ceeds the amount of Na^+ it reabsorbs.

Ninety-five percent of the neurons in the cochlear nerve are afferents originating in the inner hair cells. The cell bodies of these neurons are present in the spi-ral ganglion, which is located inside the modiolus. The remaining five percent of the neurons are efferent fi-bers that each innervate several of the more numerous outer hair cells.

Auditory Pathway

The cochlear nerve terminates tonotopically, i.e., in an orderly sequence representing the various frequen-cies in the cochlear nuclei in the medulla. The fibers of the cochlear nuclei pass to the superior olivary nuclei which receive inputs from both ears. From each olivary nucleus, fibers ascend in the lateral lemniscus to proj-ect to the ipsilateral inferior colliculus and thence to the medial geniculate body of the thalamus. From the

medial geniculate body, auditory fibers project to the primary auditory cortex. Auditory fibers decussate at several levels in the brainstem. The trapezoid body is formed by decussating auditory fibers in the medulla (Fig. 118.**4**).

The primary auditory cortex lies principally in area 41 in the anterior transverse temporal (Heschl's) gyrus, situated on the upper surface of the superior temporal gyrus along the floor of the posterior ramus of the lat-eral sulcus (Figs. 95.**6** and 100.**5**). It also extends into the lateral border of the temporal lobe, over much of the insular cortex and even into the lateral portion of the parietal operculum. The primary auditory cor-tex projects to the auditory association area (area 42) formed by the posterior transverse temporal gyrus. The rest of the superior temporal gyrus behind areas 41 and 42 forms Wernicke's area (area 22) which is concerned with the interpretation of sounds and comprehension of spoken language.

Auditory Processing

Vibration of basilar membrane

Vibrations of the oval window produced by move-ments of the stapes set up a pressure wave within the scala vestibuli. This pressure wave ultimately finds its way membrane to the round window where it causes bulging of the secondary tympanic membrane, thereby

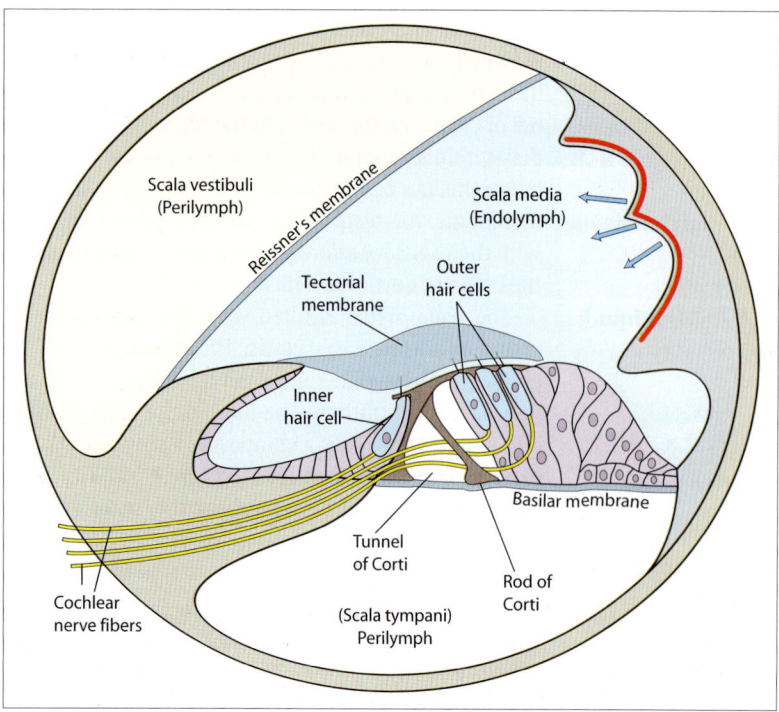

Fig. 118.**3**   The organ of Corti.

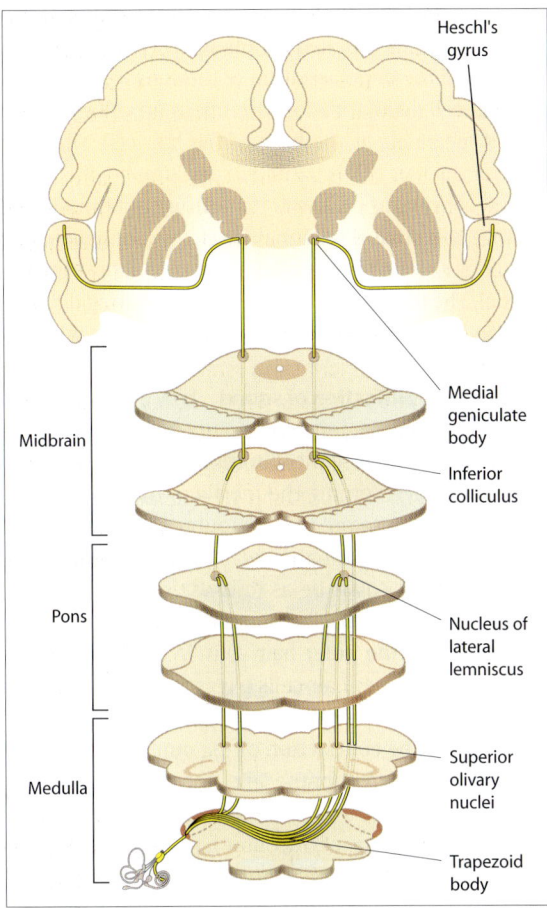

Fig. 118.**4**   Auditory pathway. Only the trapezoid decussation is shown. Decussations higher up are not shown for clarity.

membrane. This ripple has been called the *traveling wave*. The amplitude of the ripple increases till it reaches its maximum at a specific site on the membrane and then declines rapidly, somewhat like an ocean wave that gradually gains in height till it lashes on the shore. Its entire progression can be conveniently depicted by superimposing the waveforms at regular intervals into a single envelope (Fig. 118.**6**). The oscillatory characteristics of the basilar membrane are attributable to its structure: It is made of transverse fibers that are shorter at the base and longer at the apex, which make the base stiff and the apex floppy.

Auditory transduction

When the basilar membrane oscillates, a shearing force is created between the basilar membrane and the tectorial membrane. This causes the stereocilia of the hair cells to bend (Fig. 118.**7**), which leads to the production of a generator potential in the hair cells. The greater the displacement of the basilar membrane and the bending of the stereocilia, the greater is the generator potential produced, and the greater is the frequency of action potentials in the cochlear afferents issuing from the base of the hair cell. A higher frequency of action potentials is perceived as a louder sound.

Atop each hair cell, the stereocilia are arranged in rows of increasing length. The tips of adjacent stereocilia are connected to each other by fine filaments called

releasing the pressure into the middle ear. From the middle ear the pressure wave is released into the atmosphere through the Eustachian tube. The route taken by the pressure wave depends on the site where the basilar membrane oscillates in response to the wave, which in turn depends on the wave frequency. The basilar membrane does not oscillate in response to waves with frequency lower than 20 Hz and therefore these waves have to take a detour through the helicotrema to reach the round window. Waves with higher frequency make the basilar membrane oscillate and therefore are able to take a shorter route across the scala media. The highest audible frequency of 20,000 Hz makes the membrane oscillate maximally near its base while the lowest audible frequency of 20 Hz makes the membrane oscillate maximally near its apex (Fig. 118.**5**).

It might seem from the simplified account given above that the basilar membrane resonates at a specific site in response to a specific frequency. The pressure wave does not cause resonance of the basilar membrane: Rather, it sets the basilar membrane into a ripple, i.e., a transverse wave that travels along the

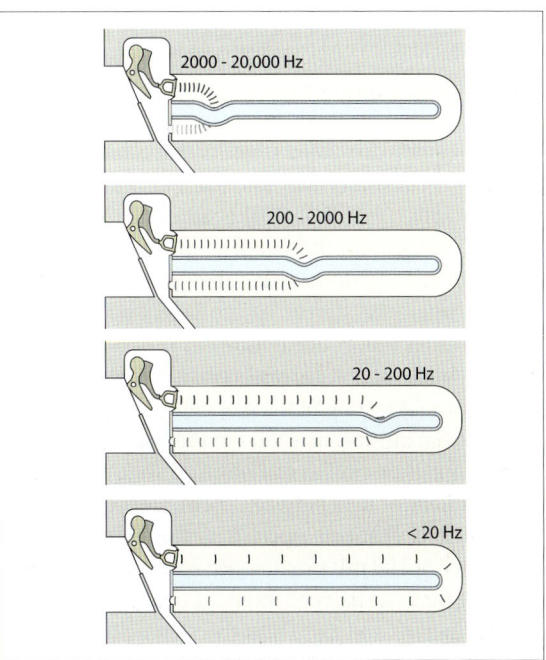

Fig. 118.**5**   The traveling wave on the basilar membrane.

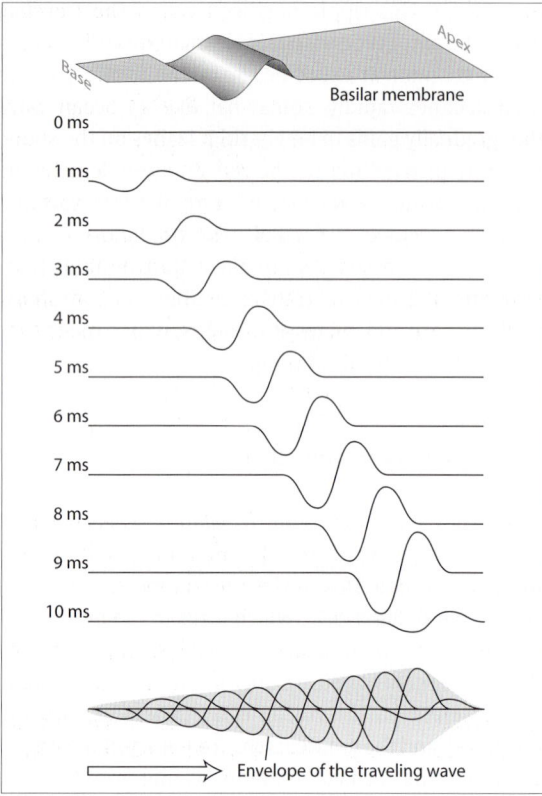

Fig. 118.**6** The traveling wave on the basilar membrane. As an example, in the first 9 milliseconds, the wave gradually gains in amplitude as it travels on the basilar membrane. In the tenth millisecond, the amplitude diminishes markedly. A convenient way of depicting all the waves across time is to represent them on the same baseline and enclose them in a common outline called the envelope.

allowing more Ca^{2+} to enter the cell. The influx of Ca^{2+} causes the release of neurotransmitter (probably gluta-mate) and the depolarization of the sensory afferents.

Hair cell depolarization also opens up voltage-sensitive K^+ channels at the base of the hair cell. Since the base of the hair cell is bathed in perilymph which has a low K^+ concentration, the K^+ ions diffuse out through the voltage-sensitive K^+ channels into the perilymph. The efflux of K^+ ions restores the hair cell to its resting potential. The Ca^{2+} that enters the hair cell during depolarization is pumped out to the exterior by a Ca^{2+} pump.

Cochlear amplification of sound Low intensity sound waves entering the inner ear do not have adequate energy in them to cause the vibration of the basilar membrane after overcoming the inertia of the cochlear fluids. It seems therefore that the outer hair cells amplify the sound that enters the cochlea, a possible mechanism of which could be as follows: The initial, feeble movement of the basilar membrane stimulates both the inner and the outer hair cells. While the depolarization of the inner hair cells triggers action potentials, the depolarization of the outer hair cells causes them to contract. The contraction of the outer hair cell pulls on the basilar membrane and thereby amplifies its movements.

The contractions of the outer hair cells are probably responsible for the evoked *otoacoustic emissions*. In response to a click sound, the ear produces a sound of

tip links. These links control the opening of mechano-sensitive cation channels in the stereocilia (Fig. 118.**8**). These channels are called *transduction channels* to distinguish them from other ion channels at the base of the hair cells. Deflection of the stereocilia toward the tallest stereocilium fully opens these transduction channels, causing depolarization of the hair cell. Conversely when the stereocilia are bent away from the tallest stereocilium, the tension in the links reduces and the transduction channels close completely, causing hyperpolarization of the hair cell.

Since the intracellular fluid of the hair cells and the endolymph have similar composition, there is no diffusion gradient of ions between them. Hence, when the transduction channels open, it is the large potential difference of 150 mV existing between the endolymph (+80 mV) and the hair cell interior (−70 mV) that drives the cations (K^+ and Ca^{2+}) into the stereocilia, depolarizing the cell. The depolarization causes voltage gated Ca^{2+} channels at the base of the hair cells to open,

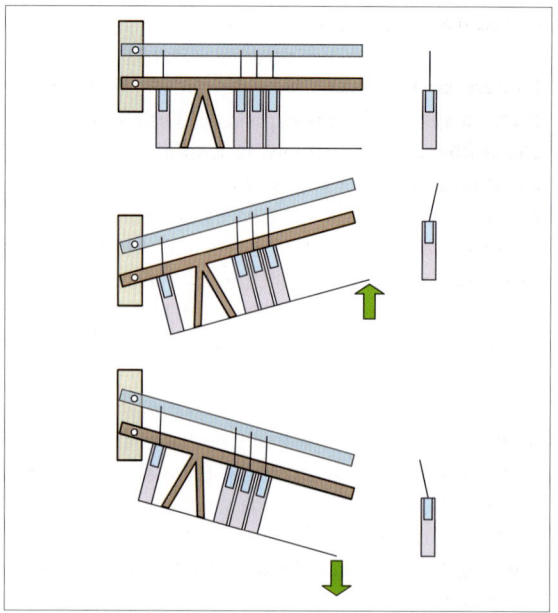

Fig. 118.**7** The vibration of the basilar membrane is associated with bending of the stereocilia due to the shearing movement between the hair cells and the tectorial membrane.

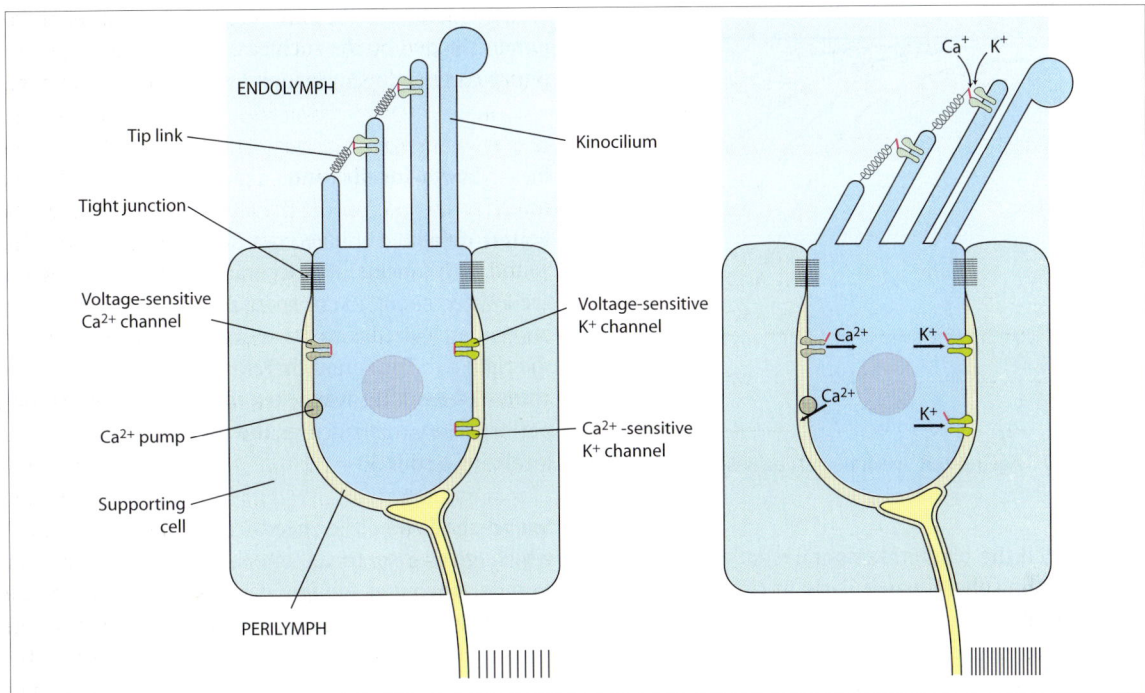

Fig. 118.8 The hair cells and auditory transduction. (*Left*) A hair cell in the resting state. (*Right*) The stereocilia are bent toward the kinocilium. The tip links pull open the mechanosensitive cation channels. The K^+ and Ca^{2+} entering the stereocilia leave through channels and pumps present in the cell body.

the same frequency after a latency of 5 to 20 ms. These emissions have been recorded in neonates as a screening test for the functional integrity of the cochlea. Otoacoustic emissions are also produced spontaneously and in neonates, they are sometimes loud enough to be audible directly.

Efferent cochlear control The sensitivity of the cochlea to specific frequencies is controlled by olivocochlear fibers which innervate the outer hair cells. It is a common experience that in a noisy radio channel, we are able to selectively hear the music or speech of our interest and ignore the rest (Fig. 118.**9**). To some extent, this is made possible by tuning our inner hair cells to selected frequencies. Such tuning is made possible by the olivocochlear fibers to the outer hair cells.

Masking The presence of one sound decreases an individual's ability to hear other sounds. This phenomenon is known as masking. It is believed to be due to the relative refractoriness of previously stimulated auditory receptors and nerve fibers to other stimuli. Masking is an important consideration in the tests for hearing (see below).

Cortical auditory processing shows the following characteristics: (1) The cortical neuronal response to acous-

Fig. 118.9 Two boys listening to the radio, each trying to concentrate on his favorite broadcast, as four different channels are mixed up. The situation signifies the role of efferent cochlear control.

tic stimuli is brief, seen mostly at the onset and at the termination of the stimulus. The phenomenon is analogous to the accentuation of contrast of a visual stimulus and helps in identifying any change in the sound quality. (2) There is tonal localization in the primary auditory cortex. Low tones are represented anterolaterally and high tones posteromedially in the auditory cortex. (3) A portion of posterior superior temporal

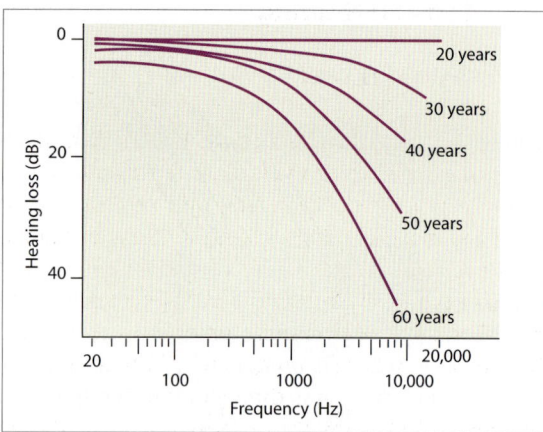

Fig. 118.**10** Age-related sensorineural deafness (presbycusis).

gyrus called the planum temporal is larger in the left cerebral hemisphere, particularly in right-handed individuals (see Fig. 100.**5**). This area is important for language-related auditory processing. The asymmetry is greater in musicians and others who have *perfect pitch*—a rare ability to tell the pitch of a sound without reference to a standard pitch.

Pitch and timbre discrimination

The brain infers the pitch of a sound from the place along the basilar membrane which responds with the maximum displacement, stimulating the hair cells located on it. There are different versions of this 'place theory', each offering a different explanation for the cause of the maximal vibration at a specific location on the basilar membrane. Thus, the resonance theory held resonance as the cause. The current theory is the traveling wave theory that posits that the sound sets up a traveling wave in the basilar membrane (Fig. 118.**6**). The wave travels from the base to the apex of the membrane. The basilar membrane is narrow and stiff at its base and becomes wider and more flexible toward the apex. This 'tunes' the membrane so that when an audible sound of a given frequency sets the basilar membrane into a wave motion, the peak amplitude of the wave occurs at a unique place on the membrane (near the base for high frequencies and near the apex for low frequencies). At this place, the hair cells are excited most intensely.

The timbre or quality of a sound depends on the various overtones that are present in addition to the fundamental frequency. The basilar membrane breaks down a sound into all its component frequencies, fundamental as well as overtones, with each frequency registering its maximum amplitude at a different place

on the membrane. The ensemble of the frequency components coded by the cochlea are identified in the auditory cortex as a single sound with a distinct timbre.

Sound localization

Sound delays Humans can localize the source of sound with considerable accuracy because the two ears are located about 15 cm apart and a sound originating on the left side has to travel an extra 15 cm to reach the right ear. The sound therefore reaches the right ear about 0.4 ms after it reaches the left. The medial superior olivary nucleus uses the time delay as a cue for localizing a sound.

Sound-shadows The head casts a sound-shadow which gives rise to an intensity difference for the sounds arriving at the two ears, especially for sounds with wavelength smaller than the dimensions of the head. The intensity differences are processed by the medial superior olivary nucleus. The sound-shadow cast by the pinnae provides a means of distinguishing sounds originating in front from the sounds originating behind the head. The convolutions of the pinnae of the ears provide information regarding the localization of sounds in the vertical plane.

Deafness

Deafness is expressed as the difference in sound pressure level (in dB) between the patient's threshold of hearing and that of a normal healthy subject. If the threshold intensity of a particular tone is 20 dB higher than that of normal subjects, it is described as 20 dB hearing loss.

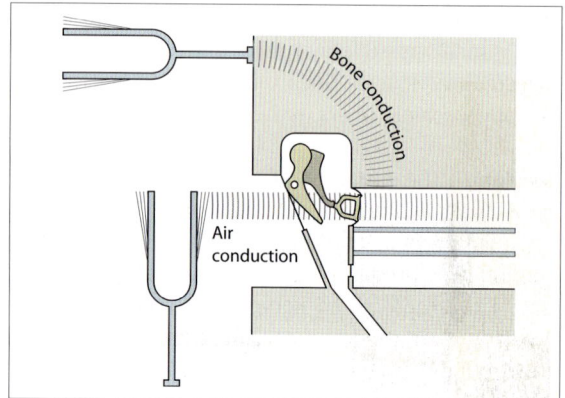

Fig. 118.**11** The paths of air-conducted and bone-conducted sound.

Conductive deafness results from a failure of the outer and middle ear to transmit sound efficiently to the inner ear. It can result from various causes, such as: (1) wax or foreign body in the external ear, (2) middle ear infection (otitis media), or (3) otosclerosis, in which the movement of the footplate of the stapes is impeded by the growth of bone around the oval window.

Sensorineural deafness results when some part of the cochlea or auditory nerve is damaged. (1) Traumatic damage to the cochlea can result from overamplified music. A special variant of traumatic damage is *boiler maker's disease* which occurs on prolonged exposure to continuous high intensity sounds of a particular frequency. Such sounds are typically produced in industrial processes. Only those hair cells that are tuned to the sound frequency are damaged. (2) Presbycusis is an age-related sensorineural deafness that specifically affects the high frequencies (Fig. 118.**10**). (3) Loss of hair cells can be caused by ototoxic drugs (e.g., streptomycin, neomycin, frusemide). (4) A very distressing but common cause of hearing loss is tinnitus, a continuous buzzing sound generated within the ear itself, that masks the natural sound reaching the ear.

Central deafness Because the auditory pathways are extensively crossed at all levels above the cochlear nuclei, unilateral damage to the auditory cortex does not result in deafness, although the affected person may have difficulty in localizing sounds. Extensive damage to the auditory cortex of the dominant hemisphere leads to difficulties in speech recognition. Extensive damage to the auditory cortex of the nondominant hemisphere affects recognition of timbre and the interpretation of temporal sequences of sound, which are important in music and speech.

Tests for Hearing

Tuning fork tests

Tuning fork tests provide a convenient method of determining whether deafness is of the conductive or sensorineural type. The tuning fork used should be large and made of stainless steel, so that the sound is not damped quickly. Its frequency should be 512 Hz. It should have an expanded base for application to the skull. The sound of a vibrating tuning fork can be heard in two ways: When its prongs are held in front of the ear, its sound must pass through air to the tympanic membrane and across the middle ear to reach the cochlea. Such conduction of sound is called *air conduction*. On the other hand, if the base of the vibrating tuning fork is pressed firmly against the skull, the sound is transmitted through the bones of the skull to the cochlea, bypassing the middle ear. Such conduction of sound is called *bone conduction* (Fig. 118.**11**).

Rinne's test The base of a vibrating tuning fork is placed on the mastoid process (Fig. 118.**12**). When the subject no longer hears it, it is held in air next to the ear on the same side. A normal subject is still able to hear the sound, indicating that his air conduction is better than bone conduction. Rinne's test in a normal person is said to be positive. A patient with conductive deafness has better bone conduction than air conduction and Rinne's test is said to be negative. In sensorineural deafness, air conduction remains better than bone conduction and therefore Rinne's test remains positive.

In severe unilateral sensorineural deafness, Rinne's test is false negative. Air conduction is absent but bone conduction may be good because the sound is

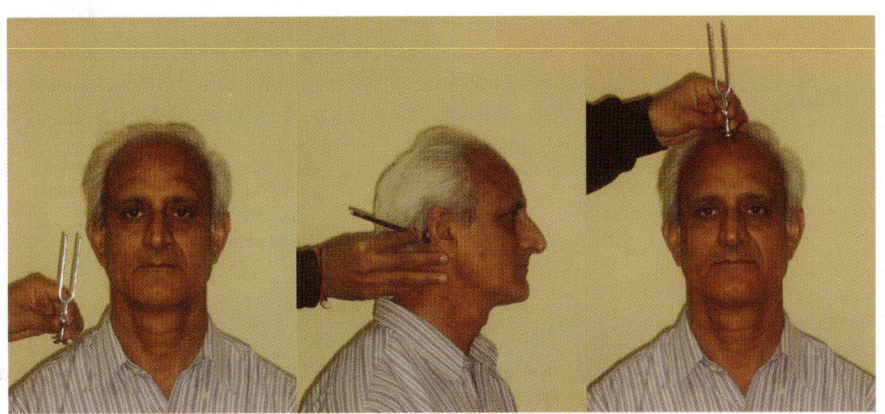

Fig. 118.**12**   The tuning fork tests for hearing.

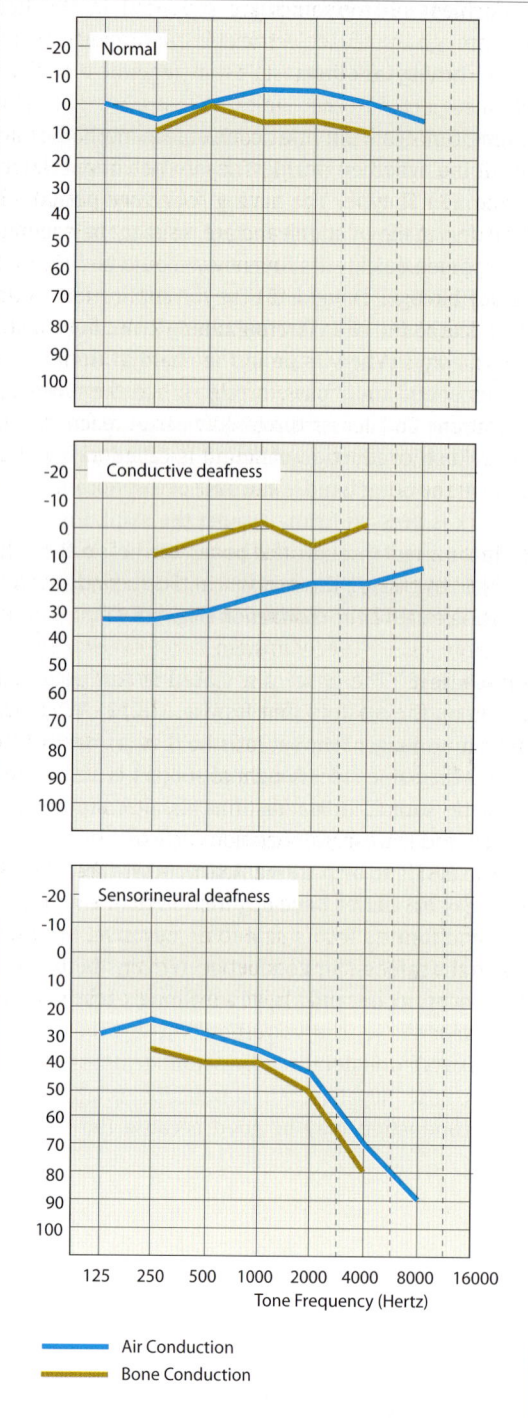

Fig. 118.**13** Audiogram in normal subjects and patients with conductive and sensorineural deafness. (*Above*) In the normal audiogram, the threshold for hearing through air conduction is nearly 0 dB and the threshold for bone conduction is slightly higher. (*Middle*) In conduction deafness, air conduction is impaired but bone conduction is normal. The threshold for air conduction is higher than bone conduction. (*Below*) In sensorineural deafness, air conduction and bone conduction are impaired with equal severity and therefore the threshold for bone conduction remains slightly higher than bone conduction.

transmitted to the opposite cochlea through the skull bones. This observation misleads the examiner into making a wrong diagnosis of conductive deafness. In this situation, Weber's test is important.

In **Weber's test**, the base of vibrating tuning fork is placed on the vertex of the skull, allowing the sound to be conducted through bone to both ears. A normal subject hears equally on both sides. If hearing is deficient in one ear, it indicates either a sensorineural deafness in that ear or a conductive deafness in the opposite ear. Bone-conducted sound is better heard in the ear with conductive deafness because of reduced background noise conducted through air. The sound is less clearly heard in the normal ear due to the masking effect of background noise.

In **Schwabach's test**, the examiner compares the bone conduction of the patient with his own, assuming that he himself has a normal hearing. The base of the vibrating tuning fork is placed first on the mastoid of the patient till he stops hearing the sound. The examiner then places the tuning fork on his own mastoid. If he still hears the sound, it indicates that the patient's bone conduction is impaired, presumably due to sensorimotor deafness. The reliability of the test is higher if the tragus of both subject and examiner are kept occluded during the test. This is called the *absolute bone conduction test* or the *modified Schwabach test*.

■ Audiometric tests

There are different types of audiometric tests like pure tone audiometry and speech audiometry. In pure tone audiometry, a pure tone audiometer delivers tones of variable frequency and intensity. Both air conduction and bone conduction are measured. For each frequency, a series of tone pips are delivered at increasing intensities (beginning from subthreshold) and the patient is instructed to signal every time he hears a sound. The frequencies tested are at octave steps, i.e., 125, 250, 500, 1000, 2000, 4000, and 8000 Hz. Each frequency is tested in the intensity range of 10–120 dB. The results are charted as an audiogram (Fig. 118.**13**).

■ Auditory evoked potentials

These are the objective tests of hearing in which the conscious response from the patient is not required. It includes electrocochleography and brainstem auditory evoked potentials.

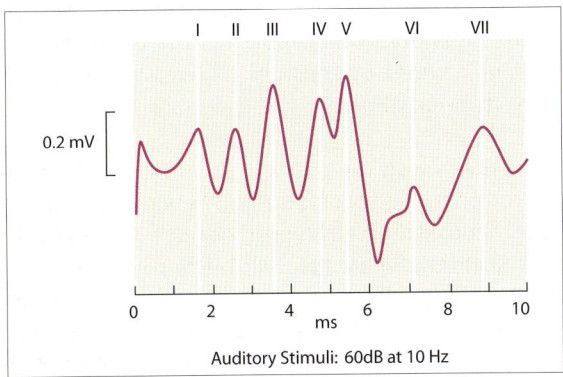

Fig. 118.**14** Brainstem auditory evoked response.

Electrocochleography measures the electrical activity generated in the cochlea in response to a click stimulus. A needle electrode is passed through the tympanic membrane and held against the promontory (an elevation on the medial wall of the middle ear, overlying the base of the cochlea). The responses obtained include the cochlear microphonics and the compound action potential of the vestibulocochlear nerve. Its main clinical use is in threshold estimation in children who cannot be tested by other means.

Cochlear microphonics develop in the hair cells of the cochlea before the development of action potentials in auditory nerve fibers. They are called 'microphonics' because when these potentials are fed into a speaker through an amplifier, the original sound is accurately reproduced. This proves that the waveform of cochlear microphonic potential is an exact replica of the sound wave that produces it.

Cochlear microphonics are produced in the hair cell through the piezoelectric effect, a property exhibited mostly by certain crystals that generate electricity when subjected to mechanical stresses. The following observations indicate that cochlear microphonics are not of biological origin. (1) They do not have a latent period. (2) They do not have a refractory period. (3) They do not show fatigue and are resistant to ischemia and hypoxia. (4) Their frequency can be unusually high for any biological signal. (5) They persist several hours after death! Cochlear microphonics do not have a significant physiological role in the hearing mechanism. At best, they might have a role in boosting receptor excitation. Cochlear microphonics have been put to clinical use for testing the integrity of the cochlea since they disappear when the hair cells are damaged.

Brainstem auditory evoked potentials   The short-latency (<10 ms) waves observed in the auditory evoked potentials (p 689) originate in the brainstem and are therefore called the brainstem auditory evoked responses (BAEP). They are also called vertex potentials because they are recorded from the vertex (Cz) with reference to one of the ear lobes. The mid-latency (10–50 ms) responses are thought to reflect activity arising in the thalamocortical radiations, the primary auditory cortex, and the early association cortex.

The BAEP waves are named I to VII (Fig. 118.**14**). Waves I, III, and V primarily represent the electrical activity from the acoustic nerve, pons, and midbrain respectively. The latencies between these three waves (the interpeak latencies) give a measure of the conduction velocities in the corresponding segments of the central auditory pathway.

The vestibular apparatus is a membranous structure consisting of three semicircular canals connected at their base to the utricle, saccule, and endolymphatic sac (Fig. 119.**1**). The utricle, saccule, and endolymphatic duct are interconnected by the utriculosaccular duct. The endolymphatic duct is the site where the endolymph is reabsorbed. It is also the site through which leucocytes and macrophages gain entry to the endolymph and clear up cellular debris in the endolymph. Damage or blockage of the endolymphatic sac causes accumulation of endolymph in the membranous labyrinth, affecting both vestibular and cochlear functions. The saccule is connected to the cochlea through the ductus reuniens.

The utricle and saccule (the otolith organs) detect linear acceleration. The semicircular canals detect rotatory acceleration. They are three in number: lateral (or horizontal), anterior (or superior), and posterior. The three canals are mutually perpendicular. The dilated end of each canal is called the ampulla and it opens in the utricle. The lateral canal is horizontal only when the head is tilted 30° forward. The left and right horizontal canals lie in the same plane and thereby constitute a functional pair. The two vertical canals (anterior and posterior) share a common origin from the utricle. From its origin to its termination, the superior canal is directed outward and forward at an angle of 45° to the sagittal plane while the posterior canal is directed outward and backward at an angle of 45° to the sagittal plane. Thus the two vertical canals are at right angles to each other. The right anterior and left posterior canals lie in parallel planes and thereby constitute a functional pair. The left anterior and right posterior canals constitute another functional pair.

The vestibular apparatus is lined internally with an epithelium. At certain small areas, the epithelial cells are specialized into hair cells bearing stereocilia. Unlike the cochlear hair cells, however, the vestibular hair cell has an additional true but nonmotile cilium known as a *kinocilium*. The structure of the hair cells and the mechanism of hair cell transduction are described on p 759 (see Fig. 118.8). Vestibular hair cells

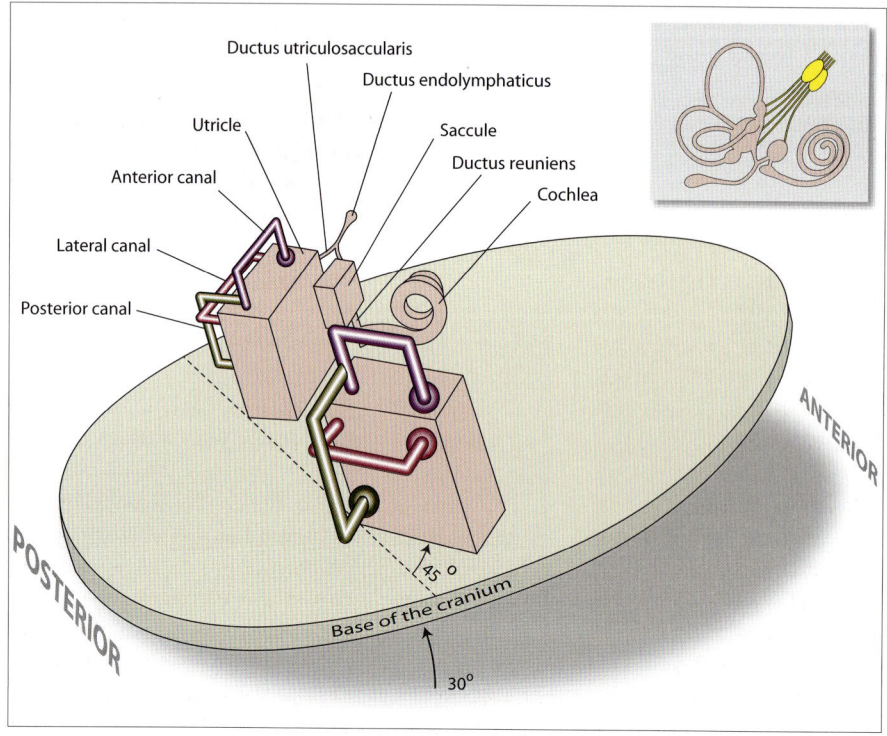

Fig. 119.**1** Schematic representation of the vestibular apparatus showing the orientation of the three semicircular canals in relation to the base of the cranium. The vestibular apparatus has been expanded for clarity. The inset shows a diagram of the vestibular apparatus.

Ductus utriculosaccularis
Ductus endolymphaticus
Utricle
Saccule
Anterior canal
Ductus reuniens
Lateral canal
Cochlea
Posterior canal
POSTERIOR
ANTERIOR
45°
Base of the cranium
30°

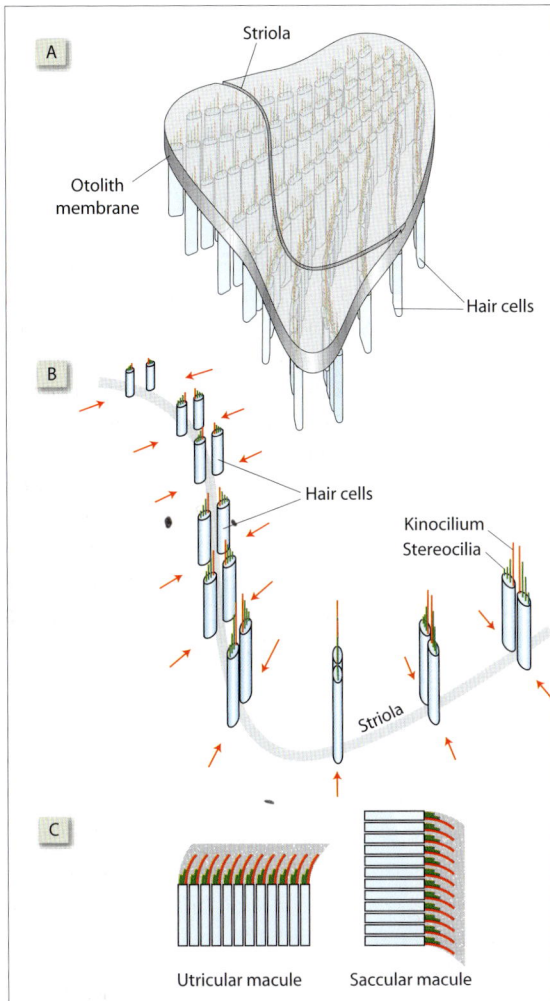

Fig. 119.**2** (A) Sensory receptors in the utricular macula. (B) Sensory receptors in the otolith organs. The arrows indicate the axis of the hair cells. (C) The macula in vertical position (left) and horizontal position (right).

membrane, owing to its higher density and consequent inertia, lags behind. This results in a shearing movement between the otolith membrane and the hair cells. The shearing movement bends the cilia and excites the hair cells, triggering action potentials at frequencies proportional to the acceleration.

The row of stereocilia and the kinocilium present on the hair cell lends to the cell a polarity (or axis) that is directed toward its kinocilium (Fig. 119.**2**A, B). A hair cell depolarizes when the stereocilia bend toward the kinocilium (i.e., along the axis) and hyperpolarizes when the stereocilia bend away from the kinocilium (i.e., opposite to the axis). If the cilia bend in a direction perpendicular to the axis, the hair cell will neither depolarize nor hyperpolarize in accordance with vectorial principles. The axes of the numerous hair cells in the macula are in different directions so that at least some of them are depolarized regardless of the direction of the shearing force. The axes of the hair cells are not randomly directed. A ridge called the striola traverses the surface of the macula. In the utricular macula, the axes of all hair cells are directed toward the striola. In the saccular macula, the axes of the hair cells are directed away from the striola. This difference in the utricular macula and saccular macula is not of much physiological significance: The important difference is that when the head is erect, the utricular macula is disposed horizontally while the saccular macula is disposed vertically. The reverse is true in the supine position (Fig. 119.**2**C). Thus regardless of the body position, the macula that is more horizontal responds more to horizontal acceleration while the macula that is more vertical responds more to vertical acceleration including gravity.

In the **semicircular canals**, the epithelium bearing the hair cells is called the crista ampullaris. It is covered with a gelatinous dome called the cupula. The cupula does not contain any crystals. The cilia of the underlying hair cells project into the cupula. At its free end the cupula is in light contact with the wall of the ampulla. The cupula therefore comes in the way of flowing endolymph and is deflected whenever the endolymph flows. In the horizontal semicircular canal, the kinocilium is located toward the utricle, while in the anterior and posterior semicircular canals, it is directed away from the utricle.

The endolymph in the horizontal canal flows in a circular motion when the head rotates about a vertical axis. When the head begins to rotate, the endolymph initially remains stationary due to inertia, resulting in its relative motion in the opposite direction. The effect occurs in both the horizontal canals simultaneously, stimulating the crista ampullaris on one side and inhibiting it on the other (Fig. 119.**3**). The reverse occurs

are structurally of two types, viz., type I and type II, but their functional difference remains unknown. The first order neurons originating from the hair cells have their cell bodies in the vestibular (Scarpa's) ganglion. The central ends of these neurons constitute the vestibular nerve and terminate in the vestibular nucleus in the brainstem (see Figs. 95.**9**, 96.**4**, and 116.**2**).

In the **otolith organs**, the epithelium bearing the hair cells is called the macula. It is covered with a gelatinous layer called the otolith membrane. Embedded in the membrane are numerous small crystals of calcium carbonate called otoconia. The cilia of the underlying hair cells project into the otolith membrane (Fig. 119.**2**). When the macula is subjected to linear acceleration, the hair cells accelerate quickly while the otolith

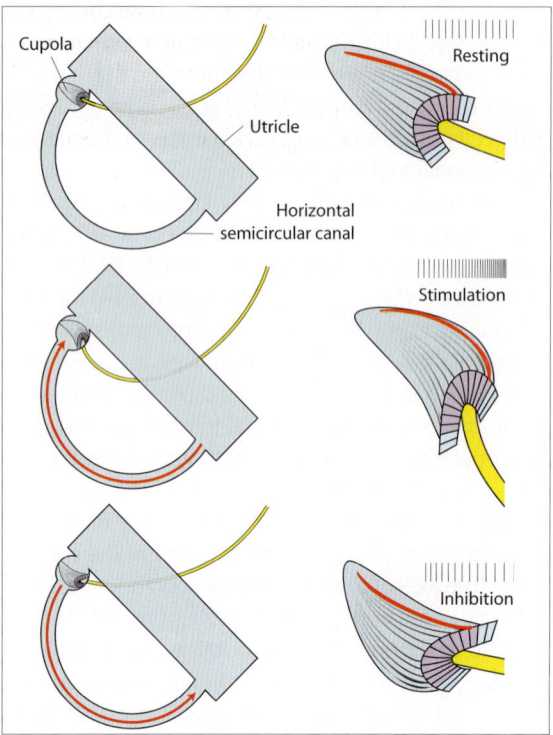

Fig. 119.**3**   Sensory receptors in the horizontal semicircular canals. Note that the kinocilium is present toward the utricle.

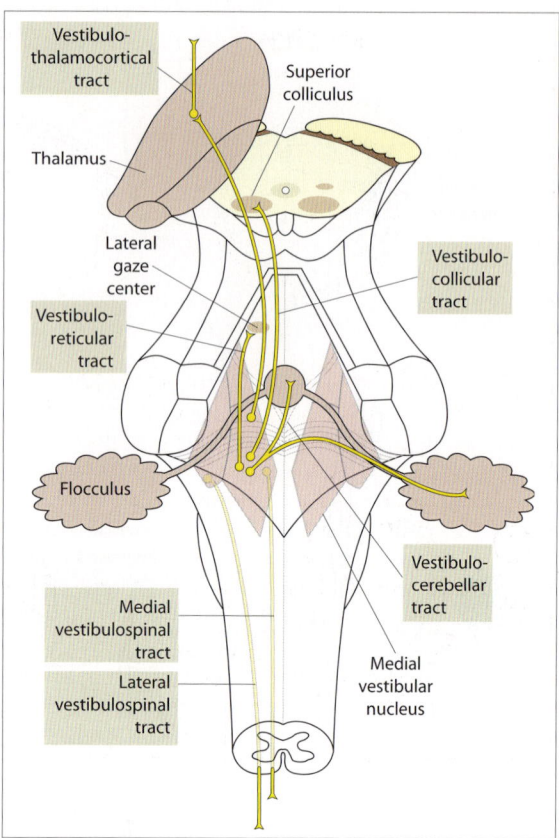

Fig. 119.**4**   Efferent connections of the vestibular nuclei

when the head suddenly stops rotating because the endolymph continues to flow for sometime due to inertia. When the head rotates to the right, the cupula bends toward the left, bending the cilia of the hair cells toward the left. Since the kinocilia in the horizontal semicircular canal are located toward the utricle, the stereocilia in the right semicircular canal move toward the kinocilia, triggering action potentials in the vestibular afferents. Simultaneously, the discharge frequency in the left vestibular afferents reduces. The combined pattern of discharge in the two vestibular nerves (right stimulated and left inhibited) results in the conscious interpretation of head rotation to right.

▪ Vestibular functions

Afferents from the vestibular apparatus terminate in the vestibular nuclei in the upper medulla. The vestibular nuclei have vestibulocollicular, vestibuloreticular, vestibulospinal, vestibulocerebellar, and vestibulothalamocortical connections with the superior colliculus, reticular formation, spinal cord, flocculonodular lobe of the cerebellum, and thalamus, respectively (Fig. 119.**4**). There are two vestibulospinal tracts, lateral and medial which descend respectively from the lateral and medial vestibular nuclei and are present in the anterior

funiculus of the spinal cord. The *medial vestibulospinal tract* descends only to the cervical and upper thoracic segments of the spinal cord and controls only the muscles of the neck and upper limbs. The *lateral vestibulospinal tract* extends down to the sacral segments and controls the muscles of trunk and lower limbs.

The central connections of the vestibular nuclei coordinate the various vestibular functions, which are as follows: (1) The vestibular apparatus stabilizes the eye in space when the head is moving or jerking, thereby preventing any blurring of vision. This function is enabled by the VOR and optokinetic reflexes. The pathway of these reflexes include the *vestibuloreticular* fibers to the gaze centers and the *vestibulocerebellar* fibers to the flocculonodular lobe (see Fig. 116.**2**). (2) When the head moves, the vestibular apparatus readjusts the posture to stabilize it against gravity. This function is enabled by the tonic labyrinthine reflexes which involve the *vestibulospinal* tracts. Simultaneously, the direction of the gaze is also readjusted through appropriate saccadic movements: the reflex pathway includes the *vestibulocollicular* fibers. (3) The vestibular apparatus also provides us with conscious senses of head tilt and head rotation. This is enabled by the *vestibulothalamocortical* fibers which reach the cerebral

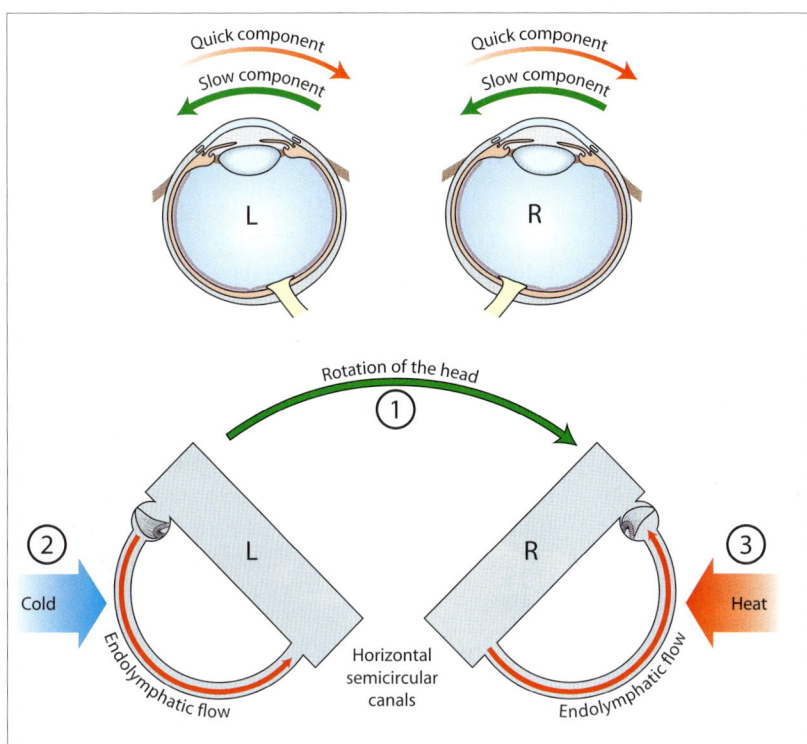

Fig. 119.**5**   Slow rotation of the eye to the left with nystagmus to the right is caused by (1) rotation of the head to the right, (2) cold saline in the left ear, or (3) warm saline in the right ear.

cortex after relaying in the ventrolateral and ventroposterior thalamic nuclei.

Vestibular function tests   The functional integrity of the vestibular apparatus is tested clinically by eliciting the VOR. The test is also one of the several tests that is used to establish whether a comatose patient is brain dead. There are two ways in which the VOR is elicited in the laboratory (Fig. 119.**5**). In the *postrotatory nystagmus test*, the patient is seated on a Barany chair, rotated ten times in 20 seconds and then abruptly stopped. The presence of postrotatory nystagmus for about 30 seconds indicates a normal vestibular function, at least on one side. In the *caloric nystagmus test*, the auditory canal is rinsed with warm water after tilting the head 60° backward (if the subject is seated) or 30° forward (if the subject is supine) so that the horizontal canals are in the vertical plane. The heat from the auditory canal is conducted to the horizontal semicircular canal and produces convection currents in the endolymph, resulting in nystagmus to the same side. Irrigating the auditory canal with cold saline results in caloric nystagmus to the other side[1]. When there is unilateral damage to the vestibular apparatus, the caloric test cannot be elicited on the affected side.

▪ Vestibular disorders

Motion sickness occurs due to excessive and repeated stimulation of the vestibular apparatus. This usually happens due to rapid and repeated change in motion while traveling. The person has nausea, vomiting, sweating, headache, and disorientation. It can be prevented by avoiding greasy and bulky food before travel and by taking antiemetic drugs. Motion sickness may be due to conflicts in visual and vestibular signals. Surgical removal of the flocculonodular lobe, which receives both optokinetic and vestibular signals, reduces or eliminates motion sickness.

Meniere's disease is associated with endolymphatic hydrops, an abnormal distension of the membranous labyrinth due to an increase in endolymph volume. The defect probably lies in the activity of the secretory cells in the membranous labyrinth and the endolymphatic sac. The disease is associated with attacks of dizziness, vertigo, and vestibular nystagmus, which are frequently accompanied by nausea and vomiting. There is loss of sensitivity to low frequency sounds. The condition is relieved by implantation of a small tube into the swollen endolymphatic sac.

---

[1]   A mnemonic that helps to remember the direction of caloric nystagmus is ACTH (Away Cold, Towards Hot).

Olfaction

Olfactory mucosa   Olfactory stimuli are detected when odorant molecules come in contact with the olfactory mucosa, a small patch of sensory epithelium located in the roof of the nasal cavity near the septum. Because of its location high in the nasal cavity, the olfactory mucosa is not directly exposed to the flow of inspired air entering the nose. Odorant molecules come in contact with the olfactory mucosa only when the airflow in the nasal cavity is turbulent. Sniffing makes the airflow turbulent and therefore helps in olfaction.

The olfactory mucosa contains olfactory receptor cells, supporting cells, basal lamina cells, and Bowman's cells. The basal lamina cells contain rapidly dividing progenitor cells that replace the aged olfactory receptor cells. The Bowman's cells secrete the mucus that forms a layer on the olfactory mucosa. Unlike the mucus layer on the adjacent nasal epithelium which is kept moving by rhythmic ciliary motion, the mucus layer on the olfactory epithelium is stagnant. Odorant molecules must pass through this stagnant mucus layer to stimulate the olfactory receptor cells. Certain odorant-binding proteins present in the olfactory mucosa help in transferring the odorant molecules to the receptors. The olfactory mucosa is innervated mainly by the olfactory nerve and partly by the trigeminal nerve. The irritative character of some odors like those of ammonia, menthol, and peppermint result from stimulation of the trigeminal nerve.

Odorant molecules must dissolve in the mucous layer lining the nose before they can come in contact with the olfactory receptors. Hence, an odorant molecule must be: (1) volatile, so as to diffuse in air and be transported into the nose, (2) water-soluble, to be able to penetrate the watery mucous layer to reach the receptor cell membrane, and (3) lipid-soluble, to penetrate the cell membranes of the olfactory receptor cells for stimulating them. Most odorant molecules are small, having fewer than 20 carbon molecules.

Olfactory pathway

The olfactory receptor cell has a dendrite that is modified into the olfactory rod, and an axon that terminates in the olfactory bulb after passing through the cribriform plate as the olfactory nerve fibers. The olfactory rod bears several cilia that project into the mucus layer.

The olfactory bulb contains the mitral cells, the dendrites of which synapse with the olfactory nerve ending to form the olfactory glomerulus. The axons of the mitral cell form the lateral and medial olfactory stria that terminate in the olfactory cortex located in the pyriform area. Olfactory fibers also project to the orbitofrontal cortex (involved in conscious discrimination of odors), amygdala (involved in emotional responses to olfactory stimuli), and the entorhinal cortex (involved in olfactory memory).

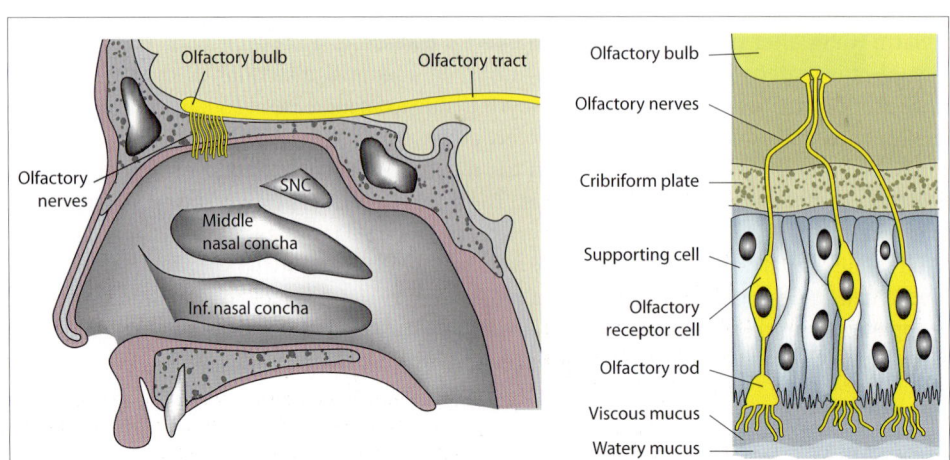

Olfactory bulb

Olfactory tract

Olfactory nerves

SNC

Middle nasal concha

Inf. nasal concha

Olfactory bulb

Olfactory nerves

Cribriform plate

Supporting cell

Olfactory receptor cell

Olfactory rod

Viscous mucus

Watery mucus

Fig. 120.**1** (*Left*) Olfactory nerves, bulb, and tract. (*Right*) Olfactory epithelium, showing the olfactory receptor cells with their olfactory rods.

Olfactory processing

Olfactory transduction   The adsorption of odorant molecules to the plasma membrane of the cilia of the receptor cells activates group-IIA hormonal mechanisms, resulting in the opening of Cl⁻ channels. The E_{Cl} of the cell is less negative than the resting potential of the olfactory cell and therefore the opening of the Cl⁻ channels depolarizes the cell. Olfactory sensation adapts very rapidly with continued exposure to an odorant. The adaptation occurs partly in the receptor cells and partly in the central neurons.

Neural encoding of different odors   An odorant substance usually stimulates a large number of olfactory receptors. No two odorant substances stimulate the same set of olfactory receptors, although individual receptors can respond to more than one odorant substance. Thus it seems that the brain interprets the odor from the pattern of receptor cells that are stimulated. An analogy can be drawn here with the auditory mechanism in which an endless variety of sound quality can be appreciated from the spatial pattern of basilar membrane vibration.

The electrical response recorded from the olfactory mucosa in response to an olfactory stimulus is called an electroolfactogram. The amplitude of the electroolfactogram increases when the intensity of the stimulus increases. The electroolfactogram is useful in the diagnosis of anosmia (absence of olfactory sense), hyposmia (diminished olfactory sense), and dysosmia (distorted olfactory sense).

Gustation

Gustatory (taste) stimuli are detected by taste receptors within the tongue, mouth, and pharynx. Taste must be distinguished from flavor, which includes the olfactory, tactile, and thermal attributes of food in addition to taste.

Sapid substances

Sapid (taste-producing) substances must dissolve in the saliva before they can stimulate the taste receptors. All taste sensations result from various combinations of four basic tastes. A sweet taste is produced by various classes of organic molecules, including sugars, glycols, and aldehydes. Saccharin (o-sulfobenzimide) and aspartame (amino acid) are sweet substances with low calorific content. The tip of the tongue is the area most sensitive to sweet stimuli. A bitter taste is produced by

alkaloids such as quinine and caffeine. The back of the tongue is the area most sensitive to bitter stimuli. A salty taste is produced by the anions of ionizable salts. The anterior half of each side of the tongue is the area most sensitive to salty stimuli. A sour taste is produced by acids. The posterior half of each side of the tongue is the area most sensitive to sour stimuli.

The tongue

Gustatory papillae   There are four types of papillae on the tongue, three of which have gustatory functions. The fourth type, called the filiform or mechanical papillae, play a role in breaking up food particles in the mouth. The gustatory papillae may be fungiform,

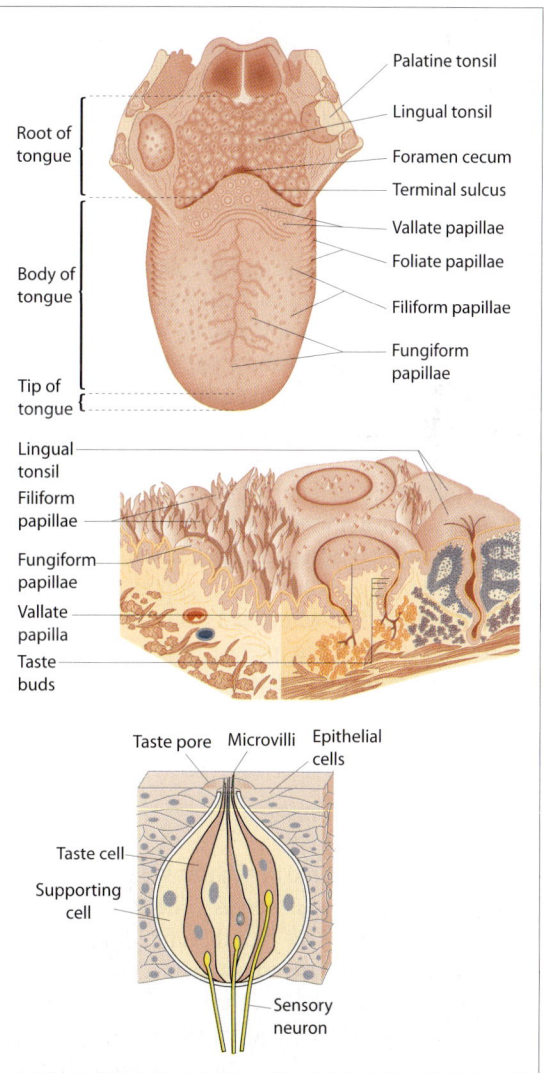

Fig. 120.2   (*Above*) The tongue, showing the location of the different types of papillae. (*Middle*) Structure of the different types of papillae. (*Below*) Structure of a taste bud.

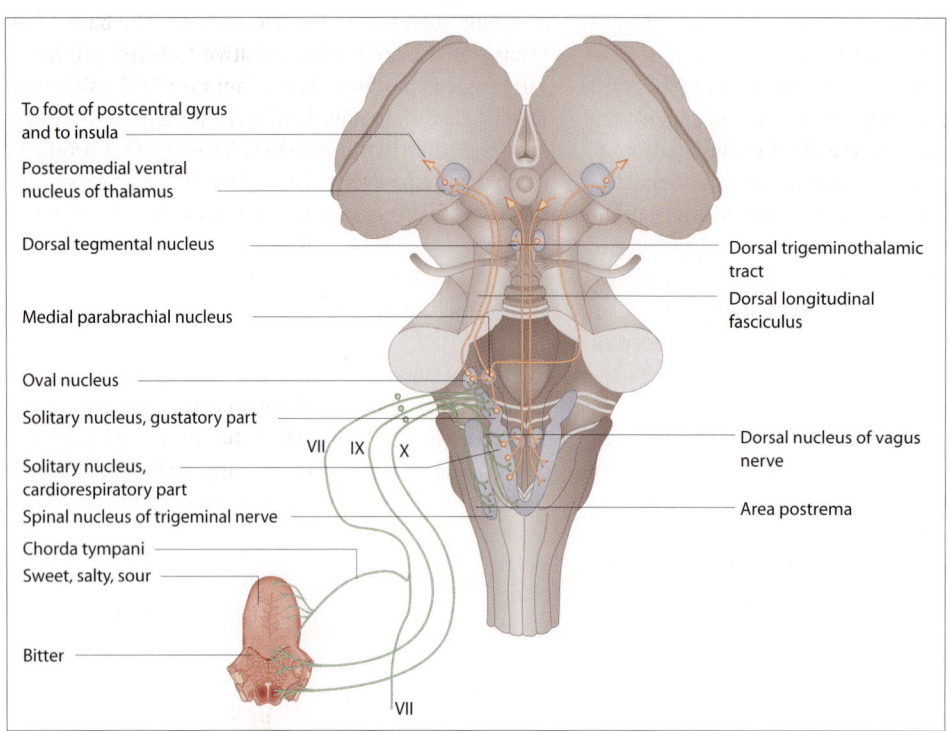

Fig. 120.**3**   Gustatory pathways.

To foot of postcentral gyrus and to insula

Posteromedial ventral nucleus of thalamus

Dorsal tegmental nucleus

Medial parabrachial nucleus

Oval nucleus

Solitary nucleus, gustatory part

Solitary nucleus, cardiorespiratory part

Spinal nucleus of trigeminal nerve

Chorda tympani

Sweet, salty, sour

Bitter

VII   IX   X

Dorsal trigeminothalamic tract

Dorsal longitudinal fasciculus

Dorsal nucleus of vagus nerve

Area postrema

VII

circumvallate or foliate. Fungiform papillae are located in the anterior two-thirds of the tongue. Circumvallate papillae are arranged in a V-shaped row on the posterior part of the tongue. Foliate papillae are located on the lateral border of the tongue anterior to the circumvallate papillae (Fig. 120.**2**).

Taste buds are located on the sides of the tongue papillae. They are also present on the hard and soft palate, epiglottis, and in the pharynx. Each taste bud is a cluster of about 50 taste cells and several supporting and basal cells. Each taste bud contains a taste pore that allows substances to reach the interior of the taste bud. The taste receptors are located on the microvilli that project from the taste cells into the taste pores (Fig. 120.**2**).

■   Gustatory pathway

Taste cells are innervated by branches of the facial, glossopharyngeal, and vagus nerves. (The tactile and temperature receptors of the mouth, tongue, and pharynx are innervated by the trigeminal nerve.) The taste buds in the anterior two-thirds of the tongue are innervated by lingual branches of the facial nerve. The cell bodies are located in the geniculate ganglion. The taste buds in the posterior third of the tongue are innervated by the glossopharyngeal nerve. The cell bodies lie in the superior and inferior ganglia of this nerve. Taste receptors in

the pharyngeal part of the tongue and on the hard palate, soft palate, and epiglottis are innervated by fibers of the vagus nerve (Fig. 120.**3**).

Gustatory fibers of the facial, glossopharyngeal, and vagus nerves terminate in the nucleus of tractus solitarius (NTS). Second order neurons originating in the NTS cross the midline and ascend in the medial lemniscus to terminate in the ventroposterior medial nucleus (VPM) of the thalamus. The third order gustatory neurons originate from the VPM and terminate at the border between the anterior insula and its frontoparietal operculum, near the foot of the postcentral gyrus.

■   Gustatory processing

Gustatory transduction   There are different mechanisms of sensory transduction for different taste substances. Sweet-tasting substances depolarize taste cells by closing K^+ channels through the group-IIA hormonal mechanism. These K^+ channels are cAMP-dependent. Bitter-tasting substances activate the group-IIC hormonal mechanism, causing a rise in intracellular Ca^{2+} that triggers neurotransmitter release. Salty-tasting substances depolarize taste cells by activating an amiloride-sensitive Na^+ channel. Sour-tasting substances (e.g., citric acid) depolarize taste cells by raising the intracellular H^+ ion concentration, which causes closure of K^+ channels.

Neural encoding of different tastes   Each nerve fiber of the gustatory nerves responds to more than one taste stimulus. However, each fiber responds best to one of the four primary taste qualities. Therefore, the coding of a gustatory sensation, like that of the olfactory system, seems to depend on the pattern of nerve fibers activated by a particular stimulus. The intensity discrimination of taste, like that of odor, is quite poor.

Certain sapid substances, like miraculin, leave an after-effect. After tasting miraculin, acids taste sweet. There are also genetic differences in taste perception, as revealed by the dichotomy with regard to the taste of phenylthiocarbamide (PTC). About 70% Caucasians find PTC to be sour while the rest find it tasteless. Abnormalities of taste include ageusia (loss of taste sensation), hypogeusia (diminished taste sensation), and dysgeusia (distorted taste sensation).

Photo Acknowledgments

The following figures are reproduced/redrawn/adapted, with permission, from other sources.

■ From Thieme Publications

Figs. 4.2, 8.5 Reproduced from Kuehnel W. **Color Atlas of Cytology, Histology, and Microscopic Anatomy**. Stuttgart–New York: Thieme; 2003.

Fig. 92.6 Reproduced from Passarge E. **Color Atlas of Genetics**. Stuttgart–New York: Thieme; 2007.

Figs. 29.1, 29.3, 30.1, 30.2, 33.3, 33.5, 33.6 Reproduced from Theml N, Diem H, Haferlach T. **Color Atlas of Hematology**. Stuttgart–New York: Thieme; 2004.

Fig. 31.2 Reproduced from Burmester G-R, Pezzutto A. **Color Atlas of Immunology**. Stuttgart-New York: Thieme; 2006.

Figs. 103.20, 105.2 Reproduced from Rohkamm R. **Color Atlas of Neurology**. Stuttgart–New York: Thieme; 2006.

Figs. 56.3, 56.4, 56.5, 76.4, 77.4, 84.4, 99.4 Reproduced from Riede U-N, Werner M. **Color Atlas of Pathology**. Stuttgart–New York: Thieme; 2005.

Figs. 8.10, 49.19 Reproduced from Despopoulos A, Silbernagl. **Color Atlas of Physiology**. Stuttgart–New York: Thieme; 2006.

Figs. 45.7, 107.1, 107.3, 108.2, 120.2, 102.4 Reproduced from Mumenthaler M, Mattle H, Taub E. **Fundamentals of Neurology**. Stuttgart–New York: Thieme; 2006.

Figs. 76.1, 76.3, 84.5 Reproduced from Eastman G, Wald C, Crossin J. **Getting Started in Radiology**. 2006, Thieme Stuttgart.

Figs. 1.17, 8.2, 9.1, 9.7, 38.6, 38.8B, 38.9, 38.13 96.10. Reproduced from Schuenke M, et al. **Thieme Atlas of Anatomy**. *General Anatomy and Musculoskeletal System*. Stuttgart–New York: Thieme; 2006.

Figs. 34.1, 36.1, 38.2, 44.1, 46.1, 47.4, 48.2, 49.1 57.1, 70.3, 71.1, 71.3, 72.1, 83.1, 84.2 84.6, 85.1, 87.1, 89.5, 90.4, 91.1 91.2, 91.4, 91.9, 93.2, 101.1 Reproduced from Schuenke M, et al. **Thieme Atlas of Anatomy**. *Neck and Internal Organs*. Stuttgart–New York: Thieme; 2006.

■ From Other Publications

Fig. 10.6A Redrawn, with permission of Wiley-Liss, Inc., a subsidiary of John Wiley & Sons, Inc., from Curtis HJ, Cole KS: Membrane resting and action potentials from the squid giant axon. J Cell Comp Physiol, 1942,19:139.

Fig. 11.4 Redrawn from Hodgkin AL, Katz B: The electrical activity of the giant axon of the squid. J Physiol (Lond), 1949,109:245.

Fig. 11.6 Redrawn from Mortimer JT: Motor prosthesis. In Brooks VB, editor: Handbook of Physiology. ed 2, Washington,1981, Motor control, American Physiological Society.

Fig. 12.6A Redrawn from Ackerman MJ, Clapham DE: Ion channels: Basic science and clinical disease. N Eng J Med 1997; 336:1575.

Fig. 13.4 Redrawn from Rushton WAH: A theory of the effects of fiber size in medullated nerve. J Physiol (Lond), 1951,115:116.

Fig. 16.1 Redrawn from Jewell BR, Wilkie DR: The mechanical properties of relaxing muscle. J Physiol (Lond), 1960,152:30.

Fig. 16.3B Redrawn from Gordon AM, Huxley AF, Julian FJ: The variation in isometric tension with sarcomere length in vertebrate muscle fibers. J Physiol (Lond), 1960,152:30.

Fig. 30.3, A, B. Redrawn, with kind permission of Springer Science and Business Media, from Faucett DW: A textbook of histology, ed 12, New York, 1994, Chapman & Hall.

Fig. 38.18 Redrawn from O'Brien ET, O'Malley K: British Medical Journal, 1979, 2(6197), 1048.

Fig. 42.6 Redrawn from O'Brien ET, O'Malley K: BMJ, 1979, 2(6198), 1126.

Fig. 45.5 *Courtesy* Dr. M Abid Geelani, Professor of Cardiothoracic Surgery, GB Pant Hospital, Maulana Azad Medical College, New Delhi.

Fig. 49.9 Redrawn from Rahn H, Otis AB, Chadwick LE, Fenn WO: The pressure-volume diagram of the thorax and lung. Am J Physiol, 1946,146:170.

Fig. 49.15 Adapted from Sherman TF: A simple analog of lung mechanics. Am J Physiol, 1993,265:S33.

Fig. 52.3 Redrawn from Sircar SS: The oxygen-carrying flask: A representational transform of oxygen dissociation curve of hemoglobin. Am J Physiol, 1994,266:S34.

Figs. 54.1, 58.3, 80.8, 80.9, 80.10, 82.5, 84.2, 92.5 Adapted from Ganong WF, Review of Medical Physiology. Boston. McGraw Hill; 2005

Fig. 55.2 *Courtesy* Lt. Col. (Dr.) DJ Chakrabarty, Army Medical Corps.

Fig. 77.2 Redrawn from WHO/FCH/CAH/Integrated management of childhood illness.

Index

A

A band 62*f*, 101
Abdominal paradox 314
ABO agglutinogens **162–163**, 197
Absolute bone conduction test 764
Accessory optic pathway **741–743**, 751
Acclimatization 361
Accommodation
  of membrane channels 83, 85
  of the eye 11–13, 727–729, **753–755**
  of the urinary bladder 429
ACE *see* Acetylcholinesterase
Acetylcholine 39, 42–43, 64–65, 71, **96–99**, 106, 128, 131*f*–132, 262, 267, 302, 436, 440–441, 444, 456, 499*t*, 534, 632, 636, **639**
Acetylcholinesterase 64, 96, 298
Achalasia 97, **460**
Achlorhydria 154, 160, 458
Acid–base balance 138, 354, **398–403**
Acidosis 110, 281–282, 287, 298, 344, 358*f*, 354, 356, 384, 386*t*–387, **398–404**, 406, 418, 420, 422, 425, 495, 532, 541–542
Acoustic reflex 757
Acquired immunodeficiency syndrome 186, **191–192**
Acromegaly 508, **509**, 640
ACTH *see* Adrenocorticotropic hormone
Actin filament 32*f*, 33, 59, 61, **62**, 63–64, 100*f*
Actinin 61, 63*t*
Action potential
  in cardiac muscle 130–132
  in neuron 82, 84
  in smooth muscle 125–129, 130–134
  propagation of 92, 95
Activity-dependent synaptic modification 638
Adaptation of receptors 649
Adaptive motor learning 607, 610, 669, 710–711, 742
ADCC *see* Antibody-dependent cell-mediated cytotoxicity
Addison's disease 406*t*, **532**

Adenohypophysis 505*f*, **506**, 507
Adequate stimulus 648–649
ADH *see* Antidiuretic hormone
Adjuvant 191
Adrenal insufficiency 532–533
Adrenergic receptors 128, 265–266, 279, 303, 311, 514, 535–536, 537, 632, 636
Adrenocorticotropic hormone 484, 487*f*, 499*t*, 506, 507*f*, **527–533**, 534–536, 583, 682*t*
Adrenogenital syndrome 527
After-discharge 673
Afterload 107*f*, 229, **234**–236, 238, 245, 273, 274, 279, 283, 287, 296–298, 329, 366
Agonist muscle 607
Agraphia 715–716
Agrin 71
Air embolism **249**, 253, 356, 360
Akinesia 618–619, 669
Albinism 506
Albumin 140
Albuminuria 380
Aldosterone 265, 275, 278, 281*f*, 382, 383, 386, 393, 398, 404, 405, **411–412**, 423, 436, 499*t*, 525–529, 531, 532, 550, 551, 580, 699
Alkaline phosphatase 138*t*, 139*t*, 496, 517, 518
Alkalosis 143, 354, 356, 362, 384, 386*t*, **399–403**, 406, 521, 579
Allocortex 621
All-or-none law **80**, 84–85, 88, **104**, 107, 134, 209
Alpha block 688
Alpha motor neuron 118*f*, 609, 660–666, 670–672
Alpha–gamma coactivation **664**, 671
Alveolar stabilization 325–326
Alveolar ventilation 334, 336, **337**, 339, 341, 355
Amacrine cells 57, 640, **735***f*
Ambiguus nucleus 267, 268, 269*f*,*t* 271, 272, 600–602*t*, 641, 682
Amenorrhea 531, 553, 574, 581, 584

Amidopyridines 97
Amino acids
  absorption of 450–452,
  essential 140, 470
  metabolism 480*f*, 483*f*, 487*f*, 491*f*, 495, 508, 530, 538*f*, 539, 541*f*
  neurotransmitters 637, 653
  nutritional requirement 143
  renal handling of 380*f*, 381, 384*f*, 407, 419, 422
  synthesis of 50–51
  transport of 41
Amnesia 529
  anterograde 629, 708
  retrograde 694, 708
  source 627, 707
AMPA receptors **637**, 643, 644
Amygdala 394, 436, 588, 591–592, 612, 615, **626–629**, 640, 652, 681, 682, 684, 704, 709–710, 770
Anabolic state 488, **491**, 495
Analgesia 456, 654
Androgen-binding protein 548, 553
Androstenedione 525–527, 547–549, 553, 578
Androsterone 528, 547
Anelectrotonic potential 83
Anemia 141–143, **144–145**, 149, 154, 157–160, 164, 167, 177, 180, 186, 188, 192, 204, 206, 236, 252, 355, 356, 421–422, 442, 465, 473–476 580, 599
Angiotensin *see* Renin–angiotensin system
Angle of His 441
Angular gyrus 626, 716
Anion gap **402**, 422–423, 542
Anismus 464
Anisocytosis 139, **141**, 145
Anrep effect 235
Ankyrin 37, 144
Anode-break excitation 83
Anomia 715, 717
ANP *see* Atrial natriuretic peptide
Antagonist muscle 107, **121–122**, 607, 619, 665, 671, 673

Anterolateral pathway 651–653, 652*f*
Antibodies, incomplete 162
Antibody-dependent cell-mediated cytotoxicity 185, **198–199**
Anticoagulants 164, 174, 176–177, **178***f*
Antidiuretic hormone 265, 271, 277, 361, 381–382, 385, 388, 391, 393–394, 396, 408, **410–412**, 419, 422, 499*t*, 504, 506, 530, 614–615
Antigravity muscles 604, 666, **673**, 692
Antigravity suit 254
Antiport **39, 41**, 133, 383–384, 398, 408, 436, 439, 454
Antithrombin 166, 173, **174**, 421
Antithyroid drugs 513, **516**
Antitrypsin 140, 313, 365
Apex beat **228**, 369, 580
Aphasia 715–716
Apneustic breathing 349–350
Apolactoferrin 180
Apoproteins 140, 154
Apoptosis 32, 44, **47–48**, 67, 187, 201, 519, 557, 572, 584
Appetite 397, 471, 628, 641, **703–706**
Apraxia 667–668
Aquaporin 410
Aqueous humor 721, **722**, 724–727
Aqueous tension 19, 307
Arcuate nucleus 505, 584, 640, 705
Area postrema 295, 411, 457, 460, 591, 595*f*, 641, 772
Argyll–Robertson pupil 755
Arneth count 179
Arrhythmias 210, **219–225**, 272, 283, 287, 516
Arteriole, structure of 246
Arterioluminal vessels 285
Arteriovenous anastomosis 236, 238, 248, 300
Artery, structure of 245–246
Asphyxia 244–245*f*, 326, 351, **356**
Assistant mover 121
Association nuclei 613
Astereognosis 659
Astigmatism 730
Astrocyte 60, 295
Asymbolia for pain 656
Ataxia
    cerebellar **610–611**, 658, 675
    sensory 658, 670, **675**
    in thalamic syndrome 614
Atelectasis 312, 326, 350, **367**, 369
Athetosis 614, 618, **620**
Atkins diet 470

Atransferrinemia 156
Atrial natriuretic peptide 265, 271, **412**, 501
Atrial septal defect 245–246
Atrioventricular block 215, 218, **219**, 220, 223
Atrioventricular dissociation **219**, 229
Atrioventricular node 209–210, 215, **219**, 221–223, 267, 285, 633*t*
Attention allocation system 627
Attention, selective 639, 655
Auditory pathway 755, **758–759**, 763, 765
Auditory transduction 759, 761
Auditory tube 756
Auerbach's plexus 433
Auscultatory gap 256
Autocrine action 264, 499
Autoimmune disease 160, 186, 188, 191, 192*t*, **193**, 465, 515, 540, 553
Automaticity **130, 209**, 218, 221, 230, 267, 619
Autonomic dysreflexia 428
Autonomic nervous system 534, 614–615, **630–634**, 681
Autoreceptors 632, 641, 682
Autoregulation
    of blood flow 262, 275*f*, 286, 290, 301, 302*f*, 303
    of GFR 380–381
    of iodine uptake 513
Auxotonic contraction 120
Axial streaming 251
Axon hillock 57, 68, 642*f*–643
Axon reflex **95***f*, 300, 649, 650
Axonotemesis 69–70
Axoplasmic transport 59–60
Azotemia 373, 422
Azurophilic granule **179**–180, 182, 184–185, 204

B

B lymphocyte 185–187, 195–197, 203
B zone 608–609
Babinski sign 598*t*, **677**, 678
Bainbridge reflex 270–272
Balanced diet 478, 703
Ballistic movement 610, 664, **669**, 670
Ballistocardiography 5, 234
Band-3 37*f*, 144
Baroreceptor 267*f*–**268**
Baroreceptor reflex 210, **269**, 270, 271*t*, 272, 274, 275, 280
Barr body 571

Barrington's reflex 427
Basal body 33, 554
Basal electric rhythm **437–438**, 453
Basal metabolic rate 477, **489–491**, 514–516, 700
Basilar membrane 14–15, **756–762**
Basket cell 605–607, 628–629
Basophil 138*t*, 179, **182–183**, 193, 202–204, 506, 512, 529
Bell–Magendie's law 598
Bell's phenomenon 740
Bereitschaft potential 668
Beriberi 236, 242, 473*t*
Bernoulli's principle 10, 238, 249, 254–255, 260
Bezold–Jarisch reflex **286**, 291
Bicarbonate (renal) reabsorption 328, **399**
Bile acid **455**, 469, 474, 514
Bile juice **443–445**, 452, 458
Bile salts 254, **444–445**, 446, 448, 463, 465
Bilirubin 143, **149–150**, 151–153, 164, 295, 417–418, 443, 455, 465, 495–496
Biliverdin 150, 443
Binocular vision 746, **749–750**, 755
Biofeedback 632, 711, **714**
Biot's breathing 357
Bipolar cells **735***f*, 737, 743, **744***f*, 745
Bipolar disorder 683
Blastocyst 567–569, 576
Bleeding time 173, **175**, 474*t*
Blind spot 627, **732–733**
Blindsight 742
Blood 138–139
Blood groups 161–165
Blood pressure 273–274
    in exercise 274
    regulation of 274–277
Blood urea nitrogen 138, 373, 422
Blood volume 138
    effect on cardiac output 237, 241
    in pregnancy 580
    pulmonary 296
    regulation of 392–393
Blood–brain barrier 66, 151, 164, 257, 263, 290, **294–295**, 353, 362, 619, 702
Blood–CSF barrier 295
Blood–testis barrier 553
Blood urea nitrogen 373, 422
BMI *see* body mass index
BMR *see* basal metabolic rate

Body fluids 23, 137, 392
Body mass index 476
Body pH, regulation of 354, **401–402**
Bohr effect 148, **344–345, 347**, 400
Bohr equation 336
Bone marrow 44, 143, 145, 149, 155–156, 179, 184–186, 202, **205–206**, 257
Botulinum toxin 97, 460
Botzinger complex 348
Bowman's capsule **373–376**, 379–380
Bowman's cell 770
Bowman's membrane 721, 723
Bradykinesia 618–619
Bradykinin **262–263**, 268, **300–301**, 313, 649
Brain electrical activity mapping 689
Brainstem 599–604
Breath sounds 369–370
Breath-holding 356, 359–360
Breathing, mechanics of 314–325
Broca's area 622, 625, 716
Brodmann areas **621–622**, 626, 745, 747
Bromocriptine 507, 509, 581, 640
Bronchial asthma 182–183, **363–365**, 533
Brown fat **698–699**
Brown–Sequard syndrome **598–599**
Bruckner's reflex 13
Bucket handle movement 314, 315*f*
Buffer nerves **268–269**
Buffers
  in blood 400
  urinary 398–399
BUN *see* blood urea nitrogen
Bundle–branch block 214, **218–221**
Bundle of His **209–210**, 215, 219, 222, 633*t*
Bundle of Kent 222, 223*f*
Burst–forming units 203

C

Calcitonin 409, 457, 499, 505, 510, 517, **519–522**
Calcitriol 423, 517, **520–524**, 529
Calcium pools of the body 519–520
Calcium absorption of 522
Calcium-induced calcium release 127, 133
Calmodulin 64, 125, 168–169, 174, **501–503**, 637, 643–644
Calsequestrin 102

Canal of Schlemm 721–722
Capacitation **565**, 569
Capgras syndrome 681
Capillary exchange 257–260
Capillary, structure of 257
Carbaminohemoglobin 148, 345–346
Carbohydrates
  assimilation of 491
  digestion of 436, 445–446, 450
  intestinal absorption of 450
Carbon dioxide transport
  in blood 45–347
Carbon monoxide 148, 150, 326, 340, 341, 343, 355–356, 636
Carboxyhemoglobin 148, 341, **355**
Carcinoid syndrome 465
Cardiac cycle 226–231
Cardiac failure 241–242
Cardiac index 323
Cardiac muscle 60*t*, **61**, 71, **130–134**
Cardiac output 232–241
Cardiac waves 273
Cardioinhibitory center 268
Carotid massage 210, 270
Carrier–mediated transport 39–41
Casts in urine 418–419, 421
Catabolic state 419, 488, **493**, 495
Catalepsy, cataplexy, catatonia 684
Catch-up growth 492
Catecholamines 128, 133, 143, 218, 235, 263, **265–266**, 278–279, 356, 400, 410, 484*f*, 487, 514, 516, 530–531, **534–536**, 537, 581, 633*t*, 698
  cardiovascular effects 265–266
  metabolic effects 535–536
  as neurotransmitter 636, **639–641**, 682*t*, 701
Catelectrotonic potential 83
Caudate nucleus 588, 598*f*, 591–592, 616–617, 620, 669, 684, 743, 753
CDH effect 346
Cell adhesion molecule **35**, 37, 67, 167, 181
Cell membrane 36–39
Central terminal of Wilson 211–212
Central venous pressure **234**–242, 252, 255, 268, 270–272, 274–275, 283–284, 303, 305, 393*f*, 411*f*, 514, 580
Centriole 33, 44–46, 57, 554
Centrosome 33
Cerebellectomy 609, 666
Cerebellum 605–611
Cerebral circulation 289–290

Cerebral hemisphere 587, 621–623, 709, 717, 762
Cerebrocerebellum 615, **609–610**, 669, 711, 717
Cerebrospinal fluid **290–295**, 362
Cerruloplasmin 139, 182
Cervical effacement, ripening 582
Charcot–Leyden crystals 182–183
Charcot–Marie–Tooth disease 35
Chemical thermogenesis 698–699
Chemoreceptor 645–646*f*
  respiratory 352–353
Chemoreceptor reflex 268, **271–272**, 274, 280
Chemoreceptor trigger zone **460–461**, 640
Cheyne–Stokes breathing 14, 28, 283, **357**
Chloride shift 346
Chloridorrhea 464
Cholagogue 444
Cholecystokinin 303, 438, 439, 444–447, **456–459**, 465, 499*t*, 500, 537–539, 704, 705
Choleretic 443–444, 457
Chorea 620, 638
Choroid plexus 257, 292, 295
Choroid layer of eye 13, 721–722, 735–736
Chromaffin cell 352, 534, 536, 630
Chromaffin granules 534
Chromatic aberration 730
Chromicity of red cells 141
Chromosomal nondisjunction 45, 47, 573
Chronaxie 83–84
Chronic bronchitis 364*t*–366
Chronic obstructive airway disorders 297–298, 317, 356, **365–366**, 369
Chronotropic effect 133, 267
Chyle 451
Chylomicron 451, 487, **491–492**, 542
Cilia 33, 312*f*
Ciliary body 721–722, 753
Cineradiography 233, 453
Cingulate circuit, anterior 612, 617, 683
Cingulate cortex, anterior 681
Cingulate gyrus 427, 628, 656, 681–682
Circadian rhythm 441, 612, **614–615**, 683, 692, 742
Circulation time 254, 357

Circulatory shock   255–266, 268, 275, **280–284**, 302, 355, 357, 530, 533

Circumventricular organs   **295**

Climbing fibers   606–607

Clonus   **665**, 677, 699

Clot retraction   167, 175–176

Clotting time   176

CNS ischemic response   274*t*, 275, **280**, 281

Coagulation of blood   167, **168–172**, 173–178, 495–496

Cochlea   14–15, **756–766**

Coefficient of oxygen utilization   363–644

Coitus   **564**, 569, 581

Coitus interruptus   570

Cold vasodilatation   300

Colliculopulvinar pathway   741–741

Colon   453–454

Colony-forming unit   202–203

Color blindness   748–749

Color vision   747–748

Compensatory pause   223

Competitive inhibition   **39–40**

Complement system   43, 140, 149, 162, 165, 181, 185, 191, 194, 198*f*, 199–**200**

Complex cells   742, 745–747

Compliance
   arterial   236
   arteriolar   272
   of intrafusal fiber   660, 662
   of lung–thorax   314, 317–326, 331, 335, 337–338, 350, 366–367
   of pulmonary venule   297
   venous   236–237, 272
   ventricular   228–229

Compound action potential   **117–118**, 765

COMT   **535**, 639–640

Concentric contraction   120, 122

Conditioning   354, 628, 681, 694, **710–714**

Conductive system of
   heart   209–210

Congenital adrenal hyperplasia   527–528

Congential anemia of newborn   164

Conjugate ocular movements   602–603, 625, **750**, 753

Connexins, connexons   34–35

Conn's syndrome   532

Consolidation of lungs   **367**, 369, 370

Constipation   453, **463–464**, 475, 515

Constitutive pathway   31

Contact junction   65

Contraceptives   568–570

Contraction remainder   102, **106**, 126

Control systems   25–28

Coomb's test   162

Cori cycle   493*f*, 494

Cornea   721

Coronary circulation   285–288

Corpus callosum   587–588, 590*f*, 616*f*, 622, 656, 668, 709

Corpus striatum   616, 620

Corticobulbar tract   602–603, 622, 626, 670

Corticospinal tract   597*f*–**600*f***, 609–610, 622, 624, 626, 665, 670–671, 677–678

Corticosteroid therapy   183, 419, **532–533**

Corticostriate fibers   622

Corticotropin releasing hormone   499*t*, 505, 528–529, 583, 614, 704

Cortisol–binding globulin   551

Cotransport *see* symport

Cough   275, 291, 294, 309, **311–313**, 333, 350, 365–367, 425, 603

Countertransport *see* antiport

Countercurrent exchanger   303, 374, 388, **390–391**, 554–555

Countercurrent multiplier   382, 385, **388–390**

Coup injury   5, 292

Cranial nerves   292, 591, 593*f*, 596*f*, **599–600*f*, 602**, 631, 460

C–reactive proteins   140

Creatine   31, 102–103, 494

Cretinism   515

Critical closing pressure   252

Critical flicker fusion frequency   742

Cross–bridge cycling
   in skeletal muscle   100–101, 103, 113
   in smooth muscle   125–126
   in cardiac muscle   133, 287

Crossed extensor reflex   671–674

Cross–matching of blood   163–164

Crypts of Lieberkuhn   433, 449

CSF *see* Cerebrospinal fluid

Cumulus oophorus   **556*f***, 566, 568–569

Cuneocerebellar tract   600, 608

Current of injury   215–218

Cushing's reflex   290, 294

Cushing's syndrome   177, 182–183, 278, 400, 406, **531–532**, 706

Cyanacobalamin   **158–159**, 461, 471*f*

Cystinuria   451

Cystometry   425, 427–429

Cytotoxic T cell   201

Cytotrophoblast   568*f*, 578

D

D cells   439, **440**, 441, 458*f*, 499, 537, **540**

Dale's vasomotor reversal   266

Dark adaptation   736, **739–740**

Dark bands   101

Davenport diagram   402

Dead spaces   **336–337**, 339, 359

Deafness   515, **762–764**

Decapacitation   **562**, 565, 566, 569

Decerebrate rigidity   **666**, 673

Decompression sickness   356, **360**

Decorticate rigidity   **666**, 675–676

Defecation reflex   454

Defibrination syndrome   176, **178**

Dehydroepiandrosterone   525–529, 532, 547, 549, 578–579

Delta wave
   in ECG   222
   in EEG   686–687, 693

Delusions   **679**, 684

Demyelination   **95**, 160, 473*t*

Dendritic spines   57, 643

Denervation hypersensitivity   **71**, 218, 428, 535, 665, 677

Dentate gyrus   588, 592*f*, **627–629*f***, 709

Dentate nucleus   594, **609–610**, 669

Deoxyribonucleic acid   29, 31, 44–46, **49–53**

Dermatome   599, **650–651**, 658

Descemet's membrane   475, **721**, 723*f*

Desmosome   33–34

DHEA *see* Dehydroepiandrosterone

DHP receptor *see* Dihydropyridine receptor

DHT *see* Dehydrotestosterone

Diabetes insipidus   **410**, 417, 419, 422

Diabetes mellitus   148, 192*t*, 344, 406, 411, 421, 450, 475, 487, 495, 508, **540–543**, 570, 573, 580, 703, 725, 748

Dialysis   174, 422, **424**

Diapedesis   35, **181**, 529

Diarrhea   464–465

Diastasis   227–229

DIC *see* Disseminated intravascular coagulation

Dietary fibers   464, **475**

Dietary requirements   577–579

Dieter's nucleus   608–609, 665

Diffuse junction   65

Diffusion capacity   239, **341–342**, 353, 355–356, 361*t*, 363, 366, 367

Diffusion trapping   409

Diffusion-limited transport   259, 340–341

Dihydropyridine receptor   102

Dihydrotestosterone   525, 547–548, 553*f*, 572, 574*t*, 575

Diopteric power   11, 727–729

2,3 Diphosphoglycerate   142*f*, 143, **148**, 164

Dipole   **17–18**, 210, 211, 685, 688

Disjugate ocular movements   602, **750**

Disseminated intravascular coagulation 178

Diuretics   242, 361, 381, 383–385, **386–387**, 400*t*, 406*t*, 408–410, 541

DNA cloning   52

Doctrine of specific nerve energies   657

Dopamine   265–266, 284, 352, 461, 507, 534*f*, 581, 616–620, 629, 632*t*, **639–641**, 683–684

Doppler method   233, 254

Dorsal columns   426, 597–599, 602, 608, **651–654**, 658, 675

Dorsal horn gating   653–654

Dorsal longitudinal fasciculus   615, 772*f*

Dorsolateral prefrontal circuit   616, 627, 753

Dorsomedial nucleus   614, 627

Double Bohr effect   347

Down-regulation   535, 541

Drowning   326, **356–357**

Duchenne's muscular dystrophy   63

Dumping syndrome   461

Dust cell   185, 311–313

Dwarfism   509

Dye-dilution technique   233, 254

Dynamic airway compression   **320–321**, 365–366

Dynein   32–33, 60

Dynorphin   636, 642, 654

Dyslexia   716

Dyspnea   241, 325, 335, 360, 366*t*

Dyspneic index   335

Dystroglycan   63

Dystrophin   61, 63

E

Ear   756–758

Ebner's gland   436

Eccentric contraction   120–122

ECG *see* Electrocardiogram

ECG artifact   689–690*f*

Edema   260
    cerebral   290, 361, 533
    pulmonary   241, 282, **297–298**, 361

Edinger–Westphal nucleus   **754–755**

EEG *see* Electroencephalogram

EEG snchronization   95, 613, 638, 685, **688**, 695

Einthoven's triangle   3, 211–213

Ejection fraction   **232**–233, 235

Ejection time, left ventricular   230–231, 515–516

Electric synapse   95, 437, 635

Electrocardiogram   132, **214–225**, 227*f*, 228, 406, 580, 693*f*

Electrocardiography   3, 18, **210–214**

Electrocochleography   764–765

Electroencephalogram   685–691

Electroencephalography   18, 685

Electromechanical coupling   127, 129

Electromechanical systole   231

Electromyogram   115–117

Electromyography   114–117

Electron transport chain   473*t*, 40–482

Electroneurography   117–119

Electrooculogram   690, 693, **739–740**

Electrophoresis   **24**, 52, 419

Electroretinogram   739

Electrotonic conduction   90–96

Electrotonic potential   82–83

Embden–Meyerhof pathway   142–143, **482**

Emboliform nucleus   594*f*, 605–606

Embolism   249, 253, 283, 356, 360, 570

Emesis *see* Vomiting

EMG *see* Electromyogram

Emotions   301, 427, 615, 626–628, 641, **679–684**, 708*f*, 709

Emphysema   **365–366**, 369*t*

Emulsification of fats   443, **445**–446

Endocytosis   31, **41–42**

Endolymph   751, 757–761, 766–769

Endomysium   70

Endoneurium   59, 66–69

Endoplasmic reticulum   29–31, 57, 59, 61

Endorphin   351, 507, 642, 654

Endothelin   166, 262–264, 463, 582

Endplate noise   116

Endplate potential   96–97, 99

Endplate zone   108

Enterochromaffin cells   440

Enterogastric reflexes   438–439, 441

Enterogastrone   438, 441, 458–459

Enterohepatic circulation   150, 152, 444, 463

Enteropathy   140, 261, 463, 465

Entorhinal cortex   **626–628**, 709, 770

EOG *see* electrooculogram

Eosinophil   138*t*, 179, **182–183**, 202–203, 364, 529,

Ephaptic conduction   60, 95, 126, 128, 130, 134, 209, 635

Epicritic sensations   645–647

Epilepsy   687, 689, **691**

Epileptiform discharge   687

Epimysium   70

Epinephrine   132, **265–266**, 301, 354, 457, 461, 474, 487–488, 491, 493–495, 493–495, **535–537**, 539–540, 634*t*, 639, 641, 680, 699*t*

Epineurium   66

Epithalamus   588, 612

EPP *see* End plate potential

EPSP *see* Excitatory postsynaptic potential

Equal pressure point theory   322

Equilibrium potential   74–75

ERG *see* Electroretinogram

Erythroblastosis fetalis   164

Erythrocyte   141–145

Erythrocytic sedimentation rate   **138–139**, 580

Erythropoiesis   **143**, 154, 156–157, 202–**204**, 206, 373, 514, 548

Erythropoietin   **143**, 145, 203, 361, 442, 499, 548

Estradiol   526–528, 547–551, 553, 558–560, 573, 577, 579

Estriol   527*f*, 547*f*–549, 577, 579

Estrone   525–527, 547–549, 553, 577, 579

Estrogens   129, 278, 499*t*, 507, 519, 525–526*f*, 528, 533, 537, **549–551**, 547*f*
    in menstrual cycle   558–560
    in oral contraceptives   569–570
    placental   576–583
    in puberty   552–557

Etiocholanone   528, 547

Eustachian tube   **756**, 759

Event–related potentials   **690**, 717

Evoked potentials   **689–690**, 764–765

Excitation–contraction coupling   **102**, 105, 127, 129, 132–133

Excitatory postsynaptic potential   81, 632, 635, 637–638, **642**, 643, 688

Exercise

airway resistance in   316

blood count changes in   179–180, 182

blood pressure in   274

body temperature in   697

bronchoconstriction in   311

cardiac output in   238–239

cardiovascular changes in   272

catabolism in   494

coefficient of O_2 utilization in   343

diffusion capacity of lungs in   342

hyperkalemia in   406

J reflex in   350–351

effect on muscle growth   70

oxygen dissociation curve in   343–345

energy source for   102–103

grading of   494

hyperventilation in   354

metabolic events in   493f–494

muscle blood flow in   301–302

muscle volume changes in   261

respiratory quotient in   489

transient proteinuria in   417–418

oxygen debt in   493f

pulmonary blood flow in   297

static and dynamic   274

Exner center   716

Exocytosis   31, 41–42

Exophthalmos   516

Expiratory pressure   323

Exteroceptors   **645–647**, 677

Extrafusal fibers   660–666

Extrapyramidal tracts   670, 677

Extrapyramidal symptoms   684

Extrastriate cortex   742–743

Extrasystole   223–225

F

Facultative reabsorption   381, 383

Fahreus–Lindqvist effect   251

Fallot's tetralogy   **245–246**, 355

Fanconi syndrome   422

Far point   727–729

Fasciculation potential   **116–117**, 677–678

Fast twitch muscle   105, 425

Fastigial nucleus   594, 608–609

Fat, dietary   469

absorption of   451

assimilation of   491f–492

digestion of   436, 446, 450

Fatty acids

oxidation of   485–486

synthesis of   484–485

Fatty liver   471, **495**

Fear conditioning   628, 681, 710, **713f**

Fecal fat test   447

Feces   445

Feeding center   442, **703–705**

Ferguson reflex   583

Ferning of cervical mucus   559–560

Ferritin   **154–156**, 206

Fertilization   566

Fetal circulation   **243–244**, 576, 578–579

Fetal hemoglobin   **147**, 344, 149

Fetoplacental unit   576–578

Fetus, exchange of gases in   274

Fever   701–702

Fibrillation   **117**, 220–221, **223–225**, 228, 287, 356, 677, 702

Fibrin   167–172, 174–176, 178, 185

Fibrinogen degradation products   175

Fibrinolysis   **167**, 171, 173–176, 178, 561

Fibrinonectin   166, 514

Fick's law of diffusion   20

Fick principle   **232**, 254, 288–289, 415

FIGLU test   158

Filopodia   66–67

Filtration coefficient   **258**, 380, 416

Final common pathway   670

Firing level   **80–81**, 83, **86**, 92, 94, 97, 128, 672, 689

First breath   244, 245f, **351**

Fixation of visual target   731, 732f, **750–751**

Flagella   **33**, 181, 554, 565, 569

Flow rate   9–10, 16–17

of blood   233, 249–250, 254, 259f, 260f, 263f, 297

of air   321, 332–335, 364

of urine   391, 398, 419, 428, 532

of glandular secretions   435–436, 444

Flow velocity   249–250, 254, 262

Flow-limited transport   259, 340–341, 385

Fluid and electrolyte balance   392–397

Fluid mosaic   37

Flutter   220, 221, 223–225

Foam cell   185

Folate trap *see* methyl trap

Folic acid   **157**, **158t**, 452, 455, 474t

Follicle-stimulating hormone   499t, 506, 508, 548–**551**, 553–555, 557–560, 569, 584

Folliculogenesis   552, 556

Follow-up servo mechanism   664, 670

Force–velocity relationship   112–113

Fornix

vaginal   565

hippocampal   587–588, 591f, 592f, 614–615, 628, 682f

Fourier transform   15

Fovea centralis   726, 732, **735–736**, 741

Fragment-1.2   171

Free-water clearance   **414–415**, 419, 424, 529

F–response   118

FRC *see* Functional residual capacity

Frey's syndrome   634

Frohlich's syndrome   706

Frontal eye field   616, 622, **625**, 743f, 752–753, 755

Fructosuria   418

FSH *see* Follicle-stimulating hormone

Functional residual capacity   318–319, 322, **328–331**, 357, 363, 366

Fusiform muscle   120–121

G

G cells   439, **440**, 441, 456–458, 540

G6PD deficiency   **144**, 145, 182

Galactokinesis   581, 583

Galactopoiesis   581, 583

Galactorrhea   581, 640

Galactosuria   418

Galanin   705

Gallstones   443, 464, **465**, 580

Galvanic skin resistance   **632**, 682t

Gametogenesis   47, **552–555**, 557, 573

Gamma motor neuron   350–351, 604, 609, 660, 663–666

Gamma–amino butyric acid   605, 613, 616–618, 621, 628–629, **636–639**, 642, 644, 683, 688–689, 695

Ganglion blockers   632

Ganglion (retinal) cells   67, 735–736, 739, 741–744, 746, 754

Gap junction   34–35, 60, 95, 517, 582–583, 635

Gastric acid output   442

Gastric acid secretion 439–442

Gastric emptying 438–439

Gastric inhibitory peptide 438–441, 456–459, 537–538

Gastrin 303, 439–442, 454, **456–459**, 461, 463, 499t, 521, 537–540

Gastrinoma 456, 461, 463

Gastrin-releasing polypeptide 440–441, 444, **456–458**, 636, 704

Gastrocephalic axis 456

Gastrocolic reflex 453–454

Gastroesophageal reflux 460

Gastroileal reflex 449

Gel-solin 139–140

Generator potential 649, 755

Geniculostriate pathway **741–743**, 755

G-force 6, 253

GFR *see* Glomerular filtration rate

GFR, single nephron 423

Ghrelin 705

Giant cell 85, 191

GIP *see* Gastric inhibitory peptide

Gibbs–Donnan equilibrium 20–21

Gigantism 509

Glaucoma 723–725

Globose nucleus 594f, 605–606, 605f

Globulin *see* plasma proteins

Globus pallidus 519f, 613, 616, 618–620, 638, 684

Glomerular filtration 379–381
measurement of 416

Glomerular membrane **375–376**, 380, 412

Glomerulonephritis 194, 417–418, **421–422**

Glomerulotubular balance **385**, 392

Glucagon 441, 456–459, 484f, 487f, 488, 493f–495, 499t, 505, 521, 530f–531, 535–538, **539–540**, 541, 634t, 704

Glucocorticoids 284, 507, **529–533**, 534, 704, 723

Gluconeogenesis 483

Glucose, renal handling of 407

Glucuronic acid 150, 512, 528

GLUT 292, 470, 535, 538

Glutamate
as buffer 399
as neutrotransmitter 268, 605, 617–618, 620–621, **635–638**, 642t, 644, 653, 684, 711, 737–743, 744, 760
in urea cycle 487

Glutamate excitotoxicity 620, 637–638

Glycine 50t, 147, 157–158, 446, 496, 604, 636–637, **639**, 642, 695

Glycogenesis 483–484, 490f, 495, 501, 530, 538

Glycogenolysis **484**, 493f, 494–495, 501, 514, 545–546, 549, 634t

Glycolysis 103, 143, 304, 344, 480, **482**, 484, 486–487, 490f, 491–495, 508, 514, 536f, 538–539, 542, 564, 722

Glycosuria, renal 418, **421–422**, 543

Glycosylated hemoglobin 149

GnRH *see* Gonnadotropin releasing hormone

Goiter 476, 513–515

Goitrogens **513–515**

Golgi apparatus 29, **30f**, 32, 68–69, 376, 410, 554

Golgi bottle neuron 662, 670

Golgi tendon organ 268, 619, 645, 446, 651, 660, **662–663**, 670

Gonadotropin releasing hormone 505, 548

G-protein 132, 181, 263–264, **500–501**, 502f, 649, 737

Graafian follicle 48, 556–557, 559

Graded potentials 58, 84

Granule cell 605–607, 621, 627–629

Granulocyte 179–183

Granulomatous disease 182

Granulosa cell 548–551, 556–558, 560, 566

Grasping reflex 742

Gravindex test 579

Gravity 5–6
effect on circulation 238, 241, 249, **252–253**, 296, 399
effect on alveolar ventilation 337–340
effect on posture 673, 675–668

Growth hormone 499t, 502, 506, **507–509**, 540, 705

GRP *see* Gastrin-releasing polypeptide

GSR *see* Galvanic skin resistance

G-suit 254

Guanylin 458

Gustation 771–773

Gustatory cortical areas 626, 772

H

H band 101

H. pylori 461

Habituation 710–712

Hair cells 645–646, 757–763, 765–768

Haldane effect 347

Hallucination 638, **656**, 684, 691

Hapten **189–190**, 193, 199

Haptoglobin 139–140, 153

Hartnup disease 451

Harvard step test 239

Haustral shuttling 453

Hb1c *see* Glycosylated hemoglobin

HCG *see* Human chorionic gonadotropin

HCS *see* Human chorionic somatom-mammotropin

HDL *see* plasma lipoproteins

Heart block *see* Atrioventricular block

Heart rate 215, 230, 232, 239, 265f–266, 268–274, 269f, 288, 290, 494b, 632–633t

Heart sounds 226–227, **229–231**

Heat stroke, exhaustion 702

Hebbian potentiation 643, 712

Helium–dilution technique **329**, 338

Helper T cell 189, 191

Hematocrit **142**, 252, 416

Hematuria 177, **417**, 421

Hemianopia 509, 741, **742**

Hemiballismus 618, 620

Hemicholinium 98

Hemidesmosome 34

Hemiplegia 588, **602**, 622, 677, 717

Hemochromatosis 154, **156–157**

Hemoglobin 146–153

Hemoglobinuria 153, 417

Hemolysis 143–144, 149, 151, 163–165, 357, 406, 418

Hemolytic anemia 144–145, 149, 157, 188, 192t, 204, 465, 475

Hemolytic disease of newborn 145, **164**, 194

Hemophilia 176–177

Hemopoiesis **202–205**, 529

Hemosiderin 153–154, **156–157**, 206

Hemosiderinuria 153

Hemostasis 138

Henderson–Hasselbalch equation 23

Henneman principle **108–109**, 662

Heparin 166, **173–174**, 176, 183, 376, 495

Hepatic circulation 303–305

Hepatic encephalopathy 455

Hepatolenticular degeneration 475

Heriditary spherocytosis 144–145

Hering–Breuer reflexes 350

Hermaphroditism   127, 528, **574–575**

Herring bodies   410, 615

High molecular weight kininogen   169, 263

High pressure nervous syndrome   360

Hippocampus   588–589, 591–592, 612, 615, **627–629**, 632, 640, 643, 656, 659, 681–682, 684, 686, 709

Hirschsprung's disease   463

Histamine   179, 183, 194, 258, 261–263, 283–284, 300, 311, 313, 439–440, 441–442, 455, 460, 529, 581, 614, 639, 641, 649–650, 695

Histiocyte   185

HMP shunt pathway   143, **492**, 495, 722

HMWK *see* High molecular weight kininogen

Hollander insulin test   442

Homeostasis   25

Homocysteinuria   160

Hopping reaction   676

Hormone action, mechanism of   499–503

Horner's syndrome   632

Horripilation   99

Housekeeper waves   438, 458

HPNS *see* High pressure nervous syndrome

H-reflex   118–119

Human chorionic gonadotropin   499t, 548, 560, **576**, 577, 579

Human chorionic somatommam-motropin   499t, 577

Human chorionic thyrotropin   579

Hunger   438, 542, 683, **703**

Hunger contractions   438

Hyaline membrane disease   **326**, 722

Hydrocephalus   291, **294**

Hydrocholeretic   444

Hydrogen breath test   455

Hydrops fetalis   164

Hydroxyproline   179, 508, 518, 524, 530

Hyperaldosteronism   278, 400t, 406t, 419, 531–**532**

Hyperalgesia   264, 648–**649**, **653**

Hyperammonemia   455

Hyperbaric oxygen therapy   343, **355–356**

Hyperbilirubinemia   **150–153**, 418, 495

Hypercapnia   353–354, **356–357**, 359, 366, 686

Hypercolumns   746

Hypergammaglobinemia   140, 296

Hyperkalemia   **164**, 214, 398, **406**, 412, 422, 531–532, 538–539, 543

Hyperlipidemia   421, 531, 541–542

Hypermetropia   **728–729**, 734

Hyperparathyroidism   496, **523–524**

Hyperpnea   350, 356, 368

Hyperproteinemia   140

Hypersensitivity   179, 182–183, 193f, **194**, 201

Hypertension   277–279

Hyperthermia   697, 700–702

Hyperventilation   290, 332, 351–356, 358–359, 361–363, 367–368, 400–401, 579, 687, 691

Hypnic myoclonia   696

Hypoalbuminemia   421, 423, 471

Hypocapnia   291, 354, **356**, 367, 579

Hypohidrosis   634

Hypokinesia   618–619

Hypoparathyroidism   523

Hypophysiotropic hormones   504–506

Hypoproteinemia   **140**, 261, 367, 380, 381, 465

Hypotension   241, 270, **280**, 281f, 283, 286, 289, 291, 300, 350, 367, 461–462, 525, 530–532, 536, 632

Hypothalamic hormones   504–507

Hypothalamo–hypophyseal axis   615

Hypothalamus   614–615

Hypothermia   283, 530, 686, 700, **702**

Hypoxia   354–355

Hysteresis of lung compliance   8, **320–321**, 324–325

I

I band   62, 101

Icterus *see* Jaundice

Icterus neonatorum gravis   164

Idioventricular rhythm   134, **219**

IGF *see* Insulin–like growth factor

Ileus   406, 462

Illusion   614, 655, **656**, 657

Immunity   189–201

Immunoglobulins   35, 149, 161–165, 181, 183, 185, 190f,t, 191, 194–195, 197, **198–200**, 421, 451

Implantation   192, 550t, 551, 560, **566**–570, 576, 581

Indifferent electrode   211, 685

Inferior temporal cortex   628, 681, 708, 742, **746–747**

Infundibular nucleus   581, 614

Infundibulum   410, 505, 507, 553, 565, 615, 640

Inhibitory postsynaptic potential   632, 637–638, **642**, 643, 688

Initial segment   **57**, 94, 642, **643**, 648

Initial spike   642–643

Innervation ratio   108

Inotropic receptors   636–638, 641, 644

Inotropic effect   **133**, 235, 266–268, 284, 637

Insertion activity   115–116

Inspiratory pressure   323

Insulin   41, 406, 442, 450, 456–459, 461, 484f, 487f–488, 491–495, 499t, 502–505, 504f, 532, 535–536, **537–539**, 703, 705

Insulin-like growth factor   492, 499f, 502, **507–508**, 538

Insulin therapy   543

Integral protein   **36–38**, 144, 500, 501

Integrin   35, 67, 167–168, 180–182, 517, 566, 569

Intercellular junction   33–**34**, 38, 257–258, 553

Interkinesis   46–47

Interleukin   201, 203

Intermediate filament   **32–34**, 57, 63

Internal capsule   588, 591f, 600, 613, 616, 622, 655

Interoceptors   645–646

Interpositus nucleus   605–606, 608–610

Intersegmental tracts   598

Interstitial cells of Cajal   437

Interstitial lung disease   341, 353, 355–356, 363, **367**

Intestinal adaptation   463

Intestinal bacteria   150, 177, **455**, 513, 636

Intestinal obstruction   392, 396, **462**

Intestinointestinal reflex   449

Intrafusal fibers   651t, 660, 661–664

Intraocular pressure   722–725

Intrapleural pressure   299, 317–321, **322–323**, 324, 338, 368–369

Intrapulmonary pressure   317–319, **323–324**

Intrathoracic pressure   235, 237, 240, 260, 270, 291, 296, 312, 314, **322–323**, 327, 359, 366, 368

Inulin   137, 380–381, 381f, 414–416, 419

Inverse-stretch reflex   619, **662–663**, 671

Iodine escape   513
Ion channel, specificity of   42–43
Ion channel, types of   43
IPSP *see* Inhibitory postsynaptic
    potential
Iridectomy   724
Iris   722
Iron cycle   155
Iron, intestinal absorption of   154–156
Iron–deficiency anemia   **144–145**, 154,
    177
Isometric contraction   100, 103–**105**,
    107, 111–**113**, **120**–121, 274
Isosthenuria   417, 419, 423
Isotonic contraction   100, 104–106,
    112–113, 120
Itch, sensation of   648

J

J point   214–215, **218**
Jaundice   **150–152**, 164, 177, 465, 474*t*,
    514
Jod–Basedow effect   513
J–receptor   325, **350–351**
J–reflex   351
Jugular venous pressure   229, 241, 253*f*
Jugular venous pulse   229, 252
Junctional folds   64, 71
Juxtaglomerular apparatus   374, 376
Juxtaglomerular cells   265, **374, 376**,
    410, 634
Juxtamedullary nephrons   374–376

K

Kallikrein   169–170, 175, **263**, 313, 436
Kartagener's syndrome   33
Karyopyknotic index   559
Kayser–Fleischer ring   472, 475
Ketoacidosis   344, 354, 400, 406, 418,
    495, 541*t*, **542**
Ketogenesis   480, 483, 508
Ketone bodies   470, **486–487**, 495, 508
Ketonuria   418
Ketosis   494, 542
Ketosteroids   527–528, 531, 547
Kety method   254, 289
Kinesin   32, 59
Kinetic automatism   618–619
Kinins   200, 261, **263**
Kinocilium   761, 766–768
Klinefelter syndrome   47, 573–574
Kluver–Bucy syndrome   **629**, 704
Korotkoff sounds   10, 252, **256**

Korsakoff's psychosis   473*t*, **629**
Krause's end bulbs   647
Krause's gland   723
Krebs cycle   103, 143, 344, 471*f*, **480–483**, 486–487, 494, 637
Kupffer cell   184–185
Kwashiorkor   **470–471**, 509

L

Lacis cells   374, 376–377
Lacrimal apparatus   723
Lactation   418, 509, 513, 521, 524, 581,
    583–584
    dietary requirements in   478
Lactose intolerance   455, 464
Lambert–Eaton syndrome   99
Lamellopodia   66–67
Laminin   35, 63, 67, 71
Language   623, 626, 708, **715–717**,
    758, 762
Laplace's law   **9**, 248–249, 325–326,
    329, 366, 429
Latch mechanism   113, **125–126**
Latent period   105, 113, 190, 765
Lateral geniculate nucleus   67, 594*f*,
    612–615, 695, **741**, 743–745, 752,
    754–755
Lateral inhibition   **654–655**, 671, 743,
    744*f*
Lateral nucleus   614*f*
Latrotoxin   97
Laxatives   454, 464, 474
LDL *see* plasma lipoproteins
Learning   607, 610, 617, 628, 668–669,
    704, 707–713, 716, 742
Lemniscus
    medial   599, 651–652, 652*f*, 772
    lateral   599, 602, 758, 759*f*
    spinal   652, 652*f*, 658, 658*f*
    trigeminal   601*t*, 602*t*, 652, 653*f*
Lengthening contraction   120
Length–servo mechanism   664
Length–tension relationship   **106–107**,
    112–113, 126, 130, 235, 323, 325,
    366
Lens   721–722
Leptin   531, 552, **704–706**
Leukopoiesis   157, 204
Leukocyte *see* granulocyte, agranulo-
    cyte
Leukotriene   181–183, 264
Lever system of skeleton   **123–124**,
    316*f*

Leydig cells   547–548, 553–554,
    571–572
LH *see* Luteinizing hormone
Light adaptation   736, 738, 740
Light bands   61, 101
Light reflex   754–755
Lipase, lingual   436
Lipogenesis   473*t*, **486–487**, 492, 514,
    538–539, 704
Lipolysis   **487**, 492–495, 501, 514, 530–531, 536, 539–540, 542, 634*t*, 704
Liver   495–496
Local anesthetics   43, **87–88**, 94, 653
Local current   90–93, 95
Local potential   58, **83–84**, 649
Locomotion   608, **676–677**
Locus ceruleus   427, 591, 595*f*, 621,
    641, 654*f*, 682*t*, 695
Longitudinal tubule   **61**–62, 101–102,
    133, 166
Long-loop stretch reflex   624, 660,
    662, 673, 675
Long-term depression   **643–644**
Long-term potentiation   637,
    643–644, 653
Lower esophageal sphincter   97,
    434–435, 460, 599
Lower motor neuron   348, 597–598,
    601*t*, 602*t*, 663, **670**, 677–678
Lower motor neuron lesion   598, 663,
    677
Lumbar puncture   292–294
Lung volumes and capacities   328–335
Lung-irritant receptors   350
Lungs, nonrespiratory functions of   311
Luteal cell   548–551, 560
Luteal-placental shift   560, **576, 577**
Luteinizing hormone   499, 506, 547–551, 553, 557, 558–560, 569, 584
Lymph node   154, 184–**187**, 191, 193,
    202, 260–261, 529
Lymph, composition of   260
Lymphagogue   260
Lymphatic circulation   260
Lymphoblast   185, 196, 203–204, 451
Lymphocyte   185–186
Lymphoid tissue   **186–188**, 191, 198,
    202, 204, 492–493
Lysosome   29, 31–32, 41–42, 156, 166,
    512, 554

M

M (magnocellular) cells   742, 747
M (microfold) cells   451

Macrophage *see* Monocyte phagocytic system

Macula adherens   34

Macula densa   374f, 376–377, 381, 410

Macula lutea   735, 741–742, 745, 767

Magnocellular neurosecretory system   504, 507

Magnocellular visual pathway   716, **742**, 745, 752

Magnocellular part of red nucleus   609–610

Malabsorption syndrome   462–463

Mamillary body/nucleus   591–592, 614f, 615, 628, 641, 682f, 701

Mammillothalamic tract   615

Mammogenesis   581, 583

Mania   679, 682–683

Manometry   254, 273 *See* also sphygmomanometry

MAO *see* Monoamine oxidase

MAO inhibitor   694–695

Marasmus   470–471

Marcus–Gunn pupil   755

Masking   761, 764

Mass contraction   453–454

Mast cell   **183**, 194, 198–200, 264, 311, 495, 529, 649, 650f

Maximum breathing capacity   334

MCH, MCHC, MCV *see* Red cell indices

Mean electrical axis of the heart   211–212, 212f, 213, **215–217**, 217f, 220

Mean systemic filling pressure   226, 229, 236–**237**, 240, 275f

Medial geniculate body   601t, 612, 758–759

Median eminence   295, 505–506, 540, 615

Megacolon   463

Megakaryocyte   166, 202–206

Megaloblastic anemia   141–**145**, 158–160, 206, 474t

Meiotonic contraction   120

Meiosis   **45–47**, 555–557, 573

Meissner's corpuscles   646f, 647f, 651

Meissner's plexus   433

Melancholic depression   641, 679, **682**, 706

Melanin   417, 506, 705

Melanin concentrating hormone   705

Melanocyte stimulating hormone   506, 704

Membrane channel   39, **42–43**, 82–83

Membrane currents   **75–76**, 87, 90–91

Membrane transport   31, 37, **39**

Memory, immunological   190

Memory cell   195–196

Memory and learning   707–714

Memory field   627

Menaquinone   177

Menarche   552–553, 558, 575

Meniere's disease   769

Menopause   558

MEPP *see* miniature end plate potential

Merkel's disk receptors   646f, 647f

Mesangial cells   263, 374f, **376–377**, 380, 411–412, 502

Mesenteric circulation   303, 305

Mesocortical dopamineric pathway   684

Mesolimbic dopamine system   616, 629, 684

MET   490, 494

Metabotropic receptors   637–638, 641, 644

Metallothioneins   472

Metamyelocyte   180, 203–204

Metastatic calcification   423, 517

Metathalamus   588, 612

Methemoglobin   143, 146, **148**, 153, 355

Methotrexate   157

Methyl trap   157–159, 158f

Micelles   445, 451

Microfilament   **32–34**, 41, 57, 66

Microfold cells   742, 747

Microglia   60, 185

Microsome   29, 150

Microtubule   **32f, 34**, 44–45, 57, 59–60, 166, 168, 377, 554

Micturition   291, **425–428**, 625

Middle temporal area   742, 747, 752

Middle temporal gyrus   589f, 621f

Migraine   641

Migrating motor complex   **437–438**, 449

Milieu interieur   25, 373

Mineral nutrients   472–475

Mineralocorticoid escape   412, 532

Mineralocorticoids   525, 527–529, **531–532**

Miniature endplate potential   97, 116

Miosis   633t, 634, 725, 755

Mitochondria   29–31, 30t

Mitosis   33, **44–45**, 47

Mitral cells   770

Mittelschmerz   560

MMC *see* Migrating motor complex

Monge's disease   362

Monoamine oxidase   **535**, 636, 640f, 641

Monoblast   203–204

Monocyte   138t, 166, 182, **184–185**, 187–189, 192, 204

Mononuclear phagocytic system   184–185

Monosynaptic reflex   660–662

Monro–Kellie doctrine   290

Morphine   642, 654

Morula   567–568

Mossy cells   628–629

Mossy fibers   606–607, 627–628, 643–644

Mossy pathway   627, 629

Motilin   438, 458

Motion sickness   461, 656, 769

Motor areas   427, **624–625**, 668–669, 678, 710

Motor end plate   **64**, 71, 96–99, 108, 115

Motor recruitment   **108**, 116, 662, 672

Motor unit   70–71, 104, **106–109**, 109t, 115–117, 662, 678, 693

Motor unit potential   **115–116**, 693

Motor unit territory   108

Mountain sickness   361–362

Mouth-to-mouth breathing   326–327

M-response   118

MSH *see* Melanocyte stimulating hormone

Mucokinesis   33, 311–312

Muller maneuver   323

Mullerian ducts   572, 575

Multiple myeloma   140, 418–419

Murmurs   10, 145, **229f**, 230, 252

Muscarinic receptors   128, 440t, **632**, 639

Muscle action
  components of   121
  insufficiency of   122
  laws of 121

Muscle contractility   107, 133

Muscle contraction
  mechanism of   100–103
  effect of temperature on   106

Muscle degeneration   71

Muscle excursion   120

Muscle fatigue   106

Muscle fibers, types of   60–61, 105

Muscle growth   70–71

Muscle heat   103

Muscle pump   **237–238**, 248f, 302

Muscle regeneration   71

Muscle spindle   660–661

Muscle tension   107*f*, 111–112, 120
Muscle twitch   104, 106
Muscles, red and white   105, 110, **301**
Muscles, tonic and phasic   301
Mutism   617, 684
Myasthenia gravis   39, 97–**99**, 186, 191, 192*t*
Myelination   **58–59**, **92–93**, 514–515, 677
Myelocyte   179, 204
Myeloid-erythroid ratio   206
Myeloperoxidase   180–182
Myoblast   70
Myocardial contractility   133, **235**, 249, 272, 283, 287–288
Myocardial infarction   174–175, 180, **216–217**, 241, 279, 283, **286–288**
Myocardial ischemia   214, 230, 241, 279, 281, **285–288**
Myocardial oxygen demand   241, **285–286**
Myoglobin   105, 109, 301, **345**, 361*t*
Myoneural cleft   **64**, 96, 98–99
Myopia   728–729, 734
Myosatellite cell   70–71
Myosin filament   **62–63**, 100, 125
Myotome   598–599
Myotube   70–71
Myxedema   490, **515**

N

Naloxone   642
Narcolepsy   694, **696**, 705
Narcosis   356, 360
Natriuresis   **277**, 285–286, 412, 422–423
Natriuretic hormone   396, **412**
Near point   13, 727–729
Near reflex   729, 750, **755**
Nebulin   61, 63
Negative pressure breathing   326
Negative stretch reflex   660
Neglect   626, 659*f*
Neocortex   588, **621**, 627–628, 684, 704
Neologism   715
Neostriatum   616
Nephritic syndrome   421
Nephron, structure of   375–377
Nephrotic syndrome   140, 261, 411, 418–419, **421**
Nernst potential   **20**, 74–75, 644
Nerve conduction failure   93–95
Nerve conduction velocity   91, 117–118

Nerve degeneration   67
Nerve growth factor   67
Nerve injury   59, 69, 658
Nerve regeneration   68
Neuregulin   72
Neurofibrillary tangle   57
Neurofilament   33, **57**, 60, 67–68
Neuroglia   59–**60**, 658
Neurohypophysis   506–**507**, 615
Neuromuscular junction   64
Neuromuscular transmission   96–99
Neuromuscular blocker   98–99
Neurons, types of   59
Neuropathic bladders   428
Neuropeptide Y   267, 705
Neuropeptides   **636–637**, 642, 653, 704–705
Neuropraxia   **69**, 94
Neurotemesis   69–70
Neurotensin   458
Neurotransmitters   636–641
Neurotrophins   67
Neutralizer   121
Neutropenia   145, **180**, 188
Neutrophil   35, 138*t*, 145, **179–183**, 189, 202–203, 293, 365, 529, 560, 571
Nicotinic receptors   632
Night blindness   474, 737
Nigrostriatal pathway   618, 640
Nissl body   57, 68–69
Nitric oxide   262–264, 311, 313, 366, 373, 499*t*, 501, 564, 636, 644
Nitrites in urine   262, 355, 418
Nitrogen balance   488, 508, 552
Nitrogen narcosis   360
Nitrogen-washout technique   329, 330
NMDA receptors   637–638, 643–644, 684, 711–712
Nociceptors   264, 286, 645–650, 653, 670
Nocturnal enuresis   696
Noncompetitive inhibition   39–40
Nonconstitutive pathway   31
Nonspecific nuclei   613
Norepinephrine   60, 65, **262**, 265–267, 278, 302, 313, 351, 436*f*, 473*t*, 530, 533–537, 632, 634*t*, 636, 639–**641**, 654*f*, 682–683, 699*t*, 705
Normoblast   156, 158, **204–205**
Northern blotting   52
Nucleic acids, digestion of   446, 450
Nucleolus   29, 57

Nucleus   29
Nucleus accumbens   612, 616, 627, 629, 682, 683
Nucleus basilis of Meynert   603, **639**
Nucleus reticularis pontis   682, 695
Nucleus of tractus solitarius   267*f*–269, 348, 460, **601***t*–**602***t*, 641, 772
Nyctalopia   737
Nystagmus   611, 740, **751–752**, 769

O

Obesity   331, 366–367, 476–477, 488, **493**, 507, 531, 540, 573, 703–704, 706
Obligatory tubular reabsorption   381
Obsessive-compulsive neurosis   617, 627, 641, **683**
Occlusion, in reflex arc   672
Ocular dominance columns   **746**, 749
Oculomotor circuit   616–617, 643, 753
Off-center cell   744–745
Olfaction   626, 645, **770**
Olfactory cortical areas   626, 772
Oligodendrocyte   58–60
Oliguria   361, 417, 421–422
Olivary nucleus   595*f*, 596, 597*f*, 600, 606, 608*f*, 610, 711, 742, 758, 759*f*, 762
Olivocochlear fibers   761
On-center cell   744
Oogenesis   45, **555**, 571, 573
Ophthalmoscopy   13, 733
Opioid peptides   642
Opponent color coding   748
Opsonization   181, **198–200**
Optical axis   726–727
Optokinetic reflex   **751–752**, 768
Oral ammonium chloride test   419
Oral metering of food   705
Oral water-loading test   419
Orbitofrontal circuit, lateral   617, 627, 683
Orbitofrontal cortex   **681–682**, 770
Orexin   705
Organ of Corti   757–758
Organelle   29
Organic acids and bases renal handling of   408–409
Orgasm   564
Orientation columns   746
Oscillations   13, 26–28
Osmolality   22
Osmolar clearance   **414**, 424

788 Index

Osmolarity 22
  of body fluids 137–138,
    regulation of 389–391
Osmoreceptors 277, **393–394**, 396,
  410, 439, 446*f*
Osmosis 21–23, 284, 346, 482, 721
Osmotic coefficient **21**, 22
Osmotic fragility of red cells **143–144**,
  149
Osmotic pressure **22**, 139, 141, 258,
  380, 394, 396
Osteitis fibrosa cystica 424, **519**, 524
Osteoblasts **517**–520, 522, 550
Osteocalcin **517**–518
Osteoclasts 185, **517**–522
Osteocytes 508, **517**, 518*f*
Osteocytic osteolysis 517
Osteomalacia 524, 574*t*, **519**–520, 524
Osteoporosis 474*t*, **519**, 530, 531, 548
Osteosclerosis 424, **519**
Otoacoustic emissions 760–761
Otolith organ 645–646, 676, 766–**767**
Ovulation 174, 551–552, **556–560**,
  563, 565–567, 569–570, 579, 584,
  697
Ovulation, tests for 560
Oxygen debt 302, 493*f*
Oxygen dissociation curve **343–344**,
  400
Oxygen therapy 284, **355–356**
Oxygen toxicity 354, **356**, 360
Oxygen transport in blood 343–345
Oxyntic cells 159*f*, **439**, 540
Oxytocin 499*t*, 504, 507, 550,
  581–584, 614–615

P

P (parvocellular) cells **742**, 746, 748
Pacemaker potential 126, 128,
  131–132, 209–210, 225, 338, 437,
  619
Pacemaker, artificial 225
Pacinian corpuscles 645–**647**,
  649–650
Packed cell volume 142
Pain 18*t*, 426, 604, 613–614, 627,
  641–642, 646–647, 649, **651–656**,
  658, 681
Paleostriatum 616
Pancreatic function tests 446–448
Pancreatic juice 438, **444–446**, 450,
  452, 457–458
Pancreatic polypeptide 537, 540
Pancreozymin 456

Papez circuit 612–615, 628, **681–682**
Paracrine action **262**–264, 440, 456,
  499, 540, 548, 553, 572
Parahippocampal gyrus 621, **627–629**,
  681, 709
Parallel elastic component 111
Paralysis agitans 618
Paraphasia 715
Paraplegia 599
Parasympathetic nervous
  system 267–268, **630–631**
Parathyroid hormone 423–424, 505,
  517, **520–523**, 537
Paraventricular nucleus 393, 410,
  504–505, 507, 581, **614**–615, 682,
  704–705
Parietooccipitotemporal association
  area 623*f*, 626
Parkinson's disease 427, **618–619**,
  640, 665, 669
Parolfactory gyrus 589*f*, 627
Partial pressure 19–20
Partial thromboplastin time 176
Parturition 581–583
Parvocellular neurosecretory system
  504–505, 540, 548
Patch clamp technique 87–88
Patent ductus arteriosus 229*f*, 230,
  245–246, 264
PCV *see* Hematocrit
Peak expiratory flow rate **332–335**,
  363–364, 366
PEFR *see* Peak expiratory flow rate
Pellegra 473
Pennate muscle 120–121
Pentose phosphate pathway 143–144,
  450, **482**
Peptic cells 439
Peptic ulcer 264, 442, **461**, 529
Peptide YY 458–459
Peptides, renal handling of 407
Percussion note 368–369
Perforant pathway **627–629**, 709
Periglomerular cells 640
Perilymph 757–671
Perimetry 731
Perimysium 70
Perineurium 66, 69
Periodic breathing 357–358
Peripheral protein **36–38**, 42, 44, 163
Peripheral resistance **236**, 238,
  240–242, 250, 252, 265–266, 270*f*,
  271*t*, 273–275, 278, 280–281, 283,
  302–303,411*f*, 580

Perirhinal area 627
Peristalsis 434, 438, 449, 453, 460,
  462–463
Periventricular system 614–615
Permissive action 530–531, 534
Pernicious anemia **160**, 192*t*, 442,
  474*t*, **599**
Peroxisome 29–31
Perspiration 392, 394, **699–700**
Peyer's patches 186, 451
Phagocytosis 181, 189
Phantom limb 614, 646, 653, 657
Phenolphthalein test 294
Phenylethylamines 636, 639–640
Pheochromocytoma 278, 536
Phosphate balance of the body 521
Phosphene 648
Phospholamban 132–133
Photopic vision 736*t*, 738
Photoreceptors 645–646*f*, 648–649,
  732, **735–739**, 743, 748
Phototransduction 735–739
Phylloquinone 177
Physical efficiency index 239
Piebaldism 506
Pineal body 612
Pinocytosis 41–42
Pitch discrimination 14–15, **762**
Pituitary gland 505–509
Place cells 628
Placenta 576–579
Placing reaction 673, 675
Plasma lipoproteins (HDL, LDL, VLDL)
  41–42, 262, 455, 469, 487,
  491*f*–**492**, 495, 514, 538*f*, 539, 542
Plasma proteins 139–140
Plasma skimming 251
Plasmapheresis 99, 140
Plasmin 167, 173, **174**, 175, 518–519
Plasminogen activator 167, **174**–175,
  518–519, 553
Plasminogen activator inhibitor 167,
  175
Plasticity
  of smooth muscle **125–126**, 275,
    429
  synaptic 607, **643–644**, 656, 669,
    711
Platelet factor-4 **166**, 181
Platelet-derived growth factor 166, 181
Platelets 166–168
Plethysmography **254**, 329, **330**
Pleural circulation 296–297, **299**

Pleural effusion **367**, 369–370

Pneumocyte **310**, 313, 325

Pneumothorax 283, 350, **368–370**

Poikilocytosis 139, **141**, 145

Poikilothermia 702

Poiseuille–Hagen law **9–10**, 249–253, 275

Polymerase chain reaction 52–53

Polymorphism 52, 139

Polyphagia 541–542, 704

Polyspermy, blocking of 566–567, 569

POMC *see* Proopiomelanocortin

Pontocerebellum 609

Portal system 16–17

  adrenal 534

  hypothalamo-hypophyseal 505–506, 615

  hepatic 241, 282, **304–305**

Positive pressure breathing 326–327

Positive supporting reaction 673

Postabsorptive phase 488, 490–492

Posterior parietal cortex 626, **656**, 707–708, 742, 753

Postsynaptic potentials 632, 637, **642–643**, 688

Posttetanic potentiation 643

Posttranslational modification 30

Post–traumatic stress disorder 683

Posture, control of 673–676

Potassium balance of the body 405–406

Potassium

  intestinal absorption of 452

  renal handling of 404

Pre-Botzinger complex 348

Pre-ejection systole 231

Prefrontal association area 622–623, **626–627**

Prefrontal leucotomy 627

Pregnancy 576–580

  dietary requirements in 478

Prekallikrein 169–170, 175*f*

Preload **107**, 229, 234–236, 274, 287

Premotor area 612–613, 617, **624–626**, 668–669

Preoptic nucleus 614, **701–702**

Preparatory-set activity 667

Prepotent reflex 283, 300

Presbyopia 728–729

Presynaptic facilitation **643–644**, 712

Presynaptic inhibition 638, **643–644**

Priming 710–711

Progenitor cell 203, 770

Progestagens 129, 144, 192, 435, 464–465, 499*t*, 507, 526*f*, 527, 531, 537, 547*f*, 548–**551**, 558–559*f*, 561, 565, 568, 570, 576–583, 577*f*

Progesterone *see* progestagens

Prokinetics 640–641

Prolactin 499, 506, 509, 577, **581**, 583–584, 640

Promonocyte 185, **203–204**

Pronase 43, 87

Pronormoblast 204

Proopiomelanocortin 704

Proprioceptors **645**, 646*f*, 655, 656*f*, 660, 677

Prosopagnosia 681, 647

Prostacyclin **173*f*, 174**, 262, 264, 410, 469

Prostaglandin 183, **264*f*, 265**, 313, 440*t*, 441, 461, 469, 529, 564–565, 581–583, 649, 702

Protein C **173**–175, 177–178

Protein losing enteropathy 140, 261

Protein S 173–**174**, 477

Protein-energy malnutrition 470–471

Protein 470

  absorption of 450–451

  assimilation of 491

  digestion of 442, 453, 450

Proteinuria 411, **417–419**, 421

Prothrombin time **176**, 496

Protodiastole 227–228

Protopathic sensations 645–646

Pseudohypoparathyroidism 523

Pseudopuberty 552

Psychic blindness 628, **704**

Pteroylglutamic acid 157–158

Ptosis 574, 632

Pubarche 552

Puberty 47, 186, 492–493, 514, 526, 548, 550, **552–553**, 555–556, 558, 571, 574, 575

Pulmonary alveolar macrophage 185, **311**, 313, 155

Pulmonary circulation 349, 356

Pulmonary function tests 363

Pulmonary stretch receptors 349, 356

Pulmonary ventilation *see* Respiratory minute ventilation

Pulmonary wedge pressure 255, 296

Pulse pressure 265–266, 269–270, 271*t*, **273–274**, 297*t*, 516, 580*f*

Pulvinar **612–613**, 656, 742, 743*f*, 745

Pump-handle movement 314–315

Punctate sensation 648

Pure pyramidal lesion 677–678

Purkinje cell **605–607**, 644

Purkinje shift 738

Purkinje system 209–210, 633*t*

Purpura 165–177, **175**, 192*f*

Pyramidal cells **621–622**, 626–628, 678, 688

Pyramidal tract 600*f*

Pyrexia 701–702

Q

Quadriplegia 599

Queckenstedts' test 294

R

Rami communicantes 630

Ramp movement 664, **669–670**

Reactive hyperemia 300, 302

Readiness potential 668–669

Receptive relaxation 438

Receptor field 744–745

Reciprocal inhibition 670–671

Red cell indices 142

Red eye reflex 13

Red muscle **105**, 301, 345

Red nucleus 600, 605, **609–610**

Reduced eye of Listing 726

Reentrant circuits 218, **221–223**

Referred pain 658

Reflexes, spinal 671–673

Refractive errors of the eye 728–730

Regulated pathway 31

Regulatory control 25

Reissner's membrane 756–758

Relaxin 110, 427, 454, 502, 548, 576–577, **581**

Relay nuclei 603, 613, 650, 655, 688, 695

Renal blood flow 254, 278, 361, **375**, 379*f*, 380–382, 391, 411, **416**, 423

Renal clearance 414–416

Renal failure 180, 279, 283, 397, 400, 400*t*, 402, 406, 417–419, **421–424**, 541

Renal function tests 417, **419–420**

Renal osteodystrophy 422–424

Renal oxygen consumption 375

Renal plasma flow **415**–416, 529

Renal threshold **413–414**, 419, 421, 422*f*, 541

Renal tubular acidosis 400*t*, 406*t*, 420, **422**

Renin-angiotensin system   241, 265, 274*t*, 275, **410**, 532

Repulsive axon guidance signals   67

Rescue reactions   673, 675

Residual volume   320*f*, **328–329**, 331, 366, 334*f*

Respiratory burst   181, 482

Respiratory distress syndrome   326

Respiratory membrane   310, **340–342**, 353, 355–356

Respiratory minute volume   334

Respiratory pump   337

Respiratory quotient   337, 488–489

Respiratory resistance   316

Respiratory rhythm   **348–349**, 353

Respiratory tract   33, 186*t*, **309–310**, 312*f*

Respiratory waves   273

Resting length of muscle fiber   105, **106–107**, 111–113, 125–126, 323

Resting membrane potential   73, 75–78

Restrictive lung disorders   333, 356, 363, **366–367**

Rete testis   **562–563**, 572

Reticular activating system *see* Reticular formation

Reticular formation   267, 434, 460, **603–605**, 607–609, 612, 615, 619, 639, 652, 665–666, 670, 677, 750, 768

Reticular thalamic nucleus   613*f*, 688–689

Reticulocyte   143–145, 156, **204**

Reticuloendothelial system   184

Reticulospinal tracts   427, 597–598, 600, **604**, 665, 670, 677

Retina   735–736

Retinal   474*t*, 736–738*f*

Retinitis pigmentosa   736, 739

Retinohypothalamic pathway   612, 615, **741–742**

Retinol   474*t*, 737–738*f*

Retinoscopy   13, 733–734

Reverberating circuits   673

Rexed's laminae   **597**, 670

Reynolds number   **10**, 145, 252

Rheobase   83–84

Rhesus agglutinogens   161–164

Ribonucleic acid   29–30, 39–49, **50–51**

Ribosome   29–30, 49, 51, 57, 131, 204

Rickets   424, 474, 496, **519**, 524

Righting reflexes   673, **676**

Rigidity   149, 609, 680, 690, 665–666, 673, 677

Rigor mortis   102

Rinne's test   763

Rouleaux formation   138–139, **141**, 252

Ruffini's end organs   646–647

Ryanodine (RYR) receptor   **102**, 133, 701

S

S/L ratio   632

Saccadic movements   616, 626, 692, 742, 751–**752**, 768

Saccule   766

Sacoplasmic reticulum   31, **61**, 103, 105–106, 109*t*, 110, 127, 130, 287

Safety factor
  of nerve conduction   93–95
  of neuromuscular transmission   97

Saliva   163, 199, 263, 268, **434–436**, 532, 771

Salivation center   436

Salt craving   393–**394**, 397, 703

Saltatory conduction   92

Salt–wasting nephropathy   423

Sarcoglycan   63

Sarcomere   **61**–63, 100–101, 105–107, 120, 125, 130

Sarcospan   63

Satiety   457, 457*t*, 542, 552, 614, **703–705**

Saxitoxin   87, 89

Scala tympani, vestibuli   756*f*, 757, 758*f*

Schaffer's collateral pathway   **628**, 638, 640–641, 656, 679, **684**

Schilling test   159, 448

Schizophrenia   627, 638, 640–641, 656, 679, **684**

Schwabach's test   764

Schwann cell   **58–60**, 64, 66, 68–70, 647b

Scotopic vision   736, 738

SCUBA   359

Scurvy   177, 474*t*

Secondary sexual characteristics   548, 550, **552**

Secretin   438, 440, 441*f*, 442, 444–447, **456–459**, 539

Segmental propulsion   453

Segmentation contractions   453

Seizures   390, 401, 529, 612, 628, 638, 689, **691**, 709

Selectin   **35**, 181

Semen, composition of   562–565

Semicircular canals   751, **766***f*–769

Seminiferous tubules   **553–555**, 563, 573

Sensitization
  immunological   185, 191
  neural   649–650, 653, **710–712**

Sensorimotor transformation   668*f*, 669

Sensory acuity   356, **648**

Sensory intensity, coding of   649

Sensory pathways   **602**, 614, **650**, 652

Sensory projection   657–658

Series elastic component   63, 101, **111**

Serotonin   138, 167, 262, 313, 461, 465, 473, 529, 581, 636, 639, 641*f*, 644, 649, 654, 682–683, 695, 705

Servo control   25

SGLT   407, 450

Sham feeding   705

Sham rage   680

Sheath of Schwann   58

Shift leukocytosis   180

Shivering   692, **698–699**, 701–702

Shock, circulatory   280–284

Short bowel syndrome   463

SIADH *see* Syndrome of inappropriate ADH secretion

Sickle cell anemia   142, 144–145, **149**

Sick–sinus syndrome   219

Siderocytes   156

SIF cells   640

Siggaard–Anderson nomogram   403

Simple (visual) cells   745–746

Sinoatrial node   209

Sinus arrhythmia   210, 272

Sinusoids, myocardial   285

Size principle   108

Sjogren's syndrome   465

Skiascopy   733

Skin, circulation in   300–301

Skinfold thickness   776–777

Skinner box   714

Sleep   692–696
  breathing during   357–358

Slow twitch muscle   105

Smell *see* Olfaction

Smooth endoplasmic reticulum   **29**, **31**, 57, 61

Smooth muscle   60*t*, 61, 63–65, 125–129

Smooth–pursuit movement   607, **751–752**

Sneeze 312–313
Snellen chart 12, 730–732
Sodium balance of the body 392
Sodium channel inactivation 85
Sodium sulfate infusion test **420**, 422
Sodium, intestinal absorption of 452
    renal handling of 383–386
Solubility product 517, 521
Soma-dendritic spike 642–643
Somatomedin *see* Insulin-like growth
    factor
Somatostatin 439–**440**, 441*f*, 456, 458,
    499, 507*f*, 521, 537, **540***f*, 614, 704
Somatotropin *see* Growth hormone
Somesthetic corical area 621,
    625–626
Somnambulism 696
Sorbitol pathway 542, 722
Sound localization 762
Source amnesia 627, **707**
Southern blotting 52
Space constant **92–93**, 643
Spastic paralysis 598*t*, **677**
Spastic pelvic floor syndrome 464
Spasticity 428, 619, 639, **665**–666,
    677
Spatial learning 628
Specific nuclei 613
Spectrin 37, 39, 144
Sperm 33, 482, **554–556**, 562–571,
    573, 581
Spermatogenesis 45, 499, 548, 552,
    554–555
Spermiogenesis 554
Spherical aberration 730
Spherocyte **141**, 144, 188
Spherocytosis 144
Sphygmomanometry 252, **255–256**,
    283
Spike potential **80**, 128, 438
Spike trigger zone 57
Spinal cord 596–599
    subacute combined degeneration of
    599
    transection of **599**, 647, 676, 686,
    695
Spinal nerve root 650
Spinal shock 599, 665, 677
Spinnbarkeit 559–561
Spinocerebellar tracts 597*f*, 598*f*, 600,
    605, **608**, 611, 645, 677
Spinocerebellum 605, 608–610
Spinohypothalamic tract 652
Spinomesencephalic tract 652

Spinoreticulothalamic tract 612, **652**,
    655
Spinothalamic pathway 597*f*–598,
    624, 649, **651**–653, 652*f*, 655, 688,
    689*f*
Spinoreticulocerebellar 598*t*, **608**
Spinoreticuloolivary 608*f*
Spirometry 328–329, 333
Splanchnic circulation 303–305
Splay **413–414**, 419, 421–422
Spleen 187–188
Sprue 157, 160, 451, **463**
Squint 97, 726–727, **753**
SRY gene 571
Stabilizer **121–122**, 315
Staircase effect **133–134**, 235
Starling forces **257**–258, 298–299,
    379–381, 385, 392, 410
Steatorrhea 462–463, 465
Stellate cell 374, 605, 621, 626
Stem cell 143, 183, 186, 189, 197, **202–
    203**, 205, 517, 554
Stereocilia 757, 759–761, 766–768
Stereognosis 598, 598*t*, **646**, 655–656,
    669
Stereopsis 746, 749
Steroid myopathy 530
Stokes–Adams syndrome 219
Stool examination 447
Strabismus 753
Strap muscle 120–121
Strength–duration curve 83
Stress 530
Stretch reflex 660–662
Stria terminalis 588, 592, 592*f*, 615,
    629, 682
Stria vascularis 757–758
Striate arteries 588, 622
Striate area 589*f*, 626
Striatopallidal pathway 617
Striatum **616**–618, 620, 638, 683, 710
Striola 767
Stroke volume 227–228, **232**,
    234–236, 238, 265–266, 270–271,
    273–274, 580*f*
Strychnine 639
Subclavian steal 17, 251, 291, **291***f*
Subfornical organ 295, 393, **411**
Subiculum 627–628
Subliminal fringe 671–672
Substance P 261, 618, **649–650***f*
Substantia gelatinosa **597**, 642, 653,
    654*f*

Substantia nigra 616–618, 638–639,
    653
Subthalamic nucleus 269, 272, 614*f*,
    616, 617–618, 620
Subthalamus 588, 590*f*, 612, 619–620
Succus entericus **449**, 452*f*
Summation
    of muscle twitches **105**, 133
    of potentials 134, **642***f*, **643**, 748
Supplementary eye field 625
Supplementary motor area 612, 617,
    624, 668
Supplementary somesthetic area 626
Suprachiasmatic nucleus 393, 612,
    614*f*, 615, 683, 742
Supramarginal gyrus 621*f*, 626
Supranuclear lesion 602–603
Supraoptic nucleus 393, 410,
    504–505*f*, 507, 581, **614**–615
Surface tension 9
    alveolar 325–326
Surfactant 311–313, 316, 321,
    325–326, 340
Sweat gland 263, **301**, 632, 634*t*, 639*f*
Sympathetic adrenergic failure 632
Sympathetic discharge
    effect on the kidneys 382
Sympathetic nervous system 267, **630**,
    631*f*
Sympathetic vasodilator fiber 267,
    301–**302**, 632
Symport **39–41**, 76, 384–386, 404*f*,
    407–410, 451–452, 465, 511, 522,
    563, 637
Synapse 64, 635
Synapse en passant 64–65
Synaptic plasticity 607, **643**, 644, 711
Synaptic vesicles 59, 64
Synaptogenesis 71
Syncope 219, 270, 283, 289–**291**, 302
Syndrome of inappropriate ADH
    secretion 410
Synergist muscle 121
Syntitiotrophoblast **576**, 568*f*
Syntrophin 63
Syringomyelia 658
Systolic multihaustral propulsion 453

T

T lymphocyte **185–187**, 195–198, 201,
    209, 203*f*
Tabes dorsalis **658**, 675
Tachycardia 219–225

Tachypnea 283, 350, **368**, 701

Tactile localization **645**, 655–657

Tactile receptors 647

Takayama test 139

Tamm–Horsfall protein 377, **417–418**

Tancytes 295

Taste 771–773

Tay–Sach disease 32

Tear film 721, 723*f*

Teichman test 139

Temperature (body) regulation of 97–701

TENS *see* Transcutaneous electrical nerve stimulation

TEPD *see* Transepithelial potential difference

Terminal button 64, 96, 99

Testicular feminizing syndrome 575

Testis determining factor 571

Testosterone 499, 525–528, **547–548**, 549–555, 570–578

Tetanus toxin 60, 636, 639

Tetanus, of muscle contraction 103, **105–106**

Tetraethylammonium 87

Tetrodotoxin 87

Thalamic pain 614, 655

Thalamic syndrome 614, 655

Thalamus 612–614, 655

Thalassemia 141, **149**

Thecal cell **548–549**, 550, 556–560

Thelarche 552

Thermoneutral zone 300, **700**

Thermoreceptors 645–646, **648–649**, 697, **700**

Thirst 265, 275, 277, 280–281, 382–397, **393**, 411–412, 541, 604, 683

Thorndike's paradigm 714

Thoroughfare channel **246–247**, 257, 302

Throbocytopenia 145, **167**, 175–176, 188

Thrombasthenia 176–177

Thrombin 170–176

Thrombin time 176

Thrombocyte *see* platelet

Thrombocytosis 167

Thrombomodulin 173–174

Thromboplastin 169, **172**, 176

Thromboplastin generation test 176

Thrombopoiesis 204–205

Thrombospondin 166, 168

Thrombosthenin 166

Thromboxane 168–169, 173, 262, 264, 469

Thrombotic disorders 173, 177–178, 281, 294, 421, 551

Thundberg's illusion 655

Thymus 99, **186–187**, 191–192, 197, 529

Thymus-independent antigen 197

Thyroid function tests 516

Thyroid hormone 143, 344, 484*f*, 487*f*, 490, 492*f*, 500, **510–516**, 583

Thyroid storm 516

Thyrotoxicosis 186, 236, 242, 490, 513, **516**

Tidal volume 323*f*, **328–329**, 331–332, 334–337, 349–350, 357, 359–363, 400, 452

Tight junction **34***f*, 257, 295, 377, 383–384, 386, 398, 439, 553, 563, 736, 757, 761*f*

Timbre discrimination 762

Tinel sign 69–70

Tip links 760–761

Titin 61, 63

Tonic labyrinthine reflexes 674–675

Tonic neck reflexes 666, 673–675

Tonicity **22–23**, 393, 396, 402, 446

Total iron binding capacity 154–155

Total lung capacity 320, 328*f*, 329, 331, 333, 357, 363, 366*t*, 368

Touch 118*t*, 351, 599*f*, 601*t*, 608, 613, 625, **645–647**, 648, 651–658

Tourniquet test 658

Trabeculectomy 724–725

Tracheobronchial tree 309–310

Transcription 50–51, 500*f*

Transcutaneous electrical nerve stimulation 653

Transducin 737

Transepithelial potential difference **377***f*–**378**, 383–386, 398, 404, 409, 411

Transferrin 139*t*, **154–156**, 421, 472, 558

Transfusion of blood 157, **163–164**, 194

Transit time
  colonic 453
  pulmonary **298**, 340

Translation 30, **51***f*, 580*f*

Transport maximum for glucose **413**, 415

Transverse (sarcoplasmic) tubule **61–62**, 102

Traube–Hering waves 273

Tremor 360, 516, **611**, 614, 618–**619**

Trendelenberg position 284

Trichromatic coding 746, 748

Trichromats 749

Trigeminal lemniscus 601*t*, 602*t*, **652**, **653***f*

Triolein breath test 447

Tripeptide hydrolysis test 447

Triple response to trauma **300**, 649

Trophoblast 192, **567–568**, 576

Tropomyosin **61–63**, 100–101, 125, 130, 133

Troponin 61, **62–63**, 100, 101, 125, 130, 133

Tuberoinfundibular tract 615

Tubular cells 377

Tubular transport 381–382

Tubulin 32–33, 60

Tubuloglomerular feedback 381

Tuning fork test 763–764

Turner syndrome 573–574

Two-point discrimination **646**, 647–648, 651

U

Unconscious senses 645

Uniport **39**, 41, 383–384

Unregulated pathway 31

Unstirred layer 451

Upper motor neuron 348, 599, 603, 663, **670**, 677–678

Upper motor neuron lesion 599, 663, **677– 688***t*

Up-regulation 535

Urea cycle 487

Urea, renal handling of 407–408

Uric acid, renal handling of 408

Urinary bladder 425–429

Uriniferous tubule 373, 375–376

Urobilin 150*f*, 417*t*, **455**

Urobilinogen **150**, 152, 418, 455

Utilization time 83*f*, 84

Utricle 766–768

Uveal tract 722

V

Valsalva maneuver 235, **270**, 275, 285, 291, 296, 323, 454, 632

Valsalva ratio 632

Van den Berg test 151

Van't Hoff equation 21

Vasoactive intestinal peptide   265, 267, 311, 436, 438–441*f*, 444, 449, 456–**458**, 457*t*, 464–465, 564, 581, 636

Vasomotor center   **268**, 272, 280–283, 291, 300, 302, 393, 603

Vasomotor failure   281

Vasopressin   177, **265**, **410**, 702

Veins, structure of   247–248

Venous pooling   **253–254**, 270–271

Venous pump   234, 236–**237**, 302

Ventilation-perfusion ratio   **339–340**, 338*f*, 355–356, 366

Ventral tegmental area   629, 682–**683**

Ventricles, cerebral   60, 294, **588**, **590***f*

Ventricular septal defect   229*f*, 230, **245**, **246***f*, 287

Ventromedial frontal cortex   681

Ventromedial nucleus   614, 703

Venules   247

Vergences   603, **750–751**

Versions   **750–751**, 762

Vestibular apparatus   766–768

Vestibular area in cerebral cortex   626

Vestibular nucleus   601, 607–608, 665, 772–773, 751, 767–**768***f*

Vestibulocerebellar tract   601*t*, 611, 768*f*

Vestibulocerebellum   605, **607**, 742–743, 751–753

Vestibuloocular reflex   26, 607, 611, 711, 742, **751–752**

Vestibuloreticular tract   601–602, 668, **768***f*

Vestibulospinal tract   597*f*, 598*t*, 600, 601*t*, 665, 670, **768***f*

Vestibulothalamocortical tract   601*t*, 602, **768**

VIP *see* Vasoactive intestinal peptide

Viscosity   **9–10**, 139, 141, 145, 149, 249, 251–252, 283, 320–321, 375

Visual acuity   12, **730–732**, 736, 742

Visual agnosia   629, **707**

Visual angle   **12–13**, 730, 732

Visual areas   **626**, 745, **623***f*, 742, **745–747**

Visual axis   **726–727**, 731, 733

Visual cycle   736–738

Vital capacity   322, **329**, 331–335, 339, 364, 366, 368

Vitamin K-dependent proteins   174, **177**, 178*f*

Vitamins   452, 472, **473***t***–474***t*

Vitamin B$_{12}$ *see* cyanacobalamine

Vitamin D   524

Vitiligo   506

VLDL *see* plasma lipoproteins

Vocal fremitus   370

Vocal resonance   368–370, 369*t*

Voltage clamp technique   86–87

Volume receptor   **268**, 271, 277, 393–394, 396, 410

Vomiting   283, 354–355, **460–462**

von Willebrand factor   140, 166, **168**

Von Willebrand's disease   176–177

VPRC *see* Hematocrit

W

Wallerian degeneration   **67–68**, 70

Water balance of the body   381, 392, 412

Water
  intestinal absorption of   452
  tubular absorption of   381

Water-deprivation test   419

Water diuresis   394, 419

Watery diarrhea syndrome   465

WDR neurons   650*f*, 652

Weber–Fechner law   14, 109, **657**

Weber's test   764

Wenckebach phenomenon   219–220

Wernicke's area   622–623, 626, **716**, 758

Wernicke's aphasia   715–716

Western blotting   52

White muscle   110, 301

Wilson's disease   475, 620

Winding up   653

Windkessel effect   245

Withdrawal bleeding   561, 569

Withdrawal reflex   **671–672**, 674, 711–712

Wolff–Chaikoff effect   513

Wolff–Parkinson–White syndrome   222

Wolffian ducts   48, 572, 573*f*, 575

Word blindness   715

Word deafness   715–716

Work
  of breathing   8, **324–325**, 360, 366–367
  of ventricular contraction   8, **248–249**

Writing center   625

X

Xerophthalamia   723

Xerostomia   465

Z

Z-line   61, 101

Zollinger–Ellison syndrome   442, 441

Zona occludens   **34**, 377

Zona pellucida   556, 566–569

Zonule   721–722, 753–754

Zona adherens   34, 735*f*